Wills' Mineral Processing Technology

Wills' Mineral Processing Technology

An Introduction to the Practical Aspects of Ore Treatment and Mineral Recovery

Eighth Edition

Barry A. Wills

Senior Partner, MEI, UK

James A. Finch, FRSC, FCIM, P.Eng.

Gerald G. Hatch Chair in Mining and Metallurgical Engineering,
Department of Mining and Materials Engineering,
McGill University, Montréal, Canada

AMSTERDAM • BOSTON • HEIDELBERG • LONDON
NEW YORK • OXFORD • PARIS • SAN DIEGO
SAN FRANCISCO • SINGAPORE • SYDNEY • TOKYO
Butterworth-Heinemann is an imprint of Elsevier

Butterworth-Heinemann is an imprint of Elsevier
The Boulevard, Langford Lane, Kidlington, Oxford OX5 1GB, UK
225 Wyman Street, Waltham, MA 02451, USA

First published 1979
Second edition 1981
Third edition 1985
Fourth edition 1988
Fifth edition 1992
Sixth edition 1997
Seventh edition 2006
Eighth edition 2016

ISBN: 978-0-08-097053-0

British Library Cataloguing-in-Publication Data
A catalogue record for this book is available from the British Library

Library of Congress Cataloging-in-Publication Data
A catalog record for this book is available from the Library of Congress

For Information on all Butterworth-Heinemann publications
visit our website at http://store.elsevier.com/

Contents

Preface

The email came as a surprise: would I be interested in editing the eighth edition of the classic *Wills' Mineral Processing Technology*? The book is the most-widely used English-language textbook on the subject and the SME's 'readers' choice', making it both an honor to be asked and a daunting prospect. My old colleague, the late Dr. Rao, had always encouraged me to write another book, but after *Column Flotation* I vowed "never again". Nevertheless, after a career of more than 40 years, I was looking for new challenges, and with several day's contemplation, I agreed. This turned out to be the largest single professional task I have undertaken, consuming most of the last two years, and on several occasions I came to question my sanity.

It was at that point that I seriously started to read whole chapters of the seventh edition, as opposed to the odd sections I had consulted in the past. You have to admire that Barry Wills wrote the first book, and the following 5 editions, alone. As Tim Napier-Munn noted in the Preface to the seventh edition, a team is needed now, and for this edition the team, which included experts and recently graduated students, is recognized in the "Acknowledgments". But what was the plan for the team?

Elsevier had hinted that the artwork could do with an upgrade, and we set about re-drawing the originals that were retained, and adding new artwork based on my teaching experience. By having students in the team, we aimed to have illustrations that addressed the usual student questions.

The next part of the plan was to use, as much as possible, the chapter structure that Barry had devised, but to number the sub-sections, as, according to the students, it was sometimes difficult to know where you were. A team member read a chapter and made suggestions, in some cases taking on the task of re-writing sections. All the chapters were finalized by me, including adding more example calculations, and then passed by Barry for a final review. I appreciate the confidence Barry had that the book was in 'good hands'. Chapter 3 became unwieldy, and we decided to divide the contents between two chapters. To avoid altering the numbering of the established chapters, a Chapter 17 was inserted. Elsevier Americanized the English, although a few "sulphides" snuck through. This also meant that my "tonnes" became "tons" and the reader should keep that in mind when working through the example calculations.

I am proud of the final product, and grateful to Barry and Elsevier for giving me the chance to be part of this enterprise.

James A. Finch
July 2015

Acknowledgments

Two people made the book possible and worked with me (JAF) throughout and thus deserve to be noted first: **Dr. Jarrett Quinn** who handled much of the correspondence, as well as managing most of Chapters 6, 8, and 14 as well as the first draft of Chapter 3 and the "Flotation Machine" section of Chapter 12; and **Dr. Yue Hua Tan** who contributed all the artwork, redrawing the originals and creating the new, and organized the supporting documentation and spreadsheets. My sincere thanks to both individuals, who never doubted we would finish.

It was apparent from the start that assistance from experts in the various aspects covered by the book was going to be required for it to be brought up to date. Most of my requests for assistance were enthusiastically met and knowing the time and effort involved I am most grateful for the help, and the book has benefitted accordingly. The contributions were edited to bring some conformity to style, and any errors thus incurred are mine alone. In alphabetical order the contributors are:

Bittner, J, Dr (Separation Technologies): Chapter 13, review of material

Bouchard, J, Prof (Université Laval): Chapter 12, "Control" section

Boucher, D (PhD candidate, McGill University): Chapter 17, "Machine Design" section

Brissette, M (Corem): Chapters 5 and 7, editing and adding new material

Bulled, D (SGS Canada): Chapter 17, "Geometallurgy" section

Cappuccitti, F (Flottec): Chapter 12, editing the "Collector" section; Chapter 15, supplying chemical data

Cunningham, R (Met-Chem): Chapter 7, "Stirred Milling" section

Demers, I, Prof (Université du Québec en Abitibi-Témiscamingue): Chapter 16

Doll, A (Alex G Doll Consulting): Chapters 5 and 7, editing and adding new material

Emad, M, Dr (Post-doctoral student, McGill University): Chapter 2

Flament, F (Triple Point Technology): Chapter 3, "Mass Balancing" section

Gilroy, T (student, McGill University): Chapter 4, first draft

Grammatikopoulos, T (SGS Canada): Chapter 17, "Applied Mineralogy" section

Hart, B, Dr (Surface Science Western): Chapter 12, "Diagnostic Surface Analysis" section

Jordens, A (PhD candidate, McGill University): Chapter 13

Krishnamoorthy, N, Dr (Research assistant, McGill University): Chapter 15, first draft

Lotter, N, Dr (XPS Consulting & Testwork Services): Chapter 12, "High Confidence Flotation Testing" section

Major, K (KMW Consulting): Chapters 6 and 8, initial review

Maldonado, M, Prof (Universidad de Santiago): Chapter 3, "Control" section

Marcuson, S, Dr (Marcuson and Associates): Chapter 1, "Sustainability" section

McIvor, R, Dr (Metcom Technologies): Chapters 5, 7, and 9, initial review

Mitri, H, Prof (McGill University): Chapter 2, review of material

Morrell, S, Dr (SMC Testing): Chapter 5, review of material

Nesset, J, Dr (NesseTech Consulting Services): Chapter 2, "Self-Heating" section; Chapter 3, "Sampling" section; Chapter 17, "Design of Experiments" section

O'Keefe, C, Dr (CiDRA): Chapter 3, contribution to "On-Line Analysis" section

Pax, R, Dr (RAP Innovation and Development): Chapter 3, "Mineral/Phase Analysis" section

Robben, C, Dr (TOMRA Sorting): Chapter 14, review of material

Schaffer, M (Portage Technologies): Chapters 6 and 7, "Control" sections

Singh, N (Research assistant, McGill University): Chapter 1, first draft
Sosa, C, Dr (SGS Canada): Chapter 17, "Geometallurgy" section
Sovechles, J (PhD candidate, McGill University): Chapter 9
Smart, R, Prof (University of South Australia): Chapter 12, "Diagnostic Surface Analysis" section
Waters, K, Prof (McGill University): Chapters 10 and 11, initial review
Williams-Jones, A, Prof (McGill University): Appendices I and II, update

The book could not have been completed without financial and logistical support. Thanks go to the Natural Sciences and Engineering Research Council of Canada (NSERC) for the funding to support the McGill team; and to McGill University for the time to devote to the book and the use of the facilities.

Last but never least, JAF would like to thank his wife, Lois, of 42 years for her support with what grew to occupy most of my time over the past 24 months.

Chapter 1

Introduction

1.1 MINERALS

The forms in which metals are found in the crust of the earth and as seabed deposits depend on their reactivity with their environment, particularly with oxygen, sulfur, and carbon dioxide. Gold and platinum metals are found principally in the *native* or metallic form. Silver, copper, and mercury are found native as well as in the form of sulfides, carbonates, and chlorides. The more reactive metals are always in compound form, such as the oxides and sulfides of iron and the oxides and silicates of aluminum and beryllium. These naturally occurring compounds are known as *minerals*, most of which have been given names according to their composition (e.g., galena—lead sulfide, PbS; cassiterite—tin oxide, SnO_2).

Minerals by definition are natural inorganic substances possessing definite chemical compositions and atomic structures. Some flexibility, however, is allowed in this definition. Many minerals exhibit *isomorphism*, where substitution of atoms within the crystal structure by similar atoms takes place without affecting the atomic structure. The mineral olivine, for example, has the chemical composition $(Mg,Fe)_2SiO_4$, but the ratio of Mg atoms to Fe atoms varies. The total number of Mg and Fe atoms in all olivines, however, has the same ratio to that of the Si and O atoms. Minerals can also exhibit *polymorphism*, different minerals having the same chemical composition, but markedly different physical properties due to a difference in crystal structure. Thus, the two minerals graphite and diamond have exactly the same composition, being composed entirely of carbon atoms, but have widely different properties due to the arrangement of the carbon atoms within the crystal lattice.

The term "mineral" is often used in a much more extended sense to include anything of economic value that is extracted from the earth. Thus, coal, chalk, clay, and granite do not come within the definition of a mineral, although details of their production are usually included in national figures for mineral production. Such materials are, in fact, *rocks*, which are not homogeneous in chemical and physical composition, as are minerals, but generally consist of a variety of minerals and form large parts of the earth's crust. For instance, granite, which is one of the most abundant *igneous* rocks, that is, a rock formed by cooling of molten material, or magma, within the earth's crust, is composed of three main mineral constituents: feldspar, quartz, and mica. These three mineral components occur in varying proportions in different parts of the same granite mass.

Coals are a group of bedded rocks formed by the accumulation of vegetable matter. Most coal-seams were formed over 300 million years ago by the decomposition of vegetable matter from the dense tropical forests which covered certain areas of the earth. During the early formation of the coal-seams, the rotting vegetation formed thick beds of *peat*, an unconsolidated product of the decomposition of vegetation, found in marshes and bogs. This later became overlain with shales, sandstones, mud, and silt, and under the action of the increasing pressure, temperature and time, the peat-beds became altered, or *metamorphosed*, to produce the sedimentary rock known as coal. The degree of alteration is known as the *rank* of the coal, with the lowest ranks (lignite or brown coal) showing little alteration, while the highest rank (anthracite) is almost pure graphite (carbon).

While metal content in an ore is typically quoted as percent metal, it is important to remember that the metal is contained in a mineral (e.g., tin in SnO_2). Depending on the circumstances it may be necessary to convert from metal to mineral, or vice versa. The conversion is illustrated in the following two examples (Examples 1.1 and 1.2).

The same element may occur in more than one mineral and the calculation becomes a little more involved.

1.2 ABUNDANCE OF MINERALS

The price of metals is governed mainly by supply and demand. Supply includes both newly mined and recycled metal, and recycling is now a significant component of the lifecycle of some metals—about 60% of lead supply comes from recycled sources. There have been many prophets of doom over the years pessimistically predicting the imminent exhaustion of mineral supplies, the most

Wills' Mineral Processing Technology.

Example 1.1

Given a tin concentration of 2.00% in an ore, what is the concentration of cassiterite (SnO_2)?

Solution

Step 1: What is the Sn content of SnO_2?
Molar mass of Sn (M_{Sn}) 118.71 g mol^{-1}
Molar mass of O (M_O) 15.99 g mol^{-1}

$$\%\text{Sn in SnO}_2 = \frac{M_{Sn}}{M_{Sn} + 2 \times M_O} = \frac{118.71}{118.71 + 2 \times 15.99} = 78.8\%$$

Step 2: Convert Sn concentration to SnO_2

$$\frac{2.00\%\text{Sn}}{78.8\%\text{Sn in SnO}_2} = 2.54\% \text{ SnO}_2$$

Example 1.2

A sample contains three phases, chalcopyrite ($CuFeS_2$), pyrite (FeS_2), and non-sulfides (containing no Cu or Fe). If the Cu concentration is 22.5% and the Fe concentration is 25.6%, what is the concentration of pyrite and of the non-sulfides?

Solution

Note, Fe occurs in two minerals which is the source of complication. The solution, in this case, is to calculate first the % chalcopyrite using the %Cu data in a similar manner to the calculation in Example 1.1 (Step 1), and then to calculate the %Fe contributed by the Fe in the chalcopyrite (Step 2) from which %Fe associated with pyrite can be calculated (Step 3).

Molar masses (g mol^{-1}): Cu 63.54; Fe 55.85; S 32.06
Step 1: Convert Cu to chalcopyrite (Cp)

$$\%\text{Cp} = 22.5\% \left[\frac{63.54 + 55.85 + (2 \times 32.06)}{63.54}\right] = 65.0\%$$

Step 2: Determine %Fe in Cp

$$\%\text{Fe in Cp} = 65\% \left[\frac{55.85}{63.54 + 55.85 + (2 \times 32.06)}\right] = 19.8\%$$

Step 3: Determine %Fe associated with pyrite (Py)

$$\%\text{Fe in Py} = 25.6 - 19.8 = 5.8\%$$

Step 4: Convert Fe to Py (answer to first question)

$$\%\text{Py} = 5.8\% \left[\frac{55.85 + (2 \times 32.06)}{55.85}\right] = 12.5\%$$

Step 5: Determine % non-sulfides (answer to second question)

$$\%\text{non-sulfides} = 100 - (\%\text{Cp} + \%\text{Py}) = 100 - (65.0 + 12.5) = 22.5\%$$

extreme perhaps being the "Limits to Growth" report to the Club of Rome in 1972, which forecast that gold would run out in 1981, zinc in 1990, and oil by 1992 (Meadows et al., 1972). Mouat (2011) offers some insights as to the past and future of mining.

In fact, major advances in productivity and technology throughout the twentieth century greatly increased both the resource base and the supply of newly mined metals, through geological discovery and reductions in the cost of production. These advances actually drove down metal prices in real terms, which reduced the profitability of mining companies and had a damaging effect on economies heavily dependent on resource extraction, particularly those in Africa and South America. This in turn drove further improvements in productivity and technology. Clearly mineral resources are finite, but supply and demand will generally balance in such a way that if supplies decline or demand increases, the price will increase, which will motivate the search for new deposits, or technology to render marginal deposits economic, or even substitution by other materials. Gold is an exception, its price having not changed much in real terms since the sixteenth century, due mainly to its use as a monetary instrument and a store of wealth.

Estimates of the crustal abundances of metals are given in Table 1.1 (Taylor, 1964), together with the amounts of some of the most useful metals, to a depth of 3.5 km (Tan and Chi-Lung, 1970).

The abundance of metals in the oceans is related to some extent to the crustal abundances, since they have come from the weathering of the crustal rocks, but superimposed upon this are the effects of acid rainwaters on mineral leaching processes; thus, the metal availability from seawater shown in Table 1.2 (Tan and Chi-Lung, 1970) does not follow precisely that of the crustal abundance. The seabed may become a viable source of minerals in the future. Manganese nodules have been known since the beginning of the nineteenth century (Mukherjee et al., 2004), and mineral-rich hydrothermal vents have been discovered (Scott, 2001). Mining will eventually extend to space as well.

It can be seen from Table 1.1 that eight elements account for over 99% of the earth's crust: 74.6% is silicon and oxygen, and only three of the industrially important metals (aluminum, iron, and magnesium) are present in amounts above 2%. All the other useful metals occur in amounts below 0.1%; copper, for example, which is the most important non-ferrous metal, occurring only to the extent of 0.0055%. It is interesting to note that the so-called common metals, zinc and lead, are less plentiful than the rare-earth metals (cerium, thorium, etc.).

TABLE 1.1 Abundance of Metal in the Earth's Crust

Element	Abundance (%)	Amt. in Top 3.5 km (tons)	Element	Abundance (%)	Amt. in Top 3.5 km (tons)
(Oxygen)	46.4		Vanadium	0.014	10^{14}–10^{15}
Silicon	28.2		Chromium	0.010	
Aluminum	8.2	10^{16}–10^{18}	Nickel	0.0075	
Iron	5.6		Zinc	0.0070	
Calcium	4.1		Copper	0.0055	10^{13}–10^{14}
Sodium	2.4		Cobalt	0.0025	
Magnesium	2.3	10^{16}–10^{18}	Lead	0.0013	
Potassium	2.1		Uranium	0.00027	
Titanium	0.57	10^{15}–10^{16}	Tin	0.00020	
Manganese	0.095		Tungsten	0.00015	10^{11}–10^{13}
Barium	0.043		Mercury	$8 \times 10 \times^{-6}$	
Strontium	0.038		Silver	7×10^{-6}	
Rare earths	0.023		Gold	$<5 \times 10^{-6}$	$<10^{11}$
Zirconium	0.017	10^{14}–10^{16}	Platinum metals	$<5 \times 10^{-6}$	

TABLE 1.2 Abundance of Metal in the Oceans

Element	Abundance (tons)	Element	Abundance (tons)
Magnesium	10^{15}–10^{16}	Vanadium Titanium	10^{9}–10^{10}
Silicon	10^{12}–10^{13}		
Aluminium Iron Molybdenum Zinc	10^{10}–10^{11}	Cobalt Silver Tungsten	10^{12}–10^{13}
Tin Uranium Copper Nickel	10^{9}–10^{10}	Chromium Gold Zirconium Platinum	$<10^{8}$

1.3 DEPOSITS AND ORES

It is immediately apparent that if the minerals containing important metals were uniformly distributed throughout the earth, they would be so thinly dispersed that their economic extraction would be impossible. However, the occurrence of minerals in nature is regulated by the geological conditions throughout the life of the mineral. A particular mineral may be found mainly in association with one rock type (for example, cassiterite mainly associates with granite rocks) or may be found associated with both igneous and *sedimentary* rocks (i.e., those produced by the deposition of material arising from the mechanical and chemical weathering of earlier rocks by water, ice, and chemical decay). Thus, when granite is weathered, cassiterite may be transported and redeposited as an *alluvial* deposit. Besides these surface processes, mineral deposits are also created due to magmatic, hydrothermal, and other geological events (Ridley, 2013).

Due to the action of these many natural agencies, mineral deposits are frequently found in sufficient concentrations to enable the metals to be profitably recovered; that is, the deposit becomes an *ore*. Most ores are mixtures of extractable minerals and extraneous nonvaluable material described as *gangue*. They are frequently classed according to the nature of the valuable mineral. Thus, in *native* ores the metal is present in the elementary form; *sulfide* ores contain the metal as sulfides, and in *oxidized* ores the valuable mineral may be present as oxide, sulfate, silicate, carbonate, or some hydrated form of these. *Complex* ores are those containing profitable amounts of more than one valuable mineral. Metallic minerals are often found in certain associations within which they may occur as mixtures of a wide range of grain sizes or as single-phase solid solutions or compounds. Galena and sphalerite, for example, are commonly associated, as, to a lesser extent, are copper sulfide minerals and sphalerite. Pyrite is almost always associated with these minerals as a sulfide gangue.

There are several classifications of a deposit, which from an investment point of view it is important to understand: *mineral resources* are potentially valuable and are further classified in order of increasing confidence into *inferred*, *indicated*, and *measured* resources; *mineral (ore) reserves* are known to be economically (and legally) feasible for extraction and are further classified, in order of increasing confidence, into *probable* and *proved* reserves.

1.4 METALLIC AND NONMETALLIC ORES

Ores of economic value can be classed as metallic or nonmetallic, according to the use of the mineral. Certain minerals may be mined and processed for more than one purpose. In one category, the mineral may be a metal ore, that is, when it is used to prepare the metal, as when bauxite (hydrated aluminum oxide) is used to make aluminum. The alternative is for the compound to be classified as a nonmetallic ore, that is, when bauxite or natural aluminum oxide is used to make material for refractory bricks or abrasives.

Many nonmetallic ore minerals associate with metallic ore minerals (Appendixes I and II) and are mined and processed together. For example, galena, the main source of lead, sometimes associates with fluorite (CaF_2) and barytes ($BaSO_4$), both important nonmetallic minerals.

Diamond ores have the lowest grade of all mined ores. One of the richest mines in terms of diamond content, Argyle (in Western Australia) enjoyed grades as high as 2 ppm in its early life. The lowest grade deposits mined in Africa have been as low as 0.01 ppm. Diamond deposits are mined mainly for their gem quality stones which have the highest value, with the low-value industrial quality stones being essentially a by-product: most industrial diamond is now produced synthetically.

1.5 THE NEED FOR MINERAL PROCESSING

"As-mined" or "run-of-mine" ore consists of valuable minerals and gangue. Mineral processing, also known as *ore dressing, ore beneficiation, mineral dressing,* or *milling*, follows mining and prepares the ore for extraction of the valuable metal in the case of metallic ores, or to produce a commercial end product as in the case of minerals such as potash (soluble salts of potassium) and coal. Mineral processing comprises two principal steps: *size reduction* to *liberate* the grains of valuable mineral (or *paymineral*) from gangue minerals, and physical separation of the particles of valuable minerals from the gangue, to produce an enriched portion, or *concentrate*, containing most of the valuable minerals, and a discard, or *tailing* (*tailings* or *tails*), containing predominantly the gangue minerals. The importance of mineral processing is today taken for granted, but it is interesting to reflect that little more than a century ago, ore concentration was often a fairly crude operation, involving relatively simple density-based and hand-sorting techniques. The twentieth century saw the development of mineral processing as an important profession in its own right, and certainly without it the concentration of many ores, and particularly the metalliferous ores, would be hopelessly uneconomic (Wills and Atkinson, 1991).

It has been predicted that the importance of mineral processing of metallic ores may decline as the physical processes utilized are replaced by the hydro- and pyrometallurgical routes used by the extractive metallurgist (Gilchrist, 1989), because higher recoveries are obtained by some chemical methods. This may apply when the useful mineral is very finely disseminated in the ore and adequate liberation from the gangue is not possible, in which case a combination of chemical and mineral processing techniques may be advantageous, as is the case with some highly complex deposits of copper, lead, zinc, and precious metals (Gray, 1984; Barbery, 1986). Heap leaching of gold and oxidized copper ores are examples where mineral processing is largely by-passed, providing only size reduction to expose the minerals. *In-situ* leaching is used increasingly for the recovery of uranium and bitumen from their ores. An exciting possibility is using plants to concentrate metals sufficiently for chemical extraction. Known as *phytomining* or *agro-mining*, it has shown particular promise for nickel (Moskvitch, 2014). For most ores, however, concentration of metals for subsequent extraction is best accomplished by mineral processing methods that are inexpensive, and their use is readily justified on economic grounds.

The two fundamental operations in mineral processing are, therefore, *liberation* or release of the valuable minerals from the gangue, and *concentration*, the separation of these values from the gangue.

1.6 LIBERATION

Liberation of the valuable minerals from the gangue is accomplished by *size reduction* or *comminution*, which involves crushing and grinding to such a size that the product is a mixture of relatively clean particles of mineral and gangue, that is, the ore minerals are *liberated* or *free*. An objective of comminution is liberation at the coarsest possible particle size. If such an aim is achieved, then not only is energy saved but also by reducing the amount of fines produced any subsequent separation stages become easier and cheaper to operate. If high-grade solid products are required, then good liberation is essential; however, for subsequent hydrometallurgical processes, such as leaching, it may only be necessary to expose the required mineral.

Grinding is often the greatest energy consumer, accounting for up to 50% of a concentrator's energy consumption (Radziszewski, 2013). As it is this process which achieves liberation of values from gangue, it is also the process that is essential for efficient separation of the minerals. In order to produce clean concentrates with little contamination with gangue minerals, it is often necessary to grind the ore to a fine size (<100 μm). Fine grinding increases energy costs and can lead to the production of very fine difficult to treat "slime" particles which may be lost into the tailings, or even discarded before the concentration process. Grinding therefore becomes a compromise between producing clean (high-grade) concentrates, operating costs, and losses of fine minerals. If the ore is low grade, and the minerals have very small grain size and are disseminated through the rock, then grinding energy costs and fines losses can be high.

In practice, complete liberation is seldom achieved, even if the ore is ground down to less than the grain size of the desired minerals. This is illustrated by Figure 1.1, which shows a lump of ore containing a grain of valuable mineral with a breakage pattern superimposed that divides the lump into cubic particles of identical volume (for simplicity) and of a size below that of the mineral grain. It can be judged that each particle produced containing mineral also contains a portion of gangue. Complete liberation has not been attained, but the bulk of the major mineral—the gangue—has, however, been liberated from the valuable mineral.

The particles of "locked" (or "composite") mineral and gangue are known as *middlings*, and further liberation from this fraction can only be achieved by further comminution. The "degree of liberation" refers to the percentage of the mineral occurring as free particles in the broken ore in relation to the total mineral content in locked and free particles. Liberation can be high if there are weak boundaries between mineral and gangue particles, which is often the case with ores composed mainly of rock forming minerals, particularly sedimentary minerals. This is sometimes referred to as "liberation by detachment." Usually, however, the adhesion between mineral and gangue is strong and during comminution the various constituents are cleft across the grain boundaries; that is, breakage is random. Random breakage produces a significant amount of middlings. Approaches to increasing the degree of liberation involve directing the breaking stresses at the mineral grain boundaries, so that the rock can be broken without breaking the mineral grains (Wills and Atkinson, 1993). For example, microwaves can be used, which cause differential heating among the constituent minerals and thus create stress fractures at grain boundaries (Kingman et al., 2004).

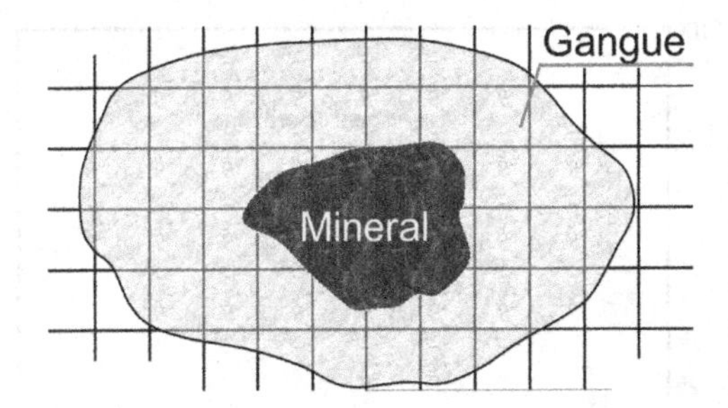

FIGURE 1.1 Mineral locked with gangue and illustrating effect of breakage on liberation.

Many researchers have tried to quantify (model) the degree of liberation (Barbery, 1991; King, 2012). These models, however, suffer from many unrealistic assumptions that must be made with respect to the grain structure of the minerals in the ore. For this reason liberation models have not found much practical application. However, some fresh approaches by Gay (2004a,b) have demonstrated that there may yet be a useful role for such models.

Figure 1.2 shows predictions using the simple liberation model based on random breakage derived by Gaudin (1939), but which is sufficient to introduce an important practical point. The degree (fraction) of liberation is given as a function of the particle size to grain size ratio and illustrates that to achieve high liberation, say 75%, the particle size has to be much smaller than the grain size, in this case ca. 1/10th the size. So, for example, if the grain size is 1 mm then size reduction must produce a particle size at least 0.1 mm (100 μm) or less, and if the grain size is 0.1 mm the particle size should be 10 μm or less. This result helps understand the fine size required from the comminution process. For example, in processing base metal sulfides a target grind size of 100 μm for adequate liberation was common in the 1960s but the finer grained ores of today may require a target grind size of 10 μm, which in turn has driven the development of new grinding technologies.

The quantification of liberation is now routine using the dedicated scanning electron microscope systems, for

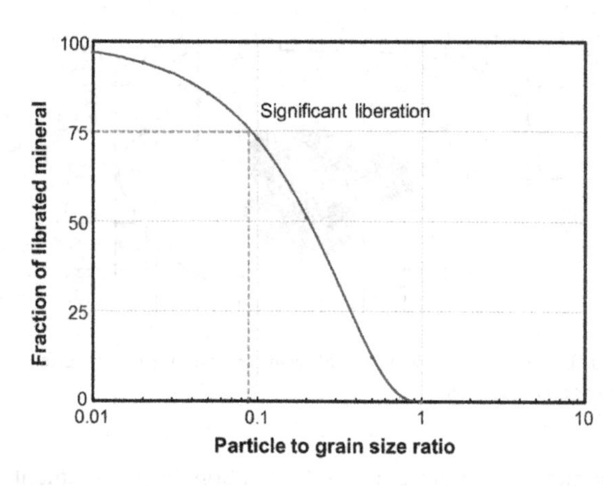

FIGURE 1.2 Degree of liberation as a function of the particle to grain ratio *(Derived from model of Gaudin (1939) for the least abundant mineral).*

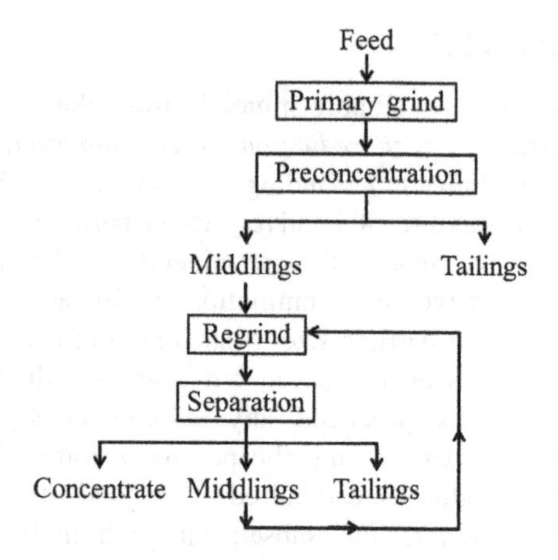

FIGURE 1.4 Flowsheet for process utilizing two-stage separation.

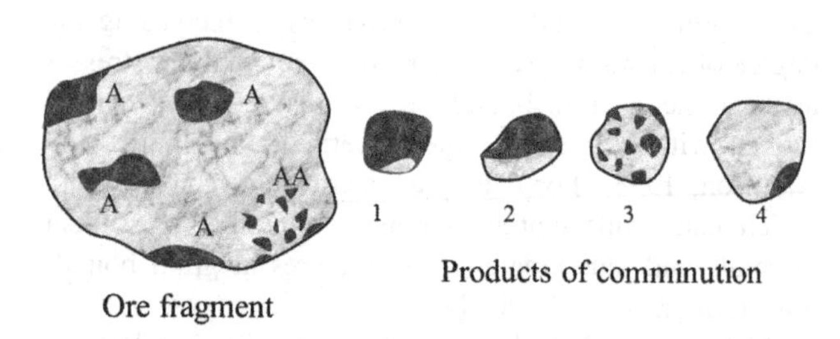

FIGURE 1.3 Example cross sections of ore particles.

example, and concentrators are increasingly using such systems to monitor the liberation in their processes.

It should be noted that a high degree of liberation is not always necessary and may be undesirable in some cases. For instance, it is possible to achieve a high recovery of values by gravity and magnetic separation even though the valuable minerals are completely enclosed by gangue, and hence the degree of liberation of the values is zero. As long as a pronounced density or magnetic susceptibility difference is apparent between the locked particles and the free gangue particles, the separation is possible. On the other hand, flotation requires at least a surface of the valuable mineral to be exposed.

Figure 1.3 is a cross section through a typical ore particle and illustrates the liberation dilemma often facing the mineral processor. Regions A represent valuable mineral, and region AA is rich in valuable mineral but is highly intergrown with the gangue mineral. Comminution produces a range of fragments, ranging from fully liberated mineral and gangue particles, to those illustrated. Particles of type 1 are rich in mineral (are high-grade particles) and are classed as concentrate as they have an acceptably low-level degree of locking with the gangue to still make a saleable concentrate grade. Particles of type 4 would be classed as tailings, the small amount of mineral present representing an acceptable loss of mineral. Particles of types 2 and 3 would probably be classed as middlings, although the degree of regrinding needed to promote liberation of mineral from particle 3 would be greater than in particle 2. In practice, ores are ground to an optimum grind size, determined by laboratory and pilot scale testwork, to produce an economic degree of liberation. The concentration process is then designed to produce a concentrate consisting predominantly of valuable mineral, with an accepted degree of locking with the gangue minerals, and a middlings fraction, which may require further grinding to promote optimum release of the minerals. The tailings should be mainly composed of gangue minerals.

During the grinding of a low-grade ore, the bulk of the gangue minerals is often liberated at a relatively coarse size (see Figure 1.1). In certain circumstances it may be economic to grind to a size much coarser than the optimum in order to produce, in the subsequent concentration process, a large middlings fraction and tailings which can be discarded at a coarse grain size (Figure 1.4). The middlings fraction can then be reground to produce a feed to the final concentration process. This method discards most of the coarse gangue early in the process, thus considerably reducing grinding costs.

An intimate knowledge of the mineralogical assembly of the ore ("texture") is essential if efficient processing is to be carried out. *Process mineralogy* or *applied mineralogy* thus becomes an important tool for the mineral processor.

Texture refers to the size, dissemination, association, and shape of the minerals within the ore. Processing of the minerals should always be considered in the context of the mineralogy of the ore in order to predict grinding and concentration requirements, feasible concentrate grades, and potential difficulties of separation (Guerney et al., 2003; Baum et al., 2004; Hoal et al., 2009; Evans et al., 2011; Lotter, 2011; Smythe et al., 2013).

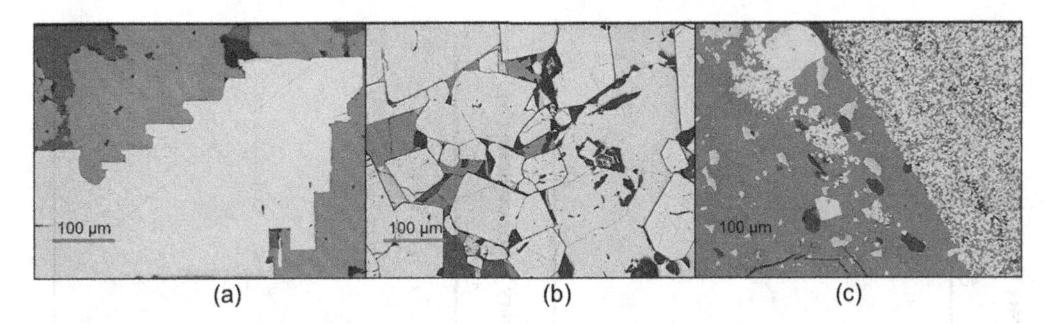

FIGURE 1.5 Micrographs showing range in ore texture: (a) galena/sphalerite (Pine Point, Canada), (b) sphalerite/chalcopyrite (Geco, Canada), and (c) galena/sphalerite/pyrite (Mt Isa, Australia) *(Courtesy Giovanni Di Prisco).*

Microscopic analysis of ores, concentrates, and tailings yields much valuable information regarding the efficiency of the liberation and concentration processes. Figure 1.5 shows examples of increasing ore complexity, from simple ("free milling" ore, Figure 1.5(a)) to fine-grained inter-grown texture (Figure 1.5(c)). Microscopic analysis is particularly useful in troubleshooting problems that arise from inadequate liberation. Conventional optical microscopes can be used for the examination of thin and polished sections of mineral samples, and in mineral sands applications the simple binocular microscope is a practical tool. However, it is now increasingly common to employ quantitative automated mineral analysis using scanning electron microscopy, such as the Mineral Liberation Analyser (MLA) (Gu, 2003), the QEMSCAN (Gottlieb et al., 2000), and the Tescan Integrated Mineral Analyser (TIMA) (Motl et al., 2012), which scan polished sections to give 2D information, and X-ray microcomputed tomography (micro CT) that allows for 3D visualization of particulates (Lin et al., 2013).

1.7 CONCENTRATION

After the valuable mineral particles have been liberated, they must be separated from the gangue particles. This is done by exploiting the physical properties of the different minerals. The most important physical properties which are used to concentrate ores are:

1. Optical. This is often called *sorting*, which used to be done by hand but is now mostly accomplished by machine (Chapter 14).
2. Density. *Gravity concentration*, a technology with its roots in antiquity, is based on the differential movement of mineral particles in water due to their different density and hydraulic properties. The method has seen development of a range of gravity concentrating devices, and the potential to treat dry to reduce reliance on often scarce water resources (Chapter 10). In *dense medium separation*, particles sink or float in a dense liquid or (more usually) an artificial dense suspension. It is widely used in coal beneficiation, iron ore and diamond processing, and in the preconcentration of some metalliferous ores (Chapter 11).
3. Surface properties. *Froth flotation* (or simply "flotation"), which is the most versatile method of concentration, is effected by the attachment of the mineral particles to air bubbles within an agitated pulp. By adjusting the "chemistry" of the pulp by adding various chemical reagents, it is possible to make the valuable minerals water-repellant (*hydrophobic*) and the gangue minerals water-avid (*hydrophilic*). This results in separation by transfer of the valuable minerals to the bubbles which rise to form froth on the surface of the pulp (Chapter 12).
4. Magnetic susceptibility. *Low-intensity magnetic separators* can be used to concentrate strongly magnetic minerals such as magnetite (Fe_3O_4) and pyrrhotite (Fe_7S_8), while *high-intensity magnetic separators* are used to recover weakly magnetic minerals. Magnetic separation is an important process in the beneficiation of iron ores and finds application in the processing of nonmetallic minerals, such as those found in mineral sand deposits (Chapter 13).
5. Electrical conductivity. *Electrostatic separation* can be used to separate conducting minerals from nonconducting minerals. Theoretically this method represents the "universal" concentrating method; virtually all minerals show some difference in conductivity and it should be possible to separate almost any two by this process. However, the method has fairly limited application, and its greatest use is in separating some of the minerals found in heavy sands from beach or stream placers. Minerals must be completely dry and the humidity of the surrounding air must be regulated, since most of the electron movement in dielectrics takes place on the surface and a film of moisture can change the behavior completely. The low capacity of economically sized units is stimulating developments to overcome (Chapter 13).

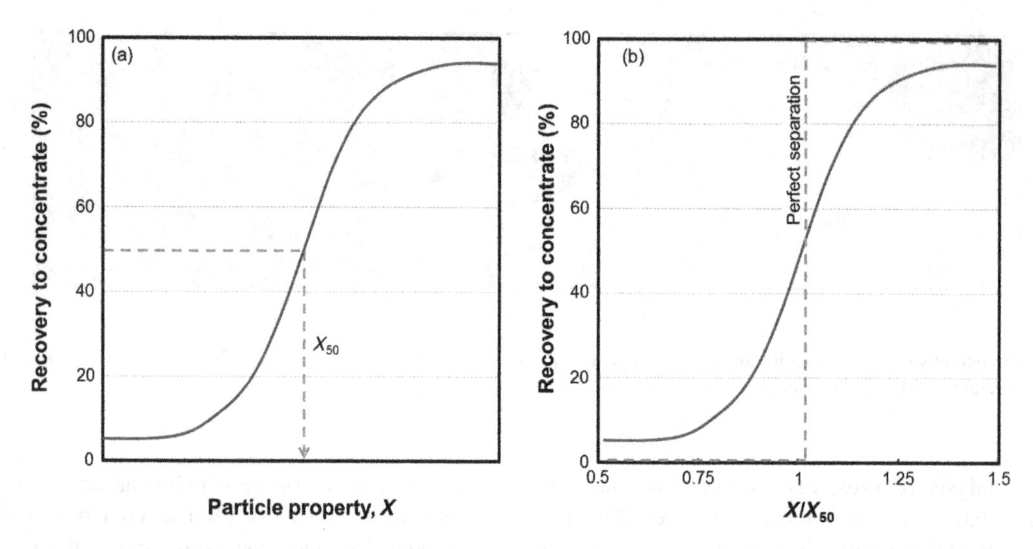

FIGURE 1.6 (a) General representation of a physical (mineral) separation process: mineral property can be density, magnetic susceptibility, hydrophobicity, size, etc., and (b) same as (a) but with property made dimensionless.

A general way to show separation is to represent as a recovery to one stream, usually the concentrate, as a function of some mineral property, variously called an *efficiency*, *performance*, or *partition* curve, as illustrated in Figure 1.6(a). The property can be density, magnetic susceptibility, some measure of hydrophobicity, or particle size (in size separation devices). The plot can be made dimensionless by dividing the property X by X_{50}, the property corresponding to 50% recovery (Figure 1.6(b)). This is a *normalized* or *reduced efficiency* curve. Treating X_{50} as the target property for separation then the ideal or perfect separation is the dashed line in Figure 1.6(b) passing through $X/X_{50} = 1$.

The size of particle is an important consideration in mineral separation. Figure 1.7 shows the general size range of efficient separation of the concentration processes introduced above. It is evident that all these physical-based techniques fail as the particle size reduces. Extending the particle size range drives innovation.

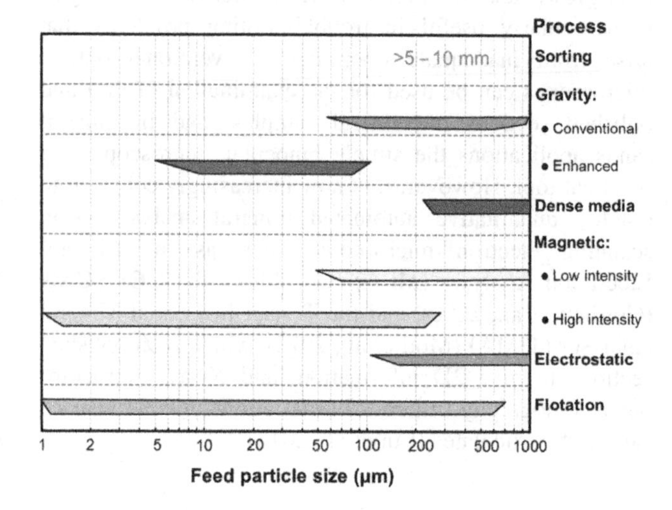

FIGURE 1.7 Effective range of application of selected mineral separation techniques *(Adapted from Mills (1978)).*

In many cases, a combination of two or more separation techniques is necessary to concentrate an ore economically. Gravity separation, for instance, may be used to reject a major portion of the gangue, as it is a relatively cheap process. It may not, however, have the selectivity to produce the final clean concentrate. Gravity concentrates therefore often need further upgrading by more expensive techniques, such as flotation. Magnetic separation can be integrated with flotation—for example, to reject pyrrhotite in processing some Ni-sulfide ores.

Ores which are very difficult to treat (*refractory*), due to fine dissemination of the minerals, complex mineralogy, or both, have driven technological advances. An example is the zinc–lead–silver deposit at McArthur River, in Australia. Discovered in 1955, it is one of the world's largest zinc–lead deposits, but for 35 years it resisted attempts to find an economic processing route due to the very fine grained texture of the ore. However, the development of the proprietary IsaMill fine grinding technology (Pease, 2005) by the mine's owners Mount Isa Mines, together with an appropriate flotation circuit, allowed the ore to be processed and the mine was finally opened in 1995. The concentrator makes a *bulk* (i.e., combined) zinc–lead concentrate with a very fine product size of 80% (by weight) finer than 7 μm.

Chemical methods can be used to alter mineralogy, allowing the low-cost mineral processing methods to be applied to refractory ores (Iwasaki and Prasad, 1989). For instance, nonmagnetic iron oxides can be roasted in a weakly reducing atmosphere to produce magnetite. In Vale's matte separation process mineral processing (comminution and flotation) is used to separate Ni–Cu matte into separate Cu- and Ni-concentrates which are sent for metal extraction (Damjanovic and Goode, 2000).

Some refractory copper ores containing sulfide and oxidized minerals have been pretreated hydrometallurgically to enhance flotation performance. In the Leach-Precipitation-Flotation process, developed in the years 1929–1934 by the Miami Copper Co., USA, the oxidized minerals are dissolved in sulfuric acid, after which the copper in solution is precipitated as *cement copper* by the addition of metallic iron. The cement copper and acid-insoluble sulfide minerals are then recovered by flotation. This process, with several variations, has been used at a number of American copper concentrators. A more widely used method of enhancing the flotation performance of oxidized ores is to allow the surface to react with sodium sulfide. This "sulfidization" process modifies the flotation response of the mineral causing it to behave, in effect, as a pseudo-sulfide (Chapter 12).

Developments in biotechnology are being exploited in hydrometallurgical operations, particularly in the bacterial oxidation of sulfide gold ores and concentrates (Brierley and Brierley, 2001; Hansford and Vargas, 2001). There is evidence to suggest that certain microorganisms could be used to enhance the performance of conventional mineral processing techniques (Smith et al., 1991). It has been established that some bacteria will act as pyrite depressants in coal flotation, and work has shown that certain organisms can aid flotation in other ways (e.g., Botero et al., 2008). Microorganisms have the potential to profoundly change future industrial flotation practice.

Extremely fine mineral dissemination leads to high energy costs in comminution and losses to tailings due to the generation of difficult-to-treat fine particles. Much research has been directed at minimizing fine mineral losses, either by developing methods of enhancing mineral liberation, thus minimizing the amount of comminution needed, or by increasing the efficiency of conventional physical separation processes, by the use of innovative machines or by optimizing the performance of existing ones. Several methods have been proposed to increase the apparent size of fine particles, by causing them to come together and aggregate. *Selective flocculation* of certain minerals in suspension, followed by separation of the aggregates from the dispersion, has been achieved on a variety of ore-types at laboratory scale, but plant application is limited (Chapter 12).

1.8 REPRESENTING MINERAL PROCESSING SYSTEMS: THE FLOWSHEET

The flowsheet shows diagrammatically the sequence of operations in the plant. In its simplest form it can be presented as a block diagram in which all operations of similar character are grouped (Figure 1.8). In this case, "comminution" deals with all crushing and grinding. The next block, "separation," groups the various treatments incident to production of concentrate and tailing. The third, "product handling," covers the shipment of concentrates and disposal of tailings.

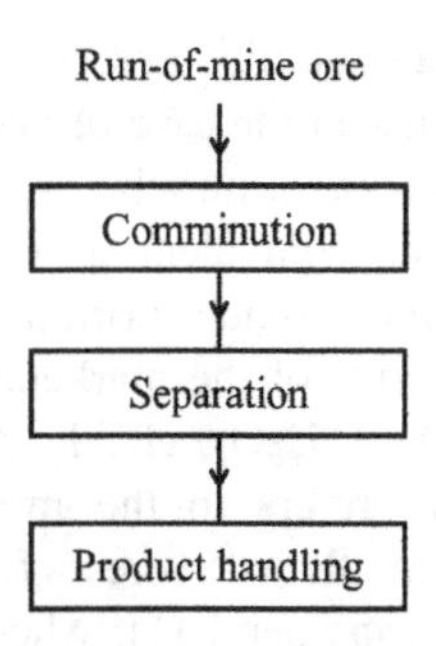

FIGURE 1.8 Simple block flowsheet.

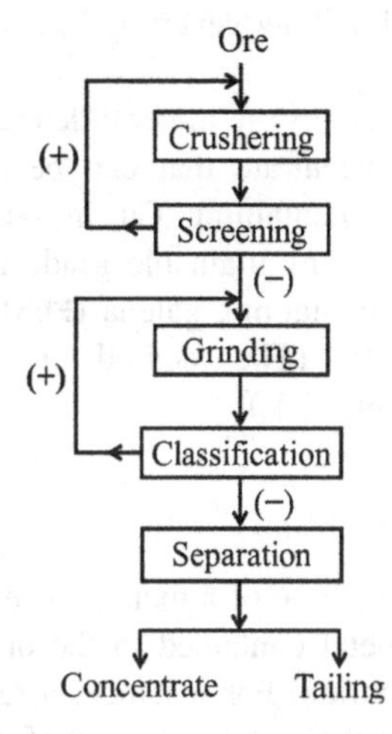

FIGURE 1.9 Line flowsheet: (+) indicates oversized material returned for further treatment and (−) undersized material, which is allowed to proceed to the next stage.

Expanding, a simple line flowsheet (Figure 1.9) can be sufficient and can include details of machines, settings, rates, etc. Most flowsheets today use symbols to represent the unit operations. Example flowsheets are given in many of the chapters, with varying degrees of sophistication.

1.9 MEASURES OF SEPARATION

The object of mineral processing, regardless of the methods used, is always the same: to separate the minerals with the values in the concentrates, and the gangue in the tailings. The two most common measures of the separation are grade and recovery.

1.9.1 Grade

The *grade*, or *assay*, refers to the content of the marketable commodity in any stream, such as the feed and concentrate. In metallic ores, the percent metal is often quoted, although in the case of very low-grade ores, such as gold, metal content may be expressed as parts per million (ppm), or its equivalent grams per ton ($g\,t^{-1}$). Some metals are sold in oxide form, and hence the grade may be quoted in terms of the marketable oxide content: for example, $\%WO_3$, $\%U_3O_8$, etc. In nonmetallic operations, grade usually refers to the mineral content: for example, $\%CaF_2$ in fluorite ores. Diamond ores are usually graded in *carats* per 100 t, where 1 carat is 0.2 g. Coal is graded according to its *ash* content, that is, the amount of incombustible mineral present within the coal. Most coal burned in power stations ("steaming coal") has an ash content between 15% and 20%, whereas "coking coal" used in steel making generally has an ash content of less than 10%.

The metal content of the mineral determines the maximum grade of concentrate that can be produced. Thus processing an ore containing Cu in only chalcopyrite ($CuFeS_2$) the maximum attainable grade is 34.6%, while processing an ore containing galena (PbS), the maximum Pb grade is 86.6%. (The method of calculation was explained in Example 1.1.)

1.9.2 Recovery

The *recovery*, in the case of a metallic ore, is the percentage of the total metal contained in the ore that is recovered to the concentrate. For instance, a recovery of 90% means that 90% of the metal in the ore (feed) is recovered in the concentrate and 10% is lost in ("rejected" to) the tailings. The recovery, when dealing with nonmetallic ores, refers to the percentage of the total mineral contained in the ore that is recovered into the concentrate. In terms of the usual symbols recovery R is given by:

$$R = \frac{C\,c}{F\,f} \tag{1.1}$$

where C is weight of concentrate (or more precisely flowrate, e.g., $t\,h^{-1}$), F the weight of feed, c the grade (assay) of metal or mineral in the concentrate, and f the grade of metal/mineral in the feed. Provided the metal occurs in only one mineral, metal recovery is the same as the recovery of the associated mineral. (Example 1.2 shows how to deal with situations where an element resides in more than one mineral.) Metal assays are usually given as % but it is often easier to manipulate formulas if the assays are given as fractions (e.g., 10% becomes 0.1).

Related measures to grade and recovery include: the *ratio of concentration*, the ratio of the weight of the feed (or *heads*) to the weight of the concentrate (i.e., *F/C*); *weight recovery* (or *mass* or *solids recovery*) also known as *yield* is the inverse of the ratio of concentration, that is, the ratio of the weight of concentrate to weight of feed (*C/F*); *enrichment ratio*, the ratio of the grade of the concentrate to the grade of the feed (*c/f*). They are all used as measures of *metallurgical efficiency*.

1.9.3 Grade–Recovery Relationship

The grade of concentrate and recovery are the most common measures of metallurgical efficiency, and in order to evaluate a given operation it is necessary to know both. For example, it is possible to obtain a very high grade of concentrate (and ratio of concentration) by simply picking a few lumps of pure galena from a lead ore, but the recovery would be very low. On the other hand, a concentrating process might show a recovery of 99% of the metal, but it might also put 60% of the gangue minerals in the concentrate. It is, of course, possible to obtain 100% recovery by not concentrating the ore at all.

Grade of concentrate and recovery are generally inversely related: as recovery increases grade decreases and vice versa. If an attempt is made to attain a very high-grade concentrate, the tailings assays are higher and the recovery is low. If high recovery of metal is aimed for, there will be more gangue in the concentrate and the grade of concentrate and ratio of concentration will decrease. It is impossible to give representative values of recoveries and ratios of concentration. A concentration ratio of 2 to 1 might be satisfactory for certain high-grade nonmetallic ores, but a ratio of 50 to 1 might be considered too low for a low-grade copper ore, and ratios of concentration of several million to one are common with diamond ores. The aim of milling operations is to maintain the values of ratio of concentration and recovery as high as possible, all factors being considered.

Ultimately the separation is limited by the composition of the particles being separated and this underpins the inverse relationship. To illustrate the impact of particle composition, consider the six particle array in Figure 1.10 is separated in an ideal separator, that is, a

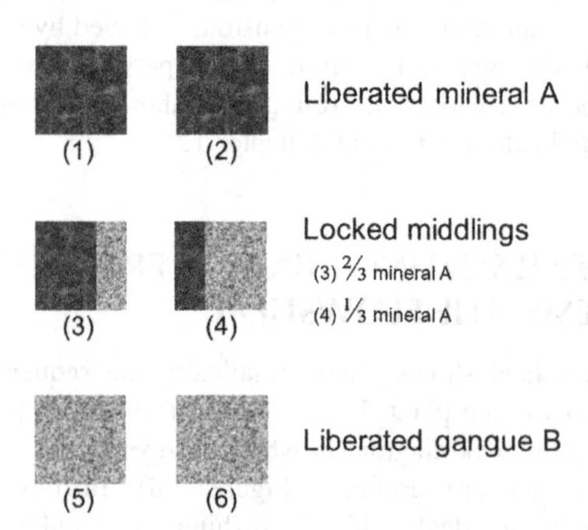

FIGURE 1.10 Example particle assemblage fed to perfect separator.

TABLE 1.3 Recovery of Particles in Figure 1.10 by Perfect Separator

Particle Recovered	1	2	3	4	5	6
Concentrate grade (%)	100 *(1/1)*[a]	100 *(2/2)*	89 *(2.66/3)*	75 *(3/4)*	60 *(3/5)*	50 *(3/6)*
Recovery (%)	33 *(1/3)*	66 *(2/3)*	89 *(2.66/3)*	100 *(3/3)*	100 *(3/3)*	100 *(3/3)*

[a] *Figure in parenthesis is the ratio of the mass of particles.*

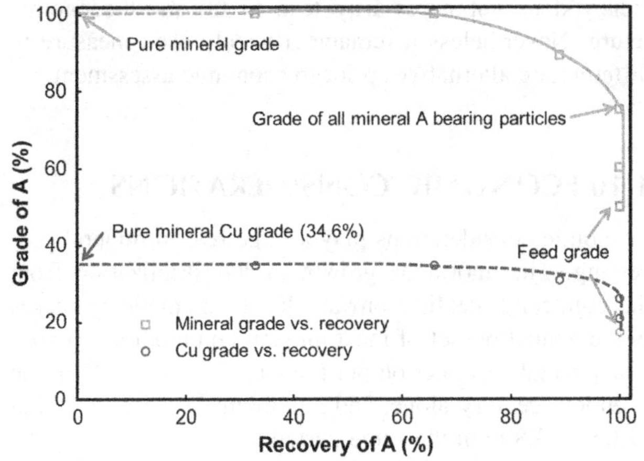

FIGURE 1.11 Grade–recovery curve for perfect separation of the six-particle assemblage in Figure 1.10 showing the inverse relationship (Cu grades based on chalcopyrite, $CuFeS_2$).

separator that recovers particles sequentially based on valuable mineral content. The six particles represent three mineral (A) equivalents and three gangue (B) equivalents (all of equal weight), thus the perfect separator would recover the particles in order 1 through 6. The grade of A and the recovery of A after each particle is separated into the concentrate is given in Table 1.3 and plotted in Figure 1.11 (which also shows the conversion from mineral grade to metal grade assuming the valuable mineral in this case is chalcopyrite, $CuFeS_2$). The inverse relationship is a consequence of the distribution of particle composition. The figure includes reference to other useful features of the grade–recovery relationship: the grade of all mineral A-containing particles, the pure mineral grade approached as recovery goes to zero, and the feed grade, which corresponds to recovery of all particles.

The grade–recovery corresponding to perfect separation is known as the *mineralogically-limited* or *liberation-limited* grade–recovery. For simple two component mixtures with well differentiated density, it is possible to approach perfect separation in the laboratory using a sequence of "heavy" liquids (e.g., some organics or concentrated salt solutions) of increasing density (Chapter 11). The liberation-limited curve can be generated from mineralogical data, essentially following the calculations used to generate Figure 1.11. The liberation-limited grade–recovery curve is used to compare against the actual operation and determine how far away it is from the theoretical maximum separation.

The grade–recovery curve is the most common representation of metallurgical performance. When assessing the effect of a process change, the resulting grade–recovery curves should be compared, not individual points, and differences between curves should be subjected to tests of statistical significance. To construct the curve, the data need to be collected in increments, such as increments of density or increments of flotation time, and then assembled into cumulative recovery and cumulative grade. There are other methods of presenting the data, such as recovery versus total solids recovery, and recovery of mineral A versus recovery of mineral B. Together these are sometimes referred to as "separability curves" and examples will be found in various chapters.

1.9.4 A Measure of Technical Separation Efficiency

There have been many attempts to combine recovery and concentrate grade into a single index defining the metallurgical efficiency of the separation. These have been reviewed by Schulz (1970), who proposed the following definition:

$$\text{Separation efficiency (SE)} = R_m - R_g \tag{1.2}$$

where R_m is the recovery of the valuable mineral and R_g is the recovery of the gangue into the concentrate.

Calculation of SE will be illustrated using Eq. (1.1). This equation applies equally well to the recovery of gangue, provided we know the gangue assays. To find the gangue assays, we must first convert the metal assay to mineral assay (see Example 1.1). For a concentrate metal grade of c if the metal content in the mineral is m then the mineral content is c/m and thus the gangue assay of the concentrate is given by (assays in fractions):

$$g = 1 - \frac{c}{m} \tag{1.3}$$

This calculation, of course, applies to any stream, not just the concentrate, and thus in the feed if f is the metal assay the mineral assay is f/m. The recovery of gangue (R_g) is therefore

$$R_g = \frac{C\left(1-\frac{c}{m}\right)}{F\left(1-\frac{f}{m}\right)} \rightarrow \frac{C(m-c)}{F(m-f)} \tag{1.4}$$

and thus the separation efficiency is:

$$\begin{aligned} SE &= \frac{Cc}{Ff} - \frac{C(m-c)}{F(m-f)} \rightarrow \frac{C}{F}\left(\frac{c}{f} - \frac{(m-c)}{(m-f)}\right) \\ &= \frac{C\,m\,(c-f)}{F\,f\,(m-f)} \end{aligned} \tag{1.5}$$

Example 1.3 illustrates the calculations. The concept can be extended to consider separation between any pair of minerals, A and B, such as in the separation of lead from zinc.

Example 1.3

A tin concentrator treats a feed containing 1% tin, and three possible combinations of concentrate grade and recovery are:

High grade	63% tin at 62% recovery
Medium grade	42% tin at 72% recovery
Low grade	21% tin at 78% recovery

Determine which of these combinations of grade and recovery produce the highest separation efficiency.

Solution

Assuming that the tin is totally contained in the mineral cassiterite (SnO_2) then $m = 78.6\%$ (0.786), and we can complete the calculation:

Case	$\frac{C}{F}$ (from Eq. (1.1))	SE (Eq. (1.5))
High grade	$0.62 = \frac{C}{F}\left(\frac{0.63}{0.01}\right)$	$SE = 9.841E^{-3}\left(\frac{0.63}{0.01} - \frac{0.786-0.63}{0.786-0.01}\right)$
	$\frac{C}{F} = 9.841E^{-3}$	$= 61.8\%$
Medium grade	$0.72 = \frac{C}{F}\left(\frac{0.42}{0.01}\right)$	$SE = 1.714E^{-2}\left(\frac{0.42}{0.01} - \frac{0.786-0.42}{0.786-0.01}\right)$
	$\frac{C}{F} = 1.714E^{-2}$	$= 71.2\%$
Low grade	$0.78 = \frac{C}{F}\left(\frac{0.21}{0.01}\right)$	$SE = 3.714E^{-2}\left(\frac{0.21}{0.01} - \frac{0.786-0.21}{0.786-0.01}\right)$
	$\frac{C}{F} = 3.714E^{-2}$	$= 75.2\%$

The answer to which gives the highest separation efficiency, therefore, is the low-grade case.

The concept of separation efficiency is examined in Appendix III, based on Jowett (1975) and Jowett and Sutherland (1985), where it is shown that defining separation efficiency as separation achieved relative to perfect separation yields Eq. (1.5).

Although separation efficiency can be useful in comparing the performance of different operating conditions, it takes no account of economic factors, and is sometimes referred to as the "technical separation efficiency." As will become apparent, a high value of separation efficiency does not necessarily lead to the most economic return. Nevertheless it remains a widely used measure to differentiate alternatives prior to economic assessment.

1.10 ECONOMIC CONSIDERATIONS

Economic considerations play a large role in mineral processing. The enormous growth of industrialization from the eighteenth century onward led to dramatic increases in the annual output of most mineral commodities, particularly metals. Copper output grew by a factor of 27 in the twentieth century alone, and aluminum by an astonishing factor of 3,800 in the same period. Figure 1.12 shows the world production of aluminum, copper, and zinc for the period 1900–2012 (USGS, 2014).

All of these metals suffered to a greater or lesser extent when the Organization of Petroleum Exporting Countries (OPEC) quadrupled the price of oil in 1973–1974, ending the great postwar industrial boom. The situation worsened in 1979–1981, when the Iranian revolution and then the Iran–Iraq war forced the price of oil up from $13 to nearly $40 a barrel, plunging the world into another and deeper recession. While in the mid-1980s a glut in the world's oil supply as North Sea oil production grew cut the price to below $15 in 1986, Iraq's invasion of Kuwait in 1990 pushed the price up

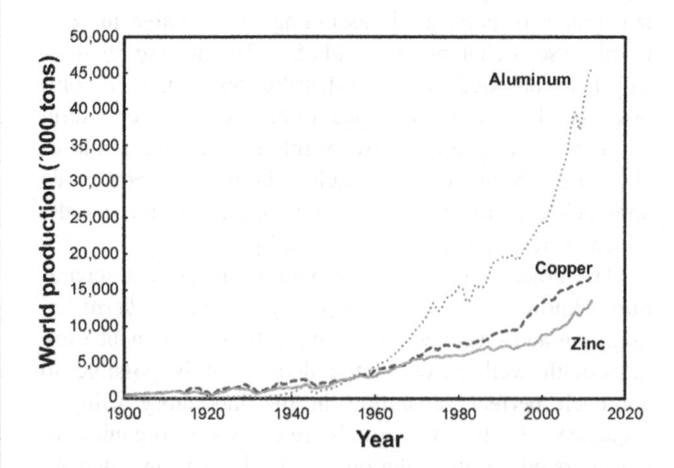

FIGURE 1.12 World production of aluminum, copper, and zinc for the period 1900–2012.

again, to a peak of $42 in October, although by then 20% of the world's energy was being provided by natural gas.

In 1999, overproduction and the Asian economic crisis depressed oil prices to as low as $10 a barrel from where it climbed steadily to a record $147 a barrel in 2008, driven by demand from the now surging Asian economies, particularly China. Turmoil in oil producing regions continued in the twenty-first century from the invasion of Iraq (2003) to the "Arab Spring" (start date 2009) and the festering stand-off with Russia over the Ukraine (2014). Over the past three years, the oil price fluctuated around $100 a barrel, apparently inured against this turmoil. In the last half of 2014, however, the price dropped precipitously to less than $50 a barrel as the world moved to a surplus of oil driven by decreased growth rate in countries like China, and a remarkable, and unexpected, increase in shale oil production in the United States. The decision by OPEC not to reduce production has accelerated the price decline. This will drive out the high-cost producers, including production from shale oil that needs about $60 a barrel to be profitable, and the price will climb.

These fluctuations in oil prices impact mining, due to their influence both on the world economy and thus the demand for metals, and directly on the energy costs of mining and processing. Metal and mineral commodities are thus likewise subject to cycles in price. Figure 1.13 shows the commodity price index and identifies the recent "super-cycle" starting about 2003, and the decline following that of oil during 2014. These "boom and bust" cycles are characterized by overoptimistic forecasts as prices rise (reference to the "super-cycle") and dire warnings about exporting countries suffering from the "natural resource disease" on the price downslide.

The cycles spur innovation. To reduce dependence on OPEC, oil production turned to "nonconventional" sources, such as the oil sands in Canada. A radical nonconventional resource, methane chlorates, appeared set to make Siberia the next Saudi Arabia and although interest has waned recently these deposits remain part of our energy future.

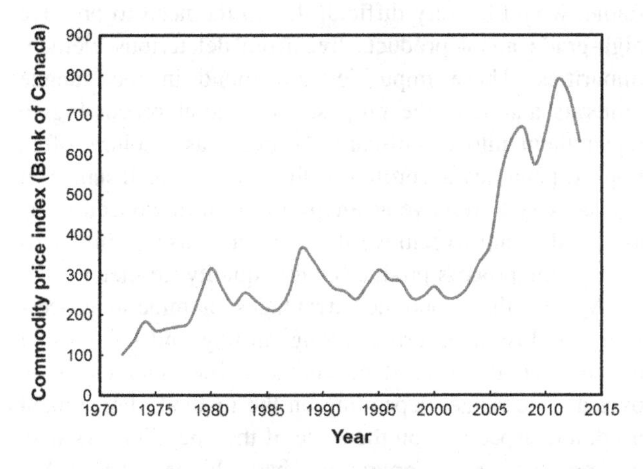

FIGURE 1.13 Commodity price index (Bank of Canada) representing the cost of nine commodities (potash, aluminum, gold, nickel, iron, copper, silver, zinc, and lead).

The most dramatic new nonconventional source, however, comes from the shale oil deposits in the United States. The vast volumes of natural gas in shale formations in the United States were well known by the 1990s and attempts to crack open the shale in vertical wells by injecting sand, water, and chemicals, known as *hydraulic fracturing* or "fracking," were underway but had proved unprofitable. The breakthrough innovation was to drill horizontally rather than vertically and thus expose thousands of meters of gas-bearing shale rather than just a few tens of meters (Hefner, 2014). The same technology also released reservoirs of oil in shale and other "tight," that is, impermeable, rock formations. As a consequence, the United States could become the world's largest oil producer by the end of the decade, a revolution with impacts both economic and geopolitical. While other large such shale formations are known throughout the world, environmental concerns over groundwater contamination has, for the moment, cooled the spread of the technology.

The cycles stimulate substitution. In the case of oil, electricity generation turned to natural gas (including liquefied natural gas, LNG) and reconsidered nuclear, while sustaining a large role for coal which remains in plentiful supply and is more generously distributed among the advanced economies. Solar-, wind-, and tidal-generated sources are starting to contribute to the electric power grid. Subsidies, carbon taxing, and cap-and-trade (emissions trading) initiatives spur these "green" alternatives. For transport, we are witnessing a resurgence of interest in electric-powered vehicles. Another energy innovation is "cogeneration," the capture and distribution of waste heat to heat buildings. Mining operations are using this concept, for example, by tapping the natural heating underground and examining the potential to recover the heat generated in comminution (Radziszewski, 2013). The demand to limit greenhouse gas emissions to combat climate change will drive substitution of carbon-based energy sources.

Certain metals and minerals likewise face competition from substitutes and some face additional factors. Avoiding "conflict minerals" influences the choice of supply source. Some companies make an advert of not using commodities seen as environmentally harmful. Lifecycle analysis may influence which commodities are used and in what combinations in order to facilitate recycling.

While commodity prices have gone up in current dollars since 2000 (Figure 1.13), the capital cost of mining projects has increased by 200–300% (Thomas et al., 2014) over the same period. This is due to increased equipment and construction costs (competition for materials and labor), increased environmental regulations, and the added complexity of constructing mine sites in remote areas. While the recent high metal prices attracted interest in previously uneconomic deposits, the project cost escalation has put pressure on the mines to optimize performance and has caused investors to be wary.

1.10.1 Contained Value

Every ton of material in the deposit has a certain contained value that is dependent on the metal content and current price of the contained metal. For instance, at a copper price of £2,000 t^{-1} and a molybdenum price of £18 kg^{-1}, a deposit containing 1% copper and 0.015% molybdenum has a contained value of more than £22 t^{-1}. To be processed economically, the cost of processing an ore must be less than the contained value.

1.10.2 Processing Costs

Mining is a major cost, and this can vary from only a few pence per ton of ore to well over £50 t^{-1}. High-tonnage operations are cheaper in terms of operating costs but have higher initial capital costs. These capital costs are paid off over a number of years, so that high-tonnage operations can only be justified for the treatment of deposits large enough to allow this. Small ore bodies are worked on a smaller scale to reduce overall capital costs, but capital and operating costs per ton are correspondingly higher (Ottley, 1991).

Alluvial mining is the cheapest method and, if on a large scale, can be used to mine ores of low contained value due to low grade or low metal price, or both. For instance, in S.E. Asia, tin ores containing as little as 0.01% Sn are mined by alluvial methods. These ores have a contained value of less than £1 t^{-1}, but very low processing costs allow them to be economically worked.

High-tonnage open-pit and underground block-caving methods are used to treat ores of low contained value, such as low-grade copper ores. Where the ore must be mined selectively, however, as is the case with underground vein-type deposits, mining methods become very expensive, and can only be justified on ores of high contained value. An underground selective mining cost of £30 t^{-1} would obviously be hopelessly uneconomic on a tin ore of alluvial grade, but may be economic on a hard-rock ore containing 1.5% tin, with a contained value of around £50 t^{-1} ore.

In order to produce metals, the ore minerals must be broken down by the action of heat (pyrometallurgy), solvents (hydrometallurgy), or electricity (electrometallurgy), either alone or in combination. The most common method is the pyrometallurgical process of smelting. These chemical methods consume large quantities of energy. The smelting of 1 t of copper ore, for instance, consumes in the region of 1,500–2,000 kWh of electrical energy, which at a cost of say 5 p kWh^{-1} is around £85 t^{-1}, well above the contained value of most copper ores. In addition, smelters are often remote from the mine site, and thus the cost of transport to the site must be considered.

The essential economic purpose of mineral processing is to reduce the bulk of the ore which must be transported to and processed by the smelter, by using relatively cheap, low-energy physical methods to separate the valuable minerals from the gangue minerals. This enrichment process considerably increases the contained value of the ore to allow economic transportation and smelting. Mineral processing is usually carried out at the mine site, the plant being referred to as a *mill* or *concentrator*.

Compared with chemical methods, the physical methods used in mineral processing consume relatively small amounts of energy. For instance, to upgrade a copper ore from 1% to 25% metal would use in the region of 20–50 kWh t^{-1}. The corresponding reduction in weight of around 25:1 proportionally lowers transport costs and reduces smelter energy consumption to around 60–80 kWh in relation to a ton of mined ore. It is important to realize that, although the physical methods are relatively low energy consumers, the reduction in bulk lowers smelter energy consumption to the order of that used in mineral processing. It is significant that as ore grades decline, the energy used in mineral processing becomes an important factor in deciding whether the deposit is viable to exploit or not.

Mineral processing reduces not only smelter energy costs but also smelter metal losses, due to the production of less metal-bearing slag. (Some smelters include a slag mineral processing stage to recover values and recycle to the smelter.) Although technically feasible, the smelting of low-grade ores, apart from being economically unjustifiable, would be very difficult due to the need to produce high-grade metal products free from deleterious element impurities. These impurities are found in the gangue minerals and it is the purpose of mineral processing to reject them into the discard (tailings), as smelters often impose penalties according to their level. For instance, it is necessary to remove arsenopyrite from tin concentrates, as it is difficult to remove the contained arsenic in smelting and the process produces a low-quality tin metal.

Against the economic advantages of mineral processing, the losses occurred during milling and the cost of milling operations must be charged. The latter can vary over a wide range, depending on the method of treatment used, and especially on the scale of the operation. As with mining, large-scale operations have higher capital but lower operating costs (particularly labor and energy) than small-scale operations.

Losses to tailings are one of the most important factors in deciding whether a deposit is viable or not. Losses will depend on the ore mineralogy and dissemination of the minerals, and on the technology available to achieve efficient concentration. Thus, the development of flotation allowed the exploitation of the vast low-grade porphyry copper deposits which were previously uneconomic to treat (Lynch et al., 2007). Similarly, the introduction of solvent extraction enabled Nchanga Consolidated Copper Mines in Zambia to treat 9 Mt per year of flotation tailings, to produce 80,000 t of finished copper from what was previously regarded as waste (Anon., 1979).

In many cases not only is it necessary to separate valuable from gangue minerals, but also to separate valuable minerals from each other. For instance, complex sulfide ores containing economic amounts of copper, lead, and zinc usually require separate concentrates of the minerals of each of these metals. The provision of clean concentrates, with little contamination with associated metals, is not always economically feasible, and this leads to another source of loss other than direct tailing loss. A metal which reports to the "wrong" concentrate may be difficult, or economically impossible, to recover and never achieves its potential valuation. Lead, for example, is essentially irrecoverable in copper concentrates and is often penalized as an impurity by the copper smelter. The treatment of such *polymetallic* base metal ores, therefore, presents one of the great challenges to the mineral processor.

Mineral processing operations are often a compromise between improvements in metallurgical efficiency and milling costs. This is particularly true with ores of low contained value, where low milling costs are essential and cheap unit processes are necessary, particularly in the early stages, where the volume of material treated is high. With such low-value ores, improvements in metallurgical efficiency by the use of more expensive methods or reagents cannot always be justified. Conversely, high metallurgical efficiency is usually of most importance with ores of high contained value and expensive high-efficiency processes can often be justified on these ores.

Apart from processing costs and losses, other costs which must be taken into account are indirect costs such as ancillary services—power supply, water, roads, tailings disposal—which will depend much on the size and location of the deposit, as well as taxes, royalty payments, investment requirements, research and development, medical and safety costs, etc.

1.10.3 Milling Costs

As discussed, the balance between milling costs and metal losses is crucial, particularly with low-grade ores, and because of this most mills keep detailed accounts of operating and maintenance costs, broken down into various subdivisions, such as labor, supplies, energy, etc. for the various areas of the plant. This type of analysis is used to identify high-cost areas where improvements in performance would be most beneficial. It is impossible to give typical operating costs for milling operations, as these vary considerably from mine to mine, and particularly from country to country, depending on local costs of energy, labor, water, supplies, etc.

TABLE 1.4 Relative Costs for a 100,000 t d^{-1} Copper Concentrator

Item	Cost (%)
Crushing	2.8
Grinding	47.0
Flotation	16.2
Thickening	3.5
Filtration	2.8
Tailings	5.1
Reagents	0.5
Pipeline	1.4
Water	8.0
Laboratory	1.5
Maintenance support	0.8
Management support	1.6
Administration	0.6
Other expenses	8.1
Total	100

Table 1.4 is an approximate breakdown of costs for a 100,000 t d^{-1} copper concentrator. Note the dominance of grinding, due mainly to power requirements.

1.10.4 Tailings Reprocessing and Recycling

Mill tailings which still contain valuable components constitute a potential future resource. New or improved technologies can allow the value contained in tailings, which was lost in earlier processing, to be recovered, or commodities considered waste in the past to become valuable in a new economic order. Reducing or eliminating tailings dumps by retreating them also reduces the environmental impact of the waste.

The cost of tailings retreatment is sometimes lower than that of processing the original ore, because much of the expense has already been met, particularly in mining and comminution. There are many tailings retreatment plants in a variety of applications around the world. The East Rand Gold and Uranium Company closed its operations in 2005 after 28 years having retreated over 870 Mt of the iconic gold dumps of Johannesburg, significantly modifying the skyline of the Golden City and producing 250 t of gold in the process. Also in 2005, underground mining in Kimberley closed, leaving a tailings dump retreatment operation as the only source of diamond production in the Diamond City. Some platinum producers in South Africa now operate tailings retreatment plants for the recovery of platinum group metals (PGMs), and also chromite as a by-product from the chrome-rich UG2 Reef. The tailings of the historic Timmins gold mining area of Canada are likewise being reprocessed.

Although these products, particularly gold, tend to dominate the list of tailings retreatment operations because of the value of the product, there are others, both operating and being considered as potential major sources of particular commodities. For example: coal has been recovered from tailings in Australia (Clark, 1997), uranium is recovered from copper tailings by the Uranium Corporation of India, and copper has been recovered from the Bwana Mkubwa tailings in Zambia, using solvent extraction and electrowinning. The Kolwezi Tailings project in the DRC (Democratic Republic of Congo) that recovered oxide copper and cobalt from the tailings of 50 years of copper mining ran from 2004 to 2009. Phytomining, the use of plants to accumulate metals, could be a low-cost way to detoxify tailings (and other sites) and recover metals. Methods of resource recovery from metallurgical wastes are described by Rao (2006).

Recovery from tailings is a form of recycling. The reprocessing of industrial scrap and domestic waste for metal recycling is a growing economic, and environmental, activity. "Urban ore" is a reference to forgotten supplies of metals that lie in and under city streets. Recovery of metals from electronic scrap is one example; another is recovery of PGMs that accumulate in road dust as car catalytic converters wear (Ravilious, 2013). Many mineral separation techniques are applicable to processing urban ores but tapping them is held back by their being so widely spread in the urban environment. The principles of recycling are comprehensively reviewed by Reuter et al. (2005).

1.10.5 Net Smelter Return and Economic Efficiency

Since the purpose of mineral processing is to increase the economic value of the ore, the importance of the grade–recovery relationship is in determining the most economic combination of grade and recovery that will produce the greatest financial return per ton of ore treated in the plant. This will depend primarily on the current price of the valuable product, transportation costs to the smelter, refinery, or other further treatment plant, and the cost of such further treatment, the latter being very dependent on the grade of concentrate supplied. A high-grade concentrate will incur lower smelting costs, but the associated lower recovery means lower returns of final product. A low-grade concentrate may achieve greater recovery of the values, but incur greater smelting and transportation costs due to the included gangue minerals. Also of importance are impurities in the concentrate which may be penalized by the smelter, although precious metals may produce a bonus.

The net return from the smelter (NSR) can be calculated for any grade–recovery combination from:

$$\text{NSR} = \text{Payment for contained metal} - (\text{Smelter charges} + \text{Transport costs}) \tag{1.6}$$

This is summarized in Figure 1.14, which shows that the highest value of NSR is produced at an optimum concentrate grade. It is essential that the mill achieves a concentrate grade that is as close as possible to this target grade. Although the effect of moving slightly away from the optimum may only be of the order of a few pence per ton treated, this can amount to very large financial losses, particularly on high-capacity plants treating thousands of tons per day. Changes in metal price, smelter terms, etc. obviously affect the NSR versus concentrate grade relationship and the value of the optimum concentrate grade. For instance, if the metal price increases, then the optimum grade will be lower, allowing higher recoveries to be attained (Figure 1.15).

It is evident that the terms agreed between the concentrator and the smelter are of paramount importance in the

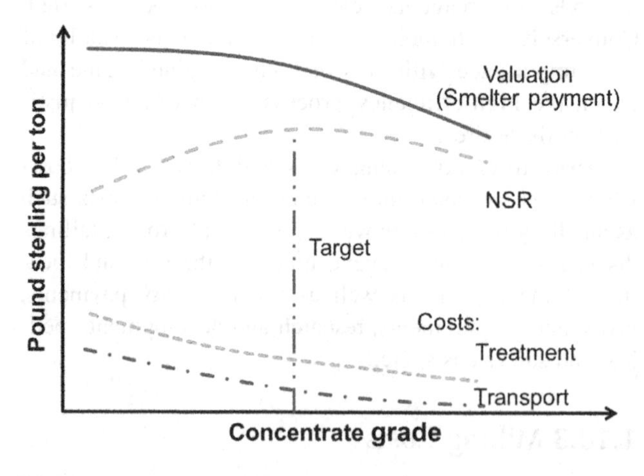

FIGURE 1.14 Variation of payment and charges (costs) with concentrate grade.

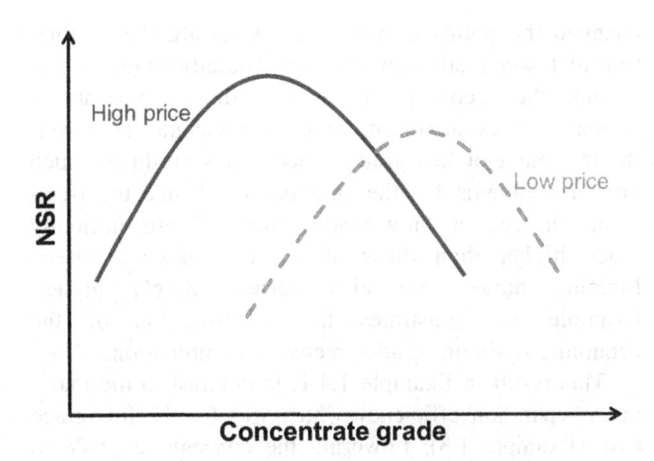

FIGURE 1.15 Effect of metal price on NSR–grade relationship.

TABLE 1.5 Simplified Tin Smelter Contract

Material	Tin concentrates, assaying no less than 15% Sn, to be free from deleterious impurities not stated and to contain sufficient moisture as to evolve no dust when unloaded at our works.
Quantity	Total production of concentrates.
Valuation	Tin, less 1 unit per dry ton of concentrates, at the lowest of the official LME prices.
Pricing	On the seventh market day after completion of arrival of each sampling lot into our works.
Treatment charge	£385 per dry ton of concentrates.
Moisture	£24 t^{-1} of moisture.
Penalties	Arsenic £40 per unit per ton.
Lot charge	£175 per lot sampled of less than 17 t.
Delivery	Free to our works in regular quantities, loose on a tipping lorry (truck) or in any other manner acceptable to both parties.

economics of mining and milling operations. Such *smelter contracts* are usually fairly complex. Concentrates are sold under contract to "custom smelters" at prices based on quotations on metal markets such as the London Metal Exchange (LME). The smelter, having processed the concentrates, disposes of the finished metal to the consumers. The proportion of the "free market" price of the metal received by the mine is determined by the terms of the contract negotiated between mine and smelter, and these terms can vary widely. Table 1.5 summarizes a typical low-grade smelter contract for the purchase of tin concentrates. As is usual in many contracts, one assay unit (1%) is deducted from the concentrate assay in assessing the value of the concentrates, and arsenic present in the concentrate (in this case) is penalized. The concentrate assay is of prime importance in determining the valuation, and the value of the assay is usually agreed on the result of independent sampling and assaying performed by the mine and smelter. The assays are compared, and if the difference is no more than an agreed value, the mean of the two results may be taken as the agreed assay. In the case of a greater difference, an "umpire" sample is assayed at an independent laboratory. This umpire assay may be used as the agreed assay, or the mean of this assay and that of the party which is nearer to the umpire assay may be chosen.

The use of smelter contracts, and the importance of the by-products and changing metal prices, can be seen by briefly examining the economics of processing two base metals—tin and copper—whose fortunes have fluctuated over the years for markedly different reasons.

1.10.6 Case Study: Economics of Tin Processing

Tin constitutes an interesting case study in the vagaries of commodity prices and how they impact the mineral industry and its technologies. Almost half the world's supply of tin in the mid-nineteenth century was mined in south-west England, but by the end of the 1870s Britain's premium position was lost, with the emergence of Malaysia as the leading producer and the discovery of rich deposits in Australia. By the end of the century, only nine mines of any consequence remained in Britain, where 300 had flourished 30 years earlier. From alluvial or secondary deposits, principally from South-East Asia, comes 80% of mined tin. Unlike copper, zinc, and lead, production of tin has not risen dramatically over the years and has rarely exceeded 250,000 t per annum.

The real price of tin spent most of the first half of the twentieth century in a relatively narrow band between US $10,000 and US$15,000 t^{-1} (1998$), with some excursions (Figure 1.16). From 1956 its price was regulated by a series of international agreements between producers and consumers under the auspices of the International Tin Council (ITC), which mirrored the highly successful policy of De Beers in controlling the gem diamond trade. Price stability was sought through selling from the ITC's huge stockpiles when the price rose and buying into the stockpile when the price fell.

From the mid-1970s, however, the price of tin was driven artificially higher at a time of world recession, a toxic combination of expanding production and falling consumption, the latter due mainly to the increasing use of aluminum in making cans, rather than tin-plated steel.

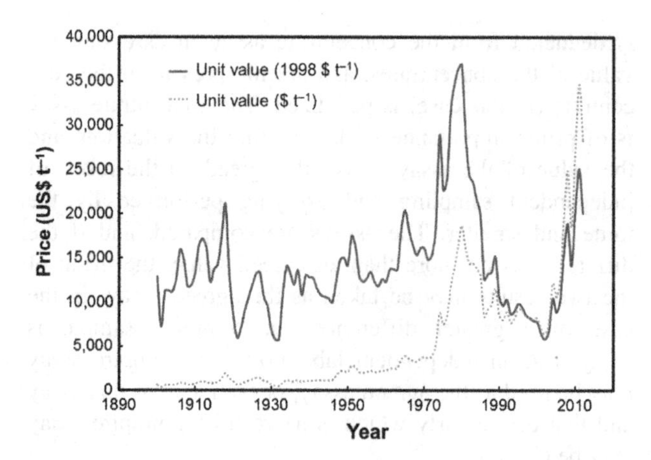

FIGURE 1.16 Tin price 1900–2012 (USGS, 2014).

Although the ITC imposed restrictions on the amount of tin that could be produced by its member countries, the reason for the inflating tin price was that the price of tin was fixed by the Malaysian dollar, while the buffer stock manager's dealings on the LME were financed in sterling. The Malaysian dollar was tied to the American dollar, which strengthened markedly between 1982 and 1984, having the effect of increasing the price of tin in London simply because of the exchange rate. However, the American dollar began to weaken in early 1985, taking the Malaysian dollar with it, and effectively reducing the LME tin price from its historic peak. In October 1985, the buffer stock manager announced that the ITC could no longer finance the purchase of tin to prop up the price, as it had run out of funds, owing millions of pounds to the LME traders. This announcement caused near panic, the tin price fell to £8,140 t^{-1} and the LME halted all further dealings. In 1986, many of the world's tin mines were forced to close due to the depressed tin price, and prices continued to fall in subsequent years, rising again in concert with other metals during the "super-cycle." While the following discussion relates to tin processing prior to the collapse, including prices and costs, the same principles can be applied to producing any metal-bearing mineral commodity at any particular period including the present day.

It is fairly easy to produce concentrates containing over 70% tin (i.e., over 90% cassiterite) from alluvial ores, such as those worked in South-East Asia. Such concentrates present little problem in smelting and hence treatment charges are relatively low. Production of high-grade concentrates also incurs relatively low freight charges, which is important if the smelter is remote. For these reasons it has been traditional in the past for hard-rock, lode tin concentrators to produce high-grade concentrates, but high tin prices and the development of profitable low-grade smelting processes changed the policy of many mines toward the production of lower-grade concentrates. The advantage of this is that the recovery of tin into the concentrate is increased, thus increasing smelter payments. However, the treatment of low-grade concentrates produces much greater problems for the smelter, and hence the treatment charges at "low-grade smelters" are normally much higher than those at the high-grade smelters. Freight charges are also correspondingly higher. Example 1.4 illustrates the identification of the economic optimum grade–recovery combination.

This result in Example 1.4 is in contrast to the maximum separation efficiency which was for the low-grade case (Example 1.3). Lowering the concentrate grade to 21% tin, in order to increase recovery, increased the separation efficiency, but adversely affected the economic return from the smelter, the increased charges being greater than the increase in revenue from the metal.

In terms of contained value, the ore, at free market price, has £85 worth of tin per ton ((1%/100) × £8,500 t^{-1} (of tin)); thus even at the economic optimum combination, the mine realizes only 62% of the ore value in payments received (£52.80/£85).

The situation may alter, however, if the metal price changes appreciably. If the tin price falls and the terms of the smelter contract remain the same, then the mine profits will suffer due to the reduction in payments. Rarely does a smelter share the risks of changing metal price, as it performs a service role, changes in smelter terms being made more on the basis of changing smelter costs rather than metal price. The mine does, however, reap the benefits of increasing metal price.

At a tin price of £6,500 t^{-1}, the NSR per ton of ore from the low-grade smelter treating the 42% tin concentrate is £38.75, while the return from the high-grade smelter, treating a 63% Sn concentrate, is £38.96. Although this is a difference of only £0.21 t^{-1} of ore, to a small 500 t d^{-1} tin concentrator this change in policy from relatively low- to high-grade concentrate, together with the subsequent change in concentrate market, would expect to increase the revenue by £0.21 × 500 × 365 = £38,325 per annum. The concentrator management must always be prepared to change its policies, both metallurgical and marketing, if maximum returns are to be made, although generation of a reliable grade–recovery relationship is often difficult due to the complexity of operation of lode tin concentrators and variations in feed characteristics.

It is, of course, necessary to deduct the costs of mining and processing from the NSR in order to deduce the *profit* achieved by the mine. Some of these costs will be indirect, such as salaries, administration, research and development, medical and safety, as well as direct costs, such as operating and maintenance, supplies and energy. The breakdown of milling costs varies significantly from mine

Example 1.4

Suppose that a tin concentrator treats a feed containing 1% tin, and that three possible combinations of concentrate grade and recovery are those in Example 1.3:

High grade	63% tin at 62% recovery
Medium grade	42% tin at 72% recovery
Low grade	21% tin at 78% recovery

Using the low-grade smelter terms set out in the contract in Table 1.5, and assuming that the concentrates are free of arsenic, and that the cost of transportation to the smelter is £20 t^{-1} of dry concentrate, what is the combination giving the economic optimum, that is, maximum NSR?

Solution

As a common basis we use 1 t of ore (feed). For each condition it is possible to calculate the amount of concentrate that can be made from 1 t of ore and from this, the smelter payment and charges for each combination of grade and recovery can be calculated.

The calculation of the amount of concentrate follows from the definition of recovery (Eq. (1.1)):

$$R = \frac{C\,c}{F\,f};\quad F = 1 \rightarrow C = \frac{(1\times f)\times R}{c}$$

For a tin price of £8,500 t^{-1} (of tin), and a treatment charge of £385 t^{-1} (of concentrate) the NSR for the three cases is calculated in the table below:

Case	Weight of Concentrate (kg t^{-1})	Smelter Payment[a]	Charge		NSR
			Treatment	Transport	
High grade	$\frac{(1t\times 1\%)\times 62\%}{63\%} = 9.84$ kg	$\frac{P\times 9.84\times(63-1)}{100{,}000} = £51.86$	$\frac{9.84\times 385}{1{,}000} = £3.79$	£0.20	£47.88
Medium grade	$\frac{(1t\times 1\%)\times 72\%}{42\%} = 17.14$ kg	$\frac{P\times 17.14\times(42-1)}{100{,}000} = £59.73$	$\frac{17.14\times 385}{1{,}000} = £6.59$	£0.34	£52.80
Low grade	$\frac{(1t\times 1\%)\times 78\%}{21\%} = 37.14$ kg	$\frac{P\times 37.14\times(21-1)}{100{,}000} = £63.14$	$\frac{37.14\times 385}{1{,}000} = £14.30$	£0.74	£48.10

[a]*Note: units of 10 are to convert from %.*

The optimum combination is thus the second case

to mine, depending on the size and complexity of the operations (Table 1.4 is one example breakdown). Mines with large ore reserves tend to have high throughputs, and so although the capital outlay is higher, the operating and labor costs tend to be much lower than those on smaller plants, such as those treating lode tin ores. Mining costs also vary considerably and are much higher for underground than for open-pit operations.

If mining and milling costs of £40 and £8, respectively, per ton of ore are typical of underground tin operations, then it can be seen that at a tin price of £8,500, the mine producing a concentrate of 42% tin, which is sold to a low-grade smelter, makes a profit of £52.80 − 48 = £4.80 t^{-1} of ore. It is also clear that if the tin price falls to £6,500 t^{-1}, the mine loses £48 − 38.96 = £9.04 for every ton of ore treated.

The mine profit per ton of ore treated can be illustrated by considering "contained values." For the 72% recovery case (medium-grade case in Example 1.4) the contained value in the concentrate is £85 × 0.72 = £61.20, and thus the contained value lost in the tailings is $23.80 (£85 − £61.20). Since the smelter payment is £52.80 the effective cost of transport and smelting is £61.20 − 52.80 = £8.40. Thus the mine profit can be summarized as follows:

$$\text{Contained value of ore} - (\text{costs} + \text{losses})$$

which for the 72% recovery case is:

$$£(85 - (8.4 + 40 + 8 + 23.8) = £4.80\ t^{-1}.$$

The breakdown of revenue and costs in this manner is summarized in Figure 1.17.

In terms of effective cost of production, since 1 t of ore produces 0.0072 t of tin in concentrates, and the free market value of this contained metal is £61.20 and the profit is £4.80, the total effective cost of producing 1 t of tin in concentrates is £(61.20 − 4.80)/0.0072 = £7,833.

The importance of metal losses in tailings is shown clearly in Figure 1.17. With ore of relatively high contained value, the recovery is often more important than the cost of promoting that recovery. Hence relatively high-cost unit processes can be justified if significant improvements in recovery are possible, and efforts to improve recoveries should always be made. For instance, suppose the concentrator, maintaining a concentrate grade of 42% tin, improves the recovery by 1%, i.e., to 73%, with no change in actual operating costs. The NSR will be £53.53 t^{-1} of ore and after deducting mining and milling costs, the profit realized by the mine will be £5.53 t^{-1} of ore. Since 1 t of ore now produces 0.0073 t of tin, having a contained value of £62.05, the cost of producing 1 t of tin in concentrates is thereby reduced to £(62.05 − 5.53)/0.0073 = £7,742.

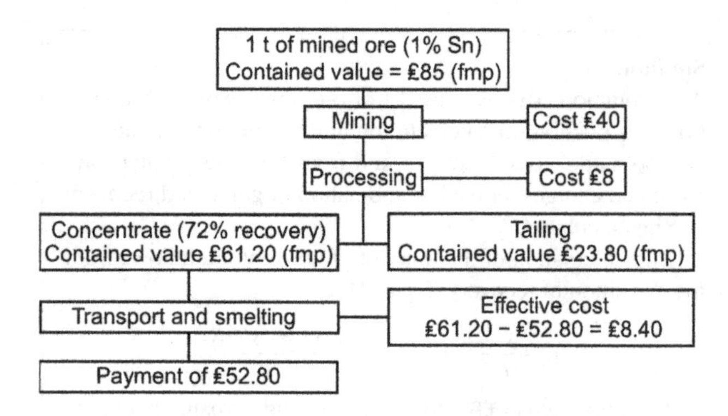

FIGURE 1.17 Breakdown of revenues and costs for treatment of lode tin (fmp = free market price).

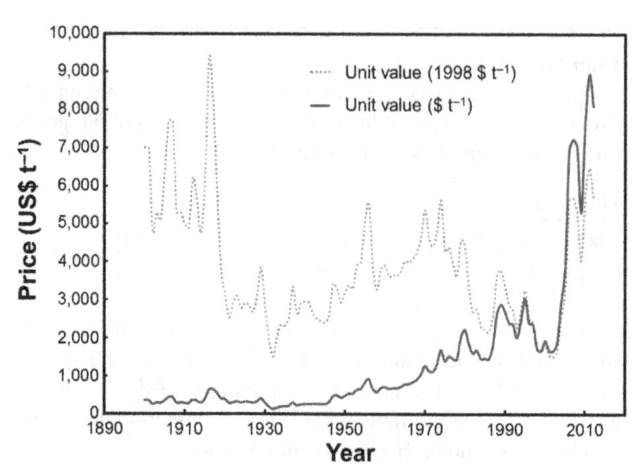

FIGURE 1.18 Copper price 1900–2012 *(USGS, 2014)*.

Due to the high processing costs and losses, hard-rock tin mines, such as those in Cornwall and Bolivia, had the highest production costs, being above £7,500 t^{-1} of ore in 1985 (for example). Alluvial operations, such as those in Malaysia, Thailand, and Indonesia, have lower production costs (around £6,000 t^{-1} in 1985). Although these ores have much lower contained values (only about £1–2 t^{-1}), mining and processing costs, particularly on the large dredging operations, are low, as are smelting costs and losses, due to the high concentrate grades and recoveries produced. In 1985, the alluvial mines in Brazil produced the world's cheapest tin, having production costs of only about £2,200 t^{-1} of ore (Anon., 1985a).

1.10.7 Case Study: Economics of Copper Processing

In 1835, the United Kingdom was the world's largest copper producer, mining around 15,000 t per annum, just below half the world production. This leading position was held until the mid-1860s when the copper mines of Devon and Cornwall became exhausted and the great flood of American copper began to make itself felt. The United States produced about 10,000 t in 1867, but by 1900 was producing over 250,000 t per annum. This output had increased to 1,000,000 t per annum in the mid-1950s, by which time Chile and Zambia had also become major producers. World annual production now exceeds 15,000,000 t (Figure 1.12).

Figure 1.18 shows that the price of copper in real terms grew steadily from about 1930 until the period of the oil shocks of the mid-1970s and then declined steeply until the early twenty-first century, the average real price in 2002 being lower than that at any time in the twentieth century. The pressure on costs was correspondingly high, and the lower cost operators such as those in Chile had more capacity to survive than the high-cost producers such as some of those in the United States. However, world demand, particularly from emerging economies such as China, drove the price strongly after 2002 and by 2010 it had recovered to about US$9000 t^{-1} (in dollars of the day) before now sliding back along with oil and other commodities.

The move to large-scale operations (Chile's Minerara Escondida's two concentrators had a total capacity of 230,000 t d^{-1} in 2003), along with improvements in technology and operating efficiencies, have kept the major companies in the copper-producing business. In some cases by-products are important revenue earners. BHP Billiton's Olympic Dam produces gold and silver as well as its main products of copper and uranium, and Rio Tinto's Kennecott Utah Copper is also a significant molybdenum producer.

TABLE 1.6 Simplified Copper Smelter Contract

Payments	
Copper:	Deduct from the agreed copper assay 1 unit, and pay for the remainder at the LME price for higher-grade copper.
Silver:	If over 30 g t^{-1} pay for the agreed silver content at 90% of the LME silver price.
Gold:	If over 1 g t^{-1} pay for the agreed gold content at 95% of the LME gold price.
Deductions	
Treatment charge: £30 per dry ton of concentrates	
Refining charge: £115 t^{-1} of payable copper	

A typical smelter contract for copper concentrates is summarized in Table 1.6. As in the case of the tin example, the principles illustrated can be applied to current prices and costs.

Consider a porphyry copper mine treating an ore containing 0.6% Cu to produce a concentrate containing 25% Cu, at 85% recovery. This represents a concentrate production of 20.4 kg t^{-1} of ore treated. Therefore, at a copper price of £980 t^{-1}:

$$\text{Payment for copper} = £980 \times \frac{20.4}{1000} \times 0.24 = £4.80$$

$$\text{Treatment charge} = £30 \times \frac{20.4}{1,000} = £0.61$$

$$\text{Refining charge} = £115 \times \frac{20.4}{1,000} \times 0.24 = £0.56$$

Assuming a transport cost of £20 t^{-1} of concentrate, the total deductions are £(0.61 + 0.56 + 0.41) = £1.58, and the NSR per ton of ore treated is thus £(4.80 − 1.58) = £3.22.

As mining, milling and other costs must be deducted from this figure, it is apparent that only those mines with very low operating costs can have any hope of profiting from such low-grade operations at the quoted copper price. Assuming a typical large open-pit mining cost of £1.25 t^{-1} of ore, a milling cost of £2 t^{-1} and indirect costs of £2 t^{-1}, the mine will lose £2.03 for every ton of ore treated. The breakdown of costs and revenues, following the example for tin (Figure 1.17), is given in Figure 1.19.

As each ton of ore produces 0.0051 t of copper in concentrates, with a free market value of £5.00, the total effective production costs are £(5.00 + 2.03)/0.0051 t = £1,378 t^{-1} of copper in concentrates. However, if the ore contains appreciable by-products, the effective production costs are reduced. Assuming the concentrate contains 25 g t^{-1} of gold and 70 g t^{-1} of silver, then the payment for gold, at a LME price of £230/troy oz (1 troy oz = 31.1035 g) is:

$$\frac{20.4}{1000} \times \frac{25}{31.1035} \times 0.95 \times 230 = £3.58$$

and the payment for silver at a LME price of £4.5 per troy oz

$$\frac{20.4}{1000} \times \frac{70}{31.1035} \times 0.95 \times 4.5 = £0.19$$

The net smelter return is thus increased to £6.99 t^{-1} of ore, and the mine makes a profit of £1.7 t^{-1} of ore treated. The effective cost of producing 1 t of copper is thus reduced to £(5.00 − 1.74)/0.0051 = £639.22.

By-products are thus extremely important in the economics of copper production, particularly for very low-grade operations. In this example, 42% of the mine's revenue is from gold. This compares with the contributions to revenue realized at Bougainville Copper Ltd (Sassos, 1983).

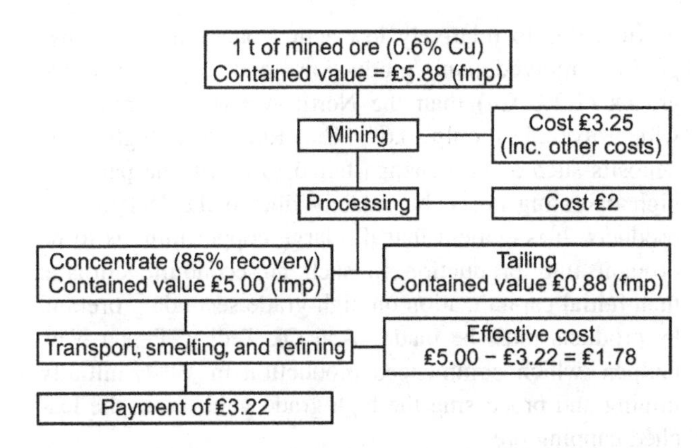

FIGURE 1.19 Breakdown of costs and revenues for treatment of typical porphyry copper ore.

TABLE 1.7 Effective Costs at World's Leading Copper Mines in 1985

Mine	Country	Effective Cost of Processed Cu (£ t^{-1})
Chuquicamata	Chile	589
El Teniente	Chile	622
Bougainville	Papua New Guinea	664
Palabora	South Africa	725
Andina	Chile	755
Cuajone	Peru	876
El Salvador	Chile	906
Toquepala	Peru	1,012
Inspiration	USA	1,148
San Manuel	USA	1,163
Morenci	USA	1,193
Twin Buttes	USA	1,208
Utah/Bingham	USA	1,329
Nchanga	Zambia	1,374
Gecamines	Zaire	1,374

Table 1.7 lists estimated effective costs per ton of copper processed in 1985 at some of the world's major copper mines, at a copper price of £980 t^{-1} (Anon., 1985b).

It is evident that, apart from Bougainville (now closed), which had a high gold content, and Palabora, a large open-pit operation with numerous heavy mineral by-products, the only economic copper mines in 1985 were the large South American porphyries. Those mines profited due to relatively low actual operating costs, by-product molybdenum production, and higher average grades (1.2% Cu) than the North American porphyries, which averaged only 0.6% Cu. Relatively high-grade deposits such as at Nchanga failed to profit due partly to high operating costs, but mainly due to the lack of by-products. It is evident that if a large copper mine is to be brought into production in such an economic climate, then initial capitalization on high-grade secondary ore and by-products must be made, as at Ok Tedi in Papua New Guinea, which commenced production in 1984, initially mining and processing the high-grade gold ore in the leached capping ore.

Since the profit margin involved in the processing of modern copper ores is usually small, continual efforts must be made to try to reduce milling costs and metal losses. This need has driven the increase in processing rates, many over 100,000 t ore per day, and the corresponding increase in size of unit operations, especially grinding and flotation equipment. Even relatively small increases in return per ton can have a significant effect, due to the very large tonnages that are often treated. There is, therefore, a constant search for improved flowsheets and flotation reagents.

Figure 1.19 shows that in the example quoted, the contained value in the flotation tailings is £0.88 t^{-1} treated ore. The concentrate contains copper to the value of £5.00, but the smelter payment is £3.22. Therefore, the mine realizes only 64.4% of the free market value of copper in the concentrate. On this basis, the actual metal loss into the tailings is only about £0.57 t^{-1} of ore. This is relatively small compared with milling costs, and an increase in recovery of 0.5% would raise the net smelter return by only £0.01. Nevertheless, this can be significant; to a mine treating 50,000 t d^{-1}, this is an increase in revenue of £500 d^{-1}, which is extra profit, providing that it is not offset by any increased milling costs. For example, improved recovery in flotation may be possible by the use of a more effective reagent or by increasing the dosage of an existing reagent, but if the increased reagent cost is greater than the increase in smelter return, then the action is not justified.

This balance between milling cost and metallurgical efficiency is critical on a concentrator treating an ore of low contained value, where it is crucial that milling costs be as low as possible. Reagent costs are typically around 10% of the milling costs on a large copper mine, but energy costs may contribute well over 25% of these costs. Grinding is by far the greatest energy consumer and this process undoubtedly has the greatest influence on metallurgical efficiency. Grinding is essential for the liberation of the minerals in the ore, but it should not be carried out any finer than is justified economically. Not only is fine grinding energy intensive, but it also leads to increased media (grinding steel) consumption costs. Grinding steel often contributes as much as, if not more than, the total mill energy cost, and the quality of grinding medium used often warrants special study. Figure 1.20 shows the effect of fineness of grind on NSR and grinding costs for a typical low-grade copper ore. Although flotation recovery, and hence NSR, increases with fineness of grind, it is evident that there is no economic benefit in grinding finer than a certain grind size (80% passing size). Even this fineness may be beyond the economic limit because of the additional capital cost of the grinding equipment required to achieve it.

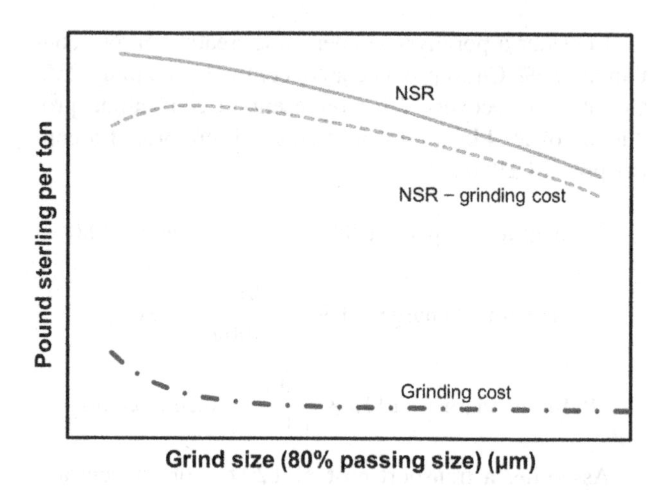

FIGURE 1.20 Effect of fineness of grind on net smelter return and grinding costs.

1.10.8 Economic Efficiency

It is apparent that a certain combination of grade and recovery produces the highest economic return under certain conditions of metal price, smelter terms, etc. *Economic efficiency* compares the *actual* NSR per ton of ore milled with the *theoretical* return, thus taking into account all the financial implications. The theoretical return is the maximum possible return that could be achieved, assuming "perfect milling," that is, complete separation of the valuable mineral into the concentrate, with all the gangue reporting to tailings. Using economic efficiency, plant efficiencies can be compared even during periods of fluctuating market conditions (Example 1.5).

In recent years, attempts have been made to optimize the performance of some concentrators by controlling plant conditions to achieve maximum economic efficiency (see Chapters 3 and 12). A dilemma often facing the metallurgist on a complex flotation circuit producing more than one concentrate is: how much contamination of one concentrate by the mineral that should report to the other concentrate can be tolerated? For instance, on a plant producing separate copper and zinc concentrates, copper is always present in the zinc concentrate, as is zinc in the copper concentrate. Metals misplaced into the wrong concentrate are rarely paid for by the specialist smelter and are sometimes penalized. There is, therefore, an optimum "degree of contamination" that can be tolerated. The most important reagent controlling this factor is often the depressant (Chapter 12), which, in this example, inhibits flotation of the zinc minerals. Increase in the addition of this reagent not only produces a cleaner copper concentrate but also tends to reduce copper recovery into this concentrate, as it also has some depressing effect on the copper minerals. The depressed copper minerals are likely to report to the zinc concentrate, so the addition rate of depressant needs to be carefully monitored and controlled to produce an optimum compromise. This should occur when the economic efficiency is maximized (Example 1.6).

Example 1.5

Calculate the economic efficiency of a tin concentrator, treating an ore grading 1% tin producing a concentrate grading 42% tin at 72% recovery, under the conditions of the smelter contract shown in Table 1.5. The cost of transportation to the smelter is £20 t^{-1} of concentrate. Assume a tin price of £8,500 t^{-1}.

Solution

It was shown in Example 1.4 that this concentrate would realize a net smelter return of £52.80. Assuming perfect milling, 100% recovery of the tin would be achieved into a concentrate grading 78.6% tin (i.e., pure cassiterite), thus the calculations proceed as follows:

The weight of concentrate produced from 1 t of feed = 12.72 kg.

Therefore, transport cost $= £20 \times \frac{12.72}{1,000} = £0.25$

Treatment charge $= £385 \times \frac{12.72}{1,000} = £4.90$

Valuation $= £12.72 \times (78.6 - 1) \times \frac{8500}{100,000} = £83.90$

Therefore, the "perfect milling" net smelter return = £(83.90 − 4.90 − 0.25) = £78.75, and thus:

Economic efficiency = 100 × 52.80/78.75 = 67.0%

Example 1.6

The following assay data were collected from a copper–zinc concentrator:

		Feed	Cu concentrate	Zn concentrate
Assay	Cu (%)	0.7	24.6	0.4
	Zn (%)	1.94	3.40	49.7

Mass balance calculation (Chapter 3) showed that 2.6% of the feed reported to the copper concentrate and 3.5% to the zinc concentrate.

Calculate the overall economic efficiency under the following simplified smelter terms:

Copper:
Copper price: £1,000 t^{-1}
Smelter payment: 90% of Cu content
Smelter treatment charge: £30 t^{-1} of concentrate
Transport cost: £20 t^{-1} of concentrate

Zinc:
Zinc price: £400 t^{-1}
Smelter payment: 85% of zinc content
Smelter treatment charge: £100 t^{-1} of concentrate
Transport cost: £20 t^{-1} of concentrate

Solution

1. Assuming perfect milling
   a. Copper
   Assuming that all the copper is contained in the mineral chalcopyrite, then maximum copper grade is 34.6% Cu (pure chalcopyrite, $CuFeS_2$).
   If C is weight of copper concentrate per 1,000 kg of feed, then for 100% recovery of copper into this concentrate:

$$100 = \frac{34.6 \times C \times 100}{0.7 \times 1,000}, C = 20.2(\text{kg})$$

   Transport cost = £20 × 20.2/1,000 = £0.40
   Treatment cost = £30 × 20.2/1,000 = £0.61
   Revenue = £20.2 × 0.346 × 1,000 × 0.9/1,000 = £6.29
   Therefore, NSR for copper concentrate = £5.28 t^{-1} of ore.

   b. Zinc
   Assuming that all the zinc is contained in the mineral sphalerite, maximum zinc grade is 67.1% (assuming sphalerite is ZnS).
   If Z is weight of zinc concentrate per 1,000 kg of feed, then for 100% recovery of zinc into this concentrate:

$$100 = \frac{67.1 \times Z \times 100}{1,000 \times 1.94}, Z = 28.9(\text{kg})$$

   Transport cost = £20 × 28.9/1,000 = £0.58
   Treatment cost = £100 × 28.9/1,000 = £2.89
   Revenue = £28.9 × 0.671 × 0.85 × 400/1,000 = £6.59
   Therefore, NSR for zinc concentrate = £3.12 t^{-1} of ore.
   Total NSR for perfect milling = £(5.28 + 3.12) = £8.40 t^{-1}
2. Actual milling
   Similar calculations give:
   Net copper smelter return = £4.46 t^{-1} ore
   Net zinc smelter return = £1.71 t^{-1} ore
   Total net smelter return = £6.17 t^{-1} ore
   Therefore, overall economic efficiency = 100 × 6.17/8.40 = 73.5%.

1.11 SUSTAINABILITY

While making a profit is essential to sustaining any operation, *sustainability* has taken on more aspects in recent years, something that increasingly impacts the mining industry.

The concept of *Sustainable Development* arose from the environmental movement of the 1970s and was created as a compromise between environmentalists who argued that further growth and development was untenable due to depletion of resources and destructive pollution, and proponents of growth who argued that growth is required to prevent developed countries from stagnating and to allow underdeveloped countries to improve. Thus, the principles of sustainable development, that is, growth within the capacity that the earth could absorb and remain substantially unchanged, were embraced by industry. The 1987 Brundtland report, commissioned by the UN, defined sustainable development as "development that meets the needs of the present without compromising the ability of future generations to meet their own needs" (Brundtland, 1987; Anon., 2000).

Sustainable development concepts have had major impact on mining operations and projects; for example, the many safety and worker health regulations continuously promulgated since the 1960s, as well as environmental regulations controlling and limiting emissions to air, land, and sea. In fact, meeting these regulations has been a main driver of technology and operational improvement projects (van Berkel and Narayanaswamy, 2004; Marcuson, 2007).

Achieving sustainable development requires balancing economic, environmental, and social factors (Figure 1.21). Sustainability attempts to account for the entire value of ecosystems and divides this value into three segments: direct, the value that can be generated by the animals, plants, and other resources; indirect, the value generated by items such as erosion control, water purification, and pollination; and intangible, the value to humans derived from beauty and religious/spiritual significance. Before the genesis of sustainability, economic evaluations concentrated on direct benefits undervaluing or ignoring indirect benefits. However, these indirect benefits may have significant economic value that is difficult to quantify and is only realized after the fact.

The mining enterprise is directly connected to and dependent upon the earth. Mining has been a crucible for sustainability activities, especially when dealing with greenfield sites inhabited by indigenous people. Algie (2002) and Twigge-Molecey (2004) highlighted six fundamental features of mining that account for this:

- Metal and mineral resources are nonrenewable.
- Economic mineralization often occurs in remote underdeveloped areas that are high conservation zones rich in biodiversity with many sites of cultural importance to the indigenous inhabitants. The mining company becomes responsible for developing essential infrastructure. Mining activities irreversibly alter both the natural and social environments.
- The mining sector is diverse in size, scope, and responsibility. It comprises government and private organizations, super-major corporations and junior miners, and exploration companies.
- The mining industry has a poor legacy.
- The risks and hazards in production, use, and disposal are not well understood by the public and are often poorly communicated.
- In contrast to manufacturing, which primarily involves physical change, processing of metals involves chemical change, which inherently is more polluting.

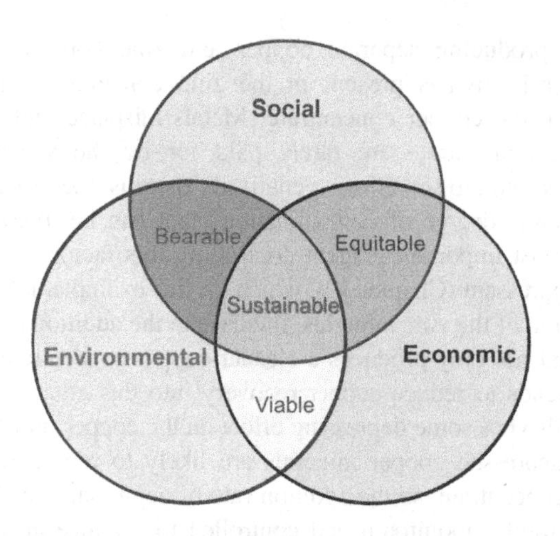

FIGURE 1.21 Sustainable development: balancing economic, environmental, and social demands.

In a mining endeavor, a large proportion of the benefits accrue globally while major environmental effects are felt locally. Indigenous people are closely tied to the local environment but remote from the global market exacerbating negative effects. Moreover, modern mining is not labor intensive and projects in remote locations generate jobs for only a small proportion of the aboriginal population. This may lead to an unhealthy division of wealth and status within the aboriginal community (Martin, 2005).

Sustainable development principles have dramatically impacted new projects. Extensive monitoring of baseline environmental conditions and pre-engineering studies to investigate likely impact on both the natural and human environments are mandated (Figure 1.22). While these activities certainly identify and reduce errors made during project implementation, they exponentially increase the amount of time required for completing projects. These delays may stretch project duration across several price cycles. During this time, low prices erode confidence in

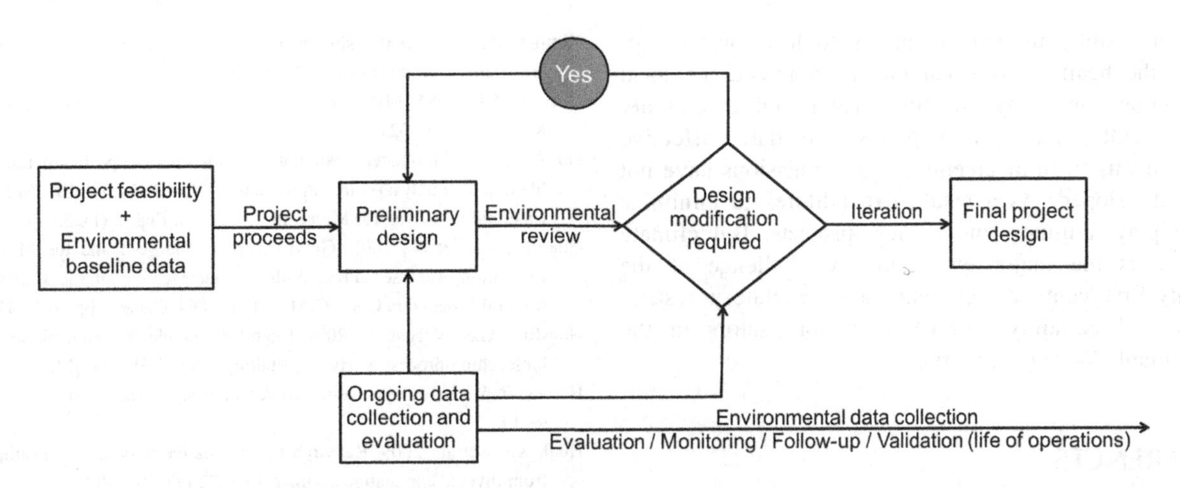

FIGURE 1.22 Simplified schematic of one aspect of the iterative environmental assessment process.

the worth of the project; project "champions" leave organizations, either voluntarily or involuntarily; projects are stopped and reevaluated. All of these result in more delays and additional cost.

Sustainable development has irreversibly altered the relationship between First Nations and industry (Marcuson et al., 2009). Fifty years ago, aboriginal peoples were typically ignored when creating a new project. Today, after years of advocacy, protests, and demonstrations, they have gained a place at the bargaining table, have input into the environmental permitting process, and frequently possess veto power over a new project. As a result, *Impact Benefit Agreements* (IBAs), designed to compensate the local population for damage to their environment and to create training, educational, and employment opportunities, are negotiated. These have succeeded in allocating a greater portion of the global benefits to the local population and in transferring social costs to the bottom line of the project. But this process has its challenges—multiparty negotiations between a company, government(s), and aboriginals take time and add costs. Moreover, governments in developing countries and aboriginal communities have different levels of sophistication in business and corporate matters, and they may have unrealistic expectations, and may be subject to exploitation by internal and external individuals and groups. Additionally, the terms of IBAs are frequently held in close confidence and thereby immune to external examination and regulation.

A modern mining project of reasonable size requires a minimum capital outlay of $1b and has an operating life of 25+ years. Costs of $4–6b are common, and time horizons of 10+ years from project initiation to successful start-up are not unusual. Achieving sustainable financing to cover these costs and extended times has necessitated globalization and consolidation of the industry.

However, even mega-corporations have not successfully modulated the historic cyclic nature of the mining business. This cyclic nature presents existential challenges to those charged with sustaining mining corporations. Net Present Value, greatly influenced by price assumptions and heavily weighted to income during the first 10 years, is a major financial metric for investment decisions. Thus, while a key financial indicator is short term focused, both project duration and asset life are long, and a fundamental mismatch exists, further complicating decision making. Sequential delays between commodity price increases, decisions to increase supply, and project completion create a sluggish, fundamentally unstable system. Finally, due to high capital costs and relatively low incremental operating costs there is hesitancy to close operations when prices fall. When coupled with political and social pressure to maintain employment, operations may be maintained even when taking losses adding more fuel to the boom and bust cycle.

Faced with this situation and compelled by investors demanding immediate returns, mining companies cut costs in areas that will immediately improve the bottom line. This means deferring capital projects, negotiating delays in environmental improvement projects, reducing support for universities, and curtailing exploration and technology initiatives. While these endeavors support the long-term economic sustainability of the enterprise, curtailment and concomitant loss of expertise further exacerbate the cyclical nature of the industry.

Through regulatory and economic forces, sustainable development has become a revolutionary driver of the mining industry. These forces are increasing (Starke, 2002). Developing countries are demanding a greater portion of the economic benefits. Social and environmental sustainability become inextricably linked in the realm of water utilization and waste disposal. These present visible, immediate threats to local water supplies, the health of the natural environment, the local economies, and ways of life. Water and community health will remain flash points. To date, effective mechanisms to limit greenhouse gas emissions have not been developed. As a result, expenditures to minimize these play a minor role in new projects. But climate change is the major environmental challenge of the twenty-first century; economic and regulatory restrictions will certainly emerge as major factors in the Sustainable Development portfolio.

REFERENCES

Algie, S.H., 2002. Global materials flows in minerals processing. Green Processing 2002. AusIMM, Carlton, Australia, pp. 39–48.

Anon., 1979. Nchanga consolidated copper mines. Eng. Min. J. 180 (Nov.), 150.

Anon., 1985a. Tin-paying the price. Min. J.(Dec), 477–479.

Anon., 1985b. Room for improvement at Chuquicamata. Min. J. (Sep), 249.

Anon., 2000. A Guide to World Resources 2000–2001: People and Ecosystems: The Fraying Web of Life. United Nations Development Programme, United Nations Environment Programme, World Bank and World Resources Institute, World Resources Institute, Washington, DC, USA, pp. 1–12.

Barbery, G., 1986. Complex sulphide ores-processing options. In: Wills, B.A., Barley, R.W. (Eds.), Mineral Processing at a Crossroads: Problems and Prospects, 117. Springer, Netherlands, pp. 157–194.

Barbery, G., 1991. Mineral Liberation, Measurement, Simulation and Practical Use in Mineral Processing. Les Editions GB, Quebec, Canada.

Baum, W., et al., 2004. Process mineralogy—a new generation for ore characterisation and plant optimisation. 2003 SME Annual Meeting and Exhibit, Denver, CO, USA, Preprint 04-012: 1–5.

Botero, A.E.C., et al., 2008. Surface chemistry fundamentals of biosorption of *Rhodococcus opacus* and its effect in calcite and magnesite flotation. Miner. Eng. 21 (1), 83–92.

Brierley, J.A., Brierley, C.L., 2001. Present and future commercial applications of biohydrometallurgy. Hydrometallurgy. 59 (2-3), 233–239.

Brundtland, G. (Ed.), 1987. Our Common Future: World Commission on Environment and Development. Oxford University Press, Oxford, England.

Clark, K., 1997. The business of fine coal tailings recovery. The Australian Coal Review. CSIRO Division of Coal and Energy Technology, Sydney, Australia, *3*: pp. 24–27.

Damjanovic, B., Goode, J.R. (Eds.), 2000. Canadian Milling Practice, Special vol. 49. CIM, Montréal, QC, Canada.

Evans, C.L., et al., 2011. Application of process mineralogy as a tool in sustainable processing. Miner. Eng. 24 (12), 1242–1248.

Gaudin, A.M., 1939. Principles of Mineral Dressing. McGraw-Hill Book Company, Inc., London, England.

Gay, S.L., 2004a. A liberation model for comminution based on probability theory. Miner. Eng. 17 (4), 525–534.

Gay, S.L., 2004b. Simple texture-based liberation modelling of ores. Miner. Eng. 17 (11-12), 1209–1216.

Gilchrist, J.D., 1989. Extraction Metallurgy. third ed. Pergamon Press, Oxford, UK.

Gottlieb, P., et al., 2000. Using quantitative electron microscopy for process mineral applications. JOM. 52 (4), 24–25.

Gray, P.M.J., 1984. Metallurgy of the complex sulphide ores. Mining Mag. Oct, 315–321.

Gu, Y., 2003. Automated scanning electron microscope based mineral liberation analysis—an introduction to JKMRC/FEI Mineral Liberation Analyser. J. Miner. Mat. Charact. Eng. 2 (1), 33–41.

Guerney, P.J., et al., 2003. Gravity recoverable gold and the Mineral Liberation Analyser. Proc. 35th Annual Meeting of the Canadian Mineral Processors Conf. CIM, Ottawa, ON, Canada, pp. 401–416.

Hansford, G.S., Vargas, T., 2001. Chemical and electrochemical basis of bioleaching processes. Hydrometallurgy. 59 (2-3), 135–145.

Hefner, R.A., 2014. The United States of gas. Foreign Affairs. 93 (3), 9–14.

Hoal, K.O., et al., 2009. Research in quantitative mineralogy: examples from diverse applications. Miner. Eng. 22 (4), 402–408.

Iwasaki, I., Prasad, M.S., 1989. Processing techniques for difficult-to-treat ores by combining chemical metallurgy and mineral processing. Miner. Process. Extr. Metall. Rev. 4 (3-4), 241–276.

Jowett, A., 1975. Formulae for the technical efficiency of mineral separations. Int. J. Miner. Process. 2 (4), 287–301.

Jowett, A., Sutherland, D.N., 1985. Some theoretical aspects of optimizing complex mineral separation systems. Int. J. Miner. Process. 14 (2), 85–109.

King, R.P., 2012. In: Schneider, C.L, King, E.A. (Eds.), Modeling and Simulation of Mineral Processing Systems, second ed. SME, Englewood, CO, USA.

Kingman, S.W., et al., 2004. Recent developments in microwave-assisted comminution. Int. J. Miner. Process. 74 (1-4), 71–83.

Lin, C.L., et al., 2013. Characterization of rare-earth resources at Mountain Pass, CA using high-resolution X-ray microtomography (HRXMT). Miner. Metall. Process. 30 (1), 10–17.

Lotter, N.O., 2011. Modern process mineralogy: an integrated multi-disciplined approach to flowsheeting. Miner. Eng. 24 (12), 1229–1237.

Lynch, A.J., et al., 2007. History of flotation technology. In: Fuerstenau, M.C., et al., (Eds.), Froth Flotation: A Century of Innovation. SME, Littleton, CO, USA, pp. 65–94.

Marcuson, S.W., 2007. SO_2 Abatement from Copper Smelting Operations: A 40 Year Perspective, vol. 3 (Book 1), The Carlos Diaz Symposium on Pyrometallurgy, Copper 2007, pp. 39–62.

Marcuson, S.W., et al., 2009. Sustainability in nickel projects: 50 years of experience at Vale Inco, Pyrometallurgy of Nickel and Cobalt 2009. Proceedings of the 48th Annual Conference of Metallurgists of CIM, Sudbury, ON, Canada, pp. 641–658.

Martin, D.F., 2005. Enhancing and measuring social sustainability by the minerals industry: a case study of Australian aboriginal people. Sustainable Development Indicators in the Mineral Industry, Aachen International Mining Symp. Verlag Gluckauf, Essen, Germany, pp. 663–679.

Meadows, D.H., et al., 1972. The Limits to Growth: A Report for the Club of Rome's Project on the Predicament of Mankind. Universal Books, New York, NY, USA.

Mills, C., 1978. Process design, scale-up and plant design for gravity concentration. In: Mular, A.L., Bhappu, R.B. (Eds.), Mineral Processing Plant Design. AIMME, New York, NY, USA, pp. 404–426. (Chapter 18).

Moskvitch, K., 2014. Field of dreams: the plants that love heavy metal. New Scientist. 221 (2961), 46–49.

Motl, D., et al., 2012. Advanced scanning mode for automated precious metal search in SEM (abstract only). Annual Meeting of the Nordic Microscopy Society SCANDEM 2012, University of Bergen, Norway, p. 24.

Mouat, J., 2011. Canada's mining industry and recent social, economic and business trends. In: Kapusta, J., et al., (Eds.), *The Canadian Metallurgical & Materials Landscape* 1960 to 2011. Met Soc/CIM, Westmount, Québec, Canada, pp. 19–26.

Mukherjee, A., et al., 2004. Recent developments in processing ocean manganese nodules—a critical review. Miner. Process. Extr. Metall. Rev. 25 (2), 91–127.

Ottley, D.J., 1991. Small capacity processing plants. Mining Mag. 165 (Nov.), 316–323.

Pease, J.D., 2005. Fine grinding as enabling technology—The IsaMill. Proceedings of the Sixth Annual Crushing and Grinding Conference, Perth, WA, Australia, pp. 1–21.

Radziszewski, P., 2013. Energy recovery potential in comminution processes. Miner. Eng. 46-47, 83–88.

Rao, S.R., 2006. Resource Recovery and Recycling from Metallurgical Wastes. Elsevier, Amsterdam.

Ravilious, K., 2013. Digging downtown: the hunt for urban gold. New Scientist. 218 (2919), 40–43.

Reuter, M.A., et al., 2005. In: Wills, B.A. (Ed.), The Metrics of Material and Metal Ecology: Harmonizing the Resource, Technology and Environmental Cycles. Developments in Mineral Processing. Elsevier, Amsterdam.

Ridley, J., 2013. Ore Deposit Geology. first ed. Cambridge University Press, Cambridge, UK.

Sassos, M.P., 1983. Bougainville copper: tropical copper with a golden bloom feeds a 130,000-mt/d concentrator. Eng. Min. J. Oct, 56–61.

Schulz, N.F., 1970. Separation efficiency. Trans. SME/AIME. 247 (Mar.), 81–87.

Scott, S.D., 2001. Deep ocean mining. Geosci. Canada. 28 (2), 87–96.

Smith, R.W., et al., 1991. Mineral bioprocessing and the future. Miner. Eng. 4 (7-11), 1127–1141.

Smythe, D.M., et al., 2013. Rare earth element deportment studies utilising QEMSCAN technology. Miner. Eng. 52, 52–61.

Starke, J. (Ed.), 2002. Breaking New Ground: Mining, Minerals and Sustainable Development: The Report of the MMSD Project. Mining, Minerals, and Sustainable Development Project, Earthscan Publications Ltd., London, England.

Tan, L., Chi-Lung, Y., 1970. Abundance of the chemical elements in the Earth's crust. Int. Geol. Rev. 12 (7), 778–786.

Taylor, S.R., 1964. Abundance of chemical elements in the continental crust: a new table. Geochim. Cosmochim. Acta. 28 (8), 1273–1285.

Thomas, K.G., et al., 2014. Project execution & cost escalation in the mining industry. Proc. 46th Annual Meeting of the Canadian Mineral Processors Conf. CIM, Ottawa, Canada, pp. 105–114.

Twigge-Molecey, C., 2004. Approaches to plant design for sustainability. Green Processing 2004. AusIMM, Fremantle, WA, Australia, pp. 47–52.

USGS, 2014. <http://minerals.usgs.gov/> (accessed Aug. 2014.).

van Berkel, R., Narayanaswamy, V., 2004. Sustainability as a framework for innovation in minerals processing. Green Processing 2004. AusIMM, Fremantle, WA, Australia, pp. 197–205.

Wills, B.A., Atkinson, K., 1991. The development of minerals engineering in the 20th century. Miner. Eng. 4 (7-11), 643–652.

Wills, B.A., Atkinson, K., 1993. Some observations on the fracture and liberation of mineral assemblies. Miner. Eng. 6 (7), 697–706.

Chapter 2

Ore Handling

2.1 INTRODUCTION

Ore handling is a key function in mining and mineral processing, which may account for 30–60% of the total delivered price of raw materials. It covers the processes of transportation, storage, feeding, and washing of the ore *en route* to, or during, the various stages of treatment in the mill.

Since the physical state of ores *in situ* may range from friable, or even sandy material, to monolithic deposits with the hardness of granite, the methods of mining and provisions for the handling of freshly excavated material will vary widely. Ore that has been well fragmented can be transported by trucks, belts, or even by sluicing, but large lumps of hard ore may need secondary blasting. Developments in nonelectric millisecond delay detonators and plastic explosives have resulted in more controllable primary breakage and easier fragmentation of occasional overly-large lumps. At the same time, crushers have become larger and lumps up to 2 m in size can now be fed into some primary units.

Ores are by and large heterogeneous in nature. The largest lumps blasted from an open pit operation may be over 1.5 m in size. The fragmented ore from a blast is loaded directly into trucks, holding up to 400 t of ore in some cases, and is transported directly to the primary crushers. Storage of such ore is not always practicable, due to its wide particle size range which causes segregation during storage, the fines working their way down through the voids between the larger particles. Extremely coarse ore is sometimes difficult to start moving once it has been stopped. Sophisticated storage and feed mechanisms are therefore often dispensed with, the trucks depositing their loads directly on the grizzly feeding the primary crusher.

The operating cycle of an underground mine is complex. Drilling and blasting are often performed in one or two shifts; the blasted ore is hoisted to the surface during the next couple of shifts. The ore is transported through the passes via chutes and tramways and is loaded into skips, holding as much as 30 t of blasted ore, to be hoisted to the surface. Large boulders are often broken up underground by primary rock breakers in order to facilitate loading and handling at this stage. The ore, on arrival at the surface, having undergone some initial crushing, is easier to handle than that from an open pit mine. The storage and feeding is usually easier, and indeed essential, due to the intermittent arrival of skips at the surface.

2.2 THE REMOVAL OF HARMFUL MATERIALS

Ore entering the mill from the mine (*run-of-mine ore)* normally contains a small proportion of material which is potentially harmful to the mill equipment and processes. For instance, large pieces of iron and steel broken off from mine machinery can jam in the crushers. Wood is a major problem in many mills as it is ground into a fine pulp and causes choking or blocking of screens, flotation cell ports, etc. Wood pulp may also consume flotation reagents by absorption, which reduces mineral floatability. Clays and slimes adhering to the ore are also harmful as they hinder screening, filtration, and thickening, and again may consume flotation reagents.

All these *tramp materials* must be removed as far as possible at an early stage in treatment. Removal by hand (hand sorting) from conveyor belts has declined with the development of mechanized methods of dealing with large tonnages, but it is still used when plentiful cheap labor is available.

Skips, ore bins, and mill equipment can be protected from large pieces of "tramp" iron and steel, such as rockbolts and wire meshes, by electromagnets suspended over conveyor belts (*guard magnets*) (Figure 2.1). The magnets are generally installed downstream of the primary crusher to protect skips and ore bins. These powerful electromagnets can pick up large pieces of iron and steel travelling over the belt. They may operate continuously (as shown) or be stationary and, at intervals, are swung away from the belt and unloaded when the magnetic field is removed. Guard magnets, however, cannot be used to remove tramp iron from magnetic ores, such as those containing magnetite, nor will they remove nonferrous

Wills' Mineral Processing Technology.

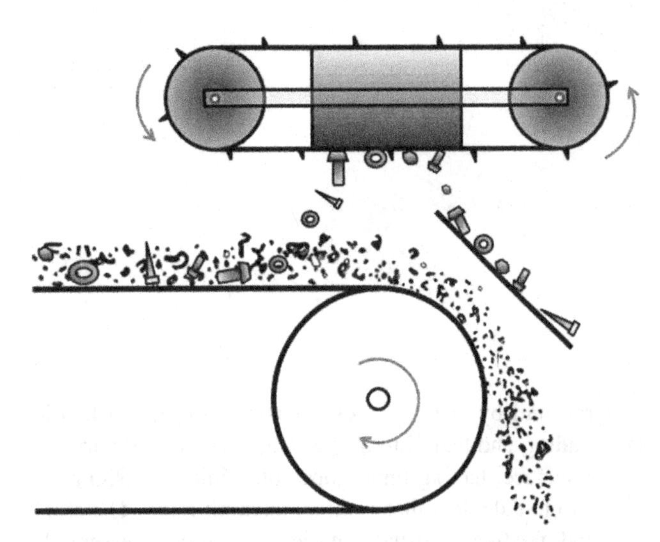

FIGURE 2.1 Conveyor guard magnet (see also Chapter 13).

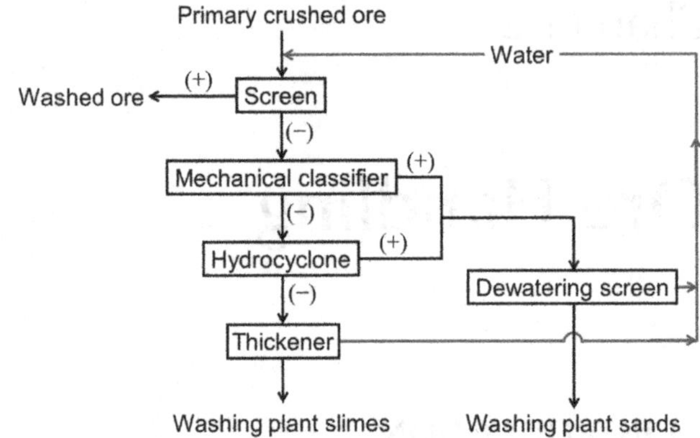

FIGURE 2.2 Typical washing plant flowsheet.

metals or nonmagnetic steels from the belt. Metal detectors, which measure the electrical conductivity of the material being conveyed, can be fitted over or around conveyor belts. The electrical conductivity of ores is much lower than that of metals, and fluctuations in electrical conductivity in the conveyed material can be detected by measuring the change that tramp metal causes in a given electromagnetic field. When a metal object triggers an alarm, the belt automatically stops and the object can be removed. With nonmagnetic ores it is advantageous to precede the metal detector with a guard magnet, which will remove the ferromagnetic tramp metals and thus minimize belt stoppages.

Large pieces of wood that have been flattened by passage through a primary crusher can be removed by passing the ore feed over a vibrating scalping screen. Here, the apertures of the screen are slightly larger than the maximum size of particle in the crusher discharge, allowing the ore to fall through the apertures and the flattened wood particles to ride over the screen and be collected separately. (On cold nights the collected wood might find a use.)

Wood can be further removed from the pulp discharge from the grinding mills by passing the pulp through a fine screen. Again, while the ore particles pass through the apertures, the wood collects on top of the screen and can be periodically removed.

Washing of run-of-mine ore can be carried out to facilitate sorting (Chapter 14) by removing obscuring dirt from the surfaces of the ore particles. However, washing to remove very fine material, or *slimes*, of little or no value is more important.

Washing is normally performed after primary crushing as the ore is then of a suitable size to be passed over washing screens. It should always precede secondary crushing as slimes severely interfere with this stage. The ore is passed through high-pressure jets of water on mechanically vibrated screens. The screen undersize product is usually directed to the grinding mills and thus the screen apertures are usually of similar size to the particles in the feed to the grinding mills.

Ore washing is sometimes assisted by adding *scrubbers* in the circuit. Scrubbers are designed to clean crushed ore, sand, and gravel, but they can also upgrade an ore by removing soft rock by attrition. Scrubbers are self-aligning, steel trunnions supported on flanged railroad type bearings, and driven by a saddle drive chain.

In the circuit shown in Figure 2.2, material passing over the screen, that is, washed ore, is transported to the secondary crushers. Material passing through the screens is classified into coarse and fine fractions by a mechanical classifier or hydrocyclone (Chapter 9), or both. It may be beneficial to classify initially in a mechanical classifier as this is more able to smooth out fluctuations in flow than is the hydrocyclone and it is better suited to handling coarse material.

The coarse product from the classifier, designated "washing plant sands," is either routed direct to the grinding mills or is dewatered over vibrating screens before being sent to mill storage. A considerable load, therefore, is taken off the dry crushing section.

The fine product from classification, that is, the "slimes," may be partially dewatered in shallow large diameter settling tanks known as thickeners (Chapter 15), and the thickened pulp is either pumped to tailings disposal or, if containing values, pumped direct to the concentration process, thus removing load from the grinding section. In Figure 2.2, the thickener overflows are used to feed the high-pressure washing sprays on the screens. Water conservation in this manner is practiced in most mills.

Wood pulp may again be a problem in the above circuit, as it will tend to float in the thickener, and will choke the water spray nozzles unless it is removed by retention on a fine screen.

2.3 ORE TRANSPORTATION

In a mineral processing plant, operating at the rate of 400,000 t d^{-1}, this is equivalent to about 28 t of solid per minute, requiring up to 75 m^3 min^{-1} of water. It is therefore important to operate with the minimum upward or horizontal movement and with the maximum practicable pulp density in all of those stages subsequent to the addition of water to the system. The basic philosophy requires maximum use of gravity and continuous movement over the shortest possible distances between processing units.

Dry ore can be moved through chutes, provided they are of sufficient slope to allow easy sliding and sharp turns are avoided. Clean solids slide easily on a 15–25° steel-faced slope, but for most ores, a 45–55° working slope is used. The ore may be difficult to control if the slope is too steep.

The belt conveyor system is the most effective and widely used method of handling loose bulk materials in mining and mineral processing industries. In a belt conveyor system, the belt is a flexible and flat loop, mounted over two pulleys, one of which is connected to a drive to provide motion in one direction. The belt is tensioned sufficiently to ensure good grip with the drive pulley, and is supported by a structural frame, with idlers or slider bed in between the pulleys. Belts today have capacities up to 40,000 t h^{-1} (Alspaugh, 2008) and single flight lengths exceeding 15,000 m, with feasible speeds of up to 10 m s^{-1}.

The standard rubber conveyor belt has a foundation, termed a *carcass*, of sufficient strength to withstand the belt tension, impact, and strains due to loading. This foundation can be single-ply or multi-ply (Figure 2.3) and is made of cotton, nylon, leather, plastic, steel fabric, or steel cord. The foundation is bound together with a rubber matrix and completely covered with a layer of vulcanized rubber. The type of vulcanized rubber cover may vary depending upon the ore properties and operational conditions (e.g., abrasiveness of ore, powder or lump material, temperature) (Ray, 2008).

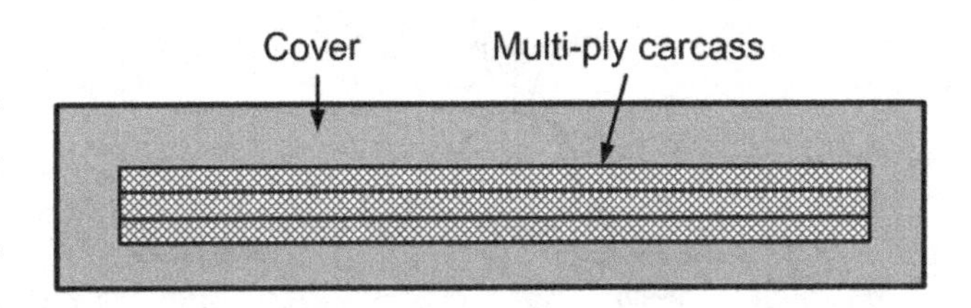

FIGURE 2.3 Construction of a multi-ply belt (in cross section).

The carrying capacity of the belt is increased by passing it over *troughing* idlers. These are support rollers set normal to the travel of the belt and inclined upward from the center so as to raise the edges and give it a trough-like profile. There may be three or five in a set and they will be rubber-coated under a loading point, so as to reduce the wear and damage from impact. Spacing along the belt is at the maximum interval that avoids excessive sag. The return belt is supported by horizontal straight idlers that overlap the belt by a few inches at each side. The idler dimensions and troughing angle are laid down in BIS in IS 8598:1987(2). The diameters of carrying and return idlers range from ca. 63 to 219 mm. Idler length may vary from 100 to 2,200 mm. Controlling factors for idler selection may include unit weight, lump size, and belt speed. The length of an idler is proportional to its diameter. The troughing idler sets are installed with a troughing angle ranging 15–50°. Idler spacing on the loaded side is a function of unit weight of material and belt width (Ray, 2008).

To control belt wandering (lateral movement), either training idlers are installed normal to the direction of the belt motion or wing idlers are installed at a forward angle. Sensors and electronic devices are also used to detect wandering and keep the belt on track (Anon., 2014).

The pulleys of belt conveyors are manufactured from rolled steel plates or from cast iron. The pulley drum is keyed to the steel shaft and the finished dimensions are machined. Generally, crowning is done to the pulley face to decrease belt wandering at the pulley. The length of the pulley is 100–200 mm more than the belt width (Ray, 2008). Inducing motion without slipping requires good contact between the belt and the drive pulley. This may not be possible with a single 180° turn over a pulley (wrap angle, α), and some form of "snubbed pulley" drive (Figure 2.4(a)) or "tandem" drive arrangement (Figure 2.4 (b) and (c)) may be more effective. The friction can also be provided by embedded grooves or covering the pulleys with rubber, polyurethane, or ceramic layer of thickness 6–12 mm. The layer can also be engraved with patterned grooves for increased friction and better drainage when dealing with wet material (Ray, 2008).

The belt system must incorporate some form of tensioning device, also known as belt take-up load, to adjust the belt for stretch and shrinkage and thus prevent undue sag between idlers, and slip at the drive pulley. The take-up device also removes sag from the belt by developing tensile stress in the belt. In most mills, gravity-operated arrangements are used, which adjust the tension continuously (Figure 2.5). A screw type take-up loading system can be used instead of gravity for tensioning the conveyor (Ray, 2008). Hydraulics have also been used extensively, and when more refined belt-tension control is required, especially in starting and stopping long conveyors, load-cell-controlled electrical tensioning devices are used.

Advances in control technology have enhanced the reliability of belt systems by making possible a high

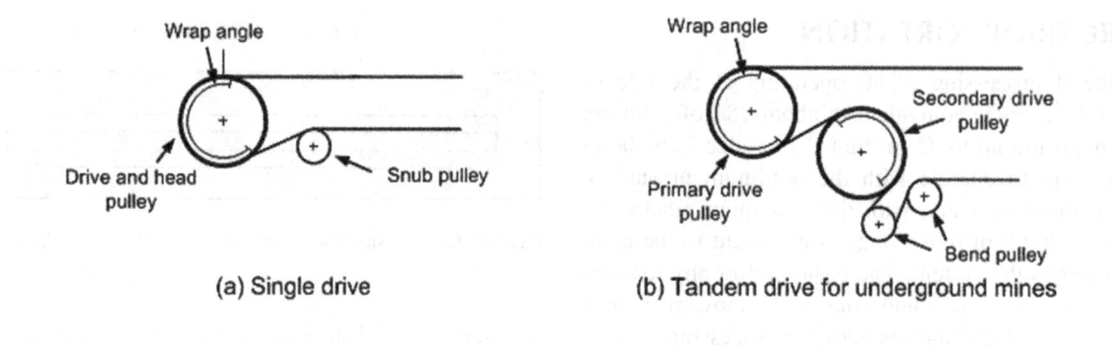

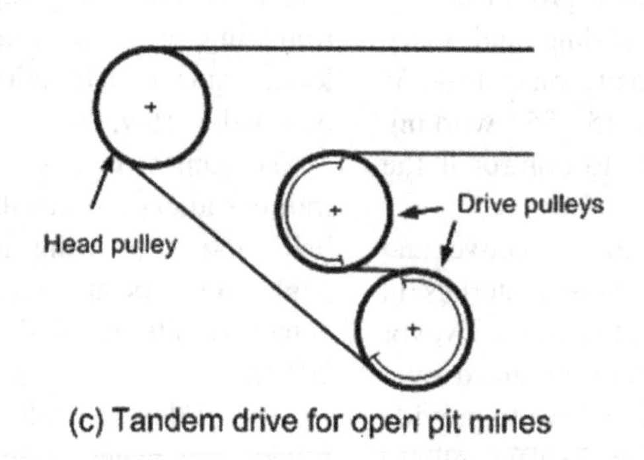

FIGURE 2.4 Conveyor-belt drive arrangements.

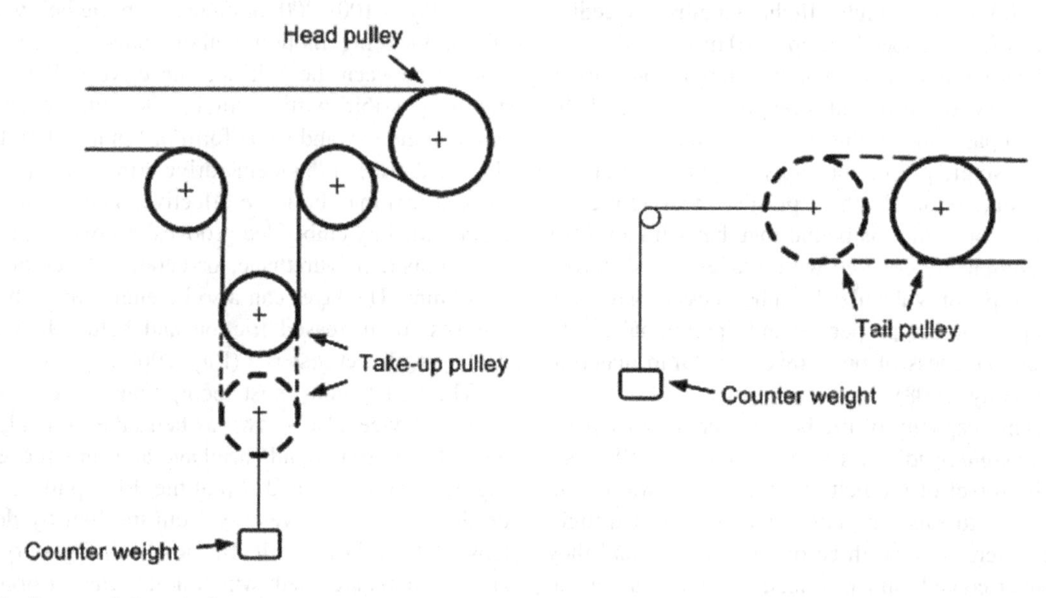

FIGURE 2.5 Conveyor-belt tensioning systems.

degree of fail-safe automation. A series of belts should incorporate an interlock system such that failure of any particular belt will automatically stop preceding belts. Interlock with devices being fed by the belt is important for the same reasons. It should not be possible to shut down any machine in the system without arresting the feed to the machine at the same time and, similarly, motor failure should lead to the automatic tripping (stopping) of all preceding belts and machines. Similarly, the belt start up sequence of conveyor systems is fixed such that the

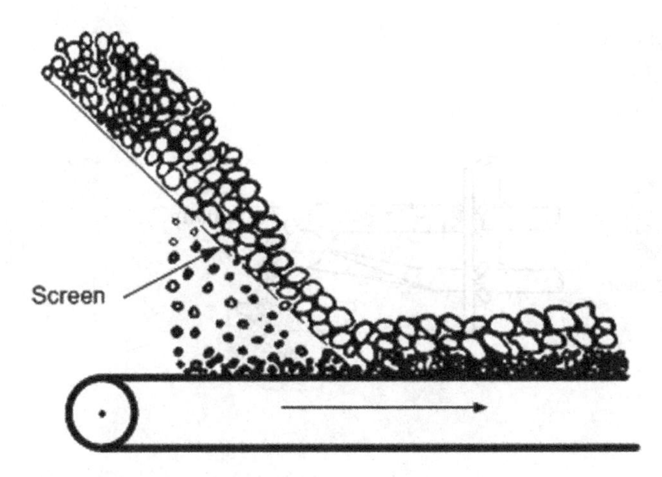

FIGURE 2.6 Belt-loading system.

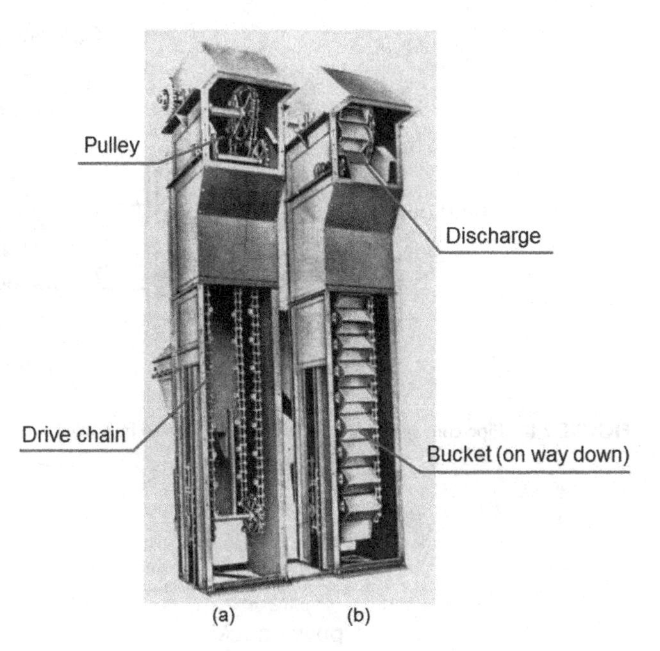

FIGURE 2.7 Gravity bucket elevator: (a) rear view without buckets, and (b) return (empty) buckets.

last conveyor should start first followed by the second to last and so on. Sophisticated electrical, pneumatic, and hydraulic circuits have been widely employed to replace all but a few manual operations.

Several methods can be used to minimize loading shock on the belt. A typical arrangement is shown in Figure 2.6 where the fines are screened onto the belt first and provide a cushion for the larger pieces of rock. The impact loading on a belt can also be reduced by increasing the number of idlers at the loading points; such idlers are called impact idlers.

Feed chutes must be designed to deliver the bulk of the material to the center of the belt at a velocity close to that of the belt. Side boards or skirt plates with a length of 2–3 m are installed to guide the material on the belt and help reduce dust. Ideally the speed of material being placed should be equal to the belt speed, but in practice this condition is seldom achieved, particularly with wet sand or sticky materials. Where conditions will allow, the angle of the chute should be as great as possible, thereby providing sufficient velocity to the material in order to match the belt speed. The chute angle with the belt is adjusted until the correct speed of flow is obtained. Higher chute angles may produce impact loading on the belt. Material, particularly when heavy or lumpy, should never be allowed to strike the belt vertically. Baffles in transfer chutes, to guide material flow, are now often remotely controlled by hydraulic cylinders. Feed chutes are sometimes installed with hydraulically operated regulator gates for better control.

The conveyor may discharge at the head pulley, or the load may be removed before the head pulley is reached. The most satisfactory device for achieving this is a *tripper*. This is an arrangement of pulleys in a frame by which the belt is raised and doubled back so as to give it a localized discharge point. The frame is usually mounted on wheels, running on tracks, so that the load can be delivered at several points, over a long bin or into several bins. The discharge chute on the tripper can deliver to one or both sides of the belt. The tripper may be moved manually, by head and tail ropes from a reversible hoisting drum, or by a motor. It may be automatic, moving backward and forward under power from the belt drive. A plough can also be used to discharge the material. A plough is a v-shaped, rubber tipped blade extending along the width of the conveyor at an angle of 60°. A troughed conveyor is made flat by passing over a slider, at such a discharge point. Cameras allow the level of bin filling to be monitored.

Shuttle belts are reversible self-contained conveyor units mounted on carriages, which permit them to be moved lengthwise to discharge to either side of the feed point. The range of distribution is approximately twice the length of the conveyor. They are often preferred to trippers for permanent storage systems because they require less head room and, being without reverse bends, are much easier on the belt.

Belt conveyors can operate at a maximum angle of 10–18° depending on the material being conveyed. Beyond the range of recommended angle of incline, the material may slide down the belt, or it may topple on itself. Where space limitation does not permit the installation of a regular belt conveyor, steep angle conveying can be installed which includes gravity bucket elevators (Figure 2.7), molded cleat belts, fin type belts, pocket belts, totally enclosed belts, and sandwich belts.

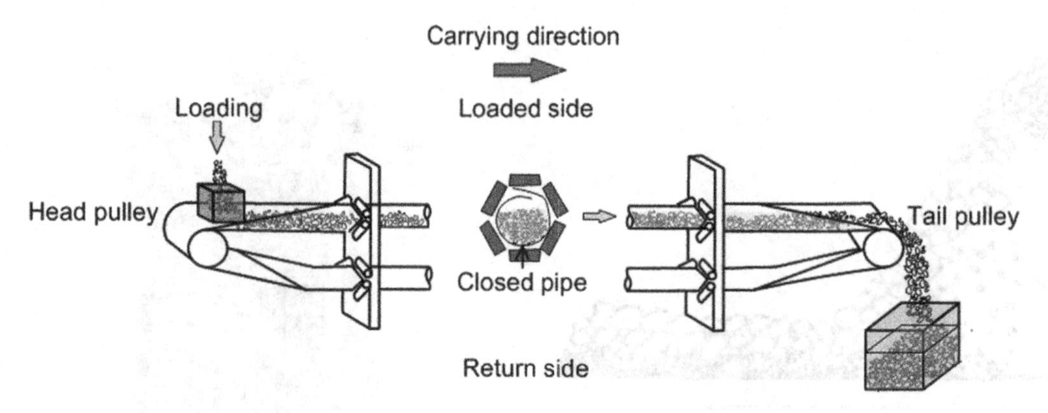

FIGURE 2.8 Pipe conveyor; note the return side can also be loaded.

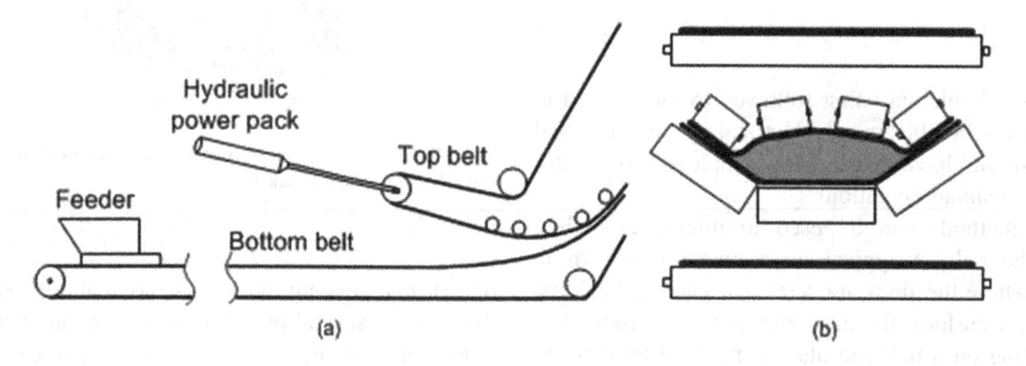

FIGURE 2.9 Sandwich conveyor: (a) side view, and (b) sectional view.

All these methods give only low handling rates with both horizontal conveying and elevating of the material. The gravity bucket elevators consist of a continuous line of buckets attached by pins to two endless roller chains running on tracks and driven by sprockets. The buckets are pivoted so that they always remain in an upright position and are dumped by means of a ramp placed to engage a shoe on the bucket, thus turning it into the dumping position.

Pipe conveyors were developed in Japan and are now being marketed throughout the world. This belt conveyor, after being loaded, is transformed into a pipe by arranging idlers. Five or six idlers are used to achieve the belt wrapping. The conveyor is unwrapped at the discharge point using idler arrangement and the belt is then passed over the head pulley in the conventional manner (Figure 2.8). The pipe conveyors are clean and environment friendly, especially when it comes to transporting hazardous material. This conveyor is compact and does not require a hood. It can efficiently negotiate horizontal curves with a short turning radius (McGuire, 2009).

Sandwich conveyor systems can be used to transport solids at steep inclines from 30° to 90°. The material being transported is "sandwiched" between two belts that hold the material in position and prevent it from sliding, rolling back or leaking back down the conveyor even after the conveyor has stopped or tripped. As pressure is applied to material to hold it in place, it is important that the material has a reasonable internal friction angle. Pressure is applied through roller arrangements, special lead-filled belts, belt weights, and tension in the belt (Figure 2.9). The advantage of sandwich belt conveyors is that they can transport material at steep angles at similar speeds to conventional belt conveyors (Walker, 2012). These belts can achieve high capacity and high lift. They can also accommodate large lumps. Belt cleaners can be used for cleaning the belts. The disadvantages include extra mechanical components and maintenance. These belts cannot elevate fine materials effectively (Anon., 2014).

Screw conveyors, also termed auger conveyors, are another means of transporting dry or damp particles. The material is pushed along a trough by the rotation of a helix, which is mounted on a central shaft. The action of the screw conveyor allows for virtually any degree of mixing of different materials and allows for the

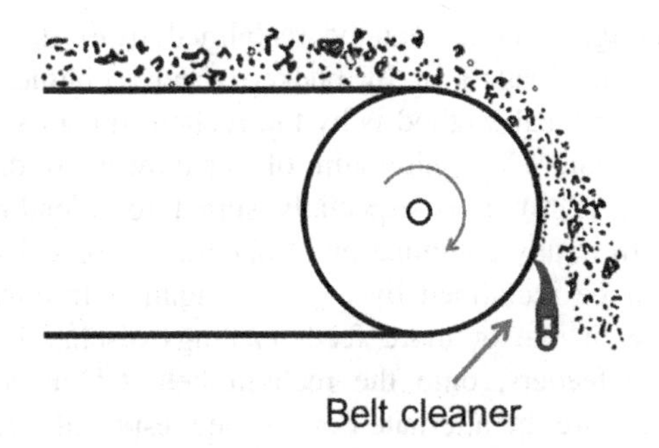

FIGURE 2.10 Conveyor belt cleaner.

transportation of material on any incline from the horizontal to vertical. The screw conveyors are also capable of moving nonfree flowing and semi-solid materials (slurries) as well (Bolat and Boğoçlu, 2012; Patel et al., 2013). The main limitation of screw conveyors is the feed particle size and capacity, which has a maximum of ca. 300 m^3 h^{-1} (Perry and Green, 1997).

Belt cleaners and washing systems are installed when handling sticky material (Figure 2.10). A fraction of the sticky material clings to the belt conveyor surface, which must be removed. Residual sticky material on the belt if not removed is carried back by the belt on the return side and may fall off at different points along the belt causing maintenance and housekeeping problems. The carry-back material also causes excessive wear, build up on return idlers, and misalignment and damage to the belt due to the accumulation of material. Belt cleaners and washing systems are generally installed near the discharge point (Anon., 2014).

Hydraulic transport of the ore stream normally takes over from dry transportation when ore is mixed with water, which is the grinding stage in most mills. Pulp may be made to flow through open *launders* by gravity in some cases. Launders, also termed flumes, are gently sloping troughs of rectangular, triangular, or semicircular section, in which the solid is carried in suspension, or by sliding or rolling. In mills, only fine sized ores (−5 mm) are transported. The slope must increase with particle size, with the solid content of the suspension, and with specific gravity of the solid. The effect of depth of water is complex: if the particles are carried in suspension, a deep launder is advantageous because the rate of solid transport is increased. If the particles are carried by rolling, a deep flow may be disadvantageous.

In plants of any size, the pulp is moved through piping via centrifugal pumps. Pipelines should be as straight as possible to prevent abrasion at bends. Abrasion can be reduced by using liners inside the pipelines. The use of oversize pipe is to be avoided whenever slow motion might allow the solids to settle and hence choke the pipe. The factors involved in pipeline design and installation are complex and include the solid–liquid ratio, the average pulp density, the density of the solid constituents, the size analysis and particle shape, and the fluid viscosity (Loretto and Laker, 1980).

Centrifugal pumps are cheap in capital cost and maintenance and occupy little space (Wilson, 1981; Pearse, 1985). Single-stage pumps are normally used, lifting up to 30 m and in extreme cases 100 m. The merits of centrifugal pumps include no drive seals, very little friction is produced in the pump, and almost no heat is transferred from the motor. Also, the centrifugal pump is not prone to breakage due to the coupling arrangement of the pump and motor (Wilson et al., 2006; Gülich, 2010). Their main disadvantage is the high velocity produced within the impeller chamber, which may result in serious wear of the impeller and chamber itself, especially when coarse sand is being pumped.

2.4 ORE STORAGE

The necessity for storage arises from the fact that different parts of the operation of mining and milling are performed at different rates, some being intermittent and others continuous, some being subject to frequent interruption for repair and others being essentially batch processes. Thus, unless reservoirs for material are provided between the different steps, the whole operation is rendered spasmodic and, consequently, uneconomical. Ore storage is a continuous operation that runs 24 h a day and 7 days a week. The type and location of the material storage depends primarily on the feeding system. The ore storage facility is also used for blending different ore grades from various sources.

The amount of storage necessary depends on the equipment of the plant as a whole, its method of operation, and the frequency and duration of regular and unexpected shutdowns of individual units.

For various reasons, at most mines, ore is hoisted for only a part of each day. On the other hand, grinding and concentration circuits are most efficient when running continuously. Mine operations are more subject to unexpected interruption than mill operations, and coarse-crushing machines are more subject to clogging and breakage than fine crushers, grinding mills, and concentration equipment. Consequently, both the mine and the coarse-ore plant should have a greater hourly capacity than the fine crushing and grinding plants, and storage reservoirs should be provided between them. Ordinary mine shutdowns, expected or unexpected, will not generally exceed a 24 h duration, and ordinary coarse-crushing plant repairs can be made within an equal period if a good supply of spare parts is kept on hand.

Therefore, if a 24 h supply of ore that has passed the coarse-crushing plant is kept in reserve ahead of the mill proper, the mill can be kept running independent of shutdowns of less than a 24 h duration in the mine and coarse-crushing plant. It is wise to provide for a similar mill shutdown and, in order to do this, the reservoir between coarse-crushing plant and mill must contain at all times unfilled space capable of holding a day's tonnage from the mine. This is not economically possible, however, with many of the modern very large mills; there is a trend now to design such mills with smaller storage reservoirs, often supplying less than a two-shift supply of ore, the philosophy being that storage does not *do* anything to the ore, and can, in some cases, has an adverse effect by allowing the ore to oxidize. Unstable sulfides must be treated with minimum delay, the worst case scenario being *self-heating* with its attendant production and environmental problems (Section 2.6). Wet ore cannot be exposed to extreme cold as it will freeze and become difficult to move.

Storage has the advantage of allowing blending of different ores so as to provide a consistent feed to the mill. Both tripper and shuttle conveyors can be used to blend the material into the storage reservoir. If the units shuttle back and forth along the pile, the materials are layered and mix when reclaimed. If the units form separate piles for each quality of ore, a blend can be achieved by combining the flow from selected feeders onto a reclaim conveyor.

Depending on the nature of the material treated, storage is accomplished in stockpiles, bins, or tanks. Stockpiles are often used to store coarse ore of low value outdoors. In designing stockpiles, it is merely necessary to know the angle of repose of the ore, the volume occupied by the broken ore, and the tonnage. The s-tockpile must be safe and stable with respect to thermal conductivity, geomechanics, drainage, dust, and any radiation emission. The shape of a stockpile can be conical or elongated. The conical shape provides the greatest capacity per unit area, thus reduces the plant footprint. Material blending from a stockpile can be achieved with any shape but the most effective blending can be achieved with elongated shape.

Although material can be reclaimed from stockpiles by front-end loaders or by bucket-wheel reclaimers, the most economical method is by the reclaim tunnel system, since it requires a minimum of manpower to operate (Dietiker, 1980). It is especially suited for blending by feeding from any combination of openings. Conical stockpiles can be reclaimed by a tunnel running through the center, with one or more feed openings discharging via gates, or feeders, onto the reclaim belt. Chain scraper reclaimers are the alternate device used, especially for the conical stock pile. The amount of reclaimable material, or the *live storage*, is about 20–25% of the total (Figure 2.11). Elongated stockpiles are reclaimed in a similar manner, the live storage being 30–35% of the total (Figure 2.12).

For continuous feeding of crushed ore to the grinding section, feed bins are used for transfer of the coarse material from belts and rail and road trucks. They are made of wood, concrete, or steel. They must be easy to fill and must allow a steady fall of the ore through to the discharge gates with no "hanging up" of material or opportunity for it to segregate into coarse and fine fractions. The discharge must be adequate and drawn from several alternative points if the bin is large. Flat-bottom bins cannot be emptied completely and retain a substantial tonnage of dead rock. This, however, provides a cushion to protect the bottom from wear, and such bins are easy to construct. This type of bin, however, should not be used with easily oxidized ore, which might age dangerously and mix with the fresh ore supply. Bins with sloping bottoms are better in such cases.

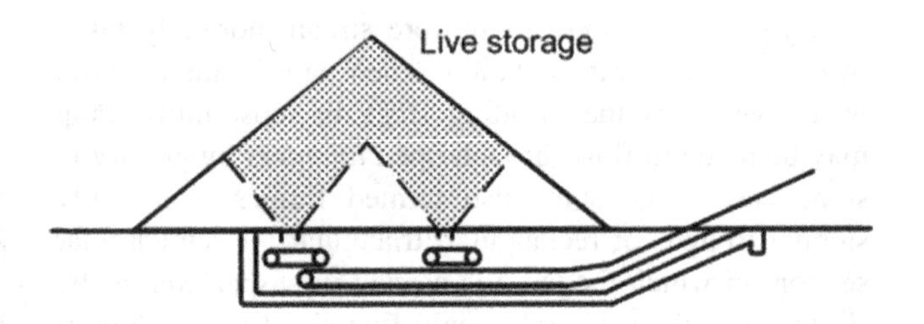

FIGURE 2.11 Reclamation from conical stock pile.

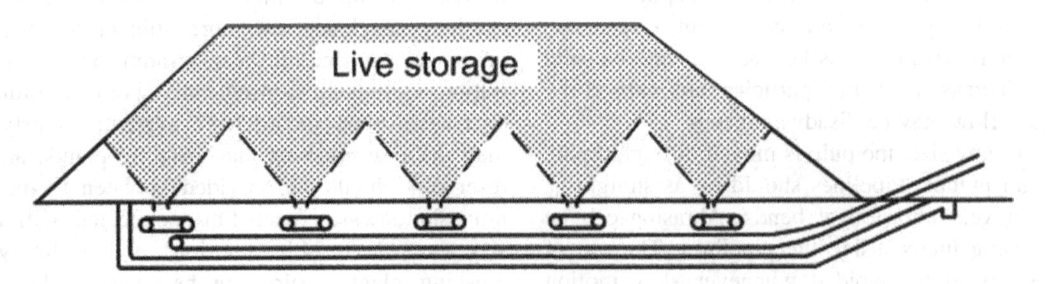

FIGURE 2.12 Reclamation from elongated stock pile.

Pulp storage on a large scale is not as easy as dry ore storage. Conditioning tanks are used for storing suspensions of fine particles to provide time for chemical reactions to proceed. These tanks must be agitated continuously, not only to provide mixing but also to prevent settlement and choking up. Surge tanks are placed in the pulp flow-line when it is necessary to smooth out small operating variations of feed rate. Their content can be agitated by stirring, by blowing in air, or by circulation through a pump.

2.5 FEEDING

Feeders are necessary whenever it is desired to deliver a uniform stream of dry or moist ore, since such ore will not flow evenly from a storage reservoir of any kind through a gate, except when regulated by some type of mechanism.

Feeding is essentially a conveying operation in which the distance travelled is short and in which close regulation of the rate of passage is required. Where succeeding operations are at the same rate, it is unnecessary to interpose feeders. Where, however, principal operations are interrupted by a storage step, it is necessary to provide a feeder. Feeders also reduce wear and tear, abrasion, and segregation. They also help in dust control and reduce material spillage. Feeder design must consider desired flow rates, delivery of a stable flow rate, the feeding direction, and particle size range of feed to be handled (Roberts, 2001).

A typical feeder consists of a small bin, which may be an integral part of a large bin, with a gate and a suitable conveyor. The feeder bin is fed by chutes, delivering ore under gravity. Feeders of many types have been designed, notably apron, belt, chain, roller, rotary, revolving disc, drum, drag scraper, screw, vane, reciprocating plate, table, and vibrating feeders. Sometimes feeders are not used and instead feeding is achieved by chutes only. Factors like type of material to be handled, the storage method, and feed rate govern the type of feeder (Anon., 2014).

In the primary crushing stage, the ore is normally crushed as soon as possible after its arrival. Many underground mines have primary crushers underground to reduce ore size and improve hoisting efficiency. Skips, trucks, and other handling vehicles are intermittent in arrival, whereas the crushing section, once started, calls for steady feed. Surge bins provide a convenient holding arrangement able to receive all the intermittent loads and to feed them steadily through gates at controllable rates. The chain-feeder (Figure 2.13) is sometimes used for smooth control of bin discharge.

The chain-feeder consists of a curtain of heavy loops of chain, lying on the ore at the outfall of the bin at approximately the angle of repose. The rate of feed is controlled automatically or manually by the chain sprocket drive such that when the loops of chain move, the ore on which they rest begins to slide.

Primary crushers depend for normal operation on the fact that broken rock contains a certain voidage (space between particles). If all the feed goes to a jaw crusher without a preliminary removal of fines, there can be danger of choking when there has been segregation of coarse and fine material in the bin. Such fines could pass through the upper zones of the crusher and drop into the finalizing zone and fill the voids. Should the bulk arriving at any level exceed that departing, it is as though an attempt is being made to compress solid rock. This choking, that is, packing of the crushing chamber (or "bogging"), is just as serious as tramp iron in the crusher and likewise can cause major damage. It is common practice, therefore, to "scalp" the feed to the crusher, heavy-duty screens known as *grizzlies* normally preceding the crushers and removing fines.

Primary crusher feeds, which scalp and feed in one operation, have been developed, such as the vibrating grizzly feeder (Chapter 8). The elliptical bar feeder (Figure 2.14) consists of elliptical bars of steel which form the bottom of a receiving hopper and are set with the long axes of the ellipses in alternate vertical and horizontal positions. Material is dumped directly onto the bars, which rotate in the same direction, all at the same time, so that the spacing remains constant. As one turns down, the succeeding one turns up, imparting a rocking, tumbling motion to the load. This works to loosen the fines, which sift through the load directly on to a conveyor belt, while the oversize is moved forward to deliver

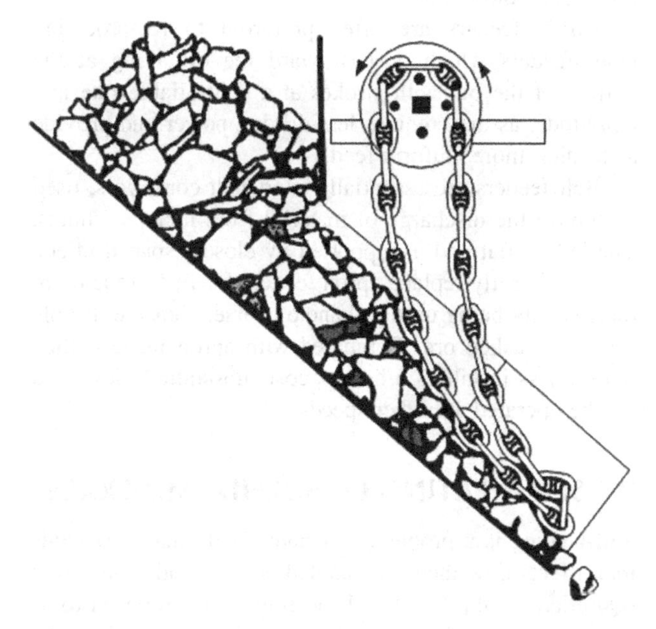

FIGURE 2.13 Side view of a chain-feeder.

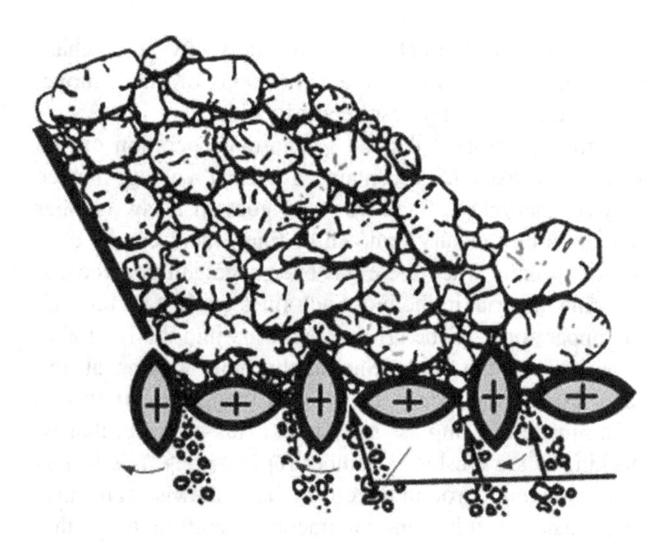

FIGURE 2.14 Cross section of elliptical bar feeder.

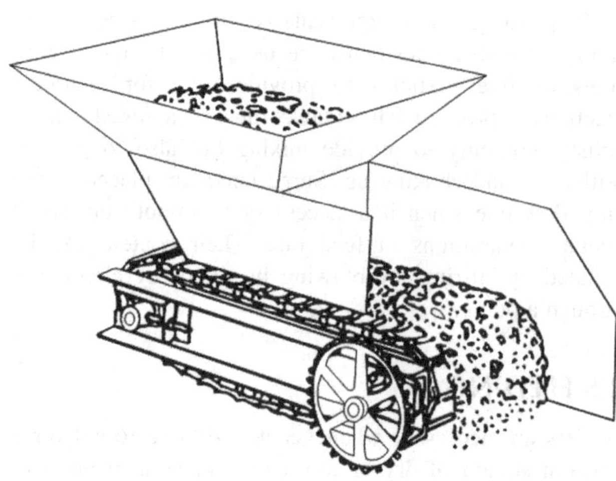

FIGURE 2.15 Apron feeder.

to the crusher. This type of feeder is probably better suited to handling high clay or wet materials such as laterite, rather than hard, abrasive ores.

The apron feeder (Figure 2.15) is one of the most widely used feeders for handling coarse ore, especially jaw crusher feed. The overlapping metal plates or pans mounted on strands of conveyor chains convey the material (Anon., 2014). It is ruggedly constructed, consisting of a series of high carbon or manganese steel pans, bolted to strands of heavy-duty chain, which run on steel sprockets. The rate of discharge is controlled by varying the speed or the height of the ribbon of ore by means of an adjustable gate. It can handle abrasive, heavy, and lumpy materials (Anon., 2014).

Apron feeders are often preferred to reciprocating plate feeders which push forward the ore lying at the bottom of the bin with strokes at a controllable rate and amplitude, as they require less driving power and provide a steadier, more uniform feed.

Belt feeders are essentially short belt conveyors, used to control the discharge of material from inclined chutes. The belt is flat and is supported by closely spaced idlers. They frequently replace apron feeders for fine ore and are increasingly being used to handle coarse, abrasive, friable primary crushed ore. Compared with apron feeders, they require less installation height, cost substantially less, and can be operated at higher speeds.

2.6 SELF-HEATING OF SULFIDE MINERALS

Self-heating is a problem associated with many materials that affects how they are handled, stored, and transported (Quintiere et al., 2012). Self-heating is also referred to as *spontaneous heating* and *pyrophoric behavior* and results when the rate of heat generation (due to oxidation) exceeds the rate of heat dissipation. In the minerals industry, environmental effects of self-heating for coals are well documented, from the production of toxic fumes (CO, NO_x, SO_2) and greenhouse gases (CH_4, CO_2), to the contamination of runoff water (Kim, 2007; Stracher, 2007). There is also growing concern over self-heating of sulfides as regulations for shipping tighten (Anon., 2011).

Many base metals occur in nature as mineral sulfides, a form which has made their extraction, concentration, and conversion into metals a challenge, but a challenge that has been successfully met by technologies such as flotation, leaching, and autogenous smelting. The propensity of sulfur-containing materials to oxidize is largely the reason for the successful extraction of these metals, as well as the source of some of the associated problems of base metal processing. These problems include acid rock (acid mine) drainage (see Chapter 16), dust explosions, and self-heating of ores, concentrates, waste rock, tailings, and mine paste fill.

Heating may occur when the sulfide material is contained or piled in sufficient quantity (i.e., reducing the heat dissipation rate), with both oxygen (air) and some moisture present (ca. 3–8% by weight). If conditions are favorable, and this includes long storage times, presence of fine particles, high relative humidity, and temperatures exceeding 30°C, heating can proceed beyond 100°C, at which point SO_2 gas begins to evolve and may continue to drive temperatures well in excess of 400°C (Rosenblum et al., 2001).

Figure 2.16 shows examples of waste rock (a) and concentrate (b) that have heated beyond 100°C leading to evolution of SO_2. "Hot muck" underground at the Sullivan lead–zinc mine in British Columbia, Canada,

FIGURE 2.16 Examples of self-heating of sulfides: (a) steam (and SO_2) emanating from sulfide ore waste dump, and (b) high temperature and SO_2 emanating from stockpiled copper sulfide concentrate *((a) Courtesy T. Krolak; (b) courtesy F. Rosenblum).*

made the cover of the *CIM Bulletin* magazine (June 1977). That mine first reported issues with self-heating of ore in 1926 (O'Brien and Banks, 1926), illustrating that the issues associated with sulfide self-heating have been around for considerable time. The sinking of the N.Y.K. Line's SS Bokuyo Maru in 1939 was attributed to spontaneous combustion of copper concentrate (Kirshenbaum, 1968). The consequences are rarely so dramatic but can result in significant storage and transportation issues that may threaten infrastructure and the workplace environment.

Dealing with materials that have the potential to self-heat requires an understanding of the material reactivity and a proactive risk management approach (Rosenblum, et al., 2001). A variety of single-stage testing methods are in use for different materials with potential for self-heating (e.g., coal, wood chips, powdered milk). However, mineral sulfides require a two-stage assessment: one that mimics weathering (i.e., oxidation) at near ambient conditions and where elemental sulfur is created, followed by a higher temperature stage above 100°C to assess the impact of the weathering stage and where the elemental sulfur is oxidized to form SO_2 (Rosenblum et al., 2014).

It is thought that the reactions governing self-heating are electrochemical as well as thermodynamic in origin (Payant et al., 2012; Somot and Finch, 2010). Pure sulfide minerals do not readily self-heat, the exception being pyrrhotite ($Fe_{1-x}S$), likely due to its nonstochiometric excess of sulfur. Payant et al. (2012) have reported that a difference in the electrochemical rest potential between minerals in a binary mixture needs to exceed 0.2 V in order for self-heating to proceed. From Table 2.1, this means the pyrite–galena mix will self-heat, and that pyrite will accelerate self-heating of pyrrhotite, as observed experimentally (Payant et al., 2012). (See Chapter 12 for discussion of electrochemical effects.)

Mitigation strategies used to control the risk of self-heating include controlling pyrrhotite content to below 10 %wt, monitoring for hot-spots with infrared thermal detectors, blending any hot material with cooler material, "blanketing" with CO_2 (in ships' holds, and storage sheds), drying to below 1 %wt moisture, and sealing with plastic (e.g., shipping concentrate in tote bags) to eliminate oxygen. The addition of various chemical agents to act as oxidation inhibitors is reportedly also practiced.

TABLE 2.1 Rest Potential Values of Some Sulfide Minerals

Mineral	Formula[a]	Rest Potential (vs. SHE) (V)
Pyrite	FeS_2	0.66
Chalcopyrite	$CuFeS_2$	0.56
Sphalerite	ZnS	0.46
Pentlandite	$(Fe,Ni)_9S_8$	0.35
Pyrrhotite	$Fe_{(1-x)}S$	0.31
Galena	PbS	0.28

[a]*Nominal formula, natural samples can vary.*
Source: From Payant et al. (2012).

REFERENCES

Alspaugh, M., 2008. Bulk Material Handling by Conveyor Belt 7. SME, Littleton, CO., USA.

Anon, 2011. The International Maritime Solid Bulk Cargoes (IMSBC) Code adopted by Resolution MSC. 268(85) (Marine Safety Committee). DNV Managing Risk, UK.

Anon. (2014). Belt conveyors for bulk materials. Report by members of CEMA Engineering Conference, Conveyor Equipment Manufacturers Association (CEMA), CBI Pub. Co., Boston, MA, USA.

Bolat, B., Boğoçlu, M.E. (2012). Increasing of screw conveyor capacity. Proceedings of the 16th International Research/Expert Conference, Trends in the Development of Machinery and Associated Tech., TMT, Dubai, UAE, pp. 515–518.

Dietiker, F.D., 1980. Belt conveyor selection and stockpiling and reclaiming applications. In: Mular, A.L., Bhappu, R.B. (Eds.), Mineral Processing Plant Design, second ed. SME, New York, NY, USA, pp. 618–635. (Chapter 30).

Gülich, J.F., 2010. Centrifugal Pumps. Second ed. Springer, New York, NY, USA.

Kim, A.G., 2007. Greenhouse gases generated in underground coal-mine fires. Rev. Eng. Geol. 18, 1–13.

Kirshenbaum, N.W., 1968. Transport and Handling of Sulphide Concentrates: Problems and Possible Improvements. Second ed. Technomic Publishing Company Inc., Pennsylvania, PA, USA.

Loretto, J.C., Laker, E.T., 1980. Process piping and slurry transportation. In: Mular, A.L., Bhappu, R.B. (Eds.), Mineral Processing Plant Design, second ed. SME, New York, NY, USA, pp. 679–702. (Chapter 33).

McGuire, P.M., 2009. Conveyors: Application, Selection, and Integration. Industrial Innovation Series. Taylor & Francis Group, CRC Press, UK.

O'Brien, M.M., Banks, H.R., 1926. The Sullivan mine and concentrator. CIM Bull. 126, 1214–1235.

Patel, J.N., et al., 2013. Productivity improvement of screw conveyor by modified design. Int. J. Emerging Tech. Adv. Eng. 3 (1), 492–496.

Payant, R., et al., 2012. Galvanic interaction in self-heating of sulphide mixtures. CIM J. 3 (3), 169–177.

Pearse, G., 1985. Pumps for the minerals industry. Mining Mag. Apr, 299–313.

Perry, R.H., Green, D.W., 1997. Perry's Chemical Engineers' Handbook. seventh ed. McGraw-Hill, New York, NY, USA.

Quintiere, J.G., et al. (2012). Spontaneous ignition in fire investigation. U.S. Department of Justice, National Critical Justice Reference Service Library, NCJ Document No: 239046.

Ray, S., 2008. Introduction to Materials Handling. New Age International (P) Ltd., New Delhi, India.

Roberts, A.W., 2001. Recent developments in feeder design and performance. In: Levy, A., Kalman, H. (Eds.), Handbook of Conveying and Handling of Particulate Solids, Vol.10. Elsevier, Amsterdam, New York, NY, USA, pp. 211–223.

Rosenblum, F., et al., 2001. Evaluation and control of self-heating in sulphide concentrates. CIM Bull. 94 (1056), 92–99.

Rosenblum, F., et al., 2014. Review of self-heating testing methodologies. Proceedings of the 46th Annual Meeting Canadian Mineral Processors Conference, CIM, Ottawa, Canada, pp. 67–89.

Somot, S., Finch, J.A., 2010. Possible role of hydrogen sulphide gas in self-heating of pyrrhotite-rich materials. Miner. Eng. 23 (2), 104–110.

Stracher, G.B., 2007. Geology of Coal Fires: Case Studies from Around the World. Geological Society of America, Boulder, CO, USA, 279.

Walker, S.C., 2012. Mine Winding and Transport. Elsevier, New York, NY, USA.

Wilson, G., 1981. Selecting centrifugal slurry pumps to resist abrasive wear. Mining Eng., SME, USA. 33, 1323–1327.

Wilson, K.C., et al., 2006. Slurry Transport Using Centrifugal Pumps. third ed. Springer, New York, NY, USA.

Chapter 3

Sampling, Control, and Mass Balancing

3.1 INTRODUCTION

This chapter deals with the collection, analysis, and use of process data. Collection of reliable data is the science of sampling. The collected samples are analyzed for some quality, metal content (assay), particle size, etc. and the data are used for process control and metallurgical accounting.

Computer control of mineral processing plants requires continuous measurement of such parameters. The development of real-time on-line sensors, such as flowmeters, density gauges, and chemical and particle size analyzers, has made important contributions to the rapid developments in this field since the early 1970s, as has the increasing availability and reliability of cheap microprocessors.

Metallurgical accounting is an essential requirement for all mineral processing operations. It is used to determine the distribution and grade of the values in the various products of a concentrator in order to assess the economic efficiency of the plant (Chapter 1). The same accounting methods are used in plant testing to make decisions about the operation. To execute metallurgical accounting requires mass balancing and data reconciliation techniques to enhance the integrity of the data.

3.2 SAMPLING

Sampling the process is often not given the consideration or level of effort that it deserves: in fact, the experience is that plants are often designed and built with inadequate planning for subsequent sampling, this in spite of the importance of the resulting data on metallurgical accounting, process control or plant testing and trials (Holmes, 1991, 2004). A simple example serves to illustrate the point. A typical porphyry copper concentrator may treat 100,000 tons of feed per day. The composite daily sub-sample of feed sent for assay will be of the order of 1 gram. This represents 1 part in 100 billion and it needs to be representative to within better than ±5%. This is no trivial task. Clearly the sampling protocol needs to produce samples that are as representative (free of bias, i.e., accurate) and with the required degree of precision (confidence limits) for an acceptable level of cost and effort. This is not straightforward as the particles to be sampled are not homogenous, and there are variations over time (or space) within the sampling stream as well. Sampling theory has its basis in probability, and a good knowledge of applied statistics and the sources of sampling error are required by those assigned the task of establishing and validating the sampling protocol.

3.2.1 Sampling Basics: What the Metallurgist Needs to Know

In order to understand the sources of error that may occur when sampling the process, reference will be made to probability theory and classical statistics—the *Central Limit Theorem* (CLT)—which form the basis for modern sampling theory. Additionally important to understanding the source of sampling error is the *additive nature of variance*, that is, that the total variance of a system of independent components is the sum of the individual variances of the components of that system. In other words, errors (standard deviation of measurements) do not cancel out; they are cumulative in terms of their variance. In general sampling terms this may be expressed as (Merks, 1985):

$$\sigma^2_{\text{overall}} = \sigma^2_{\text{composition}} + \sigma^2_{\text{distribution}} + \sigma^2_{\text{preparation}} + \sigma^2_{\text{analysis}} \quad (3.1)$$

where $\sigma^2_{\text{composition}}$, $\sigma^2_{\text{distribution}}$, $\sigma^2_{\text{preparation}}$, and $\sigma^2_{\text{analysis}}$ are the respective variances associated with error due to: i) heterogeneity among individual particles in the immediate vicinity of the sampling instrument, ii) variation in space (large volume being sampled) or over the time period being sampled, iii) the stages of preparation of a suitable sub-sample for analysis, and iv) the elemental or sizing analysis of the final sub-sample itself. This representation is convenient for most purposes, but total error may be subdivided even further, as addressed by Gy (1982), Pitard (1993), and Minnitt et al. (2007) with up to ten sources of individual sampling error identified. For the purposes of the discussion here, however, we will use the four sources identified above.

Some appropriate terminology aids in the discussion. The *sampling unit* is the entire population of particles being sampled in a given volume or over time; the *sampling increment* is that individual sample being taken at a point in time or space; a *sample* is made up of multiple sampling increments or can also be a generic term for any subset of the overall sampling unit; a *sub-sample* refers to a sample that has been divided into a smaller subset usually to facilitate processing at the next stage. It is important to recognize that the true value or *mean* of the sampling unit (i.e., entire population of particles) can never be known, there will always be some error associated with the samples obtained to estimate the true mean. This is because, as noted, the individual sampled particles are not homogeneous, that is, they vary in composition. This notion of a minimum variance that is always present is referred to as the *fundamental sampling error.* Good sampling protocol will be designed to minimize the overall sampling variance as defined in Eq. (3.1), in a sense, to get as close as practical to the fundamental sampling error which itself cannot be eliminated.

One of the pioneers of sampling theory was Visman (1972) who contributed a more useful version of Eq. (3.1):

$$\sigma_t^2 = \frac{\sigma_c^2}{m_i n} + \frac{\sigma_d^2}{n} + \sigma_p^2 + \frac{\sigma_a^2}{l} \quad (3.2)$$

where σ_t^2, σ_c^2, σ_d^2, σ_p^2, σ_a^2 = variance: total, composition, distribution, preparation, and analysis; m_i = mass of sample increment; n = number of sampling increments; and l = number of assays per sample.

It is evident from Eq. (3.2) that the error associated with composition can be reduced by increasing both the mass and the number of sampling increments, that the error associated with distribution can be reduced by increasing the number of increments but not their mass; and that assay error can be reduced by increasing the number of assays per sample. We will now proceed to discuss each of the components of the overall sampling variance in more detail.

Composition Variance

Also known as composition heterogeneity, it is dependent on factors that relate to the individual particles collected in an increment, such as particle top size, size distribution, shape, density and proportion of mineral of interest, degree of liberation, and sample size, and is largely the basis for Pierre Gy's well-known fundamental equations and *Theory of Sampling* (Gy, 1967, 1982). Gy's equation for establishing the fundamental sampling error will be detailed later in this section. To illustrate the importance

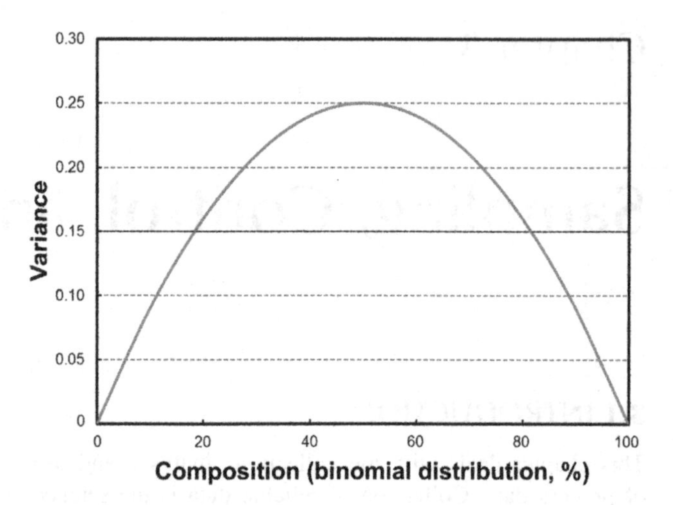

FIGURE 3.1 Composition variance as a function of mineral proportion p in a two component system (Eq. (3.3)).

of relative mineral proportions in composition variance, consider a particle comprised of two components, p and q. The composition variance can be described as:

$$\sigma^2_{\text{composition}} \propto pq \text{ or } \sigma^2_{\text{composition}} \propto p(1-p) \quad (3.3)$$

Figure 3.1 shows the composition variance as a function of mineral proportion p over the range 0 to 100%. Variance is at a maximum when the proportion of p in particles is at 50% and at a minimum as it approaches either 0% or 100%. Clearly, composition variance will be minimized the closer we get to individual mineral particle liberation. Achieving (at or near) liberation is therefore an important element of sampling protocols, particularly in the sub-sampling stages for coarser materials.

This is an important factor that needs to be understood about the system that is being sampled. The micrographs in Chapter 1 Figure 1.5 illustrate that various mineralogy textures will produce very different liberation sizes.

Selecting the appropriate sample size is also key to minimizing composition variance as noted in Eq. (3.2), and there are many examples of underestimating and overestimating precious metal content that can be traced back to sample sizes that were too small. Based on a computational simulation, the following two examples illustrate the problems associated with sample size (Example 3.1 and 3.2).

A program of sampling plus mineralogical and liberation data are important components of establishing the appropriate sampling protocol in order to minimize composition variance. We will now examine the second main source of overall sampling error, the distribution variance.

Distribution Variance

Also known as distribution heterogeneity, or *segregation*, this occurs on the scale of the sampling unit, be that in

Example 3.1

Determine an appropriate sample size for mill feed (minus 12 mm) having a valuable mineral content of 50% (e.g., iron ore).

Solution

From the simulation, sample increment sizes ranging from 10 to 10,000 g were taken 100 times to generate statistical data that are free from the effect of sampling frequency. Table ex 3.1 and Figure ex 3.1-S illustrate that total error (expressed as relative standard deviation, σ, or coefficient of variation) reaches a limiting value of about 5% when the sample size is 5 kg or greater. In a sense, this value approaches the fundamental sampling error. The pitfall of selecting a sample size that is too small is clearly evident from the table and figure. Coarser materials require a significantly larger sample size in order to minimize composition error.

The sampling of materials having low valuable mineral content (e.g., parts per million) also increases the risk of large errors if sampling increments are too small. This becomes even more pronounced if there are large density differences between valuable mineral and gangue.

TABLE EX 3.1 Effect of Sample Increment Size on Sampling Error (expressed as relative standard deviation)

Increment Size[a] (g)	Mean Assay (%)	Number of Assays within ±2.5%	Relative Standard Deviation (%)
10	46.7	14	88.55
100	49.7	24	45.6
500	50.35	37	18.38
1,000	50.08	74	14.8
2,500	50.18	86	9.94
3,500	49.82	93	7.09
5,000	50.12	98	5.1
10,000	49.97	99	5.01

[a]Each of 100 sample increments true mean = 50%

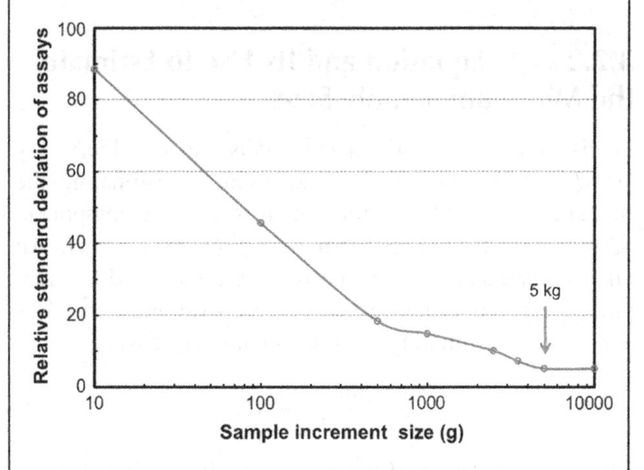

FIGURE EX 3.1-S A plot of Table ex 3.1 data showing the reduction in sampling error as sampling increment size is increased. A 5 kg sample size is indicated as suitable.

Example 3.2

Consider a 500 g test sample contains 25 particles of free (i.e., liberated) gold. Sub-samples of 15, 30, and 60 g are repeatedly selected to generate frequency distributions of the number of gold particles in each of the sub-sample sets.

Solution

Note that, on average, the 15 g sub-sample set should contain 0.75 gold particles; the 30 g set should contain 1.5 gold particles and the 60 g set, 3 gold particles. The resulting frequency distributions are presented in Table ex 3.2. It is clear that too small a sub-sample size (15 and 30 g) results in a skewed frequency distribution and a corresponding high probability of significantly underestimating or overestimating the true gold content. The table shows that for a 15 g

TABLE EX 3.2 Frequency Distribution Data of Gold Particle Content for 15, 30 and 60 g Sub-sample Sets

Predicted Number of Au Particles in Sample	Frequency Distribution (%)		
	Incremental Sample Size		
	60 g	30 g	15 g
0	5	22.3	47.2
1	14.9	33.5	35.4
2	22.4	25.1	13.3
3	22.4	12.6	3.3
4	16.8	4.7	0.6
5	10.1	1.4	0.1
6	5	0.4	
7	2.2	0.1	
8	0.8		
9	0.3		
10	0.1		

Data from Merks, Sampling and Statistics, geostats.com website sample size there is an almost 50% chance of having zero gold content (47.2%) and a similarly high probability (approaching 50%) of significantly overestimating the gold content. This risk of seriously under- or over-estimating gold content due to insufficient sample size is sometimes referred to as the *nugget effect*. (Example taken from J. Merks, Sampling and Statistics, website: geostatscam.com).

space (e.g., a stockpile or process vessel) or over time (e.g., a 24-hour mill feed composite). Clearly there will be variation in samples taken within the sampling unit and so a single *grab* or *point* sample has virtually zero probability of being representative of the entire population of particles. Multiple samples are therefore required. A well-designed sampling *protocol* is needed to minimize distribution error, including sampling bias. Such a protocol should be based, as much as is practical, on the *ideal sampling model* which states that:

- All *strata* of the volume to be sampled are included in the sampling increment; for example, cut the entire stream or flow of a conveyor or pipe, sample top to bottom in a railcar or stockpile
- Sample increment collected is proportional to the mass throughput, that is, weighted average
- Randomized sequence of sampling to avoid any natural frequency within the sampling unit
- All particles have an equal probability of being sampled

The ideal sampling model, sometimes referred to as the *equi-probable* or *probabilistic sampling model*, is often not completely achievable but every attempt within reason should be made to implement a sampling protocol that adheres to it as closely as possible. Sampling a conveyor belt or a flowing pipe across the full width of the flow at a discharge point using a sampling device that does not exclude any particles, at random time intervals and in proportion to the mass flow, is a preferred protocol that achieves close to the ideal sampling model. This can be considered as a linear or *one-dimensional* model of sampling. An example of a *two-dimensional* model would be a railcar containing concentrate where sampling increments are collected top to bottom by auger or pipe on a grid system (randomized if possible). An example of a *three-dimensional* model would be a large stockpile or slurry tank where top to bottom sampling increments are not possible and where significant segregation is likely. Point sampling on a volumetric grid basis or surface sampling of a pile as it is being constructed or dismantled can be used in three-dimensional cases, but such a protocol would deviate significantly from the ideal sampling model. It is far better to turn these three-dimensional situations into one-dimensional situations by sampling when the stockpile or tank volume is being transferred or pumped to another location.

It is evident that, in order to establish a proper sampling protocol that minimizes composition and distribution error, information is needed about the total size of, and segregation present in, the sampling unit, the size distribution and mineralogy of the particles, approximate concentrations of the elements/minerals of interest, and the level of precision (i.e., acceptable error) required from the sampling.

Preparation Variance and Analysis Variance

Equation (3.1) indicates that variances due to preparation and analysis also contribute to the overall variance. Merks (1985) has noted that variance due to preparation and analysis is typically much smaller than for distribution and composition. Illustrating the point for 1 kg sample increments of coal analyzed for ash content, Merks (1985) shows what are typical proportions of error: composition, 54%; distribution, 35%; preparation, 9%; and analysis, 2%. Minnitt et al. (2007) provide estimates of 50–100%, 10–20%, and 0.1–4% for composition, distribution and preparation/analytical error, respectively, as a portion of total error similar to Merks.

Estimates of $\sigma^2_{\text{analysis}}$ have been obtained by performing multiple assays on a sub-sample, while $\sigma^2_{\text{preparation}}$ can be estimated by collecting, preparing and assaying multiple paired sub-sets of the original sample to yield an estimate of the combined variance of preparation and analysis, and then subtracting away the $\sigma^2_{\text{analysis}}$ component previously established. Merks (2010) provides detailed descriptions of how to determine individual components of variance by the method of *interleaved samples*.

3.2.2 Gy's Equation and Its Use to Estimate the Minimum Sample Size

In his pioneering work in the 1950s and 60s Pierre Gy (1967, 1982) established an equation for estimating the fundamental sampling error, which is the only component of the overall error that can be estimated *a priori*. All other components of error need to be established by sampling the given system. Expressed as a variance, σ_f^2, Gy's equation for estimating the fundamental error is:

$$\sigma_f^2 = \frac{(M - m_s)}{M m_s} C d^3 \tag{3.4}$$

where M = the mass of the entire sampling unit or lot (g); m_s = the mass of the sample taken from M (g); d = nominal size of the particle in the sample, typically taken as the 95% passing size, d_{95}, for the distribution (cm); and C = constant for the given mineral assemblage in M given by:

$$C = f\,g\,l\,m$$

where f is the shape factor relating volume to the diameter of the particles, g the particle size distribution factor, l the liberation factor, and m the mineralogical composition factor for a 2-component system. The factors in C are estimated as follows:

f: The value of $f = 0.5$ except for gold ores when $f = 0.2$.

g: Using 95% and 5% passing sizes to define a size distribution, the following values for g can be selected:

$g = 0.25$ for $d_{95}/d_5 > 4$
$g = 0.5$ for $d_{95}/d_5 = 2 \rightarrow 4$
$g = 0.75$ for $d_{95}/d_5 < 2$
$g = 1$ for $d_{95}/d_5 = 1$

l: Values of l can be estimated from the table below for corresponding values of d_{95}/L, where L is the liberation size (cm), or can be calculated from the expression:

$$l = \left(\frac{L}{d_{95}}\right)^{0.5}$$

d_{95}/L	<1	1–4	4–10	10–40	40–100	100–400	>400
L	1	0.8	0.4	0.2	0.1	0.05	0.02

m: The m can be calculated from the expression:

$$m = \frac{1-a}{a}\left[(1-a)r + at\right]$$

where r and t are the mean densities of the valuable mineral and gangue minerals respectively, and a is the fractional average mineral content of the material being sampled. The value of a could be determined by assaying a number of samples of the material. Modern electron microscopes can measure most of these properties directly. For low grade ores, such as gold, m can be approximated by mineral density (g cm^{-3})/grade (ppm). For a more detailed description of Gy's sampling constant (C), see Lyman (1986) and Minnitt et al. (2007).

In the limit where $M \gg m_s$, which is almost always the case, Eq. (3.4) simplifies to:

$$\sigma_f^2 = \frac{Cd^3}{m_s} \tag{3.5}$$

Note that nm_i, the number of sample increments × increment mass, can be substituted for m_s. Note also that particle size, being cubed, is the more important contributor to fundamental error than is the sample mass.

Use can be made of Eq. (3.5) to calculate the sample size (m_s) required to minimize fundamental error provided the level of precision (σ^2) we are looking for is specified. The re-arranged relationship then becomes:

$$m_s = \frac{Cd^3}{\sigma^2} \tag{3.6}$$

As noted, total error is comprised of additional components (to fundamental sampling error) so a rule of thumb is to at least double the sample mass indicated by the Gy relationship. The literature continues to debate the applicability of Gy's formula (François-Bongarçon and Gy, 2002a; Geelhoed, 2011), suggesting that the fundamental sampling error solution may be overestimated and that the exponent in the liberation factor (l) relationship should not be 0.5 but a variable factor between 0 and 3 depending on the material being sampled. Gold ores would typically have an exponent of 1.5 (Minnitt et al., 2007). Nevertheless, Gy's equation has proven itself a valuable tool in the sampling of mineral streams and continues to evolve. An example calculation is used to illustrate the use of Gy's formula (Example 3.3).

If instead of the crushing product as in Example 3.3 the sampling takes place from the pulp stream after grinding to the liberation size of the ore, then $d_{95} = 0.015$ cm and $d_5 = 0.005$ (assuming classification has given fairly narrow size distribution) and the various factors become $f = 0.5$ (unchanged), $g = 0.5$, $l = 1$, and $m = 117.2$ (unchanged) giving a constant $C = 29.31$ g cm^{-3}. The new calculation for $m_s = 29.31 \times 0.015^3/0.01^2 = 0.989$ g. The advantages of performing the sampling at closer to the liberation size are clearly evident. This advantage is not lost on those preparing samples for assaying where size reduction methods such as grinders and pulverizers are used prior to splitting samples into smaller fractions for assay. See Johnson (2010) for an additional example of the use of Gy's method.

3.2.3 Sampling Surveys

These are often conducted within the plant or process to assess metallurgical performance for benchmarking, comparison, or other process improvement-related purposes. The data collected in such sampling campaigns will, typically, be subjected to *mass balancing* and *data reconciliation* techniques (discussed later in this Chapter) in order to provide a *balanced* data set of mass, chemical and mineralogical elements. The reconciliation techniques (e.g., Bilmat™ software) make small adjustments to the raw data, based on error estimates provided, to yield values that are statistically better estimates of the true values. In spite of such adjustment procedures, it is important that every effort be made to collect representative data, since these procedures will not "correct" data poorly collected. Proper sampling lies at the heart of process measurement

Example 3.3

Consider a lead ore, assaying ~5% Pb, which must be routinely sampled on crusher product for assay to a 95% confidence level of ±0.1% Pb. Assume the top size of the ore is 2.5 cm and the lower size is estimated as 0.1 cm, and that the galena is essentially liberated from the quartz gangue at a particle size of 150 μm.

Solution

σ: The required precision, σ, in relative terms, is calculated from:

$$2\sigma(\sim 95\%\text{ confidence level}) = 0.1\%/5\% = 0.02,\text{ or}$$
$$\sigma = 0.01(\text{i.e., }1\%\text{ error}).$$

C:

f: Since this is not a gold ore the shape factor f is taken as 0.5.

g: Since $d_{95} = 2.5$ cm and $d_5 = 0.1$ cm, the ratio $d_{95}/d_5 = 2.5/0.1 = 25$ giving the particle size distribution factor $g = 0.25$, a wide distribution.

m: Assuming a galena s.g. of 7.6 and gangue s.g. of 2.65, and that galena is stoichiometrically PbS (86.6% Pb), then the ore is composed of 5.8% PbS giving a = 0.058, r = 7.6, and t = 2.65, resulting in a mineralogical composition factor m = 117.8 g cm^{-3}

l: The liberation factor $l = (0.015/2.5)^{0.5} = 0.0775$

The overall constant C becomes = $f\,g\,l\,m$

$$= 0.5 \times 0.25 \times 0.0775 \times 117.8$$
$$= 1.135(\text{g cm}^{-3})$$

Thus the required sample mass becomes

$$M_s = Cd^3/\sigma^2$$
$$= 1.135 \times 2.5^3/0.01^2$$
$$= 177.4(\text{kg})$$

In practice, therefore, about 350 kg of ore would have to be sampled in order to give the required degree of confidence, and to allow for other errors associated with distribution, preparation and assaying. Clearly, further size reduction of this coarse sample and secondary sampling would be required prior to assaying.

and to achieve this survey sampling needs to be (as much as possible):

- consistent (reproducible)
- coherent (in = out)
- unbiased
- with redundancy (extra data for the data reconciliation step)

Extra process measurement and sampling locations (i.e., more than the minimum) and multiple element, mineral or size assays will provide data redundancy and improve the quality of the reconciled data.

How many "cuts" to include in the sample or sampling increment, and for how long to conduct the sampling survey, are important considerations. Use is again made of the CLT, which states that the variance of the mean of n measurements is n times smaller than that of a single measurement. This can be re-stated as;

$$CV_2 = CV_1\sqrt{\frac{n_1}{n_2}} \tag{3.7}$$

where CV refers to the coefficient of variation (or the relative standard deviation; i.e., the standard deviation relative to the mean) and n is the number of sample "cuts" making up the composite.

Equation (3.7) is plotted in Figure 3.2 as the % reduction in CV of the mean versus the number of sample cuts (n) in the composite. It is clear that collecting more than ~8 samples does little to improve the precision (CV) of the composite. However, a minimum of at least 5 is recommended.

Sampling surveys need to be conducted for a sufficient length of time to allow all elements within the process volume (i.e., sampling unit) an equal chance of being sampled (the equi-probable sampling model). Defining the mean retention time of the process volume τ as:

$$\tau = \frac{\text{Sampling unit volume }(V)}{\text{volumertic flowrate }(Q)} \tag{3.8}$$

and, dimensionless time θ as:

$$\theta = \frac{\text{time}}{\tau} \tag{3.9}$$

use is then be made of the tanks-in-series model for N perfectly mixed reactors (covered in Chapter 12) to generate the exit frequency distribution or *residence time*

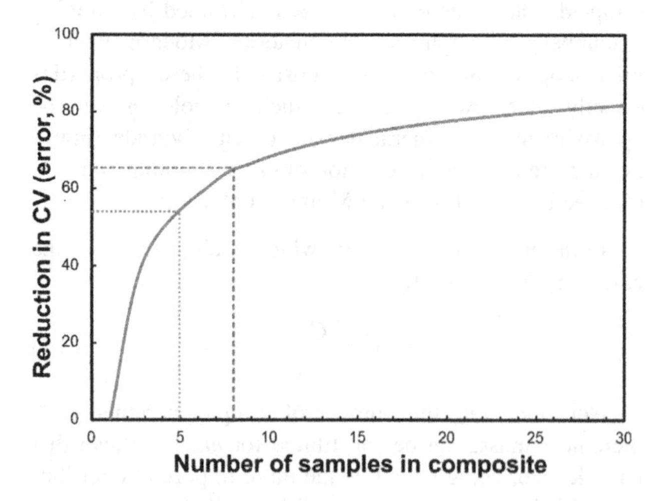

FIGURE 3.2 The % reduction in CV (rel std dev) of the mean for a composite sample as the number of sample cuts is increased based on Eq. (3.7) with $n_1 = 1$ (i.e., %Red = $100(1-(1/n_2)^{0.5})$).

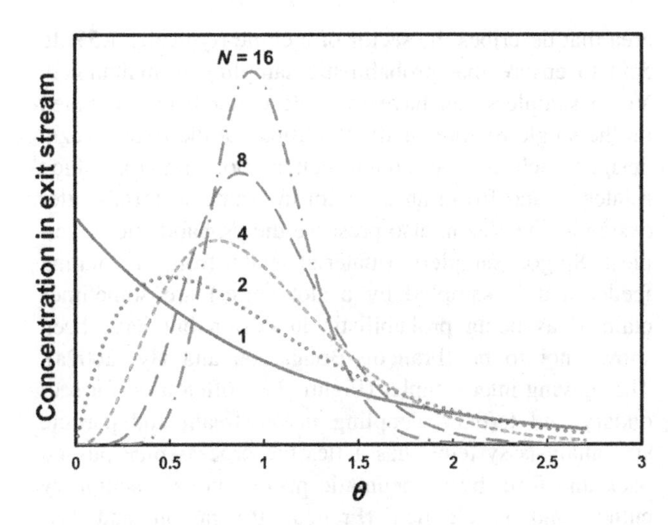

FIGURE 3.3 The residence time frequency distribution (RTD) as a function of N tanks-in-series. Note that by time = 3θ all material entering at $\theta = 0$ has fully exited the system.

distribution (RTD) for the process. To establish the value of θ needed to ensure that all elements have passed the sampling points (process volume), Figure 3.3 has been constructed showing the RTD for increasing N tanks in series. The figure indicates that for all values of N, all elements have exited the process volume by $\theta = 3$. Good sampling protocol should therefore conduct sampling at randomized time intervals (i.e., recall the ideal sampling model) over a period of 3 mean retention times. A corollary is that following a process change, subsequent sampling should also wait an equivalent $\theta = 3$ retention times before re-sampling the circuit. Figure 3.3 does indicate that, with only a small reduction in probability, a sampling time of $\theta = 2$ to 2.5 could be adequate if that improves the logistics of performing the testwork.

Once a sampling survey has been completed, a check of the process *data historian* should be made to ensure that reasonably steady state conditions have prevailed during the sampling period for important process variables such as volumetric and mass flows, pulp density, chemical additions and key on-stream analyzer (OSA) measurements. Cao and Rhinehart (1995) provide a useful method for establishing if steady state process conditions have been maintained.

Sampling surveys are not all successful as there are many uncontrolled variables impacting a process, and as many as 30% to 50% may need to be rejected and rerun. It is far better to do so before time and resources have been committed to sample preparation and data analysis on a poor test run.

Rules of Thumb

Several guidelines have evolved to aid in sampling process streams (Merks, 1985; François-Bongarçon and Gy, 2002b) some being:

- The sample size collected should be at least 1000 times the mass of the largest individual particle; this may encounter a practical limit for material with a very large top size.
- The slot on the sampler should be at right angles to the material flow and at least 3x the width of the largest particle (minimum width 10 mm).
- The sampler cross-cutting speed should not exceed 0.6 m s^{-1} unless the largest particles are <1 mm in which case the speed can be up to double this. Some ISO standards permit cutter speeds up to 1.5 m s^{-1} provided the opening slot width is significantly greater than 3x the diameter of the largest particle. Constant sample cutter speed is critically important, even if done manually.
- Sampler cutting edges need to be knife edge in design so that impacting particles are not biased toward either side.

Installed sampling systems such as those required for metallurgical accounting, on-stream analysis or bulk materials handling need to be properly designed in terms of sample size and sampling frequency. The concepts of *variograms*, *time-series analysis* and statistical measures such as *F ratios* and *paired t-tests* become important tools when selecting sampling frequency, sample increment size, and validation of sampling protocols. Further discussion is beyond this introductory text and the reader is referred to Merks (1985, 2002, 2010), Pitard (1993), Holmes (1991, 2004) and Gy (1982). In addition to the *primary* sampling device, a sampling system will often require a *secondary* and even a *tertiary* level of sub-sampling in order to deliver a suitably sized stream for preparation and analysis.

Proper sampling protocols go beyond simply good practice. There are international standards in place for the sampling of bulk materials such as coal, iron ore, precious metals and metalliferous concentrates, established by the International Standards Organization (ISO, website iso.com) and others (AS (Australia), ASTM international, and JIS (Japan)). Other regulations have also been put in place for the sampling and reporting of mineral deposits by various national industry organizations to safeguard against fraud and to protect investors.

3.2.4 Sampling Equipment

Recalling the ideal sampling model previously introduced, the concept that every particle or fluid element should have an equi-probable chance of being sampled requires that sampling equipment design adheres to the same principle. Those that do are referred to as equi-probable or *probabilistic* samplers, and those that do not, and hence will introduce bias and increased variance to the measurement data, are deemed as *non-probabilistic*. There are situations where non-probabilistic samplers are chosen for convenience or cost considerations, or on "less-important" streams within the plant. Such cases need to be recognized and the limitations carefully assessed. Critical sample streams such as process and plant feeds, final concentrates and tailings that are used for accounting purposes should always use probabilistic samplers.

Probabilistic Samplers

Figure 3.4 illustrates some of the critical design features of probabilistic samplers: cutting the full process steam at right angles to the flow Figure 3.4(a), and the knife-edged design cutter blades Figure 3.4(b).

Linear samplers are the most common and preferred devices for intermittent sampling of solids and pulp streams at discharge points such as conveyor head pulleys and ends of pipe. Examples are shown in Figure 3.5 and 3.6. Installations can be substantial in terms of infrastructure for large process streams if enclosures and sample handling systems are required. Such primary samplers often feed secondary and even tertiary devices to reduce the mass of sample collected. In cases where small flows are involved (e.g., secondary sampling), Vezin style rotary samplers such as the one shown in Figure 3.7 are often used. The Vezin design has a cutter cross sectional area that describes the sector of a circle (typically 1.5% to 5%) to ensure that probabilistic sampling is maintained. Vezin samplers can have multiple cutter heads mounted on the single rotating shaft. Variations of the rotary Vezin design, such as the arcual cutter (not shown), which rotates to and fro in an arc with the same sectorial-cutter design as the Vezin, also preserve the probabilistic principles. Spigot samplers (material exits from a rotating feeder and is sampled by a slot cutter) are sometimes claimed as being probabilistic in design but have been shown not to be (François-Bongarçon and Gy, 2002a). The moving inlet sampler (Figure 3.8), often used for secondary and tertiary sampling in on-stream and particle size analysis systems, has a flexible process pipe pushed back and forth by a pneumatic piston above a stationary cutter, and is claimed (François-Bongarçon and Gy, 2002a) to not be fully probabilistic, although this may depend on design and the ability of the cutter to fully meet the flow at a right angle.

Non-Probabilistic Samplers

Non-probabilistic designs have found use in less critical applications where some amount of measurement bias is deemed acceptable. This includes, as examples, on-stream analyzer (OSA) and particle size analyzer (PSA) installations and belt sampling of solids for moisture content. They violate the ideal sampling model in that they do not cut the entire stream, so not every particle or fluid element has an equal chance of being sampled. Strong proponents of proper sampling practice such as François-Bongarçon and Gy (2002a) deem that any sampling

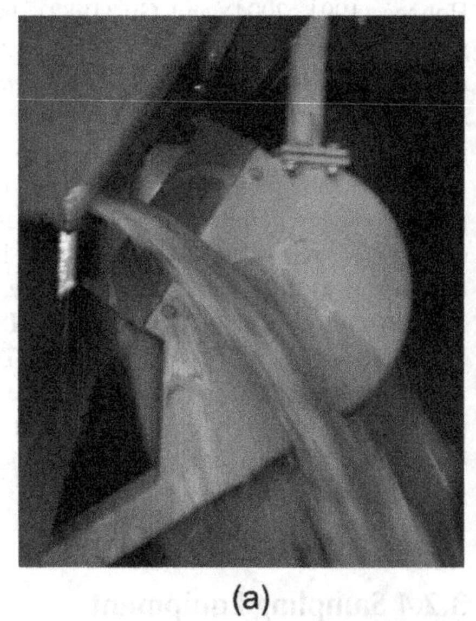
(a)

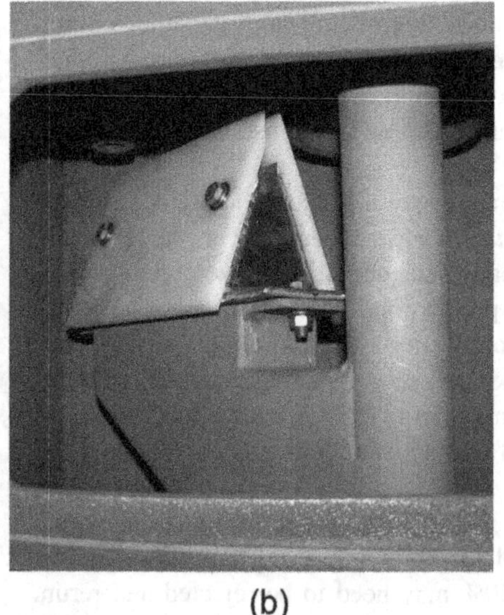
(b)

FIGURE 3.4 Features of probabilistic samplers: (a) Cutting the full process stream at right angles (linear sampler), and (b) Replaceable, non-adjustable knife-edge cutter blades (Vezin sampler) *(Courtesy Heath and Sherwood).*

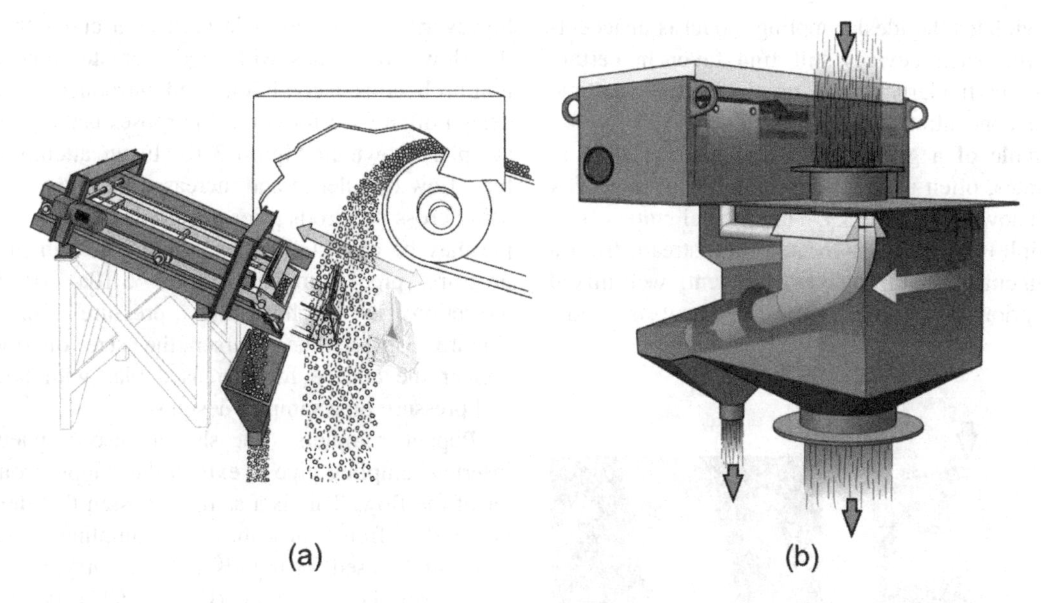

FIGURE 3.5 Example of linear sampler for solids: (a) Conveyor transfer/discharge point, and (b) Vertically discharging slurry pipe *(Courtesy Heath and Sherwood).*

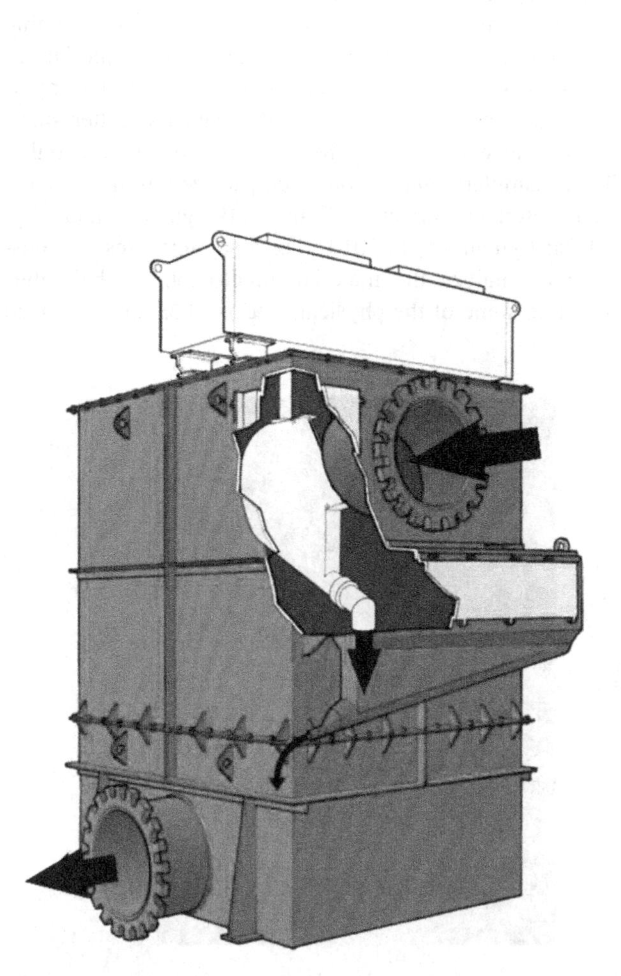

FIGURE 3.6 Example of a linear sampler for a horizontally discharging slurry pipe *(Courtesy Heath and Sherwood).*

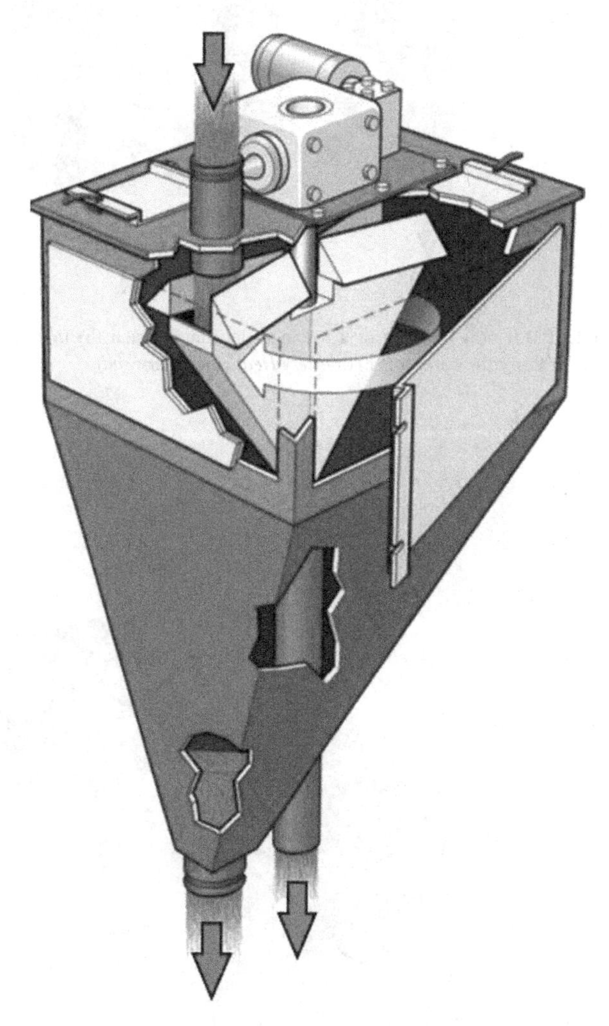

FIGURE 3.7 Example of a two-cutter Vezin sampler showing the sector design of the cutters *(Courtesy Heath and Sherwood).*

device that violates the ideal sampling model is unacceptable; however, such devices still find favor in certain applications, particularly those involving fine particle sizes such as concentrates and tailings.

An example of a gravity sampler for slurry flow in horizontal lines, often used for concentrates feeding OSA systems, is shown in Figure 3.9. The internal cutters (single or multiple) are typically located downstream from a flume arrangement that promotes turbulent, well-mixed conditions prior to sampling. The argument is made, however, that any obstacle such as a cutter that changes the flow streamlines will always create some degree of non-probabilistic conditions and introduce bias. Another design often used for control purposes is the pressure pipe sampler shown in Figure 3.10. In an attempt to reduce bias, flow turbulence and increased mixing are introduced using cross-pipe rods prior to sample extraction. The sample may be extracted either straight through in a vertical pressure pipe sampler (Figure 3.10a) or from the Y-section in a horizontal pressure pipe sampler (Figure 3.10b). The coarser the size distribution the greater the chance for increased bias with both gravity and pressure pipe sampler designs.

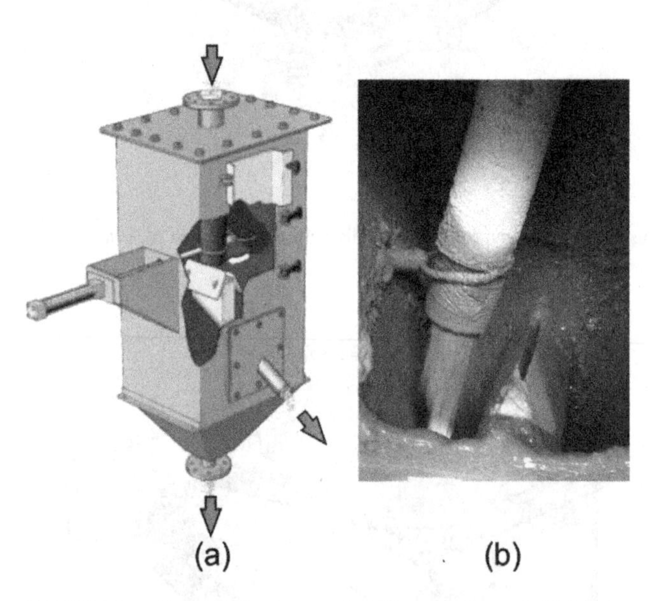

FIGURE 3.8 (a) Example of a moving inlet sampler, and (b) Interior view of slurry flow and cutter *(Courtesy Heath and Sherwood).*

Poppet samplers (not shown) use a pneumatically inserted sampling pipe to extract the sample from the center of the flow. This is a sampler design that deviates significantly from probabilistic sampling and should therefore be used for liquids and not slurry, if at all.

A cross-belt sampler (Figure 3.11) is designed to mimic stationary reference sampling of a conveyor belt, a probabilistic and very useful technique for bias testing and plant surveys. This belt sampling, however, is difficult to achieve with an automated sampler swinging across a moving conveyor, particularly if the material has fractions of very coarse or very fine material. Small particles may remain on the belt underneath the cutter sides, while coarse lumps may be pushed aside preferentially. These samplers find use on leach pad feed, mill feeds and final product streams (François-Bongarçon and Gy, 2002a; Lyman, et al., 2010). For assay purposes it is possible to analyze the material directly on the belt, thus avoiding some of the physical aspects of sampling to feed

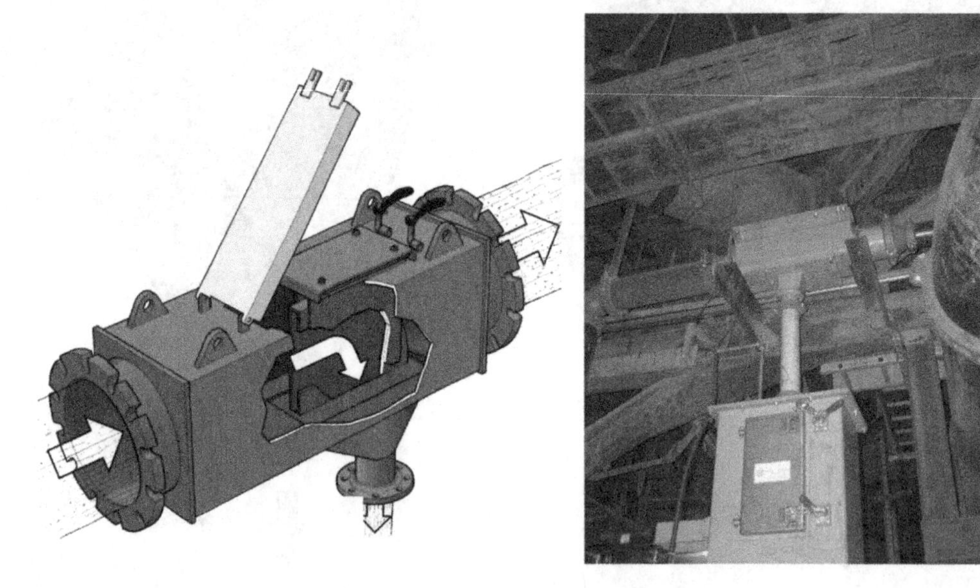

FIGURE 3.9 Example of gravity slot sampler. Photo shows the primary gravity sampler feeding a secondary sampler preparing to feed an OSA *(Courtesy Heath and Sherwood).*

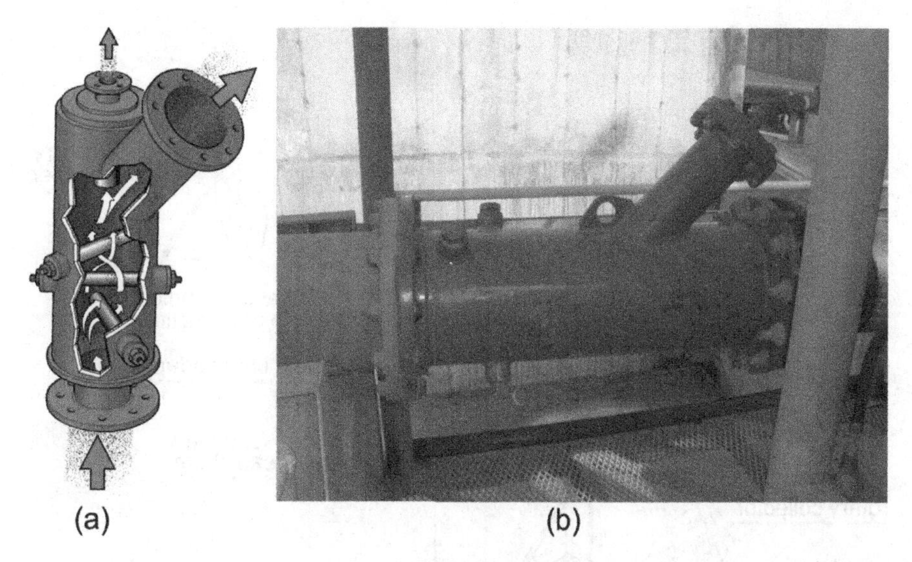

FIGURE 3.10 Example of pressure pipe sampler: (a) Sample extraction from straight pipe section on a vertical pressure pipe sampler, and (b) Sample extraction from Y-section on a horizontal pressure pipe sampler (Note the internal rods to promote increased mixing prior to sampling) *(Courtesy Heath and Sherwood).*

FIGURE 3.11 Example of a cross belt sampler. Photo is from an iron ore installation *(Courtesy Heath and Sherwood).*

an OSA system (e.g., Geoscan-M, Elemental analyzer, Scantech) (see Section 3.3).

Sampling systems for pulp and solids will typically require primary, secondary and possibly tertiary sampling stages. Figure 3.12 and Figure 3.13 show examples of such sampling systems. Once installed, testing should be conducted to validate that the system is bias-free.

Manual or automatic (e.g., using small pumps) in-pulp sampling is a non-probabilistic process and should not be used for metallurgical accounting purposes, but is sometimes necessary for plant surveys and pulp density sampling. An example of a manual in-pulp device is shown in Figure 3.14 and is available in various lengths to accommodate vessel size. One option for reducing random error associated with grab sampling using these samplers is to increase the number of "cuts", however this does not eliminate the inherent bias of non-probabilistic samplers. An example of a stop-belt manual reference sampler is shown in Figure 3.15. These provide probabilistic samples when used carefully and are very useful for bias testing of automatic samplers and for plant surveys.

Traditional methods for manually extracting a sample increment, such as grab sampling, use of a riffle box, and coning, are inherently non-probabilistic and should be avoided. The caveat being that they may be acceptable if the method is used to treat the entire lot of material,

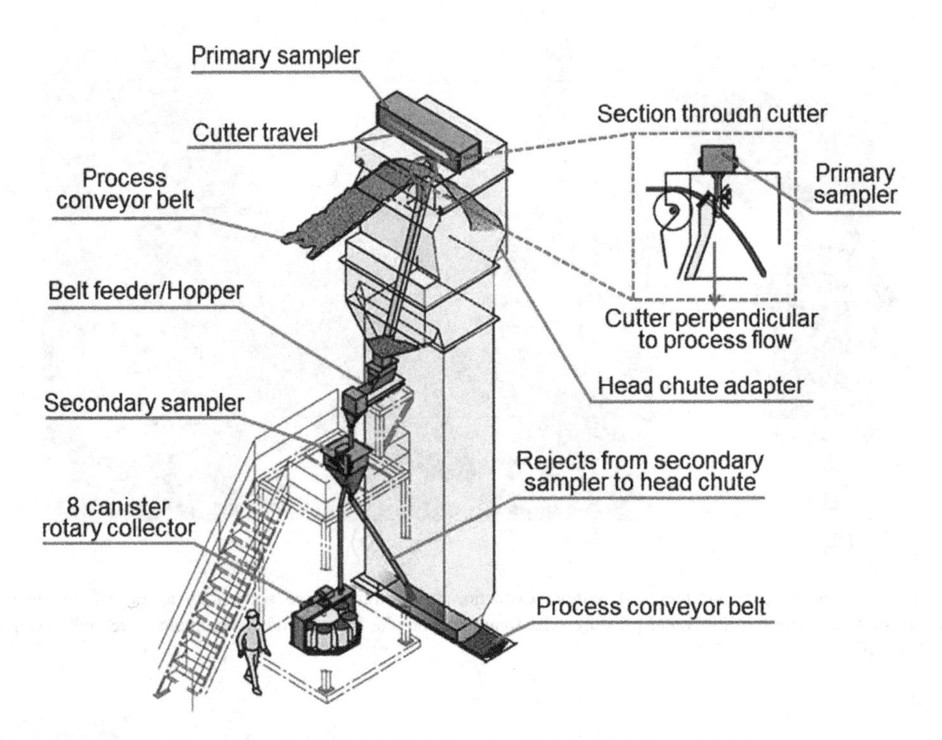

FIGURE 3.12 Solids sampling system: Primary and secondary sampling followed by a rotary collector *(Courtesy Heath and Sherwood).*

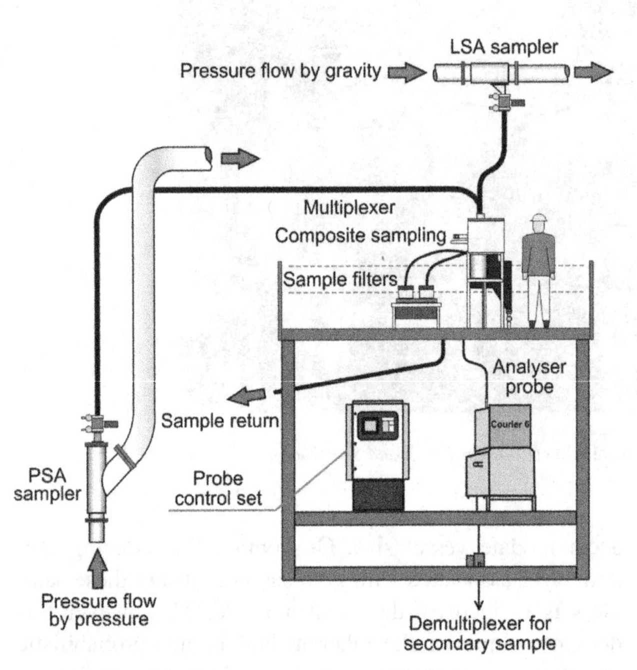

FIGURE 3.13 On-stream analysis pulp sampling system (primary and secondary sampling) *(Courtesy Outotec).*

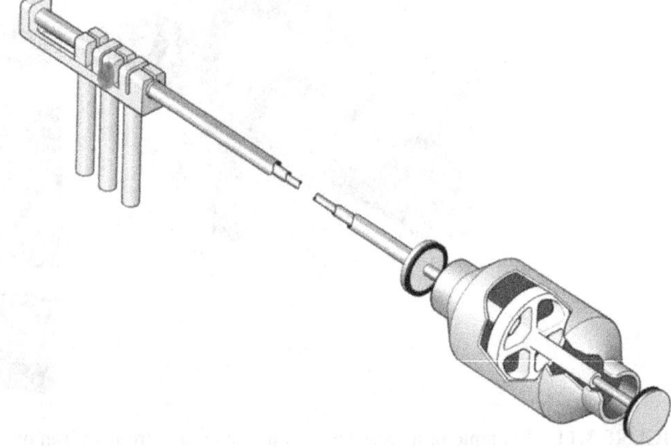

FIGURE 3.14 Example of a manual in-pulp sampler. The handles open the top and bottom valves independently to allow for entry and discharge of the sample *(Courtesy Heath and Sherwood).*

dividing it into two equal sub-lots each time, and repeating the process the required number of times to obtain the final mass needed for analysis. Riffling is inherently biased if the two outside slots lead to the same sample side, or the device feeding the riffle is not exactly the same width as the combined width of all slots or the feeding is slow and uneven. Care must be taken with these methods to avoid segregation leading to bias. The recommended method for dividing samples is to use a rotary table riffle or splitter, as shown in Figure 3.16. The close proximity and radial design of the containers, and constant speed of the table and feeder, ensure that equiprobable splitting is achieved.

FIGURE 3.15 Images of a stop-belt reference sampler. A perfect fit to the shape of the belt is critical (used with permission, Minnitt, 2014, J. S. Afr. Inst. Min. Metall, copyright SAIMM).

FIGURE 3.16 Rotary table splitter *(Courtesy TM Engineering Ltd.)*.

3.3 ON-LINE ANALYSIS

On-line analysis enables a change of quality to be detected and corrected rapidly and continuously, obviating the delays involved in off-line laboratory testing. This method also frees skilled staff for more productive work than testing of routine samples. The field has been comprehensively reviewed elsewhere (Kawatra and Cooper, 1986; Braden et al., 2002; Shean and Cilliers, 2011).

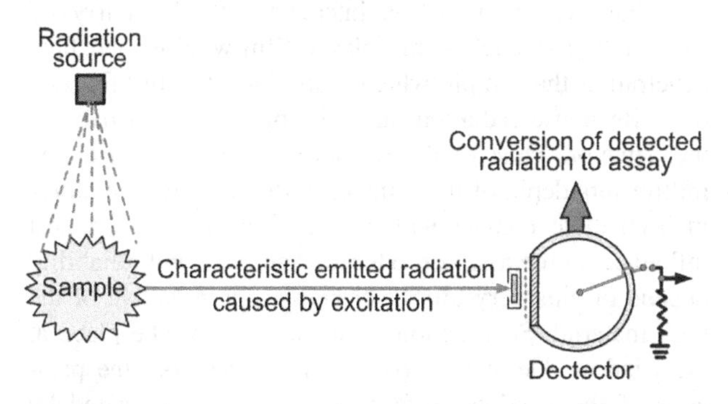

FIGURE 3.17 Principle of on-line chemical analysis.

3.3.1 On-line Element Analysis

Basically, on-line element analysis consists of a source of radiation that is absorbed by the sample and causes it to give off fluorescent response radiation characteristic of each element. This enters a detector that generates a quantitative output signal as a result of measuring the characteristic radiation of one element from the sample. Through calibration, the detector output signal is used to obtain an assay value that can be used for process control (Figure 3.17).

On-stream Analysis

The benefits of continuous analysis of process streams in mineral processing plants led to the development in the early 1960s of devices for X-ray fluorescence (XRF) analysis of flowing slurry streams. The technique exploits the emission of characteristic "secondary" (or fluorescent) X-rays generated when an element is excited by high energy X-rays or gamma rays. From a calibration the intensity of the emission gives element content.

Two methods of on-line X-ray fluorescence analysis are centralized X-ray (on-stream) and in-stream probe systems. Centralized on-stream analysis employs a single high-energy excitation source for analysis of several slurry samples delivered to a central location where the equipment is installed. In-stream analysis employs sensors installed in, or near, the slurry stream, and sample excitation is carried out with convenient low-energy sources, usually radioactive isotopes (Bergeron and Lee, 1983; Toop et al., 1984). This overcomes the problem of obtaining and transporting representative samples of slurry to the analyzer. The excitation sources are packaged with a detector in a compact device called a probe.

One of the major problems in on-stream X-ray analysis is ensuring that the samples presented to the radiation are representative of the bulk, and that the response radiation is obtained from a representative fraction of this sample. The exciting radiation interacts with the slurry by first passing through a thin plastic film window and then penetrating the sample which is in contact with this window. Response radiation takes the opposite path back to the detector. Most of the radiation is absorbed in a few millimeters depth of the sample, so that the layer of slurry in immediate contact with the window has the greatest influence on the assays produced. Accuracy and reliability depend on this very thin layer being representative of the bulk material. Segregation in the slurry can take place at the window due to flow patterns resulting from the presence of the window surface. This can be eliminated by the use of high turbulence flow cells in the centralized X-ray analyzer.

Centralized analysis is usually installed in large plants, requiring continuous monitoring of many different pulp streams, whereas smaller plants with fewer streams may incorporate probes on the basis of lower capital investment.

Perhaps the most widely known centralized analyzer is the one developed by Outotec, the Courier 300 system (Leskinen et al., 1973). In this system a continuous sample flow is taken from each process slurry to be analyzed. The Courier 300 has now been superseded by the Courier SL series. For example, the Courier 6 SL (Figure 3.13) can handle up to 24 sample streams, with typical installations handling 12–18 streams. The primary sampler directs a part of the process stream to the multiplexer for secondary sampling. The unit combines high-performance wavelength and energy dispersive X-ray fluorescence methods and has an automatic reference measurement for instrument stability and self-diagnostics. The built-in calibration sampler is used by the operator to take a representative and repeatable sample from the measured slurry for comparative laboratory assays.

The measurement sequence is fully programmable. Critical streams can be measured more frequently and more measurement time can be used for the low grade tailings streams. The switching time between samples is used for internal reference measurements, which are used for monitoring and automatic drift compensation.

The probe in-stream system uses an isotope as excitation source. These probes can be single element or multi-element (up to 8, plus % solids). Accuracy is improved by using a well-stirred tank of slurry (analysis zone). Combined with solid state cryogenic (liquid N_2) detectors, these probes are competitive in accuracy with traditional X-ray generator systems and can be multiplexed to analyze more than one sample stream. Such devices were pioneered by AMDEL and are now marketed by Thermo Gamma-Metrics. Figure 3.18 shows the TGM in-stream "AnStat" system – a dedicated analyzer with sampling system (Boyd, 2005).

X-ray fluorescence analysis is not effective on light elements. The Outotec Courier 8 SL uses laser-induced breakdown spectroscopy (LIBS) to measure both light and heavy elements. Since light elements are often associated with gangue this analyzer permits impurity content to be tracked.

On-belt Analysis

Using a technique known as prompt gamma neutron activation analysis (PGNAA), the GEOSCAN-M (M for minerals applications) is an on-belt, non-contact elemental analysis system for monitoring bulk materials (Figure 3.19) (units specifically designed for the cement and coal industries are also available). By early 2015, over 50 units were operating in minerals applications in iron ore, manganese, copper, zinc-lead and phosphate (Arena and McTiernan, 2011; Matthews and du Toit, 2011; Patel, 2014). A Cf-252 source placed under the belt generates neutrons that pass through the material and interact with the element nuclei to release gamma ray emissions characteristic of the elements. Detectors above the belt produce spectral data, that is, intensity versus energy level. Input from a belt scale provides tonnage weighted averaging of results over time and measurement reporting increments are typically every 2 to 5 minutes. A moisture (TBM) monitor is incorporated to correct for moisture content and allow for elemental analysis on a dry solids basis. Moisture is measured using the microwave transmission technique. A generator below the belt creates a microwave field and free moisture is measured by the phase shift and attenuation.

Laser induced breakdown spectroscopy (LIBS) is also used for on-belt analysis, supplied by Laser Distance Spectrometry (LDS, www.laser-distance-spectrometry.com). In this instrument a laser is directed to the material on the belt and characteristic emissions are detected and converted to element analysis. An advantage is that the analysis

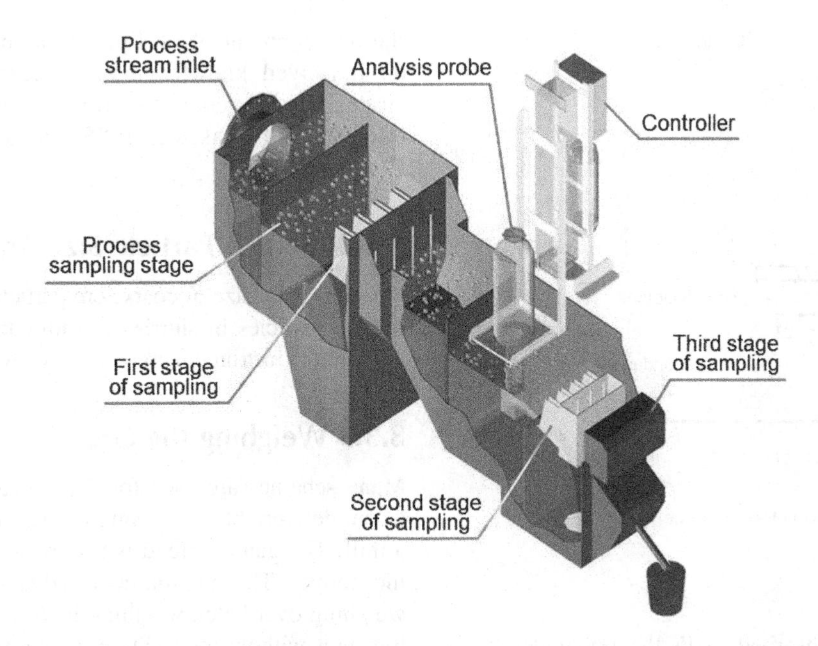

FIGURE 3.18 The Thermo Gamma-Metrics AnStat in-stream analysis probe and sampler *(Courtesy Thermo Electron Corporation).*

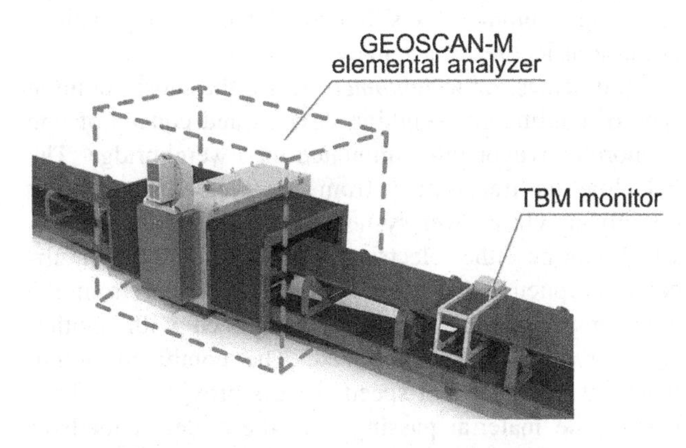

FIGURE 3.19 The GEOSCAN-M and TBM moisture monitor on-belt analyzers *(Courtesy Scantech).*

includes light elements (Li, Be, etc.). The MAYA M-2010 has been installed in a variety of mining applications.

3.3.2 On-stream Mineral Phase Analysis

Rather than elemental analysis, there is a case to measure actual mineral content as it is the minerals that are being separated. The analysis of phase can be broadly divided into two types: 1. elemental analysis coupled with mineralogical knowledge and mass balancing; and 2. direct determination using diffraction or spectroscopy. The first is essentially adapting the techniques described in Section 3.3.1; the electron beam techniques such as QEMSCAN and MLA (Chapters 1 and 17) use this approach and on-line capability may be possible.

Of the second category X-ray diffraction (XRD) is the common approach and on-line applications are being explored (Anon., 2011). XRD requires that the minerals be crystalline and not too small (>100 nm), and quantitative XRD is problematic due to matrix effects. Spectroscopic techniques include Raman, near infra-red (NIR), nuclear magnetic resonance, and Mössbauer. Their advantage is that the phases do not have to be crystalline. Both NIR and Raman are available as on-line instruments (BlueCube systems). Mössbauer spectroscopy (MS) is particularly useful as a number of mineral phases contain Fe and Fe MS data can be obtained at room temperature. Combining XRD and MS can result in simplifying phase identification (Saensunon et al., 2008). Making such units on-line will be a challenge, however.

The various technologies being developed to identify phases in feeds to ore sorters are another source of on-line measurements. For example, Lessard et al. (2014) describe a dual-energy X-ray transmission method. X-ray transmission depends on the atomic density of the element such that denser elements adsorb more (transmit less). Thus phases containing high density elements can be differentiated from phases containing low density elements by the magnitude of the transmitted intensity. By using two (dual) X-ray energies it is possible to compensate for differences in the volume that is being detected.

3.3.3 On-stream Ash Analysis

The operating principle is based on the concept that when a material is subjected to irradiation by X-rays, a portion

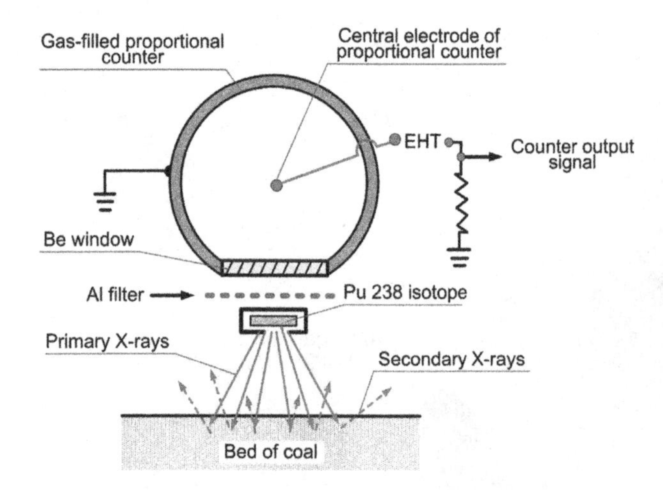

FIGURE 3.20 Sensor system of ash monitor.

of this radiation is absorbed, with the remainder being reflected. The radiation absorbed by elements of low atomic number (carbon and hydrogen) is lower than that absorbed by elements of high atomic number (silicon, aluminum, iron), which form the ash in coal, so the variation of absorption coefficient with atomic number can be directly applied to the ash determination.

A number of analyzers have been designed, and a typical one is shown in Figure 3.20. A representative sample of coal is collected and crushed, and fed as a continuous stream into the presentation unit of the monitor, where it is compressed into a compact uniform bed of coal, with a smooth surface and uniform density. The surface is irradiated with X-rays from a Pu-238 isotope and the radiation is absorbed or back-scattered in proportion to the elemental composition of the sample, the back-scattered radiation being measured by a proportional counter. At the low-energy level of the Pu-238 isotope (15–17 keV), the iron is excited to produce fluorescent X-rays that can be filtered, counted, and compensated. The proportional counter simultaneously detects the back-scattered and fluorescent X-rays after they have passed through an aluminum filter. This filter absorbs the fluorescent X-rays preferentially, its thickness preselected to suit the iron content and its variation. The proportional counter converts the radiation to electrical pulses that are then amplified and counted in the electronic unit. The count-rate of these pulses is converted to a voltage which is displayed and is also available for process control.

A key sensor required for the development of an effective method of controlling coal flotation is one which can measure the ash content of coal slurries. Production units have been manufactured and installed in coal preparation plants (Kawatra, 1985), and due to the development of on-line ash monitors, coupled with an improved knowledge of process behavior, control strategies for coal flotation have been developed (Herbst and Bascur, 1985; Salama et al., 1985; Clarkson, 1986).

3.3.4 On-line Particle Size Analysis

Measuring the size of coarse ore particles on conveyor belts or fine particles in slurries can now be done on-line with appropriate instrumentation. This is covered in Chapter 4.

3.3.5 Weighing the Ore

Many schemes are used for determination of the tonnage of ore delivered to or passing through different sections of a mill. The general trend is toward weighing materials on the move. The predominant advantage of continuous weighing over batch weighing is its ability to handle large tonnages without interrupting the material flow. The accuracy and reliability of continuous-weighing equipment have improved greatly over the years. However, static weighing equipment is still used for many applications because of its greater accuracy.

Belt scales, or *weightometers*, are the most common type of continuous-weighing devices and consist of one or more conveyor idlers mounted on a weighbridge. The belt load is transmitted from the weighbridge either directly or via a lever system to a load-sensing device, which can be either electrically, mechanically, hydraulically, or pneumatically activated. The signal from the load-sensing device is usually combined with another signal representing belt speed. The combined output from the load and belt-speed sensors provides the flowrate of the material passing over the scale. A totalizer can integrate the flowrate signal with time, and the total tonnage carried over the belt scale can be registered on a digital read-out. Accuracy is normally 1-2% of the full-scale capacity.

It is the dry solids flowrate that is required. Moisture can be measured automatically (see Section 3.3.1). To check, samples for moisture determination are frequently taken from the end of a conveyor belt after material has passed over the weighing device. The samples are immediately weighed wet, dried at a suitable temperature until all hygroscopic (free) water is driven off, and then weighed again. The difference in weight represents moisture and is expressed as:

$$\%\text{moisture} = \frac{\text{Wet weight} - \text{Dry weight}}{\text{Wet weight}} \tag{3.10}$$

The drying temperature should not be so high that breakdown of the minerals, either physically or chemically, occurs. Sulfide minerals are particularly prone to lose sulfur dioxide if overheated; samples should not be dried at temperatures above 105°C.

Periodic testing of the weightometer can be made either by passing known weights over it or by causing a length of heavy roller chain to trail from an anchorage over the suspended section while the empty belt is running.

Many concentrators use one master weight only, and in the case of a weightometer this will probably be located at some convenient point between the crushing and the grinding sections. The conveyor feeding the fine-ore bins is often selected, as this normally contains the total ore feed of the plant.

Weighing the concentrates is usually carried out after dewatering, before the material leaves the plant. Weighbridges can be used for material in wagons, trucks, or ore cars. They may require the services of an operator, who balances the load on the scale beam and notes the weight on a suitable form. After tipping the load, the tare (empty) weight of the truck must be determined. This method gives results within 0.5% error, assuming that the operator has balanced the load carefully and noted the result accurately. With recording scales, the operator merely balances the load, then turns a screw which automatically records the weight. Modern scales weigh a train of ore automatically as it passes over the platform, which removes the chance of human error except for occasional standardization. Sampling must be carried out at the same time for moisture determination. Assay samples should be taken, whenever possible, from the moving stream of material, as described earlier, before loading the material into the truck.

Tailings weights are rarely, if ever, measured. They are calculated from the difference in feed and concentrate weights. Accurate sampling of tailings is essential. Mass balance techniques are now routinely used to estimate flows (see Section 3.6).

3.3.6 Mass Flowrate

By combining slurry flowrate and % solids (*mass-flow integration*), a continuous recording of dry tonnage of material from pulp streams is obtained. The *mass-flow unit* consists of a flowmeter and a density gauge often fitted to a vertical section of a pipeline carrying the upward-flowing stream. The principal flowmeters are the magnetic-, ultrasonic-, and array-based.

Magnetic Flowmeter

The operating principle of the magnetic flowmeter (Figure 3.21) is based on Faraday's law of electromagnetic induction, which states that the voltage induced in any conductor as it moves across a magnetic field is

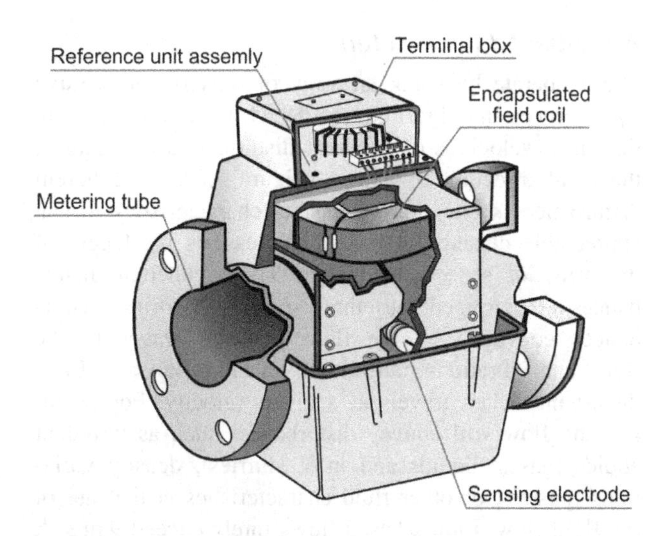

FIGURE 3.21 Magnetic flowmeter.

proportional to the velocity of the conductor. Thus, providing the pulp completely fills the pipeline, its velocity will be directly proportional to the flowrate. Generally, most aqueous solutions are adequately conductive for the unit and, as the liquid flows through the metering tube and cuts through the magnetic field, an electromotive force (emf) is induced in the liquid and is detected by two small measuring electrodes fitted virtually flush with the bore of the tube. The flowrate is then recorded on a chart or continuously on an integrator. The coil windings are excited by a single-phase AC mains supply and are arranged around the tube to provide a uniform magnetic field across the bore. The unit has the advantage that there is no obstruction to flow, pulps and aggressive liquids can be handled, and it is immune to variations in density, viscosity, pH, pressure, or temperature. A further development is the DC magnetic flowmeter, which uses a pulsed or square excitation for better stability and reduced zero error.

Ultrasonic Flowmeters

Two types of ultrasonic flowmeters are in common use. The first relies on reflection of an ultrasonic signal by discontinuities (particles or bubbles) into a transmitter/receiver ultrasonic transducer. The reflected signal exhibits a change in frequency due to the Doppler Effect that is proportional to the flow velocity; these instruments are commonly called "Doppler flow meters". As the transducer can be attached to the outside of a suitable pipe section, these meters can be portable.

The second type of meter uses timed pulses across a diagonal path. These meters depend only on geometry and timing accuracy. Hence they can offer high precision with minimal calibration.

Array-based Flowmeters

These operate by using an array of sensors and passive sonar processing algorithms to detect, track, and measure the mean velocities of coherent disturbances traveling in the axial direction of a pipe (Figure 3.22). A coherent disturbance is one that retains its characteristics for an appreciable distance, which in this case is the length of the array of sensors or longer. These coherent disturbances are grouped into three major categories: disturbances conveyed by the flow, acoustic waves in the fluid, and vibrations transmitted via the pipe walls. Each disturbance class travels at a given velocity. For example, the flow will convey disturbances such as turbulent eddies (gases, liquids and most slurries), density variations (pastes), or other fluid characteristics at the rate of the fluid flow. Liquid-based flows rarely exceed 9 m s^{-1}. Acoustic waves in the fluid will typically have a minimum velocity of 75 m s^{-1} and a maximum velocity of 1500 m s^{-1}. The third group, pipe vibrations, travels at velocities that are several times greater than the acoustic waves. Thus each disturbance class may be clearly segregated based on its velocity.

The velocities are determined as follows: as each group of disturbances pass under a sensor in the array, a signal unique to that group of disturbances is detected by the sensor. In turbulent flow, this signal is generated by the turbulent eddies which exert a miniscule stress on the interior of the pipe wall as they pass under the sensor location. The stress strains the pipe wall and thus the sensor element which is tightly coupled to the pipe wall. Each sensor element converts this strain into a unique electrical signal that is characteristic of the size and energy of the coherent disturbance. By tracking the unique signal of each group of coherent disturbances through the array of sensors, the time for their passage through the array can be determined. The array length is fixed, therefore the passage time is inversely proportional to the velocity. Taking this velocity and the known inner diameter of the pipe, the flow meter calculates and outputs the volumetric flowrate.

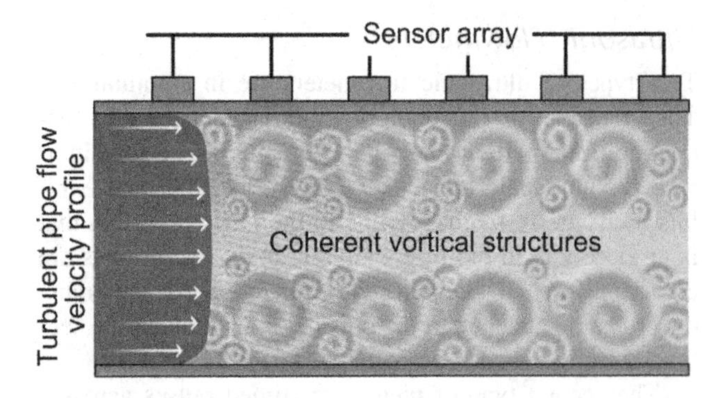

FIGURE 3.22 Array-based flow meter *(Courtesy CiDRA Mineral Processing).*

This technology accurately measures multiphase fluids (any combination of liquid, solids and gas bubbles), on almost any pipe material including multilayer pipes, in the presence of scale buildup, with conductive or non-conductive fluids, and with magnetic or non-magnetic solids. In addition, the velocity of the acoustic waves, which is also measured, can be used to determine the gas void fraction of bubbles in the pipe. The gas void fraction is the total volume occupied by gas bubbles, typically air bubbles, divided by the interior volume of the pipe. When gas bubbles are present, this measurement is necessary to provide a true non-aerated volumetric flowrate and to compensate for the effect of entrained gas bubbles on density measurements from Coriolis meters or nuclear density meters. This principle can be used to determine void fraction, or *gas holdup*, in flotation systems (Chapter 12).

Slurry Density

The density of the slurry is measured automatically and continuously in the nucleonic density gauge (Figure 3.23) by using a radioactive source. The gamma rays produced by this source pass through the pipe walls and the slurry at an intensity that is inversely proportional to the pulp density. The rays are detected by a high-efficiency ionization chamber and the electrical signal output is recorded directly as pulp density. Fully digital gauges using scintillation detectors are now in common use. The instrument must be calibrated initially "on-stream" using conventional laboratory methods of density analysis from samples withdrawn from the line.

The mass-flow unit integrates the rate of flow provided by the flowmeter and the pulp density to yield a continuous record of tonnage of dry solids passing through the pipe, given that the specific gravity of the solids comprising the ore stream is known. The method offers a reliable means of weighing the ore stream and removes chance of operator error and errors due to moisture sampling. Another advantage is that accurate sampling points, such as poppet valves (Section 3.2.4), can be incorporated at the same location as the mass-flow unit. Mass-flow integrators are less practicable, however, with concentrate pulps, especially after flotation, as the pulp contains many air bubbles, which lead to erroneous values of flowrate and density. Inducing cyclonic flow of slurry can be used to remove bubbles in some cases.

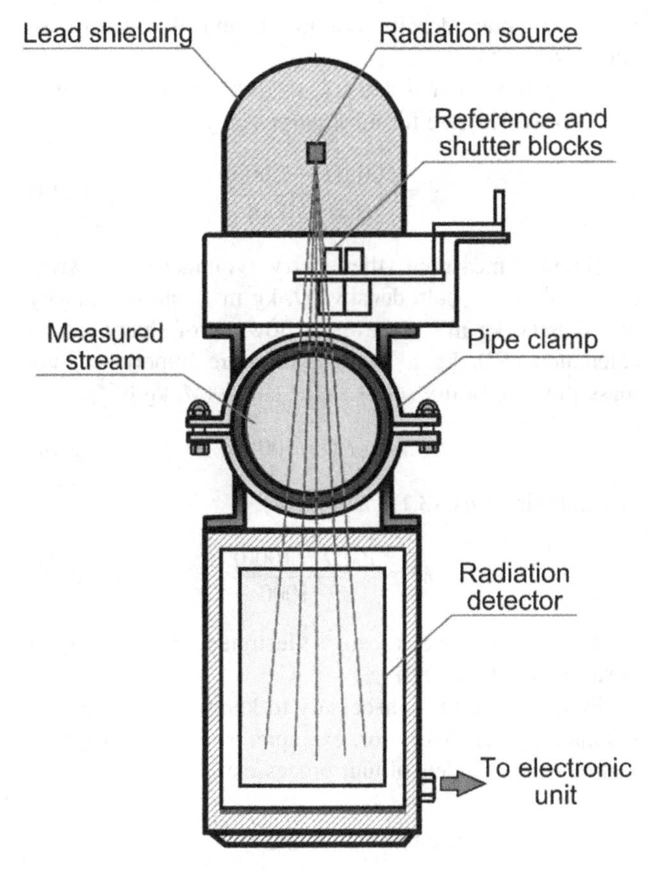

FIGURE 3.23 Nucleonic density gauge.

3.4 SLURRY STREAMS: SOME TYPICAL CALCULATIONS

Volumetric Flowrate

From the grinding stage onward, most mineral processing operations are carried out on slurry streams, the water and solids mixture being transported through the circuit via pumps and pipelines. As far as the mineral processor is concerned, the water is acting as a transport medium, such that the *weight* of slurry flowing through the plant is of little consequence. What is of importance is the *volume* of slurry flowing, as this will affect residence times in unit processes. For the purposes of metallurgical accounting, the weight of dry solids contained within the slurry is important.

If the volumetric flowrate is not excessive, it can be measured by diverting the stream of pulp into a suitable container for a measured period of time. The ratio of volume collected to time gives the flowrate of pulp. This method is ideal for most laboratory and pilot scale operations, but is impractical for large-scale operations, where it is usually necessary to measure the flowrate by online instrumentation.

Volumetric flowrate is important in calculating retention times in processes. For instance, if 120 $m^3\,h^{-1}$ of slurry is fed to a flotation conditioning tank of volume 20 m^3, then, on average, the retention time of particles in the tank will be (Eq. (3.8)):

$$\text{Retention time} = \frac{\text{Tank volume}}{\text{flowrate}} = \frac{20\ \text{m}^3}{\frac{120\ \text{m}^3}{1\ \text{h}} \times \frac{1\ \text{h}}{60\ \text{min}}} = 10\ \text{min}$$

Retention time calculations sometimes have to take other factors into account. For example, to calculate the retention time in a flotation cell the volume occupied by the air (the gas holdup) needs to be deducted from the tank volume (see Chapter 12); and in the case of retention time in a ball mill the volume occupied by the slurry in the mill is required (the slurry holdup).

Slurry Density and % Solids

Slurry, or pulp, density is most easily measured in terms of weight of pulp per unit volume. Typical units are kg m^{-3} and t m^{-3}, the latter having the same numerical value as specific gravity, which is sometimes useful to remember. As before, on flowstreams of significant size, this is usually measured continuously by on-line instrumentation.

Small flowstreams can be diverted into a container of known volume, which is then weighed to give slurry density directly. This is probably the most common method for routine assessment of plant performance, and is facilitated by using a density can of known volume which, when filled, is weighed on a specially graduated balance giving direct reading of pulp density.

The composition of a slurry is often quoted as the % solids by weight (100−% moisture), and can be determined by sampling the slurry, weighing, drying and reweighing, and comparing wet and dry weights (Eq. (3.10)). This is time-consuming, however, and most routine methods for computation of % solids require knowledge of the density of the solids in the slurry. There are a number of methods used to measure this, each method having pros and cons. For most purposes the use of a standard density bottle has been found to be a cheap and, if used with care, accurate method. A 25- or 50-ml bottle can be used, and the following procedure adopted:

1. Wash the density bottle with acetone to remove traces of grease.
2. Dry at about 40°C.

3. After cooling, weigh the bottle and stopper on a precision analytical balance, and record the weight, $M1$.
4. Thoroughly dry the sample to remove all moisture.
5. Add about 5–10 g of sample to the bottle and reweigh. Record the weight, $M2$.
6. Add double distilled water to the bottle until half-full. If appreciable "slimes" (minus 45 μm particles) are present in the sample, there may be a problem in wetting the mineral surfaces. This may also occur with certain hydrophobic mineral species, and can lead to false low density readings. The effect may be reduced by adding one drop of wetting agent, which is insufficient to significantly affect the density of water. For solids with extreme wettability problems, an organic liquid such as toluene can be substituted for water.
7. Place the density bottle in a desiccator to remove air entrained within the sample. This stage is essential to prevent a low reading. Evacuate the vessel for at least 2 min.
8. Remove the density bottle from the desiccator, and top up with double distilled water (do not insert stopper at this stage).
9. When close to the balance, insert the stopper and allow it to fall into the neck of the bottle under its own weight. Check that water has been displaced through the stopper, and wipe off excess water from the bottle. Record the weight, $M3$.
10. Wash the sample out of the bottle.
11. Refill the bottle with double distilled water, and repeat procedure 9. Record the weight, $M4$.
12. Record the temperature of the water used, as temperature correction is essential for accurate results.

The density of the solids (s, kg m^{-3}) is given by:

$$s = \frac{M2 - M1}{(M4 - M1) - (M3 - M2)} \times D_f \tag{3.11}$$

where D_f = density of fluid used (kg m^{-3}).

Knowing the densities of the pulp and dry solids, the % solids by weight can be calculated. Since pulp density is mass of slurry divided by volume of slurry, then for unit mass of slurry of x % solids by weight, the volume of solids is $x/100s$ and volume of water is $(100-x)/100\,W$ then (the 100's compensating for x in percent):

$$D = \frac{1}{\frac{x}{100\,s} + \frac{(100-x)}{100\,W}} \tag{3.12a}$$

or:

$$1 = \frac{x\,D}{100s} + (100-x)\frac{D}{100W} \tag{3.12b}$$

where D = pulp density (kg m^{-3}), and W = density of water (kg m^{-3}).

Assigning water a density of 1000 kg m^{-3}, which is sufficiently accurate for most purposes, gives:

$$x = \frac{100\,s(D-1000)}{D(s-1000)} \tag{3.13}$$

Having measured the slurry volumetric flowrate (F, m^3 h^{-1}), the pulp density (D, kg m^3), and the density of solids (s, kg m^{-3}), the mass flowrate of slurry can be calculated (FD, kg h^{-1}), and, of more importance, the mass flowrate of dry solids in the slurry (M, kg h^{-1}):

$$M = FDx/100 \tag{3.14}$$

or combining Eqs. (3.13) and (3.14):

$$M = \frac{Fs(D-1000)}{s-1000} \tag{3.15}$$

The computations are illustrated by examples (Examples 3.4 and 3.5).

In some cases it is necessary to know the % solids by volume, a parameter, for example, sometimes used in mathematical models of unit processes:

$$\%\text{solids by volume} = \frac{xD}{s} \tag{3.16}$$

Also of use in milling calculations is the ratio of the weight of water to the weight of solids in the slurry, or the *dilution ratio*. This is defined as:

$$\text{Dilution ratio} = \frac{100-x}{x} \tag{3.17}$$

This is particularly important as the product of dilution ratio and weight of solids in the pulp is equal to the weight of water in the pulp (see also Section 3.7) (Examples 3.6 and 3.7).

3.5 AUTOMATIC CONTROL IN MINERAL PROCESSING

Control engineering in mineral processing continues to grow as a result of more demanding conditions such as low grade ores, economic changes (including reduced tolerance to risk), and ever more stringent environmental regulations, among others. These have motivated technological developments on several fronts such as:

a. Advances in robust sensor technology. On-line sensors such as flowmeters, density gauges, and particle size analyzers have been successfully used in grinding circuit control. Machine vision, a non-invasive

Example 3.4

A slurry stream containing quartz is diverted into a 1-liter density can. The time taken to fill the can is measured as 7 s. The pulp density is measured by means of a calibrated balance, and is found to be 1400 kg m^{-3}. Calculate the % solids by weight, and the mass flowrate of quartz within the slurry.

Solution

The density of quartz is 2650 kg m^{-3}. Therefore, from Eq. (3.13), % solids by weight:

$$x = \frac{100 \times 2650 \times (1400 - 1000)}{1400 \times (2650 - 1000)}$$

$$= 45.9\%$$

The volumetric flowrate:

$$F = \frac{1\ \text{L}}{7\ \text{s}} \times \frac{3600\ \text{s}}{1\ \text{h}} \times \frac{1\ \text{m}^3}{1000\ \text{L}} = 0.51(\text{m}^3\text{h}^{-1})$$

Therefore, mass flowrate:

$$M = \frac{0.51\ \text{m}^3}{1\ \text{h}} \times \frac{1400\ \text{kg slurry}}{1\ \text{m}^3\ \text{slurry}} \times \frac{45.9\ \text{kg solids}}{100\ \text{kg slurry}}$$

$$= 330.5(\text{kg h}^{-1})$$

Example 3.5

A pump is fed by two slurry streams. Stream 1 has a flowrate of 5.0 m^3 h^{-1} and contains 40% solids by weight. Stream 2 has a flowrate of 3.4 m^3 h^{-1} and contains 55% solids by weight. Calculate the tonnage of dry solids pumped per hour (density of solids is 3000 kg m^{-3}.)

Solution

Slurry stream 1 has a flowrate of 5.0 m^3 h^{-1} and contains 40% solids. Therefore, from a re-arranged form of Eq. (3.13):

$$D = \frac{1000 \times 100\, s}{s(100 - x) + 1000\, x}$$

thus:

$$D = \frac{1000 \times 100 \times 3000}{(3000 \times (100 - 40)) + (1000 \times 40)} = 1364(\text{kg m}^{-3})$$

Therefore, from Eq. (3.15), the mass flowrate of solids in slurry stream 1:

$$= \frac{5.0 \times 3000 \times (1364 - 1000)}{(3000 - 1000)}$$

$$= 2.73(\text{t h}^{-1})$$

Slurry stream 2 has a flowrate of 3.4 m^3 h^{-1} and contains 55% solids. Using the same equations as above, the pulp density of the stream = 1579 kg m^{-3}. Therefore, from Eq. (3.14), the mass flowrate of solids in slurry stream 2 = 1.82 t h^{-1}. The tonnage of dry solids pumped is thus:

$$2.73 + 1.82 = 4.55(\text{t h}^{-1})$$

Example 3.6

A flotation plant treats 500 t of solids per hour. The feed pulp, containing 40% solids by weight, is conditioned for 5 min with reagents before being pumped to flotation. Calculate the volume of conditioning tank required. (Density of solids is 2700 kg m^{-3}.)

Solution

The volumetric flowrate of *solids* in the slurry stream:

$$\frac{\text{mass flowrate}}{\text{density}} = \frac{\frac{500\ \text{t}}{1\ \text{h}}}{\frac{2700\ \text{kg}}{1\ \text{m}^3} \times \frac{1\ \text{t}}{1000\ \text{kg}}} = 185.2(m^3\ \text{h}^{-1})$$

The mass flowrate of water in the slurry stream

$$= \text{mass flowrate of solids} \times \text{dilution ratio}$$

$$= 500 \times (100 - 40)/40$$

$$= 750(\text{t h}^{-1})$$

Therefore, the volumetric flowrate of water is 750 m^3 h^{-1}. The volumetric flowrate of slurry = 750 + 185.2

$$= 935.2(\text{m}^3\text{h}^{-1})$$

Therefore, for a nominal retention time of 5 min, the volume of conditioning tank should be:

$$935.2 \times 5/60 = 77.9(\text{m}^3)$$

Example 3.7

Calculate the % solids content of the slurry pumped from the sump in Example 3.5.

Solution

The mass flowrate of solids in slurry stream 1 is: 2.73 t h^{-1}

The slurry contains 40% solids, hence the mass flowrate of water:

$$= 2.73 \times 60/40$$

$$= 4.10(\text{t h}^{-1})$$

Similarly, the mass flowrate of water in slurry stream 2:

$$= 1.82 \times 45/55$$

$$= 1.49(\text{t h}^{-1})$$

Total slurry weight pumped:

$$= 2.73 + 4.10 + 1.82 + 1.49$$

$$= 10.14(\text{t h}^{-1})$$

Therefore, % solids by weight:

$$= 4.55 \times 100/10.14$$

$$= 44.9\%$$

technology, has been successfully implemented for monitoring and control of mineral processing plants (Duchesne, 2010; Aldrich et al., 2010; Janse Van Vuuren et al., 2011; Kistner et al., 2013). Some commercial vision systems are VisioRock/Froth (Metso Minerals), FrothMaster (Outotec) and JKFrothCam (JK-Tech Pty Ltd).

Other important sensors are pH meters, level and pressure transducers, all of which provide a signal related to the measurement of the particular process variable. This allows the final control elements, such as servo valves, variable speed motors, and pumps, to manipulate the process variables based on signals from the controllers. These sensors and final control elements are used in many industries besides the minerals industry, and are described elsewhere (Edwards et al., 2002; Seborg et al., 2010).

b. Advances in microprocessor and computer technology. These have led to the development of more powerful Distributed Control Systems (DCS) equipped with user friendly software applications that have facilitated process supervision and implementation of advanced control strategies. In addition, DCS capabilities combined with the development of intelligent sensors have allowed integration of automation tasks such as sensor configuration and fault detection.

c. More thorough knowledge of process behavior. This has led to more reliable mathematical models of various important process units that can be used to evaluate control strategies under different simulated conditions (Mular, 1989; Napier-Munn and Lynch, 1992; Burgos and Concha, 2005; Yianatos et al., 2012).

d. Increasing use of large units, notably large grinding mills and flotation cells. This has reduced the amount of instrumentation and the number of control loops to be implemented. At the same time, however, this has increased the demands on process control as poor performance of any of these large pieces of equipment will have a significant detrimental impact on overall process performance.

Financial models have been developed for the calculation of costs and benefits of the installation of automatic control systems (Bauer et al., 2007; Bauer and Craig, 2008). Benefits reported include energy savings, increased metallurgical efficiency and throughput, and decreased consumption of reagents, as well as increased process stability (Chang and Bruno, 1983; Flintoff et al., 1991; Thwaites, 2007).

The concepts, terminology and practice of process control in mineral processing have been comprehensively reviewed by Ulsoy and Sastry (1981), Edwards et al. (2002) and Hodouin (2011). Some general principles are introduced here, with specific control applications being reviewed in some later chapters.

3.5.1 Hierarchical Multilayer Control System

The overall objective of a process control system is to maximize economic profit while respecting a set of constraints such as safe operation, environmental regulations, equipment limitations, and product specifications. The complexity of mineral processing plants, which exhibit multiple interacting nonlinear processes affected by unmeasured disturbances, makes the problem of optimizing the whole plant operation cumbersome, if not intractable. In order to tackle this problem a *hierarchical control system*, as depicted in Figure 3.24, has been proposed. The objective of this control system structure is to decompose the global problem into simpler, structured subtasks handled by dedicated control layers following the divide-and-conquer strategy. Each control layer performs specific tasks at a different timescale, which provides a functional and temporal decomposition (Brdys and Tatjewski, 2005).

3.5.2 Instrumentation Layer

At the lowest level of the hierarchical control system is the *instrumentation layer* which is in charge of providing information to the upper control layers about the process status through the use of devices such as sensors and transmitters. At the same time, this layer provides direct access to the process through final control elements (actuators) such as control valves and pumps. Intelligent instrumentation equipped with networking capabilities has allowed the implementation of networks of instruments facilitating installation, commissioning and troubleshooting while reducing wiring costs (Caro, 2009). Examples of industrial automation or fieldbus networks are Profibus-PA, Foundation Fieldbus™, Ethernet/IP and Modbus Plus.

3.5.3 Regulatory Control Layer

The *regulatory control layer* is implemented in control hardware such as a DCS, PLC (programmable logic controllers) or stand-alone single station field devices. In this control layer, multiple single-input single-output (SISO) control loops are implemented to maintain local variables such as speed, flow or level at their target values in spite of the effect of fast acting disturbances such as pressure and flow variations. In order to achieve a desired regulation performance, the controller executes an algorithm at a high rate. As an example of regulatory

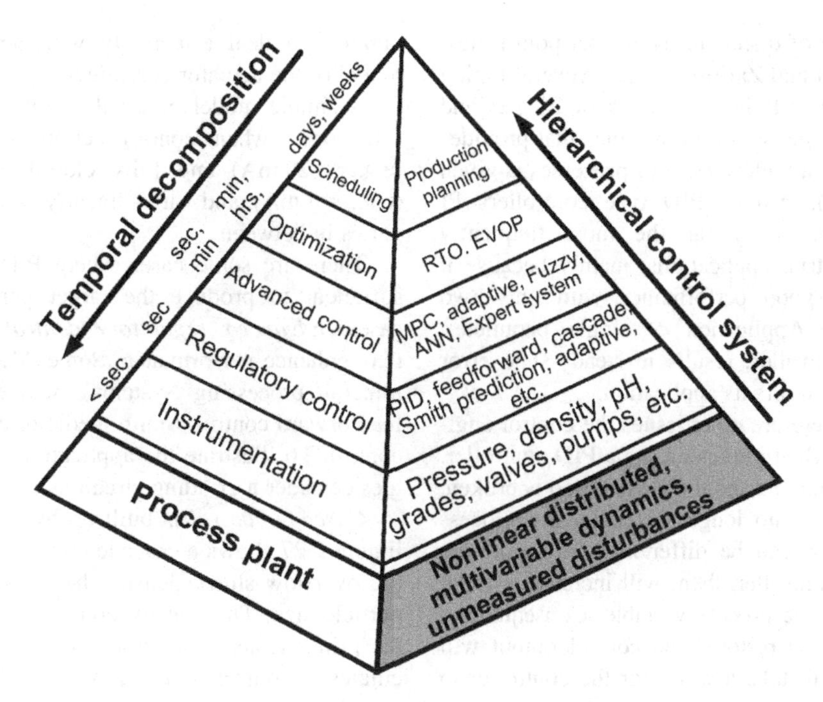

FIGURE 3.24 Hierarchical multilayer control system.

control, consider Figure 3.25, which illustrates a froth depth (level) control system for a flotation column. The objective is to regulate the froth depth at its set-point by manipulating the valve opening on the tailing stream to counter the effect of disturbances such as feed and air rate variations. To that end, a level transmitter (LT) senses the froth depth (process variable, PV) based on changes in some physical variables, for example, pressure variations by using a differential pressure sensor, and transmits an electrical signal (e.g., 4–20 mA) proportional to froth depth. The local level controller (LC) compares the desired froth depth (set-point, SP) to the actual measured PV and generates an electric signal (control output, CO) that drives the valve. This mechanism is known as *feedback* or *closed-loop* control.

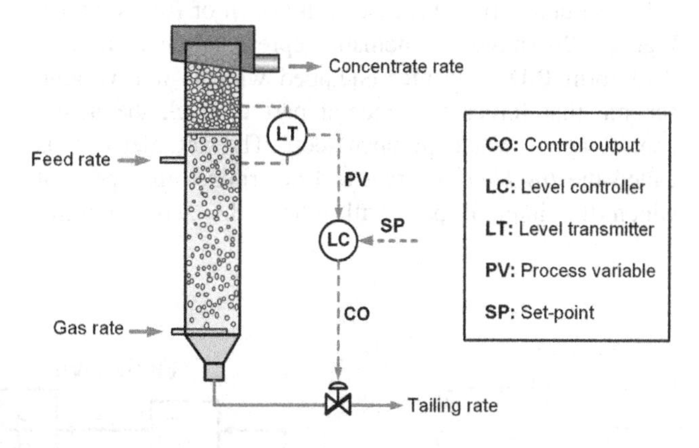

FIGURE 3.25 Feedback control of froth depth in a flotation column.

The most widely used regulatory control algorithm is the Proportional plus Integral plus Derivative (PID) controller. It can be found as a standard function block in DCS and PLC systems and also in its stand-alone single station version. There are several types of algorithms available. The standard (also known as non-interacting) algorithm has the following form:

$$CO(t) = \underbrace{K_C e(t)}_{\text{Proportional}} + \underbrace{\frac{K_C}{T_i}\int_0^t e(\tau)d\tau}_{\text{Integral}} + \underbrace{K_C T_d \frac{d\,e(t)}{dt}}_{\text{Derivative}} \quad (3.18)$$

where CO = controller output; K_c = controller gain; $e(t)$ = deviation of process variable from set-point (tracking error); T_i = integral time constant; T_d = derivative time constant; t = time. The process of selecting the parameters of a PID controller, namely K_c, T_i and T_d for the standard form, is known as *tuning*. Although the search for the appropriate PID parameters can be done by trial-and-error, this approach is only acceptable for simple processes having fast dynamics. To design an effective controller, a process model, performance measure, and signal characterization (type of disturbances and set-point) must be considered (Morari and Zafiriou, 1989). Several tuning rules have been proposed since the work of Ziegler and Nichols (1942). A simple set of tuning rules that provides good performance for a wide variety of processes is given by Skogestad (2003). Among PID type controllers PI (Proportional-Integral) is by far the most frequently encountered in industrial applications, mainly because it provides reasonably good performance with only two parameters (K_c, T_i). Application of single parameter, Proportional-only controller, results in steady-state error (offset) which largely limits its application.

Besides tuning, there are other issues the control engineer must deal with when implementing a PID controller. For example, if for some reason the closed-loop is broken, then the controller will no longer act upon the process and the tracking error can be different from zero. The integral part of the controller, then, will increase without having any effect on the process variable. Consequently, when the closed-loop is restored the control output will be too large and it will take a while for the controller to recover. This issue is known as *integral windup*.

There are situations when a closed-loop system becomes open-loop: (1) the controller is switched to manual mode, and (2) the final control element, such as a valve, saturates (i.e., becomes fully open or fully closed). Figure 3.26 shows a schematic representation of a standard form PID controller equipped with an anti-windup scheme that forces the integral part to track the actual control action when in open loop. The parameter T_t is called the *tracking constant* and determines the speed at which the integral part will follow the actual control output. To deal effectively with actuator saturation, a model of the actuator is required.

A simple model of a valve, for instance, is that it is fully open when control action is larger than 100% (e.g., >20 mA) and fully closed when less than 0% (e.g., <4 mA), and varies linearly when the control action varies in between.

There are some cases where PID controllers are not sufficient to produce the target performance. In those cases, *advanced regulatory control* (ARC) techniques may enhance performance. Some ARC strategies used in mineral processing systems include: cascade control, feedforward control, Smith predictor control, and adaptive control. To illustrate the application of these ARC strategies consider a grinding circuit as in Figure 3.27.

Cascade control is built up by nesting control loops. Figure 3.27 shows a cascade control for the regulation of the overflow slurry density, here used as a surrogate of particle size. The density controller (DC) in this case is the primary, also known as the master controller. It calculates a control action based on the deviation of the actual slurry density measurement from its set-point. However, rather than directly acting upon the final control element (valve), it provides a set-point to the inner water flowrate controller FC (secondary or slave controller). This technique has proven useful in attenuating the effect of disturbances occurring in the inner loop (e.g., water pressure variations) on the primary variable (density), especially when the outer process exhibits a transport delay. Moreover, when the inner loop is faster than the outer control loop then the tuning procedure can be carried sequentially starting by tuning the inner

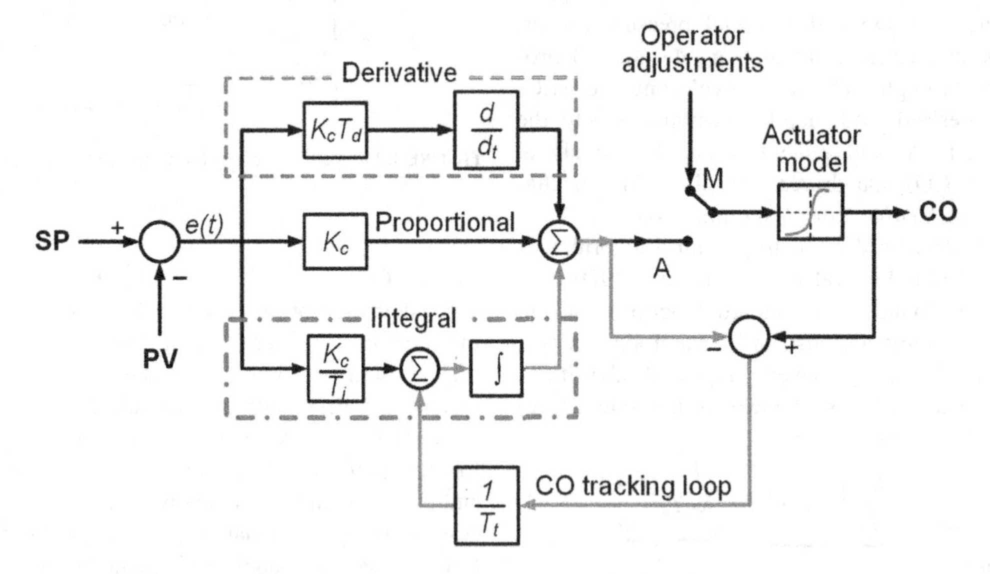

FIGURE 3.26 PID control scheme with anti-windup scheme.

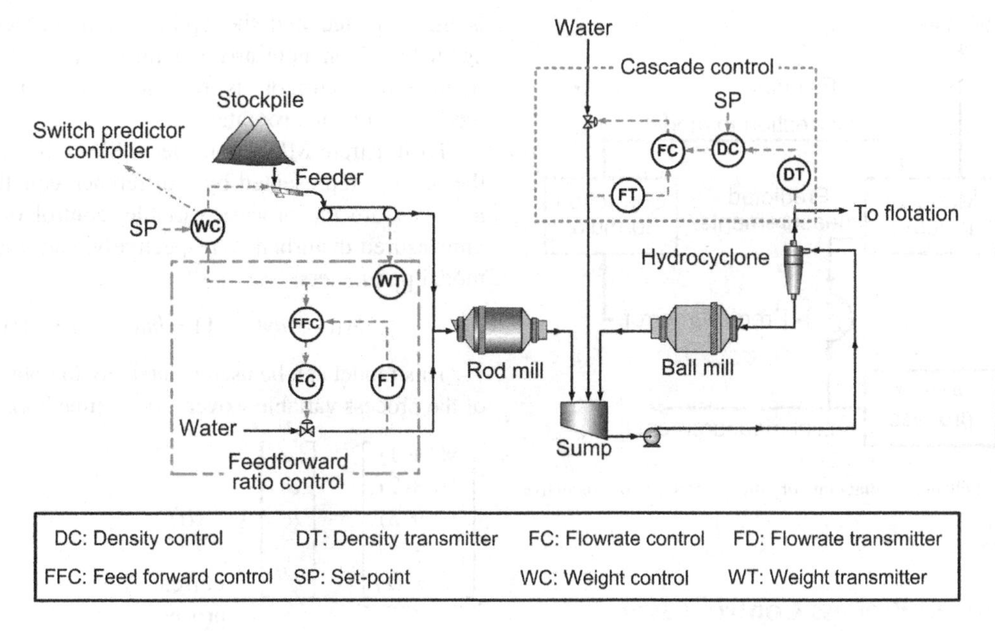

FIGURE 3.27 Control strategy of a grinding section *(Adapted from Nuñez et al., 2009).*

controller first (with outer control loop off) and subsequently tuning the primary controller neglecting the inner loop dynamics. Inner feedback control also reduces the effect of actuator nonlinearities, such as valve sticking, and dead band and nonlinear flow characteristics, on the primary control performance. Note that the whole hierarchical control system of Figure 3.27 is actually a cascade control system.

A disadvantage of feedback control is that it compensates for disturbances once they have already affected the process variable, that is, once an error has been produced. *Feedforward control* takes anticipatory corrective actions to compensate for a major disturbance at the expense of having an extra sensor to measure it. The control engineer has to assess whether the investment in extra instrumentation will be repaid with better regulation.

A typical application of a type of feedforward control is *ratio control.* Figure 3.27 shows a feedforward ratio control that aims to maintain the solid content (holdup) in the rod mill. Depending on the solid mass flowrate entering the mill measured by a weightometer (WT), a feedforward ratio controller acts by changing the set-point for the water flowrate control loop accordingly.

Processes that exhibit large transport delays relative to their dynamics can seriously limit the achievable performance of any feedback control, regardless of the complexity of the controller. In those cases, PID control performance can be improved by augmenting with a predictor, that is, a mathematical model of the process without the transport delay. This strategy is known as "Smith predictor controller" and it has been successfully applied to improve control performance of solid mass flowrate for SAG mills when manipulating the feed to a conveyor belt (Sbarbaro et al., 2005). In this case the transport delay is a function of the length and speed of the conveyor belt.

Changes in raw material characteristics and/or changes in the operating conditions mean any mathematical model used for controller synthesis may no longer reflect the actual process dynamic behavior. In those cases a control system that modifies controller parameters based on the mismatch between process model and actual process behavior is known as adaptive control (Hodouin and Najim, 1992). This type of "self-tuning" control adapts not only to load-dependent dynamic changes, but also to time-related and/or random dynamic characteristic changes (Rajamani and Hales, 1987; Thornton, 1991; Duarte et al., 1998). The measurements coming from the actual process are compared to the measurements predicted by the model, and the difference is used as the basis for the correction to the model. This correction alters the parameters and states (i.e., a set of variables that determines the state of the process) in the model so as to make the predictions better match those of the actual process. The basis of parameter and state estimation is the recursive least squares algorithm (Ljung, 1999), where the estimator updates its estimates continuously with time as each input arrives, rather than collecting all the information together and processing in a single batch (Figure 3.28).

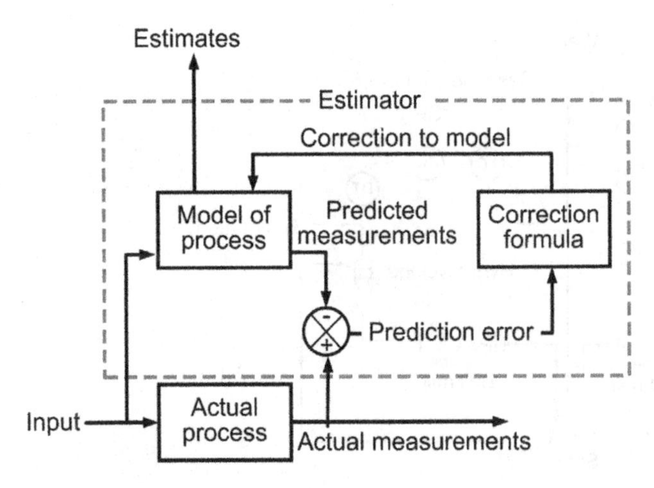

FIGURE 3.28 Schematic diagram of the operation of recursive estimation.

3.5.4 Advanced Process Control Layer

The control strategy for the grinding section shown in Figure 3.27 where multiple independent local SISO control loops were implemented is known as *decentralized control*. When the process exhibits severe interaction this decentralized strategy strongly limits the achievable performance and more advanced control strategies are required. A control system that takes interactions explicitly into account is known as centralized *multivariable control* and it is commonly implemented at the advanced process control layer. This control layer is in charge of keeping secondary variables related to process performance at their optimum, determined by the upper optimization layer, while dealing with interactions and constraints in a dynamic fashion. Receding horizon model based predictive control (MPC) provides a methodology to deal with this type of problem in a systematic way. The main cost associated with MPC is the identification and maintenance of good mathematical models. However, user friendly MPC packages implemented in DCS systems have greatly facilitated these tasks and thus the application of MPC technology in industrial processes. Examples of MPC software packages commercially available are Connoiseur (Invensys), Delta V Predict MPC (Emerson), cpmPlus Expert Optimizer (ABB), BrainWave (Andritz), Profit® Controller (Honeywell) and StarCS MPC (Mintek).

MPC translates a control problem into a dynamic optimization problem over a finite future horizon. A discrete mathematical model is used to generate predictions of the future evolution of the process variable based on current information, future control actions, and assumptions about the future behavior of disturbances. The solution to this optimization problem is a sequence of control moves. Due to the uncertainty in the process model and the disturbances, only the first calculated optimal control move is implemented and the optimization problem is solved again based on new arriving information. This way, the optimization horizon is receding and thus a feedback mechanism is incorporated.

To illustrate MPC, consider a SISO discrete model of the process represented by a difference equation where y, u, and d are the process variable, control output and an unmeasured disturbance, respectively, and a and b are the model parameters:

$$y(t) = ay(t-1) + bu(t-1) + d(t) \tag{3.19}$$

This model can be used recursively to obtain predictions of the process variable y over a finite time horizon H.

$$\underbrace{\begin{bmatrix} y(t+1) \\ y(t+2) \\ y(t+3) \\ \vdots \\ y(t+H) \end{bmatrix}}_{\text{process variable predictions}} = \begin{bmatrix} a \\ a^2 \\ a^3 \\ \vdots \\ a^H \end{bmatrix} \cdot \underbrace{y(t)}_{\text{current process variable}}$$
$$+ \begin{bmatrix} b & 0 & 0 & 0 & 0 \\ ab & b & 0 & 0 & 0 \\ a^2b & ab & b & 0 & 0 \\ \vdots & \vdots & \vdots & \vdots & \vdots \\ a^{H-1}b & a^{H-2}b & \cdots & ab & b \end{bmatrix} \cdot \underbrace{\begin{bmatrix} u(t) \\ u(t+1) \\ u(t+2) \\ \vdots \\ u(t+H-1) \end{bmatrix}}_{\text{actual and future control moves}} + \underbrace{\begin{bmatrix} d(t+1) \\ d(t+2) \\ d(t+3) \\ \vdots \\ d(t+H) \end{bmatrix}}_{\text{future disturbances}} \tag{3.20}$$

An optimization problem can then be formulated: find the sequence of future control moves $u(t)$, $u(t+1)$, ..., $u(t+H-1)$ that minimizes the deviations of the process output from its reference trajectory while satisfying multiple constraints such as limits on actuator range/rate of change, output constraints, etc. Note that minimization of the control moves is commonly incorporated in the objective function to be minimized, that is:

$$\underset{u(t),u(t+1),\ldots u(t+H-1)}{\text{minimize}} \sum_{i=1}^{H} (r(t+j) - y(t+j))^2 + \lambda \cdot \sum_{i=1}^{H} \Delta u(t+j)^2 \tag{3.21}$$

where r is the reference trajectory over the time horizon H, and λ an adjustable parameter that weights control moves with respect to tracking errors.

Note that the solution to the optimization problem will be a function of the future values of the disturbance, which are not known in advance. There are different strategies to deal with this issue, a simple way being to assume that the disturbance is constant over the time horizon and takes the same value as its current value $d(t)$, which can be estimated from the difference between the actual measured process variable $y_{meas}(t)$ and that provided by the model:

$$\hat{d}(t) = y_{meas}(t) - y(t) \tag{3.22}$$

where $\hat{d}(t)$ stands for disturbance estimation. Figure 3.29 illustrates the MPC strategy. Some examples of MPC applications in mineral processing systems include grinding control (Ramasamy et al., 2005; Jonas, 2008; Steyn and Sandrock, 2013), and flotation control (Cortes et al., 2008).

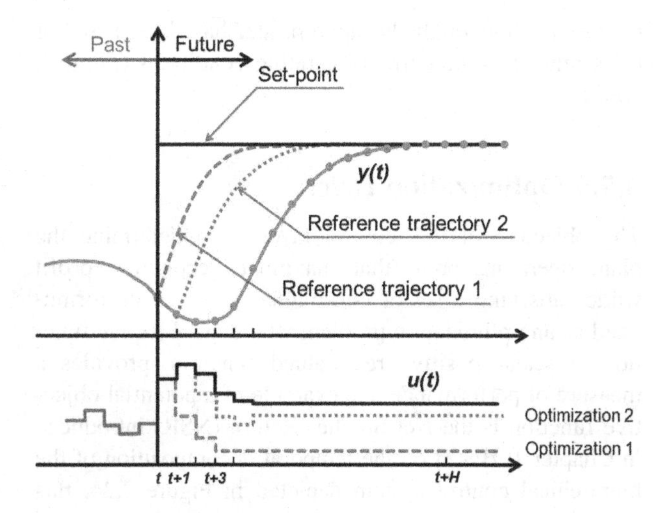

FIGURE 3.29 Receding MPC strategy.

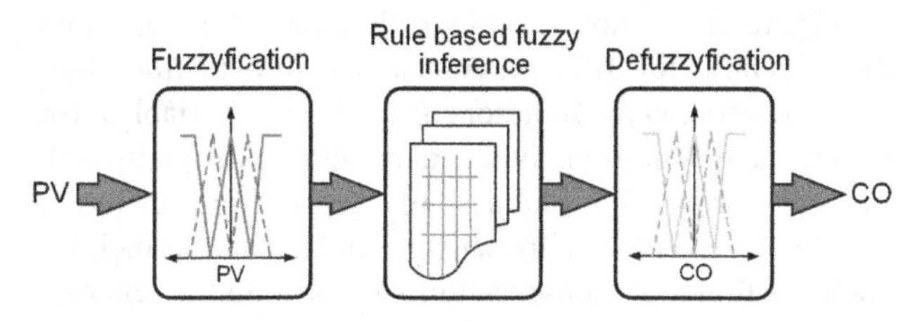

FIGURE 3.30 Block diagram of a Fuzzy inference system.

Expert Systems

In cases where the identification and maintenance of good mathematical models for the application of MPC technology become unwieldy, expert systems, which do not require an algorithmic mathematical model, are an alternative (Ruel, 2014). Expert systems derive from research into artificial intelligence (AI) and are computer programs that emulate the reasoning of human expertise in a narrowly focused domain (Laguitton and Leung, 1989; Bearman and Milne, 1992). Essentially, they are computer systems that achieve high levels of performance in tasks for which humans would require years of special education, training, or experience. Human operators often make "rule-of-thumb" educated guesses that have come to be known as *heuristics*. As the expert knowledge is commonly expressed with some degree of uncertainty, ambiguity or even contradiction *fuzzy-logic* forms an integral part of expert systems.

Fuzzy-logic theory involves the development of an ordered set of fuzzy conditional statements, which provide an approximate description of a control strategy, such that modification and refinement of the controller can be performed without the need for special technical skills. These statements are of the form: if *X is A* then *Y is B*, where *X* and *Y* are process variables and *A* and *B* represent linguistic expressions such "low" and "high". All the conditional rules taken together constitute a fuzzy decisional algorithm for controlling the plant.

Figure 3.30 illustrates a fuzzy inference-based expert system. It consists of three stages: fuzzyfication, rule based fuzzy inference, and defuzzification. In the fuzzification stage, the operational span of each variable, that is, the range or the *universe of discourse* as known in AI, is covered by fuzzy sets represented by membership functions labeled with English-like linguistic expressions. Membership functions are usually of triangular, trapezoidal or Gaussian shape. Then, any measurement can be evaluated to determine its degree of membership to each fuzzy set. As a result, a fuzzy sentence like: thickener underflow density is 70% "a little high", 30% "medium high" and 0% "very low" can be obtained. Then, a set of rules of the form IF-THEN is generated in a process called *knowledge acquisition phase*. An example of a fuzzy rule for thickener control is: if underflow density is "a little high" then flocculant addition change is "small decrease". Finally, the last stage of this fuzzy-inference system (defuzzification) converts a fuzzy action such as small decrease of flocculant addition into a crisp value (e.g., 10% decrease).

Harris and Meech (1987) have presented an excellent introduction to Fuzzy Control, showing how it has been applied to a simulation model of a secondary crushing plant. The simplicity of the method allows for its application to systems such as grinding circuits (Hales and Ynchausti, 1997), flotation plants (Osorio et al., 1999), and thickening (Shoenbrunn and Toronto, 1999).

Control systems based on rules with fuzzy logic support have steadily become more ambitious. Van der Spuy et al. (2003) propose an "on-line expert" which will provide both operator training and assistance. Expert systems have also proven useful for selecting which advanced control method might be appropriate, based on a set of rules related to the current operating conditions (Flintoff, 2002).

3.5.5 Optimization Layer

The objective of this control layer is to determine the plant operating point that maximizes economic profit while satisfying a set of constraints. This can be formulated as an optimization problem where an objective function, a scalar positive real-valued function, provides a measure of performance. An example of a potential objective function is the Net Smelter Return (NSR) introduced in Chapter 1. Based on the temporal decomposition of the hierarchical control system depicted in Figure 3.24, this layer assumes the process operates in steady-state and therefore utilizes a stationary nonlinear model of the process to determine the optimum operating conditions.

Figure 3.31 shows a block diagram of a real-time optimization (RTO) layer. A steady-state detection algorithm continuously monitors the process variables for steady-state condition. When this condition is confirmed, the measured data, subject to errors from different sources, are adjusted to satisfy fundamental principles such as those of conservation of mass and/or energy. This process is known as *data reconciliation* (Hodouin, 2011) (see Section 3.6). The reconciled data are then used to update the process model parameters including unmeasured external inputs in order for the model to represent the actual process condition as accurately as possible.

To deal with complexity, plant optimization is usually broken down into multiple staged local optimizations coordinated by a supervisory system. For example,

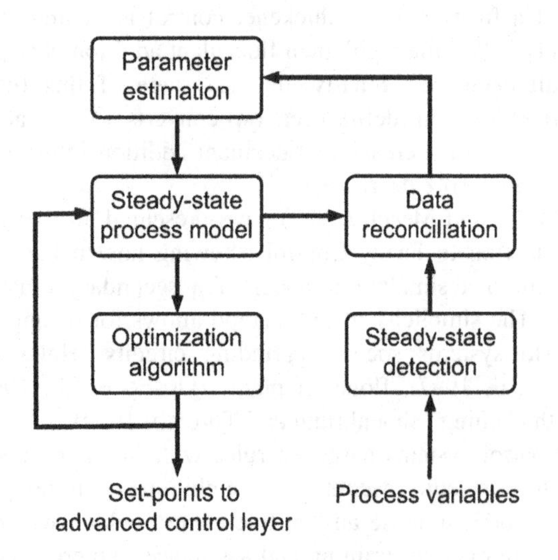

FIGURE 3.31 RTO layer components.

optimization of the grinding-flotation section can be decomposed into two optimization problems with specific objective functions and interfacing constraints (McIvor and Finch, 1991). A good example of RTO was reported by Holdsworth et al. (2002) for the Greens Creek silver–lead–zinc mine in Alaska. The highly instrumented concentrator carried out frequent on-line mass balances based on the last 2 h of operation. The balanced data were used to calibrate a simple flotation model. The operator could then assess strategies for "moving" metal between low-value bulk concentrate and high-value lead or zinc concentrates with the aid of a simple on-line NSR model. Another application of the RTO-based linear programming technique was proposed to maximize throughput while maintaining product fineness at its target value for a rod-ball grinding section (Lestage et al., 2002). Other optimizing control systems in mineral processing can be found in (Herbst and Bascur, 1984; McKee and Thornton, 1986; Hodouin, 2011).

If an accurate model of the process is not available, a model-free optimization strategy based on evolutionary operation EVOP (Box and Draper, 1998) can be considered. EVOP optimization is based on control action to either continue or reverse the direction of movement of process efficiency by manipulating the set-point of the controller to achieve some higher level of performance. The controlled variables are altered according to a predetermined strategy, and the effect on the process efficiency is computed on-line. If the efficiency is increased as a result of the change, then the next change is made in the same direction, otherwise the direction is reversed; eventually the efficiency should converge to the optimum (Krstev and Golomeov, 1998). Hulthén and Evertsson (2011) implemented an automatic EVOP algorithm that varied eccentric speed on a cone crusher. Although the algorithm pointed in the optimal direction it was found to be too slow to cope with short-term variations such as changes in the feed.

There are cases where primary variables related to economic performance are not available or cannot be measured frequently enough for optimizing control purposes. This can arise for multiple reasons: for example, high cost instrumentation, long sampling measurement delays (e.g., on-stream analyzer, lab analysis), or because the primary variable is a calculated variable (e.g., recovery). In those cases the missing variable can be inferred from a mathematical model using secondary (surrogate) variables that are correlated to the primary variable and measured on-line in real time. These models are known as *soft or virtual sensors* and their use in a control system is called *inferential control*.

Models relating secondary to primary variables can be classified into three categories, depending whether the model structure and parameters have a physical meaning.

While phenomenological models are derived from first principles, black box models, as their name suggests, provide no insight into the phenomena governing the process. Gray models combine both phenomenological models with empirical relationships. Artificial Neural Networks (ANN), a general class of parametric nonlinear mapping between inputs and outputs, belong to the black box type of model and have been successfully used for developing soft-sensors, due to their ability to learn complex input-output patterns. Some applications of soft-sensor based ANN in mineral processing are particle size estimation of hydrocylone overflow (Sbarbaro et al., 2008), particle size distribution from a conveyor belt using image processing (Ko and Shang, 2011), SAG mill power predictions (Hales and Ynchausti, 1992), and concentrate grade estimation in flotation (González et al., 2003). Soft-sensors based on gray-box modeling through the use of Kalman filter and its extended version have been reported to infer SAG mill parameters (Herbst and Pate, 1996; Apelt et al., 2002) and frother concentration in flotation systems (Maldonado et al., 2010). Statistical methods such as latent variable methods (Principal Component Analysis (PCA), Partial Least Squares (PLS)) have also been implemented to estimate process variables from analysis of digital images. Some examples are the estimation of froth grade and froth properties in flotation systems (Duchesne, 2010; Kistner et al., 2013). González (2010) provides an extensive list of soft-sensing applications in mineral processing.

3.5.6 The Control Room

Control computers are housed in dedicated rooms. It is recognized that many of the systems used within the control room need to be addressed ergonomically using a human, or a user-centered design approach: Figure 3.32 shows a design adhering to several of the ISO 11064 "Control Room Ergonomic Design" standards.

Remote monitoring and control are used in several industries, including some examples in mining: remote operation of mining equipment, such as haul trucks, for example, is standard at many sites. Arguably the main driver of remote "command centers" is the retention of highly skilled workforce in an urban setting. Integrated operations (iOps, https://www.youtube.com/watch?v = qDrObBfv9q8) seeks to integrate the best talents, local and remote, to meet operational challenges. Flintoff et al. (2014) offer some observations on these developments as they relate to mining.

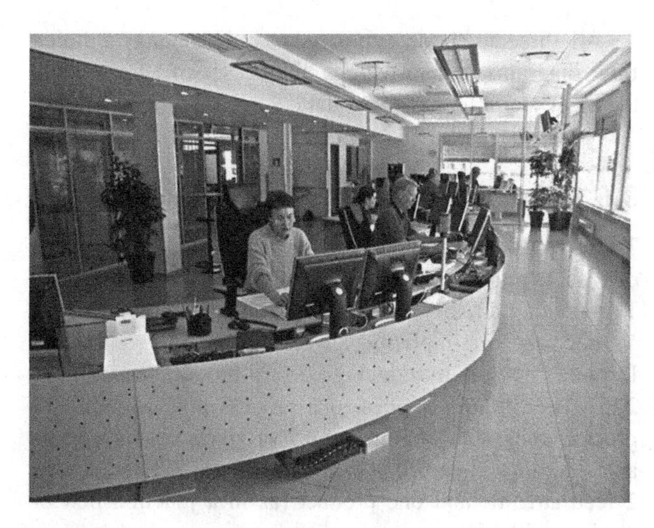

FIGURE 3.32 Nikkelverk Refinery (Glencore Nickel) – Site Central Control Room (July 2004) *(Courtesy P. Thwaites, (XPS Glencore) and R. Løkling, (Glencore Nickel – Nikkelverk, Kristiansand)).*

3.6 MASS BALANCING METHODS

Mass balancing or accounting for phase masses and component mass fractions by phase is critical to many responsibilities of the metallurgical team of any mineral or metal processing plant: troubleshooting/improving/assessing plant performance, monitoring/controlling plant operation, accounting/reporting metal production, and stock movements.

Today, data acquisition and mass balancing are increasingly computerized and automated. Mass balancing techniques have evolved accordingly from the well-known 2-product formula to advanced computerized statistical algorithms that can handle the many situations and issues a metallurgist may have to address when performing mass balance calculations.

The evolution is particularly significant in mass balancing for metal and production accounting and reporting needs. With tighter and more stringent accounting regulations, guidelines and good practices in metal accounting have been generated and published (AMIRA, 2007; Jansen, 2009; Lachance et al., 2012). Statistical reconciliation of excess (*redundant*) data is nowadays the norm in mass balancing (Reid et al., 1982) and the 2-product formula does not fit any longer with best practices due to its limitations.

This section of Chapter 3 is an introduction to mass balancing and metal accounting, summarizing years of evolution from the basic well-known 2-product formula to computerized advanced statistical techniques. While the n-product formula was used for many years, its limitations and the progress of the information technology sector have made possible the use of more advanced methods. Three of them are reviewed here:

- The node imbalance minimization method.
- The 2-step least squares minimization method.
- The generalized least squares minimization method.

All three methods apply to mass balancing complex flow diagrams and provide a unique set of estimates in the presence of an excess of data.

3.6.1 The *n*-product Formula

The *n*-product formula applies to a process unit (or group of process units) with 1 feed and *n* products, or *n* feeds and 1 product, or a total of $n+1$ feed and product streams.

For sake of simplicity, let us start with the 2-product formula before introducing the *n*-product case. The 2-product formula applies to a process unit or group of process units with one feed and 2 products (a concentrate and a tailings stream, for example (Figure 3.33)), or with 2 feed streams and one product (as in a junction box or mixing unit).

If W_F, W_C and W_T are, respectively, the mass flowrates of feed, concentrate and tailings, then, assuming steady state, the (solids) mass conservation equation is:

$$W_F = W_C + W_T \tag{3.23}$$

Assuming x_F, x_C, and x_T are the assay values of a metal of interest, then a second (or component) mass conservation equation is:

$$W_F x_F = W_C x_C + W_T x_T \tag{3.24}$$

Now assuming W_F, x_F, x_C, and x_T are measured, it follows from the previous 2 equations that:

$$W_C = \frac{x_F - x_T}{x_C - x_T} W_F \text{ and } W_T = \frac{x_C - x_F}{x_C - x_T} W_F \tag{3.25}$$

and hence, the mass flowrates of concentrate and tailings are both estimable and the 6 values, W_F, W_C, W_T, x_F, x_C, and x_T, all balance together. Equation (3.25) is known as the 2-product formula.

From the above results, several performance indicators are typically calculated.

The *solid split* (or its inverse, the *ratio of concentration*), which states how much of the feed solid reports to the concentrate stream:

$$\frac{W_C}{W_F} = \frac{x_F - x_T}{x_C - x_T} \tag{3.26}$$

The metal *recovery*, R, which gives how much of the metal contained in the feed has been recovered in the concentrate stream (Chapter 1) and is a measure of the plant metallurgical performance:

$$R = \frac{W_C x_C}{W_F x_F} = \frac{x_F - x_T}{x_C - x_T} \times \frac{x_C}{x_F} \tag{3.27}$$

The simplicity of the method and resulting equations and the limited amount of data required for providing information on the process performance and efficiency account for the success of the 2-product formula over the years (Example 3.8).

Components other than metal assays can be used, including particle size and percent solids data (see Section 3.7).

By extension of the 2-product formula, the *n*-product formula works for a process unit or group of process units with one feed and *n* product streams (or any combination of $n+1$ streams around the process unit). Assuming the mass flowrate of the feed stream is known, the *n*-product formula provides a way of calculating the mass flowrates of the *n* product streams, provided metal assays are known on each of the $n+1$ streams. The mass conservation equations are:

$$W_F = \sum_{i=1}^{n} W_{P_i} \tag{3.28}$$

and

$$W_F x_F^j = \sum_{i=1}^{n} W_{P_i} x_{P_i}^j \tag{3.29}$$

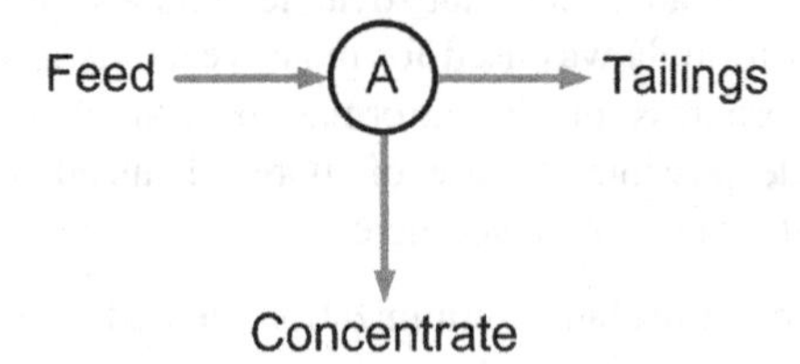

FIGURE 3.33 A simple 2-product process unit.

Example 3.8

The feed to a Zn flotation plant assays 3.93% Zn. The concentrate produced assays 52.07% Zn and the tailings 0.49% Zn. Calculate the solid split, ratio of concentration, enrichment ratio, and Zn recovery.

Solution

The solid split (Eq. (3.26)) is:

$$\frac{W_C}{W_F} = \frac{x_F - x_T}{x_C - x_T} = \frac{3.93 - 0.49}{52.07 - 0.49} = 0.067 \text{ or } 6.7\%$$

The ratio of concentration is the inverse of the solid split:

$$\frac{W_F}{W_C} = \frac{x_C - x_T}{x_F - x_T} = \frac{52.07 - 0.49}{3.93 - 0.49} = 15.0$$

The enrichment ratio (x_C/x_F) is:

$$\frac{x_C}{x_F} = \frac{52.07}{3.93} = 13.25$$

The Zn recovery (Eq. (3.27)) is:

$$R = \frac{W_C x_C}{W_F x_F} = \frac{x_F - x_T}{x_C - x_T} \times \frac{x_C}{x_F} = \frac{3.93 - 0.49}{52.07 - 0.49} \times \frac{52.07}{3.93}$$

$$= 0.884 \text{ or } 88.4\%$$

To be solvable the number of required assayed metals depends on n, the number of product streams. There are n unknown mass flowrates and hence n mass conservation equations are needed. Since there is always one equation available that expresses the mass conservation of solids (Eq. (3.28)), then $n-1$ additional equations are needed (equations of the form of Eq. (3.29)) and therefore $n-1$ metals need be assayed on the $n+1$ streams. Table 3.1 summarizes the n-product formula requirements in terms of assayed metals.

While the 2-product formula remains simple, the formula complexity increases very quickly with n, which explains why its use has been limited to $n=4$. Mathematically speaking, there exists a matrix formulation to the problem that can easily be computerized.

Sensitivity Analysis

The 2-product formula calculates the values of W_C and W_T, the concentrate and tailings streams mass flowrates, knowing W_F, x_F, x_C, and x_T respectively the feed stream mass flowrate and the metal assay of interest on each of the 3 streams. The measured values of W_F, x_F, x_C, and x_T are tainted with measurement errors. The measurement errors are propagated through the calculations on to the values of W_C and W_T. The following explains how to estimate V_C and V_T, the variances of W_C and W_T, knowing V_F, V_f, V_c, and V_t, the variances of W_F, x_F, x_C, and x_T.

Assuming covariance terms can be neglected, the variance of a function is obtained from its first derivatives:

$$V_{f(x,y)} = \left(\frac{\delta f}{\delta x}\right)^2 V_x + \left(\frac{\delta f}{\delta y}\right)^2 V_y \tag{3.30}$$

Applying Eq. (3.30) to the 2-product formula (Eq. (3.25)) gives (all assays and variances as fractions):

$$V_C = \left(\frac{W_F}{x_C - x_T}\right)^2 V_f + \left(\frac{(x_C - x_F)\,W_F}{(x_C - x_T)^2}\right)^2 V_t + \left(\frac{(x_F - x_T)W_F}{(x_C - x_T)^2}\right)^2 V_c + \left(\frac{x_F - x_T}{x_C - x_T}\right)^2 V_F \tag{3.31}$$

TABLE 3.1 Requirements for n-product Formula

Number of Product Streams	Number of Required Assayed Metals	Number of Samples
2	1	3
3	2	4
4	3	5
n	$n-1$	$n+1$

$$V_T = \left(\frac{W_F}{x_C - x_T}\right)^2 V_f + \left(\frac{(x_C - x_F)W_F}{(x_C - x_T)^2}\right)^2 V_t + \left(\frac{(x_F - x_T)W_F}{(x_C - x_T)^2}\right)^2 V_c + \left(\frac{x_C - x_F}{x_C - x_T}\right)^2 V_F \tag{3.32}$$

In the above equations, the difference $x_C - x_T$ is present in the denominator of each term of the equations. It follows that the calculated variance increases when the $x_C - x_T$ difference value decreases and hence when the assayed metal is not well separated in the process unit.

Applying Eq. (3.30) to the recovery formula (Eq. (3.27)) gives (again, reminding, all assays and variances as fractions):

$$V_R = \frac{x_C^2 x_T^2}{x_F^2 (x_C - x_T)^4}\left[\frac{(x_C - x_T)^2}{x_F^2} V_f + \frac{(x_F - x_T)^2}{x_C^2} V_c + \frac{(x_C - x_F)^2}{x_T^2} V_t\right] \tag{3.33}$$

This equation clearly shows that the calculated recovery variance is also strongly dependent on the process unit separation efficiency for the metal of interest (Example 3.9).

Example 3.9

The streams in the Zn flotation plant (Example 3.8) are also assayed for their content in Cu and Fe. The laboratory reports the measured values in Table ex 3.9.

Calculate the recoveries of zinc and copper and provide the expression of their variances as functions of the variances of Zn and Cu assays on all 3 streams.

Solution

The recoveries of Zn and Cu are calculated using Eq. (3.27):

For Zn: $R = \frac{W_C x_C}{W_F x_F} = \frac{(3.93 - 0.49) \times 52.07}{(52.07 - 0.49) \times 3.93} = 0.884$ or 88.4%

For Cu: $R = \frac{W_C x_C}{W_F x_F} = \frac{(0.16 - 0.14) \times 0.66}{(0.66 - 0.14) \times 0.16} = 0.159$ or 15.9%

and their variances are provided by Eq. (3.33):

For Zn:

$$V_R = 0.059547 \times [172.2573 V_f + 0.004365 V_c + 9652.06 V_t] = 10.25733 V_f + 0.00026 V_c + 574.7472 V_t$$

For Cu:

$$V_R = 45613.21 \times [10.5625 V_f + 0.000918 V_c + 12.7551 V_t] = 481789.5 V_f + 41.88541 V_c + 581801.1 V_t$$

TABLE EX 3.9 Zinc Flotation Plant Metal Assays (%)

	Feed	Concentrate	Tailings
Zn	3.93	52.07	0.49
Cu	0.16	0.66	0.14
Fe	11.57	14.67	13.09

From this numerical example, it is evident that: a) the stream samples that contribute most to the recovery variance are those with the lower assay values (feed and tailings); b) the variance of the least separated metal (Cu) is much higher than the variance of the most separated metal (Zn); and c) when using the 2-product formula, the accuracy of the recovery value comes from the feed and tailings stream assay accuracies rather than from the concentrate. There is a challenge here since these are not necessarily the easiest samples to collect. Worked examples estimating the standard deviation on recovery are given in Example 3.19.

Error Propagation

The n-product formula enables the computation of non-measured mass flowrates from metal assays and from the feed stream flowrate. The n-product formula does not take into account measurement errors; and measurement errors, when present, are propagated onto the calculated mass flowrates.

Example 3.9 provides a good example of error propagations. Table 3.2 displays the calculated values of the solid split using the Zn assay, the Cu assay, and the Fe assay.

Should the measured assays be free of measurement errors, the solid split values would be identical since, in theory (and reality), there exists only one solid split value. Obviously, the Fe assay is erroneous on at least one of the 3 streams, but it is impossible to identify which one. If each of the Zn and the Cu assays are taken individually, it is not possible to ascertain that the Zn assay or the Cu assay suffers error on one of the 3 samples. It is only because at least 2 metals have been assayed that we can ascertain the presence of measurement errors.

Excess of Data

Example 3.9 data introduces a second limitation of the n-product formula: there is no provision on how to deal with an excess of data. And due to the unavoidable presence of measurement errors, it is not good practice to keep only just sufficient data to apply the n-product formula even when the data set includes the metals best separated in the process unit.

TABLE 3.2 Error Propagation Example

	Feed	Concentrate	Tailings
Zn	1.0	6.7	93.3
Cu	1.0	4.5	95.5
Fe	1.0	−96.2	196.2

Since an excess of data reveals measurement errors, there is a need for a mass balancing method which can deal with data in excess.

More than One Process Unit

Another limitation of the n-product formula is that it deals with only one process unit or group of process units. The formula does not apply to more complex problems.

In the process flow diagram of Figure 3.34, the feed flowrate is measured and Cu has been assayed on each of the 6 stream samples. The question is: can we apply the n-product formula to estimate the non-measured flowrates and complete the mass balance?

Applying the 2-product formula on process unit A, it is possible to calculate the flowrate of each of the 2 outgoing streams. Considering that the flowrate of the feed stream to process unit B is now known, the 2-product formula provides values of the 2 outgoing stream flowrates of process unit B.

As a result of the above calculations, we now have an excess of data for process unit C, since the only unknown flowrate is for the outgoing stream. Consequently the 2-product formula fails to complete the mass balance calculation.

It is worth noticing that selecting any other path in the applications of the 2-product formula would have ended up in the same situation: an excess of data preventing the completion of the mass balance calculation. There are various ways to address this kind of issue and to complete the mass balance, but the point is that the n-product formula is not sufficient.

No Correction of Measurement Errors

The n-product formula closes the mass balance and provides values for the non-measured flowrates; hence it

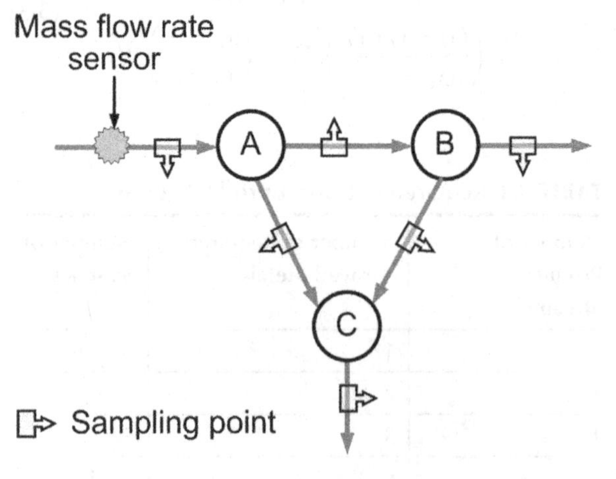

FIGURE 3.34 Flow diagram example.

gives access to key performance indicators such as the solid split and the metal recovery.

We have seen, however, that the n-product formula propagates measurement errors onto the calculated flowrate values. Another drawback is that it does not correct the measurement errors. In the example of Table 3.2 and without any indication about the uncertainties associated with Zn and Cu assays, there is no reason to trust more the solid split value calculated with the Zn assays than with the Cu assays. Obviously there is something very wrong with the Fe assays, but there is no way to identify which of the feed, concentrate or tailings stream sample is culprit.

For all the above reasons, the n-product formula cannot be considered any longer as a mass balancing method. The following section presents a more sound approach to mass balancing in presence of measurement errors.

3.6.2 Node Imbalance Minimization

In the Node Imbalance Minimization method, the mass conservation equations are written considering the deviation to rigorous mass conservation due to measurement errors (Finch and Matwijenko, 1977; Lynch, 1977; Tipman et al., 1978; Klimpel, 1979). For the simple case of a process unit with one feed and two products, the mass conservation equations are:

$$W_F - W_C - W_T = I_W \tag{3.34}$$

$$W_F x_F - W_C x_C - W_T x_T = I_x \tag{3.35}$$

where the W's stand for the mass flowrates and the x's for the metal assays; I_W and I_x are the node imbalances (errors) for the mass of solids and the mass of metal of interest, respectively. Additional equations can be written for each assayed metal of interest and/or each process unit.

In a given mass balance, the total number of mass conservation equations is therefore a function of the number of process units and metals assayed. For the flow diagram of Figure 3.34 and assuming three metals have been assayed, the total number of mass conservation equations would be: 3 process units × (1 solid equation + 3 assay equations) = 12 equations. Assuming that the feed flowrate is known, then the total number of unknown is 5: the solid mass flowrates of the other streams.

Typically there are more equations than unknowns and, as a consequence, there is not a unique solution that satisfies all equations. The best mass flowrate estimates, $\hat{W}_k$, are those which minimize the following criterion:

$$S = \sum_j (I_W^j)^2 + \sum_i \sum_j (I_{x_i}^j)^2 \tag{3.36}$$

In the above equation, i stands for an assayed metal and j for a process unit.

The calculation of these best estimates requires the calculation of the derivatives of S for each of the unknowns, that is, the non-measured solid mass flowrates. The best estimates are those unknown variable values for which the derivatives are all together equal to zero:

$$\frac{\delta S}{\delta W_K} = 0 \tag{3.37}$$

It is worth noting that with the obtained best estimates the mass conservation equations are not rigorously verified; essentially, the node imbalances are only minimized. By experience, the estimates obtained by this method are as good as the measured values used for the calculation are themselves (Example 3.10).

Example 3.10

For the example data set in Table ex 3.9, apply the Node Imbalance Minimization method to estimate the solid split value and compare it with the 2-product formula estimates.

Solution

In this example there is one process unit and three metals assayed; hence, there are four mass conservation equations and hence four imbalances. The criterion S can be written as:

$$\begin{aligned} S = {} & W_F^2 + W_C^2 + W_T^2 - 2W_F W_C - 2W_F W_T + 2W_C W_T \\ & + W_F^2 \sum_{i=1}^{3} x_{Fi}^2 + W_C^2 \sum_{i=1}^{3} x_{Ci}^2 + W_T^2 \sum_{i=1}^{3} x_{Ti}^2 \\ & - 2W_F W_C \sum_{i=1}^{3} x_{Fi} x_{Ci} - 2W_F W_T \sum_{i=1}^{3} x_{Fi} x_{Ti} \\ & + 2W_C W_T \sum_{i=1}^{3} x_{Ci} x_{Ti} \end{aligned} \tag{3.38}$$

where i stands for the assayed metal.

The derivatives of S for W_C and W_T are then:

$$\begin{aligned} \frac{\delta S}{\delta W_C} = 2 \Bigg(& W_C - W_F + W_T + W_C \sum_{i=1}^{3} x_{Ci}^2 \\ & - W_F \sum_{i=1}^{3} x_{Fi} x_{Ci} + W_T \sum_{i=1}^{3} x_{Ci} x_{Ti} \Bigg) = 0 \end{aligned} \tag{3.39}$$

$$\begin{aligned} \frac{\delta S}{\delta W_T} = 2 \Bigg(& W_T - W_F + W_C + W_T \sum_{i=1}^{3} x_{Ti}^2 \\ & - W_F \sum_{i=1}^{3} x_{Fi} x_{Ti} + W_C \sum_{i=1}^{3} x_{Ci} x_{Ti} \Bigg) = 0 \end{aligned} \tag{3.40}$$

Now, the solid split value can be obtained from the above 2 equations:

$$\frac{W_T}{W_F} = \frac{\left(1 + \sum_{i=1}^{3} x_{Fi} x_{Ti}\right)\left(1 + \sum_{i=1}^{3} x_{Ci}^2\right) - \left(1 + \sum_{i=1}^{3} x_{Fi} x_{Ci}\right)\left(1 + \sum_{i=1}^{3} x_{Ci} x_{Ti}\right)}{\left(1 + \sum_{i=1}^{3} x_{Ti}^2\right)\left(1 + \sum_{i=1}^{3} x_{Ci}^2\right) - \left(1 + \sum_{i=1}^{3} x_{Ci} x_{Ti}\right)^2} \tag{3.41}$$

and $\dfrac{W_C}{W_F} = 1 - \dfrac{W_T}{W_F}$

Using the data set of Table ex 3.9, Table ex 3.10 is derived.

The results of Table ex 3.10 show that the method is sensitive to the accuracy of the metal assays. Indeed, the Fe assay is bad as demonstrated by the application of the 2-product formula (see Table 3.2).

TABLE EX 3.10 Node Imbalance Example

	$\frac{W_T}{W_F}$ (%)	$\frac{W_C}{W_F}$ (%)
Using Zn, Cu and Fe	80.8	19.1
Using Zn and Cu	93.1	6.87
Using Zn only	93.3	6.67

Limitations

The node imbalance minimization method solves two issues with the *n*-product formula:

- It handles data in excess
- It works for more than one process unit at a time

However, the method exhibits 2 limitations:

- It is sensitive to measurement errors as can be seen in Example 3.10
- It only provides estimates of the non-measured solid flowrates; measured flowrate values and assay values are not adjusted

An extension of the method aims to decrease the influence of bad measurements. The criterion node imbalance terms are weighted according to the presence or absence of gross measurement errors.

3.6.3 Two-step Least Squares Minimization

The 2-step least squares minimization method proposes a way to adjust and correct metal assays for the measurement errors that affect them (Wiegel, 1972; Mular, 1979). It is a compromise between the limitations of the node imbalance minimization method (Section 3.6.2) and the complexity of the *generalized least squares minimization* method (Section 3.6.4).

In the node imbalance method, the deviations to the mass conservation of solids and of metals are considered all together. The obtained estimated flowrate values minimize the deviations to all the mass conservation equations.

In the 2-step least squares method, flowrate values that rigorously verify the mass conservation equations of solids and, at the time, minimize the deviations to the mass conservation equations of metals, are first estimated. In a second step, corrected values of metal assays that rigorously verify the mass conservation equations of metals are estimated.

To achieve, taking Eqs. (3.23) and (3.24) as an example, the mass conservation equations are re-written as:

$$\frac{W_T}{W_F} = 1 - \frac{W_C}{W_F} \tag{3.42}$$

$$x_F = \frac{W_C}{W_F} x_C + \left(1 - \frac{W_C}{W_F}\right) x_T \tag{3.43}$$

and the node imbalance equations for metals become:

$$x_F - \frac{W_C}{W_F} x_C - \left(1 - \frac{W_C}{W_F}\right) x_T = I_x \tag{3.44}$$

or:

$$(x_F - x_T) - \frac{W_C}{W_F}(x_C - x_T) = I_x \tag{3.45}$$

The general problem can now be stated as: find the best estimate of $\frac{W_C}{W_F}$ which minimizes the sum of squared imbalances I_x for all nodes and assayed metals:

$$S = \sum_j \sum_i (i^i_{x_j})^2 \tag{3.46}$$

By comparison with the node imbalance minimization method, the criterion only contains the node imbalances for metals and the search variables are a set of independent relative solid flowrates. Similar to the node imbalance minimization method, the calculation of the best estimates of the independent relative solid flowrates requires the calculation of the derivatives of S for each of the unknowns, that is, the independent relative solid flowrates. The best estimates are those values for which the derivatives are all together equal to zero. Due to the way the node imbalances for metals are written (Eq. (3.45)), the mass conservation of relative solid flowrates is rigorously verified for the estimated relative solid flowrate values.

Having determined $\frac{W_C}{W_F}$, assuming the feed solid flowrate W_F is measured, it is possible to estimate W_C and W_T. All these flowrate values rigorously verify their mass conservation equations.

In a second step, it is possible to adjust the metal assay measured values (x) to values ($\hat{x}$) that verify their own mass conservation equations as well. Let us call r_x the required adjustment for each metal assay value, then it follows that:

$$r_x = x - \hat{x} \tag{3.47}$$

and since:

$$(\hat{x}_F - \hat{x}_T) - \frac{W_C}{W_F}(\hat{x}_C - \hat{x}_T) = 0 \tag{3.48}$$

it follows by difference with Eq. (3.45) that:

$$I_x - (r_{x_F} - r_{x_T}) - \frac{W_C}{W_F}(r_{x_C} - r_{x_T}) = 0 \tag{3.49}$$

The problem is now to find the values of r_x which minimize the following least squares criterion:

$$S = \sum_i \sum_j (r^i_{x_j})^2 \tag{3.50}$$

under the equality constraint K of Eq. (3.49). In Eq. (3.50), i stands for the number of streams (or samples) and j for the number of assayed metals. The problem is best solved using the Lagrange technique where the criterion becomes:

$$L = \sum_i \sum_j (r^i_{x_j})^2 + \sum_k \lambda_k K_k \tag{3.51}$$

where k is the number of nodes (or mass conservation equations), and λ_k is the Lagrange coefficient of equality constraint K_k.

The problem is solved by calculating the derivatives of L for each of the unknown, that is, $r^i_{x_j}$ and λ_i. Criterion L is minimal when all its derivatives are equal to zero, which leads to a set of equations to be solved. Solving the set of equations provides the best estimates of the $r^i_{x_j}$ values, that is, the adjustments to the measured assay values which will make the adjusted assay values coherent from a mass conservation point of view (Example 3.11).

Limitations

The 2-step least squares minimization method has the advantage over the previous methods of providing adjusted assay values that verify the mass conservation equations. The method remains simple and can easily be programmed (e.g., using Excel Solver) for simple flow diagrams.

Although presenting significant improvements over the previous methods, the 2-step method is not mathematically optimal since the mass flowrates are estimated from measured and therefore erroneous metal assay values. Measurement errors are directly propagated to the estimated flowrate values. In other words, the reliability of the estimated flowrate values depends strongly on the reliability of the measured assay values. The equations could, however, be modified to include weighting factors with the objective of decreasing the influence of poor assays.

3.6.4 Generalized Least Squares Minimization

The generalized least squares minimization method proposes a way to estimate flowrate values and adjust metal assay values in a single step, all together at the same time (Smith and Ichiyen, 1973; Hodouin and Everell, 1980). There is also no limitation on the number of mass conservation equations.

The mass conservation equations are written for the theoretical values of the process variables. The process variable measured values carry measurement errors and hence do not verify the mass conservation equations. It is assumed that the measured values are unbiased, uncorrelated to the other measured values and belong to Normal distributions, $N(\mu,\sigma)$. The balances can be expressed as follows:

$$W_F^* = W_C^* + W_T^* \tag{3.52}$$

$$W_F^* x_F^* = W_C^* x_C^* + W_T^* x_T^* \tag{3.53}$$

where * denotes the theoretical value of the variable. Since the theoretical values are not known, the objective is to find the best estimates of the theoretical values. With the statistical assumptions made, the maximum likelihood estimates are those which minimize the following generalized least square criterion:

$$S = \sum_i \frac{(W_j - \hat{W}_j)^2}{\sigma^2_{W_j}} + \sum_j \sum_i \frac{(x_{ij} - \hat{x}_{ij})^2}{\sigma^2_{x_{ij}}} \tag{3.54}$$

where the ^ denotes the best estimate value and σ^2 is the variance of the measured value.

It is evident that the best estimates must obey the mass conservation equations while minimizing the generalized least squares criterion S. Therefore, the problem consists in minimizing a least squares criterion under a set of equality constraints, the mass conservation equations. Such a problem is solved by minimizing a Lagrangian of the form:

$$L = \sum_i \frac{(W_j - \hat{W}_j)^2}{\sigma^2_{W_j}} + \sum_j \sum_i \frac{(x_{ij} - \hat{x}_{ij})^2}{\sigma^2_{x_{ij}}} + \sum_k \lambda_k K_k \tag{3.55}$$

where λ_k are the Lagrange coefficients, and K_k the mass conservation equations.

As seen previously, the problem is solved by calculating the derivatives of L for each of the unknowns, that is, the $\hat{W}$, $\hat{x}$ and the λ. Criterion L is minimal when all its derivatives are equal to zero, which leads to a set of equations to be solved. Solving the set of equations provides the best estimate values $\hat{W}$ and $\hat{x}$, which verify the mass conservation equations. It is important to note that even non-measured variables can be estimated, since they appear in the mass conservation equations of the Lagrangian, as long as there is the necessary redundancy.

In Eq. (3.55), the variance of the measured values, σ^2, are weighting factors. An accurate measured value being associated with low variance cannot be markedly adjusted without significantly impacting the whole criterion value. In

Example 3.11

For the example data set of Table ex 3.9 apply the 2-step least squares minimization method to estimate the solid split value and compare it with the 2-product formula and node imbalance estimates. Calculate the adjusted metal assay values and compare with the measured values.

Solution

In the present case, Eq. (3.46) is:

$$S = \sum_j \sum_i (l^i_{x_j})^2 = \sum_{j=1}^{3} \left[(x_{F_j} - x_{T_j}) - \frac{W_C}{W_F}(x_{C_j} - x_{T_j}) \right]^2$$

where j stands for Zn, Cu and Fe. Deriving S for the unknown $w_C = \frac{W_C}{W_F}$ gives:

$$\frac{\delta S}{\delta w_C} = 2 \sum_{j=1}^{3} \left[(x_{F_j} - x_{T_j}) - w_C(x_{C_j} - x_{T_j}) \right] \left(x_{C_j} - x_{T_j} \right) = 0$$

The solution of which is:

$$w_C = \frac{W_C}{W_F} = \frac{\sum_{j=1}^{3} (x_{F_j} - x_{T_j})(x_{C_j} - x_{T_j})}{\sum_{j=1}^{3} (x_{C_j} - x_{T_j})^2}$$

Using the data set of Table ex 3.9, Table ex 3.11a is derived.

TABLE EX 3.11a Two-step Least Squares Minimization Example

	$\frac{W_T}{W_F}$ (%)	$\frac{W_C}{W_F}$ (%)
Using Zn, Cu and Fe	93.4	6.57
Using Zn and Cu	93.3	6.67
Using Zn only	93.3	6.67

The results of Table ex 3.11a show that the method is much less sensitive to the accuracy of the metal assays than the results of the node imbalance method shown in Table ex 3.10. Having determined the solid split, the next step is to adjust the metal assay values to make them coherent with the calculated solid split and coherent from a mass conservation point of view.

In the present case, Eq. (3.51) is:

$$L = \sum_{i=1}^{3} (r^i_{x_F})^2 + (r^i_{x_C})^2 + (r^i_{x_T})^2 + \sum_{i=1}^{3} \lambda_i \left[l^i_x - (r^i_{x_F} - r^i_{x_T}) - w_C(r^i_{x_C} - r^i_{x_T}) \right]$$

and the derivatives for each variable are:

$$\frac{\delta L}{d\, r^i_{x_F}} = 2r^i_{x_F} - \lambda_i = 0$$

$$\frac{\delta L}{d\, r^i_{x_F}} = 2r^i_{x_F} + \lambda_i w_C = 0$$

$$\frac{\delta L}{d\, r^i_{x_T}} = 2r^i_{x_F} + \lambda_i(1 - w_C) = 0$$

$$\frac{\delta L}{d\lambda_i} = l^i_x - (r^i_{x_F} - r^i_{x_T}) - w_C(r^i_{x_C} - r^i_{x_T}) = 0$$

Solving the system of equations leads to:

$$r^i_{x_F} = \frac{l^i_x}{1 + w_C^2 + (1 - w_C)^2}$$

$$r^i_{x_C} = \frac{-\, l^i_x w_C}{1 + w_C^2 + (1 - w_C)^2}$$

$$r^i_{x_T} = \frac{-\, l^i_x(1 - w_C)}{1 + w_C^2 + (1 - w_C)^2}$$

and the numerical solution is:

	l^i_x	Adjustments $r^i_{x_j}$			Adjusted values $\hat{x}^i_j$		
		Feed	Concentrate	Tailings	Feed	Concentrate	Tailings
Zn	0.049885	0.049885	0.049885	0.049885	3.903426	52.0717466	0.514828
Cu	−0.01418	−0.01418	−0.01418	−0.01418	0.167552	0.65950362	0.132944
Fe	−1.62385	−1.62385	−1.62385	−1.62385	12.43504	14.6131448	12.28181

Table ex 3.11b shows a comparison of the measured assays (Table ex 3.9) with the adjusted ones where adjustment = (measured − adjusted)/measured as a percent.

TABLE EX 3.11b Adjustment Distribution (%)

	Feed	Concentrate	Tailings
Zn	0.68	0.00	−5.07
Cu	−4.72	0.08	5.04
Fe	−7.48	0.39	6.17

Note the concentrate stream assays are barely adjusted by comparison with the other 2 stream assays.

contrast, a bad measured value associated to a high standard deviation can be significantly adjusted without impacting the whole criterion value. Hence, adjustments, small or large, will be made in accordance to the confidence we have in the measured value whenever possible considering the mass conservation constraints that must always be verified by the adjusted values (Example 3.12).

While Example 3.12 presents the calculations in detail for a simple case, there exists a more elegant and generic mathematical solution that can be programmed. The solution uses matrices and requires knowledge in matrix algebra.

Example 3.12

For the example data set of Table ex 3.9 apply the generalized least squares minimization method to estimate the solid split value and compare it with the 2-step least squares minimization estimate. Calculate the adjusted metal assay values and compare with the measured values.

Solution

In the present case, the Lagrangian is:

$$L = \sum_{j=1}^{3}\sum_{i=1}^{3}\frac{(x_{ij}-\hat{x}_{ij})^2}{\sigma^2_{x_{ij}}} + \sum_{i=1}^{3}\lambda_i K_i$$

There is no measured mass flowrate, there are 3 stream samples with 3 metal assays each and there are 3 mass conservation equations:

$$\hat{x}_{F_i} = \frac{\hat{W}_C}{\hat{W}_F}\hat{x}_{C_i} + \left(1 - \frac{\hat{W}_C}{\hat{W}_F}\right)\hat{x}_{T_i}$$

where i stands for Zn, Cu, or Fe, and the problem variables are: λ_i, $\hat{x}_{ij}$, and $\hat{w}_C = \frac{\hat{W}_C}{\hat{W}_F}$.

The Lagrangian derivatives are:

$$\frac{\delta L}{d\hat{w}_C} = -\sum_{i=1}^{3}(\hat{x}_{C_i} - \hat{x}_{T_i}) = 0$$

$$\frac{\delta L}{d\lambda_i} = \hat{x}_{F_i} - \hat{w}_C\hat{x}_{C_i} - (1 - \hat{w}_C)\hat{x}_{T_i} = 0$$

$$\frac{\delta L}{d\hat{x}_{F_i}} = -2\frac{(x_{F_i} - \hat{x}_{F_i})}{\sigma^2_{x_{F_i}}} + \lambda_i$$

$$\frac{\delta L}{d\hat{x}_{C_i}} = -2\frac{(x_{C_i} - \hat{x}_{C_i})}{\sigma^2_{x_{C_i}}} - \lambda_i w_C = 0$$

$$\frac{\delta L}{d\hat{x}_{T_i}} = -2\frac{(x_{T_i} - \hat{x}_{T_i})}{\sigma^2_{x_{T_i}}} - \lambda_i(1 - w_C) = 0$$

where i stands for Zn, Cu or Fe, for a total of 13 equations and 13 unknowns. It can be shown that once the value of $\hat{w}_C$ is known then there is an analytical solution for $\hat{x}_{ij}$:

$$\hat{x}_{F_i} = x_{F_i} - \frac{x_{F_i} - x_{C_i}\hat{w}_C - x_{T_i}(1 - \hat{w}_C)}{\frac{1}{\sigma^2_{x_{F_i}}} + \frac{\hat{w}_C^2}{\sigma^2_{x_{C_i}}} + \frac{(1-\hat{w}_C)^2}{\sigma^2_{x_{T_i}}}} \times \frac{1}{\sigma^2_{x_{F_i}}}$$

$$\hat{x}_{C_i} = x_{C_i} + \frac{x_{F_i} - x_{C_i}\hat{w}_C - x_{T_i}(1 - \hat{w}_C)}{\frac{1}{\sigma^2_{x_{F_i}}} + \frac{\hat{w}_C^2}{\sigma^2_{x_{C_i}}} + \frac{(1-\hat{w}_C)^2}{\sigma^2_{x_{T_i}}}} \times \frac{\hat{w}_C}{\sigma^2_{x_{C_i}}}$$

$$\hat{x}_{T_i} = x_{T_i} + \frac{x_{F_i} - x_{C_i}\hat{w}_C - x_{T_i}(1 - \hat{w}_C)}{\frac{1}{\sigma^2_{x_{F_i}}} + \frac{\hat{w}_C^2}{\sigma^2_{x_{C_i}}} + \frac{(1-\hat{w}_C)^2}{\sigma^2_{x_{T_i}}}} \times \frac{1 - \hat{w}_C}{\sigma^2_{x_{T_i}}}$$

where i stands for Zn, Cu or Fe.

Therefore, it is only required to find the best value of $\hat{w}_C$ and the algorithm consists in:

1. Assuming an initial value for $\hat{w}_C$
2. Calculating the $\hat{x}_{ij}$ values
3. Calculating the criterion S of Eq. (3.54)
4. Iterating on the value of $\hat{w}_C$ to minimize the value of S

The solution is given in Table ex 3.12a for variance values all equal to 1 for simplicity. The initial solid split $\hat{w}_C$ has been estimated at 6.67% (based on result in Example 3.11).

Table ex 3.12b shows a comparison of the measured assays with the adjusted ones. The obtained results are similar to the ones obtained with the previous method.

TABLE EX 3.12a Generalized Least Squares Minimization Example

	Feed	Concentrate	Tailings
Zn	3.89	52.1	0.527
Cu	0.167	0.660	0.133
Fe	12.4	14.6	12.3

TABLE EX 3.12b Adjustment Distribution (%)

	Feed	Concentrate	Tailings
Zn (%)	1.00	0.00	−7.47
Cu (%)	−4.64	0.07	4.96
Fe (%)	−7.47	0.38	6.17

3.6.5 Mass Balance Models

The formulation of the generalized least squares minimization method enables resolving complex mass balance problems. The Lagrange criterion consists of 2 types of terms: the weighted adjustments and the mass conservation constraints. The criterion can easily be extended to include various types of measurements and mass conservation equations.

The whole set of mass conservation equations that apply to a given data set and mass balance problem is called the *mass balance model*. Depending on the performed measurements and sample analyses, the following types of mass conservation equations may typically apply in mineral processing plants:

- Conservation of slurry mass flowrates
- Conservation of solid phase mass flowrates
- Conservation of liquid phase mass flowrates (for a water balance, for instance)
- Conservation of solid to liquid ratios
- Conservation of components in the solid phase (e.g., particle size, liberation class)
- Conservation of components in the liquid phase (in leaching plants, for instance)

There are also additional mass conservation constraints such as:

- The completeness constraint of size or density distributions (the sum of all mass fractions must be equal to 1 for each size distribution)
- Metal assays by size fraction of size distributions
- The coherency constraints between the reconstituted metal assays from assay-by-sizes and the sample metal assays

All these constraint types can be handled and processed by the Lagrangian of the generalized least squares minimization algorithm. As mentioned, all the necessary mass conservation equations for a given data set constitute the mass balance model.

It is convenient to define the mass balance model using the concept of networks. Indeed, the structure and the number of required mass conservation equations depend on the process flow diagram and the measurement types. Therefore, it is convenient to define a network type by process variable type, knowing that the structure of mass conservation equations for solid flowrates (Eq. (3.23)) is different from that of metal assays (Eq. (3.24)), for example.

Not only does a network type mean a mass conservation equation type, but it also expresses the flow of the mass of interest (solids, liquids, metals...).

For the 2-product process unit of Figure 3.33, assuming 3 metals (Zn, Cu, and Fe) have been assayed on each stream, then Figure 3.35 shows the 2 networks that can be developed, one for each equation type (Eqs. (3.23) and (3.24)).

For the more complex flow diagram of Figure 3.34, assuming again 3 metals (Zn, Cu, and Fe) have been assayed on each of the 6 streams, then Figure 3.36 shows the 2 networks that can be developed.

Assuming a fourth metal, Au for example, has been assayed on the main Feed, main Tailings and main Concentrate streams only, then a third network should be developed for Au as shown in Figure 3.37.

Networks are conveniently represented using a matrix (Cutting, 1976). In a network matrix, an incoming stream is represented with a + 1, an outgoing stream with a −1 and any other stream with a 0. For the networks of Figure 3.35 and Figure 3.36, the network matrices are, respectively:

$$M = \begin{bmatrix} 1 & -1 & -1 \end{bmatrix} \qquad (3.56)$$

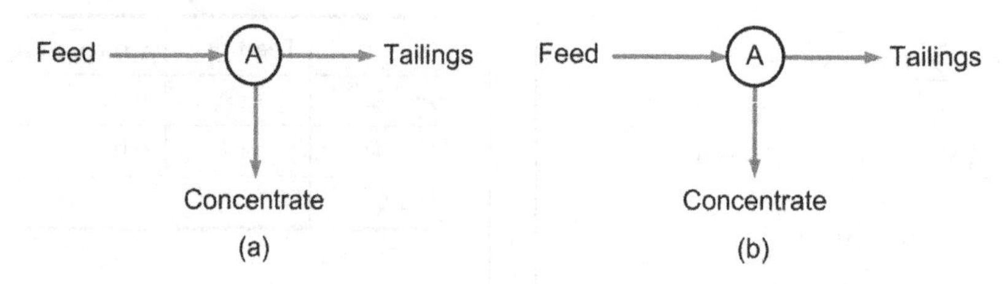

FIGURE 3.35 Mass conservation networks for the flow diagram of Figure 3.33: (a) For solids, and (b) For metal (in this case they are the same).

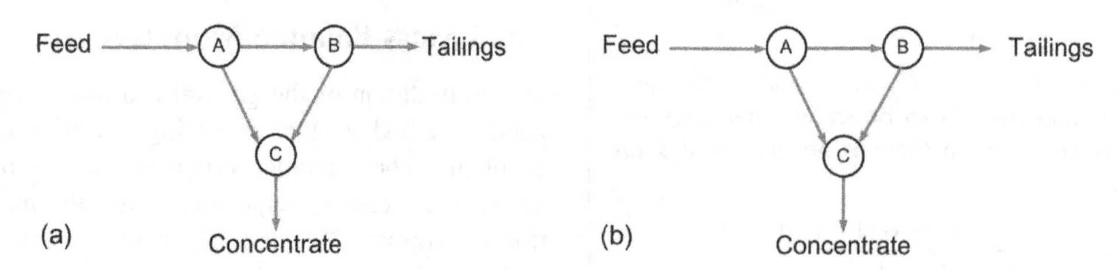

FIGURE 3.36 Mass conservation networks for the flow diagram of Figure 3.34: (a) For solids, and (b) For metal (again, they are the same in this case).

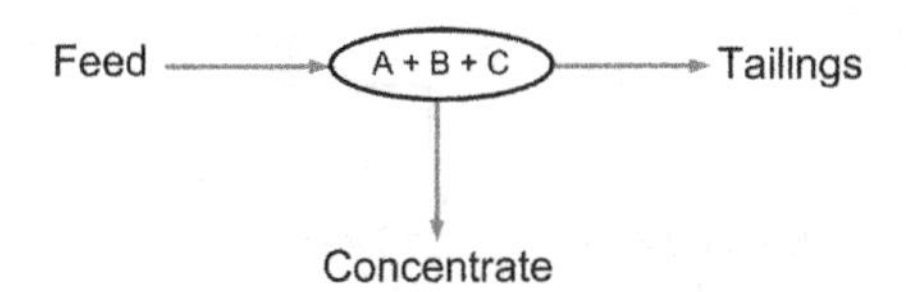

FIGURE 3.37 Mass conservation networks for Au in the flow diagram of Figure 3.34.

and

$$M = \begin{bmatrix} 1 & -1 & -1 & 0 & 0 & 0 \\ 0 & 1 & 0 & -1 & -1 & 0 \\ 0 & 0 & 1 & 0 & 1 & -1 \end{bmatrix} \begin{matrix} \text{for node A} \\ \text{for node B} \\ \text{for node C} \end{matrix} \tag{3.57}$$

Each row represents a node; each column stands for a stream. Equations (3.56) and (3.57) expressed in a matrix form become:

$$M_W W = 0 \tag{3.58}$$

$$M_x \overline{W}_x X_i = 0 \tag{3.59}$$

where M_W and M_x are the network matrices for the solid network and the metal assay network, W is the column matrix of solid mass flowrates, $\overline{W}_x$ the diagonal matrix of the solid mass flowrates used in the metal assay network, and X_i the column matrix of metal assay i on each stream of the metal assay network (i stands for Zn, Cu and Fe) (Example 3.13).

3.6.6 Error Models

In the generalized least square minimization method, weighting factors prevent large adjustments of trusted measurements and, on the contrary, facilitate large adjustments of poorly measured process variables. Setting the weighting factors is the main challenge of the method (Almasy and Mah, 1984; Chen et al., 1997; Darouach et al., 1989; Keller et al., 1992; Narasimhan and Jordache, 2000; Blanchette et al., 2011).

Assuming a measure obeys to a Normal distribution, the distribution variance is a measure of the confidence we have in the measure itself. Hence, a measure we trust exhibits a small variance, and the variance increases when the confidence decreases. The variance should not be confused with the process variable variation, which also includes the process variation: the variance is the representation of the measurement error only.

Measurement errors come from various sources and are of different kinds. Theories and practical guidelines have been developed to understand the phenomenon and hence render possible the minimization of such errors as discussed in Section 3.2. A brief recap of measurement errors as they pertain to mass balancing will help drive the message home.

Errors fall under two main categories: systematic errors or biases, and random errors. Systematic errors are difficult to detect and identify. First, they can only be suspected over time by nature and definition. Second, a

Example 3.13

A survey campaign is being designed around the grinding circuit in Figure ex 3.13. For the location of flow meters and samplers, and for the analysis types performed on the samples, develop the mass conservation networks for each data type and mass conservation equation type that apply.

Solution

With the slurry feed flowrate to the grinding circuit being measured, water addition stream flowrates being measured and % solids in slurry being measured, the slurry and water mass balances can be calculated and associated measured values adjusted. The mass conservation equations that apply obey to the Slurry and Water Mass Conservation Networks are given in Figure ex 3.13-S.

The mass conservation equations for the dry solid obey the same network except water addition streams are removed.

Mass size fractions are not conserved through grinding devices (SAG mill and ball mills). Therefore there is no mass conservation equation for size fractions around the grinding devices and consequently there is no node for grinding devices in mass size fraction networks. The same reasoning applies to Cu-by-size fractions.

The (overall) Cu on stream samples is conserved through grinding. Since Cu has been assayed only on the SAG mill feed and the cyclone overflows feeding the flotation circuit, the mass conservation network for Cu on stream samples consists of only one node.

Assuming eleven sieves have been used for measuring the size distributions, then the total system of mass conservation equations consist in:

- Seven mass conservation equations for slurry flowrate variables
- Seven mass conservation equations for % solids variables
- Seven mass conservation equations for water flowrate variables
- Seven mass conservation equations for solid flowrate variables
- Forty eight (4 nodes × 12 mass fractions) mass conservation equations for size fractions
- Forty eight mass conservation equations for Cu-by-size fractions
- One mass conservation equation for Cu

for a total of 135 equations and 316 variables.

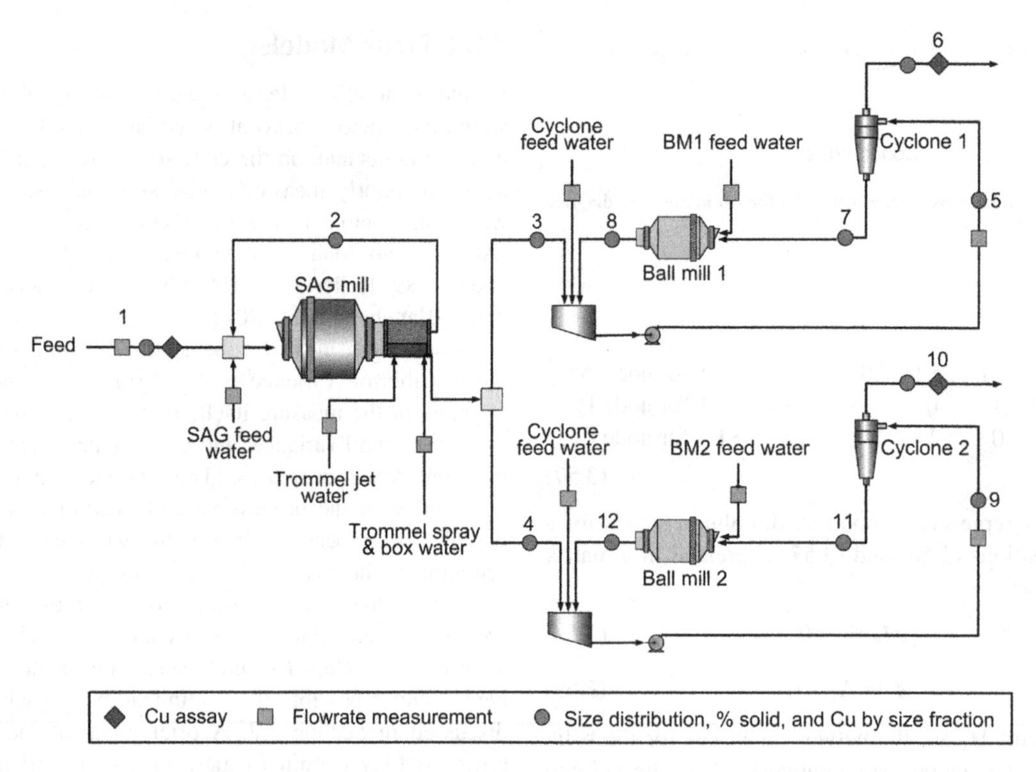

FIGURE EX 3.13 Grinding circuit showing sampling points.

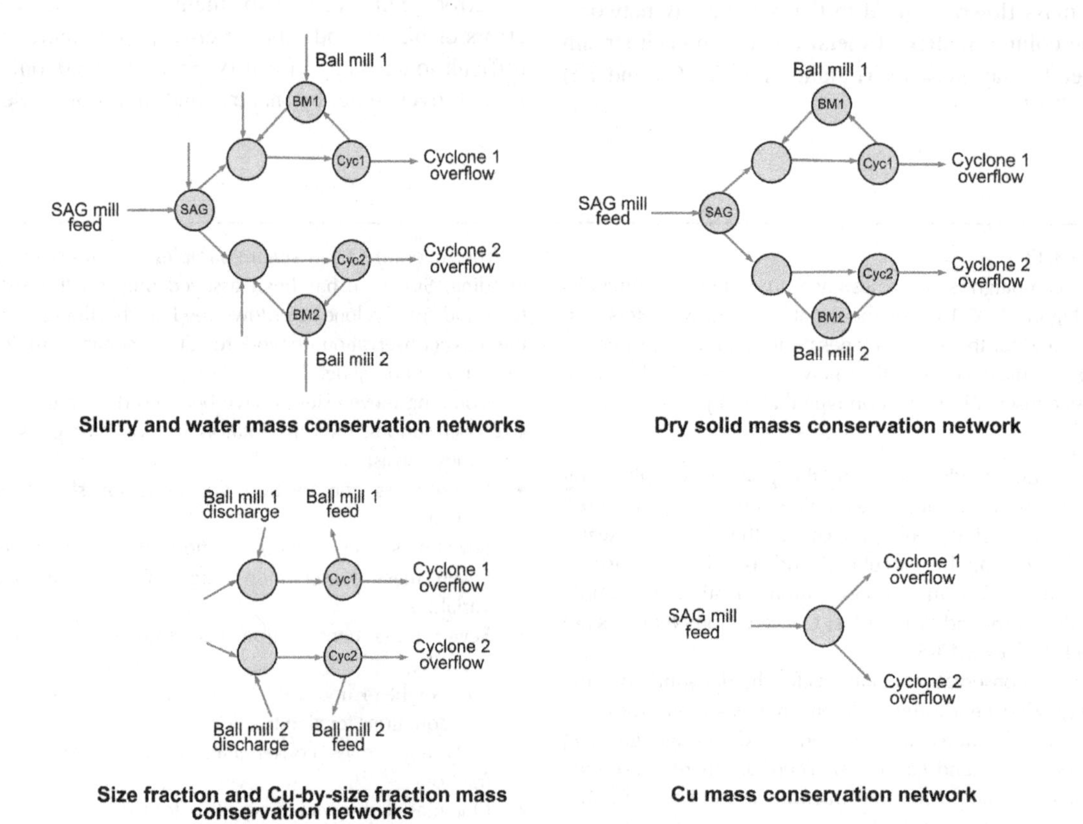

FIGURE EX 3.13-S The networks corresponding to Figure ex 3.13

source of comparison must be available. That is achievable to some extent through data reconciliation (Berton and Hodouin, 2000). A bias can be suspected if a measure over time is systematically adjusted to a lower (or higher) value. Once a bias is detected and identified, the bias source and the bias itself should be eliminated. Remember, the generalized least square method is a statistical method for data reconciliation by mass balancing and bias is a deterministic error, not a statistical variable, and is therefore a non-desired disturbance in the statistical data processing.

Measurement errors can also be characterized by their amplitude, small and gross errors, and by the frequency distribution of the measure signal over time: low frequency, high frequency or cyclical. Although there is no definition for a gross error, gross errors are statistically barely probable. As such, if a gross error is suspected it should be corrected before the statistical adjustment is performed. Typically, gross errors have a human source or result from a malfunctioning instrument.

Measurement errors originate from two main sources: sampling errors, and analysis errors. Sampling errors mainly result from the heterogeneity of the material to sample and the difficulty in collecting a representative sample; analysis errors result from the difficulty to analyze sample compositions. The lack of efficiency of the instruments used for measuring process variables also contributes to measurement errors. Analysis errors can be measured in the laboratory and therefore the variance of the analysis error is estimable. Sampling errors are tricky to estimate and one can only typically classify collected samples from the easiest to the more difficult to collect. Assuming systematic and gross errors have been eliminated, then purely random measurement errors can be modeled using 2 functions: one that represents the sampling error and the other one that represents the analysis error. The easiest model is the so-called multiplicative error model where the analysis error variance (σ_{AE}^2) is multiplied by a factor (k_{SE}) representing the sample representativeness or sampling error: the higher the factor value, the less representative the sample.

$$\sigma_{ME}^2 = k_{SE}\sigma_{AE}^2 \tag{3.60}$$

The multiplicative error model can serve as a basis to develop more advanced error models.

3.6.7 Sensitivity Analysis

The remaining unknown, once a mass balance is obtained for a given set of measured process variables and measurement errors, is: how trustable are the mass balance results or is it possible to determine a confidence interval for each variable estimated by mass balance? That is the objective of the sensitivity analysis.

Assuming systematic and gross errors have been eliminated, and measurement errors are random and obey to Normal distributions, there are two main ways to determine the variance of the estimated variable values (stream flowrates and their composition): Monte Carlo simulation, and error propagation calculation. Each method has its own advantages and disadvantages; each method provides an estimate of the variance of the adjusted or estimated variables.

In the Monte Carlo simulation approach (Laguitton, 1985), data sets are generated by disturbing the mass balance results according to the assumed error model. Each new data set is statistically reconciled and the statistical properties of the reconciled data sets determined.

In the error propagation approach (Flament et al., 1986; Hodouin et al., 1989), the equations through which measurement errors have to be propagated are complex and therefore it is preferable to linearize the equations around a stationary point. Obviously, the obtained mass balance is a good and valid stationary point. The set of linearized equations is then solved to determine the variable variances. This method enables the calculation of covariance values, a valuable feature for calculating confidence intervals around key performance indicators, for instance.

From a set of measured process variables and their variances, therefore, it is possible to determine a set of reconciled values and their variances:

$$\left\{ \begin{matrix} \left(W_j, \sigma_{W_j}^2\right) \\ \left(x_{ij}, \sigma_{x_{ij}}^2\right) \end{matrix} \right\} \Rightarrow \left\{ \begin{matrix} \left(\hat{W}_j, \sigma_{\hat{W}_j}^2\right) \\ \left(\hat{x}_{ij}, \sigma_{\hat{x}_{ij}}^2\right) \end{matrix} \right\} \tag{3.61}$$

It can be demonstrated that the variance of the adjusted values is less than the variance of the measured values. While the variance of the measured values contributes the most to the variance of the adjusted values, the reduction in the variance values results from the statistical information provided. The data redundancy and the topology of the mass conservation networks are the main contributors to the statistical content of the information provided for the mass balance calculation.

3.6.8 Estimability and Redundancy Analysis

We have seen with the n-product formula that $n-1$ components need be analyzed on the $n+1$ streams around the process unit to estimate the n product stream flowrates, assuming the feed stream flowrate is known (or taken as unity to give relative flowrates). The n-product formula is the solution to a set of n equations with n unknowns.

In such a case, there are just enough data to calculate the unknowns: no excess of data, that is, no data redundancy, and the unknowns are mathematically estimable. A definition of redundancy is therefore:

A measure is redundant if the variable value remains estimable should the measured value become unavailable.

It follows from the definition that the measured value is itself an estimate of the variable value. Other estimates can be obtained by calculation using other measured variables and mass conservation equations.

With the *n*-product formula, determining data redundancy and estimability is easy. In complex mass balances, determining data redundancy and estimability is quite tricky (Frew, 1983; Lachance and Flament, 2011). Estimability and redundancy cannot be determined just by the number of equations for the number of unknowns. It is not unusual to observe global redundancy with local lack of estimability (Example 3.14).

What are the factors influencing estimability and redundancy? Obviously, from Example 3.14, the number of measured variables is a strong factor, but it is not the only one. The network topology is also a factor. In the case of Example 3.14, if one of the product streams of unit B was recycled to unit A, then, with two metals analyzed on each stream, all the metal analyses would be redundant and adjustable.

It is worth noting that analyzing additional stream components does not necessarily increase redundancy and enable estimability. A component that is not separated in the process unit has the same or almost same concentration in each stream sample. The associated mass conservation equation is then similar (collinear) to the mass conservation of solids preventing the whole set of equations to be solved.

Estimability and redundancy analyses are complex analyses that are better performed with mathematical algorithm. However, the mathematics is too complex to be presented here.

Example 3.14

In the following mass balance problem (Figure ex 3.14), determine the estimability and redundancy of each variable entering the mass balance considering the Feed stream mass flowrate is measured and:

1. One metal has been analyzed on each sample;
2. Two metals have been analyzed on each sample.

Solution

When one metal is analyzed on each of the 6 streams:

There are 4 equations (2 for the mass flowrates and 2 for the metal assay) and 5 unknowns. The system is globally underdetermined since there are not enough equations for the number of unknowns. Around node A, the 2-product formula can be applied and therefore the mass flowrates of unit A products can be estimated. However, around node B, a 3-product formula cannot be applied since only one metal assay has been performed. Unit B product stream flowrates are not estimable. Metal assays are not redundant and cannot be adjusted by statistical data reconciliation.

When two metals are analyzed on each of the 6 streams:

There are a total of 6 equations and 5 unknowns. The system is therefore globally redundant and estimable *a priori*. Around node A, the 2-product formula can be applied with each metal assay and therefore the two product stream mass flowrates are estimable and the metal analyses on the 3 streams around unit A are redundant and adjustable. Around node B, the 3-product formula can be applied and therefore the 3 product stream mass flowrates are estimable now that the mass flowrate of the feed stream to unit B is known. However, the metal analyses are not redundant and not adjustable.

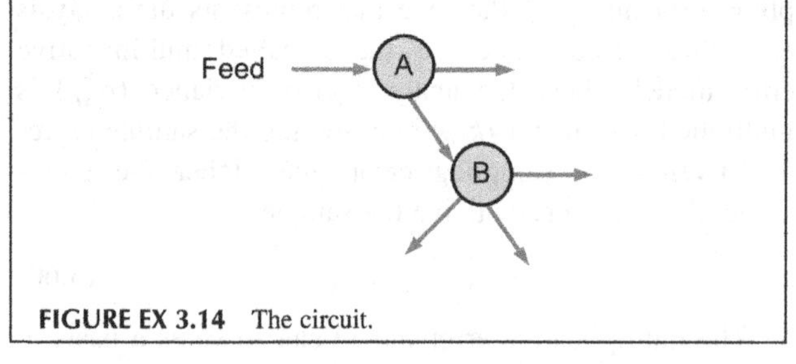

FIGURE EX 3.14 The circuit.

3.6.9 Mass Balancing Computer Programs

It might be tempting to use spreadsheets to compute mass balances. Indeed, spreadsheets offer most of the features required to easily develop and solve a mass balance problem. However, spreadsheets are error prone, expensive to troubleshoot and maintain over time, and become very quickly limited to fulfill the needs and requirements of complex mass balances and their statistics (Panko, 2008).

There exist several providers of computer programs for mass balance calculations for a variety of prices. While all the features of an advanced solution may not be required for a given type of application (research, process survey, modeling and simulation project, on-line mass balancing for automatic control needs, rigorous production and metallurgical accounting), a good computer program should offer:

- An easy way to define, list and select mass conservation equations
- An easy way to define error models

- Visualization and validation tools to quickly detect gross errors
- A sensitivity analysis tool
- An estimability and redundancy analysis tool
- Statistical tools to validate the error models against the adjustments made

Furthermore, depending on the mass balance types and application, the program should provide:

- Support for complete analyses (e.g., size distributions)
- Support for assay-by-size fractions
- Connectivity to external systems (archiving databases, Laboratory Information Management Systems) for automated data acquisition
- Automated detection and removal of gross errors
- Automated configuration of mass conservation models
- A reporting tool
- A dedicated database to support ancillary requirements
- Trend analysis
- Multi-mass balance capabilities

Additional information can be found in Crowe (1996), Romagnoli and Sánchez (1999) and Morrison (2008).

3.6.10 Metallurgical Balance Statement

One principal output of the sampling, assaying, mass balancing/data reconciliation exercise is the *metallurgical balance*, a statement of performance over a given period: a shift, a day, a week, etc. Assuming that the reconciled data in Example 3.12 correspond to a day when throughput was 25,650 t (measured by weightometer and corrected for moisture), the metallurgical balance would look something like in Table 3.3.

The calculations in the Wt (weight) column is from the solid (mass) split using Eq. (3.25). It should be evident using the reconciled data in Table ex 3.12a that regardless of which metal assay is selected to perform the calculation that the result for solid split is the same (including using Fe), giving $W_C/W_F = 0.065$. (In checking you will note that it is important to retain a large number of significant figures to avoid "rounding errors", rounding at the finish to create the values given in Table 3.3.) From the solid split the tonnage to concentrate is calculated by multiplying by the feed solids flowrate (i.e., $25{,}650 \times 0.065 = 1{,}674.1$), which, in turn, gives the tonnage to tailings. The recovery to concentrate is given using Eq. (3.27), and thus the recovery (loss) to tails is known; together recovery and loss are referred to as "distribution" in the table. The information in the table would be used, for instance, to compare performance between time periods, and to calculate Net Smelter Return (Chapter 1).

TABLE 3.3 Metallurgical Balance for XY Zn Concentrator for Day Z

Product	Wt (t)	Grade (%)		Distribution (%)	
		Zn	Cu	Cu	Zn
Feed	25,650.0	3.89	0.17	100	100
Zn conc	1,674.1	52.07	0.66	87.35	25.71
Tailings	23,975.9	0.53	0.13	12.65	74.29

3.7 EXAMPLE MASS BALANCE CALCULATIONS

Mass balancing based on metal assays has been described, and used to illustrate data reconciliation. As noted, components other than metal (or mineral) can be used for mass balancing. The use of particle size and per cent solids is illustrated; this is done without the associated data reconciliation, but remember, this is always necessary for accurate work.

3.7.1 Use of Particle Size

Many units, such as hydrocyclones and gravity separators, produce a degree of size separation and the particle size data can be used for mass balancing (Example 3.15).

Example 3.15 is an example of node imbalance minimization; it provides, for example, the initial value for the generalized least squares minimization. This graphical approach can be used whenever there is "excess" component data; in Example 3.9 it could have been used.

Example 3.15 uses the cyclone as the node. A second node is the sump: this is an example of 2 inputs (fresh feed and ball mill discharge) and one output (cyclone feed). This gives another mass balance (Example 3.16).

In Chapter 9 we return to this grinding circuit example using adjusted data to determine the cyclone partition curve.

3.7.2 Use of Percent Solids

A unit that gives a large difference in %solids between streams is a thickener (Examples 3.17); another is the hydrocyclone (Example 3.18).

3.7.3 Illustration of Sensitivity of Recovery Calculation

See Example 3.19.

Example 3.15

The streams around the hydrocyclone in the circuit (Figure ex 3.15) were sampled and dried for particle size analysis with the results in Table ex 3.15. Determine the solids split to underflow and the circulating load.

Solution

Solids split: We can set up the calculation in the following way, first recognizing the two balances:

Solids: $F = O + U$

Particle size: $F f = O o + U u$

where *F*, *O*, *U* refer to solids (dry) flowrate of feed, overflow and underflow, and *f*, *o*, *u* refer to weight fraction (or %) in each size class. Since the question asks for solids split to underflow, *U*/*F*, we can use the solids balance to substitute for $O\ (=F - U)$ in the particle size balance. After gathering terms and re-arranging we arrive at:

$$\frac{U}{F} = \frac{(f-o)}{(u-o)}$$

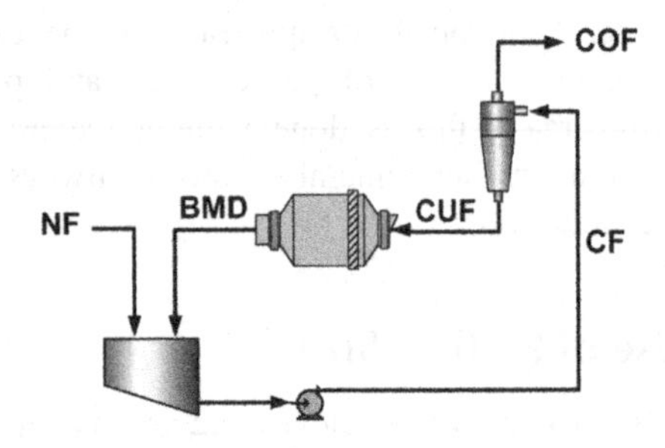

FIGURE EX 3.15 Grinding circuit, ball mill closed with a hydrocyclone.

While this may seem that we could simply apply the two-product formula, it is always advisable to set up the solution starting from the basic balances. Attempting to remember formulae is an invitation to error.

The estimates of *U*/*F* derived from each size class are included in the table. This is another example where errors in the data produce uncertainty in the mass balance. Note that the two estimates that are far from the others, for the 105 μm and 74 μm size fractions, correspond to situations where there is little difference in data between the three streams, making the calculation numerically unstable (if all three were equal no solution can be found).

By re-arranging we can write:

$$(f-o) = \frac{U}{F}(u-o)$$

This has the form of a linear equation passing through the origin ($y = mx$) where the slope is solids split *U*/*F*. Figure ex 3.15-S shows the resulting plot and includes the solids split resulting from linear regression using the function in Excel.

The plot agrees with expectation and the solids split is therefore:

$$\frac{U}{F} = 0.616$$

Circulating load: While this is encountered and defined later (e.g., Chapter 7), it is sufficient here to note it is given by the flowrate returning to the ball mill (*U*) divided by the fresh feed rate to the circuit (*N*); that is, and observing that $N = O$ (at steady state), we can write:

TABLE EX 3.15 Size Distribution Data Obtained on Cyclone Feed, Overflow and Underflow

Size* (μm)	CF	COF	CUF	f-o	u-o	(f−o)/(u−o)
+592	7.69	0.01	12.84	7.68	12.83	0.599
−592 + 419	4.69	0.34	7.18	4.35	6.84	0.636
−419 + 296	6.68	0.7	9.96	5.98	9.26	0.646
−296 + 209	7.03	2.61	9.32	4.42	6.71	0.659
−209 + 148	11.29	7.63	13.6	3.66	5.97	0.613
−148 + 105	13.62	13.55	14.91	0.07	1.36	0.051
−105 + 74	11.39	11.52	10.42	−0.13	−1.1	0.118
−74 + 53	9.81	15	7.19	−5.19	−7.81	0.665
−53 + 37	5.65	9.37	3.32	−3.72	−6.05	0.615
−37	22.15	39.27	11.26	−17.12	−28.01	0.611
	100	**100**	**100**			

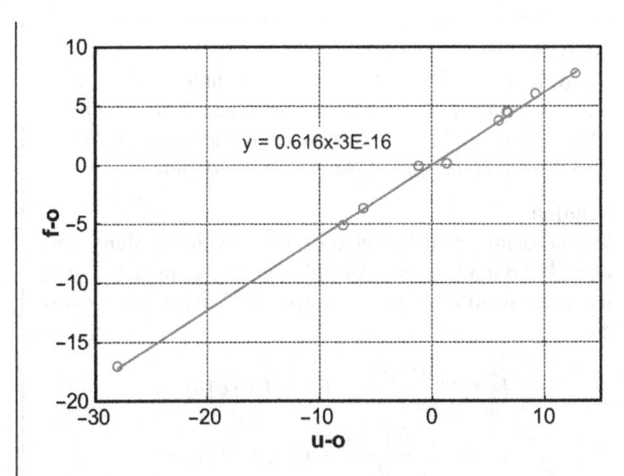

FIGURE EX 3.15-S Plot of (f-o) vs. (u-o) to determine solids split.

$$CL = \frac{U}{O}$$

We could approach the problem in the same way as we did to solve for U/F (by substituting for F in this case). More directly CL is given by noting that:

$$U = 0.616\ F \quad \text{and thus} \quad O = 0.384\ F$$

giving:

$$\frac{U}{O} = \frac{0.616}{0.384} = 1.60$$

Thus the circulating load is 1.60, often quoted as a percentage, 160%. We will see that this is a typical value for a closed ball mill-cyclone circuit (Chapters 7 and 9).

Example 3.16

At the same time as the streams around the cyclone were sampled, the streams around the sump were also sampled, and sized. The measurement results for two size classes, +592 μm and −37 μm are given in the Table ex 3.16.

Determine the circulating load.

Solution

The mass balance equations for the sump are:

Solids: $N + B = F$

Particle size: $N\,n + B\,b = F\,f$

and $CL = \frac{B}{N}$

Combining equations to eliminate F we deduce that:

$$CL = \frac{n - f}{f - b}$$

The estimates of CL are included in the table. Clearly, there is error, the result for −37 μm even being negative. After reconciliation the adjusted data and mass balance are given below:

592	19.42	1.44	8.27	11.15	6.83	1.633
− 37	20.66	21.61	21.25	− 0.59	− 0.36	1.639

Clearly now the data do agree more closely with the result in Example 3.15. By experience, mass balancing around the sump is less reliable than around the cyclone.

TABLE EX 3.16 Raw (unadjusted) Size Distribution Data for New Feed, Ball Mill Discharge, and Cyclone Feed

	n	b	f	n-f	f-b	(n−f)/(f−b)
592	19.86	2.16	7.69	12.17	5.53	2.201
−37	19.84	20.27	22.15	−2.31	1.88	−1.229

Example 3.19 raises a question as to how to estimate the variance. In this case a relative standard deviation (standard deviation divided by the mean) was taken and assumed equal for all assays. The standard deviation could be broken down into that associated with sampling and that associated with assaying, or some other "error model". Regardless, there is some questioning of the common use of relative standard deviation as the absolute standard deviation decreases as the assay value decreases and some would argue this is not realistic. Alternative error models which avoid this have been suggested (Morrison, 2010), but none appear to be universally accepted.

Example 3.17
Feed to a thickener is 30% solids by weight and the underflow is 65% by weight. Calculate the recovery of water to the overflow $R_{w/o}$ assuming no solids report to the overflow.

Solution
We can approach the problem in two ways: 1, starting with a balance on the slurry, and 2, using dilution ratios. Using F, O, U to represent feed, overflow and underflow flowrates, f, o, u to represent % solids (as a fraction), and subscripts sl, s and w to represent slurry, solids, and water we can write the appropriate balances.

Balances based on slurry: We can write three balances:
Slurry: $F_{sl} = O_{sl} + U_{sl}$
Solids: $F_{sl}f_s = O_{sl}o_s + U_{sl}u_s$
Water: $F_{sl}(1 - f_s) = O_{sl}(1 - o_s) + U_{sl}(1 - u_s)$

and thus water recovery will be given by:

$$R_{w/o} = \frac{O_{sl}}{F_{sl}}\frac{1-o_s}{1-f_s} \rightarrow \frac{O_{sl}}{F_{sl}} \times \frac{1}{1-f_s}$$

Since U_{sl} does not appear in the recovery equation it can be substituted by $(F_{sl} - O_{sl})$ from the slurry balance into the water balance (this is effectively combining the two equations). In this manner the ratio O_{sl}/F_{sl} is obtained from the water balance equation:

$$\frac{O_{sl}}{F_{sl}} = \frac{u_s - f_s}{u_s}$$

and thus $R_{w/o}$ is determined:

$$R_{w/o} = \frac{u_s - f_s}{u_s(1-f_s)}$$

Balances based on dilution ratios: Using dilution ratios (Section 3.4) the water mass flowrate can be obtained, as illustrated using the feed stream as an example:
Feed solids: $F_s = F_{sl}f_s$
Feed water: $F_w = F_{sl}(1 - f_s)$
From the ratio: $F_w = F_s\frac{1-f_s}{f_s}$

Analogous expressions for water flowrate in the other streams can be written. It will be noticed that the expression for O_w in this example is not determinable (as O_s and o_s are zero). Thus the $R_{w/o}$ is determined from:

$$R_{w/o} = 1 - \frac{U_w}{F_w}$$

After substituting by the dilution ratios and re-arranging we obtain:

$$R_{w/o} = \frac{u_s - f_s}{u_s(1-f_s)}$$

Thus the two approaches give the same result, as must be the case.
Solving we obtain:

$$R_{w/o} = \frac{0.65 - 0.30}{0.65 \times (1 - 0.3)}$$

$$R_{w/o} = 0.769 \text{ or } 77\%$$

Example 3.18
Referring to Example 3.15, at the same time % solids data were determined on each stream: feed, 55%; overflow, 41%; and underflow, 70%. Determine the split of solids to underflow and the split of water to the underflow.

Solution
The problem could be set up starting with the slurry balance, but it lends itself to the dilution ratio approach. Using the same symbolism as in Example 3.17 the dilution ratios are:

$$F_w = F_s\frac{100 - f_s}{f_s} = F_sf_s' = F_s(0.818)$$

$$O_w = O_s\frac{100 - o_s}{o_s} = O_so_s' = O_s(1.439)$$

$$U_w = U_s\frac{100 - u_s}{u_s} = U_su_s' = U_s(0.429)$$

Solids split: The two balances are:
Solids: $F_s = O_s + U_s$
Water: $F_sf_s' = O_so_s' + U_su_s'$

giving:
Solids split: $R_{s/u} = \frac{U_s}{F_s}$

By combining equations to eliminate O_s we derive:

$$R_{s/u} = \frac{f_s' - o_s'}{u_s' - o_s'} = \frac{0.818 - 1.439}{0.429 - 1.439}$$

$$R_{s/u} = 0.614 (\text{or } 61.4\%)$$

This compares well with the estimate in Example 3.15, which is often not the case using raw, that is, unadjusted data; the % solids data could be included in mass balance/data reconciliation, and must be if use is to be made of stream values derived from those data.

Water split: The water split is:

$$R_{w/u} = \frac{U_w}{F_w} = \frac{U_su_s'}{F_sf_s'}$$

$$R_{w/u} = 0.614 \times \frac{0.429}{0.818}$$

$$R_{w/u} = 0.322 (\text{or } 32.2\%)$$

The water split to underflow, also referred to as water recovery to underflow, is an important parameter in modeling the performance of a cyclone (Chapter 9).

Example 3.19
A concentrator treats a feed of 2.0% metal. Compare the 95% confidence interval on the recovery for the following two conditions:

a. Producing concentrate grading 40% metal and a tailings of 0.3% metal.

b. Producing concentrate grading 2.2% metal and a tailings of 1.3% metal.

Take that the relative standard deviation on the metal assays is 5% in both cases.

Solution a)

Recovery:

From Eq. (3.27): $R = 85.6\%$

Standard deviation on R:

Solving for V_R (Eq. (3.33)) gives:

$$V_R = 0.14492(394.023V_f + 0.001806V_c + 16044.44V_t)$$

To insert the values of V_f, V_c, and V_t the relative standard deviations must be changed to absolute values, σ (remembering the Eq. (3.33) is written in terms of fractions):

$\sigma_f = 5\%$ of $2.0\% = 0.1\%$ (or 0.001);

$\sigma_c = 5\%$ of $40\% = 2\%$ (or 0.02);

and $\sigma_t = 5\%$ of $0.3\% = 0.015\%$ (or 0.00015)

Substituting gives (recalling that $V = \sigma^2$):

$V_R = 0.14492\ (0.000394 + 7.225\text{E}{-10} + 0.000361)$

$V_R = 0.00011$

That is: $\sigma_R = 0.01046$ (or 1.05%)

The standard deviation on the recovery is $\pm 1.05\%$ or an approximate 95% confidence interval of $2 \times 1.05\%$ or 2.1%. The recovery is best reported, therefore, as:

$$R = 85.6 \pm 2.1\%$$

Solution b)

Recovery:

From Eq. (3.27): $R = 85.6\%$ (i.e., the same)

Standard deviation of R:

Solving for V_R (Eq. (3.33)):

$$V_R = 3.11675(0.2025\ V_f + 0.10124\ V_c + 0.02367\ V_t)$$

To insert the values of V_f, V_c, and V_t the relative standard deviations must be changed to absolute values, σ:

$\sigma_f = 5\%$ of $2.0\% = 0.1\%$ (or 0.001);

$\sigma_c = 5\%$ of $2.2\% = 0.11\%$ (or 0.0011);

and $\sigma_t = 5\%$ of $1.3\% = 0.065\%$ (or 0.00065)

Substituting gives (recalling that $V = \sigma^2$):

$V_R = 3.11675(2.025\text{E} - 07 + 1.225\text{E} - 07 + 1\text{E} - 08)$

$V_R = 0.01044$

That is:

$$\sigma_R = 0.102(\text{or } 10.2\%)$$

And the 95% confidence interval is:

$$R = 85.6 \pm 20.4\%$$

The much higher uncertainty in the Example 3.19 (b) compared to a) again illustrates the greater uncertainty in the value of the recovery when the metal (or any component) is not well separated.

REFERENCES

Aldrich, C., et al., 2010. On-line monitoring and control of froth flotation systems with machine vision: a review. Int. J. Miner. Process. 96 (1-4), 1–13.

Almasy, G.A., Mah, R.S.H., 1984. Estimation of measurement error variances from process data. Ind. Eng. Chem. Process Des. Dev. 23 (4), 779–784.

AMIRA, 2007. P754: Metal Accounting - Code of Practice and Guidelines, Release 3.

Anon., 2011. Using x-ray diffraction to solve mineral processing problems – CSIRO Process Science and Engineering, Oct 2011.

Apelt, T.A., et al., 2002. Inferential measurement of SAG mill parameters II: state estimation. Miner. Eng. 15 (12), 1043–1053.

Arena, T., McTiernan, J., 2011. On-belt analysis at Sepon copper operation. Proc. Metallurgical Plant Design and Operating Strategies (MetPlant 2011). AusIMM, Perth, WA, Australia, pp. 527–535.

Bauer, M., Craig, I.K., 2008. Economic assessment of advanced process control – A survey and framework. J. Process Control. 18 (1), 2–18.

Bauer, M., et al., 2007. A profit index for assessing the benefits of process control. Ind. Eng. Chem. Res. 46 (17), 5614–5623.

Bearman, R.A., Milne, R.W., 1992. Expert systems: opportunities in the minerals industry. Miner. Eng. 5 (10-12), 1307–1323.

Bergeron, A., Lee, D.J., 1983. A practical approach to on-stream analysis. Min. Mag. Oct., 257–261.

Berton, A., Hodouin, D., 2000. Statistical detection of gross error in material balance calculations. In: Herbst, J.A. (Ed.), Control 2000: Mineral and Metallurgical Processing. SME, Littleton, CO, USA, pp. 23–33. (Chapter 3).

Blanchette, G., et al., 2011. Estimation errors for gold inventory in large carbon-in-leach tanks. Proc. 43th Annual Meeting of the Canadian Mineral Processors Conf. CIM, Ottawa, ON, Canada, pp. 477–490.

Box, G.E.P., Draper, N.R., 1998. Evolutionary Operation: A Statistical Method for Process Improvement. Classics Edn. John Wiley and Sons, New York, NY, USA.

Boyd, R., 2005. An innovative and practical approach to sampling of slurries for metallurgical accounting. <www.thermo.com.cn/Resources/201007/27101513615.doc>.

Braden, T.F., et al., 2002. On-line composition analysis of mineral slurries. In: Mular, A.L., et al., (Eds.), Mineral Processing Plant Design, Practice and Control, vol. 2. SME, Littleton, CO, USA, pp. 2020–2047.

Brdys, M., Tatjewski, P., 2005. Iterative Algorithms for Multilayer Optimizing Control. Imperial College Press.

Burgos, R., Concha, F., 2005. Further development of software for the design and simulation of industrial thickeners. Chem. Eng. J. 111 (2-3), 135–144.

Cao, S., Rhinehart, R.R., 1995. An efficient method for on-line identification of steady state. J. Process Control. 5 (6), 363–374.

Caro, D., 2009. Automation Network Selection: A Reference Manual. second ed. Publisher: ISA (The International Society of Automation).

Chang, J.W., Bruno, S.J., 1983. Process control system in the mining industry. World Min. 36 (May), 37–41.

Chen, J., et al., 1997. Robust estimation of measurement error variance/covariance from process sampling data. Comput. Chem. Eng. 21 (6), 593–600.

Clarkson, C.J., 1986. The potential for automation and process control in coal preparation. Norrie, A.W., Turner, D.R. (Eds.), Proc. Automation for Mineral Resource Development: 1st IFAC Symposium, Brisbane, Queenland, Australia, pp. 247–252.

Cortes, G., et al., 2008. Rougher flotation multivariable predictive control: concentrador A-1 Division Codelco Norte. Kuyvenhoven, R., et al. (Eds.), Proc. Inertnation Mineral Processing Conf., Santiago, Chile, pp. 327–337.

Crowe, C.M., 1996. Data reconciliation – progress and challenges. J. Process Control. 6 (2-3), 89–98.

Cutting, G.W., 1976. Estimation of interlocking mass balances on complex mineral beneficiation plants. Int. J. Miner. Process. 3 (3), 207–218.

Darouach, M., et al., 1989. Maximum likelihood estimator of measurement error variances in data reconciliation, Proc. Advanced Information Processing in Automatic Control (AIPAC'89), IFAC, Nancy, France, pp. 109–112.

Duarte, M., et al., 1998. Grinding operation optimization of the CODELCO-Andina concentrator plant. Miner. Eng. 11 (12), 1119–1142.

Duchesne, C, 2010. Multivariate image analysis in mineral processing. In: Sbárbaro, D., del Villar, R. (Eds.), Advanced Control and Supervision of Mineral Processing Plants. Springer, London, pp. 85–142. (Chapter 3).

Edwards, R., et al., 2002. Strategies for instrumentation and control of grinding circuits. In: Mular, A.L., et al., (Eds.), Mineral Processing Plant Design, Practice and Control, vol. 2. SME, Littleton, CO, USA, pp. 2130–2151.

Finch, J.A., Matwijenko, O., 1977. Individual mineral behaviour in a closed-grinding circuit. CIM Bull., 70(788), 164–172.

Flament, F., et al., 1986. Propagation of measurement errors in mass balance calculation of mineral processing data, Proc. 19th International Symposium on Application of Computers and Operations Research (APCOM'86). Mineral Processing Applications I, pp. 553–562 (Chapter 50).

Flintoff, B., 2002. Introduction to process control. In: Mular, A.L., et al., (Eds.), Mineral Processing Plant Design, Practice and Control, vol. 2. SME, Littleton, CO, USA, pp. 2051–2065.

Flintoff, B., et al., 1991. Justifying control projects. In: Halbe, D.N. (Ed.), Plant Operators. SME, Littleton, CO, USA, pp. 45–53.

Flintoff, B., et al., 2014. Innovations in comminution instrumentation and control. Mineral Processing and Extractive Metallurgy: 100 Years of Innovation. SME, Englewood, CO, USA, pp. 91–116.

François-Bongarçon, D., Gy, P., 2002a. Critical aspects of sampling in mills and plants: a guide to understanding sampling audits. J. S. Afr. Inst. Min. Metall. 102 (8), 481–484.

François-Bongarçon, D., Gy, P., 2002b. The most common error in applying 'Gy's Formula' in the theory of mineral sampling, and the history of the liberation factor. J. S. Afr. Inst. Min. Metall. 102 (8), 475–480.

Frew, J.A., 1983. Computer-aided design of sampling schemes. Int. J. Miner. Process. 11 (4), 255–265.

Geelhoed, B., 2011. Is Gy's formula for the fundamental sampling error accurate? Experimental evidence. Miner. Eng. 24 (2), 169–173.

González, G.D., 2010. Soft sensing. In: Sbárbaro, D., del Villar, R. (Eds.), Advanced Control and Supervision of Mineral Processing Plants. Springer-Verlag, London Limited, UK, pp. 143–212. (Chapter 4).

González, G.D., et al., 2003. Local models for soft-sensors in a rougher flotation bank. Miner. Eng. 16 (5), 441–453.

Gy, P.M., 1967. L'Echantillonage des Minerais en Vrac (Sampling of Particulate Materials). Vol. 1. Revue de l'Industrie Minerale, Special, No. St. Etienne, France.

Gy, P.M., 1982. Sampling of Particulate Materials: Theory and Practice. second ed. Elsevier, Amsterdam.

Hales, L.B., Ynchausti, R.A., 1992. Neural networks and their use in the prediction of SAG mill power. In: Kawatra, S.K. (Ed.), Comminution - Theory and Practice. SME, Littleton, CO, USA, pp. 495–504. (Chapter 36).

Hales, L.B., Ynchausti, R.A., 1997. Expert systems in comminution – cases studies. In: Kawatra, S.K. (Ed.), Comminution Practices. SME, Littleton, CO, USA, pp. 57–68. (Chapter 8).

Harris, C.A., Meech, J.A., 1987. Fuzzy logic: a potential control technique for mineral processing. CIM Bull. 80 (905), 51–59.

Herbst, J.A., Bascur, O.A., 1984. Mineral processing control in the 1980s – realities and dreams. In: Herbst, J.A. (Ed.), Control '84 Mineral/Metallurgical Processing. Mineral Processing Control in the 1980s. SME, New York, NY, USA, pp. 197–215. (Chapter 21).

Herbst, J.A., Bascur, O.A., 1985. Alternative control strategies for coal flotation. Miner. Metall. Process. 2, 1–9.

Herbst, J.A., Pate, W.T., 1996. On-line estimation of charge volumes in semiautogenous and autogenous grinding mills, Proc. 2nd International Conference on Autogenous and Semiautogenous Grinding Technology, Vol. 2, Vancouver, BC, Canada, pp. 669–686.

Hodouin, D., 2011. Methods for automatic control, observation and optimization in mineral processing plants. J. Process Control. 21 (2), 211–225.

Hodouin, D., Everell, M.D., 1980. A hierarchical procedure for adjustment and material balancing of mineral processes data. Int. J. Miner. Process. 7 (2), 91–116.

Hodouin, D., Najim, K., 1992. Adaptive control in mineral processing. CIM Bull. 85 (965), 70–78.

Hodouin, D., et al., 1989. Reliability of material balance calculations a sensitivity approach. Miner. Eng. 2 (2), 157–169.

Holdsworth, M., et al., 2002. Optimising concentrate production at the Green's Creek Mine. Proc. SME Annual Meeting, Preprint 02-063. SME, Phoenix, AZ, USA, pp. 1–10.

Holmes, R.J., 1991. Sampling methods: problems and solutions. In: Malhotra, D., et al., (Eds.), Evaluation and Optimization of Metallurgical Performance. SME, Littleton, CO, USA, pp. 157–167. (Chapter 16).

Holmes, R.J., 2004. Correct sampling and measurement—The foundation of accurate metallurgical accounting. Chemom. Intell. Lab. Syst. 74 (1), 71–83.

Hulthén, E., Evertsson, C.M., 2011. Real-time algorithm for cone crusher control with two variables. Miner. Eng. 24 (9), 987–994.

Janse Van Vuuren, M.J., et al., 2011. Detecting changes in the operational states of hydrocyclones. Miner. Eng. 24 (14), 1532–1544.

Jansen, W., 2009. Applying best practices in metal production accounting and reconciliation. Eng. Min. J. (E&MJ). 211 (10), 68–71.

Johnson, N.W., 2010. In: Greet, C.J. (Ed.), Existing Methods for Process Analysis. Flotation Plant Optimization: A Metallurgical Guide to Identifying and Solving Problems in Flotation Plants. AusIMM, pp. 35–63.

Jonas, R., 2008. Advanced control for mineral processing: better than expert systems. Proc. 40th Annual Mineral Processors Operators Conf. CIM, Ottawa, ON, Canada, pp. 421–443.

Kawatra, S.K., 1985. The development and plant trials of an ash analyzer for control of a coal flotation circuit, Proc. 15th International Mineral Processing Cong., Cannes, France, pp. 176–188.

Kawatra, S.K., Cooper, H.R., 1986. On-stream composition analysis of mineral slurries. In: Mular, A.L., Anderson, M.A. (Eds.), Design and Installation of Concentration and Dewatering Circuits. SME Inc., Littleton, CO, USA, pp. 641–654. (Chapter 41).

Keller, J.Y., et al., 1992. Analytical estimator of measurement error variances in data reconciliation. Comput. Chem. Eng. 16 (3), 185–188.

Kistner, M., et al., 2013. Monitoring of mineral processing systems by using textural image analysis. Miner. Eng. 52, 169–177.

Klimpel, A., 1979. Estimation of weight ratios given component make-up analyses of streams, AIME Annual Meeting, New Orleans, LA, USA, Preprint No. 79-24.

Ko, Y-D., Shang, H., 2011. A neural network-based soft sensor for particle size distribution using image analysis. Powder Technol. 212 (2), 359–366.

Krstev, B., Golomeov, B., 1998. Use of evolutionary planning for optimisation of the galena flotation process, Proc. 27th International Symposium on Computer Applications in the Minerals Industries (APCOM), London, UK, pp. 487–492.

Lachance, L., Flament, F., 2011. A new ERA tool for state of the art metallurgical balance calculations. Proc. 43th Annual Meeting of the Canadian Mineral Processors Conf. CIM, Ottawa, ON., Canada, pp. 331–346.

Lachance, L., et al., 2012. Implementing best practices of metal accounting at the Strathcona Mill. Proc. 44th Annual Canadian Mineral Processors Operators Conf. CIM, Ottawa, ON, Canada, pp. 375–386.

Laguitton, D., 1985. *The SPOC Manual: Simulated Processing of Ore and Coal.* Canada Centre for Mineral and Energy Technology, Department of Energy, Mines and Resources. Canadian Government Publishing Centre, Ottawa, ON, Canada.

Laguitton, D., Leung, J., 1989. Advances in expert system applications in mineral processing. Dobby, G.S., Rao, R. (Eds.), Proc. International Symposium on Processing of Complex Ores, Halifax, NS, Canada, pp. 565–574.

Leskinen, T., et al., 1973. Performance of on-stream analysers at Outokumpu concentrators, Finland. CIM Bull. 66 (Feb.), 37–47.

Lessard, J., et al., 2014. Developments of ore sorting and its impact on mineral processing economics. Miner. Eng. 65, 88–97.

Lestage, R., et al., 2002. Constrained real-time optimization of a grinding circuit using steady-state linear programming supervisory control. Powder Technol. 124 (3), 254–263.

Ljung, L., 1999. System Identification: Theory for the User, second ed., Englewood Cliffs, NJ, USA.

Lyman, G.J., 1986. Application of Gy's sampling theory to coal: a simplified explanation and illustration of some basic aspects. Int. J. Miner. Process. 17 (1-2), 1–22.

Lyman, G., et al., 2010. Bias testing of cross belt samplers. J. S. Afr. Inst. Min. Metall. 110, 289–298.

Lynch, A.J., 1977. Mineral Crushing and Grinding Circuits: Their Simulation, Optimisation, Design and Control. Elsevier Scientific Pub. Co, Amsterdam; New York, NY, USA.

Maldonado, M., et al., 2010. On-line estimation of frother concentration for flotation processes. Can. Metall. Q. 49 (4), 435–446.

Matthews, D., du Toit, T., 2011. Real-time online analysis of iron ore, validation of material stockpiles and roll out for overall elemental balance as observed in the Khumani Iron Ore Mine, South Africa, Proc. Iron Ore Conference, Perth, WA, Australia, pp. 297–305.

McKee, D.J., Thornton, A.J., 1986. Emerging automatic control approaches in mineral processing. In: Wills, B.A., Barley, R.W. (Eds.), Mineral Processing at a Crossroads – Problems and Prospects, Vol. 117. Springer, Netherlands, pp. 117–132. , NATO ASI Series.

McIvor, R.E., Finch, J.A., 1991. A guide to interfacing of plant grinding and flotation operations. Miner. Eng. 4 (1), 9–23.

Merks, J.W., 1985. Sampling and Weighing of Bulk Solids. Trans Tech Publications, Clausthal-Zellerfeld, Germany.

Merks, J.W., 2002. Sampling in mineral processing. In: Mular, A.L., et al., (Eds.), Mineral Processing Plant Design, Practice and Control, vol. 1. SME, Littleton, CO, USA, pp. 37–62.

Merks, J.W., 2010. Spatial dependence in material sampling. In: Geelhoed, B. (Ed.), Approaches in Material Sampling. IOS Press, Amsterdam, pp. 105–149. (Chapter 6).

Minnitt, R.C.A., et al., 2007. Part 1: understanding the components of the fundamental sampling error: a key to good sampling practice. J. S. Afr. Inst. Min. Metall. 107 (8), 505–511.

Minnitt, R.C.A., 2014. Sampling in the South African minerals industry. J. S. Afr. Inst. Min. Metall. 114 (1), 63–81.

Morari, M., Zafiriou, E., 1989. Robust Process Control. Prentice Hall, Englewood Cliffs, NJ, USA.

Morrison, R.D., 2008. An Introduction to Metal Balancing and Reconciliation, JKMRC Monograph Series in Mining and Mineral Processing, No.4, The Univerisity of Queensland, Australia.

Morrison, R.D., 2010. In: Greet, C.J. (Ed.), Mass Balancing Flotation Data, Flotation Plant Optimisation. AusIMM, Carlton, Victoria, Australia, pp. 65–81., Spectrum Number 16.

Mular, A.L., 1979. Data adjustment procedures for mass balances. In: Weiss, A. (Ed.), Computer Methods for the 80's in the Minerals Industry. SME, New York, NY, USA, pp. 843–849. (Chapter 4).

Mular, A.L., 1989. Modelling, simulation and optimization of mineral processing circuits. In: Sastry, K.V.S., Fuerstenau, M.C. (Eds.), Challenges in Mineral Processing. SME, Littleton, CO, USA, pp. 323–349. (Chapter 19).

Napier-Munn, T.J., Lynch, A.J., 1992. The modelling and computer simulation of mineral treatment processes - current status and future trends. Miner. Eng. 5 (2), 143–167.

Narasimhan, S., Jordache, C., 2000. Data Reconciliation and Gross Error Detection: An Intelligent Use of Process Data. Gulf Publishing Company.

Nuñez, E., et al., 2009. Self-optimizing grinding control for maximizing throughput while maintaining cyclone overflow specifications, Proc. 41th Annual Meeting of the Canadian Mineral Processors Conf., Ottawa, ON, Canada, pp. 541–555.

Osorio, D., et al., 1999. Assessment of expert fuzzy controllers for conventional flotation plants. Miner. Eng. 12 (11), 1327–1338.

Panko, R.R., 2008. What we know about spreadsheet errors. J. Organ. End User Comput. (JOEUC). 10 (2), 15–21.

Patel, M., 2014. On-belt elemental analysis of lead-zinc ores using prompt gamma neutron activation analysis, Proc. 27th International Mineral Processing Congr. (IMPC 2014), Santiago, C1717, Chile, 1–9 (Chapter 17).

Pitard, F.F., 1993. Pierre Gy's Sampling Theory and Sampling Practice: Heterogeneity, Sampling Correctness, and Statistical Process Control. second ed. CRC Press.

Rajamani, K., Hales, L.B., 1987. Developments in adaptive control and its potential in the minerals industry. Miner. Metall. Process. 4 (2), 18–24.

Ramasamy, M., et al., 2005. Control of ball mill grinding circuit using model predictive control scheme. J. Process Control. 15 (3), 273–283.

Reid, K.J., et al., 1982. A survey of material balance computer packages in the mineral industry. In: Johnson, T.B., Barnes, R.J. (Eds.), Proc. 17th International Symposium(APCOM)-Application of Computers and Operations Research. SME, New York, NY, USA, pp. 41–62. (Chapter 6).

Romagnoli, J.A., Sánchez, M.C., 1999. Data Processing and Reconciliation for Chemical Process Operations. Process Systems Engineering, vol. 2. Academic Press.

Ruel, M., 2014. Advanced control decision tree. Proc. 46th Annual Mineral of the Canadian Mineral Processors Conf. CIM, Ottawa, ON, Canada, pp. 179–187.

Saensunon, B., et al., 2008. A combined ^{57}Fe-Mössbauer and X-ray diffraction study of the ilmenite reduction process in commercial rotary kiln. Int. J. Miner. Process. 86 (1-4), 26–32.

Salama, A.I.A., et al., 1985. Coal preparation process control. CIM Bull. 78 (881), 59–64.

Sbarbaro, D., et al., 2005. A multi-input-single-output Smith predictor for feeders control in SAG grinding plants. IEEE Trans. Control Syst. Technol. 13 (6), 1069–1075.

Sbarbaro, D., et al., 2008. Adaptive soft-sensors for on-line particle size estimation in wet grinding circuits. Control Eng. Pract. 16 (2), 171–178.

Seborg, D.E., et al., 2010. Process Dynamics and Control. third ed. John Wiley and Sons, USA.

Shean, B.J., Cilliers, J.J., 2011. A review of froth flotation control. Int. J. Miner. Process. 100 (3-4), 57–71.

Shoenbrunn, F., Toronto, T., 1999. Advanced thickener control. Advanced Process Control Applications for Industry Workshop. IEEE Industry Applications Society, Vancouver BC, Canada, 83-86.

Skogestad, S., 2003. Simple analytic rules for model reduction and PID controller tuning. J. Process Control. 13 (4), 291–309.

Smith, H.W., Ichiyen, N.M., 1973. Computer adjustment of metallurgical balances. CIM Bull. 66 (Sep.), 97–100.

Steyn, C.W., Sandrock, C., 2013. Benefits of optimisation and model predictive control on a fully autogeneous mill with variable speed. Miner. Eng. 53, 113–123.

Thornton, A.J., 1991. Cautious adaptive control of an industrial flotation circuit. Miner. Eng. 4 (12), 1227–1242.

Thwaites, P., 2007. Process control in metallurgical plants – From an Xstrata perspective. Ann. Rev. Control. 31 (2), 221–239.

Tipman, R., et al., 1978. Mass balances in mill metallurgical operations. Proc. 10th Annual Meeting of the Canadian Mineral Processors Conf. CIM, Ottawa, ON, Canada, pp. 336–356.

Toop, A., et al., 1984. Advances in in-stream analysis. In: Jones, M.J., Gill, P. (Eds.), Mineral Processing and Extractive Metallurgy: Papers Presented at the International Conference Mineral Processing and Extractive Metallurgy. IMM and Chinese Society of Metals, Kunming, Yunnan, China, pp. 187–194.

Ulsoy, A.G., Sastry, K.V.S., 1981. Principal developments in the automatic control of mineral processing systems. CIM Bull. 74 (836), 43–53.

Van der Spuy, D.V., et al., 2003. An on-line expert for mineral processing plants, Proc. 22th International Mineral Processing Cong., (IMPC), Paper 368. Cape Town, South Africa, pp. 223–231.

Visman, J., 1972. Discussion 3: a general theory of sampling. J. Mater. 7 (3), 345–350.

Wiegel, R.L., 1972. Advances in mineral processing material balances. Can. Metall. Q. 11 (2), 413–424.

Yianatos, J., et al., 2012. Modelling and simulation of rougher flotation circuits. Int. J. Miner. Process. 112-113, 63–70.

Ziegler, J.G., Nichols, N.B., 1942. Optimum settings for automatic controllers. Trans. ASME AIMME. 64, 759–768.

Chapter 4

Particle Size Analysis

4.1 INTRODUCTION

Size analysis of the various products of a concentrator is of importance in determining the quality of grinding, and in establishing the degree of liberation of the values from the gangue at various particle sizes. In the separation stage, size analysis of the products is used to determine the optimum size of the feed to the process for maximum efficiency and to determine the size range at which any losses are occurring in the plant, so that they may be reduced.

It is essential, therefore, that methods of size analysis be accurate and reliable, as important changes in plant operation may be made based on the results of these tests. Since it is often the case that only relatively small amounts of material are used in sizing tests, it is essential that the sample is representative of the bulk material and the same care should be taken over sampling for size analysis as is for assaying (Chapter 3).

4.2 PARTICLE SIZE AND SHAPE

The primary function of precision particle analysis is to obtain quantitative data about the size and size distribution of particles in the material (Bernhardt, 1994; Allen, 1997). However, exact size of an irregular particle cannot be measured. The terms "length," "breadth," "thickness," or "diameter" have little meaning because so many different values of these quantities can be determined. The size of a spherical particle is uniquely defined by its diameter. For a cube, the length along one edge is characteristic, and for other regular shapes there are equally appropriate dimensions.

For irregular particles, it is desirable to quote the size of a particle in terms of a single quantity, and the expression most often used is the "equivalent diameter." This refers to the diameter of a sphere that would behave in the same manner as the particle when submitted to some specified operation.

The assigned equivalent diameter usually depends on the method of measurement, hence the particle-sizing technique should, when possible, duplicate the type of process one wishes to control.

Several equivalent diameters are commonly encountered. For example, the *Stokes' diameter* is measured by sedimentation and elutriation techniques, the *projected area diameter* is measured microscopically, and the *sieve-aperture diameter* is measured by means of sieving. The latter refers to the diameter of a sphere equal to the width of the aperture through which the particle just passes. If the particles under test are not true spheres, and they rarely are in practice, this equivalent diameter refers only to their second largest dimension.

Recorded data from any size analysis should, where possible, be accompanied by some remarks which indicate the approximate shape of the particles. Fractal analysis can be applied to particle shape. Descriptions such as "granular" or "acicular" are usually quite adequate to convey the approximate shape of the particle in question.

Some of these terms are given below:

Acicular	Needle-shaped
Angular	Sharp-edged or having roughly polyhedral shape
Crystalline	Freely developed in a fluid medium of geometric shape
Dendritic	Having a branched crystalline shape
Fibrous	Regular or irregularly thread-like
Flaky	Plate-like
Granular	Having approximately an equidimensional irregular shape
Irregular	Lacking any symmetry
Modular	Having rounded, irregular shape
Spherical	Global shape

There is a wide range of instrumental and other methods of particle size analysis available. A short list of some of the more common methods is given in Table 4.1, together with their effective size ranges (these can vary greatly depending on the technology used), whether they can be used wet or dry and whether fractionated samples are available for later analysis.

Wills' Mineral Processing Technology.

TABLE 4.1 Some Methods of Particle Size Analysis

Method	Wet/Dry	Fractionated Sample?	Approx. Useful Size Range (μm)[a]
Test sieving	Both	Yes	5–100,000
Laser diffraction	Both	No	0.1–2,500
Optical microscopy	Dry	No	0.2–50
Electron microscopy	Dry	No	0.005–100
Elutriation (cyclosizer)	Wet	Yes	5–45
Sedimentation (gravity)	Wet	Yes	1–40
Sedimentation (centrifuge)	Wet	Yes	0.05–5

[a]*A micrometer (micron) (μm) is 10^{-6} m.*

4.3 SIEVE ANALYSIS

Test sieving is the most widely used method for particle size analysis. It covers a very wide range of particle sizes, which is important in industrial applications. So common is test sieving as a method of size analysis that particles finer than about 75 μm are often referred to as being in the "sub-sieve" range, although modern sieving methods allow sizing to be carried out down to about 5 μm.

Sieve (or screen) analysis is one of the oldest methods of size analysis and is accomplished by passing a known weight of sample material through successively finer sieves and weighing the amount collected on each sieve to determine the percentage weight in each size fraction. Sieving is carried out with wet or dry materials and the sieves are usually agitated to expose all the particles to the openings. Sieving, when applied to irregularly shaped particles, is complicated by the fact that a particle with a size near that of the nominal aperture of the test sieve may pass only when presented in a favorable orientation. As there is inevitably a variation in the size of sieve apertures due to irregularity of weaving, prolonged sieving will cause the larger apertures to exert an unduly large effect on the sieve analysis. Given time, every particle small enough could find its way through a very few such holes. The procedure is also complicated in many cases by the presence of "near-size" particles which cause "blinding," or obstruction of the sieve apertures, and reduce the effective area of the sieving medium. Blinding is most serious with test sieves of very small aperture size.

The process of sieving may be divided into two stages. The first step is the elimination of particles considerably smaller than the screen apertures, which should occur fairly rapidly. The second step is the separation of the so-called "near-size" particles, which is a gradual process rarely reaching completion. Both stages require the sieve to be manipulated in such a way that all particles have opportunities to pass through the apertures, and so that any particles that blind an aperture may be removed from it. Ideally, each particle should be presented individually to an aperture, as is permitted for the largest aperture sizes, but for most sizes this is impractical.

The effectiveness of a sieving test depends on the amount of material put on the sieve (the "charge") and the type of movement imparted to the sieve.

A comprehensive account of sampling techniques for sieving is given in BS 1017-1 (Anon., 1989a). Basically, if the charge is too large, the bed of material will be too deep to allow each particle a chance to meet an aperture in the most favorable position for sieving in a reasonable time. The charge, therefore, is limited by a requirement for the maximum amount of material retained at the end of sieving appropriate to the aperture size. On the other hand, the sample must contain enough particles to be representative of the bulk, so a minimum size of sample is specified. In some cases, the sample will have to be subdivided into a number of charges if the requirements for preventing overloading of the sieves are to be satisfied. In some cases, air is blown through the sieves to decrease testing time and decrease the amount of blinding occurring, a technique referred to as *air jet sieving*.

4.3.1 Test Sieves

Test sieves are designated by the nominal aperture size, which is the nominal central separation of opposite sides of a square aperture or the nominal diameter of a round aperture. A variety of sieve aperture ranges are used, the most popular being the following: the German Standard, DIN 4188; ASTM standard, E11; the American Tyler series; the French series, AFNOR; and the British Standard, BS 1796.

Woven-wire sieves were originally designated by a mesh number, which referred to the number of wires per inch, which is the same as the number of square apertures per square inch. This has the serious disadvantage that the same mesh number on the various standard ranges corresponds to different aperture sizes depending on the thickness of wire used in the woven-wire cloth. Sieves are now designated by aperture size, which gives the user directly the information needed.

Since some workers and the older literature still refer to sieve sizes in terms of mesh number, Table 4.2 lists mesh numbers for the British Standards series against

TABLE 4.2 BSS 1796 Wire-mesh Sieves

Mesh Number	Nominal Aperture Size (μm)	Mesh Number	Nominal Aperture Size (μm)
3	5,600	36	425
3.5	4,750	44	355
4	4,000	52	300
5	3,350	60	250
6	2,800	72	212
7	2,360	85	180
8	2,000	100	150
10	1,700	120	125
12	1,400	150	106
14	1,180	170	90
16	1,000	200	75
18	850	240	63
22	710	300	53
25	600	350	45
30	500	400	38

nominal aperture size. A fuller comparison of several standards is given in Napier-Munn et al. (1996).

Wire-cloth screens are woven to produce nominally uniform square apertures within required tolerances (Anon., 2000a). Wire cloth in sieves with a nominal aperture of 75 μm and greater are plain woven, while those in cloths with apertures below 63 μm may be twilled (Figure 4.1).

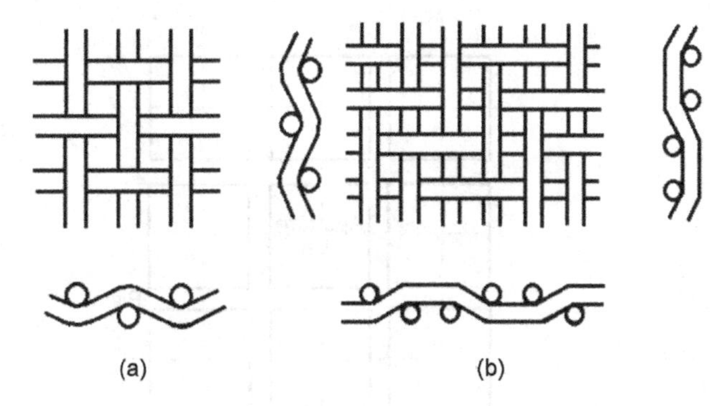

FIGURE 4.1 Weaves of wire cloth: (a) plain weave, and (b) twilled weave.

Standard test sieves are not available with aperture sizes smaller than about 20 μm. Micromesh sieves are available in aperture sizes from 2 μm to 150 μm, and are made by electroforming nickel in square and circular mesh. Another popular type is the "micro-plate sieve," which is fabricated by electroetching a nickel plate. The apertures are in the form of truncated cones with the small circle uppermost (Figure 4.2). This reduces blinding (a particle passing the upper opening falls away unhindered) but also reduces the percentage *open area*, that is, the percentage of the total area of the sieving medium occupied by the apertures.

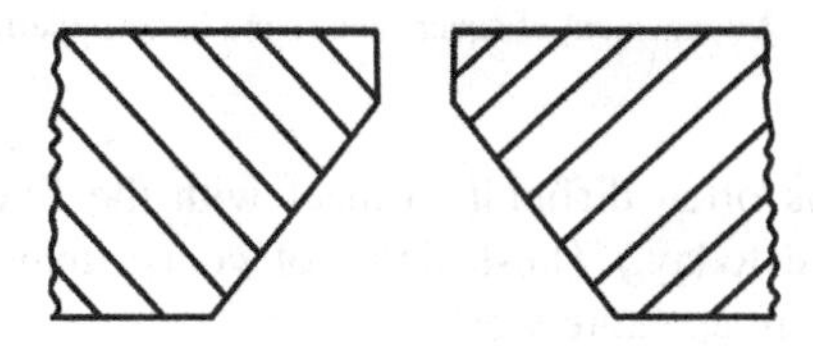
FIGURE 4.2 Cross-section of a micro-plate aperture.

Micro-sieves are used for wet or dry sieving where accuracy is required in particle size analysis down to the very fine size range (Finch and Leroux, 1982). The tolerances in these sieves are much better than those for woven-wire sieves, the aperture being guaranteed to within 2 μm of nominal size.

For aperture sizes above about 1 mm, perforated plate sieves are often used, with round or square holes (Figure 4.3). Square holes are arranged in line with the center points at the vertices of squares, while round holes are arranged with the centers at the apices of equilateral triangles (Anon., 2000b).

4.3.2 Choice of Sieve Sizes

In each of the standard series, the apertures of consecutive sieves bear a constant relationship to each other. It has long been realized that a useful sieve scale is one in which the ratio of the aperture widths of adjacent sieves is the square root of 2 ($\sqrt{2} = 1.414$). The advantage of such a scale is that the aperture areas double at each sieve, facilitating graphical presentation of results with particle size on a log scale.

Most modern sieve series are based on a fourth root of 2 ratio ($\sqrt[4]{2} = 1.189$) or, on the metric scale, a tenth root of 10 ($\sqrt[10]{10} = 1.259$), which makes possible much closer sizing of particles.

For most size analyses it is usually impracticable and unnecessary to use all the sieves in a particular series. For most purposes, alternative sieves, that is, a $\sqrt{2}$ series, are quite adequate. However, over certain size ranges of particular interest, or for accurate work the $\sqrt[4]{2}$ series may be used. Intermediate sieves should never be chosen at random, as the data obtained will be difficult to interpret.

In general, the sieve range should be chosen such that no more than about 5% of the sample is retained on the coarsest sieve, or passes the finest sieve. (The latter

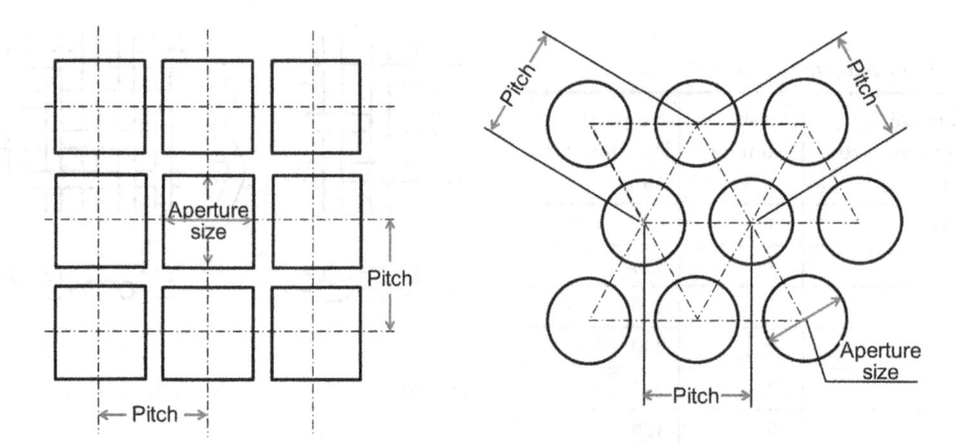

FIGURE 4.3 Arrangement of square and round holes in perforated plate sieves.

guideline is often difficult to meet with the finer grinds experienced today.) These limits of course may be lowered for more accurate work.

4.3.3 Testing Methods

The general procedures for test sieving are comprehensively covered in BS 1796 (Anon., 1989b). Machine sieving is almost universally used, as hand sieving is tedious, and its accuracy and precision depends to a large extent on the operator.

Sieves can be procured in a range of diameters, depending on the particle size and mass of material to be sieved. A common diameter for laboratory sieves is 200 mm (Figure 4.4).

The sieves chosen for the test are arranged in a stack, or *nest*, with the coarsest sieve on the top and the finest at the bottom. A tight-fitting pan or receiver is placed below the bottom sieve to receive the final undersize, and a lid is placed on top of the coarsest sieve to prevent escape of the sample.

The material to be tested is placed in the uppermost, coarsest sieve, and the nest is then placed in a sieve shaker, which vibrates the material in a vertical plane (Figure 4.5), and, on some models, a horizontal plane. The duration of screening can be controlled by an automatic timer. During the shaking, the undersize material falls through successive sieves until it is retained on a sieve having apertures which are slightly smaller than the diameter of the particles. In this way, the sample is separated into size fractions.

After the required time, the nest is taken apart and the amount of material retained on each sieve is weighed. Most of the near-mesh particles, which block the openings, can be removed by inverting the sieve and tapping the frame gently. Failing this, the underside of the gauze may be brushed gently with a soft brass wire or nylon brush. Blinding becomes more of a problem the finer the

FIGURE 4.4 No. 10 USA standard test sieve *(Courtesy Haver Tyler Corporation).*

FIGURE 4.5 W.S. TYLER® RO-TAP® test sieve shaker *(Courtesy Haver Tyler Corporation, W.S. TYLER® and RO-TAP® are registered trademarks of Haver Tyler Corporation).*

TABLE 4.3 Results of Typical Sieve Test

(1)	(2)	(3)	(4)	(5)	(6)
Sieve Size Range (μm)	Sieve Fraction		Nominal Aperture Size (μm)	Cumulative (%)	
	wt (g)	wt (%)		Undersize	Oversize
+250	0.02	0.1	250	99.9	0.1
−250 to +180	1.32	2.9	180	97.0	3.0
−180 to +125	4.23	9.5	125	87.5	12.5
−125 to +90	9.44	21.2	90	66.3	33.7
−90 to +63	13.10	29.4	63	36.9	63.1
−63 to +45	11.56	26.0	45	10.9	89.1
−45	4.87	10.9			

aperture, and brushing, even with a soft hair brush, of sieves finer than about 150 μm aperture tends to distort the individual meshes.

Wet sieving can be used on material already in the form of slurry, or it may be necessary for powders which form aggregates when dry-sieved. A full description of the techniques is given in BS 1796 (Anon., 1989b).

Water is the liquid most frequently used in wet sieving, although for materials which are water-repellent, such as coal or some sulfide ores, a wetting agent may be necessary.

The test sample may be washed down through a nest of sieves. At the completion of the test the sieves, together with the retained oversize material, are dried at a suitable low temperature and weighed.

4.3.4 Presentation of Results

There are several ways in which the results of a sieve test can be tabulated. The three most convenient methods are shown in Table 4.3 (Anon., 1989b). Table 4.3 shows for the numbered columns:

1. The sieve size ranges used in the test.
2. The weight of material in each size range. For example, 1.32 g of material passed through the 250 μm sieve, but was retained on the 180 μm sieve: the material therefore is in the size range −250 to +180 μm.
3. The weight of material in each size range expressed as a percentage of the total weight.
4. The nominal aperture sizes of the sieves used in the test.
5. The cumulative percentage of material passing through the sieves. For example, 87.5% of the material is less than 125 μm in size.
6. The cumulative percentage of material retained on the sieves.

The results of a sieving test should always be plotted graphically to assess their full significance (Napier-Munn et al., 1996).

There are many ways of recording the results of sieve analysis, the most common being that of plotting cumulative undersize (or oversize) against particle size, that is, the screen aperture. Although arithmetic graph paper can be used, it suffers from the disadvantage that points in the region of the finer aperture sizes become congested. A semi-logarithmic plot avoids this, with a linear ordinate for percentage oversize or undersize and a logarithmic abscissa for particle size. Figure 4.6 shows graphically the results of the sieve test tabulated in Table 4.3.

It is not necessary to plot both cumulative oversize (or coarser than) and undersize (finer than) curves, as they are mirror images of each other. A valuable quantity that can be determined from such curves is the "median size" of the sample. This refers to the mid-point in the size distribution, X_{50}, where 50% of the particles are smaller than this size and 50% are larger, also called the 50% *passing size*.

Size analysis is important in assessing the performance of grinding circuits. The product size is usually quoted in terms of one point on the cumulative undersize curve, this often being the 80% passing size, X_{80}. On Figure 4.6 the X_{80} is about 110 μm. In a grinding circuit it is common to refer to the 80% passing size of the feed and product as F_{80} and P_{80}, respectively, and to T_{80} as the *transfer size* between grinding units (e.g., SAG mill product going to a ball mill). Although the X_{80} does not show the overall size distribution of the material, it does facilitate routine control of the grinding circuit. For instance, if the target size is 80% −250 μm (80% finer than 250 μm), then for routine control the operator needs only screen a fraction of the mill product at one size. If

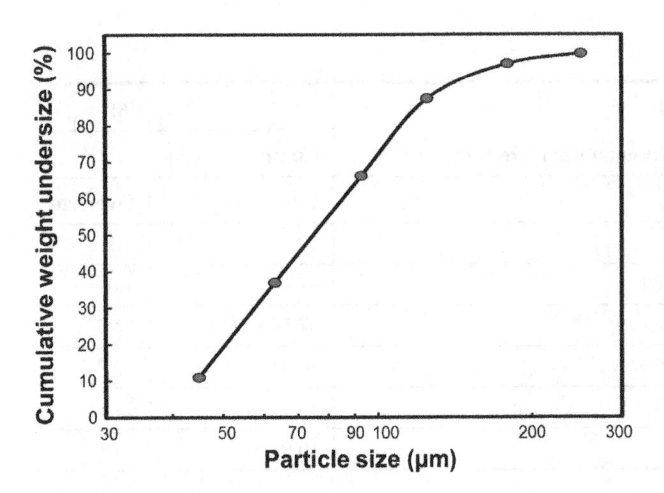

FIGURE 4.6 Sieve analysis presented as semi-log plot (Table 4.3).

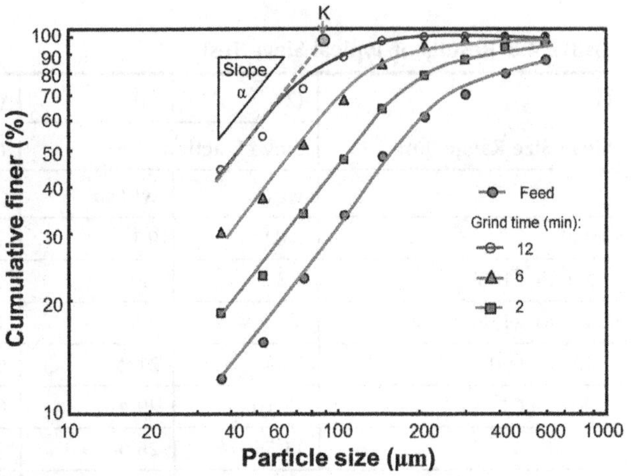

FIGURE 4.7 Example of Gates-Gaudin-Schuhmann type plot for the set of laboratory grind times in Appendix IV.

it is found that, say, 70% of the sample is −250 µm, then the product is too coarse, and control steps to remedy this can be made.

Many curves of cumulative oversize or undersize against particle size are S-shaped, leading to congested plots at the extremities of the graph. More than a dozen methods of plotting in order to proportion the ordinate are known. The two most common methods, which are often applied to comminution studies where non-uniform size distributions are obtained, are the Gates-Gaudin-Schuhmann (Schuhmann, 1940) and the Rosin–Rammler (Rosin and Rammler, 1933) methods. Both methods are derived from attempts to represent particle size distribution curves by means of equations. This results in scales which, relative to a linear scale, are expanded in some regions and contracted in others.

In the Gates-Gaudin-Schuhmann (G-G-S) method, cumulative undersize data are plotted against sieve aperture on log–log axes. This frequently leads to a linear trend from which data can be interpolated easily. The linear trend is fitted to the following:

$$\log P = \alpha \log \frac{X}{K} + \log 100 \tag{4.1}$$

which can be written as:

$$P = 100\left(\frac{X}{K}\right)^{\alpha} \tag{4.2}$$

where P is the cumulative undersize (or passing) in percent, X the particle size, K the apparent (theoretical) top size (i.e., when P = 100%) obtained by extrapolation, and α a constant sometimes referred to as the *distribution coefficient* (Austin et al., 1984). This method of plotting is routinely used in grinding studies (laboratory and plant). An example G-G-S plot is shown in Figure 4.7 for increased grinding time in a laboratory ball mill (data given in Appendix IV). The figure shows a series of roughly parallel (or *self-similar*) linear sections of the size distributions. This result, showing a consistent behavior, is quite common in grinding (and other) studies (but see Chapter 5 for exceptions) and supports the use of a single metric to describe the distribution, for example, the 80% passing size, X_{80} (Austin et al., 1984).

Self-similar distributions such as in Figure 4.7 become a unique distribution by plotting against a reduced particle size, typically X/X_{50} where X_{50} is the 50% passing size or median size. If it is known that data obtained from the material usually yield a linear plot on log–log axes, then the burden of routine analysis can be greatly relieved, as relatively few sieves will be needed to check essential features of the size distribution. Self-similarity can be quickly identified, as a plot of, say, % − 75 µm versus % +150 µm will yield a single curve if this property holds.

Plotting on a log–log scale considerably expands the region below 50% in the cumulative undersize curve, especially that below 25%. It does, however, severely contract the region above 50%, and especially above 75%, which is a disadvantage of the method (Figure 4.8).

The Rosin–Rammler method is often used for representing the results of sieve analyses performed on material which has been ground in ball mills. Such products have been found to obey the following relationship (Austin et al., 1984):

$$100 - P = 100 \exp(bX)^{n} \tag{4.3}$$

where b and n are constants.

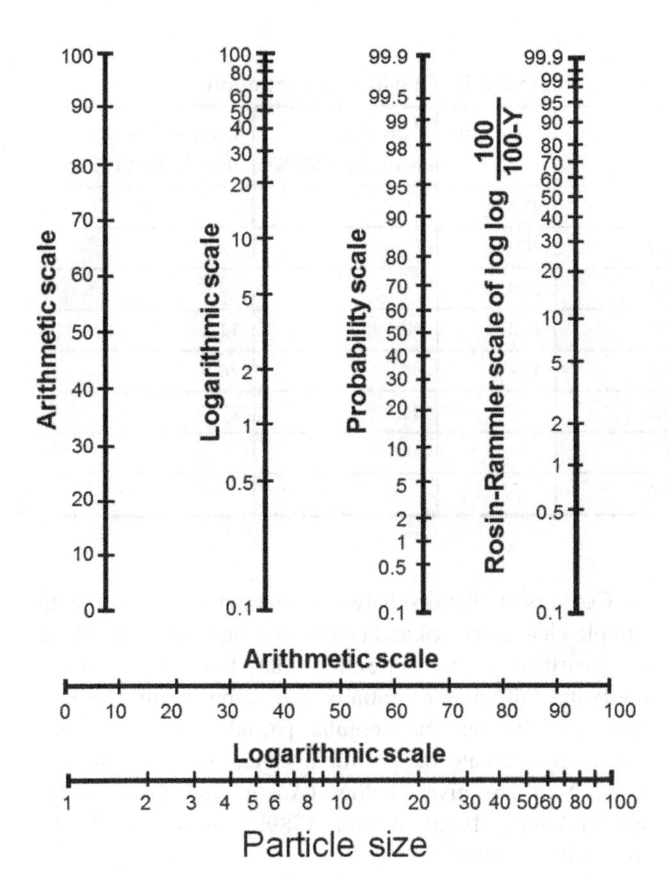

FIGURE 4.8 Comparison of scales.

This can be rewritten as:

$$\log\left[\ln\frac{100}{100-P}\right] = \log b + n \log X \tag{4.4}$$

Thus, a plot of ln [100/(100−P)] versus X on log−log axes gives a line of slope n.

In comparison with the log−log method, the Rosin−Rammler plot expands the cumulative undersize regions below 25% and above 75% (Figure 4.8) and it contracts the 30−60% region. It has been shown, however, that this contraction is insufficient to cause adverse effects (Harris, 1971). The method is tedious to plot manually unless charts having the axes divided proportionally to log [ln (100/(100−P))] and log X are used. The data can, however, be easily plotted in a spreadsheet.

The Gates-Gaudin-Schuhmann plot is often preferred to the Rosin−Rammler method in mineral processing applications, the latter being more often used in coal-preparation studies, for which it was originally developed. The two methods have been assessed by Harris (1971), who suggests that the Rosin−Rammler is the better method for mineral processing applications. The Rosin−Rammler is useful for monitoring grinding operations for highly skewed distributions, but, as noted by Allen (1997), it should be used with caution, since taking logs always apparently reduces scatter, and thus taking logs twice is not to be recommended.

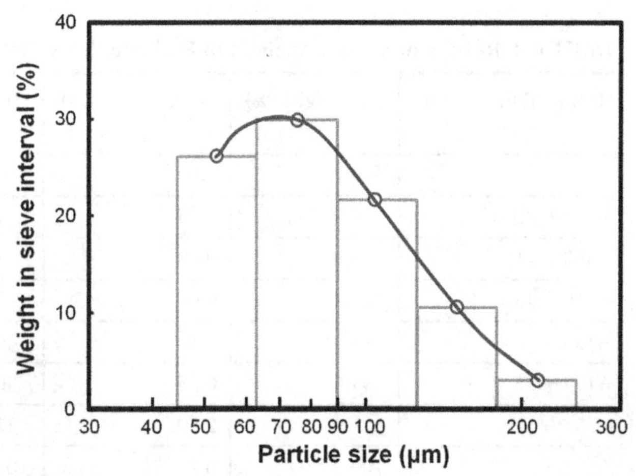

FIGURE 4.9 Fractional (frequency) distribution and histogram representation of screen analysis data.

Although cumulative size curves are used almost exclusively, the particle size distribution curve itself is sometimes more informative. Ideally, this is derived by differentiating the cumulative undersize curve and plotting the gradient of the curve obtained against particle size. In practice, the size distribution curve is obtained by plotting the retained fraction of the sieves against size. This can be as a histogram or as a frequency (fractional) curve plotted at the "average" size in between two sieve sizes. For example, material which passes a 250 μm sieve but is retained on a 180 μm sieve, may be regarded as having an arithmetic mean particle size of 215 μm or, more appropriately since the sieves are in a geometric sequence, a geometric mean ($\sqrt{(250 \times 180)}$) of 212 μm for the purpose of plotting. If the distribution is represented on a histogram, then the horizontals on the columns of the histogram join the various adjacent sieves used in the test. Unless each size increment is of equal width, however, the histogram has little value. Figure 4.9 shows the size distribution of the material in Table 4.3 represented on a frequency curve and a histogram.

Fractional curves and histograms are useful and rapid ways of visualizing the relative frequency of occurrence of the various sizes present in the material. The only numerical parameter that can be obtained from these methods is the "mode" of the distribution, that is, the most commonly occurring size.

For assessment of the metal losses in the tailings of a plant, or for preliminary evaluation of ores, assaying must be carried out on the various screen fractions. It is important, therefore, that the bulk sample satisfies the minimum sample weight requirement given by Gy's equation for the fundamental error (Chapter 3).

TABLE 4.4 Results of Screen Analysis to Evaluate the Suitability for Treatment by Gravity Concentration

(1) Size Range (μm)	(2) Weight (%)	(3) Assay (% Sn)	Distribution (% Sn)	Size (μm)	Cumulative oversize (Wt %)	Cumulative distribution (% Sn)
+422	9.7	0.02	0.9	422	9.7	0.9
−422 + 300	4.9	0.05	1.2	300	14.6	2.1
−300 + 210	10.3	0.05	2.5	210	24.9	4.6
−210 + 150	23.2	0.06	6.7	150	48.1	11.3
−150 + 124	16.4	0.12	9.5	124	64.5	20.8
−124 + 75	33.6	0.35	56.5	75	98.1	77.3
−75	1.9	2.50	22.7			
	100.0	0.21	100.0			

Table 4.4 shows the results of a screen analysis performed on an alluvial tin deposit for preliminary evaluation of its suitability for treatment by gravity concentration. Columns 1, 2, and 3 show the results of the sieve test and assays, which are evaluated in the other columns. It can be seen that the calculated overall assay for the material is 0.21% Sn, but that the bulk of the tin is present within the finer fractions. The results show that, for instance, if the material was initially screened at 210 μm and the coarse fraction discarded, then the bulk required for further processing would be reduced by 24.9%, with a loss of only 4.6% of the Sn. This may be acceptable if mineralogical analysis shows that the tin is finely disseminated in this coarse fraction, which would necessitate extensive grinding to give reasonable liberation. Heavy liquid analysis (Chapter 11) on the −210 μm fraction would determine the theoretical (i.e., liberation-limited, Chapter 1) grades and recoveries possible, but the screen analysis results also show that much of the tin (22.7%) is present in the −75 μm fraction, which constitutes only 1.9% of the total bulk of the material. This indicates that there may be difficulty in processing this material, as gravity separation techniques are not very efficient at such fine sizes (Chapters 1 and 10).

4.4 SUB-SIEVE TECHNIQUES

Sieving is rarely carried out on a routine basis below 38 μm; below this size the operation is referred to as *sub-sieving*. The most widely used methods are sedimentation, elutriation, microscopy, and laser diffraction, although other techniques are available.

There are many concepts in use for designating particle size within the sub-sieve range, and it is important to be aware of them, particularly when combining size distributions determined by different methods. It is preferable to cover the range of a single distribution with a single method, but this is not always possible.

Conversion factors between methods will vary with sample characteristics and conditions, and with size where the distributions are not self-similar. For spheres, many methods will give essentially the same result (Napier-Munn, 1985), but for irregular particles this is not so. Some approximate factors for a given characteristic size (e.g., X_{80}) are given below (Austin and Shah, 1983; Napier-Munn, 1985; Anon., 1989b)—these should be used with caution:

Conversion	Multiplying Factor
Sieve size to Stokes' diameter (sedimentation, elutriation)	0.94
Sieve size to projected area diameter (microscopy)	1.4
Sieve size to laser diffraction	1.5
Square mesh sieves to round hole sieves	1.2

4.4.1 Stokes' Equivalent Diameter

In sedimentation techniques, the material to be sized is dispersed in a fluid and allowed to settle under carefully controlled conditions; in elutriation techniques, samples are sized by allowing the dispersed material to settle against a rising fluid velocity. Both techniques separate the particles on the basis of resistance to motion in a fluid. This resistance to motion determines the terminal velocity which the particle attains as it is allowed to fall in a fluid under the influence of gravity.

For particles within the sub-sieve range, the terminal velocity is given by the equation derived by Stokes (1891):

$$v = \frac{d^2 g(\rho_s - \rho_l)}{18\,\eta} \tag{4.5}$$

where v is the terminal velocity of the particle (m s^{-1}), d the particle diameter (m), g the acceleration due to gravity (m s^{-2}), ρ_s the particle density (kg m^{-3}), ρ_f the fluid density (kg m^{-3}), and η the fluid viscosity (N s m^{-2}); ($\eta = 0.001$ N s m^{-2} for water at 20°C).

Stokes' law is derived for spherical particles; non-spherical particles will also attain a terminal velocity, but this velocity will be influenced by the shape of the particles. Nevertheless, this velocity can be substituted in the Stokes' equation to give a value of d, which can be used to characterize the particle. This value of d is referred to as the "Stokes' equivalent spherical diameter" (or simply "Stokes' diameter" or "sedimentation diameter").

Stokes' law is only valid in the region of laminar flow (Chapter 9), which sets an upper size limit to the particles that can be tested by sedimentation and elutriation methods in a given liquid. The limit is determined by the particle Reynolds number, a dimensionless quantity defined by:

$$\text{Re} = \frac{v d \rho_f}{\eta} \tag{4.6}$$

The Reynolds number should not exceed 0.2 if the error in using Stokes' law is not to exceed 5% (Anon., 2001a). In general, Stokes' law will hold for all particles below 40 μm dispersed in water; particles above this size should be removed by sieving beforehand. The lower limit may be taken as 1 μm, below which the settling times are too long, and also the effects of Brownian motion and unintentional disturbances, such as those caused by convection currents, are far more likely to produce serious errors.

4.4.2 Sedimentation Methods

Sedimentation methods are based on the measurement of the rate of settling of the powder particles uniformly dispersed in a fluid and the principle is well illustrated by the common laboratory method of "beaker decantation."

The material under test is uniformly dispersed in low concentration in water contained in a beaker or similar parallel-sided vessel. A wetting agent may need to be added to ensure complete dispersion of the particles. A syphon tube is immersed into the water to a depth of h below the water level, corresponding to about 90% of the liquid depth L.

The terminal velocity v is calculated from Stokes' law for the various sizes of particle in the material, say 35, 25, 15, and 10 μm. For a distribution, it is usual to fix d_s for the particles that are most abundant in the sample.

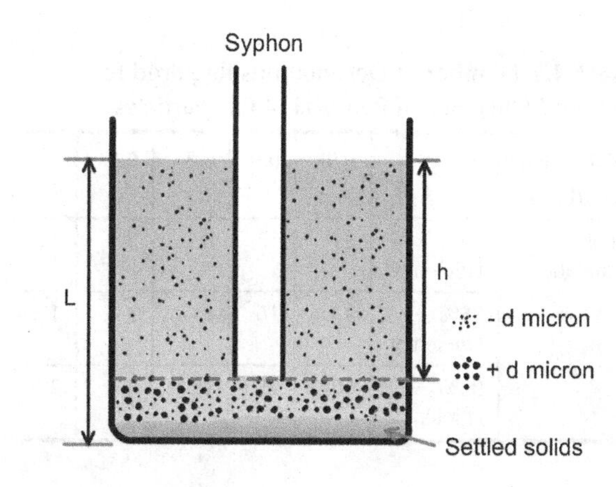

FIGURE 4.10 Beaker decantation.

The time required for a 10 μm particle to settle from the water level to the bottom of the syphon tube, distance h, is calculated ($t = h/v$). The pulp is gently stirred to disperse the particles through the whole volume of water and then it is allowed to stand for the calculated time. The water above the end of the tube is syphoned off and all particles in this water are assumed to be smaller than 10 μm diameter (Figure 4.10). However, a fraction of the −10 μm material, which commenced settling from various levels below the water level, will also be present in the material below the syphon level. In order to recover these particles, the pulp remaining must be diluted with water to the original level, and the procedure repeated until the decant liquor is essentially clear. In theory, this requires an infinite number of decantations, but in practice at least five treatments are usually sufficient, depending on the accuracy required. The settled material can be treated in a similar manner at larger separating sizes, that is, at shorter decanting times, until a target number of size fractions is obtained.

The method is simple and cheap, and has an advantage over many other sub-sieve techniques in that it produces a true fractional size analysis, that is, reasonable quantities of material in specific size ranges are collected, which can be analyzed chemically and mineralogically.

This method is, however, extremely tedious, as long settling times are required for very fine particles, and separate tests must be performed for each particle size. For instance, a 25 μm particle of quartz has a settling velocity of 0.056 cm s^{-1}, and therefore takes about 3½ min to settle 12 cm, a typical immersion depth for the syphon tube. Five separate tests to ensure a reasonably clear decant therefore require a total settling time of about 18 min. A 5 μm particle, however, has a settling velocity of 0.0022 cm s^{-1}, and therefore takes about 1½ h to settle 12 cm. The total time for evaluation of such material is thus about 8 h. A complete analysis may therefore take an operator several days.

TABLE 4.5 Number of Decantations Required for Required Efficiency of Removal of fine Particles

Relative Particle Size (d_1/d)		0.95	0.9	0.8	0.5	1
Number of decantations	90% Efficiency	25	12	6	2	1
	95% Efficiency	33	16	8	3	1
	99% Efficiency	50	25	12	4	2

Another problem is the large quantity of water, which dilutes the undersize material, due to repeated decantation.

In the system shown in Figure 4.10, after time t, all particles larger than size d have fallen to a depth below the level h. All particles of a size d_1, where $d_1 < d$, will have fallen below a level h_1 below the water level, where $h_1 < h$. The efficiency of removal of particles of size d_1 into the decant is thus:

$$\frac{h - h_1}{L}$$

since at time $t = 0$ the particles were uniformly distributed over the whole volume of liquid, corresponding to depth L, and the fraction removed into the decant is the volume above the syphon level, $h - h_1$.

Now, since $t = h/v$, and $v \propto d^2$,

$$\frac{h}{d^2} = \frac{h_1}{d_1^2}$$

Therefore, the efficiency of removal of particles of size d_1

$$= \frac{h - h(d_1/d)^2}{L} = \frac{h[1 - (d_1/d)^2]}{L} = a\left[1 - (d_1/d)^2\right] = E$$

where $a = h/L$.

If a second decantation step is performed, the amount of minus d_1 material in the dispersed suspension is $1 - E$, and the efficiency of removal of minus d_1 particles after two decantations is thus:

$$E + (1 - E)E = 2E - E^2 = 1 - [1 - E]^2$$

In general, for n decantation steps the efficiency of removal of particles of size d_1, at a separation size of d, is:

$$\begin{aligned} & 1 - [1 - E]^n \\ \text{or efficiency} = \; & 1 - \{1 - a[1 - (d_1/d)^2]\} \end{aligned} \quad (4.7)$$

Table 4.5 shows the number of decantation steps required for different efficiencies of removal of various sizes of particle expressed relative to d, the separating size, where the value of $a = 0.9$. It can be shown (Heywood, 1953) that the value of a has relatively little effect; therefore there is nothing to be gained by attempting to remove the suspension adjacent to the settled particles, thus risking disturbance and re-entrainment.

Table 4.5 shows that a large number of decantations are necessary for effective removal of particles close to the separation size, but that relatively small particles are quickly eliminated. For most purposes, unless very narrow size ranges are required, no more than about twelve decantations are necessary for the test.

A much quicker and less-tedious method of sedimentation analysis is the *Andreasen pipette technique* (Anon., 2001b).

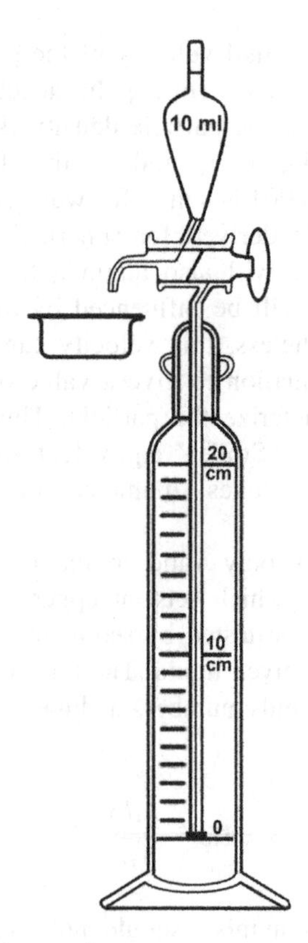

FIGURE 4.11 Andreasen pipette.

The apparatus (Figure 4.11) consists of a half-liter graduated cylindrical flask and a pipette connected to a 10 mL reservoir by means of a two-way stop-cock. The tip of the pipette is in the plane of the zero mark when the ground glass stopper is properly seated.

A 3 to 5% suspension of the sample, dispersed in the sedimentation fluid, usually water, is added to the flask. The pipette is introduced and the suspension agitated by inversion. The suspension is then allowed to settle, and at given intervals of time, samples are withdrawn by applying suction to the top of the reservoir, manipulating the two-way cock so that the sample is drawn up as far as the calibration mark on the tube above the 10 mL reservoir. The cock is then reversed, allowing the sample to drain into the collecting dish. After each sample is taken, the new liquid level is noted.

The samples are then dried and weighed, and the weights compared with the weight of material in the same volume of the original suspension.

There is a definite particle size, d, corresponding to each settling distance h and time t, and this represents the size of the largest particle that can still be present in the sample. These particle sizes are calculated from Stokes' law for the various sampling times. The weight of solids collected, w, compared with the corresponding original weight, w_o (i.e., w/w_o) then represents the fraction of the original material having a particle size smaller than d, which can be plotted on the size-analysis graph.

The Andreasen pipette method is quicker than beaker decantation, as samples are taken off successively throughout the test for increasingly finer particle sizes. For example, although 5 μm particles of quartz will take about 2½ h to settle 20 cm, once this sample is collected, all the coarser particle-size samples will have been taken, and so the complete analysis, in terms of settling times, is only as long as the settling time for the finest particles.

The disadvantage of the method is that the samples taken are each representative of the particles smaller than a particular size, which is not as valuable, for mineralogical and chemical analysis, as samples of various size ranges, as are produced by beaker decantation.

Sedimentation techniques tend to be tedious, due to the long settling times required for fine particles and the time required to dry and weigh the samples. The main difficulty, however, lies in completely dispersing the material within the suspending liquid, such that no agglomeration of particles occurs. Combinations of suitable suspending liquids and dispersing agents for various materials are given in BS ISO 13317−1 (Anon., 2001a).

Various other techniques have been developed that attempt to speed up testing. Examples of these methods, which are comprehensively reviewed by Allen (1997), include: the photo-sedimentometer, which combines gravitational settling with photo-electric measurement; and the sedimentation balance, in which the weight of material settling out onto a balance pan is recorded against time to produce a cumulative sedimentation size analysis.

4.4.3 Elutriation Techniques

Elutriation is a process of sizing particles by means of an upward current of fluid, usually water or air. The process is the reverse of gravity sedimentation, and Stokes' law applies.

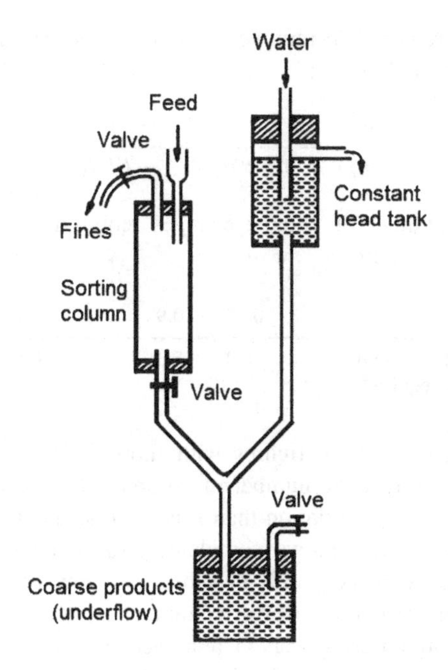

FIGURE 4.12 Simple elutriator.

All elutriators consist of one or more "sorting columns" (Figure 4.12) in which the fluid is rising at a constant velocity. Feed particles introduced into the sorting column will be separated into two fractions, according to their terminal velocities, calculated from Stokes' law.

Those particles having a terminal velocity less than that of the velocity of the fluid will report to the overflow, while those particles having a greater terminal velocity than the fluid velocity will sink to the underflow. Elutriation is carried out until there are no visible signs of further classification taking place or the rate of change in weights of the products is negligible.

An elutriator based on air is the Haultain infrasizer, which comprises six sorting columns ("cones") plus a dust collector (Price, 1962). More commonly, water is the fluid. This involves the use of much water and dilution of the undersize fraction, but it can be shown that this is not as serious as in beaker decantation. Consider a sorting column of depth h, sorting material at a separating size of d. If the upward velocity of water flow is v, then (as noted above) by Stokes' law, $v \propto d^2$.

Particles smaller than the separating size d will move upwards in the water flow at a velocity dependent on their size. Thus, particles of size d_1, where $d_1 < d$, will move upwards in the sorting column at a velocity v_1, where $v_1 \propto (d^2 - d_1^2)$.

The time required for a complete volume change in the sorting column is h/v, and the time required for particles of size d_1 to move from the bottom to the top of the sorting column is h/v_1. Therefore, the number of volume changes required to remove all particles of size d_1 from the sorting column is:

$$= \frac{h/v_1}{h/v} = \frac{d^2}{d^2 - d_1^2} = \frac{1}{1 - (d_1/d)^2}$$

and the number of volume changes required for various values of d_1/d are:

d_1/d	0.95	0.9	0.8	0.5	0.1
Number of volume changes required	10.3	5.3	2.8	1.3	1.0

Comparing these figures with those in Table 4.5, it can be seen that the number of volume changes required is far less with elutriation than it is with decantation. It is also possible to achieve complete separation by elutriation, whereas this can only be achieved in beaker decantation by an infinite number of volume changes.

Elutriation thus appears more attractive than decantation, and has certain practical advantages in that the volume changes need no operator attention. It suffers from the disadvantage, however, that the fluid velocity is not constant across the sorting column, being a minimum at the walls of the column, and a maximum at the center. The separation size is calculated from the mean volume flow, so that some coarse particles are misplaced in the overflow, and some fines are misplaced into the underflow. The fractions thus have a considerable overlap in particle size and are not sharply separated. Although the decantation method never attains 100% efficiency of separation, the lack of sharpness of the division into fractions is much less than that due to velocity variation in elutriation (Heywood, 1953).

Elutriation is limited at the coarsest end by the validity of Stokes' law, but most materials in the sub-sieve range exhibit laminar flow. At the fine end of the scale, separations become impracticable below about 10 μm, as the material tends to agglomerate, or extremely long separating times are required. Separating times can be considerably decreased by utilizing centrifugal forces. One of the most widely used methods of sub-sieve sizing in mineral processing laboratories is the Warman cyclosizer (Finch and Leroux, 1982), which is extensively used for routine testing and plant control in the size range 8–50 μm for materials of specific gravity similar to quartz (s. g. 2.7), and down to 4 μm for particles of high specific gravity, such as galena (s. g. 7.5).

The cyclosizer unit consists of five cyclones (see Chapter 9 for a full description of the principle of the hydrocyclone), arranged in series such that the overflow of one unit is the feed to the next unit (Figure 4.13).

The individual units are inverted in relation to conventional cyclone arrangements, and at the apex of each, a chamber is situated so that the discharge is effectively closed (Figure 4.14).

FIGURE 4.13 Cyclosizer *(Courtesy MARC Technologies Pty Ltd.)*.

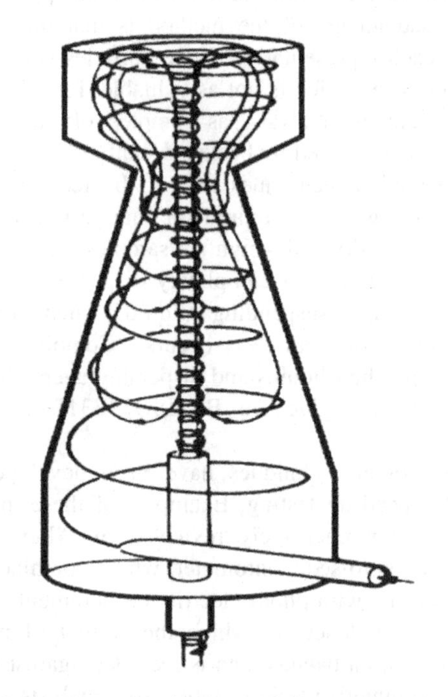

FIGURE 4.14 Flow pattern inside a cyclosizer cyclone unit.

Water is pumped through the units at a controlled rate, and a weighed sample of solids is introduced ahead of the cyclones.

The tangential entry into the cyclones induces the liquid to spin, resulting in a portion of the liquid, together with the faster-settling particles, reporting to the apex opening, while the remainder of the liquid, together with the slower settling particles, is discharged through the vortex outlet and into the next cyclone in the series. There is a successive decrease in the inlet area and vortex outlet diameter of each cyclone in the direction of the flow, resulting in a corresponding increase in inlet velocity and an increase in the centrifugal forces within the cyclone. This results in a successive decrease in the limiting particle-separation size of the cyclones.

The cyclosizer is manufactured to have definite limiting separation sizes at standard values of the operating variables, viz. water flow rate, water temperature, particle density, and elutriation time. To correct for practical operation at other levels of these variables, a set of correction graphs is provided.

Complete elutriation normally takes place after about 20 min, at which time the sized fractions are collected by discharging the contents of each apex chamber into separate beakers. The samples are dried, and weighed to determine the mass fractions of each particle size.

4.4.4 Microscopic Sizing and Image Analysis

Microscopy can be used as an absolute method of particle size analysis, since it is the only method in which individual mineral particles are observed and measured (Anon., 1993; Allen, 1997). The image of a particle seen in a microscope is two-dimensional and from this image an estimate of particle size must be made. Microscopic sizing involves comparing the projected area of a particle with the areas of reference circles, or *graticules*, of known sizes, and it is essential for meaningful results that the mean projected areas of the particles are representative of the particle size. This requires a random orientation in three dimensions of the particle on the microscope slide, which is unlikely in most cases.

The optical microscope method is applicable to particles in the size range 0.8–150 μm, and down to 0.001 μm using electron microscopy.

Basically, all microscopy methods are carried out on extremely small laboratory samples, which must be collected with great care in order to be representative of the process stream under study.

In manual optical microscopy, the dispersed particles are viewed by transmission, and the areas of the magnified images are compared with the areas of circles of known sizes inscribed on a graticule.

The relative numbers of particles are determined in each of a series of size classes. These represent the size distribution by number from which it is possible to calculate the distribution by volume and, if all the particles have the same density, the distribution by weight.

Manual analysis of microscope slides is tedious and error prone; semi-automatic and automatic systems have been developed which speed up analyses and reduce the tedium of manual methods (Allen, 1997).

The development of quantitative image analysis has made possible the rapid sizing of fine particles. Image analyzers accept samples in a variety of forms—photographs, electron micrographs, and direct viewing—and are often integrated in system software. Figure 4.15 shows the grayscale electron backscatter image of a group of mineral particles obtained with a scanning electron microscope; grains of chalcopyrite (Ch), quartz (Qtz), and epidote (Epd) are identified in the image. On the right are plotted the size distributions of the "grains" of the mineral chalcopyrite (i.e., the pieces of chalcopyrite identified by the instrument, whether liberated or not) and the "particles" in which the chalcopyrite is present. The plots are based on the analysis of several hundred thousand particles in the original sample, and are delivered automatically by the system software. Image analysis of this kind is available in many forms for the calculation of many quantities (such as size, surface area, boundary lengths)

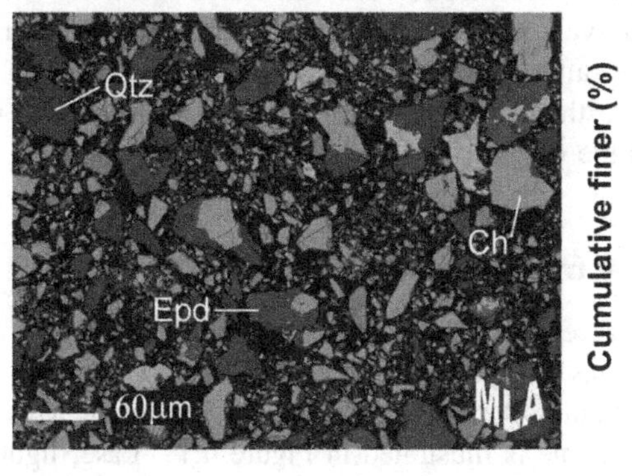

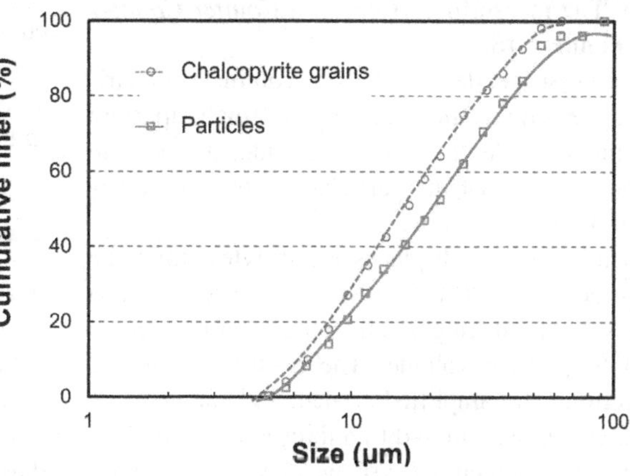

FIGURE 4.15 Using image analysis to estimate the size distributions of particles and mineral grains *(Courtesy JKMRC and JKTech Pty Ltd).*

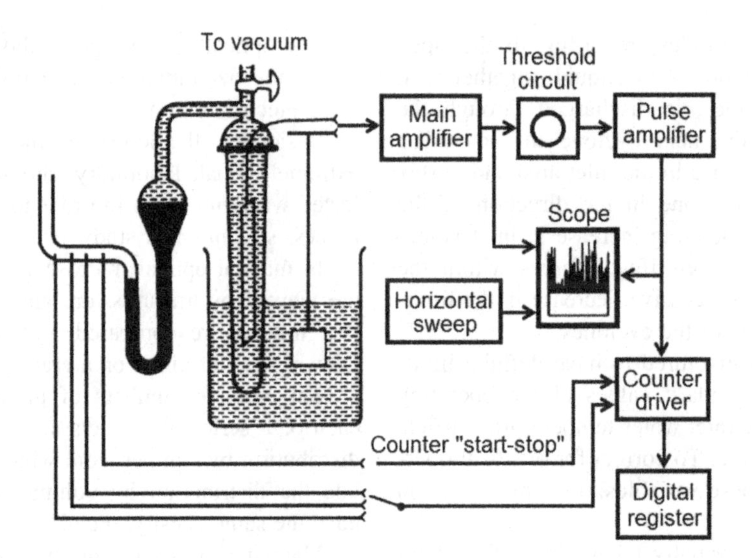

FIGURE 4.16 Principle of the Coulter Counter.

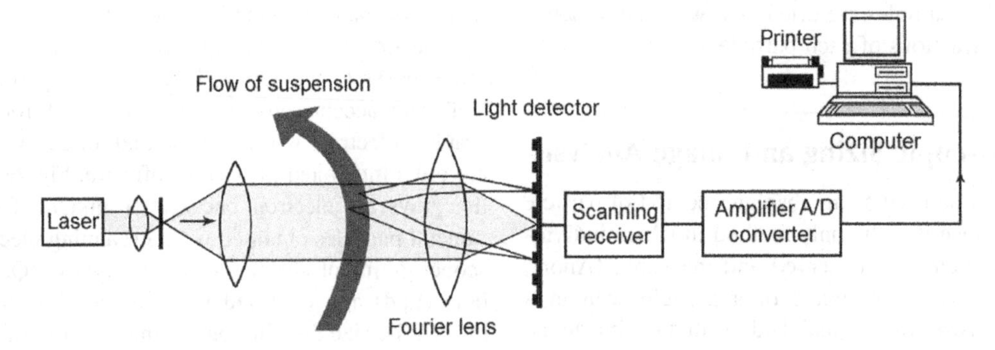

FIGURE 4.17 The laser diffraction instrument principle *(from Napier-Munn et al., 1996; Courtesy JKMRC, The University of Queensland).*

for most imaging methods, for example, optical, electron. (See also Chapter 17.)

4.4.5 Electrical Impedance Method

The Beckman Coulter Counter makes use of current changes in an electrical circuit produced by the presence of a particle. The measuring system of a *Coulter Counter* is shown in Figure 4.16.

The particles, suspended in a known volume of electrically conductive liquid, flow through a small aperture having an immersed electrode on either side; the particle concentration is such that the particles traverse the aperture substantially one at a time.

Each particle passage displaces electrolyte within the aperture, momentarily changing the resistance between the electrodes and producing a voltage pulse of magnitude proportional to particle volume. The resultant series of pulses is electronically amplified, scaled, and counted.

The amplified pulses are fed to a threshold circuit having an adjustable screen-out voltage level, and those pulses which reach or exceed this level are counted, this count representing the number of particles larger than some determinable volume proportional to the appropriate threshold setting. By taking a series of counts at various amplification and threshold settings, data are directly obtained for determining number frequency against volume, which can be used to determine the size distribution.

As the instrument measures particle volume, the equivalent diameter is calculated from that of a sphere of the same volume, which can be a more relevant size measure than some alternatives. The instrument is applicable in the range 0.4–1,200 μm.

4.4.6 Laser Diffraction Instruments

In recent years, several instruments based on the diffraction of laser light by fine particles have become available, including the Malvern Master-Sizer and the Microtrac. The principle is illustrated in Figure 4.17. Laser light is passed through a dilute suspension of the particles which circulate through an optical cell. The light is scattered by the particles, which is detected by a solid state detector which measures light intensity over a range of angles. A theory of light scattering is used to calculate the particle size distribution from the light distribution pattern, finer particles inducing more scatter than coarser particles. The early instruments used Fraunhofer theory, which is suitable for coarse particles in the approximate range 1–2,000 μm (the upper limit being imposed mainly by mechanical constraints). More recently Mie theory has been used to extend the capability down to 0.1 μm and below. Some modern instruments offer both options to cover a wide size range.

Laser diffraction instruments are fast, easy to use, and give reproducible results. However, light scattering theory does not give a definition of size that is compatible with other methods, such as sieving. In most mineral processing applications, for example, laser diffraction size distributions tend to appear coarser than those of other methods. Austin and Shah (1983) have suggested a procedure for inter-conversion of laser diffraction and sieve-size distributions, and a simple conversion can be developed by regression for materials of consistent characteristics. In addition, the results can depend on the relative refractive indices of the solid particles and liquid medium (usually, though not necessarily, water), and even particle shape. Most instruments claim to compensate for these effects, or offer calibration inputs to the user.

For these reasons, laser diffraction size analyzers should be used with caution. For routine high volume analyses in a fixed environment in which only *changes* in size distribution need to be detected, they probably have no peer. For comparison across several environments or materials, or with data obtained by other methods, care is needed in interpreting the data. These instruments do not, of course, provide fractionated samples for subsequent analysis.

4.5 ON-LINE PARTICLE SIZE ANALYSIS

The major advantage of on-line analysis systems is that by definition they do not require sampling by operations personnel, eliminating one source of error. They also provide much quicker results, in the case of some systems continuous measurement, which in turn decreases production delays and improves efficiency. For process control on-line measurement is essential.

4.5.1 Slurry Systems

While used sometimes on final concentrates, such as Fe concentrates, to determine the Blaine number (average particle size deduced from surface area), and on tailings for control of paste thickeners, for example, the prime application is on cyclone overflow for grinding circuit control (Kongas and Saloheimo, 2009). Control of the grinding circuit to produce the target particle size distribution for flotation (or other mineral separation process) at target throughput maximizes efficient use of the installed power.

FIGURE 4.18 The Thermo GammaMetrics PSM-400MPX on-line particle size analyzer *(Courtesy Thermo Electron Corporation).*

Continuous measurement of particle size in slurries has been available since 1971, the PSM (particle size monitor) system produced then by Armco Autometrics (subsequently by Svedala and now by Thermo Gamma-Metrics) having been installed in a number of mineral processing plants (Hathaway and Guthnals, 1976).

The PSM system uses ultrasound to determine particle size. This system consists of three sections: the air eliminator, the sensor section, and the electronics section. The air eliminator draws a sample from the process stream and removes entrained air bubbles (which otherwise act as particles in the measurement). The de-aerated pulp then passes between the sensors. Measurement depends on the varying absorption of ultrasonic waves in suspensions of different particle sizes. Since solids concentration also affects the absorption, two pairs of transmitters and receivers, operating at different frequencies, are employed to measure particle size and solids concentration of the pulp, the processing of this information being performed by the electronics. The Thermo GammaMetrics PSM-400MPX (Figure 4.18) handles slurries up to 60% w/w solids and outputs five size fractions simultaneously.

Other measurement principles are now in commercial form for slurries. Direct mechanical measurement of particle size between a moving and fixed ceramic tip, and

FIGURE 4.19 CiDRA's CYCLONEtrac^{SM} systems in operation: the PST (at top) and OSM *(Courtesy CiDRA Minerals Processing).*

laser diffraction systems are described by Kongas and Saloheimo (2009). Two recent additions are the CYCLONEtrac systems from CiDRA Minerals Processing (Maron et al., 2014), and the OPUS ultrasonic extinction system from Sympatec (Smith et al., 2010).

CiDRA's CYCLONEtrac PST (particle size tracking) system comprises a hardened probe that penetrates into the cyclone overflow pipe to contact the stream and effectively "listens" to the impacts of individual particles. The output is % above (or below) a given size and has been shown to compare well with sieve sizing (Maron et al., 2014). The OPUS ultrasonic extinction system (USE) transmits ultrasonic waves through a slurry that interact with the suspended particles. The detected signal is converted into a particle size distribution, the number of frequencies used giving the number of size classes measured. Applications on ores can cover a size range from 1 to 1,000 μm (Smith et al., 2010).

In addition to particles size, recent developments have included sensors to detect malfunctioning cyclones. Westendorf et al. (2015) describe the use of sensors (from Portage Technologies) on cyclone overflow and underflow piping. CiDRA's CYCLONEtrac OSM (oversize monitor) is attached to the outside of the cyclone overflow pipe and detects the acoustic signal as oversize particles ("rocks") hit the pipe (Cirulis and Russell, 2011). The systems are readily installed on individual cyclones thus permitting

FIGURE 4.20 Online visual imaging system (Split Engineering's Split-Online system) (camera is at top center) *(Courtesy Split Engineering).*

poorly operating units to be identified and changed while allowing the cyclone battery to remain in operation. Figure 4.19 shows an installation of both CiDRA systems (PST, OSM) on the overflow pipe from a cyclone.

4.5.2 On-belt Systems

Image analysis is used in sizing rocks on conveyor belts. Systems available include those supplied by Split Engineering, WipFrag, Metso, and Portage Technologies (see also Chapter 6). A fixed camera with appropriate lighting captures images of the particles on the belt. Then, software segments the images, carries out appropriate corrections, and calculates the particle size distribution. A system installed on a conveyor is shown in Figure 4.20. Figure 4.21 shows the original camera views and the segmented images for a crusher feed and product, together with the calculated size distributions. (Another example is given in Chapter 6.) Common problems with imaging systems are the inability to "see" particles under the top layer, and the difficulty of detecting fines, for which correction algorithms can be used. The advantage of imaging systems is that they are low impact, as it possible to acquire size information without sampling or interacting with the particles being measured. These systems are useful in detecting size changes in crusher circuits, and are increasingly used in measuring the feed to SAG mills for use in mill control (Chapter 6), and in mining applications to assess blast fragmentation.

There are other on-line systems available or under test. For example, CSIRO has developed a version of the ultrasonic attenuation principle, in which velocity spectrometry and gamma-ray transmission are incorporated to produce a more robust measurement in the range 0.1–1,000 μm (Coghill et al., 2002). Most new developments in on-line

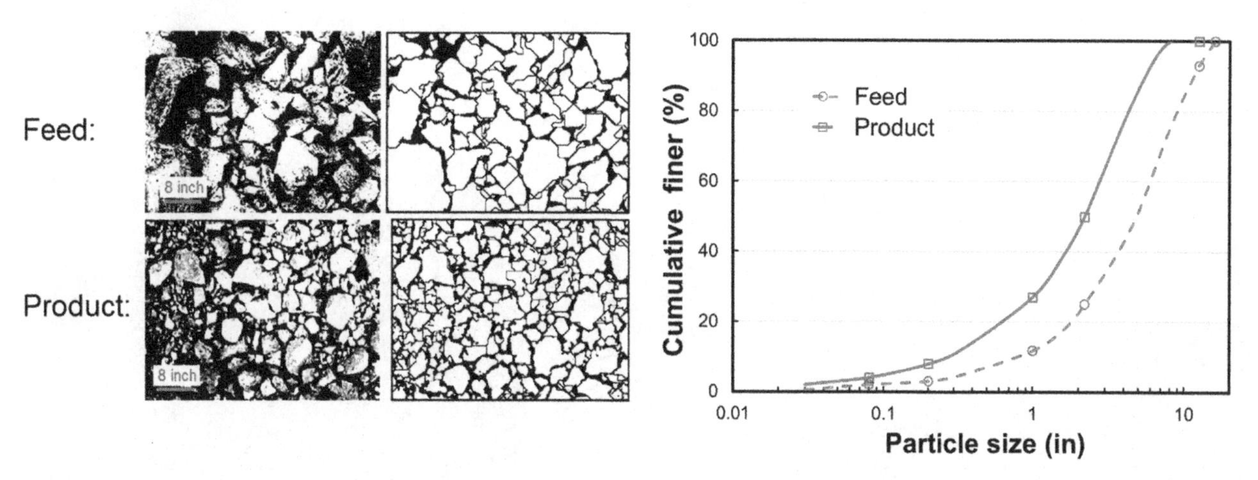

FIGURE 4.21 The original and segmented images of crusher feed and product, with the estimated size distributions (Split Engineering's Split-Online system) *(Courtesy Split Engineering)*.

analysis are modifications and improvements of previous systems, or adaptations of off-line systems for on-line usage.

REFERENCES

Allen, T., 1997. fifth ed. Particle Size Measurement, vol. 1. Chapman and Hall, London. UK.

Anon, 1989a. British Standard 1017-1: 1989. Sampling of coal and coke. Methods for sampling of coal.

Anon, 1989b. British Standard 1796-1: 1989. ISO 2591-1: 1988. Test sieving. Methods using test sieves of woven wire cloth and perforated metal plate.

Anon, 1993. British Standard 3406-4: 1993. Methods for the determination of particle size distribution. Guide to microscope and image analysis methods.

Anon, 2000a. British Standard 410-1: 2000. ISO 3310-1: 2000. Test sieves. Technical requirements and testing. Test sieves of metal wire cloth.

Anon, 2000b. British Standard 410-2: 2000. ISO 3310-2: 2000. Test sieves. Technical requirements and testing. Test sieves of perforated metal plate.

Anon, 2001a. British Standard ISO 13317-1: 2001. Determination of particle size distribution by gravitational liquid sedimentation methods. General principles and guidelines.

Anon, 2001b. British Standard ISO 13317-2: 2001. Determination of particle size distribution by gravitational liquid sedimentation methods. Fixed pipette method.

Austin, L.G., Shah, I., 1983. A method for inter-conversion of microtrac and sieve size distributions. Powder Tech. 35 (2), 271–278.

Austin, L.G., et al., 1984. Process Engineering of Size Reduction: Ball Milling. AIME, New York, NY, USA.

Bernhardt, C., 1994. Particle Size Analysis: Classification and Sedimentation Techniques, vol. 5. Chapman & Hall, London, UK.

Cirulis, D., Russell, J., 2011. Cyclone monitoring system improves operations at KUC's Copperton concentrator. Eng. Min. J (E&MJ). 212 (10), 44–49.

Coghill, P.J., et al., 2002. On-line measurement of particle size in mineral slurries. Miner. Eng. 15 (1-2), 83–90.

Finch, J.A., Leroux, M., 1982. Fine sizing by cyclosizer and micro-sieve. CIM Bull. 75 (839), 235–240.

Harris, C.C., 1971. Graphical presentation of size distribution data: An assessment of current practice. Trans. Inst. Min. Metall., Sec. C. 80, C133–C139.

Hathaway, R.E., Guthnals, D.L., 1976. The continuous measurement of particle size in fine slurry processes. CIM Bull. 766 (Feb.), 64–71.

Heywood, H., 1953. Fundamental principles of sub-sieve particle size measurement. Recent Developments in Mineral Dressing: a Symposium Arranged by the Institution. Institute of Mining and Metallurgy (Great Britain), Imperial College of Science & Technology, London, England. pp.31–58.

Kongas, M., Saloheimo, K., 2009. Online slurry particle size analyzers. In: Malhotra, D., et al., (Eds.), Recent Advances in Mineral Processing Plant Design. SME, Littleton, CO, USA, pp. 419–432.

Maron, R., et al., 2014. Low maintenance, real-time tracking of particle size in individual hydrocyclones for process optimization. Proc. 11th Encuentro International de Mantenedores de Plantas Mineras (MAPLA) Conf., Paper: BIO535, Santiago, Chile. 1–11.

Napier-Munn, T.J., 1985. The determination of the size distribution of ferrosilicon powders. Powder Tech. 42 (3), 273–276.

Napier-Munn, T.J., et al., 1996. Mineral Comminution Circuits: Their Operation and Optimisation (*Appendix 3*). Julius Kruttschnitt Mineral Research Centre (JKMRC), The University of Queensland, Brisbane, Australia.

Price, E.W., 1962. Stokes' law and the Haultain infrasizer unit. Ind. Eng. Chem. Process Des. Dev. 1 (1), 79–80.

Rosin, P., Rammler, E., 1933. The laws governing the fineness of powdered coal. J. Institute Fuel. 7, 29–36.

Schuhmann, R., Jr., 1940. Principles of comminution, I-Size distribution and surface calculations. Technical Publication. No. 1189, AIME, 1–11.

Smith, A., et al., 2010. Ultrasonic extinction for full concentration, real time particle size analysis in the mining industry. Proc. 2nd International Congress on Automation in the Mining Industry (Automining), Santiago Chile. 1–10.

Stokes, S.G.G., 1891. Mathematical and Physical Papers, vol. 3. Cambridge University Press Warehouse, London, UK.

Westendorf, M., et al., 2015. Managing cyclones: A valuable asset, the Copper Mountain case study. Mining Eng. 67 (6), 26–41.

Chapter 5

Comminution

5.1 INTRODUCTION

Because most minerals are finely disseminated and intimately associated with the gangue, they must be initially "unlocked" or "liberated" before separation can be undertaken. This is achieved by *comminution*, in which the particle size of the ore is progressively reduced until the clean particles of mineral can be separated by such methods as are available. Comminution in its earliest stages is carried out in order to make the freshly mined material easier to handle by scrapers, conveyors, and ore carriers, and in the case of quarry products to produce material of controlled particle size.

Explosives are used in mining to remove ores from their natural beds, and blasting can be regarded as the first stage in comminution. Comminution in the mineral processing plant, or "mill," takes place as a sequence of crushing and grinding processes.

Crushing reduces the particle size of run-of-mine (ROM) ore to such a level that grinding can be carried out until the mineral and gangue are substantially produced as separate particles. Crushing is accomplished by compression of the ore against rigid surfaces, or by impact against surfaces in a rigidly constrained motion path. A range of crushing machines is available (Chapter 6). Crushing is usually a dry process, and is performed in several stages, *reduction ratios* (feed size to product size) being small, ranging from three to six in each stage. The reduction ratio of a crushing stage can be defined as the ratio of maximum particle size entering to maximum particle size leaving the crusher, although other definitions are used (Chapter 6).

Conventional grinding takes place in tumbling mills where the ore is introduced into a horizontal mill where the cylindrical body of the mill is turned by a motor, causing the *mill charge* of ore and grinding media to tumble. Grinding is accomplished by impact, attrition and abrasion of the ore by the free motion of unconnected media such as steel rods, steel or ceramic balls, or coarse ore pebbles. Grinding is usually performed "wet" to provide a slurry feed to the concentration process, although dry grinding has various applications.

Primary autogenous or semi-autogenous mills are tumbling mills capable of grinding very coarse feed, thereby replacing one or two stages of crushing. There is an overlapping set of particle sizes where it is possible to crush or grind the ore. From a number of case studies, it appears that at the fine end of crushing operations, equivalent reduction can be achieved for roughly half the energy and costs required by tumbling mills (Flavel, 1978) but at the cost of a more complicated and expensive circuit that is less compact (Barratt and Sochocky, 1982; Söderlund et al., 1989) and with higher maintenance costs (Knight et al., 1989). Recent developments in crusher technology are re-addressing this energy advantage of fine crushing (Chapter 6).

A relatively new comminution device that is somewhat intermediate between fine crushing and coarse tumbling mills is the high pressure grinding rolls (HPGR). These dry crushing devices utilize two rotating rolls creating compression breakage of a particle bed, in which inter-particle breakage occurs (Schönert, 1988). Some evidence has also been reported for downstream benefits such as increased grinding efficiency and improved leachability due to microcracking (Knecht, 1994). The HPGR offers a realistic potential to markedly reduce the comminution energy requirements needed by tumbling mills. Reports have suggested the HPGR to be between 20% and 50% more efficient than conventional crushers and mills, although circuits require additional ore conveying (Doll, 2015). Pre-treatment of ball mill feed by HPGR may reduce total combined energy needed substantially, and has become standard practice in the cement industry. Several hard-rock mineral processing plants with competent ore have opted for HPGR (Amelunxen and Meadows, 2011; Burchardt et al., 2011; Morley, 2011). The HPGR technology is described in Chapter 6.

Stirred mills are now common in mineral processing for fine grinding, though they have been used in other industries for many years (Stehr and Schwedes, 1983). They represent the broad category of mills that use a stirrer to impart motion to the steel, ceramic, or other fine particle media. Both vertical and horizontal configurations exist. Compared to ball mills, stirred mills emphasize shear energy rather than impact energy, which coupled

Wills' Mineral Processing Technology.

with the fine media size, is more energy efficient for fine grinding, below, say, $P_{80} = 50\ \mu m$ (i.e., lower specific energy, kWh t^{-1}) (Stief et al., 1987; Mazzinghy et al., 2012). Stirred mills also provide higher *power intensity* (power per unit volume of mill, kW m^{-3}) than ball mills, making for more compact units (Nesset et al., 2006). Stirred mills are described in Chapter 7.

5.2 PRINCIPLES OF COMMINUTION

Most minerals are crystalline materials in which the atoms are regularly arranged in three-dimensional arrays. The configuration of atoms is determined by the size and types of physical and chemical bonds holding them together. In the crystalline lattice of minerals, these inter-atomic bonds are effective only over small distances, and can be broken if extended by a tensile stress. Such stresses may be generated by tensile or compressive loading (Figure 5.1).

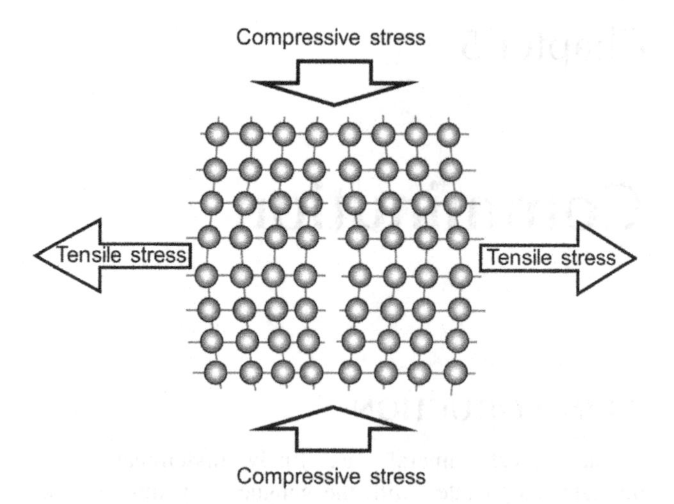

FIGURE 5.1 Strain of a crystal lattice resulting from tensile or compressive stresses.

Even when rocks are uniformly loaded, the internal stresses are not evenly distributed, as the rock consists of a variety of minerals dispersed as grains of various sizes. The distribution of stress depends upon the mechanical properties of the individual minerals, but more importantly upon the presence of cracks or flaws in the matrix, which act as sites for stress concentration (Figure 5.2).

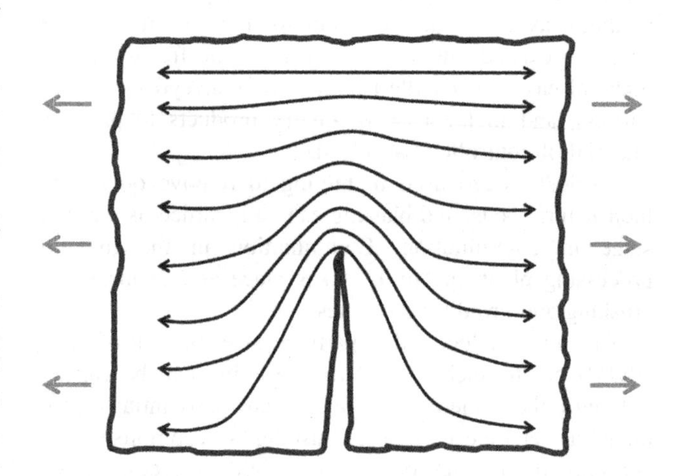

FIGURE 5.2 Stress concentration at a crack tip.

It has been shown (Inglis, 1913) that the increase in stress at such a site is proportional to the square root of the crack length perpendicular to the stress direction. Therefore, there is a critical value for the crack length at any particular level of stress at which the increased stress level at the crack tip is sufficient to break the atomic bond at that point. Such rupture of the bond will increase the crack length, thus increasing the stress concentration and causing a rapid propagation of the crack through the matrix, thus causing fracture.

Although the theories of comminution assume that the material is brittle, crystals can store energy without breaking, and release this energy when the stress is removed. Such behavior is known as *elastic*. When fracture does occur, some of the stored energy is transformed into free surface energy, which is the potential energy of atoms at the newly produced surfaces. Due to this increase in surface energy, newly formed surfaces are often more chemically active, being more amenable to the action of flotation reagents, for example. They also oxidize more readily in the case of sulfide minerals.

Griffith (1921) showed that materials fail by crack propagation when this is energetically feasible, that is, when the energy released by relaxing the strain energy is greater than the energy of the new surface produced. Brittle materials relieve the strain energy mainly by crack propagation, whereas "tough" materials can relax strain energy without crack propagation by the mechanism of plastic flow, where the atoms or molecules slide over each other and energy is consumed in distorting the shape of the material. Crack propagation can also be inhibited by encounters with other cracks or by meeting crystal boundaries. Fine-grained rocks, such as taconites, are therefore usually tougher than coarse-grained rocks.

The energy required for comminution is reduced in the presence of water, and can be further reduced by chemical additives which adsorb onto the solid (Hartley et al., 1978). This may be due to the lowering of the surface energy on adsorption providing that the surfactant can penetrate into a crack and reduce the bond strength at the crack tip before rupture.

Natural particles are irregularly shaped, and loading is not uniform but is achieved through points, or small areas, of contact. Breakage is achieved mainly by crushing, impact and attrition, and all three modes of fracture (compressive, tensile and shear) can be discerned depending

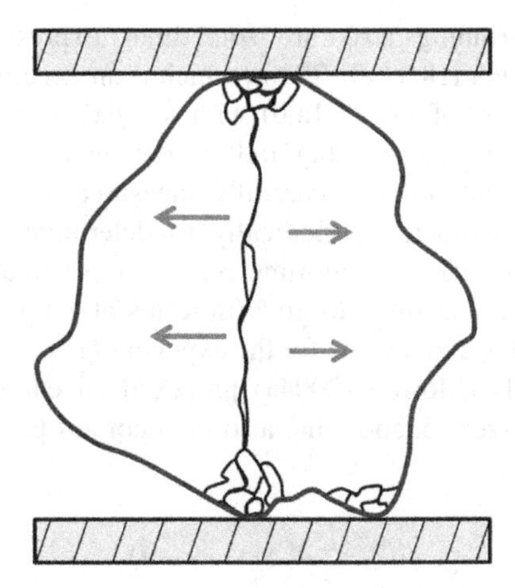

FIGURE 5.3 Fracture by crushing.

on the rock mechanics and the type of loading, that is, the comminution device.

When an irregular particle is broken by compression, or crushing, the products fall into two distinct size ranges—coarse particles resulting from the induced tensile failure, and fines from compressive failure near the points of loading, or by shear at projections (Figure 5.3). The amount of fines produced can be reduced by minimizing the area of loading, and this is often done in compressive crushing machines by using corrugated crushing surfaces (Partridge, 1978; Chapter 6).

In impact breakage, due to the rapid loading, a particle experiences a higher average stress while undergoing more strain than is necessary to achieve simple fracture, and tends to break apart rapidly, mainly by tensile failure. The products from impact breakage when failure occurs are typically linear when cumulative finer is plotted in log–log space (e.g., Figure 4.7). If impact energy is lower than the yield stress, fracture may not occur but chipping, where edges are broken away, does.

Attrition or abrasion is not strictly a breakage event but rather a surface phenomenon where shear stress causes material to abrade off. It produces much fine material, and may be undesirable depending on the comminution stage and industry sector. Attrition occurs due to particle–particle interaction and in stirred mills due to media–particle interaction as well; attrition seems to be the main mechanism in stirred mills.

5.3 COMMINUTION MODELING

Models for comminution processes fall into the three categories: empirical, phenomenological, and fundamental. Under empirical are the *energy-based models*. These are used in preliminary and detailed design, for assessing comminution efficiency, and for geometallurgy (distributing hardness data into mine block models, Chapter 17). Phenomenological models include *population balance models*, which have some underlying principles but with parameters fitted by calibration for given comminution devices. They are widely used in unit and circuit simulators. Lastly, the fundamental models include those based on discrete element method (DEM), computational fluid dynamics (CFD) and, some would argue, statistical physics (Martins and Radziszewski, 2014). They are computationally intensive, used for design and optimization of specific components of crushers, tumbling mills, and material handling systems.

5.3.1 Energy-based Comminution Models

This family of models is the oldest of the comminution models and they continue to find widespread use (Morrell, 2014a). Energy-based models assume a relationship between energy input of the comminution device and the resultant effective particle size of the product. Many rely on the feed and product size distributions being self-similar; that is, parallel when cumulative finer is plotted in log-log space (Chapter 4). The energy input is for net power, that is, after correcting for motor efficiency and drive train mechanical losses. Typically, energy is measured as kWh t^{-1} or Joules, depending on the model.

The most familiar general energy-based comminution equation is that first presented by Walker et al. (1937). In differential form it is:

$$dE = -K\,x^{-n}\,dx \tag{5.1a}$$

where dE is the increment in energy to effect an incremental decrease dx in particle size x, and n and K are constants (K also converts units). In integral form it is:

$$E = -K\int_F^P x^{-n}dx \tag{5.1b}$$

where F and P are some measure of feed and product particle size (usually a diameter). Letting $n = 2$, 1 or 3/2 gives the "three comminution laws" due to Von Rittinger, Kick, and Bond.

The oldest theory, Von Rittinger (1867), stated that the energy consumed in size reduction is proportional to the area of new surface produced. The surface area of a known weight of particles of uniform diameter is inversely proportional to the diameter, hence Von Rittinger's law equates to:

$$E_{VR} = K\left(\frac{1}{P} - \frac{1}{F}\right) \tag{5.2}$$

which is the solution of Eq. (5.1b) for $n = 2$.

The second oldest theory, Kick (1885), stated that the energy required is proportional to the reduction in volume of the particles. Solving Eq. (5.1b) for $n = 1$ gives one form of the Kick equation:

$$E_K = K\left(\ln\frac{F}{P}\right) \tag{5.3}$$

where F/P is the reduction ratio. Kick's theory is that energy requirement depends only the reduction ratio, and not on the original size of the particles.

As Lynch and Rowland (2005) note, the means to make measurements of energy and size necessary to validate the Von Rittinger and Kick models did not exist until the middle of the twentieth century when electrical motors and precision laboratory instruments became available. The literature from this period includes work by a group at the Allis Chalmers Company who were trying to calibrate Von Rittinger's equation to industrial rod mills (Bond and Maxson, 1938; Myers et al., 1947).

Often referred to as the "third theory", Bond (1952) stated that the energy input is proportional to the new crack tip length produced in particle breakage. Bond redefined his "theory" to rather be an empirical relationship in a near-final treatise (Bond, 1985). The equation is commonly written as:

$$W = 10 \times Wi\left(\frac{1}{\sqrt{P_{80}}} - \frac{1}{\sqrt{F_{80}}}\right) \tag{5.4}$$

where W is the energy input (work) in kilowatt hours per metric ton (or per short ton in Bond's original publications), Wi is the *work index* (or *Bond work index*) in kilowatt hours per metric ton, and P_{80} and F_{80} are the 80% product and feed passing sizes, in micrometers.

Solving Eq. (5.1b) for $n = 3/2$ gives the same form as Eq. (5.4) with the constant 2 K ahead of the bracket. In effect the 2 K is replaced by ($10 \times Wi$), which is convenient because Wi becomes equal to W in the case of grinding from a theoretical infinite feed size to 80% passing 100 μm. The Bond model remains the most widely used, at least for the "conventional" comminution equipment in use at the time Bond developed the model and calibrated it against industrial data. It is one reason that the 80% passing size became the common single point metric ("mean") of a particle size distribution.

A modification of Eq. (5.1a,b) was proposed by Hukki (1962), namely substituting n by a function of particle size, $f(x)$. This provoked debate over the size range that the three established models applied to. What can be agreed is that all the models predict that energy consumption will increase as product particle size (i.e., P) decreases. Typical specific energy values (in kWh t^{-1}) are (Morrell, 2014b): primary crushing (i.e., 1000-100 mm), 0.1-0.15; secondary crushing (100-10 mm), 1-1.2; coarse grinding (10-1 mm), 3-3.5; and fine grinding (1-0.1 mm), 10.

Fine grinding tests are sometimes expressed as a *signature plot* (He et al., 2010), which is an experimentally fitted version of Eq. (5.1a,b) with $n = f(x)$. A laboratory test using a fine grinding mill is conducted where the energy consumption is carefully measured and a slurry sample is extracted periodically to determine the 80% passing size. The energy-time relationship versus size is then plotted and fitted to give (in terms of Eq. (5.1a,b)) a coefficient K and a value for the exponent $f(x)$.

Recently, Morrell (2004a) proposed an energy-based model of size reduction that also incorporates particle size dependence:

$$dE = -M\ g(x)^{-f(x)}\ dx \tag{5.5}$$

where $g(x)$ is a function describing the variation in breakage properties with particle size, and M a constant related to the breakage properties of the material.

The problem that occurs when trying to solve Eq. (5.5) is the variable nature of the function $g(x)$. A pragmatic approach was to assume M is a constant over the normal range of particle sizes treated in the comminution device and leave the variation in size-by-size hardness to be taken up by $f(x)$. Morrell (2009) gives the following:

$$W = M_i 4(P^{f(P)} - F^{f(F)}) \tag{5.6}$$

where M_i is the work index parameter related to the breakage property of an ore and the type of comminution machine, W is the specific comminution energy (kWh t^{-1}), P and F are the product and feed 80% passing size (μm), and $f(x)$ is given by (Morrell, 2006):

$$f(x) = \left(0.295 + \frac{x}{10^6}\right) \tag{5.7}$$

The parameter M_i takes on different values depending on the comminution machine: M_{ia} for primary tumbling mills (AG/SAG mills) that applies above 750 μm; M_{ib} for secondary tumbling mills (e.g., ball mills) that applies below 750 μm; M_{ic} for conventional crushers; and M_{ih} for HPGRs. The values for M_{ia}, M_{ic}, and M_{ih} were developed using the SMC Test® combined with a database of operating comminution circuits. A variation of the Bond laboratory ball work index test was used to determine values of M_{ib}. This is similar to the approach Bond used in relating laboratory results to full scale machines. The methodology continues to be refined as the database expands (Morrell, 2010).

Morrell (2009) gave a worked example comparing the energy requirements for three candidate circuits to illustrate the calculations. Taking just the example for the fine particle tumbling mill serves that purpose here (Example 5.1).

Example 5.1

An SMC Test® and Bond ball mill work index test (with closing screen 150 μm) were performed on an ore sample with the following results:

$$M_{ia} = 19.4\ \text{kWh t}^{-1};\ M_{ic} = 7.2\ \text{kWh}^{-1};$$
$$M_{ih} = 13.9\ \text{kWh}^{-1};\ \ M_{ib} = 18.8\ \text{kWh t}^{-1}.$$

Calculate the specific grinding energy for the fine particle tumbling mill to give a product P_{80} of 106 μm.

Solution

From the M_i data the relevant value is $M_{ib} = 18.8$ kWh t^{-1}. Noting it is fine grinding then the feed F_{80} is taken as 750 μm. Combining Eqs. (5.6) and (5.7) and substituting the values:

$$W = 18.8 \times 4 \times (106^{-(0.295+750/1{,}000{,}000)} - 750^{-(0.295+750/1{,}000{,}000)})$$
$$= 8.4(kWh\ t^{-1})$$

5.3.2 Breakage Characterization

Using the energy-based models requires estimates of parameters that relate to the ore, variously expressed as hardness, grindability, or resistance to breakage. This has given rise to a series of procedures, summarized by Mosher and Bigg (2002) and Morrell (2014a).

Bond Tests

The most widely used parameter to measure ore hardness is the Bond work index *Wi*. Calculations involving Bond's work index are generally divided into steps with a different *Wi* determination for each size class. The *low energy crushing work index* laboratory test is conducted on ore specimens larger than 50 mm, determining the crushing work index (Wi_C, *CWi* or *IWi* (impact work index)). The *rod mill work index* laboratory test is conducted by grinding an ore sample prepared to 80% passing 12.7 mm (½ inch, the original test being developed in imperial units) to a product size of approximately 1 mm (in the original and still the standard, 14 mesh; see Chapter 4 for definition of mesh), thus determining the rod mill work index (Wi_R or *RWi*). The *ball mill work index* laboratory test is conducted by grinding an ore sample prepared to 100% passing 3.36 mm (6 mesh) to product size in the range of 45-150 μm (325-100 mesh), thus determining the ball mill work index (Wi_B or *BWi*). The work index calculations across a narrow size range are conducted using the appropriate laboratory work index determination for the material size of interest, or by chaining individual work index calculations using multiple laboratory work index determinations across a wide range of particle size.

To give a sense of the magnitude, Table 5.1 lists Bond work indices for a selection of materials. For preliminary design purposes such reference data are of some guide but measured values are required at the more advanced design stage.

TABLE 5.1 Selection of Bond Work Indices (kWh t^{-1})

Material	Work Index	Material	Work Index
Barite	4.73	Fluorspar	8.91
Bauxite	8.78	Granite	15.13
Coal	13.00	Graphite	43.56
Dolomite	11.27	Limestone	12.74
Emery	56.70	Quartzite	9.58
Ferro-silicon	10.01	Quartz	13.57

A major use of the Bond model is to select the size of tumbling mill for a given duty. (An example calculation is given in Chapter 7.) A variety of *correction factors* (*EF*) have been developed to adapt the Bond formula to situations not included in the original calibration set and to account for relative efficiency differences in certain comminution machines (Rowland, 1988). Most relevant are the EF_4 factor for coarse feed and the EF_5 factor for fine grinding that attempt to compensate for sizes ranges beyond the bulk of the original calibration data set (Bond, 1985).

The standard Bond tumbling mill tests are time-consuming, requiring locked-cycle testing. Smith and Lee (1968) used batch-type tests to arrive at the work index; however, the grindability of highly heterogeneous ores cannot be well reproduced by batch testing.

Berry and Bruce (1966) developed a comparative method of determining the hardness of an ore. The method requires the use of a reference ore of known work index. The reference ore is ground for a certain time (*T*) in a laboratory tumbling mill and an identical weight of the test ore is then ground for the same time. Since the power input to the mill is constant (*P*), the energy input ($E = P \times T$) is the same for both reference and test ore. If *r* is the reference ore and *t* the ore under test, then we can write from Bond's Eq. (5.4):

$$W_r = W_t = Wi_r\left[\frac{10}{\sqrt{P_{80r}}} - \frac{10}{\sqrt{F_{80r}}}\right] = Wi_t\left[\frac{10}{\sqrt{P_{80t}}} - \frac{10}{\sqrt{F_{80t}}}\right]$$

Therefore:

$$Wi_t = Wi_r \frac{\left[\dfrac{10}{\sqrt{P_{80r}}} - \dfrac{10}{\sqrt{F_{80r}}}\right]}{\left[\dfrac{10}{\sqrt{P_{80t}}} - \dfrac{10}{\sqrt{F_{80t}}}\right]} \tag{5.8}$$

Reasonable values for the work indices are obtained by this method as long as the reference and test ores are ground to about the same product size distribution.

Work indices have been obtained from grindability tests on different sizes of several types of equipment, using identical feed materials (Lowrison, 1974). The values of work indices obtained are indications of the efficiencies of the machines. Thus, the equipment having the highest indices, and hence the largest energy consumers, are found to be jaw and gyratory crushers and tumbling mills; intermediate consumers are impact crushers and vibration mills, and roll crushers are the smallest consumers. The smallest consumers of energy are those machines that apply a steady, continuous, compressive stress on the material.

A class of comminution equipment that does not conform to the assumption that the particle size distributions of a feed and product stream are self-similar includes autogenous mills (AG), semi-autogenous (SAG) mills and high pressure grinding rolls (HPGR). Modeling these machines with energy-based methods requires either recalibrating equations (in the case of the Bond series) or developing entirely new tests that are not confused by the non-standard particle size distributions.

Drop Weight Tests

In contrast to the Bond tests which yield a single work index value, drop weight tests are designed to produce more detailed information on the size distributions produced on breaking rocks and their relationship to input energy. This was required for comminution circuit simulations, one of the aims being the prediction of size distribution from comminution units to link to other units, such as classifiers. The JK Drop-weight Test, for example, is linked to the JK comminution models (Morrell, 2014b).

The JK Drop-weight Test is conducted on single-size particles ranging from $-63+53$ mm to $-16+13.2$ mm. Each size class is broken at three input energies and the progeny particles sized. From the size distribution the fraction less than 1/10th the original size is determined, the t_{10}. The t_{10} is related to the associated input specific energy (*Ecs*, kWh t^{-1}) by the following:

$$t_{10} = A[1 - e^{(-b.Ecs)}] \tag{5.9}$$

where A and b are dependent on the ore properties (type, size, etc.). The product $A \times b$ can be correlated with other comminution system variables: a high $A \times b$, for example, indicates material easy to break and corresponds to a low work index, *Wi*. The t_{10} can be related to other t_n values. This approach is described in detail by Napier-Munn et al. (1996).

A drawback of the drop-weight test is that usually over 60 kg of rock are required for a single test. This problem is addressed with the introduction of the SMC Test[R], noted earlier, which reduces the amount of sample required (Morrell, 2004b).

MacPherson Test

Developed by Art MacPherson while at Aerofall Mills Ltd., it uses a 450 mm dry air-swept SAG mill with an 8% ball charge. The test is run in closed circuit to steady state. At completion, the particle size distribution (PSD) is measured. Based on the PSD of feed and product and the measured power input, the *autogenous work index* (*AWi*) is determined. Two drawbacks of the test are: it is restricted to particles finer than 32 mm, and due to the low number of high energy impacts a correction for hard ores is required (Mosher and Bigg, 2002).

SPI and SGI Tests

The SPI™ (SAG Power Index, trademark of SGS Mineral Services and invented by John Starkey) and the generic SGI (SAG Grindability Index) tests are dry batch tests commonly used for ore variability characterization in SAG milling circuits. The main differences between the SPI and the SGI are: 1) sample prep for the test feed is different; 2) the SGI is run progressively with fixed timing. The two, largely interchangeable, batch tests are conducted in a 30.5 cm diameter by 10.2 cm long grinding mill charged with 5 kg of steel balls. Two kilograms of sample are crushed to 100% minus 1.9 cm and 80% minus 1.3 cm and placed in the mill. The test is run until the sample is reduced to 80% minus 1.7 mm. The test result is *M*, the time (in minutes) required to reach a P_{80} of 1.7 mm (Starkey and Dobby, 1996). This time is used to calculate the SAG mill specific energy, E_{SAG}, either via the proprietary SPI equation (Kosick and Bennett, 1999) or a published SGI equation (Amelunxen et al., 2014):

$$E_{SAG} = K\left(\frac{M}{\sqrt{P_{80}}}\right)^n f_{SAG} \tag{5.10}$$

where K and n are empirical factors, and f_{SAG} incorporates a series of calculations that estimate the influence of factors such as pebble crusher recycle load (Chapter 6), ball load, and feed size distribution. Amelunxen et al. (2014) published a set of parameters (calibrated mostly from porphyry ores in the Americas) for the SGI equation where $K = 5.9$, $n = 0.55$ and $f_{SAG} = 1.0$ for a circuit without pebble crushing or $f_{SAG} = 0.85$ for a circuit with pebble crushing.

SAGDesign Test

The Standard Autogenous Grinding Design (SAGDesign) test measures macro and micro ore hardness by means of a SAG mill test and a standard Bond ball mill work index test on SAG ground ore. Ore feed is prepared from a minimum of 10 kg of split diamond drill core samples by stage crushing the ore in a jaw crusher to 80% product passing 19 mm. The SAG test reproduces commercial SAG mill grinding conditions on 4.5 liter of ore and determines the SAG mill specific pinion energy needed to grind ore from 80% passing 152 mm to 80% passing 1.7 mm, herein referred to as macro ore hardness. The SAG mill product is then crushed to 100% passing 3.35 mm and is subjected to a standard Bond ball mill work index (S_d-BWi) grinding test to provide the total pinion energy at the specified grind size for mill design purposes. A calibration equation is used to convert the test result (total number of revolutions to reach the endpoint and the initial mass of feed ore) to the pinion energy. The SAGDesign test has been abbreviated for the needs of geometallurgy, that is, measuring the ore variability in a deposit (Brissette et al., 2014).

Bond-based AG/SAG Models

Several authors have developed autogenous and semiautogenous grinding models that are re-calibrated versions of Bond equations (e.g., Barratt, 1979; Sherman, 2011; Lane et al., 2013). These methods typically partition SAG breakage into ranges of particle sizes where one or another of the Bond work index values is used, the sum of which is then multiplied by empirical adjustment factors to give an estimate of the SAG mill (or SAG-ball mill circuit) specific energy consumption. A detailed example of a Bond-type SAG circuit sizing calculation is provided by Doll (2013).

The validity of the Bond method has been questioned in the case of coarser particle sizes in SAG milling (Millard, 2002). Laboratory testing of rocks at different sizes frequently indicates material reporting harder (higher) crushing work index values at coarser sizes. If the model is valid, then the only way this is possible is if the defects (fractures) where a rock breaks are eliminated at larger sizes, which is plainly not possible. Fortunately, Bond-based models tend to not be sensitive to coarse sized material due to the $(F_{80})^{-½}$ term being small in relation to the $(P_{80})^{-½}$ term.

5.3.3 Population Balance Models

The energy-based models, such as the Bond model, do not predict the complete product size distribution, only the 80% passing size, nor do they predict the effect of operating variables on mill circulating load, nor classification performance. The complete size distribution is required in order to simulate the behavior of the product in ancillary equipment such as screens and classifiers, and for this reason population balance models are used in the design, optimization and control of grinding circuits (Napier-Munn et al., 1996).

Population balance—or mass-size balance—describes a family of modeling techniques that involve tracking and manipulating whole (or partial) particle size distributions as they proceed through the comminution process (or, indeed, any mineral processing unit). These techniques allow grinding circuits to be simulated without the assumption that all particle size distributions have a "normal" shape (as in many energy-based models), but at a cost of being more computationally intensive and requiring more model configuration parameters.

There are various commercially available programs that perform this style of modeling (e.g., USIMpac, JK SimMet, and Moly-Cop Tools). A range of case studies can be found in the literature, covering both design and optimization of grinding circuits. Examples include Richardson (1990), Lynch and Morrell (1992), McGhee et al. (2001), and Dunne et al. (2001).

The general equation describing any comminution device is (Napier-Munn et al., 1996; King, 2012):

$$p_i = T_i f_i \tag{5.11}$$

where p_i, f_i are the fraction in size class i in product and feed, respectively, and T_i is some transfer function. The task is to determine T_i.

This modeling approach is most often applied to tumbling mills. The starting assumption is that the rate of disappearance (rate of breakage) of mass of size class i is proportional to the mass of that size fraction in the mill at time t, $f_i(t)$, that is, a first-order kinetics model:

$$\frac{df_i(t)}{dt} = -S_i f_i(t) \tag{5.12}$$

where S_i is the *specific rate of breakage* (or *selection*) *parameter* of size class i. By convention, the size classes are numbered 1 to n, from coarsest to finest (with $n+1$ being the pan fraction in a typical sieve sizing method).

To complete the balance on any size class, the breakage into and out of that class is required. This introduces a second parameter, the *breakage* (or *breakage appearance*) *parameter*, B:

$$\frac{df_i(t)}{dt} = \sum_{j=1}^{i-1} B_{ij} S_j f_j(t) - S_i f_i(t) \tag{5.13}$$

where B_{ij} describes the fraction of size j ($j < i$) that reports to size i. If we apply the balance around a single size class then the solution to Eq. (5.13) is (dropping the t):

Product out = Feed in + Breakage in − Breakage out

$$p_i = f_i + \sum_{j=1}^{i-1} B_{ij} S_j f_i - S_i f_i \text{ or } \sum_{j=1}^{i-1} B_{ij} S_j f + (1 - S_i) f_i \tag{5.14}$$

TABLE 5.2 Cumulative Mass Retained as a Function of Time (%)

Size (μm)	Time (min)						
	0	2	4	6	8	10	12
600	13.12	3.83	1.66	0.70	0.33	0.15	0.10
300	29.99	13.36	5.49	2.11	0.78	0.31	0.16
150	52.05	35.98	24.24	15.64	8.64	5.05	2.81
75	77.10	66.30	56.71	48.37	40.16	33.05	27.60
37	87.62	81.52	75.02	69.88	64.49	59.97	55.69

Provided we know the *B* and *S* parameters as a function of size, we can predict the product size distribution.

The simplest way to test the model is to consider the rate of disappearance of the top size class. Writing Eq. (5.12) for size class (sieve) 1 and assuming all particles have the same residence time (as in batch grinding) we derive:

$$\ln\left(\frac{f_1(t)}{f_1(0)}\right) = -S_1 t \tag{5.15}$$

where $f_1(0)$ and $f_1(t)$ are the initial weight and weight retained after *t* grind time, and S_1 is the rate of breakage parameter for size class 1. Table 5.2 and Figure 5.4 show the result of a laboratory batch ball-mill grinding test on a sample of Pb-Zn ore taken from a ball mill feed. Figure 5.4 shows that the rate of disappearance of "coarse" material does obey the first order rate model; the slope is *S* and we note that *S* increases with increasing particle size (usually to a maximum depending on factors such as media size to particle size). The equation could have been written in terms of fine material: the rate of appearance of fines would also follow the first-order model. This is a common finding for batch tests (Austin et al., 1984) but first order appearance of fines has also long been recognized in continuous milling (Dorr and Anable, 1934).

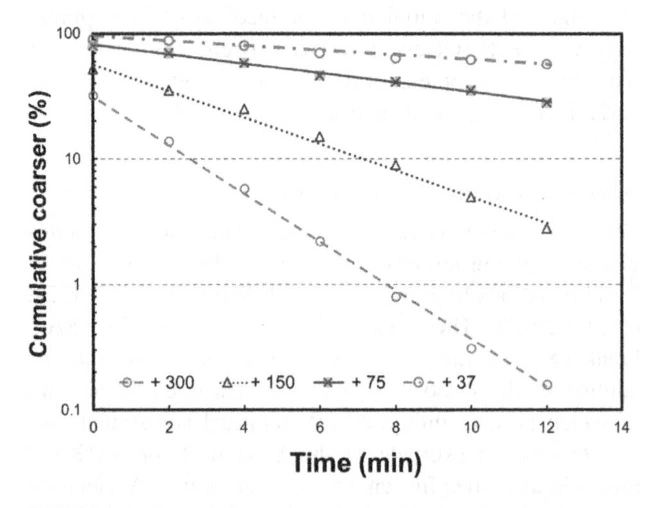

FIGURE 5.4 Test of first order grinding kinetics, the "disappearance plot" (Eq. (5.15)).

Figure 5.5 shows the general relationship for the rate of breakage parameter as a function of particle size for ball milling and SAG milling. The maximum is related to the ball size to particle size ratio passing through an optimum (at too large a particle size the balls are too small for effective breakage). The minimum is related to the "critical size" observed in AG/SAG milling where particles of this size self-break slowly and build up in the mill requiring extraction and crushing to eliminate (Chapters 6 and 7).

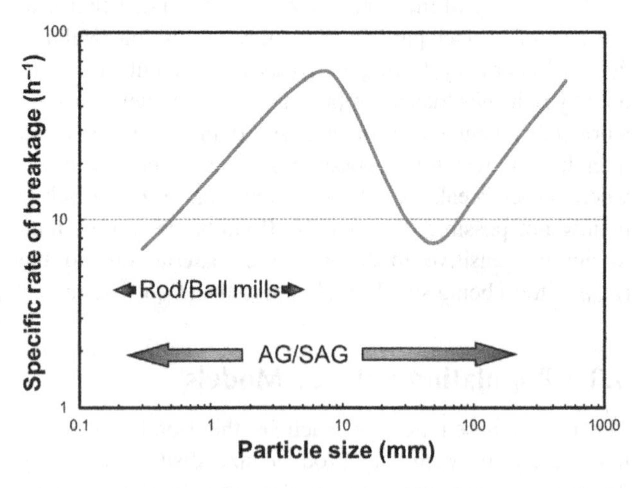

FIGURE 5.5 General result for specific rate of breakage as a function of particle size.

The model (Eq. (5.14)) can be combined with information on material transport, measured by the distribution of residence times in the mill, and mill discharge characteristics to provide a description of open-circuit grinding, which can be coupled with information concerning the classifier to produce closed-circuit grinding conditions (Napier-Munn et al., 1996). A simplification for closed circuit ball mill grinding is that transport can be approximated by plug flow (Furuya et al., 1971). In that case Eq. (5.15), the solution for batch grinding and hence also plug flow, applies. (See Chapter 12 for more discussion on material transport.) These models can only realize their full potential, however, if there are accurate methods of estimating the B and S model parameters (or functions). The complexity of the breakage environment in a tumbling mill precludes the calculation of these values from first principles, so that successful application depends on the development of techniques for the calibration of model parameters to experimental data. There are methods to estimate the selection and breakage functions simultaneously (e.g., Rajamani and Herbst, 1984). Commercial simulators often use a default breakage function, which is a drawback. Refinements to determine S and B continue to evolve (Shi and Xie, 2015).

Experimental characterization of the properties of heterogeneous ores is a challenge in population-type models. Although breakage characteristics for homogeneous materials can be determined on a small scale and used to predict large-scale performance, it is more difficult to predict the behavior of mixtures of two or more components. Furthermore, the relationship of material size reduction to subsequent processing is even more difficult to predict (e.g., using the output of population balance comminution models to predict flotation characteristics) due to the complexities of mineral release (liberation). Work has advanced on the development of grinding models that include mineral liberation in the size-reduction description (Choi et al., 1988; Herbst et al., 1988). An entropy-based multiphase approach, which models particles individually rather than the standard approach of using composite classes, is seen as an advance (Gay, 2004). The development of liberation models is essential if simulation of integrated plants is to be realized.

Linking to Energy

The mass-size balance models as written above are in the time-domain. To be more practical they need to be converted to the energy-domain. One way is by arguing that the specific rate of breakage parameter is proportional to the net specific power input to the mill charge (Herbst and Fuerstenau, 1980; King, 2012). For a batch mill this becomes:

$$S_i = S_i^E \frac{P}{M} \tag{5.16}$$

where S_i^E is the energy-specific rate of breakage parameter, P the net power drawn by the mill, and M the mass of charge in the mill excluding grinding media (i.e., just the ore). The energy-specific breakage rate is commonly given in t kWh^{-1}. For a continuous mill, the relationship is:

$$S_i \tau = S_i^E \frac{P}{F} \tag{5.17}$$

where τ is the mean retention time, and F the solids mass flow rate through the mill. Assuming plug flow, Eq. (5.17) can be substituted into Eq. (5.15) to apply to a grinding mill in closed circuit (where $t = \tau$).

Simplified Grinding Model

The result in Figure 5.5 suggests a simplified approach to grinding models by basing on disappearance of "coarse" without reference to where the products end up, that is, eliminating the B function. Given that the coarse fraction is determined by cumulating the data above a given size (see Table 5.2), the S−parameter is no longer size discrete but on a cumulative basis. Thus, the model is also referred to as the *cumulative-basis grinding model.*

Finch and Ramirez-Castro (1981) adopted the cumulative basis model in order to extend the grinding model to mineral classes, which otherwise would require the estimation of the B function for these classes, making an already difficult task essentially insurmountable. Applied to a Pb-Zn ore, they found that mineral grinding rates were similar but classification was highly dependent on mineral density, which dominated the mineral size distributions in the ball mill-cyclone circuit (see also Chapter 9). Hinde and Kalala (2009) used the cumulative-basis model for tumbling mills, stirred mills and HPGRs in assessing competing circuit arrangements. They converted to the energy-domain by adapting the approach of Herbst and Fuerstenau (1980) (Eq. (5.16)) and confirmed the model in batch grinding tests. Jankovic and Valery (2013), in essence, used the cumulative-basis (or "coarse disappearance") model in assessing closed circuit ball mill operation.

5.3.4 Fundamental Models

Use of Discrete Element Modeling (DEM) and Computational Fluid Dynamics (CFD), while becoming routine in modeling many unit operations in mineral processing (Chapter 17), has arguably been most widely applied to comminution units. The computationally-intensive technique combines detailed physical models to describe the motion of balls, rocks, and slurry and attendant breakage of particles as they are influenced by moving liners/lifters and grates. Powell and Morrison (2007) make the case that the future of comminution modeling must include these advanced techniques that the ever-increasing computational power brings into reach.

These advanced modeling techniques applied to tumbling mills are leading to improved understanding of charge dynamics, and to improved designs of mill internals. There is potential to reduce downtime and increase the efficient use of energy. The approach has been shown to reliably model breakage in tumbling mills (Carvalho and Tavares, 2010), and is in common use simulating breakage in crushers (Quist and Evertsson, 2010; Kiangi et al., 2012). Applied to stirred mills it has helped identify the importance of shear in size reduction (Radziszewski, 2013a). Incorporating slurry properties, moving to simulation of continuous operation (simulations are currently batch and the introduction of feed is a disturbance to charge motion), and direct prediction of particle breakage are some of the areas of active research.

One of the features of three-dimensional DEM simulation is the cutaway images of particle motion in the mill, an example being in Figure 5.6 for a 1.8 m diameter pilot SAG mill. Radziszewski and Allen (2014) provide a series of images of charge motion in stirred mills.

Validation of the predictions made by DEM/CFD models is critical to verifying the simulations. Visual inspection through a transparent end wall is commonly used (e.g., Maleki-Moghaddam et al., 2013). Figure 5.7 shows a good agreement between simulation and experiment for charge motion for a scaled-down SAG mill. Positron Emission Particle Tracking (PEPT, Chapter 17) is a powerful non-invasive tool for validation but is restricted by the size of the scanner to relatively small test units. Acoustic monitoring of industrial-scale units provides another non-invasive opportunity to test model predictions, for example, detecting where balls strike the shell (Pax, 2012).

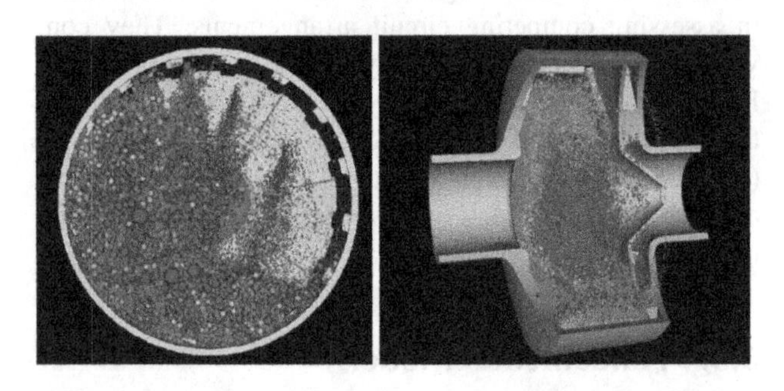

FIGURE 5.6 Example of ball and particle motion in a slice of a pilot SAG mill using three-dimensional DEM simulation *(Courtesy CSIRO (Dr. Paul Cleary))*.

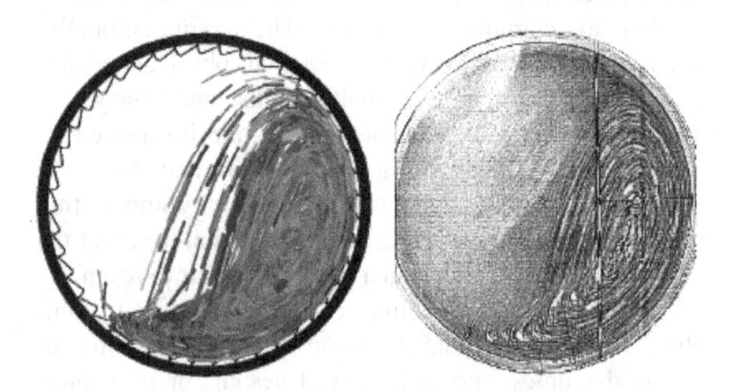

FIGURE 5.7 Comparison of three-dimensional DEM with experiment for 75% critical speed, scale model 0.6 m diameter SAG mill *(Courtesy CSIRO (Dr. Paul Cleary))*.

5.4 COMMINUTION EFFICIENCY

Comminution consumes the largest part of the energy used in mining operations, from 30 to 70% (Radziszewski, 2013b; Nadolski et al., 2014). This has consequently drawn most of the sustainability initiatives designed to reduce energy consumption in mining, including, for example, the establishment of CEEC (Coalition for Eco-efficient Comminution, www.ceecthefuture.org) and GMSG (Global Mining Standards Guidelines Group, www.globalminingstandards.org).

One approach is to ask if all the ore needs fine grinding, where the bulk of the energy is consumed. Certainly in the case of low grade ores, much of the gangue can be liberated at quite coarse size and its further size reduction would represent inefficient expenditure of comminution energy (Chapter 1). This provides an opportunity for ore sorting, for example (Chapter 14). Lessard et al. (2014) provide case studies showing the impact on comminution energy of including ore sorting on crusher products ahead of grinding. Similar in objective, the development of coarse particle flotation technologies (Chapter 12) aims to reduce the mass of material sent to the fine grinding (liberation) stage.

This still leaves the question of how efficiently the energy is used in comminution. It is common experience that significant heat is generated in grinding (in particular) and can be considered a loss in efficiency. However, it may be that heat is an inevitable consequence of breakage. Capturing the heat could even be construed as a benefit (Radziszewski, 2013b).

A fundamental approach to assessing energy efficiency is to compare input energy relative to the energy associated with the new surface created. This increase in surface energy is calculated by multiplying the area of new surface created (m^2) by the surface tension expressed as an energy ($J\ m^{-2}$). On this basis, efficiency is calculated to be as little as 1% (Lowrison, 1974). This may not be an entirely fair basis to evaluate as we suspect that some input energy goes into deforming particles and creating micro-cracks (without breakage) and that the new surface created is more energetic than the original surface, meaning the surface tension value may be underestimated. In addition, both these factors may provide side-benefits of comminution. Rather than this comparison as a basis, other measures use a comparison against a "standard."

Possible Standards

Single particle slow compressive loading is considered about the most energy efficient way to comminute.

Comparing to this basis, Fuerstenau and Abouzeid (2002) found that ball milling quartz was about 15% energy efficient. From theoretical reasoning, Tromans (2008) estimated the energy associated with breakage by slow compression and showed that relative to this value the efficiency of creating new surface area could be as high as 26%, depending on the mineral.

Nadolski et al. (2014) propose an "energy benchmarking" measure based on single particle breakage. A methodology derived from the JK Drop-weight Test is used to determine the limiting energy required for breakage, termed the "essential energy." They derive a comminution benchmark energy factor (BEF) by dividing actual energy consumed in the comminution machine by the essential energy. They make the point that the method is independent of the type of equipment, and can be used to include energy associated with material transport as well to assess competing circuit designs.

A standard already in use is the Bond work index against which the *operating work index* can be compared.

Operating Work Index Wi_O

This is obtained for a comminution device using Eq. (5.4), by measuring W (the specific energy being used, kWh t^{-1}), F_{80} and P_{80} and solving for Wi as the operating work index, Wi_O. (Note that the value of W is the power applied to the pinion shaft of the mill. Motor input power thus has to be converted to power at the mill pinion shaft output by applying corrections for electrical and mechanical losses between the power measurement point and the shell of the mill.) The ratio of laboratory determined work index to operating work index, $Wilab{:}Wi_O$, is the measure of efficiency relative to the standard: for example, if $Wilab{:}Wi_O < 1$, the unit or circuit is using more energy than predicted by the standard test, that is, it is less efficient than predicted. Values of $Wilab{:}Wi_O$ obtained from specific units can be used to assess the effect of operating variables, such as mill speed, size of grinding media, type of liner, etc. Note, this means that the Bond work index has to be measured each time the comparison is to be made. An illustration of the use of this energy efficiency calculation is provided by Rowland and McIvor (2009) (Example 5.2).

The procedure has the virtue of simplicity and the use of well recognized formulae. Field studies have shown the ratio can vary as much as ±35% from unity (some circuits will operate at greater efficiency than the Bond energy predicts). Rowland and McIvor (2009) discuss some of the limitations of the method. They note that the size distributions from AG/SAG milling can be quite unlike those from rod and ball mills on which the technique originated; and, beyond giving a measure of efficiency, it does not provide any specific indication of the causes of inefficiency. In the case of ball

Example 5.2

a. A survey of a SAG-ball mill circuit processing ore from primary crushing showed size reduction of circuit (SAG) feed F_{80} of approximately 165,000 μm to flotation circuit feed (cyclone overflow) P_{80} of 125 μm. The total specific energy input for the two milling stages was 14.6 kWh t^{-1}. Calculate the operating work index for the circuit.

b. Circuit feed samples taken at the same time were sent for Bond work index testing. The rod mill test gave *RWilab* of 14.5 kWh t^{-1} and a ball mill work index *BWilab* of 13.8 kWh t^{-1}. Accepting that the rod mill work index applies to size reduction of the circuit feed down to the rod mill test product P_{80} of 1050 μm and that the ball mill work index applies from this size to the circuit product size, calculate the standard Bond energy for the circuit.

c. Calculate the combined *Wilab* and the relative efficiency, *Wilab*:*Wio*. What do you conclude?

Solution

a. The appropriate form of Eq. (5.4) is:

$$14.6 = 10 \times Wi_o\left(\frac{1}{\sqrt{125}} - \frac{1}{\sqrt{165000}}\right)$$

$$Wi_o = 16.8\ (\text{kWh t}^{-1})$$

b. Bond energy for both size reduction stages is:

From rod mill test:

$$W = 10 \times 14.5\left(\frac{1}{\sqrt{1050}} - \frac{1}{\sqrt{165000}}\right)$$

$$W = 4.1\ (\text{kWh t}^{-1})$$

From ball mill test:

$$W = 10 \times 13.8\left(\frac{1}{\sqrt{125}} - \frac{1}{\sqrt{1050}}\right)$$

$$W = 8.1(\text{kWh t}^{-1})$$

Therefore predicted total is: $W_T = 12.2$ kWh t^{-1}

c. The combined test work index for this ore, $Wilab^C$ is given by:

$$12.2 = 10 \times Wilab^C\left(\frac{1}{\sqrt{125}} - \frac{1}{\sqrt{165000}}\right)$$

$$Wilab^C = 14.0\ (\text{kWh t}^{-1})$$

and, thus:

$$\frac{Wilab^C}{Wi_o} = \frac{14.0}{16.8} = 0.83\ (\text{or } 83\%)$$

This result indicates the circuit is only 83% efficient compared to that predicted by the Bond standard test.

mill-classifier circuits "functional performance analysis" does provide a tool to identify which of the two process units (or both) may be the source of inefficiency.

Functional Performance Analysis

McIvor (2006, 2014) has provided an intermediate level (i.e., between the simple lumped parameter work index of Bond, and the highly detailed computerized circuit modeling approach) to characterize ball mill-cyclone circuit performance.

Classification System Efficiency (CSE) is the percentage of "coarse" size (typically with reference to the P_{80}) material occupying the mill and can be calculated by taking the average of the percentage of coarse material in the mill feed and mill discharge. As this also represents the percentage of the mill energy expended on targeted, coarse size material, it is directly proportional to overall grinding circuit efficiency and production rate. Circuit performance can then be expressed in the Functional Performance Equation for ball milling circuits, which is derived as follows.

Define "fines" as any product size material, and "coarse" as that targeted to be further ground, the two typically differentiated by the circuit target P_{80}. For any grinding circuit, the production rate of new, product size material or fines (Production Rate of Fines, PRF) must equal the specific grinding rate of the coarse material (i.e., fine product generated per unit of energy applied to the coarse material) times the power applied to the coarse material:

$$\text{Circuit PRF} = \text{Power Applied to Coarse Material} \times \text{Specific Grinding Rate of Coarse Material}$$

The power applied to the coarse material is the total mill power times the fraction of coarse material in the mill, the latter already defined as CSE. Therefore:

$$\text{Circuit PRF} = \text{Mill Power} \times \text{CSE} \times \text{Specific Grinding Rate of Coarse Material}$$

A measure of the plant ball mill's grinding efficiency is the ratio of the grinding rate of the coarse material in the plant ball mill compared to the grinding rate, or "grindability," of the same coarse material (g rev^{-1}) as measured in a standardized test mill set up. That is:

$$\text{Mill Grinding Eff.} = \text{Specific Grinding Rate of Coarse Material/Material Grindability}$$

Therefore, if we substitute for specific grinding rate of coarse in the previous equation, we have the Functional Performance Equation for ball milling circuits:

$$\text{Circuit PRF} = \text{Mill Power} \times \text{CSE} \times \text{Mill Grinding Efficiency} \times \text{Material Grindability}$$

This is a simple yet insightful expression of how a ball mill-cyclone circuit generates new product size material. Rate of production is a direct function of the material grindability, as well as the amount of power provided by the mill. It is also a direct function of two separate and distinct efficiencies that are in play. Each of these efficiencies is specifically related to certain physical design and operating variables, which we can manipulate. The terms in the equation are generated by a circuit survey. Thus, the Functional Performance Equation provides understanding and opportunity for plant ball mill circuit optimization.

It is also noteworthy that the complement to CSE is the fraction of mill energy being used on unnecessary further grinding of fines. Such over-grinding is often detrimental to downstream processing and thus an important motivator to achieve high CSE, even beyond its impact on grinding circuit efficiency.

REFERENCES

Amelunxen, P., Meadows, D., 2011. Not another HPGR trade-off study!. Miner. Metall. Process. 28 (1), 1–7.

Amelunxen, P., et al., 2014. The SAG grindability index test. Miner. Eng. 55, 42–51.

Austin, L.G., et al., 1984. Process Engineering of Size Reduction: Ball Milling. SME, Quinn Printing Inc., Hoboken, NJ, USA.

Barratt, D.J., 1979. Semi-autogenous grinding: a comparison with the conventional route. CIM Bull. 71 (811), 74–80.

Barratt, D.J., Sochocky, M.A., 1982. Factors which influence selection of comminution circuits. In: Mular, A.L., et al., (Eds.), Design and Installation of Comminution Circuits. SME, New York, NY, USA, pp. 1–26. (Chapter 1).

Berry, T.F., Bruce, R.M., 1966. A simple method of determining the grindability of ores. Can. Min. J. Jul., 63–65.

Bond, F.C., 1952. The third theory of comminution. Trans. AIME. 193, 484–494.

Bond, F.C., 1985. Testing and calculations. SME Mineral Processing Handbook. Weiss, N.L. (Ed.), Section 3A: General Aspects of Comminution, pp. 16–27.

Bond, F.C., Maxson, W.L., 1938. Grindability and grinding characteristics of ores. Trans. AIME.296–322.

Brissette, M., et al., 2014. Geometallurgy: new accurate testwork to meet required accuracies of mining project development. Proc. 27th International Mineral Processing Cong., (IMPC), Ch. 14, Paper C1417, Santiago, Chile, pp. 200–209.

Burchardt, E., et al., 2011. HPGR's in minerals: what do existing operations tell us for the future? *Proc. International Autogenous and Semi Autogenous Grinding Technology (SAG) Conf.*, Paper 108. CIM, Vancouver, BC, Canada, pp. 1–17.

Carvalho, R.M., Tavares, L.M., 2010. Towards high-fidelity simulation of SAG mills using a mechanistic model. Proc. 7th International Mineral Processing Seminar (Procemin 2010), Santiago, Chile, pp. 83–92.

Choi, W.Z., et al., 1988. Estimation of model parameters for liberation and size reduction. Miner. Metall. Process. 5 (Feb.), 33–39.

Doll, A.G., 2013. Technical Memorandum: SAG mill + ball mill circuit sizing. MINE331 lecture notes at UBC. <https://www.sagmilling.com/articles/12/view/?s=1>.

Doll, A.G., 2015. A simple estimation method of materials handling specific energy consumption in HPGR circuits. *Proc. 47th Canadian Mineral Processors Conf.* CIM, Ottawa, ON, Canada, pp. 3–13.

Dorr, J.V.N., Anable, A., 1934. Fine grinding and classification. Trans. AIME. 112, 161–177.

Dunne, R., et al., 2001. Design of the 40 foot diameter SAG mill installed at the Cadia gold copper mine. Proc. International Autogenous and Semi Autogenous Grinding Technology (SAG) Conf., vol. 1, Vancouver, BC, Canada, pp. 43–58.

Finch, J.A., Ramirez-Castro, J, 1981. Modelling mineral size reduction in the closed -circuit ball mill at the Pine Point Mines concentrator. Int. J. Miner. Process. 8 (1), 61–78.

Flavel, M.D., 1978. Control of crushing circuits will reduce capital and operating costs. Mining Mag. 138-139 (Mar.), 207–213.

Fuerstenau, D.W., Abouzeid, A.-Z.M., 2002. The energy efficiency of ball milling in comminution. Int. J. Miner. Process. 67 (1-4), 161–185.

Furuya, M., et al., 1971. Theoretical analysis of closed-circuit grinding system based on comminution kinetics. Ind. Eng. Chem. Process. Des. Dev. 10 (4), 449–456.

Gay, S.L., 2004. Simple texture-based liberation modelling of ores. Miner. Eng. 17 (11-12), 1209–1216.

Griffith, A.A., 1921. The phenomena of rupture and flow in solids. Phil. Trans. R Soc. 221, 163–198.

Hartley, J.N., et al., 1978. Chemical additives for ore grinding: how effective are they? Eng. Min. J. Oct., 105–111.

He, M., et al., 2010. Enhancement of energy efficiency in fine grinding of copper sulfide minerals using a pilot-scale stirred media mill – Isamill. Proc. 25th International Mineral Processing Cong., (IMPC), Brisbane, Queensland, Australia, pp. 791–799.

Herbst, J.A., Fuerstenau, D.W., 1980. Scale-up procedure for continuous grinding mill design using population balance models. Int. J. Miner. Process. 7 (1), 1–31.

Herbst, J.A., et al., 1988. Development of a multicomponent-multisize liberation model. Miner. Eng. 1 (2), 97–111.

Hinde, A.L., Kalala, J.T., 2009. The application of a simplified approach to modelling tumbling mills, stirred media mills and HPGR's. Miner. Eng. 22 (7-8), 633–641.

Hukki, R.T., 1962. Proposal for a Solomonic settlement between the theories of Von Rittinger, kick and bond. Trans. AIME. 220, 403–408.

Inglis, C.E., 1913. Stresses in a plate due to the presence of cracks and sharp corners. Trans. Royal Inst. Naval Arch. 60, 219–241.

Jankovic, A., Valery, W., 2013. Closed circuit ball mill – basics revisited. Miner. Eng. 43-44, 148–153.

Kiangi, K., et al., 2012. DEM investigations into the processing challenges faced by a primary gyratory crusher installed in Nkomati Mine. Proc. 26th International Mineral Processing Cong., (IMPC), New Delhi, India, pp. 2429–2442.

Kick, F., 1885. Das Gesetz der Proportionalen Widerstände und Seine Anwendungen: Nebst Versuchen über das Verhalten Verschiedener Materialien bei Gleichen Formänderungen Sowohl unter der Presse als dem Schlagwerk. A. Felix, Leipzig.

King, R.P., 2012. Modeling and Simulation of Mineral Processing Systems. second ed. SME, Englewood, CO, USA, Elsevier.

Knecht, J., 1994. High-pressure grinding rolls - a tool to optimize treatment of refractory and oxide gold ores. *Proc. 5th Mill Operators Conf.* AusIMM, Roxby Downs, Australia, pp. 51–59.

Knight, D.A., et al., 1989. Comminution circuit comparison – Conventional vs. semi-autogenous. Proc. International Autogenous and Semi Autogenous Grinding Technology (SAG) Conf., Vancouver, BC, Canada, pp. 217–224.

Kosick, G., Bennett, C., 1999. The value of orebody power requirement profiles for SAG circuit design. Proc. 31st Annual Meeting of The Canadian Mineral Processors Conf. CIM, Ottawa, ON, Canada, pp. 241–253.

Lane, G., et al., 2013. Power-based comminution calculations using Ausgrind. Proc. 10th International Mineral Processing Conf., (Procemin 2013), Santiago, Chile, pp. 85–96.

Lessard, J., et al., 2014. Development of ore sorting and its impact on mineral processing. Miner. Eng. 65, 88–97.

Lowrison, G.C., 1974. Crushing and Grinding. Butterworths, London.

Lynch, A.J., Morrell, S., 1992. The understanding of comminution and classification and its practical application in plant design and operation. In: Kawatra, S.K. (Ed.), Comminution: Theory and Practice. AIME, Littleton, CO, USA, pp. 405–426. (Chapter 30).

Lynch, A.J., Rowland, C.A., 2005. The History of Grinding. SME, Littleton, CO, USA.

Maleki-Moghaddam, M., et al., 2013. A method to predict shape and trajectory of charge in industrial mills. Miner. Eng. 46-47, 157–166.

Martins, S., Radziszewski, P., 2014. Trumbling mills and the theory of large deviations. Proc. 27th International Mineral Processing Cong., (IMPC). Ch. 10, Paper C1009, Santiago, Chile.

Mazzinghy, D., et al., 2012. Predicting the size distribution in the product and the power requirements of a pilot scale vertimill. Proc. 9th International Mineral Processing Conf., (Procemin 2012), Santiago, Chile, pp. 412–420.

McGhee, S., et al., 2001. SAG feed pre-crushing at ASARCO's ray concentrator: development, implementation and evaluation. Proc. International Autogenous and Semi Autogenous Grinding Technology (SAG) Conf., vol. 1, Vancouver, BC, Canada, pp. 234–247.

McIvor, R.E., 2006. Industrial validation of the functional performance equation for ball milling and pebble milling circuits. Mining Eng. 58 (11), 47–51.

McIvor, R.E., 2014. Plant performance improvements using the grinding circuit "classification system efficiency. Mining Eng. 66 (9), 72–76.

Millard, M., 2002. The use of comminution testwork results in SAG mill design. Proc. Metallurgical Plant Design and Operation Strategies. AusIMM, Sydney, NSW, Australia, pp. 56–71.

Morley, C.T., 2011. HPGR trade-off studies and how to avoid them. Proc. International Autogenous and Semi Autogenous Grinding Technology (SAG) Conf., Paper 170, Vancouver, BC, Canada, pp. 1–23.

Morrell, S., 2004a. An alternative energy-size relationship to that proposed by Bond for the design and optimization of grinding circuits. Int. J. Miner. Process. 74 (1-4), 133–141.

Morrell, S., 2004b. Predicting the specific energy of autogenous and semi autogenous mills from small diameter drill core samples. Miner. Eng. 17 (3), 447–451.

Morrell, S., 2006. Rock characterisation for high pressure grinding rolls circuit design. Proc. International Autogenous and Semi Autogenous Grinding Technology (SAG) Conf., vol. 4, Vancouver, BC, Canada, pp. 267–278.

Morrell, S., 2009. Predicting the overall specific energy requirement of crushing, high pressure grinding roll and tumbling mill circuits. Miner. Eng. 22 (6), 544–549.

Morrell, S., 2010. Predicting the specific energy required for size reduction of relatively coarse feeds in conventional crushers and high pressure grinding rolls. Miner. Eng. 23 (2), 151–153.

Morrell, S., 2014a. Innovations in comminution modelling and ore characterisation. In: Anderson, C.G., et al., (Eds.), Mineral Processing and Extractive Metallurgy: 100 years of Innovation. SME, Englewood, CO, USA, pp. 74–83.

Morrell, S., 2014b. Personal communication.

Mosher, J., Bigg, A., 2002. Bench-scale and pilot plant tests for comminution circuit design, Mineral Processing Plant Design, Practice and Control, vol. 1. SME, Littleton, CO, USA, pp. 123–135.

Myers, J.F., et al., 1947. Rod milling - plant and laboratory data. Mining Technology. AIME, Technical publication: No. 2175, pp. 1–11.

Nadolski, S., et al., 2014. An energy benchmarking model for mineral comminution. Miner. Eng. 65 (15), 178–186.

Napier-Munn, T.J., et al., 1996. Mineral Comminution Circuits: Their Operation and Optimisation, Julius Kruttschnitt Mineral Research Centre (JKMRC), University of Queensland, Brisbane, Queensland, Australia.

Nesset, J.E., et al., 2006. Assessing the performance and efficiency of fine grinding technologies. *Proc. 38th Annual Meeting of The Canadian Mineral Processors Conf.* CIM, Ottawa, ON, Canada, pp. 283–309.

Partridge, A.C., 1978. Principles of comminution. Mine Quarry. 7 (288), 70–73.

Pax, R.A., 2012. Determining mill operational efficiencies using non-contact acoustic emissions from microphone arrays. Proc. 11th Mill Operators Conference. AusIMM, Melbourne, Australia, pp. 119–126.

Powell, M.S., Morrison, R.D., 2007. The future of comminution modelling. Int. J. Miner. Process. 84 (1-4), 228–239.

Quist, J., Evertsson, C., 2010. Application of discrete element method for simulating feeding conditions and size reduction in cone crushers. Proc. 25th International Mineral Processing Cong., (IMPC), Brisbane, Queensland, Australia, pp. 3337–3347.

Radziszewski, P., 2013a. Assessing the stirred mill design space. Miner. Eng. 41, 9–16.

Radziszewski, P., 2013b. Energy recovery potential in comminution processes. Miner. Eng. 46-47, 83–88.

Radziszewski, P, Allen, J., 2014. Towards a better understanding of stirred milling technologies: estimating power consumption and energy use. *Proc. 46th Annual Meeting of The Canadian Mineral Processors Conf.* CIM, Ottawa, ON, Canada, pp. 55–66.

Rajamani, K., Herbst, J.A., 1984. Simultaneous estimation of selection and breakage functions from batch and continuous grinding data. Trans. Inst. Min. Metall. Sect. B. 93 (6), C74–C85.

Richardson, J.M., 1990. Computer simulation and optimisation of mineral processing plants.three case studies. In: Rajamani, R.K., Herbst, J.A. (Eds.), Proc. Control 90-Mineral and Metallurgical Processing. SME, pp. 233–244.

Rowland Jr., C.A., 1988. Using the Bond Work Index to measure operating comminution efficiency. Miner. Metall. Process. 15 (4), 31–36.

Rowland Jr., C.A., McIvor, R.E., 2009. The Bond standard for comminution efficiency. In: Malhorta, D., et al., (Eds.), Recent Advances in Mineral Processing Plant Design. SME, Littleton, CO, USA, pp. 328–331.

Schönert, K., 1988. A first survey of grinding with high-compression roller mills. Int. J. Miner. Process. 22 (1-4), 401–412.

Sherman, M., 2011. Bond is back! Proc. International Autogenous and Semi Autogenous Grinding Technology (SAG) Conf., Paper No. 017, Vancouver, BC, Canada, pp. 1–13.

Shi, F., Xie, W., 2015. A specific energy-based size reduction model for batch grinding ball mill. Miner. Eng. 70, 130–140.

Smith, R.W., Lee, K.H., 1968. A comparison of data from Bond type simulated closed-circuit and batch type grindability tests. Trans. AIME SME. 241, 91–99.

Söderlund, A., et al., 1989. Autogenous vs. conventional grinding influence on metallurgical results. Proc. Autogenous and Semi-Autogenous Grinding Technology (SAG) Conf., Vol. 1, Vancouver, BC, Canada, pp. 187–198.

Starkey, J., Dobby, G., 1996. Application of the Minnovex SAG power index at five Canadian SAG plants. Proc. Autogenous and Semi-Autogenous Grinding Technology (SAG) Conf., Vancouver, BC, Canada, pp. 345–360.

Stehr, N., Schwedes, J., 1983. Investigation of the grinding behaviour of a stirred ball mill. German Chem. Eng. 6, 337–343.

Stief, D.E., et al., 1987. Tower mill and its application to fine grinding. Miner. Metall. Process. 4 (1), 45–50.

Tromans, D., 2008. Mineral comminution: energy efficiency considerations. Miner. Eng. 21 (8), 613–620.

Von Rittinger, P.R., 1867. Lehrbuch der Aufbereitungs Kunde. Ernst and Korn, Berlin, German.

Walker, W.H., et al., 1937. Principles of Chemical Engineering, New York, NY, USA.

Chapter 6

Crushers

6.1 INTRODUCTION

Crushing is the first mechanical stage in the process of comminution in which a principal objective is the liberation of the valuable minerals from the gangue. Crushing is typically a dry operation that is performed in two- or three-stages (i.e., primary, secondary, tertiary crushing). Lumps of run-of-mine ore as large as 1.5 m across are reduced in the primary crushing stage to 10–20 cm in heavy-duty machines.

In most operations, the primary crushing schedule is the same as the mining schedule. When primary crushing is performed underground it is normally a responsibility of the mining department; for primary crushing at the surface it is customary for the mining department to deliver the ore to the crusher and for the mineral processing department to crush and handle the ore from this point through the successive ore-processing stages. Primary crushers are commonly designed to operate 75% of the available time, mainly due to interruptions caused by insufficient crusher feed and mechanical delays (Lewis et al., 1976; McQuiston and Shoemaker, 1978; Major, 2002).

Primary crusher product from most metalliferous ores can be crushed and screened satisfactorily, and subsequent crushing consists of one or two size-reduction stages with appropriate crushers and screens (Major, 2002). In three-stage circuits, ore is reclaimed from ore storage with secondary crushing product typically ranging from 3.7–5.0 cm, and tertiary crushing further reducing the ore to ca. 0.5–2 cm in diameter. The product size is determined by the size of the opening at the discharge, called the *set* or *setting*. The *reduction ratio* is the ratio of feed size to product size, often with reference to the 80% passing size, that is, reduction ratio $= F_{80}/P_{80}$. If the ore tends to be slippery and tough, the tertiary crushing stage may be substituted by coarse grinding. On the other hand, more than three size-reduction stages may be required if the ore is extra-hard, or in special cases where it is important to minimize the production of fines (Major, 2009).

Vibrating screens are sometimes placed ahead of secondary or tertiary crushers to remove undersize material (i.e., *scalp* the feed), thereby increasing the capacity of the crushing plant. Undersize material tends to pack the voids between large particles in the crushing chamber, and can choke the crusher, causing damage, because the packed mass of rock is unable to swell in volume as it is broken.

Crushing may be in open- or closed-circuit, depending on the required product size distribution. Two basic crushing flowsheets are shown in Figure 6.1: (a) the older style ("traditional") 3-stage crushing circuit ahead of a rod mill, and (b) the more modern open-circuit primary crushing prior to SAG milling with crushing and recycling of "critical size" material (see Chapters 5 and 7 for discussion on "critical size"). Both flowsheets show the primary crusher is in open circuit. Figure 6.1(a) shows the secondary crusher operating in open-circuit while the tertiary crusher is closed with the screen undersize feeding the rod mill. In open-circuit crushing there is no recycle of crusher product to the feed. Open-circuits may include *scalping* ahead of the crushers with undersize material from the screen being combined with the crusher product, which is then routed to the next operation. If the crusher is producing rod mill (or ball mill) feed, it is good practice to use final stage closed-circuit crushing in which the undersize from the screen feeds the mill. The crusher product is returned to the screen so that any over-size material is recirculated. To meet the tonnage and product size requirements there may be more than one secondary and tertiary crusher operating in *parallel* with the feed split between the units. The various circuits are illustrated by Major (2002).

One of the main reasons for closing the circuit is the greater flexibility given to the crushing plant as a whole. The crusher can be operated at a wider setting if necessary, thus altering the size distribution of the product, and by making a selective cut on the screen, the finished product can be adjusted to give the required specification. There is the added factor that if the material is wet or sticky (potentially due to climatic conditions) it is possible to open the setting of the crusher to prevent the possibility of packing, and by this means the throughput of the machine is increased, which will compensate for the additional circulating load. Closed-circuit

Wills' Mineral Processing Technology.

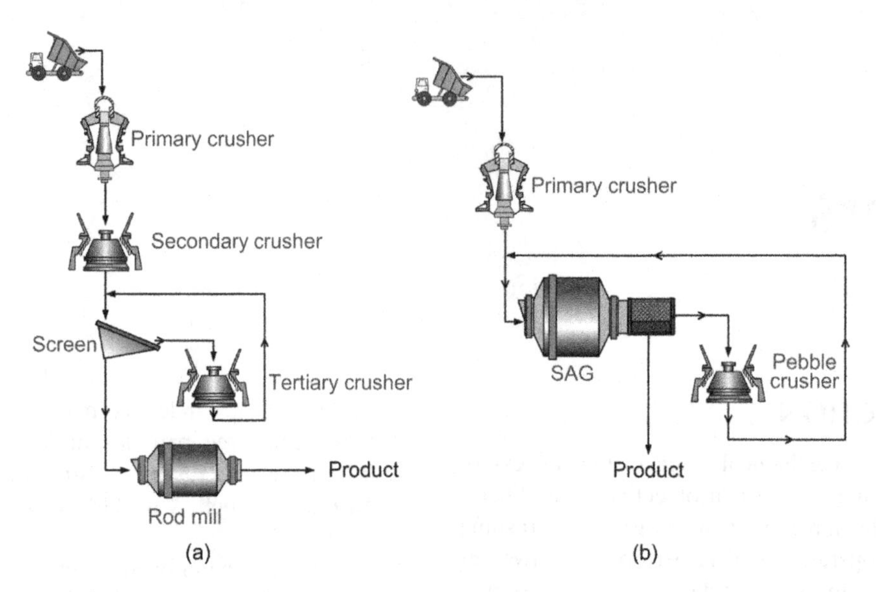

FIGURE 6.1 Example crushing flowsheets feeding a: (a) rod mill, and (b) SAG mill with a "recycle" (pebble) crusher to size reduce "critical size" material.

operation also allows compensation for wear that takes place on liners, and generally gives greater freedom to meet changes in requirements from the plant.

Surge bins precede the primary crusher to receive dumped loads from ships or trucks and should have enough storage capacity to maintain a steady feed to the crusher. In most mills the crushing plant does not run for 24 h a day, as hoisting and transport of ore is usually carried out on two shifts only, the other shift being used for drilling and blasting. The crushing section must therefore have a greater hourly capacity than the rest of the plant, which is run continuously. Crushed ore surge capacity is generally included in the flowsheet to ensure continuous supply to the grinding circuit. The obvious question is, why not have similar storage capacity before the crushers and run this section continuously also? Apart from the fact that it is cheaper in terms of power consumption to crush at off-peak hours, large storage bins are expensive, so it is uneconomic to have bins at both the crushing and grinding stages. It is not practical to store large quantities of run-of-mine (ROM) ore, as it consists of a large range of particle sizes and the small ones move down in the pile and fill the voids. This packed mass is difficult to move after it has settled. ROM ore should therefore be kept moving as much as possible, and surge bins should have sufficient capacity only to even out the flow to the crusher.

6.2 PRIMARY CRUSHERS

Primary crushers are heavy-duty machines, used to reduce ROM ore down to a size suitable for transport and for feeding the secondary crushers or AG/SAG mills. The units are always operated in open circuit, with or without heavy-duty scalping screens (*grizzlies*). There are two main types of primary crushers in metalliferous operations: jaw and gyratory crushers, although impact crushers have limited use as primaries and will be considered separately. Scalping is typically associated with jaw crusher circuits and can be included to maximize throughput.

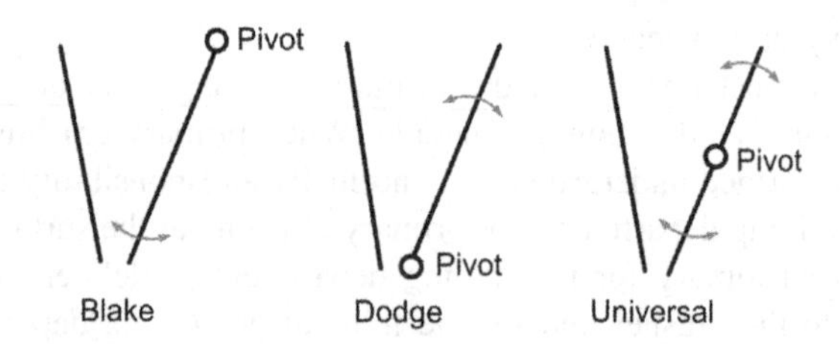

FIGURE 6.2 Jaw-crusher types.

6.2.1 Jaw Crushers

The Blake crusher was patented by E.W. Blake in 1858 and variations in detail on the basic form are found in most of the jaw crushers used today. The patent states that the stone breaker "consists of a pair of jaws, one fixed and the other movable, between which the stones are to be broken." The jaws are set at an acute angle with one jaw pivoting so as to swing relative to the other fixed jaw. Material fed into the jaws is repetitively *nipped* and released to fall further into the crushing chamber until the discharge aperture.

Jaw crushers are classified by the method of pivoting the swing jaw (Figure 6.2). In the *Blake crusher*, the jaw is pivoted at the top and thus has a fixed receiving area and a variable discharge opening. In the *Dodge crusher*, the jaw is pivoted at the bottom, giving it a variable feed area but fixed delivery area. The Dodge crusher is

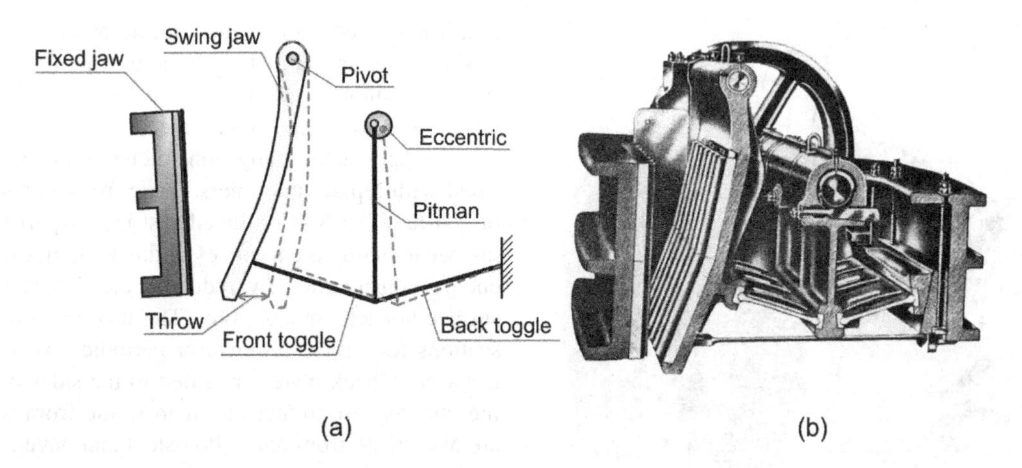

FIGURE 6.3 Double-toggle jaw crusher: (a) functional diagram, and (b) cross section.

restricted to laboratory use, where close sizing is required. The *Universal crusher* is pivoted in an intermediate position, and thus has a variable delivery and receiving area.

There are two forms of the Blake crusher: *double toggle and single toggle*.

In *double-toggle Blake crushers*, the oscillating movement of the swinging jaw is effected by vertical movement of the pitman, which moves up and down under the influence of the eccentric (Figure 6.3). The back toggle plate causes the pitman to move sideways as it is pushed upward. This motion is transferred to the front toggle plate, which in turn causes the swing jaw to close on the fixed jaw, this minimum separation distance being the *closed set* (or *closed side setting*). Similarly, downward movement of the pitman allows the swing jaw to open, defining the *open set* (or *open side setting*).

The important features of the machine are:

1. Since the jaw is pivoted from above, it moves a *minimum* distance at the entry point and a *maximum* distance at the delivery. This maximum distance is called the *throw* of the crusher, that is, the difference between the open side and closed side settings.
2. The horizontal displacement of the swing jaw is greatest at the bottom of the pitman cycle and diminishes steadily through the rising half of the cycle as the angle between the pitman and the back toggle plate becomes less acute.
3. The crushing force is *least* at the start of the cycle, when the angle between the toggles is most acute, and is strongest at the top of the cycle, when full power is delivered over a reduced travel of the jaw.

Figure 6.3 shows a cross section through a double-toggle jaw crusher. Jaw crushers are rated according to their receiving area, that is, the *gape*, which is the distance between the jaws at the feed opening, and the *width* of the plates. For example, a 1,220 × 1,830 mm crusher has a *gape* of 1,220 mm and a *width* of 1,830 mm.

Consider a large piece of rock falling into the mouth of the crusher. It is nipped by the jaws, which are moving relative to each other at a rate depending on the size of the machine (and which usually varies inversely with the size). Basically, time must be given for the rock broken at each "bite" to fall to a new position before being nipped again. The ore falls until it is arrested. The swing jaw closes on it, quickly at first, and then more slowly with increasing power toward the end of the stroke. The fragments fall to a new arrest point as the jaws open and are gripped and crushed again. During each "bite" of the jaws, the rock swells in volume due to the creation of voids between the particles. Since the ore is also falling into a gradually reducing cross sectional area of the crushing chamber, choking of the crusher would soon occur if it were not for the increasing amplitude of swing toward the discharge end of the crusher. This accelerates the material through the crusher, allowing it to discharge at a rate sufficient to leave space for material entering from above. This is termed *arrested* or *free* crushing as opposed to *choked crushing*, which occurs when the volume of material arriving at a particular cross section is greater than that leaving. In arrested crushing, crushing is by the jaws only, whereas in choked crushing, particles break one other. This *inter-particle comminution* can lead to excessive production of fines, and if choking is severe can damage the crusher.

The discharge size of material from the crusher is controlled by the open side set, which is the maximum opening of the jaws at the discharge end. This can be adjusted by using toggle plates of the required length. The back pillow into which the back toggle plate bears can be adjusted to compensate for jaw wear. A number of manufacturers offer jaw setting by hydraulic jacking, and some fit electro-mechanical systems, which allow remote control (Anon., 1981).

A feature of all jaw crushers is the heavy fly-wheel (seen at the back of Figure 6.3(b)) attached to the drive,

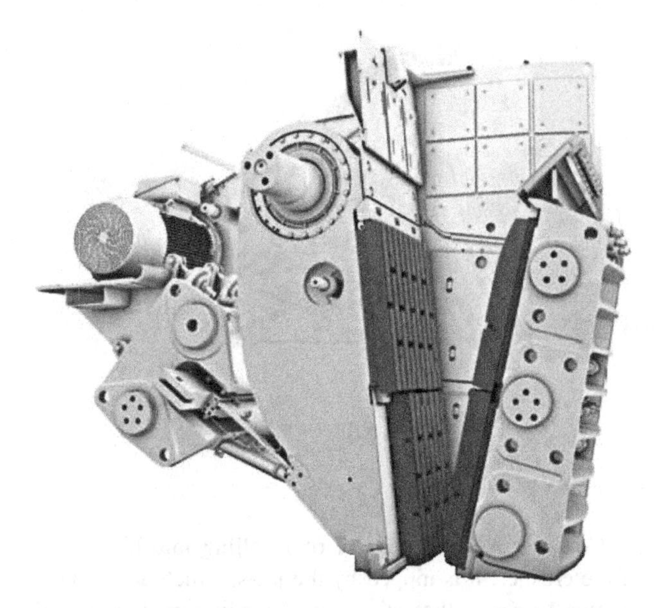

FIGURE 6.4 Cross section of single-toggle Nordberg® C Series™ jaw crusher *(Courtesy Metso)*.

which is necessary to store energy on the idling half of the stroke and deliver it on the crushing half. Since the jaw crusher works on half-cycle only, it is limited in capacity for its weight and size. Due to the alternate loading and release of stress, jaw crushers must be rugged and require strong foundations to accommodate the vibrations.

In *single-toggle jaw crushers* (Figure 6.4) the swing jaw is suspended on the eccentric shaft, which allows a lighter, more compact design than with the double-toggle machine. The motion of the swing jaw also differs from that of the double-toggle design. Not only does the swing jaw move toward the fixed jaw under the action of the toggle plate, but it also moves vertically as the eccentric rotates. This elliptical jaw motion assists in pushing rock through the crushing chamber. The single-toggle machine therefore has a somewhat higher capacity than the double-toggle machine of the same gape. The eccentric movement, however, increases the rate of wear on the jaw plates. Direct attachment of the swing jaw to the eccentric imposes a high degree of strain on the drive shaft, and as such maintenance costs tend to be higher than with the double-toggle machine.

Double-toggle machines are usually used on tough, hard and abrasive material. This being said, single-toggle crusher do see use (primarily in Europe) for heavy-duty work on tough taconite ores, and it is often choke fed, since the jaw movement tends to make it self-feeding.

Jaw-crusher Construction

Jaw crushers are heavy-duty machines and hence must be robustly constructed. The main frame is often made from cast iron or steel, connected with tie-bolts. It is commonly made in sections so that it can be transported underground for installation. Modern jaw crushers may have a main frame of welded mild steel plate.

The jaws are usually constructed from cast steel and fitted with replaceable liners, made from manganese steel, or "Ni-hard," a Ni-Cr alloyed cast iron. Apart from reducing wear, hard liners are essential to minimize crushing energy consumption by reducing the deformation of the surface at each contact point. The jaw plates are bolted in sections for simple removal or periodic reversal to equalize wear. Cheek plates are fitted to the sides of the crushing chamber to protect the main frame from wear. These are also made from hard alloy steel and have similar lives to the jaw plates. The jaw plates may be smooth, but are often corrugated, the latter being preferred for hard, abrasive ores. Patterns on the working surface of the crushing members also influence capacity, especially at small settings. The corrugated profile is claimed to perform compound crushing by compression, tension, and shearing. Conventional smooth crushing plates tend to perform crushing by compression only, though irregular particles under compression loading might still break in tension. Since rocks are around 10 times weaker in tension than compression, power consumption and wear costs should be lower with corrugated profiles. Regardless, some type of pattern is desirable for the jaw plate surface in a jaw crusher, partly to reduce the risk of undesired large flakes easily slipping through the straight opening, and partly to reduce the contact surface when crushing flaky blocks. In several installations, a slight wave shape has proved successful. The angle between the jaws is usually less than 26°, as the use of a larger angle causes particle to slip (i.e., not be nipped), which reduces capacity and increases wear.

In order to overcome problems of choking near the discharge of the crusher, which is possible if fines are present in the feed, curved plates are sometimes used. The lower end of the swing jaw is concave, whereas the opposite lower half of the fixed jaw is convex. This allows a more gradual reduction in size as the material nears the exit, minimizing the chance of packing. Less wear is also reported on the jaw plates, since the material is distributed over a larger area.

The speed of jaw crushers varies inversely with the size, and usually lies in the range of 100–350 rpm. The main criterion in determining the optimum speed is that particles must be given sufficient time to move down the crusher throat into a new position before being nipped again.

The throw (maximum amplitude of swing of the jaw) is determined by the type of material being crushed and is usually adjusted by changing the eccentric. It varies from 1 to 7 cm depending on the machine size, and is highest for tough, plastic material and lowest for hard, brittle ore.

The greater the throw the less danger of choking, as material is removed more quickly. This is offset by the fact that a large throw tends to produce more fines, which inhibits arrested crushing. Large throws also impart higher working stresses to the machine.

In all crushers, provision must be made for avoiding damage that could result from uncrushable material entering the chamber. Many jaw crushers are protected from such "tramp" material (often metal objects) by a weak line of rivets on one of the toggle plates, although automatic trip-out devices are now common. Certain designs incorporate automatic overload protection based on hydraulic cylinders between the fixed jaw and the frame. In the event of excessive pressure caused by an overload, the jaw is allowed to open, normal gap conditions being reasserted after clearance of the blockage. This allows a full crusher to be started under load (Anon., 1981). The use of "guard" magnets to remove tramp metal ahead of the crusher is also common (Chapters 2 and 13).

Jaw crushers are supplied in sizes up to 1,600 mm (gape) × 1,900 mm (width). For coarse crushing application (closed set ~ 300 mm), capacities range up to ca. 1,200 t h^{-1}. However, Lewis et al. (1976) estimated that the economic advantage of using a jaw crusher over a gyratory diminishes at crushing rates above 545 t h^{-1}, and above 725 t h^{-1} jaw crushers cannot compete.

6.2.2 Gyratory Crushers

Gyratory crushers are principally used in surface-crushing plants. The gyratory crusher (Figure 6.5) consists essentially of a long spindle, carrying a hard steel conical grinding element, the *head*, seated in an eccentric sleeve. The spindle is suspended from a "spider" and, as it rotates, normally between 85 and 150 rpm, it sweeps out a conical path within the fixed crushing chamber, or shell, due to the gyratory action of the eccentric. As in the jaw crusher, maximum movement of the head occurs near the discharge. This tends to relieve the choking due to swelling. The gyratory crusher is a good example of arrested crushing. The spindle is free to turn on its axis in the eccentric sleeve, so that during crushing the lumps are compressed between the rotating head and the top shell segments, and abrasive action in a horizontal direction is negligible.

With a gyratory crusher, at any cross section there are in effect two sets of jaws opening and shutting like jaw crushers. In fact, the gyratory crusher can be regarded as an infinitely large number of jaw crushers each of infinitely small width, and, as consequence, the same terms gape, set, and throw, have identical meaning in the case of the gyratory crusher. Since the gyratory, unlike the jaw crusher, crushes on full cycle, it has a higher capacity than a jaw crusher of the same gape,

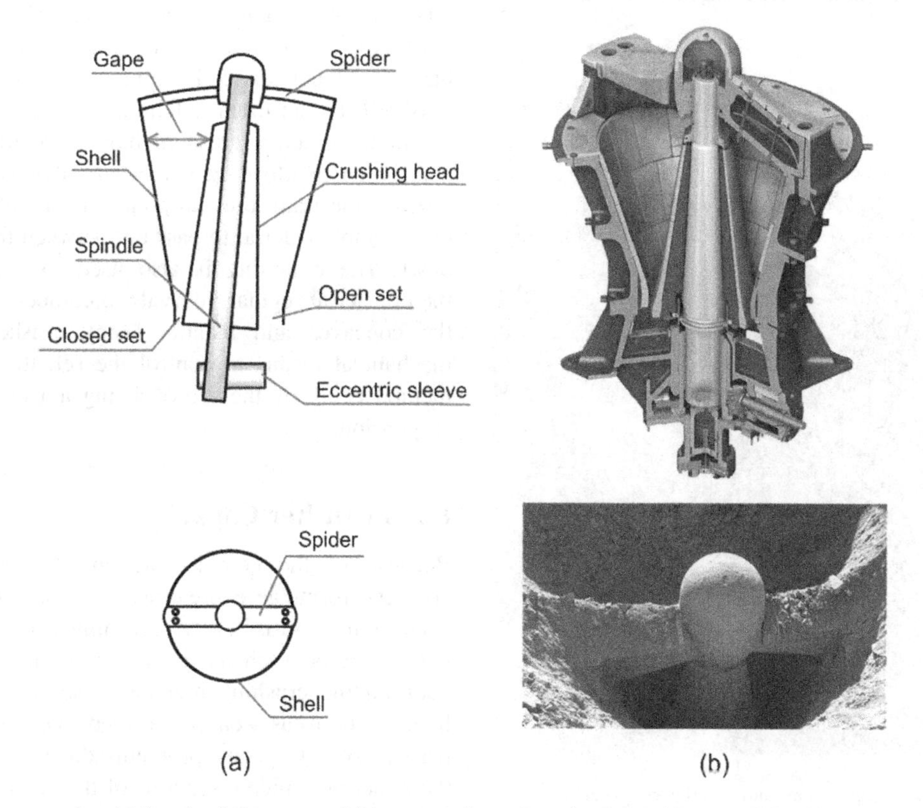

FIGURE 6.5 Gyratory crusher: (a) functional diagrams, and (b) cross section and overhead view *(Courtesy FLSmidth)*.

roughly by a factor of 2.5–3, and is usually favored in plants handling large throughputs: in mines with crushing rates above 900 t h^{-1}, gyratory crushers are always selected.

Gyratory crushers are identified by the size of the gape and the size of the mantle at the discharge. They range in size up to ca. 1,600 mm × 2,900 mm (gape × mantle diameter) with power consumption as high as 1,200 kW and capable of crushing up to ca. 10,000 t h^{-1} at a discharge open side setting up to 240 mm.

Large gyratories typically dispense with expensive feeding mechanisms and are often fed direct from trucks (Figure 6.6). They can be operated with the head buried in feed. Although excessive fines may have to be "scalped" from the feed (more common in jaw crushing circuits), the trend in large-capacity plants with gyratory crushers is to dispense with grizzlies. This reduces capital cost of the installation and reduces the height from which the ore must fall into the crusher, thus minimizing damage to the spider. Choked crushing is encouraged to some extent as rock-to-rock crushing in primary stages reduces the rock-to-steel crushing required in the secondary crushers, thus reducing wear (McQuiston and Shoemaker, 1978). Choke feeding of a gyratory crusher has been claimed beneficial when the crusher is followed by SAG mills, as their throughput is sensitive to the mill feed size (Simkus and Dance, 1998). Operating crushers under choke feeding conditions gives more even wear and longer crusher life.

FIGURE 6.6 Truck dumping ore into a primary gyratory crusher *(Courtesy Sandvik)*.

The last ten years or so have seen advances in increased installed power which, without increasing crusher size, has increased capacity (Erikson, 2014). Other innovations have been in serviceability, and measurement of wear components (Erikson, 2014).

Gyratory Crusher Construction

The outer shell of the crusher is constructed from heavy steel casting or welded steel plate, with at least one constructional joint, the bottom part taking the drive shaft for the head, the top, and lower shells providing the crushing chamber. If the spindle is carried on a suspended bearing, as in most primary gyratories, then the spider carrying the bearing forms a joint across the reinforced alloyed white cast-iron (Ni-hard) liners or *concaves*. In smaller crushers, the concave is one continuous ring bolted to the shell. Large machines use sectionalized concaves, called *staves*, which are wedge-shaped, and either rest on a ring fitted between the upper and the lower shell, or are bolted to the shell. The concaves are backed with some soft filler material, such as white metal, zinc, or plastic cement, which ensures even seating against the steel *bowl*.

The head consists of the steel forgings, which make up the spindle. The head is protected by a *mantle* (usually of manganese steel) fastened to the head by means of nuts on threads which are pitched so as to be self-tightening during operation. The mantle is typically backed with zinc, plastic cement, or epoxy resin. The vertical profile is often bell-shaped to assist the crushing of material that has a tendency to choke. Figure 6.7 shows a gyratory crusher head during installation.

Some gyratory crushers have a hydraulic mounting and, when overloading occurs, a valve is tripped which releases the fluid, thus dropping the spindle and allowing the "tramp" material to pass out between the head and the bowl. The mounting is also used to adjust the set of the crusher at regular intervals to compensate for wear on the concaves and mantle. Many crushers use simple mechanical means to control the set, the most common method being by the use of a ring nut on the main shaft suspension.

6.2.3 Crusher Capacity

Because of the complex action of jaw and gyratory crushers, formulae expressing their capacities have never been entirely satisfactory. Crushing capacity depends on many factors, such as the angle of nip (i.e., the angle between the crushing members), stroke, speed, and the liner material, as well as the feed material and its initial particle size. Capacity problems do not usually occur in the upper and middle sections of the crushing cavity, providing the angle of nip is not too great. It is normally the

FIGURE 6.7 Crusher head during installation *(Courtesy FLSmidth)*.

discharge zone, the narrowest section of the crushing chamber, which determines the crushing capacity.

Broman (1984) describes the development of simple models for optimizing the performance of jaw and gyratory crushers. The volumetric capacity (Q, $m^3\ h^{-1}$) of a jaw crusher is expressed as:

$$Q = BSs \cdot \cot a \cdot k \cdot 60n \tag{6.1}$$

where B = inner width of crusher (m); S = open side setting (m); s = throw (m); a = angle of nip; n = speed of crusher (rpm); and k is a material constant which varies with the characteristics of the crushed material, the feeding method, liner type, etc., normally having values between 1.5 and 2.

For gyratory crushers, the corresponding formula is:

$$Q = (D - S)\pi Ss \cdot \cot a \cdot k \cdot 60n \tag{6.2}$$

where D = diameter of the head mantle at the discharge point (m), and k, the material constant, normally varying between 2 and 3.

6.2.4 Selection of a Jaw or Gyratory Crusher

As noted, in deciding whether a jaw or a gyratory crusher should be used, the main factor is the maximum size of ore which the crusher will be required to handle and the throughput required. Gyratory crushers are, in general, used where high capacity is required. Jaw crushers tend to be used where the crusher gape is more important than the capacity. For instance, if it is required to crush material of a certain maximum diameter, then a gyratory having the required gape would have a capacity about three times that of a jaw crusher of the same gape. If high capacity is required, then a gyratory is the answer. If, however, a large gape is needed but not capacity, then the jaw crusher will probably be more economical, as it is a smaller machine and the gyratory would be running idle most of the time. A guiding relationship was that given by Taggart (1945): if $t\ h^{-1} < 161.7 \times (\text{gape in } m^2)$, use a jaw crusher; conversely, if the tonnage is greater than this value, use a gyratory crusher.

There are some secondary considerations. The capital and maintenance costs of a jaw crusher are slightly less than those of the gyratory but they may be offset by the installation costs, which are lower for a gyratory, since it occupies about two-thirds the volume and has about two-thirds the weight of a jaw crusher of the same capacity. The circular crushing chamber allows for a more compact design with a larger proportion of the total volume being accounted for by the crushing chamber. Jaw-crusher foundations need to be much more rugged than those of the gyratory, due to the alternating working stresses.

In some cases, the self-feeding capability of the gyratory compared with the jaw results in a capital cost saving, as expensive feeding devices, such as the heavy-duty chain feeders (Chapter 2), may be eliminated. In other cases, the jaw crusher has found favor, due to the ease with which it can be shipped in sections to remote locations and for installation underground.

The type of material being crushed may also determine the crusher used. Jaw crushers perform better than gyratories on clay or plastic materials due to their greater throw. Gyratories have been found to be particularly suitable for hard, abrasive material, and they tend to give a more cubic product than jaw crushers if the feed is laminated or "slabby."

6.3 SECONDARY/TERTIARY CRUSHERS

Secondary crushers are lighter than the heavy-duty, rugged primary machines. The bulk of secondary/tertiary crushing of metalliferous ores is performed by cone crushers. Since they take the primary crushed ore as feed, the maximum feed size will normally be less than 15 cm in diameter and, because most of the harmful constituents in the ore, such as tramp metal, wood, clays, and slimes have already been removed, it is much easier to handle. Similarly, the transportation and feeding arrangements serving the crushers do not need to be as rugged as in the primary stage. Secondary/tertiary crushers also operate with dry feeds, and their purpose is to reduce the ore to a size suitable for grinding (Figure 6.1(a)). Tertiary crushers

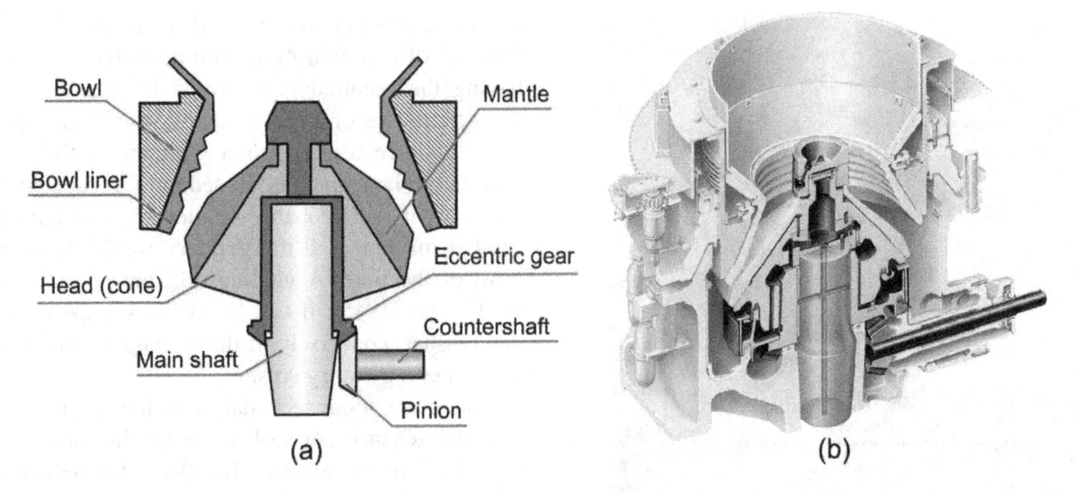

FIGURE 6.8 Cone crusher: (a) functional diagram, and (b) cross section *(Courtesy Metso).*

are, to all intents and purposes, of the same design as secondaries, except that they have a closer set.

6.3.1 Cone Crushers

The cone crusher is a modified gyratory crusher, and accordingly many of the same terms including gape, set, and throw, apply. The essential difference is that the shorter spindle of the cone crusher is not suspended, as in the gyratory, but is supported in a curved, universal bearing below the gyratory head or *cone* (Figure 6.8). Major suppliers of cone crushers include Metso (GP, HP, and MP Series), FLSmidth (Raptor® Models) and Sandvik (CS- and CH-series, developed from the Allis Chalmers Hydrocone Series), who manufacture a variety of machines for coarse and fine crushing applications.

Power is transmitted from the source to the countershaft through a V-belt or direct drive. The countershaft has a bevel pinion pressed and keyed to it, and drives the gear on the eccentric assembly. The eccentric has a tapered, offset bore and provides the means whereby the head and main shaft follow an eccentric path during each cycle of rotation.

Since a large gape is not required, the crushing shell or bowl flares outwards, which allows for the swell of broken ore by providing an increasing cross sectional (annular) area toward the discharge end. The cone crusher is therefore an excellent arrested crusher. The flare of the bowl allows a much greater head angle than in the gyratory crusher, while retaining the same angle between the crushing members. This gives the cone crusher a high capacity, since the capacity of gyratory crushers is roughly proportional to the diameter of the head.

The head is protected by a replaceable mantle, which is held in place by a large locking nut threaded onto a collar bolted to the top of the head. The mantle is backed with plastic cement, or zinc, or more recently with an epoxy resin.

Unlike a gyratory crusher, which is identified by the dimensions of the feed opening and the mantle diameter, a cone crusher is rated by the diameter of the cone mantle. Cone crushers are also identified by the power draw: for example, the Metso MP1000 refers to 1,000 HP (746 kW), and the FLSmidth XL2000 refers to 2,000 HP (1,491 kW). As with gyratories, one advance has been to increase power draw (Erikson, 2014). The largest unit is the Metso MP2500 recently installed at First Quantum Minerals' Sentinel mine in Zambia with capacity of 3,000 to 4,500 t h^{-1} (Anon., 2014). The increase in cone crusher capacity means they better match the capacity of the primary gyratories, simplifying circuits, which in the past could see several secondary and especially tertiary cone crushers in parallel (Major, 2002).

The throw of cone crushers can be up to five times that of primary crushers, which must withstand heavier working stresses. They are also operated at much higher speeds (700–1,000 rpm). The material passing through the crusher is subjected to a series of hammer-like blows rather than being gradually compressed as by the slowly moving head of the gyratory crusher.

The high-speed action allows particles to flow freely through the crusher, and the wide travel of the head creates a large opening between it and the bowl when in the fully open position. This permits the crushed fines to be rapidly discharged, making room for additional feed. The fast discharge and non-choking characteristics of the cone crusher allow a reduction ratio in the range 3:1–7:1, but this can be higher in some cases.

The classic Symons cone crusher heads were produced in two forms depending on the application: the standard (Figure 6.9(a)) for normal secondary crushing and the short-

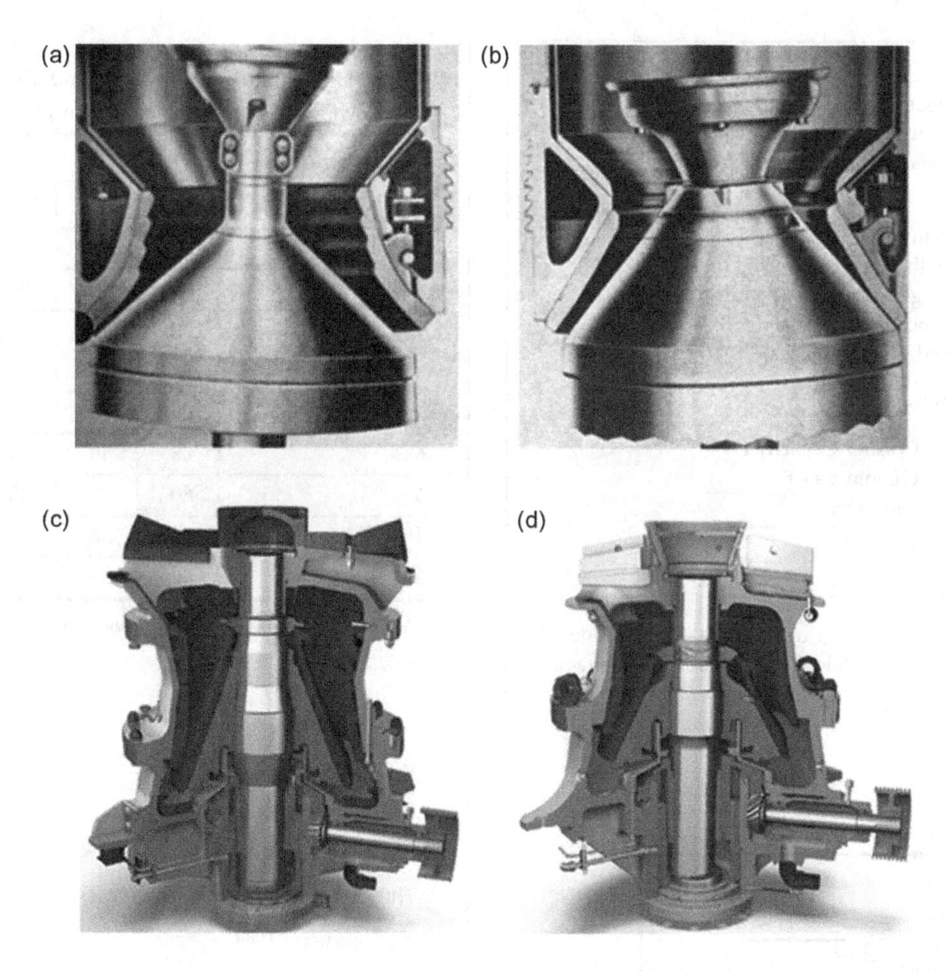

FIGURE 6.9 Cone crushers: original (a) standard, (b) short-head, and modern (c) secondary, and (d) tertiary cone crushers *((c) and (d) Courtesy Metso).*

head (Figure 6.9(b)) for fine, or tertiary duty. They differ mainly in the shape of their crushing chambers. The standard cone has "stepped" liners, which allow a coarser feed than in the short-head, and delivers a product varying from ca. 0.5 to 6 cm. The short-head has a steeper head angle than the standard, which helps to prevent choking from the much finer material being handled. It also has a narrower feed opening and a longer parallel section at the discharge, and delivers a product of ca. 0.3–2.0 cm. Contemporary secondary (Figure 6.9(c)) and tertiary (Figure 6.9(d)) cone crusher designs have evolved and geometry varies depending on the ore type, feed rate, and feed and product particle size.

The parallel section between the liners at the discharge is a feature of all cone crushers and is incorporated to maintain a close control on product size. Material passing through the parallel zone receives more than one impact from the crushing members. This tends to give a product size dictated by the closed set rather than the open set. The distributing plate on the top of the cone helps to centralize the feed, distributing it at a uniform rate to the entire crushing chamber. An important feature of the crusher is that the bowl is held down either by an annular arrangement of springs or by a hydraulic mechanism. These allow the bowl to yield if uncrushable "tramp" material enters the crushing chamber, so permitting the offending object to pass. (The presence of such tramp material is readily identified by the noise and vibration caused on its passage through the chamber.) If the springs are continually activated, as may happen with ores containing many tough particles, oversize material will be allowed to escape from the crusher. This is one of the reasons for using closed-circuit crushing in the final stages. It may be necessary to choose a screen for the circuit that has apertures slightly larger than the set of the crusher (Example 6.1). This is to reduce the tendency for very tough particles, which are slightly oversize, to "spring" the crusher, causing an accumulation of such particles in the closed-circuit and a build-up of pressure in the crushing throat.

The set on the crusher can easily be changed, or adjusted for liner wear, by screwing the bowl up or down by means of a capstan and chain arrangement or by adjusting the hydraulic setting, which allows the operator to change settings even if the equipment is operating under maximum load (Anon., 1985). To close the setting,

Example 6.1
For the circuit Figure ex 6.1 and the given data:
1. Which cone crusher would you select?
2. What is the feed rate to the screen?

Given data:
Fresh feed to the circuit (N) = 200 t h^{-1}
Fraction finer than screen opening (n) = 20%
Screen opening (aperture) = 13 mm
Screen efficiency (E_U, see Chapter 8) = 90%
Crusher closed side setting (c.s.s) = 10 mm

Crusher options:

Crusher Model	Crusher Capacity t h^{-1} (10 mm c.s.s.)
Model #1	140–175
Model #2	175–220
Model #3	260–335

Note: In Figure ex 6.1, upper case variables denote solids flow rates, t h^{-1}, and lower case variables the fraction less than the screen opening, % (Table ex 6.1).

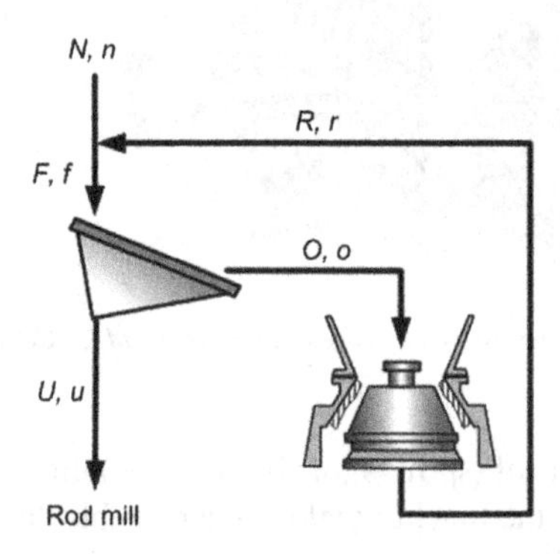

FIGURE EX 6.1 Typical closed circuit final crushing stage.

Solution
Part 1
Solve for circulating load, R/U as follows:
Mass balances:
Across circuit $N = U$
Across crusher $O = R$
Around screen $F = O + U$

$$Ff = Oo + U(\text{assume } u = 1)$$

$$E_U = \frac{U}{Ff}$$

At feed node to screen $F = N + R = U + R$

$$Ff = Nn + Rr = Un + Rr$$

By dividing by U $\frac{Ff}{U} = n + \frac{Rr}{U}$
substituting $Ff = U/E_U$ $\frac{1}{E} = n + \frac{R}{U}r$

TABLE EX 6.1 Crusher Product Gradation Table (wt% passing particle size for given c.s.s.)[a]

Particle size, mm	C.S.S., mm			
	8	10	13	16
20	100	100	100	90
13	100	**90**	80	64
10	100	80	64	54
8	93	68	55	45
7	74	60	50	40
6	68	50	45	35
5	60	45	38	30
4	50	38	32	24
3	40	30	26	18

[a] *Based on Metso Crushing and Screening Handbook, 5th Edn.*

then, re-arranging $\frac{R}{U} = \frac{1}{r}\left(\frac{1}{E_U} - n\right)$
From the crusher product gradation table provided:

$r = 0.9(90\%)$(i.e., % − 13 mm for c.s.s. = 10 mm)

Thus solving for R (=O) where U (=N) = 200 t h^{-1}:

$$\frac{R}{200} = \frac{1}{0.9}\left(\frac{1}{0.9} - 0.2\right)$$

$R = 202.5$ t h^{-1}

From the crusher capacity data, choice is model #2.
Part 2
Feed rate to the screen $F = N + R = 402.5$ t h^{-1}
This illustrates the approach to crusher sizing (as well as giving an exercise in identifying the balances). The approach is described in detail in Mular and Jergensen (1982) and Mular et al. (2002). Crusher suppliers, of course, have propriety methods of selecting and sizing units for a given duty.

the operator opens a valve which pumps hydraulic oil to the cylinder supporting the crusher head. To open the setting, another valve is opened, allowing the oil to flow out of the cylinder. Efficiency is enhanced through automatic tramp iron clearing and reset. When tramp iron enters the crushing chamber, the crushing head will be forced down, causing hydraulic oil to flow into the accumulator. When the tramp iron has passed from the chamber, nitrogen pressure forces the hydraulic oil from the accumulator back into the supporting hydraulic cylinder, thus restoring the original setting.

Wet Crushing

In 1988 Nordberg Inc. introduced wet tertiary cone crushing at a Brazilian lead-zinc mine (Karra, 1990). The *Water Flush* technology (now supplied by Metso) uses a cone crusher incorporating special seals, internal components, and lubricants to handle the large flow of water, which is added to the crusher to produce a product slurry containing 30–50% solids by weight, that can be fed directly to ball mills. The flushing action permits tighter closed side settings (Major, 2009). Such technology has potential for the crushing of sticky ores, for improving productivity in existing circuits, and for developing more cost-effective circuits. However, the presence of water during crushing can increase the liner wear rates, depending on the application.

Wear

Maintenance of the wear components in both gyratory and cone crushers is one of the major operating costs. Wear monitoring is possible using a Faro Arm (Figure 6.10), which is a portable coordinate measurement machine. Ultrasonic profiling is also used. A more advanced system using a laser scanner tool to profile the mantle and concave produces a 3D image of the crushing chamber (Erikson, 2014). Some of the benefits of the liner profiling systems include: improved prediction of mantle and concave liner replacement; identifying asymmetric and high wear areas; measurement of open and closed side settings; and quantifying wear life with competing liner alloys.

FIGURE 6.10 Faro Arm *(FARO, FAROARM, and the Faro Blue color are registered trademarks of FARO Technologies Inc. © 2006 FARO Technologies Inc. All Rights Reserved).*

6.3.2 The Gyradisc® Crusher

The Gyradisc® crusher is a specialized form of cone crusher, used for producing finer material, which has found application in the quarrying industry (primarily sand and gravel) for the production of large quantities of sand at economic cost.

The main modification to the conventional cone crusher is a shorter crushing member, with the lower one being at a flatter angle (Figure 6.11). Crushing is by *inter-particle comminution* by the impact and attrition of a multi-layered mass of particles.

The angle of the lower member is less than the angle of repose of the ore, so that when at rest the material does not slide. Transfer through the crushing zone is by movement of the head. Each time the lower member moves away from the upper, material enters the attrition chamber from the surge load above. When reduction begins, material is picked up by the lower member and is moved outward. Due to the slope, it is carried to an advanced position and caught between the crushing members.

The length of stroke and the timing are such that after the initial stroke the lower member is withdrawn faster than the previously crushed material falls by gravity. This permits the lower member to recede and return to strike the previously crushed mass as it is falling, thus scattering it so that a new alignment of particles is obtained prior to another impact. At each withdrawal of the head, the void is filled by particles from the surge chamber.

At no time does single-layer crushing occur. Crushing is by particle on particle, so that the setting of the crusher is not as directly related to the size of product as it is on the cone crusher. When used in open circuit, the Gyradisc® will produce a product of chippings from about 1 cm downwards, of good cubic shape, with a satisfactory amount of sand, which obviates the use of blending and re-handling. In closed circuit, they are used to

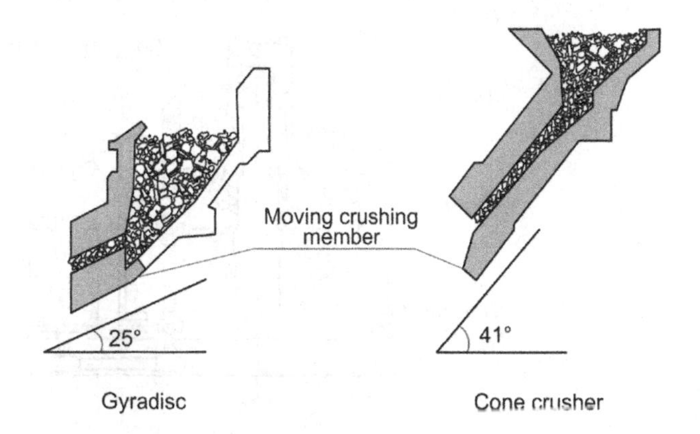

FIGURE 6.11 Comparison of the Gyradisc and a conventional cone crusher *(Adapted from Nordberg, n.d.).*

produce large quantities of sand. They may be used in open circuit on clean metalliferous ores with no primary slimes to produce an excellent ball-mill feed. Feeds of less than 19 mm may be crushed to about 3 mm (Lewis et al., 1976).

6.3.3 The Rhodax® Crusher

The Rhodax® crusher is another specialized form of cone crusher, referred to as an inertial cone crusher. Developed by FCB (now Fives fcb) Research Center in France, the Rhodax® crusher is claimed to offer process advantages over conventional cone crushers and is based on inter-particle compression crushing. It consists of a frame supporting a cone and a mobile ring, and a set of rigid links forming ties between the two parts (Figure 6.12). The frame is supported on elastic suspensions isolating the environment from dynamic stresses created by the crushing action. It contains a central shaft fixed on a structure. A grinding cone is mounted on this shaft and is free to rotate. A sliding sleeve on this shaft is used to adjust the vertical position of the cone and therefore the setting, making it simple to compensate for wear. The ring structure is connected to the frame by a set of tie rods. The ring and the cone are made of wear resistant steel.

One set of synchronized unbalanced masses transmits a known and controlled crushing force to the ring when the masses rotate. This fragmentation force stays constant even if the feed varies, or an unbreakable object enters the crushing chamber. The Rhodax® is claimed to achieve reduction ratios varying from 4 to more than 30 in open circuit. The relative positions of the unbalanced masses can be changed if required, so the value of the crushing force can thus be remotely controlled. As feed particles enter the fragmentation chamber, they slowly advance between the cone and the moving ring. These parts are subjected to horizontal circular translation movements and move toward and away from each other at a given point.

During the approach phase, materials are subjected to compression, up to 10–50 MPa. During the separation phase, fragmented materials pass further down in the chamber until the next compression cycle. The number of cycles is typically 4 to 5 before discharge. During these cycles the cone rolls on a bed of stressed material a few millimeters thick, with a rotation speed of a 10–20 rpm. This rotation is actually an epicyclical movement, due to the lack of sliding friction between the cone and the feed material. The unbalanced masses rotate at 100–300 rpm. The following three parameters can be adjusted on the Rhodax® crusher: the gap between the cone and the ring, the total static moment of unbalanced masses, and the rotation speed of these unbalanced masses.

The combination of the latter two parameters enables the operator to fix the required fragmentation force easily and quickly. Two series of machines have been developed on this basis, one for the production of aggregates (maximum pressure on the material bed between 10 and 25 MPa), and the other for feeds to grinding (25–50 MPa maximum pressure on the material bed). Given the design of the machine (relative displacement of two non-imposed wear surfaces), the product size distribution is independent of the gap and wear. These are distinct advantages over conventional cone crushers, which suffer problems with the variable product quality caused by wear.

6.3.4 A Development in Fine Crushing

IMP Technologies Pty. Ltd. has recently tested a pilot-scale super fine crusher that operates on dry ore and is envisaged as a possible alternative to fine or ultra-fine grinding circuits (Kelsey and Kelly, 2014). The unit includes a rotating compression chamber and an internal gyrating mandrel (Figure 6.13). Material is fed into the compression chamber and builds until the gyratory motion of the mandrel is engaged. Axial displacement of the compression chamber and the gyratory motion of

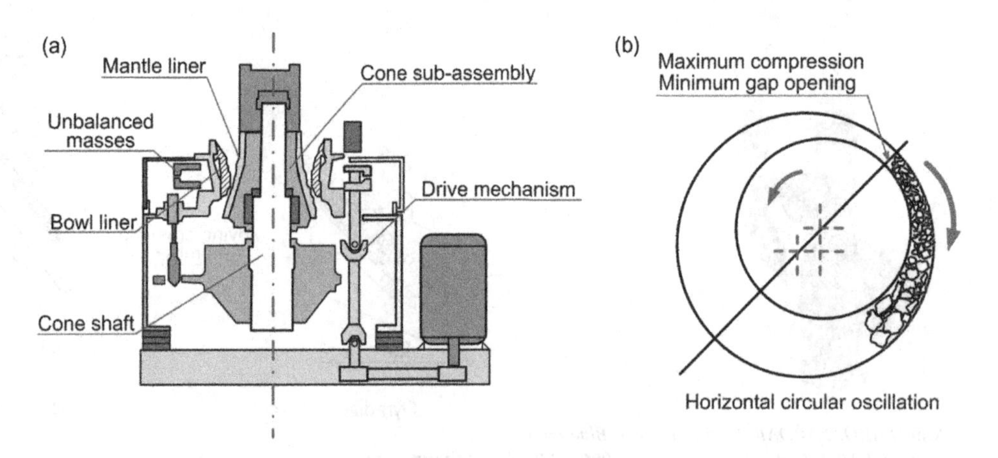

FIGURE 6.12 (a) Schematic of the Rhodax crusher, and (b) principle of operation *(Adapted from Portal, 2007).*

the mandrel result in fine grinding of the feed material. In one example, a feed F_{80} of 300 μm was reduced to P_{80} of 8 μm, estimated to be the equivalent to two stages of grinding. This development is the latest in a resurgence in crushing technology resulting from the competition of AG/SAG milling and the demands for increased comminution energy efficiency.

6.3.5 Roll Crushers

Although not widely used in the minerals industry, roll crushers can be effective in handling friable, sticky, frozen, and less abrasive feeds, such as limestone, coal, chalk, gypsum, phosphate, and soft iron ores.

Roll crusher operation is fairly straightforward: the standard spring rolls consist of two horizontal cylinders that revolve toward each other (Figure 6.14(a)). The gap (closest distance between the rolls) is determined by shims which cause the spring-loaded roll to be held back from the fixed roll. Unlike jaw and gyratory crushers, where reduction is progressive by repeated nipping action as the material passes down to the discharge, the crushing process in rolls is one of single pressure.

FIGURE 6.13 The pilot-scale IMP super fine crusher *(Courtesy IMP Technologies Pty. Ltd.)*.

Roll crushers are also manufactured with only one rotating cylinder (Figure 6.14(b)), which revolves toward a fixed plate. Other roll crushers use three, four, or six cylinders, although machines with more than two rolls are rare today. In some crushers the diameters and speeds of the rolls may differ. The rolls may be gear driven, but this limits the distance adjustment between the rolls. Modern rolls are driven by V-belts from separate motors.

The disadvantage of roll crushers is that, in order for reasonable reduction ratios to be achieved, very large rolls are required in relation to the size of the feed particles. They therefore have the highest capital cost of all crushers for a given throughput and reduction ratio.

The action of a roll crusher, compared to the other crushers, is amenable to a level of analysis. Consider a spherical particle of radius r, being crushed by a pair of rolls of radius R, the gap between the rolls being $2a$ (Figure 6.15). If μ is the coefficient of friction between the rolls and the particle, θ is the angle formed by the tangents to the roll surfaces at their points of contact with the particle (the angle of nip), and C is the compressive force exerted by the rolls acting from the roll centers through the particle center, then for a particle to be just gripped by the rolls, equating vertically, we derive:

$$C \sin\left(\frac{\theta}{2}\right) = \mu C \cos\left(\frac{\theta}{2}\right) \quad (6.3)$$

Therefore,

$$\mu = \tan\left(\frac{\theta}{2}\right) \quad (6.4)$$

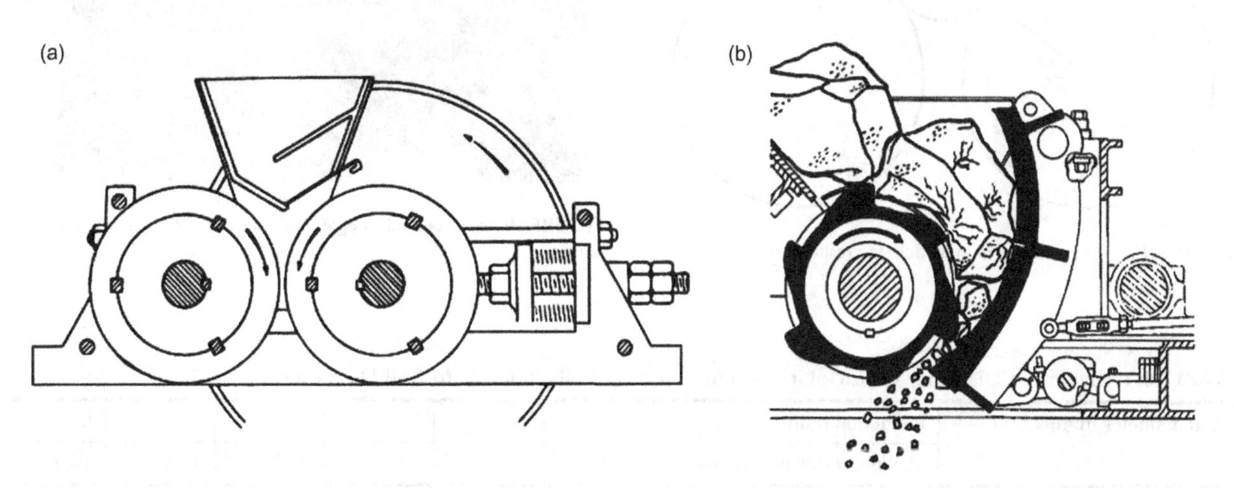

FIGURE 6.14 Roll crushers: (a) double, and (b) single roll crusher *((b) Courtesy TerraSource Global)*.

The coefficient of friction between steel and most ore particles is in the range 0.2–0.3, so that the value of the angle of nip θ should never exceed about 30°, or the particle will slip. It should also be noted that the value of the coefficient of friction decreases with speed, so that the speed of the rolls depends on the angle of nip, and the type of material being crushed. The larger the angle of nip (i.e., the coarser the feed), the slower the peripheral speed needs to be to allow the particle to be nipped. For smaller angles of nip (finer feeds), the roll speed can be increased, thereby increasing the capacity. Peripheral speeds vary between about 1 m s^{-1} for small rolls, up to about 15 m s^{-1} for the largest sizes of 1,800 mm diameter upwards.

The value of the coefficient of friction between a particle and moving rolls can be calculated from:

$$\mu_k = \left[\frac{1 + 1.12v}{1 + 6v}\right]\mu \quad (6.5)$$

where μ_k is the kinetic coefficient of friction and v the peripheral velocity of the rolls (m s^{-1}). From Figure 6.15:

$$\cos\left(\frac{\theta}{2}\right) = \frac{R + a}{R + r} \quad (6.6)$$

Equation 6.6 can be used to determine the maximum size of rock gripped in relation to roll diameter and the reduction ratio (r/a) required. Table 6.1 gives example values for 1,000 mm roll diameter where the angle of nip should be less than 20° in order for the particles to be gripped (in most practical cases the angle of nip should not exceed about 25°).

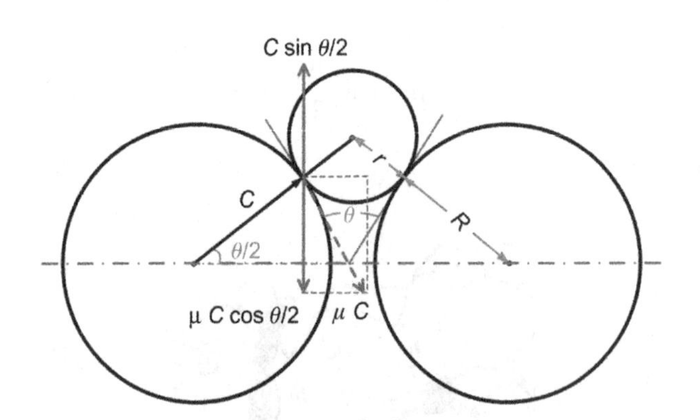

FIGURE 6.15 Forces acting on a particle in crushing rolls.

Unless very large diameter rolls are used, the angle of nip limits the reduction ratio of the crusher, and since reduction ratios greater than 4:1 are rare, a flowsheet may require coarse crushing rolls to be followed by fine rolls.

Smooth-surfaced rolls are usually used for fine crushing, whereas coarse crushing is often performed in rolls having corrugated surfaces, or with stub teeth arranged to present a chequered surface pattern. "Sledging" or "slugger" rolls have a series of intermeshing teeth, or slugs, protruding from the roll surfaces. These dig into the rock so that the action is a combination of compression and ripping, and large pieces in relation to the roll diameter can be handled. Toothed crushing rolls (Figure 6.16) are typically used for coarse crushing of soft or sticky iron ores, friable limestone or coal, where rolls of ca. 1 m diameter are used to crush material of top size of ca. 400 mm.

Wear on the roll surfaces is high and they often have a manganese steel tire, which can be replaced when worn. The feed must be spread uniformly over the whole width of the rolls in order to give even wear. One simple method is to use a flat feed belt of the same width as the rolls.

Since there is no provision for the swelling of broken ore in the crushing chamber, roll crushers must be "starvation fed" if they are to be prevented from choking. Although the floating roll should only yield to an uncrushable body, choked crushing causes so much pressure that the springs are continually activated during

FIGURE 6.16 Toothed single roll crusher *(Courtesy McLanahan Corporation).*

TABLE 6.1 Maximum Diameter of Rock Gripped in Crushing Rolls Relative to Roll Diameter

Roll diameter (mm) 1,000	Reduction ratio	2	3	4	5	6
	Maximum size of rock gripped (mm)	30.9	23.0	20.4	19.1	18.3

crushing, and some oversize escapes. Rolls should therefore be used in closed circuit with screens. Choked crushing also causes inter-particle comminution, which leads to the production of material finer than the gap of the crusher.

The capacity of the rolls can be calculated in terms of the ribbon of material that will pass the space between the rolls. Thus theoretical capacity (Q, kg h^{-1}) is equal to:

$$Q = 188.5NDWsd \tag{6.7}$$

where N is the speed of rolls (rpm), D the roll diameter (m), W the roll width (m), s the density of feed material (kg m^{-3}), and d the distance between the rolls (m).

In practice, allowing for voids between the particles, loss of speed in gripping the feed, etc., the capacity is usually about 25% of the theoretical.

6.4 HIGH PRESSURE GRINDING ROLLS

The pressure exerted on the feed particles in conventional roll crushers is in the range 10–30 MPa. During the 1970–80s, work by Prof. Schönert of the Technical University of Clausthal, Germany led to the development of the *High-Compression Roller Mill*, which utilized forces from 50–150 MPa (Schönert, 1979, 1988; McIvor, 1997). The units are now commonly termed *High Pressure Grinding Rolls* (HPGR) and employ a fixed and movable roll to crush material. A hydraulic pressure system acts on pistons that press the movable roll against a material bed (density >70% solids by volume) fed to the rolls (Figure 6.17). The roll gap can be adjusted depending on the feed particle size and application.

The first commercial installation began operation in 1985 to grind cement clinker (McIvor, 1997). Since then, HPGR technology has seen wide use in the cement, limestone and diamond industries and has recently been implemented in hard rock metalliferous operations (Morley, 2010). Researchers have noted that the high pressure exerted on the particle bed produces a high proportion of fines and particles with micro-cracks and improved mineral liberation, which can be advantageous for subsequent comminution or metallurgical processes (Esna-Ashari and Kellerwessel, 1988; Clarke and Wills, 1989; Knecht, 1994; Watson and Brooks, 1994; Daniel and Morrell, 2004).

Unlike conventional crushers or tumbling mills, which employ impact and attrition to break particles, the HPGR employs inter-particle crushing in the bed, and as such the particle packing properties of the feed material play a role in determining breakage. Feed that is scalped (fines removed) prior to the HPGR is termed "truncated." The removal of fines impacts HPGR operation, as coarse particles tend to have a greater impact on roll wear, and the elimination of fines creates a less compact bed, which reduces the inter-particle breakage action. To ensure proper bed, formation the units should be choke fed the entire length of the rolls.

The HPGR product typically comprises fines and portions of compacted cake referred to as "flakes" (Figure 6.18). Depending on flake competency, the product may require subsequent deagglomeration (ranging from a mild pre-soaking to modest attritioning) to release the fines (van der Meer and Grunedken, 2010). It has been shown that the specific energy consumption for compression and ball mill-deagglomeration is

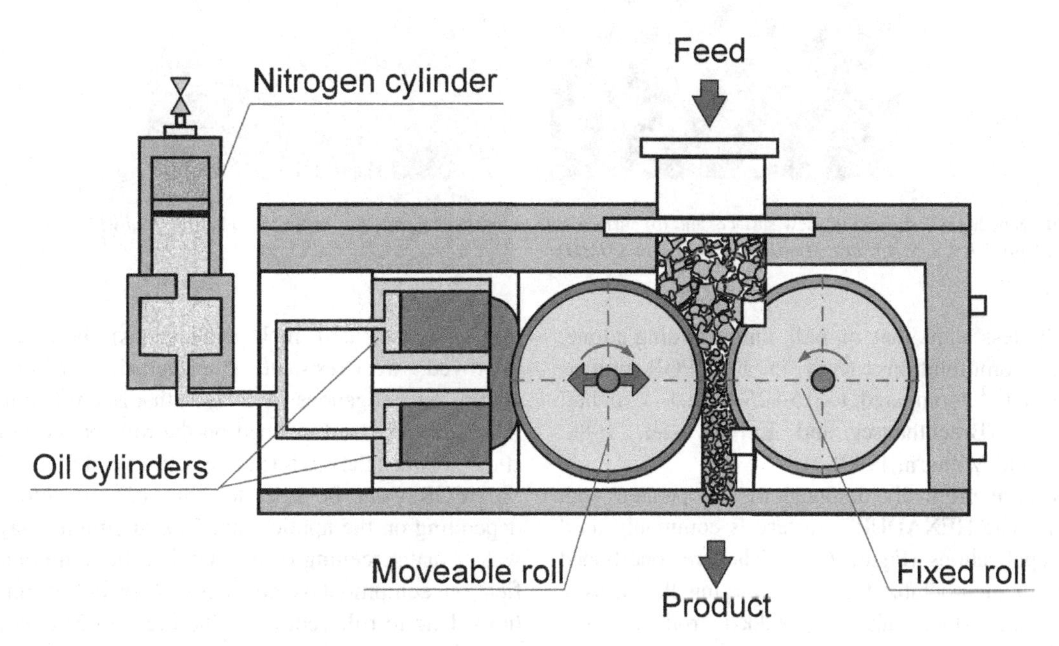

FIGURE 6.17 High pressure grinding rolls *(Adapted from Napier-Munn et al., 1996; Courtesy JKMRC and The University of Queensland).*

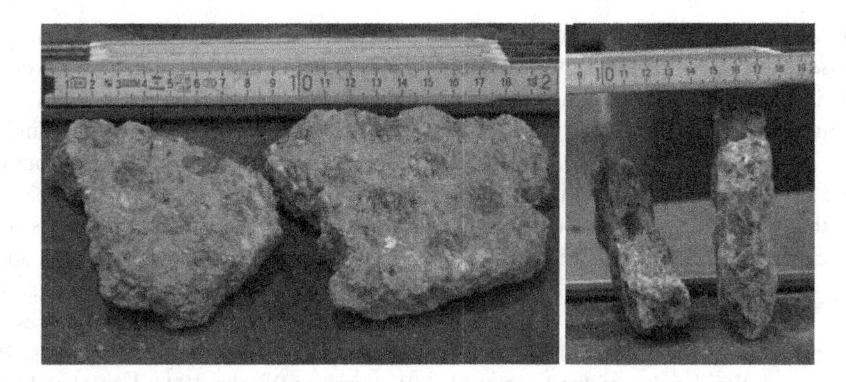

FIGURE 6.18 Example flake product from an HPGR *(Used with permission, van der Meer and Gruendken, 2010, Minerals Engineering, copyright Elsevier).*

FIGURE 6.19 HEXADUR® (a) standard roll surface, (b) pre-conditioned surface, and (c) surface after use retaining autogenous wear layer *(Courtesy Köppern, Germany).*

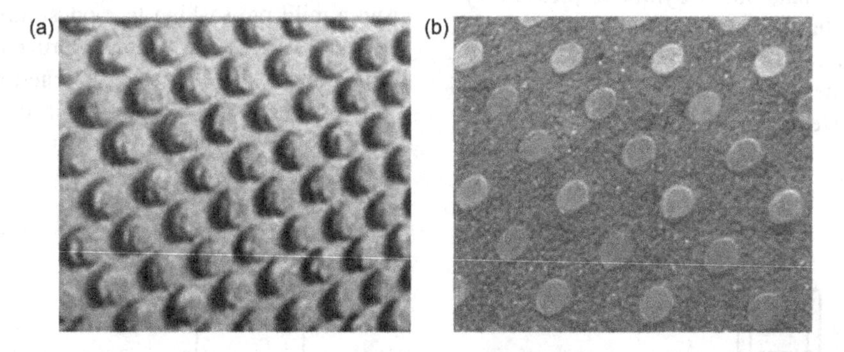

FIGURE 6.20 Studded roll surface: (a) new surface, and (b) surface with autogenous wear layer between studs *((b) Used with permission, van der Meer and Maphosa, 2012, J. S. Afr. Inst. Min. Metall, copyright SAIMM).*

considerably less than that of ball mill grinding alone. The typical comminution energy in an HPGR unit is 2.5–3.5 kWh t^{-1}, compared to 15–25 kWh t^{-1} in ball mill grinding (Brachthauser and Kellerwessel, 1988; Schwechten and Milburn, 1990).

HPGRs were originally designed to be operated with smooth rolls. The HEXADUR® surface is commonly used in cement applications (Figure 6.19). The pre-conditioned surface design incorporates tiles with varying thicknesses, which enhances feed intake. Studded roll surfaces (Figure 6.20) have become standard in the new designs (especially in hard rock applications), because of their improved wear-resistant characteristics. Most surfaces employ an autogenous wear layer, that is, crushed feed material is captured and retained on the roll surface in the interstices between the studs (Figure 6.19(c) and Figure 6.20(b)).

HPGRs can be operated in open- or closed-circuit depending on the application. Closed circuits may employ wet or dry screening or air classification, although classification equipment is not a necessity in certain applications. Due to roll geometry, the press force exerted at the roll edges is less than in the center, resulting in a coarser

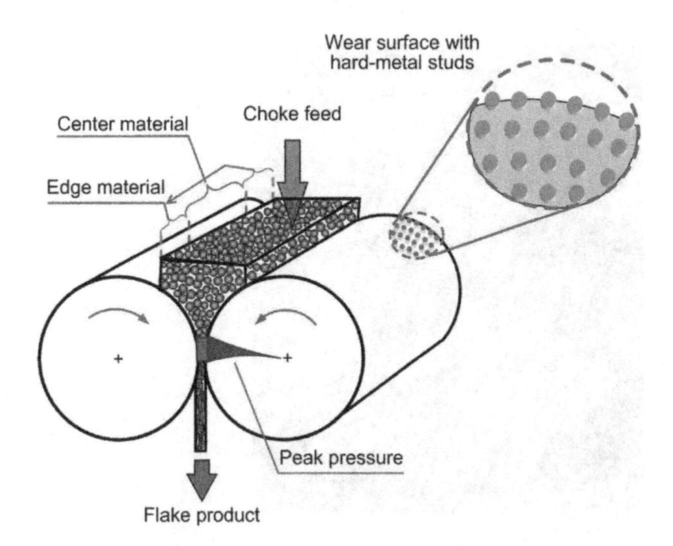

FIGURE 6.21 Diagram of HPGR operation *(Adapted from van der Meer and Gruendken, 2010).*

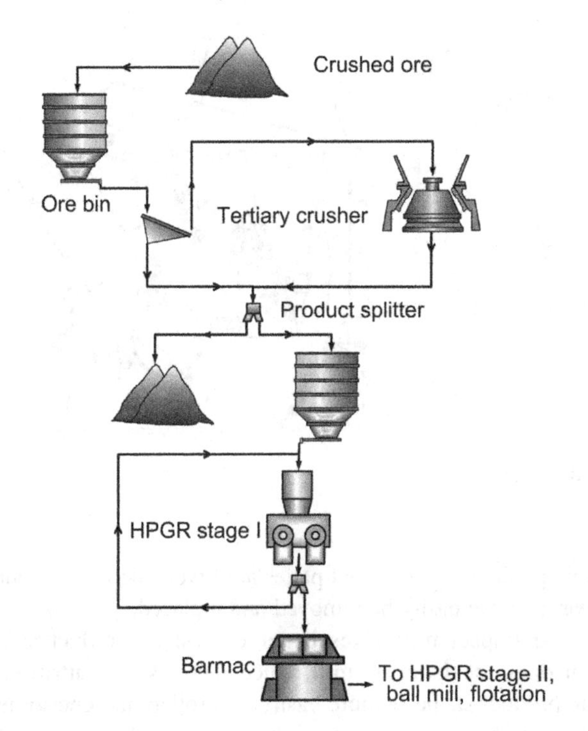

FIGURE 6.22 Example of an HPGR with "edge" recycle in a crushing circuit *(Adapted from van der Meer and Gruendken, 2012).*

edge product (Figure 6.21). Splitters can be used to separate the edge product for recycle (van der Meer and Grunedken, 2010). Figure 6.22 shows an example of edge material recycle using a splitter in a crushing application at a copper flotation plant.

6.5 IMPACT CRUSHERS

Impact crushers (e.g., hammer mills and impact mills) employ sharp blows applied at high speed to free-falling rocks where comminution is by impact rather than compression. The moving parts are "beaters," which transfer some of their kinetic energy to the ore particles upon contact. Internal stresses created in the particles are often large enough to cause them to shatter. These forces are increased by causing the particles to impact upon an anvil or breaker plate.

There is an important difference between the states of materials crushed by pressure and by impact. There are internal stresses in material broken by pressure that can later cause cracking. Impact causes immediate fracture with no residual stresses. This stress-free condition is particularly valuable in stone used for brick-making, building, and roadmaking, in which binding agents (e.g., tar) are subsequently added. Impact crushers, therefore, have a wider use in the quarrying industry than in the metal-mining industry. They may give trouble-free crushing on ores that tend to be plastic and pack when the crushing forces are applied slowly, as is the case in jaw and gyratory crushers. These types of ore tend to be brittle when the crushing force is applied instantaneously by impact crushers (Lewis et al., 1976).

Impact crushers are also favored in the quarry industry because of the improved product shape. Cone crushers tend to produce more elongated particles because of their ability to pass through the chamber unbroken. In an impact crusher, all particles are subjected to impact and the elongated particles, having a lower strength due to their thinner cross section, would be broken (Ramos et al., 1994; Kojovic and Bearman, 1997).

6.5.1 Hammer Mills

Figure 6.23(a) shows the cross section of a typical hammer mill. The hammers (Figure 6.23(b)) are made from manganese steel or nodular cast iron containing chromium carbide, which is extremely abrasion resistant. The breaker plates are made of the same material.

The hammers are pivoted so as to move out of the path of oversize material (or tramp metal) entering the crushing chamber. Pivoted (swing) hammers exert less force than they would if rigidly attached, so they tend to be used on smaller impact crushers or for crushing soft material. The exit from the mill is perforated, so that material that is not broken to the required size is retained and swept up again by the rotor for further impacting. There may also be an exit chute for oversize material which is swept past the screen bars. Certain design configurations include a central discharge chute (an opening in the screen) and others exclude the screen, depending on the application.

The hammer mill is designed to give the particles velocities of the order of that of the hammers. Fracture is either due to impact with the hammers or to the

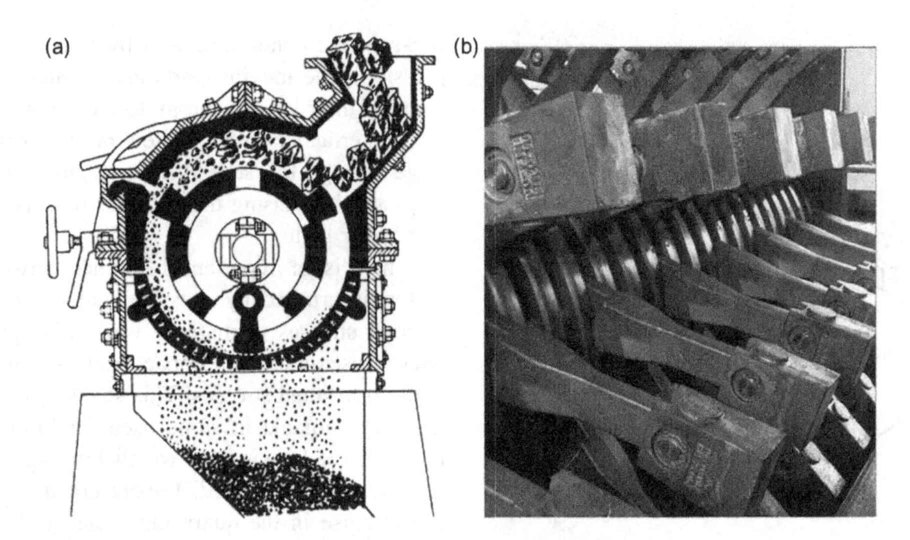

FIGURE 6.23 (a) Diagram of a hammer mill, and (b) close-up of hammers *((b) Courtesy Sandvik).*

subsequent impact with the casing or grid. Since the particles are given high velocities, much of the size reduction is by attrition (i.e., particle on particle breakage), and this leads to little control on product size and a much higher proportion of fines than with compressive crushers.

The hammers can weigh over 100 kg and can work on feed up to 20 cm. The speed of the rotor varies between 500 and 3,000 rpm. Due to the high rate of wear on these machines (wear can be taken up by moving the hammers on the pins) they are limited in use to relatively non-abrasive materials. They have extensive use in limestone quarrying and in the crushing of coal. A great advantage in quarrying is the fact that they produce a relatively cubic product.

A model of the swing hammer mill has been developed for coal applications (Shi et al., 2003). The model is able to predict the product size distribution and power draw for given hammer mill configurations (breaker gap, under-screen orientation, screen aperture) and operating conditions (feed rate, feed size distribution, and breakage characteristics).

6.5.2 Impact Mills

For coarser crushing, the fixed hammer *impact mill* is often used (Figure 6.24). In these machines the material falls tangentially onto a rotor, running at 250–500 rpm, receiving a glancing impulse, which sends it spinning toward the impact plates. The velocity imparted is deliberately restricted to a fraction of the velocity of the rotor to avoid high stress and probable failure of the rotor bearings.

The fractured pieces that can pass between the clearances of the rotor and breaker plate enter a second chamber created by another breaker plate, where the clearance is smaller, and then into a third smaller chamber. The *grinding path* is designed to reduce flakiness and to produce cubic particles. The impact plates are reversible to even out wear, and can easily be removed and replaced.

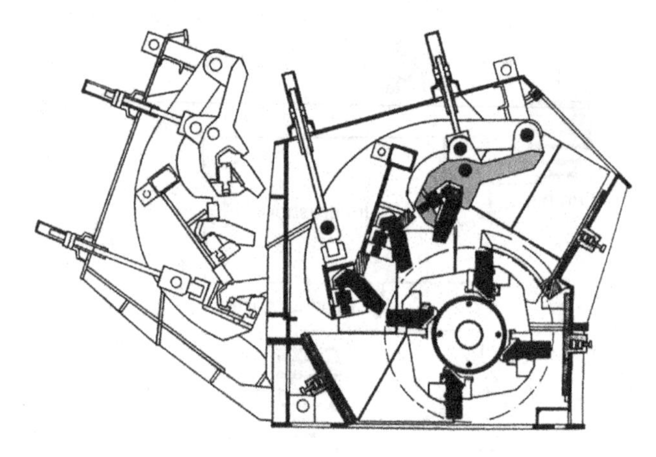

FIGURE 6.24 Impact mill.

The impact mill gives better control of product size than does the hammer mill, since there is less attrition. The product shape is more easily controlled and energy is saved by the removal of particles once they have reached the size required.

Large impact crushers will reduce 1.5 m top size ROM ore to 20 cm, at capacities of around $1500\ \text{t h}^{-1}$, although units with capacities of $3000\ \text{t h}^{-1}$ have been manufactured. Since they depend on high velocities for crushing, wear is greater than for jaw or gyratory crushers. Hence impact crushers are not recommended for use on ores containing over 15% silica (Lewis et al., 1976). However, they are a good choice for primary crushing when high reduction ratios are required (the ratio can be as high as 40:1) and the ore is relatively non-abrasive.

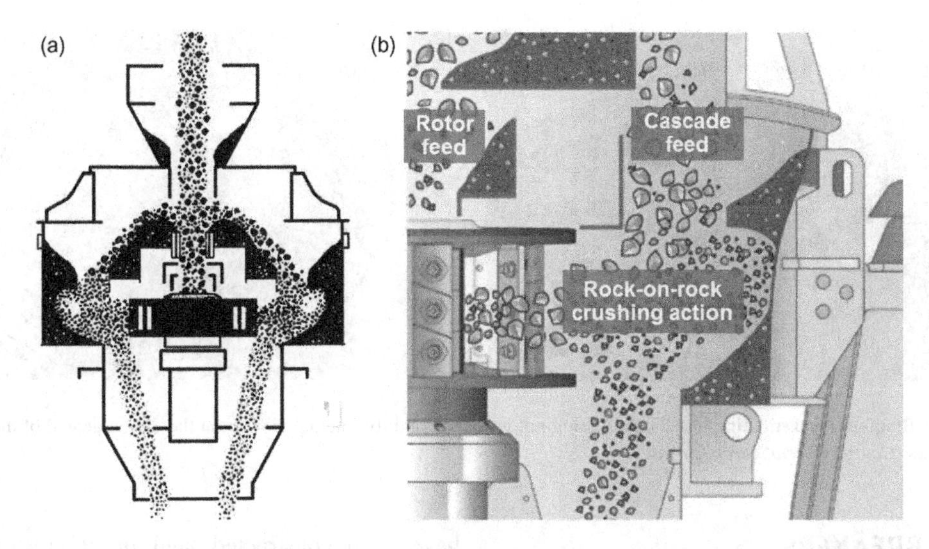

FIGURE 6.25 (a) Cross section of Barmac VSI Crusher, and (b) rock on rock crushing action in the chamber *(Courtesy Metso)*.

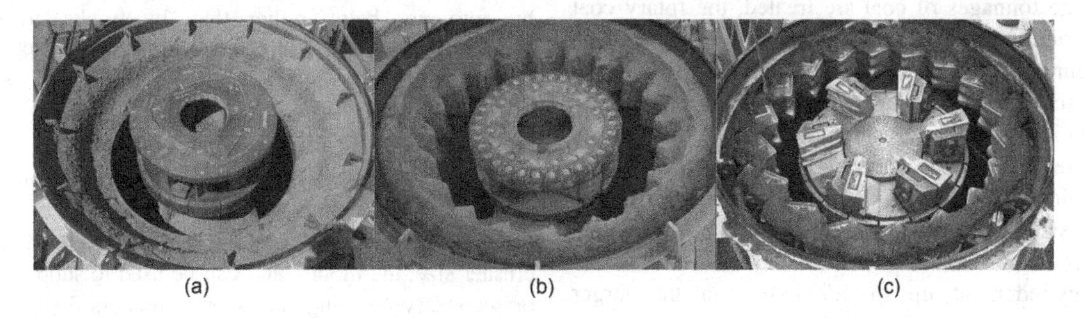

FIGURE 6.26 Example Canica VSI crusher internal chamber configurations: (a) Rock on Rock, (b) Rock on Anvil, and (c) Shoe and Anvil *(Courtesy Terex® Mineral Processing Systems)*.

6.5.3 Vertical Shaft Impact (VSI) Crushers

Barmac Vertical Shaft Impact Crusher

Developed in New Zealand in the late 1960s, over the years it has been marketed by several companies (Tidco, Svedala, Allis Engineering, and now Metso) under various names (e.g., *duopactor*). The crusher is finding application in the concrete industry (Rodriguez, 1990). The mill combines impact crushing, high-intensity grinding, and multi-particle pulverizing, and as such, is best suited in the tertiary crushing or primary grinding stage, producing products in the 0.06–12 mm size range. It can handle feeds of up to 650 t h^{-1} at a top size of over 50 mm. Figure 6.22 shows a Barmac in a circuit; Figure 6.25 is a cross-section and illustration of the crushing action.

The basic comminution principle employed involves acceleration of particles within a special ore-lined rotor revolving at high speed. A portion of the feed enters the rotor, while the remainder cascades to the crushing chamber. Breakage commences when rock enters the rotor, and is thrown centrifugally, achieving exit velocities up to 90 m s^{-1}. The rotor continuously discharges into a highly turbulent particle "cloud" contained within the crushing chamber, where reduction occurs primarily by rock-on-rock impact, attrition, and abrasion.

Canica Vertical Shaft Impact Crusher

This crusher developed by Jaques (now Terex® Mineral Processing Solutions) has several internal chamber configurations available depending on the abrasiveness of the ore. Examples include the Rock on Rock, Rock on Anvil and Shoe and Anvil configurations (Figure 6.26). These units typically operate with 5 to 6 steel impellers or hammers, with a ring of thin anvils. Rock is hit or accelerated to impact on the anvils, after which the broken fragments freefall into the discharge chute and onto a product conveyor belt. This impact size reduction process was modeled by Kojovic (1996) and Djordjevic et al. (2003) using rotor dimensions and speed, and rock breakage characteristics measured in the laboratory. The model was also extended to the Barmac crushers (Napier-Munn et al., 1996).

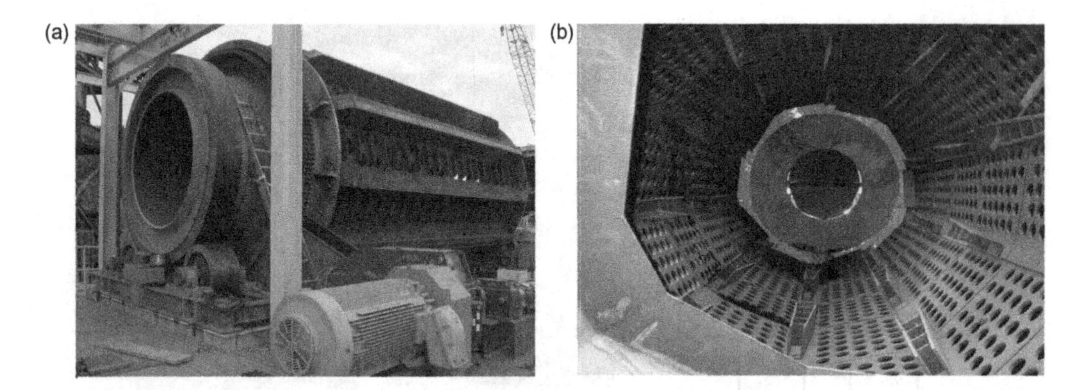

FIGURE 6.27 (a) A Bradford breaker during installation (casing not installed), and (b) internal view from the discharge end of the breaker showing screen plates and lifters *(Courtesy TerraSource Global)*.

6.6 ROTARY BREAKERS

Where large tonnages of coal are treated, the rotary coal breaker (commonly termed the Bradford breaker) can be used (Figure 6.27(a)). This is similar to the cylindrical trommel screen (Chapter 8), consisting of a cylinder 1.8–4.5 m in diameter and length of about 1½ to 2½ times the diameter, revolving at a speed of about 10–18 rpm. The machine is massively constructed, with perforated walls, the size of the perforations being the size to which the coal is to be broken. The ROM coal is fed into the rotating cylinder, at up to 1,500 t h^{-1} in the larger machines. The machine utilizes differential breakage, the coal being much more friable than the associated stones and shales, and trash such as wood or steel from the mine. The resulting small particles of coal fall through the holes, while the larger lumps of coal are transported by longitudinal lifters (Figure 6.27(b)) within the cylinder until they reach a point where they slide off the lifters and fall to the bottom of the cylinder, breaking by their own impact, and fall through the holes. The lifters are inclined to give the coal a forward motion through the breaker. Shale and stone do not break as easily, and are usually discharged from the end of the breaker, which thus cleans the coal to a certain degree and, as the broken coal is quickly removed from the breaker, produces few coal fines. Although the rotary breaker is an expensive piece of equipment, maintenance costs are low, and it produces positive control of top size product.

Esterle et al. (1996) reviewed the work on modeling of rotary breakers. The work was based at three open pit coal mines in Central Queensland, Australia, where 3 m diameter breakers were handling ROM coal.

6.7 CRUSHING CIRCUITS AND CONTROL

Efforts continue to improve crusher energy efficiency and to reduce capital and operating costs. Larger crushers have been constructed, and *in-pit* crushing units have been used, which allow relatively cheap ore transportation by conveyor belts, rather than by trucks, to a fixed crushing station (Griesshaber, 1983; Frizzel, 1985; Utley, 2009). The in-pit units are either fixed plants at the pit edge or semi- or fully-mobile units in the pit. A mobile crusher is a completely self-contained unit, mounted on a frame that is moved by means of a transport mechanism either in the open pit as mining progresses or through different mineral processing plants as required. Depending on the crusher size, the mobile unit can be used to support primary or secondary crushing stages or to process the critical size pebbles from a SAG mill. Semi-mobile units can include gyratories, being the crusher of choice for throughputs over 2,500 t h^{-1}. Fully-mobile units typically use jaw, hammer, or roll crushers, fed directly or by apron feeders, at rates of up to 1,000 t h^{-1}.

Crushing plants may be housed with the rest of the milling plant but today are often a separate facility, one reason being to better control dust. Two possible flowsheets were illustrated in Figure 6.1 to provide feed to rod (or ball mill) or to an AG/SAG mill. In some cases, the crushing circuit is designed not only to produce mill feed, but also to provide media for autogenous grinding (Wills, 1983). Crushing plants are characterized by extensive use of conveyors and the energy for transport is a consideration in selecting the type of circuit.

Two possible flowsheets for a crushing plant producing ball mill feed are shown in Figure 6.28. The circuit in Figure 6.28(a) is a "conventional" design and is typical in that the secondary feed is scalped and the secondary product is screened and conveyed to a storage bin, rather than feeding the tertiary crushers directly. The intermediate bins allow good mixing of the secondary screen oversize with the circulating load, and regulation of the tertiary crusher feed, providing more efficient crushing. Note the tertiary circuit is an example of crushers operating in parallel. The circuit is adaptable to

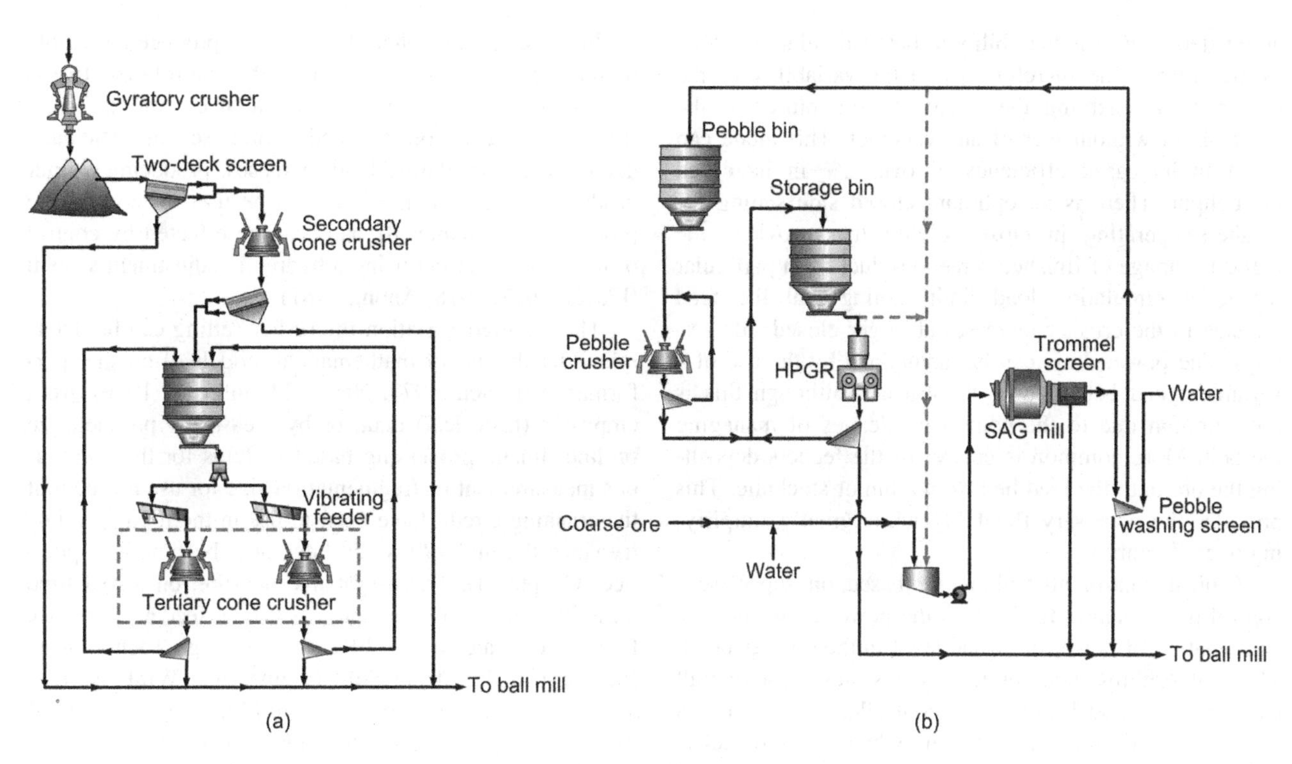

FIGURE 6.28 (a) Three-stage crushing circuit for ball mill feed, and (b) crushing circuit including HPGR *((a) Adapted from Motz, 1978; (b) Adapted from Rosario et al., 2009).*

automatic feed control to maintain maximum power utilization (Mollick, 1980). Figure 6.28(b) shows a circuit including an HPGR stage. Rosario et al. (2009) provide a selection of possible crushing circuits that plant designers can contemplate.

Advances in instrumentation and process control hardware have made the implementation of computer control more common in crushing circuits. Instrumentation includes ore level detectors, oil flow sensors, power measurement devices, belt scales, variable speed belt drives and feeders, blocked chute detectors, and particle size measurement devices (Horst and Enochs, 1980; Flintoff et al., 2014). An early example of the importance of automatic control is the crushing plant at Mount Isa in Australia, where the output increased by over 15% after controls were introduced (Manlapig and Watsford, 1983).

Supervisory control systems are not usually applied to primary crushers, the instrumentation basically being used to protect them. Thus lubrication flow indicators and bearing temperature detectors may be fitted, together with high and low level alarms in the chamber under the crusher. Vision systems are increasingly being used on both the feed and discharge of primary crushers to guide when changes in the discharge setting are required. Additionally, various systems, including vision, are being explored to detect metal (steel bars, etc.) in the truck dump to the crushers. Steel can tear a conveyor belt leading to downtime and often is undetected as it is hidden beneath ore on the conveyor after the crusher.

The operating and process control objectives for secondary and tertiary crushing circuits differ from one plant to the next. Typically, the main objective is to maximize crusher throughput at some specified product size which often requires ensuring, as best as possible, choke feed in the crusher and a properly selected closed side setting. Due to an increase in power cost and availability at a number of operations, the efficient use of power is becoming a more dominant issue. Numerous variables affect the performance of a crusher, but only three—ore feed rate, crusher set, and, in some cases, feed size—can be adjusted (with the exception of a water flush crusher or HPGR where additional degrees of freedom exist).

Lynch (1977) has described case studies of automatic control systems for various applications. When the purpose of the crushing plant is to produce feed for the grinding circuit, the most important objective of the control system is to ensure a supply of crushed ore at the rate required by the grinding plant. The fineness of the crusher product is maintained by the selection of screens of the appropriate aperture in the final closed circuit loop.

The most effective way of maximizing throughput is to maintain the highest possible crusher power draw, and this has been used to control many plants. A benefit of automated control is the ability to better regulate the feed to the crusher and therefore lessen the variability of the power draw, enabling the system to run closer to the power limit without fear of an overshoot. This alone can result in increased efficiency of over 2% in increased throughput. There is an optimum closed side setting for crushers operating in closed circuit that provides the highest tonnage of finished screen product for a particular power or circulating load limit, noting that the feed tonnage to the crusher increases at larger closed side settings. The power draw can be maintained by the use of a variable speed belt feeding the crusher, although this is not common due to the control challenges of managing the belt. More common is control of the feeders depositing the ore onto the feed belt from a bin or stockpile. This provides the necessary flexibility while greatly simplifying overall control.

Typical control algorithms are based on supervisory control that manages feed versus the power draw and/or a combination of power draw and level in the crusher bowl. These algorithms incorporate expert systems for overall strategy and model predictive controllers to provide a predictable feed rate. In situations where the size delivered from individual feeders is known, another degree of freedom for control is added, the incorporation of feeder bias and selection. Uneven feed from the bin or stockpile combined with the long time delays inherent in the distance between the feed stock and crusher, reduce the efficacy of traditional control and have driven the industry to embrace more advanced expert, model-based and multivariable control. At the same time, the software and technology for these solutions has become more accessible to the plants, resulting in robust solutions.

Operations under choked conditions also require sensing of upper and lower levels of feed in the crusher by mechanical, nuclear, sonic, vision, or proximity switches. Operation at high power draw (choked conditions) leads to increased fines production, such that if the increased throughput provided by the control system cannot be accommodated by the grinding plant, then the higher average power draw can be used to produce a finer product. In most cases, high throughput increases screen loading, which decreases screening efficiency, particularly for the particles close to the screen aperture size. This has the effect of reducing the effective "cut-size" of the screen, producing a finer product (see also Chapter 8). Thus a possible control scheme during periods of excess closed circuit crushing capacity or reduced throughput requirement is to increase the circulating load by reducing the number of screens used, leading to a finer product. The implementation of this type of control loop requires accurate knowledge of the behavior of the plant under various conditions.

In those circuits where the crushers produce a saleable product (e.g., road-stone quarries), the control objective is usually to maximize the production of certain size fractions from each ton of feed. Since screen efficiency decreases as circulating load increases, producing a finer product size, circulating load can be used to control the product size (Chapter 8). This can be effected by control of the crusher setting using a hydraulic adjustment system (Flavel, 1977, 1978; Anon., 1981).

The required variation in crusher setting can be determined by the use of mathematical models of crusher performance (Lynch, 1977; Napier-Munn et al., 1996), from empirical (historical) data, or by measuring product size on-line. Image processing based systems for the continuous measurement of fragmentation size for use throughout the crushing circuit have been in use in the mining industry since the mid-1990s and have now become best practice (Chapter 4). These systems measure, on a real time basis, the size of the ore on a belt or a feeder. Currently four systems are in use: PRC from Portage Technologies Inc., Split-Online from Split Engineering, WipFrag from WipWare Inc., and VisoRock from Metso. An example of the screen capture from a moving conveyor belt is shown in Figure 6.29 (see also Chapter 4).

Additional loops are normally required in crushing circuits to control levels in surge bins between different stages. For instance, the crusher product surge bins can be monitored such that at high level feed is increased to draw down the bins.

The importance of primary crusher control on SAG mill performance at Highland Valley Copper was well recognized, and through the use of image analysis, HVC was able to quantify the effect, and thereby regulate crusher product size through a combination of feed rate and setting control (Dance, 2001). Figure 6.30 illustrates the effect of primary crusher product size on the SAG mill throughput. Tracking the crusher product through the stockpile network, as the amount of medium size material (50–125 mm) increased, the amount of this material in the feed to the SAG mill increased, as measured 24 hours later (the +24 hours key in the figure). This size material constitutes SAG mill critical size in this operation and, as expected, as the amount fed to the SAG mill increased the tonnage decreased, for one of the SAG mills from 2,000 to 1,800 $t\,h^{-1}$. This change in amount of medium size material was caused by an increase in the amount in the feed to the primary crusher, resulting from a period of higher energy blasting, which reduced the amount of + 125 mm. Because in this operation the medium size material passed through the crusher virtually unchanged, the increase was reflected in the crusher product. It is necessary to understand these interacting factors to effect control and maximize the throughput of the circuit.

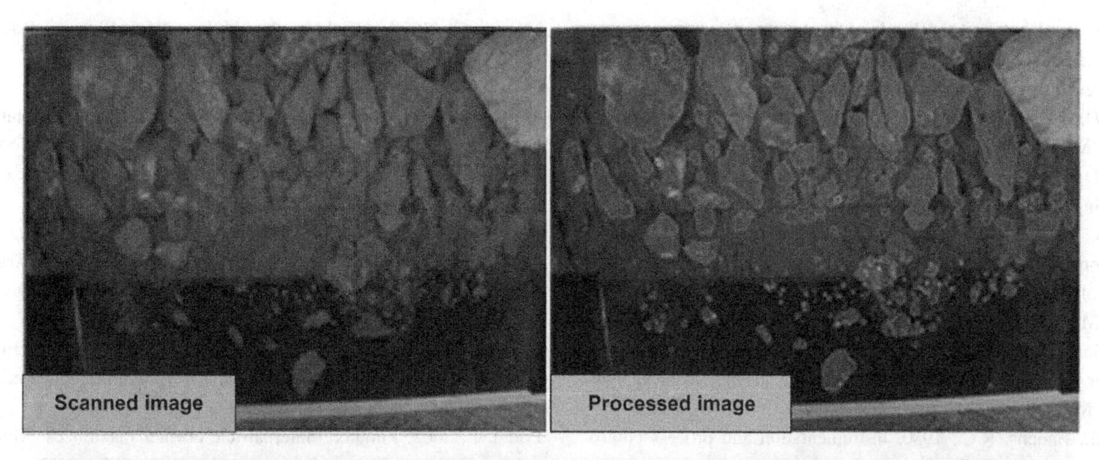

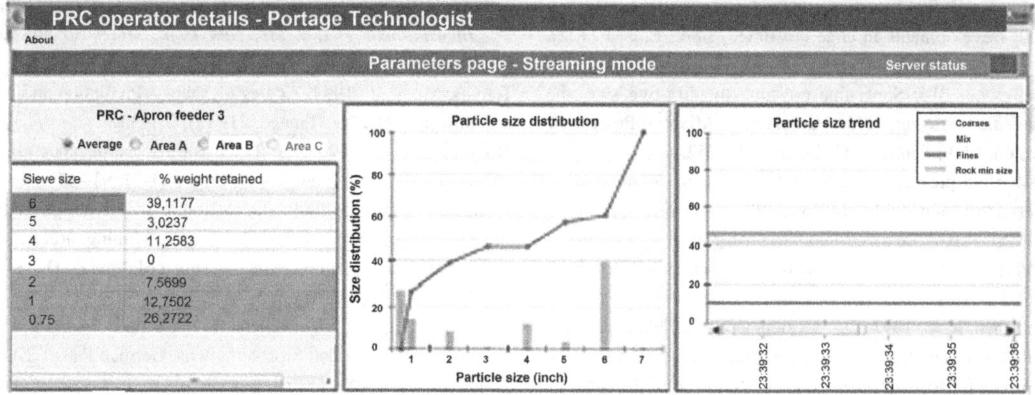

FIGURE 6.29 Screen capture of Portage PRC® *(Courtesy Portage Technologies Inc.).*

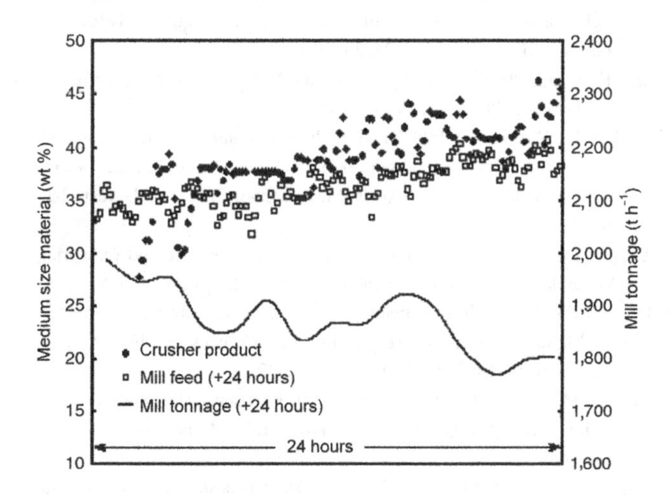

FIGURE 6.30 Effect of increasing medium size (50–125 mm) fraction in crusher product (and mill feed) on mill throughput (tonnage) *(Courtesy Teck, Highland Valley Copper).*

REFERENCES

Anon., 1981. Crushers. Mining Mag. 144-145 (Aug.), 94–113.

Anon., 1985. Rugged roller-bearing crusher. Mining Mag. 153 (Sept.), 240–240.

Anon., 2014. World's largest cone crushers go into service at African mine. Eng. Min. J. 215 (12), 102–104.

Blake, E.W., 1858. Machine for Crushing Stone. US Patent No. US20542.

Brachthauser, M., Kellerwessel, H., 1988. High pressure comminution with roller presses in mineral processing. Forssberg, E. (Ed.), Proc. 16th International Mineral Processing Cong., (IMPC), Stockholm, Sweden, pp. 209–219.

Broman, J., 1984. Optimising capacity and economy in jaw and gyratory crushers. Eng. Min. J. 185 (6), 69–71.

Clarke, A.J., Wills, B.A., 1989. Technical note enhancement of cassiterite liberation by high pressure roller comminution. Miner. Eng. 2 (2), 259–262.

Dance, A., 2001. The importance of primary crushing in mill feed size optimisation. Proc. International Autogenous and Semiautogenous Grinding Technology (SAG) Conf., vol. 2, Vancouver, BC, Canada, pp. 270–281.

Daniel, M.J., Morrell, S., 2004. HPGR model verification and scale-up. Miner. Eng. 17 (11-12), 1149–1161.

Djordjevic, N., et al., 2003. Applying discrete element modelling to vertical and horizontal shaft impact crushers. Miner. Eng. 16 (10), 983–991.

Erikson, M.T., 2014. Innovations in comminution equipment: high pressure grinding rolls, semi-autogenous grinding, ball mills, and regrind mills. In: Anderson, C.G., et al., (Eds.), Mineral Processing and Extractive Metallurgy: 100 Years of Innovation. SME, Englewood, CO, USA, pp. 65–76.

Esna-Ashari, M., Kellerwessel, H., 1988. Interparticle crushing of gold ore improves leaching. Randol Gold Forum 1988, Scottsdale, AZ, USA, pp. 141–146.

Esterle, J.S., et al., 1996. Coal breakage modeling: a tool for managing fines generation. Howarth, H. et al. (Eds.), Proc. Mining Technology Conference, Perth, WA, Australia. pp. 211–228.

Flavel, M.D., 1977. Scientific methods to design crushing and screening plants. Mining Eng. 29 (7), 65–70.

Flavel, M.D., 1978. Control of crushing circuits will reduce capital and operating costs. Mining Mag. 138 (3), 207–213.

Flintoff, B., et al., 2014. Innovations in comminution instrumentation and control. Mineral Processing and Extractive Metallurgy: 100 Years of Innovation. SME, Englewood, CO, USA, pp. 91-116.

Frizzel, E.M., 1985. Technical note: mobile in-pit crushing - product of evolutionary change. Min. Eng. 37 (6), 578–580.

Griesshaber, H.E., 1983. Crushing and grinding: design considerations. World Min. 36 (10), 41–43.

Horst, W.E., Enochs, R.C., 1980. Instrumentation and process control. Eng. Min. J. 181 (6), 70–95.

Karra, V., 1990. Developments in cone crushing. Miner. Eng. 3 (122), 75–81.

Kelsey, C., Kelly, J., 2014. Super-fine crushing to ultra-fine size, the "IMP" super-fine crusher. 27th International Mineral Processing Cong., (IMPC), Ch 9. Santiago, Chile, pp. 239–252.

Knecht, J., 1994. High-pressure grinding rolls - a tool to optimize treatment of refractory and oxide gold ores. *Proc.* 5th *Mill Operators Conf.* AusIMM, Roxby Downs, Melbourne, Australia, pp. 51–59.

Kojovic, T., 1996. Vertical shaft impactors: predicting performance. Quarry Aus. J. 4 (6), 35–39.

Kojovic, T., Bearman, R.A., 1997. The development of a flakiness model for the prediction of crusher product shape. Proc. 41st Annual Institute of Quarrying Conf., Brisbane, Queenland, Australia, pp. 135–148.

Lewis, F.M., et al., 1976. Comminution: a guide to size-reduction system design. Mining Eng. 28 (9), 29–34.

Lynch, A.J., 1977. Mineral Crushing and Grinding Circuits: Their Simulation, Optimisation, Design, and Control. Elsevier Scientific Pub. Co, Amsterdam.

Major, K., 2002. Types and characteristics of crushing equipment and circuit flowsheets. In: Mular, A.L., et al., (Eds.), Mineral Processing Plant Design, Practice, and Control, vol. 1. SME, Littleton, CO, USA, pp. 566–583.

Major, K., 2009. Factors influencing the selection and sizing of crushers. In: Malhotra, D., et al., (Eds.), Recent Advances in Mineral Processing Plant Design. SME, Englewood, CO, USA, pp. 356–360.

Manlapig, E.V., Watsford, R.M.S., 1983. Computer control of the lead-zinc concentrator crushing plant operations of Mount Isa Mines Ltd. Proc. 4th IFAC Symposium on Automation in Mining, Mineral and Metal Processing (MMM 83). Helsinki University of Technology, Helsinki, Finland, pp. 435–445.

McIvor, R.E., 1997. High pressure grinding rolls - a review. Comminution Practices. SME, Littleton, CO, USA (Chapter 13), pp. 95–98.

McQuiston, F.W., Shoemaker, R.S., 1978. Primary Crushing Plant Design. SME, Port City Press, MD, USA.

Mollick, L., 1980. Crushing. Eng. Min. J. 181 (1-6), 96–103.

Morley, C., 2010. HPGR-FAQ HPGR: transaction paper. J. S. Afr. Inst. Min. Metall. 110 (3), 107–115.

Motz, J.C., 1978. Crushing. In: Mular, A.L., Bhappu, R.B. (Eds.), Mineral Processing Plant Design, second ed. SME, New York, NY, USA, pp. 203–238. (Chapter 11).

Mular, A.L., Jergensen, G.V.II (Eds.), 1982. Design and Installation of Comminution Circuits. SME, New York, NY; Port City Press, MD, USA.

Mular, A.L. (Ed.), 2002. Mineral Processing Plant Design, Practice, and Control. SME, Littleton, CO, USA.

Napier-Munn, et al., 1996. *Mineral Comminution circuits - Their Operation and Optimisation* (Appendix 3). Julius Kruttschnitt Mineral Research Centre (JKMRC), The University of Queensland, Brisbane, Queensland, Australia.

Nordberg, n.d. Nordberg Gyradisc Sales Literature. <www.crushers.co.uk/download/37/nordberg-gyradisc-sales-literature> viewed at Dec., 2014.

Portal, J., 2007. Rhodax® interparticle crusher maximizes chloride slag production with a minimum generation of fines. *Proc. The* 6th *International Heavy Minerals Conf. 'Back to Basics'.* SAIMM, Johannesburg, South Africa, pp. 63–68.

Ramos, M., et al., 1994. Aggregate shape - Prediction and control during crushing. Quarry Manage. 21 (11), 23–30.

Rodriguez, D.E., 1990. The Tidco Barmac autogenous crushing mill-A circuit design primer. Miner. Eng. 3 (1-2), 53–65.

Rosario, P.P., et al., 2009. Recent trends in the design of comminution circuits for high tonnage hard rock Miming. Recent Advances in Mineral Processing Plant Design. Malhptra, D., et al., (Eds.), Littleton, CO, USA, pp. 347–355.

Schönert, K., 1979. Verfahren zar Fein-und Feinstzerkleinerung von Materialen Sproden Stoffverhaltens. German Patent 2708053.

Schönert, K., 1988. A first survey of grinding with high-compression roller mills. Int. J. Miner. Process. 22 (1-4), 401–412.

Schwechten, D., Milburn, G.H., 1990. Experiences in dry grinding with high compression roller mills for end product quality below 20 microns. Miner. Eng. 3 (1-2), 23–34.

Shi, F., et al., 2003. An energy-based model for swing hammer mills. Int. J. Miner. Process. 71 (1-4), 147–166.

Simkus, R., Dance, A., 1998. Tracking hardness and size: measuring and monitoring ROM ore properties at Highland Valley copper. Proc. Mine to Mill 1998 Conf. AusIMM, Brisbane, Queensland, Australia, pp. 113–119.

Taggart, A.F., 1945. Handbook of Mineral Dressing: Ore and Industrial Minerals. Wiley, & Sons., Chapman & Hall, Ltd, London, UK.

Utley, R.W., 2009. In-pit crushing - the move to continuous mining. In: Malhotra, D., et al., (Eds.), Recent Advances in Mineral Plant Design. SME, Littleton, CO, USA, pp. 332–339.

van der Meer, F.P., Grunedken, A., 2010. Flowsheet considerations for optimal use of high pressure grinding rolls. Miner. Eng. 23 (9), 663–669.

van der Meer, F.P., Maphosa, W., 2012. High pressure grinding moving ahead in copper, iron, and gold processing. J. S. Afr. Inst. Min. Metall. 112, 637–647.

Watson, S., Brooks, M., 1994. KCGM evaluation of high pressure grinding roll technology. *Proc.* 5th *Mill Operators' Conf.* AusIMM, Roxby Downs, SA, Australia, pp. 69-83.

Wills, B.A., 1983. Pyhasalmi and Vihanti concentrators. Mining Mag. 149 (3), 176–185.

Chapter 7

Grinding Mills

7.1 INTRODUCTION

Grinding is the last stage in the comminution process where particles are reduced in size by a combination of impact and abrasion, either dry, or more commonly, in suspension in water. It is performed in cylindrical steel vessels that contain a charge of loose crushing bodies—the grinding medium—which is free to move inside the mill, thus comminuting the ore particles. According to the ways by which motion is imparted to the charge, grinding mills are generally classified into two types: *tumbling mills* and *stirred mills*. In tumbling mills, the mill shell is rotated and motion is imparted to the charge via the mill shell. The grinding medium may be steel rods, balls, or rock itself. Media ball sizes, for example, range from about 20 mm for fine grinding to 150 mm for coarse grinding. Tumbling mills are typically employed in the mineral industry for primary grinding (i.e., stage immediately after crushing), in which particles between 5 and 250 mm are reduced in size to between 25 and 300 μm. In stirred mills, the mill shell is stationary mounted either horizontally or vertically and motion is imparted to the charge by the movement of an internal stirrer. Grinding media (25 mm or less) inside the mill are agitated or rotated by the stirrer, which typically comprises a central shaft to which are attached screws, pins, or discs of various designs. Stirred mills find application in regrinding, fine (15–40 μm) and ultrafine (<15 μm) grinding.

All ores have an economic optimum particle size which maximizes the difference between net smelter return (NSR) and grinding costs (Chapter 1): too coarse a grind and the inadequate liberation limits recovery (and thus revenue) in the separation stage; too fine a grind and grinding costs exceed any increment in recovery (and may even reduce recovery depending on the separation process). The optimum grind size will depend on many factors, including the extent to which the values are dispersed in the gangue, and the subsequent separation process to be used. It is the purpose of the grinding section to exercise close control on this product size and, for this reason, correct grinding is often said to be the key to good mineral processing.

Grinding costs are driven by energy and steel (media, liners, etc.) consumption; grinding is the most energy-intensive operation in mineral processing. On a survey of the energy consumed in a number of Canadian copper concentrators it was shown that the average energy consumption in kWh t^{-1} was 2.2 for crushing, 11.6 for grinding, and 2.6 for flotation (Joe, 1979). It can be shown, using Bond's equation (Equation 5.4), that 19% extra energy must be consumed in grinding one screen size finer on a $\sqrt{2}$ screen series. A strategy to reduce grinding energy consumption in low grade high tonnage flotation plants is to employ a coarse primary grind and regrind the rougher flotation concentrate, which represents a much lower tonnage. This has driven development of coarse flotation technology (see Chapter 12), and energy efficient regrinding (see later in this chapter).

Although tumbling mills have been developed to a high degree of mechanical efficiency and reliability, their energy efficiency (conversion of energy delivered by the mill into broken material) remains an area of debate. Breakage of the ore is mostly the result of repeated, random impact and abrasion, events which break liberated as well as unliberated particles. At present there is no practical way that these impacts can be directed at the interfaces between the mineral grains, which would produce optimum liberation (liberation by detachment), although various ideas have been postulated (Wills and Atkinson, 1993). Assisted breakage through application of such technologies as microwave heating (Jones et al., 2006) and high voltage pulsing (van der Wielen et al. 2014) aims to increase liberation and reduce overall grinding energy consumption by creating multiple internal stresses and micro-cracks, that is weakening the ore prior to tumble milling.

Although the economic degree of liberation is the principal purpose of grinding in mineral processing, grinding is sometimes used to increase mineral surface area. Production of some industrial minerals such as talc involves size reduction to meet customer requirements, and iron ore concentrates are reground to produce pelletizer feed. Where grinding is followed by hydrometallurgical methods of extraction, as in gold-ore processing, a high surface area increases cyanide leaching rate

Wills' Mineral Processing Technology.

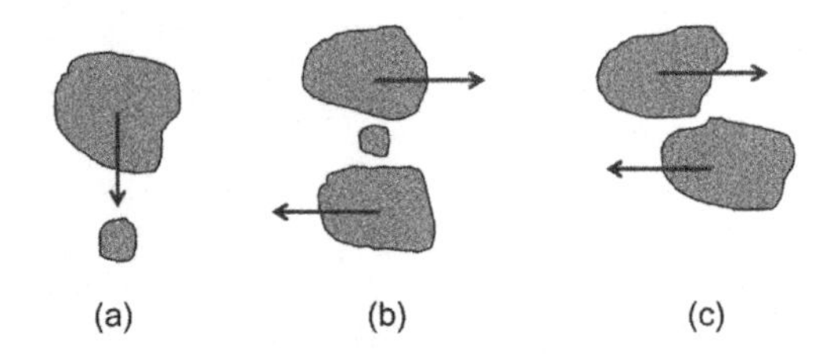

FIGURE 7.1 Mechanisms of breakage: (a) Impact or compression, (b) Chipping or attrition, and (c) Abrasion.

and in such cases extensive grinding may not be a disadvantage, as the increase in energy consumption may be offset by the increased gold recovery.

Grinding within a tumbling mill is influenced by the size, quantity, the type of motion, and the spaces between the individual pieces of the medium in the mill. As opposed to crushing, which takes place between relatively rigid surfaces, grinding is a random process. The degree of grinding of an ore particle depends on the probability of the ore entering a zone between the medium entities (balls, etc.) and the probability of some breakage event occurrence after entry. Grinding can be achieved by several mechanisms, including: impact or compression due to sudden forces applied almost normally to the particle surface; chipping or attrition due to steady forces that break the matrix of a particle; and abrasion due to forces acting parallel to and along the surfaces (i.e., shear) (Figure 7.1). These mechanisms distort the particles and change their shape beyond certain limits determined by their degree of elasticity, which causes them to break.

Grinding is usually performed wet, although dry grinding is used in applications such as cement production or when a material is simultaneously ground and dried. Water conservation efforts may see an increase in dry grinding in the future. In the mill, the mixture of medium, ore, and water, known as the *mill charge*, is intimately mixed, the medium comminuting the particles by any of the above methods.

Apart from laboratory testing, grinding in mineral processing is a continuous process, material being fed at a controlled rate into one end of the mill and discharging at the other end after a suitable dwell (residence) time. Control of product size is exercised by the properties of the mill charge, the nature of the ore feed, and the type of mill and circuit used.

7.2 TUMBLING MILLS

Tumbling mills are of three basic types: rod, ball, and autogenous/semi-autogenous (AG/SAG), based on the type of grinding media (SAG mills are AG mills with some grinding balls). Structurally, each type of mill consists of a horizontal cylindrical shell, provided with

FIGURE 7.2 Tumbling mill.

renewable wearing liners and a charge of grinding medium. The drum is supported so as to rotate on its axis on hollow *trunnions* attached to the end walls (Figure 7.2). The diameter of the mill determines the impact that can be exerted by the medium on the ore particles and, in general, the larger the feed size the larger the mill diameter needs to be. The length of the mill, in conjunction with the diameter, determines the volume, and hence the capacity of the mill.

Material is usually fed to the mill continuously through one end trunnion, the ground product leaving via the other trunnion, although in certain applications the product may leave the mill through a number of ports spaced around the periphery of the shell. All types of mill can be used for wet or dry grinding by modification of feed and discharge equipment.

7.2.1 Motion of the Charge

The distinctive feature of tumbling mills is the use of loose crushing bodies, which are large, hard, and heavy in relation to the ore particles, but small in relation to the volume of the mill, and which occupy (including voids) slightly less than half the volume of the mill.

Due to the rotation and friction of the mill shell, the grinding medium is lifted along the rising side of the mill until a position of dynamic equilibrium is reached (the *shoulder*), when the bodies cascade and cataract down the free surface of the other bodies, about a dead zone where little movement occurs, down to the *toe* of the mill charge (Figure 7.3).

The driving force of the mill is transmitted via the liner to the charge. The speed at which a mill is run and the liner design governs the motion and thus nature of the product and the amount of wear on the shell liners. For instance, a practical knowledge of the trajectories followed by the steel balls in a mill determines the speed at which it must be run in order that the descending balls shall fall on to the toe of the charge, and not on to the liner, which could lead to liner damage. Simulation of charge motion can be used to identify such potential problems (Powell et al., 2011), and acoustic monitoring can give indication of where ball impact is occurring (Pax, 2012).

At relatively low speeds, or with smooth liners, the medium tends to roll down to the toe of the mill and essentially abrasive comminution occurs. This *cascading* leads to finer grinding and increased liner wear. At higher speeds the medium is projected clear of the charge to describe a series of parabolas before landing on the toe of the charge. This *cataracting* leads to comminution by impact and a coarser end product with reduced liner wear. At the *critical speed* of the mill *centrifuging* occurs and the medium is carried around in an essentially fixed position against the shell.

In traveling around inside the mill, the medium (and the large ore pieces) follows a path which has two parts: the lifting section near to the shell liners, which is circular, and the drop back to the toe of the mill charge, which is parabolic (Figure 7.4(a)).

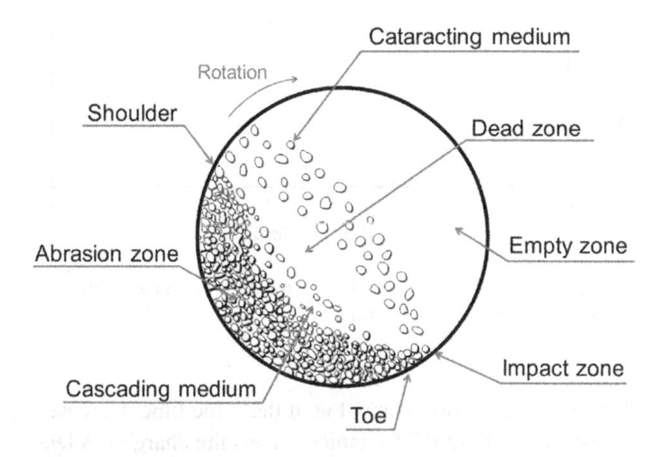

FIGURE 7.3 Motion of charge in a tumbling mill.

Consider a ball (or rod) of radius r meters, which is lifted up the shell of a mill of radius R meters, revolving at N rev min^{-1}. The ball abandons its circular path for a parabolic path at point P (Figure 7.4(b)), when the weight of the ball is just balanced by the centrifugal force, that is when:

$$\frac{mV^2}{R-r} = mg\cos\alpha \tag{7.1}$$

where m is the mass of the ball (kg), V the linear velocity of the ball (m s^{-1}), and g the acceleration due to gravity (m s^{-2}).

Since V is related to N by the following:

$$V = \frac{2\pi N(R-r)}{60} \tag{7.2}$$

Then:

$$\cos\alpha = \frac{4\pi^2N^2(R-r)}{60^2g} = \frac{0.0011\,N^2(D-d)}{2} \tag{7.3}$$

where D is the mill diameter and d the ball diameter in meters.

The critical speed of the mill occurs when $\alpha = 0$, that is, the medium abandons its circular path at the highest vertical point. At this point, $\cos\alpha = 1$.

Therefore:

$$N_c = \frac{42.3}{\sqrt{D-d}}\,\text{rev min}^{-1} \tag{7.4}$$

where N_c is the critical speed of the mill. Equation (7.4) assumes that there is no slip between the medium and the shell liner.

Mills are driven, in practice, at speeds of 50–90% of critical speed. The speed of rotation of the mill influences the power draw through two effects: the value of N and the shift in the center of gravity with speed. The center of gravity first starts to shift away from the center of the mill (to the right in Figure 7.4(a)) as the speed of rotation increases, causing the torque exerted by the charge to

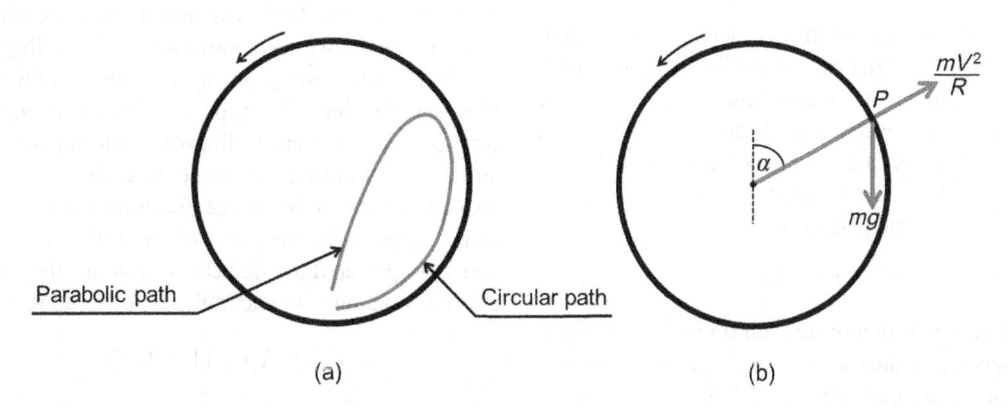

FIGURE 7.4 (a) Trajectory of grinding medium in tumbling mill, and (b) Forces acting on the medium.

increase and draw more power (see Section 7.2.2). But, as critical speed is reached, the center of gravity moves toward the center of the mill as more and more of the material is held against the shell throughout the cycle, causing power draw to decrease. Since grinding effort is related to grinding energy, there is little increase in efficiency (i.e., delivered kWh t^{-1}) above about 40–50% of the critical speed. It is also essential that the cataracting medium should fall well inside the mill charge and not directly onto the liner, thus excessively increasing steel consumption.

At the toe of the load the descending liner continuously underruns the churning mass, and moves some of it into the main mill charge. The medium and ore particles in contact with the liners are held with more firmness than the rest of the charge due to the extra weight bearing down on them. The larger the ore particle, rod, or ball, the less likely it is to be carried to the breakaway point by the liners. The cataracting effect should thus be applied in terms of the medium of largest diameter.

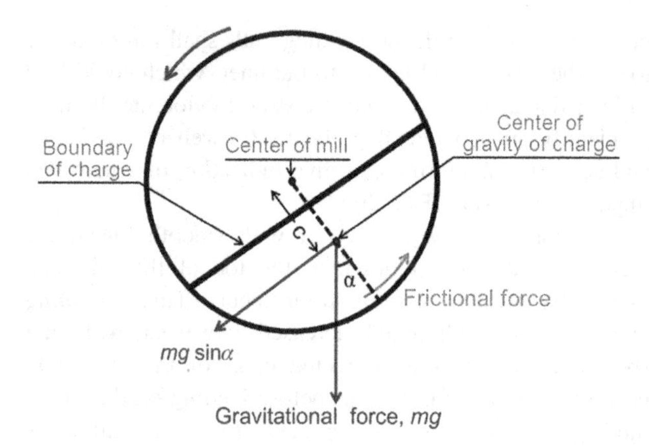

FIGURE 7.5 Idealized depiction of charge to set up power draw problem (m is mass of charge).

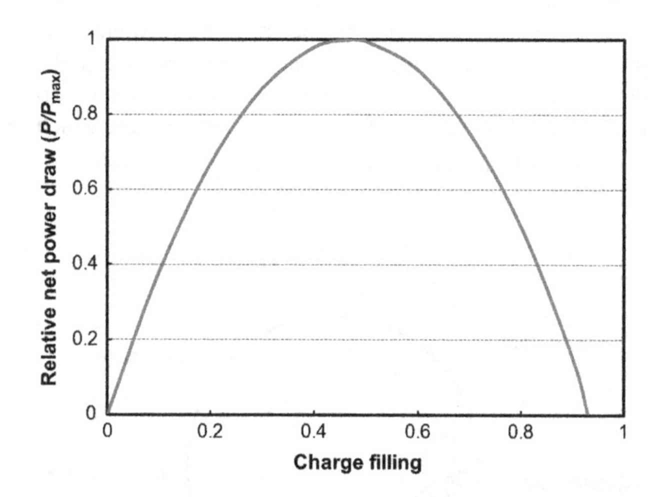

FIGURE 7.6 Effect of charge filling on net power draw relative to maximum net power draw (A = 1.065).

7.2.2 Power Draw

The gross power drawn by a rotating mill is the sum of the no-load power (to account for frictional and mechanical losses in the drive power) and net power drawn by the charge. It is the latter, P_{net}, that is considered here, starting with the ball mill.

Austin et al. (1984) calculated net power by considering the energy to lift a single ball, summing over the total number of balls and multiplying by the number of times the balls are lifted per second, given by the speed (N, revs per second) of the mill. Hogg and Fuerstenau (1972) treated the problem as one of rotation of the center of mass of the charge about the center of the mill (Figure 7.5); that is, power is given by the torque times the angular speed.

The end result in both approaches is similar and the power model has the form:

$$P_{net} = KD^{2.5}L\rho_{ap}\alpha\beta \tag{7.5}$$

where K is a calibration constant that depends on the mill type, D is the mill diameter inside the liners, L the effective length (allowing for conical ends, etc.), ρ_{ap}, the apparent density of the charge (steel, plus ore plus voids), and α and β are factors to account for fractional filling of the charge and mill rotational speed, respectively.

For charge filling, the general form of α is:

$$\alpha = J(1 - AJ) \tag{7.6}$$

where J is the charge filling of the mill (media and ore plus voids) and suggested values of A range from 1.03 to 1.065. The effect of increasing charge filling is evident in Figure 7.6, which shows the P/P_{max} calculated for $A = 1.065$. Increasing J increases the mass of the charge m, which increases power draw, but at the same time decreases the distance c (Figure 7.5), which means the charge has less lift distance. These offsetting effects on the power draw mean P_{net} passes through a maximum, in the case of $A = 1.065$ at $J \sim 0.47$. This result indicates why tumbling mills are rarely operated above about 45% filling.

As regards mill speed, up to about a critical speed of 80%, which represents a practical range in most cases, β increases approximately linearly with fraction of critical speed, N_c. (As discussed above, as critical speed is reached the centrifuging of the charge means the center of mass of the charge approaches the center of the mill, that is, c decreases and thus power draw decreases.) Putting this together, the general equation to predict mill net power draw is:

$$P_{net} = K\rho_{ap}J(1 - AJ)D^{2.5}LN_c \tag{7.7}$$

In ball milling, where the steel media dominates the mass of the charge, the apparent density term does not need to consider the ore. This is not the case with AG/SAG milling. The apparent density for the SAG case (i.e., with some steel balls) is calculated as follows:

$$\rho_{ap} = \frac{J_b(1-f_v)\rho_b + J_r(1-f_v)\rho_r + (J_b+J_r)f_v U \rho_{sl}}{J_b(1-f_v) + J_r(1-f_v) + (J_b+J_r)f_v U} \quad (7.8)$$

where J_b is the filling of balls, J_r the filling of rock (rock = fraction of ore acting as grinding media), f_v fractional voidage (same for balls and ore), U fractional filling of the voids by non-rock particles, and ρ_{sl} slurry density given by:

$$\rho_{sl} = \theta_p \rho_p + (1 - \theta_p) \quad (7.9)$$

where θ_p is fractional volumetric solids content in the slurry, ρ_p solids density, and taking water density as 1 t m^{-3}. Example 7.1 illustrates the impact of charge density.

The example shows that compared to all ball grinding media (Part (b)), the apparent density in the SAG case (Part (a)) has decreased by ~58% (=(5.19 − 3.0)/5.19), which corresponds to the decrease in power draw. One response is in the design of the SAG mill, to increase mill diameter, taking advantage of the $D^{2.5}$ term in Eq. (7.5) giving the characteristic high D/L ratio ("pancake shape") of the AG/SAG mill (at least outside of South Africa, where L was also increased). Experience, however, suggests that replacing balls with rock is not just a matter of a decrease in apparent density that can be offset by increasing D, but that balls are more effective in transmitting power into grinding action (size reduction). This has led to a steady increase in ball load in SAG mills (Sepúlveda, 2001).

Example 7.1

Part a). Calculate the apparent density of charge in a SAG mill for the following conditions: $J_b = 0.1$, $J_r = 0.3$, $f_v = 0.4$, $\rho_b = 7.85$ t m^{-3}, $\rho_r = 3$ t m^{-3}, $\theta_p = 0.1$.

Solution

$$\rho_{ap} = \frac{0.1\times(1-0.4)\times 7.85 + 0.3\times(1-0.4)\times 3 + 0.4\times 0.4\times 1.2}{0.1\times(1-0.4) + 0.3\times(1-0.4) + 0.4\times 0.4\times 1}$$

$$\rho_{av} = 3.0\ (\text{t m}^{-3})$$

Part b). Determine the effect of replacing rock by steel.

Solution

If the rocks are replaced with balls (i.e., $J_b = 0.4$, $J_r = 0$) then,

$$\rho_{ap} = 5.19\ (\text{t m}^{-3})$$

FIGURE 7.7 Tube mill end and trunnion.

7.2.3 Construction of Mills

Shell

Mill shells are designed to sustain impact and heavy loading, and are constructed from rolled mild steel plates connected together. Holes are drilled to take the bolts for holding the liners. For attachment of the trunnion heads, heavy flanges of fabricated or cast steel are usually welded or bolted to the ends of the plate shells, planed with parallel faces that are grooved to receive a corresponding spigot on the head, and drilled for bolting to the head.

Mill Ends

Also known as trunnion heads, they may be of nodular or gray cast iron for diameters less than about 1 m; larger heads are constructed from cast steel, which is relatively light and can be welded. The heads are ribbed for reinforcement and may be flat, slightly conical, or dished. They are machined and drilled to fit shell flanges (Figure 7.7).

Trunnions and Bearings

The trunnions are made from cast iron or steel and are spigoted and bolted to the end plates, although in small mills they may be integral with the end plates (see also Anon., 1990). They are highly polished to reduce bearing friction. Most trunnion bearings are rigid high-grade iron castings with 120–180° lining of white metal in the bearing area, surrounded by a fabricated mild steel housing, which is bolted into the concrete foundations (Figure 7.8).

The bearings in smaller mills may be grease lubricated, but oil lubrication is favored in large mills, via motor-driven oil pumps. The effectiveness of normal lubrication protection is reduced when the mill is shut down for any length of time, and many mills are fitted

FIGURE 7.8 180° oil-lubricated trunnion bearing.

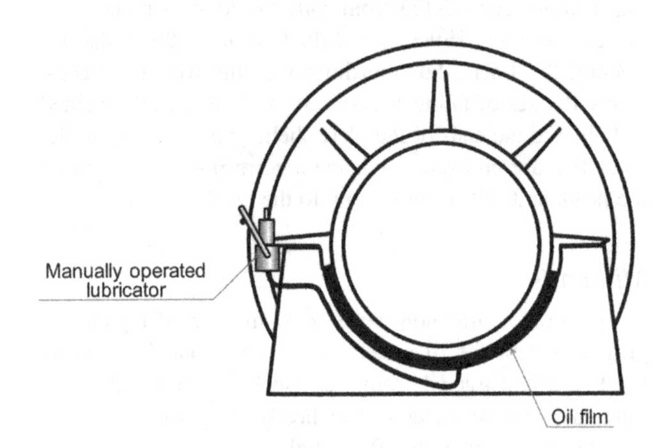

FIGURE 7.9 Hydraulic starting lubricator.

with manually operated hydraulic starting lubricators, which force oil between the trunnion and trunnion bearing, preventing friction damage to the bearing surface on starting by re-establishing the protecting film of oil (Figure 7.9).

Drive

Tumbling mills are most commonly rotated by a pinion meshing with a girth ring bolted to one end of the machine (Figure 7.10). The pinion shaft is either coupled directly or via a clutch to the output shaft of a slow-speed synchronous motor, or to the output shaft of a motor-driven helical or double helical gear reducer. In some mills, electrical

FIGURE 7.10 Gear/pinion assembly on ball mill.

thyristors and/or DC motors are used to give variable speed control. Very large mills driven by girth gears require two motors, each driving separate pinions, with a complex load sharing system balancing the torque generated by the two motors. (See also Knecht, 1990.)

The larger the mill, the greater are the stresses between the shells and heads and the trunnions and heads. In the early 1970s, maintenance problems related to the application of gear and pinion and large speed reducer drives on dry grinding cement mills of long length drove operators to seek an alternative drive design. As a result, a number of gearless drive (ring motor) cement mills were installed and the technology became relatively common in the European cement industry.

The gearless drive design features motor rotor elements bolted to a mill shell, a stationary stator assembly surrounding the rotor elements, and electronics converting the incoming current from 50/60 Hz to about 1 Hz. The mill shell actually becomes the rotating element of a large low speed synchronous motor. Mill speed is varied by changing the frequency of the current to the motor, allowing adjustments to the mill power draw as ore grindability changes.

The gearless drive design was not applied to the mills in the mineral industry until 1981 when the then-world's largest ball mill, 6.5 m diameter and 9.65 m long driven by a 8.1 MW motor, was installed at Sydvaranger in Norway (Meintrup and Kleiner, 1982). A gearless drive SAG mill, 12 m diameter and 6.1 m length (belly inside liners) with a motor power of more than 20 MW, went into operation at Newcrest Mining's Cadia Hill gold and copper mine in Australia, with a throughput of over 2,000 t h^{-1} (Dunne et al., 2001). Motor designs capable of 35 MW have been reported (van de Vijfeijken, 2010).

The major advantages of the gearless drive include: variable speed capacity, removal of limits of design power, high drive efficiency, low maintenance requirements, and less floor space for installation.

Liners

The internal working faces of mills consist of renewable liners, which must withstand impact, be wear-resistant, and promote the most favorable motion of the charge. Rod mill ends have plain flat liners, slightly coned to encourage the self-centering and straight-line action of rods. They are made usually from manganese or chrome-molybdenum steels, having high impact strength. Ball mill ends usually have ribs to lift the charge with the mill rotation. The ribs prevent excessive slipping and increase liner life. They can be made from white cast iron, alloyed with nickel (Ni-hard), other wear-resistant materials, and rubber (Durman, 1988). Trunnion liners are designed for each application and can be conical, plain, with advancing or retarding spirals. They are manufactured from hard cast iron or cast alloy steel, a rubber lining often being bonded to the inner surface for increased life.

Shell liners have an endless variety of lifter shapes. Smooth linings result in much abrasion, and hence a fine grind, but with associated high metal wear. The liners are therefore generally shaped to provide lifting action and to add impact and crushing. From a survey (Wei and Craig, 2009), the most common shapes were wave, rib, step, and Osborn (Figure 7.11). The liners are attached to the mill shell and ends by forged steel countersunk liner bolts.

Rod mill liners are generally of alloyed steel or cast iron, and of the wave type, although Ni-hard step liners may be used with rods up to 4 cm in diameter. Lorain liners consist of high carbon rolled steel plates held in place by manganese or hard alloy steel lifter bars. Ball mill liners may be made of hard cast iron when balls of up to 5 cm in diameter are used, but otherwise cast manganese steel, cast chromium steel, or Ni-hard are used.

Efforts to prolong liner life are constantly being made. With the lost production cost associated with shut-downs for replacing liners, the trend is toward selecting liners that have the best service life and least relining downtime (Orford and Strah, 2006).

Rubber liners and lifters have supplanted steel at some operations, particularly in ball mills. They have been found to be longer lasting, easier and faster to install, and their use results in a significant reduction of noise level. In primary grinding applications with severe grinding forces, the higher wear rate of rubber tends to inhibit its use. Rubber lining may have drawbacks in processes requiring the addition of flotation reagents directly into the mill, or temperatures exceeding 80°C. They are also thicker than their steel counterparts, which reduces mill capacity, a potentially important factor in small mills. There are also important differences in design aspects between steel and rubber linings (Moller and Brough, 1989). A combination of rubber lifter bars with steel inserts embedded in the face, the steel providing the wear resistance and the rubber backing cushioning the impacts, is a compromise design (Moller, 1990).

A different concept is the magnetic liner. Magnets keep the lining in contact with the steel shell and the end plates without using bolts, while the ball "scats" in the charge and any magnetic minerals are attracted to the liner to form a 30–40 mm protective layer, which is continuously renewed as it wears.

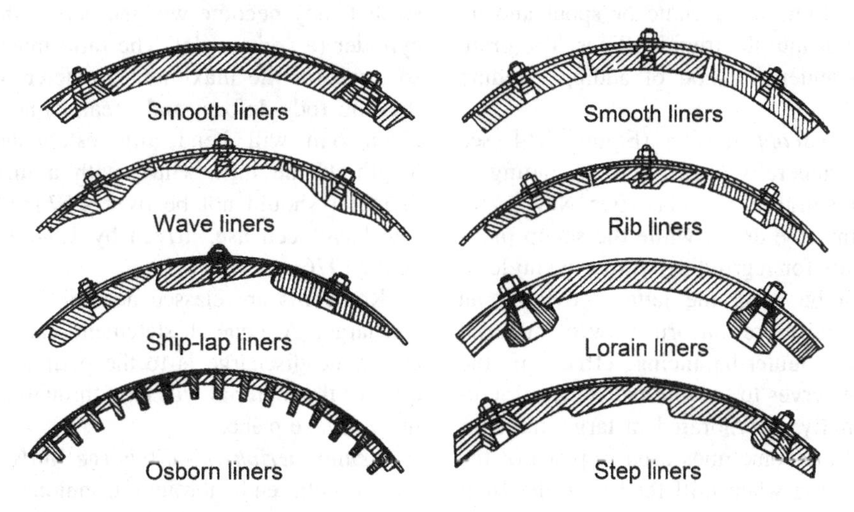

FIGURE 7.11 Mill shell liners.

FIGURE 7.12 Spout feeder.

FIGURE 7.13 Drum feeder on ball mill.

Mill Feeders

The type of feeding arrangement used on the mill depends on whether the grinding is done in open or closed circuit and whether it is done wet or dry. The size and rate of feed are also considerations. Dry mills are usually fed by some sort of vibratory feeder. Three types of feeder are in use in wet-grinding mills. The simplest form is the *spout feeder* (Figure 7.12), consisting of a cylindrical or elliptical chute supported independently of the mill, and projecting directly into the trunnion liner. Material is fed by gravity through the spout to the mills. They are often used for feeding rod mills operating in open circuit or mills in closed circuit with hydrocyclone classifiers.

An alternative to a spout feeder, *drum feeders* (Figure 7.13) may be used when headroom is limited. The mill feed enters the drum via a chute or spout and an internal spiral carries it into the trunnion liner. The drum also provides a convenient method of adding grinding balls to a mill.

Combination drum-scoop feeders (Figure 7.14, see also Figure 7.2) are generally used for wet grinding in closed circuit with a spiral or rake classifier. New material is fed directly into the drum, while the scoop picks up the classifier sands for regrinding. Either a single or a double scoop can be used, the latter providing an increased feed rate and more uniform flow of material into the mill; the counter-balancing effect of the double-scoop design serves to smooth out power fluctuation and it is normally incorporated in large-diameter mills. *Scoop feeders* are sometimes used in place of the drum-scoop combination when mill feed is in the fine-size range.

7.2.4 Types of Tumbling Mill

Rod Mills

These mills can be considered as either fine crushers or coarse grinding machines. They are capable of taking feed as large as 50 mm and making a product as fine as 300 μm, reduction ratios (F_{80}:P_{80}) normally being in the range 15:1–20:1. They are often preferred to fine crushing machines when the ore is "clayey" or damp, thus tending to choke crushers. They are still to be found in older plants but are often replaced upon modernization (e.g., Leung et al., 1992) and are now relatively rare (Wei and Craig, 2009).

The distinctive feature of a rod mill is that the length of the cylindrical shell is between 1.5 and 2.5 times its diameter (Figure 7.15). This ratio is important because the rods, which are only a few centimeters shorter than the length of the shell, must be prevented from turning so that they become wedged across the diameter of the cylinder (a *rod tangle*). The ratio must not, however, be so large for the maximum diameter of the shell in use that the rods deform and break. Since rods longer than about 6 m will bend, this establishes the maximum length of the mill. Thus, with a mill 6.4 m long the diameter should not be over 4.57 m. Rod mills of this size have seen use, driven by 1,640 kW motors (Lewis et al., 1976).

Rod mills are classed according to the nature of the discharge. A general statement can be made that the closer the discharge is to the periphery of the shell, the quicker the material will pass through and less overgrinding will take place.

Center peripheral discharge mills (Figure 7.16) are fed at both ends through trunnions and discharge the ground product through circumferential ports at the center of the shell. The short path and steep gradient give a coarse grind with a minimum of fines, but the reduction ratio is limited. This mill can be used for wet or dry grinding and has found its greatest use in the preparation of specification sands, where high tonnage rates and an extremely coarse product are required.

End peripheral discharge mills (Figure 7.17) are fed at one end through the trunnion, discharging the ground product from the other end of the mill by means of several peripheral apertures into a close-fitting circumferential chute. This type of mill is used mainly for dry and damp grinding, where moderately coarse products are involved.

The most widely used type of rod mill in the mining industry is the *trunnion overflow* (Figure 7.18), in which the feed is introduced through one trunnion and discharges through the other. This type of mill is used only for wet grinding and its principal function is to convert crushing-plant product into ball mill feed. A flow gradient is provided by making the discharge (overflow) trunnion diameter 10–20 cm larger than that of the feed opening. The discharge trunnion is often fitted with a spiral (*trommel*) screen to remove tramp material.

FIGURE 7.14 Drum-scoop feeder.

FIGURE 7.15 Rod mill.

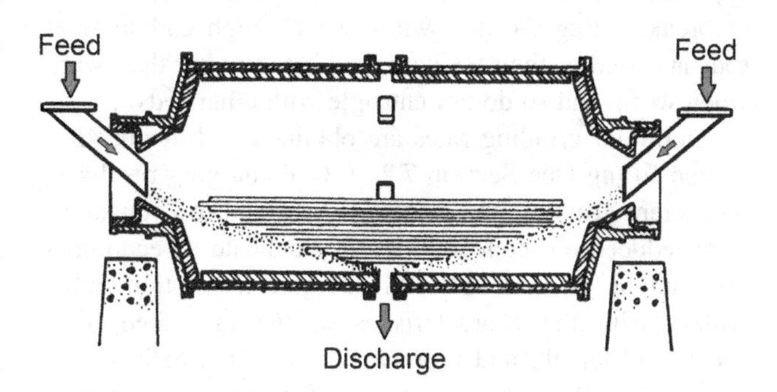

FIGURE 7.16 Central peripheral discharge mill.

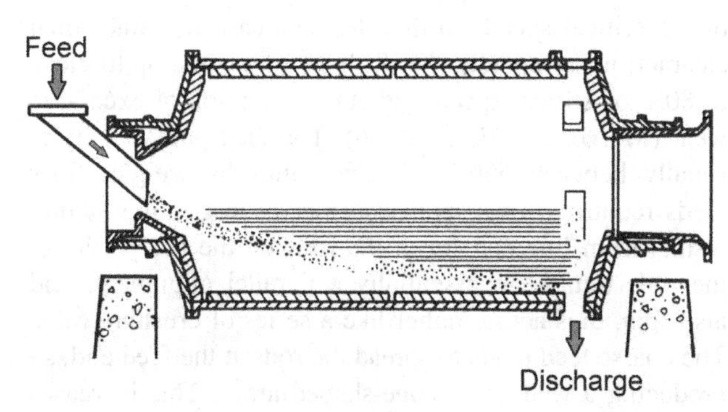

FIGURE 7.17 End peripheral discharge mill.

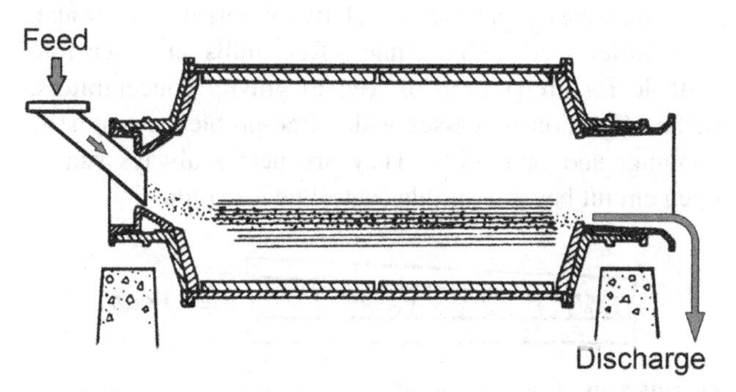

FIGURE 7.18 Overflow mill.

Rod mills are charged initially with a selection of rods of assorted diameters, the proportion of each size being calculated to approximate to a seasoned or equilibrium charge. A seasoned charge will contain rods of varying diameters, ranging from fresh replacements to those that have worn down to such a size as to warrant removal. Actual diameters in use range from 25 to 150 mm. The largest diameter should be no greater than that required to break the largest particle in the feed. A coarse feed or product normally requires larger rods. Provided the large feed particles are broken, the smaller the rods the larger is the total surface area and hence the greater is the size reduction. Generally, rods should be removed when they are worn down to about 25 mm in diameter or less, depending on the application, as small ones tend to bend or break, filling the mill with "scats". High carbon steel rods are used as they are hard, and break rather than warp when worn, and so do not entangle with other rods.

Optimum grinding rates are obtained with about 45% charge filling (see Section 7.2.2). Overcharging results in inefficient grinding and increased liner and rod consumption; reduced rod consumption may indicate an economic optimum at less than 45% filling. Rod consumption varies widely with the characteristics of the mill feed, mill speed, rod length, and product size; it is normally in the range 0.1–1.0 kg of steel per ton of ore for wet grinding, being less for dry grinding.

Rod mills are normally run at between 50% and 65% of the critical speed, so that the rods cascade rather than cataract; many operating mills have been sped up to close to 80% of critical speed without any reports of excessive wear (McIvor and Finch, 1986). The feed pulp density is usually between 65% and 85% solids by weight, finer feeds requiring lower pulp densities. The grinding action results from line contact of the rods on the ore particles; the rods tumble in essentially a parallel alignment, and also spin, thus acting rather like a series of crushing rolls. The coarse feed tends to spread the rods at the feed end, so producing a wedge- or cone-shaped array. This increases the tendency for grinding to take place preferentially on the larger particles, thereby producing a minimum amount of extremely fine material (Figure 7.19). This selective grinding gives a product of relatively narrow size range, with little oversize or slimes. Rod mills are therefore suitable for preparation of feed to gravity concentrators, certain flotation processes with slime problems, magnetic cobbing, and ball mills. They are nearly always run in open circuit because of this controlled size reduction.

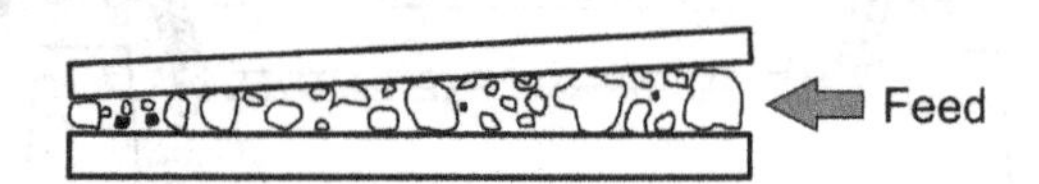

FIGURE 7.19 Preferential grinding action of rods on coarse material.

Ball Mills

The final stages of primary comminution are performed in tumbling mills using steel balls as the grinding medium, and so are designated "ball mills." A typical ball mill setup is shown in Figure 7.20.

Since balls have a greater surface area per unit weight than rods, they are better suited for fine grinding. The term ball mill is restricted to those having a length to diameter ratio of 2 to 1 and less. Ball mills in which the length to diameter ratio is between 3 and 5 are designated *tube mills*. The latter are sometimes divided into several longitudinal compartments, each having a different charge composition; the charges can be steel balls or rods, or pebbles, and they are often used dry to grind cement clinker, gypsum, and phosphate. Tube mills having only one compartment and a charge of hard, screened ore particles as the grinding medium are known as *pebble mills*. They have the advantage over ball mills when iron contamination needs to be avoided. Since the weight of pebbles per unit volume is about 35–55% of that of steel balls, and as the power input is directly proportional to charge weight, the power draw and capacity of pebble mills are correspondingly lower (see Section 7.2.2). Thus in a given grinding circuit, for a certain feed rate, a pebble mill would be much larger than a ball mill, with correspondingly higher capital cost. The increment in capital cost might be justified by reduction in operating cost attributed to the lower cost of the grinding medium, provided this is not offset by higher energy cost per ton of finished product (Lewis et al., 1976).

Ball mills are also classified by the nature of the discharge. They may be simple trunnion overflow mills, operated in open or closed circuit, or *grate discharge* (low-level discharge) mills. The latter type is fitted with discharge grates between the cylindrical mill body and the discharge trunnion. The pulp can flow freely through the openings in the grate and is then lifted up to the level of the discharge trunnion (Figure 7.21). These mills have a lower pulp level than overflow mills, thus reducing the residence time of particles in the mill. Consequently, little overgrinding takes place and the product contains a large fraction of coarse material, which is returned to the mill by some form of classifying device. Closed-circuit grinding, with high circulating loads, produces a closely sized end product and a high output per unit volume compared with open circuit grinding. Grate discharge mills usually take a coarser feed than overflow mills and are not required to grind so finely, the main reason being that with many small balls forming the charge, the grate open area plugs very quickly. The trunnion overflow mill is the simplest to operate and is used for most ball mill applications, especially for fine grinding and regrinding. Energy consumption is said to be about 15%

FIGURE 7.20. Ball mill in operation: 7.9 m (26 ft) diameter ball mills (power rating 16.4 MW each) at Tintaya Antapaccay *(Courtesy Antapaccay Perú)*.

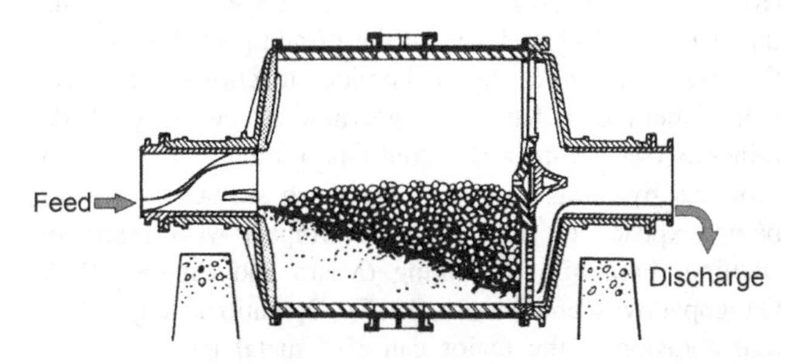

FIGURE 7.21 Grate discharge mill.

less than that of a grate discharge mill of the same size, although the grinding efficiencies of the two mills are similar (Lewis et al., 1976).

The trend in recent years has been to use fewer comminution machines per grinding line with the result that units have increased considerably in size and thus capacity. For example, in the 1980s, the largest operating ball mill was 5.5 m in diameter by 7.3 m in length driven by a 4 MW motor. Today the largest ball mills in operation are over 8 m in diameter utilizing gearless (wrap-around) motors with power output of more than 16 MW.

Grinding in a ball mill is effected by point contact of balls and ore particles and, given time, any degree of fineness can be achieved. The process is completely random—the probability of a fine particle being struck by a ball is the same as that of a coarse particle. The product from an open-circuit ball mill therefore exhibits a wide range of particle size, and overgrinding of at least some of the ore can be a problem. Closed-circuit grinding in mills providing low residence time for the particles is almost always used in the last stages to overcome this.

Several factors influence the efficiency of ball mill grinding. The pulp density of the feed should be as high as possible, consistent with ease of flow through the mill. It is essential that the balls are coated with a layer of ore; too dilute a pulp increases metal-to-metal contact, giving increased steel consumption and reduced efficiency. Ball mills should operate between 65% and 80% solids by weight, depending on the ore. The viscosity of the pulp increases with the fineness of the particles, therefore fine-grinding circuits may need lower pulp densities. The major factors affecting the pulp rheology and its effects on grinding circuits have been discussed by a number of researchers (Klimpel, 1984; Kawatra and Eisele, 1988; Moys, 1989; Shi and Napier Munn, 2002; He et al., 2004; Pax, 2012). It was found that not only the viscosity of the pulp but also the rheological type, Newtonian or non-Newtonian, would affect ball milling performance. An extreme case of high viscosity slurry is grinding fibrous ore particles where mills have to operate at low %solids.

The efficiency of grinding depends on the surface area of the grinding medium. Thus, balls should be as small as possible and the charge should be graded such that the largest balls are just heavy enough to grind the largest and hardest particles in the feed. A seasoned charge will consist of a wide range of ball sizes and new balls added to the mill are usually of the largest size required. Undersize balls leave the mill with the ore product and can be removed by passing the discharge through trommel screens or by magnets mounted near the discharge trunnion. Various formulae have been proposed for the required ratio of ball size to ore size, none of which is entirely satisfactory. The correct sizes are often determined by trial and error, primary grinding usually requiring a graded charge of 10-5 cm diameter balls, and secondary grinding requiring 5-2 cm. Concha et al. (1988) have developed a method to calculate ball mill charge by using a grinding circuit simulator with a model of ball wear in a tumbling mill.

Segregation of the ball charge within the mill is achieved in the *Hardinge mill* (Figure 7.22). The conventional drum shape is modified by fitting a conical section, the angle of the cone being about 30°. Due to the centrifugal force generated, the balls are segregated so that the largest are at the feed end of the cone, that is, the largest diameter and greatest centrifugal force, and the smallest are at the discharge. By this means, a regular gradation of ball size and of size reduction is produced.

There are also non-spherical shapes of grinding media such as Doering Cylpebs. The Cylpebs are slightly tapered cylindrical grinding media with length equaling diameter, and all the edges being rounded (Figure 7.23). They are available in sizes from 85 mm × 85 mm to 8 mm × 8 mm. Because of their geometry, these grinding media have greater surface area than balls of the same mass. For instance, Cylpebs with a diameter to height ratio of unity have 14.5% more surface area than balls. It was thus supposed that these grinding media would produce more fines in a grinding mill than balls. However, in laboratory tests it was found that the comparison depended on what was held constant between the two media types: mass, size distribution, or surface area (Shi, 2004).

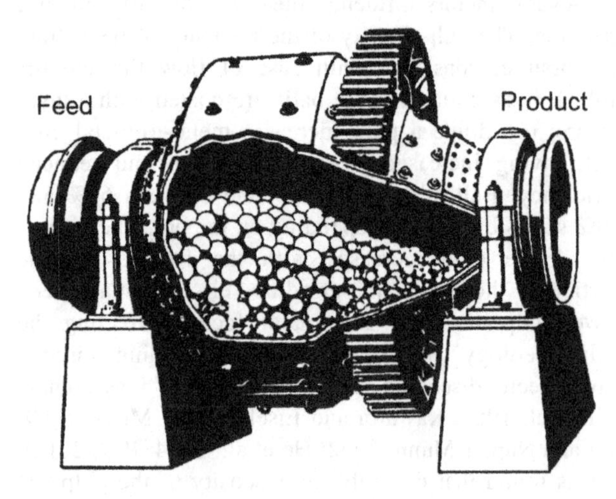

FIGURE 7.22 Hardinge mill.

Grinding balls are usually made of forged or rolled high-carbon or alloy steel, cast alloy steel and some of white iron (Rajagopal and Iwasaki, 1992). Consumption varies between 0.1 and as much as 1 kg t^{-1} of ore depending on hardness of ore, fineness of grind, and medium quality. Medium consumption can be a high proportion, sometimes as much as 40% of the total milling cost, so is an area that often warrants special attention. Good quality grinding media may be more expensive, but may be economic due to lower wear rates. Finer grinding may lead to improved metallurgical efficiency, but at the expense of higher grinding energy and media consumption. Therefore, particularly with ore of low value, where milling costs are crucial, the economic limit of grinding has to be carefully assessed (Chapter 1).

As the medium consumption contributes significantly to the total milling cost, great effort has been expended in the study of medium wear. Three wear mechanisms are generally recognized: abrasion, corrosion, and impact (Rajagopal and Iwasaki, 1992). Abrasion refers to the direct removal of metal from the grinding media surface. Corrosion refers to electrochemical reactions with oxygen, sometimes enhanced by galvanic effects with sulfide minerals (see Chapter 12); continual removal of oxidation products by abrasion means the fresh surface is always being exposed to such reactions. Impact wear refers to spalling, breaking, or flaking (Misra and Finnie, 1980; Gangopadhyay and Moore, 1987). Operational data show that abrasion is the major cause of metal loss in coarse (primary) grinding, with more corrosion occurring in finer regrinding (Dunn, 1982; Dodd et al., 1985). Corrosion products (iron oxy-hydroxides) can be detrimental to flotation and has led to the use of "inert" media, including high chrome balls, but also ceramic and other non-iron media (see Chapter 12).

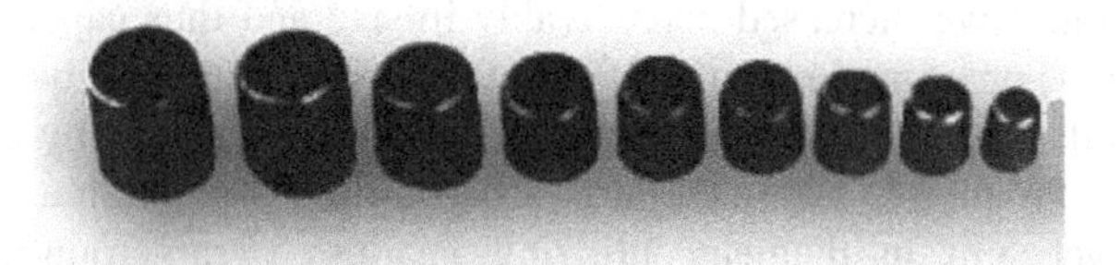

FIGURE 7.23 Doering Cylpebs used as grinding media *(Courtesy JKMRC and JKTech Pty Ltd).*

Attempts have been extended to predict media wear by developing a model incorporating the abrasive, corrosive, and impact wear mechanisms (Radziszewski, 2002). An example is the following empirical ball wear model for forged carbon steel based on fitting data from 46 Peruvian mills to ore hardness, pulp acidity and particle top size given by Rabanal and Guzmán (2013):

$$\Omega_E d^R = K_d^E = 1.36\left[\left(\frac{A_i - 0.05}{0.20}\right)^{0.166}\left(\frac{F_{80}}{5000}\right)^{0.069}\left(\frac{pH}{10}\right)^{-0.243}\right] \tag{7.10}$$

where Ω_E is the energy-corrected ball wear rate, g per kWh, d^R is the diameter of the largest balls in the mill (the recharge size), mm, K_d^E is the linear wear rate of balls, μm per kWh t^{-1}, Ai is the Bond abrasion index determined in a laboratory test (unitless), F_{80} is the feed 80% passing size of the ore, μm, and pH is the water acid/base measurement of the mill pulp.

The charge filling in a ball mill is about 30–45% of the internal volume of the mill, about 40% of this being void space. As discussed in Section 7.2.2, the reason is the power draw passes through a maximum at about a filling of 45%. Given the cost of media it may be economic to operate at the lower end of the filling range, even at the expense of a loss in energy efficiency, the same point noted for rod mills.

Ball mills are often operated at higher speeds than rod mills, so that the larger balls cataract and impact on the ore particles. The work input to a mill increases in proportion to the speed, and ball mills are run at as high a speed as is possible without centrifuging. Normally this is 70–80% of the critical speed.

Autogenous/Semi-autogenous Mills

The highest throughput grinding circuits in the mining industry use autogenous grinding (AG) or semi-autogenous grinding (SAG) mills. An AG mill is a tumbling mill that uses the ore itself as grinding media. The ore must contain sufficient competent pieces to act as grinding media and preferably be high specific gravity (s.g.) which, for example, favors AG milling of iron ores (s.g. 4 vs. 2.7 for high silicate ores). A SAG mill is an autogenous mill that uses steel balls in addition to the natural grinding media. A typical setup is shown in Figure 7.24. The ball charges in SAG mills have generally been most effective in the range of 4–15% of the mill volume, including voids. As noted in Section 7.2.2 there is a case to increase ball charge to increase power draw. In South African practice SAG mills can have a ball charge as high as 35% of mill volume (Powell et al., 2001).

The first paper describing ore as grinding media was delivered to the American Institute of Mining and

FIGURE 7.24 SAG mill in operation: 12.2 m (40 ft) diameter SAG mill (power rating 24 MW) at Tintaya Antapaccay *(Courtesy Antapaccay Perú).*

Metallurgical Engineers in 1908. In the early 1930s, Alvah Hadsel installed the Hadsel crushing and pulverizing machine, which was later improved by the Hardinge Company and was called the Hardinge Hadsel Mill. Early deployment of AG mills can be attributed primarily to a need in the iron ore industry to economically process large quantities of ore in the late 1950s. The high density of iron ores compared to many others probably favored AG milling. Non-ferrous operations (mostly copper and gold) utilized AG milling to a lesser extent before recognizing that SAG milling was better able to handle a variety of ore types. By December 2000, at least 1075 commercial AG/SAG mills had been sold worldwide with a total installed power exceeding 2.7 GW. SAG mills installed in Chile now regularly exceed 100,000 tons per day treated in a single mill.

AG/SAG mills usually replace the final two stages of crushing (secondary and tertiary) and rod milling of the traditional circuit (Chapter 6). This gives some advantages over the traditional circuit: lower capital cost, ability to treat a wide range of ore type including sticky and clayey feeds, relatively simple flowsheets, large size of available equipment, lower manpower requirements, and reduced steel consumption (noting that these mills replace some crushing/rod milling stages). The use of AG/SAG milling has grown to the point where many existing plants are retrofitting them, whilst new plants rarely consider a design that does not include them. This may not continue in the future, as new technologies such as fine crushing, high pressure grinding rolls, and ultra fine grinding (stirred milling) offer alternative flowsheet options (Erikson, 2014).

A schematic of various sections of the autogenous grinding mill is shown in Figure 7.25. AG or SAG mills are defined by the aspect ratio of the mill shell design and the product discharge mechanism. The aspect ratio is defined as the ratio of diameter to length. Aspect ratios generally fall into three groups: high aspect ratio mills where the diameter is 1.5–3 times of the length, "square mills" where the diameter is approximately equal to the length, and low aspect ratio mills where the length is 1.5–3 times that of the diameter. Although Scandinavian and South African practice favors low aspect ratio AG/SAG mills, in North America and Australia the mills are distinguished by high aspect ratios. The largest SAG mills are up to 12 m in diameter by 6.1 m length (belly inside liners), driven by a motor power of more than 20 MW. Many high aspect ratio mills, particularly the larger diameter units, have conical rather than flat ends. Attention should therefore be given to the mill length definition.

Inside the milling chamber the mill is lined with wearing plates held by lifter bars bolted to the shell. Lifter bars are essential to reduce slippage of the mill load, which causes rapid wear of the liners and also impairs the grinding action. The shape and geometry of the lifter bars, particularly the height and the face angle, have a significant influence on milling performance and wear rates.

The grate, as shown in Figure 7.25, is used to hold back the grinding media and large rock pieces and allow fine particles and slurry to flow through. Shapes of the grate apertures can be square, round, or slotted, with size varying from 10 to 40 mm. In some installations there are large holes varying from 40 to 100 mm in the grate, designated as *pebble ports*, which allow pebbles to be extracted and crushed (to ca. 12–20 mm) prior to recirculation to the mill. The total open area of the grate is ca. 2–12% of the mill cross-sectional area.

The slurry of particles smaller than the grate apertures discharges into the pulp lifter chambers. The pulp lifters, which are radially arranged, as shown in Figure 7.25, rotate with the mill and lift the slurry into the discharge trunnion and out of the mill. Each pulp lifter chamber is emptied before its next cycle to create a gradient across the grate for slurry transportation from the milling chamber into the pulp lifter chambers.

There are two types of pulp lifter design: radial and curved (also known as spiral, refer to Figure 7.26). The radial (or straight type) is more common in the mineral processing industry.

While the major portion of the slurry passing through the grate is discharged from the mill via the discharge trunnion, a proportion of the slurry in the pulp lifter

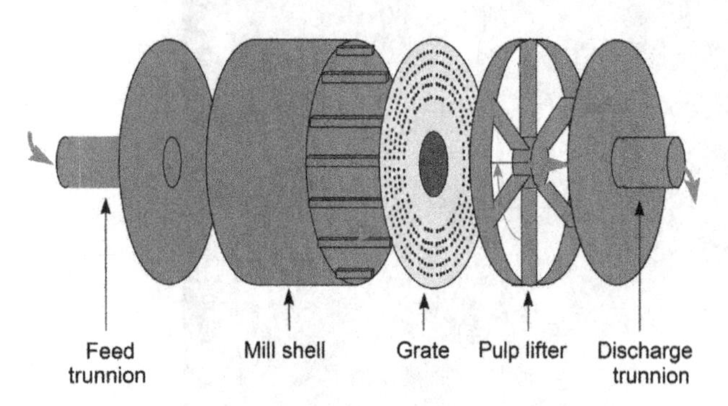

FIGURE 7.25 Schematic of various sections of an AG mill *(Courtesy Latchireddi, 2002).*

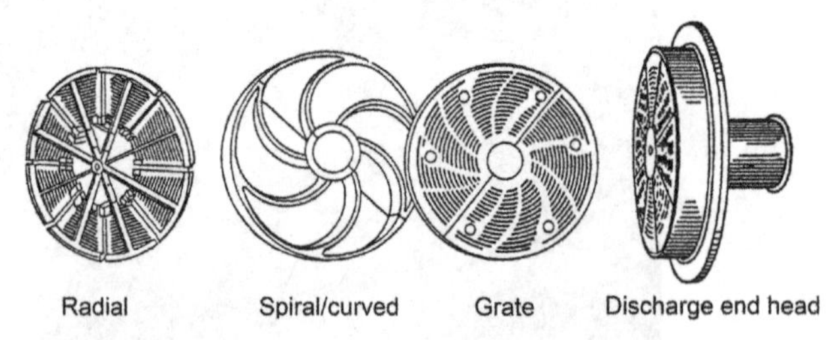

FIGURE 7.26 Conventional designs of pulp lifter *(from Taggart, 1945; Courtesy John Wiley and Sons).*

chamber flows back into the mill as the mill rotates. This flow-back process often leads to higher slurry holdup inside the mill, and may sometimes contribute to the occurrence of "slurry pooling", which has adverse effects on the grinding performance (Morrell and Kojovic, 1996). To improve the slurry transportation efficiency of the pulp lifters, particularly where slurry pooling limits capacity, a new concept of a Twin Chamber Pulp Lifter was developed (Latchireddi and Morrell, 2003, 2006) and first tested in a bauxite plant (Alcoa's 7.7 m diameter mill) in Western Australia (Morrell and Latchireddi, 2000). In this design, the slurry first enters the section exposed to the grate, the transition chamber, and then flows into the lower section, the collection chamber, which is not exposed to the grate. This mechanism prevents the pulp from flowing backward into the mill, which can significantly increase the capacity of the mill. The Turbo Pulp Lifter from Outotec is a similar concept. DEM analysis has helped refine lifter design to minimize pulp flow back. With curved pulp lifters the rotation of the mill cannot be reversed (used to extend liner life); in consequence asymmetric liner profiles are employed designed to resist wear on the leading side.

AG/SAG milling may be wet or dry. Dry AG mills were once common in iron ore grinding. They have more workplace environmental problems (dust), do not handle materials containing clay well, and are more difficult to control than wet mills. In certain talc and mica grinding applications dry SAG mills are operated.

AG/SAG mills can handle feed ore as large as 200 mm, normally the product of the primary crusher or the run-of-mine ore, and achieve a product of 0.1 mm in one stage. The particle size distribution of the product depends on the characteristics and structure of the ore being ground. The main mechanism of comminution in AG/SAG mills is considered to be abrasion and impact. Due to the relatively gentle comminution action, fractures in rock composed of strong equidimensional mineral grains in a weaker matrix are principally at the grain or crystal boundaries. Thus the product sizing is predominantly around the region of grain or crystal size. This is generally desirable for subsequent mineral separation as the wanted minerals are liberated with minimal overgrinding, and the grains keep their original prismatic shape more intact. A pilot plant study was made of the liberation characteristics of a nickel sulfide ore with full autogenous and semi-autogenous milling conditions. The mill products were sized and assayed and then analyzed by QEMSEM. Evidence indicated that selective breakage was occurring in both cases, leading to preferential liberation of sulfides (Wright et al., 1991).

Investigations have shown that ores ground autogenously may float faster than if ground with steel media (Forssberg et al., 1993). Grinding with steel media can suppress the floatability of minerals, due to release of iron oxidation products into the slurry (Chapter 12).

In comparison with high aspect ratio mills, SAG mills in the gold mines of South Africa (locally known as "run-of-mine" mills, or ROM mills) are low aspect ratio (largest mills are 4.88 m diameter and up to 12.2 m in length), and are normally operated at high ball charge (up to 35%), high total filling (up to 45%), and high speeds (up to 90% critical). They are often operated in a single stage of grinding to produce final product of 75–80% passing 75 μm from a feed top size around 200 mm, but the mill throughput is relatively low. The initial cost of low aspect ratio AG/SAG mills is less than high aspect ratio mills, but they consume more power per ton of product. The development of the different operating practice in South Africa occurred for historical rather than operational reasons. The ROM mills evolved from pebble tube mills used for fine secondary grinding that were converted to primary mills by directing the full run-of-mine feed to them.

The influence of feed size and hardness on AG or SAG mill operation is more significant than that on rod mill or ball mill operation. In rod or ball mills, the mass of the media accounts for approximately 80% of the total mass of the charge and dominates both the power draw and the grinding performance of the mills (see Section 7.2.2). In SAG mills a significant proportion (or all of it, in AG mills) of the grinding media derives from the feed ore. Any change in the feed size distribution and hardness will therefore result in a change of the breakage characteristics, and the mill charge level will be changed, which affects the mill power draw. As a result, the measured AG/SAG mill power draw often varies widely with time. This is one of the significant differences in operation between the AG/SAG mill and the rod/ball mill, power draw of the latter being relatively stable. In response to the variation in feed size and hardness, the mill feed rate has to be changed. At BHP-Billiton OK Tedi Mine (Papua New Guinea), the ore hardness varied between 5 and 16 kWh t^{-1}, and the throughput to the 9.8 m by 4.3 m (7.5 MW) SAG mill could vary between 700 and 3,000 t h^{-1} (Sloan et al., 2001). The effects of feed size and hardness on AG/SAG mill operation have been discussed (Bouajila et al., 2001; Hart et al., 2001; Morrell and Valery, 2001).

The AG and SAG mills respond in a different manner to feed size changes. In an AG mill sufficient numbers of large rocks need to be provided to maintain a high enough frequency of breakage collisions. In general, AG mill performance is better with coarser feeds, up to 200 mm. In SAG mills, however, ball charge tends to dominate rock breakage, and the contribution of rock grinding media will decrease (see Section 7.2.2). The coarser feed rocks provide less of a grinding media role and will instead provide a rock burden that requires to be ground. By reducing the feed size in these circumstances, the grinding burden will be reduced (Napier-Munn et al., 1996).

Since the feed size and hardness exert an important influence on AG/SAG mill performance, there are incentives to take account of the mine to mill operation through controlling blasting practices, mining methods, run-of-mine stockpiling, partial or fully secondary crushing, and selective pre-screening of AG/SAG mill feed. The mine-to-mill exercise has resulted in significant benefits to mining companies in terms of improving AG/SAG mill throughput, energy consumption, and grinding product size distribution (Scott et al., 2002).

The effect of feed characteristics on AG/SAG mills also differs from that on high pressure grinding rolls (Chapter 6). HPGRs are close to constant tonnage units: changes in feed characteristics will affect the power draw of the HPGR but tonnage remains fairly consistent compared to AG/SAG mills. This provides for greater circuit stability, and is a contributing reason to the growing interest in using HPGR technology (Lichter, 2014).

7.2.5 Motor Selection for Tumbling Mills

The choice of motor is dictated by the amount of power required, and whether or not the mill requires a variable speed capability. The efficiency of a particular drive dictates the economics of each motor type.

Three types of electric motors are commonly used to drive large tumbling mills:

Gearless

The largest mills, any requiring over 18 MW of power output, are almost always driven by gearless drives where the *rotor* of the motor is mounted to an external circumferential flange of the mill and the stationary *stator* of the motor is mounted securely to a foundation wrapped around the mill flange. Because of the large stator structure, these drives are sometimes called *wrap-around* or *ring* motors. Gearless drives operate at the same speed that the mill is turning and require a sophisticated drive control system that includes an inherent variable speed capability.

Synchronous

Widely used in North and South America, the synchronous motor is mounted beside one end of a mill and is connected to the mill via gears. The motor drives a *pinion*, which is a small geared shaft that connects to the *bull gear* that wraps around the mill circumference on a flange at one end of the mill. Synchronous motors derive their name from the way that the motor speed synchronizes with the waveform of the electric alternating current (AC) phases. This means the motor operates at a fixed speed in its simplest form. A maximum of two motors can be connected to the mill's bull gear via separate pinions, called a *twin-pinion* arrangement, developed for large mills (>5.5 m).

Wound-rotor Induction

Common in Australia and Africa, the wound-rotor motor also uses a gear system to transmit energy to the mill. It operates at a higher motor speed than the synchronous motor and needs an additional mechanical gear-box to reduce the motor shaft speed to the pinion speed needed to drive the mill. A slip energy recovery system is commonly used to permit modest speed adjustments when the motor is operating near its design speed. Wound-rotor induction motors can also be in a twin-pinion arrangement.

It is often desirable to adjust the speed at which a tumbling (primarily AG/SAG) mill operates; in such a circumstance, the mill drive system design must include a mechanism to achieve a variable speed capability (Barratt et al., 1999). Both the synchronous and wound-rotor induction motors can be made fully variable speed if the electric AC waveforms delivered to them by the drive electrical system are modified. A variety of technologies can be used to adapt the AC waveform. Two common variable speed drives for synchronous motors are the *cycloconverter* and the *load commutating inverter* (Grandy et al., 2002; von Ow, 2009).

The choice of motor and drive type is dictated first by the motor size (von Ow, 2009):

- Single-pinion drives are available up to 9 MW
- Twin-pinion arrangements of synchronous and wound-rotor induction motors are available in sizes ranging from 18 MW down to 6 MW
- Gearless drive systems are available in sizes ranging from 30 MW down to 12 MW

The electrical efficiency of fixed speed synchronous motors is greater than fixed speed wound-rotor induction motors, but the initial cost is higher. The gearless motor has a higher efficiency than a variable speed synchronous drive, which has a higher efficiency than a fully variable speed wound-rotor induction drive. The initial capital cost, however, and installation time favor the wound-rotor induction motor, followed by the synchronous motor, with the gearless being the most expensive motor with the longest installation time. A financial and plant power grid evaluation is often used to choose between motor options.

7.2.6 Sizing Tumbling Mills

Tumbling mills are rated by power rather than capacity, since the capacity is determined by many factors, such as the grindability, determined by laboratory testing (Chapter 5), and the reduction in size required.

Rod and Ball Mills

The specific energy needed for a certain required capacity may be estimated by the use of Bond's equation, written here as:

$$W = 10Wi\left(\frac{1}{\sqrt{P_{80}}} - \frac{1}{\sqrt{F_{80}}}\right)EF_x \quad (7.11)$$

where W is the specific energy consumption of the mill, kWh t^{-1}; Wi is the work index measured in a laboratory mill, kWh t^{-1}; F_{80} and P_{80} are, respectively, the circuit feed and product 80% passing sizes, μm; and EF_x is the product of the Rowland efficiency factors applicable to a grinding stage. The Rowland efficiency factors dependent on the size of mill, size and type of media, type of grinding circuit, etc. (Rowland and Kjos, 1980; Rowland, 1982; King, 2012). The power required (in kW) is then given by W × T, where T is the throughput tonnage (t h^{-1}) (see Example 7.2).

AG/SAG Mills

Extensive testing is being conducted for sizing autogenous or semi-autogenous mills because of the lack of equivalent methodology as represented in the Bond approach for rod and ball mills. Pilot scale testing of ore samples, more specifically for an autogenous mill where the grinding medium is also the material to be ground and consequently a variable itself, has become a necessity in assessing the feasibility of AG/SAG milling, predicting the energy requirement, flowsheet, and product size (Rowland, 1987; Mular and Agar, 1989; Mosher and Bigg, 2001).

The use of energy-based or mathematical modeling and simulation can help reduce the cost of the pilot tests by narrowing the choice of processing route at the pre-feasibility stage. This approach includes collection of a large quantity of representative ore samples that are typically drill cores or samples taken throughout the ore body. The rock breakage characterization data are obtained through standard laboratory tests such as the drop-weight impact and tumbling tests (Napier-Munn et al., 1996), SAGDesign tests (Starkey et al., 2006), the MacPherson autogenous mill work index test (MacPherson, 1989; Mosher and Bigg, 2001) and more recently, the Rotary Breakage Tester (Shi et al., 2009). The breakage characteristics of the ore are compared with extensive databases of ore types and plant performance, which provides an indication of the ore's relative strength and amenability to processing in various circuit configurations. Following pilot scale tests, computer simulation software such as JKSimMet (Napier-Munn et al., 1996) is used to scale-up to the full-size plant. There are other computer software available for plant simulation, such as MODSIM developed by King and co-workers (Ford and King, 1984), USIM-PAC (Evans et al., 1979), and CEET (Comminution Economic Evaluation Tool) (Dobby et al., 2001; Starkey et al., 2001). Even in the absence of pilot data, energy-based models or simulation can still provide reasonably accurate prediction of full-scale plant performance.

Example 7.2

Using the Bond equation, select the size of ball mill in closed circuit with a cyclone for the following conditions: $F_{80} = 600$ μm, $P_{80} = 110$ μm, $Wi = 10.5$ kW t^{-1}, T = 150 t h^{-1}.

Solution

Note that the P_{80} is the circuit product (i.e., cyclone overflow) and circulating load is assumed to be 250% (Rowlands, 1982). Solving for W, and ignoring efficiency factors (for simplicity) then:

$$W = 10 \times 10.5 \times \left(\frac{1}{\sqrt{110}} - \frac{1}{\sqrt{600}}\right) = 5.72 \text{ (kWh t}^{-1}\text{)}$$

Therefore the power draw P is;

$$P = 5.72 \times 150 = 859 \text{ (kW)}$$

The size of mill is then selected from either a calibrated power model (Section 7.2.2) or, more generally, from manufacturer's data, such as in the table below (Table ex 7.2).

The mill (4.12 m × 3.96 m) meets the target power.

TABLE EX 7.2 Ball Mill Specifications (overflow discharge, 40% charge filling: see Rowlands, 1982)

Diameter (m)	Length (m)	Ball Size (mm)	Mill Speed (% critical)	Charge Weight (t)	Mill Power (kW)
3.96	3.96	50	71.7	82.8	843
4.12	3.96	64	71.7	89.4	945
4.27	4.27	64	70.7	104	1093

The cost of a thorough pilot testing or standard testwork program can be prohibitive, particularly for deposits with highly variable ore types. Instead of collecting large representative samples, a large number of small quantity drill core samples, representatives of the different geological domains of the deposit, are collected throughout the ore body to be characterized by simpler tests such as the SPI test (SAG Power Index) or abbreviated versions of the standard laboratory tests, such the SMC (Morrell, 2004), the RBT Lite (JKTech Brochure) or the SVT (Brissette et al., 2014). These tests are recognized to be geometallurgical tests (Chapter 17). Geometallurgy methodology is increasingly used as a standard in both design and production forecasting of AG/SAG circuits worldwide, giving a much better understanding of the ore response to milling for any deposit.

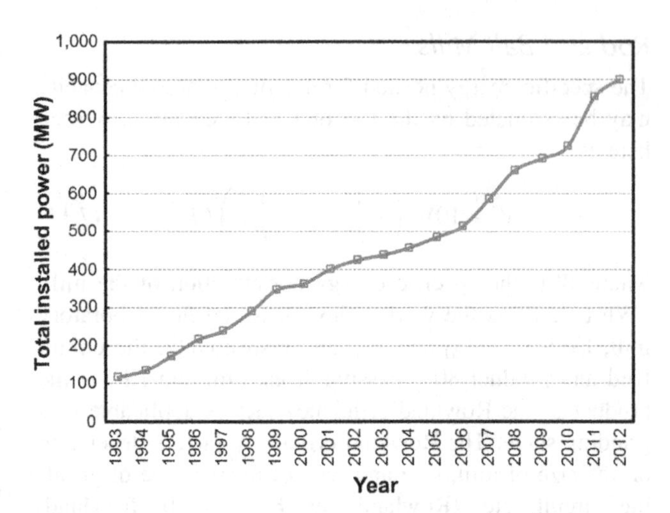

FIGURE 7.27 Increase in installed power for fine grinding applications (all fine grinding technologies: more than 1500 units in 2012 worldwide) *(Courtesy M. Brissette).*

7.3 STIRRED MILLS

The concept of stirred milling dates back to about 1928 (Stehr, 1988) when the idea of using "an agitator and spherical grinding media" was presented. In 1948 DuPont introduced the "sand mill" for pigment grinding. Stirred milling found applications in multiple fine to ultrafine grinding applications from pharmaceuticals to industrial minerals. Starting about two decades ago, stirred milling has become more prevalent in milling applications, corresponding to the increase in processing of more complex fine-grained ores demanding liberation grinds of 10 μm and less (Underle et al., 1997; Ellis and Gao, 2002).

The common stirred mills in mineral processing are of two types: those that operate at low stirrer speed where gravity plays a role, the Metso Vertimill®, and the Eirich TowerMill®; and high-speed stirred mills which fluidize the pulp, the Metso Stirred Media Detritor (SMD®), the Xstrata IsaMill®, the FLSmidth VXP® Mill and, since 2012, the Outotec High Intensity Grinding Mill (HIGMill®). The Figure 7.27 shows the increase in installed power of fine grinding technologies in the past 20 years. The application is mainly in regrinding to increase liberation and concentrate grade.

Stirred mills differ from tumbling mills in how grinding energy is transferred to the material being ground. Tumbling mills use both impact and shear (abrasion/attrition) energy in roughly equal measure (Radziszewski and Allen, 2014), while stirred mills use predominately shear energy. For fine grinding, shear is more effective and stirred mills are more energy efficient (use less energy per ton) than tumbling mills when the product P_{80} is less than about 100 μm (Nesset et al., 2006; Mazzinghy et al., 2012). One simple explanation for this higher energy efficiency is that media impacts (i.e., balls colliding) tend to expel fine particles with the slurry rather than nip and break them ("splash out" effect). While initial applications were in regrinding stirred mills have attracted interest for use in primary grinding circuits.

7.3.1 Power Draw and Stress Intensity

Power Draw

From a review of existing power models for stirred mills, Radziszewski and Allen (2014) proposed that if all grinding is accomplished via shear then grinding power can be represented by the following:

$$P_\tau = \mu\omega^2 V_\tau \tag{7.12}$$

where P_τ is the shear power (W), μ the viscosity of the mill contents (N.s m^{-2}) (for which a relationship was given), ω the angular velocity (rad s^{-1}), and V_τ the "shear volume" (referring to all the shear surface pairs between an impeller and the mill chamber). With appropriate calibration the predicted power matched the measured power for the data of Gao et al. (1996) and Jankovic (1998).

Stress Intensity

This is a measure of the pressure acting on the particles and is fixed during the comminution process as the conditions—grinding media size and density, stirrer speed and slurry density—are normally constant during operation. In the fluidized slurry mills the stress derives mainly from the action of the media, while for lower stirrer speed mills gravity (the weight of the charge) also contributes stress. Expressions for stress intensity have been developed by Kwade et al. (1996) and Jankovic (2001). There is a relationship between stress intensity and product size and an optimum exists that gives maximum size reduction (lowest product P_{80}) and thus is the most energy efficient condition (Jankovic, 2001). As with tumbling mills, the ratio of media size to particle size is of particular significance in optimizing energy consumption (Jankovic, 2003).

Mill orientation	Horizontal		Vertical
	Pin	Disc	Screw (Auger)
Impeller shape			
	Smooth	Disc	Pin
Mill shell surface (liner)			
Impeller speed	Low	Medium	High

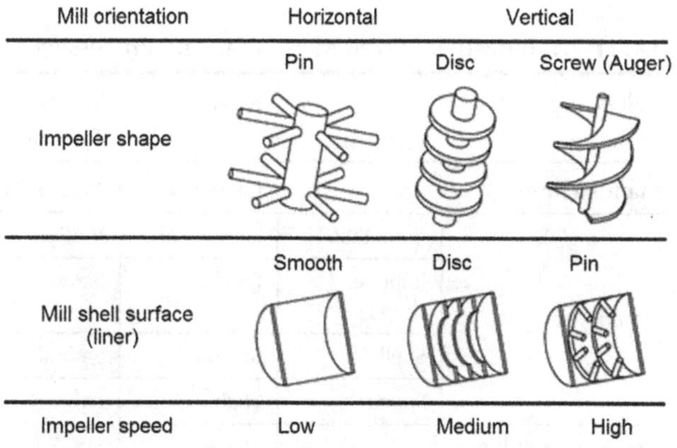

FIGURE 7.28 Stirred milling technology classification. *(Courtesy R. Cunningham).*

7.3.2 Sizing Stirred Mills

Specific grinding energy (kWh t^{-1}) is the key parameter for sizing industrial units. The Bond method, successfully used to size ball mills, for example, is not applicable to stirred milling. Development of testing methodologies is still continuing and in some cases there can be discrepancies in scale-up of 100–300% (Larson et al., 2011). The three most common bench techniques for sizing a stirred grinding mill are the Levin Test, the Metso Jar Ball Mill Test, and the Signature Plot.

Levin Test

This test requires "baseline" data from an operating mill (i.e., feed size, product size, mill tonnage, and total grinding energy consumption). A sample of the baseline reference material is used to calibrate the lab scale mill. The test sample is ground to the same product size as the baseline reference and the number of revolutions required to achieve the target size is recorded. Once calibrated, the energy delivered per revolution is assumed to be the same. Any future material's specific energy requirements can be estimated by recording the number of revolutions required to achieve a particular target size. The challenge of the method is obtaining high quality baseline data.

Metso Jar Ball Mill Test

This method can be carried out wet or dry in a tumbling grinding mill (8″ D × 10″ L). Grinding is carried out over several grinding periods (e.g., 10 min and 20 min). Both product size and energy consumption are recorded for each subsequent grinding period. The measured energy is multiplied by 0.65 (VTM factor) and 1.10 (SMD factor) to predict the energy required by a Vertimill® and Stirred Media Detritor, respectively.

Signature Plot Technique

The signature plot takes a 20 kg sample at 50 wt% solids and runs the sample through a test mill. At fixed periods, with a known energy expenditure, a sample is taken to determine the particle size distribution. (See also Chapter 5.) Typically, the method takes either 12 data points or as many until a reduction in size is no longer taking place. This technique has been used for sizing both the Xstrata IsaMill® and the FLSmidth VXP® mill.

Pilot testing of a stirred mill can normally be done with a scale model of the milling technology. Fluidized mills (such as the IsaMill®) have been found to scale up directly to industrial units as long at the test units are 1 L in volume or greater (Karbstein et al., 1996). This is attributed to the fact that the stress intensity responsible for breakage is due to the high speed stirring. Technologies that employ both lower stirring speeds and gravitational stress intensity to induce grinding (such as the Vertimill®) need larger pilot units in order to ensure that the significant gravity stress component is being reproduced.

7.3.3 Mill Types

Stirred milling technologies can be classified by four properties: the orientation of the mill, the shape of the impeller, the mill shell surface, and the speed of the impeller (Figure 7.28).

All mills currently on the market can be described by the combinations in Figure 7.28. For example, a Vertimill® would be a vertical, screw impeller, smooth shell liner, and low speed mill, while an IsaMill® would be a horizontal, disc impeller, smooth surface and high speed mill. Table 7.1 summarizes the features of the available stirred mills. A mill's orientation tends to be more a plant design factor as horizontal units require more footprint, while vertical units need more head room. The disc and pin impellers are currently associated with the medium to high speed (i.e., the fluidized) technologies, while the screw impeller is associated with the low speed technologies. The selection of a non-smooth liner would be in cases where one wanted to create more shear volume (Radziszewski, 2013).

TowerMill®/Vertimill®

The predecessor of today's Vertimill® is the TowerMill, which was introduced in 1953 by the Nichitsu Mining

TABLE 7.1 Industrial Stirred Mills in Mineral Processing

Mill Name		IsaMill	Vertmill/ TowerMill	Stirred Media Detritor	VXP Mill	HIGMill
Manufacturer		Glencore	Metso/Eirich	Mesto	FLSmidth	Outotec
Description	Orientation	Horizontal	Vertical	Vertical	Vertical	Vertical
	Impeller shape	Disc	Screw	Pin	Disc	Disc
	Shell	Smooth	Smooth	Smooth	Smooth	Disc
	Speed	High	Low	Medium	Medium	Medium
Density % Solids ($v\ v^{-1}$)		10–30	30–50	10–30	10–30	10–30
Impeller tip speed ($m\ s^{-1}$)		19–23	<3	3–8	10–12	8–12
Power intensity ($kW\ m^{-3}$)		300–1,000	20–40	50–100	240–765	100–300
Typical size range (μm)	Feed (F_{80})	300–70	6,000–800	300–70	300–70	300–70
	Product (P_{80})	<10	20	<10	<10	<10
Grinding media size (mm)		1–3	12–38	1–8	1.5–3	1–6
Largest unit	Model	50,000 L	VTM6000/ ETM1500	SMD1100	VXP10000	HIG5000
	Power	8 MW	4.5 MW/ 1.1 MW	1.1 MW	3 MW	5 MW

Courtesy R. Cunningham.

Industry Co., Ltd. By 1954 a new company, Nippon Funsaiki, was founded to exclusively produce the Abrasive Toushiki (steeple-like) Crusher. Hereafter it was named "Tower Mill" and in 1965 the Japan Tower Mill Co., Ltd. was established. In 1983, Kubota Ironworks Co. purchased the Japan Tower Mill Co. and supplied the technology as Kubota Tower mills. The latest "owner" of the tower mill technology is Eirich. The Koppers Company, Inc. adopted the new fine grinding technology in the early 1980s, after which the Tower Mill was manufactured by MPSI under a license agreement. In 1991, the license expired and Svedala Industries, Inc. (now Metso Minerals Ltd.) obtained all rights to the technology, except the name, which was changed to Vertimill®.

A schematic of a tower mill is shown in Figure 7.29. Steel balls or pebbles are placed in a vertical grinding chamber in which an internal screw flight provides medium agitation. The feed enters at the top, with mill water, and is reduced in size by attrition and abrasion as it falls, the finely ground particles being carried upward by pumped liquid and overflowing to a classifier. Oversize particles are returned to the bottom of the chamber. An alternative configuration is to feed the mill by the bottom.

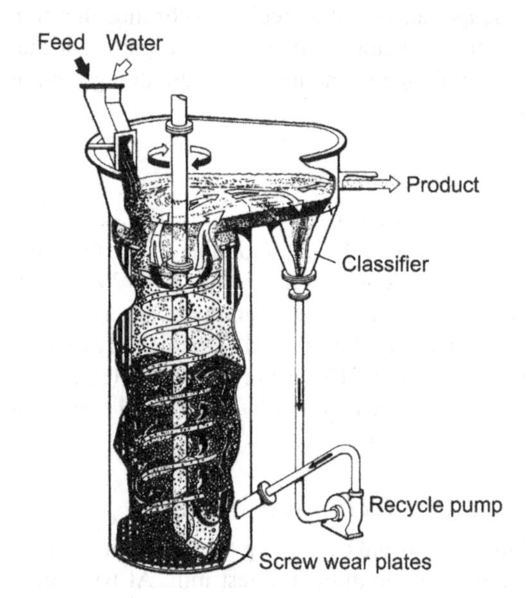

FIGURE 7.29 Tower mill.

Stirred Media Detritor (SMD®)

English China Clays developed an attrition stirred "sand mill" during the 1960s and in 1969 the first production units were installed in a kaolin plant. Currently, ECC, now Imerys, operate more than 200 attrition sand mills in

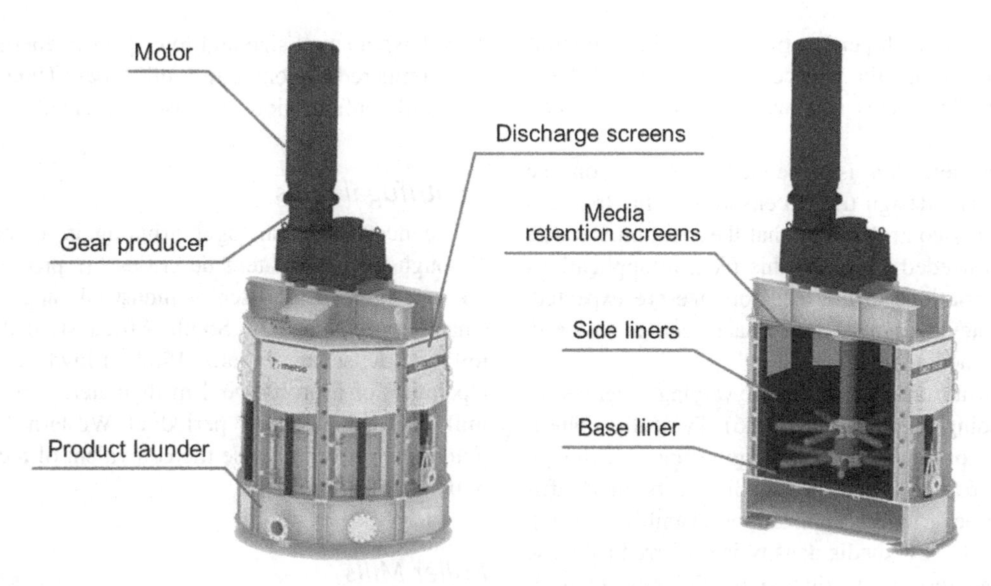

FIGURE 7.30 The stirred media detritor *(Courtesy Metso Minerals).*

their kaolin and calcium carbonate plants around the world. In 1996, Svedala and ECC signed a license agreement enabling attrition sand mills to be supplied for the Century Zinc Project in Australia. In the following year, this license was expanded, enabling Svedala (now Metso Minerals Ltd) to manufacture and supply the Stirred Media Detritor (SMD®) globally for all applications other than the white pigment industry.

Figure 7.30 shows an SMD®. Normally, natural silica or ceramic is used as the grinding media. Grinding media are added through a pneumatic feed port or the manual feed chute located on top of the mill. Feed slurry enters through a port in the top of the unit.

IsaMill®

Developed by Mount Isa Mines Ltd. (now Glencore Technology) and Netszch-Feinmahltechnik GmbH in the 1990s, the IsaMill® is a large version of the Netszch horizontal stirred mill that was being used for ultrafine grinding applications in various chemical industries. To make this horizontal mill suitable for the mining industry, the engineering challenge was to expand the volume by a factor of 6.

Inside the horizontal shell is a series of rotating discs mounted on a shaft that is coupled to a motor and gearbox (Figure 7.31). The high-speed discs fluidize the media and the slurry that is continuously fed into the feed port. A patented product separator keeps the media inside the mill, allowing only the product to exit. The grinding media can include granulated slag, river sand, ceramic beads or a sized portion of the ore itself.

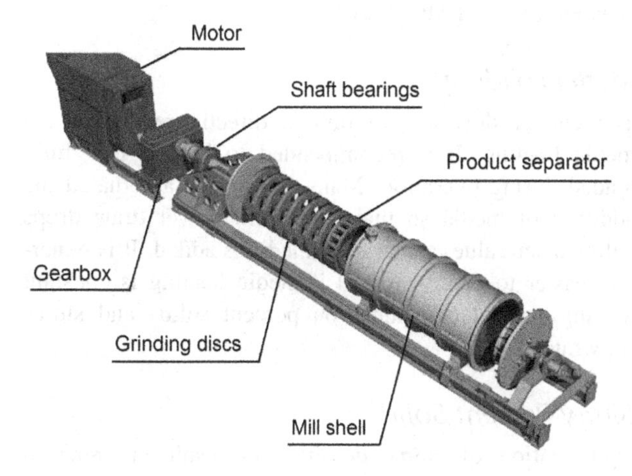

FIGURE 7.31 Schematic view of M10000 IsaMill with the mill shell removed *(Courtesy Xstrata Technology).*

VXPMill®

Originally developed in the mid-1990s for the fine pigment industry by Deswik Ltd., in 2010 Deswik and Knelson signed a partnership and the name was changed to the Knelson-Deswik mill. In 2012, Knelson became a part of the FLSmidth group and the mill received its present name: the FLSmidth VXPmill®.

HIGMill®

In 2012, Outotec launched a new fine grinding technology for the mineral processing industry. The technology has been utilized for more than 30 years in the calcium carbonate industry, but until recently, was not available for mineral processing.

7.3.4 Some Operational Points

Media

Proper media selection is key for a successful stirred milling application. A balance needs to be achieved between milling efficiency (kWh per ton fresh feed) versus operating costs (cost of media replacement per ton of fresh feed). Milling efficiencies vary with media size, density, and shape.

Media size selection is dependent normally on the feed top size. The design tries to ensure that the top feed size to the mill is controlled and that the medium is a size no larger than needed to handle this feed. In applications where large variations in the feed top size are expected, energy efficiency will need to be sacrificed in order to select a larger media size.

Media density and shape have varying degrees of effect on grinding (Nesset et al., 2006). Typically, spherical media are considered to be energy efficient, although there is evidence that rod shaped media can be more efficient in ultrafine grinding applications (Tamblyn, 2009). For vertical mills, low media density is preferred as dense media sink and results in grinding inefficiency. In horizontal mills, high density media were found to be more efficient (Xu and Mao, 2011).

Media Loading

The energy drawn by a mill is directly related to the media loading. It is recommended to keep a mill fully loaded during operation. Many mills have automated the addition of media so that when mill power draw drops below a set value, a charge of media is added. It is generally easier to control a mill if media loading is constant leaving control dependent on percent solids and slurry flow rate.

Slurry Percent Solids

Optimization of slurry density can result in gains in energy efficiency. If there is too high a water content, the number of contact events between particles and media drops, resulting in less grinding; if the solids content becomes too great, then fluidization of the mill is lost and the only media movement is that directly caused by the impeller.

7.4 OTHER GRINDING MILL TYPES

Vibratory Mills

These are available for continuous or batch grinding materials to a very fine end product. Two vibrating cylinders charged with ball or rod media are placed either directly above the other or inclined at 30° to perpendicular. Located between the two cylinders is an eccentrically supported weight driven to produce a small circular oscillation of a few millimeters, which fluidizes the mill contents and provides impulses to the material that is ground by attrition. The material, wet or dry, is fed and discharged through flexible bellow-type hoses. Vibratory mills have a small size and low specific energy consumption compared to competing mill types. Three-tube vibratory mill configurations are also available.

Centrifugal Mills

The concept of centrifugal grinding is an old one, and although an 1896 patent describes the process, it has so far not gained full-scale industrial application. The Chamber of Mines of South Africa studied centrifugal milling (Kitschen et al., 1982; Lloyd et al., 1982). Operation of a prototype 1 m diameter, 1 m long, 1 MW mill over an extended period at Western Deep Levels gold mine proved it to be the equivalent of a conventional 4 m × 6 m ball mill.

Roller Mills

These mills are often used for the dry grinding of medium soft materials of up to 4–5 mohs hardness. Above this hardness, excessive wear offsets the advantage of lower energy consumption compared with conventional mills.

Table and Roller Mills

These units are used for dry grinding of medium-hard materials such as coal, limestone, phosphate rock, and gypsum. Two or three rollers, operating against coiled springs, grind material which is fed onto the center of a rotating grinding table (Figure 7.32). Ground material spilling over the edge of the table is air-swept into a

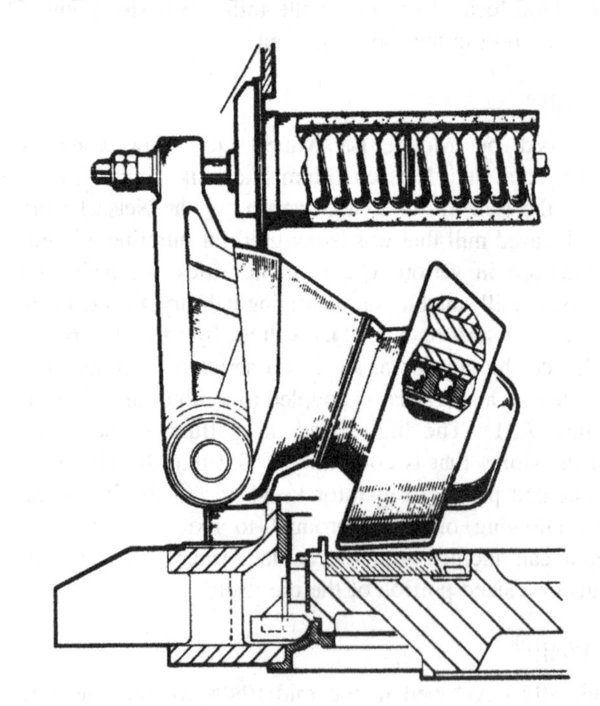

FIGURE 7.32 Section through table and roller mill.

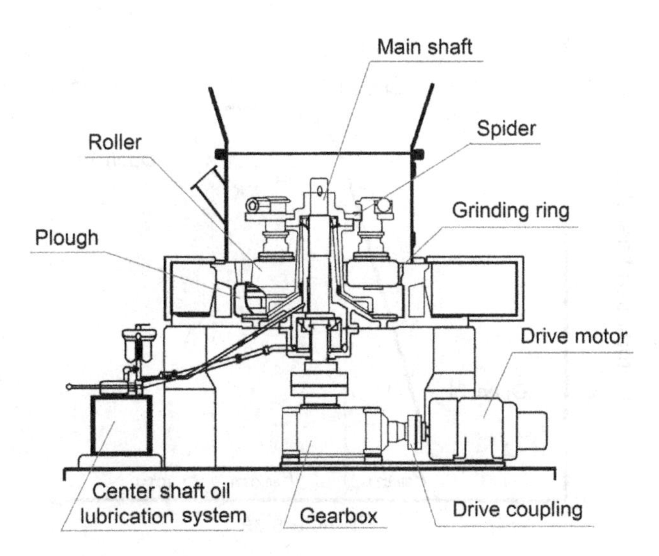

FIGURE 7.33 Section through pendulum roller mill.

classifier mounted on the mill casing, coarse particles being returned for further grinding.

Pendulum Roller Mills

Used for dry fine grinding non-metallic minerals such as barytes and limestone, material is reduced by the centrifugal action of suspended rollers running against a stationary grinding ring (Figure 7.33). The rollers are pivoted on a spider support fitted to a gear-driven shaft. Feed material falls onto the mill floor, to be scooped up by ploughs into the "angle of nip" between the rolls and the grinding ring. Ground material is air-swept from the mill into a classifier, oversize material being returned.

7.5 GRINDING CIRCUITS

7.5.1 Circuit Types

The feed can be wet or dry, depending on the subsequent process and the nature of the product. Dry grinding is necessary with some materials due to physical or chemical changes that occur if water is added. Areas with a shortage of water and the growing determination to limit water use may see increased dry grinding in the future. It can cause less wear on the liners and grinding media as there is no corrosion. There also tends to be a higher proportion of fines in the product, which may be desirable in some cases. Control of dust is an issue.

Wet grinding is generally used in mineral processing operations because of the overall economies of operation. The advantages of wet grinding include:

1. Lower power per ton of product.
2. Higher capacity per unit mill volume.

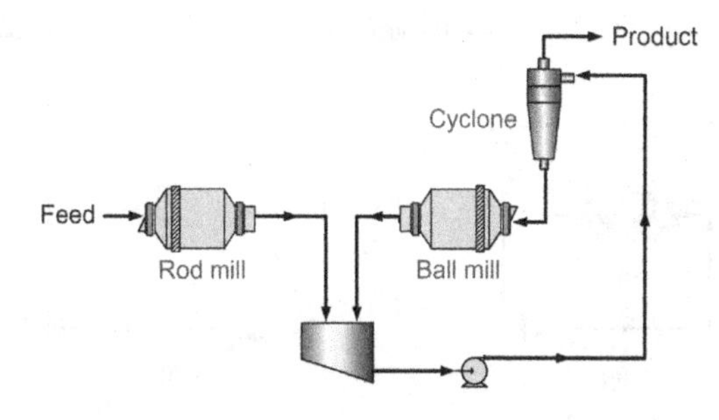

FIGURE 7.34 Rod/ball mill circuit illustrating open circuit (rod mill) and closed circuit (ball mill-hydrocyclone).

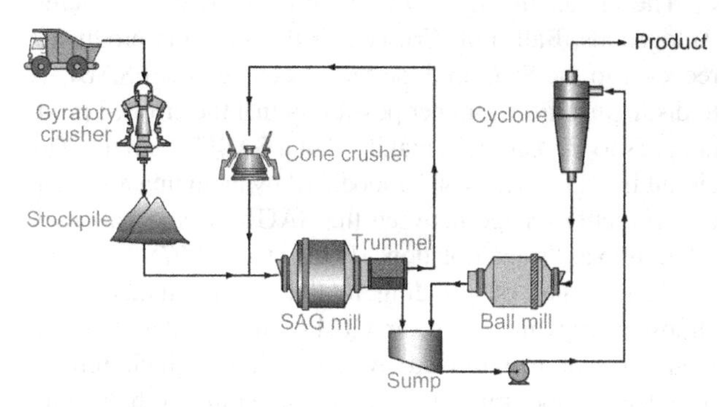

FIGURE 7.35 Semi-autogenous/ball mill circuit.

3. Makes possible the use of wet screening or classification for close product control.
4. Eliminates the dust problem.
5. Makes possible the use of simple handling and transport methods such as pumps, pipes, and launders.

The type of mill for a particular grind, and the circuit in which it is to be used, must be considered simultaneously. Circuits are divided into two broad categories, *open* and *closed* (the same terms introduced for crushing circuits, Chapters 6 and 8). Figure 7.34 shows the older style rod/ball mill arrangement, which illustrates open circuit, the rod mill, and closed circuit, the ball mill closed with a hydrocyclone (Chapter 9). Figure 7.35 shows a SAG/ball mill circuit where both grinding mills are in closed circuit, the SAG mill closed with a crusher to control the amount of "critical size" material in the circuit (see Chapters 5 and 6). In open circuit, the material is fed to the mill at a rate calculated to produce the correct product in one pass. Open circuit is sometimes used for regrinding iron concentrates for pelletization.

The first mill in the two circuits (Figures 7.34 and 7.35) is sometimes referred to as the "primary" mill, and the second mill as the "secondary". As well, the circuits may be referred to as "primary circuits" to distinguish them from regrind circuits that may exist downstream in the plant.

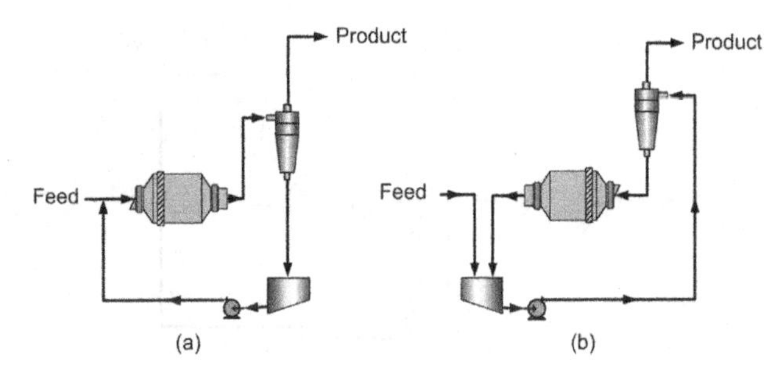

FIGURE 7.36 Closed circuit arrangements: (a) Forward classification circuit, FCC, and (b) Reverse classification circuit, RCC.

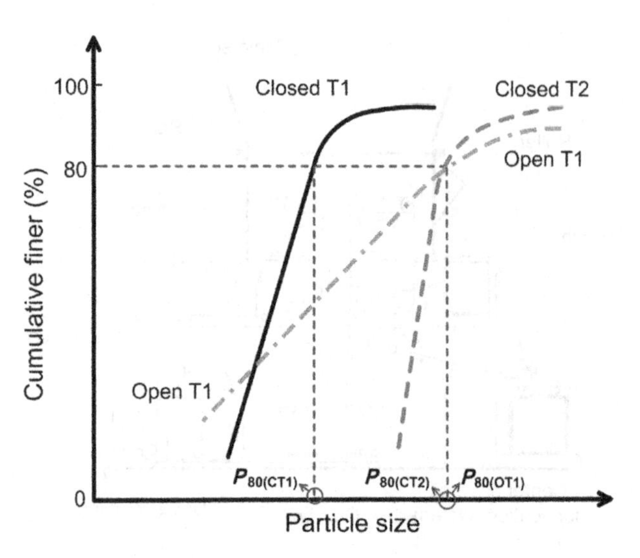

FIGURE 7.37 Effect of closing a circuit on product size distribution and circuit capacity (see text for description of the symbols).

The circuit in Figure 7.35 is termed SABC, for Semi-Autogenous, Ball mill, Crusher. In this case the product is recycled to the SAG mill, so this circuit becomes SABC-A to distinguish from another possibility that the crushed product goes to the ball mill, making that a SABC-B circuit. The circuit in Figure 7.35 can be modified by including a separation (flotation) stage between the SAG mill and ball mill, referred to as "stage flotation" (Finch et al., 2007).

Final stages of grinding in the mining industry are almost always in closed circuit, in which material of the required size is removed by a classifier, which returns oversize to the mill, which is commonly a ball mill. Figure 7.36 shows the two closed ball mill-cyclone circuit arrangements: *forward classification* (a) where the fresh feed goes directly to the mill, and *reverse classification* (b) where the fresh feed goes first to the classifier. As a rule of thumb, forward circuit classification is used if the fresh feed contains less than 30% finished product (i.e., finer than the circuit product P_{80}). The material returned to the mill by the classifier is known as the *circulating load* and is expressed as a ratio or percentage of the weight of new feed. (This same term was introduced in Chapter 6 describing the closed crusher/screen arrangement.) Circulating loads are typically 150–350%. The sizing of a ball mill is based on a mill in closed circuit with a 250% circulating load (see Example 7.2).

In closed-circuit operation, no effort is made to effect all the size reduction in a single pass. Instead, every effort is made to remove material from the circuit as soon as it reaches the required size. This has the effect of giving a narrower size distribution in the circuit product (classifier fine product) compared to open circuit, as finished product (fines) is removed quickly and avoids "over-grinding" while the coarse material is preferentially recycled and ground. The larger the circulating load the shorter the residence time in the mill and thus the greater is the content of coarse material in the mill feed. This means that a larger fraction of the grinding energy is being usefully applied to over-size material, which increases the capacity of the circuit to produce a target P_{80}. (See also Chapter 5 discussion on "functional performance analysis".) This impact on product particle size distribution (PSD) and circuit capacity is illustrated in Figure 7.37. Here the increased fineness and narrowing of the PSD of the closed circuit product compared to the open circuit product at the same tonnage T1 is depicted. The finer product P_{80} of the closed circuit at T1 (P_{80CT1}) compared to the open case (P_{80OT1}) means the tonnage can be increased (T2 > T1) to return to the original open circuit P_{80} (P_{80OT1}). In this manner, at a given P_{80} the closed circuit has higher throughput than the open circuit at the same mill power draw.

Another advantage of the closed circuit is that the mill contents become narrower in size distribution as circulating load increases and the distribution changes less from feed to discharge, making it, in principle, easier to optimize the media size to particle size, which further increases grinding energy efficiency.

The optimum circulating load for a particular circuit depends on the classifier capacity and the cost of transporting (pumping) the load to the mill. It is usually in the range of 100–350%, although it can be as high as 600%. The increase in capacity is most rapid in the first 100% of circulating load, then continues up to a limit, depending on the circuit, just before the mill "chokes" (i.e., is overloaded, identified in an overflow mill by balls sometimes exiting with the slurry). The increase in capacity can be considerable: Davis, as quoted by Gaudin (1939), showed the quantity of new −150 mesh (−105 μm) material increased by about 2.5 times as the circulating load increased from zero (i.e., open circuit) to 250%. While some current work is challenging this finding (Jankovic and Valery, 2013), it is still widely used as a guide to increasing ball mill-cyclone circuit efficiency (McIvor, 2014).

Similar considerations apply to regrind applications. Using a ball mill, it is usually closed with a cyclone. With the newer stirred mills there may be a combination of internal and external classification. Given the increasing fine size demanded and achievable with stirred mills, efficient classification becomes a challenge.

Various types of classifying devices can be used to close the circuit (Chapter 9). Mechanical classifiers are still found in some older mills, but suffer from classifying by gravitational force, which restricts their capacity when dealing with fine material. Hydrocyclones (or simply "cyclones") classify by centrifugal action, which intensifies the classification of fine particles, giving sharper separations and allowing higher circulating loads. They occupy much less floor space than mechanical classifiers of the same capacity and have lower capital and installation costs. Due to their fast response, the grinding circuit can rapidly be brought into balance if changes are made and the reduced dwell time of particles in the circuit gives less time for oxidation to occur, which may be important with sulfide minerals that are to be subsequently floated. Cyclones have, therefore, come to dominate grinding circuits preceding flotation operations. Figure 7.38 shows a ball mill/cyclone closed circuit installation.

FIGURE 7.38 Ball mill closed with cluster of hydrocyclones (above and in the background), Batu Hijau mine, Indonesia *(Courtesy JKMRC and JKTech Pty Ltd).*

The action of all classifiers in grinding circuits is dependent on the differential settling rates of particles in a fluid, which means that particles are classified not only by size but also by specific gravity. Consequently, a small dense particle may classify in a similar way to a large, low-density particle. Thus, when an ore containing a heavy valuable mineral is fed to a closed grinding circuit, this mineral is preferentially recycled compared to the low density mineral (usually gangue) and is ground finer. (This effect of density on classification is discussed further in Chapter 9.) Selective grinding of heavy sulfides in this manner prior to flotation can allow a coarser overall grind, the light, coarse gangue reporting to the classifier overflow, while the heavy valuable mineral particles are selectively ground. It can, however, pose problems with gravity and magnetic separation circuits, where particles should be as coarse and as closely sized as possible in order to achieve maximum separation efficiency.

To address the differential mineral classification due to density has contributed to the use of "double classification" (Castro et al., 2009), which re-classifies the cyclone underflow, and stimulated development of the "three-product cyclone" (Obeng and Morrell, 2003). In some cases circuits can be closed by screens rather than classifiers, which cut on the basis of size only, as discussed in Chapter 8. Combinations of classifiers and screens are another option.

Other strategies aim to take advantage of the concentration of heavy minerals in the circulating load. A flotation unit ("flash flotation") or gravity separation stage can be incorporated in the grinding circuit, e.g., on the cyclone underflow, cyclone feed or ball mill discharge. Incorporation of gravity separation is almost universal in gold circuits with a high degree of "gravity recoverable gold" (Chapter 10), and in some instances can be extended to the recovery of platinum group minerals.

Multi-stage grinding, in which ball mills are arranged in series, can be used to produce successively finer products (Figure 7.39). These circuits introduce the possibility of separation stages between the grinding units but the

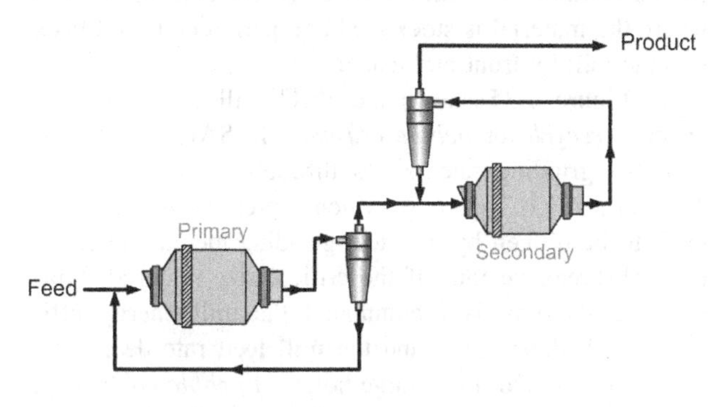

FIGURE 7.39 Two-stage ball mill grinding circuit.

trend is toward large single-stage primary ball mills, both to reduce capital and operating costs and to facilitate automatic control. The disadvantage of single-stage milling is that if a high reduction ratio is required, then relatively large balls are required for coarse feed, which may not efficiently grind the finer particles.

Rosario et al. (2009) review trends in comminution circuit design with reference to high tonnage hard rock operations. The circuits described, in addition to the ones introduced above, include: SAG mill closed with a screen (no recycle crusher); use of HPGR (high pressure grinding roll) as the recycle crusher; pre-crushing ahead of a SABC circuit; and HPGR with stirred mills.

7.5.2 Circuit Operation

Grinding circuits are fed at a controlled rate from the stockpile or bins holding the crusher plant product. There may be a number of grinding circuits in parallel, each circuit taking a definite fraction of the feed. An example is the Highland Valley Cu/Mo plant with five parallel grinding lines (Chapter 12). Parallel mill circuits increase circuit flexibility, since individual units can be shut down or the feed rate can be changed, with a manageable effect on production. Fewer mills are, however, easier to control and capital and installation costs are lower, so the number of mills must be decided at the design stage.

The high unit capacity SAG mill/ball mill circuit is dominant today and has contributed toward substantial savings in capital and operating costs, which has in turn made many low-grade, high-tonnage operations such as copper and gold ores feasible. Future circuits may see increasing use of high pressure grinding rolls (Rosas et al., 2012).

Autogenous grinding or semi-autogenous grinding mills can be operated in open or closed circuit. However, even in open circuit, a coarse classifier such as a trommel attached to the mill, or a vibrating screen can be used. The oversize material is recycled either externally or internally. In internal recycling, the coarse material is conveyed by a reverse spiral or water jet back down the center of the trommel into the mill. External recycling can be continuous, achieved by conveyor belt, or is batch where the material is stockpiled and periodically fed back into the mill by front-end loader.

In Figure 7.35 shows the SAG mill closed with a crusher (*recycle* or *pebble crusher*). In SAG mill operation, the grinding rate passes through a minimum at a "critical size" (Chapter 5), which represents material too large to be broken by the steel grinding media, but has a low self-breakage rate. If the critical size material, typically 25–50 mm, is accumulated the mill energy efficiency will deteriorate, and the mill feed rate decreases. As a solution, additional large holes, or *pebble ports* (e.g., 40–100 mm), are cut into the mill grate, allowing coarse material to exit the mill. The crusher in closed circuit is then used to reduce the size of the critical size material and return it to the mill. As the pebble ports also allow steel balls to exit, a steel removal system (such as a guard magnet, Chapters 2 and 13) must be installed to prevent them from entering the crusher. (Because of this requirement, closing a SAG mill with a crusher is not used in magnetic iron ore grinding circuits.) This circuit configuration is common as it usually produces a significant increase in throughput and energy efficiency due to the removal of the critical size material.

An example SABC-A circuit is the Cadia Hill Gold Mine, New South Wales, Australia (Dunne et al., 2001). The project economics study indicated a single grinding line. The circuit comprises a SAG mill, 12 m diameter by 6.1 m length (belly inside liners, the effective grinding volume), two pebble crushers, and two ball mills in parallel closed with cyclones. The SAG mill is fitted with a 20 MW gearless drive motor with bi-directional rotational capacity. (Reversing direction evens out wear on liners with symmetrical profile and prolongs operating time.) The SAG mill was designed to treat 2,065 t h^{-1} of ore at a ball charge of 8% volume, total filling of 25% volume, and an operating mill speed of 74% of critical. The mill is fitted with 80 mm grates with total grate open area of 7.66 m^2 (Hart et al., 2001). A 4.5 m diameter by 5.2 m long trommel screens the discharge product at a cut size of ca. 12 mm. Material less than 12 mm falls into a cyclone feed sump, where it is combined with discharge from the ball mills. Oversize pebbles from the trommel are conveyed to a surge bin of 735 t capacity, adjacent to the pebble crushers. Two cone crushers with a closed side set of 12–16 mm are used to crush the pebbles with a designed product P_{80} of 12 mm and an expected total recycle pebble rate of 725 t h^{-1}. The crushed pebbles fall directly onto the SAG mill feed belt and return to the SAG mill.

SAG mill product feeds two parallel ball mills of 6.6 m × 11.1 m (internal diameter × length), each with a 9.7 MW twin pinion drive. The ball mills are operated at a ball charge volume of 30–32% and 78.5% critical speed. The SAG mill trommel undersize is combined with the ball mills' discharge and pumped to two parallel packs (clusters) of twelve 660 mm diameter cyclones. The cyclone underflow from each line reports to a ball mill, while the cyclone overflow is directed to the flotation circuit. The designed ball milling circuit product is 80% passing 150 μm.

Several large tonnage copper porphyry plants in Chile use an open-circuit SAG configuration where the pebble crusher product is directed to the ball mills (SABC-B circuit). The original grinding circuit at Los Bronces is an example: the pebbles generated in the two SAG mills are crushed in a satellite pebble crushing plant, and then are conveyed to the three ball mills (Moglía and Grunwald, 2008).

7.5.3 Control of the Grinding Circuit

The purpose of milling is to reduce the size of the ore particles to the point where economic liberation of the valuable minerals takes place. Thus, it is essential that not only should a milling circuit process a certain tonnage of ore per day, but should also yield a product that is of known and controllable particle size. The principal variables that can affect control are: changes in the feed rate and the circulating load, size distribution and hardness of the ore, and the rate of water addition to the milling circuit. Operational factors are also important, such as to add grinding media into the mill or clear a plugged cyclone. For the purposes of stabilizing control, feed rate, densities, and circulating loads may be maintained at optimal values that would produce the required product (typically known through experience); however, this objective fails when disturbances cause deviations from target values. Fluctuation in feed size and hardness are probably the most significant factors disturbing the balance of a grinding circuit. Such fluctuations may arise from differences in composition, mineralization, particle size of the feed, and texture of the ore from different parts of the mine, from changes in mining methods (e.g., blast pattern), from changes in the crusher settings, often due to wear, and from damage to screens in the crushing circuit. Minor fluctuations in ore characteristics can be smoothed out by ore blending prior to feeding to the crusher. Ore storage increases smoothing out of variations, provided that ore segregation does not take place in the coarse-ore stockpile or storage bins. The amount of storage capacity available depends on the nature of the ore (such as its tendency to oxidize) and on the mill economics (real estate, capital budget, etc.).

Increase in feed size or hardness produces a coarser mill product unless the feed rate is proportionally reduced. Conversely, a decrease in feed size or hardness may allow an increase in mill throughput. A coarser mill product results in a greater circulating load from the classifier, thus increasing the volumetric flow rate. Since the product size from a hydrocyclone (type of classifier) is affected by flow rate (Chapter 9), the size distribution of the product will change. Control of the circulating load is therefore crucial in control of product particle size. A high circulating load at a fixed product size means lower energy consumption, but excessive circulating loads may result in the mill, classifier, conveyor, or pump overloading, hence there is an upper limit to mill feed rate. Determination of circulating load from a sampling campaign was illustrated in Chapter 3, Example 3.15.

Grinding is the most energy intensive section in a mineral processing plant, and the product from grinding has a major impact on subsequent separation processes, hence close control is required. Automatic control of milling circuits is one of the most advanced and successful areas of the application of process control in the minerals industry (Herbst et al., 1988). Control of grinding circuits is reviewed by Wei and Craig (2009) and Flintoff et al. (2014).

Instrumentation

Typical instrumentation is illustrated in Figure 7.40 for a primary (AG/SAG or rod) mill-secondary ball mill/cyclone circuit. Control of the feed rate is essential, and devices

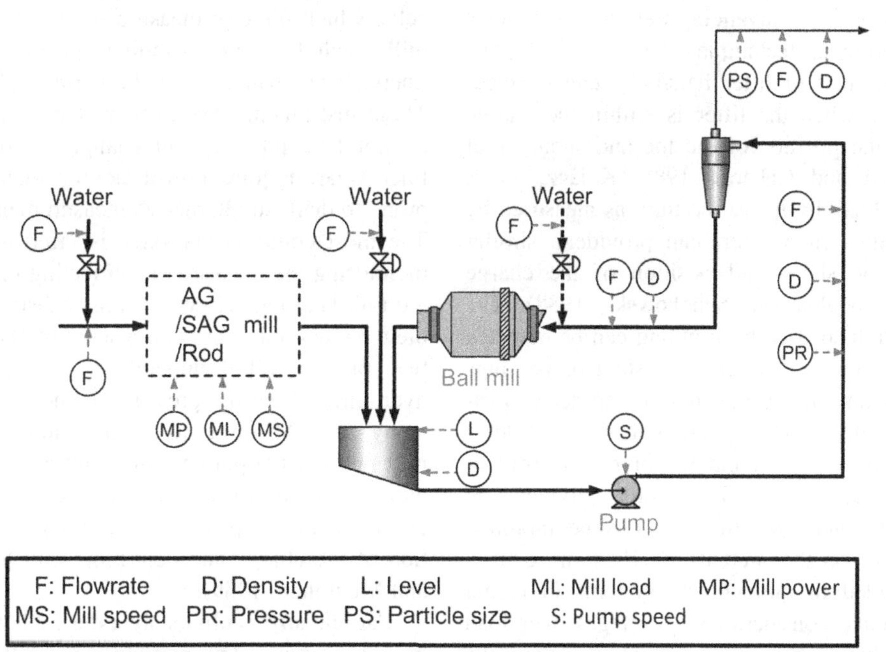

FIGURE 7.40 Grinding circuit control variables.

such as variable speed apron feeders or feeder belts, in conjunction with weightometers, are often used. Slurry density is measured mainly by nuclear density gauges, and flow rates by magnetic flowmeters or, increasingly, CiDRA's SONARtrac® technology (Chapter 3). Sump level is commonly indicated by bubble tube, capacitance-type detectors, or other electronic devices.

Newer mills are often equipped with variable speed drives, allowing mill rotational speed to be adjusted as an additional (and important) degree of freedom. Control of the grinding media charge can be made by continuous monitoring of the mill power consumption. A fall in power consumption to a certain level necessitates charging of fresh grinding media. Continuous power monitoring finds its greatest use in autogenous grinding, the power drawn being used as a measure of mill load and a constraint on mill feed rate. AG/SAG mills with crushers in pebble recirculation loops can utilize the crusher operating time or closed side setting (Chapter 6) to control the size of crushed pebbles returning to the mill, which in turn affects mill load and circulating load.

Arguably, the most important parameter required to provide good control of mill performance (especially AG/SAG mills) is the load (or amount of charge) inside the mill. Volumetric load (the percentage of mill internal volume taken up by ore and grinding media) is the key variable as it affects how the charge behaves inside the mill (e.g., shoulder and toe positions, Figure 7.3), but cannot currently be measured online, so it must be inferred from other, measurable parameters. Two categories of instruments are available: "off-shell" stationary sensors and "on-shell" sensors attached to the shell (Tano, 2005). On-shell load monitoring is an advancing field and includes a variety of measurement techniques. The strain experienced by the mill shell or lifter bars/bolts can be measured to determine when the lifter is within the charge volume and thus the position of the toe and shoulder of the charge (Herbst and Gabardi, 1988; Kolacz, 1997; Tano, 2005). Conductivity inside the mill, as measured by two probes mounted on a lifter, can provide a similar measurement of the shape and position of the charge (Moys, 1988; Vermeulen and Schakowski, 1988). The mechanical vibrations caused by grinding can be detected by accelerometers attached to the mill shell or bearings and analyzed to determine charge motion and detect various operational conditions (Campbell et al., 2001, 2006).

Acoustic systems use a single sensor or an array of microphones to detect the sound vibrations produced by charge movement inside the mill and can be mounted close to the mill or on the shell itself. These have been increasingly included in mill control systems in recent years. They can have considerable advantages over load cells (Pax, 2001) due to their ease of maintenance and the additional process information they provide. One system is to use two off-shell microphones, one placed above, and the other placed below the normal level of impact of the charge on the mill shell. As the mill load increases, the point of impact moves toward the upper microphone, and conversely down toward the lower microphone as it decreases. By comparing the output from the two microphones, it becomes possible to determine whether the charge level is rising or falling. This information is correlated with the power draw of the mill and used to calculate the rate of addition of the new feed. It was found that the sound level changed ahead of detectable changes in the mill bearing pressure, power, torque, motor current, and mill weight. Therefore the plant took the sound reading as an indicator of changes in SAG load for milling circuit control. Other systems adjust mill speed in response to acoustic information to control the mill load and ensure maximum grinding efficiency by moving the point of impact to the toe of the charge (Pax et al., 2014). Several commercial systems are available to analyze the sound produced by the mill and detect factors such as ball-on-liner impacts, charge geometry (toe and shoulder position), water holdup or specific information on the behavior of balls and ore, and is used to infer the mill volumetric loading (La Rosa et al., 2008; Pax, 2011, 2012; Pax et al., 2014; Shuen et al., 2014). Sound-based systems tend to be more successful in the long term as a tool to detect steel-on-liner collisions and therefore indicate when the mill should be slowed to protect the liners rather than a real time indicator of load. This is primarily due to calibration drift and the influence of unmeasured acoustic disturbances in the concentrator.

Measurement of mill weight has mainly been by load cells, which directly measure the mass of material in the mill (including ore, grinding media, water, shell, and liners), from which the volumetric load can be inferred. These instruments, however, rely on accurate calibration, do not take into account changes in rod/ball charge or liner wear, require downtime for maintenance and are prone to drift, so alternative measurements are often used. The most common approach is inferring mill weight by monitoring the pressure of lubricating oil flowing through the mill bearings. However, this measure can be susceptible to signal noise and is impacted by the physical properties of the oil (temperature, viscosity) and system hydraulics. Bearing pressure is nonetheless utilized in many AG/SAG mills as an indicator of the load in the mill and is incorporated into control systems. Other systems use the power draw (more specifically the amplitude and frequency of power oscillations) to calculate the position of the charge and determine mill load (Koivistoinen and Miettunen, 1989).

Particle size is obviously a desirable measurement in a grinding circuit. The particle size of the circuit product (cyclone overflow) can be inferred by the use of mathematical models (Chapter 9) but it is preferable to measure directly by the use of on-line monitors, which recent developments are making increasingly robust (Chapter 4). As noted by Flintoff et al. (2014), on-line measurement of liberation would be a notable breakthrough for grinding circuit control.

Mill feed particle size is significant for AG/SAG load management. Monitoring of the feed to the mill by image analysis, for example, provides a measure of the particle size distribution (Chapter 4). Felsher et al. (2015) describe stabilization of SAG power draw by adjusting apron feeder ratios to target higher fines fraction in the SAG feed.

Upset conditions in cyclone operations, such as roping and plugging, have a significant impact on the plant stability and performance, but they often go undetected for extended periods of time due to lack of instrumentation. There are technologies and sensing methods today for monitoring cyclone performance (Chapter 4). When integrated with the plant control system, "offending" cyclones can be automatically switched off so that corrective action is taken within seconds rather than many minutes.

Pumpbox control plays a significant role in stability of the circuit. Many mineral processing plants run their cyclone feed pumpboxes using traditional level control, where the pump speed adjusts to maintain a constant level in the pumpbox. The issue with this philosophy is that a continuously varying flowrate to the cyclopac leads to variability in cyclone pressure and cut-point, which negatively impacts recovery in downstream operations. Pumpbox level management strategy, focused around stabilizing the flowrate to the cyclones, is designed to allow the level in the cyclone feed pumpbox to vary while maintaining the pump speed (Westendorf et al., 2015). The result is more consistent grind size to downstream operations as well as stabilized circulating load within the grinding circuit. Reductions of variability in excess of 50% have been reported.

Control Objectives/strategies

In implementing instrumentation and process control for grinding circuits, the control objective must first be defined, which may be to:

1. Maintain a constant feed rate within a limited range of product size.
2. Maintain a constant product size at maximum throughput.
3. Maximize production per unit time in conjunction with downstream circuit performance (e.g., flotation recovery).

Aspects of grinding circuit control can be introduced using Figure 7.40. Of the variables shown, only the ore feed and the water addition rates can be varied independently, as the other variables depend upon and respond to changes in these items. It is therefore ore feed rate and water addition to the mill and classifier that are the major manipulated variables used to control grinding circuits. There is an important difference in the dynamic response of the circuit to changes in ore feed rate and to changes in water addition rate. Changes in ore feed rate initiate a slow progressive change to the final steady state condition, while changes in classifier water addition initiate an immediate response. Increase in water addition rate also results in a simultaneous increase in circulating load and sump level, confirming the necessity of using a large capacity sump and variable speed pump to maintain effective control.

If the strategy is constant product size at constant feed rate (objective 1), then the only manipulated variable is the classifier water, and the circuit must therefore tolerate variations in the cyclone overflow density and volumetric flow rate when ore characteristics change. Such variations have to be considered with respect to downstream operations (objective 3). Variations in circulating load will also occur. Changes in mill rotational speed provide a fast response in the circuit because the mill discharge rate will be rapidly affected; however, this can lead to instability in the circulating load, hence speed control loops are generally tuned to be slow-acting to ensure stability. Additionally, mill speed has a direct relationship to power draw, meaning that in an overload or high-power situation, increasing the speed of the mill will cause a further increase in power before the mill will begin to unload.

In many applications, the control objective is maximum throughput at constant particle size (objective 2), which allows manipulation of both ore feed rate and classifier water addition rate. Allowing for the fact that the circuit has a limited capacity, this objective can be stated as a fixed product particle size set-point at a circulating load set-point corresponding to a value just below the maximum tonnage constraint. The circulating load is either calculated from measured values, measured directly, or inferred.

Since both ore feed rate and classifier water addition rate can be varied independently, two control strategies are available for objective 2. In the first, product size is controlled by ore feed rate and the circulating load by classifier water addition rate. In the second, product size is controlled by the classifier water addition rate and circulating load by the ore feed rate. The choice of control strategy depends on which of the control loops, the particle size loop or the circulating load loop, is required to respond faster. This depends on many factors, such as the ability of the grinding and concentration circuits to handle variations in flow, the sensitivity of the concentration process to deviations from the optimum particle size, time lags within the grinding circuit, and the number of grinding stages. If the particle size response must be fast, then this loop is controlled by the classifier water, whereas if a fast mill throughput response is more important, then the product size is controlled by the ore feed rate.

It is important to consider the impact of unmeasured disturbances within the grinding circuit, such as ore hardness, liner, lifter, and grate condition, media charge and holdup in the mill. The importance of these parameters makes traditional control difficult, resulting in the need for more advanced techniques.

A common problem in the operation of a milling circuit, particularly when controlling particle size, is that the control actions are often constrained by physical or operational limitations. This problem is particularly relevant in the case of multivariable control, in which control of the process depends critically on the correct combination of several control actions. The saturation of one of these control actions may result in partial loss of control, or even instability of the process. In order to deal with this problem, a method has been developed, in which the multivariable control scheme includes a robust "limit algorithm", thus retaining effective stabilizing control of the milling process (Hulbert et al., 1990; Craig et al., 1992; Metzner, 1993; Galán et al., 2002). Alternatively, an expert system is incorporated to "manage" the controllers and modify the action based on the situation in the plant.

Expert systems are an accepted form of control for complex grinding circuits. Expert systems are computer programs that compile the process knowledge and rules employed by the control room operators into a heuristic model (Chapter 3). They allow continuous monitoring of the grinding circuit, applying operational best practice consistently. They require little additional investment and infrastructure and can operate from a desktop computer. Several commercial systems are in use.

Model-predictive control has its origins in work done in the 1970s (Lynch, 1977) and gained ground with the advent of more reliable models and more powerful computing platforms (Herbst et al., 1993; Evans, 2001; Galán et al., 2002; Lestage et al., 2002; Muller et al., 2003; Schroder et al., 2003; Yutronic and Toro, 2010). This approach involves development of mathematical models of a unit or a circuit based on existing knowledge, operating mechanisms, and experimental data. Such models often incorporate a number of model parameters that need to be calibrated by experimental data from a specific unit or circuit. Once the model parameters are determined and the ore breakage characteristics are known, the models can predict the performance of the unit or the circuit in response to changes in the feed condition or the operational conditions, either steady-state performance or the dynamic response. The model can also be used in conjunction with an expert system (Felsher et al., 2015) whereby the model predictive controller provides a fundamental foundation for stabilization and the expert system uses heuristics for optimization, providing a framework for the controllers to work within.

REFERENCES

Anon., 1990. Modern tube mill design for the mineral industry-Part I. Mining Mag. 163 (Jul.), 20–27.

Austin, L.G., et al., 1984. Process Engineering of Size Reduction: Ball Milling. SME, Quinn Printing Inc., Hoboken, NJ, USA.

Barratt, D.J., et al., 1999. SAG milling design trends, comparative economics, mill sizes and drives. CIM Bull. 92 (1026-1035), 62–66.

Bouajila, A., et al., 2001. The impact of feed size analysis on the autogenous grinding mill, Proc. SAG 2001 Conf., Vol. 2. Vancouver, Canada, pp. 317–330.

Brissette, M., et al., 2014. Geometallurgy: new accurate testwork to meet required accuracies of mining project development, Proc. 27th International Mineral Processing Congr., Ch. 14, Paper C1417. Santiago, Chile.

Campbell, J., et al., 2001. SAG mill monitoring using surface vibrations. Barret, D.J., et al. (Eds.), Proc. International Autogenous and Semiautogenous Grinding Technology (SAG) Conf., Vol. 2, Vancouver, Canada, pp. 373–385.

Campbell, J.J., et al., 2006. The development of an on-line surface vibration monitoring system for AG/SAG mills, Proc. 4th International Conference on Autogenous and Semiautogenous Grinding Technology (SAG) Conf., Vancouver, BC, Canada, Vol. 3, pp. 326–336.

Castro, E., et al., 2009. Maximum classification through double classification using the ReCyclone. In: Malhotra, D., et al., (Eds.), Recent Advances in Mineral Processing Plant Design. SME, Englewood, CO, USA, pp. 444–454.

Concha, F., et al., 1988. Optimization of the ball charge in a tumbling mill. Forssberg, E. (Ed.), Proc. 16th International Mineral Processing Congr., Stockholm, Sweden, pp. 147–156.

Craig, I.K., et al., 1992. Optimised multivariable control of an industrial run-of-mine milling circuit. J. S. Afr. Inst. Min. Metall. 92 (6), 169–176.

Dobby, G., et al., 2001. Advances in SAG circuit design and simulation applied to the mine block model, Proc. of SAG 2001 Conf., Vol. 4. Vancouver, Canada, pp. 221–234.

Dodd, J., et al., 1985. Relative importance of abrasion and corrosion in metal loss in ball milling. Miner. Metall. Process. 2 (Nov.), 212–216.

Dunn, D.J., 1982. Selection of liners and grinding media for comminution circuits-case studies. In: Mular, A.L., Jergensen, G.V. (Eds.), Design and Installation of Comminution Circuits. SME, New York, NY, USA, pp. 973–985 (Chapter 59).

Dunne, R., et al., 2001. Design of the 40 foot diameter SAG mill installed at the Cadia Gold Copper Mine, Proc. of SAG 2001 Conf., Vol.1, Vancouver, Canada, pp. 43–58.

Durman, R.W., 1988. Progress in abrasion-resistant materials for use in comminution processes. Int. J. Min. Process. 22 (1-4), 381–399.

Ellis, S., Gao, M., 2002. The development of ultrafine grinding at KCGM. SME Annual Meeting, Phoenix, AZ, USA, Preprint 02-072. pp. 1–6.

Erikson, M.T., 2014. Innovations in comminution equipment: crushers, high pressure grinding rolls, semi-autogenous grinding, ball mills and regrind mills. In: Anderson, C.G., et al., (Eds.), Mineral Processing and Extractive Metallurgy: 100 Years of Innovation. SME, Englewood, CO, USA, pp. 65–76.

Evans, G., 2001. A new method for determining charge mass in AG/ SAG mills, Proc. International Conference on Autogenous and Semiautogenous Grinding Technology (SAG) Conf., Vol. 2, Vancouver, BC., Canada, pp. 331–345.

Evans, L.B., et al., 1979. ASPEN: an advanced system for process engineering. Comput. Chem. Eng. 3 (1-4), 319–327.

Felsher, D., et al., 2015. Sustainable optimisation of the Bell Creek SAG Mill. Proc. 47th Annual Meeting of the Canadian Mineral Processors Conf. CIM, Ottawa, ON, Canada, pp. 54–65.

Finch, J.A., et al., 2007. Iron control in mineral processing. Proc. 39th Annual Meeting of the Canadian Mineral Processors Conf.. CIM, Ottawa, ON, Canada, pp. 365–386.

Flintoff, B., et al., 2014. Innovations in comminution instrumentation and control. In: Anderson, C.G., et al., (Eds.), Mineral Processing and Extractive Metallurgy: 100 years of Innovation. SME, Littleton, CO, USA, pp. 91–116.

Ford, M.A., King, R.P., 1984. The simulation of ore-dressing plants. Int. J. Miner. Process. 12 (4), 285–304.

Forssberg, E.K.S., et al., 1993. Influence of grinding method on complex sulphide ore flotation: a pilot plant study. Int. J. Miner. Process. 3-4, 157–175.

Galán, O., et al., 2002. Robust control of a SAG mill. Powder Technol. 124 (3), 264–271.

Gangopadhyay, A.K., Moore, J.J., 1987. Effect of impact on the grinding media and mill liner in a large semi-autogenous mill. Wear. 114 (2), 249–260.

Gao, M.W., et al., 1996. Power predictions for a pilot scale stirred ball mill. Int. J. Miner. Process. 44-45, 641–652.

Gaudin, A.M., 1939. Principles of Mineral Dressing. McGraw-Hill Book Company, New York, NY, USA and London, UK.

Grandy, G.A., et al., 2002. Selection and evaluation of grinding mill drives. Mular, A. L., et al. (Eds.), Mineral Processing Plant Design, Practice, and Control. Vol. 1, Vancouver, Canada, pp. 819–839.

Hart, S., et al., 2001. Optimisation of the Cadia Hill SAG mill circuit, Proc. of SAG 2001 Conf., Vol. 1, Vancouver, Canada, pp. 11–30.

He, M., et al., 2004. Slurry rheology in wet ultrafine grinding of industrial mineral: a review. Powder Technol. 147 (1-3), 94–112.

Herbst, J.A., Gabardi, T.L., 1988. Closed loop media charging of mills based on a smart sensor system. Sommer, G. (Ed.), Proc. IFAC Applied Measurements in Mineral and Metallurgical Processing Conf., Transvaal, South Africa, pp. 17–23.

Herbst, J.A., et al., 1988. Optimal control of comminution operations. Int. J. Miner. Process. 22 (1-4), 275–296.

Herbst, J.A., et al., 1993. A model-based methodology for steady state and dynamic optimisation of autogenous and semiautogenous grinding mills, Proc. 18th International Mineral Processing Cong., (IMPC), Sydney, Australia, pp. 519–527.

Hogg, R., Fuerstenau, D.W., 1972. Power relationships for tumbling mills. Trans. SME-AIME. 252, 418–432.

Hulbert, D.G., et al., 1990. Multivariable control of a run-of-mine milling circuit. J. S Afr. Inst. Min. Metall. 90 (3), 173–181.

Jankovic, A., 1998. Mathematical Modeling of Stirred Mills. PhD thesis. University of Queensland. Australia.

Jankovic, A., 2001. Media stress intensity analysis for vertical stirred mills. Miner. Eng. 14 (10), 1177–1186.

Jankovic, A., 2003. Variables affecting the fine grinding of minerals using stirred mills. Miner. Eng. 16 (4), 337–345.

Jankovic, A., Valery, W., 2013. Closed circuit ball milling-basics revisited. Miner. Eng. 43-44, 148–153.

Joe, E.G., 1979. Energy consumption in Canadian mills. CIM Bull. 72 (801-806), 147–151.

Jones, D.A., et al., 2006. Understanding microwave assisted breakage. Miner. Eng. 18 (7), 659–669.

Karbstein, H., et al., 1996. Scale-up for grinding in stirred ball mills. Aufbereitungs-Technick. 37 (10), 469–479.

Kawatra, S.K., Eisele, T.C., 1988. Rheological effects in grinding circuits. Int. J. Miner. Process. 22 (1-4), 251–259.

King, R.P., 2012. Modeling and Simulation of Mineral Processing Systems. second ed. SME, Englewood, CO, USA, Elsevier.

Kitschen, L.P., et al., 1982. The centrifugal mill: experience with a new grinding system and its applications. CIM Bull. 75 (845), 72.

Klimpel, R.R., 1984. Influence of material breakage properties and associated slurry rheology on breakage rates in the wet grinding of coal/ores in tumbling media mills. Part. Sci. Technol. 2 (2), 147–156.

Knecht, J., 1990. Modern tube mill design for the mineral industry: part II. Mining Mag. 163 (Oct.), 264–269.

Koivistoinen, P., Miettunen, J., 1989. The effect of mill lining on the power draw of a grinding mill and its utilisation in grinding control, Proc. 1st International Conference on Autogenous and Semiautogenous Grinding Technology (SAG) Conf., Vol. 2, Vancouver, BC, Canada, pp. 687–695.

Kolacz, J., 1997. Measurement system of the mill charge in grinding ball mill circuits. Miner. Eng. 10 (12), 1329–1338.

Kwade, A., et al., 1996. Motion and stress intensity of grinding beads in a stirred media mill. Part2: stress intensity and its effect on comminution. Powder Technol. 86 (1), 69–76.

La Rosa, D., et al., 2008. The use of acoustics to improve load estimation in the Cannington AG mill. Proc. Metallurgical Plant Design and Operating Strategies (MetPlant). AusIMM, Perth, WA, Australia, pp. 105–116.

Larson, M., et al., 2011. Regrind mills: challenges of scaleup. SME Annual Meeting, Denver, CO, USA, Preprint 11-130.

Latchireddi, S., 2002. Modelling the Performance of Grates and Pulp Lifters in Autogenous and Semi-autogenous Mills. PhD Thesis. The University of Queensland. Australia.

Latchireddi, S., Morrell, S., 2003. Slurry flow in mills: grate-only discharge mechanism (Parts 1 and Part 2). Miner. Eng. 16 (7), 625–633, 635–642.

Latchireddi, S., Morrell, S., 2006. Slurry flow in mills with TCPL: an efficient pulp lifter for ag/sag mills. Int. J. Miner. Process. 79 (3), 174–187.

Lestage, R., et al., 2002. Constrained real-time optimization of a grinding circuit using steady-state linear programming supervisory control. Powder Technol. 124 (3), 254–263.

Leung, K., et al., 1992. Decision of Mount Isa Mines Limited to change to autogenous grinding. In: Kowatra, S.K. (Ed.), Comminution - Theory and Practice. SME, Littleton, CO, USA, pp. 331–338 (Chapter 24).

Lewis, F.M., et al., 1976. Comminution: a guide to size reduction system design. Mining Eng. 28 (Sept.), 29–35.

Lichter, J., 2014. The future of comminution. In: Anderson, C.G., et al., (Eds.), Mineral Processing and Extractive Metallurgy: 100 Years of Innovation. SME, Littleton, CO, USA, pp. 129–144.

Lloyd, P.J.D., et al., 1982. Centrifugal grinding on a commercial scale. Eng. Min. J. 183 (12), 49–54.

Lynch, A.J., 1977. Mineral Crushing and Grinding Circuits: Their Simulation, Optimisation and Control. Elsevier, Amsterdam, Netherlands.

MacPherson, A.R., 1989. Autogenous grinding, 1987-update. CIM Bull. 82 (921), 75–82.

Mazzinghy, D.B., et al., 2012. Predicting the size distribution in the product and the power requirements of a pilot scale Vertimill. Procemin2012. Santiago, Chile, pp. 412–420.

McIvor, R.E., 2014. Plant performance improvements using the grinding circuit "classification system efficiency". Mining Eng. 66 (9), 72–76.

McIvor, R.E., Finch, J.A., 1986. The effects of design and operating variables on rod mill performance. CIM Bull. 79 (895), 39–46.

Meintrup, W., Kleiner, F., 1982. World's largest ore grinder without gears. Mining Eng. 34 (Sep.), 1328–1331.

Metzner, G., 1993. Multivariable and optimising mill control--- the South African experience, Proc. 18th International Mineral Processing Congr. (IMPC), Sydney, Australia, pp. 293–300.

Misra, A., Finnie, I., 1980. A classification of three-body abrasive wear and design of a new tester. Wear. 60 (1), 111–121.

Moglía, M., Grunwald, V., 2008. Advanced control implementation in the SAG grinding plant, Anglo American Chile, Los Bronces division, Proc. 5th International Mineral Processing Seminar (Procemin 2008), Santiago, Chile, pp. 3360–367 (Chapter 6).

Moller, J., 1990. The best of two worlds: a new concept in primary grinding wear protection. Miner. Eng. 3 (1-2), 221–226.

Moller, T.K., Brough, R., 1989. Optimizing the performance of a rubber-lined mill. Min. Eng. 41 (8), 849–853.

Morrell, S., 2004. Predicting the specific energy of autogenous and semi-autogenous mills from small diameter drill core samples. Miner. Eng. 17 (3), 447–451.

Morrell, S., Kojovic, T., 1996. The influence of slurry transport on the power draw of autogenous and semi-autogenous mills, Proc. 2nd International Conference on Autogenous and Semi-autogenous Grinding Technology, Vancouver, Canada, pp. 378–389.

Morrell, S., Latchireddi, S., 2000. The operation and interaction of grates and pulp lifter in autogenous and semi-autogenous mills. Proc. 7th Mill Operators Conf.. AusIMM, Kalgoorlie, Australia, pp. 13–22.

Morrell, S., Valery, W., 2001. Influence of feed size on AG/SAG mill performance, Proc. SAG 2001 Conf., Vol.1, Vancouver, Canada, pp. 203–214.

Mosher, J., Bigg, T., 2001. SAG mill test methodology for design and optimisation, Proc. SAG 2001 Conf., Vol. 1, Vancouver, Canada, pp. 348–361.

Moys, M.H., 1988. Measurement of slurry properties and load behaviour in grinding mills. Sommer, G. (Ed.), Proc. IFAC Applied Measurements in Mineral and Metallurgical Processing, Transvaal, South Africa, pp. 3–9.

Moys, M.H. (1989). Slurry rheology: the key to a further advance in grinding mill control, Proc. 1st International Conference On:Autogenous and Semiautogenous Grinding Technology, Vancouver, Canada, pp. 713–727.

Mular, A.L., Agar, G.E. (Eds.), 1989. Advances in Autogenous and Semiautogenous Grinding Technology. University of British Columbia, Vancouver, Canada.

Muller, B., et al., 2003. Model predictive control in the minerals processing industry, Proc. 22nd International Mineral Processing Cong., Cape Town, South Africa, pp. 1692–1702.

Napier-Munn, T.J., et al., 1996. Mineral Comminution Circuits: Their Operation and Optimisation, Julius Kruttschnitt Mineral Research Centre (JKMRC), Brisbane, Australia.

Nesset, J.E., et al., 2006. Assessing the performance and efficiency of fine grinding technologies. Proc. 38th Annual Meeting of The Canadian Mineral Processors Conf.. CIM, Ottawa, ON, Canada, pp. 283–309.

Obeng, D.P., Morrell, S., 2003. The JK three-product cyclone-performance and potential applications. Int. J. Miner. Process. 69 (1-4), 129–142.

Orford, I.W., Strah, L.M., 2006. Optimizing SAG mill liner changes through procedures and liner design, Proc. SAG 2006 Conf., Vol. 3, Vancouver, Canada, pp. 30–58.

Pax, R.A., 2001. Non-contact acoustic measurement of in-mill variables of SAG mills, Proc. International Conference on Autogenous and Semiautogenous Grinding Technology Conf., Vol. 2, Vancouver, Canada, pp. 386–393.

Pax, R.A., 2011. Non-contact acoustic measurement of dynamic in-mill processes for SAG/AG mills. Proc. Metallurgical Plant Design and Operating Strategies (MetPlant) Conf.. AusIMM, Perth, WA, Australia, pp. 163–175.

Pax, R.A., 2012. Determining mill operational efficiencies using non-contact acoustic emissions from microphone arrays. Proc. 11th Mill Operators Conference. AusIMM, Melbourne, Australia, pp. 119–126.

Pax, R., et al., 2014. Implementation of acoustic arrays for semi-autogenous grinding mill operation. Proc. 12th Mill Operators' Conf.. AusIMM, Townsville, QLD, Australia, pp. 273–282.

Powell, M.S., et al., 2001. Developments in the understanding of South African style SAG mills. Miner. Eng. 14 (10), 1143–1153.

Powell, M.S., et al., 2011. DEM modelling of liner evolution and its influence on grinding rate in ball mills. Miner. Eng. 24 (3-4), 341–351.

Rabanal, C., Guzmán, L., 2013. Empirical correlation for estimating grinding media consumption, Proc. 10th International Mineral Processing Conf. (Procemin 2013), Santiago, Chile, pp. 125–135.

Radziszewski, P., 2002. Exploring total media wear. Miner. Eng. 15 (12), 1073–1087.

Radziszewski, P., 2013. Assessing the stirred mill design space. Miner. Eng. 41, 9–16.

Radziszewski, P., Allen, J., 2014. Towards a better understanding of stirred milling technologies - estimating power consumption and energy use, Proc. 46th Annual Meeting of The Canadian Mineral Processors Conf., Ottawa, ON, Canada, pp. 55–66.

Rajagopal, V., Iwasaki, I., 1992. Grinding media selection criteria for wear resistance and flotation performance. In: Kowatra, S.K. (Ed.), Comminution - Theory and Practice. SME, Littleton, CO, USA, pp. 181–200 (Chapter 14).

Rosario, P.P., et al., 2009. Recent trends in the design of comminution circuits for high tonnage hard rock mining. In: Malhotra, D., et al., (Eds.), Recent Advances in Mineral Processing Plant Design. SME, Littleton, CO, USA, pp. 347–355.

Rosas, J., et al., 2012. Update of Chilean developments in mineral processing technology. Proc. 44th Annual Mineral Processors Conference Conf.. CIM, Ottawa, ON, Canada, pp. 3-12.

Rowland, C.A., 1982. Selection of rod mills, ball mills, pebble mills, and regrind mills. In: Mular, A.L., Jergensen, G.V. (Eds.), Design and Installation of Comminution Circuits. SME, New York, NY, USA, pp. 393–438 (Chapter 23).

Rowland, C.A., 1987. New developments in the selection of comminution circuits. Eng. Min. J. 188 (2), 34–38.

Rowland, C.A., Kjos, D.M., 1980. Rod and ball mills. In: Mular, A.L., Bhappu, R.B. (Eds.), Mineral Processing Plant Design. SME, New York, NY, USA, pp. 239–278 (Chapter 12).

Schroder, A.J., et al., 2003. On-line dynamic simulation of milling operations, Proc. Copper 2003, Vol. 3. Santiago, Chile, pp. 27-44.

Scott, A., et al., 2002. Tracking and quantifying value from "mine to mill" improvement. Proc. Value Tracking Symp.. AusIMM, Brisbane, Australia, pp. 77–84.

Sepúlveda, J.E., 2001. A phenomenological model of semiautogeneous grinding processes in a Moly-Cop tools environment, Proc. SAG 2001 Conf., Vol. 4, Vancouver, Canada, pp. 301–315.

Shi, F., 2004. Comparison of grinding media: cylpebs versus balls. Miner. Eng. 17 (11-12), 1259–1268.

Shi, F., Napier-Munn, T.J., 2002. Effects of slurry rheology on industrial grinding performance. Int. J. Min. Proc. 65 (3-4), 125–140.

Shi, F., et al., 2009. Development of a rapid particle breakage characterisation device – the JKRBT. Miner. Eng. 22 (7-8), 602–612.

Shuen, D., et al., 2014. The benefits of using SmartEar™ at Pueblo Viejo. Proc. 12th Mill Operators' Conf.. AusIMM, Townsville, Australia, pp. 297–304.

Sloan, R., et al., 2001. Expert systems on SAG circuits: three comparative case studies, Proc. SAG 2001 Conf., Vol 2, Vancouver, Canada, pp. 346-357.

Starkey, J., et al., 2001. Design of the Agnico-Eagle Laronde Division SAG mill, Proc. SAG 2001 Conf., Vol. 3, Vancouver, Canada, pp. 165–178.

Starkey, J., et al., 2006. SAG Design testing-what it is and why it works, Proc. SAG 2006 Conf., Vol. 4, Vancouver, Canada, 240–254.

Stehr, N., 1988. Recent developments in stirred ball milling. Int. J. Miner. Process. 22 (1-4), 431–444.

Tano, K., 2005. Continuous Monitoring of Mineral Processes with Special Focus on Tumbling Mills – A Multivariate Approach. Doctoral Thesis. Luleå University of Technology, Sweden.

Taggart, A.F., 1945. Handbook of Mineral Dressing, Ore and Industrial Minerals. John Wiley, & Sons., Chapman & Hall, Ltd, London, UK.

Tamblyn, R.J., 2009. Analysis of Energy Requirements in Stirred Media Mills. PhD thesis. University of Birmingham, UK.

Underle, U., et al., 1997. Stirred mill technology for regrinding McArthur River and Mount Isa zinc/lead ores, Proc. 20th International Mineral Processing Congr., Aachen, Germany, pp. 71–78.

van de Vijfeijken, M., 2010. Mills and GMDs. Int. Mining. Oct., 30–31.

van der Wielen, K., et al., 2014. High voltage breakage: a review of theory and applications, Proc. 27th International Mineral Processing Congr., (IMPC), Santiago, Chile, pp. 78–86 (Chapter 9).

Vermeulen, L.A., Schakowski, F.A., 1988. Estimation, by use of a conductivity bolt, of the charge volume and other milling parameters in an operating mill. Sommer, G., (Ed.), Proc. IFAC Applied Measurements in Mineral and Metallurgical Processing Conf., Transvaal, South Africa, pp. 11–15.

von Ow, T.R., 2009. Molienda-alto consumo de energía y altos requerimientos de mantención, cómo superar los desafíos mediante el uso de convertidores de frecuencia, Proc. de MAPLA 2009, Antofagasta, Chile, pp. 207–218.

Wei, D., Craig, I.K., 2009. Grinding mill circuits: a survey of control and economic concerns. Int. J. Miner. Process. 90 (1-4), 56–66.

Westendorf, M., et al., 2015. Managing cyclones – a valuable asset: the Copper Mountain case study. SME Annual Meeting, Denver, CO, USA, preprint 15-140.

Wills, B.A., Atkinson, K., 1993. Some observations on the fracture and liberation of mineral assemblies. Miner. Eng. 6 (7), 697–706.

Wright, P., et al., 1991. A liberation study of autogenous and SAG mills. Proc. 4th Mill Operators Conf.. AusIMM, Burnie, Tas., Australia, pp. 171–174.

Xu, X., Mao, S., 2011. Select a proper density ceramic media for fine and ultrafine fluidized stirred milling. <http://www.chemcobeads.com> Downloaded 30th Nov. 2011.

Yutronic, N.I., Toro, R., 2010. Design and implementation of advanced automatic control strategy based on dynamic models for high capacity SAG mill. Proc. 42nd Annual Meeting of The Canadian Mineral Processors Conf.. CIM, Ottawa, ON, Canada, pp. 461–472.

Chapter 8

Industrial Screening

8.1 INTRODUCTION

Industrial screening is extensively used for size separations from 300 mm down to roughly 40 μm, although the efficiency decreases rapidly with fineness. Dry screening is generally limited to material above ca. 5 mm in size, while wet screening down to ca. 250 μm is common. Although there are screen types that are capable of efficient size separations down to 40 μm, sizing below 250 μm is more commonly undertaken by classification (Chapter 9). Selection between screening and classification is influenced by the fact that finer separations demand large areas of screening surface and therefore can be expensive compared to classification for high-throughput applications.

The types of screening equipment are many and varied. Likewise, there are a wide range of screening objectives. The main purposes in the minerals industry are:

a. *Sizing* or *classifying*: to separate particles by size, usually to provide a downstream unit process with the particle size range suited to that unit operation.
b. *Scalping*: to remove the coarsest size fractions in the feed material, usually so that they can be crushed or removed from the process.
c. *Grading*: to prepare a number of products within specified size ranges. Examples include quarrying and iron ore processing, where the final product size is an important part of the specification.
d. *Media recovery*: for washing magnetic media from ore in dense medium circuits; or to retain grinding media inside grinding mills.
e. *Dewatering*: to drain free moisture from a wet sand slurry.
f. *De-sliming* or *de-dusting*: to remove fine material, generally below 0.5 mm, from a wet or dry feed.
g. *Trash removal*: usually to remove coarse wood fibers or tramp material from a slurry stream.

8.2 SCREEN PERFORMANCE

In its simplest form, the screen is a surface having many apertures, or holes, usually with uniform dimensions. Particles presented to a screen surface either pass through or are retained, according to whether the particles are smaller or larger than the governing dimensions of the apertures. The efficiency of screening is determined by the degree of separation of the material into size fractions above and/or below the aperture size.

Efficiency Formulae

These can be derived from mass balances around a screen, Figure 8.1, where:

$F(\mathrm{t\,h^{-1}})$ = mass flow rate of feed material to the screen.
$O(\mathrm{t\,h^{-1}})$ = mass flow rate of the coarse product stream (oversize or overflow stream).
$U(\mathrm{t\,h^{-1}})$ = mass flow rate of the fine product stream (undersize or underflow stream).
f = mass fraction of feed material finer than a defined cut-point (or cut-size) (e.g., the screen aperture size).
o = mass fraction of material finer than the aperture size in the coarse product.
u = mass fraction of material finer than the aperture size in the fine product.

The mass fractions f, o, and u can be determined by sieve analysis on a representative sample of each stream on a laboratory screen of the same aperture size as the industrial screen.

At steady state, the overall solids mass balance on the screen is:

$$F = O + U \tag{8.1}$$

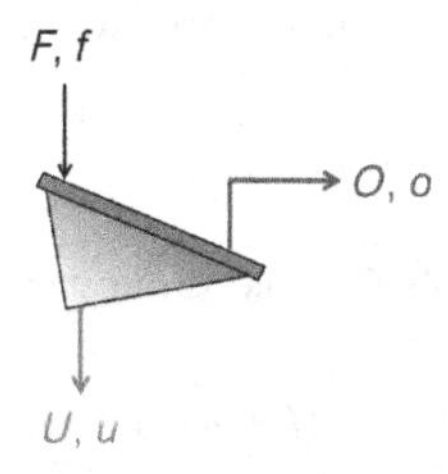

FIGURE 8.1 Mass balance on a screen.

and the mass balance of the material finer than the screen aperture size is:

$$Ff = Oo + Uu \quad (8.2)$$

Several methods of defining screen performance exist (Colman and Tyler, 1980; Nichols, 1982; Bothwell and Mular, 2002; Valine and Wennen, 2002). Commonly, the fine product stream is the important one and efficiency is defined by the recovery of finished product (material less than cut-size) to the fine (underflow) stream, E_U:

$$E_U = \frac{Uu}{Ff} \quad (8.3)$$

Equation (8.3) represents the actual mass of screen undersize that reports to the underflow, compared to the amount that should report theoretically, Ff. Solving for U/F by rearranging Eq. (8.1) and substituting into Eq. (8.2), the efficiency can be expressed in terms of the measured undersize mass fractions in each stream:

$$O = F - U$$

$$Ff = (F - U)o + Uu$$

$$\frac{U}{F} = \frac{f - o}{u - o} \quad (8.4)$$

$$E_U = \frac{(f - o)u}{(u - o)f} \quad (8.5)$$

In most cases (and if there are no broken or deformed apertures), the amount of coarse material in the underflow is usually negligible and a simplification is to assume $u = 1$ (i.e., all particles (100%) reporting to the fine product stream are below the screen aperture size), in which case the efficiency reduces to (f, o as fractions):

$$E_U = \frac{U}{Ff} = \frac{f - o}{f(1 - o)} \quad (8.6)$$

This is the efficiency equation used in Example 6.1 in Chapter 6 to calculate the flows around a closed crusher/screen circuit.

If, rather than the fine product, the coarse product is of more interest, a second definition of efficiency is recovery of oversize to the overflow, E_O:

$$E_O = \frac{O(1 - o)}{F(1 - f)} \quad (8.7)$$

or, substituting for O/F in the same way as for U/F

$$E_O = \frac{(f - u)(1 - o)}{(o - u)(1 - f)} \quad (8.8)$$

Example 8.1

What is the % − 12.7 mm in the screen underflow for 25.4 mm aperture screen if the feed to the screen is 50% − 25.4 mm and 25% − 12.7 mm given the screen efficiency is 85%?

Solution

To solve we need the efficiency of recovery of finished product to the underflow, Eq. (8.6). From this definition the mass flow rate of underflow U is

$$U = E_U \times F \times f = 0.85 \times F \times 0.5$$

Assuming all-12.7 mm material in feed reports to underflow (reasonable):

$$\% - 12.7\text{ mm in underflow} = \frac{F \times 0.25}{F \times 0.85 \times 0.5} = 0.588 \text{ or } 58.8\%$$

Example 8.1 illustrates the use of the screen efficiency formulae.

Formulae such as these are acceptable for assessing the efficiency of a screen under different conditions, operating on the same feed. They do not, however, give an absolute value of the efficiency, as no allowance is made for the difficulty of the separation. A feed composed mainly of particles of a size near to that of the screen aperture—"near-size" (or "near-mesh") material—presents a more difficult separation than a feed composed mainly of very coarse and very fine particles relative to the screen aperture.

Efficiency and Circulating Load

Circulation of material occurs in several parts of a mineral processing flowsheet, in grinding and flotation circuits, for example, as well as the crushing stage. In the present context, the circulating load (C) is the mass of coarse material returned from the screen to the crusher relative to the circuit final product (or fresh feed to the circuit), often quoted as a percentage. Figure 8.2 shows two closed circuit arrangements. Circuit (a) was considered in Chapter 6 (Example 6.1), and circuit (b) is an alternative. The symbols have the same meaning as before. The relationship of circulating load to screen efficiency for circuit (a) was derived in Example 6.1, namely (where all factors are as fractions):

$$C = \frac{R}{U} = \frac{1}{r}\left(\frac{1}{E_U} - n\right) \quad (8.9)$$

In circuit (b), at steady state Eqs. (8.1–8.3) apply and the circulating load ratio $C = O/U$ can be related to E_U as follows:

From Eq. (8.1) dividing through by U:

$$C = \frac{O}{U} = \frac{F}{U} - 1$$

and from the definition of E_U we can substitute for F $(=U/E_U f$, assuming $u=1)$, giving:

$$C = \frac{1}{E_U f} - 1 \quad (8.10)$$

The circulating load as a function of screen efficiency for the two circuits is shown in Figure 8.3. The circulating load increases with decreasing screen efficiency and as crusher product coarsens (f or r decreases), which is related to the crusher set (specifically the closed side setting, c.s.s.). For circuit (a) C also increases as the fresh feed coarsens (n decreases), which is likely coming from another crusher. In this manner, the circulating load can be related to crusher settings.

Efficiency or Partition Curve

For a screen this is drawn by plotting the partition coefficient, defined as the fraction (percentage) of the feed reporting to the oversize product, against the geometric mean size on a logarithmic scale. (For particles in the range, say, −8.0 +6.3 mm, the geometric mean size is $\sqrt{(8 \times 6.3)} = 7.1$ mm.) Figure 8.4 shows ideal and real partition curves (see also Chapter 9).

The separation size, or *cut-point*, is obtained at 50% recovery to oversize product, or 50% probability (i.e., the size of particle which has equal probability of reporting to the undersize or oversize product). The cut-point is always less than the size of the aperture.

The efficiency of separation is assessed from the steepness of the partition curve (see Chapter 9). The efficiency curve is used for simulation and design purposes (Napier-Munn et al., 1996; King, 2012).

Separation Efficiency

The efficiency definitions above refer to the recovery of finished product to either stream, for example, E_U is recovery of finished product (material less than cut-size, or undersize) to the underflow stream. We could consider a separation efficiency as introduced in Chapter 1 (Eq. (1.2)), which is recovery of target material minus recovery of nontarget material. With respect to a screen, for the underflow we could consider the difference in recovery of undersize to the underflow minus recovery of oversize to the underflow, and an analogous statement could be made for the overflow, that is, recovery of oversize to the overflow minus recovery of undersize to the overflow. The appropriate equations can be derived, and this definition of screen efficiency may be of more use than the standard definitions in certain cases. The same could be applied to any particle size classifier, such as a cyclone.

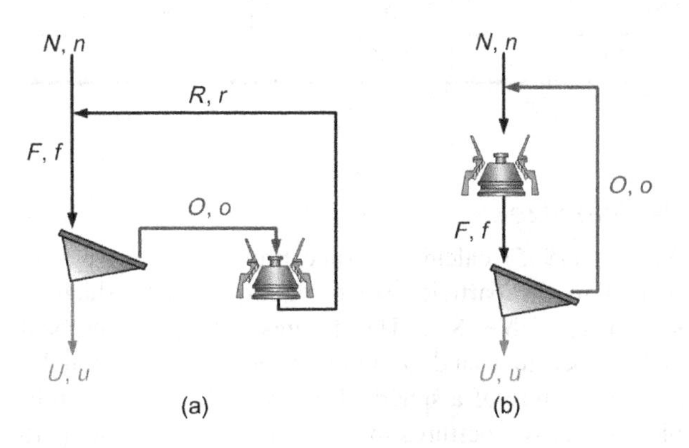

FIGURE 8.2 Flow around a crusher closed with a screen: (a) fresh feed to screen, and (b) fresh feed to crusher.

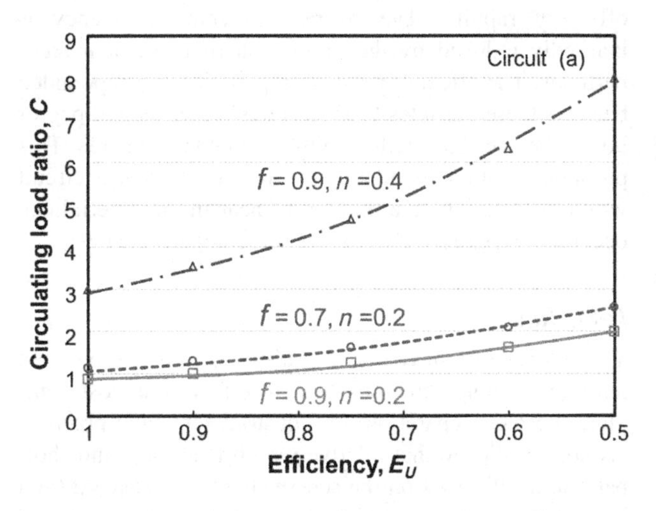

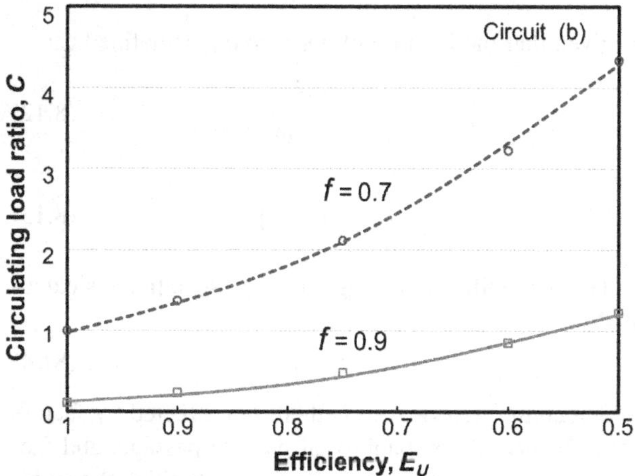

FIGURE 8.3 Circulating load ratio as a function of screen efficiency (E_U) and crusher product size for the two circuits in Figure 8.2.

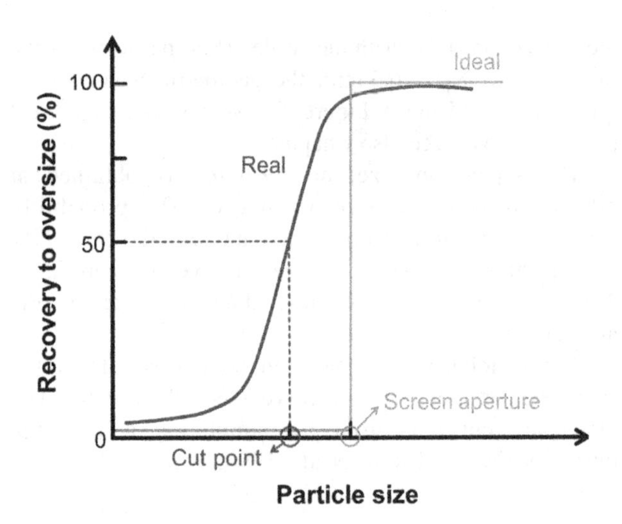

FIGURE 8.4 Ideal and real partition curves.

8.3 FACTORS AFFECTING SCREEN PERFORMANCE

Screen effectiveness must always be coupled with capacity as it is often possible, by the use of a low feed rate and a very long screening time, to effect an almost complete separation. At a given capacity, the effectiveness depends on the nature of the screening operation, that is, on the overall chance of a particle passing through the screen once it has reached it.

The process of screening is frequently described as a series of probabilistic events, where particles are presented to a screening surface many times, and on each presentation there exists a given probability that a particle of a given size will pass. In its simplest form, the probability of passage for a single spherical particle size d passing a square aperture with a size x bordered by a wire diameter w in a single event is given by Gaudin (1939):

$$p = \left(\frac{x - d}{x + w}\right)^2 \tag{8.11}$$

or, given that the fraction of *open area* f_o is defined as:

$$f_o = \frac{x^2}{(x+w)^2} \tag{8.12}$$

$$p = f_o\left(1 - \frac{d}{x}\right)^2 \tag{8.13}$$

The probability of passage for n presentations is calculated by:

$$p' = (1-p)^n \tag{8.14}$$

Screening performance is therefore affected by factors that influence the probability of particle passage, and factors that influence the number of opportunities the particles are given to pass through the screen mesh.

TABLE 8.1 Probability of Passage

Ratio of Particle to Aperture Size	Chance of Passage per 1,000	Number of Apertures Required in Path
0.001	998	1
0.01	980	2
0.1	810	2
0.2	640	2
0.3	490	2
0.4	360	3
0.5	250	4
0.6	140	7
0.7	82	12
0.8	40	25
0.9	9.8	100
0.95	2.0	500
0.99	0.1	10^4
0.999	0.001	10^6

Particle Size

Taggart (1945) calculates some probabilities of passage related to the particle size using Eq. (8.14), which are shown in Table 8.1. The figures relate the probable chance per thousand of unrestricted passage through a square aperture of a spherical particle and give the probable number of apertures in series in the path of the particle necessary to ensure its passage through the screen.

It can be seen from Table 8.1 that as the particle size approaches that of the aperture, the chance of passage falls off very rapidly. The overall screening efficiency is markedly reduced by the proportion of these *near-mesh* particles. The effect of near-mesh particles is compounded because these particles tend to "plug" (also termed peg or blind) the apertures, reducing the available open area. This problem is often found on screens run in closed circuit with crushers, where a buildup of near-mesh material can occur and progressively reduces screening efficiency.

Feed Rate

The principle of sieve sizing analysis is to use a low feed mass and a long screening time to effect an almost complete (perfect) separation. In industrial screening practice, economics dictate that relatively high feed rates and short particle dwell times on the screen should be used. At high feed rates, a bed of material is presented to the screen, and fines must travel to the bottom of the particle bed before they have an opportunity to pass through the screen surface. The net effect is reduced efficiency. High capacity and high efficiency are often opposing requirements for any given separation, and a compromise is necessary.

Screen Angle

Equation (8.13) assumes that the particle approaches the aperture perpendicular to the aperture. If a particle approaches the aperture at a shallow angle, it will "see" a narrower effective aperture dimension and near-mesh particles are less likely to pass. The slope of the screening surface affects the angle at which particles are presented to the screen apertures. Some screens utilize this effect to achieve separations significantly finer than the screen aperture. For example, sieve bends (see Section 8.5.3) cut at approximately half the aperture size. Where screening efficiency is important, horizontal screens are selected.

The screen angle also affects the speed at which particles are conveyed along the screen, and therefore the dwell time on the screen and the number of opportunities particles have of passing the screen surface. Banana screens (see Section 8.5.1) incorporate a variable-angle slope which allows for increased throughput.

Particle Shape

Most granular materials processed on screens are nonspherical. While spherical particles pass with equal probability in any orientation, irregular-shaped near-mesh particles must orient in an attitude that permits them to pass. Elongated and slabby particles will present a small cross section for passage in some orientations and a large cross section in others. The extreme particle shapes therefore have a low screening efficiency: Mica, for instance, screens poorly on square aperture screens, its flat, platelike crystals tending to "ride" over the screen apertures.

Open Area

The chance of passing through the aperture is proportional to the percentage of open area in the screen material (Eq. (8.13)), which is defined as the ratio of the net area of the apertures to the whole area of the screening surface (Eq. (8.12)). The smaller the area occupied by the screen deck construction material, the greater the chance of a particle reaching an aperture.

Open area generally decreases with the fineness of the screen aperture. In order to increase the open area of a fine screen, very thin and fragile wires or deck construction must be used. This fragility and the low-throughput capacity are the main reasons for classifiers replacing screens at fine aperture sizes. Advances in screen design continue to be made and examples of cyclones being replaced by screens are reported (Barkhuysen, 2009; Valine et al., 2009).

Vibration

Screens are vibrated in order to throw particles off the screening surface so that they can again be presented to the screen, and to convey the particles along the screen. The right type of vibration also induces stratification of the feed material (Figure 8.5), which allows the fines to work through the layer of particles to the screen surface while causing larger particles to rise to the top. Stratification tends to increase the rate of passage in the middle section of the screen (Soldinger, 1999).

The vibration must be sufficient to prevent pegging and blinding. However, excessive vibration intensity will cause particles to bounce from the screen deck and be thrown so far from the surface that there are very few effective presentations to the screen surface. Higher vibration rates can, in general, be used with higher feed rates, as the deeper bed of material has a "cushioning" effect that inhibits particle bounce.

Vibration can be characterized by the vibration frequency, v cycles per second, and amplitude, a meters.

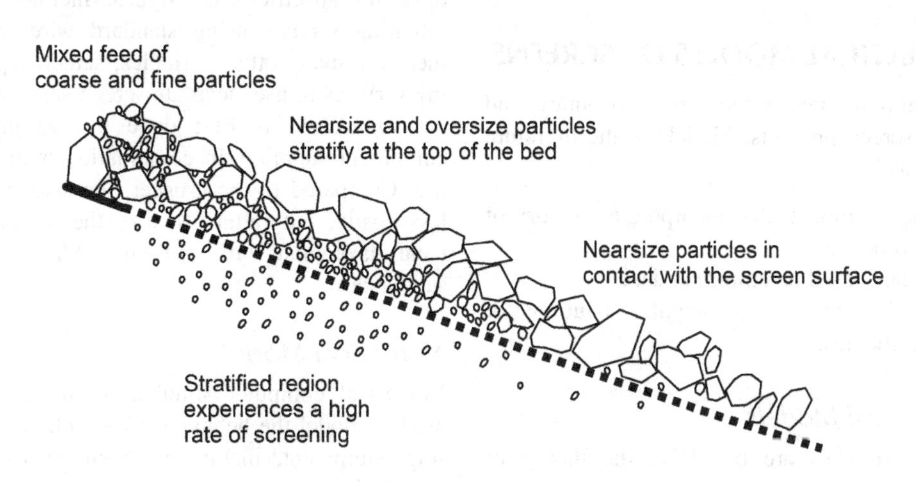

FIGURE 8.5 Stratification of particles on a screen *(Courtesy JKMRC and JKTech Pty Ltd).*

The term "stroke" is commonly used and refers to the peak-to-peak amplitude, or $2a$. Generally, screening at larger apertures is performed using larger amplitudes and lower frequencies; whereas for fine apertures, small amplitudes and high frequencies are preferred. The intensity of vibration is defined by the vibration g-force, Γ:

$$\Gamma = \frac{a(2\pi v)^2}{9.81} \tag{8.15}$$

Vibrating screens typically operate with a vibration force of between 3 and 7 times the gravitational acceleration, or 3–7 G. Vibrations are induced by mechanical exciters driven by electric motors, or electrical solenoids in the case of high-frequency screens. The power required is small compared to other unit operations within the concentrator and is approximately proportional to the loaded mass of the screen.

Moisture

The amount of surface moisture present in the feed has a marked effect on screening efficiency, as does the presence of clays and other sticky materials. Damp feeds screen poorly as they tend to agglomerate and "blind" the screen apertures. As a rule, dry screening at less than around 5 mm aperture size must be performed on perfectly material, unless special measures are taken to prevent blinding. These measures may include using heated decks to break the surface tension of water between the screen wire and particles, ball-decks (a wire cage containing balls directly below the screening surface) to impart additional vibration to the underside of the screen cloth, or the use of nonblinding screen cloth weaves.

Wet screening allows finer sizes to be processed efficiently down to 250 μm and less. Adherent fines are washed off large particles, and the screen is cleaned by the flow of pulp and additional water sprays.

8.4 MATHEMATICAL MODELS OF SCREENS

Screen models aim to predict the size distribution and flow rate of the screen products. Models in the literature can be classified as:

1. phenomenological models that incorporate a theory of the screening process;
2. empirical models based on empirical data;
3. numerical models based on computer solutions of Newtonian mechanics.

Phenomenological Models

Phenomenological models are based on the theory of particle passage through a screening surface. The two dominant theories are *probabilistic*, treating the process as a series of probabilistic events, and *kinetic*, treating the process as one or more kinetic rate processes. These modeling approaches are comprehensively reviewed by Napier-Munn et al. (1996) and King (2012). They have been incorporated into simulators that help guide screen selection and operation.

Empirical Models

Empirical or *capacity* models aim to predict the required area of screen and are frequently used by screen manufacturers. There are a number of formulations of these models of the general form (Bothwell and Mular, 2002):

$$\text{Area required} = \frac{\text{Screen feed rate (t h}^{-1})}{C \times F_1 \times F_2 \times F_3 \times \cdots \times F_n} \tag{8.16}$$

where F_1 to F_n are correction factors and C is the base screen capacity in t h^{-1} per unit area when the other factors are 1.

One set of correction factors: account for the feed size distribution, namely: the quantity of oversize (material larger than the aperture), half-size (material less than half the aperture size), and near-size (material between 75% and 125% of the aperture size). Other factors include: the density of material being screened; whether the screen is a top deck or a lower deck on a multideck screen; the open area of the screen cloth; whether square or slotted apertures are used; whether wet-screening is employed; and the desired screening efficiency.

The values of the base capacity and for each of the factors are given in the form of tables or charts (e.g., Nichols, 1982). Karra (1979) converted the data into equation form to simplify use in spreadsheet calculations. A screen area calculation is given in Example 8.2.

While these capacity-based calculations are popular, the information should be treated as a guide only (Olsen and Coombe, 2003). The calculations have been developed for specific screen-types: inclined circular stroke vibrating screens using standard wire-mesh screen. As there are many other variables, screen types, and screening surfaces in use, accurate screen selection for a particular application is best done by seeking advice from equipment suppliers in combination with pilot-scale testing. Compared to the crusher, the screen is considerably less costly, but failure to size the screen correctly will mean throughput will not be met. Flavel (1982) describes such a case.

Numerical Models

Numerical computer simulations are being increasingly used to model the behavior of particles in various processing equipment, including screens (Cleary, 2003). It is expected that numerical simulation techniques such as the

Example 8.2
Size the screen in the crusher-screen closed circuit, Example 6.1.

Solution
A simplified screen area calculation is:

$$A = \frac{F}{C \times M \times K}$$

where F is feed rate to screen ($t\,h^{-1}$), C the base capacity ($t\,h^{-1}\,m^{-2}$), M is the "oversize" correction factor, and K is the "half-size" correction factor. The base capacity is empirically determined for a feed 25% coarser than the screen opening and 40% finer than half the screen opening (i.e., $M=1$ and $K=1$, respectively, under these feed conditions). As feeds vary from this size distribution corrections factors, M and K, and introduced. Some values of C, K, and M are given (Nichols, 1982); recall that the screen opening (aperture "a") in Example 6.1 was 13 mm.

Opening, *a* (mm)	Capacity, *C* ($t\,h^{-1}\,m^{-2}$)	% + *a*	*M*	% − 0.5 *a*	*K*
13	41.5	40	1.10	25	0.7
19	51.5	45	1.13	30	0.8
25	59.0	50	1.18	35	0.9

For $a = 13$ mm, from the capacity data

$$C = 41.5\ (t\,h^{-1} m^{-2})$$

From Example 6.1, the feed rate, F, to the screen is:

$$F = 402.5\ (t\,h^{-1})$$

M: require % + a (% +13 mm)
We can get to this by calculating f which is the % − 13 mm

$$f = Ff/F = 222.2/402.5 = 0.552 \text{ or } 55.2\%$$

thus

$$\% + 13\text{ mm} = 100 - 55.2 = 44.8\%$$

and

$$M = 1.13 \text{ (close enough)}$$

K: require % − 0.5 a (% − 6.5 mm)
To calculate need % − 6.5 mm in fresh feed (N) and recycle stream (R).
In N assume that % − 6.5 mm = 10%
In R, from crusher gradation table (c.s.s. = 10 mm) % − 6.5 mm ~ 55%
The calculation for % − 6.5 mm (x) in screen feed F is then

$$402.5 \times x = 200 \times (0.1) + 202.5 \times (0.55)$$

thus

$$x = 0.326 \text{ (or 32.6\%)}$$

and

$$K = 0.82 \text{ (by interpolation)}$$

The calculated screen area A is then

$$A = \frac{402.5}{41 \times 1.13 \times 0.82} = 10.5\ m^2$$

Applying a safety factor:

$$A \sim 12\ m^2$$

Depending on availability, a 2.44 m by 5.5 m single deck screen meets the area requirement.

Discrete Element Method will gain wider application in the modeling of industrial screens and assist in the design and optimization of new screening machines.

8.5 SCREEN TYPES

There are numerous types of industrial screens available. The dominant type is the vibrating screen, of which there are many subtypes in use for coarse and fine-screening applications.

8.5.1 Vibrating Screens

Vibrating screens are the most important and versatile screening machines for mineral processing applications (Crissman, 1986). The success of the vibrating screen has made many older screen types obsolete in the minerals industry, including shaking and reciprocating screens, details of which can be found in Taggart (1945). Vibrating screens have a rectangular screening surface with feed and oversize discharge at opposite ends. They perform size separations from 300 mm down to 45 μm and they are used in a variety of sizing, grading, scalping, dewatering, wet screening, and washing applications.

Vibrating screens of most types can be manufactured with more than one screening deck. On multiple-deck systems, the feed is introduced to the top coarse screen, the undersize falling through to the lower screen decks, thus producing a range of sized fractions from a single screen.

Inclined or Circular Motion Screens

A vertical, circular or elliptical vibration is induced mechanically by the rotation of unbalanced weights or flywheels attached usually to a single drive shaft (see Section 8.5.2) (Figure 8.6). The amplitude of throw can be adjusted by adding or removing weight elements bolted to the flywheels. The rotation direction can be contra-flow or in-flow. Contra-flow slows the matcrial more and permits more efficient separation, whereas in-flow permits a greater throughput. Single-shaft screens

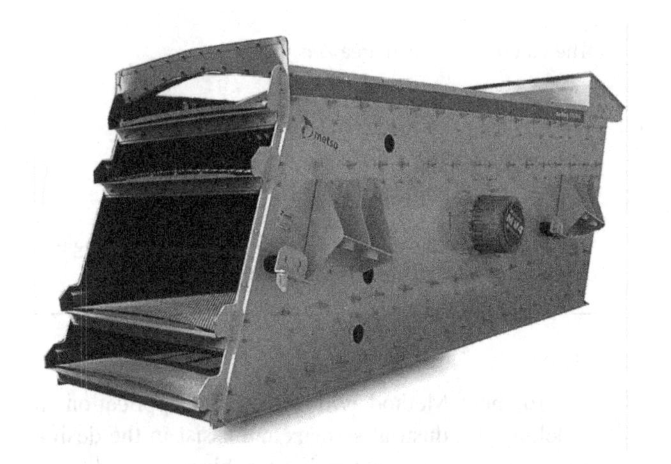

FIGURE 8.6 Inclined four-deck vibrating screen *(Courtesy Metso Minerals)*.

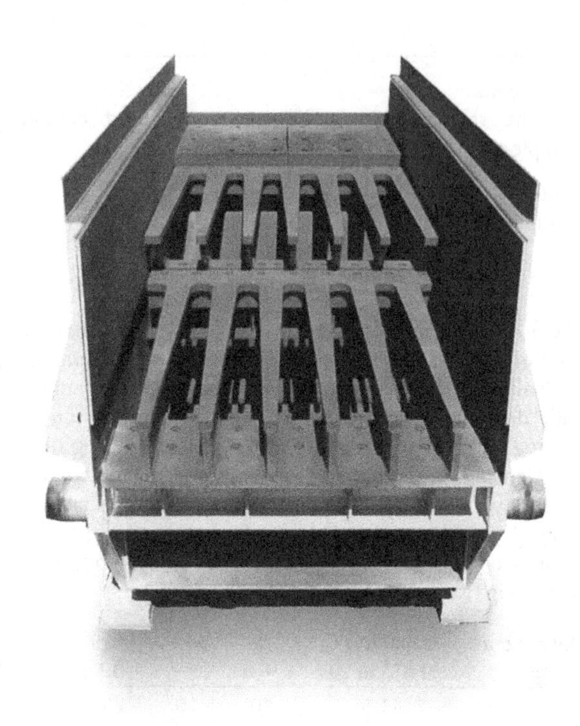

FIGURE 8.7 Vibrating grizzly screen *(Courtesy Metso Minerals)*.

must be installed on a slope, usually between 15° and 28°, to permit flow of material along the screen.

Grizzly Screens

Very coarse material is usually screened on an inclined screen called a grizzly screen. Grizzlies are characterized by parallel steel bars or rails (Figure 8.7) set at a fixed distance apart and installed in line with the flow of ore. The gap between grizzly bars is usually greater than 50 mm and can be as large as 300 mm, with feed topsize as large as 1 m. Vibrating grizzlies are usually inclined at an angle of around 20° and have a circular-throw mechanism (Section 8.5.2). The capacity of the largest machines exceeds 5,000 t h^{-1}.

The most common use of grizzlies in mineral processing is for sizing the feed to primary and secondary crushers. If a crusher has a 100 mm setting, then feed can be passed over a grizzly with a 100 mm gap in order to reduce the load on the crusher.

The bars are typically made from wear-resistant manganese steel and are usually tapered to create gaps that become wider toward the discharge end of the screen to prevent rocks from wedging between the bars. Domed or peaked profiles on the tops of the bars give added wear protection and prevent undersized rocks from "riding" along the bars and being misplaced.

Horizontal, Low-Head, or Linear Vibrating Screens

As shown in Figure 8.8, they have a horizontal or near-horizontal screening surface, and therefore need less headroom than inclined screens. Horizontal screens must be vibrated with a linear or an elliptical vibration produced by a double or triple-shaft vibrator (Section 8.5.2). The accuracy of particle sizing on horizontal screens is superior to that on inclined screens; however, because gravity does not assist the transport of material along the screen they have lower capacity than inclined screens (Krause, 2005). Horizontal screens are used in sizing applications where screening efficiency is critical, and in drain-and-rinse screens in heavy medium circuits (Chapter 11).

Resonance Screens

These is a type of horizontal screen consisting of a screen frame connected by rubber buffers to a dynamically balanced frame having a natural resonance frequency which is the same as that of the vibrating screen body. The vibration energy imparted to the screen frame is stored up in the balancing frame and reimparted to the screen frame on the return stroke. The energy losses are reduced to a minimum, and the sharp return motion produced by the resonant action imparts a lively action to the deck and promotes good screening.

Dewatering Screens

This type of vibrating screen receives a thick slurry feed and produces a drained sand product. Dewatering screens are often installed with a slight up-hill incline to ensure that water does not flow over with the product. A thick bed of particles forms, trapping particles finer than the screen aperture.

FIGURE 8.8 Horizontal screen *(Courtesy Schenk Australia).*

Banana or Multislope Screens

Becoming popular in high-tonnage sizing applications where both efficiency and capacity are important, banana screens (Figure 8.9) typically have a variable slope of around 40–30° at the feed end of the screen, reducing to around 0–15° in increments of 3.5–5° (Beerkircher, 1997). Banana screens are usually designed with a linear-stroke vibrator (Section 8.5.2).

The steep sections of the screen cause the feed material to flow rapidly at the feed end of the screen. The resulting thin bed of particles stratifies more quickly and therefore has a faster screening rate for the very fine material than would be possible on a slower moving thick bed. Toward the discharge end of the screen, the slope decreases to slow down the remaining material, enabling more efficient screening of the near-size material. The capacity of banana screens is reported to be up to three or four times that of conventional vibrating screens (Meinel, 1998).

Modular Screens

Units such as the *OmniScreen* (Figure 8.10) consist of two or more independent screen modules arranged in series, effectively making a large screen from a number of smaller units. A key advantage of this arrangement is that each screen module can be separately configured with a unique screen slope, screen surface type, vibration stroke, and frequency. This allows screening performance to be optimized separately on different sections of the screen. The individual screen sections. being smaller and lighter, are mechanically more robust compared with a single screen with an equivalent total area. Modular screens are frequently installed in a multislope configuration.

Mogensen Sizer

This is a vibrating screen exploiting the principle that particles smaller than the aperture statistically require a certain number of presentations to the screen in order to pass (refer to Table 8.1). The Mogensen Sizer (Figure 8.11) consists of a system of oscillating and sloping screens of decreasing aperture size, the smallest of which has a mesh size up to twice the size of the desired separation size (Hansen, 2000). This arrangement allows particles very much finer than the screen apertures to pass through quickly while causing larger particles to be rejected by one of the screen surfaces.

A thin layer of particles on each screen surface is maintained, enabling high capacity such that a particular screening duty can be met with a machine occupying less floor space than a conventional screen, and blinding and

FIGURE 8.9 Banana screen *(Courtesy Schenk Australia).*

FIGURE 8.10 Omni screen *(Courtesy Omni Crushing and Screening).*

wear are reduced. The units typically range in size from 0.5 to 3.0 m wide and can contain up to six decks.

High-Frequency Screens

Efficient screening of fine particles requires a vibration with small amplitude and high frequency. Frequencies up to 3,600 rpm are used to separate down to 100 μm, compared with vibrating screens for coarser applications that are vibrated at around 700–1,200 rpm. The vibration of the screening surface can be created by electric motors or with electrical solenoids. In the case of the *Tyler H-series* (or *Hum-mer*) screens, the vibrators are mounted above and connected by rods directly to the screening surface so that energy is not wasted in vibrating the entire screen body.

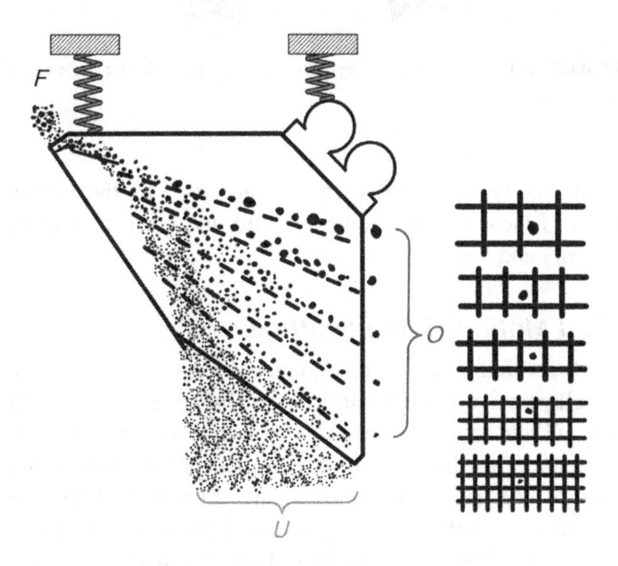

FIGURE 8.11 Mogensen sizer separating feed, *F*, into oversize, *O*, and undersize, *U*, streams *(Adapted from Hansen (2000))*.

Derrick Stack Sizer®

Widely used in wet screening applications, the Stack Sizer® comprises up to five individual screen decks positioned one above the other operating in parallel (Figure 8.12). The "stacked" design allows for high-capacity units in a small footprint. The flow distributor (Flo-Divider™) splits the feed stream evenly to the individual polyurethane screen decks (openings down to 45 μm) where feeders distribute the stream across the entire width (up to 6 m) of each screen. Dual vibratory motors provide uniform linear motion to all screen decks. The undersize and oversize streams are individually combined and exit toward the bottom of the Stack Sizer®. Repulp sprays and trays are an optional addition in between screen sections, which allow for increased screen efficiency.

By classifying by size-only, screens compared to hydrocyclones, give a sharper separation with multidensity feeds (for example, in Pb–Zn operations), and reduce overgrinding of the dense minerals. Valine et al. (2009) document several concentrators in base metals, phosphate, and iron ore operations that replaced hydrocyclones with Stack Sizers® in closing ball mill circuits. An example is the Minera Cerro Lingo (Peru) operation, which produces copper, lead, and zinc concentrates, where 26-in. (66 cm) diameter hydrocyclones in the ball mill circuit were replaced with four Stack Sizers. The result was a decrease in the circulating load from 260% to 108% and 14% increase in circuit throughput.

8.5.2 Vibration Modes

Circular Motion (Single-Shaft) Screens

When the shaft of an inclined screen is located precisely at the screen's center of gravity, the entire screen body

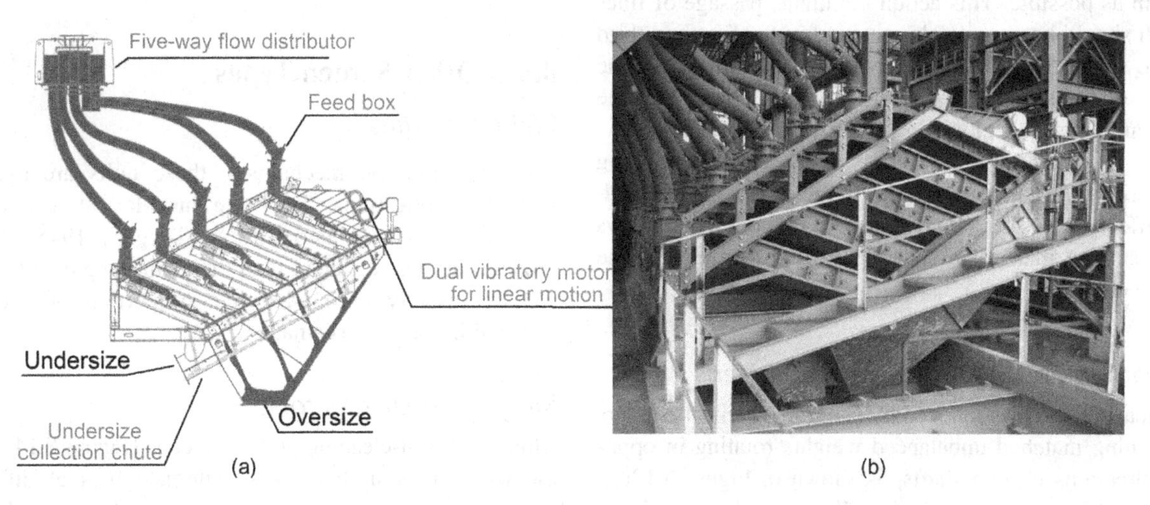

FIGURE 8.12 The Derrick Stack Sizer®: (a) diagram, and (b) in operation *(Copyright © 2014 Derrick Corporation. All rights reserved. This picture is used with the permission of Derrick Corporation)*.

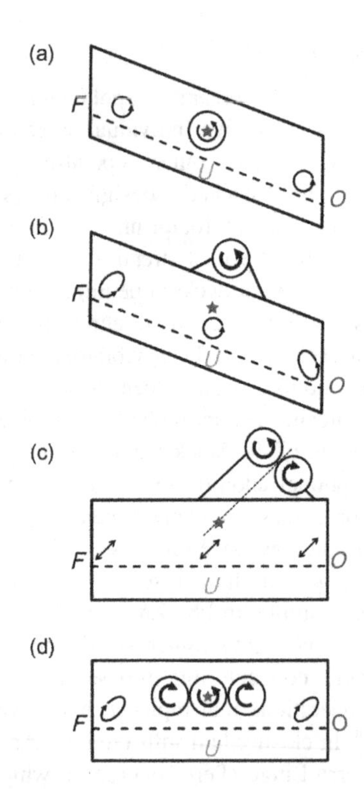

FIGURE 8.13 Vibration patterns generated by various exciter designs. The star represents the location of the screen's center of gravity *(Courtesy JKMRC and JKTech Pty Ltd).*

vibrates with a circular vibration pattern (Figure 8.13(a)). Occasionally, the shaft is installed above or below the center of gravity, as in the system shown in Figure 8.13(b). This placement results in an elliptical motion, slanting forward at the feed end, a circular motion at the center, and an elliptical motion, slanting backward at the discharge end. Forward motion at the feed end serves to move oversize material rapidly out of the feed zone to keep the bed as thin as possible. This action facilitates passage of fines which should be completely removed in the first one-third of the screen length. As the oversize bed thins, near the center of the screen the motion gradually changes to the circular pattern to slow down the rate of travel of the solids. At the discharge end, the oversize and remaining near-size materials are subjected to the increasingly retarding effect of the backward elliptical motion. This allows the near-size material more time to find openings in the screen cloth.

Linear-Vibration (Double-Shaft) Screens

A linear vibration is induced by using mechanical exciters containing matched unbalanced weights rotating in opposite directions on two shafts, as shown in Figure 8.13(c). Linear stroke screens can be installed on a slope, horizontally or even on a small up-hill incline. The angle of stroke is typically between 30° and 60° to the screen deck. Linear-vibration exciters are used on horizontal screens and banana screens.

FIGURE 8.14 Self-cleaning grizzly attached to a feeder. *(Courtesy Metso Minerals)*

Oval Motion (Triple-Shaft) Screens

A three-shaft exciter design can be used to generate an elliptical vibratory motion, as shown in Figure 8.13 (d), which can also be used on horizontal and banana screens. The three shafts are connected by gears and one of the shafts is driven. The elliptical motion is claimed to offer the efficiency benefit of a linear vibrating screen with the tumbling action of a circular motion screen. Higher capacities and increased efficiencies are claimed over either linear or circular motion machines.

8.5.3 Other Screen Types

Static Grizzlies

With no vibration mechanism, these units are used in scalping applications. They are installed at a slope of 35–50° to assist material flow (Taggart, 1945). Static grizzlies are less efficient than their vibrating counterparts and are usually used when the proportion of oversize material in the feed is small.

Mogensen Divergators

Along with self-cleaning grizzly screens (Figure 8.14), they use round bars in two rows—alternate bars at different angles, and fixed at one end to prevent the possibility of blinding. Divergators are used for coarse separations

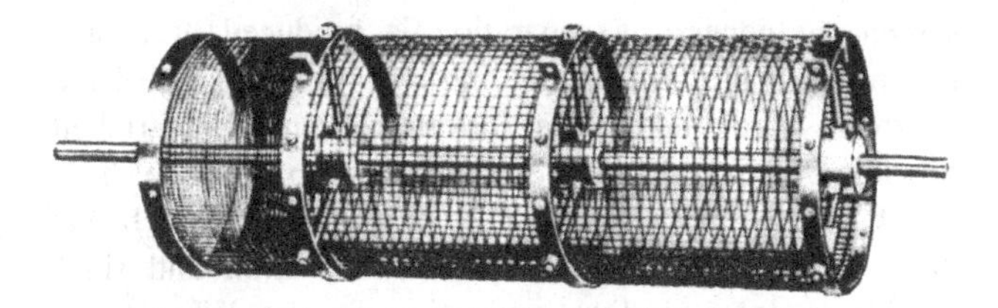

FIGURE 8.15 Trommel screen.

between 25 and 400 mm. Divergators are used in grizzly scalping duties and in chutes to direct the fine material onto the conveyor first to cushion the impact from coarser lumps.

Trommels

These revolving screens are one of the oldest screening devices (Figure 8.15), comprising a cylindrical screen typically rotating at between 35% and 45% critical speed. Trommels are installed on a small angle to the horizontal or use a series of internal baffles to transport material along the cylinder. Trommels can be made to deliver several sized products by using trommel screens in series from finest to coarsest such as the one shown; or using concentric trommels with the coarsest mesh being innermost. Trommels can handle material from 55 mm down to 6 mm, and even smaller sizes can be handled under wet screening conditions. Although trommels are typically cheaper, vibration-free, and mechanically robust, they generally have lower capacities than vibrating screens since only part of the screen surface is in use at any one time, and they can be more prone to blinding.

Trommels remain widely used in some screening duties, including aggregate screening plants and the screening of mill discharge streams. Tumbling mill (AG, SAG, rod, and ball mills) discharge streams usually pass through a trommel screen attached to the mill outlet to prevent grinding media *scats* from reaching subsequent processing equipment, and the case of AG/SAG mills, to extract pebbles to send to crushing (Chapters 6 and 7). Trommels are also used for wet-scrubbing ores, such as bauxite.

Rotaspiral

Introduced in 2001 by Particle Separation Systems, it is a trommel-like device designed for fine screening between 1,000 and 75 μm. The drum contains an internal spiral to move the material through the screen. Water sprays are used to fluidize the screen bed and wash the screen surface. The Rotaspiral can also be used in a dewatering duty.

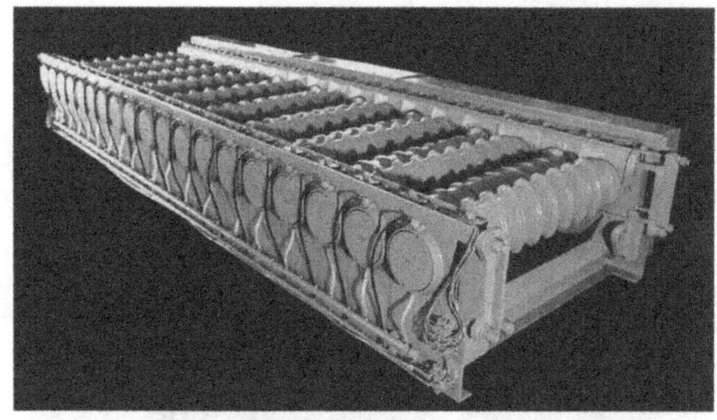

FIGURE 8.16 Roller screen *(Courtesy Metso Minerals).*

Bradford Breaker

A variation of the trommel screen, it is employed in the coal industry (see also Chapter 6). It serves a dual function of breaking coal, usually to between 75 and 100 mm, and separating the harder shale, rock, tramp metal, and wood contaminants into the oversize. Bradford breakers are operated at between 60% and 70% critical speed (see Chapter 7).

Roller Screens

Used for screening applications from 3 to 300 mm (Clifford, 1999), roller screens (Figure 8.16) use a series of parallel driven rolls (circular, elliptical, or profiled) or discs to transport oversize while allowing fines to fall through the gaps between the rolls or discs. They offer advantages of high capacity, low noise levels, require little head-room, subject the material to little impact, and permit screening of very sticky materials.

Flip-Flow Screen

The concept used in the Liwell "Flip-flow" screens is a system of flexible screen panels that are alternately stretched and relaxed to impart motion to the screen bed instead of relying only on mechanical vibration of the screen body. The throwing action can generate forces of up to 50 G on the screen surface, preventing material from blinding the apertures. The screen body may be static or subjected to accelerations in the range 2–4 G (Kingsford, 1991).

Flip-flow screens can be used for separations ranging from 0.5 up to 50 mm and for feed rates up to 800 $t\,h^{-1}$. Flip-flow screens are particularly suited for fine separations of damp material that cannot be screened efficiently on conventional vibrating screens (Meinel, 1998). Both roller and flip-flow screens are now finding application in coal preparation plants where damp feeds blind standard vibrating screens (Luttrell, 2014).

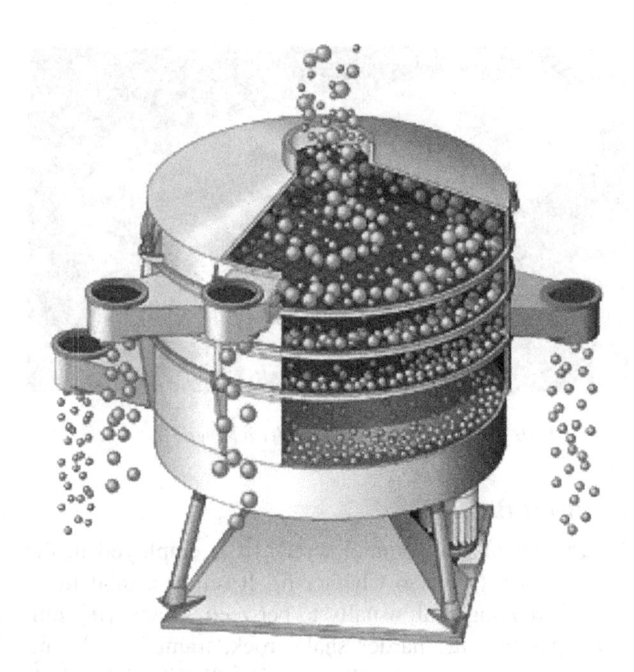

FIGURE 8.17 Gyratory screen.

Circular, Gyratory, or Tumbler Screens

As shown in Figure 8.17, they impart a combined gyratory and vertical motion and are used for fine-screening applications, wet or dry, down to 40 μm, including laboratory use. The basic components consist of a nest of sieves up to around 2.7 m in diameter supported on a table that is mounted on springs on a base. Suspended from beneath the table is a motor with double-shaft extensions, which drives eccentric weights and in doing so effects horizontal gyratory motion. Vertical motion is imparted by the bottom weights, which swing the mobile mass about its center of gravity, producing a circular tipping motion to the screen, the top weights producing the horizontal gyratory motion. Ball trays and ultrasonic devices may be fitted below the screen surfaces to reduce blinding. Circular screens are often configured to produce multiple size fractions. These types of screens are typically used in low-capacity applications, one example being the dewatering of carbon in gold plants.

Sieve Bends

Along with *inclined flat screens*, sieve bends are used for dewatering and fine screening applications. The sieve bend has a curved screen composed of horizontal wedge bars, whereas flat screens are installed on a slope of between 45° and 60°. Feed slurry enters the upper surface of the screen tangentially and flows down the surface in a direction perpendicular to the openings between the wedge bars. As the stream of slurry passes each opening a thin layer is peeled off and directed to the underside of the screen. In general, a separation is produced at a size roughly half the bar spacing and so little plugging of the apertures should take place. Separation can be undertaken down to 50 μm and screen capacities are up to 180 $m^3 h^{-1}$.

One of the most important applications for sieve bends is in draining water from the feed to drain and rinse screens in dense medium separation circuits. When treating abrasive materials sieve bends will require regular reversal of the screen surface as the leading edge of the apertures will lose their sharpness over time.

Sieve bends and inclined wedge-wire screens are sometimes installed with mechanical devices to periodically vibrate or rap the screen surface in order to remove blinded particles.

Linear Screen

Developed by Delkor (now Tenova Delkor), this screen is predominantly used for removing wood chips and fiber from the ore stream feeding carbon-in-pulp systems, and for the recovery of loaded carbon in gold CIP circuits (Walker, 2014). The machine (Figure 8.18) comprises a synthetic monofilament screen cloth supported on rollers and driven by a head pulley coupled to a variable speed drive unit. Mesh sizes in use are typically around 500 μm. Dilute slurry enters through a distributor on to the moving cloth. The undersize drains through the cloth by gravity and is collected in the underpan. The oversize material retained on the screen is discharged at the drive pulley, and any adhering material is washed from the screen cloth using water sprays. As the screen is not vibrated, linear screens are quiet and the energy consumption is much less than that required for vibrating screens.

Pansep Screen

It has a similar principle to the linear screen, but rather than a continuous screen surface, the deck is divided into a series of pans that move in a manner similar to a conveyor (Figure 8.19). The base of each pan consists of a tensioned wire screen mesh, permitting finer cut-points than on linear screens. Cut-points in the range 45–600 μm are possible. Screening occurs both on the top of the "conveyor" motion and on the bottom, giving high screening capacity for the occupied foot print. As well, a deck cleaning action is provided by continually reversing the screening direction (Buisman and Reyneke 2000; Mohanty, 2003). Panels are washed twice each rotation.

8.5.4 Screening Surfaces

There are many types of screening surface available. The size and shape of the apertures, the proportion of open area, the material properties of the screening surface, and

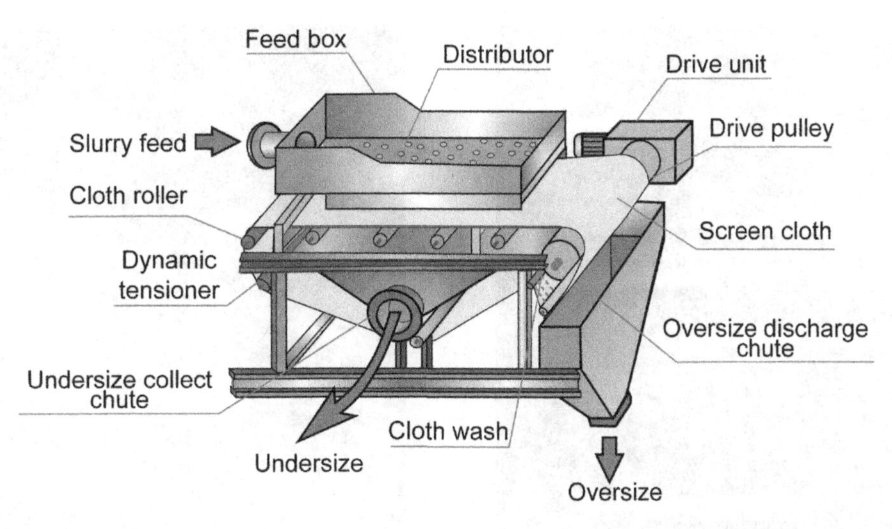

FIGURE 8.18 Linear screen *(Courtesy Delkor).*

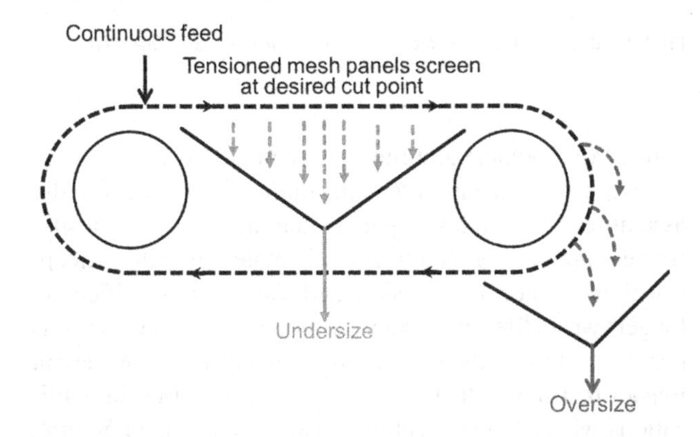

FIGURE 8.19 Principle of the Pansep screen *(Adapted from Buisman and Reyneke (2000)).*

FIGURE 8.20 Bolt-in screening surface.

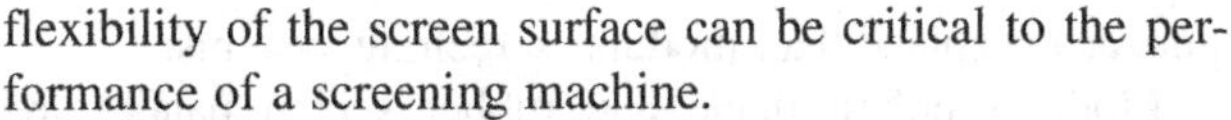

flexibility of the screen surface can be critical to the performance of a screening machine.

Screening surfaces are usually manufactured from steel, rubber, or polyurethane and can be classified according to how they are fixed to the screen. Bolt-in, tensioned, and modular fixing systems are used on industrial vibrating screens.

Bolt-In Screening Surfaces

Screening surfaces for screening duties with particles larger than around 50 mm frequently consist of large sheets of punched, laser-cut, or plasma-cut steel plate, often sandwiched with a polyurethane or rubber wear surface to maximize wear life. These sheets are rigid and are bolted to the screen (Figure 8.20). Curved sections of screens of this type are also commonly used on trommels.

These screening surfaces are available with custom-designed aperture shapes and sizes. Apertures usually have a tapered profile, becoming wider with depth, thereby reducing the propensity of particles pegging in the aperture.

Tensioned Screening Surfaces

These screen surfaces consist of cloths that are stretched taut, either between the sides of the screen (cross tensioned) or along the length of the screen (end tensioned). Maintaining the correct tension in the screen cloth is essential to ensure screening efficiency and to prevent premature failure of the screening surface. Tensioned screens are available in various wire weaves as well as polyurethane and rubber mats.

Woven-Wire Cloth

Usually constructed from steel or stainless steel, these traditional screen surfaces remain popular. Wire cloths are

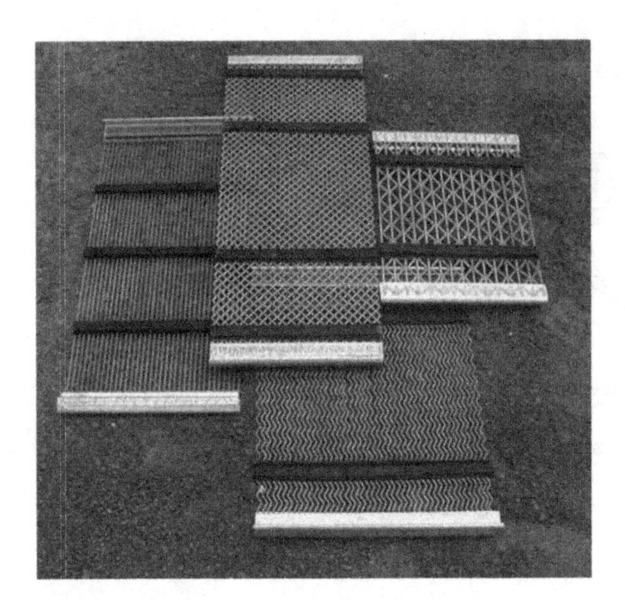

FIGURE 8.21 Various types of self-cleaning wire mesh.

FIGURE 8.22 Modular screen panels *(Courtesy Metso Minerals).*

the cheapest screening surfaces, have a high open area, and are comparatively light. The high open area generally allows a screen to be smaller than a screen with modular panels for the same capacity duty. In relatively light screening duties, therefore, wire-tensioned screens are often preferred. Increasing the wire thickness increases their strength, but decreases open area and hence capacity.

Various types of square and rectangular weaves are available. Rectangular screen apertures have a greater open area than square-mesh screens of the same wire diameter. The wire diameter chosen depends on the nature of the work and the capacity required. Fine screens can have the same or greater open areas than coarse screens, but the wires used must be thinner and hence are more fragile.

"Self-Cleaning" Wire

Traditionally, blinding problems have been countered by using wire with long-slotted apertures or no cross-wires at all ("piano-wire"), but at the cost of lower screening efficiency. Self-cleaning wire (Figure 8.21) is a variation on this, having wires that are crimped to form "apertures", but individual wires are free to vibrate and therefore have a high resistance to blinding and pegging. Screening accuracy can be close to that of conventional woven wire mesh; and they have a longer wear life, justifying their higher initial cost. There are three main types of self-cleaning weave: diamond, triangle, and wave or zig-zag shaped apertures. The triangle and diamond weaves give a more efficient separation.

Tensioned Rubber and Polyurethane Mats

Interchangeable with tensioned wire cloths, these mats are usually reinforced with internal steel cables or synthetic cords. Polyurethane and rubber screen decks (Figure 8.22) are popular in harsh screening duties. They are usually assembled in modules or panels that are fixed onto a sub-frame (*modular screening*). Both materials offer exceptional resistance to abrasion and can have significantly longer wear life than steel, although the open area is generally lower than wire. Rubber also has excellent impact resistance; therefore rubber is often used in applications where top size can be greater than around 50 mm. Polyurethane is generally preferred in wet screening applications. Aggregate producers prefer tensioned mats because they must be able to make frequent deck changes to produce different specifications, and tensioned mats are quicker to replace than modular screening systems.

Modular polyurethane and rubber screen panels are typically 305 mm × 305 mm, 610 mm × 305 mm, or similar, in size. The edges of the panel usually contain a rigid steel internal frame to give the panel strength. Panel systems allow for rapid replacement of the deck. Different panel types and aperture sizes can be installed at different positions along the screen to address high wear areas and to optimize any given screening task. Modular screens do not require tensioning and retensioning and damaged sections of the screen can be replaced *in situ*. Polyurethane and rubber screens are also quieter, and the more flexible apertures reduce blinding compared with steel wire cloths.

Square, rectangular, and slot apertures are the most commonly used aperture shapes. Rectangular and slot apertures can be in-flow (usual for sizing applications), cross-flow (usual for dewatering applications), or diagonal. With slabby particles rectangular and slot apertures provide greater open area, throughput, resistance to pegging, and efficiency compared with square apertures. Other aperture shapes include circles, hexagons, octagons, rhomboids, and teardrops. Combinations of shapes and configurations are also possible. Circular apertures are considered to give the most accurate cut, but are more prone to pegging. Slotted, teardrop, and more complex aperture shapes are used where blinding or pegging can be a problem. Apertures are tapered, being wider at the bottom than the top, to ensure that a particle that has passed through the aperture at the deck surface can fall freely to undersize.

Modular Wire and Wedge Wire Panels

These have much greater open area compared with modular polyurethane screens. These wire panels consist of a polyurethane or rubber fixing system molded around a woven-wire or wedge-wire screening surface.

REFERENCES

Barkhuysen, N.J., 2009. Implementation strategies to improve mill capacity and efficiency through classification by particle size only, with cases studies. Proc. of the Fifth Southern African Base Metals Conference, SAIMM, Kasane, Botswana, pp. 101–113.

Beerkircher, G., 1997. Banana screen technology. In: Kawatra, S.K. (Ed.), Comminution Practices. SME, Littleton, CO, USA, pp. 37–40.

Bothwell, M.A., Mular, A.L., 2002. Coarse screening. In: Mular, A.L., et al., (Eds.), Mineral Processing Plant Design, Practice and Control, vol. 1. SME, Littleton, CO, USA, pp. 894–916.

Buisman, R., Reyneke, K., 2000. Fine coal screening using the new Pansep screen. Proceedings of the 17th International Coal Preparation Confernece, Lexington, KY, USA, pp. 71–85.

Cleary, P.W., 2003. DEM as a tool for design and optimisation of mineral processing equipment. Proceedings of the 22th International Mineral Processing Congress (IMPC), Cape Town, South Africa, pp. 1648–1657.

Clifford, D., 1999. Screening for profit. Mining Mag. May, 236–248.

Colman, K.G., Tyler, W.S., 1980. Selection guidelines for size and type of vibrating screens in ore crushing plants. In: Mular, A.L., Bhappu, R.B. (Eds.), Mineral Processing Plant Design, second ed. SME, New York, NY, USA, pp. 341–361.

Crissman, H., 1986. Vibrating screen selection. Pit and Quarry: Part I, 78(June), 39–44; Part II, 79(Nov.), 46–50.

Flavel, M.D., 1982. Selection and sizing of crushers. In: Mular, A.L., Jergensen, G.V. (Eds.), Design and Installation of Comminution Circuits. SME, Littleton, CO, USA, pp. 343–386.

Gaudin, A.M., 1939. Principles of Mineral Dressing. McGraw-Hill Book Company, New York, NY, USA and London, UK.

Hansen, H., 2000. Fundamentals and further development of sizer technology. Aufbereitungs Tech. 41 (7), 325–329.

Karra, V.K., 1979. Development of a model for predicting the screening performance of a vibrating screen. CIM Bull. 72 (801-806), 167–171.

King, R.P., 2012. Modeling and Simulation of Mineral Processing Systems. second ed. SME, Englewood, CO, USA.

Kingsford, G.R., 1991. The evaluation of a non-blinding screen for screening iron ore fines. Proceedings of the Fourth Mill Operators Conference, Burnie, Tasmania, pp. 25–29.

Krause, M., 2005. Horizontal versus inclined screens. Quarry (Mar.).26–27.

Luttrell, G.H., 2014. Innovations in coal processing. In: Anderson, C.G., et al., (Eds.), Mineral Processing and Extractive Metallurgy: 100 Years of Innovation. SME, Englewood, CO, USA, pp. 277–296.

Meinel, A., 1998. Classification of fine, medium-sized and coarse particles on shaking screens. Aufbereitungs Tech. 39 (7), 317–327.

Mohanty, M.K., 2003. Fin coal screening performance enhancement using the Pansep screen. Int. J. Miner. Process. 69 (1-4), 205–220.

Napier-Munn, T.J., et al., 1996. Mineral Comminution Circuits—Their Operation and Optimisation. Julius Kruttschnitt Mineral Research Centre (JKMRC), University of Queensland, , Brisbane, Australia.

Nichols, J.P., 1982. Selection and sizing of screens. In: Mular, A.L, Jergensen, G.V (Eds.), Design and Installation of Comminution Circuits. SME, New York, NY, USA, pp. 509–522.

Olsen, P., Coombe, A., 2003. Is screening a science or art? Quarry. 11 (8), 20–25.

Soldinger, M., 1999. Interrelation of the stratification and passage in the screening process. Miner. Eng. 12 (5), 497–516.

Taggart, A.F., 1945. Handbook of Mineral Dressing, Ore and Industrial Minerals. John Wiley & Sons., Chapman & Hall, Ltd, London, UK.

Valine, S.B., Wennen, J.E., 2002. Fine screening in mineral processing operations. In: Mular, A.L., et al., (Eds.), Mineral Processing Plant Design, Practice, and Control, vol. 1. SME, Littleton, CO, USA, pp. 917–928.

Valine, S.B., et al., 2009. Fine sizing with the Derrick® Stack Sizer™ screen. In: Malhotra, D., et al., (Eds.), Recent Advances in Mineral Processing Plant Design. SME, Littleton, CO, USA, pp. 433–443.

Walker, S., 2014. On deck: screens for all processes. Eng. Mining J. 215 (5), 66–71.

Chapter 9

Classification

9.1 INTRODUCTION

Classification, as defined by Heiskanen (1993), is a method of separating mixtures of minerals into two or more products on the basis of the velocity with which the particles fall through a fluid medium. The carrying fluid can be a liquid or a gas. In mineral processing, this fluid is usually water, and wet classification is generally applied to mineral particles that are considered too fine (<200 μm) to be sorted efficiently by screening. As such, this chapter will only discuss wet classification. A description of the historical development of both wet and dry classification is given by Lynch and Rowland (2005).

Classifiers are nearly always used to close the final stage of grinding and so strongly influence the performance of these circuits (Chapter 7). Since the velocity of particles in a fluid medium is dependent not only on the size, but also on the specific gravity and shape of the particles, the principles of classification are also important in mineral separations utilizing gravity concentrators (Chapter 10).

9.2 PRINCIPLES OF CLASSIFICATION

9.2.1 Force Balance

When a solid particle falls freely in a vacuum, there is no resistance to the particle's motion. Therefore, if it is subjected to a constant acceleration, such as gravity, its velocity increases indefinitely, independent of size and density. Thus, a lump of lead and a feather fall at exactly the same rate in a vacuum.

In a viscous medium, such as air or water, there is resistance to this movement and this resistance increases with velocity. When equilibrium is reached between the gravitational force and the resistant force from the fluid, the body reaches its *terminal velocity* and thereafter falls at a uniform rate.

The nature of the resistance, or *drag force*, depends on the velocity of the descent. At low velocities, motion is smooth because the layer of fluid in contact with the body moves with it, while the fluid a short distance away is motionless. Between these two positions is a zone of intense shear in the fluid all around the descending particle. Effectively, all resistance to motion is due to the shear forces, or the viscosity of the fluid, and is hence called *viscous resistance*. At high velocities the main resistance is due to the displacement of fluid by the body, with the viscous resistance being relatively small; this is known as *turbulent resistance*. Whether viscous or turbulent resistance dominates, the acceleration of particles in a fluid rapidly decreases and the terminal velocity is reached relatively quickly.

A particle accelerates according to Newton's well-known equation where $\sum \vec{F}$ is the net force acting on a particle, m the mass of the particle, and $\vec{a}$ the acceleration of the particle:

$$\sum \vec{F} = m\vec{a} \tag{9.1}$$

As mass, a combination of a particle's size and density, is a factor on the particle acceleration, it is common for fine high-density material, such as gold or galena, to be misclassified and report to the same product with the coarser low density particles. This occurrence will be further discussed in Section 9.2.4.

The classification process involves the balancing of the accelerating (gravitational, centrifugal, etc.) and opposing (drag, etc.) forces acting upon particles, so that the resulting net force has a different direction for fine and coarse particles. Classifiers are designed and operated so that the absolute velocities, resulting from the total net force, cause particles to be carried into separable products. Forces acting upon particles can include:

- Gravitational or electrostatic field force
- Inertial force, centrifugal force, and Coriolis force (only in rotational systems)
- Drag force
- Pressure gradient force, buoyancy force
- Basset history force
- Particle–particle interaction force

An example of a classifier is a *sorting column*, in which a fluid is rising at a uniform rate (Figure 9.1). Particles introduced into the sorting column either sink or rise according to whether their terminal velocities, a result

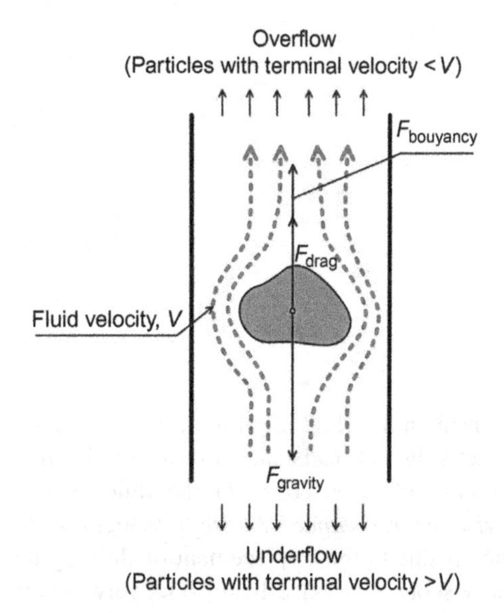

FIGURE 9.1 Balance of forces on a particle in a sorting column.

of the net force, are greater or smaller than the upward velocity of the fluid. The sorting column therefore separates the feed into two products—an *overflow* consisting of particles with terminal velocities smaller than the velocity of the fluid and an *underflow* or *spigot product* containing particles with terminal velocities greater than the rising velocity.

9.2.2 Free Settling

Free settling refers to the sinking of particles in a volume of fluid which is large with respect to the total volume of particles, hence particle–particle contact is negligible. For well-dispersed pulps, free settling dominates when the percentage by weight of solids is less than about 15% (Taggart, 1945).

Consider a spherical particle of diameter d and density ρ_s falling under gravity in a viscous fluid of density ρ_f under free-settling conditions, that is, ideally in a fluid of infinite size. The particle is acted upon by three forces: a gravitational force acting downward (taken as positive direction), an upward buoyant force due to the displaced fluid, and a drag force D acting upward (see Figure 9.1). Following Newton's law of motion in Eq. (9.1), the equation of motion of the particle is therefore:

$$\sum \vec{F} = m\vec{a}$$

$$F_{\text{Gravity}} - F_{\text{Buoyancy}} - F_{\text{Drag}} = m\frac{dx}{dt} \qquad (9.2)$$

$$mg - m'g - D = m\frac{dx}{dt}$$

where m is the mass of the particle, m' the mass of the displaced fluid, x the particle velocity, and g the acceleration due to gravity. When terminal velocity is reached, acceleration (dx/dt) is equal to zero, and hence:

$$D = g(m - m') \qquad (9.3)$$

Therefore, using the volume and density of a sphere:

$$D = \left(\frac{\pi}{6}\right) g\, d^3(\rho_s - \rho_f) \qquad (9.4)$$

Stokes (1891) assumed that the drag force on a spherical particle was entirely due to viscous resistance and deduced the expression:

$$D = 3\pi d\eta v \qquad (9.5)$$

where η is the fluid viscosity and v the terminal velocity.

Hence, substituting in Eq. (9.4) we derive:

$$3\pi d\eta v = \left(\frac{\pi}{6}\right) g\, d^3(\rho_s - \rho_f)$$

or solving for the terminal velocity

$$v = \frac{g\, d^2(\rho_s - \rho_f)}{18\,\eta} \qquad (9.6)$$

This expression is known as *Stokes' law*.

Newton assumed that the drag force was entirely due to turbulent resistance, and deduced:

$$D = 0.055\,\pi\, d^2 v^2 \rho_f \qquad (9.7)$$

Substituting in Eq. (9.4) gives:

$$v = \left[\frac{3\, g\, d(\rho_s - \rho_f)}{\rho_f}\right]^{1/2} \qquad (9.8)$$

This is *Newton's law* for turbulent resistance.

The range for which Stokes' law and Newton's law are valid is determined by the dimensionless Reynolds number (Chapter 4). For Reynolds numbers below 1, Stokes' law is applicable. This represents, for quartz particles settling in water, particles below about 60 μm in diameter. For higher Reynolds numbers, over 1,000, Newton's law should be used (particles larger than about 0.5 cm in diameter). There is, therefore, an intermediate range of Reynolds numbers (and particle sizes), which corresponds to the range in which most wet classification is performed, in which neither law fits experimental data. In this range there are a number of empirical equations that can be used to estimate the terminal velocity, some of which can be found in Heiskanen (1993).

Stokes' law (Eq. (9.6)) for a particular fluid can be simplified to:

$$v = k_1\, d^2(\rho_s - \rho_f) \qquad (9.9)$$

and Newton's law (Eq. (9.8)) can be simplified to:

$$v = k_2[d(\rho_s - \rho_f)]^{1/2} \qquad (9.10)$$

where k_1 and k_2 are constants, and $(\rho_s - \rho_f)$ is known as the *effective density* of a particle of density ρ_s in a fluid of density ρ_f.

9.2.3 Hindered Settling

As the proportion of solids in the pulp increases above 15%, which is common in almost all mineral classification units, the effect of particle–particle contact becomes more apparent and the falling rate of the particles begins to decrease. The system begins to behave as a heavy liquid whose density is that of the pulp rather than that of the carrier liquid; *hindered-settling* conditions now prevail. Because of the high density and viscosity of the slurry through which a particle must fall, in a separation by hindered settling the resistance to fall is mainly due to the turbulence created (Swanson, 1989). The interactions between particles themselves, and with the fluid, are complex and cannot be easily modeled. However, a modified form of Newton's law can be used to determine the approximate falling rate of the particles, in which ρ_p is the *pulp density:*

$$v = k_2[d(\rho_s - \rho_p)]^{1/2} \tag{9.11}$$

9.2.4 Effect of Density on Separation Efficiency

The aforementioned laws show that the terminal velocity of a particle in a particular fluid is a function of the particle size and density. It can be concluded that:

1. If two particles have the same density, then the particle with the larger diameter has the higher terminal velocity.
2. If two particles have the same diameter, then the heavier (higher density) particle has the higher terminal velocity.

As the feed to most industrial classification devices will contain particles with varying densities, particles will not be classified based on size alone. Consider two mineral particles of densities ρ_a and ρ_b and diameters d_a and d_b, respectively, falling in a fluid of density ρ_f at exactly the same settling rate (their terminal velocity). Hence, from Stokes' law (Eq. (9.9)), for fine particles:

$$d_a^2(\rho_a - \rho_f) = d_b^2(\rho_b - \rho_f)$$

or

$$\frac{d_a}{d_b} = \left(\frac{\rho_b - \rho_f}{\rho_a - \rho_f}\right)^{1/2} \tag{9.12}$$

This expression is known as the *free-settling ratio* of the two minerals, that is, the ratio of particle size required for the two minerals to fall at equal rates.

Similarly, from Newton's law (Eq. (9.8)), the free-settling ratio of large particles is:

$$\frac{d_a}{d_b} = \frac{\rho_b - \rho_f}{\rho_a - \rho_f} \tag{9.13}$$

Example 9.1

Consider a binary mixture of galena (specific gravity 7.5) and quartz (s.g. 2.65) particles being classified in water. Determine the free-settling ratio (a) from Stokes' law and (b) from Newton's law. What do you conclude?

Solution

a. For small particles, obeying Stokes' law, the free-settling ratio (Eq. (9.12)) is:

$$\frac{d_a}{d_b} = \left(\frac{7.5 - 1}{2.65 - 1}\right)^{1/2} = 1.99$$

That is, a particle of galena will settle at the same rate as a particle of quartz which has a diameter 1.99 times larger than the galena particle.

b. For particles obeying Newton's law, the free-settling ratio (Eq. (9.13)) is:

$$\frac{d_a}{d_b} = \frac{7.5 - 1}{2.65 - 1} = 3.94$$

Conclusion: The free-settling ratio is larger for coarse particles obeying Newton's law than for fine particles obeying Stokes' law.

The general expression for free-settling ratio can be deduced from Eqs. (9.12) and (9.13) as:

$$\frac{d_a}{d_b} = \left(\frac{\rho_b - \rho_f}{\rho_a - \rho_f}\right)^n \tag{9.14}$$

where $n = 0.5$ for small particles obeying Stokes' law and $n = 1$ for large particles obeying Newton's law. The value of n lies in the range 0.5–1 for particles in the intermediate size range of 50 μm to 0.5 cm (Example 9.1).

The result in Example 9.1 means that the density difference between the particles has a more pronounced effect on classification at coarser size ranges. This is important where gravity concentration is being utilized. Over-grinding of the ore must be avoided, such that particles are fed to the separator in as coarse a state as possible, so that a rapid separation can be made, exploiting the enhanced effect of specific gravity difference. Since the enhanced gravity effect also means that fine heavy minerals are more likely to be recycled and overground in conventional ball mill-classifier circuits, it is preferable where possible to use open circuit rod mills for the primary coarse grind feeding a gravity circuit.

When considering hindered settling, the lower the density of the particle, the greater is the effect of the reduction of the effective density ($\rho_s - \rho_f$). This then leads to a greater reduction in falling velocity. Similarly, the larger the particle, the greater is the reduction in falling rate as the pulp density increases. This is important in classifier design; in effect, hindered-settling reduces the effect of size, while increasing the effect of density on classification. This is illustrated by considering a mixture of quartz and galena particles settling in a pulp of density 1.5. The *hindered-settling ratio* can be derived from Eq. (9.11) as:

$$\frac{d_a}{d_b} = \frac{\rho_b - \rho_p}{\rho_a - \rho_p} \tag{9.15}$$

Therefore, in this system:

$$\frac{d_a}{d_b} = \frac{7.5 - 1.5}{2.65 - 1.5} = 5.22$$

A particle of galena will thus fall in the pulp at the same rate as a particle of quartz, which has a diameter 5.22 times as large. This compares with the free-settling ratio, calculated as 3.94 for turbulent resistance (Example 9.1).

The hindered-settling ratio is always greater than the free-settling ratio, and the denser the pulp, the greater is the ratio of the diameter of equal settling particles. For quartz and galena, the greatest hindered-settling ratio that we can attain practically is about 7.5. Hindered-settling classifiers are used to increase the effect of density on the separation, whereas free-settling classifiers use relatively dilute suspensions to increase the effect of size on the separation (Figure 9.2). Relatively dense slurries are fed to certain gravity concentrators, particularly those treating heavy alluvial sands. This allows high tonnages to be treated and enhances the effect of specific gravity difference on the separation. The efficiency of separation, however, may be reduced since the viscosity of a slurry increases with density. For separations involving feeds with a high proportion of particles close to the required density of separation, lower slurry densities may be necessary, even though the density difference effect is reduced.

As the pulp density increases, a point is reached where each mineral particle is covered only with a thin film of water. This condition is known as a *quicksand*, and because of surface tension, the mixture is a perfect suspension and does not tend to separate. The solids are in a condition of *full teeter*, which means that each grain is free to move, but is unable to do so without colliding with other grains and as a result stays in place. The mass acts as a viscous liquid and can be penetrated by solids with a higher specific gravity than that of the mass, which will then move at a velocity impeded by the viscosity of the mass.

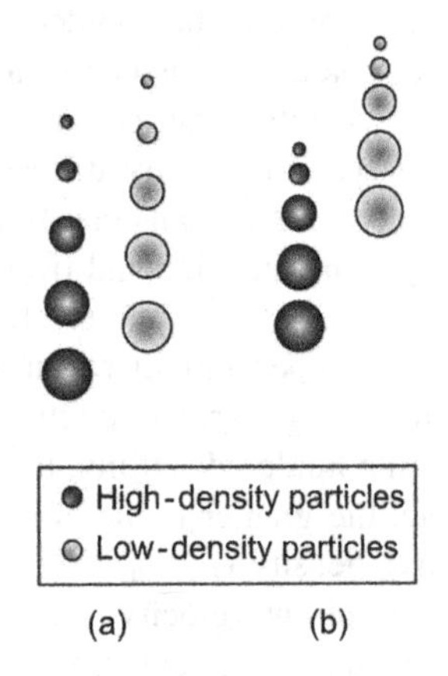

FIGURE 9.2 Classification by: (a) free settling, and (b) hindered settling.

A condition of teeter can be produced in a classifier sorting column by putting a constriction in the column, either by tapering the column or by inserting a grid into the base. Such hindered-settling sorting columns are known as *teeter chambers*. Due to the constriction, the velocity of the introduced water current is greatest at the bottom of the column. A particle falls until it reaches a point where its falling velocity equals that of the rising current. The particle can now fall no further. Many particles reach this condition, and as a result, a mass of particles becomes trapped above the constriction and pressure builds up in the mass. Particles move upward along the path of least resistance, which is usually the center of the column, until they reach a region of lower pressure at or near the top of the settled mass; here, under conditions in which they previously fell, they fall again. As particles from the bottom rise at the center, those from the sides fall into the resulting void. A general circulation is established, the particles being said to *teeter*. The constant jostling of teetering particles has a scouring effect which removes any entrained or adhering slimes particles, which then leave the teeter chamber and pass out through the classifier overflow. Clean separations can therefore be made in such classifiers. The teeter column principle is also exploited in some coarse particle flotation cells (Chapter 12).

The analysis in this section has assumed, directly or implicitly, that the particles are spherical. While this is never really true, the resulting broken particle shapes are often close enough to spherical for the general findings above to apply. The obvious exceptions are ores containing flakey or fibrous mineral such as talc, mica, and some serpentines. Shape in those cases will influence the classification behavior.

9.2.5 Effect of Classifier Operation on Grinding Circuit Behavior

Classifiers are widely employed in closed-circuit grinding operations to enhance the size reduction efficiency. A typical ball mill-classifier arrangement is seen in Figure 7.38 (Chapter 7). The benefits of classification can include: improved comminution efficiency; improved product (classifier overflow) quality; and greater control of the circulating load to avoid overloading the circuit (Chapter 7). The improvement in efficiency of the grinding circuit is seen as either a reduction in energy consumption or increase in throughput (capacity). The main increase in efficiency is due to the reduction of overgrinding. By removing the finished product size particles from the circuit they are not subject to unnecessary further grinding (overgrinding), which is a waste of comminution energy. This, combined with the recycling of unfinished (oversize) particles, results in the circuit product (classifier overflow) having a narrower size distribution than is the case for open circuit grinding. This narrow size distribution and restricted amount of excessively fine material benefits downstream mineral separation processes. Other benefits come from reduced particle–particle contact cushioning from fines in the grinding mill and less misplaced coarse material in the overflow, which would reduce downstream efficiencies. Therefore, classifier performance is critical to the optimal running of a mineral processing plant.

9.3 TYPES OF CLASSIFIERS

Although they can be categorized by many features, the most important is the force field applied to the unit: either gravitational or centrifugal. Centrifugal classifiers have gained widespread use as classifying equipment for many different types of ore. In comparison, gravitational classifiers, due to their low efficiencies at small particle sizes (<70 μm), have limited use as classifiers and are only found in older plants or in some specialized cases. Table 9.1 outlines the key differences between the two types of classifiers and further highlights the benefits of centrifugal classifiers. Many types of classifiers have been designed and built, and only some common ones will be introduced. A more comprehensive guide to the major types of classification equipment used in mineral processing can be found elsewhere (Anon., 1984; Heiskanen, 1993; Lynch and Rowland, 2005).

TABLE 9.1 A Comparison of Key Parameters for Centrifugal and Gravitational Classifiers

Item	Centrifugal Classifiers	Gravitational Classifiers
Capacity	High	Low
Cut-size	Fine—Coarse	Coarse
Capacity/cut-size dependency	Yes	No
Energy consumption	High (feed pressure)	Low
Initial investment	Low	High
Footprint	Small	Large

9.4 CENTRIFUGAL CLASSIFIERS—THE HYDROCYCLONE

The hydrocyclone, commonly abbreviated to just cyclone, is a continuously operating classifying device that utilizes centrifugal force to accelerate the settling rate of particles. It is one of the most important devices in the minerals industry, its main use in mineral processing being as a classifier, which has proved extremely efficient at fine separation sizes. Apart from their use in closed-circuit grinding, cyclones have found many other uses, such as de-sliming, de-gritting, and thickening (dewatering). The reasons for this include their simplicity, low investment cost, versatility, and high capacity relative to unit size.

Hydrocyclones are available in a wide range of sizes, depending on the application, varying from 2.5 m in diameter down to 10 mm. This corresponds to cut-sizes of 300 μm down to 1.5 μm, with feed pressures varying from 20 to about 200 kPa (3–30 psi). The respective flowrates vary from 2 $m^3\ s^{-1}$ in a large unit to $2.5 \times 10^{-5}\ m^3\ s^{-1}$ in a small unit. Units can be installed on simple supports as single units or in clusters ("cyclopacs") (Heiskanen, 1993).

9.4.1 Basic Design and Operation

A typical hydrocyclone (Figure 9.3) consists of a conically shaped vessel, open at its *apex* (also known as *spigot* or underflow), joined to a cylindrical section, which has a tangential feed inlet. The top of the cylindrical section is closed with a plate through which passes an axially mounted overflow pipe. The pipe is extended into the body of the cyclone by a short, removable section known as the *vortex finder*. The vortex finder forces the feed to travel downward, which prevents short-circuiting of feed directly into the overflow. The impact of these design parameters on performance is discussed further in Section 9.4.5.

The feed is introduced under pressure through the tangential entry, which imparts a swirling motion to the pulp. This generates a vortex in the cyclone, with a low-pressure zone along the vertical axis. An air core develops along the axis, normally connected to the atmosphere through the apex opening, but in part created by dissolved air coming out of solution in the zone of low pressure.

The conventional understanding is that particles within the hydrocyclone's flow pattern are subjected to two opposing forces: an outward acting centrifugal force and an inwardly acting drag (Figure 9.4). The centrifugal force developed accelerates the settling rate of the particles, thereby separating particles according to size and specific gravity (and shape). Faster settling particles move to the wall of the cyclone, where the velocity is lowest, and migrate down to the apex opening. Due to the action of the drag force, the slower-settling particles move

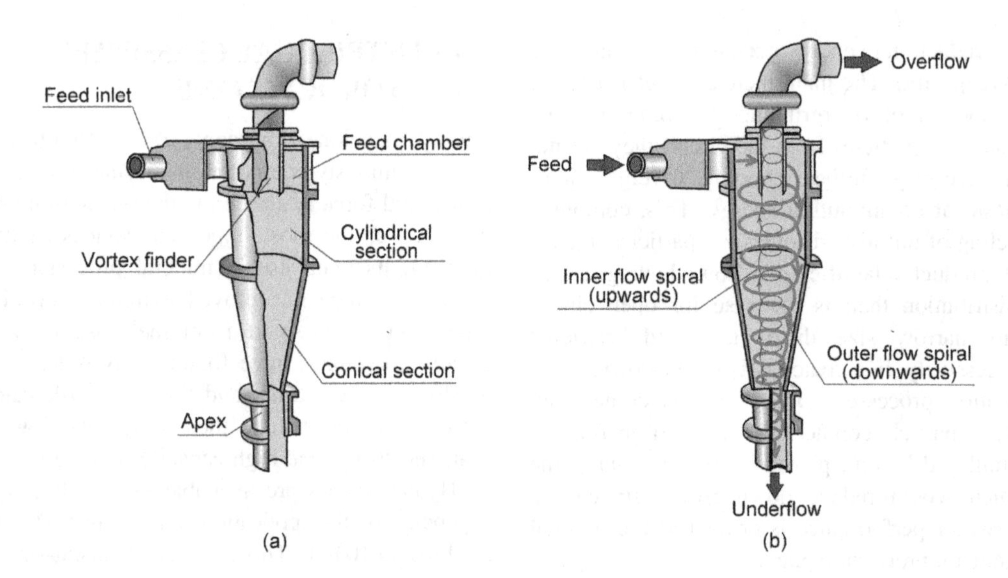

FIGURE 9.3 A hydrocyclone showing: (a) main components, and (b) principal flows *(Adapted from Napier-Munn et al. (1996); courtesy JKMRC, The University of Queensland).*

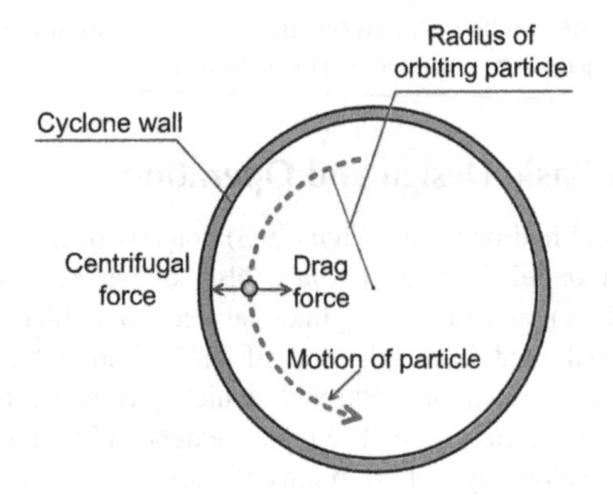

FIGURE 9.4 Forces acting on an orbiting particle in the hydrocyclone.

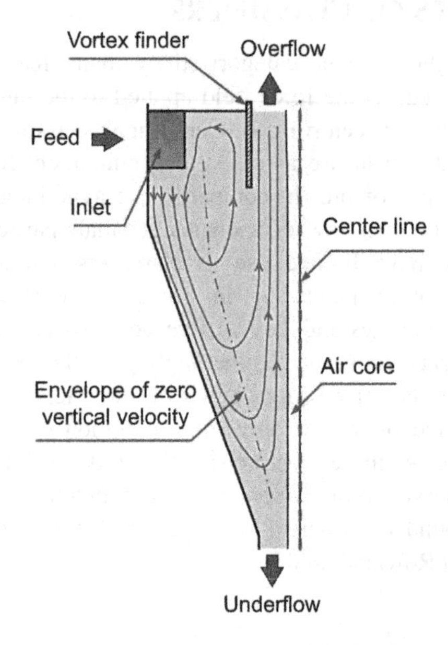

FIGURE 9.5 Distribution of the vertical components of velocity in a hydrocyclone.

toward the zone of low pressure along the axis and are carried upward through the vortex finder to the overflow.

The existence of an outer region of downward flow and an inner region of upward flow implies a position at which there is no vertical velocity. This applies throughout the greater part of the cyclone body, and an envelope of zero vertical velocity should exist throughout the body of the cyclone (Figure 9.5). Particles thrown outside the envelope of zero vertical velocity by the greater centrifugal force exit via the underflow, while particles swept to the center by the greater drag force leave in the overflow. Particles lying on the envelope of zero velocity are acted upon by equal centrifugal and drag forces and have an equal chance of reporting either to the underflow or overflow. This concept should be remembered when considering the cut-point described later.

Experimental work by Renner and Cohen (1978) has shown that classification does not take place throughout the whole body of the cyclone. Using a high-speed probe, samples were taken from several selected positions within a 150-mm diameter cyclone and were subjected to size analysis. The results showed that the interior of the cyclone may be divided into four regions that contain distinctively different size distributions (Figure 9.6).

Essentially unclassified feed exists in a narrow region, A, adjacent to the cylinder wall and roof of the cyclone. Region B occupies a large part of the cone of the cyclone and contains fully classified coarse material, that is, the size distribution is practically uniform and resembles that of the underflow (coarse) product. Similarly, fully classified fine material is contained in region C, a narrow

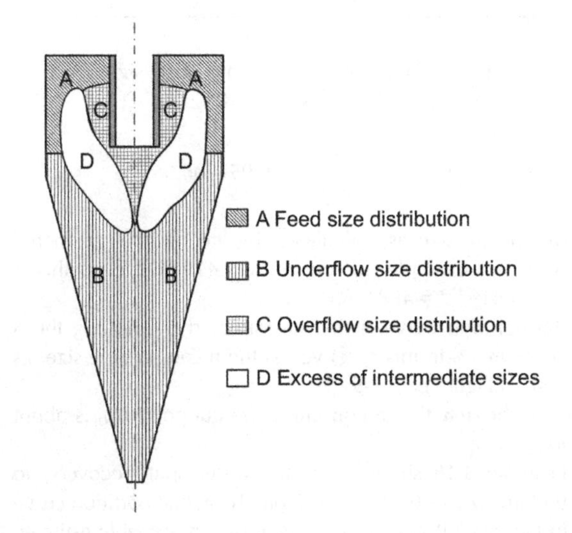

FIGURE 9.6 Regions of similar size distribution within cyclone (Renner and Cohen, 1978).

region surrounding the vortex finder and extending below the latter along the cyclone axis. Only in the toroid-shaped region D does classification appear to be taking place. Across this region, size fractions are radially distributed, so that decreasing sizes show maxima at decreasing radial distances from the axis. These results, however, were taken with a cyclone running at low pressure, so the region D may be larger in production units.

9.4.2 Characterization of Cyclone Efficiency

The Partition Curve

The most common method of representing classifier efficiency is by a *partition curve* (also known as a *performance, efficiency*, or *selectivity* curve). The curve relates the weight fraction of each particle size in the feed which reports to the underflow to the particle size. It is the same as the partition curve introduced for screening (Chapter 8). The *cut-point*, or separation size, is defined as the size for which 50% of the particles in the feed report to the underflow, that is, particles of this size have an equal chance of going either with the overflow or underflow (Svarovsky and Thew, 1992). This point is usually referred to as the d_{50} size.

It is observed in constructing the partition curve for a cyclone that the partition value does not appear to approach zero as particle size reduces, but rather approaches some constant value. To explain, Kelsall (1953) suggested that solids of all sizes are entrained in the coarse product liquid, bypassing classification in direct proportion to the fraction of feed water reporting to the underflow. For example, if the feed contains 16 t h^{-1} of material of a certain size, and 12 t h^{-1} reports to the underflow, then the percentage of this size reporting to the underflow, and plotted on the partition curve, is 75%. However, if, say, 25% of the feed water reports to the underflow, then 25% of the feed material will short-circuit with it; therefore, 4 t h^{-1} of the size fraction will short-circuit to the underflow, and only 8 t h^{-1} leave in the underflow due to classification. The corrected recovery of the size fraction is thus:

$$100 \times \frac{12-4}{16-4} = 67\%$$

The uncorrected, or actual, partition curve can therefore be corrected by utilizing Eq. (9.16):

$$C = \frac{S - R_{w/u}}{1 - R_{w/u}} \tag{9.16}$$

where C is the corrected mass fraction of a particular size reporting to underflow, S the actual mass fraction of a particular size reporting to the underflow, and $R_{w/u}$ the fraction of the feed liquid that is recovered in the coarse product (underflow) stream, which defines the *bypass fraction*. The corrected curve thus describes particles recovered to the underflow by true classification and introduces a corrected cut-size, d_{50c}. Kelsall's assumption has been questioned, and Flintoff et al. (1987) reviewed some of the arguments. However, the Kelsall correction has the advantages of simplicity, utility, and familiarity through long use.

The construction of the *actual* and *corrected* partition curves can be illustrated by means of an example. The calculations are easily performed in a spreadsheet (Example 9.2).

Modern literature, including manufacturer's data, usually quotes the corrected d_{50} with the subscript "c" dropped. Some care in reading the literature is therefore required. To avoid confusion, the designation d_{50c} will be retained.

Although not demonstrated here, the partition curve is often plotted against d/d_{50}, that is, making the size axis nondimensional—a reduced or normalized size—giving rise to a *reduced partition* curve.

The exponential form of the curves has led to several fitting models (Napier-Munn et al., 1996). One of the most common (Plitt, 1976) is:

$$C = 1 - \exp(-K\, d^m) \tag{9.17}$$

where K is a constant, d the mean particle size, and m the "sharpness" of separation. By introducing $C = 0.5$ at $d = d_{50c}$ and rearranging Eq. (9.17) gives K:

$$K = \frac{\ln 0.5}{d_{50c}^m} = \frac{0.693}{d_{50c}^m}$$

thus:

$$C = 1 - \exp\left[-0.693\left(\frac{d}{d_{50c}}\right)^m\right] \tag{9.18a}$$

Example 9.2

a. Determine the actual partition curve and actual d_{50} for the cyclone data in Example 3.15.

b. Determine the corrected partition curve and corrected d_{50c}.

As a reminder, the grinding circuit in Example 3.15 is

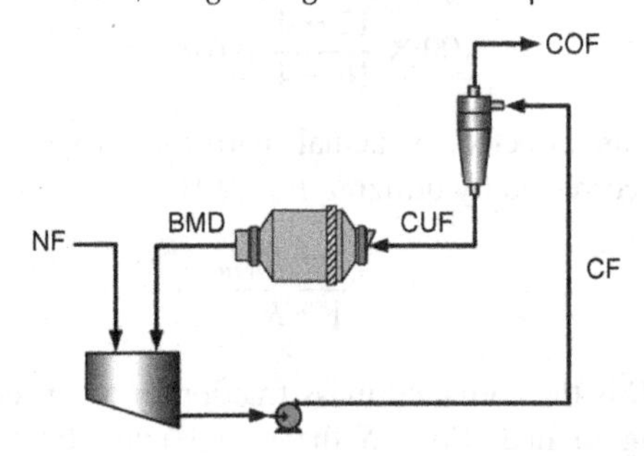

Solution

a. Using the same symbolism as in Example 3.15 then the partition value S is given by:

$$S = \frac{U u}{F f}$$

From the size distribution data using the generalized least squares minimization procedure, the solids split (recovery) to underflow (U/F) is 0.619. From this, data reconciliation was executed and the adjusted size distribution data are given in Table ex 9.2 along with the actual partition (S) and the corrected partition (C) values.

Specimen calculation of S and mean particle size:
For the $-592+419$ μm size class:

$$S = 0.619 \times \frac{7.51}{4.86} = 0.957 \text{ or } 95.7\%$$

The mean size is calculated by taking the geometric mean; for example, for the $-592+419$ size class this is $(592 \times 419)^{0.5} = 498$ μm.

The partition curve is then constructed by plotting the S values (as % in this case) versus the mean particle size, as done in Figure ex 9.2-S.

From the actual partition curve the cut-point d_{50} is about 90 μm.

b. Example 3.18 showed that the water split (recovery) to underflow was 0.32 (32%). From the actual partition curve in Figure ex 9.2-S this appears to be a reasonable estimate of the bypass fraction. Using this bypass value the corrected partition C values are computed and the corrected partition curve constructed which is included in the figure.

Specimen calculation:
For the $-209+148$ μm size class:

$$C = \frac{0.743 - 0.32}{1 - 0.32} = 0.622 \quad \text{or } 62.2\%$$

From the corrected partition curve the d_{50c} is about 120 μm.

TABLE EX 9.2 Adjusted Size Distribution Data and Partition Values

Size Interval (μm)	Mean Size (μm)	CF	COF	CUF	Partition, S	Corrected, C
+592		8.27	0.23	13.21	98.9	98.3
−592 + 419	498	4.86	0.54	7.51	95.7	93.6
−419 + 296	352	6.77	0.94	10.35	94.6	92.1
−296 + 209	249	7.04	2.81	9.64	84.8	77.6
−209 + 148	176	11.38	7.67	13.66	74.3	62.2
−148 + 105	125	13.91	13.21	14.35	63.9	46.9
−105 + 74	88	11.09	11.7	10.71	59.8	40.9
−74 + 53	63	9.86	14.78	6.83	42.9	16.0
−53 + 37	44	5.57	9.33	3.26	36.2	6.2
−37		21.25	38.79	10.48	30.5	−2.2
		100	100	100		

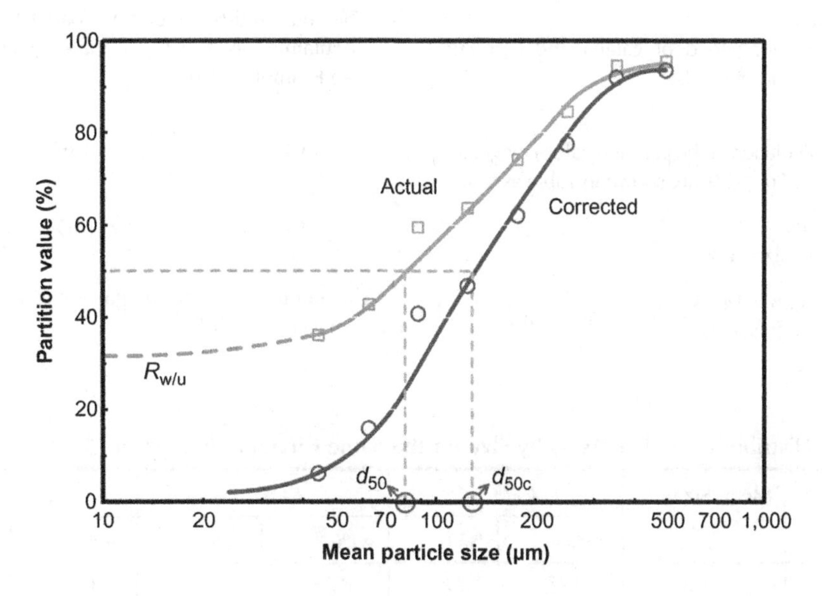

FIGURE EX 9.2-S Actual and corrected partition curves for data in the table.

and:

$$S = C(1 - R_{w/u}) + R_{w/u} \tag{9.18b}$$

The normalized size is evident in Eq. (9.18a). By fitting data to Eq. (9.18a), the two unknowns, d_{50c} and m, can be estimated. For instance, for Example 9.2 the fitted values (rounded) are: $d_{50c} = 125$ μm and $m = 1.7$.

Sharpness of Cut

The value of m in Eq. (9.18a) is one measure of the sharpness of the separation. The value is determined by the slope of the central section of the partition curve; the closer the slope is to vertical, the higher the value of m and the greater the classification efficiency. Perfect classification would give $m = \infty$; but in reality m values are rarely above 3. The slope of the curve can be also expressed by taking the points at which 75% and 25% of the feed particles report to the underflow. These are the d_{75} and d_{25} sizes, respectively. The sharpness of separation, or the so-called *imperfection*, I, is then given by (where the actual or corrected d_{50} may be used):

$$I = \frac{d_{75} - d_{25}}{2d_{50}} \tag{9.19}$$

Multidensity Feeds

Inspection of the partition curve in Example 9.2 (Figure ex 9.2-S) shows a deviation at about 100 μm which we chose to ignore by passing a smooth curve through the data, implying, perhaps, some experimental uncertainty. In fact, the deviation is real and is created by classifying feeds with minerals of different specific gravity. In this case the feed is a Pb–Zn ore with galena (s.g. 7.5), sphalerite (4.0), pyrite (5.0), and a calcite/dolomite non-sulfide gangue, NSG (2.85). The individual mineral classification curves can be generated from the metal assay on each size fraction using the following:

$$S_{i,M} = \frac{Uu_i u_{i,M}}{Ff_i f_{i,M}} \tag{9.20}$$

where subscript i refers to size class i (which we did not need to specify before) and subscript i,M refers to metal (or mineral) assay M in size class i. The calculations are illustrated using an example (Example 9.3).

From Example 9.3 it is evident that the high-density galena preferentially reports to the underflow. The minerals, in fact, classify according to their density. Treating the elemental assay curves to be analogous with their associated minerals has the implication that they are liberated, free minerals. This is not entirely true, after all, the purpose of the grinding circuit is to liberate, so in the cyclone feed liberation is unlikely to be complete. It is better to refer to mineral-by-size curves, which does not imply liberated minerals. The analysis can be extended to determine the corrected partition curves and the corresponding m and $d_{50c,M}$ values, which we can then use as a first approximation for the associated minerals. In doing so we find the following values: $m = 2.2$ for all minerals, and $d_{50c,M}$ equals 37 μm for galena, 66 μm for pyrite; 89 μm for sphalerite, and 201 μm for NSG. The m value

Example 9.3

a. Determine the actual partition values for Pb (galena) from the data in the table.

b. Determine the circulating load of galena and compare with the circulating load of solids.

Solution

a. The table and the calculation of the partition values for galena ($S_{i,\mathrm{Pb}}$) calculated using Eq. (9.20) are shown in Table ex 9.3.

Specimen calculation:
For the $-53+37$ μm size class:

$$S_{i,\mathrm{Pb}} = 0.619 \times \frac{3.26 \times 14.64}{5.57 \times 6.77} = 0.783 \text{ or } 78.3\%$$

The actual partition curves for all minerals are given in Figure ex 9.3-S.

b. The total solids circulating load CL_{Tot} and total galena circulating load $CL_{Tot,Pb}$ are given by, respectively (see Example 3.15):

$$CL_{Tot} = \frac{U}{O} = \frac{0.619}{0.381} = 1.63$$

$$CL_{Tot,Pb} = \frac{U\,u_{Tot,Pb}}{O\,o_{Tot,Pb}} = 1.63 \times \frac{5.91}{1.66} = 5.78$$

The circulating load of galena is thus 3.56 (5.78/1.63) times that of the solids.

TABLE EX 9.3 Size Distribution and Pb Assay by Size for the Same Circuit as in Example 9.2

Size Interval (μm)	Mean Size	CF		COF		CUF		$S_{i,\mathrm{Pb}}$
		wt%	%Pb	wt%	%Pb	wt%	%Pb	
+592		8.27	1.33	0.23	1.41	13.21	1.32	98.1
−592 + 419	498	4.86	1.02	0.54	1.3	7.51	1	93.8
−419 + 296	352	6.77	1.73	0.94	−0.16	10.35	1.84	100.7
−296 + 209	249	7.04	2.55	2.81	−0.12	9.64	3.02	100.4
−209 + 148	176	11.38	3.19	7.67	0.25	13.66	4.21	98.1
−148 + 105	125	13.91	3.83	13.21	0.22	14.35	5.87	97.9
−105 + 74	88	11.09	8.14	11.7	0.35	10.71	13.36	98.1
−74 + 53	63	9.86	9.21	14.78	1.02	6.83	20.07	93.4
−53 + 37	44	5.57	6.77	9.33	2.29	3.26	14.64	78.3
−37		21.25	3.55	38.79	3.09	10.48	4.59	39.5
Head		100	4.29	100	1.66	100	5.91	85.3

(Note, the "head" row gives the total Pb assay for the stream; and that data reconciliation was not constrained to zero on the Pb assay giving a couple of small negative values in the COF column and consequently $S_{i,\mathrm{Pb}}$ values slightly larger than 100%.)

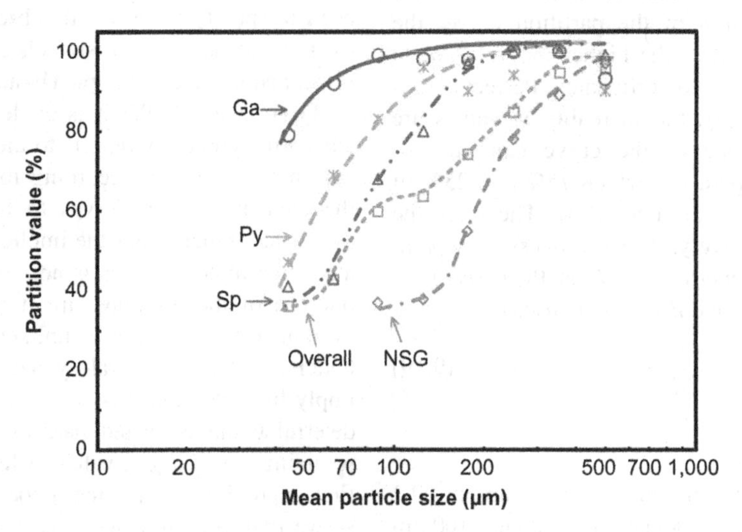

FIGURE EX 9.3-S Actual partition curves for minerals and total solids (overall).

shows the sharpness of separation is higher for the minerals than for the total solids ($m = 1.7$, Example 9.2). By inspection, the $d_{50c,M}$ values correspond well with the hindered settling expression, from Eq. (9.15). For example, the ratio $d_{50c,NSG}$: $d_{50c,Ga}$ is 5.4, which is close to the value calculated for the hindered settling of essentially this pair of minerals. It is this density effect, concentrating heavy minerals in the circulating stream, which gives the opportunity for recovery from this stream (or other streams inside the circuit), using gravity separation devices or flash flotation.

Unusual Partition Curves

"Unusual" refers to deviations from the simple exponential form, which can range from inflections to the curve displaying a minimum and maximum (Laplante and Finch, 1984; Kawatra and Eisele, 2006). The partition curve in Example 9.2 illustrates the phenomena: the inflection at about 100 μm was initially ignored in favor of a smooth curve, suggesting, perhaps, experimental error. Example 9.3 shows that the inflection in the total solids curve is because it is the sum of the mineral partition curves (weighted for the mineral content); below 100 μm the NSG is no longer being classified and the solids curve shifts over to reflect the continuing classification of the denser minerals. The "unusual" nature of the curve is even more evident in series cyclones where the downstream (secondary) cyclone receives overflow from the primary cyclone, which now comprises fine dense mineral particles and coarse light mineral particles.

The impact of the component minerals is also evident in the m values: that for the mineral curves is higher than for the solids, and the correspondence to the exponential model, Eq. (9.18a), is better for the minerals than for the solids. It should be remembered when determining m that a low value may not indicate performance that needs improving, as the component minerals may be comparatively quite sharply classified.

Apart from density, the shape of the particles in the feed is also a factor in separation. Flakey particles such as mica often reporting to the overflow, even though they may be relatively coarse. Wet classification of asbestos feeds can reveal similar unusual overall partition curves as for density, this time the result of the shape: fibrous asbestos minerals, and the granular rock particles.

Another "unusual" feature of the partition curve sometimes reported that is not connected to a density (or shape) effect is a tendency for the curve to bend up toward the fine end (Del Villar and Finch, 1992). Called a "fish-hook" by virtue of its shape, it is well recognized in mechanical air classifiers (Austin et al., 1984). Connected to the bypass, the notion is that the fraction reporting to the underflow is size-dependent, that the finer the particles the closer they split in the same proportion as the water, thus bending the partition curve upward (the fish-hook) at the finer particle sizes to intercept the Y-axis at $R_{w/u}$. Whether the fish-hook is real or not continues to attract a surprising amount of literature (Bourgeois and Majumder, 2013; Nageswararao, 2014).

Cyclone Overflow Size Distribution

Although partition curves are useful in assessing and modeling classifier performance, the minerals engineer is usually more interested in knowing fineness of grind (i.e., cyclone overflow particle size distribution) than the cyclone cut-size. Simple relationships between fineness of grind and the partition curve of a hydrocyclone have been developed by Arterburn (1982) and Kawatra and Seitz (1985).

Figure 9.7(a) shows the evolution of size distribution of the solids from the feed to product streams for the

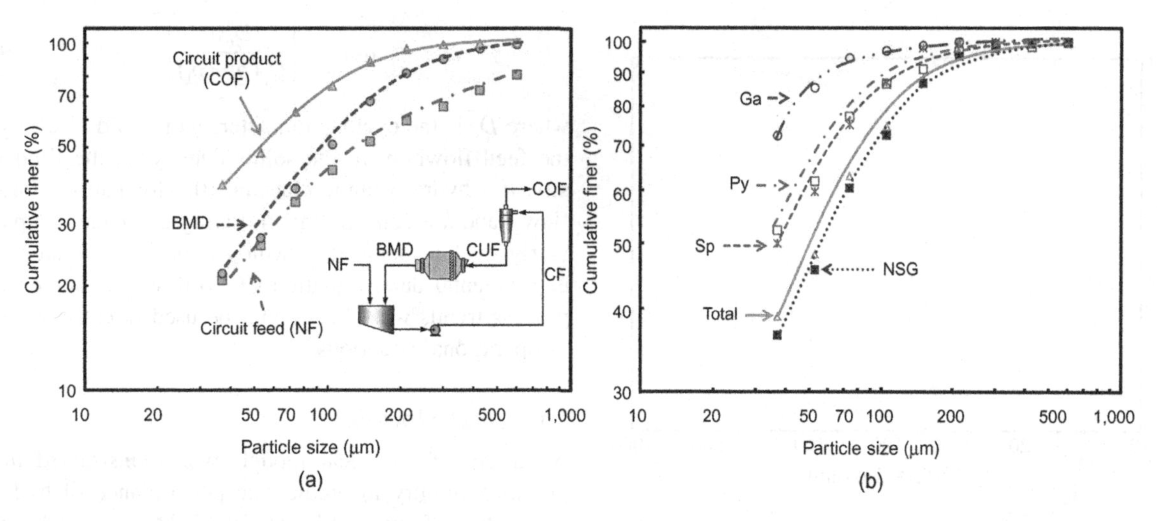

FIGURE 9.7 (a) Evolution of size distribution through closed grinding circuit (Example 9.2), and (b) size distribution of minerals in the overflow product from the same circuit.

grinding circuit in Example 9.2; and Figure 9.7(b) shows the size distribution of the solids and the minerals in the cyclone overflow for the same circuit. The latter shows the much finer size distribution of the galena ($P_{80,Pb} \sim 45\ \mu m$) compared to the solids ($P_{80} \sim 120\ \mu m$) and the NSG ($P_{80,NSG} \sim 130\ \mu m$) due to the high circulating load, and thus additional grinding, of the galena. It is this observation which gives rise to the argument introduced in Chapter 7 that the grinding circuit treating a high-density mineral component can be operated at a coarse solids P_{80}, as the high-density mineral will be automatically ground finer, and the grind instead could be made the P_{80} of the target mineral for recovery.

This density effect in cyclones (or any classification device) is often taken to be a reason to consider screening which is density independent. This is a good point at which to compare cyclones with screens.

9.4.3 Hydrocyclones Versus Screens

Hydrocyclones have come to dominate classification when dealing with fine particle sizes in closed grinding circuits (<200 μm). However, recent developments in screen technology (Chapter 8) have renewed interest in using screens in grinding circuits. Screens separate on the basis of size and are not directly influenced by the density spread in the feed minerals. This can be an advantage. Screens also do not have a bypass fraction, and as Example 9.2 has shown, bypass can be quite large (over 30% in that case). Figure 9.8 shows an example of the difference in partition curve for cyclones and screens. The data is from the El Brocal concentrator in Peru with evaluations before and after the hydrocyclones were replaced with a Derrick Stack Sizer® (see Chapter 8) in the grinding circuit (Dündar et al., 2014). Consistent with expectation, compared to the cyclone the screen had a sharper separation (slope of curve is higher) and little bypass. An increase in grinding circuit capacity was reported due to higher breakage rates after implementing the screen. This was attributed to the elimination of the bypass, reducing the amount of fine material sent back to the grinding mills which tends to cushion particle–particle impacts.

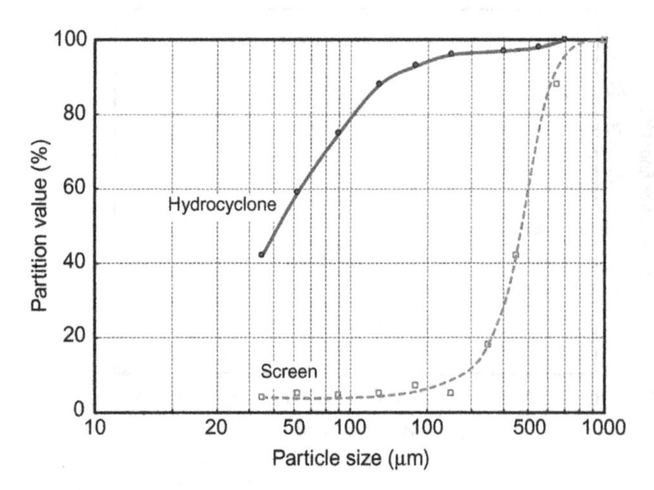

FIGURE 9.8 Partition curves for cyclones and screens in the grinding circuit at El Brocal concentrator. *(Adapted from Dündar et al. (2014))*

Changeover is not one way, however: a recent example is a switch from screen to cyclone, to take advantage of the additional size reduction of the denser payminerals (Sasseville, 2015).

9.4.4 Mathematical Models of Hydrocyclone Performance

A variety of hydrocyclone models have been proposed to estimate the key relationships between operating and geometrical variables for use in design and optimization, with some success. These include empirical models calibrated against experimental data, as well as semi-empirical models based on equilibrium orbit theory, residence time, and turbulent flow theory. Progress is also being made in using computational fluid dynamics to model hydrocyclones from first principles (e.g., Brennan et al., 2003; Nowakowski et al., 2004; Narasimha et al., 2005). All the models commonly used in practice are still essentially empirical in nature.

Bradley model

Bradley's seminal book (1965) listed eight equations for the cut-size, and the number has increased significantly since then. Bradley's equation based on the equilibrium orbit hypothesis (Figure 9.4) was:

$$d_{50} = k\left[\frac{D_c^3\eta}{Q_f(\rho_s - \rho_l)}\right]^n \tag{9.21}$$

where D_c is the cyclone diameter, η the fluid viscosity, Q_f the feed flowrate, ρ_s the solids density, ρ_l the fluid density, n a hydrodynamic constant (0.5 for particle laminar flow), and k a constant incorporating other factors, particularly cyclone geometry, which must be estimated from experimental data. Equation (9.21) describes some of the process trends well, but cannot be used directly in design or operational situations.

Empirical Models

A variety of empirical models were constructed in the previous century to predict the performance of hydrocyclones (e.g., Leith and Licht, 1972). More recently, models have been proposed by, among others, Nageswararao et al. (2004) and Kraipech et al. (2006). The most widely used empirical models are probably those of Plitt (1976) and its later modified form (Flintoff et al., 1987), and Nageswararao (1995). These models, based on a phenomenological description of the process with numerical constants determined from large databases, are described in Napier-Munn et al. (1996) and were reviewed and compared by Nageswararao et al. (2004).

Plitt's modified model for the corrected cut-size d_{50c} in micrometers is:

$$d_{50c} = \frac{F_1 39.7 D_c^{0.46} D_i^{0.6} D_o^{1.21} \eta^{0.5} \exp(0.06 C_v)}{D_u^{0.71} h^{0.38} Q_f^{0.45} \left(\frac{\rho_s - 1}{1.6}\right)^k} \quad (9.22)$$

where D_c, D_i, D_o, and D_u are the inside diameters of the cyclone, inlet, vortex finder, and apex, respectively (cm), η the liquid viscosity (cP), C_v the feed solids volume concentration (%), h the distance between apex and end of vortex finder (cm), k a hydrodynamic exponent to be estimated from data (default value for laminar flow 0.5), Q_f the feed flowrate (l min^{-1}), and ρ_s the solids density (g cm^{-3}). Note that for noncircular inlets, $D_i = \sqrt{4A/\pi}$ where A is the cross-sectional area of the inlet (cm^2).

The equation for the volumetric flowrate of slurry to the cyclone Q_f is:

$$Q_f = \frac{F_2 P^{0.56} D_c^{0.21} D_i^{0.53} h^{0.16} (D_u^2 + D_o^2)^{0.49}}{\exp(0.0031 C_v)} \quad (9.23)$$

where P is the pressure drop across the cyclone in kilopascals (1 psi = 6.895 kPa). The F_1 and F_2 in Eqs. (9.22) and (9.23) are material-specific constants that must be determined from tests with the feed material concerned. Plitt also reports equations for the flow split between underflow and overflow, and for the sharpness of separation parameter m in the corrected partition curve.

Nageswararao's model includes correlations for corrected cut-size, pressure-flowrate, and flow split, though not sharpness of separation. It also requires the estimate of feed-specific constants from data, though first approximations can be obtained from libraries of previous case studies. This requirement for feed-specific calibration emphasizes the important effect that feed conditions have on hydrocyclone performance.

Asomah and Napier-Munn (1997) reported an empirical model that incorporates the angle of inclination of the cyclone, as well as explicitly the slurry viscosity, but this has not yet been validated in the large-scale use which has been enjoyed by the Plitt and Nageswararao models.

A useful general approximation for the flowrate in a hydrocyclone is

$$Q \approx 9.5 \times 10^{-3} D_c^2 \sqrt{P} \quad (9.24)$$

Flowrate and pressure drop together define the useful work done in the cyclone:

$$\text{Power} = \frac{PQ}{3,600} \text{kW} \quad (9.25)$$

where Q is the flowrate (m^3 h^{-1}); P the pressure drop (kPa); and D_c the cyclone diameter (cm). The power can be used as a first approximation to size the pump motor, making allowances for head losses and pump efficiency.

These models are easy to incorporate in spreadsheets, and so are particularly useful in process design and optimization using dedicated computer simulators such as JKSimMet (Napier-Munn et al., 1996) and MODSIM (King, 2012), or the flowsheet simulator Limn (Hand and Wiseman, 2010). They can also be used as a virtual instrument or "soft sensor" (Morrison and Freeman, 1990; Smith and Swartz, 1999), inferring cyclone product size from geometry and operating variables as an alternative to using an online particle size analyzer (Chapter 4).

Computational Models

With advances in computer hardware and software, considerable progress has been made in the fundamental modeling of hydrocyclones. The multiphase flow within a hydrocyclone consists of solid particles, which are dispersed throughout the fluid, generally water. In addition, an air core is present. Such multiphase flows can be studied using a combination of Computational Fluid Dynamics (CFD) and Discrete Element Method (DEM) techniques (see Chapter 17).

The correct choice of the turbulence, multiphase (air–water interaction), particle drag, and contact models are essential for the successful modeling of the hydrocyclone. The highly turbulent swirling flows, along with the complexity of the air core and the relatively high feed percent solids, incur large computational effort. Choices are therefore made based on a combination of accuracy and computational expense. Studies have included effects of short-circuiting flow, motion of different particle sizes, the surging phenomena, and the "fish-hook" effect as well as particle–particle, particle–fluid, and particle–wall interactions in both dense medium cyclones (Chapter 11) and hydrocyclones.

The data required to validate numerical models can be obtained through methods such as particle image velocimetry and laser Doppler velocimetry. These track the average velocity distributions of particles, as opposed to Lagrangian tracking, where individual particles of a dispersed phase can be tracked in space and time. Wang et al. (2008) used a high-speed camera to record the motion of a particle with density 1,140 kg m^{-3} in water in a hydrocyclone. The two-dimensional particle paths obtained showed that local or instantaneous instabilities in the flow field can have a major effect on the particle trajectory, and hence on the separation performance of the hydrocyclone. Lagrangian tracking includes the use of positron emission particle tracking (PEPT), which is a more recent development in process engineering. PEPT locates a point-like positron emitter by cross-triangulation (see Chapter 17). Chang et al. (2011) studied the flow of a particle through a hydrocyclone using PEPT with an ^{18}F radioactive tracer and could track the particle in the cyclone with an accuracy of 0.2 mm ms^{-1}.

9.4.5 Operating and Geometric Factors Affecting Cyclone Performance

The empirical models and scale-up correlations, tempered by experience, are helpful in summarizing the effects of operating and design variables on cyclone performance. The following process trends generally hold true:

Cut-size (inversely related to solids recovery)

- Increases with cyclone diameter
- Increases with feed solids concentration and/or viscosity
- Decreases with flowrate
- Increases with small apex or large vortex finder
- Increases with cyclone inclination to vertical

Classification efficiency

- Increases with correct cyclone size selection
- Decreases with feed solids concentration and/or viscosity
- Increases by limiting water to underflow
- Increases with certain geometries

Flow split of water to underflow

- Increases with larger apex or smaller vortex finder
- Decreases with flowrate
- Decreases with inclined cyclones (especially low pressure)
- Increases with feed solids concentration and/or viscosity

Flowrate

- Increases with pressure
- Increases with cyclone diameter
- Decreases (at a given pressure) with feed solids concentration and/or viscosity

Since the operating variables have an important effect on the cyclone performance, it is necessary to avoid fluctuations, such as in flowrate and pressure drop, during operation. Pump surging should be eliminated either by automatic control of level in the sump, or by a self-regulating sump, and adequate surge capacity should be installed to eliminate flowrate fluctuations.

The feed flowrate and the pressure drop across the cyclone are related (Eq. (9.22)). The pressure drop is required to enable design of the pumping system for a given capacity or to determine the capacity for a given installation. Usually the pressure drop is determined from a feed-pressure gauge located on the inlet line some distance upstream from the cyclone. Within limits, an increase in feed flowrate will improve fine particle classification efficiency by increasing the centrifugal force on the particles. All other variables being constant, this can only be achieved by an increase in pressure and a corresponding increase in power, since this is directly related to the product of pressure drop and capacity. Since increase in feed rate, or pressure drop, increases the centrifugal force, finer particles are carried to the underflow, and d_{50} is decreased, but the change has to be large to have a significant effect. Figure 9.9 shows the effect of pressure on the capacity and cut-point of cyclones.

The effect of increase in feed pulp density is complex, as the effective pulp viscosity and degree of hindered settling is increased within the cyclone. The sharpness of the separation decreases with increasing pulp density and the cut-point rises due to the greater resistance to the swirling motion within the cyclone, which reduces the effective pressure drop. Separation at finer sizes can only be achieved with feeds of low solids content and large pressure drop. Normally, the feed concentration is no greater than about 30% solids by weight, but for closed-circuit grinding operations, where relatively coarse separations are often required, high feed concentrations of up to 60% solids by weight are often used, combined with low-pressure drops, often less than 10 psi (68.9 kPa). Figure 9.10 shows that feed concentration has an important effect on the cut-size at high pulp densities.

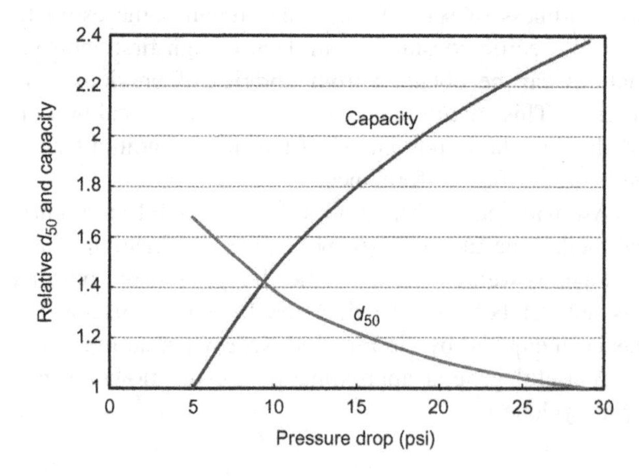

FIGURE 9.9 Effect of pressure on capacity and cut-point of hydrocyclone.

In practice, the cut-point is mainly controlled by the cyclone design variables, such as those of the inlet, vortex-finder, and apex openings, and most cyclones are designed such that these are easily changed.

The area of the inlet determines the entrance velocity and an increase in area increases the flowrate. The geometry of the feed inlet is also important. In most cyclones the shape of the entry is developed from circular cross section to rectangular cross section at the entrance to the cylindrical section of the cyclone. This helps to "spread" the flow along the wall of the chamber. The inlet is normally tangential, but involuted feed entries are also common (Figure 9.11). Involuted entries are said to minimize turbulence and reduce wear. Such design differences are reflected in proprietary cyclone developments such as Weir Warman's CAVEX® and Krebs' gMAX® units.

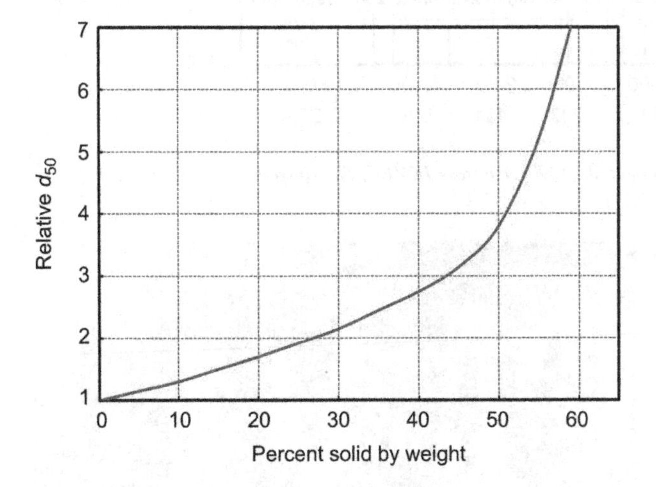

FIGURE 9.10 Effect of solids concentration on cut-point of hydrocyclone.

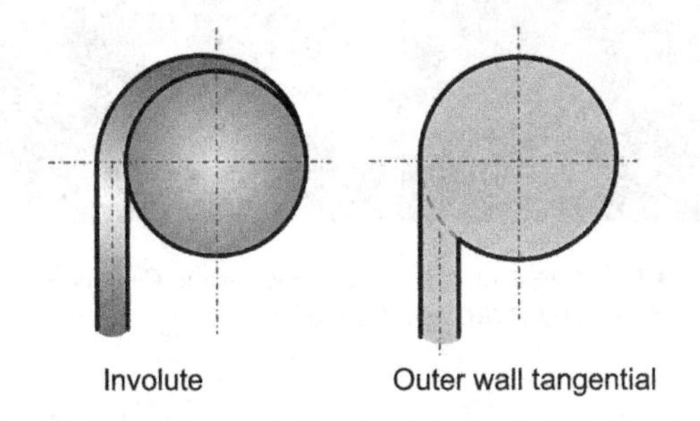

FIGURE 9.11 Involute and tangential feed entries.

The diameter of the vortex finder is an important variable. At a given pressure drop across the cyclone, an increase in the diameter of the vortex finder will result in a coarser cut-point and an increase in capacity.

The size of the apex opening determines the underflow density and must be large enough to discharge the coarse solids that are being separated by the cyclone. The orifice must also permit the entry of air along the axis of the cyclone to establish the air vortex. Cyclones should be operated at the highest possible underflow density, since unclassified material (the bypass fraction) leaves the underflow in proportion to the fraction of feed water leaving via the underflow. Under correct operating conditions, the discharge should form a hollow cone spray with a 20–30° included angle (Figure 9.12(a)). Air can then enter the cyclone, the classified coarse particles will discharge freely, and solids concentrations greater than 50% by weight can be achieved. Too small apex opening can lead to the condition known as "roping" (Figure 9.12(b)), where an extremely thick pulp stream of the same diameter as the apex is formed, and the air vortex may be lost, the separation efficiency will fall, and oversize material will discharge through the vortex finder. (This condition is sometimes encouraged where a very high underflow solids concentration is required, but is otherwise deleterious: the impact of this "tramp" oversize on downstream flotation operations can be quite dramatic (Bahar et al., 2014)). Too large an apex orifice results in the larger hollow cone pattern seen in Figure 9.12(c). The underflow will be excessively dilute and the additional water will carry unclassified fine solids that would otherwise report to the

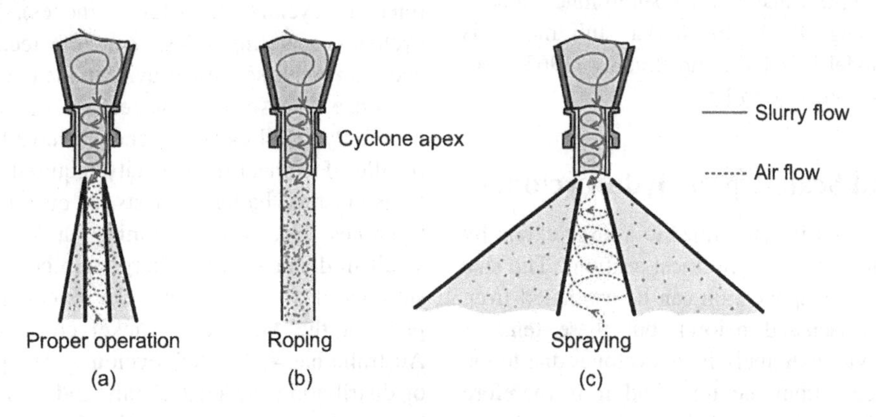

FIGURE 9.12 Nature of underflow discharge: (a) correct apex size—proper operation, (b) apex too small—"roping" leading to loss of air core, and (c) apex too large—"spraying" which leads to lower sharpness of separation.

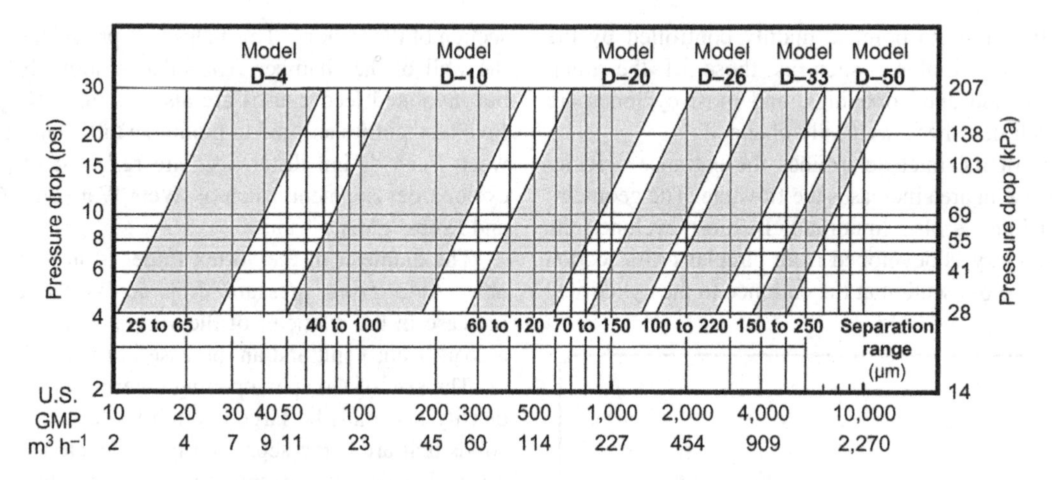

FIGURE 9.13 Hydrocyclone performance chart (Krebs) *(From Napier-Munn et al. (1996); courtesy JKMRC, The University of Queensland).*

overflow. The state of operation of a cyclone is important to optimize both grinding efficiency and downstream separation processes, and online sensors are now being incorporated to identify malfunctioning cyclones (Cirulis and Russell, 2011; Westendorf et al., 2015).

Some investigators have concluded that the cyclone diameter has no effect on the cut-point and that for geometrically similar cyclones the efficiency curve is a function only of the feed material characteristics. The inlet and outlet diameters are the critical design variables, the cyclone diameter merely being the size of the housing required to accommodate these apertures (Lynch et al., 1974, 1975; Rao et al., 1976). This is true where the inlet and outlet diameters are essentially proxies for cyclone diameter for geometrically similar cyclones. However, from theoretical considerations, it is the cyclone diameter that controls the radius of orbit and thus the centrifugal force acting on the particles. As there is a strong interdependence between the aperture sizes and cyclone diameter, it is difficult to distinguish the true effect, and Plitt (1976) concluded that the cyclone diameter has an independent effect on separation size.

For geometrically similar cyclones at constant flowrate, $d_{50} \propto \text{diameter}^x$, but the value of x is open to much debate. The value of x using the Krebs–Mular–Jull model is 1.875, for Plitt's model it is 1.18, and Bradley (1965) concluded that x varies from 1.36 to 1.52.

9.4.6 Sizing and Scale-Up of Hydrocyclones

In practice, the cut-point is determined to a large extent by the cyclone size (diameter of cylindrical section). The size required for a particular application can be estimated from empirical models (discussed below), but these tend to become unreliable with extremely large cyclones due to the increased turbulence within the unit, and it is therefore more common to choose the required model by referring to manufacturers' charts, which show capacity and separation

FIGURE 9.14 A nest of 150 mm cyclones at the Century Zinc mine, Australia *(Courtesy JKMRC and JKTech Pty Ltd).*

size range in terms of cyclone size. A typical performance chart is shown in Figure 9.13 (where the *D* designation refers to cyclone diameter in inches). This is for Krebs cyclones, operating at less than 30% feed solids by weight, and with solids specific gravity in the range 2.5–3.2.

Since fine separations require small cyclones, which have only small capacity, enough have to be connected in parallel if to meet the capacity required; these are referred to as clusters, batteries, nests, or *cyclopacs* (Figure 9.14). Cyclones used for de-sliming duties are usually very small in diameter, and a large number may be required if substantial flowrates must be handled. The de-sliming plant at the Mr. Keith Nickel concentrator in Western Australia has 4,000 such cyclones. The practical problems of distributing the feed evenly and minimizing blockages have been largely overcome by the use of Mozley cyclone assemblies. A 16 × 51 mm (16 × 2 in.) assembly is shown

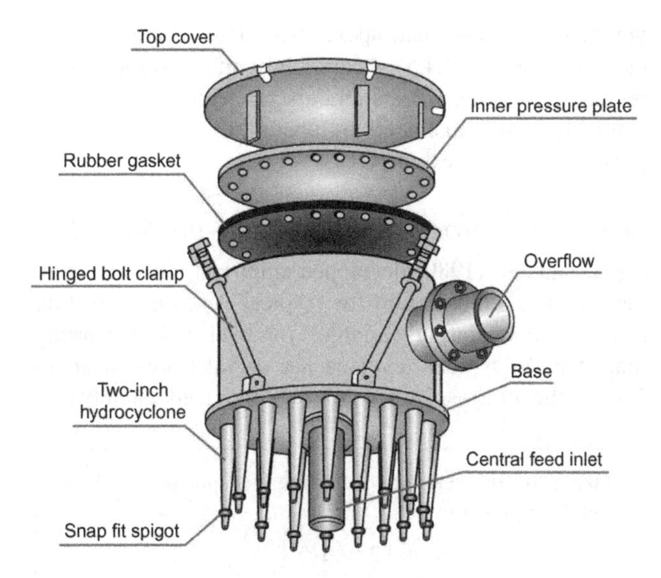

FIGURE 9.15 Mozley 16 × 2-in. cyclone assembly.

FIGURE 9.16 Interior of Mozley 16 × 2-in. cyclone assembly.

in Figure 9.15. The feed is introduced into a housing via a central inlet at pressures of up to 50 psi (344.8 kPa). The housing contains 16 2-in. cyclones, which have interchangeable spigot and vortex finder caps for precise control of cut-size and underflow density. The feed is forced through a trash screen and into each cyclone without the need for separate distributing ports (Figure 9.16). The overflow from each cyclone leaves via the inner pressure plate and leaves the housing through the single overflow pipe on the side. The assembly design reduces maintenance, the removal of the top cover allowing easy access so that individual cyclones can be removed (for repair or replacement) without disconnecting feed or overflow pipework.

Since separation at large particle size requires large diameter cyclones with consequent high capacities, in some cases, where coarse separations are required, cyclones cannot be utilized, as the plant throughput is not high enough. This is often a problem in pilot plants where simulation of the full-size plant cannot be achieved, as scaling down the size of cyclones to allow for smaller capacity also reduces the cut-point produced.

Scale-Up of Hydrocyclones

A preliminary scale-up from a known situation (e.g., a laboratory or pilot plant test) to the unknown (e.g., a full production installation) can be done via the basic relationships between cut-size, cyclone diameter, flowrate, and pressure drop. These are:

$$\frac{d_{50c_2}}{d_{50c_1}} = \left(\frac{D_{c2}}{D_{c1}}\right)^{n_1}\left(\frac{Q_1}{Q_2}\right)^{n_2} = \left(\frac{D_{c2}}{D_{c1}}\right)^{n_3}\left(\frac{P_1}{P_2}\right)^{n_4} \tag{9.26}$$

and

$$\frac{P_1}{P_2} \cong \left(\frac{Q_1}{Q_2}\right)^{n_5}\left(\frac{D_{c2}}{D_{c1}}\right)^{n_6} \tag{9.27}$$

where P is pressure drop, Q the flowrate, D_c the cyclone diameter, the subscripts 1 and 2 indicate the known and scale-up applications respectively, and n_{1-6} are constants which are a function of the flow conditions. The theoretical values (for dilute slurries and laminar flow in small cyclones) are: $n_1 = 1.5$, $n_2 = 0.5$, $n_3 = 0.5$, $n_4 = 0.25$, $n_5 = 2.0$, and $n_6 = 4.0$. The constants to be used in practice will depend on conditions and which model is favored. In particular, at a given flowrate high feed solids concentrations will substantially influence both cut-size (increase) and pressure drop (decrease). There is no general consensus, but in most applications the following values will give more realistic predictions: $n_1 = 1.54$, $n_2 = 0.43$, $n_3 = 0.72$, $n_4 = 0.22$, $n_5 = 2.0$, and $n_6 = 3.76$.

These relationships tell us that the diameter, flowrate, and pressure must be considered together. For example, cut-size cannot be scaled purely on cyclone diameter, as a new diameter will bring either a new flowrate or pressure or both. For example, if it is desired to scale to a larger cyclone at the same cut-size, then $d_{50c_1} = d_{50c_2}$ and $D_{c2} = D_{c1}(P_2/P_1)^{n_4/n_3}$.

Classification efficiency can sometimes be improved by arranging several cyclones in series to retreat overflow, underflow, or both. Svarovsky and Thew (1984) have pointed out that if N cyclones with identical classification curves are arranged in series, each treating the overflow of the previous one, then the overall recovery of size d to the combined coarse product, $R_{d(\mathrm{T})}$, is given by:

$$R_{d(\mathrm{T})} = 1 - (1 - R_d)^N \tag{9.28}$$

where R_d = recovery of size d in one cyclone.

Sizing of Hydrocyclones—Arterburn Technique

Arterburn (1982) published a method, based on the performance of a "typical" Krebs cyclone, which allows for the cyclone size (i.e., diameter) to be estimated for a given application. Using a series of empirical and semi-empirical relationships, the method relates the overflow P_{80} to the cut-point, d_{50c}, of a "base" cyclone, which is related to cyclone size. An example calculation illustrates the procedure (Example 9.4).

The cyclone size is the main choice for preliminary circuit design purposes. Variables, such as diameter of vortex finder, inlet, and apex, also affect separation (discussed in Section 9.4.5). Accordingly, most cyclones have replaceable vortex finders and apexes with different sizes available, and adjustments by operations will be made to provide the final design.

Sizing of Hydrocyclones—Mular–Jull Model

Mular and Jull (1980) developed empirical formulae from the graphical information for "typical" cyclones, relating d_{50c} to the operating variables for cyclones of varying diameter. A "typical" cyclone has an inlet area of about 7% of the cross-sectional area of the feed chamber, a

Example 9.4

Select (a) the size and (b) number of cyclones for a ball mill circuit for the following conditions:

- Target overflow particle size is 80% passing 90 μm (i.e., $P_{80} = 90\ \mu m$).
- The pressure drop across the cyclone is 50 kPa.
- Slurry feed rate to the cyclone is 1,000 $m^3\ h^{-1}$.
- Slurry is 28% by volume of solids specific gravity 3.1 (55% solids by weight).

Solution

a. Size of cyclone (D_c)

Step 1: Estimate the cut-size for the application.

Arterburn gives a relationship between cut-size for the application and the P_{80}:

$$d_{50c}(\text{appl}) = 1.25 \times P_{80}$$

$$d_{50c}(\text{appl}) = 112.5\ \mu m$$

Step 2: Determine a generic or base cut-size, d_{50c} (base) from the application, d_{50c} (appl):

$$d_{50c}(\text{base}) = \frac{d_{50}(\text{appl})}{C_1 \times C_2 \times C_3}$$

where the Cs are correction factors. The relationship changes the specific application to a cyclone operating under base conditions, which is therefore independent of the specific application operating parameters. (Olson and Turner (2002) introduce some additional correction factors reflective of cyclone design.)

The first correction term (C_1) is for the influence of the concentration of solids in the feed slurry:

$$C_1 = \left(\frac{53 - V}{53}\right)^{-1.43}$$

where V is the % solids by volume.

$$C_1 = \left(\frac{53 - 28}{53}\right)^{-1.43}$$

$$C_1 = 2.93$$

The second correction (C_2) is for the influence of pressure drop (ΔP, kPa) across the cyclone, measured by taking the difference between the feed pressure and the overflow pressure:

$$C_2 = 3.27 \cdot (\Delta P)^{-0.28}$$

$$C_2 = 3.27 \times 50^{-0.28}$$

$$C_2 = 1.09$$

The final correction (C_3) is for the effect of solids specific gravity:

$$C_3 = \left(\frac{1.65}{\rho_s - \rho_l}\right)^{0.5}$$

where ρ_s and ρ_l are the specific gravity of solid and liquid (usually taken as water, $\rho_l = 1$). (Arterburn uses Stokes' law for particles of different densities but equal terminal velocities (Eq. (9.12)), with the reference being a quartz particle of specific gravity 2.65 in water.)

$$C_3 = \left(\frac{1.65}{3.1 - 1}\right)^{0.5}$$

$$C_3 = 0.89$$

Thus, the d_{50c}(base) is:

$$d_{50c}(\text{base}) = \frac{112.5}{2.93 \times 1.09 \times 0.89}$$

$$d_{50c}(\text{base}) = 40\ \text{cm}$$

Step 3: Estimate size (diameter) of cyclone from d_{50c} (base) using:

$$D_c = \left(\frac{d_{50c}(\text{base})}{2.84}\right)^{1.51}$$

$$D_c = 54\ \text{cm}$$

The cyclone sizes in Figure 9.13 are in inches; from those shown the closest, but larger, selection is 26 in. (D-26). From the pressure drop-capacity chart (Figure 9.13) a D-26 unit is suited to the target separation as the d_{50c} (appl) range is 70–150 μm.

b. Number of cyclones

From Figure 9.13 a D-26 cyclone can treat about 300 $m^3\ h^{-1}$. Thus to handle the target volumetric flowrate of 1,000 $m^3\ h^{-1}$ we need (1,000/300), that is, four cyclones.

vortex finder of diameter 35–40% of the cyclone diameter, and an apex diameter normally not less than 25% of the vortex-finder diameter.

The equation for the cyclone cut-point is:

$$d_{50c} = \frac{0.77 D_c^{1.875} \exp(-0.301 + 0.0945V - 0.00356V^2 + 0.0000684V^3)}{Q^{0.6}(S-1)^{0.5}} \quad (9.29)$$

Equations such as these have been used in computer-controlled grinding circuits to infer cut-points from measured data, but their use in this respect is declining with the increased use of online particle size monitors (Chapter 4). Their value, however, remains in the design and optimization of circuits by the use of computer simulation, which greatly reduces the cost of assessing circuit options.

Sizing of Hydrocyclones—Simulation Packages

A complement to the above methods of sizing cyclones is to use a simulation package such as those mentioned earlier (JKSimMet, MODSIM, or Limn). These packages incorporate empirical cyclone models such as those by Plitt and Nageswararao and can be used for optimizing processing circuits incorporating cyclones (Morrison and Morrell, 1998).

9.5 GRAVITATIONAL CLASSIFIERS

Gravitational classifiers are best suited for coarser classification and are often used as dewatering and washing equipment. They are simple to operate and have low energy requirements, but capital outlay is relatively high compared to cyclones. Gravitational classifiers can be further categorized into two broad groups, depending on the direction of flow of the carrying current: if the fluid movement is horizontal and forms an angle with the particle trajectory, the classification is called *sedimentation* classification; if the fluid movement and particle settling directions are opposite, the classification is called *hydraulic* or *counter flow*. Sedimentation, or horizontal current, classifiers are essentially of the free-settling type and accentuate the sizing function. On the other hand, hydraulic, or vertical current, classifiers are usually hindered-settling types and so increase the effect of density on the separation.

9.5.1 Sedimentation Classifiers

Nonmechanical Sedimentation Classifiers

As the simplest form of classifier, there is little attempt to do more than separate the solids from the liquid,

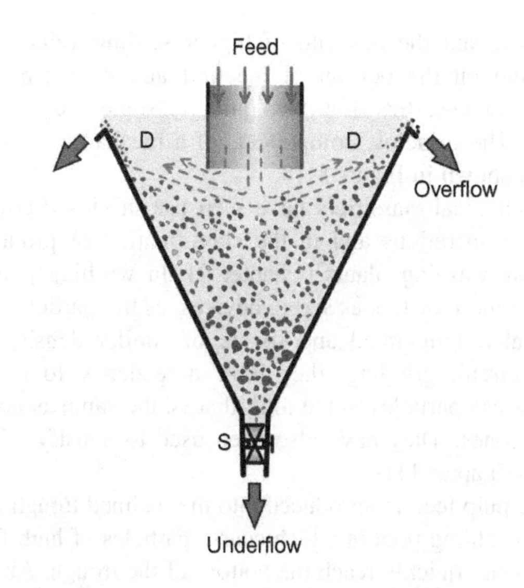

FIGURE 9.17 Nonmechanical sedimentation classifier: settling cone operation.

and as such they are sometimes used as dewatering units in small-scale operations. Therefore, they are not suitable for fine classification or if a high separation efficiency is required. They are often used in the aggregate industry to de-slime coarse sand products. The principle of the settling cone is shown in Figure 9.17. The pulp is fed into the tank as a distributed stream, with the spigot discharge valve, S, initially closed. When the tank is full, overflow of water and slimes commences, and a bed of settled sand builds until it reaches the level shown. The spigot valve is now opened and sand is discharged at a rate equal to that of the input. Classification by horizontal current action takes place radially across zone D from the feed pipe to the overflow lip. The main difficulty in operation of such a device is the balancing of the sand discharge and deposition rates; it is virtually impossible to maintain a regular discharge of sand through an open pipe under the influence of gravity. Many different designs of cone have been introduced to overcome this problem (Taggart, 1945).

In the "Floatex" separator, which is essentially a hindered-settling classifier over a dewatering cone, automatic control of the coarse lower discharge is governed by the specific gravity of the teeter column. The use of the machine as a de-sliming unit and in upgrading coal and mica, as well as its possible application in closed-circuit classification of metalliferous ores, is discussed by Littler (1986).

Mechanical Sedimentation Classifiers

This term describes classifiers in which the material of lower settling velocity is carried away in a liquid overflow, and the material of higher settling velocity is deposited on the bottom of the unit and is transported upward against the flow of liquid by some mechanical means. The principle components of a mechanical classifier are shown in Figure 9.18.

Mechanical classifiers have seen use in closed-circuit grinding operations and in the classification of products from ore-washing plants (Chapter 2). In washing plants, they act more or less as sizing devices, as the particles are essentially unliberated and so are of similar density. In closed-circuit grinding, they have a tendency to return small dense particles to the mill, that is, the same as noted for cyclones. They have also been used to densify dense media (Chapter 11).

The pulp feed is introduced into the inclined trough and forms a settling pool in which coarse particles of high falling velocity quickly reach the bottom of the trough. Above this, coarse sand is a quicksand zone where hindered settling takes place. The depth and shape of this zone depends on the classifier action and on the feed pulp density. Above the quicksand is a zone of essentially free-settling material, comprising a stream of pulp flowing horizontally across the top of the quicksand zone from the feed inlet to the overflow weir, where the fines are removed.

The settled sands are conveyed up the inclined trough by a mechanical rake or by a helical screw. The conveying mechanism also serves to keep fine particles in suspension in the pool by gentle agitation and when the sands leave the pool they are slowly turned over by the raking action, thus releasing entrained slimes and water, increasing the efficiency of the separation. Washing sprays are often directed on the emergent sands to wash the released slimes back into the pool.

The rake classifier (Figure 9.18(a)) uses rakes actuated by an eccentric motion, which causes them to dip into the settled material and to move it up the incline for a short distance. The rakes are then withdrawn, and return to the starting-point, where the cycle is repeated. The settled material is thus slowly moved up the incline to the discharge. In the *duplex* type, one set of rakes is moving up, while the other set returns; simplex and quadruplex machines are also made, in which there are one or four raking assemblies, respectively.

In spiral classifiers (Figure 9.18(b); Figure 9.19), a continuously revolving spiral moves the sands up the slope. They can be operated at steeper slopes than the rake classifier, in which the sands tend to slip back when the rakes are removed. Steeper slopes aid the drainage of sands, giving a cleaner, drier product. Agitation in the pool is less than in the rake classifier, which is important in separations of very fine material.

The size at which the separation is made and the quality of the separation depend on a number of factors. Increasing the feed rate increases the horizontal carrying velocity and thus increases the size of the particles leaving in the overflow. The feed should not be introduced directly into the pool, as this causes agitation and releases coarse material from the hindered-settling zone, which may report to the overflow. The feed stream should be slowed down by spreading it on an apron, partially submerged in the pool, and sloped toward the sand discharge end, so that most of the kinetic energy is absorbed in the part of the pool furthest from the overflow.

The speed of the rakes or spiral determines the degree of agitation of the pulp and the tonnage rate of sand

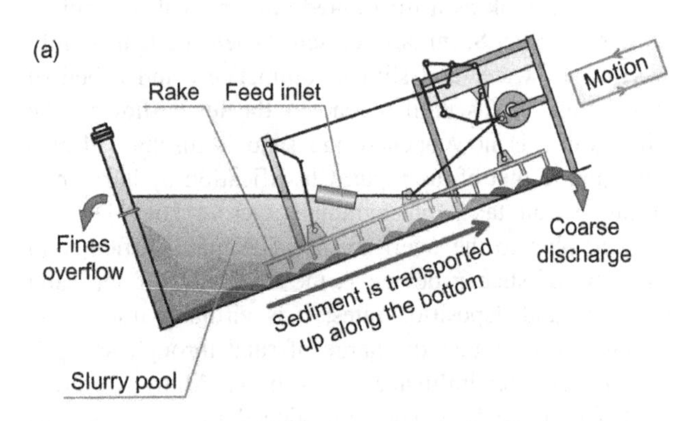

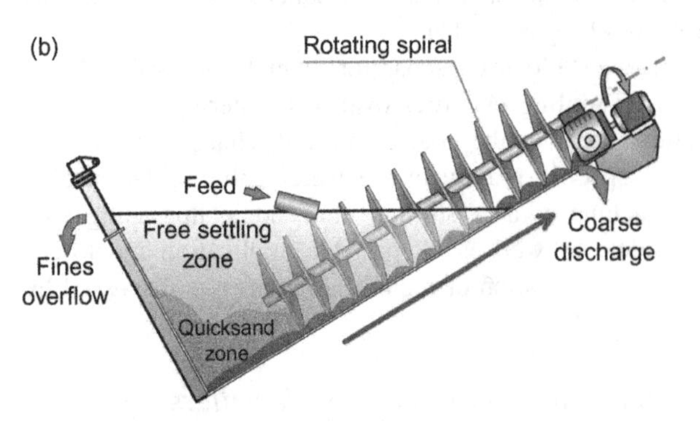

FIGURE 9.18 Principle of mechanical classifier: (a) rake classifier, and (b) spiral classifier.

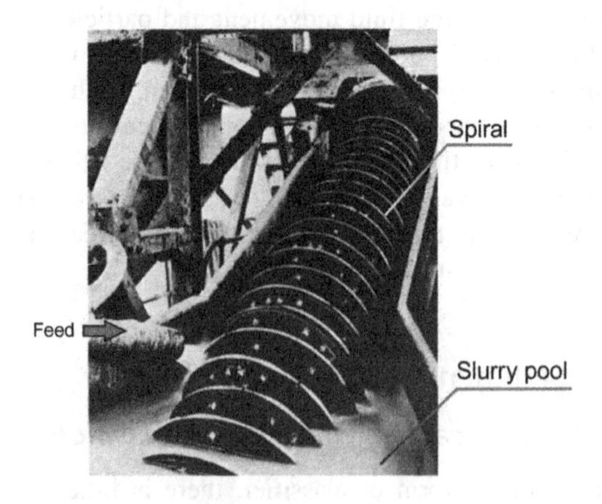

FIGURE 9.19 Spiral classifier.

removal. For coarse separations, a high degree of agitation may be necessary to keep the coarse particles in suspension in the pool, whereas for finer separations, less agitation, and thus lower raking speeds, are required. It is essential, however, that the speed is high enough to transport the sands up the slope.

The height of the overflow weir is an operating variable in some mechanical classifiers. Increasing the weir height increases the pool volume, and hence allows more settling time and decreases the surface agitation, thus reducing the pulp density at overflow level, where the final separation is made. High weirs are thus used for fine separations.

Dilution of the pulp is the most important variable in the operation of mechanical classifiers. In closed-circuit grinding operations, ball mills rarely discharge at less than 65% solids by weight, whereas mechanical classifiers never operate at more than about 50% solids. Water to control dilution is added in the feed launder or onto the sand near the "V" of the pool. Water addition determines the settling rate of the particles. Increased dilution reduces the density of the weir overflow product, and increases free settling, allowing finer particles to settle out of the influence of the horizontal current. Therefore finer separations are produced, provided that the overflow pulp density is above a value known as the *critical dilution*, which is normally about 10% solids. Below this density, the effect of increasing rising velocity with dilution becomes more important than the increase in particle settling rates produced by decrease of pulp density. The overflow therefore becomes coarser with increasing dilution (Figure 9.20). In mineral processing applications, however, very rarely is the overflow density less than the critical dilution.

One of the major disadvantages of the mechanical classifier is its inability to produce overflows of fine particle size at reasonable pulp densities. To produce fine particle separations, the pulp may have to be diluted to such an extent to increase particle settling rates that the overflow becomes too dilute for subsequent operations. It may therefore require thickening before mineral separation (concentration) can take place efficiently. This is undesirable as, apart from the capital cost and floor space of the thickener, oxidation of sulfide particles may occur in the thickener, which may affect subsequent processes, especially froth flotation.

9.5.2 Hydraulic Classifiers

Hydraulic classifiers are characterized by the use of water additional to that of the feed pulp, introduced so that its direction of flow opposes that of the settling particles. They normally consist of a series of sorting columns through which, in each column, a vertical current of water is rising and particles are settling out (Figure 9.21(a)).

The rising currents are graded from a relatively high velocity in the first sorting column, to a relatively low velocity in the last, so that a series of underflow (spigot)

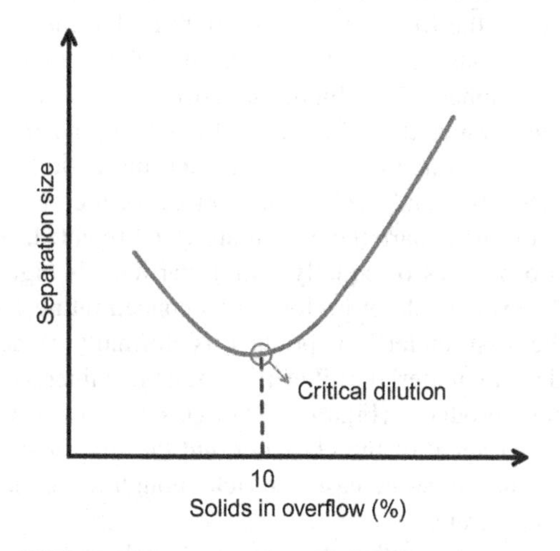

FIGURE 9.20 Effect of dilution of overflow on mechanical classifier separation.

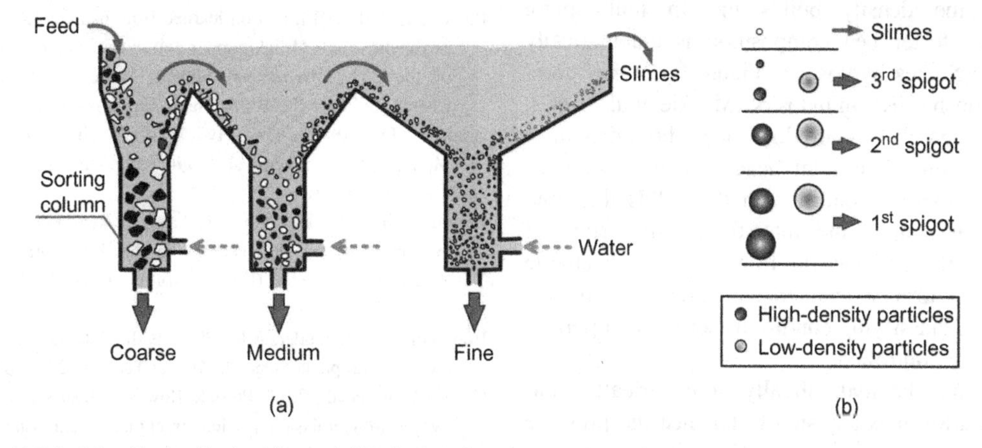

FIGURE 9.21 (a) Principle of hydraulic classifier, and (b) spigot products.

products can be obtained, with the coarser, denser particles in the first underflow and progressively finer products in the subsequent underflows (Figure 9.21(b)). The finest fraction (slimes) overflows the final sorting column. The size of each successive vessel is increased, partly because the amount of liquid to be handled includes all the water used for classifying in the previous vessels and partly because it is desired to reduce, in stages, the surface velocity of the fluid flowing from one vessel to the next.

Hydraulic classifiers may be free-settling or hindered-settling types. The free-settling types, however, are rarely used; they are simple and have high capacities, but are inefficient in sizing and sorting. They are characterized by the fact that each sorting column is of the same cross-sectional area throughout its length.

The greatest use for hydraulic classifiers in the mineral industry is for sorting the feed to certain gravity concentration processes so that the size effect can be suppressed and the density effect enhanced (Chapter 10). Such classifiers are of the hindered-settling type. These differ from the free-settling classifiers in that the sorting column is constricted at the bottom in order to produce a teeter chamber. The hindered-settling classifier uses much less water than the free-settling type and is more selective in its action, due to the scouring action in the teeter chamber, and the buoyancy effect of the pulp, as a whole, on those particles which are to be rejected. Since the ratio of sizes of equally falling particles is high, the classifier is capable of performing a concentrating effect, and the first underflow product is normally richer in high-density material (often the valuable mineral) than the other products (Figure 9.21). This is known as the *added increment* of the classifier and the first underflow product may in some cases be rich enough to be classed as a concentrate.

During classification the teeter bed tends to grow, as it is easier for particles to become entrapped in the bed rather than leave it. This tends to alter the character of the spigot discharge, as the density builds up. In multi-spigot *hydrosizers*, the teeter bed composition is automatically controlled. The Stokes hydrosizer (Figure 9.22) was common in the Cornish (UK) tin industry (Mackie et al., 1987).

Each teeter chamber is provided at its bottom with a supply of water under constant head, which is used for maintaining a teetering condition in the solids that find their way down against the interstitial rising flow of water. Each teeter chamber is fitted with a discharge spigot that is, in turn, connected to a pressure-sensitive valve so that the classifying conditions set by the operator can be accurately controlled.

The valve may be hydraulically or electrically controlled; in operation it is adjusted to balance the pressure set up by the teetering material. The concentration of solids in a particular compartment can be held steady in spite of the normal variations in feed rate taking place from time to time. The rate of discharge from each spigot will, of course, change in sympathy with these variations, but since these changing tendencies are always being balanced by the valve, the discharge will take place at a nearly constant density. For a quartz sand this is usually about 65% solids by weight, but is higher for heavier minerals.

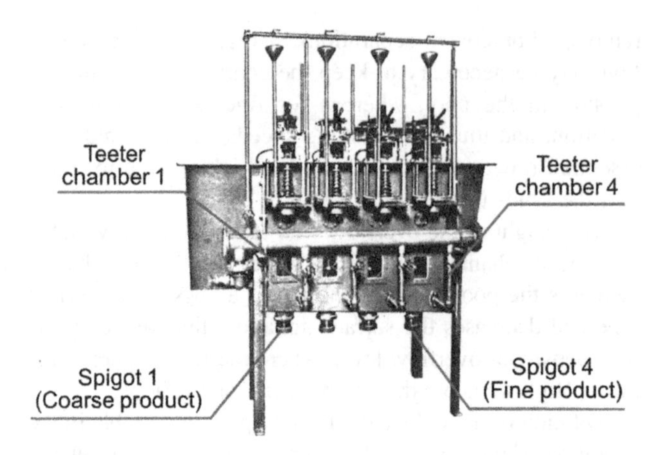

FIGURE 9.22 Stokes multi-spigot hydrosizer.

REFERENCES

Anon, 1984. Classifiers Part 2: some of the major manufacturers of classification equipment used in mineral processing. Mining Mag. 151 (Jul.), 40–44.

Arterburn, R.A., 1982. Sizing, and selection of hydrocyclones. In: Mular, A. L., Jergensen, G.V. (Eds.), Design and Installation of Comminution Circuits, vol. 1. AIME, New York, NY, USA, pp. 597–607. (Chapter 32).

Asomah, A.K., Napier-Munn, T.J., 1997. An empirical model of hydrocyclones incorporating angle of cyclone inclination. Miner. Eng. 10 (3), 339–347.

Austin, L.G., et al., 1984. Process Engineering of Size Reduction: Ball Milling. SME, New York, NY, USA.

Bahar, A., et al., 2014. Lesson learned from using column flotation cells as roughers: the Miduk copper concentrator plant case. Proceedings of the 27th International Mineral Processing Congress (IMPC), Chapter 6, Paper C0610, Santiago, Chile, pp. 1–12.

Bradley, D., 1965. The Hydrocyclone: International Series of Monographs in Chemical Engineering. first ed. Pergamon Press Ltd., Oxford, UK.

Brennan, M.S., et al., 2003. Towards a new understanding of the cyclone separator. Proceedings of the 22nd International Mineral Processing Congress (IMPC), vol. 1, Cape Town, South Africa, pp. 378–385.

Bourgeois, F., Majumder, A.K., 2013. Is the fish-hook effect in hydrocyclones a real phenomenon?. Powder Technol. 237, 367–375.

Chang, Y.-F., et al., 2011. Particle flow in a hydrocyclone investigated by positron emission particle tracking. Chem. Eng. Sci. 66 (18), 4203–4211.

Cirulis, D., Russell, J., 2011. Cyclone monitoring system improves operations at KUC's Copperton concentrator. Eng. Min. J. 212 (10), 44–49.

Del Villar, R., Finch, J.A., 1992. Modelling the cyclone performance with a size dependent entertainment factor. Miner. Eng. 5 (6), 661–669.

Dündar, H., et al., 2014. Screens and cyclones in closed grinding circuits. Proceedings of the 27th International Mineral Processing Congress (IMPC), Chapter 16, Paper C1607, Santiago, Chile, pp. 1–11.

Flintoff, B.C., et al., 1987. Cyclone modelling: a review of present technology. CIM Bull. 80 (905), 39–50.

Hand, P., Wiseman, D., 2010. Addressing the envelope. J. S. Afr. Inst. Min. Metall. 110, 365–370.

Heiskanen, K., 1993. Particle Classification. Chapman & Hall, London, UK.

Kawatra, S.K., Eisele, T.C., 2006. Causes and significance of inflections in hydrocyclone efficiency curves. In: Kawatra, S.K. (Ed.), Advances in Comminution. SME, Littleton, CO, USA, pp. 131–147.

Kawatra, S.K., Seitz, R.A., 1985. Technical note: calculating the particle size distribution in a hydrocyclone overflow product for simulation purposes. Miner. Metall. Process. 2 (Aug.), 152–154.

Kelsall, D.F., 1953. A further study of the hydraulic cyclone. Chem. Eng. Sci. 2 (6), 254–272.

King, R.P., 2012. Modeling and Simulation of Mineral Processing Systems. second ed. SME, Englewood, CO, USA.

Kraipech, W., et al., 2006. The performance of the empirical models on industrial hydrocyclone design. Int. J. Miner. Process. 80, 100–115.

Laplante, A.R., Finch, J.A., 1984. The origin of unusual cyclone performance curves. Int. J. Miner. Process. 13 (1), 1–11.

Leith, D., Licht, W., 1972. The collection efficiency of cyclone type particle collector: a new theoretical approach. AIChE Symp. Series (Air-1971). 68 (126), 196–206.

Littler, A., 1986. Automatic hindered-settling classifier for hydraulic sizing and mineral beneficiation. Trans. Inst. Min. Metall., Sec. C. 95, C133–C138.

Lynch, A.J., Rowland, C.A., 2005. The History of Grinding. SME, Littleton, CO, USA.

Lynch, A.J., et al., 1974. The influence of hydrocyclone diameter on reduced-efficiency curves. Int. J. Miner. Process. 1 (2), 173–181.

Lynch, A.J., et al., 1975. The influence of design and operating variables on the capacities of hydrocyclone classifiers. Int. J. Miner. Process. 2 (1), 29–37.

Mackie, R.I., et al., 1987. Mathematical model of the Stokes hydrosizer. Trans. Inst. Min. Metall., Sec. C. 96, C130–C136.

Morrison, R.D., Freeman, N., 1990. Grinding control development at ZC Mines. Proc. AusIMM. 295 (2), 45–49.

Morrison, R.D., Morrell, S., 1998. Comparison of comminution circuit energy efficiency using simulation. Miner. Metall. Process. 15 (4), 22–25.

Mular, A.L., Jull, N.A., 1980. The selection of cyclone classifiers, pumps and pump boxes for grinding circuits, Mineral Processing Plant Design. second ed. AIMME, New York, NY, USA, pp. 376–403.

Nageswararao, K., 1995. Technical note: a generalised model for hydrocyclone classifiers. Proc. AusIMM. 300 (2), 21-21.

Nageswararao, K., 2014. Comment on: 'Is the fish-hook effect in hydrocyclones a real phenomenon?' by F. Bourgeois and A.K Majumder [Powder Technology 237 (2013) 367-375]. Powder Technol. 262, 194–197.

Nageswararao, K., et al., 2004. Two empirical hydrocyclone models revisited. Miner. Eng. 17 (5), 671–687.

Napier-Munn, T.J., et al., 1996. Mineral Comminution Circuits: Their Operation and Optimisation. Chapter 12, Julius Kruttschnitt Mineral Research Centre (JKMRC), The University of Queensland, Brisbane, Australia.

Narasimha, M., et al., 2005. CFD modelling of hydrocyclone—prediction of cut-size. Int. J. Miner. Process. 75 (1-2), 53–68.

Nowakowski, A.F., et al., 2004. Application of CFD to modelling of the flow in hydrocyclones. Is this a realizable option or still a research challenge? Miner. Eng. 17 (5), 661–669.

Olson, T.J., Turner, P.A., 2002. Hydrocyclone selection for plant design. In: Mular, A.L., et al., (Eds.), Mineral Processing Plant Design, Practice, and Control, vol. 1. Vancouver, BC, Canada, pp. 880–893.

Plitt, L.R., 1976. A mathematical model of the hydrocyclone classifier. CIM Bull. 69 (Dec.), 114–123.

Rao, T.C., et al., 1976. Influence of feed inlet diameter on the hydrocyclone behaviour. Int. J. Miner. Process. 3 (4), 357–363.

Renner, V.G., Cohen, H.E., 1978. Measurement and interpretation of size distribution of particles within a hydrocyclone. Trans. Inst. Min. Metall., Sec. C. 87 (June), C139–C145.

Sasseville, Y., 2015. Study on the impact of using cyclones rather than screens in the grinding circuit at Niobec Mine. Proceedings of the 47th Annual Meeting of The Canadian Mineral Processors Conference, Ottawa, ON, Canada, pp. 118–126.

Smith, V.C., Swartz, C.L.E., 1999. Development of a hydrocyclone product size soft-sensor. In: Hodouin, D. et al., (Eds.), Proceedings of the International Symposium on Control and Optimization in Minerals, Metals and Materials Processing. 38th Annual Conference of Metallurgists of CIM: Gateway to the 21st Century. Quebec City, QC, Canada, pp. 59–70.

Stokes, G.G., 1891. Mathematical and Physical Papers, vol. 3. Cambridge University Press.

Svarovsky, L., Thew, M.T., 1992. Hydrocyclones: Analysis and Applications. Springer Science + Business Media, LLC, Technical Communications (Publishing) Ltd., Letchworth, England.

Swanson, V.F., 1989. Free and hindered settling. Miner. Metall. Process. 6 (Nov.), 190–196.

Taggart, A.F., 1945. Handbook of Mineral Dressing, Ore and Industrial Minerals. John Wiley & Sons, Chapman & Hall, Ltd, London, UK.

Wang, Zh.-B., et al., 2008. Experimental investigation of the motion trajectory of solid particles inside the hydrocyclone by a Lagrange method. Chem. Eng. J. 138 (1-3), 1–9.

Westendorf, M., et al., 2015. Managing cyclones: A valuable asset, the Copper Mountain case study. Min. Eng. 67 (6), 26–41.

Chapter 10

Gravity Concentration

10.1 INTRODUCTION

Gravity concentration is the separation of minerals based upon the difference in density. Techniques of gravity concentration have been around for millennia. Some believe that the legend of the Golden Fleece from Homer's Odyssey was based upon a method of gold recovery, which was to place an animal hide (such as a sheep's fleece) in a stream containing alluvial gold; the dense gold particles would become trapped in the fleece and be recovered (Agricola, 1556). In the various gold rushes of the nineteenth century, many prospectors used gold panning as a means to make their fortune—another ancient method of gravity concentration. Gravity concentration, or density-based separation methods, declined in importance in the first half of the twentieth century due to the development of froth flotation (Chapter 12), which allowed for the selective treatment of low-grade complex ores. They remain, however, the main concentrating methods for iron and tungsten ores and are used extensively for treating tin ores, coal, gold, beach sands, and many industrial minerals.

In recent years, mining companies have reevaluated gravity systems due to increasing costs of flotation reagents, the relative simplicity of gravity processes, and the fact that they produce comparatively little environmental impact. Modern gravity techniques have proved efficient for concentration of minerals having particle sizes down to the 50 μm range and, when coupled with improved pumping technology and instrumentation, have been incorporated in high-capacity plants (Holland-Batt, 1998). In many cases a high proportion of the mineral in an ore body can be preconcentrated effectively by gravity separation systems; the amount of reagents and energy used can be cut significantly when the more expensive methods are restricted to the processing of a gravity concentrate. Gravity separation at coarse sizes (as soon as liberation is achieved) can also have significant advantages for later treatment stages, due to decreased surface area, more efficient dewatering, and the absence of adsorbed chemicals, which could interfere with further processing.

10.2 PRINCIPLES OF GRAVITY CONCENTRATION

Gravity concentration methods separate minerals of different specific gravity by their relative movement in response to gravity and one or more other forces, the latter often being the resistance to motion offered by a viscous fluid, such as water or air.

It is essential for effective separation that a marked density difference exists between the mineral and the gangue. Some idea of the type of separation possible can be gained from the *concentration criterion*, $\Delta\rho$:

$$\Delta\rho = \frac{\rho_h - \rho_f}{\rho_l - \rho_f} \quad (10.1)$$

where ρ_h is the density of the heavy mineral, ρ_l is the density of the light mineral, and ρ_f is the density of the fluid medium.

In general terms, if the quotient has a magnitude greater than 2.5, then gravity separation is relatively easy, with the efficiency of separation decreasing as the value of the quotient decreases. To give an example, if gold is being separated from quartz using water as the carrier fluid, the concentration criterion is 11.1 (density of gold being 19,300 kg m^{-3}; quartz being 2,650 kg m^{-3}), which is why panning for gold has been so successful.

The motion of a particle in a fluid is dependent not only on its specific gravity, but also on its size (Chapter 9)—large particles will be affected more than smaller ones. The efficiency of gravity processes therefore increases with particle size, and the particles should be sufficiently coarse to move in accordance with Newton's law (Eq. (9.8)). Particles small enough that their movement is dominated mainly by surface friction respond relatively poorly to commercial high-capacity gravity methods. In practice, close size control of feeds to gravity processes is required to reduce the size effect and make the relative motion of the particles specific gravity-dependent. The incorporation of enhanced gravity concentrators, which impart additional centrifugal acceleration to the particles (Section 10.4), have been utilized in order to overcome some of the drawbacks of fine particle

Wills' Mineral Processing Technology.

TABLE 10.1 Dependence on Concentration Criterion (Eq. (10.1)) for Separations

Concentration Criterion	Separation?	Useful for?
2.5	Relatively easy	To 75 μm
1.75–2.5	Possible	To 150 μm
1.5–1.75	Difficult	To 1.7 mm
1.25–1.5	Very difficult	
<1.25	Not possible	

processing. Table 10.1 gives the relative ease of separating minerals using gravity techniques, based upon the particle size and concentration criterion (Anon., 2011).

10.3 GRAVITATIONAL CONCENTRATORS

Many machines have been designed and built to effect separation of minerals by gravity (Burt, 1985). The dense medium separation (DMS) process is used to preconcentrate crushed material prior to grinding and will be considered separately in the next chapter (Chapter 11). Design and optimization of gravity circuits is discussed by Wells (1991) and innovations in gravity separation are reviewed by Honaker et al. (2014).

It is essential for the efficient operation of all gravity separators that the feed is carefully prepared. Grinding is particularly important to provide particles of adequate liberation; successive regrinding of middlings is required in most operations. Primary grinding should be performed where possible in open-circuit rod mills, but if fine grinding is required, closed-circuit ball milling should be used, preferably with screens closing the circuits rather than hydrocyclones, in order to reduce selective overgrinding of heavy friable valuable minerals (see also Chapters 8 and 9). Other methods of comminution, as described in Chapters 6 and 7, such as semi-autogeneous grinding mills and high-pressure grinding rolls, may find application in preparing gravity feeds.

Gravity separators are sensitive to the presence of *slimes* (ultrafine particles), which increase the viscosity of the slurry and hence reduce the sharpness of separation, and obscure visual cutpoints for operators. It has been common practice to remove particles less than about 10 μm from the feed (i.e., de-slime), and divert this fraction to the tailings, which can incur considerable loss of values. De-sliming is often achieved by the use of hydrocyclones, although hydraulic classifiers may be preferable in some cases since the high shear forces produced in hydrocyclones tend to cause degradation of friable minerals (and create more loss of slimed values).

The feed to jigs and spirals should, if possible, be screened before separation takes place, each fraction being treated separately. In most cases, however, removal of the oversize by screening, in conjunction with desliming, is adequate. Processes which employ flowing-film separation, such as shaking tables, should always be preceded by good hydraulic classification in multi-spigot hydrosizers (Chapter 9).

Although most slurry transportation is achieved by centrifugal pumps and pipelines, as much as possible use should be made of natural gravity flow; many old gravity concentrators were built on hillsides to achieve this. Reduction of slurry pumping to a minimum not only lowers energy consumption, but also reduces slimes production in the circuit. To minimize degradation of friable minerals, pumping velocities should be as low as possible, consistent with maintaining the solids in suspension.

One of the most important aspects of gravity circuit operations is correct water balance within the plant. Almost all gravity concentrators have an optimum feed pulp density, and relatively little deviation from this density causes a rapid decline in efficiency. Accurate pulp density control is therefore essential, and this is most important on the raw feed. Automatic density control should be used where possible, and the best way of achieving this is by the use of nucleonic density gauges (Chapter 3) controlling the water addition to the new feed. Although such instrumentation is expensive, it is usually economic in the long term. Control of pulp density within the circuit can be made by the use of settling cones (Chapter 9) preceding the gravity device. These thicken the pulp, but the overflow often contains solids, and should be directed to a central large sump or thickener. For substantial increase in pulp density, hydrocyclones or thickeners may be used. The latter are the more expensive, but produce less particle degradation and also provide substantial surge capacity. It is usually necessary to recycle water in most plants, so adequate thickener or cyclone capacity should be provided, and slimes build-up in the recycled water must be minimized.

If the ore contains an appreciable amount of sulfide minerals then, because they are relatively dense and tend to report with the "heavy" product, they need to be removed. If the primary grind is finer than ca. 300 μm, the sulfides should be removed by froth flotation prior to gravity concentration. If the primary grind is too coarse for effective sulfide flotation, then the gravity concentrate must be reground prior to removal of the sulfides. The sulfide flotation tailing is then usually cleaned by further gravity concentration.

The final gravity concentrate often needs cleaning by magnetic separation, leaching, or some other method, in order to remove certain mineral contaminants. For instance, at the South Crofty tin mine in Cornwall, the gravity concentrate was subjected to cleaning by magnetic separators, which removed wolframite from the cassiterite product.

10.3.1 Jigs

Jigging is one of the oldest methods of gravity concentration, yet the basic principles have only recently been understood. A mathematical model developed by Jonkers et al. (2002) predicts jig performance on a size-by-density basis; Mishra and Mehrotra (1998) developed discrete element method models of particle motion in a jig; and Mishra and Mehrotra (2001) and Xia et al. (2007) have developed a computational fluid dynamics model of coal stratification in a jig.

In the jig, the separation of minerals of different specific gravity is accomplished in a particle bed, which is fluidized by a pulsating current of water, producing stratification based upon density. The bed may be a specific mineral added to and retained in the jig, called *ragging*, composed of a certain density and shape through which the dense particles penetrate and the light particles pass over the top. The aim is to dilate the bed and to control the dilation so that the heavier, smaller particles penetrate the interstices of the bed and the larger high-specific gravity particles fall under a condition similar to hindered settling (Lyman, 1992).

On the pulsion stroke the bed is normally lifted as a mass, then as the velocity decreases it tends to dilate, the bottom particles falling first until the whole bed is loosened. On the suction stroke it then closes slowly again, and this is repeated at every stroke. Fine particles tend to pass through the interstices after the large ones have become immobile. The motion can be obtained either by using a fixed sieve jig, and pulsating the water, or by employing a moving sieve, as in the simple hand-jig shown in Figure 10.1.

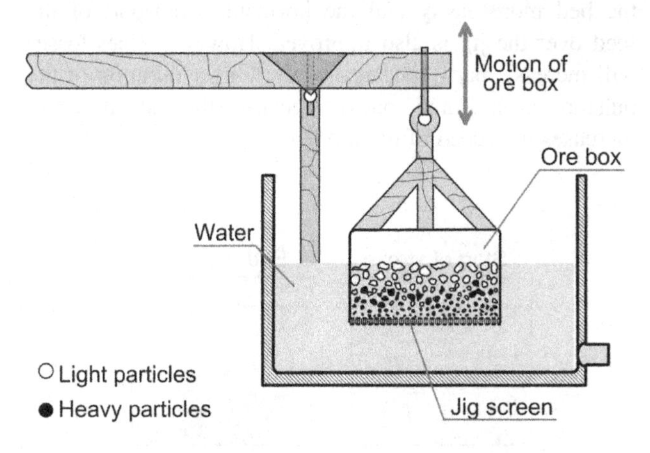

FIGURE 10.1 Jigging action: by moving box up and down the particles segregate by density *(Adapted from Priester et al. (1993))*.

The jig is normally used to concentrate relatively coarse material and, if the feed is fairly close-sized (e.g., 3–10 mm), it is not difficult to achieve good separation of a fairly narrow specific gravity range in minerals in the feed (e.g., fluorite, s.g. 3.2, from quartz, s.g. 2.7). When the specific gravity difference is large, good concentration is possible with a wider size range. Many large jig circuits are still operated in coal, cassiterite, tungsten, gold, barytes, and iron ore concentrators. They have a relatively high unit capacity on classified feed and can achieve good recovery of values down to 150 μm and acceptable recoveries often down to 75 μm. High proportions of fine sand and slime interfere with performance and the fines content should be controlled to provide optimum bed conditions.

Jigging Action

It was shown in Chapter 9 that the equation of motion of a particle settling in a viscous fluid is:

$$m\frac{dx}{dt} = mg - m'g - D \tag{10.2}$$

where m is the mass of the mineral particle, dx/dt is the acceleration, g is the acceleration due to gravity, m' is the mass of displaced fluid, and D is the fluid resistance due to the particle movement.

At the beginning of the particle movement, since the velocity x is very small, D can be ignored, as it is a function of velocity.

Therefore

$$\frac{dx}{dt} = \left(\frac{m - m'}{m}\right)g \tag{10.3}$$

and since the particle and the displaced fluid are of equal volume,

$$\frac{dx}{dt} = \left(\frac{\rho_s - \rho_f}{\rho_s}\right)g \tag{10.4}$$

where ρ_s and ρ_f are the respective densities of the solid and the fluid.

The initial acceleration of the mineral grains is thus independent of size and dependent only on the densities of the solid and the fluid. Theoretically, if the duration of fall is short enough and the repetition of fall frequent enough, the total distance travelled by the particles will be affected more by the differential initial acceleration, and therefore by density, than by their terminal velocities, and therefore by size. In other words, to separate small heavy mineral particles from large light particles, a short jigging cycle is necessary. Although relatively short, fast strokes are used to separate fine minerals, more control and better stratification can be achieved by using longer, slower strokes, especially with the coarser particle sizes.

It is therefore good practice to screen the feed to jigs into different size ranges and treat these separately. The effect of differential initial acceleration is shown in Figure 10.2.

If the mineral particles are examined after a longer time, they will have attained their terminal velocities and will be moving at a rate dependent on their specific gravity and size. Since the bed is really a loosely packed mass with interstitial water providing a very thick suspension of high density, hindered-settling conditions prevail, and the settling ratio of heavy to light minerals is higher than that for free settling (Chapter 9). Figure 10.3 shows the effect of hindered settling on the separation.

The upward flow can be adjusted so that it overcomes the downward velocity of the fine light particles and carries them away, thus achieving separation. It can be increased further so that only large heavy particles settle, but it is apparent that it will not be possible to separate the small heavy and large light particles of similar terminal velocity.

Hindered settling has a marked effect on the separation of coarse minerals, for which longer, slower strokes should be used, although in practice, with coarser feeds, it is unlikely that the larger particles have sufficient time to reach their terminal velocities.

At the end of a pulsion stroke, as the bed begins to compact, the larger particles interlock, with the smaller particles still moving downward through the interstices under the influence of gravity. The fine particles may not settle as rapidly during this *consolidation trickling* phase (Figure 10.4) as during the initial acceleration or suspension, but if consolidation trickling can be made to last long enough, the effect, especially in the recovery of the fine heavy minerals, can be considerable.

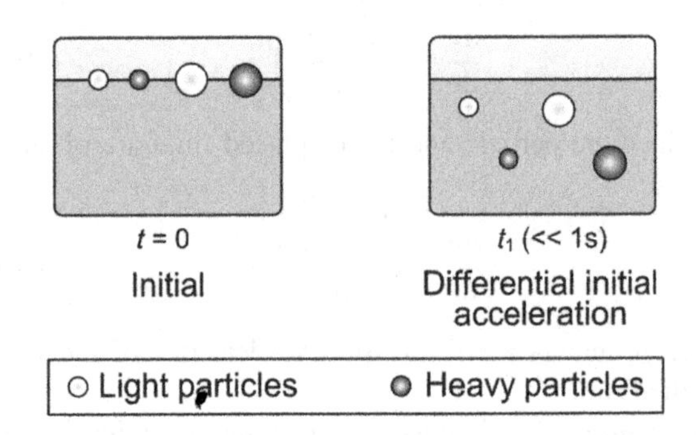

FIGURE 10.2 Differential initial acceleration.

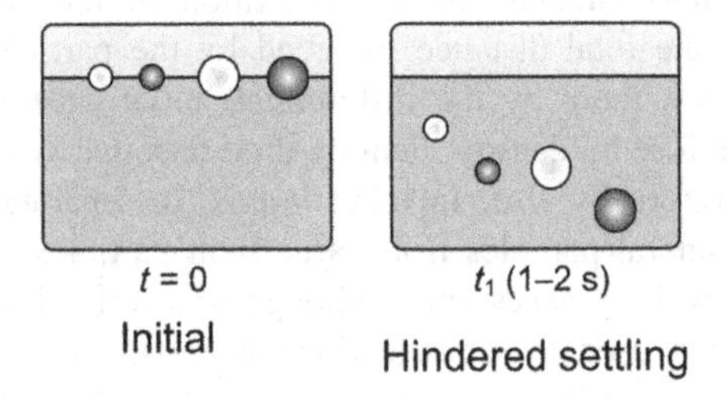

FIGURE 10.3 Hindered settling.

Figure 10.5 shows an idealized jigging process by the described phenomena.

In the jig the pulsating water currents are caused by a piston having a movement that is a harmonic waveform (Figure 10.6). The vertical speed of flow through the bed is proportional to the speed of the piston. When this speed is greatest, the speed of flow through the bed is also greatest (Figure 10.7).

The upward speed of flow increases after point A, the beginning of the cycle. As the speed increases, the particles will be loosened and the bed will be forced open, or dilated. At, say, point B, the particles are in the phase of hindered settling in an upward flow, and since the speed of flow from B to C still increases, the fine particles are pushed upward by the flow. The chance of them being carried along with the top flow into the low-density product (often the tailings) is then at its greatest. In the vicinity of D, first the coarser particles and later the remaining fine particles will fall back. Due to the combination of initial acceleration and hindered settling, it is mainly the coarser particles that will lie at the bottom of the bed.

At E, the point of transition between the pulsion and the suction stroke, the bed will be compacted. Consolidation trickling can now occur to a limited extent. In a closely sized ore, the heavy particles can now penetrate only with difficulty through the bed and may be lost to the low-density stream. Severe compaction of the bed can be reduced by the addition of hutch water, a constant volume of water, which creates a constant upward flow through the bed. This flow, coupled with the varying flow caused by the piston, is shown in Figure 10.8. Thus suction is reduced by hutch-water addition and is reduced in duration; by adding a large quantity of water, the suction may be entirely eliminated. The coarse ore then penetrates the bed more easily and the horizontal transport of the feed over the jig is also improved. However, fines losses will increase, partly because of the longer duration of the pulsion stroke, and partly because the added water increases the speed of the top flow.

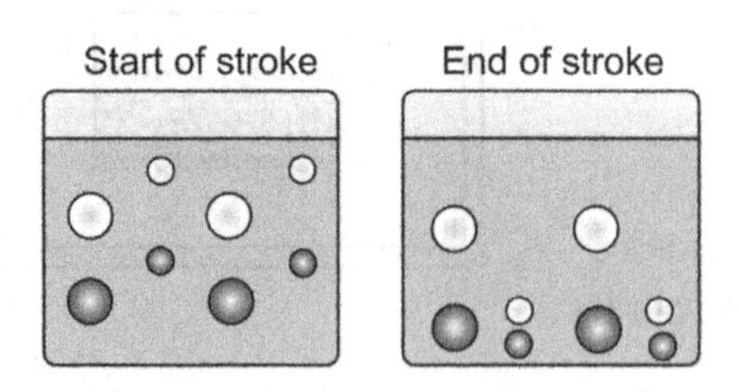

FIGURE 10.4 Consolidation trickling.

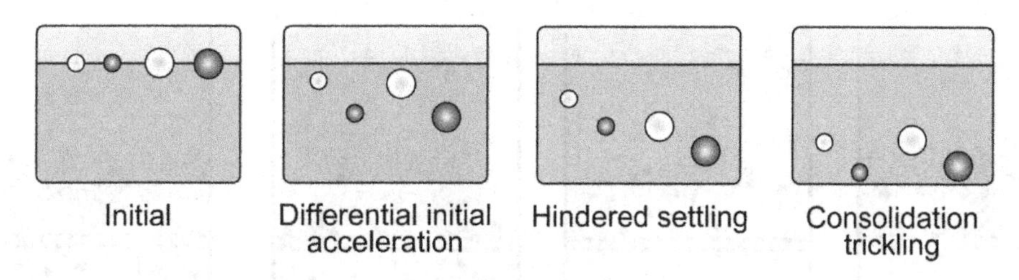

FIGURE 10.5 Ideal jigging process.

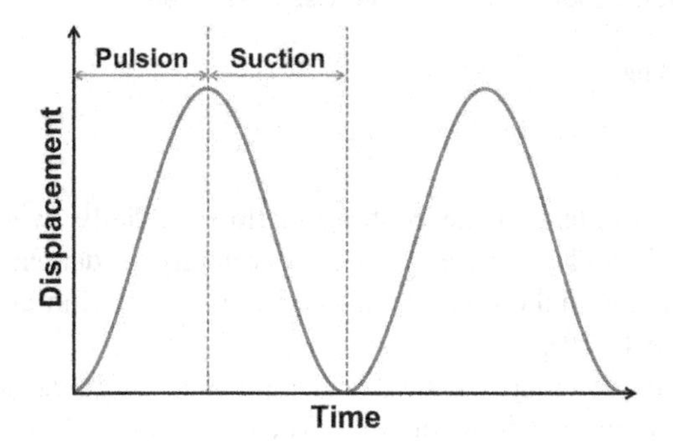

FIGURE 10.6 Movement of the piston in a jig.

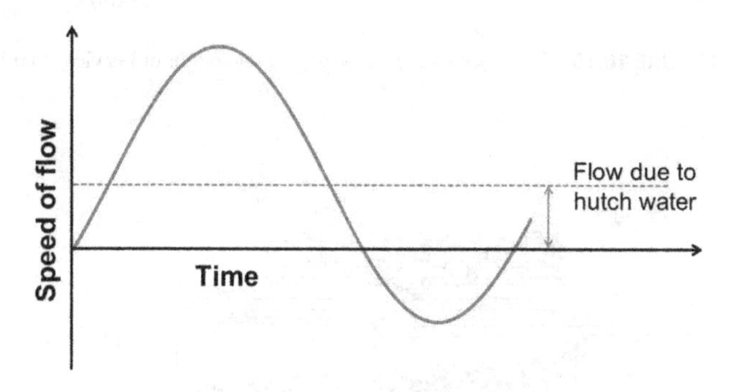

FIGURE 10.8 Effect of hutch water on flow through bed.

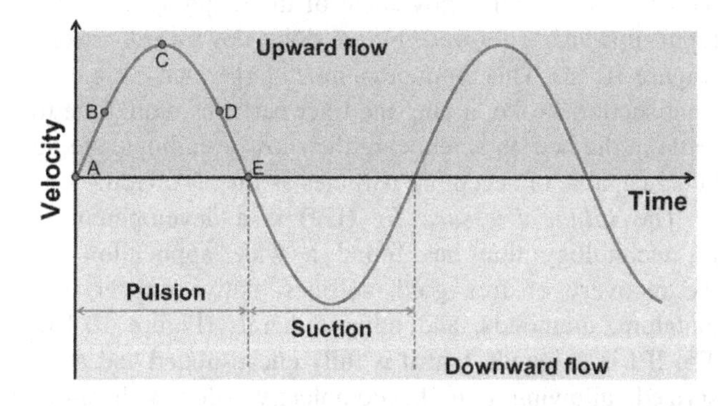

FIGURE 10.7 Speed of flow through bed during jig cycle.

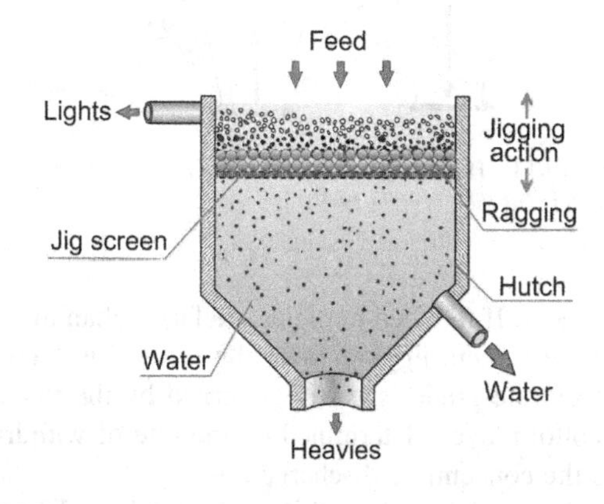

FIGURE 10.9 Basic jig construction.

Types of Jig

Essentially, the jig is an open tank filled with water, with a horizontal jig screen at the top supporting the jig bed, and provided with a spigot in the bottom, or hutch compartment, for "heavies" removal (Figure 10.9). Current types of jig are reviewed by Cope (2000). The jig bed consists of a layer of coarse, heavy particles, the ragging, placed on the jig screen on to which the slurry is fed. The feed flows across the ragging and the separation takes place in the jig bed (i.e., in the ragging), so that particles with a high specific gravity penetrate through the ragging and then pass the screen to be drawn off as the heavy product, while the light particles are carried away by the cross-flow to form the light product. The type of ragging material, particle density, size, and shape, are important factors. The harmonic motion produced by the eccentric drive is supplemented by a large amount of continuously supplied hutch water, which enhances the upward and diminishes the downward velocity of the water (Figure 10.8). The combination of actions produces the segregation depicted in Figure 10.10.

One of the oldest types is the *Harz jig* (Figure 10.11) in which the plunger moves up and down vertically in a separate compartment. Up to four successive compartments are placed in series in the hutch. A high-grade heavy product is produced in the first compartment, successively lower grades being produced in the other compartments, and the light product overflowing the final

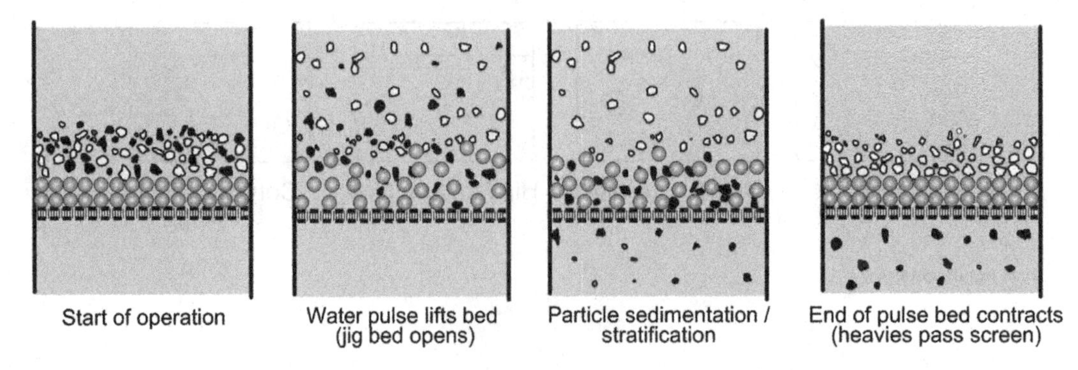

FIGURE 10.10 Sequence of events leading to separation of heavies from lights in a jig.

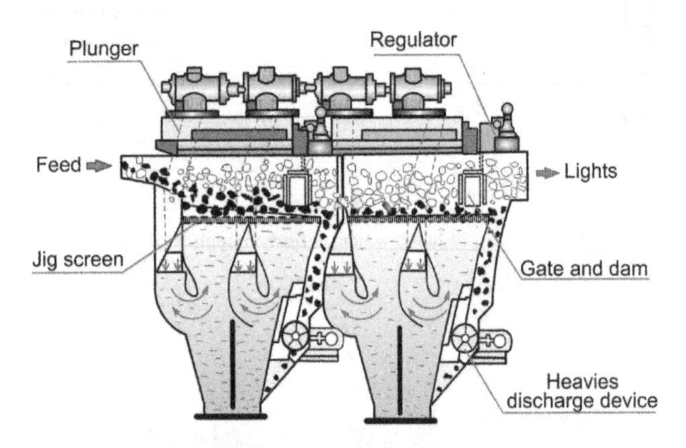

FIGURE 10.11 Harz jig *(Adapted from The Great Soviet Encyclopedia, 3rd Edn. (1970–1979))*.

compartment. If the feed particles are larger than the apertures of the screen, jigging "over the screen" is used, and the concentrate grade is partly governed by the thickness of the bottom layer, determined by the rate of withdrawal through the concentrate discharge port.

The *Denver mineral jig* (Figure 10.12) is widely used, especially for removing heavy minerals from closed grinding circuits, thus preventing over-grinding (Chapters 7 and 9). The rotary water valve can be adjusted so as to open at any desired part of the jig cycle, synchronization between the valve and the plungers being achieved by a rubber timing belt. By suitable adjustment of the valve, any desired variation can be achieved, from complete neutralization of the suction stroke with hydraulic water to a full balance between suction and pulsion.

Conventional mineral jigs consist of square or rectangular tanks, sometimes combined to form two, three, or four cells in series. In order to compensate for the increase in cross-flow velocity over the jig bed, caused by the addition of hutch water, trapezoidal-shaped jigs were developed. By arranging these as sectors of a circle, the modular *circular*, or *radial*, jig was introduced, in which the feed enters in the center and flows radially over the jig bed (and thus the cross-flow velocity is decreasing) toward the light product discharge at the circumference (Figure 10.13).

The advantage of the circular jig is its large capacity. Since their development in 1970, *IHC Radial Jigs* were installed on most newly built tin dredges in Malaysia and Thailand. In the IHC jig, the harmonic motion of the conventional eccentric-driven jig is replaced by an asymmetrical "saw-tooth" movement of the diaphragm, with a rapid upward, followed by a slow downward, stroke (Figure 10.14). This produces a much larger and more constant suction stroke, giving the finer particles more time to settle in the bed, thus reducing their loss to tailings, the jig being capable of accepting particles as fine as 60 μm.

The *InLine Pressure Jig* (IPJ) is a development in jig technology that has found a wide application for the recovery of free gold, sulfides, native copper, tin/tantalum, diamonds, and other minerals (Figure 10.15). The IPJ is unique in that it is fully encapsulated and pressurized, allowing it to be completely filled with slurry (Gray, 1997). It combines a circular bed with a vertically pulsed screen. Length of stroke and pulsation frequency, as well as screen aperture, can all be altered to suit the application. IPJs are typically installed in grinding circuits, where their low water requirements allow operators to treat the full circulating load, maximizing recovery of liberated values. Both heavy and light products are discharged under pressure.

Jigs are widely used in coal cleaning (also referred to as "coal washing") and are preferred to the more expensive DMS when the coal has relatively little middlings, or "near-gravity" material. No feed preparation is required, as is necessary with DMS, and for coals that are easily washed, that is, those consisting predominantly of liberated coal and denser "rock" particles, the lack of close density control is not a disadvantage.

Two types of air-pulsated jig—*Baum* and *Batac*—are used in the coal industry. The standard Baum jig

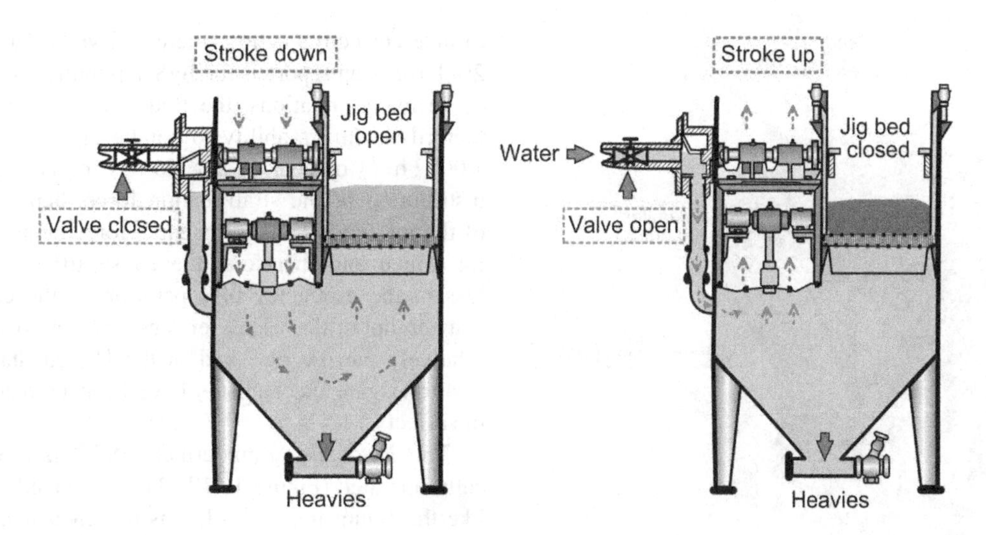

FIGURE 10.12 Denver mineral jig.

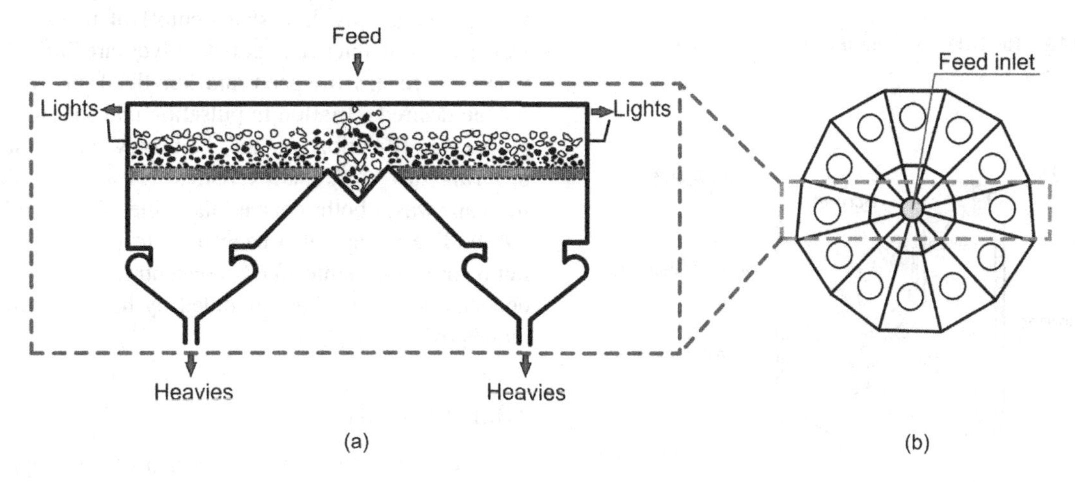

FIGURE 10.13 (a) Cross section through a circular jig, and (b) radial jig up to 12 modules.

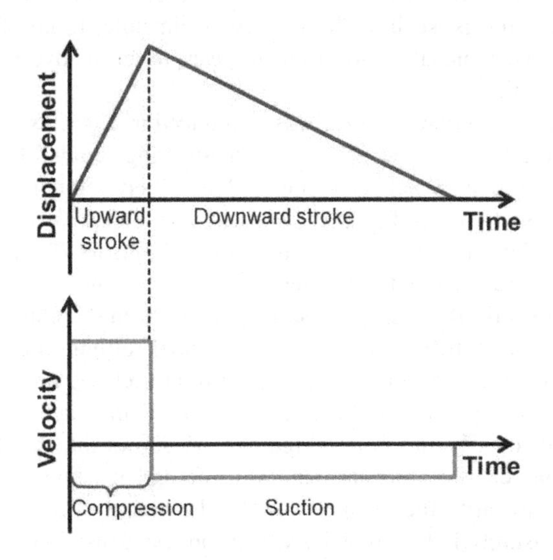

FIGURE 10.14 IHC jig drive characteristics.

(Figure 10.16), with some design modifications (Green, 1984), has been used for nearly 100 years and is still the dominant device.

Air under pressure is forced into a large air chamber on one side of the jig vessel, causing pulsation and suction to the jig water, which in turn causes pulsation and suction through the screen plates upon which the raw coal is fed, thus causing stratification. Various methods are used to continuously separate the *refuse* (heavy non-coal matter) from the lighter coal product, and all the modern Baum jigs are fitted with some form of automatic refuse extraction. One form of control incorporates a float immersed in the bed of material. The float is suitably weighted to settle on the dense layer of refuse moving across the screen plates. An increase in the depth of refuse raises the float, which automatically controls the refuse discharge, either by adjusting the height of a moving gate,

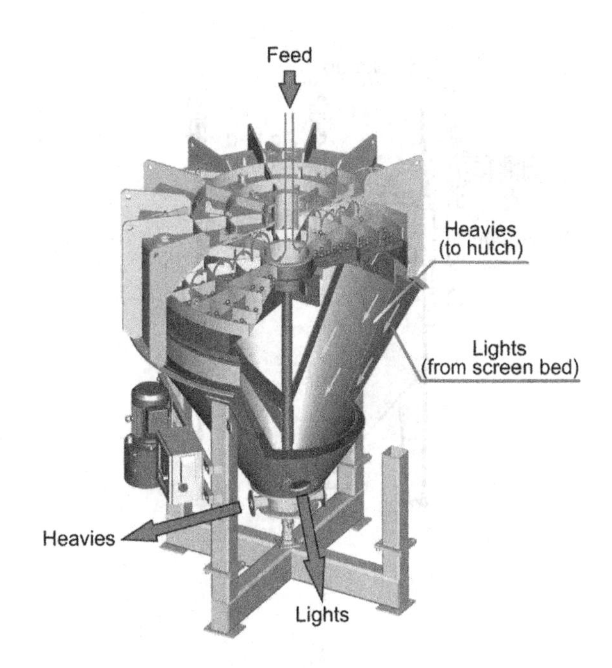

FIGURE 10.15 The Gekko Systems IPG *(Courtesy Gekko Systems)*.

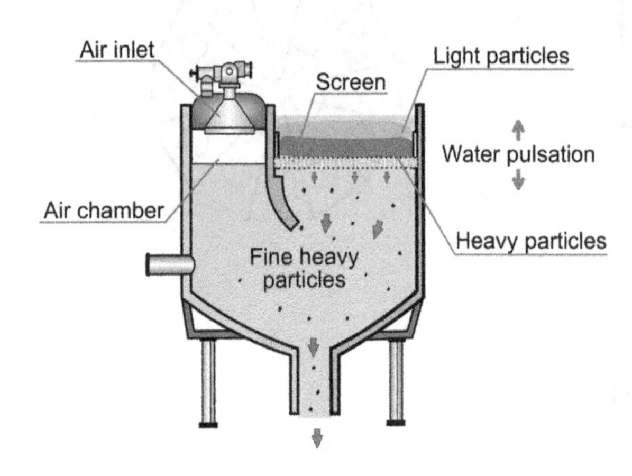

FIGURE 10.16 Baum jig.

or by controlling the pulsating water, which lifts the rejects over a fixed weir plate (Wallace, 1979). This system is reported to respond quickly and accurately.

It is now commonplace for an automatic control system to determine the variations in refuse bed thickness by measuring the differences in water pressure under the screen plates arising from the resistance offered to pulsation. The JigScan control system (developed at the Julius Kruttschnitt Mineral Research Centre) measures bed conditions and pulse velocity many times within the pulse using pressure sensing and nucleonic technology (Loveday and Jonkers, 2002). Evidence of a change in the pulse is an indicator of a problem, allowing the operator to take corrective action. Increased yields of greater than 2% have been reported for JigScan-controlled jigs.

In many situations, the Baum jig still performs satisfactorily, with its ability to handle large tonnages (up to 1,000 t h^{-1}) of coal of a wide size range. However, the distribution of the stratification force, being on one side of the jig, tends to cause unequal force along the width of jig screen and therefore uneven stratification and some loss in the efficiency of separation of the coal from its heavier impurities. This tendency is not so important in relatively narrow jigs, and in the United States multiple float and gate mechanisms have been used to counteract the effects.

The *Batac jig* (Zimmerman, 1975) is also pneumatically operated (Figure 10.17), but has no side air chamber like the Baum jig. Instead, it is designed with a series of multiple air chambers, usually two to a cell, extending under the jig for its full width, thus giving uniform air distribution. The jig uses electronically controlled air valves which provide a sharp cutoff of the air input and exhaust. Both inlet and outlet valves are infinitely variable with regard to speed and length of stroke, allowing for the desired variation in pulsation and suction by which proper stratification of the bed may be achieved for differing raw coal characteristics. As a result, the Batac jig can wash both coarse and fine sizes well (Chen, 1980). The jig has also been used to produce high-grade lump ore and sinter-feed concentrates from iron ore deposits that cannot be upgraded by heavy-medium techniques (Miller, 1991).

10.3.2 Spirals

Spiral concentrators have found many varied applications in mineral processing, but perhaps their most extensive application has been in the treatment of heavy mineral sand deposits, such as those carrying ilmenite, rutile, zircon, and monazite, and in recent years in the recovery of fine coal.

The *Humphreys spiral* was introduced in 1943, its first commercial use being on chrome-bearing sands. It is composed of a helical conduit of modified semicircular cross section. Feed pulp of between 15% and 45% solids by weight and in the size range from 3 mm to 75 μm is introduced at the top of the spiral. As it flows spirally downward, the particles stratify due to the combined effect of centrifugal force, the differential settling rates of the particles, and the effect of interstitial trickling through the flowing particle bed. The result of this action is depicted in Figure 10.18. Figure 10.19 shows the stratification across a spiral trough, with the darker heavy mineral toward the center, with the band becoming increasingly lighter radially where the less dense material flows.

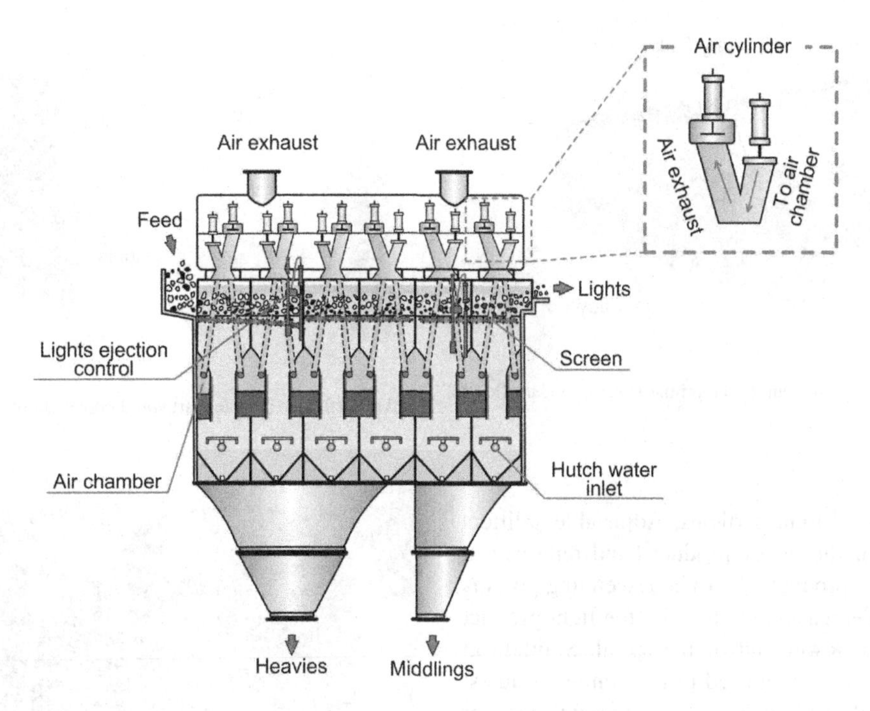

FIGURE 10.17 Batac jig.

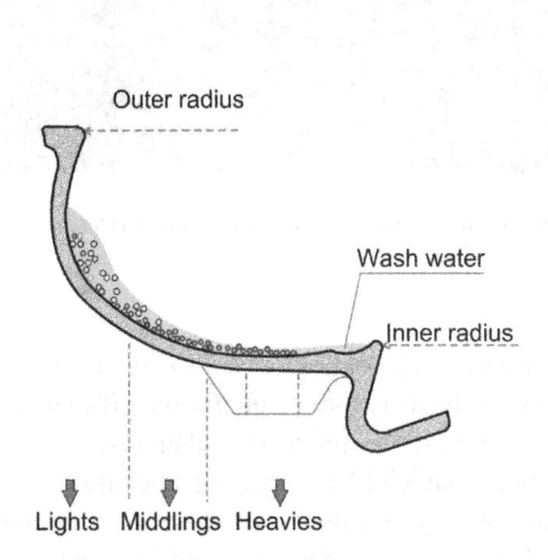

FIGURE 10.18 Cross section of spiral stream.

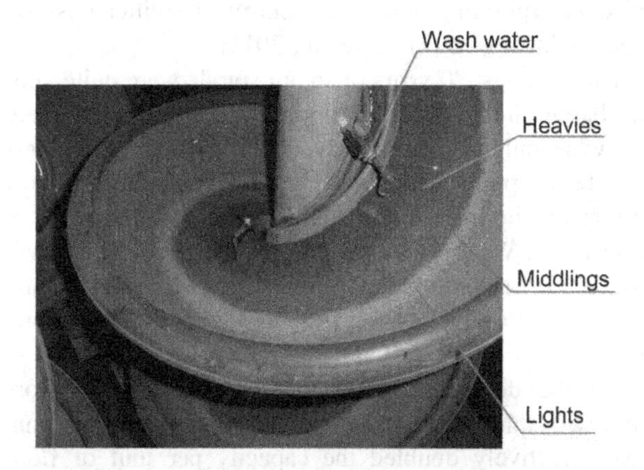

FIGURE 10.19 Example of stratification across a spiral trough, with dense, dark material in the center and lighter, less dense extending out radially *(Courtesy Multotec)*.

These mechanisms are complex, being much influenced by the slurry density and particle size. Mills (1980) reported that the main separation effect is due to hindered settling, with the largest, densest particles reporting preferentially to the band that forms along the inner edge of the stream. Bonsu (1983), however, reported that the net effect is reverse classification, the smaller, denser particles preferentially entering this band.

Determination of size-by-size recovery curves of spiral concentrators has shown that both fine and coarse dense particles are lost to the light product, the loss of coarse particles being attributed to the Bagnold force (Bazin et al., 2014). Some of the complexities of the spiral concentrator operation arise from the fact that there is not one flow pattern, but rather two: a primary flow down the spiral and a secondary flow across the trough flowing outward at the top of the stream and inward at the bottom (Figure 10.20) (Holland-Batt and Holtham, 1991).

Ports for the removal of the higher specific gravity particles are located at the lowest points in the cross section. Wash water is added at the inner edge of the stream and flows outwardly across the concentrate band to aid in

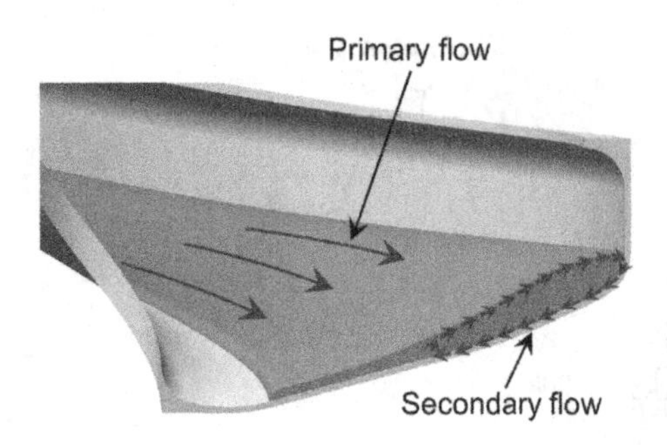

FIGURE 10.20 Schematic diagram of the primary and secondary flows in a spiral concentrator.

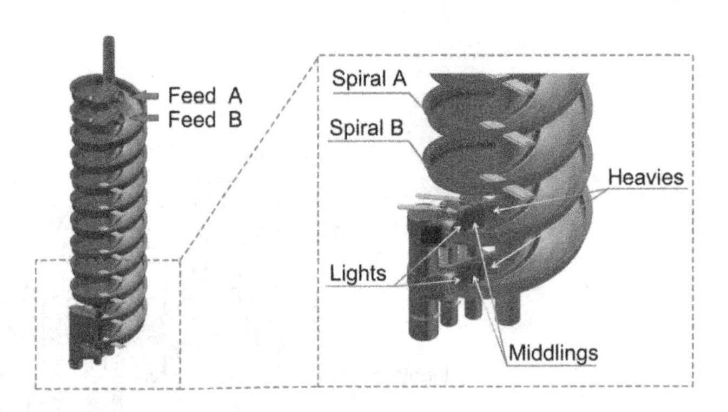

FIGURE 10.21 Double-start spiral concentrator *(Courtesy Multotec)*.

flushing out entrapped light particles. Adjustable splitters control the width of the heavy product band removed at the ports. The heavy product taken via descending ports is of a progressively decreasing grade, with the light product discharged from the lower end of the spiral. Splitters at the end of the spiral are often used to give three products: heavies, lights, and middlings, giving possibilities for recycle and retreatment in other spirals (e.g., rougher-cleaner spiral combination) or to feed other separation units. Incorporating automatic control of splitter position is being developed (Zhang et al., 2012).

Until the last 20 years or so, all spirals were quite similar, based on the original Humphreys design. Today there is a wide range of designs available. Two developments have been spirals with only one heavy product take-off port at the bottom of the spiral, and the elimination of wash water. Wash waterless-spirals reportedly offer lower cost, easier operation, and simplified maintenance, and have been installed at several gold and tin processing plants.

Another development, double-start spiral concentrators with two spirals integrated around a common column, have effectively doubled the capacity per unit of floor space (Figure 10.21). At Mount Wright in Canada, 4,300 double-start spirals have been used to upgrade specular hematite ore at 6,900 t h^{-1} at 86% recovery (Hyma and Meech, 1989). Figure 10.22 shows an installation of double-start spirals.

One of the most important developments in fine coal washing was the introduction in the 1980s of spiral separators specifically designed for coal. It is common practice to separate coal down to 0.5 mm using dense medium cyclones (Chapter 11), and below this by froth flotation. Spiral circuits have been installed to process the size range that is least effectively treated by these two methods, typically 0.1–2 mm (Honaker et al., 2008).

A notable innovation in fine coal processing is the incorporation of multistage separators and circuitry, specifically recycling middling streams (Luttrell, 2014). Both

FIGURE 10.22 Installation of double-start spirals *(Courtesy Multotec)*.

theoretical and field studies have shown that single-stage spirals have relatively poor separation efficiencies, as a compromise has to be made to either discard middlings and sacrifice coal yield or accept some middlings and a lower quality coal product. To address this, two-stage compound spirals have been designed in which clean coal and middlings are retreated in a second stage of spirals and the middlings from this spiral are recycled to the first spiral. This is essentially a rougher-cleaner closed-circuit configuration and it can be shown that his will give a higher separation efficiency than a single-stage separator (Section 10.8). The separation efficiencies achievable rival those of dense medium separators.

Some of the developments in spiral technology are the result of modeling efforts. Davies et al. (1991) reviewed the development of spiral models and described the mechanism of separation and the effects of operating parameters. A semi-empirical mathematical model of the spiral has been developed by Holland-Batt (1989). Holland-Batt (1995) discussed design aspects, such as the pitch of the trough and the trough shape. A detailed CFD model of fluid flow in a spiral has been developed and validated by Matthews et al. (1998). In Chapter 17, an example of particle tracking along a spiral used to verify model predictions is illustrated.

Spirals are made with slopes of varying steepness, the angle depending on the specific gravity of separation. Shallow angles are used, for example, to separate coal from shale, while steeper angles are used for heavy mineral–quartz separations. The steepest angles are used to separate heavy minerals from heavy waste minerals, for example, zircon (s.g. 4.7) from kyanite and staurolite (s.g. 3.6). Capacity ranges from 1 to 3 $t\ h^{-1}$ on low slope spirals to about double this for the steeper units. Spiral length is usually five or more turns for roughing duty and three turns in some cleaning units. Because treatment by spiral separators involves a multiplicity of units, the separation efficiency is very sensitive to the pulp distribution system employed. Lack of uniformity in feeding results in substantial falls in operating efficiency and can lead to severe losses in recovery. This is especially true with coal spirals (Holland-Batt, 1994).

10.3.3 Shaking Tables

When a film of water flows over a flat, inclined surface, the water closest to the surface is retarded by the friction of the water absorbed on the surface and the velocity increases toward the water surface. If mineral particles are introduced into the film, small particles will not move as rapidly as large particles, since they will be submerged in the slower-moving portion of the film. Particles of high specific gravity will move more slowly than lighter particles, and so a lateral displacement of the material will be produced (Figure 10.23).

The flowing film effectively separates coarse light particles from small dense particles, and this mechanism is exploited to some extent in the shaking-table concentrator (Figure 10.24), which is perhaps the most metallurgically efficient form of gravity concentrator, being used to treat small, more difficult flow-streams, and to produce finished concentrates from the products of other forms of gravity system.

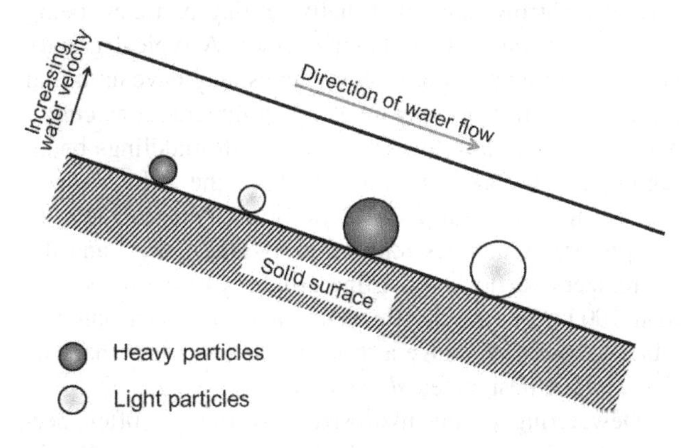

FIGURE 10.23 Action in a flowing film.

The shaking table consists of a slightly inclined deck, on to which feed, at about 25% solids by weight, is introduced at the feed box and is distributed across the table by the combination of table motion and flow of water (wash water). Wash water is distributed along the length of the feed side, and the table is vibrated longitudinally, using a slow forward stroke and a rapid return, which causes the mineral particles to "crawl" along the deck parallel to the direction of motion. The minerals are thus subjected to two forces, that due to the table motion and that, at right angles to it, due to the flowing film of water. The net effect is that the particles move diagonally across the deck from the feed end and, since the effect of the flowing film depends on the size and density of the particles, they will fan out on the table, the smaller, denser particles riding highest toward the concentrate launder at the far end, while the larger lighter particles are washed into the tailings launder, which runs along the length of the table. An adjustable splitter at the concentrate end is often used to separate this product into two fractions—a high-grade concentrate (heavy product) and a middlings fraction.

Although true flowing film concentration requires a single layer of feed, in practice, a multilayered feed is introduced onto the table, enabling much larger tonnages to be processed. Vertical stratification due to shaking action takes place behind the *riffles*, which generally run parallel with the long axis of the table and are tapered from a maximum height on the feed side, till they die out near the opposite side, part of which is left smooth. In the protected pockets behind the riffles, the particles stratify so that the finest and heaviest particles are at the bottom and the coarsest and lightest particles are at the top (Figure 10.25). Layers of particles are moved across the riffles by the crowding action of new feed and by the flowing film of wash water. Due to the taper of the riffles, progressively finer sized and higher density particles are continuously being brought into contact with the flowing film of water that tops the riffles. Final concentration takes place at the unriffled area at the end of the deck, where the layer of material at this stage is usually only one or two particles deep.

The significance of the many design and operating variables and their interactions have been reviewed by Sivamohan and Forssberg (1985a), and the development of a mathematical model of a shaking table is described by Manser et al. (1991). The separation on a shaking table is controlled by a number of operating variables, such as wash water, feed pulp density, deck slope, amplitude, and feed rate, and the importance of these variables in the model development is discussed.

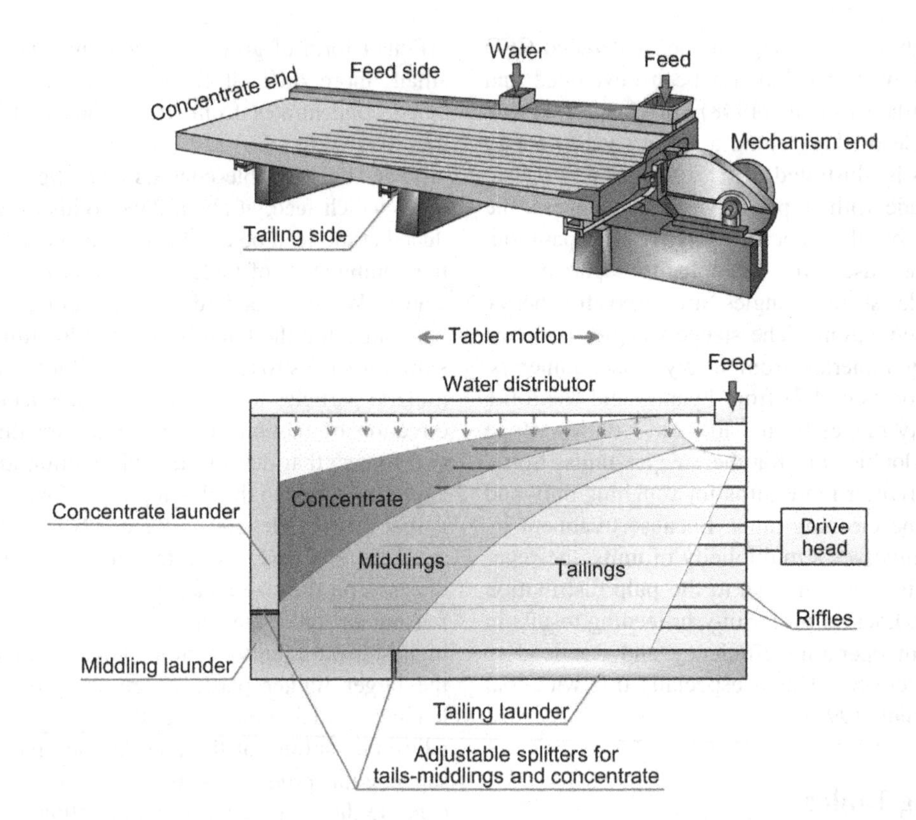

FIGURE 10.24 Shaking table (concentrate refers to heavy product, etc.).

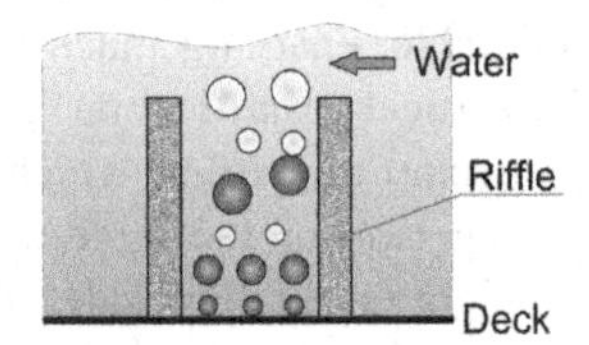

FIGURE 10.25 Vertical stratification between riffles.

Other factors, including particle shape and the type of deck, play an important part in table separations. Flat particles, such as mica, although light, do not roll easily across the deck in the water film; such particles cling to the deck and are carried down to the concentrate discharge. Likewise, spherical dense particles may move easily in the film toward the tailings launder.

The table decks are usually constructed of wood, lined with materials with a high coefficient of friction, such as linoleum, rubber, and plastics. Decks made from fibreglass are also used which, although more expensive, are extremely hard wearing. The riffles on such decks are incorporated as part of the mold.

Particle size plays an important role in table separations; as the range of sizes in a table feed increases, the efficiency of separation decreases. If a table feed is made up of a wide range of particle sizes, some of these sizes will be cleaned inefficiently. The middlings produced may not be "true middlings," that is, locked particles of associated mineral and gangue, but relatively coarse dense particles and fine light particles. If these particles are returned to the grinding circuit, together with the true middlings, then they will be needlessly reground.

Since the shaking table effectively separates coarse light from fine dense particles, it is common practice to classify the feed, since classifiers put such particles into the same product, on the basis of their equal settling rates. In order to feed as narrow a size range as possible onto the table, classification is usually performed in multi-spigot hydrosizers (Chapter 9), each spigot product, comprising a narrow range of equally settling particles, being fed to a separate set of shaking tables. A typical gravity concentrator employing shaking tables may have an initial grind in rod mills to liberate as much mineral at as coarse a size as possible to aid separation, with middlings being reground before returning to the hydrosizer. Tables operating on feed sizes in the range 3 mm to 100 μm are sometimes referred to as *sand tables*, and the hydrosizer overflow, consisting primarily of particles finer than 100 μm, is usually thickened and then distributed to tables whose decks have a series of planes, rather than riffles, and are designated *slime tables*.

Dewatering of the hydrosizer overflow is often performed by hydrocyclones, which also remove particles in the overflow smaller than about 10 μm, which will not separate efficiently by gravity methods due to their extremely slow settling rates.

Successive stages of regrinding are a feature of many gravity concentrators. The mineral is separated at all stages in as coarse a state as possible in order to achieve reasonably fast separation and hence high throughputs.

The capacity of a table varies according to size of feed particles and the concentration criteria. Tables can handle up to 2 $t\,h^{-1}$ of 1.5 mm sand and perhaps 1 $t\,h^{-1}$ of fine sand. On 100–150 μm feed materials, table capacities may be as low as 0.5 $t\,h^{-1}$. On coal feeds, however, which are often tabled at sizes of up to 15 mm, much higher capacities are common. A normal 5 mm raw coal feed can be tabled with high efficiency at 12.5 $t\,h^{-1}$ per deck, while tonnages as high as 15 $t\,h^{-1}$ per deck are not uncommon when the feed top size is 15 mm. The introduction of double and triple-deck units has improved the area/capacity ratio at the expense of some flexibility and control.

Separation can be influenced by the length of stroke, which can be altered by means of a hand-wheel on the vibrator, or head motion, and by the reciprocating speed. The length of stroke usually varies within the range of 10–25 mm or more, the speed being in the range 240–325 strokes per minute. Generally, a fine feed requires a higher speed and shorter stroke that increases in speed as it goes forward until it is jerked to a halt before being sharply reversed, allowing the particles to slide forward during most of the backward stroke due to their built-up momentum.

The quantity of water used in the feed pulp varies, but for ore-tables normal feed dilution is 20–25% solids by weight, while for coal tables pulps of 33–40% solids are used. In addition to the water in the feed pulp, clear water flows over the table for final concentrate cleaning. This varies from a few liters to almost 100 $L\,min^{-1}$ according to the nature of the feed material.

Tables slope from the feed to the tailings (light product) discharge side and the correct angle of incline is obtained by means of a handwheel. In most cases the line of separation is clearly visible on the table, so this adjustment is easily made.

The table is slightly elevated along the line of motion from the feed end to the concentrate end. The moderate slope, which the high-density particles climb more readily than the low-density minerals, greatly improves the separation, allowing much sharper cuts to be made between concentrate, middlings, and tailings. The correct amount of end elevation varies with feed size and is greatest for the coarsest and highest specific gravity feeds. The end elevation should never be less than the taper of the riffles, otherwise there is a tendency for water to flow out toward the riffle tips rather than across the riffles. Normal end elevations in ore tabling range from a maximum of 90 mm for a very heavy, coarse sand, to as little as 6 mm for an extremely fine feed.

Ore-concentrating tables are used primarily for the concentration of minerals of tin, iron, tungsten, tantalum, mica, barium, titanium, zirconium, and, to a lesser extent, gold, silver, thorium, and uranium. Tables are now being used in the recycling of electronic scrap to recover precious metals.

Duplex Concentrator

This machine was originally developed for the recovery of tin from low-grade feeds, but has a wider application in the recovery of tungsten, tantalum, gold, chromite, and platinum from fine feeds (Pearl et al., 1991). Two decks are used alternately to provide continuous feeding, the feed slurry being fed onto one of the decks, the lower density minerals running off into the discharge launder, while the heavy minerals remain on the deck. The deck is washed with water after a preset time, in order to remove the gangue minerals, after which the deck is tilted and the concentrate is washed off. One table is always concentrating, while the other is being washed or is discharging concentrates. The concentrator has a capacity of up to 5 $t\,h^{-1}$ of −100 μm feed producing enrichment ratios of between 20 and 500 and is available with various sizes and numbers of decks.

Mozley Laboratory Separator

This flowing film device, which uses orbital shear, is now used in many mineral processing laboratories and is designed to treat small samples (100 g) of ore, allowing a relatively unskilled operator to obtain information for a recovery grade curve within a very short time (Anon., 1980).

10.4 CENTRIFUGAL CONCENTRATORS

In an attempt to recover fine particles using gravity concentration methods, devices have been developed to make use of centrifugal force. The ability to change the apparent gravitational field is a major departure in the recovery of fine minerals.

Kelsey Centrifugal Jig

The Kelsey centrifugal jig (KCJ) takes a conventional jig and spins it in a centrifuge. The main operating variables adjusted to control processing different types of feed are: centrifugal acceleration, ragging material, and feed size distribution. The 16 hutch J1800 KCJ can treat over 100 $t\,h^{-1}$, depending on the application. The use of a J650 KCJ in tin recovery is described by Beniuk et al. (1994).

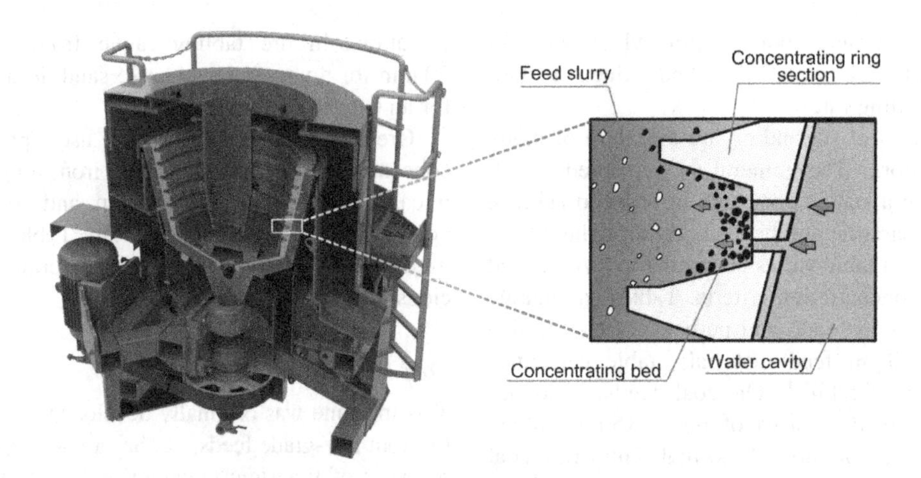

FIGURE 10.26 Knelson concentrator cutaway; and action inside the riffle *(Courtesy FLSmidth).*

Other non-jig centrifugal separators have also been developed. An applied gravitational acceleration, such as that imparted by a rapidly rotating bowl, will increase the force on fine particles, allowing for easier separation based upon differences in density. Exploiting this principle, enhanced gravity, or centrifugal, concentrators were developed initially to process fine gold ores but now are applied to other minerals.

Knelson Concentrator

This is a compact batch centrifugal separator with an active fluidized bed to capture heavy minerals (Knelson, 1992; Knelson and Jones, 1994) (Figure 10.26). A centrifugal force up to 60 times that of gravity acts on the particles, trapping denser particles in a series of rings (riffles) located in the machine, while the low-density particles are flushed out. Unit capacities range from laboratory scale to $150\,t\,h^{-1}$ for particles ranging in size from $10\,\mu m$ to a maximum of 6 mm. It is generally used for feeds in which the dense component to be recovered is a very small fraction of the total material, less than $500\,g\,t^{-1}$ (0.05% by weight).

Feed slurry is introduced through a stationary feed tube and into the concentrate cone. When the slurry reaches the bottom of the cone it is forced outward and up the cone wall under the influence of centrifugal force. Fluidization water is introduced into the concentrate cone through a series of fluidization holes (see inset in Figure 10.26). The slurry fills each ring to capacity to create a concentrating bed, with compaction of the bed prevented by the fluidization water. The flow of water that is injected into the rings is controlled to achieve optimum bed fluidization. High-specific gravity particles are captured and retained in the concentrating cone; the high-density material may also substitute for the low-density material that was previously in the riffles—made possible by the fluidization of the bed. When the concentrate cycle is complete, concentrates are flushed from the cone into the concentrate launder. Under normal operating conditions, this automated procedure is achieved in less than 2 min in a secure environment.

The units have seen a steady improvement in design from the original Manual Discharge (MD), to Centre-Discharge (CD), to Extended Duty (XD), and the Quantum Series. The first installation was in the grinding circuit at Camchib Mine, Chibougamau, Quebec, Canada, in 1987 (Nesset, 2011).

Falcon Concentrator

Another spinning batch concentrator (Figure 10.27), it is designed principally for the recovery of free gold in grinding circuit classifier underflows where, again, a very small (<1%) mass pull to concentrate is required. The feed first flows up the sides of a cone-shaped bowl, where it stratifies according to particle density before passing over a concentrate bed fluidized from behind by back-pressure (process) water. The bed retains dense particles such as gold, and lighter gangue particles are washed over the top. Periodically the feed is stopped, the bed rinsed to remove any remaining lights and is then flushed out as the heavy product. Rinsing/flushing frequency, which is under automatic control, is determined from grade and recovery requirements.

The units come in several designs, the Semi-Batch (SB), Ultrafine (UF), and i-Con, designed for small scale and artisanal miners. The first installation was at the Blackdome Gold Mine, British Columbia, Canada, in 1986 (Nesset, 2011).

These two batch centrifugal concentrators have been widely applied in the recovery of gold, platinum, silver, mercury, and native copper; continuous versions are also

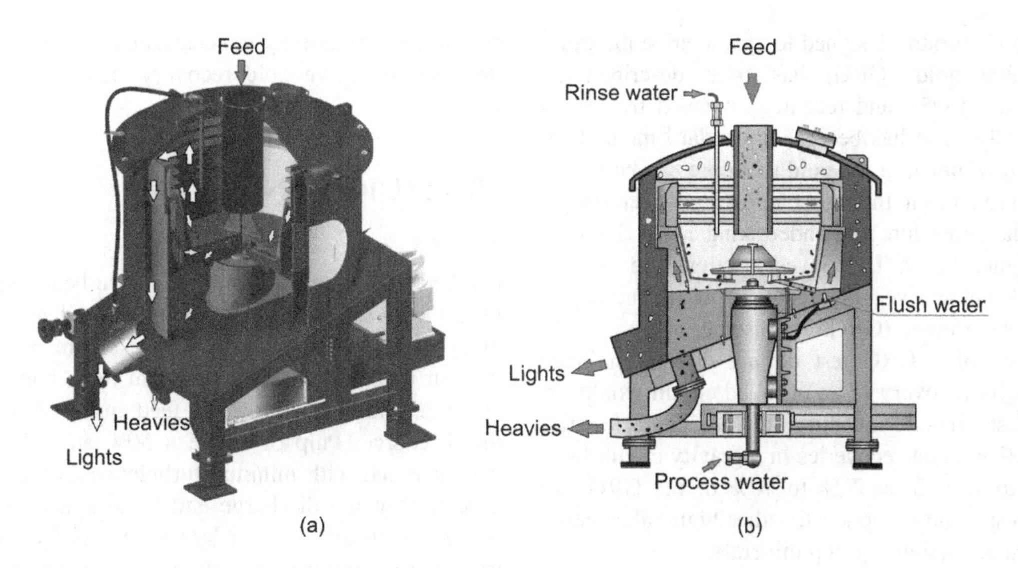

FIGURE 10.27 Falcon SB Concentrator: (a) cutaway view, and (b) flow of feed and products.

operational, the Knelson Continuous Variable Discharge (CVD) and the Falcon Continuous (C) (Klein et al., 2010; Nesset, 2011).

Multi-Gravity Separator

The principle of the multi-gravity separator (MGS) can be visualized as rolling the horizontal surface of a conventional shaking table into a drum, then rotating it so that many times the normal gravitational pull can be exerted on the mineral particles as they flow in the water layer across the surface. Figure 10.28 shows a cross section of the pilot scale MGS. The Mine Scale MGS consists of two slightly tapered open-ended drums, mounted "back to back," rotating at speeds variable between 90 and 150 rpm, enabling forces of between 5 and 15 G to be generated at the drum surfaces. A sinusoidal shake with an amplitude variable between 4 and 6 cps is superimposed on the motion of the drum, the shake imparted to one drum being balanced by the shake imparted to the other, thus balancing the whole machine. A scraper assembly is mounted within each drum on a separate concentric shaft, driven slightly faster than the drum but in the same direction. This scrapes the settled solids up the slope of the drum, during which time they are subjected to countercurrent washing before being discharged as concentrate at the open, outer, narrow end of the drum. The lower density minerals, along with the majority of the wash water, flow downstream to discharge via slots at the inner end of each drum. The MGS has been used to effect improvements in final tin concentrate grade (Turner and Hallewell, 1993).

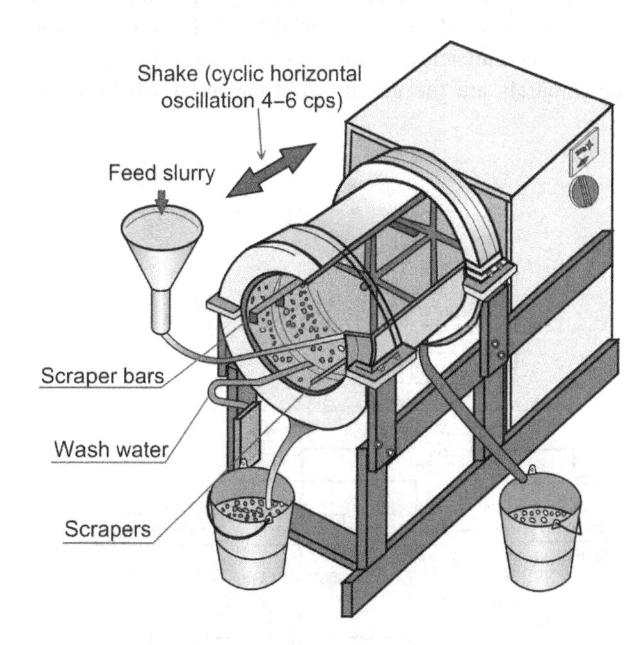

FIGURE 10.28 Pilot scale MGS.

Testing for Gravity Recoverable Gold

Although most of the gold from gold mines worldwide is recovered by dissolution in cyanide solution, a proportion of coarse (+75 μm) gold is recovered by gravity separators. It has been argued that separate treatment of the coarse gold in this way constitutes a security risk and increases costs. Gravity concentration can remain an attractive option only if it can be implemented with low capital and operating costs. A test using a laboratory centrifugal concentrator designed to characterize the gravity recoverable gold (GRG) has been described by Laplante et al. (1995), and recently reviewed by Nesset (2011). The GRG test has become a standard method for determining how much of the gold in an ore can be recovered by gravity, often through employing a centrifugal separator. The procedure for undertaking a GRG test is shown in Figure 10.29. The sieved fractions are assayed for gold and cumulative gold recovery as a function of particle size determined (GRG response curves).

The results of a GRG test do not directly indicate what the gravity recovery of an installed circuit would be. The GRG test aims to quantify the ore characteristics only. In practice, plant recoveries in a gravity circuit have been found to vary from 20% to 90% of the GRG test value. The test is now applied to other high-value dense minerals such as platinum group minerals.

Often, the gravity concentrator unit will be placed inside the closed grinding circuit, treating the hydrocyclone underflow (Chapter 7). Using GRG results, it is possible to simulate recovery by a gravity separation device, which can be used to decide on the installation and how much of the underflow to treat. In certain cases where sulfide minerals are the gravity gold carrier, flash flotation combined with gravity concentration technology provides the most effective gold recovery (Laplante and Dunne, 2002).

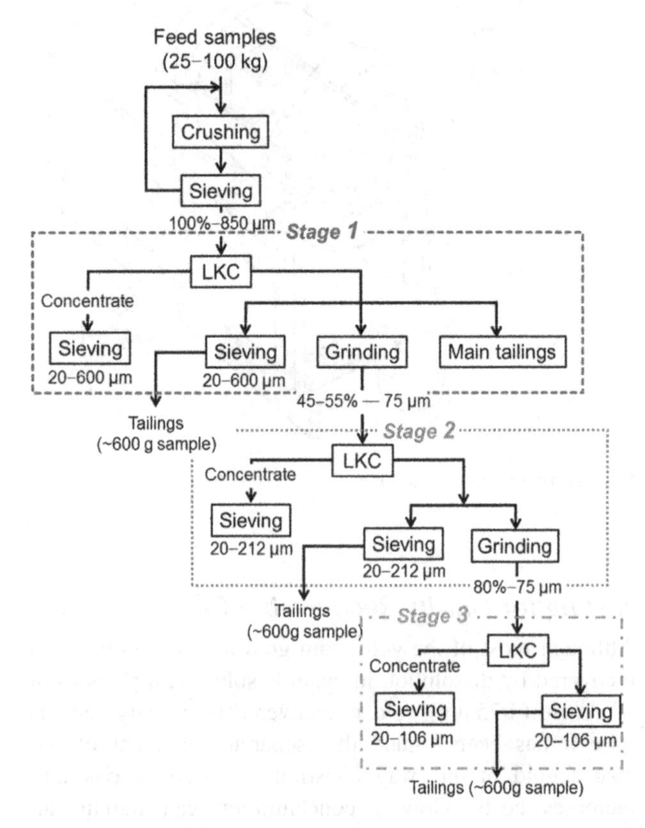

FIGURE 10.29 Flowsheet detailing the GRG test (LKC is laboratory Knelson concentrator).

10.5 SLUICES AND CONES

Sluices

Pinched sluices of various forms have been used for heavy mineral separations for centuries and are familiar in many western movies. In its simplest form (Figure 10.30), it is an inclined launder about 1 m long, narrowing from about 200 mm in width at the feed end to about 25 mm at the discharge. Pulp of between 50% and 65% solids by weight is fed with minimal turbulence and stratifies as it descends; at the discharge end these strata are separated by various means, such as by splitters, or by some type of tray (Sivamohan and Forssberg, 1985b). The fundamental basis for gravity concentration in sluices is described by Schubert (1995). The simple sluice box can be a relatively efficient gravity concentrator, provided it is correctly operated. A recent episode on the Discovery Channel showed the use of sluices in gold processing.

Reichert Cone

The Reichert cone is a wet gravity concentrating device that was designed for high-capacity applications in the early 1960s, primarily to treat titanium-bearing beach sands (Ferree, 1993). Its principle of operation is similar to the pinched sluice. Figure 10.31 shows a schematic of

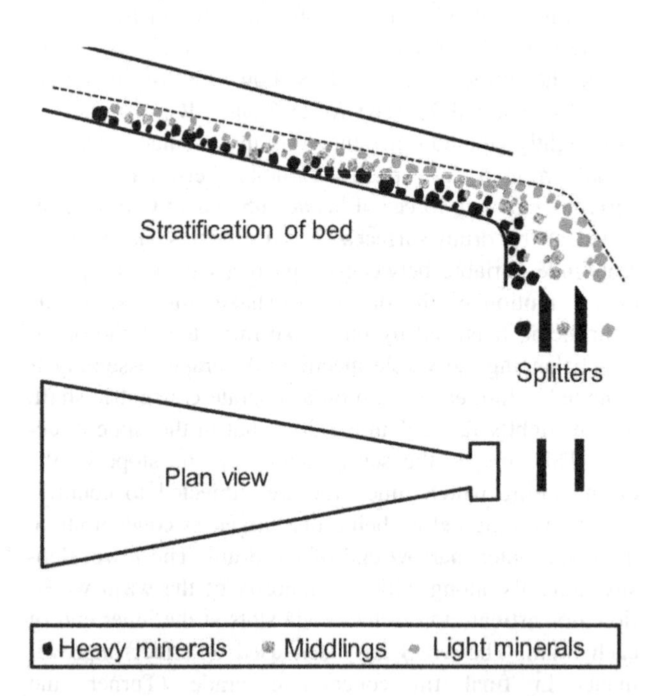

FIGURE 10.30 Pinched sluice.

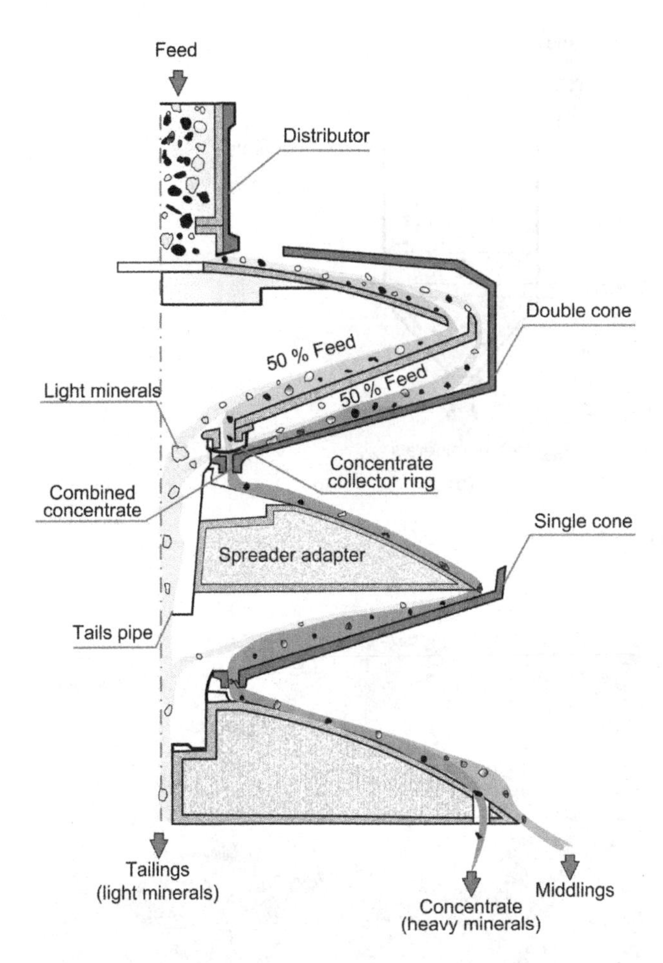

FIGURE 10.31 Schematic of the Reichert cone concentrator.

a Reichert cone, with the feed pulp being distributed evenly around the periphery of the cone. As it flows toward the center of the cone the heavy particles separate to the bottom of the film. A slot in the bottom of the concentrating cone removes this heavy product (usually the concentrate); the part of the film flowing over the slot is the light product (tailings). The efficiency of this separation process is relatively low and is repeated a number of times within a single machine to achieve effective performance. Inability to observe the separation is also a disadvantage (Honaker et al., 2014).

The success of cone circuits in the Australian mineral sand industry led to their application in other fields. At one point, Palabora Mining Co. in South Africa used 68 Reichert cones to treat 34,000 t d^{-1} of flotation tailings. However, the separation efficiency was always lower than spirals and as the design and efficiency of spiral concentrators improved, especially with the addition of double-start spirals, they started to retake the section of the processing industry that cones had initially gained. Today, there are relatively few circuits that include Reichert cones.

10.6 FLUIDIZED BED SEPARATORS

Fluidized bed separators (FBS), also known as teetered-bed or hindered-bed separators, have been in mineral processing plants for over a century. Initially used for size separation (Chapter 9), FBS units can be operated to provide efficient density-based separation in the particle size range 1–0.15 mm by exploiting hindered settling and the autogeneous dense medium naturally provided by fine high-density particles in the feed.

A typical arrangement is to feed slurry into the vessel and let the particles descend against an upward current of water. The upward velocity is set to match the settling velocity of the finest fraction of the dense particles, resulting in accumulation to form the fluidized bed. The bed level is monitored, and underflow discharge controlled to remove the heavy product at a rate dependent on the mass of heavies in the incoming feed. The lights cannot penetrate the bed and report as overflow.

Units available today include the Stokes classifier (see also Chapter 9), Lewis hydrosizer, Linatex hydrosizer, Allflux separator, and Hydrosort (Honaker et al., 2014). Some positive features of FBS units include efficient separation at flowrates up to 20 t h^{-1} m^{-2}, the capability to adjust to variations in feed characteristics, and general simplicity of operation. There is a need for close control of the top size and for clean fluidization water to avoid plugging the injection system.

There have been some significant advances in design. Noting that entering feed slurry could cause some disruption to the teeter bed, Mankosa and Luttrell (1999) developed a feed system to gently introduce the feed across the top of an FBS unit, now known as the CrossFlow™ Separator.

Another development was to combine a fluidization chamber with an upper chamber comprising a system of parallel inclined channels (Galvin et al., 2002). This unit is the Reflux Classifier™, now well established in the coal industry (Bethell, 2012).

CrossFlow™ Separator

In this device, rather than the feed entering the teeter bed, a tangential feed inlet, which increases in area to the full width of the separator to reduce input turbulence, directs the feed slurry across the top of the chamber, leaving chamber contents largely undisturbed (Figure 10.32). The upward velocity in the separator is thus constant and because the feed does not directly enter the teeter bed, variations in feed characteristics have little impact on the separation performance. A baffle plate at the discharge end of the feed inlet prevents short-circuiting of solids into the floats product.

An additional improvement was to include a slotted plate above a series of bars carrying large diameter holes

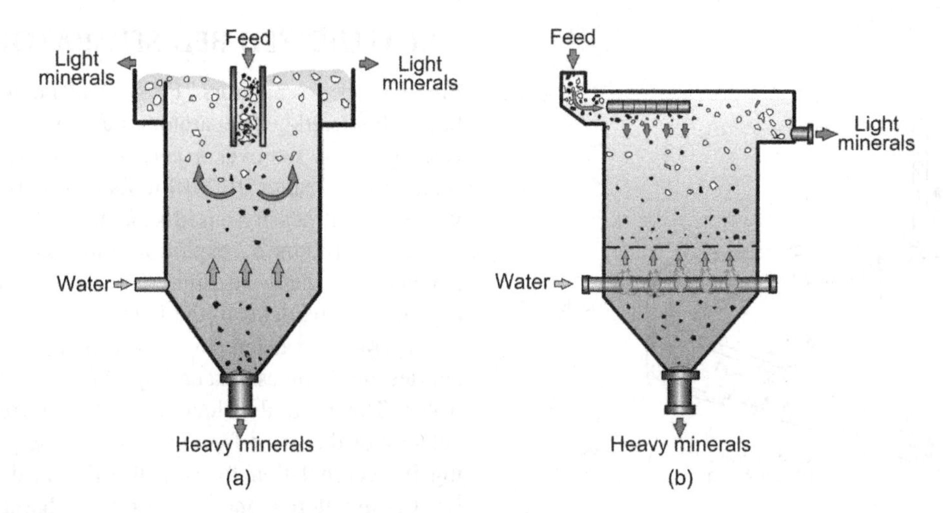

FIGURE 10.32 Comparison of (a) traditional, and (b) CrossFlow separators.

(>12.5 mm) through which the fluidization water is injected. In this arrangement, the holes are simply used to introduce the water while the slotted plate acts to distribute the water, a combination that reduces the problem of plugging faced by the prior system of distribution piping.

The new design has increased separation efficiency and throughput, which combine to reduce operating costs compared to the traditional designs. Honaker et al. (2014) report that tests have been conducted on a mineral sands application and a unit has been installed in a coal plant.

Reflux Classifier™

The Reflux Classifier is a system of parallel inclined channels above a fluidization chamber (Galvin et al., 2002; Galvin, 2012). Using closely spaced channels promotes laminar flow ("laminar-shear mechanism"), which results in fine dense particles segregating and sliding downward back to the fluidization chamber while a broad size range of light particles are transported upward to the overflow (Galvin et al., 2010). This is a version of lamella technology (see lamellae thickener, Chapter 15). An inverted version, that is, with the inlined channels at the bottom, is being developed into a flotation machine (Chapter 12, Figure 12.92).

Ghosh et al. (2012) report on an installation of two Reflux Classifiers treating 1−0.5 mm coal at $110\,\mathrm{t\,h^{-1}}$ with excellent ash reduction. Pilot tests have shown potential for use in processing fine iron ore (Amariei et al., 2014). Figure 10.33 shows a recent installation on a coal processing plant in Australia.

FIGURE 10.33 Reflux classifiers used in coal processing at site in Australia *(Courtesy FLSmidth).*

10.7 DRY PROCESSING

Honaker et al. (2014) review the historical development of using air as the medium to achieve gravity-based separations. The initial prime application was in coal preparation, peaking in tons treated in the United States around the mid-1960s. The pneumatic technologies follow the same basic mechanisms described for wet processing. With demand to reduce water usage growing, these technologies are worth reexamining.

Pneumatic Tables

These were initially the most common pneumatic gravity devices. They use the same throwing motion as their wet counterparts to move the feed along a flat riffled deck, while blowing air continuously up through a porous bed. The stratification produced is somewhat different from that of wet tables. Whereas in wet tabling the particle size increases and the density decreases from the top of the concentrate band to the tailings, on an air table both particle size and density decrease from the top down, the coarsest particles in the middlings band having the lowest density. Pneumatic tabling is therefore similar in effect to hydraulic classification. They are commonly used in combination with wet tables to clean zircon concentrates, one of the products obtained from heavy mineral sand deposits. Such concentrates are often contaminated with small amounts of fine quartz, which can effectively be separated from the coarse zircon particles by air tabling. Some fine zircon may be lost in the tailings and can be recovered by treatment on wet shaking tables. Recent testwork into air tabling for coal is detailed by Honaker et al. (2008) and Gupta et al. (2012). A modified air-table, the FGX Dry Separator, from China is modeled by Akbari et al. (2012).

Air Jigs

Gaudin (1939) describes devices common in the early part of the last century. Two modern descendants are the Stump jig and the Allair jig (Honaker et al., 2014). These units employ a constant air flow through a jig bed supported on a fixed screen in order to open the bed and allow stratification. Bed level is sensed and used to control the discharge rate in proportion to the amount of material to be rejected. There are several installations around the world, mostly for coal cleaning. Commercial units, for example, can treat up to 60 t h^{-1} of 75–12 mm coal (Honaker et al., 2014).

Other Pneumatic-Based Devices

Various units have been modified to operate dry including: the Reflux Classifier (Macpherson and Galvin, 2010), Knelson concentrator (Greenwood et al., 2013; Kökkılıç et al., 2015), Reichert cone (Rotich et al., 2013), and fluidized bed devices (Franks et al., 2013). The latter devices overlap with DMS. For example, applications in coal employ a dense medium of air and magnetite, the air dense medium fluidized bed (ADMFB) process. The first commercial ADMFB installation was a 50 t h^{-1} unit in China in 1994 (Honaker et al., 2014). Various ADMFB processes are reviewed by Sahu et al. (2009).

10.8 SINGLE-STAGE UNITS AND CIRCUITS

Single Versus Two Stages of Spirals

As noted, one of the innovations in coal processing is to incorporate circuits. By making some simplifying assumptions, it is possible to deduce an analytical solution for a circuit, as Luttrell (2002) demonstrates based on the work of Meloy (1983). (The same approach is used quite extensively in Chapter 12, Section 12.11.) Figure 10.34(a) shows two stages of spiral in a possible coal application: the rougher gives a final discard heavy product (refuse) and light product that is sent to a cleaner stage which gives the final light product (i.e., clean coal) with the heavy product (middlings) being recycled to the rougher. For this rougher-cleaner circuit the solution for the circuit recovery R_{circ} is given by a mass balance across the dashed box.

Letting the feed to the rougher be X (mass units per unit time) then:

$$F = XR_cR_r + X(1 - R_r) \tag{10.5}$$

and thus circuit recovery is:

$$R_{circ} = \frac{R_cR_r}{R_cR_r + (1 - R_r)} \tag{10.6}$$

For each unit there is a partition curve, that is, recovery to specified concentrate as a function of particle density. Figure 10.35 shows a generic partition curve in reduced form (Chapter 1), where the abscissae is density divided by the density corresponding to 50% recovery, that is, ρ/ρ_{50}. (Note the values of ρ/ρ_{50} decrease from left

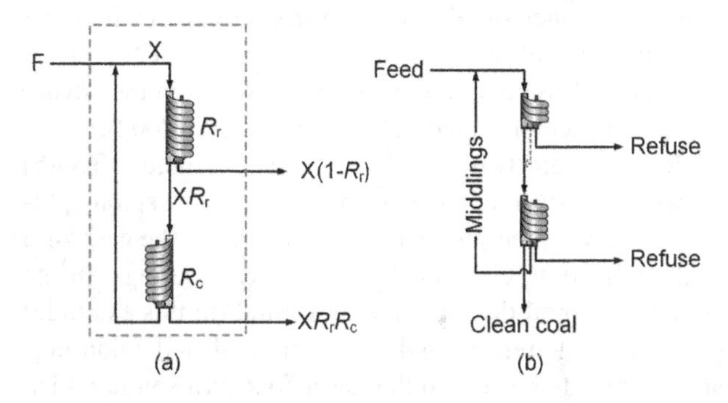

FIGURE 10.34 (a) Two stages of spirals in rougher-cleaner spiral combination, and (b) a two-stage compound spiral.

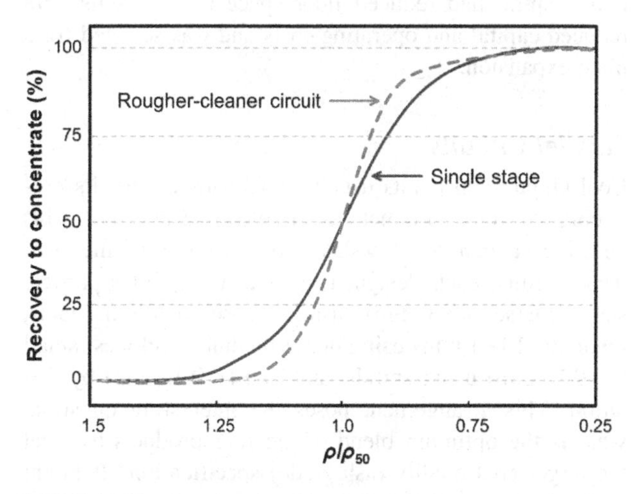

FIGURE 10.35 Generic reduced form of partition curve, showing slope for rougher-cleaner circuit with middlings recycle (Figure 10.34(a)), dashed line, is superior to the single unit, solid line.

to right, indicating the concentrate is the light product, e.g., clean coal.) The slope at any point is the sharpness of the separation (the ability to separate between density classes). Assuming the same recovery point for both units, we can differentiate to solve for the sharpness of separation (slope) of the circuit, namely:

$$\frac{dR_{circ}}{dR} = \frac{2R - R^2}{(R^2 - R + 1)^2} \quad (10.7)$$

A convenient point for comparison is the slope at $\rho/\rho_{50} = 1$, or $R = 0.5$, which upon substituting in Eq. (10.7) gives $dR_{circ}/dR = 1.33$; in other words, the slope at $\rho/\rho_{50} = 1$ for the circuit is 1.33 times that for the single unit. This increase in sharpness is illustrated in Figure 10.35 (dashed line). A variety of circuits can be analyzed using this approach (Noble and Luttrell, 2014).

The circuit, in effect, compensates for deficiencies in the stage separation. When the stage separation efficiency is high, such as in DMS (Chapter 11), circuits offer less benefit. But this is rather the exception and is why circuits are widely used in mineral processing, for example, rougher-cleaner-recleaner spiral circuits in iron ore processing, and the wide range of flotation circuits is evident in Chapter 12. Circuits are less common in coal processing, but the advantage can be demonstrated (Bethell and Arnold, 2002).

Rather than two stages of spirals, Figure 10.34(b) shows a two-stage compound spiral. Typically, spiral splitters will give three products, for example, in the case of a coal application: clean coal, refuse, and middlings. In the compound spiral the top three-turn spiral (in this example) produces final refuse, and the clean coal and middlings are combined and sent to the lower four-turn spiral, which produces another final refuse, final clean coal, and a middlings, which is recycled. Bethell and Arnold (2002) report that compared to two stages of spirals, the compound spiral had reduced floor space requirements with reduced capital and operating costs and was selected for a plant expansion.

Parallel Circuits

Coal preparation plants usually have parallel circuits producing a final product as a blend. Luttrell (2014) describes a generic flowsheet comprising four independent circuits, each designed to treat a particular particle size: coarse (>10 mm) using dense medium baths, medium (10–1 mm) using dense medium cyclones, small (1–0.15 mm) using spirals, and fine (<0.15 mm) by flotation. This arrangement poses the interesting question: what is the optimum blend of the four products to meet the target coal quality (ash grade) specification? It might seem that controlling all four to produce the same ash content would be the answer. However as Luttrell (2002) shows, the answer is to blend when each circuit has achieved the same target *incremental* ash grade. (The increment in grade can be understood using the grade-gradient plot in Appendix III; the tangent to the operating line is the incremental grade.) In this manner the yield of coal product will be maximized at the target ash grade for the blend; that is, in this respect the process is optimized. Since coal feeds are basically a two-density mineral situation (coal and ash minerals), then the density of composite (locked) coal-ash mineral particles will exactly reflect the composition, that is, the ash grade. This means that the density cutpoint for each circuit should be the same so that the increment of product from each at the cutpoint density has the same increment in density and thus the same increment in ash grade. These considerations are of less import in most ore processing plants, which usually comprise one flowsheet producing final concentrate, but whenever more than one independent product is being combined to produce final concentrate this same question of the optimized blend arises.

REFERENCES

Agricola, G., 1556. De Re Metallica. Translated by Hoover, H.C., Hoover, L.H., Dover Publications, Inc. New York, NY, USA, 1950.

Akbari, H., et al., 2012. Application of neural network for modeling the coal cleaning performance of the FGX dry separator. In: Young, C.A., Luttrell, G.H. (Eds.), Separation Technologies for Minerals, Coal, and Earth Resources. SME, Englewood, CO, USA, pp. 189–197.

Amariei, D., et al., 2014. The use of a Reflux Classifier for iron ores: assessment of fine particles recovery at pilot scale. Miner. Eng. 62, 66–73.

Anon, 1980. Laboratory separator modification improves recovery of coarse grained heavy minerals. Mining Mag. 142-143 (Aug.), 158–161.

Anon, 2011. Basics in Mineral Processing. eighth ed. Metso Corporation.

Bazin, C., et al., 2014. Simulation of an iron ore concentration circuit using mineral size recovery curves of industrial spirals. Proceedings of the 46th Annual Meeting of the Canadian Mineral Processors Conference, CIM, Ottawa, ON, Canada, pp. 387–402.

Beniuk, V.G., et al., 1994. Centrifugal jigging of gravity concentrate and tailing at Renison Limited. Miner. Eng. 7 (5-6), 577–589.

Bethell, P.J., 2012. Dealing with the challenges facing global coal preparation. In: Klima, M.S., et al., (Eds.), Challenges in Fine Coal Processing, Dewatering, and Disposal. SME, Englewood, CO, USA, pp. 33–45.

Bethell, P.J., Arnold, B.J., 2002. Comparing a two-stage spiral to two stages of spirals for fine coal preparation. In: Honaker, R.Q., Forrest, W.R. (Eds.), Advances in Gravity Concentration. SME, Littleton, CO, USA, pp. 107–114.

Bonsu, A.K., 1983. Influence of Pulp Density and Particle Size on Spiral Concentration Efficiency. Thesis (M. Phil.), Camborne School of Mines, University of New South Wales (UNSW), Australia.

Burt, R.O., 1985. *Gravity Concentration Technology*. (Development in Mineral Processing Series, vol. 5. Elsevier, Amsterdam, The Netherlands.

Chen, W.L., 1980. Batac jig cleaning in five U.S. plants. Mining Eng. 32 (9), 1346–1350.

Cope, L.W., 2000. Jigs: the forgotten machine. Eng. Min. J. 201 (8), 30–34.

Davies, P.O.J., et al., 1991. Recent developments in spiral design, construction and application. Miner. Eng. 4 (3-4), 437–456.

Ferree, T.J., 1993. Application of MDL Reichert cone and spiral concentrators for the separation of heavy minerals. CIM Bull. 86 (975), 35–39.

Franks, G.V., et al., 2013. Copper ore density separations by float/sink in a dry sand fluidised bed dense medium. Int. J. Miner. Process. 121 (10), 12–20.

Galvin, K.P., 2012. Development of the Reflux Classifier. In: Klima, M. S., et al., (Eds.), Challenges in Fine Coal Processing, Dewatering, and Disposal. SME, Englewood, CO, USA, pp. 159–186.

Galvin, K.P., et al., 2002. Pilot plant trial of the reflux classifier. Miner. Eng. 15 (1-2), 19–25.

Galvin, K.P., et al., 2010. Application of closely spaced inclined channels in gravity separation of fine particles. Miner. Eng. 23 (4), 326–338.

Gaudin, A.M., 1939. Principles of Mineral Dressing. McGraw-Hill Book Company, Inc., London, England.

Ghosh, T., et al., 2012. Performance evaluation and optimization of a full-scale Reflux classifier. Coal Prep. Soc. Amer. J. 11 (2), 24–33.

Gray, A.H., 1997. InLine pressure jig—an exciting, low cost technology with significant operational benefits in gravity separation of minerals. Proceedings of the AusIMM Annual Conference, AusIMM, Ballarat, VIC, pp. 259–266.

Green, P., 1984. Designers improve jig efficiency. Coal Age. 89 (1), 50–53.

Greenwood, M., et al., 2013. The potential for dry processing using a Knelson concentrator. Miner. Eng. 45, 44–46.

Gupta, N., et al., 2012. Application of air table technology for cleaning Indian coal. In: Young, C.A., Luttrell, G.H. (Eds.), Separation Technologies for Minerals, Coal, and Earth Resources. SME, Englewood, CO, USA, pp. 199–210.

Holland-Batt, A.B., 1989. Spiral separation: theory and simulation. Trans. Inst. Min. Metall. Sec. C. 98 (Jan.-Apr.), C46–C60.

Holland-Batt, A.B., 1994. The effect of feed rate on the performance of coal spirals. Coal Preparation. 14 (3-4), 199–222.

Holland-Batt, A.B., 1995. Some design considerations for spiral separators. Miner. Eng. 8 (11), 1381–1395.

Holland-Batt, A.B., 1998. Gravity separation: a revitalized technology. Mining Eng. 50 (Sep), 43–48.

Holland-Batt, A.B., Holtham, P.N., 1991. Particle and fluid motion on spiral separators. Miner. Eng. 4 (3-4), 457–482.

Honaker, R.Q., et al., 2008. Upgrading coal using a pneumatic density-based separator. Int. J. Coal Prep. Utils. 28 (1), 51–67.

Honaker, R., et al., 2014. Density-based separation innovations in coal and minerals processing applications. In: Anderson, C.G., et al., (Eds.), Mineral Processing and Extractive Metallurgy: 100 Years of Innovation. SME, Englewood, CO, USA, pp. 243–264.

Hyma, D.B., Meech, J.A., 1989. Preliminary tests to improve the iron recovery from the -212 micron fraction of new spiral feed at Quebec Cartier Mining Company. Miner. Eng. 2 (4), 481–488.

Jonkers, A., et al., 2002. Advances in modelling of stratification in jigs. Proceedings of the 13th International Coal Preparation Congress, vol. 1, Jonannesburg, South Africa, pp. 266–276.

Klein, B., et al., 2010. A hybrid flotation-gravity circuit for improved metal recovery. Int. J. Miner. Process. 94 (3-4), 159–165.

Knelson, B., 1992. The Knelson concentrator. Metamorphosis from crude beginning to sophisticated world wide acceptance. Miner. Eng. 5 (10-12), 1091–1097.

Knelson, B., Jones, R., 1994. "A new generation of Knelson concentrators" a totally secure system goes on line. Miner. Eng. 7 (2-3), 201–207.

Kökkılıç, O., et al., 2015. A design of experiments investigation into dry separation using a Knelson concentrator. Miner. Eng. 72, 73–86.

Laplante, A., Dunne, R.C., 2002. The gravity recoverable gold test and flash flotation. Proceedings of the 34th Annual Meeting of the Canadian Mineral Processors Conference, Ottawa, ON, Canada, pp. 105–124.

Laplante, A.R., et al., 1995. Predicting gravity separation gold recoveries. Miner. Metall. Process. 12 (2), 74–79.

Loveday, G., Jonkers, A., 2002. The Apic jig and the JigScan controller take the guesswork out of jigging. Proceedings of the 14th International Coal Preparation Congress and Exhibition, Johannesburg, South Africa, pp. 247–251.

Luttrell, G.H., 2002. Density separation: are we really making use of existing process engineering knowledge? In: Honaker, R.Q., Forrest, W.R. (Eds.), Advances in Gravity Concentration. SME, Littleton, CO, USA, pp. 1–16.

Luttrell, G.H., 2014. Innovations in coal processing. In: Anderson, C.G., et al., (Eds.), Mineral Processing and Extractive Metallurgy: 100 Years of Innovation. SME, Englewood, CO, USA, pp. 277–296.

Lyman, G.J., 1992. Review of jigging principles and control. Coal Preparation. 11 (3-4), 145–165.

Macpherson, S.A., Galvin, K.P., 2010. The effect of vibration on dry coal beneficiation in the Reflux classifier. Int. J. Coal Prep. Util. 30 (6), 283–294.

Mankosa, M.J., Luttrell, G.H., 1999. Hindered-Bed Separator Device and Method. Patent No. US 6264040 B1. United States Patent Office.

Manser, R.J., et al., 1991. The shaking table concentrator: the influence of operating conditions and table parameters on mineral separation—the development of a mathematical model for normal operating conditions. Miner. Eng. 4 (3-4), 369–381.

Matthews, B.W., et al., 1998. Fluid and particulate flow on spiral concentrators: computational simulation and validation. Appl. Math. Model. 22 (12), 965–979.

Meloy, T.P., 1983. Analysis and optimization of mineral processing and coal-cleaning circuits—circuit analysis. Int. J. Miner. Process. 10 (1), 61–80.

Miller, D.J., 1991. Design and operating experience with the Goldsworthy Mining Limited BATAC jig and spiral separator iron ore beneficiation plant. Miner. Eng. 4 (3-4), 411–435.

Mills, C., 1980. Process design, scale-up and plant design for gravity concentration. In: Mular, A.L., Bhappu, R.B. (Eds.), Mineral Processing Plant Design, second ed. AIMME, New York, NY, pp. 404–426. (Chapter 18).

Mishra, B.K., Mehrotra, S.P., 1998. Modelling of particle stratification in jigs by the discrete element method. Miner. Eng. 11 (6), 511–522.

Mishra, B.K., Mehrotra, S.P., 2001. A jig model based on the discrete element method and its experimental validation. Int. J. Miner. Process. 63 (4), 177–189.

Nesset, J.E., 2011. Significant Canadian developments in mineral processing technology—1961 to 2011. In: Kapusta, J., et al., (Eds.), The Canadian Metallurgical & Materials Landscape 1960 to 2011. MetSoc, CIM, Westmount, Montreal, Quebec, Canada, pp. 241–293.

Noble, A., Luttrell, G.H., 2014. The matrix reduction algorithm for solving separation circuits. Miner. Eng. 64, 97–108.

Pearl, M., et al., 1991. A mathematical model of the duplex concentrator. Miner. Eng. 4 (3-4), 347–354.

Priester, M., et al., 1993. *Tools for Mining: Techniques and Processes for Small Scale Mining*. Informatica International, Incorporated (Publisher) Braunschweig, Lower Saxony, Germany.

Rotich, N., et al., 2013. Modeling and simulation of gravitational solid-solid separation for optimum performance. Powder Technol. 239, 337–347.

Sahu, A.K., et al., 2009. Development of air dense medium fluidised bed technology for dry beneficiation of coal—a review. Int. J. Coal Prep. Util. 29 (4), 216–2241.

Schubert, H., 1995. On the fundamentals of gravity concentration in sluices and spirals. Aufbereitungs Technik. 36 (11), 497–505.

Sivamohan, R., Forssberg, E., 1985a. Principles of tabling. Int. J. Miner. Process. 15 (4), 281–295.

Sivamohan, R., Forssberg, E., 1985b. Principles of sluicing. Int. J. Miner. Process. 15 (3), 157–171.

Turner, J.W.G., Hallewell, M.P., 1993. Process improvements for fine cassiterite recovery at Wheal Jane. Miner. Eng. 6 (8-10), 817–829.

Wallace, W.M., 1979. Electronically controlled Baum jig washing. Mine & Quarry. 8 (7), 43–45.

Wells, A., 1991. Some experiences in the design and optimisation of fine gravity concentration circuits. Miner. Eng. 4 (3-4), 383–398.

Xia, Y., et al., 2007. CFD simulation of fine coal segregation and stratification in jigs. Int. J. Miner. Process. 82 (3), 164–176.

Zhang, B., et al., 2012. Development of an automatic control system for spiral concentrator; phase 1. In: Young, C.A., Luttrell, G.H. (Eds.), Separation Technologies for Minerals, Coal, and Earth Resources. SME, Englewood, CO, USA, pp. 497–508.

Zimmerman, R.E., 1975. Performance of the Batac jig for cleaning fine and coarse coal sizes. Proceedings of the AIME Annual Meeting, AIME, Dallas, TX. Preprint 74-F-18: 1–25.

Chapter 11

Dense Medium Separation (DMS)

11.1 INTRODUCTION

Dense medium separation (DMS) is also known as heavy medium separation (HMS) or the sink-and-float process. It has two principal applications: the preconcentration of minerals, that is, the rejection of gangue prior to grinding for final liberation, and in coal preparation to produce a commercially graded end-product, that is, clean coal being separated from the heavier shale or high-ash coal. The history of the process, innovations, and failures are reviewed by Napier-Munn et al. (2014).

In principle, it is the simplest of all gravity processes and has long been a standard laboratory method for separating minerals of different specific gravity. Heavy liquids of suitable density are used, so that those minerals less dense ("lighter") than the liquid float, while those denser ("heavier") than it sink (Figure 11.1). Rather than quoting a density value, it is common to refer to specific gravity (s.g.), "relative density" (RD), or simply "density."

Since most of heavy liquids are expensive or toxic, the dense medium used in industrial separations is a suspension of particles of some dense solid in water, which behaves as a heavy liquid.

The process offers some advantages over other gravity processes. It has the ability to make sharp separations at any required density, with a high degree of efficiency, even in the presence of high percentages of near-density material (or near-gravity material, i.e., material close to the desired density of separation). The density of separation can be closely controlled, within a RD of ± 0.005, and can be maintained under normal conditions for indefinite periods. The separating density can be changed as required and fairly quickly, to meet varying requirements. The process is, however, rather expensive, mainly due to the ancillary equipment needed to clean and recycle the medium, and the cost of the medium itself.

For preconcentration, DMS is applicable to any ore in which, after a suitable degree of liberation by comminution, there is enough difference in specific gravity between the particles to separate those which will repay the cost of further treatment from those which will not. The process is most widely applied when the density difference occurs at a coarse particle size, for example, after crushing, as separation efficiency decreases with size due to the slower rate of settling of the particles. Particles should preferably be larger than about 4 mm in diameter, in which case separation can be effective on a difference in specific gravity of 0.1 or less. If the values are finely disseminated throughout the host rock, then a suitable density difference between the crushed particles cannot be developed by coarse crushing.

Providing that a density difference exists, there is no upper size limit except that determined by the ability of the plant to handle the material. Separation down to 500 μm, and less, can be facilitated by the use of centrifugal separators.

Preconcentration is most often performed on metalliferous ores that are associated with relatively light country rock, such as silicates and carbonates. Lead–zinc (galena–sphalerite) ores can be candidates, examples being the operations at Mount Isa (Queensland, Australia) and the Sullivan concentrator (British Columbia, Canada (now closed)). In some of the Cornish tin ores, the cassiterite is found in lodes with some degree of banded structure in which it is associated with other high-specific-gravity minerals such as the sulfides of iron, arsenic, and copper, as well as iron oxides. The lode fragments containing these minerals therefore have a greater density than the siliceous waste and allow early separation.

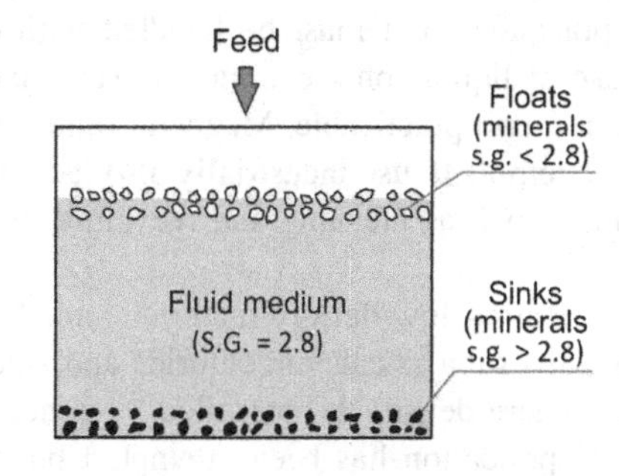

FIGURE 11.1 Principle of DMS (fluid medium s.g. = 2.8 assumed for illustration).

Wills' Mineral Processing Technology.

11.2 THE DENSE MEDIUM

11.2.1 Liquids

Heavy liquids have wide use in the laboratory for the appraisal of gravity-separation techniques on ores. Heavy liquid testing may be performed to determine the feasibility of DMS on a particular ore and to determine the economic separating density, or it may be used to assess the efficiency of an existing dense medium circuit by carrying out tests on the sink and float products. The aim is to separate the ore samples into a series of fractions according to density, establishing the relationship between the high- and the low-specific-gravity minerals (see Section 11.6).

Tetrabromoethane, having a specific gravity of 2.96, is commonly used and may be diluted with white spirit or carbon tetrachloride (s.g. 1.58) to give a range of densities below 2.96.

Bromoform (s.g. 2.89) may be mixed with carbon tetrachloride to give densities in the range 1.58–2.89. For densities up to 3.3, diiodomethane is useful, diluted as required with triethyl orthophosphate. Aqueous solutions of sodium polytungstate have certain advantages over organic liquids, such as being virtually nonvolatile, nontoxic, and of lower viscosity, and densities of up to 3.1 can easily be achieved (Plewinsky and Kamps, 1984).

For higher density separations, Clerici solution (thallium formate–thallium malonate solution) allows separation at densities up to specific gravity 4.2 at 20°C or 5.0 at 90°C. Separations of up to specific gravity 18 can be achieved by the use of *magneto-hydrostatics*, that is, the utilization of the supplementary weighting force produced in a solution of a paramagnetic salt or ferrofluid when situated in a magnetic field gradient. This type of separation is applicable primarily to nonmagnetic minerals with a lower limiting particle size of about 50 μm (Parsonage, 1980; Domenico et al., 1994). Lin et al. (1995) describe a modification to the Franz Isodynamic Separator (Chapter 13) for use with magnetic fluids.

Many heavy liquids give off toxic fumes and must be used with adequate ventilation: the Clerici liquids are extremely poisonous and must be handled with extreme care. The use of liquids on a commercial scale has therefore not been found practicable. Magnetic fluids avoid the toxicity but attempts to use industrially also face problems of practicality, such as cleaning and recycling the expensive fluids.

For fractionating low-density materials, notably coals, solutions of salts such as calcium chloride and zinc sulfate can be used where density is controlled by concentration. Commercial application has been attempted but the problems encountered reclaiming the salts for recycle have proven difficult to surmount.

11.2.2 Suspensions

Below a concentration of about 15% by volume, finely ground suspensions in water behave essentially as simple Newtonian fluids. Above this concentration, however, the suspension becomes non-Newtonian and a certain minimum stress, or yield stress, has to be applied before shear will occur and the movement of a particle can commence. Thus, small particles, or those close to the medium density, are unable to overcome the resistance offered by the medium before movement can be achieved. This can be solved to some extent either by increasing the shearing forces on the particles or by decreasing the apparent viscosity of the suspension. The shearing force may be increased by substituting centrifugal force for gravity. The viscous effect may be decreased by agitating the medium, which causes elements of liquid to be sheared relative to each other. In practice, the medium is never static, as motion is imparted to it by paddles, air, etc., and also by the sinking material itself. All these factors, by reducing the yield stress, tend to bring the parting or separating density as close as possible to the density of the medium in the bath.

In order to produce a stable suspension of sufficiently high density, with a reasonably low viscosity, it is necessary to use fine, high-specific-gravity solid particles, agitation being necessary to maintain the suspension and to lower the apparent viscosity. The solids comprising the medium must be hard, with no tendency to slime, as degradation increases the apparent viscosity by increasing the surface area of the medium. The medium must be easily removed from the mineral surfaces by washing and must be easily recoverable from the fine-ore particles washed from the surfaces. It must not be affected by the constituents of the ore and must resist chemical attack, such as corrosion.

For ore preconcentration, galena was initially used as the medium and, when pure, it can give a bath specific gravity of about 4. Above this level, ore separation is slowed down by the viscous resistance. Froth flotation, which is an expensive process, was used to clean the contaminated medium, but the main disadvantage is that galena is fairly soft and tends to slime easily, and it also has a tendency to oxidize, which impairs the flotation efficiency.

The most widely used medium for metalliferous ores is now ferrosilicon, while magnetite is used in coal preparation. Recovery of medium in both cases is by magnetic separation.

Ferrosilicon (s.g. 6.7–6.9) is an alloy of iron and silicon which should contain not less than 82% Fe and 15–16% Si (Collins et al., 1974). If the silicon content is less than 15%, the alloy will tend to corrode, while if it is more than 16% the magnetic susceptibility and density will be greatly reduced. Losses of ferrosilicon from a dense medium circuit vary widely, from as little as 0.1 to more than

2.5 kg t^{-1} of ore treated, the losses, apart from spillages, mainly occurring in magnetic separation and by the adhesion of medium to ore particles. Corrosion usually accounts for relatively small losses and can be effectively prevented by maintaining the ferrosilicon in its passive state. This is normally achieved by atmospheric oxygen diffusing into the medium or by the addition of small quantities of sodium nitrite (Stewart and Guerney, 1998).

Milled ferrosilicon is produced in a range of size distributions, from 30% to 95%-45 μm, the finer grades being used for finer ores and centrifugal separators. The coarser, lower viscosity grades can achieve medium densities up to about 3.3. Atomized ferrosilicon consists of rounded particles, which produce media of lower viscosity and can be used to achieve densities up to 3.8 (Napier-Munn et al., 2014).

Magnetite (s.g. ca. 5) is used in coal washing as separation densities are not as high as needed for metalliferous ores. Medium densities are up to 2.3 but work on spheroidized magnetite aims to reach bath densities up to 2.8 (Napier-Munn et al., 2014).

11.3 SEPARATING VESSELS

Several types of separating vessel are in use, and these may be classified into gravitational ("static-baths") and centrifugal ("dynamic") vessels. There is an extensive literature on the performance of these processes, and mathematical models are being developed, which can be used for circuit design and simulation purposes (King, 2012).

11.3.1 Gravitational Vessels

Gravitational units comprise some form of vessel into which the feed and medium are introduced and the floats are removed by paddles or merely by overflow. Removal of the sinks is the most difficult part of separator design. The aim is to discharge the sinks particles without removing sufficient of the medium to cause disturbing downward currents in the vessel. They are largely restricted to treat feeds coarser than ca. 5 mm in diameter. There are a wide range of gravitational devices (Leonard, 1991; Davis, 1992) and just a selection is described here.

Wemco Cone Separator (Figure 11.2)

This unit is widely used for ore treatment, having a relatively high sinks capacity. The cone, which has a diameter of up to 6 m, accommodates feed particles of up to 10 cm in diameter, with capacities up to 500 t h^{-1}.

The feed is introduced on to the surface of the medium by free-fall, which allows it to plunge several centimeters into the medium. Gentle agitation by rakes mounted on the central shaft (stirring mechanism) helps keep the medium in suspension. The float fraction simply overflows a weir, while the sinks are removed by pump (Figure 11.2(a)) or by external or internal air lift (Figure 11.2(b)).

Drum Separators (Figure 11.3)

These are built in several sizes, up to 4.3 m diameter by 6 m long, with capacities of up to 450 t h^{-1} and treating feed particles of up to 30 cm in diameter. Separation is accomplished by the continuous removal of the sink product through the action of lifters fixed to the inside of the rotating drum. The lifters empty into the sinks launder when they pass the horizontal position. The float product overflows a weir at the opposite end of the drum from the feed chute. Longitudinal partitions separate the float surface from the sink-discharge action of the revolving lifters.

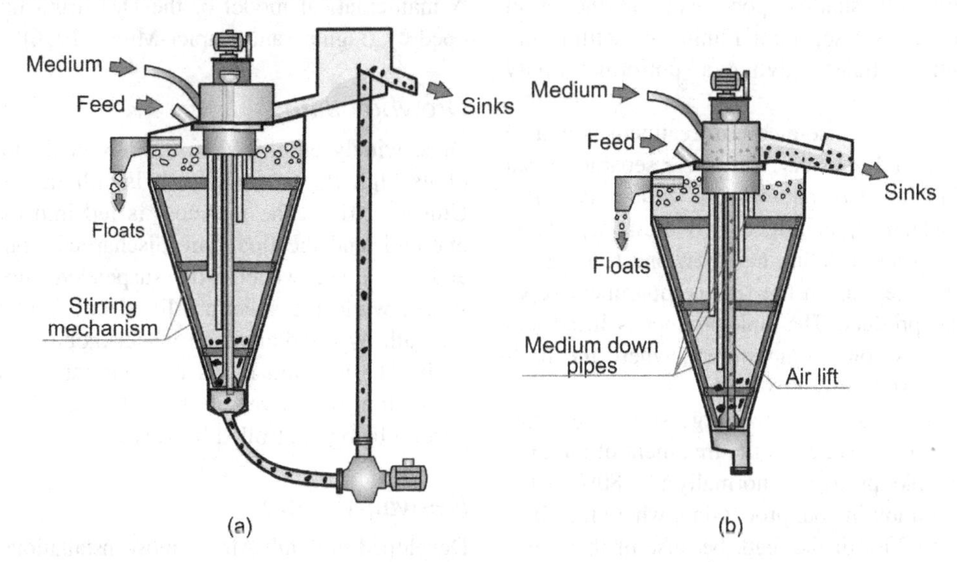

FIGURE 11.2 Wemco cone separator: (a) with torque-flow-pump sinks removal, and (b) with compressed-air sinks removal.

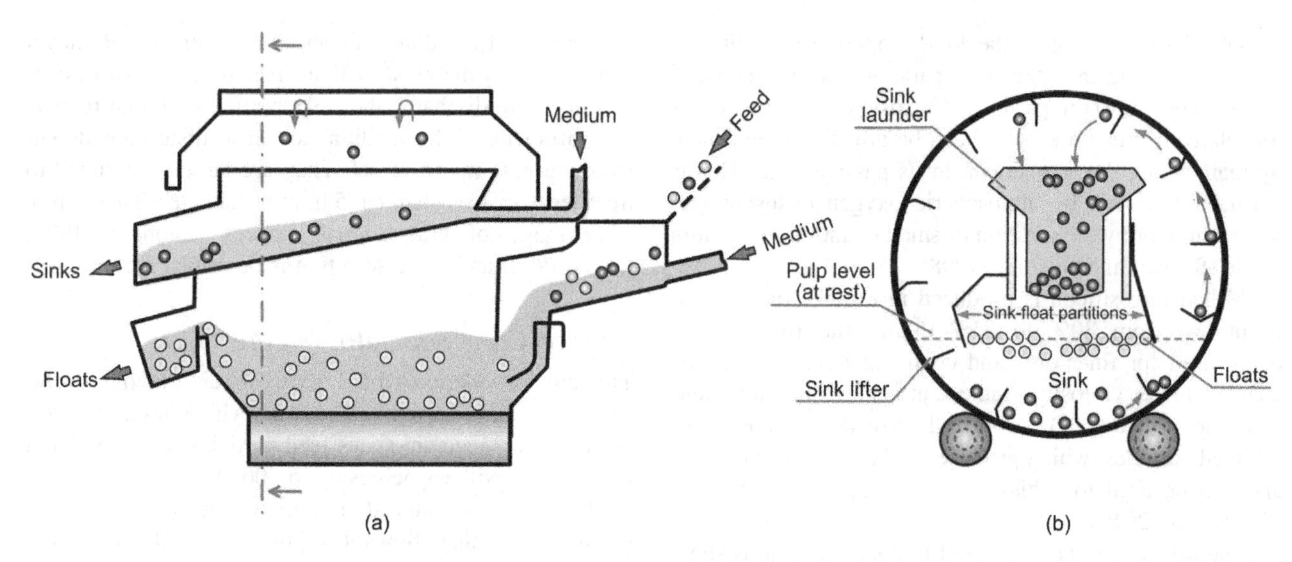

FIGURE 11.3 Dense media (DM) drum separator: (a) side view, and (b) end view.

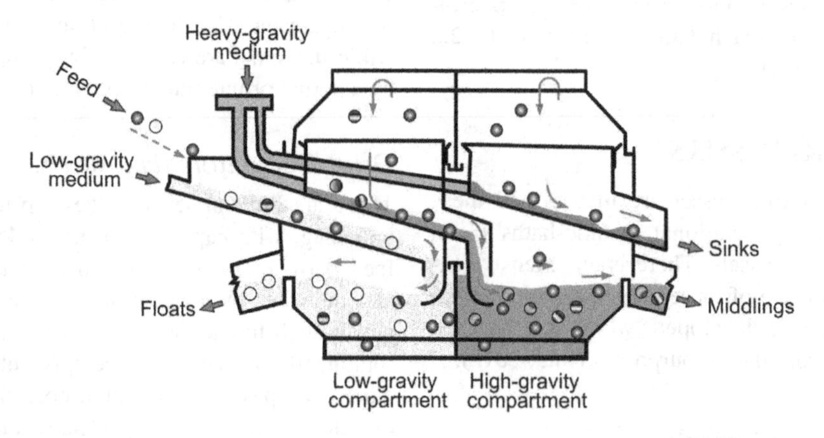

FIGURE 11.4 Two-compartment DM drum separator.

The comparatively shallow pool depth in the drum compared with the cone separator minimizes settling out of the medium particles, giving a uniform gravity throughout the drum.

Where single-stage dense-medium treatment is unable to produce the desired recovery, two-stage separation can be achieved in the *two-compartment drum separator* (Figure 11.4), which is, in effect, two drum separators mounted integrally and rotating together, one feeding the other. The lighter medium in the first compartment separates a true float product. The sink product is lifted and conveyed into the second compartment, where the middlings and the true sinks are separated.

Although drum separators have large sinks capacities and are inherently more suited to the treatment of metallic ores, where the sinks product is normally 60–80% of the feed, they are common in coal processing, where the sinks product is only 5–20% of the feed, because of their simplicity, reliability, and relatively small maintenance needs. A mathematical model of the DM drum has been developed by Baguley and Napier-Munn (1996).

Drewboy Bath

Once widely employed in the UK coal industry because of its high floats capacity, it is still in use (Cebeci and Ulusoy, 2013). The raw coal is fed into the separator at one end, and the floats are discharged from the opposite end by a star-wheel with suspended rubber, or chain straps, while the sinks are lifted out from the bottom of the bath by a radial-vaned wheel mounted on an inclined shaft. The medium is fed into the bath at two points—at the bottom of the vessel and with the raw coal—the proportion being controlled by valves.

Norwalt Washer

Developed in South Africa, most installations are still to be found in that country. Raw coal is introduced into the center of the annular separating vessel, which is provided with stirring arms. The floats are carried round by the stirrers and are discharged over a weir on the other side of the vessel, being carried out of the vessel by the medium flow. The heavies sink to the bottom of the vessel and are moved along by scrapers attached to the bottom of the stirring arms and are discharged via a hole in the bottom of the bath into a sealed elevator, either of the wheel or bucket type, which continuously removes the sinks product.

11.3.2 Centrifugal Separators

Cyclonic dense medium separators have now become widely used in the treatment of ores and coal. They provide a high centrifugal force and a low viscosity in the medium, enabling much finer separations to be achieved than in gravitational separators. Feed to these devices is typically de-slimed at about 0.5 mm, to avoid contamination of the medium with slimes and to simplify medium recovery. A finer medium is required than with gravitational vessels, to avoid fluid instability. Much work has been carried out to extend the range of particle size treated by centrifugal separators. This is particularly the case in coal preparation plants, where advantages to be gained are elimination of de-sliming screens and reduced need for flotation of the screen undersize, as well as more accurate separation of fine coal. Froth flotation has little effect on sulfur reduction, whereas pyrite can be removed, and oxidized coal can be treated by DMS. Work has shown that good separations can be achieved for coal particles as fine as 0.1 mm, but below this size separation efficiency decreases rapidly. Tests on a lead–zinc ore have shown that good separations can be achieved down to 0.16 mm using a centrifugal separator (Ruff, 1984). These, and similar results elsewhere, together with the progress made in automatic control of medium consistency, add to the growing evidence that DMS can be considered for finer material than had been thought economical or practical until recently. As the energy requirement for grinding, flotation, and dewatering is often up to 10 times that required for DMS, a steady increase of fines preconcentration DMS plants is likely.

Dense Medium Cyclones (DMC)

By far the most widely used centrifugal DM separator is the cyclone (DMC) (Figure 11.5) whose principle of operation is similar to that of the conventional hydrocyclone (Chapter 9). Cyclones typically treat ores and coal in the range 0.5–40 mm. Cyclones up to 1 m in diameter for coal preparation were introduced in the 1990s, and units up to 1.4 m diameter and capable of throughputs of over 250 t h^{-1} treating feed particles up to 75–90 mm are now common in the coal industry (Luttrell, 2014). Osborne (2010) has documented the decrease in circuit complexity that has accompanied this increase in unit size. The larger DMC units treating coarser sizes may obviate the requirement for static-bath vessels, and the need for fewer units minimizes differences in cut-point densities and surges that are common in banks of smaller units.

FIGURE 11.5 Cast iron dense medium cyclones used in the coal and diamond industry *(Courtesy Multotec)*.

DMC sizes have lagged in metalliferous operations, the largest being 0.8 m diameter, but confer similar advantages as experience at Glencore's lead–zinc plant at Mount Isa has shown (Napier-Munn et al., 2009).

The feed is suspended in the medium and introduced tangentially to the cyclone either via a pump or it is gravity-fed. Gravity feeding requires a taller and therefore more expensive building, but achieves a more consistent flow and less pump wear and feed degradation. The dense material (reject in the case of coal, product in the case of iron ore, for example) is centrifuged to the cyclone wall and exits at the apex. The light product "floats" to the vertical flow around the axis and exits via the vortex finder. In a DMC, there is a difference in density at various points. Figure 11.6 shows a rough indication of the density variations in a DMC containing medium only (Bekker, 2014). The figure is constructed assuming a density cut-point at a RD (RD_{50}) of 2.85 (RD_{50} refers the density of a particle that has a 50:50 chance of reporting to either floats or sinks, see later). Figure 11.6 is an idealized representation, as in reality there are density gradients radially across the cyclone as well, which the mathematical and computational models of DMCs are showing. Mathematical models of the DMC for coal were developed by King and Juckes (1988); and for minerals by Scott and Napier-Munn (1992). More recently, computational fluid dynamics models of DMCs have been developed, revealing further detail on the flows inside the device (Kuang et al., 2014).

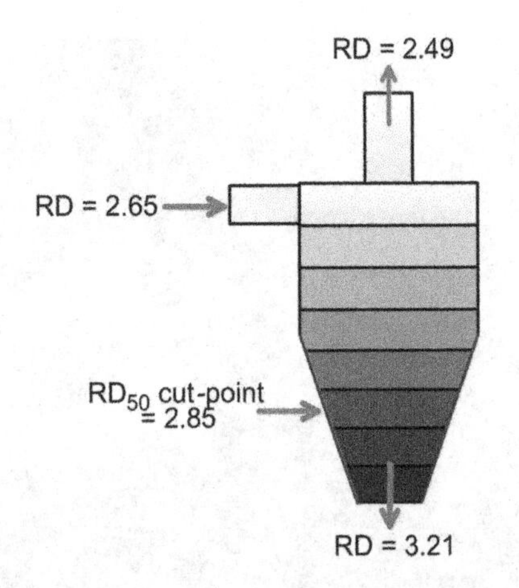

FIGURE 11.6 Density gradients inside a dense medium cyclone *(Adapted from Bekker, 2014)*.

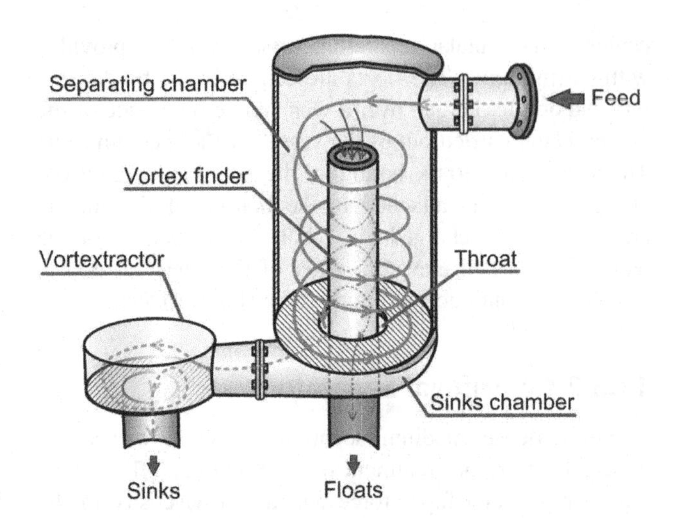

FIGURE 11.7 Vorsyl separator.

In general, DMCs have a cone angle of 20°, with manufacturers generally staying with one type of cone angle, as there has been shown to be no real benefit achieved by altering it (Bekker, 2014).

Water-Only Cyclones

Particles below ca. 0.5–1 mm are generally too fine for the drainage and washing screens used as part of the circuit to recover/recycle DM (see Section 11.4), and particles in the range 0.2–1 mm are therefore processed by water-based gravity techniques. In the coal industry, the most common such device is the spiral concentrator (Chapter 10) but water-only cyclones are also used (Luttrell, 2014). They separate coal from rock within a self-generated (autogenous) dense medium derived from fine fraction of the heavy minerals in the feed (similar in concept to the fluidized bed separators, Chapter 10). Modern units have a wide angle conical bottom to emphasize density separation and suppress size effects.

Vorsyl Separator (Figure 11.7)

Developed in the 1960s at the British Coal Mining Research and Development Establishment for processing 50–5 mm sized feeds at up to 120 t h^{-1}, the unit continues to be used (Banerjee et al., 2003; Majumder et al., 2009). The feed to the separator, consisting of de-slimed raw coal, together with the separating medium of magnetite, is introduced tangentially, or more recently by an involute entry (see Chapter 9), at the top of the separating chamber, under pressure. Material of specific gravity less than that of the medium passes into the clean coal outlet via the vortex finder, while the near-density material and the heavier shale particles move to the wall of the vessel due to the centrifugal acceleration induced. The particles move in a spiral path down the chamber toward the base of the vessel where the drag caused by the proximity of the orifice plate reduces the tangential velocity and creates a strong inward flow toward the throat. This carries the shale, and near-density material, through zones of high centrifugal force, where a final precise separation is achieved. The shale, and a proportion of the medium, discharge through the throat into the shallow shale chamber, which is provided with a tangential outlet, and is connected by a short duct to a second shallow chamber known as the vortextractor. This is also a cylindrical vessel with a tangential inlet for the medium and reject and an axial outlet. An inward spiral flow to the outlet is induced, which dissipates the inlet pressure energy and permits the use of a large outlet nozzle without the passing of an excessive quantity of medium.

LARCODEMS (Large Coal Dense Medium Separator)

This was developed to treat a wide size range of coal (−100 mm) at high capacity in one vessel (Shah, 1987). The unit (Figure 11.8) consists of a cylindrical chamber which is inclined at approximately 30° to the horizontal. Medium at the required RD is introduced under pressure, either by pump or static head, into the involute tangential inlet at the lower end. At the top end of the vessel is another involute tangential outlet connected to a vortextractor. Raw coal of 0.5–100 mm is fed into the separator by a chute connected to the top end, the clean coal being removed through the bottom outlet. High RD particles pass rapidly to the separator wall and are removed through the top involute outlet and the vortextractor.

The first installation of the device was in the 250 t h^{-1} coal preparation plant at Point of Ayr Colliery in the United Kingdom (Lane, 1987). In addition to coal processing, the LARCODEMS has found application in concentrating iron ore (for example, a 1.2 m LARCODEMS is

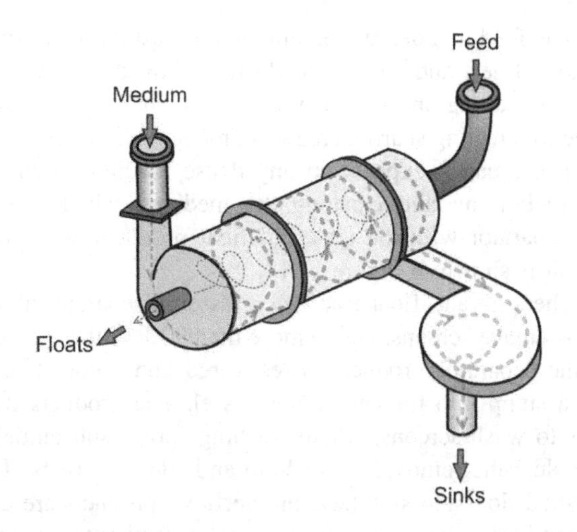

FIGURE 11.8 LARCODEMS separator.

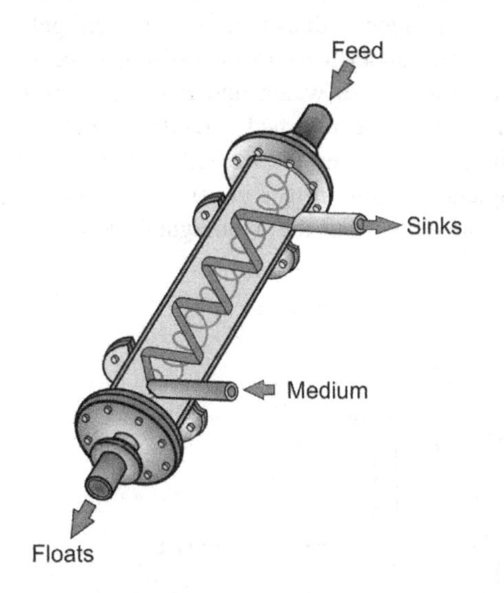

FIGURE 11.9 Dyna Whirlpool separator.

used in Kumba's iron ore concentrator at Sishen in South Africa to treat up to 800 t h^{-1} of $-90+6$ mm feeds (Napier-Munn et al., 2014)), and in recycling, notably of plastics (Pascoe and Hou, 1999; Richard et al., 2011).

Dyna Whirlpool Separator (Figure 11.9)

Developed in the United States, this device is similar to the LARCODEMS and is used for treating fine coal, particularly in the Southern Hemisphere, as well as diamonds, fluorspar, tin, and lead–zinc ores, in the size range 0.5–30 mm (Wills and Lewis, 1980).

It consists of a cylinder of predetermined length having identical tangential inlet and outlet sections at either end. The unit is operated in an inclined position and medium of the required density is pumped under pressure

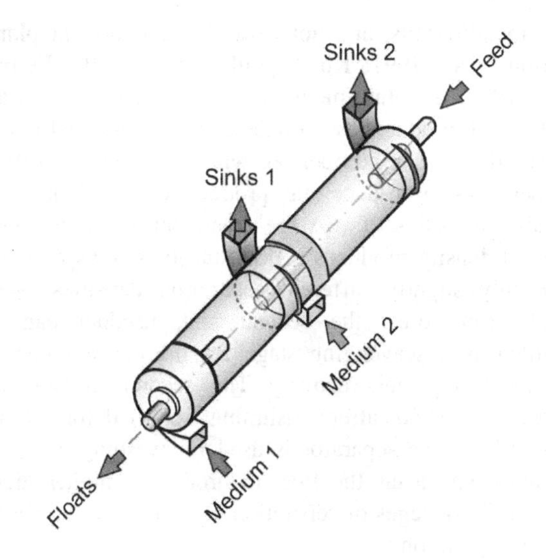

FIGURE 11.10 Tri-Flo separator.

into the lower outlet. The rotating medium creates a vortex throughout the length of the unit and leaves via the upper tangential discharge and the lower vortex outlet tube. Raw feed entering the upper vortex tube is sluiced into the unit by a small quantity of medium and a rotational motion is quickly imparted by the open vortex. Float material passes down the vortex and does not contact the outer walls of the unit, thus greatly reducing wear. The floats are discharged from the lower vortex outlet tube. The heavy particles (sinks) of the feed penetrate the rising medium toward the outer wall of the unit and are discharged with medium through the sink discharge pipe. Since the sinks discharge is close to the feed inlet, the sinks are removed from the unit almost immediately, again reducing wear considerably. Only near-density particles, which are separated further along the unit, actually come into contact with the main cylindrical body. The tangential sink discharge outlet is connected to a flexible sink hose and the height of this hose may be used to adjust back pressure to finely control the cut-point.

The capacity of the separator can be as high as 100 t h^{-1}, and it has some advantages over the DM cyclone. Apart from the reduced wear, which not only decreases maintenance costs but also maintains performance of the unit, operating costs are lower, since only the medium is pumped. The unit has a higher sinks capacity and can accept large fluctuations in sink/float ratios (Hacioglu and Turner, 1985).

Tri-Flo Separator (Figure 11.10)

This can be regarded as two Dyna Whirlpool separators joined in series and has been installed in a number of coal, metalliferous, and nonmetallic ore treatment plants (Burton et al., 1991; Kitsikopoulos et al., 1991; Ferrara et al., 1994). Involute medium inlets and sink outlets are used, which produce less turbulence than tangential inlets.

The device can be operated with two media of differing densities to produce sink products of individual controllable densities. Two-stage treatment using a single medium density produces a float and two sinks products with only slightly different separation densities. With metalliferous ores, the second sink product can be regarded as a scavenging stage for the dense minerals, thus increasing their recovery. This second product may be recrushed, and, after de-sliming, returned for retreatment. Where the separator is used for washing coal, the second stage cleans the float to produce a higher grade product. Two stages of separation also increase the sharpness of separation.

11.4 DMS CIRCUITS

Although the separating vessel is the most important element of a DMS process, it is only one part of a relatively complex circuit. Other equipment is required to prepare the feed, and to recover, clean, and recirculate the medium (Symonds and Malbon, 2002).

The feed to a dense medium circuit must be screened to remove fines, and slimes should be removed by washing, thus alleviating any tendency that such slime content may have for creating sharp increases in medium viscosity.

The greatest expense in any dense medium circuit is for reclaiming and cleaning the medium, which leaves the separator with the sink and float products. A typical circuit is shown in Figure 11.11.

The sink and float fractions pass onto separate vibrating drainage screens, where more than 90% of the medium in the separator products is recovered and pumped back via a sump into the separating vessel. The products then pass to wash screens, where washing sprays substantially complete the removal of medium and adhering fines. The finished float and sink (screen overflow) products are discharged from the screens for disposal or further treatment.

The underflows from the drainage screens are combined and a fraction reports to the main medium sump and the remainder is densified by a centrifugal or spiral densifier. The underflows from washing screens, consisting of medium, wash water, and fines, are too dilute and contaminated to be returned directly as medium to the separating vessel. They are treated (together in this case) by magnetic separation to recover the magnetic ferrosilicon or magnetite from the nonmagnetic fines, which also

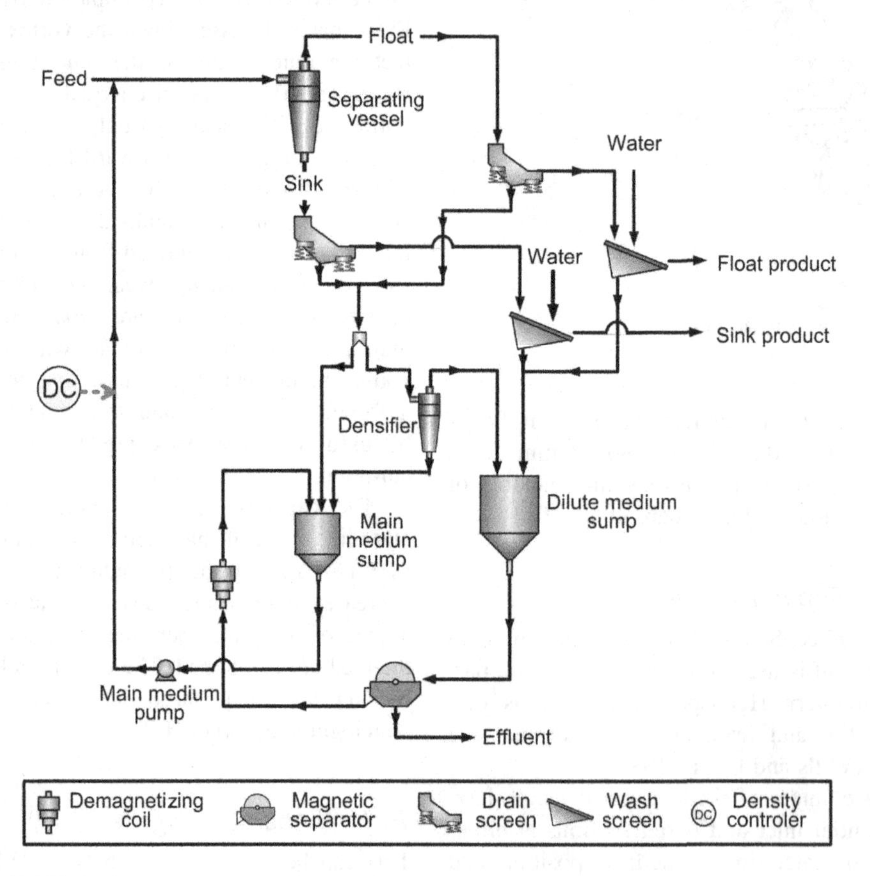

FIGURE 11.11 Typical DMS circuit.

densifies the medium. The densified medium is directed to the main medium sump passing via a demagnetizing coil to ensure a nonflocculated, uniform suspension in the separating vessel.

Most large DMS plants include automatic control of the feed medium density. This is done by densifying sufficient medium to cause the medium density to rise, measuring the feed density with a gamma attenuation gauge, and using the signal to adjust the amount of water added to the medium to return it to the correct density.

The major costs in DMS are power (for pumping) and medium consumption. Medium losses can account for 10–35% of total costs. They are principally due to adhesion to products and losses from the magnetic separators, though the proportions will depend on the size and porosity of the ore, the characteristics of the medium solids, and the plant design (Napier-Munn et al., 1995). Losses increase for fine or porous ore, fine media, and high operating densities.

Correct sizing and selection of equipment, together with correct choice of design parameters, such as rinsing water volumes, are essential. As effluent water always contains some entrained medium, the more of this that can be recycled back to the plant the better (Dardis, 1987). Careful attention should also be paid to the quality of the medium used, Williams and Kelsall (1992) having shown that certain ferrosilicon powders are more prone to mechanical degradation and corrosion than others.

Medium rheology is critical to efficient operation of dense medium systems (Napier-Munn, 1990), although the effects of viscosity are difficult to quantify (Reeves, 1990; Dunglison et al., 2000). Management of viscosity includes selecting the correct medium specifications, minimizing operating density, and minimizing the content of clays and other fine contaminants (Napier-Munn and Scott, 1990). If the amount of fines in the circuit reaches a high proportion due, say, to inefficient screening of the feed, it may be necessary to divert an increased amount of medium into the cleaning circuit. Many circuits have such a provision, allowing medium from the draining screen to be diverted into the washing screen undersize sump.

11.5 EXAMPLE DMS APPLICATIONS

The most important use of DMS is in coal preparation, where a relatively simple separation removes the low-ash coal (clean coal) from the heavier high-ash discard and associated shales and sandstones. DMS is preferred to the cheaper jigs when washing coals with a relatively large proportion of middlings, or near-density material, since the separating density can be controlled at much closer limits.

Luttrell (2014) gives a generic flowsheet for a modern US coal processing plant with four independent circuits treating different size fractions: coarse size (+10 mm) using dense medium (i.e., static) vessels; medium size (−10 + 1 mm) using dense medium cyclones; small size (−1 + 0.15 mm) using spirals; and fine size (−0.15 mm) using flotation. British coals, in general, are relatively easy to wash, and jigs are used in many cases. Where DMS is preferred, drum and Drewboys separators are most widely used for the coarser fractions, with DM cyclones and Vorsyl separators being preferred for the fines. DMS is essential with most Southern Hemisphere coals, where a high middlings fraction is present. This is especially so with the large, low-grade coal deposits found in the former South African Transvaal province. Drums and Norwalt baths are the most common separators utilized to wash such coals, with DM cyclones and Dyna Whirlpools being used to treat the finer fractions.

At the Landau Colliery in the Transvaal (operated by Anglo Coal), a two-density operation is carried out to produce two saleable products. After preliminary screening of the run-of-mine coal, the coarse (+7 mm) fraction is washed in Norwalt bath separators, utilizing magnetite as the medium to give a separating density of 1.6. The sinks product from this operation, consisting predominantly of sand and shales, is discarded, and the floats product is routed to Norwalt baths operating at a lower density of 1.4. This separation stage produces a low-ash floats product, containing about 7.5% ash, which is used for metallurgical coke production, and a sinks product, which is the process middlings, containing about 15% ash, which is used as power-station fuel. The fine (0.5–7 mm) fraction is treated in a similar two-stage manner using Dyna Whirlpool separators.

In metalliferous mining, DMS is used in the preconcentration of lead–zinc ores, in which the disseminated sulfide minerals often associate together as replacement bandings in the light country rock, such that marked specific gravity differences between particles crushed to fairly coarse sizes can be exploited.

A dense medium plant was incorporated into the lead–zinc circuit at Mount Isa Mines Ltd., Australia, in 1982 in order to increase the plant throughput by 50%. The ore, containing roughly 6.5% lead, 6.5% zinc, and 200 ppm silver, consists of galena, sphalerite, pyrite, and other sulfides finely disseminated in distinct bands in quartz and dolomite. Liberation of the ore into particles which are either sulfide-rich or predominantly gangue begins at around −50 mm and becomes substantial below 18 mm.

The plant treats about 800 $t\,h^{-1}$ of material, in the size range 1.7–13 mm by DM cyclones, at a separating RD of 3.05, to reject 30–35% of the run-of-mine ore as tailings, with 96–97% recoveries of lead, zinc, and silver to the preconcentrate. The preconcentrate has a 25% lower Bond Work Index and is less abrasive because the lower specific gravity hard siliceous material mostly reports to the rejects. (The grinding circuit product (cyclone overflow) will also be finer after installation of DMS simply due to the removal of the low-specific-gravity mineral fraction that classifies at a coarser size, as discussed in Chapter 9.) The rejects are used as a cheap source of fill for underground operations. The plant is extensively instrumented, the process control strategy being described by Munro et al. (1982).

DMS is also used to preconcentrate tin and tungsten ores, and nonmetallic ores such as fluorite and barite. It is important in the preconcentration of diamond ores, prior to recovery of the diamonds by electronic sorting (Chapter 14) or grease-tabling (Rylatt and Popplewell, 1999). Diamonds are the lowest grade of all ores mined, and concentration ratios of several million to one must be achieved. DMS produces an initial enrichment of the ore in the order of 100–1,000 to 1 by making use of the fact that diamonds have a fairly high specific gravity (3.5) and are relatively easily liberated from the ore, since they are loosely held in the parent rock. Gravitational and centrifugal separators are utilized, with ferrosilicon as the medium, and separating densities between 2.6 and 3.0. Clays in the ore sometimes present a problem by increasing the medium viscosity, thus reducing separating efficiency and the recovery of diamonds to the sinks.

Upgrading low-grade iron ores for blast furnace feed sometimes uses DMS. Both gravity and centrifugal separators are employed, and in some cases the medium density can exceed 4 (Myburgh, 2002).

11.6 LABORATORY HEAVY LIQUID TESTS

Laboratory testing may be performed on ores to assess the suitability of DMS (and other gravity methods) and to determine the economic separating density.

Liquids covering a range of densities in incremental steps are prepared, and the representative sample of crushed ore is introduced into the liquid of highest density. The floats product is removed and washed and placed in the liquid of next lower density, whose float product is then transferred to the next lower density and so on. The sinks product is finally drained, washed, and dried, and then weighed, together with the final floats product, to give the density distribution of the sample by weight (Figure 11.12).

Care should be taken when evaluating ores of fine particle size that sufficient time is given for the particles to settle into the appropriate fraction. Centrifuging is often carried out on fine materials to reduce the settling time, but this should be done carefully, as there is a tendency for the floats to become entrained in the sinks fraction. Unsatisfactory results are often obtained with porous materials, such as magnesite ores, due to the entrainment of liquid in the pores, which changes the apparent density of the particles.

After assaying the fractions for metal content, the distribution of material and metal in the density fractions of the sample can be tabulated. Table 11.1 shows such a distribution from tests performed on a tin ore. The computations are easily accomplished in a spreadsheet. Columns 2, 3, and 4 are self-explanatory. Column 5 is an intermediate calculation step, referred to as "Sn units", and is the product of % wt and %Sn (i.e., column 2 × column 4). Column 6, %Sn Distribution, is then computed by dividing each row in column 5 by the sum of units, 111.90; column 7 is then obtained by cumulating the rows in column 6. Knowing the sum of units gives a back-calculated value of the feed (or head) assay, in this case 1.12% Sn (i.e., sum of units divided by sum of increment weights, 111.90/100 as a percent).

It can be seen from columns 3 and 7 of the table that if a separation density of 2.75 was chosen, then 68.48% of the material, being lighter than 2.75, would be discarded as a float product, and only 3.78% of the tin would be lost in this fraction. Conversely, 96.22% of the tin would be recovered into the sink product (i.e., 100 − 3.78), which accounts for 31.52% of the original total feed weight. From this information we can quickly calculate the grade of Sn in the sinks by using the definition of recovery (Chapter 1):

$$R = \frac{C\,c}{F\,f} \tag{11.1}$$

where $R = 96.22\%$, C/F (weight recovery, or yield) = 31.52%, and $f = 1.12\%$, hence, solving for c, the tin grade in the sinks product, we find $c = 3.41\%$ Sn. The analogous calculation can be used to determine the Sn grade in the discard (light) product, t, which gives $t = 0.062\%$.

The choice of *optimum* separating density must be made on economic grounds. In the example shown in Table 11.1, the economic impact of rejecting 68.48% of the feed to DMS on downstream performance must be assessed. The smaller throughput will lower grinding and

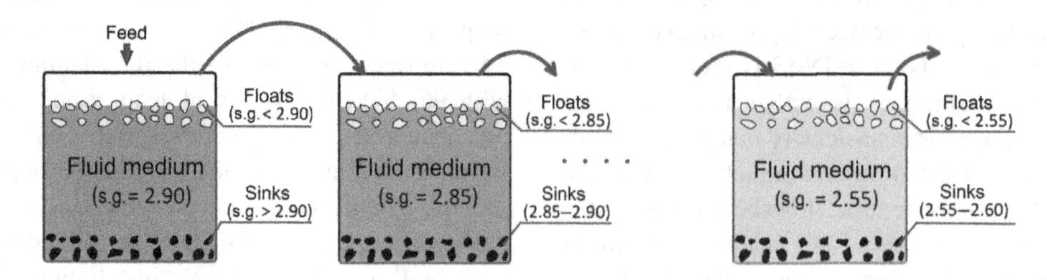

FIGURE 11.12 Heavy liquid testing.

concentration operating costs, the impact on grinding energy and steel costs often being particularly high. Against these savings, the cost of operating the DMS plant and the impact of losing 3.78% of the run-of-mine tin to floats must be considered. The amount of *recoverable* tin in this fraction has to be estimated, together with the subsequent loss in smelter revenue. If this loss is lower than the saving in overall milling costs, then DMS is economic. The optimum density is that which maximizes the difference between overall reduction in milling costs per ton of run-of-mine ore and loss in smelter revenue. Schena et al. (1990) have analyzed the economic choice of separating density.

Heavy liquid tests are important in coal preparation to determine the required density of separation and the expected yield of coal of the required ash content. The "ash" content refers to the amount of incombustible material in the coal. Since coal is lighter than the contained minerals, the higher the density of separation the higher is the *yield* (Chapter 1):

$$\text{yield} = \frac{\text{weight of coal floats product} \times 100\%}{\text{total feed weight}} \qquad (11.2)$$

but the higher is the ash content. The ash content of each density fraction from heavy liquid testing is determined by taking about 1 g of the fraction, placing it in a cold well-ventilated furnace, and slowly raising the temperature to 815°C, maintaining the sample at this temperature until constant weight is obtained. The residue is cooled and then weighed. The ash content is the mass of ash expressed as a percentage of the initial sample weight taken.

Table 11.2 shows the results of heavy liquid tests performed on a coal sample. The coal was separated into the density fractions shown in column 1, and the weight fractions and ash contents are tabulated in columns 2 and 3, respectively. The weight percent of each product is multiplied by the ash content to give the ash units (column 4) (same calculation as "units" in Table 11.1).

The total floats and sinks products at the various separating densities shown in column 5 are tabulated in columns 6–11. To obtain the cumulative floats at each separation density, columns 2 and 4 are cumulated from top to bottom to give columns 6 and 7, respectively. Column 7 is then divided by column 6 to obtain the cumulative percent ash (column 8). Cumulative sink ash is obtained in essentially the same manner, except that columns 2 and 4 are cumulated from bottom to top to give columns 9 and 10, respectively. The results are plotted in Figure 11.13 as typical *washability curves.*

Suppose an ash content of 12% is required in the coal product. It can be seen from the washability curves that such a coal would be produced at a yield of 55% (cumulative percent floats), and the required density of separation is 1.465.

The difficulty of the separation in terms of operational control is dependent mainly on the amount of material present in the feed that is close to the required density of separation. For instance, if the feed were composed entirely of pure coal at specific gravity 1.3 and shale at specific gravity 2.7, then the separation would be easily carried out over a wide range of operating densities. If, however, the feed consists of appreciable middlings, and

TABLE 11.1 Heavy Liquid Test Results on Tin Ore Sample

1	2	3	4	5	6	7
Specific Gravity Fraction	% Weight		% Sn in s.g. fraction	Sn Units	% Sn	
	Incremental	Cumulative			Distribution	Cum. Distribution
−2.55	1.57	1.57	0.003	0.0047	0.004	0.004
2.55–2.60	9.22	10.79	0.04	0.37	0.33	0.33
2.60–2.65	26.11	36.90	0.04	1.04	0.93	1.27
2.65–2.70	19.67	56.57	0.04	0.79	0.70	1.97
2.70–2.75	11.91	68.48	0.17	2.02	1.81	3.78
2.75–2.80	10.92	79.40	0.34	3.71	3.32	7.10
2.80–2.85	7.87	87.27	0.37	2.91	2.60	9.70
2.85–2.90	2.55	89.82	1.30	3.32	2.96	12.66
+2.90	10.18	100.00	9.60	97.73	87.34	100.00
Total			1.12	111.90		

TABLE 11.2 Heavy liquid test results on a coal sample

1	2	3	4	5	6	7	8	9	10	11
Sp. gr. fraction	Wt %	Ash %	Ash units	Separating density	Cumulative float			Cumulative sink		
					Wt %	Ash units	Ash %	Wt %	Ash units	Ash %
−1.30	0.77	4.4	3.39	1.30	0.77	3.39	4.4	99.23	2213.76	22.3
1.30–1.32	0.73	5.6	4.09	1.32	1.50	7.48	5.0	98.50	2209.67	22.4
1.32–1.34	1.26	6.5	8.19	1.34	2.76	15.67	5.7	97.24	2201.48	22.6
1.34–1.36	4.01	7.2	28.87	1.36	6.77	44.54	6.6	93.23	2172.61	23.3
1.36–1.38	8.92	9.2	82.06	1.38	15.69	126.60	8.1	84.31	2090.55	24.8
1.38–1.40	10.33	11.0	113.63	1.40	26.02	240.23	9.2	73.98	1976.92	26.7
1.40–1.42	9.28	12.1	112.29	1.42	35.30	352.52	10.0	64.70	1864.63	28.8
1.42–1.44	9.00	14.1	126.90	1.44	44.30	479.42	10.8	55.70	1737.73	31.2
1.44–1.46	8.58	16.0	137.28	1.46	52.88	616.70	11.7	47.12	1600.45	34.0
1.46–1.48	7.79	17.9	139.44	1.48	60.67	756.14	12.5	39.33	1461.01	37.1
1.48–1.50	6.42	21.5	138.03	1.50	67.09	894.17	13.3	32.91	1322.98	40.2
+1.50	32.91	40.2	1322.98	–	100.00	2217.15	22.2	–	–	–
Total	100.0	22.2	2217.15							

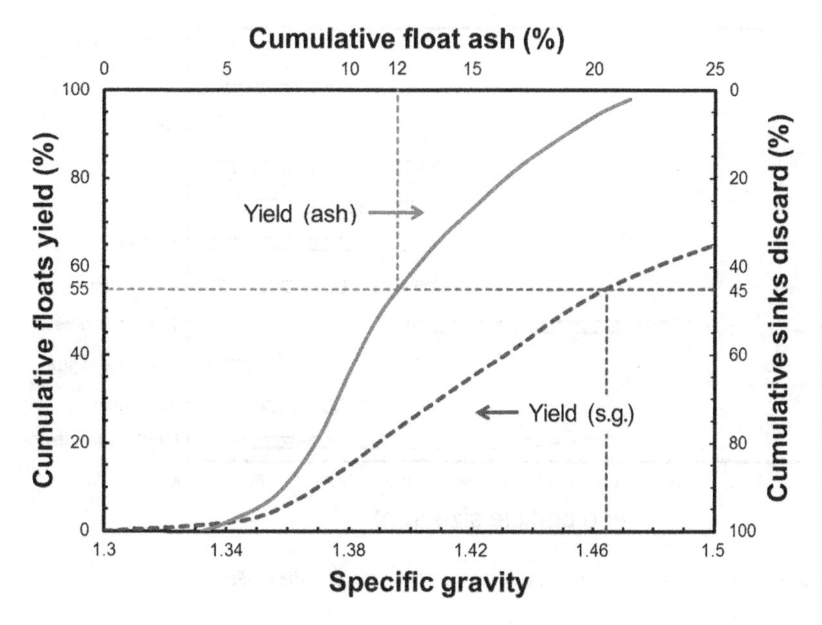

FIGURE 11.13 Typical coal washability curves.

much material present is near-density (i.e., very close to the chosen separating density), then only a small variation in this density will seriously affect the yield and ash content of the product.

The amount of near-density material present is sometimes regarded as being the weight of material in the range ±0.1 or ±0.05 of the separating RD. Separations involving feeds with less than about 7% of ±0.1 near-density material are regarded by coal preparation engineers as being fairly easy to control. Such separations are often performed in Baum jigs, as these are cheaper than dense medium plants, which require expensive media-cleaning facilities, and no feed preparation (i.e., removal of the fine particles by screening) is required. However, the density of separation in jigs is not as easy to control to fine limits, as it is in DMS, and for near-density material much above 7%, DMS is preferred.

Heavy liquid tests can be used to evaluate any ore, and combined with Table 11.3 can be used to indicate the type of separator that could effect the separation in practice (Mills, 1980).

Table 11.3 takes no account of the particle size of the material and experience is therefore required in its application to heavy liquid results, although some idea of the effective particle size range of gravity separators can be gained from Figure 11.14. The throughput of the plant must also be taken into account with respect to the type of separator chosen.

TABLE 11.3 Gravity Separation Process Depends on Amount of Near-Density Material

Wt % Within ±0.1 RD of separation	Gravity Process Recommended	Type
0–7	Almost any process	Jigs, tables, spirals
7–10	Efficient process	Sluices, cones, DMS
10–15	Efficient process with good operation	
15–25	Very efficient process with expert operation	DMS
>25	Limited to a few exceptionally efficient processes with expert operation	DMS with close control

11.7 EFFICIENCY OF DMS

Laboratory testing assumes perfect separation and, in such batch tests, conditions are indeed close to the ideal, as sufficient time can be taken to allow complete separation to take place.

In a continuous production process, however, conditions are usually far from ideal and particles can be misplaced to the wrong product for a variety of reasons. The dominant effect is that of the density distribution of the feed. Very dense or very light particles will settle through the medium and report to the appropriate product quickly, but particles of density close to that of the medium will move more slowly and may not reach the right product in the time available for the separation. In the limit, particles of density the same as, or very close to, that of the

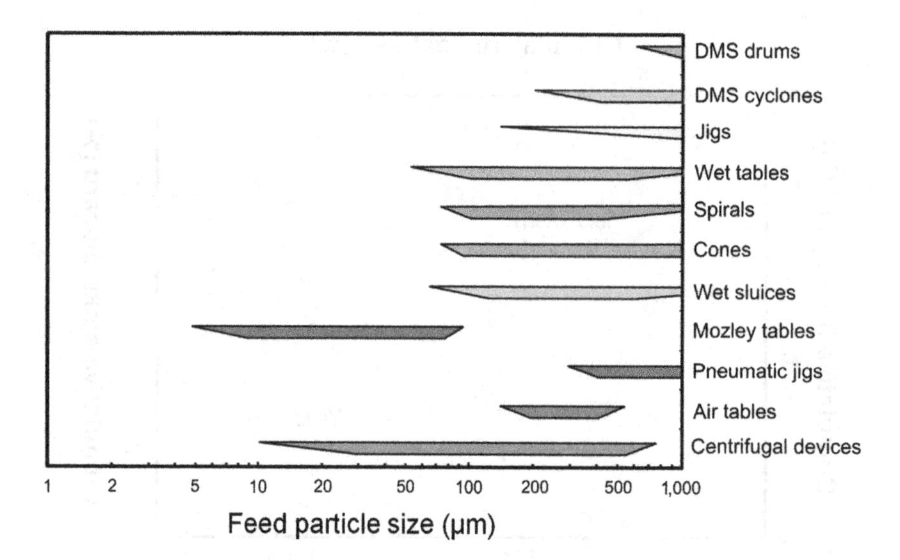

FIGURE 11.14 Effective range of gravity and dense medium devices *(Adapted from Mills (1980)).*

medium will follow the medium and divide in much the same proportion.

Other factors also play a role in determining the efficiency of separation. Fine particles generally separate less efficiently than coarse particles because of their slower settling rates. The properties of the medium, the design and condition of the separating vessel, and the feed conditions, particularly feed rate, will all influence the separation.

Partition Curve

The efficiency of separation can be represented by the slope of a partition or Tromp curve, first introduced by Tromp (1937). It describes the separating efficiency for the separator whatever the quality of the feed and can be used for estimation of performance and comparison between separators.

The partition curve relates the *partition coefficient or partition value*, that is, the percentage of the feed material of a particular specific gravity, which reports to either the sinks product (generally used for minerals) or the floats product (generally used for coal), to specific gravity (Figure 11.15). It is exactly analogous to the classification efficiency curve (Chapter 9), in which the partition coefficient is plotted against particle size rather than specific gravity.

The ideal partition curve reflects a perfect separation, in which all particles having a density higher than the separating density report to sinks and those lighter report to floats. There is no misplaced material.

The partition curve for a real separation shows that efficiency is highest for particles of density far from the operating density and decreases for particles approaching the operating density.

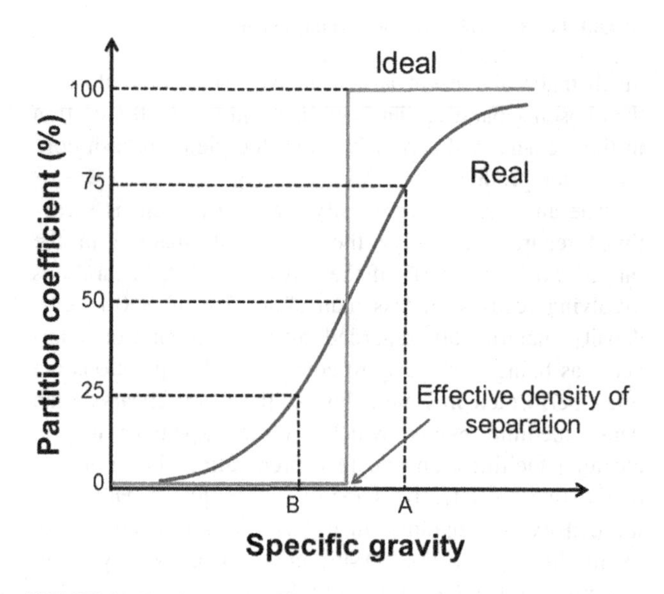

FIGURE 11.15 Partition or Tromp curve.

The area between the ideal and real curves is called the "error area" and is a measure of the degree of misplacement of particles to the wrong product.

Many partition curves give a reasonable straight-line relationship between the distribution of 25% and 75%, and the slope of the line between these distributions is used to show the efficiency of the process.

The *probable error of separation* or the *Ecart probable* (E_p) is defined as half the difference between the density where 75% is recovered to sinks and that at which 25% is recovered to sinks, that is, from Figure 11.15:

$$E_p = \frac{A - B}{2} \quad (11.3)$$

The density at which 50% of the particles report to sinks is shown as the *effective density of separation*, which may not be exactly the same as the medium density, particularly for centrifugal separators, in which the separating density is generally higher than the medium density. This density of separation is referred to as the RD_{50} or ρ_{50} where the 50 refers to 50% chance of reporting to sinks (or floats).

The lower the E_p, the nearer to vertical is the line between 25% and 75% and the more efficient is the separation. An ideal separation has a vertical line with an $E_p = 0$, whereas in practice the E_p usually lies in the range 0.01–0.10.

The E_p is not commonly used as a method of assessing the efficiency of separation in units such as tables and spirals due to the many operating variables (wash water, table slope, speed, etc.) which can affect the separation efficiency. It is, however, ideally suited to the relatively simple and reproducible DMS process. However, care should be taken in its application, as it does not reflect performance at the tails of the curve, which can be important.

Construction of Partition Curves

The partition curve for an operating dense medium vessel can be determined by sampling the sink and float products and performing heavy liquid tests to determine the amount of material in each density fraction. The range of liquid densities applied must envelope the working density of the dense medium unit. The results of heavy liquid tests on samples of floats and sinks from a vessel separating coal (floats) from shale (sinks) are shown in Table 11.4. The calculations are easily performed in a spreadsheet.

Columns 1 and 2 are the results of laboratory tests on the float and sink products, and columns 3 and 4 relate these results to the total distribution of the feed material to floats and sinks, which, in this example, is determined by directly weighing the two products over a period of time. The result of such determinations showed that 82.60% of the feed reported to the floats product (and 17.40% reported to the sinks) (see the Total row). Thus, for example, the first number in column 3 is (83.34/100) × 82.60 (=63.84) and in column 4 it is (18.15/100) × 17.40 (=3.16). The weight fraction in columns 3 and 4 can be added together to produce the reconstituted feed weight distribution in each density fraction (column 5). Column 6 gives the nominal (average) specific gravity of each density range, for example material in the density range 1.30–1.40 is assumed to have a specific gravity lying midway between these densities, that is, 1.35. Since the −1.30 specific gravity fraction and the +2.00 specific gravity fraction have no bound, no nominal density is given.

The partition coefficient (column 7) is the percentage of feed material of a certain nominal specific gravity which reports to sinks, that is:

$$\frac{\text{column 4}}{\text{column 5}} \times 100\%.$$

The partition curve can then be constructed by plotting the partition coefficient against the nominal specific gravity, from which the separation density and probable error of separation of the vessel can be determined. The plot is shown in Figure 11.16. Reading from the plot, the RD_{50} is ca. 1.52 and the E_p ca. 0.12 ((1.61 − 1.37)/2).

TABLE 11.4 Determination of Partition Coefficient for Vessel Separating Coal from Shale

	1	2	3	4	5	6	7
Specific Gravity Fraction	Analysis (wt %)		% of feed		Reconstituted Feed (%)	Nominal Specific Gravity	Partition Coefficient
	Floats	Sinks	Floats	Sinks			
−1.30	83.34	18.15	68.84	3.16	71.98	–	4.39
1.30–1.40	10.50	10.82	8.67	1.88	10.56	1.35	17.84
1.40–1.50	3.35	9.64	2.77	1.68	4.45	1.45	37.74
1.50–1.60	1.79	13.33	1.48	2.32	3.80	1.55	61.07
1.60–1.70	0.30	8.37	0.25	1.46	1.71	1.65	85.46
1.70–1.80	0.16	5.85	0.13	1.02	1.15	1.75	88.51
1.80–1.90	0.07	5.05	0.06	0.88	0.94	1.85	93.83
1.90–2.00	0.07	4.34	0.06	0.76	0.81	1.95	92.89
+2.00	0.42	24.45	0.35	4.25	4.60	–	92.46
Total	100.00	100.00	82.60	17.40	100.00		

The partition curve can also be determined by applying the mass balancing procedure explained in Chapter 3, provided the density distributions of feed, as well as sinks and floats, are available. Often the feed is difficult to sample, and thus resort is made to direct measurement of sinks and floats flowrates. It must be understood, however, that the mass balancing approach is the better way to perform the calculations as redundant data are available to execute data reconciliation. The mass balance/data reconciliation method is illustrated in determination of the partition curve for a hydrocyclone in Chapter 9.

An alternative, rapid, method of determining the partition curve of a separator is to use density tracers. Specially developed color-coded plastic tracers of known density can be fed to the process, the partitioned products being collected and hand sorted by density (color). It is then a simple matter to construct the partition curve directly by noting the proportion of each density of tracer reporting to either the sink or float product. Application of tracer methods has shown that considerable uncertainties can exist in experimentally determined Tromp curves unless an adequate number of tracers is used, and Napier-Munn (1985) presents graphs that facilitate the selection of sample size and the calculation of confidence limits. A system in operation in a US coal preparation plant uses sensitive metal detectors that automatically spot and count the number of different types of tracers passing through a stream (Chironis, 1987).

Partition curves can be used to predict the products that would be obtained if the feed or separation density were changed. The curves are specific to the vessel for which they were established and are not affected by the type of material fed to it, provided:

a. The feed size range is the same—efficiency generally decreases with decrease in size; Figure 11.17 shows typical efficiencies of gravitational separators or baths (drum, cone, etc.) and centrifugal separators (DMC, Dyna Whirlpool, etc.) versus particle size. It can be seen that, in general, below about 10 mm, centrifugal separators are better than gravitational separators.
b. The separating density is in approximately the same range—the higher the effective separating density the greater the probable error, due to the increased medium viscosity. It has been shown that the E_p is directly proportional to the separating density, all other factors being the same (Gottfried, 1978).
c. The feed rate is the same.

The partition curve for a vessel can be used to determine the amount of misplaced material that will report to the products for any particular feed material. For example, the distribution of the products from the tin ore, which was evaluated by heavy liquid tests (Table 11.1), can be determined for treatment in an operating separator. Figure 11.18 shows a partition curve for a separator having an E_p of 0.07.

The curve can be shifted slightly along the abscissa until the effective density of separation corresponds to the

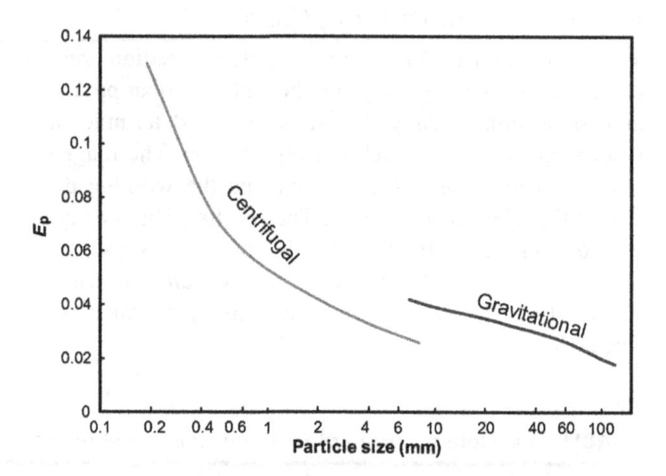

FIGURE 11.17 Effect of particle size on efficiency of DM separators.

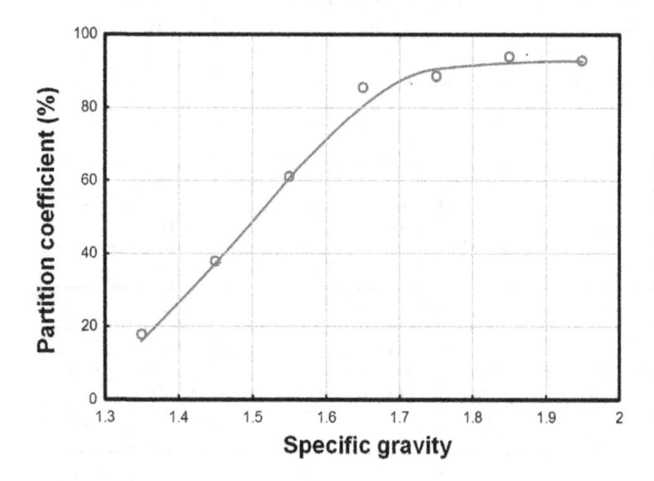

FIGURE 11.16 Partition curve for data in Table 11.4.

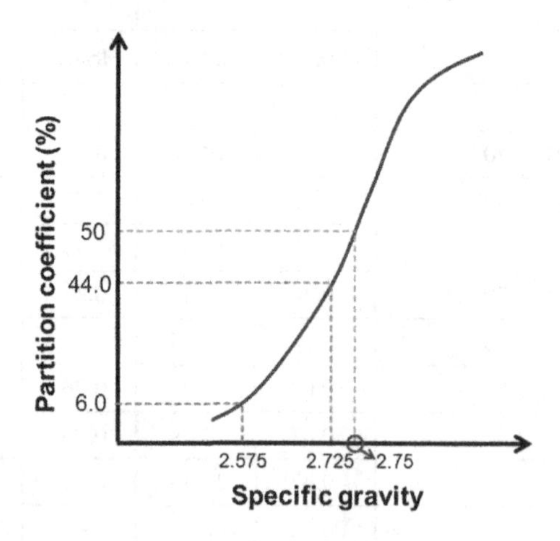

FIGURE 11.18 Partition curve for $E_p = 0.07$.

laboratory evaluated separating density of 2.75. The distribution of material to sinks and floats can now be evaluated: for example, at a nominal specific gravity of 2.725, 44.0% of the material reports to the sinks and 56.0% to the floats.

The performance is evaluated in Table 11.5. Columns 1, 2, and 3 show the results of the heavy liquid tests, which were tabulated in Table 11.1. Columns 4 and 5 are the partition values to sinks and floats, respectively, obtained from the partition curve. Column 6 = column 1 × column 4, and column 9 = column 1 × column 5. The assay of each fraction is assumed to be the same, whether or not the material reports to sinks or floats (columns 2, 7, and 10). Columns 8 and 11 are then calculated as the amount of tin reporting to sinks and floats in each fraction (columns 6 × 7 and 9 × 10) as a percentage of the total tin in the feed (sum of columns 1 × 2, i.e., 1.12).

The total distribution of the feed to sinks is the sum of all the fractions in column 6, that is, 40.26%, while the recovery of tin into the sinks is the sum of the fractions in column 8, that is, 95.29%. This compares with a distribution of 31.52% and a recovery of 96.19% of tin in the ideal separation. In terms of upgrading, the grade of tin in the sinks is now 2.65% (solving Eq. (11.1), i.e., 96.29 × 1.12/40.26) compared to the ideal of 3.42% Sn.

This method of evaluating the performance of a separator on a particular feed is tedious and is ideal for a spreadsheet, providing that the partition values for each density fraction are known. These can be represented by a suitable mathematical function. There is a large literature on the selection and application of such functions. Some are arbitrary, and others have some theoretical or heuristic justification. The key feature of the partition curve is its S-shaped character. In this it bears a passing resemblance to a number of probability distribution functions, and indeed the curve can be thought of as a statistical description of the DMS process, describing the probability with which a particle of given density reports to the sink product. Tromp himself recognized this in suggesting that the amount of misplaced material relative to a suitably transformed density scale was normally distributed, and Jowett (1986) showed that a partition curve for a process controlled by simple probability factors should have a normal distribution form.

However, many real partition curves do not behave ideally like the one illustrated in Figure 11.15. In particular, they are not asymptotic to 0 and 100%, but exhibit evidence of short-circuit flow to one or both products (e.g., Figure 11.16). Stratford and Napier-Munn (1986) identified four attributes required of a suitable function to represent the partition curve:

1. It should have natural asymptotes, preferably described by separate parameters.
2. It should be capable of exhibiting asymmetry about the separating density; that is, the differentiated form of the function should be capable of describing skewed distributions.
3. It should be mathematically continuous.
4. Its parameters should be capable of estimation by accessible methods.

A two-parameter function asymptotic to 0 and 100% is the Rosin-Rammler function, originally developed to describe size distributions (Tarjan, 1974) (see Chapter 4):

$$P_i = 100 - 100\exp\left[-\left(\frac{\rho_i}{a}\right)^m\right] \quad (11.4)$$

In this form, P_i is the partition number (feed reporting to sinks, %), ρ_i the mean density of density fraction i, and a and m the parameters of the function; m describes the steepness of the curve (high values of m indicating more efficient separations). Partition curve functions are normally expressed in terms of the *normalized* density, ρ/ρ_{50}, where ρ_{50} is the separating density (RD_{50}). The normalized curve is generally independent of cut-point and medium density, but is dependent on particle size. Inserting this normalized density into Eq. (11.4), and noting that $P = 50$ for $\rho = \rho_{50}$ ($\rho/\rho_{50} = 1$), gives:

$$P_i = 100 - 100\exp\left[-\ln 2\left(\frac{\rho_i}{\rho_{50}}\right)^m\right] \quad (11.5)$$

One of the advantages of Eq. (11.5) is that it can be linearized so that simple linear regression can be used to estimate m and ρ_{50} from experimental data:

$$\ln\left[\frac{\ln\left(\frac{100}{100-P_i}\right)}{\ln 2}\right] = m\ln\rho_i - m\ln\rho_{50} \quad (11.6)$$

(This approach is less important today with any number of curve-fitting routines available (and Excel Solver), the same point also made in Chapter 9 when curve-fitting cyclone partition curves.)

Gottfried (1978) proposed a related function, the Weibull function, with additional parameters to account for the fact that the curves do not always reach the 0 and 100% asymptotes due to short-circuit flow:

$$P_i = 100 - 100\left[f_0 + c\exp\left(-\frac{\left(\frac{\rho_i}{\rho_{50}} - x_0\right)^a}{b}\right)\right] \quad (11.7)$$

The six parameters of the function (c, f_0, ρ_{50}, x_0, a, and b) are not independent, so by the argument of Eq. (11.5), x_0 can be expressed as:

$$x_0 = 1 - \left[b\ln\left(\frac{c}{0.5 - f_0}\right)\right]^{\frac{1}{a}} \quad (11.8)$$

TABLE 11.5 Separation of Tin Ore Evaluation

Specific Gravity Fraction	Nominal S. G.	Feed			Partition Value (%)		Predicted Sinks			Predicted Floats		
		1	2	3	4	5	6	7	8	9	10	11
		Wt %	% Sn	% Sn Dist.	Sinks	Floats	Wt %	% Sn	% Sn Dist.	Wt %	% Sn	% Sn Dist.
−2.55	–	1.57	0.003	0.004	0.0	100.0	0.00	0.003	0.00	1.57	0.003	0.04
2.55–2.60	2.575	9.22	0.04	0.33	6.0	94.0	0.55	0.04	0.02	8.67	0.04	0.31
2.60–2.65	2.625	26.11	0.04	0.93	13.5	86.5	3.52	0.04	0.13	22.59	0.04	0.80
2.65–2.70	2.675	19.67	0.04	0.70	27.0	73.0	5.31	0.04	0.19	14.35	0.04	0.51
2.70–2.75	2.725	11.91	0.17	1.81	44.0	56.0	5.24	0.17	0.80	6.67	0.17	1.01
2.75–2.80	2.775	10.92	0.34	3.32	63.0	37.0	6.88	0.34	2.09	4.04	0.34	1.23
2.80–2.85	2.825	7.87	0.37	2.60	79.5	20.5	6.26	0.37	2.07	1.61	0.37	0.53
2.85–2.90	2.875	2.55	1.30	2.96	90.5	9.5	2.32	1.30	2.68	0.24	1.30	0.28
+2.90	–	10.18	9.60	87.34	100.00	0.00	10.18	9.60	87.31	0.00	9.60	0.00
Total		100.00	1.12	100.00			40.26	2.65	95.29	59.74	0.09	4.71

In this version of the function, representing percentage of feed to sinks, f_0 is the proportion of high-density material misplaced to floats, and $1-(c+f_0)$ is the proportion of low-density material misplaced to sinks, so that $c+f_0 \leq 1$. The curve therefore varies from a minimum of $100[1-(c+f_0)]$ to a maximum of $100(1-f_0)$.

The parameters of Eq. (11.8) have to be determined by nonlinear estimation. First approximations of c, f_0, and ρ_{50} can be obtained from the curve itself.

King and Juckes (1988) used Whiten's classification function (Lynch, 1977) with two additional parameters to describe the short-circuit flows or by-pass:

$$P_i = \beta + (1-\alpha-\beta)\left[\frac{\exp\left(\frac{b\rho_i}{\rho_{50}}\right)-1}{\exp\left(\frac{b\rho_i}{\rho_{50}}\right)+\exp(b)-2}\right] \quad (11.9)$$

Here, for P_i is the proportion to underflow, α the fraction of feed which short-circuits to overflow, and β the fraction of feed which short-circuits to underflow; b is an efficiency parameter, with high values of b indicating high efficiency. Again, the function is nonlinear in the parameters.

The E_p can be predicted from these functions by substitution for ρ_{75} and ρ_{25}. Scott and Napier-Munn (1992) showed that for efficient separations (low E_p) without short-circuiting, the partition curve could be approximated by:

$$P_i = \frac{1}{1+\exp\left(\frac{\ln 3(\rho_{50}-\rho_i)}{E_p}\right)} \quad (11.10)$$

Organic Efficiency

This term is often used to express the efficiency of coal preparation plants. It is defined as the ratio (normally expressed as a percentage) between the actual yield of a desired product and the theoretical possible yield at the same ash content.

For instance, if the coal, whose washability data are plotted in Figure 11.13, produced an operating yield of 51% at an ash content of 12%, then, since the theoretical yield at this ash content is 55%, the organic efficiency is equal to 51/55 or 92.7%.

Organic efficiency cannot be used to compare the efficiencies of different plants, as it is a dependent criterion, and is much influenced by the washability of the coal. It is possible, for example, to obtain a high organic efficiency on a coal containing little near-density material, even when the separating efficiency, as measured by partition data, is quite poor.

REFERENCES

Baguley, P.J., Napier-Munn, T.J., 1996. Mathematical model of the dense medium drum. Trans. Inst. Min. Metall. Sec. C. 105-106, C1–C8.

Banerjee, P.K., et al., 2003. A plant comparison of the vorsyl separator and dense medium cyclone in the treatment of Indian coals. Int. J. Miner. Process. 69 (1-4), 101–114.

Bekker, E., 2014. DMC basics: a holistic view. Proceedings of the 11th DMS Powders Conference, Mount Grace, South Africa.

Burton, M.W.A., et al., 1991. The economic impact of modern dense medium systems. Miner. Eng. 4 (3-4), 225–243.

Cebeci, Y., Ulusoy, U., 2013. An optimization study of yield for a coal washing plant from Zonguldak region. Fuel Process. Technol. 115, 110–114.

Chironis, N.P., 1987. On-line coal-tracing system improves cleaning efficiencies. Coal Age. 92 (Mar.), 44–47.

Collins, B., et al., 1974. The production, properties and selection of ferrosilicon powders for heavy medium separation. J. S. Afr. Inst. Min. Metall. 75 (5), 103–119.

Dardis, K.A., 1987. The design and operation of heavy medium recovery circuits for improved medium recovery. Proceedings of Dense Medium Operator's Conference. AusIMM, Brisbane, Australia, pp. 157–184.

Davis, J.J., 1992. Cleaning coarse and small coal—dense medium processes. In: Swanson, A.R., Partridge, A.C. (Eds.), Advanced Coal Preparation Monograph Series, vol. 3. Australian Coal Preparation Society, Broadmeadow, NSW, Australia, part 8.

Domenico, J.A., et al., 1994. Magstream® as a heavy liquid separation alternative for mineral sands exploration. SME Annual Meeting. SME, Albuquerque, NM, USA, Preprint 94–262.

Dunglison, M.E., et al., 2000. The rheology of ferrosilicon dense medium suspensions. Miner. Process. Extr. Metall. Rev. 20 (1), 183–196.

Ferrara, G., et al., 1994. Tri-Flo: a multistage high-sharpness DMS process with new applications. Miner. Metall. Process. 11 (2), 63–73.

Gottfried, B.S., 1978. A generalisation of distribution data for characterizing the performance of float-sink coal cleaning devices. Int. J. Miner. Process. 5 (1), 1–20.

Hacioglu, E., Turner, J.F., 1985. A study of the Dyna Whirlpool, Proceedings of 15th International Mineral Processing Congress, vol. 1. Cannes, France, pp. 244–257.

Jowett, A., 1986. An appraisal of partition curves for coal-cleaning processes. Int. J. Miner. Process. 16 (1-2), 75–95.

King, R.P., 2012. In: Schneider, C.L., King, E.A. (Eds.), Modeling and Simulation of Mineral Processing Systems, second ed. SME, Englewood, CO, USA.

King, R.P., Juckes, A.H., 1988. Performance of a dense medium cyclone when beneficiating fine coal. Coal Preparation. 5 (3-4), 185–210.

Kitsikopoulos, H., et al., 1991. Industrial operation of the first two-density three-stage dense medium separator processing chromite ores. Proceedings of the 17th International Mineral Processing Congress (IMPC), vol. 3. Dresden, Freiberg, Germany, pp. 55–66.

Kuang, Sh., et al., 2014. CFD modeling and analysis of the multiphase flow and performance of dense medium cyclones. Miner. Eng. 62, 43–54.

Lane, D.E., 1987. Point of Ayr Colliery. Mining Mag. 157 (Sept.), 226–237.

Leonard, J.W. (Ed.), 1991. Coal Preparation. SME, Littleton, CO, USA.

Lin, D., et al., 1995. Batch magnetohydrostatic separations in a modified Frantz Separator. Miner. Eng. 8 (3), 283–292.

Luttrell, G.H., 2014. Innovations in coal processing. In: Anderson, C.G., et al., (Eds.), Mineral Processing and Extractive Metallurgy: 100 Years of Innovation. SME, Englewood, CO, USA, pp. 277–296.

Lynch, A.J., 1977. Mineral Crushing and Grinding Circuits: Their Simulation, Optimisation, Design and Control. Elsevier Scientific Publishing Company, Amesterdam, The Netherlands.

Majumder, A.K., et al., 2009. Applicability of a dense-medium cyclone and Vorsyl Separator for upgrading non-coking coal fines for use as a blast furnace injection fuel. Int. J. Coal Prep. Utils. 29 (1), 23–33.

Mills, C., 1980. Process design, scale-up and plant design for gravity concentration. In: Mular, A.L., Bhappu, R.B. (Eds.), Mineral Processing Plant Design, second ed. AIMME, New York, NY, USA, pp. 404–426 (Chapter 18).

Munro, P.D., et al., 1982. The design, construction and commissioning of a heavy medium plant of silver-lead-zinc ore treatment-Mount Isa Mines Limited. Proceedings of the 14th International Mineral Processing Congress, Toronto, ON, Canada, pp. VI-6.1–VI-6.20.

Myburgh, H.A., 2002. The influence of the quality of ferrosilicon on the rheology of dense medium and the ability to reach higher densities. Proceedings of Iron Ore 2002. AusIMM, Perth, WA, USA, pp. 313–317.

Napier-Munn, T.J., 1985. Use of density tracers for determination of the Tromp curve for gravity separation processes. Trans. Inst. Min. Metall. 94 (Mar.), C45–C51.

Napier-Munn, T.J., 1990. The effect of dense medium viscosity on separation efficiency. Coal Preparation. 8 (3-4), 145–165.

Napier-Munn, T.J., Scott, I.A., 1990. The effect of demagnetisation and ore contamination on the viscosity of the medium in a dense medium cyclone plant. Miner. Eng. 3 (6), 607–613.

Napier-Munn, T.J., et al., 1995. Some causes of medium loss in dense medium plants. Miner. Eng. 8 (6), 659–678.

Napier-Munn, T.J., et al., 2009. Advances in dense-medium cyclone plant design. Proceedings of the 10th Mill Operation' Conference. AusIMM, Adelaide, SA, Australia, pp. 53–61.

Napier-Munn, T.J., et al., 2014. Innovations in dense medium separation technology. In: Anderson, C.G., et al., (Eds.), Mineral Processing and Extractive Metallurgy: 100 Years of Innovation. SME, Englewood, CO, USA, pp. 265–276.

Osborne, D., 2010. Value of R&D in coal preparation development. In: Honaker, R.Q. (Ed.), Proceedings of the 16th International Coal Preparation Congress. SME, Lexington, KY, USA, pp. 845–858.

Parsonage, P., 1980. Factors that influence performance of pilot-plant paramagnetic liquid separator for dense particle fractionation. Trans. Inst. Min. Metall. Sec. C. 89 (Dec.), C166–C173.

Pascoe, R.D., Hou, Y.Y., 1999. Investigation of the importance of particle shape and surface wettability on the separation of plastics in a LARCODEMS separator. Miner. Eng. 12 (4), 423–431.

Plewinsky, B., Kamps, R., 1984. Sodium metatungstate, a new medium for binary and ternary density gradient centrifugation. Die Makromolekulare Chemie. 185 (7), 1429–1439.

Reeves, T.J., 1990. On-line viscosity measurement under industrial conditions. Coal Prep. 8 (3-4), 135–144.

Richard, G.M., et al., 2011. Optimization of the recovery of plastics for recycling by density media separation cyclones. Resour. Conserv. Recy. 55 (4), 472–482.

Ruff, H.J., 1984. New developments in dynamic dense-medium systems. Mine Quarry. 13 (Dec.), 24–28.

Rylatt, M.G., Popplewell, G.M., 1999. Diamond processing at Ekati in Canada. Mining Eng. 51 (2), 19–25.

Schena, G.D., et al., 1990. Pre-concentration by dense-medium separation—an economic evaluation. Trans. Inst. Min. Metall. Sec. C. 99-100 (Jan.-Apr.), C21–C31.

Scott, I.A., Napier-Munn, T.J., 1992. Dense medium cyclone model based on the pivot phenomenon. Trans. Inst. Min. Metall. 101 (Jan.–Apr.), C61–C76.

Shah, C.L., 1987. A new centrifugal dense medium separator for treating 250 t/h of coal sized up to 100 mm. In: Wood, P. (Ed.), Proceedings of the Third International Conference on Hydrocyclones. BHRA the Fluid Engineering Center, Oxford, England, pp. 91–100.

Stewart, K.J., Guerney, P.J., 1998. Detection and prevention of ferrosilicon corrosion in dense medium plants. Proceedings of the Sixth Mill Operators' Conference. AusIMM, Madang, Papua New Guinea, pp. 177–183.

Stratford, K.J., Napier-Munn, T.J., 1986. Functions for the mathematical representation of the partition curve for dense medium cyclones. Proceedings of the 19th Application of Computers & Operations Research in the Mineral Industry (APCOM). SME, Pittsburgh, PA, USA, pp. 719–728.

Symonds, D.F., Malbon, S., 2002. Sizing and selection of heavy media equipment: design and layout. In: Mular, Halbe, Barratt (Eds.), Mineral Processing Plant Design, Practice and Control, vol. 1. SME, Littleton, CO, USA, pp. 1011–1032.

Tarjan, G., 1974. Application of distribution functions to partition curves. Int. J. Miner. Process. 1 (3), 261–264.

Tromp, K.F., 1937. New methods of computing the washability of coals. Glückauf. 73, 125–131.

Williams, R.A., Kelsall, G.H., 1992. Degradation of ferrosilicon media in dense medium separation circuits. Miner. Eng. 5 (1), 57–77.

Wills, B.A., Lewis, P.A., 1980. Applications of the Dyna Whirlpool in the minerals industry. Mining Mag. 143 (3), 255–257.

Chapter 12

Froth Flotation

12.1 INTRODUCTION

Flotation is undoubtedly the most important and versatile mineral separation technique, and both its use and application are continually being expanded to treat greater tonnages and to cover new areas. Recently celebrating its first centenary (Fuerstenau et al., 2007), flotation has permitted the mining of low-grade and complex ore bodies which would have otherwise been regarded as uneconomic. In earlier practice, the tailings of many gravity plants were of a higher grade than the ore treated in many modern flotation plants. Warranting its own history (Lynch et al., 2010), at least one technology historian has described flotation as "perhaps the greatest single metallurgical improvement of the modern era" (Mouat, 1996).

Initially developed to treat the sulfide minerals of copper, lead, and zinc, flotation has expanded to include nickel, platinum- and gold-hosting sulfides, and to nonsulfide minerals including oxides such as hematite and cassiterite, and nonmetallic minerals such as fluorite, talc, phosphates, potash, and energy (fuel) minerals, fine coal and bitumen (Rao and Leja, 2004). Flotation now finds application outside the mining industry, deinking recycled paper pulp (Hardie et al., 1998) and deoiling oil refinery effluents (Rawlins, 2009), for example.

12.2 PRINCIPLES OF FLOTATION

Flotation is a separation process that exploits natural and induced differences in surface properties of the minerals, whether the surface is readily wetted by water, that is, is *hydrophilic*, or repels water, that is, is *hydrophobic*. If hydrophobic the mineral particle can attach to air bubbles and be floated. The system is complex, involving three phases (solids, water, and air) and the interaction of chemical and physical variables. The chemical variables aim to control the transition between the hydrophilic and hydrophobic state. Physical variables include those resulting from properties of the ore, such as particle size and composition (liberation), and machine-derived factors such as air rate and bubble size. Klimpel (1984) represented the interaction as a triangle: chemistry, ore, and machine. The mix of physics and chemistry often sees flotation described as a physicochemical based process.

The subject has been reviewed comprehensively by a number of authors, among the more recent being Crozier (1992), Harris et al. (2002), Johnson and Munro (2002), Fuerstenau and Somasundaran (2003), Nguyen and Schulze (2004), Rao and Leja (2004), Fuerstenau et al. (2007), and Bulatovic (2007, 2010).

The process of material being recovered by flotation from the pulp comprises three mechanisms:

1. Selective attachment to air bubbles (or "true flotation").
2. Entrainment in the water which passes through the froth.
3. Physical entrapment between particles in the froth attached to air bubbles (often referred to as "aggregation").

The attachment of valuable minerals to air bubbles is the most important mechanism and represents the majority of particles that are recovered to the concentrate. Although true flotation is the dominant mechanism for the recovery of valuable mineral, the separation efficiency (SE) between the valuable mineral and gangue is also dependent on the degree of entrainment and physical entrapment. Unlike true flotation, which is selective to the mineral surface properties, both gangue and valuable minerals alike can be recovered by entrainment and entrapment. Drainage of these minerals occurs in the froth phase and controlling the stability of this phase is important to achieve an adequate separation. In industrial practice, entrainment of unwanted gangue can be common and hence a single flotation stage is uncommon. Often, several stages of flotation (to form "circuits") are required to reach an economically acceptable quality of valuable mineral in the final product.

True flotation exploits the differences in surface properties of particles of various minerals. After treatment with reagents, such differences in surface properties between the minerals within the flotation pulp become apparent and an air bubble is able to attach to a particle, and lift it (i.e., float it) to the water surface. This means the density of the *bubble–particle aggregate* is less than the density of the surrounding pulp.

Figure 12.1 illustrates the principle of true flotation in a mechanical flotation cell. The agitator provides enough turbulence in the pulp phase to promote collision of particles and bubbles, which results in the attachment of hydrophobic particles to bubbles (forming bubble–particle aggregates) and their transport into the froth phase for recovery.

The process can only be applied to relatively fine particles, because if they are too large, either the adhesion between the particle and the bubble will be less than the particle weight and the bubble will therefore drop its load, or the bubble–particle aggregate density will exceed that of the pulp. There is an optimum size range for flotation (Trahar and Warren, 1976; Crawford and Ralston, 1988; Finch and Dobby, 1990), and continual research to expand the size range (Jameson, 2010; Wyslouzil et al., 2013).

In flotation concentration, the valuable mineral is usually transferred to the froth, or float fraction, leaving the gangue in the pulp or tailings. This is *direct flotation* and the opposite is *reverse flotation*, in which the gangue is separated into the float fraction.

A function of the froth phase is to retain the collected particles and transport them to the overflow. Another function is to enhance the overall selectivity of the flotation process. The froth achieves this by allowing entrained material to drain while preferentially retaining the attached material. This increases the concentrate grade while limiting the reduction in recovery of valuables. The relationship between recovery and grade is a trade-off that needs to be managed according to operational constraints.

The mineral particles can only attach to the air bubbles if they are to some extent hydrophobic. Having reached the surface of the pulp, the air bubbles only continue to support the mineral particles if they can form a stable froth, otherwise they will burst and drop the mineral particles. To achieve these conditions, it is necessary to use the numerous chemical compounds known as *flotation reagents*.

The forces tending to hold a particle to a bubble are shown in Figure 12.2(a), with the common depiction of the contact angle (by convention measured in the liquid) in Figure 12.2(b). The tensile forces lead to the development of an angle between the mineral surface and the bubble surface. At equilibrium:

$$\gamma_{s/a} = \gamma_{s/w} + \gamma_{w/a} \cos\theta \tag{12.1}$$

where $\gamma_{s/a}$, $\gamma_{s/w}$, and $\gamma_{w/a}$ are the surface tensions (or energies) between solid and air, solid and water, and water and air, respectively, and θ is the contact angle between the mineral surface and the bubble.

The force required to break the particle–bubble interface is called the *work of adhesion*, $W_{s/a}$, and is equal to the work required to separate the solid–air interface and produce separate air–water and solid–water interfaces:

$$W_{s/a} = \gamma_{w/a} + \gamma_{s/w} - \gamma_{s/a} \tag{12.2}$$

Combining with Eq. (12.1) gives:

$$W_{s/a} = \gamma_{w/a}(1 - \cos\theta) \tag{12.3}$$

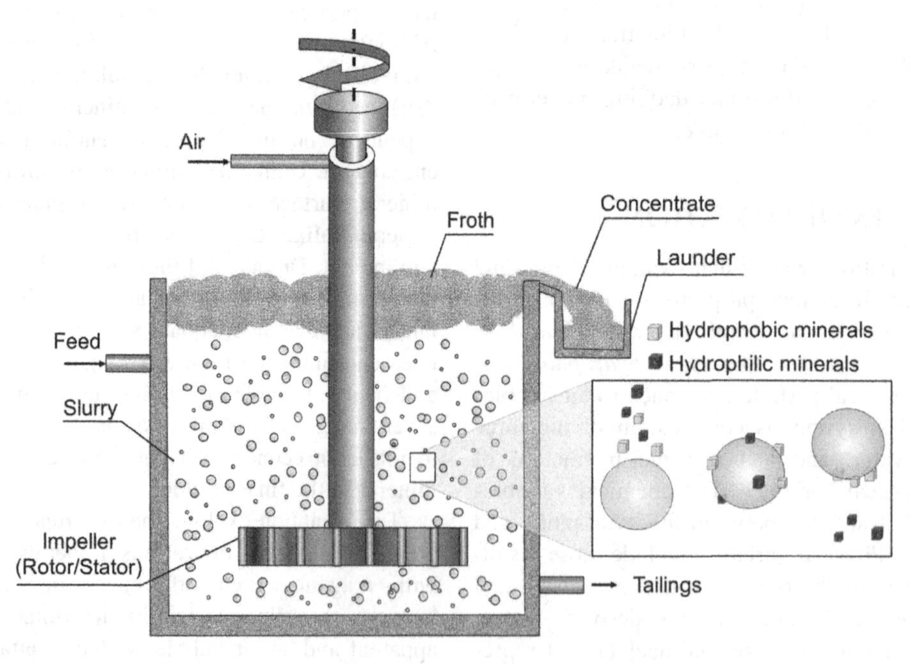

FIGURE 12.1 Principle of froth flotation.

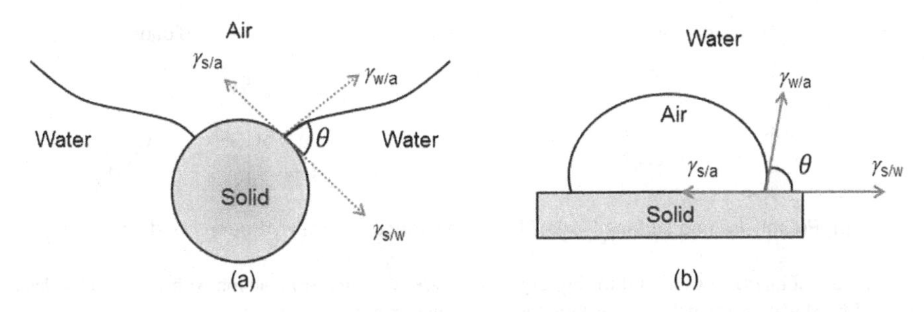

FIGURE 12.2 (a) Particle attached to bubble, and (b) classic representation of contact angle and surface tension forces.

It can be seen that the greater the contact angle the greater is the work of adhesion between particle and bubble and the more resilient the system is against disruptive forces. The hydrophobicity of a mineral therefore increases with the contact angle; minerals with a high contact angle are said to be *aerophilic*, that is, they have a higher affinity for air than for water. The terms hydrophobicity and floatability are often used interchangeably. Hydrophobicity is a necessary condition, but floatability incorporates other particle properties, such as particle size, that affect amenability to flotation.

When a particle and bubble collide in the pulp attachment is not instantaneous but requires time, referred to as *induction time*. Induction time is associated with the properties of the thin water film that separates the particle and bubble just before attachment. For a hydrophobic surface, induction time is short, a few milliseconds, and if it is less than the time the particle and bubble are in contact, attachment is successful and the particle is floated. With a hydrophilic surface the induction time is large and exceeds the particle–bubble contact time. Both contact angle and induction time are used to characterize surface properties related to flotation (Chau et al., 2009; Gu et al., 2003)

Most minerals are not water-repellent in their natural state and flotation reagents must be added to the pulp. The most important reagents are the *collectors*, which adsorb on mineral surfaces, rendering them hydrophobic and facilitating bubble attachment. *Regulators* are used to control the flotation process; these either activate or depress mineral attachment to air bubbles and are also used to control particle dispersion and the pH of the system. *Frothers* help produce the fine bubbles necessary to increase collision rates and to help maintain a reasonably stable froth. Reviews of flotation reagents and their application include those of Ranney (1980), Crozier (1984), Somasundaran and Sivakumar (1988), Ahmed and Jameson (1989), Suttill (1991), Adkins and Pearse (1992), Nagaraj (1994), Buckley and Woods (1997), Ralston et al. (2001), Fuerstenau and Somasundaran (2003), Nagaraj and Ravishankar (2007), and Bulatovic (2007, 2010).

12.3 CLASSIFICATION OF MINERALS

All minerals are classified into *polar* or *nonpolar types* according to their surface characteristics. The surfaces of nonpolar minerals are characterized by relatively weak molecular bonds. The minerals are composed of covalent molecules held together by van der Waals forces, and the nonpolar surfaces do not readily attach to the water dipoles, and in consequence are naturally (inherently) hydrophobic, with contact angles between 60° and 90°. Minerals of this type, such as graphite, sulfur, molybdenite, diamond, coal, and talc, thus have high natural floatabilities. Although it is possible to float these minerals without the aid of chemical agents, it is common to increase their hydrophobicity by the addition of hydrocarbon oils. Fuel oil, for example, is widely used to increase the floatability of coal. Layered minerals, such as molybdenite and talc, break to reveal polar edges and largely nonpolar faces; on balance they are hydrophobic but tend to become less hydrophobic as the particle size reduces, as the balance of edge to face increases.

Minerals with strong covalent or ionic surface bonding are known as polar types and react strongly with water molecules; these minerals are naturally hydrophilic.

Hydrophilicity is associated with high solid surface energy and hydrophobicity with low solid surface energy, where high and low are relative to the surface energy of water ($\gamma_{w/a} \sim 72$ mJ m^{-2}). Solid surface energy can be measured using inverse gas chromatography, a technique that is being adapted to characterize the mineral surface in flotation systems (Ali et al., 2013). These measurements reveal that surfaces are heterogeneous, having a distribution of energy sites. Accordingly, any measure such as contact angle and induction time should be taken as an average value.

The nature of the interaction of a surface with water is illustrated in Figure 12.3 for quartz, a strongly hydrophilic mineral. Freshly broken, a quartz surface has unsatisfied ("dangling") Si and O bonds, which hydrolyze to form SiOH (silanol) groups, which in turn hydrogen bond with water dipoles. Whenever interaction with water to form

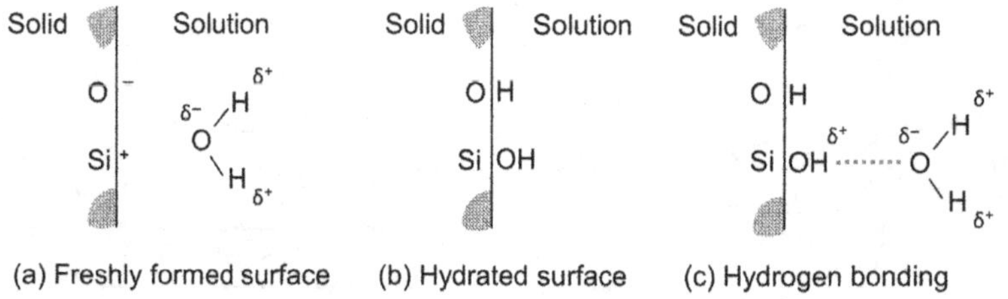

FIGURE 12.3 Stages in exposure of quartz to water: (a) freshly exposed surface of quartz, (b) reaction with water molecules to form SiOH (silanol) groups, and (c) formation of H-bond with water molecules making the surface hydrophilic.

TABLE 12.1 Classification of Polar Minerals

Group 1	Group 2	Group 3(a)	Group 4	Group 5
Galena	Barite	Cerrusite	Hematite	Zircon
Covellite	Anhydrite	Malachite	Magnetite	Willemite
Bornite	Gypsum	Azurite	Gothite	Hemimorphite
Chalcocite	Anglesite	Wulfenite	Chromite	Beryl
Chalcopyrite			Ilmenite	Feldspar
Stibnite		**Group 3(b)**	Corundum	Sillimanite
Argentite		Fluorite	Pyrolusite	Garnet
Bismuthinite		Calcite	Limonite	Quartz
Millerite		Witherite	Borax	
Cobaltite		Magnesite	Wolframite	
Arsenopyrite		Dolomite	Columbite	
Pyrite		Apatite	Tantalite	
Sphalerite		Scheelite	Rutile	
Orpiment		Smithsonite	Cassiterite	
Pentlandite		Rhodochrosite		
Realgar		Siderite		
Native Au, Pt, Ag, Cu		Monazite		

this hydrated surface is identified it indicates a hydrophilic surface.

The polar group of minerals has been subdivided into various classes depending on the magnitude of polarity (Wrobel, 1970), which increases from groups 1–5 (Table 12.1). Apart from the native metals, the minerals in group 1 are all sulfides, which are only weakly polar due to their covalent bonding, which is relatively weak compared to the ionic bonding of the silicate, carbonate, and sulfate minerals. In general, therefore, the degree of polarity, and thus surface energy and hydrophilicity, increases from sulfide minerals, through sulfates, to carbonates, halites, phosphates, etc., then to oxides–hydroxides, and, finally, silicates and quartz, which are strongly hydrophilic.

The nature of the surface is complicated by the fact that all minerals release ions to solution. Some minerals are semisoluble and some, like sylvite and halite, are soluble and processing is in their brine. In many cases, Ca and Mg species are present in the process waters.

Sulfide minerals require a little further consideration as they are reactive with oxygen dissolved in the process water. Figure 12.4 shows a series of reactions producing an "oxidized" surface comprising groups such as M-OH and sulfoxy species ($S_xO_y^{2-}$) depending on pH and oxidizing conditions. These species represent hydrophilic

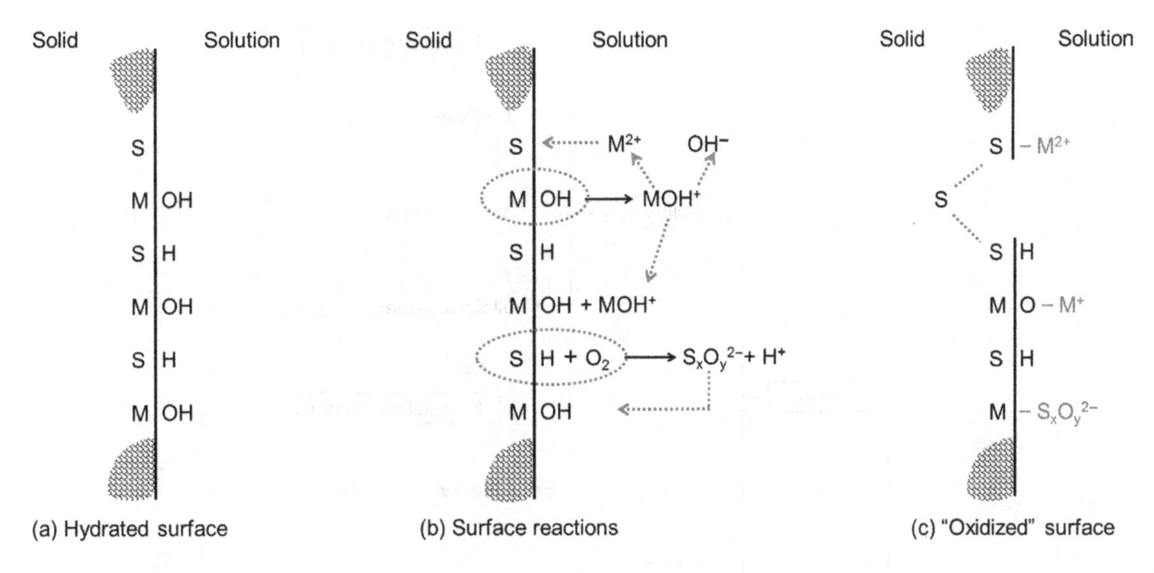

FIGURE 12.4 (a) Hydrated surface of sulfide mineral, (b) reactions with dissolved oxygen (DO) to form a variety of "oxidation" species, and (c) reaction products, most of which make surface more hydrophilic.

sites, which H-bond with water, adding to the hydrated layer and making the surface more hydrophilic. One oxidation product shown is S–S–S, a *metal deficient sulfide* or *polysulfide*, the sulfurs bonding to each other as the metal ion is removed. In contrast to the other oxidation products, this represents a hydrophobic site. These oxidation reactions and resultant species come into play in interpreting flotation response.

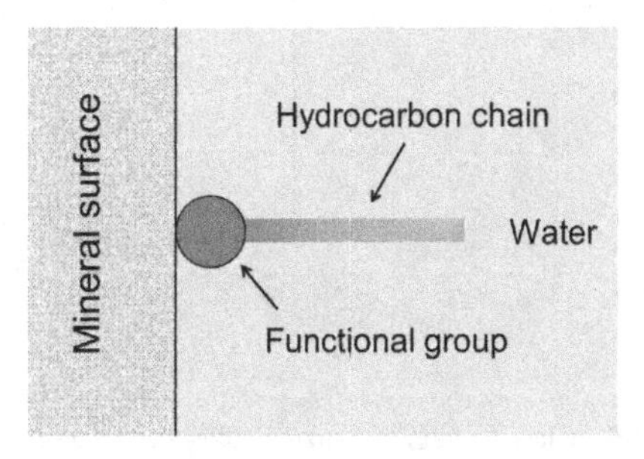

FIGURE 12.5 Adsorption of collector on mineral surface showing general orientation with hydrocarbon chain directed toward the water and making the site hydrophobic.

12.4 COLLECTORS

Hydrophobicity has to be imparted to most minerals in order to float them. To achieve this, *surfactants* known as *collectors* are added to the pulp and time is allowed for adsorption during agitation in what is known as the *conditioning period.* Collectors are organic compounds which render selected minerals water-repellent by adsorption on to the mineral surface, reducing the stability of the hydrated layer separating the mineral from the air bubble to such a level that attachment of the particle to the bubble can be made; that is, collectors reduce the induction time.

Collectors may be nonionizing compounds, which are practically insoluble and strongly hydrophobic. An example is kerosene. These collector types are used with naturally hydrophobic minerals such as coals and molybdenite to boost their floatability, sometimes referred to as *oil-assisted flotation* (Laskowski, 1992). They adsorb through hydrophobic interaction, the natural tendency of hydrophobic entities to repel water and come together. A variation is to first coat the bubbles with oil, referred to as *oily bubble flotation* (Su et al., 2006). In some situations, an insoluble collector forms (referred to as a *collector colloid*) and appears to adsorb first on the bubble and is then transferred to the mineral surface upon collision (Burdukova and Laskowski, 2009).

Soluble, ionizing collectors are the more common form and have found wide application. They are *heteropolar*, that is, the molecule contains a nonpolar hydrocarbon group (hydrocarbon chain, R) and a polar group, which may be one of a number of types. The nonpolar hydrocarbon radical has pronounced water-repellent properties, while the polar group gives the molecule its solubility. The polar group is the *functional* (or *reactive*) group and it is through this group that reaction with sites on the mineral surface occurs (i.e., adsorption). Adsorption can be chemical or physical, as discussed in Sections 12.4.3 and 12.4.4. Figure 12.5 shows the general result of adsorption, with the nonpolar hydrocarbon chain oriented toward the water, making the site hydrophobic. Surface coverage by collector could be up to a monolayer

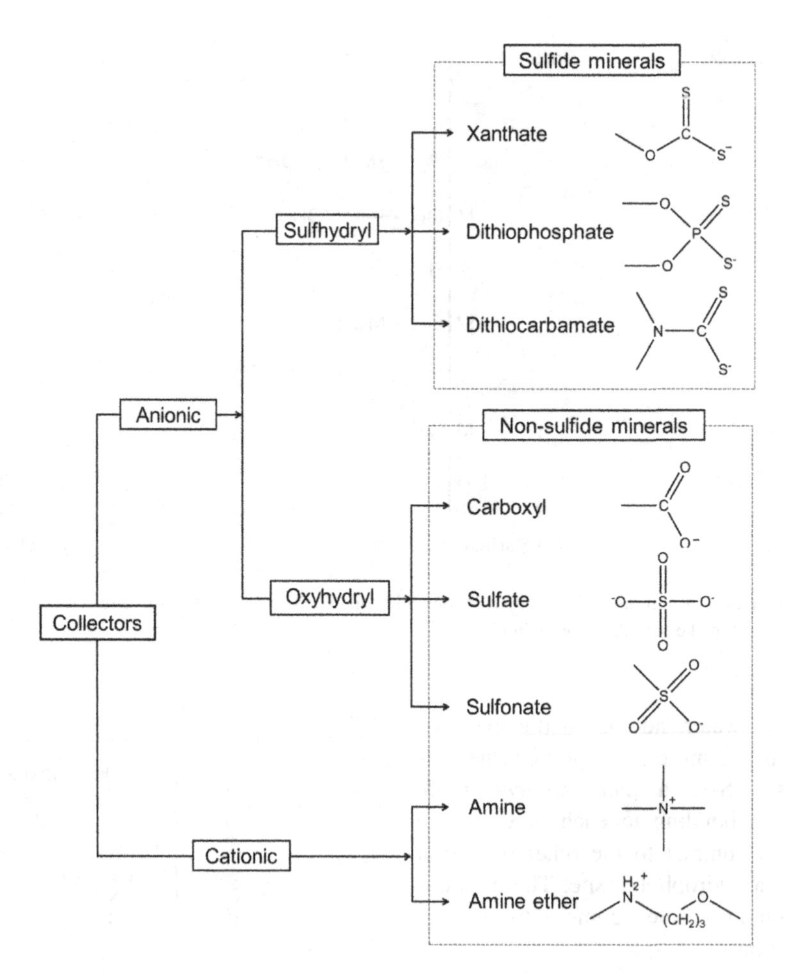

FIGURE 12.6 Classification of ionizing collectors according to type, anionic or cationic, and their application, primarily for non-sulfides or sulfides.

(or more) but is usually far less. Provided the surface hydrophobic sites overcome the hydrophilic sites, on balance the particle becomes floatable.

Ionizing collectors can be classed by the type of ion, anionic or cationic, or their major application, non-sulfide minerals or sulfide minerals. Figure 12.6 combines both classifications. Table 12.2 summarizes some of the major collectors with typical applications.

Anionic collectors consist of two types according to the structure of the polar group: *sulfhydryl* type and *oxyhydryl* type. The term sulfhydryl refers to the SH group present in undissociated form of the collector. The term *thiol* refers to carbon bonded to the SH, that is, C-SH or R-SH. Both sulfhydryl and thiol (thio) are used to describe this class of collectors. They are widely used in the flotation of sulfide minerals (Avotins et al., 1994). The other anionic type of collectors is *oxyhydryl* (referring to the OH group), and they are mainly used in non-sulfide flotation. Compared to the sulfhydryl collectors, the hydrocarbon chain is usually longer. Typically supplied as salts of Na or K, these cations play no role in the action of these anionic collectors.

Cationic collectors are based on pentavalent nitrogen, primary (fatty) amines (i.e., amines with one R radical) and ether amines being the most common (Nagaraj and Ravishankar, 2007). The hydrocarbon chain is typically C12–24, and the anion is usually a halide which plays no direct role in the collector action. They are used primarily for non-sulfides.

Other collector classes are *amphoteric* reagents and *chelating* reagents (Somasundaran and Moudgil, 1987). Amphoteric collectors possess cationic and anionic functions, depending on pH. They have been used to treat sedimentary phosphate deposits (Houot et al., 1985) and to improve the selectivity of cassiterite flotation (Broekaert et al., 1984). Chelating reagents have been suggested as flotation collectors in view of their ability to form stable compounds with selected cations. Their action with metal ions in bulk solution is well understood, the difficulty is translating this action to metal ions on a mineral surface. Whether another collector class based on nanoparticle technology emerges remains to be seen (Yang et al., 2012).

Collectors are usually used in small amounts as increased concentration, apart from the cost, tends to float

TABLE 12.2 Major Collectors and Example Applications

Collector Family		General Formula	*R* Value	Application
Sulfide minerals	Xanthate, alkyl	$R{-}O{-}C({=}S){-}S^-$	C2–C8	Mostly for Cu, Zn, Cu–Mo, Au, PGM, Ni, and oxidized minerals
	Dithiophosphate, dialkyl	$(R_1{-}O)(R_2{-}O)P({=}S){-}S^-$	C2–C6	Widely used with xanthates for PGM, Fe, Cu, Pb, Cu–Mo, Cu, Au, and complex sulfides
	Dithiocarbamate, dialkyl	$R_1R_2N{-}C({=}S){-}S^-$	C2–C6	PGM ores, Cu, Pb
	Dithiophosphinate, dialkyl	$R_1R_2P({=}S){-}S^-$	C2–C4	Widely used, especially for complex sulfides, e.g., Pd complexes with S and Se
	Thionocarbamate, dialkyl	$R_1{-}NH{-}C({=}S){-}O{-}R_2$	C2–C4	Widely used for Cu, Cu–Mo
Non-sulfide minerals	Fatty acids and their salts	$R{-}C({=}O){-}O^-$	C8–C22	Widely used for P, ZnO, CuO, Ni, Nb, Ti, Cu–Co, Sn, W
	Primary amines	$H{-}N^+(H)(H){-}R$	C12–24	Widely used, e.g., quartz, ZnO, Ti, Sn
	Amine ether	$R{-}O{-}(CH_2)_3{-}NH_2$	C6–C13	Quartz, Al silicates
	Petroleum sulfonates	$R{-}S({=}O)_2{-}O^-$	C14–C17	Widely used, e.g., Ti, W, Fe
	Hydroxamates	$R{-}C({=}O){-}N(H){-}O^-$	C5–C14	Rare earths; Sn, W, Mn, Al silicates, quartz

Source: Adapted from Nagaraj and Ravishankar (2007).

other minerals, reducing selectivity. It is always harder to eliminate a collector already adsorbed than to prevent its adsorption. Over-dosing with collector can lead to froth stability issues, ranging from immobility to collapse and can induce bubble clustering in the pulp, that is, arrays of bubbles bridged by particles, which can potentially entrap gangue.

Increasing chain length of the hydrocarbon radical increases the level of hydrophobicity imparted to the particle, that is, makes for a more powerful collector, but often one with less selectivity. Table 12.2 shows that chain length is commonly longer with collectors for non-sulfides than for sulfides. This difference can be attributed to non-sulfides being generally more strongly hydrophilic compared to sulfides (Table 12.1) which the longer hydrocarbon chain helps overcome. Chain length is limited by solubility.

It is common to add more than one collector to a flotation system. A selective collector may be used at the head of the circuit, to float the highly hydrophobic particles, after which a more powerful, but less selective one, is added to promote recovery of the slower floating particles. Mixed collectors are used. Guidelines for mixed sulfhydryl collectors are given by Lotter and Bradshaw (2010). Lee et al. (2009) used combinations of sulfhydryl (xanthate) and oxyhydryl (hydroxamate) collectors for mixed sulfide/oxide ores. Kerosene is used as an auxiliary collector with oxyhydryl collectors in non-sulfide flotation (Nagaraj and Ravishanakar, 2007).

There are hundreds of collectors that have been developed over the century of flotation and just some of the more important ones are included here. For more detail there have been some recent reviews (Nagaraj and

Ravishankar, 2007; Bulatovic, 2007, 2010; Nagaraj and Farinato, 2014a). Reagent supplier handbooks also remain a useful resource.

12.4.1 Collectors for Non-sulfide Minerals

Among the oxyhydryl collectors, fatty acids dominate in non-sulfide flotation. The salts of oleic acid, such as sodium oleate, and of linoleic acid are commonly employed. They are used for the flotation of minerals of calcium, barium, strontium, and magnesium, the carbonates of non-ferrous metals, and the soluble salts of alkali metals and alkaline earth metals (Finch and Riggs, 1986).

After fatty acids, the next largest application is amines, followed by petroleum sulfonates. Dodecylamine and other long chain alkyl amines are excellent collectors for quartz, for example, in reverse flotation from iron concentrates. Petroleum sulfonates possess similar properties to fatty acids, with less collecting power, but partly as a consequence, greater selectivity. They are used for recovering barite, celestite, fluorite, apatite, chromite, kyanite, mica, cassiterite, and scheelite (Holme, 1986).

By convention referred to as hydroxamates they are used as the acid (alkyl hydroxamic acid). They are chelating agents and typically have a carbon chain C6 to 14. The hydroxyl group and the carbonyl (C=O) group can form stable chelates with many metal cations (e.g., iron, manganese, copper, rare-earth elements) (Türkel, 2011). They are used in collection of minerals such as pyrochlore, muscovite, phosphorite, hematite, pyrolusite, rhodonite, rhodochrosite, chrysocolla, malachite, bornite, calcite, and gold and other precious metals. Hydroxamates have proved more effective than fatty acids in recovery of bastnaesite (a rare earth fluocarbonate) (Pradip and Fuerstenau, 2013), and in removal of anatase to improve brightness of kaolin clay (Yoon et al., 1992).

The oxyhydryl collectors have been used to float cassiterite, but have now been largely replaced by other reagents such as arsonic and phosphonic acids and sulfosuccinamates (Baldauf et al., 1985).

12.4.2 Collectors for Sulfide Minerals

The most widely used sulfhydryl collectors are xanthates, dithiophosphates, and the carbamates (Adkins and Pearse, 1992). The reaction with the surface is through a sulfur atom, the bonding properties being modified by neighboring atoms, especially by N and O, and other groups.

Mercaptans are the simplest thiol compounds, the ionic form having the general formula $R\text{-}S^-$. The chain length is usually C2 to C8. Mercaptans have been used as selective collectors for some refractory sulfide minerals (Shaw, 1981). Chen et al. (2010) report the use of dodecyl (C12) mercaptan as collector in flotation of auriferous pyrite and arsenopyrite.

Xanthates are the most important thiol collectors. Different from mercaptans, the hydrocarbon chain (R) is not bonded directly to the sulfur, but via an O–C linkage. They are available with alkyl chains ranging from C2 to C6 (or higher) of any isomeric type (Adkins and Pearse, 1992). The common alkyl chains are ethyl, isopropyl, isobutyl, and amyl. An example structure is given in Figure 12.7(a). Xanthates have good water solubility and stability in alkaline conditions with relatively low cost and ease of manufacture, transport, storage, and handling. However, because they interact with all sulfides (sphalerite and pyrrhotite being partial exceptions with short chain xanthates), xanthates require regulating agents in order to achieve selectivity between sulfide minerals.

Xanthates have the disadvantage of decomposing to CS_2, which poses a health hazard, and in discharge waters excess xanthate can create toxic conditions for aquatic life. This is prompting the search for "green" (i.e., environmentally friendly) alternatives to xanthate (Dong and Xu, 2011).

Dithiophosphates are the second most common thiol collectors and their use dates back to 1920 (Crozier 1991; Adkins and Pearse, 1992). An example structure is given in Figure 12.7(b). In dithiophosphate collectors pentavalent phosphorus replaces the tetravalent carbon, with now two oxygen atoms linking to their respective hydrocarbon chains. Dithiophosphates can be used alone, but are usually used in conjunction with xanthates or other collectors. They are available with alkyl chains ranging from C2 to C6 (or higher) of any isomeric type.

In dithiocarbamates nitrogen replaces oxygen in the otherwise xanthate-type structure. Compared to xanthate,

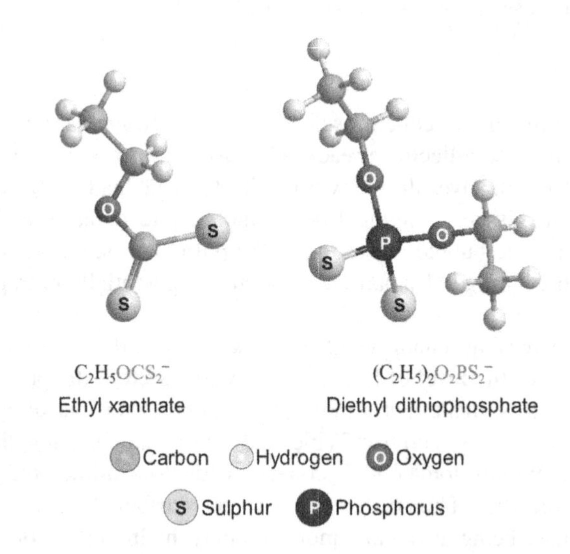

FIGURE 12.7 Structure of (a) ethyl xanthate, and (b) diethyl dithiophosphate.

both dithiophosphates and dithiocarbamates are more stable over a wider range of pH (Ngobeni and Hangone, 2013).

The structure of the collectors determines their interaction with the mineral. Oxygen is a strong electronegative atom and in xanthates withdraws electrons from the sulfur, reducing the electron density, which reduces its reactivity compared to mercaptans. The two oxygen atoms in dithiophosphates reduce the electron density on the sulfur even further, making these collectors less reactive than xanthate. In contrast, with dithiocarbamate, the nitrogen being less electronegative than oxygen makes this collector more reactive than xanthate. The more reactive the collector, in general, the less selective. Putting this together, the general sequence of selectivity is: dithiocarbamates < xanthates < dithiophosphates (Somasundaran and Moudgil, 1987). One target in identifying more selective reagents is to reduce the large consumption of lime often needed to achieve selectivity against pyrite using xanthate.

Dithiophosphinate (represented by Cytec's R3418A) is structurally similar to dithiophosphate but the R groups are directly bonded to the P atom. The lack of O or N preserves the electron density on the sulfur making this collector quite reactive. The metal complexes of dithiophosphinate are several orders of magnitude more stable than those of dithiophosphate (Güler et al., 2006).

Thionocarbamates are chelating collectors, chelation with surface metal ions taking place through two members, the C=S and N–H groups, forming stable complexes (Somasundaran and Moudgil, 1987). The Z200 reagent introduced by Dow was *O*-isopropyl *N*-ethyl thionocarbamate (referring to isopropyl being bonded to the O atom, and ethyl to the N atom). A common reference to this reagent today is isopropyl ethyl thionocarbamate (IPETC). A modified thinocarbamate, ethoxycarbonyl alkyl thionocarbamate, reacts as a six member chelate.

Table 12.3 gives examples of collector types and dosages taken from Canadian practice (Damjanović and Goode, 2000). The prevalence of xanthate is evident. Consumption is typically quoted on the basis of kg per ton ore, but dividing by the head grade gives consumption on the basis of kg per ton metal, which makes for comparison between operations. As the Table shows, collector consumption ranges from about 1–4 kg t^{-1} metal. There is some indication that the concentration in solution is also important, sometimes a factor in matching laboratory and plant conditions where pulp density (% solids) can be quite different.

Understanding the collector structure–activity relationship is probably the largest topic in basic flotation research. The target is to design collectors tailored to meet particular mineral recovery and selectivity needs (Somasundaran and Wang, 2006; Abramov, 2011; Nagaraj and Farinato, 2014a). Molecular modeling has been applied (e.g., Pradip and Rai, 2003). The challenge limiting that approach is modeling the mineral surface under flotation conditions.

12.4.3 Collector Adsorption in Non-sulfide Mineral Systems

In non-sulfide mineral flotation, the anionic oxyhydryl and cationic amines are the prime collectors. The wide range in non-sulfide mineral types, from soluble, to semisoluble to insoluble, suggests that one collector adsorption mechanism will not fit all cases. Nevertheless, the two general mechanisms, chemical and physical, apply.

Chemical adsorption, or *chemisorption*, refers to the formation of chemical bond between the collector and, usually, the metal cation in the mineral surface. The reaction falls under the general electron donor/electron acceptor model. Reaction with fatty acid (e.g., oleate) provides an example of chemisorption through formation of calcium oleate as precipitates and surface compounds (Hu et al., 1986; Rao et al., 1991). The possible surface reaction is given in Figure 12.8. In this example, the oxygen of the COO^- functional group is the electron donor, the calcium the electron acceptor.

The second mechanism, physical adsorption, is through electrostatic interaction. Collector adsorption by this mechanism is illustrated in Figure 12.9(a) for the flotation of goethite with two 12-carbon collectors, dodecyl sulfate ($R\text{-}SO_4^-$) and dodecylamine ($R\text{-}NH_3^+$). With the anionic collector flotation is successful to about pH 6 at which flotation with the cationic collector starts. The explanation for this different response lies in the fact that particles in water carry a *surface charge*, and it is through this charge that electrostatic interaction with the ionic group of the collector occurs.

The common measure of surface charge is *zeta potential* (ζ), as in Figure 12.9(b). Measurement of zeta potential is described in any standard surface chemistry text (see also Chapter 15). For goethite, the zeta potential is positive at pH less than about pH 6.7 and becomes negative at higher pH. The anionic collector (i.e., with positively charged head group) is electrostatically attracted to the goethite at pH < 6.7 and the cationic collector is electrostatically attracted at pH > 6.7, thus explaining the different flotation-pH response to the two reagents. The resulting adsorption is referred to as physical adsorption or *physisorption*, which does not result in a chemical bond.

The origin of the charge on goethite is illustrated in Figure 12.10. Considering just the Fe and O sites, by analogy with the situation described for quartz (Figure 12.3) these sites are hydrated. When the pH is low, in the case of goethite below ca. pH 6.7, the OH is removed from

TABLE 12.3 Example Collector Types and Dosage Based on Metal Tons from Canadian Practice (Damjanović and Goode, 2000)

Operation	Metals	Collector	Dosage (kg t^{-1} metal)
Strathcona Mill	Ni–Cu	PIBX	2.8 (Ni + Cu)
Clarabelle Mill	Ni–Cu	SAX	2.1 (Ni + Cu)
Thomson	Ni–Cu	SAX	1.3 (Ni + Cu)
Louvicourt Mine	Cu–Zn–Au–Ag	Aerophine 3418A	1.6 (Cu)
		PAX 51	0.5 (Cu)
		Flex 41	0.5 (Zn)
Les Mine Selbaie	Cu–Zn–Au–Ag	SIPX	1.1 (Cu)
		SP129	0.6 (Zn)
Myra Falls	Cu–Zn–Au–Ag	PAX: Aerofloat 208	0.7 (Cu + Zn)
		(73:27 blend added to both circuits)	1.5 (Cu + Zn)
Mine Bouchard-Hebert	Cu–Zn–Au–Ag	3418A (to Cu circuit)	3.5 (Cu)
		3148A (to Zn circuit)	0.7 (Zn)
		Aerofloat 208	2.1 (Cu)
Mine Langlois	Cu–Zn–Au–Ag	208 (to Cu circuit)	3.3 (Cu)
		3418 (to Cu circuit)	5.4 (Cu)
		3418 (to Zn circuit)	0.6 (Zn)
Matagami Mine	Cu–Zn–Au–Ag	Aero 3477 (to Cu circuit)	4.6 (Cu)
		Aero 3418A (to Cu circuit)	0.4 (Cu)
		Aero 3418A (to Zn circuit)	0.5 (Zn)
Brunswick Mine	Pb–Zn–Cu (Ag)	80:20 SIPX:PAX	1.8 (Pb + Zn + Cu)
		Aero 241	0.1 (Pb + Cu)
		Aerophine 5100	0.1 (Cu)
Heath Steele	Pb–Zn–Cu (Ag)	SIPX	1.0 (Pb + Zn + Cu)
		Aero 241	0.3(Pb + Zn + Cu)
		Aerophine 5100	0.2 (Pb + Zn + Cu)

Key (family type for trade name reagents): PIBX, potassium isobutyl xanthate; SAX, sodium amyl xanthate; SP129, blend of dithiophosphate and mercaptobenzothiazole plus diamine modifier; Aerophine 3418A, dithiophosphinate; PAX, potassium amyl xanthate; Flex 41, xanthate; SIBX, sodium isobutyl xanthate; Aerofloat 208, dithiophosphate; Aero 3477, dithiophosphate; Aero 241, dithiophosphate; Aerophine 5100, allyl isobutyl thionocarbamate.

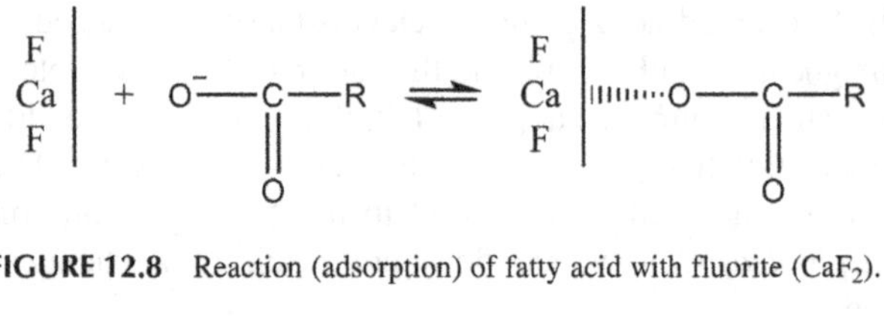

FIGURE 12.8 Reaction (adsorption) of fatty acid with fluorite (CaF_2).

some Fe sites by reaction with H^+ (to form H_2O) leaving a net positive charge, while at pH > 6.7 the H from some O sites is removed by reaction with OH^- leaving a net negative charge. For many minerals this interpretation of the effect of pH on zeta potential applies and OH^- and H^+ are referred to as *potential determining ions*. The pH when the charge is zero is the *iso-electric point* (IEP, as marked on Figure 12.9). The IEP is sometimes referred to as the *point of zero charge*, PZC.

The IEP is characteristic of the mineral and values are in the literature (Shergold, 1984; Parks, 1965); in the case of goethite, from Figure 12.9(b), the IEP is about pH 6.7, which is similar to the other major iron minerals, hematite and magnetite. If physisorption is the principal interaction mechanism, for any mineral at pH < IEP anionic collectors are effective; and at pH > IEP cationic collectors are

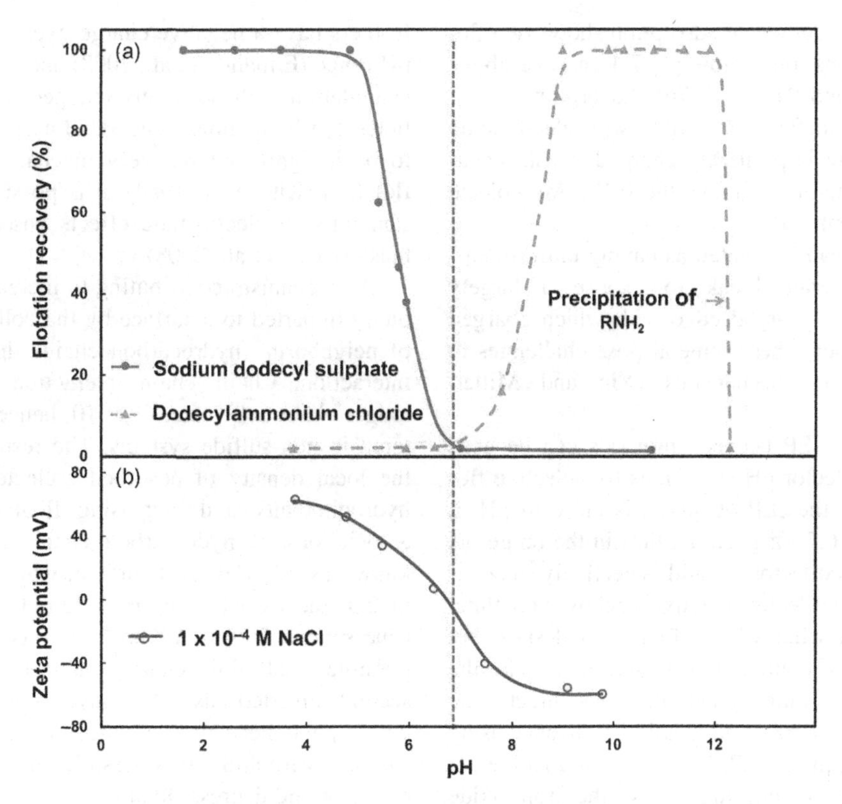

FIGURE 12.9 (a) Flotation response of goethite (FeOOH) to two collectors, anionic sulfate and cationic amine, as a function of pH (Note: the rapid decrease in flotation with amine at ca. 12 is discussed in Section 12.7), and (b) surface charge of goethite (reported as zeta potential) as function of pH (Note: reference to 10^{-4} M NaCl is as background electrolyte is to keep a constant ionic strength which would otherwise alter zeta potential) *(Adapted from Iwasaki et al. (1960)).*

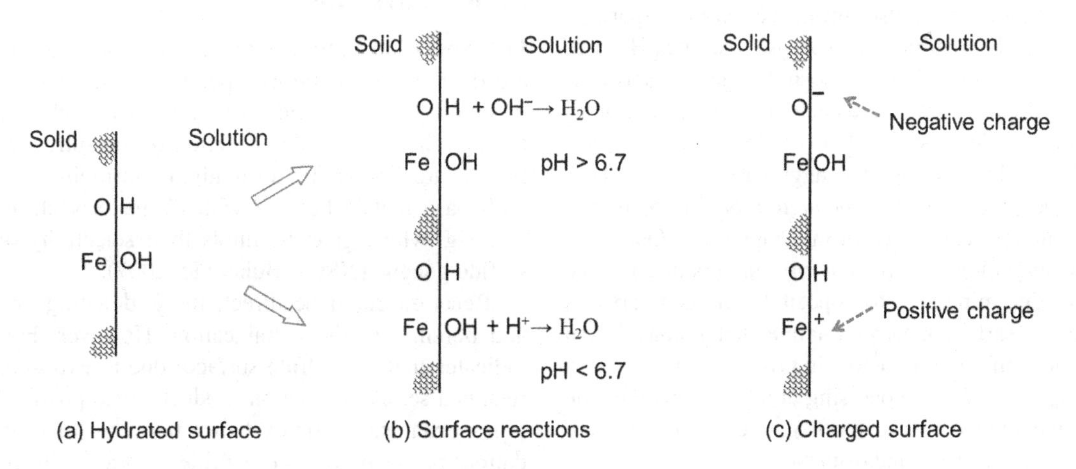

FIGURE 12.10 Origin of charge on goethite: (a) depiction of hydrated Fe and O sites, (b) reaction at low pH with H^+ and at high pH with OH^-, and (c) the resulting positive and negative charged surface sites.

effective. This is known as the *electrostatic model of flotation* (Fuerstenau, 2007).

The division in surface charge at IEP is not exact. Surface charge is a net value, consequently there are some negative sites at pH < IEP and positive sites at pH > IEP that can interact with oppositely charged collectors. This is in evidence using more hydrophobic longer chain C18 collectors in place of the C12 collectors in Figure 12.9(a), which will show greater pH overlap in the flotation response (Fuerstenau, 2007).

Both chemical and electrostatic interaction can be involved in collector adsorption. For example, pyrolusite has an IEP ~ pH 7.4 but oleate adsorption shows two flotation response regions, one below pH 7.4 and one above pH 7.4 (Fuerstenau and Palmer, 1976). Adsorption of oleate at pH below the IEP is compatible with electrostatic adsorption (pyrolusite is positively charged, oleate negatively charged), while at pH above the IEP a Mn–oleate chemical bond is identified.

Not all minerals can be treated as having uniform surfaces. Layered minerals break to expose a largely uncharged face and a charged edge with which charged collectors may interact. These mineral pose challenges in identifying interaction mechanisms (Yin and Miller, 2012).

The difference in IEP between minerals can be used as a guide to the collector-pH conditions for selective flotation. For example, the IEP of quartz is close to pH 3, compared to the pH 6.7 for goethite, thus in the range pH 3–6.7, a cationic collector should selectively recover quartz or an anionic collector selectively recover goethite. Identifying this "pH window" is often a good start, but there are a number of confounding factors that make the choice of selective flotation conditions less direct. The reverse flotation of quartz in processing iron ores using amine, rather than at pH 3–6.7, flotation is at alkaline pH with polysaccharide (starch) to depress the iron oxide minerals (Araujo et al., 2005a). Other confounding factors, including the variation in solution species with pH, are discussed in Section 12.7.

Electrostatic interactions not only contribute to adsorption of collectors, but also influence particle–particle interactions. In the goethite–quartz case over the pH range 3–6.7, the two minerals are oppositely charged and thus are electrostatically attracted to each other. Aggregation of particles due to electrostatic effects is known as *coagulation* and since in this case the aggregates are of mixed minerals, the process is *heterocoagulation*. (Coagulation and the related processes return in Chapter 15.) One form of heterocoagulation is *slime coating*, the coating of fine particles of one mineral onto oppositely charged particles of another. Clearly counterproductive, heterocoagulation is one of the confounding factors in processing iron ores in the pH range 3–6.7; by processing at alkaline pH the iron minerals and quartz both carry negative charge and are electrostatically repelled and dispersed.

Rather than adsorbing as ions, some collectors appear to come out of solution as colloidal precipitates (*collector colloids*) that interact electrostatically with particles. In potash flotation in saturated brines, one proposed mechanism is that the amine collector forms positively charged collector colloids that heterocoagulate with negatively charged sylvite (KCl) giving the selective flotation from halite (NaCl) (Yalamanchili et al., 1993).

Not only do particles carry charge, so do the bubbles. Bubbles in water alone and in the presence of most frothers have a negative charge over most of the practical pH range (Elmahdy et al., 2008) and some level of heterocoagulation with positively charged nonhydrophobic particles can be demonstrated (Uddin et al., 2012). Unlikely to be a significant recovery mechanism in conventional flotation (Kitchener, 1984), it is possible to induce flotation through electrostatic effects based on charged bubbles (Waters et al., 2008).

A mechanism contributing to increasing the hydrophobicity imparted to a surface by the collector is the linking of neighboring hydrocarbon chains through hydrophobic interaction. Chain–chain interaction occurs more with longer chain collectors, $C > 10$, hence is mainly encountered in non-sulfide systems. The result is an increase in the local density of adsorbed collector, intensifying the hydrophobicity and increasing floatability. In solution, associations of hydrocarbon chains form 3D structures known as *micelles*; at a surface, which restricts assembly in 2D, the associations are referred to as *hemi-micelles* (Fuerstenau, 2007). At too high a collector dosage it is possible that this chain–chain interaction causes a second, inverted adsorption layer to form with the molecule's polar head now exposed to the water making the surface hydrophilic. It is possible, therefore, to "overdose" collector and depress flotation.

12.4.4 Collector Adsorption in Sulfide Mineral Systems

In flotation of sulfide minerals, the principal collectors are the sulfhydryl (thiol) type. Adsorption predominantly involves chemical bond formation, that is, chemisorption, through the electron donor action of the sulfur. Thiol collectors are known to form highly insoluble precipitates with base metals but not with elements such as Si, Ca, and Mg, which give the thiols their selectivity over non-sulfide gangue (NSG) (Bulatovic, 2007).

Reaction could be direct, the S donating an electron and bonding to the metal cation. However, Figure 12.4 indicates that the sulfide surface (due to exposure to DO) retains a series of reaction products. One possible adsorption mechanism is ion exchange between some of the oxidation products on the sulfide surface, illustrated for adsorption of xanthate in Figure 12.11 (Shergold, 1984). The end result is a chemical bond between the metal cation in the surface and the sulfur of the xanthate. This mechanism goes some way to understanding the observation that, generally, increasing pH depresses flotation, the hydrophilic OH^- being in competition with the collector for adsorption sites. An important variation on the electron donor/acceptor model is where the electron acceptor is oxygen. This is the *electrochemical* or *electron transfer model of flotation* (Chander, 2003; Woods, 2010).

$$\mathrm{M}\,|\,\mathrm{OH^-} + \mathrm{S^-{-}C({=}S){-}O{-}R} \rightleftharpoons \mathrm{M}\,|\,\cdots\mathrm{S{-}C({=}S){-}O{-}R} + \mathrm{OH^-}$$

FIGURE 12.11 Ion exchange mechanism; xanthate exchanges with oxidation product (see Figure 12.4 for examples of oxidation products on a sulfide surface).

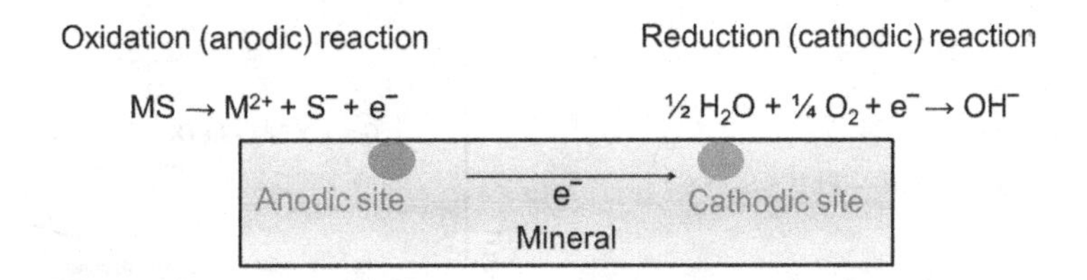

FIGURE 12.12 Depiction of a sulfide mineral−water interface hosting both anodic (oxidation) and cathodic (reduction) sites.

The electrochemical mechanism refers to one species donating electrons, the oxidizing or anodic species, and another accepting the electrons, the reducing or cathodic species. The sulfide mineral−water interface hosts both anodic and cathodic sites, as depicted in Figure 12.12. At the anodic site, the metal sulfide is oxidized and the electrons released transfer to the cathodic site, where reduction of oxygen dissolved in the water occurs to produce OH^- ions. Oxygen is the common electron acceptor in flotation systems, but on occasions other electron acceptors, for example, ferric ions, which reduce to ferrous ions, can be involved. Most sulfide minerals are semiconductors and can sustain the electron transfer process.

A way to represent the process is to use a current−potential diagram, as shown in Figure 12.13 (Woods, 2010). Here, the current associated with both the anodic and cathodic reactions is plotted against electrochemical potential: as potential is increased the anodic current increases and the cathodic current decreases. With no external potential applied, the equilibrium potential E is when the anodic current I_A equals the cathodic current I_C. This is the potential that is measured. The equilibrium potential is also referred to as the *open circuit potential*; and when measuring on a mineral alone, it is known as the *rest potential*. Measurement of rest potential and generation of the current−potential plot are described in Rao and Leja (2004) and Woods (2010).

Values of rest potential are given in Chapter 2 and others are available in the literature (Shergold, 1984; Fuerstenau, 2007). The value depends on the reference electrode in use. The common practical reference electrodes are the saturated calomel electrode (SCE, Hg/Hg_2Cl_2) and the silver/silver chloride electrode (Ag/AgCl), but the potential is often converted to the *standard hydrogen electrode* (SHE) scale, symbolized as E_h. The conversion to E_h is to add 0.24 V to the value on the SCE scale and to add 0.22 V to the value on the Ag/AgCl scale. On

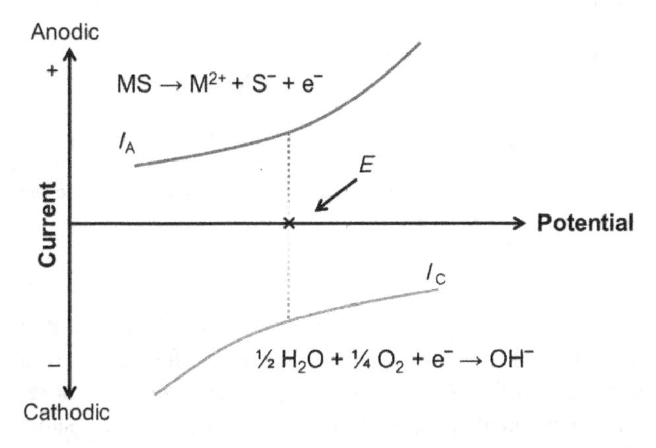

FIGURE 12.13 Current−potential diagram representing the anodic and cathodic reactions in Figure 12.12.

reading the literature, care is required to note the reference electrode. Rather than the actual value of rest potential, the relative value is often more useful (and can be remembered); the relative rest potential for some common sulfides and mild steel (representing grinding media) in decreasing order is:

Pyrite > Chalcopyrite > Sphalerite > Pentlandite > Pyrrhotite > Galena > Mild Steel

Pyrite, having the highest rest potential of the common sulfides, is sometimes referred to as the most "noble."

Adsorption of xanthate can now be understood as the anodic reaction on the mineral surface. Using galena with xanthate collector as the example system, Figure 12.14 depicts this reaction, which is represented on the current−potential diagram in Figure 12.15. The equilibrium potential of the galena−xanthate reaction is lower than the equilibrium (rest) potential of the mineral; that is, the xanthate oxidizes (loses an electron) more readily than

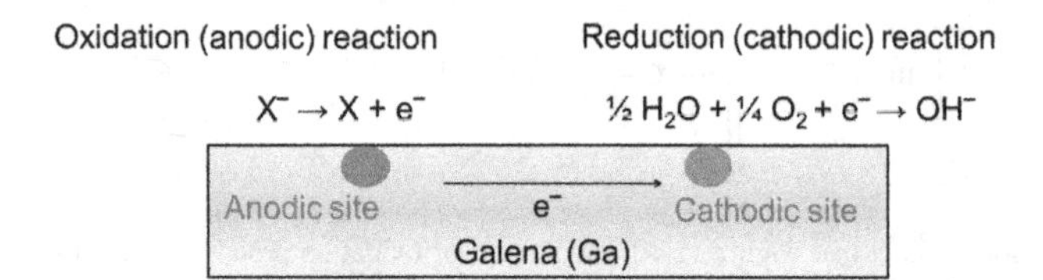

FIGURE 12.14 Adsorption of xanthate as an anodic (oxidation) reaction on the surface of galena.

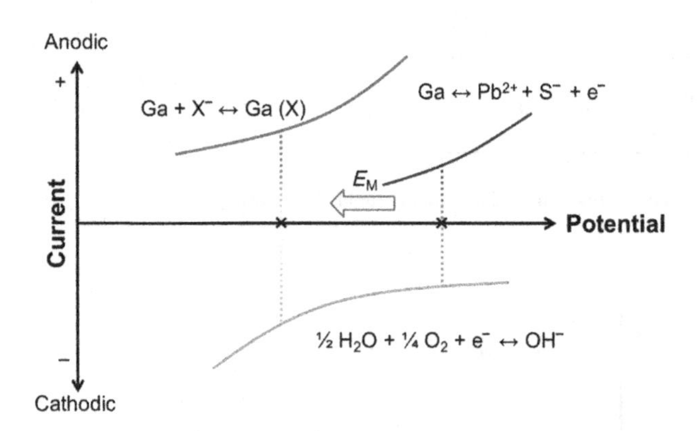

FIGURE 12.15 Current–potential diagram representing the anodic and cathodic reactions in Figure 12.14.

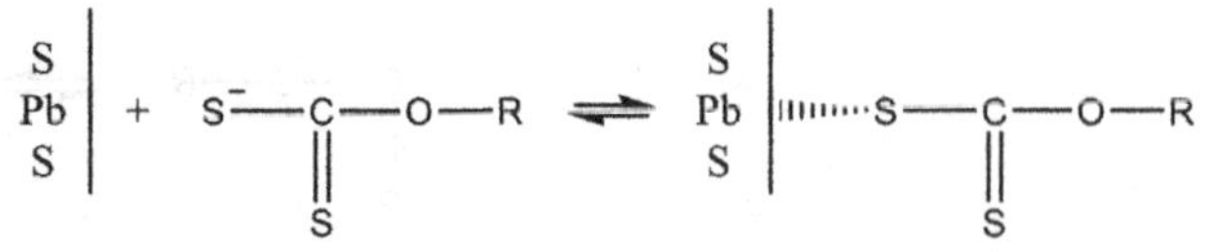

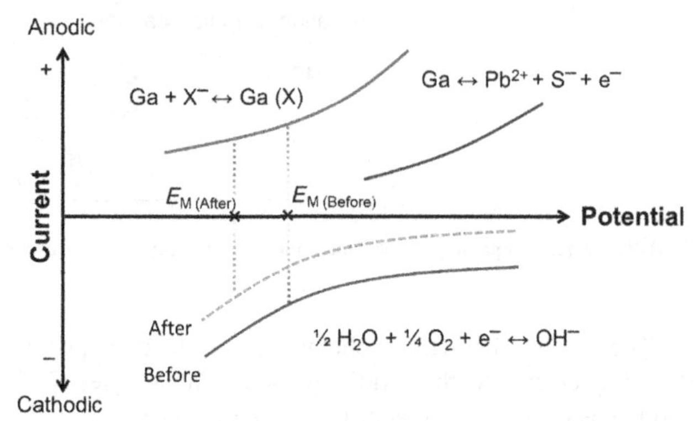

FIGURE 12.16 Result of adsorption: chemical bond formation (chemisorption) between xanthate and galena.

FIGURE 12.17 Impact of raising pH or lowering DO concentration (the "after" condition) on reducing the magnitude of the cathodic current which shifts the location of $I_A = I_C$ to lower potential; that is, the pulp potential is lowered.

does the galena. Upon adsorption, the measured potential decreases (Figure 12.15), and is now referred to as a *mixed potential* E_M, the potential when two (or more) electrochemical processes are occurring on the same electrode (mineral surface in this case). When the measurements are in a flotation slurry, the term *pulp potential* is often applied. The decrease in potential upon addition of xanthate can be detected in laboratory experiments (Labonté and Finch, 1990). The mixed potential now represents equilibrium between the anodic current associated with adsorption of xanthate and the cathodic current associated with reduction of oxygen. The reaction can be generalized as (introducing X^- to represent the xanthate ion):

Anodic reaction:

$$MS + X^- \leftrightarrow MS(X) + e^- \tag{12.4}$$

Cathodic reaction:

$$\tfrac{1}{2} H_2O + \tfrac{1}{4} O_2 + e^- \leftrightarrow OH^- \tag{12.5}$$

Xanthate adsorption is probably a combination of chemical (ion exchange) and electrochemical mechanisms. Regardless, the end result is a chemical bond between the sulfur of the xanthate and the metal atom of the mineral surface (Figure 12.16). The electrochemical model applies to all thiol reagents.

The electrochemical model explains the depressant effect on sulfides of raising pH. The increase in OH^- concentration drives the cathodic reaction (Eq. (12.5)) in reverse and thus reduces reaction with xanthate. An effect of oxygen is also identified: too low a DO concentration may reduce the cathodic reaction to the extent that the reaction with (adsorption of) xanthate is reduced and the mineral does not become hydrophobic enough to float. A decrease in DO or increase in pH has a similar effect on the mixed potential: by reducing the magnitude of the cathodic current they move the cathodic current–potential curve upward (to less negative current values), and thus the equivalence between the anodic current and cathodic current, which defines the mixed potential, moves to a lower value (Figure 12.17). At the new, lower mixed potential, note that the anodic current is less than before, indicating the reduced adsorption of xanthate.

The role of oxygen and pulp potential is illustrated in Johnson et al. (1982). They attributed poor galena recovery to low pulp potential due to oxygen consumption by the media in the grinding mill. In down-the-bank surveys, they observed both Pb grade and recovery increasing over the first few cells, an unusual feature (grade and recovery are commonly inversely related, Chapter 1), which suggested that the first cells by aerating the pulp were raising the pulp potential and increasing xanthate reaction with galena. Measuring pulp potential, Johnson et al. showed that it could be as low as −150 mV (vs. Ag/AgCl) in ball mill discharge, rising to −10 mV at the head of the flotation bank and to +150 mV mid-way down the bank.

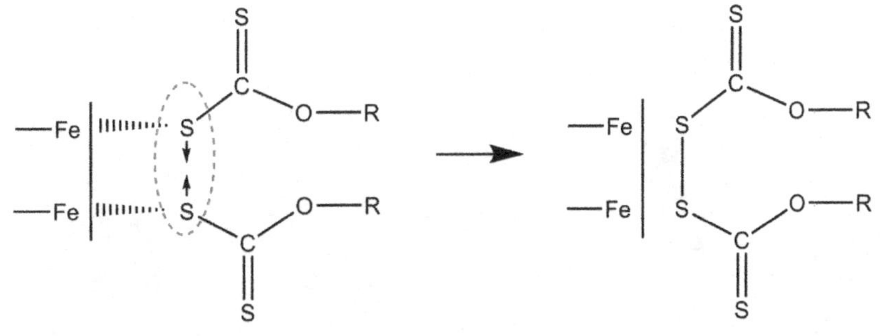

Adsorbed xanthate Dixanthogen

FIGURE 12.18 Formation of dixanthogen on surface of pyrite.

Using combinations of air and nitrogen to control potential in laboratory experiments, they demonstrated the impact on restoring galena flotation of raising DO and pulp potential.

When comparing the rest potential of sulfides it was noted that pyrite had the highest rest potential. The rest potential of pyrite actually lies above another xanthate oxidation–reduction couple, the anodic reaction to form dixanthogen (Eq. (12.6)), where again oxygen is the common electron acceptor:

$$2\,X^- \leftrightarrow X_2 + 2\,e^- \quad (12.6)$$

Formation of dixanthogen on the surface of pyrite is depicted in Figure 12.18. Being a neutral molecule it is both highly hydrophobic and poorly soluble. The presence of dixanthogen on pyrite is readily demonstrated in laboratory experiments but it is not always clear if it is the hydrophobic species in all cases of pyrite flotation.

12.5 FROTHERS

Frothers have three main functions in flotation (Klimpel and Isherwood, 1991):

- Aid formation and preservation of small bubbles.
- Reduce bubble rise velocity.
- Aid formation of froth.

Reduction in bubble size increases the number and total surface area of bubbles, which increases collision rate with particles and thus increases flotation kinetics. Reducing rise velocity increases the residence time of bubbles in the pulp which increases the number of collisions with particles and thus further increases kinetics. Formation of a froth means the bubbles do not burst when they reach the top of the pulp, which enables the collected particles to overflow as the float product.

Like collectors, frothers are a class of surfactant, this time active at the air–water interface. Most frothers are heteropolar compounds comprising a polar (i.e., hydrophilic)

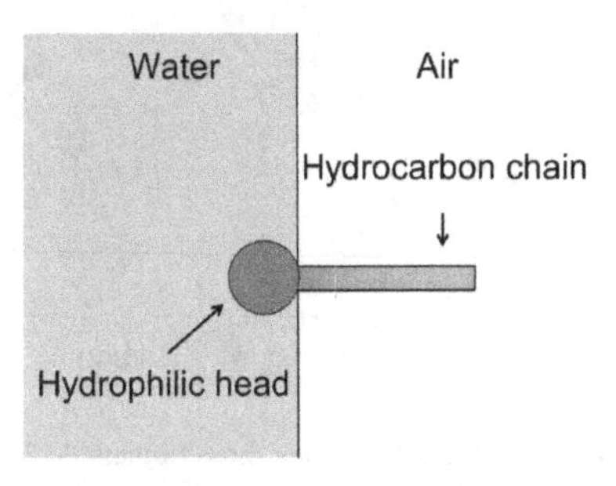

FIGURE 12.19 General structure of frother molecule and orientation at the air–water interface.

group, typically hydroxyl, and a hydrophobic hydrocarbon chain. When surface-active molecules are in water, the water dipoles combine (H-bond) readily with the polar groups, but there is practically no reaction with the nonpolar hydrocarbon group, the tendency being to force the latter into the air phase. Thus, the heteropolar structure of the frother molecule leads to its adsorption at the air–water interface, with the nonpolar groups oriented toward the air and the polar groups toward the water (Figure 12.19). Frother thus controls the properties of the air–water interface.

Initially natural oils such as pine oil were used as frothers, but their use diminished over the years. Some collecting properties of the natural oils also interfered with process selectivity (Crozier and Klimpel, 1989). Having collecting and frothing properties in the same reagent can make selective flotation difficult.

The major commercial frothers today are alcohols and polyglycols (Klimpel and Isherwood, 1991; Laskowski, 1998) with a third type, alkoxy substituted paraffins, for example, triethoxy butane, in some use. Alcohol frothers ($C_nH_{2n+1}OH$) usually contain a single hydroxyl (OH) group and are restricted to 5–7 carbons either straight or branch-chained. Shorter chain alcohols are not surface active enough and longer chain alcohols are increasingly insoluble. A common alcohol frother is MIBC, methyl

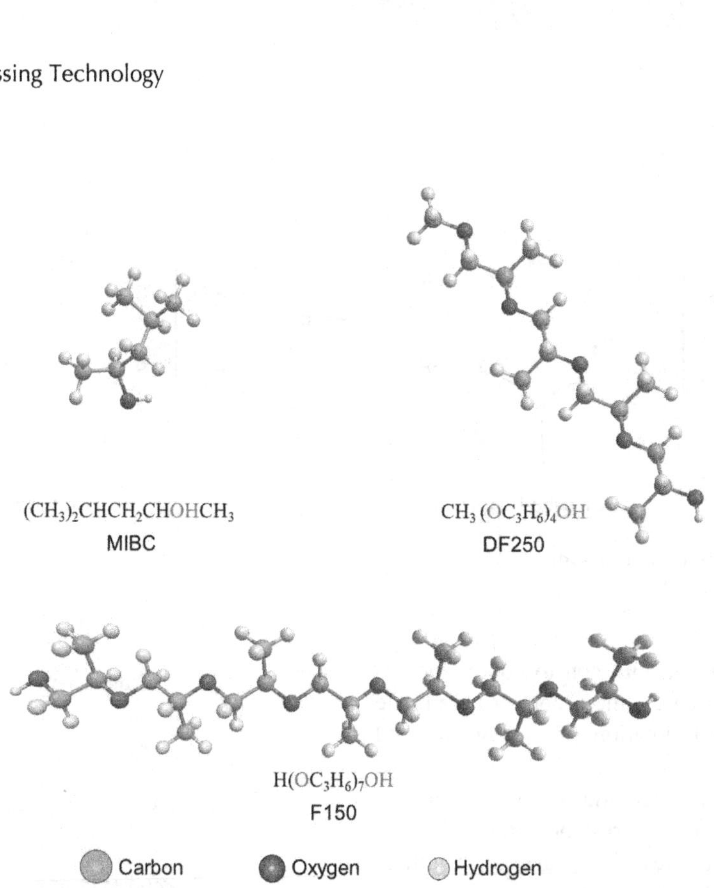

$(CH_3)_2CHCH_2CHOHCH_3$
MIBC

$CH_3(OC_3H_6)_4OH$
DF250

$H(OC_3H_6)_7OH$
F150

Carbon Oxygen Hydrogen

FIGURE 12.20 Example structures of three frothers: MIBC, DF250, and F150.

isobutyl carbinol ($(CH_3)_2CHCH_2CH(OH)CH_3$). (The name, incidentally, derives from the old nomenclature building from a base of methyl alcohol (CH_3OH) or carbinol; the modern formal chemical name is 4-methyl-2-pentanol.) Polyglycol frothers include polypropylene glycols (PPGs) ($H(OC_3H_6)_mOH$), PPG alkyl ethers ($C_nH_{2n+1}(OC_3H_6)_mOH$), and polyethylene glycol alkyl ethers ($C_nH_{2n+1}(OC_2H_4)_lOH$), which form a large class with varying molecular structure and molecular weight. The propoxy group ($PO = OC_3H_6$) and ethoxy group ($EO = OC_2H_4$) make these surfactants readily soluble in water by introducing the ether –O– linkage, which acts as another polar site. Two familiar frothers from the polypropylene family are DF250 ($CH_3(PO)_4OH$), a PPG methyl ether, and F150 ($H(PO)_7OH$), a PPG. Alternative naming is to use an acronym and include the molecular weight. For example, F150 is PPG of molecular weight 425, giving the name PPG425. Figure 12.20 shows the structure of MIBC, DF250, and F150. Rather than pure single compounds, commercial frothers are often by-products and blends.

Availability and cost are still major considerations in selection of a frother, but many operations look toward new frothers or frother blends that are tailored to the operation. Pugh (2007) estimated that MIBC and PPG ethers account for over 80% of the frothers used today in metallic ore flotation.

Frothers may be added as blends. One reason is to handle a wide particle size range. It has been observed that a single frother generally cannot float the broad particle size distribution typical of a flotation feed; in general MIBC suits finer particles, polyglycols coarser particles (Klimpel, 1995). Another argument for blends is to try to effect some independence over bubble size reduction and froth stabilization. There is some research into the bubble size and frothing characteristics of blends (Laskowski et al., 2003; Tan et al., 2005; Elmahdy and Finch, 2013). The interactions between frothers can be surprisingly complex (Zhang et al., 2012a).

12.5.1 Bubble Size Reduction

Frother action in reducing bubble size is illustrated in Figure 12.21. The images show the decreasing size and increasing number of bubbles as frother concentration increases, while the inserts show the progressive change in size distribution, from bimodal to unimodal, that is often seen (Quinn and Finch, 2012). The plot shows the Sauter mean diameter (D_{32}) as a function of frother concentration (D_{32}-C). (The Sauter mean diameter, or volume-to-surface mean bubble diameter, produces the same surface area to volume ratio as the bubble size distribution (BSD); it is the common mean size used in analysis of flotation systems. See Section 12.14.2 for more

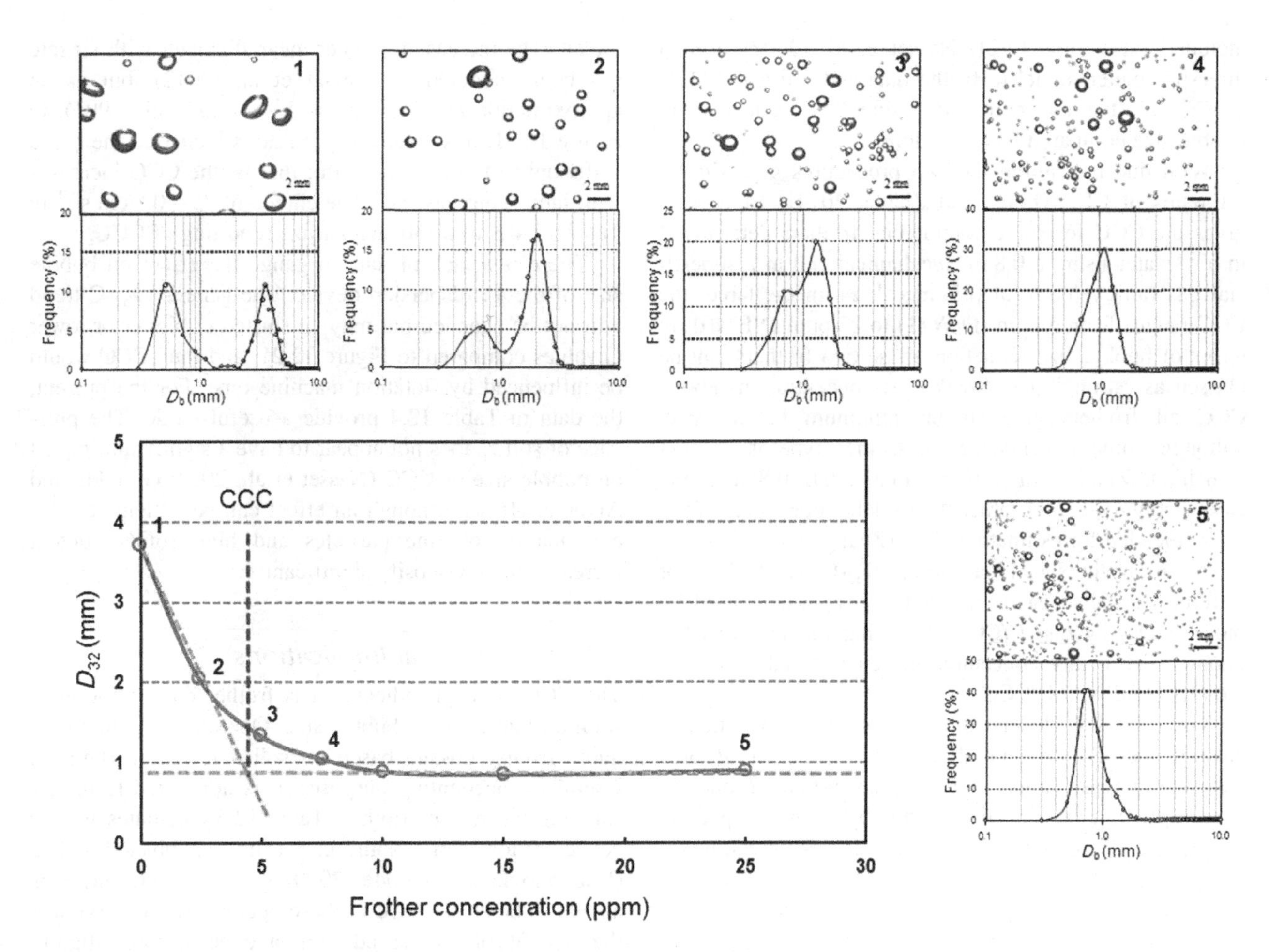

FIGURE 12.21 Reduction in bubble size (Sauter mean diameter) as a function of frother concentration illustrated with images and number frequency distribution. Included on the plot is the CCC estimation procedure of Cho and Laskowski (2002); see text (Data collected at $J_g = 0.5$ cm s^{-1} in a 0.8 m^3 mechanical cell.) *(Adapted from Nesset and Finch (2013)).*

details.) In water only the mean bubble size produced in virtually any flotation machine is at least 4 mm (diameter). As Figure 12.21 shows, addition of just a few ppm frother reduces the mean bubble size to about 1 mm or less. To give some perspective, a few ppm is equivalent to a few gram per ton of water.

To illustrate the impact of size reduction, dividing a 4 mm bubble into 1 mm bubbles increases the number of bubbles 64 times, and the total surface area 16 times, which translates to an increase in *bubble surface area flux* (BSAF) and flotation kinetics. BSAF (S_b) is given by (see also Sections 12.9.3 and 12.14.2):

$$S_b = \frac{6\,J_g}{D_{32}} \tag{12.7}$$

where J_g is superficial gas (air) velocity (volumetric air rate divided by cell cross-sectional area). For a given gas velocity, reducing the bubble size from 4 mm (0.4 cm) to 1 mm (0.1 cm) increases S_b by 4 times. The relationship between kinetics and BSAF is pursued in Section 12.9.3, but suffice it to say that achieving capacity in a flotation machine is dependent on producing small bubbles, making this arguably the key function of the frother.

The common explanation for frother's ability to reduce bubble size is that frothers retard bubble *coalescence*, that is, the merging to form larger bubbles (Harris, 1976). In this coalescence prevention interpretation the working hypothesis is that the machine produces small bubbles and the frother preserves them (Cho and Laskowski, 2002). There is some evidence that frothers also aid break-up of the air mass (Chu and Finch, 2013).

Critical Coalescence Concentration

Figure 12.21 shows the common trend observed for all frothers: a rapid initial decrease in size to a transition concentration where the minimum D_{32} is reached. Cho and Laskowski (2002) introduced the term *critical coalescence concentration* (CCC) to refer to this transition concentration. Their graphical method of estimating CCC is included on Figure 12.21. Nesset et al. (2007) used a three-parameter model to fit the data and estimated CCC as CCC95, the concentration giving 95% reduction in bubble size compared to water alone.

With due precaution the two procedures give similar estimates of CCC (Finch and Zhang, 2014). Table 12.4 gives the CCC for some commercial frothers determined in air–water using a 0.8 m^3 mechanical cell at a superficial gas (air) velocity of 0.5 cm s^{-1}. From the table, the CCC ranges from 5 ppm (PFW31) to 23 ppm (NF240); in terms of bubble size reduction, these two frothers can be classed as "strong" and "weak." At concentrations above CCC all frothers give similar minimum bubble size, although some dependence on frother type is evident (Finch and Zhang, 2014). Using a large cell (0.8 m^3 in the case of Figure 12.21) rather than a laboratory-size unit is recommended for estimating CCC (Zhang et al., 2009).

As all frothers yield a similar D_{32}-C trend, data for different frothers can be reduced to a common trend by plotting D_{32} against C/CCC. Knowing the frother CCC bubble size at any concentration can be predicted from the D_{32} trend with C/CCC.

The CCC will depend to an extent on factors such as gas rate and perhaps method of bubble generation. Bubble size will increase as gas rate increases, reflecting that the energy per unit mass of air imparted by the gas dispersion device has decreased corresponding to less break-up of the air. The increase in Sauter mean diameter with air rate has been modeled by Nesset et al. (2012), but as an approximation D_{32} increase as ${J_g}^{0.25}$ (Xu et al., 1991). In Figure 12.21, this means as gas rate is increased the curve shifts upward and to the right; that is, the CCC increases with increasing gas rate. The choice of $J_g = 0.5$ cm s^{-1} in Table 12.4 is to try to standardize reporting of CCC.

There is a lack of data to gauge the effect on bubble size of the air dispersion device. The general D_{32}-C trend will not be changed but may be shifted to higher or lower C values compared to Figure 12.21, and thus CCC would be influenced by flotation machine type. For the present, the data in Table 12.4 provide a useful guide. The presence of solids does not appear to have a significant impact on bubble size or CCC (Nesset et al., 2012; Grandon and Alvarez, 2014), although an effect can be anticipated if a combination of fine particles and high solids content increase slurry viscosity significantly.

TABLE 12.4 CCC for Some Commercial Frothers: Determined at $J_g = 0.5$ cm s^{-1} in a 0.8 m^3 Mechanical Cell

Family	Name	Supplier	CCC (ppm)
Aliphatic alcohol	FX120-01	Flottec	11
Polypropylene glycols and their ethers	DowFroth200	Dow Chemical	17
	DowFroth250		10
	DowFroth1012		6
	FX160-01	Flottec	12
	FX160-05		15
	F160		8
	F150		6
	PolyFroth W31	Huntsman	5
Polyethylene glycols and their ethers	FX120-02	Flottec	13
	NasFroth240	Nassaco	23
	NovelFrother234	Sasol	16

Notes: 1. CCC was measured as CCC95; 2. CCC values were adapted from Zhang et al. (2012a) and Finch and Zhang (2014).

CCC and Practical Implications

The CCC concept indicates it is frother concentration in solution that controls bubble size. Operating practice is to quote frother dosage based on solids feed rate (kg t^{-1}). Useful for accounting purposes, it is not suited to understanding the role of frother. Table 12.5 calculates frother concentration in solution from published data (Damjanović and Goode, 2000) showing concentrations appear to be above CCC at these operations. To maximize the benefit of frother addition a concentration slightly above CCC is the target; a concentration lower than CCC means (from Figure 12.21) any change in concentration has a large impact on bubble size which impacts kinetics. Increasing C much beyond CCC, apart from cost, risks excess water being carried to the froth, increasing entrainment (Zhang et al., 2010; Welsby, 2014). In the case of coal and other carbonaceous matter, the addition of frother may appear to greatly exceed CCC but this is because some frother is adsorbed by the solids; what remains in solution is the few ppm enough to produce the bubble size reduction (Gredelj et al., 2009). Some solids may adsorb one type of frother but not another. An example is talc, which adsorbs polyglycol frothers but not alcohols (Kuan and Finch, 2010).

Nesset et al (2012) incorporated CCC in a model to predict bubble size in mechanical flotation cells as a benchmark for operations: comparing predicted against measured D_{32} could indicate potential for further decreases in bubble size (see Section 12.14.2). Plants operating below CCC may find that increasing frother dosage is not compatible with froth stability requirements, giving excess water recovery, for example. If this is the situation it may suggest that a different frother should be considered.

TABLE 12.5 Frother Dosage at Rougher and Calculated Concentration in Solution: Data from Canadian Milling Practice

Operation	Metals	Type	Dosage (kg t^{-1})	% Solids	Concentration (ppm)
Troilus	Cu–Au–Ag	Aerofroth 65	0.014	36	25
Strathcona	Ni–Cu	DF250C	0.025	47	22
Langlois Mine	Cu–Zn–Au–Ag	MIBC	0.0082	40	12
Matagami	Cu–Zn–Au–Ag	MIBC	0.01	38	16
Thomson	Ni–Cu	MIBC	0.022	48	20

Source: From Damjanović and Goode (2000).

12.5.2 Bubble Rise Velocity Reduction

Small bubbles (<2.5 mm) rise more slowly than large bubbles and small bubbles in the presence of many solutes rise more slowly than in water only (Clift et al., 2005). Decreased bubble rise velocity means bubbles spend longer in the pulp zone of the flotation cell (increased bubble residence time), which means more time to collide with and collect particles. Over the size range 1–2.5 mm bubble rise velocity is about halved in the presence of frother; that is, bubble retention time about doubles.

Increasing bubble retention time increases the content of air in the cell, referred to as *air (gas) holdup* (ε_g or GH). The specific surface area of bubbles (surface area per unit volume, A_b) is related to GH and the Sauter mean diameter by:

$$A_b = \frac{6\,\varepsilon_g}{D_{32}} \tag{12.8}$$

In the absence of frother GH is rarely above 5% (0.05 in Eq. (12.8)), but with frother can reach 15% in mechanical cells and columns (Finch and Dobby, 1990; Dahlke et al., 2005). Combined with the decrease in bubble size from ca. 4 to 1 mm upon introducing frother, then the specific surface area increases about 12 times, potentially boosting kinetics by a comparable factor (see Section 12.9.5).

GH is recognized as an important cell operating variable (Yianatos and Henriquez, 2007). Dahlke et al. (2005) used the GH–gas rate relationship to define an operating range for a cell (see Section 12.14.2). In the newer "reactor-separator" flotation cell designs, achieving high GH to increase flotation kinetics is one of the features (see Section 12.13.3).

Compared to bubble size reduction, the mechanism of frothers slowing bubble rise is perhaps better understood, often explained by surface tension gradients forming around the bubble, which increase fluid drag (Clift et al., 2005). However, recent work has also shown that velocity is controlled by bubble shape, which may not be dependent on surface tension gradients (Tomiyama et al., 2002; Maldonado et al., 2013).

12.5.3 Froth Formation

The third principal frother function is to promote froth stability. Froth stability, while understood in general terms, has no universal definition (Farrokhpay, 2011). The coalescence inhibiting function of frother coupled with the small size of bubble generated are contributing factors, but froth stability is strongly dependent on the amount and hydrophobicity of particles in the froth (Hunter et al., 2008). MIBC, for example, gives little froth in an air–water only system but clearly can support substantial froths in practice. It is not surprising therefore that any factors influencing particle hydrophobicity, such as collector and regulating agents, can impact the froth.

The simplest explanation for the increased froth stability is that the particles attached to bubbles impede water drainage from the froth. The impact of particles is often evident down a flotation bank; the froth in the first cell can even become too loaded and "collapse", while the last cell struggles to form a froth at all. Zanin et al. (2009) note two operations achieved similar results using different froth strategies, one using a shallow froth with a "weak" frother (MIBC) and a second a deeper froth with a "strong" blended frother.

Hydrophilic particles can also influence froth stability, for example, clays that are entrained into froth (Farrokhpay and Bradshaw, 2012). Too stable a froth is not desirable as it can affect downstream operations, such as pumping and concentrate thickening. The requirement that froth not to be too stable restricts the type of surfactant that can be considered as a frother. On occasion defoaming agents may be needed.

The importance of the froth in determining flotation performance is well recognized; factors affecting the froth are reviewed by Farrokhpay (2011) and Ata (2012).

12.5.4 Frother Properties of Other Agents

The non-sulfide mineral collectors—fatty acids, alkyl sulfates, and amines—are known to exhibit frother action. In these cases, the polar group is the charged (ionic) functional group. Polymer regulating agents are other candidates with possible frother-like functions, and some inorganic salts at high concentration also mimic frother actions. Data on bubble size control of these other agents is limited but is starting to be collected (e.g., Ravichandran et al., 2013).

Atrafi et al. (2012) showed that sodium oleate produced a D_{32}-C trend similar to that in Figure 12.21 but with a much higher CCC, ca. 70 ppm, compared to commercial frothers. Most non-sulfide collectors can produce stable froth in the absence of solids, aided by the bubbles becoming charged due to the ionic group on the collector, which induces electrostatic repulsion between bubbles in the froth that retards coalescence. Sometimes used alone, it is arguably preferable not to have one reagent acting as both collector and frother to avoid compromising the two roles. The choice of frother to use with these "frothing" collectors may not be obvious, however (Espinosa-Gomez et al., 1988; El-Shall et al., 2000). In comparison, the sulfhydryl collectors probably have less frother action but the long chain members warrant testing. Polymer depressants may have frother properties; some may promote coalescence, at least in the froth (Wiese et al., 2010). In the case of bitumen flotation, the natural surfactants released in processing the ore provide the frother action (Nassif et al., 2013; Zhou et al., 2000).

Some inorganic salts are known as *coalescence inhibitors* (Craig et al., 1993) and in sufficient concentration can substitute for frother, an example being Glencore Xstrata's Raglan operation (Quinn et al., 2007). Determining CCC for a variety of salts encountered in mineral processing systems, Quinn et al. (2014) found values ranging from ca. 0.07 M ($MgSO_4$) to 0.31 M ($NaCl$). Combining the results, they estimated a critical coalescence *ionic strength* of about 0.33 M. Laboratory flotation tests on naturally hydrophobic graphite suggested that provided the same bubble size was achieved, recovery was independent of the type of salt (Alexander et al., 2012). An effect of salt on selective flotation, however, is reported (Wang and Peng, 2013).

A number of plants use high salt content process water either due to recycling or the water source, for example, bore water and seawater. Seawater has an ionic strength of about 0.7 M, which explains the fine bubble size found when this is the source of water. Using seawater is set to expand, notably in Chile (Rosas et al., 2012), and in consequence flotation using saline waters is attracting study (Castro, 2012; Castro et al., 2012; Wang and Peng, 2013). While it would seem frother is not needed to produce small bubbles in saline waters, it may be necessary to achieve a desired froth property or for some other purpose. For example, potash flotation is conducted in saturated brine and addition of frother (MIBC) in that system appears to assist the amine collector function (Burdukova and Laskowski, 2009).

12.5.5 Possible Mechanisms

As noted, a prime function of frother is to reduce bubble size, a property shared with high concentration of some salts. This shared action complicates isolating the mechanism (Finch et al., 2008). For instance, the action of frother is often associated with the property of surfactant to reduce surface tension. Most salts, however, increase surface tension. Regardless of the direction of change, both are capable of creating surface tension gradients, local variations in tension related to $\delta\gamma/\delta C$, and gradients have been linked to coalescence prevention. But the magnitudes are very different, much higher with frother than with salts due to the much lower values of C. The fact that both frothers and salts, at least the cations, hydrate and thus change the local water structure at the air–water interface could be the shared phenomenon related to coalescence prevention. A hydrated bubble surface suggests it has some level of hydrophilicity (or reduced hydrophobicity), which would contribute to the properties of the thin film that must finally rupture for particle and bubble to attach. Provided the particle is more hydrophobic than the bubble is hydrophilic, the film will rupture. Interaction between bubble and particle can be considered a form of hydrophobic interaction (Yoon, 2000).

Frothers, it has been suggested, can interact with collectors, known as *monolayer penetration* (Rao and Leja, 2004). A supporting argument derives from Eq. (12.2) which shows that high work of adhesion is favored by a low value of $\gamma_{s/a}$, the solid–air interfacial tension. Recognizing that the higher the level of adsorbed surfactant the lower will be the interfacial tension, one way to achieve this high surface concentration at the solid–air interface is for the frother layer on the bubble to intermingle with ("penetrate") the collector layer on the particle to produce the low $\gamma_{s/a}$. An interesting possibility, a weakness is that Eq. (12.2) describes an equilibrium state; the argument is thus thermodynamically based and may not be correspond to the flotation condition.

Pursuing the mechanisms by which frothers act may seem a diversion in a "practical" text but it will pay dividends in at least one area: CFD modeling (Chapter 17). This is reaching a high degree of sophistication in modeling flotation machine mechanics, leading to important advances in design of cell internals. What is missing is a model of how bubbles are formed that correctly includes the obvious large impact of frothers. As a suggestion, since the addition of frother decreases bubble size, this means an increase in bubble surface area, which corresponds to an increase in surface energy, and thus frothers could be considered as adding an energy component to the mechanical energy input to the air dispersion mechanism.

12.5.6 Frother Selection and Characterization

Selection of frother for a given duty remains largely empirical, but progress is being made developing tests to characterize frothers to aid the process. The characterization tests are designed to capture both the pulp-related functions, bubble size and rise velocity, and the froth stability function (Laskowski, 2003). Cappuccitti and Finch (2008) used froth height versus GH in air–water tests to screen candidate frothers. Tsatouhas et al. (2006) used froth half-life, the time for froth to decay to half its original height once air is turned off. Comparing water-only and pulp systems, they showed the marked effect of particles on increasing froth stability with MIBC.

Several test procedures employ a bubble column. Some care is needed as under certain combinations of fine bubble size and high air rate, bubbles arrive at the froth base faster than they burst at the top of the froth, giving uncontrolled froth buildup and no equilibrium froth height is reached. A conical section at the top of the column helps eliminate this problem as increasing froth height means increasing froth surface area, enabling the bubble burst rate to eventually match the bubble arrival rate (Cunningham and Finch, 2009).

Linking frother functions to frother structure, which aims to shorten the search for a frother with target properties, is a growing research area (Zhang et al., 2012b; Kowalczuk, 2013; Finch and Zhang, 2014; Corin and O'Connor, 2014).

12.5.7 Frother Analysis

Realizing the importance of frother to operations has led to development of an online concentration estimation technique (Maldonado et al., 2010) and to laboratory analytical techniques (Tsatouhas et al., 2006; Gélinas and Finch, 2007). Plant surveys have revealed *frother partitioning*, the loss of frother from the pulp and concentration in the froth water as a result of frother adsorption on bubbles. For alcohol frothers, partitioning appears to be minor but for some polyglycols the concentrating effect in the froth can be sufficient to disturb downstream operations (Gélinas and Finch, 2007). *Frother mapping*, the evaluation of frother distribution (deportment) around a circuit, has, among other findings, identified uneven distribution between parallel flotation banks, and remnant frother in recycle waters (Zangooi et al., 2014). Mapping helps indicate if and where frother needs to added. Frother analysis in conjunction with gas dispersion characterization (see Section 12.14.2) makes for a powerful diagnostic tool.

12.6 REGULATORS

Regulators, or modifiers, are used extensively in flotation to modify the action of the collector, either by intensifying or by reducing its water-repellent effect on the mineral surface. They thus make collector action more selective toward certain minerals. Regulators can be classed as activators, depressants, dispersants, or pH modifiers (the latter covered in Section 12.7). Recent reviews are by Fuerstenau et al. (2007), Nagaraj and Ravishankar (2007), and Bulatovic (2007, 2010).

12.6.1 Activators

These reagents alter the chemical nature of mineral surfaces so that they can react with the collector and become hydrophobic. Activators are generally soluble inorganic salts that ionize in solution, the ions then reacting with the mineral surface. Sulfide systems are considered first, the action of copper ions and sodium sulfide being described.

Copper Ions

Activation in sulfide systems is reviewed by Finkelstein (1997). The prime example is activation by copper ions. Used for a variety of sulfides (Allison and O'Connor, 2011), the classic use remains the activation of sphalerite. Sphalerite is not readily floated by a xanthate collector, for two reasons: (1) the collector product zinc xanthate is, compared to other base metal xanthates, relatively soluble and does not provide stable hydrophobic sites; and (2) because sphalerite is a poor semiconductor it does not readily support electron transfer reactions (Section 12.4.4). Floatability can be improved by the use of large quantities of long-chain xanthates, but a more efficient method is to use copper sulfate as an activator, which is readily soluble and dissociates into copper ions in solution. Activation is due to the exchange of Cu for Zn in the surface lattice of sphalerite:

$$ZnS + Cu^{2+} \leftrightarrow CuS + Zn^{2+} \quad (12.9)$$

This exchange is favored (a) because of their relative position in the electrochemical series (copper is above zinc and thus when together zinc will be oxidized to zinc ions and copper ions reduced to copper), and (b) because the sphalerite lattice can accommodate the copper ions without undue distortion. The exchange mechanism is evidenced by the fact that a mole of Cu lost from solution is replaced by a mole of Zn in solution (Fuerstenau et al., 2007). Exchange continues till the capacity to hold Cu in the sphalerite surface is reached. The addition rate in practice is anywhere between 0.2 and 1.6 g Cu per kg Zn (Damjanović and Goode, 2000; Finch et al., 2007a). That copper is held in the lattice, rather than more loosely on the surface by a physical adsorption mechanism, is confirmed by copper not being extracted by ethylene diamine tetra-acetic acid (EDTA) (Kant et al., 1994). The sphalerite acts as a "sink" for copper ions.

Sphalerites almost always contain variable iron in solid solution, the low Fe variety being pale brown and the high Fe end member, marmatite (with up to 10% Fe), being black. The presence of iron is recognized in the commonly quoted formula (Zn,Fe)S. In general, sphalerite flotation is not significantly affected by the presence of iron, although it is reported that misplacement to lead concentrates may be greater for the low iron sphalerite variety (Zieliński et al., 2000).

The presence of the copper in the sphalerite surface has two effects: (1) with xanthate the copper forms highly insoluble copper xanthate; and (2) surface conductivity increases to support the electrochemical xanthate adsorption mechanism. The anodic reaction of xanthate also appears to be associated with Cu^{2+} acting as an electron acceptor and reducing to Cu^{+} as the reaction product identified is Cu(I)-xanthate (Wang et al., 1989a,b).

The main use of copper sulfate as an activator is in the differential flotation of lead–zinc and copper–zinc ores, where after lead or copper flotation the sphalerite is activated and floated. Selectivity in the zinc flotation stage is generally against iron sulfides, pyrite, and pyrrhotite. The common approach is to first raise pH to depress the iron sulfides then add copper sulfate. The addition of copper sulfate itself, however, can promote selectivity by reducing pyrite flotation (Dichmann and Finch, 2001; Boulton et al., 2003). Experiments on sphalerite/pyrite/xanthate systems showed that in the absence of copper both sphalerite and pyrite floated, but addition of copper altered the balance of competition for xanthate in favor of the sphalerite and the pyrite became "starved" of xanthate. While the selectivity mechanism remains the subject of investigation (Chandra and Gerson, 2009), a practical outcome is in the order of reagent addition: rather than the common practice of first raising pH, adding copper sulfate first may both activate sphalerite and "starve" the pyrite of xanthate, to be followed by raising pH to complete depression of the pyrite. This $CuSO_4$–lime sequence of reagent addition has shown increased sphalerite/pyrite selectivity over the lime–$CuSO_4$ sequence in plant practice (Finch et al., 2007a). Provided copper addition rate is kept below the capacity of sphalerite to hold the copper, there appears to be little activation of the pyrite.

Copper ions are sometimes connected to *inadvertent* or *accidental activation*, the unintentional activation of other minerals, which hampers selectivity. Release of copper ions into solution from secondary copper minerals such as chalcocite (Cu_2S) can be an uncontrolled source of accidental sphalerite activation, sometimes with overwhelming effects on Cu/Zn selectivity (McTavish, 1985). On occasion, accidental activation of non-sulfide gangue by Cu ions is also suspected.

Other metals ions can cause accidental activation and hamper selectivity, notably Pb in Pb/Zn differential flotation (see Section 12.8) and possibly Ni in pentlandite/pyrrhotite separation (Finch et al., 2007a). Lead ions can sometimes be the activator of choice (Miller et al., 2006). In comparison to the Cu/sphalerite case, most metal ion activators in other systems are not held in the mineral lattice, but loosely on the surface as evidenced by their extraction using EDTA at least in the important alkaline pH range (Fuerstenau et al., 2007).

Sodium Sulfide

Oxidized minerals of lead, zinc, and copper, such as cerussite, smithsonite, azurite, and malachite, float poorly with sulfhydryl collectors due to a combination of collector loss through precipitation by heavy metal ions dissolved from the mineral lattice, and collector coatings that are weakly anchored and readily removed by particle abrasion (Fuerstenau et al., 2007). Such minerals are activated by sodium sulfide or sodium hydrosulfide by a process referred to as "sulfidization" (Fuerstenau et al., 1985; Malghan, 1986; Newell et al., 2007). Large quantities, up to 10 kg t^{-1} of such "sulfidizers," may be required, due to the relatively high solubilities of the oxidized minerals.

In the case of sodium sulfide, in solution it hydrolyzes and then dissociates:

$$Na_2S + 2\,H_2O \leftrightarrow 2\,NaOH + H_2S \qquad (12.10)$$

$$NaOH \leftrightarrow Na^+ + OH^- \qquad (12.11)$$

$$H_2S \leftrightarrow H^+ + HS^- \qquad (12.12)$$

$$HS^- \leftrightarrow H^+ + S^{2-} \qquad (12.13)$$

Since the dissociation constants of Eqs. (12.12) and (12.13) are low and that of Eq. (12.11) is high, the concentration of OH^- ions increases at a faster rate than that of H^+ ions and the pulp becomes alkaline. Maintaining alkaline pH is important to avoid formation of hazardous H_2S gas. Hydrolysis and dissociation of sodium sulfide releases OH^-, S^{2-}, and HS^- ions into solution and these react with and modify the mineral surfaces. Sulfidization causes sulfur ions to pass into the lattice of the oxidized minerals, giving them a relatively insoluble pseudo-sulfide surface coating and allowing them to be floated by sulfhydryl collectors.

The addition of sodium sulfide must be controlled, as it is also a strong general depressant for sulfide minerals. The amount required is dependent on the pulp alkalinity, as an increase in pH causes Eqs. (12.12) and (12.13) to proceed further to the right, producing more HS^- and S^{2-} ions. For this reason, sodium hydrosulfide is sometimes preferred to sodium sulfide, as it does not hydrolyze and hence increase the pH. Excess sodium sulfide also removes oxygen from the pulp by the following reaction:

$$S^{2-} + 2\,O_2 = SO_4^{2-} \tag{12.14}$$

Since DO is required in the pulp for the adsorption of sulfhydryl collectors (Section 12.4.4), flotation efficiency is reduced.

Sodium sulfide and hydrosulfide can be used in conjunction. In reprocessing weathered stockpiled Cu/Zn ore, Les mines Selbaie used a combination of NaHS to the SAG/ball mills (70/30 split) at pH 9 and Na_2S to the cleaner regrind at pH 5.5 to virtually restore the Cu and Zn metallurgy to that of normal ore (Finch et al., 2007a). Part of the problem was considered to be activation of sphalerite and pyrite by Cu ions released from the weathered ore. In addition to sulfidization, sulfide ions act to precipitate the Cu ions as Cu_2S, a process known as *sequestration*.

In processing mixed sulfide-oxidized ores, the sulfide minerals are usually floated first, followed by sulfidization of the oxidized minerals. This prevents the depression of sulfides by sodium sulfide. It has been suggested (Zhang and Poling, 1991) that the detrimental effects of residual sulfide can be eliminated by the coaddition of ammonium sulfate. Use of the relatively inexpensive ammonium sulfate appears to reduce the consumption of the more expensive sulfidizing agent while enhancing the activating effect.

Non-sulfide Systems

Because of the similarity in surface properties of many of the valuable and gangue non-sulfide minerals, they tend to respond to the same collectors and consequently the use of regulating agents is extensive (Miller et al., 2002; Nagaraj and Ravishankar, 2007). Just two examples of activation are selected as illustrations. In reverse flotation of silica from iron ores using an anionic collector, Ca ions are introduced to adsorb on the silica and alter the surface charge to positive, thus enabling collector adsorption through the electrostatic mechanism (Houot, 1983). In the flotation of feldspar from quartz with an amine (cationic) collector at acid pH, fluoride is added as activator. The fluoride selectively lowers the negative charge on the feldspar compared to the quartz, thus promoting adsorption of the cationic collector (Shergold, 1984).

12.6.2 Depressants

Depression is used to increase the selectivity of flotation by rendering certain minerals hydrophilic, thus preventing their flotation. They are key to the economic flotation of many ores.

Several mechanisms are involved alone or in combination, among them: adsorption of hydrophilic species; blocking of collector adsorption sites; removal (desorption) of activating species (*deactivation*); and removal of hydrophobic sites (desorption/destruction of adsorbed collector). The role of pH and pulp potential in controlling depression is also a factor but is largely covered elsewhere (Sections 12.7 and 12.8). The reagents involved fall into three classes (Nagaraj and Ravishankar, 2007): inorganics, small organics, and organic polymers.

12.6.3 Inorganic Depressants

These reagents see most application in sulfide systems. Among the many employed (Nagaraj and Ravishankar, 2007) five are selected as being among the most important.

Cyanide

This reagent is widely used in the selective flotation of lead–copper–zinc and copper–zinc ores to depress sphalerite, pyrite, and certain copper sulfides. Sphalerite rejection from copper concentrates is often of major concern, as zinc is a penalty element in copper smelting.

Cyanide chemistry is comprehensively reviewed by Guo et al. (2014). There are three broad categories of cyanide species: free cyanide (CN_{free}: HCN, CN^-); weak acid dissociable cyanide (CN_{WAD}: e.g., cyanide complexes with Cu, Ni, Zn); and strong acid dissociable cyanide (CN_{SAD}: e.g., complexes with Fe). Guo et al. (2014) provide thermodynamic stability constants.

Accidental activation of sphalerite by Cu^{2+} ions released from copper minerals can be countered by the addition of cyanide. In this role the cyanide could be classed as a *deactivator*. The cyanide acts by reacting with copper in solution, first reducing Cu^{2+} to Cu^+ then forming soluble cupro-cyanide complexes, $Cu(CN)_x{}^{1-x}$ (e.g., $Cu(CN)_2{}^-$), which are stable and prevent the uptake of copper by Eq. (12.9). The form of the complex depends on pH and pulp potential (Seke and Pistorius, 2006).

Cyanide needs to be added early in the circuit, for example, the grinding stage, to complex the copper ions before abstraction by sphalerite, as subsequent removal of Cu from activated sphalerite is difficult.

Cyanide is most commonly used as the sodium salt, which hydrolyzes in aqueous solution to form free alkali and relatively insoluble hydrogen cyanide:

$$NaCN + H_2O \leftrightarrow HCN + NaOH \tag{12.15}$$

The hydrogen cyanide then dissociates:

$$HCN \leftrightarrow H^+ + CN^- \quad (12.16)$$

The dissociation constant of Eq. (12.16) is very low compared with that of Eq. (12.15), so that an increase in pulp alkalinity reduces the amount of free HCN, but increases the concentration of CN^- ions. An alkaline pulp is essential, as free hydrogen cyanide is extremely dangerous. The major function of the alkali, however, is to control the concentration of cyanide ions available for formation of the copper cyanide complex:

$$2\,CN^- + Cu^+ \leftrightarrow Cu(CN)_2^- \quad (12.17)$$

Cyanide is also used in depression of pyrite. In that case formation of a surface ferric ferrocyanide complex, which both produces a hydrophilic site and blocks a possible collector adsorption site, is the suspected mechanism (Fuerstenau et al., 2007).

Apart from the reactions of cyanide with metal ions, it can react with metal xanthates to form soluble complexes, preventing xanthate adsorption on the mineral surface. The greater the solubility of the metal xanthate in cyanide (i.e., the more soluble the metal xanthate–cyanide complex), the less stable is the attachment of the collector to the mineral. It has been shown that lead xanthates have very low solubilities in cyanide, copper xanthates are fairly soluble, while the xanthates of zinc, nickel, and iron are highly soluble. On this basis iron and zinc sulfides should be readily separated from galena in processing complex ores by the use of cyanide. In the separation of chalcopyrite from sphalerite and pyrite, close control of cyanide ion concentration is needed. Sufficient cyanide should be added only to complex the heavy metal ions in solution and to solubilize the zinc and iron xanthates. Excess cyanide forms soluble complexes with the slightly less soluble copper xanthate, depressing the chalcopyrite.

The depressive effect of cyanide depends on its concentration and on the concentration of the collector and the length of its hydrocarbon chain. The longer the chain, the greater is the stability of the metal xanthate in cyanide solutions (i.e., the soluble metal xanthate–cyanide complex is harder to form) and the higher the concentration of cyanide required to depress the mineral. Relatively low concentrations of xanthates with short hydrocarbon chains are therefore preferred for selective flotation where cyanides are used as depressants.

Cyanides, being highly toxic, must be handled with care. They also have the disadvantage of being expensive and they dissolve gold and silver. If the plant includes cyanidation to recover Au from flotation concentrates and/or tails, the use of cyanide as a depressant can be accommodated as these plants include cyanide destruction. If water from the cyanidation plant is to be recycled to the flotation section then cyanide in solution needs to be controlled. As discussed in Section 12.8, under certain conditions it is possible that cyanides can cause accidental copper activation of sphalerite. Despite these disadvantages, they are widely used. For example, Umipig et al. (2012) found that cyanide was essential in processing a Cu–Zn ore to produce separate Cu- and Zn-concentrates in the face of accidental Cu activation of the sphalerite (see also Section 12.17.3). The process includes cyanide destruction and active management of accidental activation so that $CuSO_4$ addition is not required.

Zinc Sulfate

While many plants function efficiently with cyanide alone, in others an additional reagent, generally zinc sulfate, is included to ensure satisfactory depression of sphalerite. The introduction of zinc ions can counter copper ions from depositing on the sphalerite surface by shifting Eq. (12.9) to the left. Accidental activation by Pb^{2+} ions is similarly countered by adding Zn^{2+} ions; if lead activation is the principal problem zinc sulfate alone may be sufficient. More complex reactions may also occur: cyanide may react with zinc sulfate to form zinc cyanide, which is relatively insoluble and precipitates on the sphalerite surface rendering it hydrophilic and blocking collector adsorption. In an alkaline pulp, zinc hydroxide forms, which is precipitated on the sphalerite surface, again rendering it hydrophilic and preventing collector adsorption.

Sulfur Dioxide

Widely used as a galena depressant in copper–lead separation, effective sphalerite rejection is also achieved in copper cleaner and copper–lead separation circuits through acidification of pulps by injection of SO_2. The addition of SO_2 is generally considered to (a) reduce the pulp potential which retards collector adsorption (Section 12.4.4) and (b) produce a series of sulfoxy species, giving a range of hydrophilic surface reaction products. In addition, SO_2 in combination with air (oxygen) can produce strongly oxidizing conditions through formation of Caro's acid (H_2SO_5). This may result in the oxidization (destruction) of collector coatings. It does also mean that SO_2 cannot be employed when treating ores that contain secondary copper minerals, like covellite or chalcocite, since they become soluble in the presence of sulfur dioxide and the dissolved copper ions activate zinc sulfides (Konigsmann, 1985).

Sulfur dioxide does not appreciably depress chalcopyrite and other copper-bearing sulfides. In fact, adsorption of xanthate on chalcopyrite may be enhanced in the presence of SO_2, and the addition of SO_2 before xanthate results in effective sphalerite depression while increasing the floatability of chalcopyrite. The use of SO_2 in various

Swedish concentrators is discussed by Broman et al. (1985), who point out that SO_2 has the advantage over cyanide in sphalerite depression in that there is little copper depression, and no dissolution of precious metals. However, it is indicated that the use of SO_2 demands adaptation of the other reagent additions, and in some cases a change of collector type is required.

In reverse flotation of pyrite from zinc concentrate a combination of heat and SO_2 has been employed, although that practice is largely abandoned today for environmental concerns. The harsh conditions this combination implies speaks to the difficulty of deactivating sphalerite once Cu has been adsorbed (Grano, 2010).

Rather than SO_2, sulfoxy species, *sulfite* (SO_3^{2-}) and *metabisulfite* (HSO_3^-), have been used. They function by a combination of formation of hydrophilic surface products and reduction in DO and pulp potential that retards collector adsorption.

Sulfide, Hydrosulfide

More than 40% of the western world's molybdenum is produced as a by-product from porphyry copper ores. The molybdenite (MoS_2) is collected along with copper sulfides in a bulk Cu–Mo concentrate. The two minerals are then separated, almost always by depressing the copper minerals and floating the molybdenite. Sodium hydrosulfide or sodium sulfide is used most extensively, though several other inorganic compounds, such as cyanides, and Noke's reagent (a product of the reaction of sodium hydroxide and phosphorous pentasulfide) are also used (Nagaraj et al., 1986). The action is through control of pulp potential, as discussed in Section 12.8.

Phosphates

Accidental activation by lead ions can be controlled by addition of polyphosphates, $(P_nO_{3n+1})^{(n+2)-}$, an inorganic complexing agent. The shorter chain phosphates function by forming a precipitate, the longer chain members by forming soluble complexes. An example of their use is the Myra Falls Cu–Pb–Zn concentrator where increasing recovery of Zn to the Cu concentrate was traced to increasing levels of galena in the ore (0.1% Pb in the original ore rising to 0.5% Pb). Addition of monophosphate deactivated the sphalerite (Yeomans, 2008). Provided the level of Pb ions is not too high, polyphosphates can be effective and economic (Raschi and Finch, 2006).

12.6.4 Small Organic Molecule Depressants

Identified as a subgroup by Nagaraj and Ravishankar (2007), only the polyamines DETA (diethylenetriamine) and TETA (triethylenetetramine) introduced in processing Ni ores to depress pyrrhotite (Marticorena et al., 1994; Kelebek and Tukel, 1999) are considered. While the mechanism may not be fully understood, the amines' N-C-C-N structure does chelate with metal ions such as Cu and Ni that may be accidentally activating the pyrrhotite. Depression of pyroxene (a silicate) by DETA and TETA in selective flotation of pentlandite was attributed to this deactivation mechanism (Shackelton et al., 2003). In combination with sulfite ions to reduce potential and thus reaction with xanthate (even decomposing it to carbon disulfide) increases the effectiveness of polyamine depressants (Tukel and Kelebek, 2010).

12.6.5 Organic Polymer Depressants

The use of polymeric depressants has the attraction of being less hazardous than the more widely used inorganic depressants, and interest in their use continues to grow (Nagaraj and Farinato, 2014a). The general structure, shown Figure 12.22, is a hydrocarbon backbone with functional groups and OH (hydrophilic) groups. The functional groups react with sites on the target mineral surface and the OH groups confer the hydrophilicity to depress the mineral.

The functional groups can be nonionic such as OH, anionic such as COO^-, SO_3^{2-}, cationic based on N, or the composition may be amphoteric containing both anionic and cationic functional groups. The interactions with the mineral surface, while still the subject of investigation, in general follow similar mechanisms as for collectors, that is, chemical and physical interactions. Initially, natural products and their derivatives were used, but the structural variations are limited and synthetic polymers, which offer the ability to tailor the depressant, are gaining acceptance.

Natural Products

The natural products and derivatives include (Pugh, 1989; Bulatovic, 1999): starch and dextrin (polysaccharides), and carboxylmethylcellulose (CMC); tannin and quebracho; and lignosulfonates. The basic unit of starch (and dextrin) and CMC polymers is shown in Figure 12.23. Dextrins are fragmented starches of lower molecular weight (i.e., n in the Figure is smaller for dextrin compared to starch). Example uses are (Bulatovic, 2007): starch in depression of iron oxide minerals in reverse

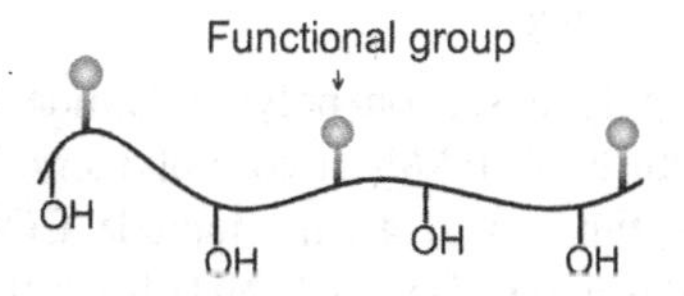

FIGURE 12.22 General structure of organic polymer depressants.

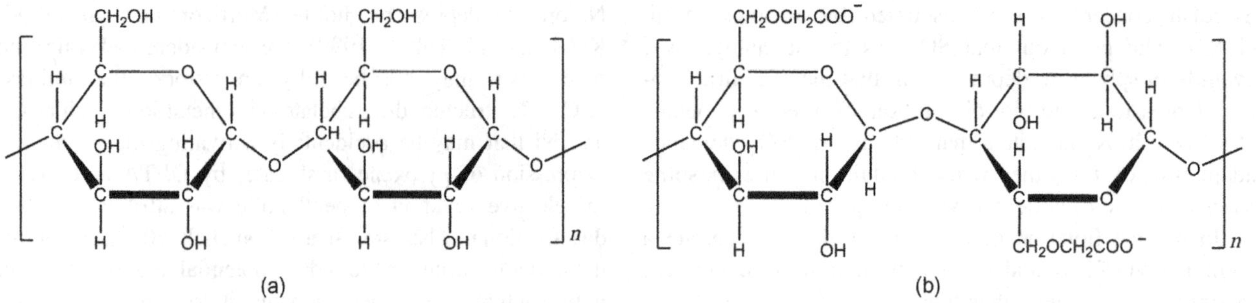

FIGURE 12.23 Basic dextrose unit structure of (a) starch and dextrin, and (b) CMC (see Pugh, 1989; Bulatovic, 1999; Trombino et al., 2012).

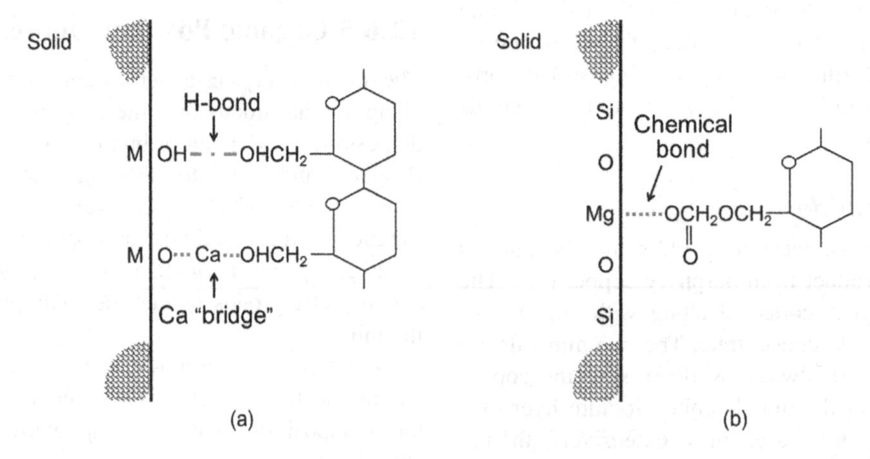

FIGURE 12.24 Possible adsorption mechanisms for (a) starch and dextrin, and (b) CMC.

flotation of silica using amine; dextrin in depression of galena in copper–lead separation; and CMC in depression of silicate minerals in processing Ni-sulfide ores and platinum group (PGM) minerals. For starches and dextrin the functional group appears to be the OH, with adsorption a combination of H-bonding with mineral surface OH groups and possible chemical bridging via Ca^{2+} and other bivalent ions, which are known to play a role (Figure 12.24(a)) (Bulatovic, 2007). In CMC the functional group is the carboxyl, COO^-, which may bond directly with surface metal atoms (Figure 12.24(b)) or again via Ca^{2+} bridging. Gums are another polysaccharide used largely in non-sulfide systems. Huang et al. (2012) have tested chitosan, a natural polyaminosaccharide, for chalcopyrite–galena separation.

Synthetic Products

These are largely based on polyacrylamide (PAM). The general structure of PAM-based polymers is shown in Figure 12.25. Functional groups include COO^-, CH_2O^- for non-sulfides and CH_2S^-, CS-NH_2 for sulfides. Targets for replacement are sulfur species (e.g., HS) used in Cu–Mo and Cu–Pb separation, which are objectionable reagents on safety and environmental grounds.

12.6.6 Dispersants

Dispersed particles represent a better condition for selective flotation than systems where the particles are aggregated. Several of the depressants also function as dispersants (Nagaraj and Ravishankar, 2007). Rendering the surface hydrophilic means water is held as a hydrated layer on the particle surface, which not only depresses the particle but also makes it less prone to aggregate (to aggregate the hydrated layer on the contacting particles has to be expelled). A reagent used primarily as a dispersant is sodium silicate. The action appears to be related to restoring high negative charge to generate electrostatic repulsion between particles, accomplished through a combination of adsorption of the silicate anion (SiO_3^{2-}) and removal of metal cations from the surface, sequestering them as insoluble metal silicates. Coating of fine particles on larger particles ("slime coating") is an example of aggregation. To avoid slime-related problems, in some non-sulfide systems feeds are deslimed by classification

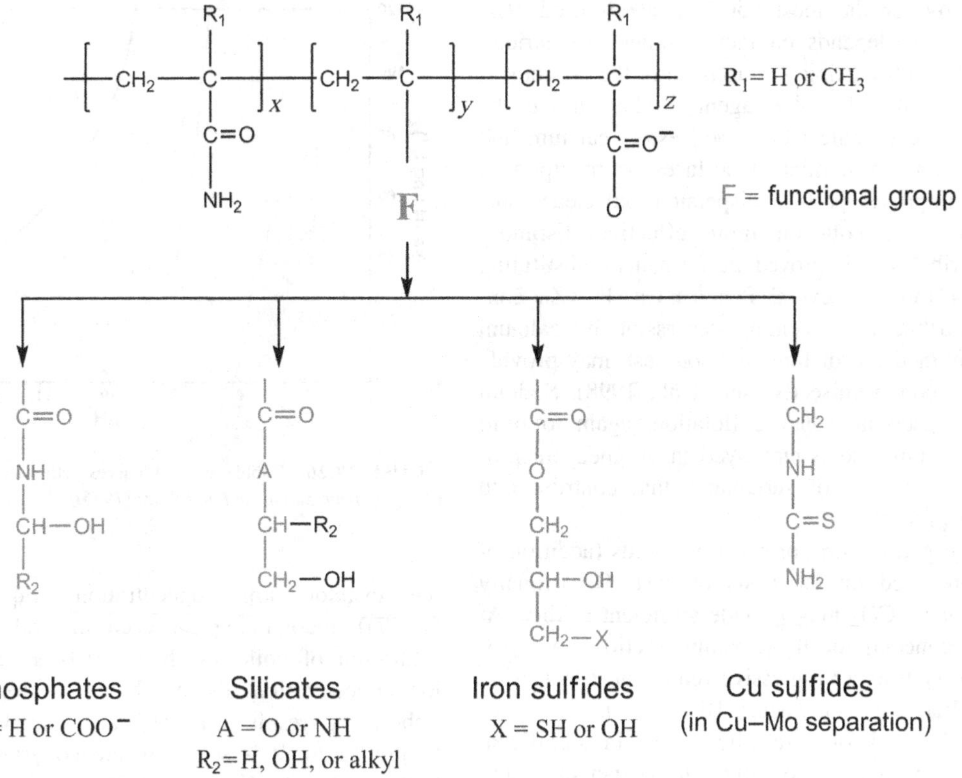

R_1 = H or CH_3

F = functional group

Phosphates
R_2 = H or COO^-

Silicates
A = O or NH
R_2 = H, OH, or alkyl

Iron sulfides
X = SH or OH

Cu sulfides
(in Cu–Mo separation)

FIGURE 12.25 General structure of PAM-based polymer depressants and target minerals *(Adapted from Nagaraj and Ravishankar (2007)).*

(e.g., cycloning, see Chapter 9) with attendant loss of values. Reducing the impact of slimes by controlled dispersion would be preferable.

Dispersants also lower pulp viscosity by reducing particle–particle interactions (Jones and Horsley, 2000; Nagaraj and Ravishankar, 2007). The role of rheology (viscosity) in flotation is reviewed by Farrokhpay (2012), who notes effects in both the pulp and froth zones. Phyllosilicate group minerals pose particular problems (Ndlovu et al., 2014), especially fibrous serpentines which can form tangled mats that entrap bubbles and are buoyed into the froth (Bhambhani et al. 2012). Processing Ni ores hosted in ultramafic rock (i.e., high in fibrous Mg, Fe silicates) is challenged by rheology. Processing fibrous ultramafic ores requires low solids percent in grinding and flotation. A combination of acid attack and grinding to disintegrate the fibers has been shown to produce a pulp more amenable to flotation (Uddin et al., 2011).

12.7 THE IMPORTANCE OF pH

It is evident from the foregoing that pulp alkalinity plays an important, though complex, role in flotation, and, in practice, selectivity in complex separations is dependent on a balance between reagent concentrations and pH. In non-sulfide mineral flotation, pH controls surface charge (zeta potential), which can control reactions with ionic collectors and certain regulating agents. Knowing the IEP for the minerals to be separated helps identify the pH window for selective separation. In sulfide mineral flotation with thiol collectors both the ion exchange and electrochemical mechanism teach that increasing pH retards adsorption of the collector and depresses the mineral. Depression occurs at a different pH for each sulfide; for the common sulfides, generally as pH is increased pyrite is depressed first, ca. pH 10, and chalcopyrite last, ca. pH 12 providing the *pH window* for selective flotation. The critical pH for sulfide depression will depend on the collector concentration. It is generally advisable to keep collector dosage low, at "starvation" levels.

pH Regulators

Flotation is generally carried out in alkaline conditions, with the added advantage that most collectors, including xanthates, are stable under these conditions, and corrosion of cells, pipework, etc., is minimized. Alkalinity is controlled by the addition of lime, sodium carbonate (soda ash), and, to a lesser extent, sodium hydroxide (caustic soda) or ammonia.

Lime is by far the most common alkali used. The choice, however, depends on factors other than strictly pH. Lime by introducing Ca ions into the system is sometimes not the desirable agent. Soda ash can be favored as the carbonate CO_3^{2-} sequesters calcium and other metal ions from mineral surfaces as precipitates, which both improves particle dispersion and cleans surfaces, making the collector more effective. Espinosa (2011) describes the improved performance substituting soda ash for lime in several Peñoles Cu–Pb–Zn concentrators attributed to avoiding depression by calcium sulfate. Combined use of lime and soda ash may provide an economic compromise (Nesset et al., 1998). Sodium hydroxide is used in iron ore flotation, again to limit effects of calcium, and is employed in bitumen ore processing to aid release of surfactants that contribute to the flotation process.

To reduce pH, sulfuric or sulfurous acids (addition of SO_2) are preferred on the basis of cost. Occasionally injecting exhaust CO_2 may provide sufficient acidity. At the Niobec concentrator the cleaning section for pyrochlore recovery uses a progressive reduction in pH using hydroflurosilicic and oxalic acids (Biss, 1984).

These pH regulators are often used in significant amounts. For example, lime, although cheaper on a unit weight basis than collectors, the overall cost in sulfide mineral flotation is roughly double that of the collector used, so significant savings can be achieved by the proper choice and use of pH regulators (Fee and Klimpel, 1986).

Lime is used in the form of milk of lime, a suspension of calcium hydroxide particles in a saturated aqueous solution. Lime or soda ash is often added to the slurry prior to flotation to precipitate heavy metal ions from solution. In this sense, the alkali may be considered a deactivator, as some heavy metal ions can activate minerals such as sphalerite and pyrite. Since the heavy metal salts precipitated by the alkali can dissociate to a limited extent and thus allow ions back into solution, cyanide is often used with the alkali to complex them.

Lime can also act as a depressant for pyrite and arsenopyrite when using xanthate collectors. Both the hydroxyl and calcium ions participate in the depressive effect by the formation of mixed films of Fe(OH), FeO (OH), $CaSO_4$, and $CaCO_3$ on the surface, so reducing the adsorption of xanthate. Lime has little such effect with copper minerals, but does depress galena to some extent. In the flotation of galena, therefore, pH control may be effected by the use of soda ash, pyrite and sphalerite being depressed by cyanide.

As was shown earlier, the effectiveness of sodium cyanide and sodium sulfide is governed to such a large extent by pH that these reagents are of scarcely any value in the absence of alkalis. Where cyanide is used as a depressant, the function of the alkali is to control

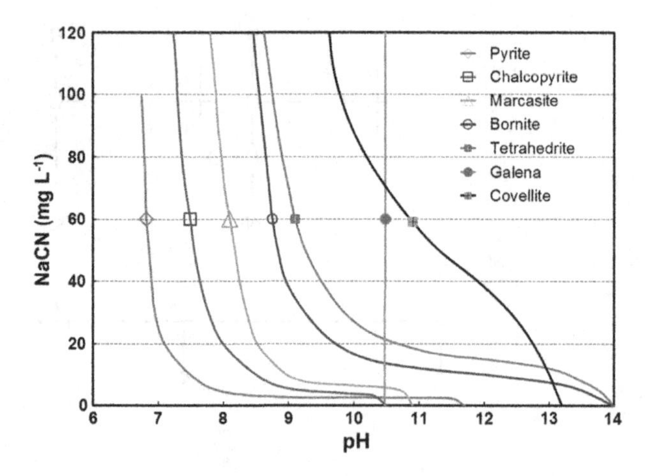

FIGURE 12.26 Bubble contact curves (ethyl xanthate = 25 mg L^{-1}) *(Adapted from Sutherland and Wark (1955)).*

the cyanide ion concentration (Eqs. (12.16) and (12.17)), recognizing for each mineral and given concentration of collector there exists a "critical" cyanide ion concentration above which flotation is impossible. Laboratory results for several minerals are given in Figure 12.26. It is seen that chalcopyrite can be floated from pyrite at pH 7.5 and 30 mg L^{-1} sodium cyanide and since, of the copper minerals, chalcopyrite lies closest to pyrite relative to the influence of alkali and cyanide, all the copper minerals will float with the chalcopyrite. Thus, by careful choice of pH and cyanide concentration, excellent separations are theoretically possible, although in practice other variables serve to make the separation more difficult. The figure shows that adsorption of xanthate by galena is uninfluenced by cyanide, the alkali alone acting as a depressant.

In addition to controlling the interaction of mineral with collector and regulating agents, pH also controls metal ion speciation in solution and in some cases collector speciation.

Metal Ion Speciation

Multivalent metals ions are *hydrolysable*, that is, they undergo the following series of reactions:

$$\begin{aligned} M^{2+} + OH^- \rightarrow M(OH)^+ + OH^- \rightarrow \\ M(OH)_2(s) + OH^- \rightarrow M(OH)_3^- \end{aligned} \quad (12.18)$$

Species distribution diagrams can be found in a variety of sources (e.g., Miller et al., 2007). The monohydroxy complex and the hydroxide precipitate are active at the mineral surface. Their interaction is revealed by measuring zeta potential, as illustrated in Figure 12.27. Depicted for silica, the same trend holds for many minerals, nonsulfides, and unoxidized sulfides. Three charge reversal

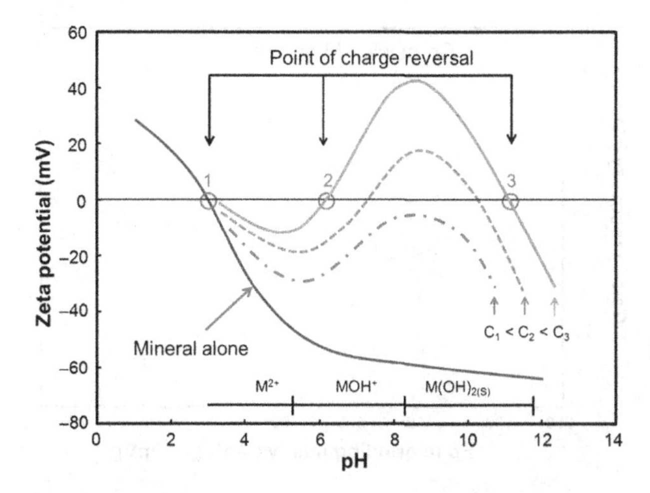

FIGURE 12.27 Zeta potential as a function of pH and concentration of hydrolysable metal ions: response based on silica but common for many minerals.

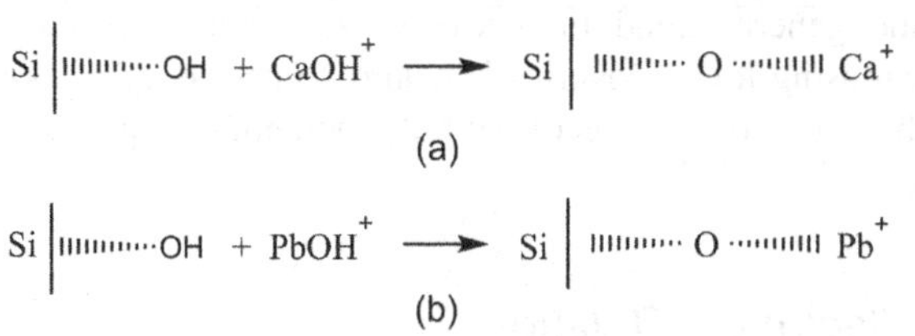

$$\mathrm{Si}\,|\cdots\mathrm{OH} + \mathrm{CaOH}^{+} \longrightarrow \mathrm{Si}\,|\cdots\mathrm{O}\cdots\mathrm{Ca}^{+}$$

(a)

$$\mathrm{Si}\,|\cdots\mathrm{OH} + \mathrm{PbOH}^{+} \longrightarrow \mathrm{Si}\,|\cdots\mathrm{O}\cdots\mathrm{Pb}^{+}$$

(b)

FIGURE 12.28 Formation of: (a) Ca^+ site, and (b) Pb^+ site as a result of interaction with the monohydroxy complex (H_2O is the other product).

pHs are shown: the first corresponds to the IEP of the mineral (there is little interaction with the free ion); the second and third reversals reflect adsorption of the monohydroxy complex and of hydroxide precipitate, the last reversal being the IEP of the hydroxide itself. An explanation for the second charge reversal through interaction with the monohydroxy complex is illustrated in Figure 12.28 for $CaOH^-$ and $PbOH^-$. The resulting Ca^+ site (Figure 12.28(a)) provides the "bridge" in Figure 12.24(a); and the Pb^+ site (Figure 12.28(b)) is sometimes considered responsible for interaction with thiol collectors, giving accidental activation.

The uptake (adsorption) of hydroxide precipitates is an example of heterocoagulation. Over a certain pH range, the particle is negatively charged and the hydroxide precipitate (colloid) is positively charged (IEP ~ pH 11 for the "hydroxide" in Figure 12.27). With sufficient coverage, the particle surface takes on the character of the precipitate, hence the third charge reversal coincides with the hydroxide IEP. In a survey of Cu–Zn concentrators, Mg- and Zn-hydroxy species were identified as promoting particle aggregation (El-Ammouri et al., 2002). Magnesium ions are considered to be the cause of molybdenite depression when floating Cu–Mo ores in seawater at about pH 10 where the hydroxide forms (Nagaraj and Farinato, 2014b; Castro et al., 2014). Figure 12.27 shows clear evidence for possible surface contamination by metal ion species that may influence flotation.

For sulfides the trend in Figure 12.27 is seen for unoxidized surfaces; most sulfides in that state have an IEP ~ pH 2–3, similar to sulfur (Mirnezami et al., 2012). When superficially oxidized, the usual situation, the buildup of oxy-hydroxide species shifts the sulfide IEP to higher pH values. The trend in Figure 12.27 is mimicked for sulfides by progressive increases in the level of oxidation.

Interestingly, the same trends in the presence of hydrolysable cations are seen in the zeta potential of bubbles, which shows the adsorption of metal ion hydrolysis products is quite general (Han et al., 2004).

Collector Speciation

The form of collector can change with pH. An extreme example is the decomposition of xanthate at low pH. In non-sulfide mineral flotation, collectors like fatty acids and fatty amines can be in molecular (undissociated) or ionic form. The reactions are:

$$\mathrm{R-COO^- + H^+ \leftrightarrow R-COOH} \tag{12.19}$$

$$\mathrm{R-NH_3^+ + OH^- \leftrightarrow R-NH_2 + H_2O} \tag{12.20}$$

In the undissociated form (to the right of the arrow), solubility is low and it is generally the ionic form that acts as a collector. There is a pH where the concentration of ionic and molecular forms is equal, the pK_{diss}: for fatty acids at pH $< pK_{diss}$ the molecular form predominates, and for amines at pH $> pK_{diss}$ the molecular form predominates. For oleate the pK_{diss} (or pK_a) = 4.95 and for dodecylamine pK_{diss} (pK_b) = 10.63 (Pugh, 1986). The pK_{diss} is thus an important parameter in the action of these collectors. For example, in Figure 12.9 (Section 12.4.3) for the amine collector, further increase in pH above ca. pH 12 results in rapid decline in flotation as the molecular form becomes dominant. Noting that the maximum flotation rate can often occur at pH near the pK_{diss} has suggested that combinations of ion and molecular forms, possibly *ion-molecular* species, can be the active collector in some instances (Pugh, 1986; Rao and Forssberg, 2007). In the fatty acid case, the complex is referred to as an *acid-soap*. Surface activity (e.g., measured by surface tension depression) is known to be maximum around the pK_{diss} (Finch and Smith, 1973; Ananthapadmanabhan and Somasundaran, 1988) thus an effect on the frother-related properties of fatty acids and amines can also be expected. Values of K_a and K_b ($pK = -\log K$) are given by Somasundaran and Wang (2006).

12.8 THE IMPORTANCE OF PULP POTENTIAL

In sulfide mineral systems the oxidation/reduction conditions play a strong role in selectivity through control of electrochemical reactions. Many of these electrochemical processes are reviewed by Bruckard et al. (2011). The pulp potential of a system is the result of all the anodic (oxidation) and cathodic (reduction) reactions taking place and is difficult to predict. That pulp potential has an impact is understood, however, and can be interpreted from the representation in Figure 12.14. In that example if the potential is too low, below that for xanthate-mineral reaction, then the collector is not adsorbed and the mineral is not floatable. Of course, this may be the desired effect, as in the addition of Na_2S to depress chalcopyrite in selective flotation of molybdenite (Poorkani and Banisi, 2005). Represented on the current–potential diagram we would see that the S^{2-} oxidation/O_2 reduction potential of the Na_2S is lower than that for the adsorption of xanthate, which lowers the pulp potential (i.e., Na_2S is a reducing agent). In other words, S^{2-} oxidation is favored over xanthate oxidation (adsorption) and the chalcopyrite is depressed. Molybdenite, being naturally hydrophobic, is not affected by the pulp potential. The action of most inorganic depressants can be interpreted electrochemically (Woods, 2010).

Rather than too low, the pulp potential may be high enough to oxidize the sulfides, producing various depressant hydrophilic oxy-hydroxide surface species. In the Noranda group preflotation aeration was common practice to depress iron sulfides (Konigsmann, et al., 1976). Sulfide mineral slurries can have a high capacity for oxygen, which can be measured by the *DO demand*, the rate of decrease in DO when the air supply is shut off (Grano, 2010). A measure of the extent of oxidation can be made by extraction with EDTA that readily solubilizes metal oxidation products, but does not attack the sulfide mineral directly (Grano, 2010). Solution analysis before and after EDTA extraction (e.g., by atomic absorption spectroscopy) reveals the metal ions that were held on the surfaces.

The effect of pulp potential, therefore, can be both to promote and depress flotation, depending on the magnitude of the potential. This is illustrated Figure 12.29 for xanthate flotation of Pb (galena) from a sample of Brunswick Mine ore. The pulp potential was controlled in the grinding mill (note reference electrode was Ag/AgCl) and as E_p was increased, first recovery increased, reached a plateau then declined. The increase in recovery is associated with increased adsorption of xanthate via the anodic (oxidation) reaction; the subsequent decrease is associated with production and buildup of oxy-hydroxide oxidation species on the galena surface. Woods (2010) gives examples for other sulfide minerals, which show the same general trend in recovery with pulp potential. In diagnosing low recoveries it would be necessary to establish where on the recovery-pulp potential response you are.

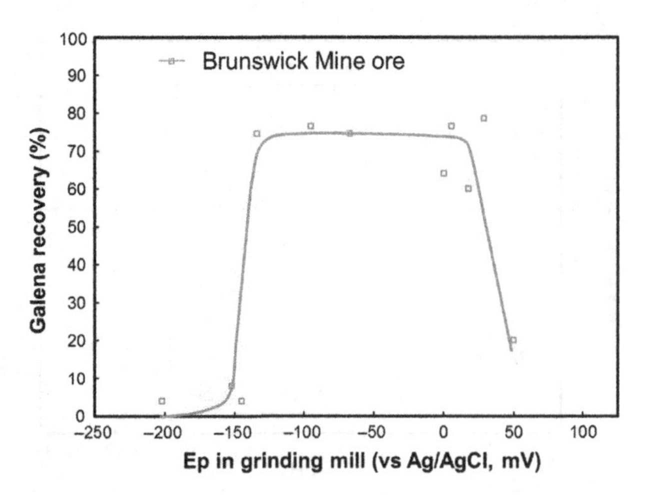

FIGURE 12.29 Flotation of galena (Pb) from a sample of Cu–Pb–Zn Brunswick Mine ore illustrating effect of pulp potential.

Collectorless Flotation

A certain level of oxidation of sulfide minerals can result in formation of metal deficient or polysulfide sites (Figure 12.4), which can make the mineral floatable without collector. Their formation is an electrochemical process, the anodic reaction being:

$$xMS \leftrightarrow M_{(x-1)}S_x + M^{2+} + 2\,e^- \qquad (12.21)$$

with the common electron acceptor again being oxygen. The reaction can continue to eventually form elemental sulfur (S_8), which is strongly hydrophobic. Referred to as *self-induced* or *collectorless flotation* it has attracted attention (Luttrell and Yoon, 1984; Chander, 1991; Ralston, 1991). Although oxygen is the usual electron acceptor there are others which could substitute, notably iron (Fe^{3+}/Fe^{2+}) ubiquitous in sulfide flotation pulps. In copper activation of sphalerite it is noted that some level of collectorless flotation is achieved, which is attributed to Cu^{2+} acting as an additional electron acceptor to oxygen and intensifying the oxidation (Lascelles and Finch, 2002). The addition of sulfide ions can also bring about collectorless flotation (Yoon, 1981). Pyrrhotite is in effect a metal deficient iron sulfide (Pratt et al., 1994) and naturally has polysulfides on the surface and often exhibits collectorless flotation.

Collectorless flotation can be demonstrated in plant practice, but there is usually little incentive to exploit as there is no easy way to determine if collectorless flotation

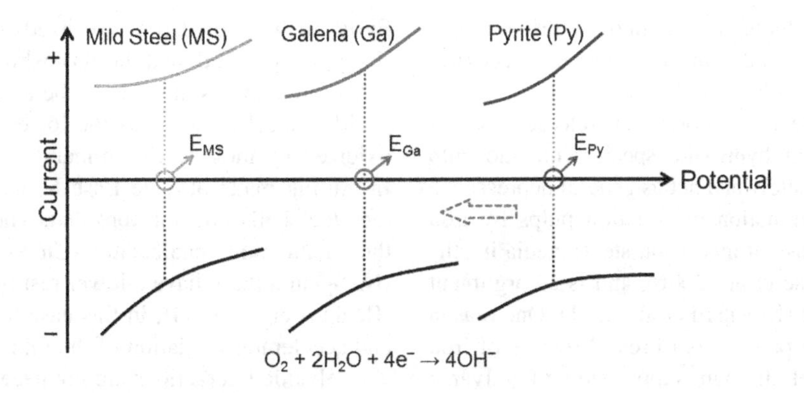

FIGURE 12.30 Current–potential diagram for the pyrite–galena–mild steel system in galvanic contact.

will be robust to ore changes and collector addition removes that concern (Leroux et al., 1994).

In the disintegration of fibers involving strong acid attack mentioned in Section 12.6.6 with respect to a novel processing route for ultramafic Ni ores, the oxidizing conditions produced by the concentrated sulfuric acid rendered the pentlandite (and pyrrhotite) collectorlessly floatable. At the same time the amount of $MgSO_4$ put into solution provided an ionic strength sufficient to replace the need for frother (see Section 12.5.4). The process, therefore, gave a technically interesting collectorless-and-frotherless flotation system.

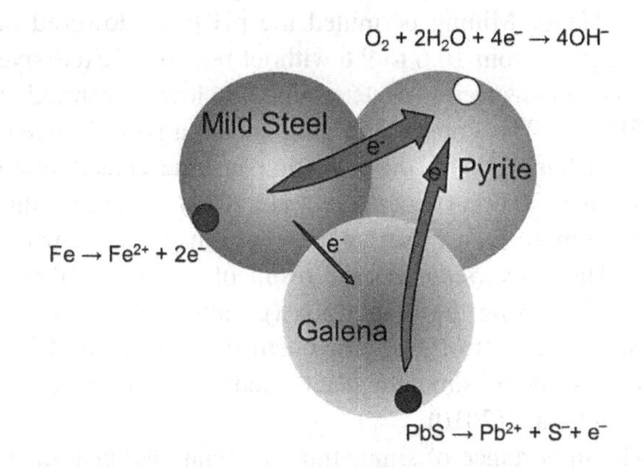

FIGURE 12.31 Physical picture of galvanic contact for the pyrite–galena–mild steel system, showing flow of electrons to pyrite (arrow width indicates relative magnitude) and release of metal ions from the steel and galena.

Galvanic Interaction

Recalling from Section 12.4.4 that sulfides have different rest potential values means that when they are in contact with each other in electrolyte (i.e., process water) it will induce *galvanic interaction*. Iwasaki and co-workers (Adam and Iwasaki, 1984; Nakazawa and Iwasaki, 1985) were among the first to consider the implications of galvanic interactions in flotation.

To illustrate galvanic interaction consider three materials in a grinding mill, in order of decreasing rest potential: pyrite, galena, and mild steel. The associated current–potential diagram is shown in Figure 12.30. In this case, each anodic reaction is matched with its own cathodic (oxygen reduction) reaction. The mixed potential at $I_A = I_C$ establishes that the highest anodic current is contributed by oxidation of the steel, and that the highest cathodic current is contributed by reduction of oxygen on the surface of pyrite, with galena oxidation/reduction reactions in between. The physical picture is illustrated in Figure 12.31: the pyrite draws electrons from both steel and galena accelerating their oxidation.

The result of the galvanic interactions depicted in Figure 12.31 includes lowering of DO (consumed by accepting electrons) and release of metal ions both with implications for flotation. The lowering of DO in the grinding mill discharge may retard reaction with collector and reduce floatability. Aeration in a bank of flotation cells may restore the DO level necessary for flotation but there may be a case for preflotation aeration. In reactor–separator flotation cells (Section 12.13.3) with short retention time, ensuring that the DO level is sufficient for flotation is an important precaution.

Oxidation promoted by galvanic interaction releases Pb and Fe ions into the pulp. (Recent work has shown that galvanic interactions, especially in the presence of pyrite, promote formation of hydrogen peroxide (H_2O_2), which further accelerates oxidation rates (Nooshabadi et al., 2013).) The released Pb^{2+} ions may accidentally activate other minerals. Acting on the suspicion that Pb^{2+} ions were activating pyrite, Brunswick Mine installed a flotation stage between the autogenous (AG) mill and the ball mill to remove galena before the further size reduction (i.e., increase in galena surface area) added to the release of more Pb^{2+} ions. The result was increased pyrite rejection from the bulk Cu–Pb concentrate (Finch et al.,

2007a). Addition of reducing agents into the grinding circuit to reduce galena oxidation was another successful option at another plant (Finch et al., 2007a).

Oxidation of the steel (i.e., corrosion) releases a series of hydrophilic iron oxy-hydroxide species into the pulp that adsorb indiscriminately and act as general depressants. Trying to limit contamination of flotation pulps by iron species has led to the use of inert (non-steel) media in stirred regrind mills (Pease et al., 2006), and is an argument in favor of AG milling (Bruckard et al., 2011). One reason raising pH is common practice is to retard release of iron species from the steel through suppression of galvanic interactions by inhibiting the reduction of oxygen on the pyrite (in the example in Figure 12.31). Replacing forged steel media by corrosion-resistant high chrome media at Ernest Henry Mining permitted the pH to be lowered in the roughers from 10.6 to 9.6 without release of excessive iron corrosion species (as determined by low Fe extraction by EDTA). The copper and gold metallurgy was essentially unchanged with the reduction in lime consumption valued at $700,000 (Kirkwood et al., 2014). Of course this needs to more than offset any increase in grinding media costs. The conversion was the result of extensive laboratory testing using the Magoteaux (grinding) Mill® which permits close control over pulp chemistry. Details on laboratory testing to simulate plant chemical conditions are given in Grano (2010).

The importance of simulating the plant chemical environment when conducting laboratory tests is illustrated by investigations into cases of apparent accidental activation of sphalerite by cyanide in Cu–Zn ores (Rao et al., 2011). At low pulp potential, copper leached into solution is held as a cupro-cyanide complex (Section 12.6.3), which is not only stable but with Cu in the +1 state means that there is no exchange with Zn^{2+} ions in the sphalerite surface lattice. If the pulp potential is raised, however, the cupro-cyanide is oxidized to cupric cyanide, a reaction that both increases leaching of copper and puts Cu in the +2 state, making exchange with Zn^{2+} ions in the sphalerite lattice now possible. The implication is that if laboratory grinding is with steel media giving low pulp potential then this accidental activation mechanism is not observed; but if the plant grinding circuit is AG then pulp potential is higher than in the lab experience and accidental activation of sphalerite may occur. Espinosa (2011) reviewing plant practices at several Cu–Zn operations reported sphalerite activation by cyanide. In tests using the Magotteaux Mill® to control pulp chemistry, he established the role of oxidation in the use of cyanide causing accidental activation of sphalerite. Alternative depressants such as sulfoxy species may have to be substituted if this cyanide activation problem is encountered.

The impact of galvanic interactions can be seen in ore stockpiles and reclaimed pillars at the end of mine life. Pyrite in weathered (i.e., oxidized) ore will still be relatively untarnished and bright, while other sulfides like sphalerite may disintegrate to the touch.

Mild steel usually has the lowest rest potential and is oxidized by most sulfide minerals. There are exceptions. In milling matte at Vale Base Metals' Sudbury operation the steel balls are not corroded. The sulfide minerals in the matte are chalcocite (Cu_2S) and heazelwoodite (Ni_3S_2) and these have a lower rest potential than the steel (Bozkurt et al., 1994); in this case the steel is "protected" and accelerates oxidation of the minerals.

Galvanic interactions are not necessarily detrimental as the preceding may imply. Considering that xanthate adsorption is an oxidation reaction, the adsorption of xanthate on galena in contact with pyrite is promoted, while at the same time the surface of pyrite hosts the formation of hydroxyl ions which generate hydrophilic (depressant) Fe oxy-hydroxide products. The galvanic contact thus promotes selective flotation of galena over pyrite. In the Galvanox™ process pyrite is deliberately added to a second sulfide to enhance leaching rates, an example where galvanic effects are exploited to advantage. There are options to include the Galvanox™ process in combination with flotation to form an integrated circuit (Dixon et al., 2007).

Use of Nitrogen

There is occasional reference to the potential for gases other than air in sulfide mineral flotation, nitrogen being the logical candidate based on cost (Johnson, 1988). The action of nitrogen substituting for air as flotation gas can be interpreted as reducing galvanic interactions. It has been shown that nitrogen promotes xanthate flotation of pyrite relative to other sulfide minerals (Martin et al., 1989; Xu et al., 1995). The postulated mechanism is that the nitrogen breaks the galvanic interaction between sphalerite and pyrite, which suppresses the formation of oxy-hydroxides on the pyrite surface. The removal of these hydrophilic species both eliminates a pyrite depressant and opens sites for adsorption of collector.

A commercial process for pyrite reverse flotation from sphalerite concentrate employs heat and SO_2 conditioning. The combination appears to first produce strong oxidizing conditions that destroy the Cu–xanthate on the sphalerite surface and ties up the Cu as an oxide, and finally leads to a low pulp potential that favors pyrite flotation. The use of nitrogen helps maintain the low potential, thus permitting less aggressive conditioning (Xu et al. 1995; Grano, 2010). In pyrite reverse flotation from Pb concentrate in Brunswick Mine's Pb upgrading circuit, heating alone (to ca. 75°) appears to be sufficient (Martin et al., 2000); this could in part be due to heating lowering the DO sufficient to suppress galvanic contact and thus prevent the oxy-hydroxide from forming on the pyrite surface.

Experiments on pyrite alone show that pyrite is depressed by nitrogen, explained by the rest potential being too low to support formation of dixanthogen (Figure 12.18). When pyrite is in combination with other sulfides this is not the case. The nature of the hydrophobic species on the pyrite at low pulp potential remains the subject of investigation (Miller et al., 2006). The situation with the other major iron sulfide mineral, pyrrhotite, is different. Using nitrogen pyrrhotite is effectively depressed in separation from pentlandite. In this case it appears that the collectorless flotation of pyrrhotite due to oxidation of the surface polysulfides to elemental sulfur is suppressed by the lowered potential with nitrogen (Finch et al., 1996, 2007a).

Practical applications of nitrogen must meet the challenge of maintaining a nitrogen atmosphere in face of the many opportunities for air (oxygen) to enter the system, via open balls mills and hydrocyclones being two examples. This ingress of oxygen has hampered commercial attempts in the promising use of nitrogen to depress pyrrhotite away from pentlandite. One possible solution is to enclose flotation cells and recycle the air which leads to an atmosphere enriched in nitrogen by taking advantage of the high oxygen demand of sulfide pulps. A small supply of nitrogen would still be needed to replace the volume of oxygen consumed to maintain ambient air pressure inside the cell. Another possibility is to take advantage of locations where a nitrogen stream is available, for example, off-gas from oxygen production. The patented N_2TEC technology implemented at Lone Tree mine used such an off-gas source of nitrogen and reported improved selective recovery of Au-bearing pyrite (Simmons, 1997). Where nitrogen is commonly used is in chalcopyrite–molybdenite separation, where it both reduces side oxidation reactions that consume the chalcopyrite depressants, NaHS/Na_2S, and helps maintain a low pulp potential, which depresses chalcopyrite (Aravena, 1987; Poorkani and Banisi, 2005).

To take advantage of pulp potential in control of mineral separations requires measurement. Measurement of E_p is more problematic than measurement of pH. Traditionally, it is measured using a noble metal (usually platinum but sometimes gold) as the sensing electrode with calomel (Hg/Hg_2Cl_2) or silver/silver chloride (Ag/AgCl) as reference electrodes. The sensing electrode needs to be inert and designed to include contributions to the pulp potential from the particles—it is the potential at the mineral/solution interface that matters in flotation. Sensing electrodes are prone to fouling and need close attention (Chander, 2003). An example of successful E_p measurement is in control of NaSH/Na_2S additions in chalcopyrite–molybdenite separation. Here the noble metal sensing electrode becomes coated with sulfur and responds to the sulfide ions in solution (Woods, 2010).

Control of potential through reagents is often not easy, the systems being buffered to change. The possibility of direct control through design of an electrochemical cell, successfully used at the laboratory scale, is an area of research for industrial application (Panayotov and Panayotova, 2013).

12.9 FLOTATION KINETICS

The engineering approach is to consider flotation as a rate process by analogy with chemical reaction kinetics (Amelunxen and Runge, 2014). The reactants in flotation are the bubble and particle and the product is the bubble–particle aggregate. From this standpoint, the rate of flotation will depend on a number of factors: concentration of particles and bubbles, frequency of collision, efficiency of particle attachment upon collision, and stability of the attachment. Making some assumptions simplifies development of a model, the key ones being: the reaction follows first order, the bubble concentration does not vary, and colliding particles can always find space on a bubble to attach. Under these conditions the rate of disappearance of particle mass ($-dW/dt$) in a flotation cell is proportional to the particle mass in the cell (W). Mathematically, this is expressed as, where k is the flotation rate constant:

$$-\mathrm{d}W/\mathrm{d}t = kW \tag{12.22}$$

The rate constant k lumps all the unmeasured variables and becomes an engineering measure of floatability; the higher the floatability the higher the rate constant, making k a widely used parameter.

To solve Eq. (12.22) requires the nature of the particle transport through the cell to be known. The simplest case is batch flotation, where all particles are treated as having the same residence time.

12.9.1 Batch Flotation

The solution is given by integrating from time 0 to t when mass in the cell decreases from W_0 to W:

$$\int_{W_0}^{W} \frac{\mathrm{d}W}{W} = k\int_0^t \mathrm{d}t \tag{12.23}$$

$$\ln\frac{W}{W_0} = kt \tag{12.24}$$

$$W = W_0 \exp(-kt) \tag{12.25}$$

Rearranging to give the more familiar expression for recovery, where $R = (W_0 - W)/W_0$ (R on a fractional basis) gives:

$$R = 1 - \exp(-kt) \tag{12.26}$$

We will return to the batch Eq. (12.26) in "flotation testing" (Section 12.12).

12.9.2 Continuous Flotation

In continuously operating units, transport behavior can be determined by a tracer test, as illustrated in Figure 12.32(a).

A tracer test is most readily performed on the liquid; for example, adding a salt (usually NaCl) and monitoring the conductivity on the discharge, or introducing a colored dye and monitoring the discharge using spectroscopy. Guidelines are given by Nesset (1988). Particle mixing can be determined by introducing a different type of particle to those in the flotation feed but of similar size and density; or, better, by radioactive tagging an element in the feed particles (Yianatos, 2007).

Transport exhibits two extremes: plug flow, and fully (or perfectly) mixed flow (Figure 12.32b). In plug flow all the particles enter and leave the cell at the same time, the time corresponding to the mean residence (retention) time. The mean residence time τ is given by the cell volume (m^3) (with allowance for volume occupied by gas) divided by the volumetric flowrate through the cell ($m^3\ min^{-1}$); the time scale can then be given as t/τ (a *reduced or normalized time*), as in Figure 12.32. Plug flow, therefore, is indicated by a (theoretical) line at $t/\tau = 1$. Fully mixed transport, on the other hand, means the tracer is immediately evenly distributed throughout the cell and thus the zero-time sample is the cell average concentration. With time the concentration decays exponentially, giving a distribution of residence times or *residence time distribution* (RTD).

The two extreme cases offer mathematical simplicity, which is why they are the usual starting point. The two solutions are:

Plug flow:

$$R = 1 - \exp(-k\tau) \tag{12.27}$$

Fully mixed:

$$R = \frac{k\tau}{1 + k\tau} \tag{12.28}$$

The plug flow solution is the same as for the batch case, as all particles have the same residence time (at least those in the tailings stream from the cell). Equation (12.28) is derived in Appendix V.

Mechanical cells and flotation columns show an RTD close to fully mixed (Yianatos, 2007; Govender et al., 2014). While mean particle residence time depends on particle mass (size and density), it is generally close enough to the liquid mean residence time (within 90%) for most practical purposes and the simpler determination of liquid residence time suffices.

RTD studies have diagnostic value. If, for example, there is significant short circuiting, the RTD will exhibit

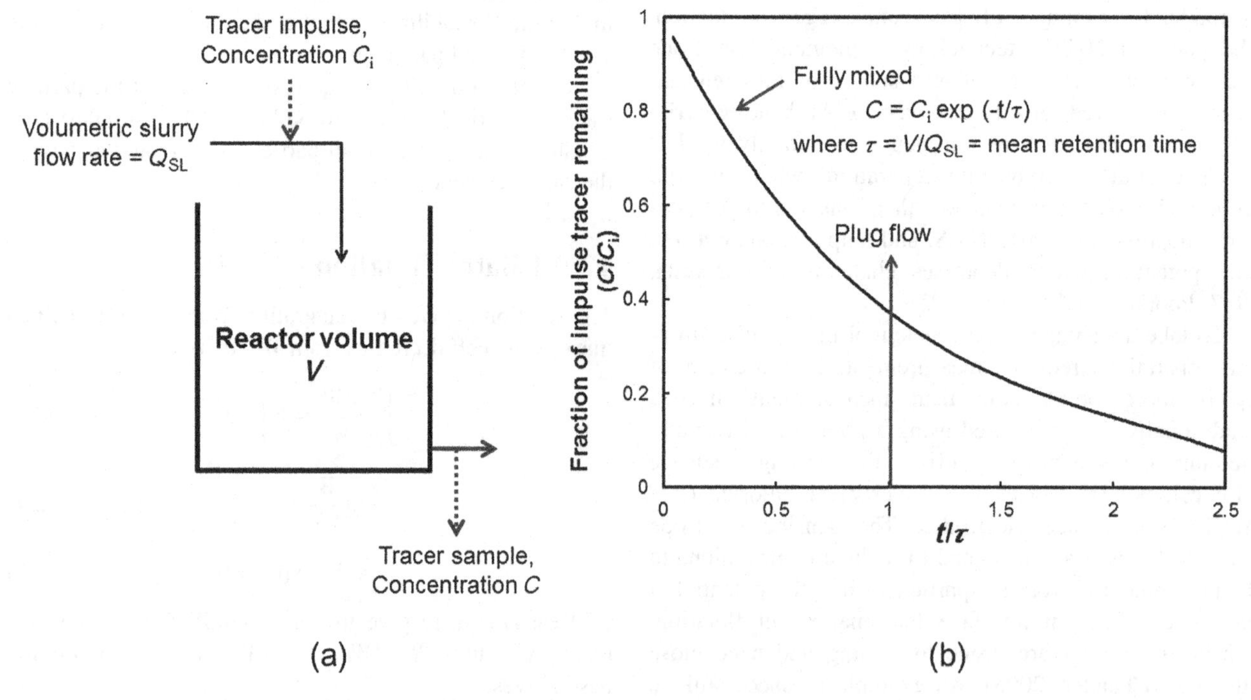

FIGURE 12.32 (a) Introduction of tracer as an impulse (concentration is based on volume of reactor.), and measurement in the discharge stream over time, and (b) the two extreme outcomes, plug flow and fully (or perfectly) mixed flow.

unusual shapes; for example, more than one maxima and/or a tail extending to long times (>2.5 t/τ). RTD studies on flotation columns baffled into vertical sections revealed pulp flowing between the sections created by uneven distribution of feed and air.

12.9.3 Components of the Rate Constant

The rate constant is a function of contributions by the machine, the particles, and the froth.

Machine

The machine is responsible for production of bubble swarms and effecting bubble–particle collision. Recent work has related the machine "factor" to the BSAF. BSAF (S_b) is the rate at which bubble surface area moves through the cell per unit of cell cross-sectional area with units $m^2\ m^{-2}\ s^{-1}$ or simply s^{-1}. Introduced in Section 12.5, to remind, it is estimated from the measurements of superficial gas velocity (the volumetric gas flowrate per unit area of cell, J_g), and the bubble size, usually taken as the Sauter mean diameter (D_{32}):

$$S_b = \frac{6\,J_g}{D_{32}} \tag{12.7}$$

Since J_g is typically quoted in cm s^{-1} D_{32} must be in cm. To give a sense of magnitude, at a typical value of $J_g = 1$ cm s^{-1} and $D_{32} = 1$ mm (0.1 cm) then $S_b = 60\ s^{-1}$.

Both J_g and D_{32} are measurable using suitable probes (see Section 12.14.2). The S_b can also be predicted using a correlation developed by Gorain et al. (1999).

Gorain et al. (1997) and Alexander et al. (2000) showed for shallow froth depths that the first order rate constant was linearly related to the BSAF. In addition, this relationship was shown to be independent of cell size and operating parameters. This independence is illustrated in Figure 12.33, which shows that the relationship measured in a pilot scale 60 L cell was essentially identical to that measured in a parallel 100 m^3 cell. Letting the proportionality constant be P, the pulp zone rate constant (k_p) can be expressed as follows:

$$k_p = P\,S_b \tag{12.29}$$

There is support for Eq. (12.29) taking a first principles approach to bubble–particle interaction (Finch, 1998), but the relationship is largely empirical.

Particles

The relationship expressed in Figure 12.33 introduces a new engineering measure of floatability, the proportionality constant P, a dimensionless quantity if k_p and S_b are in expressed in the same reciprocal time units (e.g., s^{-1} as in Figure 12.33). The value of P is a reflection of particle properties: size, composition (liberation), and hydrophobicity. While treated as a constant, P is likely an inverse function of bubble size (Yoon, 1993; Hernandez-Aguilar, 2011); nevertheless the model (Eq. (12.29)) represents an advance over using the rate constant alone in analyzing flotation systems. Based on these findings, flotation performance can be considered to arise from the interaction of a stream property—the particle floatability (P)—with parameters that characterize the operating conditions of the pulp (S_b).

Making the analogy of the bubble/particle reaction to a chemical reaction means the derived kinetic model applies to the pulp (or slurry) zone, not to the overall cell, which includes froth.

Froth

Equation (2.29) can be modified by introducing a froth zone recovery factor. Figure 12.34 shows the concept of two interacting zones, pulp and froth, with *dropback*

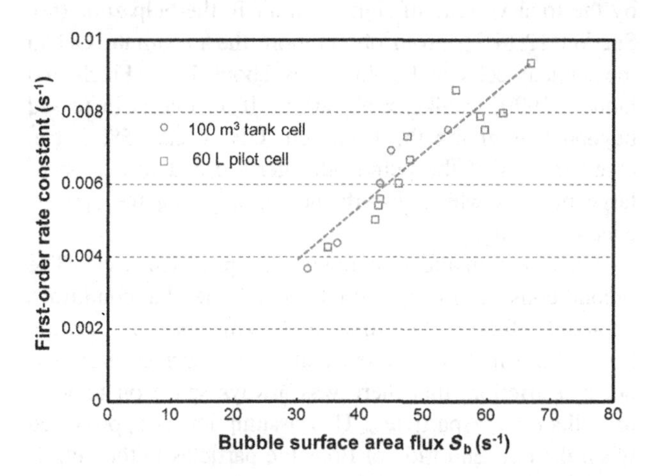

FIGURE 12.33 First-order rate constant and BSAF relationship in a 60 L pilot cell and a 100 m^3 cell *(Adapted from Alexander et al. (2000))*.

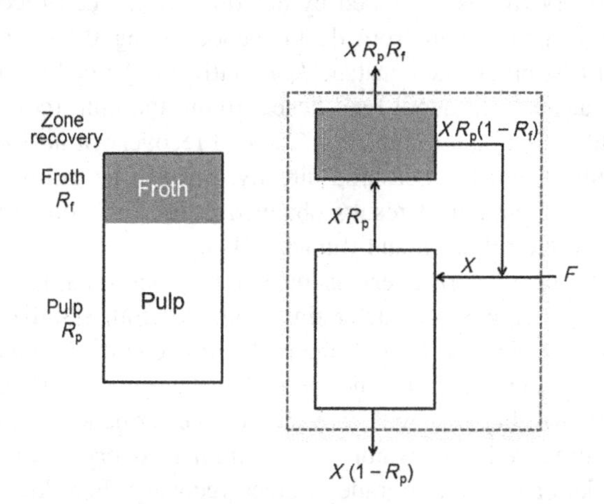

FIGURE 12.34 Flotation cell treated as interacting pulp and froth zones in derivation of overall recovery.

$(1 - R_f)$ considered to be combined with the fresh feed to the cell (Finch and Dobby, 1990). Letting the feed (dry) solids mass flowrate to the pulp zone be X and introducing R_p as pulp zone recovery and R_f as froth zone recovery, the resulting solids flowrates are as indicated on the figure. A mass balance across the dashed line block gives the incoming feed F and thus the overall flotation cell recovery R_{fc} as:

$$F = X R_p R_f + X(1 - R_f) \tag{12.30}$$

$$R_{fc} = \frac{R_p R_f}{R_p R_f + (1 - R_f)} \tag{12.31}$$

Substituting Eq. (12.28) into Eq. (12.31), the overall flotation rate constant (k_{fc}) is given by:

$$k_{fc} = k_p R_f \tag{12.32}$$

and substituting Eq. (12.29) into Eq. (12.32) gives the expression relating the overall flotation cell rate constant to the three contributing factors: particles (ore), machine, and froth.

$$k_{fc} = P S_b R_f \tag{12.33}$$

The fully mixed first-order kinetic model then becomes:

$$R_{fc} = \frac{P S_b R_f \tau}{1 + P S_b R_f \tau} \tag{12.34}$$

Techniques to quantify the froth recovery factor continue to be developed (e.g., Seaman et al., 2004; Yianatos et al., 2008; Rahman et al., 2012). However, most methods are either intrusive in the froth or subject to assumptions (e.g., no entrainment). A method initially developed for batch flotation cells by Feteris et al. (1987) was later modified by Vera et al. (1999) to determine R_f directly on industrial flotation cells. In this approach, froth zone recovery (R_f) is estimated by determining the cell recovery at a measured froth depth (hence giving the overall cell first order rate constant, k_{fc}) relative to the cell recovery at no froth depth (and hence giving the pulp (collection) zone rate constant, k_p). The cell recovery at no-froth depth cannot be measured directly, but can be estimated by extrapolation of results obtained at four or more froth depths (Amelunxen and Runge, 2014).

Froth zone recovery is related to froth stability. As was the case in characterizing frothers' froth stabilizing function (Section 12.5.6), the problem is one of definition and measurement. The notion is that there is an optimum froth stability that will give the best performance from a flotation cell, for example, maximum recovery at given grade, or maximum grade at given recovery (Farrokhpay, 2011). A promising measure of froth stability is *air recovery*, the fraction of the air delivered to the cell that overflows as unburst bubbles. Air recovery α can be calculated using the following expression:

$$\alpha = \frac{\xi \cdot v_f \cdot h \cdot w}{Q_g} \tag{12.35}$$

where Q_g is the total air flowrate into the cell, ξ the GH in the froth zone (usually assumed to be 1), v_f the overflowing froth velocity, h the froth height over the lip, and w the length of the lip where froth is overflowing. Froth velocity is measured by a camera monitoring the surface of the froth as it overflows using image processing software. A concern could be raised regarding propagation of error involved in calculation from several independent measurements (what is a typical error, e.g., 95% confidence interval, on α?) plus the fact that some air may leave with the cell tailings. Nevertheless, air recovery has been successfully used. By manipulating air rate to cells to operate at maximum (or peak) air recovery (PAR) improved bank performance was obtained, marking a new approach in flotation circuit control and optimization (Hadler and Cilliers, 2009; Hadler et al., 2010).

12.9.4 Testing the $k-S_b$ Relationship

To test Eq. (12.29) it is important not only to minimize the impact of froth (i.e., using shallow froth where $R_f \sim 1$) but also to ensure that two other considerations apply: that the pulp zone is operating in a "safe" range, and that the conditions approximate those of first order kinetics. A safe operating range refers to the air being adequately dispersed into bubbles. Yianatos and Henriquez (2007) suggest the following "safe" conditions: air velocity J_g 1–2 cm s^{-1}, D_{32} 1–1.5 mm and corresponding S_b 50–100 s^{-1}. The reason there is a safe operating range is because the pulp cannot hold more than a certain volume of air. The volume of air is measured as GH (or ε_g), the volume of air divided by the total volume of slurry and air in the pulp zone (see Section 12.14.2). From observation, the maximum GH in mechanical cells and columns is about 15% (Finch and Dobby, 1990; Dahlke et al., 2005). If air rate is increased beyond that giving the maximum GH of ca. 15% it can result in loss of the pulp/froth interface and formation of large bubbles which disturb the froth, giving the appearance of boiling.

Operating in the safe range is a practical issue. The second consideration is a fundamental one: that conditions approach those for first-order kinetics. In writing Eq. (12.22) it was assumed that bubble surface area was not a restriction, that there was always space on bubbles to collect more particles. This assumption is approached when the concentration of floatable particles in the pulp is low. The first cells in a bank, especially a cleaner bank with a high feed concentration of floatable particles, will often violate the assumption; the froth overflow solids rate from the cell is now likely determined by *bubble carrying capacity*, the maximum mass of particles that can be transported to the overflow by the bubbles (Finch and Dobby, 1990). Based on the overflow cell area typical carrying capacity is ca. $1\,\text{t}\,\text{h}^{-1}\,\text{m}^{-2}$ (Patwardhan and Honaker, 2000).

Respecting these restrictions then a linear $k-S_b$ is supported (Hernandez-Aguilar et al., 2005); otherwise, a linear $k-S_b$ relationship may not be observed, the rate constant could be independent of S_b or even decrease as S_b increases. Following the interpretation of Barbian et al. (2003), these deviations from linearity can be understood qualitatively by considering the way the particle load on the bubble changes as S_b is increased and the resulting effect on the froth: at low S_b the particle load per bubble will be high, perhaps giving too stable a froth, which does not flow readily; at a particular S_b the particle load may give an optimum froth stability; and at high S_b particle load may become too low to give adequate froth stability and froth overflow rate decreases. As froth zone recovery changes, so does the cell recovery and the derived overall rate constant.

While the importance of the froth to overall cell recovery is recognized it remains that the froth cannot deliver more mass than is delivered from the pulp. Some new cell designs aim to increase the pulp collection kinetics.

12.9.5 Modifications to Apply to Reactor–Separator Cell Designs (See Section 12.13.3)

The fully mixed kinetic flotation model expressed in Eq. (12.34) has been used primarily in analysis of mechanical cells and flotation columns. Some of the newer cell designs aimed at increasing the flotation rate fall into a category that can be described as "reactor–separator" designs. A feature of the reactor part is a high concentration of bubbles. In mechanical cells and columns, as noted, the volume fraction of gas that can be held in the pulp (GH) is limited to about 15%. In the reactor part of the reactor–separator cell designs GH can exceed 15%. For example, in the Jameson cell downcomer (the "reactor") GH can approach 60% by having the slurry and air move concurrently (Marchese et al., 1992). In addition, in these reactors the energy input goes more directly into bubble–particle collision and attachment, rather than some energy being used to circulate pulp as in mechanical cells. In recognition, the pulp flotation rate constant in Eq. (12.29) can be modified to (Williams and Crane, 1983):

$$k_p = P_g \varepsilon_g E \tag{12.36}$$

where P_g retains similar meaning (a factor associate with particle floatability, but based on GH rather than BSAF), and E is the energy directed to collision/attachment events. In consequence of the high bubble concentration, these new cell designs have much shorter retention times compared to mechanical cells and columns. For example, Harbort et al. (2003) report retention times in a Jameson cell circuit about 2–10 times shorter than in a mechanical cell circuit on the same duty; or up to one order of magnitude higher rate constant.

There is empirical support for Eq. (12.36). Hernández et al. (2003) reported a linear dependence of rate constant on GH in tests on flotation columns. Fundamentally, there is a question whether flotation rate depends on BSAF or GH. As noted in Section 12.5, from GH and bubble size the specific bubble surface area, A_b (bubble surface area per unit volume of gas), is given by Eq. (12.8):

$$A_b = \frac{6\,\varepsilon_g}{D_{32}} \tag{12.8}$$

This gives another modification to Eq. (12.29), namely:

$$k_p = P_a A_b E \tag{12.37}$$

where, again, P_a retains the meaning of floatability but now based on bubble specific surface area. Hence, k_p is either dependent on the bubble surface area passing through the cell (S_b) or the bubble surface area in the cell (A_b), a potentially important distinction, but one not resolved.

Parameters corresponding to S_b and A_b are also associated with the particles; there is a particle surface area flux passing through the cell and a particle specific surface area in the cell. Part of the science of flotation is matching these particle and bubble parameters to optimize the process, tied up in the recurring question of what is the appropriate bubble size for a given particle size that resists an easy answer but is an ambition of fundamental modeling efforts (e.g., Bloom and Heindel, 2002).

From a practical perspective, in mechanical and column cells BSAF and GH are related, for example, both increase with an increase in air rate and a decrease in bubble size (Finch et al., 2000), so whether one or the other is "fundamental" could be considered moot. Compared to BSAF, from a measurement standpoint GH has the advantage as commercial GH sensors are available (CiDRA Minerals Processing). At least one plant has used online measurement of GH in developing column flotation control strategies (Amelunxen and Rothman, 2009).

12.10 THE ROLE OF PARTICLE SIZE AND LIBERATION

12.10.1 True Flotation

As with all mineral separation processes, recovery in flotation is particle size dependent (Chapter 1). The role of

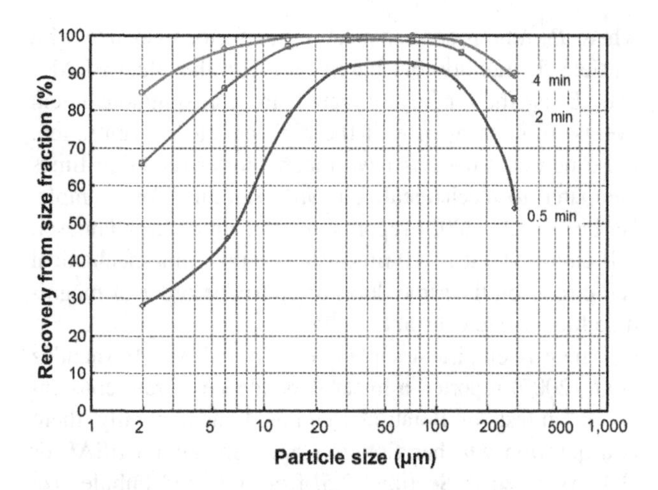

FIGURE 12.35 Typical recovery trend as a function of particle size and time *(Adapted from Trahar (1981))*.

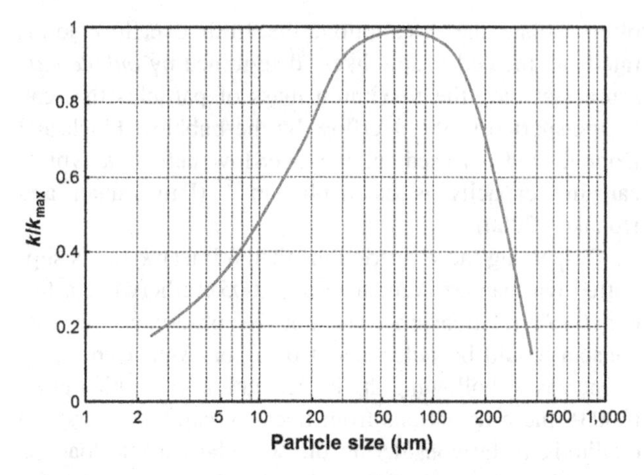

FIGURE 12.36 Trend in Figure 12.35 converted to relative rate constant (relative to maximum rate constant, k/k_{max}) as a function of particle size.

particle size is less pronounced compared to, say, gravity concentration, but analysis of recovery-by-size data can reveal opportunities for process improvement (Grano et al., 2014). With increased availability of liberation data, analysis of recovery-by-size-by-liberation data is adding an extra level of what can be called *diagnostic metallurgy*. Recovery-by-size is calculated from:

$$R_{i,m} = \frac{C\, c_i\, c_{i,m}}{F\, f_i\, f_{i,m}} \tag{12.38}$$

where $R_{i,m}$ is recovery of mineral m of size i, C/F the mass fraction reporting to concentrate, c_i, f_i the mass fraction in size interval i in concentrate and feed, and $c_{i,m}$, $f_{i,m}$ the assay of mineral m in size interval i in concentrate and feed. Data collection, therefore, requires sampling the process, sizing the samples, and assaying the sized fractions. There are the inevitable associated errors requiring data reconciliation to obtain useful information (Chapter 3).

Figure 12.35 shows a typical evolution of the recovery–size relationship as flotation time is increased. The figure reveals three size classes: an intermediate size range where recovery is high; a fine size range where recovery is lower and increases with time; and a coarse size range where recovery is again lower but less affected by time. The size divisions are not precise and will vary with mineral type. The data can be reexpressed as a rate constant versus size relationship (Figure 12.36).

Figure 12.37 illustrates three types of recovery–size relationships: one for strongly floatable mineral, a second for weakly floatable mineral, and a third for mineral recovered by entrainment. For the strongly floatable mineral there is a broad intermediate size range of high recovery, while the less floatable mineral has a similar but size-compressed relationship. Any selective flotation process will see this combination of strong and weak floatable minerals, for

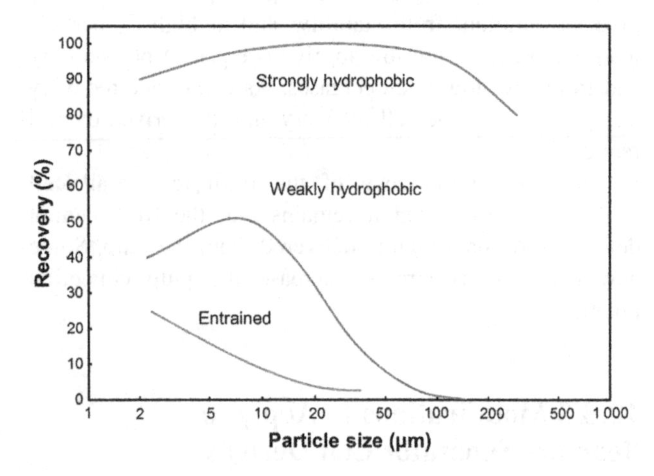

FIGURE 12.37 Illustrative recovery–size relationship for two floatable minerals, one strongly floatable, one weakly floatable, and a third mineral recovered by entrainment.

example, selective flotation of base metal sulfides from each other and from pyrite.

Fine Particles

Particle recovery by true flotation involves collision, attachment and formation of stable attachment to bubbles, often expressed as probabilities (or efficiencies). *Collision probability* is governed by the physics of the process (or "hydrodynamics"), while *attachment probability* and *stability of attachment* include a chemical component (particle hydrophobicity). The decline in recovery at the fine end is usually ascribed to low collision probability, fine particles having insufficient inertia to cross the water streamlines around the coursing bubble. Note this does not say there is a lower particle size limit to flotation, only that the rate of flotation will decrease as particle size decreases. Figure 12.36

reinforces this conclusion, showing the fines have a low but finite rate constant. (Pease et al. (2006) make the point quite forcefully that there is not a lower size limit to flotation, just slowing kinetics.)

The collision-limited explanation of low recovery of fines implies that fine particles will be strongly influenced by flotation time, which is supported in Figure 12.35, and perhaps less affected by reagent additions that alter hydrophobicity, which can be the case (Trahar, 1981; Senior et al., 1994). Chemistry is not completely absent of course: fines are subject to surface oxidation and other contamination effects (e.g., adsorption of metal ions), and fines can cause aggregation (slime coating), issues that do respond to chemistry (Grano et al., 2014).

Increasing the rate of flotation of fine particles has obvious practical benefits. The requirement is to increase exposure to bubbles in order to provide more collision events and bubble surface area to transport collected particles. In an individual cell, increasing air rate and decreasing bubble size appear to offer means to achieve this increased exposure, the combination increasing S_b (Eq. (12.7)), which predicts an increase in rate constant (and the effect may be even greater than Eq. (12.7) predicts if P is inversely related to bubble size (Yoon, 1993)). Improved fine Pt-mineral recovery by frother addition to decrease bubble size is an example of successful implementation of this approach (Hernandez-Aguilar et al., 2006).

Increasing air or decreasing bubble size, however, also increases GH and there is this noted limit, about 15%, before a cell "boils." The options to increase kinetics using air rate and bubble size in a cell are, therefore, limited. The practical option to increasing exposure to bubbles is to float for longer. In the laboratory the effect of time on increasing fine particle recovery is easy to demonstrate (Figure 12.35); in the plant, surveying cells cumulatively down a flotation bank will show the impact of increasing number of cells (i.e., time) on increasing recovery of fines (Trahar, 1981; Grano et al., 2014). The installed flotation capacity is often dictated by the slow floating fine particles, including the use of a scavenger stage to provide increased flotation time.

Rather than relying on increased exposure, the energy of interaction with the bubble could be increased to "force" the particles across the water streamlines to intercept the bubble. The positive effect of increasing energy input on fines flotation is known, which recent work using a specially designed cell to give uniform distribution of energy throughout the pulp has reinforced (Safari et al., 2013). Increased energy input can be achieved by increasing impeller speed in mechanical cells; an option sometimes exercised to increase fines recovery is to increase impeller speed down a bank of cells. The reactor–separator flotation cells aim to inject this energy more directly to the collection of fine particles, as noted in Eq. (12.36).

Coarse Particles

At the coarse end of the particle size spectrum, the decline in recovery probably includes a contribution from decreased liberation, but in terms of bubble–particle interaction events is usually attributed to decreased stability of attachment, that is, increased probability of detachment (Jameson, 2012). Dissipation of energy as fluid eddies causes the bubble–particle aggregate, such as depicted in Figure 12.1, to spin and the centrifugal force can dislodge the particle. Another factor is higher froth dropback of coarse particles (Finch and Dobby, 1990; Rahman et al., 2012). Detachment of the particle and froth dropback introduce a chemical aspect: addition of collector to increase hydrophobicity (increase contact angle in Figure 12.2) should increase stability of the bubble–particle aggregate in the pulp zone and particle retention in the froth zone and thus increase recovery. The positive impact of chemistry on the recovery of coarse particles is seen in the recovery–size data of Trahar (1981) and Senior et al. (1994). Loss of free coarse particles to tailings may indicate that altered chemistry is required.

In contrast to fine particles, high energy input is evidently detrimental to coarse particle recovery, as Safari et al. (2013) illustrate. To accommodate the different needs of fine and coarse particles some strategies have been implemented. Distributing collector along the bank is one (Bazin and Proulx, 2001). Low collector concentration at the head of the bank is sufficient for the fast floating intermediate-size particles and upon their removal further collector dosage down the bank can target recovering the slower floating coarse particles. Classifying feed into fine and coarse fractions is often discussed but there are few examples. At the Mt Keith operation the feed was first split into two size fractions then into three (Senior and Thomas, 2005), and the improved Ni recovery paid for the project within a month. The authors emphasized the benefits of understanding particle size effects. New cells combining teeter bed (hindered settling) and flotation principles are aimed expressly at enhancing coarse particle recovery (see Section 12.13.4).

The trend in the recovery vs. size shown in Figure 12.35 is qualitatively explained by the product of collision probability (that increases as particle size increases) and detachment probability (that increases as particle size increases).

Liberation

Recovery-by-size-by-liberation data are not common, a situation set to change with the increasing spread of automated mineralogy technology (Chapter 17). A set of such data is given by Welsby et al. (2010), where liberation is considered to be surface liberation. Derived from their data, Figure 12.38(a) shows rate constant as a function

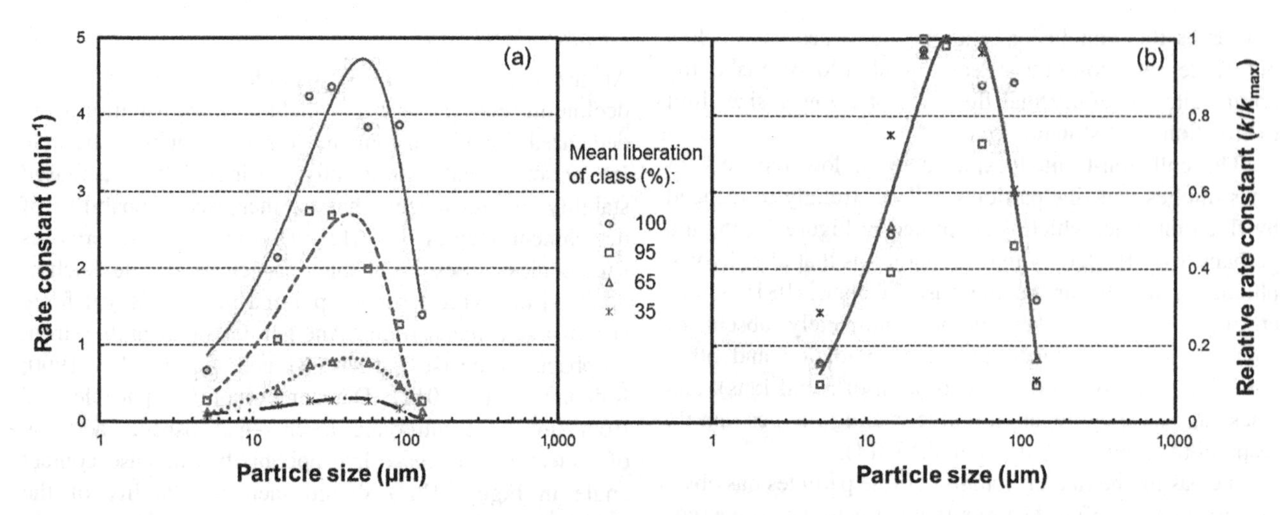

FIGURE 12.38 (a) Rate constant-by-size-by-liberation, and (b) same data but as rate constant relative to maximum rate constant showing common trend among size classes *((a) Data from Welsby et al. (2010)).*

of size and liberation. Figure 12.38(b) replots the data as rate constant relative to the maximum rate constant, k/k_{max} showing there is consistency to the rate constant–size relationship independent of liberation. Jameson (2012) noted this consistency in the data, showing the rate constant of liberation class k_x relative to the rate constant of fully liberated mineral, k_{lib} ($L = k_x/k_{lib}$), was a unique function of liberation (surface exposure, x) independent of particle size, that is, $L = \text{fn}(x)$, introduced as a *liberation function*.

In predicting the rate constant for locked (composite) particles, a first approximation is that the rate constant of the liberation class k_l is a weighted average (Nesset, 2011); for example, for a particle with three surface exposed minerals, A, B, and C the k_l is given by:

$$k_l = x_A k_A + x_B k_B + x_C k_C \tag{12.39}$$

where x is fractional surface exposure and k refers to rate constant of liberated mineral. From Eq. (12.39), a refinement of the liberation function might be to consider locked particle type, binary locked A, B particles as distinct from binary locked A, C particles, etc.

Recovery-size-liberation data offer the promise to interface comminution and flotation models (and other separation process models). The comminution step produces the size-liberation spectrum and the associated recovery (or rate constant) data for a separation unit can be used to predict the outcome (grade and recovery) from that unit. The common simplification is to assume that recovery–size data are independent of the size distribution of the feed to the separation unit and how the size distribution was produced. This simplified approach to interfacing comminution with flotation has been used over the years (Ramirez-Castro and Finch, 1980; Bazin et al., 1994; Bradshaw and Vos, 2013). A related assumption is that the size-liberation data are also independent of how the size distribution is generated. With the consistency evident in the rate constant-size-liberation relationship (Figure 12.38(b)), a way to include liberation in interfacing comminution and flotation may be forthcoming.

12.10.2 Entrainment

To increase fine particle recovery, increasing flotation time is the practical option. Increasing time incurs a penalty, however: increased entrainment recovery. The recovery–size relationship for entrained material (Figure 12.37), shows a trend distinct from the two floatable minerals, recovery increasing as particle size decreases especially below about 50 μm (Smith and Warren, 1989). Recovery of entrained particles is related to water recovery (Jowett, 1966). Water is transported by bubbles into and through the froth, and since particles are in the water (i.e., it is slurry that is actually carried by the bubbles), there is an unselective particle recovery mechanism. Entrainment recovery is a fine particle recovery mechanism as fines do not settle out of the transported water under gravity as fast as coarse particles. Johnson (1972), from industrial and lab-scale tests, established that recovery by entrainment was proportional to feed water recovery to the concentrate. From this finding the degree of entrainment was defined as the ratio of the recovery of entrained solids to the recovery of water C_i or ENT_i where subscript i refers to a particular particle class (particle type and size). An estimate of dependence on size is shown in Figure 12.39, taken from Johnson (2010), who reviewed practical aspects of entrainment.

The dependence of entrainment on water transport has driven the need to understand (model) water recovery

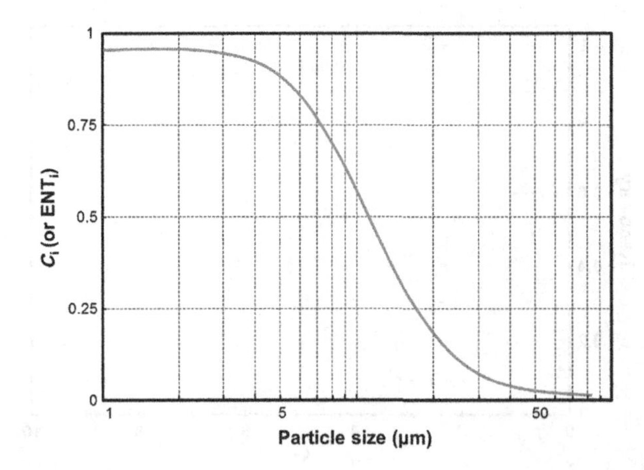

FIGURE 12.39 Entrainment factor ENT_i as a function of particle size *(Data from Johnson (2010)).*

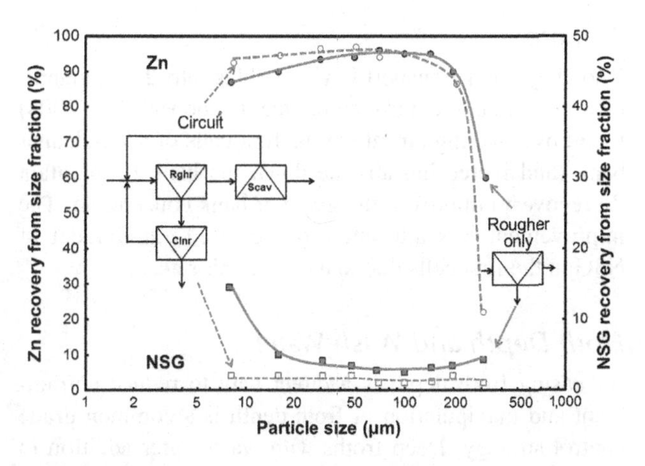

FIGURE 12.40 Example of cleaner block damping recovery of entrained non-sulfide gangue, NSG (and of scavenger increasing fine sphalerite recovery) (Note: cleaner block was four stages; unpublished data, Pine Point Mines).

mechanisms. Several water recovery models were compared by Zheng et al. (2006). A physics-based simulator, *FrothSim*, was successfully used to model entrainment in assessing ways of controlling bank performance (Smith et al., 2008).

From an operations viewpoint, the dependence means that reducing entrainment requires reducing water recovery. There are several options.

Role of Cleaner

The rougher stage contributes the initial (and often large) entrainment and the cleaners play an important role in reducing it. Figure 12.40 illustrates an example of the cleaner's effect on recovery-by-size for a floatable mineral (sphalerite) and entrained mineral (NSG). The rougher inputs significant entrained fine (<10 μm) NSG, which the combination of cleaner stages (or *cleaner block*) rejects (and at the same time the scavenger increases fine sphalerite recovery due to increased retention time, as discussed above). Cleaners do not fully eliminate entrainment; even after several stages of flotation there is evidence that operational changes can still be effective in reducing its impact (Cooper et al., 2004).

This role of cleaners in rejecting entrainment is one argument against by-passing products direct to concentrate. Candidate products include concentrates from the first cell in the rougher bank, a unit cell ahead of the roughers or a flash flotation stage inside the grinding circuit. These products might be "high grade" (e.g., exceeding final concentrate grade) but can still carry substantial entrained (−50 μm) material that would be easily rejected in a cleaner stage with little loss of value recovery. Not cleaning these products means the remaining feed to the cleaners has to be upgraded sufficiently to compensate, which is not always achievable.

Pulp Dilution

Increasing pulp dilution reduces water recovery (Johnson and Munro, 2002). It may seem counter-intuitive, but recall that water transport is via the bubbles. Diluting the feed to a flotation stage means increasing the feed rate of water, but since the bubbling rate is not changed (S_b is unchanged) then the rate of water carried to the concentrate does not change, at least to a first approximation. Since water recovery is the ratio of the concentrate water rate to feed water rate, the recovery goes down. Not necessarily practical in the roughing stage because of the volumes involved, pulp dilution to cleaner stages is widely practiced.

An illustration of the merit of cleaner feed dilution is described by Shannon et al. (1993). In that example dilution enabled the operation to take full advantage of increased liberation through regrinding by controlling entrainment of the additional fines produced. A reassignment of cleaner cell capacity retained the necessary retention time. Retrofitting pulp dilution while retaining the required retention time (e.g., by reassignment of cells) may not always be feasible and it is best to plan pulp dilution at the plant design stage.

Particles tend to stabilize froths and increase water recovery. This is one reason why flotation usually aims to float the mineral in least amount in the feed, to reduce entrainment (and entrapment) of unwanted minerals. Commonly the mineral in least amount is the valuable mineral and is floated, but in some cases such as processing iron ores, the gangue (silica) is in lesser amount and reverse flotation is practiced. In cleaner stages the floatable mineral becomes the majority and sometimes reverse flotation, for example, pyrite from base metal concentrates, is worth considering.

Air Rate

Knowing water transport is via bubbles introduces manipulating air rate to control entrainment. Cooper et al. (2004) found by lowering air rate in the first cells of a Zn-cleaner bank (and increasing air rate down the bank to maintain Zn recovery) improved the grade of bank concentrate. The improvement was attributed to decreased entrainment of NSG in the first cells due to the lower air rate.

Froth Depth and Wash Water

Increasing froth depth is a direct way to reduce entrainment and manipulation of froth depth is a common grade control strategy. Deep froths with wash water addition to control entrainment was probably the prime advantage of flotation columns when these units reemerged in the 1980s (Finch and Dobby, 1990). Wash water is added into the froth to replace the water coming from the pulp and thus eliminates the entrained particles coming with the pulp water. By replacing the water draining from the froth under gravity, the addition of wash water enables deep froths to be built. Entrainment control using wash water is successful on mechanical cells as well (Kaya et al., 1990).

Since reducing water consumption is a concern in most locations, wash water use is under some scrutiny. How to introduce wash water to minimize use while achieving the benefits remains a challenge (Neethling et al., 2006). Most operations appear to favor introducing wash water from over-head drip pans rather than submerged devices in order to facilitate visual checks. An example wash water addition system is shown in Section 12.13.2.

12.11 CELLS, BANKS, AND CIRCUITS

The purpose of this section is to provide a theoretical framework for the progression from single cells to banks to circuits that characterizes flotation plants. The emphasis is the effect on recovery and separation between floatable minerals (often referred to as "selectivity").

12.11.1 Single Cell

Recovery

Introduced in Section 12.9, recovery in a single cell has two mathematical extremes depending on material transport, namely plug flow (Eq. (12.27)) and fully mixed flow (Eq. (12.28)).

For mechanical cells and flotation columns the fully mixed approximation is usually adopted. Figure 12.41 compares recovery as a function of $k\tau$ for the plug flow and fully mixed cases: it is evident, since k is the same for both, that more time is needed in a fully mixed cell to achieve the same recovery as under plug flow. For example, for a target recovery of 90% (0.9) the fully mixed $k\tau$ is 9 (solving Eq. (12.28), see Figure 12.41) and for plug flow kt is 2.3 (Eq. (12.27)). This means retention time is 9/2.3 (i.e., 3.9) times longer in the fully mixed case compared to plug flow: for equivalent recovery this translates to 3.9 times the cell volume being required in the fully mixed case compared to plug flow.

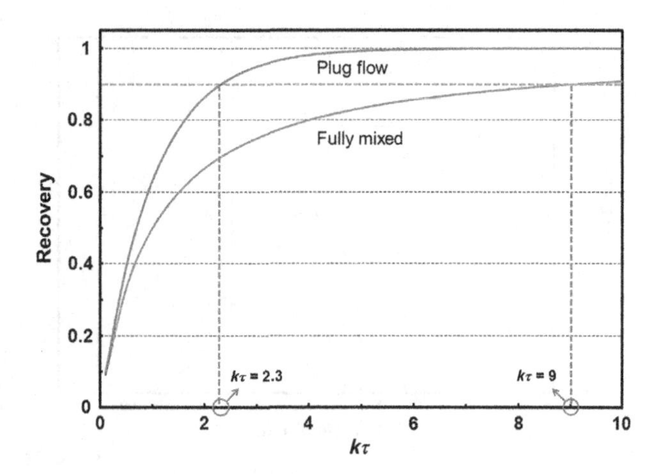

FIGURE 12.41 Recovery as a function of $k\tau$ (rate constant × time) under plug flow and fully mixed transport.

This difference in recovery between fully mixed and plug flow transport is well understood, one consequence is the need to scale up lab determined flotation time (i.e., equivalent to plug flow) to the continuous (industrial) case. A common scale up factor is 2.5 to estimate plant flotation time in sizing cells (Wood, 2002). (Increasing flotation time by 2.5 is equivalent to scaling down the rate constant by 2.5, noting that it is always the product $k\tau$ in the formulae.) Less realized perhaps is the impact of mixing degree on separation (selectivity) of floatable minerals.

Selectivity

To illustrate the impact of mixing on separation, consider two minerals A and B with rate constant $k_A = 1\ \text{min}^{-1}$ and $k_B = 0.1\ \text{min}^{-1}$ and take as the measure of efficiency the difference in recovery, that is, $SE = R_A - R_B$ (Eq. (1.2)). Solving Eqs. (12.27) and (12.28) for these k values, we find that SE in the fully mixed case is 43% (0.43) and in the plug flow case SE is 69% (0.69) (Table 12.6). The lower selectivity in the fully mixed case can be seen in Figure 12.41; the slope is the rate of change in R with $k\tau$ (i.e., $\delta R/\delta k\tau$) and it is evident that slope is higher for the plug flow case, meaning greater selectivity compared to the fully mixed case. (A convenient reference point is to compare $\delta R/\delta k\tau$ at $R = 0.5$.) In

TABLE 12.6 Comparison of SE for Fully Mixed and Plug Flow Transport at Equal Target Recovery of Mineral A = 90% (0.90) where $k_A = 1\ \text{min}^{-1}$ and $k_B = 0.1\ \text{min}^{-1}$

Fully Mixed				Plug Flow			
τ (min)	R_A	R_B	SE	t (min)	R_A	R_B	SE
9	0.90	0.47	0.43	2.3	0.90	0.21	0.69

summary, therefore, not only is more time (i.e., more cell volume) required in the fully mixed case to reach the same recovery as in plug flow but the SE is also less.

The analysis of selectivity can be addressed by writing Eq. (12.28) for both A and B and combining to eliminate τ (which is clearly the same for both minerals) to express R_B as a function of R_A. For the fully mixed case the result is:

$$R_B = \frac{1}{1 + \left(\frac{k_A}{k_B}\right)\left(\frac{1-R_A}{R_A}\right)} \tag{12.40}$$

Introducing S as the relative rate constant (k_A/k_B) Eq. (12.40) becomes:

$$R_B = \frac{1}{1 + S\left(\frac{1-R_A}{R_A}\right)} \tag{12.41}$$

Figure 12.42 shows the trend in R_B with R_A for S varying from "difficult" separation ($S = 2$) to "easy" separation ($S = 10$, equivalent to the situation above where $k_A = 1$ and $k_B = 0.1$). An example of a difficult separation is sphalerite from pyrite, where Cooper et al. (2004) reported $S \sim 2-3$. The figure shows the increasing rate of recovery of B with respect to A as recovery of A is increased.

Instead of expressing S as a relative rate constant, we could express in terms of relative floatability by expanding S in terms of the $k-S_b$ relationship (Eq. (12.33)):

$$S = \frac{k_A}{k_B} = \frac{P_A \cdot S_b \cdot R_{fA}}{P_B \cdot S_b \cdot R_{fB}} \tag{12.42}$$

Since S_b is the same for both minerals and there is evidence that $R_{fA} = R_{fB}$ for floatable minerals (Vera et al., 1999) then $S \sim P_A/P_B$.

Using SE and relative floatability, S, the following optimization problem can be addressed: to maximize SE (maximize selectivity) at target recovery. This optimization problem can be stated as follows:

Maximize

$$\text{SE} = R_A - R_B \tag{12.43}$$

by searching on R_A subject to

$$R_A = R_{A\text{target}} \quad \text{performance constraint} \tag{12.44}$$

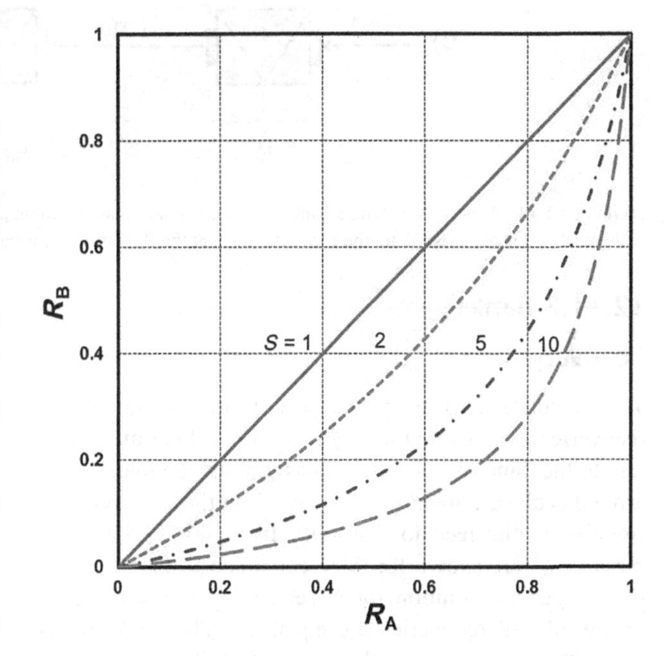

FIGURE 12.42 Recovery of mineral B as a function of mineral A and relative rate constant S *(Adapted from Maldonado et al. (2011)).*

and

$$R_B = \frac{1}{1 + S\left(\frac{1-R_A}{R_A}\right)} \quad \text{process constraint} \tag{12.41}$$

Maximizing SE is equivalent to minimizing R_B at the target R_A or maximizing grade (specifically grade of A relative to B).

Applying to a single cell it is evident that the optimization problem cannot be solved: in a single cell it is not possible to satisfy two constraints, maximum SE and target R_A (Maldonado et al., 2011).

Three limitations of the single fully mixed cell are therefore identified: the need for increased flotation residence time (i.e., cell volume) to attain the same recovery as in plug flow; the lower selectivity between floatable minerals compared to plug flow; and the inability to maximize SE at a target recovery. These limitations are addressed by placing cells in series; that is, to form a bank (or line or row).

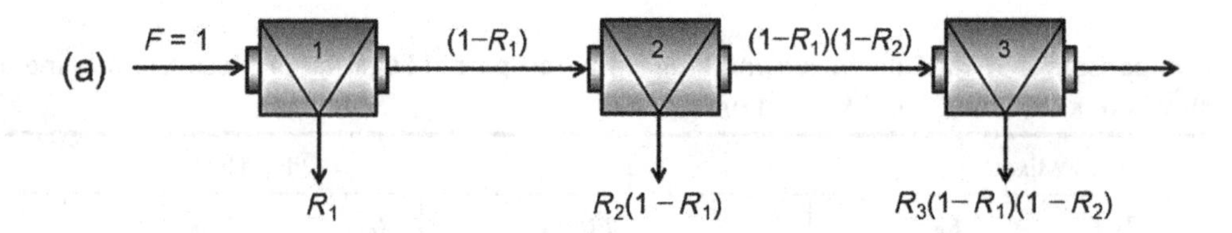

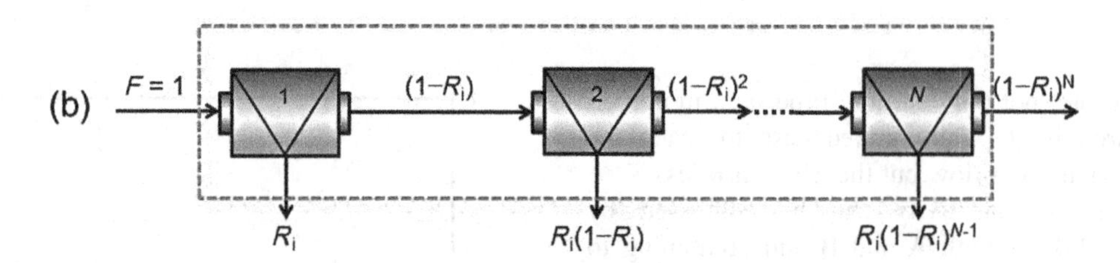

FIGURE 12.43 Mass flows along a bank based on F = unit mass flowrate and cell recovery based on feed to cell for: (a) three cells, and (b) N cells with equal cell recoveries (Note: the convention is that the float product, usually the concentrate, is drawn coming from the peak of the triangle).

12.11.2 Banks

Recovery

The calculation of bank recovery from individual cell recoveries is illustrated in Figure 12.43. Taking mass feed rate to the bank of unity ($F = 1$) as a basis, the mass flows around each cell are calculated based on the cell recovery (relative to the feed to that cell). In Figure 12.43(a) the mass flows are shown for three cells; and in Figure 12.43 (b) the general solution for N cells is given for the case where all cell recoveries are equal, R_i. The bank (total) recovery R_{bk} is given by the summation of the individual cell mass flows (since $F = 1$, the mass flows are equivalent to fractional recoveries). In case (b) where all cells have equal recovery (R_i) the bank recovery can be derived directly from a mass balance around the dashed block:

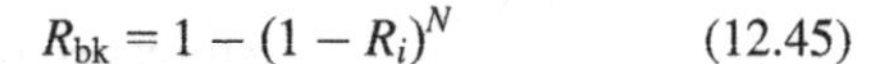

$$R_{bk} = 1 - (1 - R_i)^N \quad (12.45)$$

Assuming equal cell recovery might seem a mathematical contrivance to provide a simple solution, but a justification for equal recoveries will be given when selectivity is considered. By substituting Eq. (12.28) into Eq. (12.45) and, noting that $\tau_i = T/N$ where T is the total bank retention time, the following is derived:

$$R_{bk} = 1 - \left(\frac{N}{N+kT}\right)^N \quad (12.46)$$

Figure 12.44 illustrates the impact of N: increasing N from 1 to 4 gives a significant increase in recovery for a given kT; in the limit as $N \rightarrow \infty$, Eq. (12.46) becomes the plug flow solution, $R_{bk} = 1 - \exp(-kT)$. Recovery approaches the plug flow result as the number of cells is increased because transport approaches plug flow (Figure 12.45).

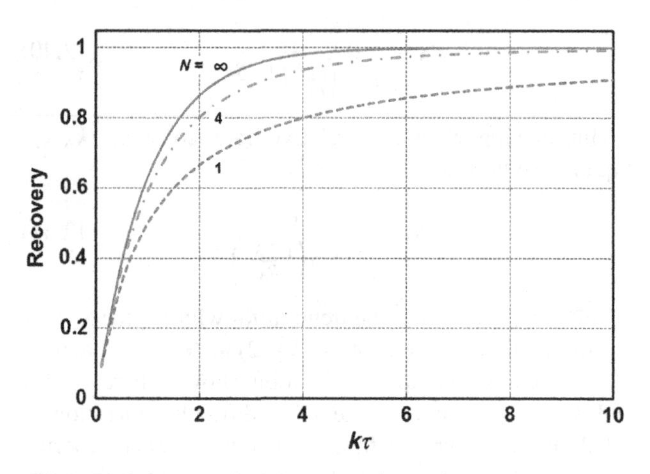

FIGURE 12.44 Bank flotation recovery as a function of $k\tau$ and number of cells N in the bank.

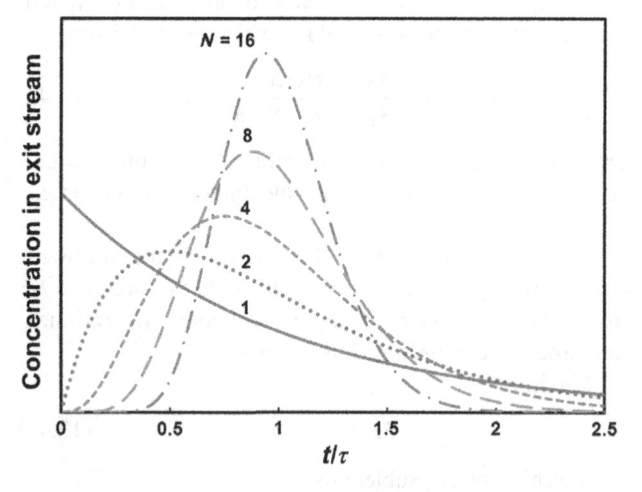

FIGURE 12.45 Bank RTD moves toward plug flow as number of cells is increased.

Banks clearly give an advantage in recovery over one large cell of equivalent total volume. A question that arises is: how many cells should the bank comprise?

The analysis above implicitly assumed the individual cells are isolated and there is no short circuiting or back mixing between cells. The modern circuit of tank cells approaches the assumption: in that case Figure 12.44 suggests a bank of four or five cells is sufficient. Some care in this interpretation is required, however. Eq. (12.46) derives from Eq. (12.45), where all cells have equal recovery, making this the basis of the calculations. Significant deviation from equal recovery invalidates the solution, potentially making the four to five cell bank too short. As a resolution, six to eight cells to a bank may offer a practical compromise (Wood, 2002). A caveat is that in practice cells are often paired in a bank and act as one unit and this may represent an effective reduction in number of cells in the bank. (Note that in some plants "bank" refers to this pairing of cells, the whole being a row or line.) With older trough style banks of cells the impact of short circuiting between cells was countered by having long banks, up to 20 or more cells (Wood, 2002).

Figure 12.44 is a general solution independent of the magnitude of k: for example, if k is low this is compensated by selecting larger cells to increase T and maintain the kT value. In practice, longer banks (more cells) are usually recommended if k is considered low (Wood, 2002).

Short banks are encouraged by the increase in the size of mechanical cells, which over the past 50 years has gone from about 6 to 600 m^3 (see Section 12.13.1). The advantage of large cells in reducing the number required and thus reducing capital and operating costs is an obvious incentive. To retain the recovery advantage of banks with large cells it is still necessary to keep a minimum number of cells in the bank, at least four, and preferably six to eight. The cost advantage of large cells is retained by reducing the number of parallel banks needed to reach plant throughput.

In constructing Figure 12.44 it was assumed that the retention time in each cell was the same, that is, each cell was the same volume. Maldonado et al. (2012) showed that for a given total bank capacity, cells of the same volume give the highest recovery; that is, equal size cells make best use of the installed volume. At the design stage there is little incentive to mix cell sizes, but there is less constraint in a plant expansion when a larger cell (usually at the head of the bank) can give the added volume perhaps more cheaply.

Selectivity

Gaining recovery for a given installed cell volume is one impact of the bank over a single cell. There is also a gain in separation between true floating minerals. Applying the optimization problem to a bank of cells, maximizing SE is solved by searching on R_{Ai} subject to the performance constraint (Eq. (12.44)) and the process constraint (Eq. (12.41)). Maldonado et al. (2011) illustrated the solution for two- and three-cell banks, which are amenable to trial-and-error solution, and went on to solve the general problem of an N-cell bank. The result is that the maximum in bank SE ($SE_{bk,\ max}$) occurs when all cells have equal recovery, referred to as a *balanced* or *flat bank recovery profile*.

The physical interpretation derives from Eq. (12.41) and Figure 12.42. The relationship between R_B and R_A means that a cell recovering more A than its neighbor incurs a penalty of incremental B recovery that is not offset by another cell under-recovering A: if no cell should recover more than its neighbors then all cells must have the same recovery. There is evidence that this flat recovery profile does offer a metallurgical advantage over other profiles (Maldonado et al., 2012).

Figure 12.46 shows the impact of number of cells in the bank on SE: SE_{bk} increases and levels off at about $N = 7$ or 8. The physical interpretation is again based on Figure 12.42: increasing the number of cells means the recovery per cell is lowered which means the slope of the $R_B - R_A$ relationship ($\delta R_B/\delta R_A$) is lowered; or in other words, less R_B is recovered per R_A recovery and thus selectivity is increased. There is evidently a potential loss in SE by having a bank shorter than about four cells, and, remember, the calculation in Figure 12.46 assumes the bank is operated with a flat recovery profile: a short bank with a nonflat profile risks even larger loss in SE than Figure 12.46 suggests. Again, we may have a practical compromise of six to eight cells in a bank. For older circuits with long banks, some up to 20 cells, balancing recovery is not really a concern; differences in recovery between so many cells will tend to be small and the profile will approach balanced.

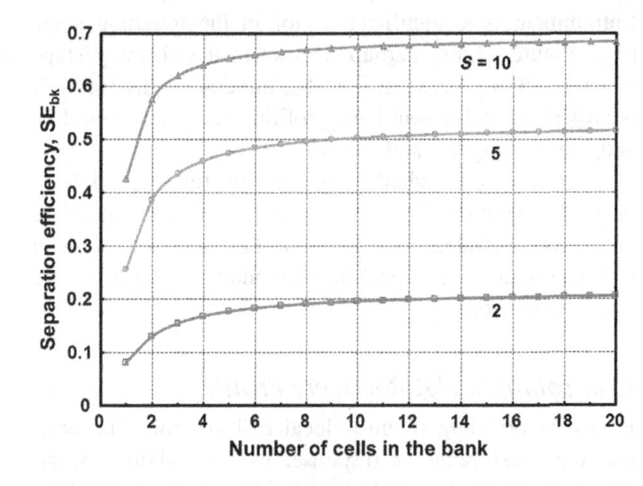

FIGURE 12.46 Bank SE as a function of number of cells: flat profile and target recovery of A 90% *(Adapted from Maldonado et al. (2011)).*

The increase in SE_{bk} as N is increased is reminiscent of the effect of N on recovery (Figure 12.44) and for the same reason: as N increases the bank approaches plug flow transport, which represents the theoretical maximum SE that can be achieved by the bank (Figure 12.46). Assuming plug flow for the bank and setting up the problem in the same fashion as in deriving Eq. (12.41), it can be shown that:

$$R_{\mathrm{Bbk}} = 1 - (1 - R_{\mathrm{Abk}})^{1/S} \tag{12.47}$$

In the limit (plug flow) setting a target bank recovery of A determines the bank recovery of B for a known S and thus the maximum bank SE can be calculated.

Note that the analysis does not consider cell volume, only recovery. Having cells of different size complicates balancing cell recoveries, as a larger cell in the bank will naturally tend to recover more than its smaller neighbors. If plant expansion (or some other reason) suggests adding cells to a bank, the effect on selectivity, along with the noted effective use of cell volume in terms of recovery, is an argument in favor of the added cell having the same volume.

The optimization problem was to find a maximum in SE for a target bank recovery. We could extract the benefit in a different way: by setting a target SE_{bk} the flat recovery profile will give the maximum bank recovery. The difficulty is knowing beforehand if the target SE_{bk} is achievable.

Singh and Finch (2014), using JKSimFloat, explored two aspects not addressed in the analytical solution: variable S and impact of entrainment. The relative rate constant S will tend to decrease down the bank as high rate constant A particles are recovered in the first cells; in the example described they showed that this variation in S did not alter the flat recovery solution. Their examination of entrainment suggested the optimum operating condition was equal mass flowrate (mass pull) from each cell (*flat mass pull profile*), but that the flat recovery profile was close in performance. Entrainment is a significant factor in the roughing stage (e.g., Figure 12.40), perhaps less so in the cleaning stage when selectivity between floatable minerals is the focus. It is possible that different bank profiling strategies should be adopted for roughers and cleaners.

A review of standard texts does not reveal guidelines on bank operation. The flat recovery profile is offered as an operating strategy: the question then is how to achieve it. Some options discussed by Maldonado et al. (2012) are briefly reviewed.

Controlling a Flat Recovery Profile

Recovery profiling requires local cell control. One way, and the most rapid in response, is manipulation of air rate. Cooper et al. (2004) found using air to redistribute recovery down a Zn-cleaner bank gave improved performance (higher bank grade at target recovery). High recovery in the first cell of a bank is common because the high proportion of fast floating mineral renders recovery "easy." Regardless of whether the flat recovery profile can be achieved, making sure the first cells do not "over-recover" is perhaps the principal practical outcome of the analysis. That observation holds whether maximizing selectivity or minimizing entrainment is the target; and is an argument that if a larger cell is to be added to a bank for some purpose that it not be made the first cell.

The Cooper et al. experience was on a cleaner bank where the first cells are probably carrying capacity limited rather than kinetically controlled. Nevertheless, there was an advantage to redistributing recovery down the bank suggesting that balancing recovery is a robust guideline.

Manipulating air rate down a bank became known as *air rate (or J_g) profiling*, part of a general strategy of *air distribution management*. An extension is *PAR profiling* (Hadler et al., 2012). In this strategy air rate to each cell in the bank is adjusted to achieve the maximum air recovery (Eq. (12.35)), which has been shown to maximize bank performance. Air distribution management to banks is gaining traction in flotation control strategies (Shean and Cilliers, 2011).

Online control of the recovery profile might be effected by monitoring *froth velocity* using froth imaging technology. If a relationship between froth velocity and recovery can be identified it would allow air to be manipulated to achieve the target froth velocity profile. As a first approach it may be easier to link froth velocity to mass pull rather than recovery, and consider the flat mass pull profile strategy identified by Singh and Finch (2014).

Variables other than air that might be considered for local cell control include froth depth (level), reagents, and impeller speed. Froth depth has been incorporated into air profiling (Gorain, 2005) and in the PAR strategy (Hadler et al., 2012). (In some control strategies froth depth is the principal manipulated local variable with air rate for fine tuning.) Distributing reagents along the bank and increasing impeller speed down the bank, both practiced to recover slow floating fractions (Section 12.10.1), have the effect of balancing recoveries. In general, reagents do not offer local control, as additions in one location will affect units downstream (and upstream through recycle). At one operation, reduced frother dosage was used to "slow down" the first cell in a bank when air rate could not be lowered sufficiently (Blonde et al., 2013).

12.11.3 Circuits

Compared to a single cell, the bank improves selectivity which approaches the theoretical maximum SE result for plug flow. A combination of flotation stages with recirculation, that is, a circuit (or network or flowsheet), enables

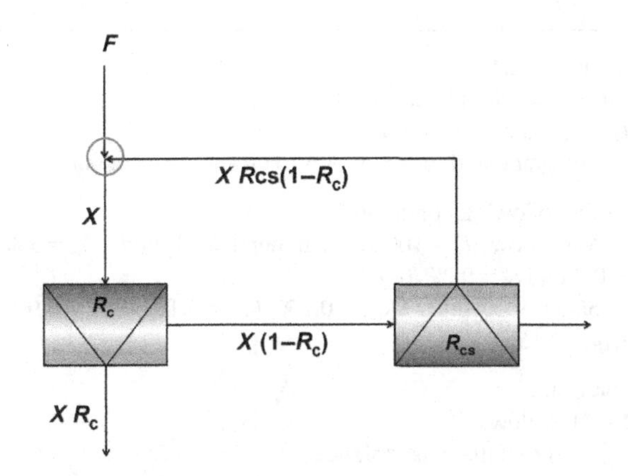

FIGURE 12.47 Mass flows around a cleaner–cleaner/scavenger circuit from which circuit recovery is calculated.

the plug flow result to be exceeded. In general, a circuit will give better separation than single-stage operation, especially for nonsharp separations (Williams and Meloy, 2007), flotation being an example nonsharp separation system as the relative rate constant S imposes a limit to selectivity.

Selectivity

A basic circuit is shown in Figure 12.47, compromising two stages, the second being a scavenger to the first (i.e., treating the tailings from the first stage). This arrangement is familiar in the rougher–scavenger arrangement, and in the cleaner–cleaner/scavenger arrangement (the intended circuit depicted in the Figure). Circuit recovery can be solved by iteration, but also in an analogous fashion to that applied to Figure 12.34, this time introducing X as the mineral mass flowrate into the cleaner stage and deriving the (mineral) mass flows around the circuit (Figure 12.47). By mass balancing at the feed node (indicated by the circle), F can be deduced and the circuit recovery R_{circ} solved:

$$F = X[1 - R_{cs}(1 - R_c)] \tag{12.48}$$

$$R_{circ} = \frac{R_c}{1 - R_{cs}\,(1 - R_c)} \tag{12.49}$$

The solution will be referred to as an "analytical" solution (an alternative name is "transfer function" (Williams and Meloy, 2007)). The impact of the circuit on SE can be judged by setting R_c and varying R_{cs} for the target mineral. Assuming the banks are long enough for plug flow to apply, the result for $S = 5$ is shown in Figure 12.48(a). The calculation proceeds as follows: for A: select $R_{c,A}$ (e.g., 0.2, 0.5, etc.), vary $R_{cs,A}$ (e.g., from 0 to 1), calculate $R_{circ,A}$ using Eq. (12.49); and for B: from $R_{c,A}$ and $R_{cs,A}$ calculate $R_{c,B}$ and $R_{cs,B}$ using Eq. (12.47) with $S = 5$, and solve for $R_{circ,B}$ using Eq. (12.49); finally, solve for $SE_{circ} = R_{circ,A} - R_{circ,B}$. The Figure shows circuit SE

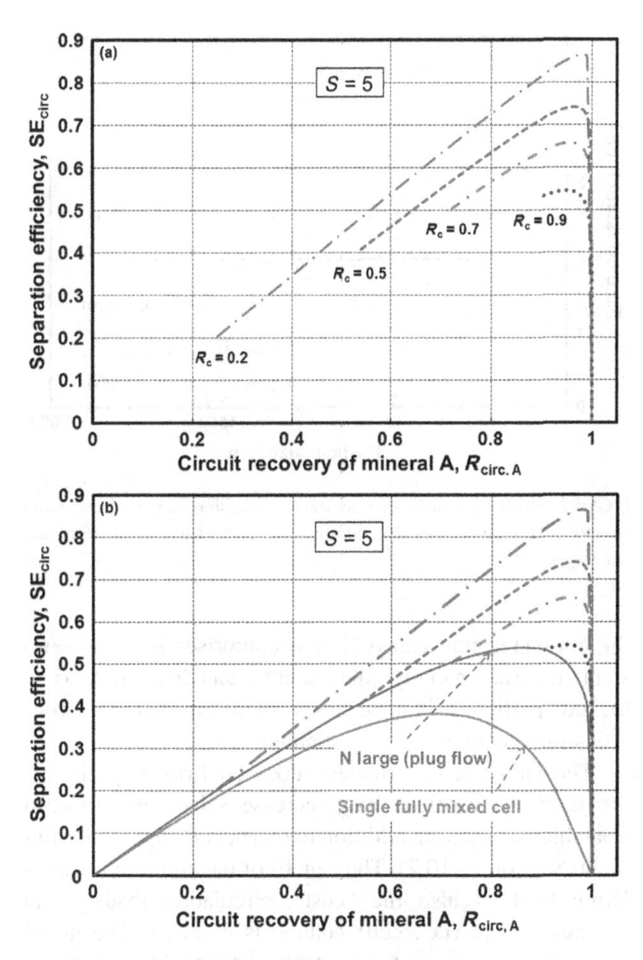

FIGURE 12.48 (a) Circuit (Figure 12.47) SE as a function of circuit recovery of A for different cleaner stage recoveries (R_c), and (b) as (a) but showing progress in SE going from single cell to bank (flat recovery profile) to circuit.

increases as R_{cs} increases. This increase in SE_{circ} is related to the increase in circulating load of A, given by:

$$CL_A = \frac{R_{cs,A}(1 - R_{c,A})}{1 - R_{cs,A}(1 - R_{c,A})} \tag{12.50}$$

The CL_A is further increased by lowering $R_{c,A}$, as Eq. (12.50) indicates, and this further increases SE_{circ} as Figure 12.48(a) indicates. Putting all this together, Figure 12.48(b) shows the advance in SE going from single cell to bank (plug flow) to the two-stage circuit.

According to Figure 12.48(b), the circuit makes a significant increase in SE. This is indeed a circuit "property", to improve selectivity. The magnitude of the impact shown, however, is exaggerated as the analysis implicitly assumed the recycled materials have the same recoverability as the fresh feed to the circuit. Figure 12.49, an example of rate constant versus size and liberation based on Figure 12.38(a), illustrates why this is not the case. The dashed horizontal lines indicate possible divisions for the cleaner and cleaner/scavenger stages; it is evident that

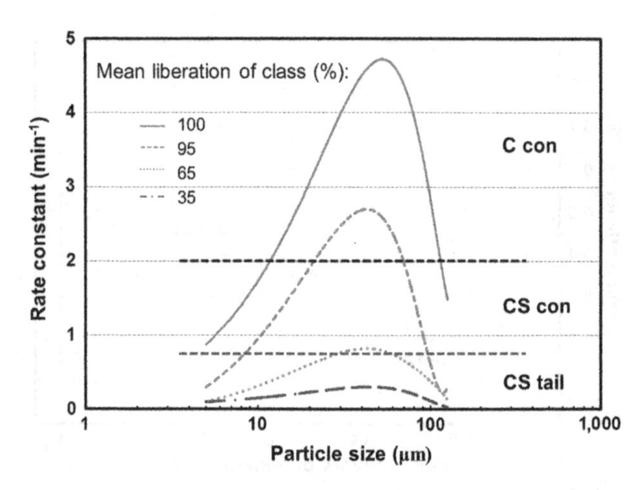

FIGURE 12.49 Example rate constant as function of particle size and composition showing possible divisions for cleaner and cleaner/scavenger stages.

the circulated material (CS con) comprises low rate constant material, namely fines, coarse and low grade (i.e., locked) particles. To take this variation in material properties into account requires simulators.

The analysis, nevertheless, does confirm why circuits are used in flotation: they increase selectivity between floatable particles (in addition to cleaners damping entrainment, Section 12.10.2). The benefit of the circuit, the circulating load, is also the "cost": circulating loads mean increased stage (i.e., cell) volume is required. The target circulating load needs to be established at the design stage.

To optimize a circuit, Agar et al. (1980) suggested using the maximum SE at each stage. Based on Jowett (1975), they showed that the maximum in bank SE coincides with the cell in the bank that produces concentrate of grade equal to the feed grade; thus this cell defines the length of the bank. This is equivalent to selecting the length of bank such that no cell is yielding concentrate of lower grade than the feed to the bank, making it a tempting logical choice. However, using SE alone does not allow for the performance constraint of reaching a target recovery, nor does the strategy say how the bank is to be operated. It was later noted that maximizing stage SE does not result in maximizing circuit SE (Jowett and Sutherland, 1985; Lauder, 1992). The solution proposed here is to set the target circuit recovery of mineral A, then select the stage recoveries that meet that target and operate the stages with a balanced recovery profile. This will result in the maximum circuit SE at the target circuit recovery of A.

Varying target A recovery will generate the predicted relationship between SE_{circ} and $R_{circ,A}$ (such as Figure 12.48) to aid reaching a decision on where to operate (which could include an economic assessment) that will then give the corresponding stage recoveries. Repeating the calculations for the other mineral components (B, etc.) and summing gives total solids flows around the circuit and coupled with estimates of the water content (% solids) slurry volume flows can be calculated and stage cell volume requirements assessed. Agar and Kipkie (1978) used this approach, employing bench tests to determine stage recovery data for the mineral components (referred to as *split factors*) which were used to estimate the flow of mineral around a circuit by an iterative procedure. An example calculation using the analytical solution for the circuit in Figure 12.47 is included below (Example 12.1). The analytical approach helps

Example 12.1

For the circuit in Figure 12.47 calculate:

1. the mass flow of A and B,
2. the grade and recovery of A to cleaner concentrate.

for the following conditions:

Mass flow (F) = 100 t h^{-1}; mineral feed grades $f_A = 5\%$ (0.05) and $f_B = 95\%$ (0.95).

Stage recoveries: $R_{cA} = 0.75$, $R_{cB} = 0.10$, $R_{csA} = 0.90$, $R_{csB} = 0.15$.

Solution

1. Mass flows.

From the mass balance:

$$F = X[1 - R_{cs}(1 - R_c)]$$

this becomes the scaling factor to determine mass flows for any stream; note, for the individual minerals $F = F\,f$ where $f = f_A$ of f_B. The table below summarizes the calculations.

Stream	Formula	Mass Flow (t h^{-1}) A	Mass Flow (t h^{-1}) B
Cleaner concentrate	$\frac{F\,f\,R_c}{[1 - R_{cs}(1 - R_c)]}$	4.84	10.98
Cleaner tails	$\frac{F\,f\,(1 - R_c)}{[1 - R_{cs}(1 - R_c)]}$	1.61	98.84
Cleaner/scavenger concentrate (circulating load)	$\frac{F\,f\,R_{cs}(1 - R_c)}{[1 - R_{cs}(1 - R_c)]}$	1.45	14.83
Cleaner feed (as a check)	$\frac{F\,f}{[1 - R_{cs}(1 - R_c)}$	6.45	109.83

The cleaner feed calculation offers a check, where cleaner feed = fresh feed + circulating load:

For A: 6.45 = 5 + 1.45 (i.e., checks); and for B: 109.83 = 95 + 14.83 (i.e., checks).

2. Grades of and recoveries to cleaner concentrate.

From the table these are readily calculated:

Grade A = 4.84/(4.84 + 10.98) = 0.3058 or 30.58%

Recovery A = 4.84/5 = 0.9677 or 96.77%

While simplified, the calculation method offers a quick assessment, and for teaching purposes provides a calculation route that does not require a simulator to complete.

TABLE 12.7 Some Common Circuit Arrangements, with Analytical Solution and Measure of Selectivity, the Slope (dR_{circ}/dR)

	Circuit	Circuit Recovery (all R Equal)	(dR_{circ}/dR)
R-S-C	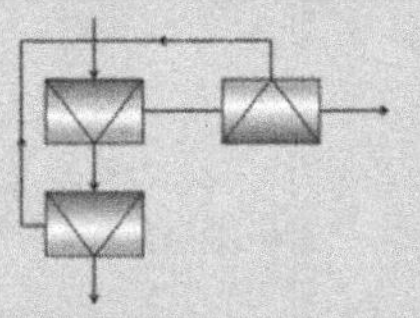	$\frac{R^2}{R^2+(1-R)^2}$	$\frac{2R(1-R)}{[R^2+(1-R)^2]^2}$
R-C-CS	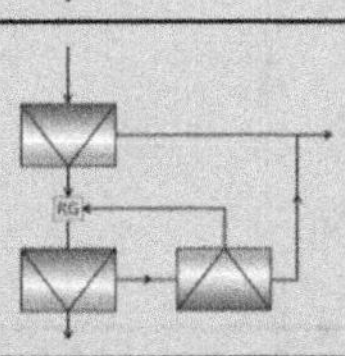	$\frac{R^2}{R+(1-R)^2}$	$\frac{2R-R^2}{[R+(1-R)^2]^2}$
R-C_3-CS	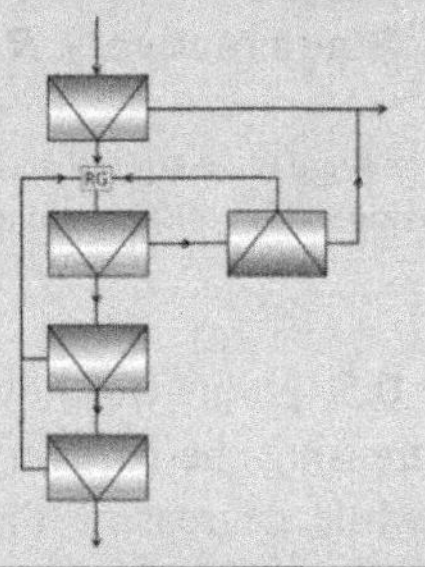	$\frac{R^4}{R^3+(1-R)^2}$	$\frac{R^6+2R^5-6R^4+4R^3}{[R^3+(1-R)^2]^2}$
R-S-C_3	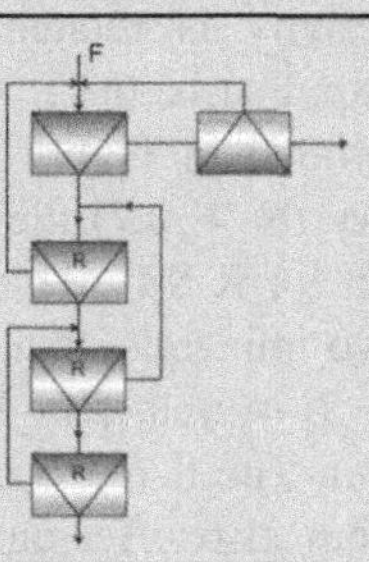	$\frac{R^4}{(R^2-R+1)(3R^2-3R+1)}$	$\frac{-2R^3(3R^3-7R^2+6R-2)}{(3R^4-6R^3+7R^2-4R+1)^2}$
$(R\text{-}C\text{-}CS)_2$	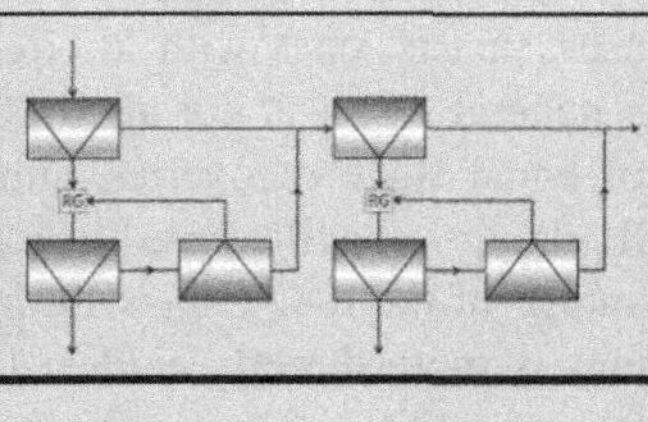	$\frac{R^2(1-R)}{R+(1-R)^2}$	$\frac{R^2(3R^5-8R^4+10R^3-5R^2-R+1)}{[R+(1-R)^2]^2}$

understand the interactions between stages, which is further illustrated later. Once the circuit is decided, control of the circulating load is key to performance.

The arrangement in Figure 12.47 represents a component of a circuit. Some circuit options are considered.

Circuit Options

Table 12.7 shows some basic circuits. Simplifying by assuming equal recovery in each stage, the analytical solution for the circuit recovery can be found and is included in the Table. (To derive, a suitable starting point is required—e.g., in the R-S-C circuit letting X be feed to rougher, and in the R-C-CS circuit letting X be feed to the cleaner—and then proceeding as for Figure 12.47). Noble and Luttrell (2014) have considered over 30 circuits using this same basic approach.

The first two circuits in Table 12.7 represent the rougher–scavenger–cleaner circuit (R-S-C) and the rougher–cleaner–cleaner/scavenger circuit (R-C-CS). (In the first circuit it is traditional to refer to "scavenger" not rougher/scavenger.) The R-S-C circuit can be treated as the conventional arrangement providing as it does some buffering against the inevitable feed changes, enabling a

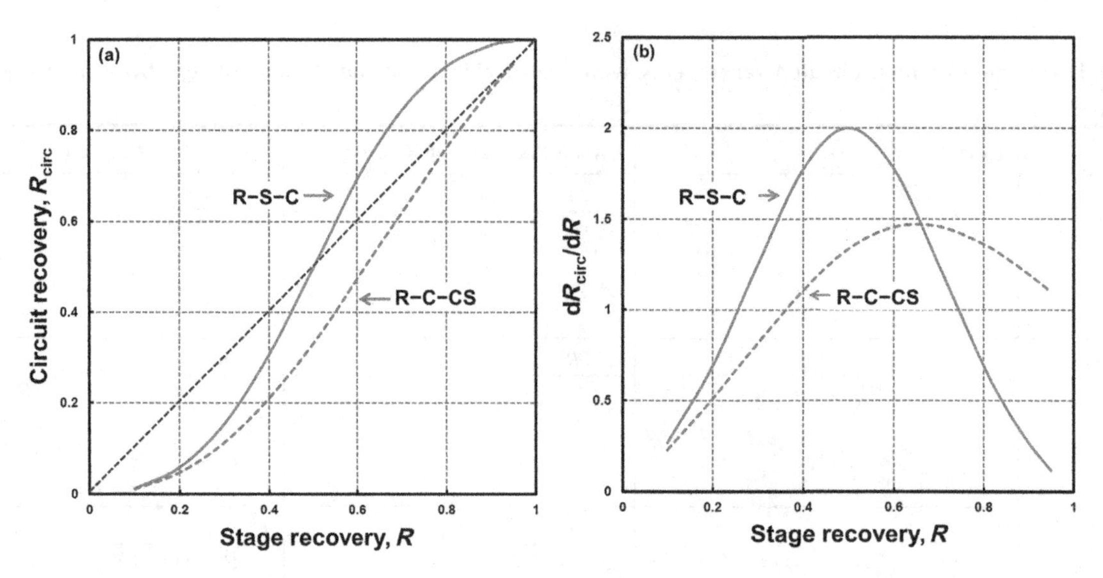

FIGURE 12.50 (a) Comparison of circuit recovery for R-S-C and R-C-CS circuits as a function of stage recovery (all stages with equal recovery, R) and compared to stage alone (the 45° dashed line), and (b) comparison of sharpness of separation (slope, $\delta R_{circ}/\delta R$) as a function of R.

level of manual supervision in preautomatic control days. The interaction between the stages, however, creates difficulties in control, as any action in one stage impacts the other stages. This made for a debate on advantages of open versus closed circuits (Lauder, 1992).

The R-C-CS circuit does partially open the circuit by isolating the rougher stage from the cleaner block (C-CS) with the CS tail being sent to final tail. The rougher function is now recovery (still at maximized SE) and the cleaner block function is grade. Independent control action appropriate to the two functions can then be implemented. In that regard the R-C-CS circuit is more "controllable" or "operable" than the R-S-C circuit. The stages may use different flotation machines, for example, flotation columns as the cleaner stage with mechanical cells in the cleaner/scavenger stage (Dobby, 2002).

It is common to regrind recycled material which will contain middling (composite) particles. Regrind is shown on the R-C-CS circuit (RG, which today would likely be a stirred mill) and could be included in the R-S-C circuit by combining the scavenger concentrate and cleaner tails. Regrind should always be considered. It may also permit a coarse primary grind if the mineral is readily floated, the lower mass of material making further grinding (regrinding) more energy efficient. Treatment of high tonnage porphyry copper ores often takes advantage of this approach.

Figure 12.50(a) compares R_{circ} for the R-S-C and R-C-CS circuits as a function of R, the stage recovery. Two features are: (1) the R_{circ} of the R-C-CS always lags that of the R-S-C and is less than the stage recovery (specifically the rougher stage) as there is no recycle to the rougher; and (2) the R-S-C circuit reduces recovery when $R < 0.5$ and enhances recovery when $R > 0.5$. This pivoting at $R = 0.5$ points to how the R-S-C circuit increases SE, increasing the recovery of target mineral A with $R_A > 0.5$ while decreasing the recovery of B with $R_B < 0.5$. Selectivity is related to the slope of the $R_{circ}-R$ relationship, that is, $(\delta R_{circ}/\delta R)$, and is included in Table 12.7 and plotted in Figure 12.50(b). Note that at high stage recovery ($R > 0.7$) the slope for R-C-CS circuit is greater than for R-S-C. This means that the selectivity between two minerals that have high and close recovery (i.e., low S) is greater in the R-C-CS circuit than in the R-S-C circuit. The R-C-CS circuit may not only be more "operable" but also offer an advantage in difficult selectivity cases, that is, ones with low relative rate constant, S. This general result is not altered by relaxing the assumption of equal stage recoveries, simply the magnitude of the SE changes (as illustrated in Figure 12.48).

This advantage of the R-C-CS circuit in difficult separation situations is more directly evident in Figure 12.51, which compares the SE of a single stage to that for the R-S-C and R-C-CS circuits under conditions of (a) "difficult" separation, $S = 2$, and (b) "easy" separation, $S = 10$. It is evident for the difficult case that at $R_A > 0.8$ the R-C-CS circuit gives higher SE_{circ} than the R-S-C circuit. Since high recovery for the target mineral, i.e., high R_A, is the target the R-C-CS circuit seems favored, the R-S-C circuit even failing to achieve the SE for a single stage at $R_A > 0.85$. For easy separations (b), the R-S-C circuit seems the better choice; this would include when separation is primarily against entrained mineral, as in the case of recovering talc, coal, or bitumen.

In the late 1980s Brunswick Mine implemented the R-C-CS type circuit, with significant impact on Zn metallurgy, increasing the selectivity of sphalerite over pyrite

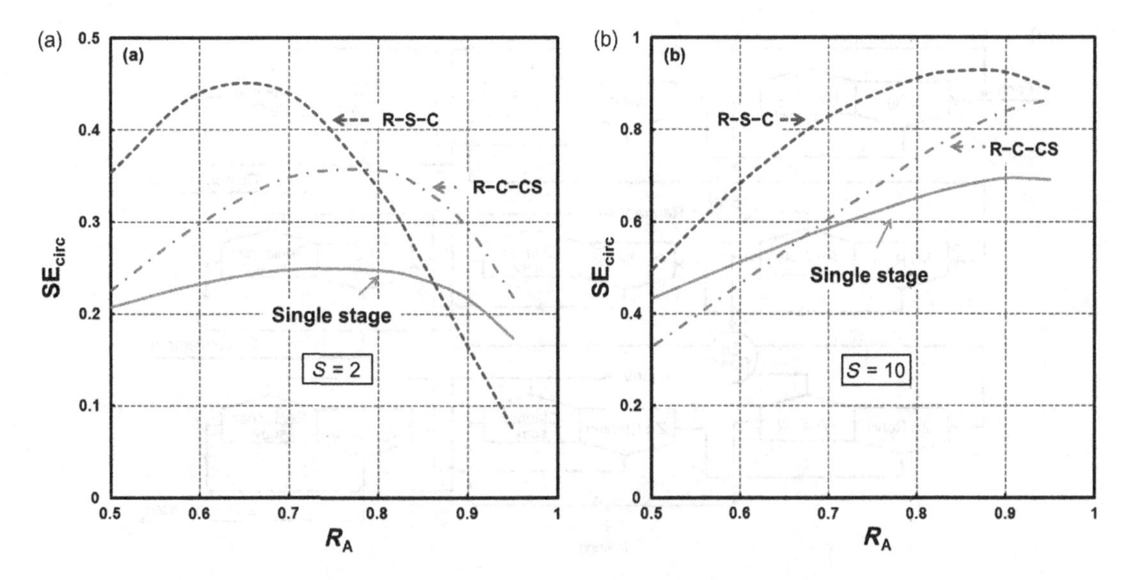

FIGURE 12.51 Circuit SE (SE_{circ}) as a function of stage recovery of A (R_A) comparing the R-S-C and R-C-CS circuits for (a) difficult separation ($S = 2$), and (b) easy separation ($S = 10$).

(Shannon et al., 1993). This separation would be classed as difficult with $S \sim 2-3$ (Cooper et al., 2004). Sending CS tail to final tailings removed material that actually decreased performance if recycled to the rougher. Including regrind (ball mill/hydrocyclone) on the feed to the cleaner block (as in Table 12.7) to increase liberation, and diluting the pulp to control entrainment of the additional fines produced (Section 12.10.2) added to the advantage of the R-C-CS circuit. The circuit modification, which was a major simplification over the prior flowsheet, included removing a thickener. Thickeners do offer surge capacity and densification of feed to subsequent stages (regrind or flotation) but in some sulfide flotation circuits they may not be desirable by giving opportunity for oxidation and introducing a time lag that makes control difficult (Bulatovic et al., 1998). The modifications at Brunswick Mine were the start of an impressive series of improvements documented by Orford et al. (2005). During the 1990s the Noranda group moved to adopt the R-C-CS circuit across operations. The R-C-CS circuit is also common in Cu–Mo plants using columns as cleaners and mechanical cells as cleaner/scavengers (Bulatovic et al., 1998).

Opening the cleaner block by sending CS tails to final tails is not a decision taken lightly. Recirculating streams can appear to contain substantial mineral values but upon opening the circuit the flow diminishes and the amount of mineral lost to final tails represented by the CS tails is minor (and, as noted, trying to recover the contained values may be counter-productive). If there is concern over potential loss represented by the CS tails the stream could be considered for alternative treatments, upgrading by pyro- or hydrometallurgical processes, such as the Galvanox™ process (Dixon et al., 2007).

The next two circuits in Table 12.7 are developments on the first two. The R-C_3-CS circuit introduces two additional cleaner stages to the R-C-CS network with cleaner 2 and cleaner 3 tails combined and recycled to the first cleaner via regrind. (Brunswick Mine had up to four cleaner stages in this arrangement.) The R-S-C_3 circuit shows the conventional countercurrent flow arrangement. In the R-S-C_3 circuit it is not easy to regrind all the recycled streams, which is a disadvantage. The analytical solution is also quite complex (to solve, incidentally, you have to start with feed to the last cleaner and work backwards through the circuit) compared to the solution for R-C_3-CS, suggesting a strong interaction between stages, which makes for more difficult control. The number of stages could get quite large: prior to replacing with flotation columns the Cu–Mo separation circuit at Mines Gaspe had up to 14 countercurrent cleaning stages of mechanical cells (Finch and Dobby, 1990). Incidentally, rather than being referred to as cleaner 1, 2, etc. the cleaner stages may have local names, such as recleaner.

The last circuit supplements the basic R-C-CS circuit by repeating it, let's call it $(\text{R-C-CS})_2$. This option allows for reagent additions into the second circuit aimed at the slow floating fractions. In that case the two concentrates will be combined. The same arrangement also serves when two minerals are to be recovered, for minerals A and B making the circuits $(\text{R-C-CS})_A$ and $(\text{R-C-CS})_B$ with separate concentrates.

Plants which concentrate two or more minerals may repeat circuits for each mineral or produce a *bulk concentrate* (e.g., a concentrate of at least two valuable minerals) then perform selective flotation on the bulk concentrate. Options are illustrated in Figure 12.52 for a Cu–Pb–Zn ore ranging from (a) "easy" milling ore

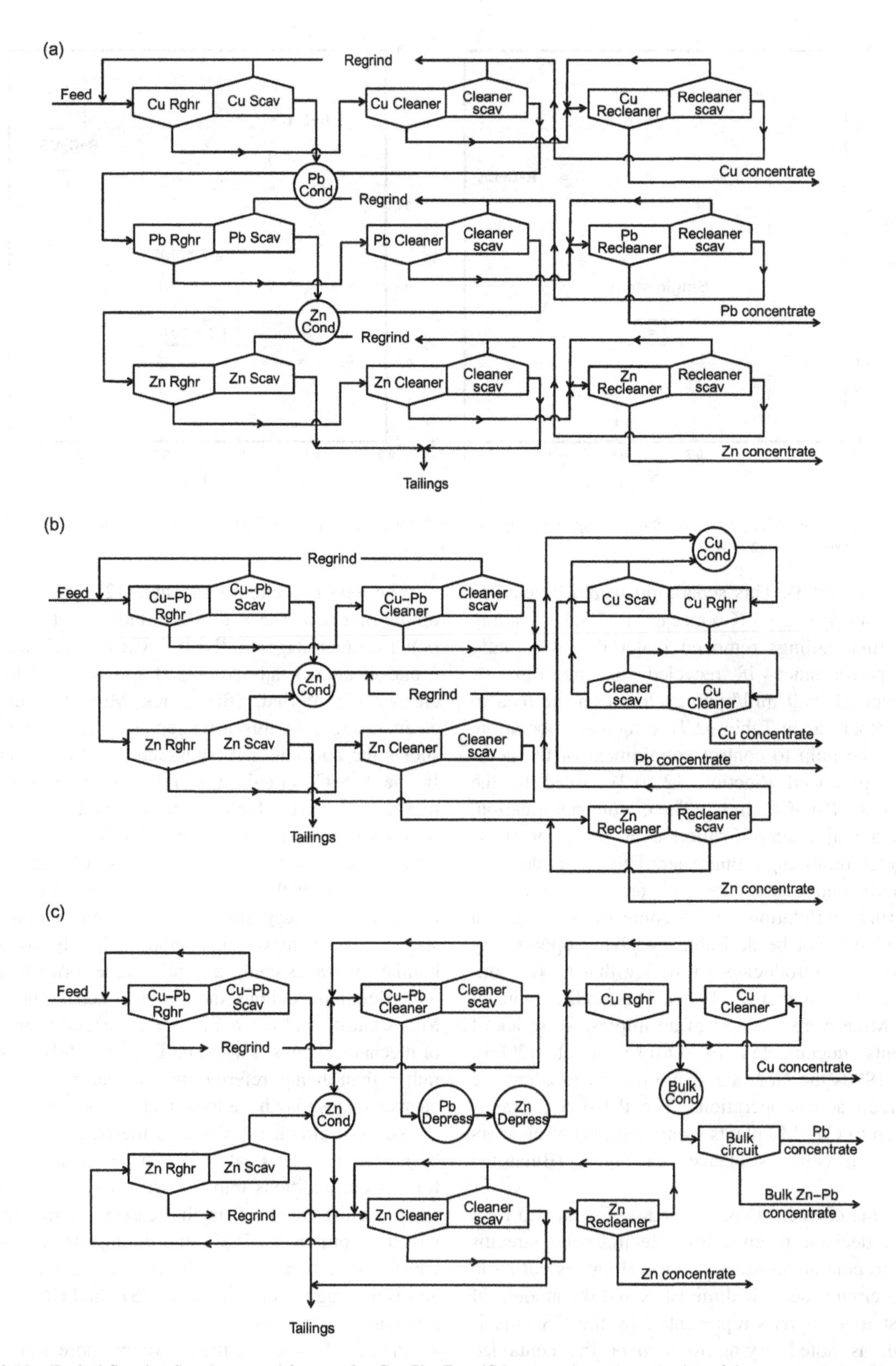

FIGURE 12.52 Typical flotation flowsheets used for complex Cu–Pb–Zn sulfide processing: (a) production of three concentrates, (b) production of bulk Cu–Pb concentrate subsequently separated to give the three concentrates, and (c) as (b) but production of a fourth concentrate, bulk Pb–Zn (Note: the circuits are a combination of R-S-C and R-C-CS types) *(Adapted from Barbery (1986)).*

where separate concentrates can be produced, to (b) more difficult ore where a bulk Cu–Pb concentrate is produced and then separated into Cu and Pb concentrates, to (c) where a bulk Zn–Pb concentrate is an additional final concentrate.

Circuit design is a combination of experience and increasingly the use of simulators, at least for the non-chemistry aspects (Herbst and Harris, 2007; Mendez et al., 2009; Ghobadi et al., 2011). While search algorithms may find optimal circuit configurations based on metallurgical performance criteria, in the end the choice has to be a circuit that is readily operated.

Circuit Flexibility

The decision having been reached to design a flotation circuit according to a certain scheme, it is necessary to provide for fluctuations in the flowrate and grade of ore to the plant.

The simplest way of smoothing out grade fluctuations and of providing a smooth flow to the flotation plant is by interposing a large agitated storage tank (agitator) between the grinding section and the flotation plant:

$$\text{Grind} \rightarrow \text{Storage Agitator} \rightarrow \text{Flotation Plant}$$

Any minor variations in grade and tonnage are smoothed out by the agitator, from which material is pumped at a controlled rate to the flotation plant. The agitator can also be used as a conditioning tank, reagents being fed directly into it. It is essential to precondition the pulp sufficiently with the reagents (including sometimes air, Section 12.8) before feeding to the flotation banks, otherwise the first few cells in the bank act as an extension of the conditioning system, and poor recoveries result.

Provision must be made to accommodate any major changes in flowrate that may occur; for example, grinding mills may have to be shut down for maintenance. This is achieved by splitting the feed into parallel banks of cells (Figure 12.53). Major reductions in flowrate below the design target can then be accommodated by shutting off the feed to the required number of banks. The optimum number of banks required will depend on the ease of control of the particular circuit. More flexibility is built into the circuit by increasing the number of banks, but the problems of controlling large numbers of banks must be taken into account. The move to very large unit processes, such as grinding mills, flotation machines, etc., in order to reduce costs and facilitate automatic control, has reduced the need for many parallel banks.

In designing each flotation bank, the number of cells required must be assessed: should a few large cells be incorporated or many small cells giving the same total capacity?

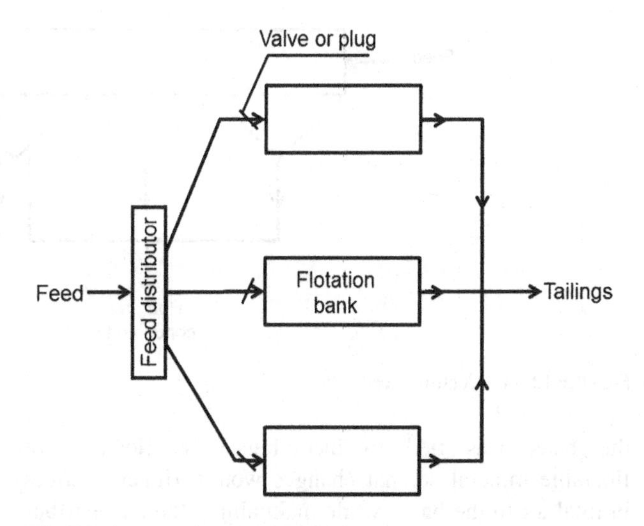

FIGURE 12.53 Parallel flotation banks.

Some theoretical considerations have been introduced (Section 12.11.2), but there is a practical aspect as well: if a small cell in a bank containing many such cells has to be shut down, then its effect on production and efficiency is not as large as that of shutting down a large cell in a bank consisting of only a few such cells.

Flexibility can include having "extra" cells in a bank. It is often suggested that the last cell in the bank normally should not be producing much overflow, thus representing reserve capacity for any increase in flowrate or grade of bank feed. This reserve capacity would have to be factored in when selecting the length of the bank (number of cells) and how to operate it, for example, trying to take advantage of recovery or mass pull profiling. If the ore grade decreases, it may be necessary to reduce the number of cells producing rougher concentrate, in order to feed the cleaners with the required grade of material. A method of adjusting the "cell split" on a bank is shown in Figure 12.54. If the bank shown has, say, 20 cells (an old-style plant), each successive four cells feeding a common launder, then by plugging outlet B, 12 cells produce rougher concentrate, the remainder producing scavenger concentrate (assuming a R-S-C type circuit). Similarly, by plugging outlet A, only eight cells produce rougher concentrate, and by leaving both outlets free, a 10–10 cell split is produced. This approach is less attractive on the shorter modern banks. Older plants may also employ double launders, and by use of *froth diverter trays* cells can send concentrate to either launder, and hence direct concentrate to different parts of the flowsheet. An example is at the North Broken Hill concentrator (Watters and Sandy, 1983).

Rather than changing the number of cells, it may be possible to adjust air (or level) to compensate for changes in mass flowrate of floatable mineral to the bank. To maintain the bank profile at Brunswick Mine, total air to

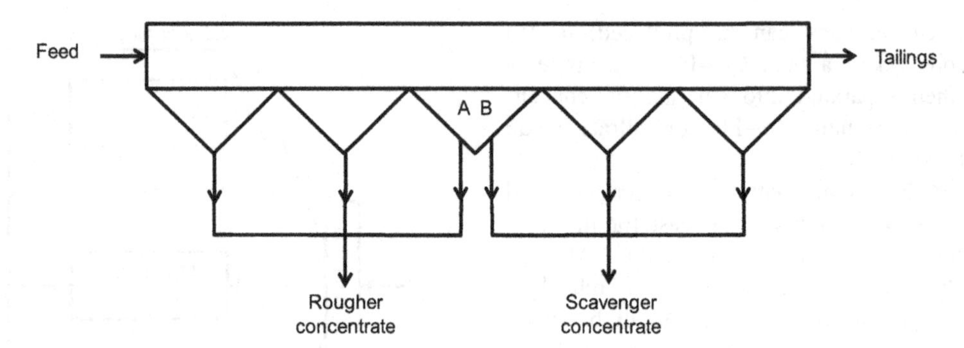

FIGURE 12.54 Control of cell split.

the bank was tied to incoming mass flowrate of floatable mineral so that changes would trigger changes in total air to the bank, while maintaining the air distribution profile (Cooper et al., 2004).

12.12 FLOTATION TESTING

To develop a flotation circuit for a specific ore, preliminary laboratory testwork is undertaken to determine the choice of reagents and the size of plant for a given throughput as well as the flowsheet and peripheral data. Flotation testing is also carried out on ores and stream samples in existing plants to improve procedures and for development of new reagents.

Increasing attention is now paid to the need to understand variation in an orebody, and a procedure for the identification, sampling, and characterization of *geometallurgical (geomet) units* (or "domains") has been successfully formulated and is now in common use (see Chapter 17).

Geometallurgical units (Lotter et al., 2003; Fragomeni et al, 2005) can be defined as an ore type or group of ore types that possess a unique set of textural and compositional properties from which it can be predicted they will have similar metallurgical performance. Sampling of an orebody based on geometallurgical units will define metallurgical variability and allow process engineers to design more robust flowsheet options. This variability can be muted when samples from different geometallurgical units are blended and tested as one sample. Rather, composite samples are created by ensuring grade and grade distributions from a specific area defining the geometallurgical unit within a resource are maintained. The method used to divide an orebody into geometallurgical units is based on a review of geological data including host rock, alteration, grain sizes, texture, structural geology, grade, mineralogy, and metal ratios with focus on characteristics that are known to affect metallurgical performance (Lotter et al., 2011). (The foregoing list is not complete and also includes hardness testing and the grade/recovery curve as characterizing parameters (e.g., Fragomeni et al., 2005).) Statistical analysis is often used to help define preliminary units. In addition, it is recommended that a variability program based on smaller samples taken from within and throughout a geometallurgical unit is completed prior to finalizing the divisions between geometallurgical units. This approach has a higher probability of capturing the full variance in composition and metallurgical behavior that can be expected from within a unit, and provides a cross check that the geometallurgical unit definition is robust.

Once a set of geomet units has been identified, sampled, and tested, the investigator is in a position to prepare a known composite using increments of each geomet unit to represent an overall plant mill feed mixture. At this point, the mixture may demonstrate interactions that either improve or deteriorate the metallurgical performance, so this last step is essential.

The purpose of sampling is to obtain an unbiased set of increments from the lot being sampled in such a manner that, when the increments are combined as a sample, the sample has the same composition as the lot that was sampled; the only difference is that the mass of the sample is less than the mass of the lot (Chapter 3). It is a definite recommendation that a proven, quantitative sampling protocol be used in the selection of increments to the sample to be used for flotation testing. In the procedure "High Confidence Flotation Testing" (Lotter, 1995a), use is made of the 50-piece experiment by Gy (1979) to obtain the sampling equation and associated minimum sample mass for a plant mill feed. This procedure is detailed elsewhere (Bartlett and Hawkins, 1987; Lotter, 1995b; Lotter et al., 2013); however, it is sufficient to remark that this approach plus use of Gy's safety line to prepare the replicate batch test charges of ore negates any argument that the material brought to the test bench in the laboratory is not representative (Lotter and Oliveira, 2013; Lotter, 1995c; Lotter and Fragomeni, 2010). An alternative sampling method validated by Lotter and Oliveira (2011) uses drill core and drill core data to sample the population physically and mathematically so as to match the parent valuable metal distribution, and tests for agreement between the parent and sample distributions using the χ^2 test (Lotter and Oliveira, 2013).

Having selected representative samples of the ore, it is necessary to prepare them for flotation testing, which involves comminution to the target particle size. Crushing must be carried out with care to avoid accidental contamination of the sample by grease or oil, or with other materials that have been previously crushed. (Even in a commercial plant, a small amount of grease or oil can temporarily upset the flotation circuit.) Samples are usually crushed with small jaw crushers or cone crushers to about 0.5 cm and then to about 1 mm with crushing rolls closed with a screen.

Storage of the crushed sample is important, since oxidation of the surfaces is to be avoided, especially with sulfide ores. Not only does oxidation inhibit collector adsorption, but it also facilitates the dissolution of heavy metal ions, which may interfere with the flotation process (Section 12.7). Sulfides should be tested as soon as possible after obtaining the sample and ore samples must be shipped in sealed drums in as coarse a state as possible. Samples should be crushed as needed during the testwork, although a better solution is to crush all the samples and to store them in an inert atmosphere.

Wet grinding of the samples should always be undertaken immediately prior to flotation testing to avoid oxidation of the liberated mineral surfaces. Batch laboratory grinding, using ball mills, produces a flotation feed with a wider size distribution than that obtained in continuous closed-circuit grinding; to minimize this, batch rod mills are often used which give products having a size distribution which approximates that obtained in closed-circuit ball mills. True simulation is never really achieved, however, as high-specific-gravity minerals in a plant grinding circuit are ground finer than the average due to recycling by the cyclone, which is a feature missed in a batch mill. It is also important to understand the effect of grinding media on flotation, making it sometimes difficult in the lab to simulate the plant chemical conditions (Section 12.8).

Predictions from laboratory tests can be improved if the mineral recovery from the batch tests is expressed as a function of mineral size rather than overall product size. The optimum mineral size can be determined and the overall size estimated to give the optimum grind size (Finch et al., 1979). This method assumes that the same fineness of the valuable mineral will give the same flotation results both from closed-circuit and batch grinding, irrespective of the differences in size distributions of the other minerals.

The potential for liberation of the minerals contained in the ore can be determined by characterizing the grain sizes of the minerals present (Chapter 17). This can be achieved by breaking the drill core samples at a relatively coarse size (typically about 600 μm) to preserve the *in situ* texture of the samples, including grain size, association, and shape. The texture can be characterized by using a scanning electron microscope configured as a mineral liberation analyzer, such as the MLA or the QEMSCAN (Chapter 17). Such an analyzer can measure the grain sizes and composition of the component minerals of the ore.

Testwork should then be carried out over a range of grinding sizes in conjunction with flotation tests in order to determine the optimum flotation feed size distribution. If the mineral is readily floatable a coarse grind may be utilized, and the subsequent concentrate regrinding to further liberate mineral before further flotation will mean a reduction in grinding costs per ton treated.

12.12.1 Small-Scale Tests

A measurement widely used in basic studies is the contact angle (Figure 12.2). The measurement can be made by attachment of a bubble on a polished surface of a mineral held in the solution (of collector, etc.) under test ("captive bubble" test). The contact angle on particles can be inferred from measuring pressure drop as a test solution is passed through a bed of particles. Most surface chemistry texts will describe the procedures. Some quasi-first principles flotation simulators use measured values of contact angle to predict grade/recovery. A related test is "bubble pickup" where a bubble is exposed to particles in the test solution and the mass collected or angle subtended on the surface by the attached particles (the two are related) is measured (Chu et al., 2014).

There are a variety of devices designed to float about 1 g of mineral particles (e.g., Partridge and Smith, 1971; Bradshaw and O'Connor, 1996). In the *Hallimond tube* technique (Figure 12.55), the mineral particles (powder) are supported on sintered glass inside the tube containing the test solution. Air (or any other gas) bubbles are introduced through the sinter and any hydrophobic mineral particles are lifted by the bubbles, which burst at the water surface, allowing the particles to fall into the

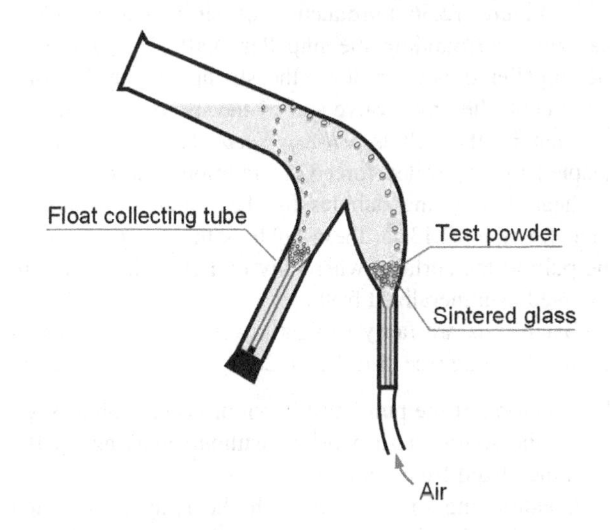

FIGURE 12.55 Hallimond tube.

collecting tube. By treating a weighed sample of mineral, the weight collected in the tube can be related to the floatability. The technique can be adapted to simple mineral mixtures for some assessment of selectivity. The Hallimond tube has the advantage that assaying is largely avoided and the procedure is straightforward (sample preparation notwithstanding), however it is doubtful whether the method, especially as conducted on single minerals, simulates industrial flotation.

On a somewhat larger scale, the EMDEE Microflot Agitator uses up to 10 g samples (Chudacek, 1991). The device is a tube half filled with test solution and subjected to controlled agitation, which produces bubbles and partitions the particles between the pulp and the resulting froth. Tested on size fractions, the partitioning results were shown to correlate well with the rate constant determined by batch flotation tests. Bradshaw and Vos (2013) have refined the technique, retaining the testing of sized fractions to develop a *Mineral Separability Indicator* (JKMSI). By characterizing the size-by-size behavior it is possible to take advantage of the consistency of the recovery–size relationship (Section 12.10.1) in predicting full-scale results. A small scale test that simulates industrial flotation has obvious merit when limited samples are available.

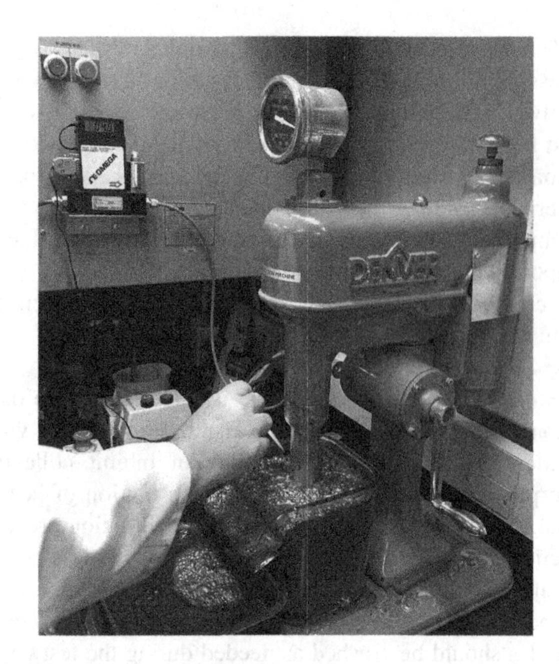

FIGURE 12.56 Laboratory flotation cell.

12.12.2 Batch Flotation Tests

The bulk of laboratory testwork is carried out in batch flotation cells usually with 500 g, 1 kg, or 2 kg samples. The cells are usually mechanically agitated, the speed of rotation of the impellers being variable. A standard design is shown in Figure 12.56; variations aiming to improve reproducibility include bottom agitated cells and automated froth removal. Bottom-driven cells provide an unhindered surface which facilitates insertion of probes (pH, etc.).

In Figure 12.56 introduction of air is via a hollow standpipe surrounding the impeller shaft. The action of the impeller draws air down the standpipe, the flowrate being controlled by a valve and by the speed of the impeller; that is, the cell is *self-aspirated*. The setup can be adapted for controlled forced air addition. The air stream is sheared into fine bubbles by the impeller (discussed later, Section 12.13.1), these bubbles then rising through the pulp to the surface, where any particles picked up are removed as mineralized froth.

Batch tests are fairly straightforward, but a few points are worth noting (see also Wood (2002) and Runge (2010)):

1. Agitation of the pulp must be vigorous enough to keep all the solids in suspension, without breaking up the mineralized froth column.
2. Conditioning of the pulp with the reagents is often required. This is a period of agitation, varying from a few seconds to several minutes, before the air is turned on, which allows the surface of the mineral particles to react with the reagents.
3. Very small quantities of frother can have marked effects, and *stage additions* are often needed to control the volume of froth. The froth depth should be between 2 and 5 cm, as very shallow froths risk losing pulp into the concentrate container. Reduction in air rate is sometimes used to limit the amount of froth produced. This should be standardized for comparative tests to prevent the introduction of another variable.
4. Test should be at low pulp density (e.g., 10% solids) to avoid overloading bubbles (one reason being this invalidates conditions for first order kinetics). (As a matter of economics, industrial separations, at least at the roughing stage, are carried out in as dense a pulp as possible consistent with good selectivity and operating conditions. The denser the pulp, the less cell volume is required, and also less reagent is required, since the effectiveness of most reagents is a function of their concentration in solution. Most commercial rougher floats are in pulps of 25–40% solids by weight, although they can be as low as 8% and as high as 55%. In cleaner stages pulps may be diluted to take advantage of increased selectivity against entrained material (Section 12.10.2).) It must be borne in mind that in batch flotation tests the pulp density varies continuously, from beginning to end, as solids are removed with the froth and water is added to maintain the cell pulp level. This continuous variation changes the concentration of reagents as well as the character of the froth.

5. As water contains dissolved chemicals that may affect flotation, water from the supply which will be used commercially should be employed, rather than, say, distilled water. Any water added should contain some frother to maintain the bubble size in the pulp.
6. Normally only very small quantities of reagent are required for batch tests. In order to give accurate control of their addition rates, they may have to be diluted. Water-soluble reagents can be added as aqueous solutions by pipette, insoluble liquid reagents by graduated dropper or hypodermic needle. Solids may either be emulsified or dissolved in organic solvents, providing the latter do not affect flotation.
7. Recovery of froth is sensitive to operator technique, which is why automated froth removal methods are recommended.
8. Batch tests tend to produce significant entrainment recovery and allowance for this in determining kinetic parameters may be warranted (e.g., Ross, 1990).
9. Most commercial flotation operations include at least one cleaning stage, in which the froth product is refloated to increase its grade, the cleaner tailings often being recycled. Since cleaner tails are not recycled in batch tests, they do not always closely simulate commercial plants. If cleaning is critical, *cycle tests* (*lock* or *locked cycle* tests) may have to be undertaken.

Locked Cycle Tests

These are multiple-step flotation tests designed to measure the effect of circulating materials to provide a better insight into the expected performance from multistage processing of an ore. To incorporate the effect of recycle streams, the tests are performed multiple times with the recycled stream from a previous interaction added in the subsequent test. Normally at least six cycles are required before the circuit reaches equilibrium and a complete material balance should be made on each cycle. Since the reagents are in solution, it is essential that liquids as well as solids recirculate, so any liquid used to adjust pulp density must be circuit liquid obtained from decantation or filtration steps. Cycle tests are very laborious to carry out, and often the test fails to reach steady state. Procedures have been developed to simulate the locked cycle test steady state result from individual batch flotation test data (Agar and Kipkie, 1978; Runge, 2010).

Two other test procedures which use multiple flotation steps are *release analysis* (Dell, 1964) and *tree flotation* (Pratten et al., 1989). Devised to determine coal washability (i.e., the liberation-limited separation of coal from ash), they serve to show how close the actual operation is to the best possible performance. In establishing the effectiveness of new flotation machines, one way is to show the separation approximates the release or tree flotation result (e.g., Adel et al., 1991).

High-Confidence Flotation Testing

This method for has become a standard in some flotation laboratories (Lotter, 1995a,b; Lotter and Fragomeni, 2010). The approach is based on the following criteria:

1. The sample of ore presented to the flotation testing laboratory must be representative. This is achieved by use of Gy's minimum sample mass and safety line models, combined with appropriate blending and subsampling protocols (Chapter 3).
2. Provided that the mass balance is maintained within certain limits across the flotation test, the *built-up* (or back-calculated) head grade should be in agreement with the sample mean grade of the assay heads within an acceptable error.
3. The use of replicate flotation tests assembled into composites, with the use of appropriate quality controls to reject outliers, instead of relying on single flotation tests for data, has a significant influence in reducing and normalizing the random errors.

The layout of the method is shown in Figure 12.57. Apart from its focus on representative sampling to produce reliable test material, which is shown as Section 1 in Figure 12.57, it has a structure of replicate flotation tests with associated quality controls that identify outlier tests and reject these at the 95% confidence level, as shown in Section 2 of the figure. The number of replicate tests required is dependent on the sample standard deviation that is typical of the ore type in question, and of the format of the flotation test. If the ore is unknown, it is best to perform a set of baseline rougher float tests and empirically obtain an estimate of the sample standard deviation.

The acceptable error in the table of 3.3% is based on experience with S. African Pt ores; the acceptable error needs to be established for the ore in question, using techniques introduced in Chapter 3.

Additionally, investigators should have a clear idea of the differences in grade and recovery that they may want to demonstrate (Napier-Munn, 2012). For base metal ores, triplicates are often suitable. However for precious metal bearing ores, or in cases where the ore is polymetallic bearing both base metal and precious metals, the number of replicates may be as high as five. In either case, the sample standard deviation and relative standard deviation (RSD) are calculated from these replicate tests for the first flotation concentrate mass. If this exceeds an RSD of 5%, the quality control triggers a check for outliers using the small data set model after Grubbs (1969). Outliers are then rejected, leaving the original set of replicate tests short.

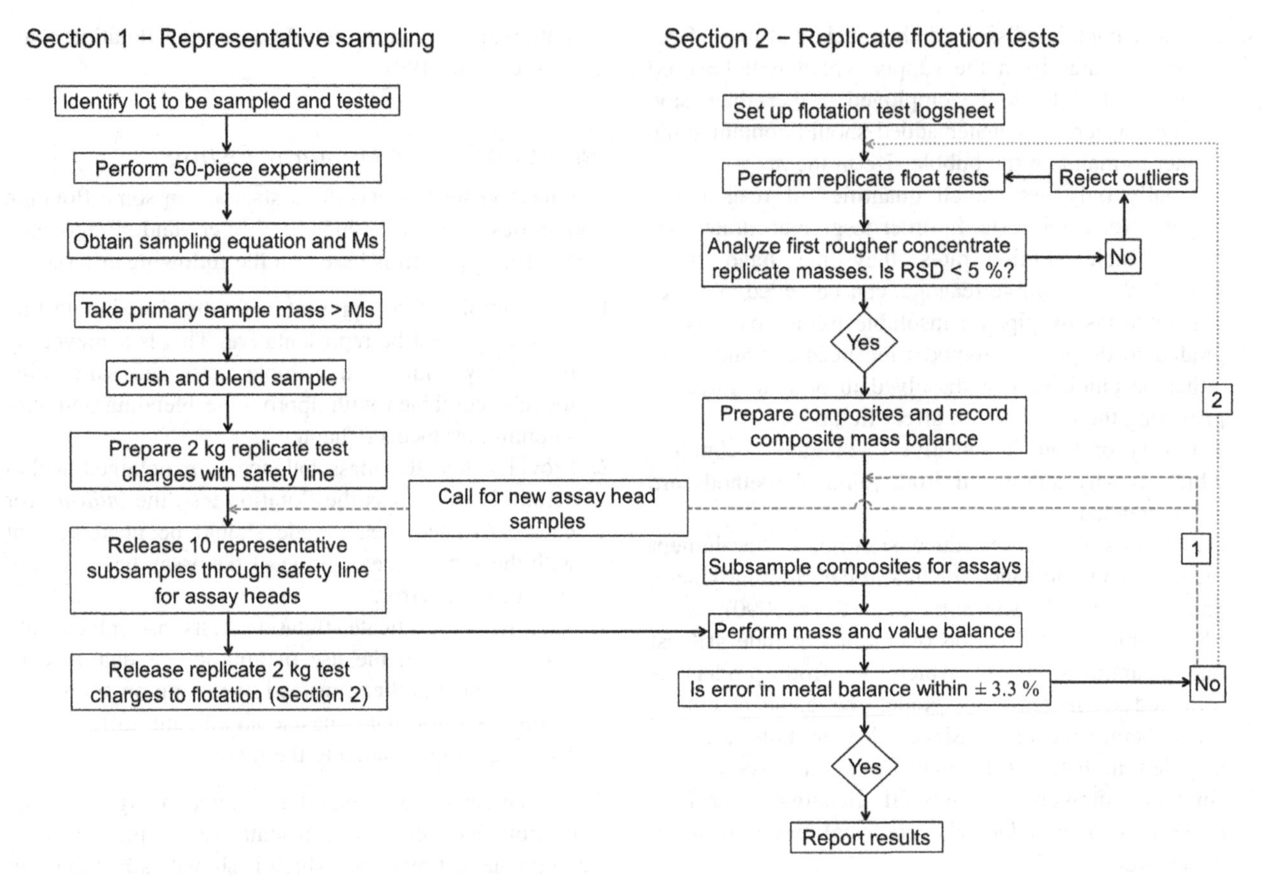

FIGURE 12.57 Schematic of activities in High-Confidence Flotation Testing.

Repeat flotation tests are then performed to replace the rejected outliers. Thereafter, acceptable replicates are taken into a set of composite concentrates and tailings for assaying and mass balancing. This amounts to the equivalent use of averages instead of singleton observations, a tenet of the Central Limit Theorem in statistics, which reduces overall random error and leaves any residual error distributed almost in the pattern of a normal distribution (Box et al., 2005). Use of this method puts the investigator in a position to compare smaller differences in grade and recovery between tests using different conditions, for example, reagent suite A versus B.

In Figure 12.57, the trial metal balance produces an independent estimate of the ore grade (the built-up head) from the weights and metal grades of the flotation test products. This built-up head value is compared to the sample mean assay head produced earlier in Section 1 of the model. The percent difference, using the assay head as the 100% basis, may trigger some repeat investigation of assays because of excessive error in the trial metal balance (i.e., in excess of ±3.3% in the case of Figure 12.57), calling for new subsamples of the assay heads and flotation test products (legend "1"). If this does not resolve the metal balance error, final resort activities indicated by legend "2" are triggered, calling for a new set of flotation tests.

The largest risk in any greenfield project is the scale-up from laboratory to operations at commissioning. The High Confidence Flotation Testing methodology has demonstrated accurate scale-up properties in the development and prediction of a flowsheet for a new orebody (Lotter and Fragomeni, 2010). For example, in the flowsheet development testwork for the Ivanhoe Kamoa copper project, which may be used as a relevant case study of this practice, the error in the copper metal balances averaged −0.3% with 80% of the individual errors falling between ±2% (Lotter et al., 2013).

Presenting Data

In addition to the common methods of presenting separation data, which include grade–recovery, recovery–mass recovery, and recovery–recovery plots (Runge, 2010), in flotation testing for scale-up time is also used a variable. Appendix VI gives a set of data on flotation of a Pb-ore collected as a function of time. The data can serve to illustrate all the data presentation methods, but here just

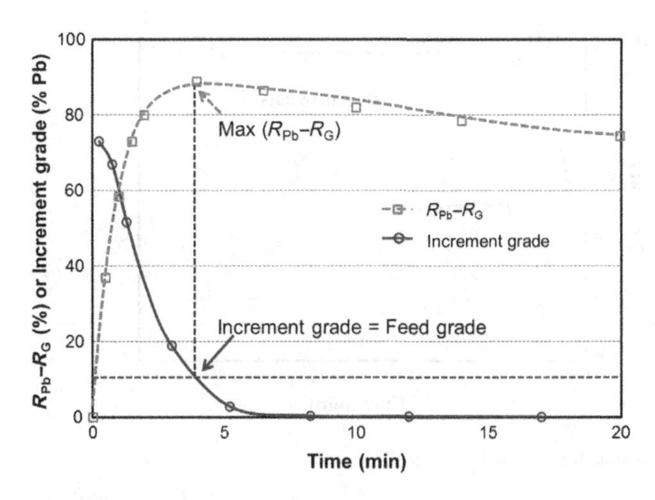

FIGURE 12.58 SE and increment grade as a function of flotation time, illustrating location (time) of maximum SE.

time-based plots are presented, first to illustrate locating the maximum SE, following Agar et al. (1980) as introduced previously (Section 12.11.3), and second to estimate the flotation rate constant.

1. Maximum SE

   This is located in two ways: at the time when the difference in recovery, in this case of galena versus the rest (lumped as one gangue G), is maximum; and at the time when the increment of grade is equal to the feed grade.

   The computations (Appendix VI) involve determining cumulative recovery of Pb (galena) and gangue G. To determine the latter, the Pb assay has to be converted to galena mineral assay (see Example 1.1) with G determined by difference, 100 − %galena. The resulting plot is shown in Figure 12.58. It is evident that the time when maximum SE occurs does correspond to the time when the increment in grade equals the feed grade. (Note that the increment of grade is plotted against the average of the time interval.) Translating to the case of a bank, it led to the notion that the optimum length of the bank would be that cell producing the same grade of concentrate as the feed (Section 12.11.3).

2. Determining flotation rate constant

   In batch flotation the assumption of first order kinetics gives the following result (model) from Section 12.9.1:

$$R = 1 - \exp(-kt) \qquad (12.26)$$

It is important that the conditions do not change while performing the test, and this includes physical conditions such as air rate as well as chemical conditions. Figure 12.59(a) illustrates the "disappearance" plot (ln $(1 - R)$ vs. t), with slope k. Taking the data up to about 80% recovery gives $k = 0.88\ \text{min}^{-1}$, which Figure 12.59(b) shows gives a reasonable fit to the recovery−time data.

Extending the disappearance plot to all the data, it is evident that a single k does not cover the full range (the reader should test this). To improve the fit has resulted in various modifications to the simple form of Eq. (12.26). Two modifications are: (1) introducing a maximum recovery less than 100% by rewriting as:

$$R = R_{\infty}[1 - \exp(-kt)] \qquad (12.51)$$

where R_{∞} refers to recovery after infinite flotation time; and (2) introducing the notion of fast and slow floating components:

$$R = \emptyset_f(1 - \exp(-k_f t) + (1 - \emptyset_f)(1 - \exp(-k_s t) \qquad (12.52)$$

where $\emptyset_f$ refers to the fraction of fast floating mineral (the fraction of slow floating mineral is then $(1 - \emptyset_f)$), and subscripts f and s refer to fast and slow. This could be expanded to consider a nonfloating component as well. These refinements are referred to as a *floatability component model* (FCM).

There are any number of curve-fitting routines to estimate the parameters in these models, including Excel Solver. For example, the best-fit parameters for model Eq. (12.51) are $R_{\infty} = 97.62\%$ and $k = 0.95\ \text{min}^{-1}$. Plotting, however, remains a useful visual check.

What is being recognized in these modifications is that the rate constant has a distribution of values, which has led to attempts to describe this distribution with mathematical functions (Weiss, 1985). Klimpel (1980) assumed a rectangular distribution, which leads to the following solution:

$$R = R_{\infty}\left[1 - \frac{1}{Kt}(1 - \exp(-Kt))\right] \qquad (12.53)$$

where K is the maximum rate constant in the distribution. Klimpel made use of this model in plant evaluations by determining the effect of variables, for example, collector, on R_{∞} and K noting sometimes K would increase but R decrease and vice versa, the so-called "R−K trade-off."

The distribution of k values derives from the effect of particle size and liberation illustrated in Figure 12.38; that is, it is "property based." It would seem best to determine the rate constant on a size-liberation basis. This represents a large experimental effort for one condition and may become prohibitive if dependence on other conditions, for example, process chemistry, is to be considered. Welsby et al. (2010) show these property-based parameters can be related to floatability components—fast, slow, and nonfloating—and thus the simpler FCM can be used. Using an FCM, it should be possible to reassign floatability components following a change in feed stream properties (such as particle size),

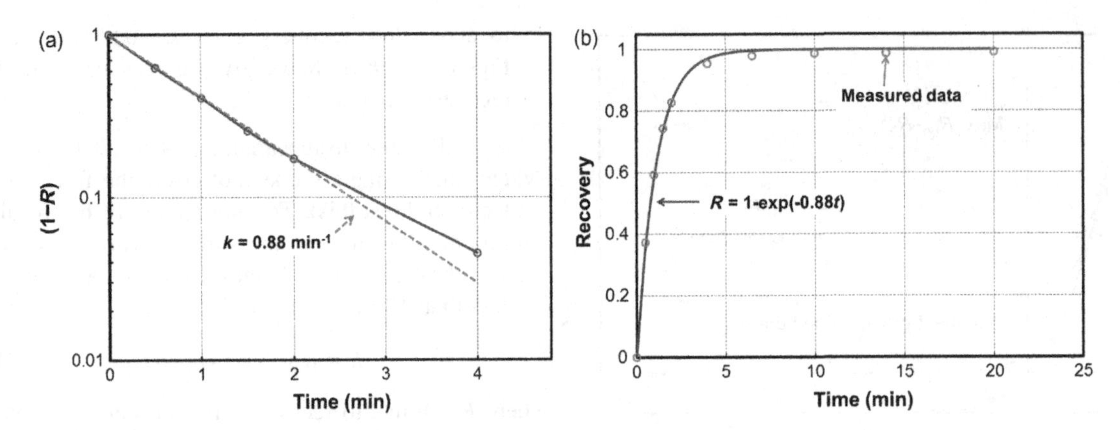

FIGURE 12.59 (a) Disappearance plot to estimate k, and (b) fit to recovery–time for data in Appendix VI.

instead of rederiving them, a large reduction in experimental effort. It must be appreciated that data fitting exercises are accompanied by a degree of uncertainty in the derived parameters that can be high (Sandoval-Zambrano and Montes-Atenas, 2012). Simulations provide trends in metallurgy rather than precise predictions.

Cell Sizing (Example 12.2)

With the target retention time established, calculation of cell volume is relatively straightforward. Based on Wood (2002), the computation is as follows:

$$N\,V_{\text{eff}} = \frac{F}{60}T\,E\,P \qquad (12.54)$$

where,

N = number of cells in bank,
V_{eff} = effective cell volume, m^3 (to allow for impeller, internal launders, etc.). If not known assume $V_{\text{eff}} = V_{\text{total}}$,
F = dry solids feed rate, metric tons per hour ($t\,h^{-1}$),
T = plant retention time (= lab test time × scale-up factor),
E = pulp expansion factor to allow for GH (% air in the pulp):

$$= \frac{100}{100 - \%\,\text{GH}} = \frac{1}{0.85}\ \text{(default)}$$

P = pulp volume per ton solids ($m^3\,t^{-1}$):

$$= \frac{1}{\text{solids density}} + \frac{100}{\text{pulp \%solids}} - 1$$

(with solids density in $t\,m^{-3}$ which is numerically equal to specific gravity).

Factors that may also need to be considered are carrying capacity, both on an area ($t\,h^{-1}\,m^{-2}$) and lip length ($t\,h^{-1}\,m^{-1}$) basis, and the gas dispersion properties lab versus plant. In the lab the 5 min flotation time represents a total amount of BSAF delivered, TBSAF (i.e., TBSAF = S_b × time × lab cell area). Although not conventionally considered, it would seem that the bank of cells should deliver at least the same TBSAF per unit mass of feed. (This does not imply that each cell delivers the same BSAF, following the discussion on bank profiling, Section 12.11.2.) In scaling up flotation columns from lab/pilot data it was possible to take gas dispersion similarity into account (Finch and Dobby, 1990).

> **Example 12.2**
> Estimate the size of cell for the following conditions related to the data for the Pb-ore (Appendix VI): feed rate 24,000 $t\,d^{-1}$ at 35% solids with ore s.g. 3.
>
> **Solution**
> F = 1,000 $t\,h^{-1}$ (assuming 100% availability, otherwise multiple by 100/% availability)
> T = 12.5 min (5 min from Figure 12.59(b), giving ~ 95% recovery) × 2.5 (scale-up factor)
> E = 1/85 (default; i.e., 15% GH assumed)
> $P = \frac{1}{3} + \frac{100}{35} - 1 = 2.19$
>
> From Eq. (12.54) $NV_{\text{eff}} = 537\ m^3$. Taking seven cells as the bank length, then $V_{\text{eff}} = 77\ m^3$ and $V_{\text{total}} = 80\ m^3$ (rounding). This cell size may be available (check with cell suppliers) otherwise another cell size is selected and N recalculated. The objective is to use the least number of cells that meet the minimum bank length requirements (Section 12.11.2).

12.12.3 Pilot Plant Testwork

Laboratory flotation tests provide the basis of design of the commercial plant. Prior to development of the plant, pilot scale testing is often carried out in order to:

1. Provide continuous operating data for design. Laboratory tests do not closely simulate commercial plants, as they are batch processes.

2. Prepare large samples of concentrate for survey by smelters, etc., in order to assess the possibility of penalties or bonuses for trace impurities.
3. Compare costs with alternative process methods.
4. Compare equipment performance.
5. Demonstrate the feasibility of the process to nontechnical investors.

Laboratory and pilot scale data should provide the optimum conditions for concentrating the ore and the effect of change of process variables. The most important data provided by testwork include:

1. The optimum grind size of the ore. This is the particle size at which the most economic recovery can be obtained. This depends not only on the grindability of the ore but also on its floatability. Some readily floatable minerals can be floated at well above the liberating size of the mineral particles, the only upper limit to size being that at which the bubbles can no longer physically lift the particles to the surface. The upper size limit is normally around 300 μm, although recent machine developments are increasing this size, for example, the HydroFloat™ (Section 12.13.4). For some ores, in particular non-sulfide ores, there is a lower limit for flotation, around 5 μm, requiring a desliming step. Like the coarse particle limit, process improvements continue to lower this value.
2. Quantity of reagents required and location of addition points.
3. Pulp density; important in determining size and number of flotation cells.
4. Flotation time; experimental data gives the time necessary to make the separation into concentrate and tailings. This depends on the particle size and the reagents used and is needed to determine the plant capacity. Comparing pilot and lab rate data should help establish the scaling factor. Efforts should be made to measure gas dispersion properties (e.g., bubble size and gas rate, see Section 12.14.2) to try to ensure similarity, as these affect the rate constant and thus the scaling up factor.
5. Pulp temperature, which affects the reaction rates. Most separations, however, are at ambient temperatures.
6. The extent of uniformity of the ore; variations in hardness, grindability, mineral content, and floatability should be investigated so that variations can be accommodated in the design, and, increasingly, to populate a geometallurgical model.
7. Corrosion and erosion qualities of the pulp; this is important in determining the materials used to construct the plant. The increasing use of seawater places a premium on establishing the impact of corrosion (Moreno et al. 2011).

FIGURE 12.60 The Floatability Characterization Test Rig *(Courtesy JKMRC and JKTech Pty Ltd).*

There are a few basic circuits (Table 12.7) with many possible variations and laboratory tests should provide data for the design of the best-suited circuit. This should be as basic as possible at this stage. Many flow schemes used in operating plants have evolved over a long period, and attempted duplication in the laboratory is often difficult and misleading. The laboratory procedures should be kept as simple as possible so that the results can be interpreted into plant operation.

A key issue in pilot plant testing is flexibility and consistency of operation. A standardized pilot plant has been developed called the *Floatability Characterization Test Rig* (FCTR). The unit, described by Rahal et al. (2000), is a fully automated pilot plant designed to move from plant to plant and characterize the floatability of each plant's ore according to standard procedures. It can be used both for testing modified circuits in existing plants and developing flowsheets for new ores. The FCTR is shown in operation in Figure 12.60.

Mini-plants have been assembled by some mining companies (e.g., Figure 12.61) and are supplied, for example, by Eriez Flotation Division.

Flotation studies at all levels, lab and pilot but especially plant scale, benefit from detailed surface analyses.

12.12.4 Diagnostic Surface Analysis

Selective flotation recovery is driven by the chemistry of the top few monolayers of the mineral surface. The surface of each individual mineral particle is a complex, distinctly nonuniform array of various species, some hydrophilic, such as oxidation products (i.e., oxy-hydroxides, oxy-sulfur species), adsorbed ions, precipitates (e.g., $CaSO_4$), and attached fine particles of other mineral phases, along with hydrophobic species, collector, and metal–collector complexes. Bubble attachment is

FIGURE 12.61 View of mini-plant *(Courtesy Manqiu Xu, Vale).*

dependent on the ratio of hydrophobic to hydrophilic surface species on individual mineral surfaces. If this ratio is too low, recovery will be low and flotation kinetics slow. This ratio varies widely between different particles of the same mineral and some nonvalue mineral phases will also have adsorbed hydrophobic collector in some form, contributing to gangue recovery (Trahar, 1976; Stowe et al., 1995; Crawford and Ralston, 1988; Piantadosi and Smart, 2002; Smart et al., 2003, 2007; Boulton et al., 2003; Malysiak et al., 2004; Shackleton et al., 2007; Muganda et al., 2012).

To optimize mineral beneficiation by flotation, a detailed evaluation of the surface chemistry of both valuable mineral (paymineral) and gangue is often of value. Ideally, the surface chemical evaluation should be along with other significant contributors to the recovery process: solution chemistry and liberation data. The approach to process improvement then becomes integrated, potentially identifying limits to recovery by linking various contributing factors, which will help identify opportunities for improvement (Gerson and Napier-Munn, 2013).

The necessary mineral surface chemical information required can essentially be obtained from two analytical techniques: time of flight secondary ion mass spectrometry (TOF-SIMS), and X-ray photoelectron spectroscopy (XPS). Two complementary techniques that are widely used with these direct surface analysis techniques are EDTA extraction and solution speciation modeling. An excellent review on a wide range of innovative surface analytical techniques, applied to the fundamental understanding of the flotation process is given by Smart et al. (2014).

The TOF-SIMS technique, being able to analyze the first few molecular layers on a mineral surface, is the most advanced tool for performing statistical analyses to evaluate the hydrophobic/hydrophilic balance (Smart et al., 2014; Chehreh Chelgani and Hart, 2014). The XPS technique (analysis of the outer 2–5 nm of the surface), when used in concert with the TOF-SIMS, is essential for identifying element speciation. The technique is more suited for single mineral and laboratory flotation systems, but has likely provided the most insight to mineral surface chemistry in relation to flotation recovery (Smart et al., 1999, 2003). To illustrate, results are briefly reviewed of one plant study where TOF-SIMS was used to identify opportunities for sphalerite recovery improvement, and from another plant study where TOF-SIMS and XPS were used to identify factors controlling pyrochlore losses.

TOF-SIMS

Brunswick Mine processes a Cu–Zn–Pb sulfide ore. In a plant survey, sphalerite was misreporting to the Cu–Pb bulk concentrate. Accidental activation by Cu and/or Pb ions was suspected. Samples were taken around the copper–lead circuit from the rougher feed (RF), the first rougher cell (A) concentrate (Con A) and tail (Tl A), and from the seventh cell in the rougher bank (F) concentrate (Con F) and tail (Tl F). The surface of 30 + sphalerite particles per sample were selected using mineral phase imaging to give reliable statistics. TOF-SIMS normalized intensities (Figure 12.62) show the range and distribution of both Cu and Pb on the surface of the sphalerite particles. The data indicate that there are several factors controlling the inadvertent activation of the sphalerite. Both Cu and Pb intensities on the surface of sphalerite in the Con A/Tl A pair and Con F are significantly higher than those in the feed (RF), suggesting dissolution/adsorption of Cu and Pb species during conditioning. For the con/tail pairs at A and F, Cu on the surface of sphalerite particles is discriminatory (there is more Cu on the cons than the tails). Lead, on the other hand, does not show any significant discrimination between the Con A and Tl A pair, whereas there is significant discrimination for Pb on sphalerite particles between the Con F and Tl F pair. The TOF-SIMS surface chemical evaluation identifies Cu activation as the primary driver for sphalerite recovery in the early portion of the rougher bank (cell A). With surface Cu adsorption and the resulting downstream decrease in pulp Cu content, the sphalerite reporting to the concentrates in the later cells is doing so in response to both Pb and Cu activation. These data fit the observations of

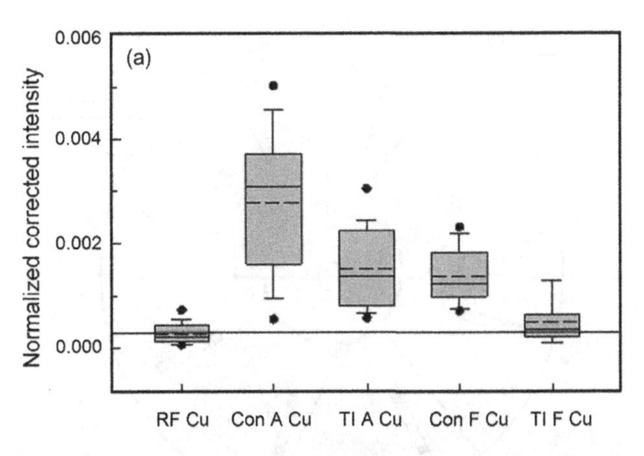

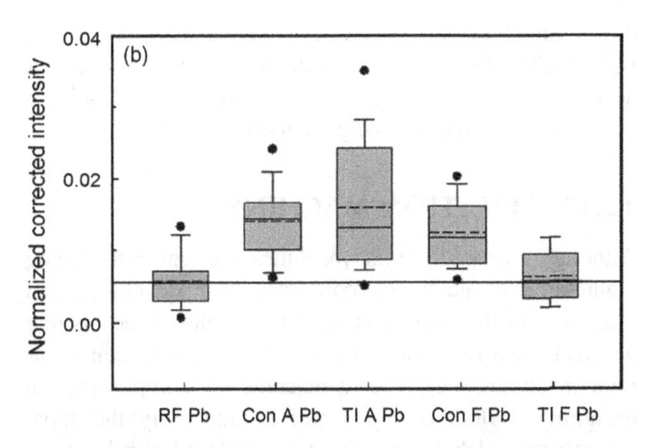

FIGURE 12.62 Vertical box plots for TOF-SIMS analysis of sphalerite surfaces from Con and Tail samples at Brunswick Mine: (a) Cu distribution, and (b) Pb distribution (Note: the solid horizontal line across the plot identifies the mean Cu or Pb content from the feed sample and is used as a baseline for comparison).

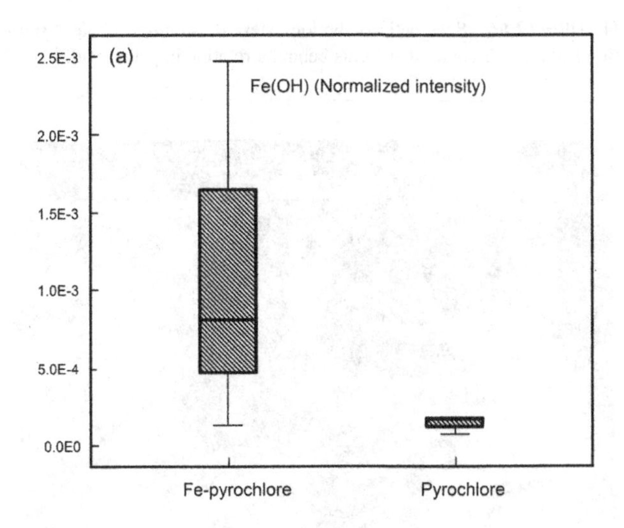

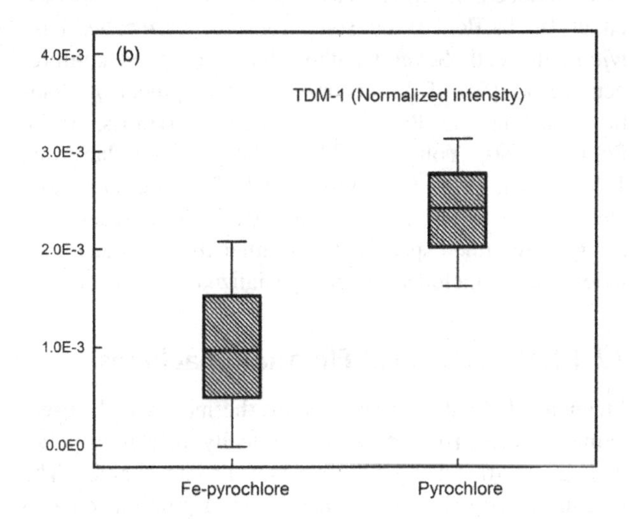

FIGURE 12.63 Vertical box plots of TOF-SIMS normalized intensity for (a) FeOH, and (b) collector (TDM-1) on the surface of high Fe (Fe-pyrochlore) and low Fe pyrochlore (pyrochlore) particles from the Niobec flotation plant.

Ralston and Healy (1980), who showed that sphalerite uptake of Cu ion occurs before that of Pb, and the observation of Sui et al. (1999) that Cu activation is favored three to four times over Pb at pH >9.

TOF-SIMS/XPS

The Niobec Mine processes a niobium ore. In a plant survey, the composition of the pyrochlore particles reporting to the tails were identified with a higher Fe content than those reporting to the concentrate. A statistical compositional analysis of some 200 + particles from the concentrate and tails samples showed that the Fe content was not related to Fe-rich inclusions, but rather the Fe occurred in the pyrochlore matrix, Fe substituting for Na or Ca. TOF-SIMS analyses identified that the pyrochlore reporting to the tail (particles with higher matrix Fe content) had a higher level of FeOH (along with other Fe-oxide species) on the surface (Figure 12.63(a)) and significantly less tallow diamine (TDM-1) collector on their surface (Figure 12.63(b)). The TOF-SIMS surface analyses established a link between mineral Fe content, the degree of surface oxidation, collector loading and the observed poor recovery of high Fe pyrochlore grains (Chehreh Chelgani et al., 2012a,b).

To verify the relationship between pyrochlore matrix Fe content and the identified surface chemical variations, a series of controlled bench conditioning tests were set up to determine the effect of accelerated oxidation on pyrochlore particles of different Fe content. XPS analysis for Fe revealed that a greater proportion of oxidative Fe species developed on the surface of the pyrochlore particles with higher matrix Fe content (Chehreh Chelgani et al., 2013). These data support the hypothesis reached from the TOF-SIMS analyses, which suggest that the development of surface oxides favors pyrochlore grains with higher Fe content, ultimately impeding collector adsorption and reducing recovery (Chehreh Chelgani et al., 2012a; Chehreh Chelgani and Hart, 2014).

12.13 FLOTATION MACHINES

Although many different machines are currently being manufactured and many more have been developed and discarded in the past, it is fair to state that three distinct groups have arisen: mechanical, column, and reactor/separator machines. The type of machine is of importance in designing a flotation plant and is frequently the topic causing most debate (Araujo et al., 2005b; Lelinski et al., 2005). Flotation machines either use air entrained by turbulent pulp addition, or more commonly air either blown in or induced, in which case the air must be dispersed either by baffles or by some form of permeable base within the cell. Several authors have compiled comprehensive reviews of the history and development of flotation machines in the minerals industry (Harris, 1976; Poling, 1980; Young, 1982; Barbery, 1984; Jameson, 1992; Finch, 1995; Laskowski, 2001; Gorain et al.; 2007; Nelson et al., 2009; Lynch et al., 2010). Some recent cell designs are aimed specifically at either fine or coarse particles and are included under "specialized" machines.

12.13.1 Mechanical Flotation Machines

Mechanical flotation machines are the most widely used, being characterized by a mechanically driven impeller that agitates the slurry and disperses air into bubbles. The machines may be self-aerated, the aspiration created by rotation of the impeller, or forced-air (supercharged), where air is introduced via an external blower. Impeller mechanisms typically comprise a *rotor* (impeller) and *stator* (or *diffuser*, i.e., baffles surrounding the rotor). As the impeller rotates, an air cavity forms in the low pressure region behind the impeller blade where bubbles are formed by a shearing action (Figure 12.64) (Crozier and Klimpel, 1989; Mavros, 1992; Schubert and Bischofberger, 1998). The stator does not have a major impact on bubble size but on the dispersion of bubbles in the cell (Harris, 1976). Machine suppliers recommend impeller speeds that allow the machine to maintain the particles in suspension and disperse the bubbles throughout the cell. Improved design of the rotor/stator mechanism, viz. to increase productivity (flotation rate) reduce power consumption, and increase life, is one of the objectives in CFD modeling.

A typical flotation bank comprises a number of such machines in series, "cell-to-cell" machines being separated by weirs between each impeller, or individual (isolated) tank cells. Historically, "open-" or "free-flow" (hog

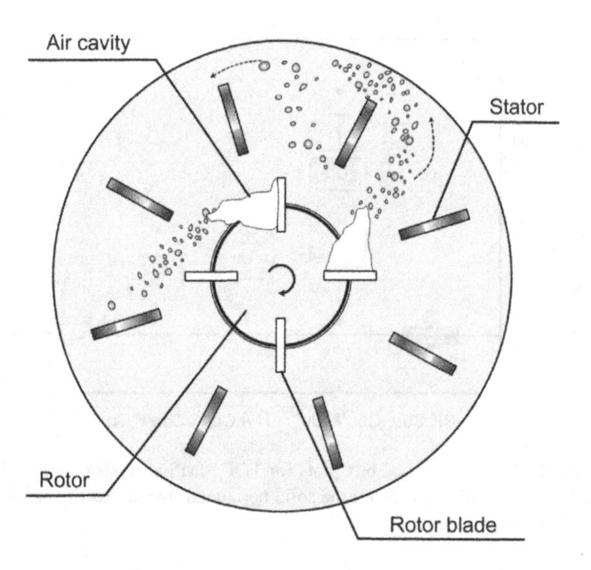

FIGURE 12.64 Schematic of bottom view illustrating bubble formation from the air cavity that forms behind a rotating impeller blade.

FIGURE 12.65 Older-style mechanical flotation cell.

trough-type) machines allowed virtually unrestricted flow of slurry down a bank.

The most pronounced trend in recent years, particularly in the flotation of base metal ores, has been the move toward larger capacity cells, with corresponding reduction in capital, maintenance, and operating costs (Murphy, 2012), particularly where automatic control is incorporated. In the mid-1960s, flotation cells were commonly 5.7 m^3 (200 ft^3) in volume or less (Figure 12.65). In the 1970s and 1980s flotation machines grew in size and typically ranged from 8.5 to 28.3 m^3 with larger cells being increasingly adopted. Manufacturers in the forefront of the industry included Denver Equipment, Galigher, Wemco, Outokumpu Oy, and Sala.

FIGURE 12.66 SuperCell (ca. 600 m^3) being fabricated *(Courtesy FLSmidth).*

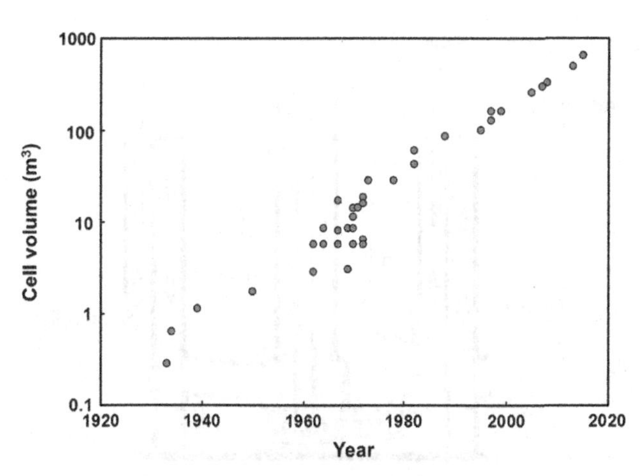

FIGURE 12.67 The trend toward larger flotation cells (Note: logarithmic scale on y-axis) *(Data from Dreyer (1976), Lynch et al. (2010), Nesset et al. (2012)).*

Relatively small cylindrical cells (tank-type cells up to $\sim 60\ m^3$) were introduced in the 1970s, but only saw widespread use in the 1990–2000s when suppliers developed the large tank cell designs. The largest installed cells are currently 300–350 m^3 (Nelson et al., 2009; Coleman and Dixon, 2010; Shen et al., 2010). One project is currently installing 500 m^3 cells (Outotec, 2014), while 600–650 m^3 cells (Figure 12.66) are being tested (Lelinski et al., 2013). The trend is clearly toward ever increasing tank sizes (Figure 12.67). Modern tank cell designs are circular (cylindrical), fitted with froth crowders, multiple froth launders, and internal discharge. The large size allows for fewer cells, but this means more accurate instrumentation is required, for example, effective level control systems.

In the 1970s most flotation machines were of the "open-flow" type, as they were better suited to high throughputs and easier to maintain than cell-to-cell types. The Denver "Sub-A" was perhaps the most well-known cell-to-cell machine, being widely used in small plants and in multistage cleaning circuits. The cells were manufactured in sizes up to 14.2 m^3. In the coal industry, users reported significant improvement in selectivity over open-flow designs. The Sub-A design was patented by Arthur W. Fahrenwald in 1922, although the machine name and general design appeared to have originated from an earlier machine produced by Minerals Separation Corporation (Hebbard, 1913; Fahrenwald, 1922; Wilkinson et al., 1926; Lynch et al., 2010), one of the pioneering companies in flotation cell design and manufacturing. The Sub-A impeller mechanism is suspended in an individual square cell separated from the adjoining cell by an adjustable weir (Figure 12.68(a)). A feed pipe carries pulp from the weir of the preceding cell to the impeller of the next cell, flow being aided by the suction of the impeller. Suction created by the impeller also draws air down the hollow standpipe surrounding the shaft and into the pulp. The air stream is sheared into fine bubbles (with the aid of frother usually) as the air and pulp are intimately mixed by the action of the rotating impeller. Directly above the impeller is a stationary hood, which prevents "sanding-up" of the impeller when the machine is shut down. Attached to the hood are four baffles which extend almost to the corners of the cell. The baffles limit pulp agitation above the impeller producing a *quiescent zone* where mineralized bubbles ascend without being subjected to excessive turbulence, which could result in particle detachment. As the bubbles cross from the pulp zone to the froth zone, they are carried upward to the overflow by the crowding action of succeeding bubbles. In some cases, removal of froth is accomplished by rotating froth paddles that aid overflow. Particles too heavy to flow over the tailings weir by-pass through sand relief ports which prevent the build-up of coarse material.

The amount of air introduced into the pulp depends on the impeller speed, which was typically in the range of 7–10 $m\ s^{-1}$ peripheral (tip) speed. More air could be introduced by increasing impeller speed but this could over agitate the pulp as well as increase impeller wear and energy consumption. In such cases, supercharging could be applied to introduce additional air down the stand-pipe by means of an external blower. There is a limit to the amount of air that can be introduced and the cell remain in a safe operating range (Sections 12.9.4 and 12.14.2).

Supercharging is required with the Denver D-R machine (Figure 12.68(b)), which ranged in size from 2.8 to 36.1 m^3, and which was developed as a result of the need for a machine to handle larger tonnages in bulk-

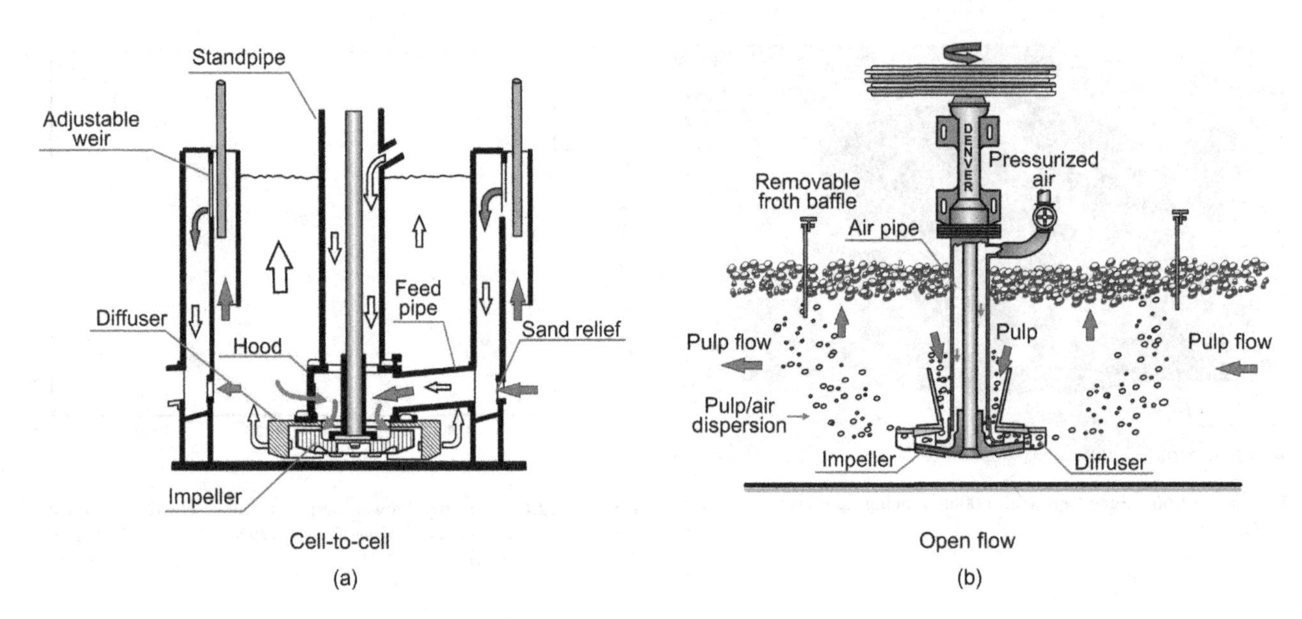

FIGURE 12.68 (a) Denver subaeration machine in cell-to-cell arrangement, and (b) D-R flotation machine in open flow arrangement.

flotation circuits. These units were characterized by the absence of intermediate partitions and weirs between cells. Individual cell feed pipes have been eliminated, and pulp flows freely through the machine without interference. The pulp level could be automatically controlled by a single tailings weir at the end of the trough. Bank operation is relatively simple and the need for operator attention is minimized.

Perhaps the best known of the forced-air machines is the Galigher Agitair (Sorensen, 1982) (Figure 12.69). This system, again, offers a straight-line flow of pulp through a row of cells, flow being produced by a gravity head. In each compartment, which may be up to 42.5 m^3 in volume, is a separate impeller rotating in a stationary baffle system. Air is blown into the pulp through the hollow standpipe surrounding the impeller shaft, and is sheared into fine bubbles, the volume of air being controlled separately for each compartment. Pulp depth is controlled by means of weir bars or dart valves at the discharge end of the bank, while the depth of froth in each cell can be controlled by varying the number and size of froth weir bars provided for each cell.

The original Fagergren machines (patented ca. 1920) were self-aerated and employed a horizontal impeller system (Margetts and Fagergren, 1920) similar to other flotation machines at the time (Lynch et al., 2010). The design evolved to a vertical agitator with draft-tube and false bottom for slurry recirculation (Figure 12.70) in the 1930s and Western Machinery Co. (WEMCO) began selling the machines which gained popularity during the 1950s under the name WEMCO Fagergren. The 1 + 1 rotor–disperser assembly design was introduced in the late 1960s (1 + 1 refers to the rotor length being equal to its diameter). The WEMCO design (Figure 12.70) continues to be the most widely used self-aerating mechanism and is currently being supplied in SuperCells up to 600 m^3. The modern designs comprise banks of individual tank cells. Feed slurry enters at the base of the cell from a feed tank (for the first cell) or from the preceding cell. Pulp passing through each cell is drawn upward into the rotor by the suction created by rotation. The suction also draws air down the standpipe. The air is thoroughly mixed with the pulp and small bubbles are formed by the disperser—a stationary, ribbed, perforated band encompassing the rotor—by abruptly diverting the swirling motion of the pulp. The air rate to the cell is influenced by impeller speed, slurry density, and level.

Dorr-Oliver designed a flotation cell in the early 1980s in response to the demand for larger cells (ca. 40 m^3). The pump-type impeller design removed the need for extensive tank baffling. The Dorr-Oliver patent (Lawrence et al., 1984) describes a U-shaped tank with rotor–stator assembly with a six-bladed impeller being curvilinear and of parabolic shape (Figure 12.71). Air flows down the standpipe surrounding the impeller shaft and into the gas chamber via an aperture in the impeller top plate. The cells are currently provided in rectangular, U-shaped bottom and cylindrical options ranging in size from 5 to 330 m^3 (Nelson et al., 2009).

Figure 12.72 shows the modern tank cell designs with baffles, radial launders, froth crowders and examples of: (a) the WEMCO mechanism with external discharge and (b) Dorr-Oliver mechanism with internal discharge systems.

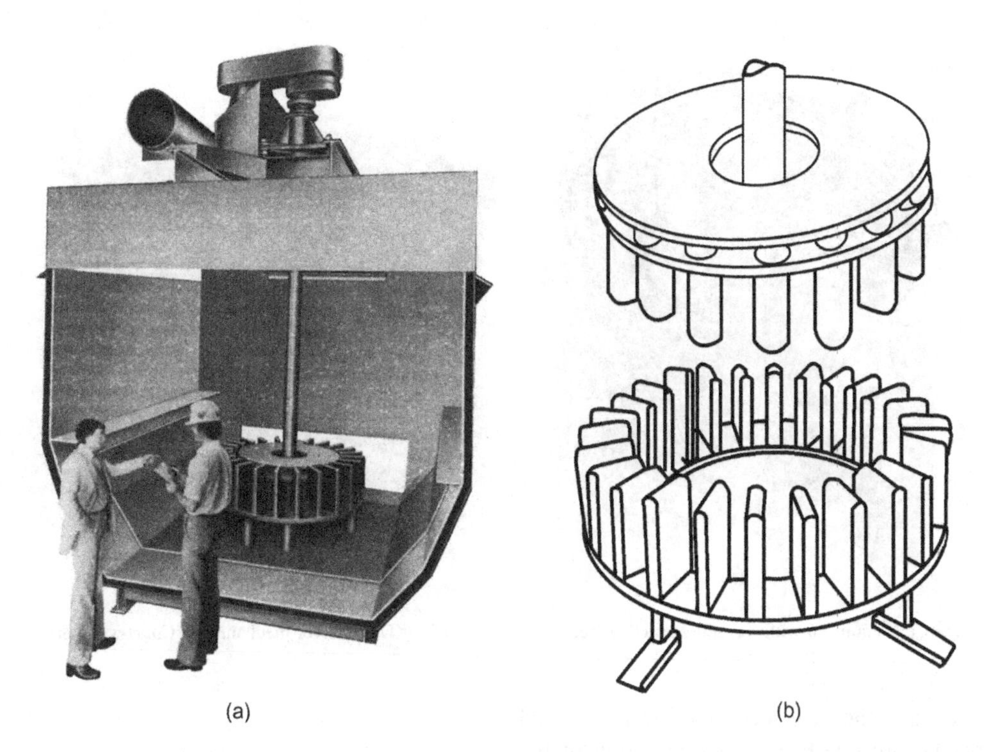

FIGURE 12.69 (a) Galigher 42.5 m^3 Agitair flotation machine, and (b) an example Agitair mechanism *(Adapted from Gorain et al. (1997)).*

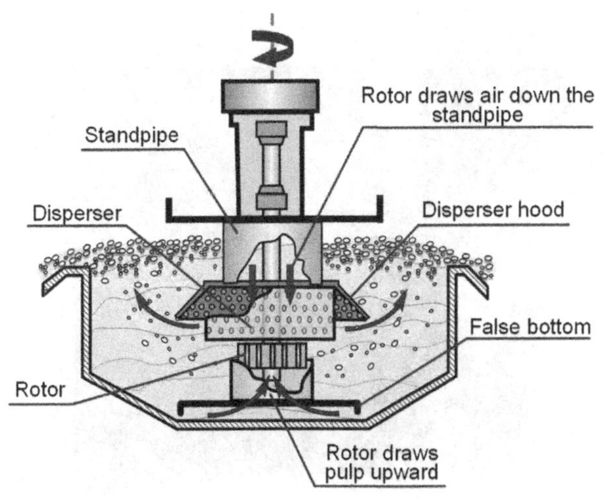

FIGURE 12.70 Diagram of the Wemco Fagergren cell.

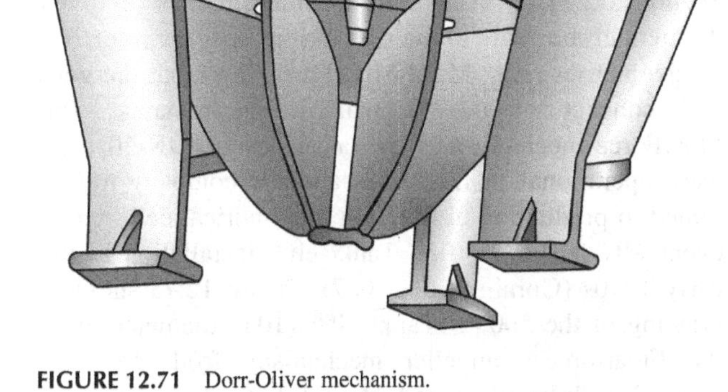

FIGURE 12.71 Dorr-Oliver mechanism.

Outotec (formerly Outokumpu Oy) has operated several base metal mines and concentrators in Finland and elsewhere and is well known for its mineral processing equipment, including the OK flotation cells. The OK cell designs originated in the 1970s and are now provided ranging in size from 0.5–38 m^3 and are supplied in two types: OK-U (rounded-bottom tank to minimize sanding with internal transversal launders) and OK-R (flat-bottom tank with one-sided longitudinal launders). The OK impeller consists of a number of vertical slots which taper downward, the top of the impeller being closed by a horizontal disc. As the impeller rotates, slurry is accelerated in the slots and expelled near the point of maximum diameter. Air is blown down the shaft and the slurry and air flows are brought into contact in the rotor–stator clearance, the aerated slurry then exiting the mechanism into the surrounding cell volume. The slurry flow is replaced by fresh slurry, which enters the slots near their base, where the diameter and peripheral speed are less. Thus the impeller acts as a pump, drawing in slurry at the base of the cell, and expelling it outward. The tank cell design and the rotor design minimize short-circuiting, as pulp flow is towards the

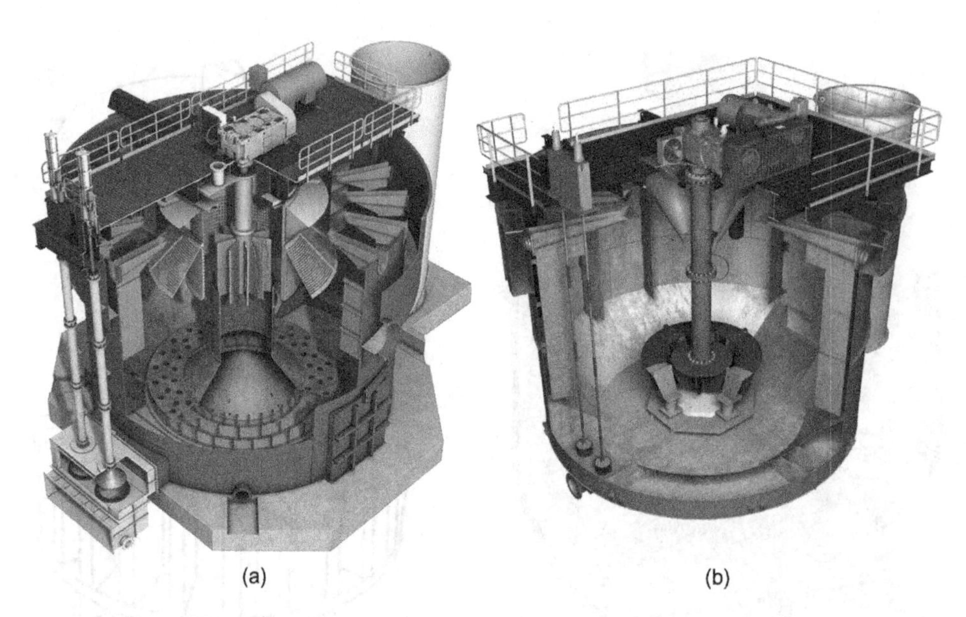

FIGURE 12.72 SuperCells equipped with: (a) self-aerated Wemco, and (b) forced-air Dorr-Oliver mechanisms *(Courtesy FLSmidth).*

bottom of the cell and the new feed entering is directed towards the mechanism due to the suction action of the rotor. It is because of this feature eliminating short circuiting that banks containing only two large cells are now in use in many of the world's concentrators (Niitti and Tarvainen, 1982) (arguments notwithstanding in favor of a minimum length of bank of about seven cells, Section 12.11.2). The OK mechanism has evolved throughout the years to include various specific rotor/stator geometries (e.g., Multi-Mix, Free-Flow), but the general concept of the original design remains. The FloatForce mechanism is the latest design. OK-50 cells were operational in the 1980s and the company transitioned to production of the TC-60 cylindrical cell, which eventually led to the first TankCell® installation in the early 1990s (Gorain et al., 2007). Figure 12.73 shows a drawing of the 500 m^3 TankCell® (10 m diameter) with the Floatforce® impeller mechanism, froth crowder, internal froth launders, and internal dart valve discharge.

The RCS (Reactor Cell System) machine (Figure 12.74(a)) is a tank cell design developed during the 1990s which employs the DV™ (Deep Vane) impeller mechanism and is available in sizes ranging from 0.8 (pilot-scale) to 200 m^3. The cells can be employed as roughers, scavengers or cleaners. RCS machines are equipped with a feed box, peripheral and/or double internal crossflow launders for concentrate removal and internal downflow discharge control dart valves, which are used for level control. A shelf baffle is located at the cell wall to create a quiescent zone above the impeller region. Figure 12.74(b) shows a bank of RCS machines in operation. A review of the development of the RCS machine is given by Gorain et al. (2007).

FIGURE 12.73 TankCell® e500 equipped with the FloatForce® mechanism *(Courtesy Outotec).*

12.13.2 Flotation Columns

The 1980–1990s saw the increasing use of flotation columns in the minerals (including coal) industry. The main advantages of columns include improved separation performance, particularly on fine materials, low capital and operational cost, less floor space demand, and adaptability to automatic control. The ability to build deep froths, coupled with the use of froth washing, make columns

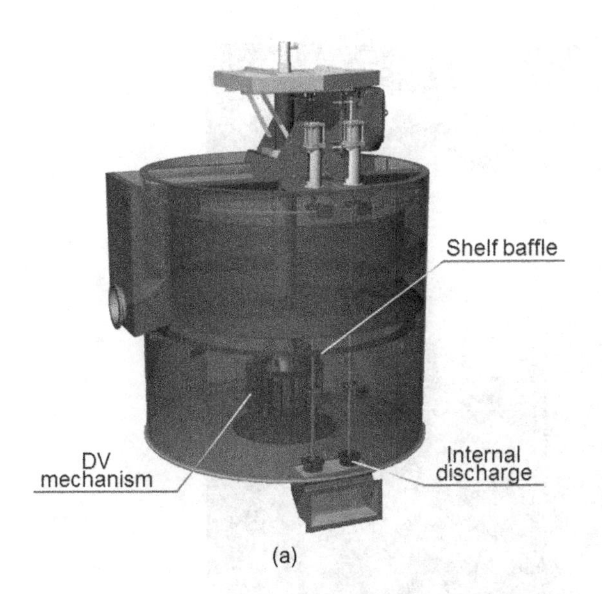

(a)

(b)

FIGURE 12.74 (a) Diagram showing RCS machine internals, and (b) bank of RCS machines in operation *(Courtesy Metso).*

attractive as final cleaners where the production of high-grade concentrates is required.

A schematic and industrial flotation column are shown in Figure 12.75. The column consists of two distinct zones: the *collection zone* (also referred to as the recovery or pulp zone) and the froth zone (also referred to as *washing* or *cleaning zone*). The collection zone is below the feed point where particles are suspended in the descending water phase and contact a rising swarm of air bubbles produced by a sparging system in the column base (Murdock and Wyslouzil, 1991). Floatable particles collide with and adhere to the bubbles and are transported to the froth zone above the feed point. Nonfloatable material is removed from the base of the column (often the tailings). Gangue particles that are entrained with water into the froth and others loosely attached to bubbles are washed back into the collection zone, hence reducing contamination of the concentrate. The wash water also ensures downward liquid flow in all parts of the column, preventing bulk flow of feed material into the concentrate. Wash water systems can be mounted above the froth or in the froth zone. Above-mounting is typically preferred as systems inside the froth are prone to plugging and are impossible to visually inspect, although certain operators prefer in-froth systems (Amelunxen and Sandoval, 2013). Figure 12.76 shows two types of wash water systems: irrigation pipes and a drip tray arrangement.

Modern flotation columns were developed in Canada in the early 1960s (Boutin and Wheeler, 1967) and were first used for cleaning molybdenum concentrates. Two column flotation units were installed in the molybdenum circuit at Mines Gaspé, Canada, in 1980, and excellent results were reported (Cienski and Coffin, 1981). The units replaced the multistage cleaner banks of mechanical cells. Since then, many of the copper–molybdenum producers have installed columns for molybdenum cleaning, and their use has been expanded into the roughing, scavenging, and cleaning of a variety of ore types, in many parts of the world. The late 1980s and early 1990s saw a focus on column flotation research, with two international conferences (Sastry, 1988; Agar et al., 1991), a monograph (Finch and Dobby, 1990), and many scientific articles (Araujo et al., 2005b) dedicated to the subject.

Flotation columns come in a range of sizes from 1 to 13 m high (Finch, 1995), with diameters of up to ca. 3.5 m (round or square, the former being more popular), the importance of height/diameter ratio having been discussed by several authors (Yianatos et al., 1988; Finch and Dobby, 1990). Instrumentation and some degree of automatic control is a necessity for column operation. Several groups have reviewed current and potential methods for column control (Bergh and Yianatos, 2003; Bouchard et al., 2005, 2009).

Bubbles are introduced into the column via a sparger. Initially, all spargers were internal devices (i.e., bubbles were produced inside the column), which injected air into the column through porous media, typically perforated rubber or filter cloth. The internal devices were prone to wear and fouling (plugging) by particles. Jetting spargers and external bubble generation devices (e.g., Microcel™) are now prevalent in the industry. Both systems allow for individual sparger isolation, which enables online replacement and maintenance. Certain systems allow water or slurry addition to the sparger, which has been shown to enhance fine bubble production (Dobby and Finch, 1991). In recent years, popular designs include the SlamJet®, SparJet®, Microcel™, and Cavitation tube (CavTube™).

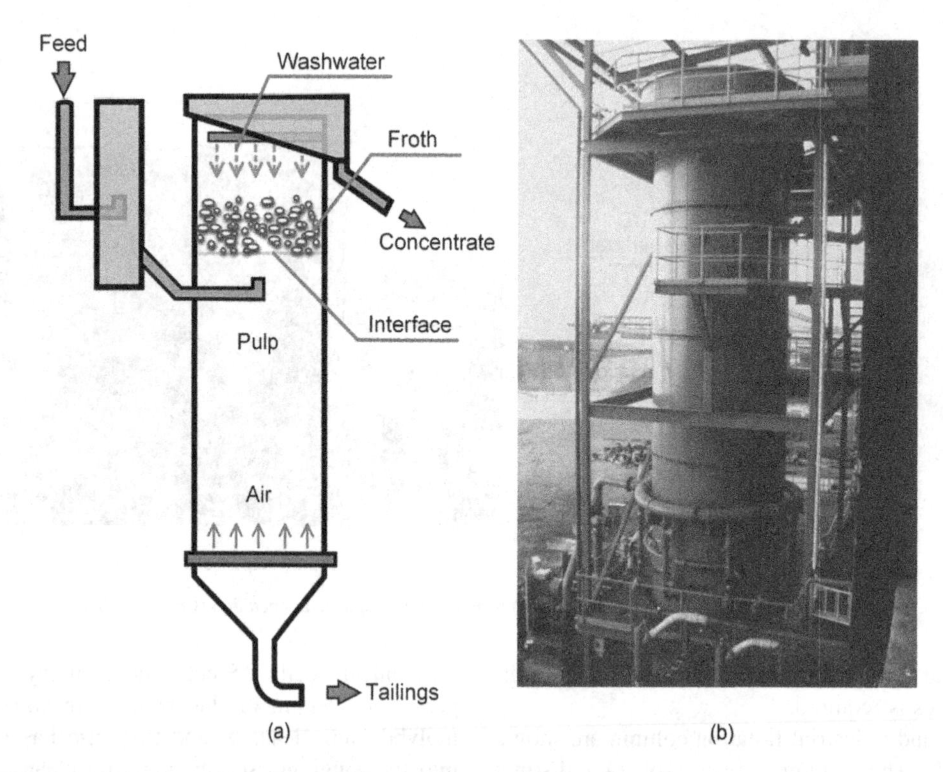

FIGURE 12.75 Flotation column: (a) schematic, and (b) in operation *(Courtesy Metso).*

FIGURE 12.76 Wash water systems: (a) irrigation pipes, and (b) drip tray *(Courtesy Aminpro Ltd.).*

Jetting spargers force high pressure air (30–100 psig) through a tube with an orifice typically ca. 1 mm in diameter (Dobby, 2002). Laboratory (air–water) testing of jetting spargers suggest two bubble production mechanisms (Bailey et al., 2005), shear along the jet surface that is responsive to frother dosage, and turbulent breakup at the end of the jet that produces large bubbles (ca. 5–10 mm) and is not responsive to frother.

An example of an internal jetting sparger is the SlamJet® (Figure 12.77), which is a gas injection tube fitted with an air-actuated flowrate control and shut-off mechanism. The Sparjet (not shown) is a more recent design that includes a wear-protected nozzle with integrated needle-valve, which helps reduce fouling.

The external CISA/Microcel™ system (Figure 12.78) pumps slurry, drawn from the column, through an in-line contactor, where it mixes with and shears air into a bubble dispersion that is injected into the column (Brake and Eldridge, 1996). The external manifold system used to recycle slurry through the in-line contactor can be seen at the base of the column (the ring structure) in Figure 12.75(b).

Figure 12.79(a) shows an isometric (cutaway) drawing and an image of the cavitation tube (CavTube™) in operation. A slurry/gas mixture is fed to the CavTube™ where a pressure drop is established due to the internal geometry, which results in hydrodynamic cavitation to produce fine bubbles. The process may result in bubble nucleation and growth on the particle surface, augmenting bubble–particle collision as the collection mechanism. Reviews of the cavitation process and its role in flotation processes can be found elsewhere (Zhou et al., 1994, 1997, 2009). Frother addition assists the process of fine bubble production within the device. Similar to the CISA system, the CavTube™ is an external sparging system that requires slurry recirculation via a slurry manifold (Figure 12.79(b)).

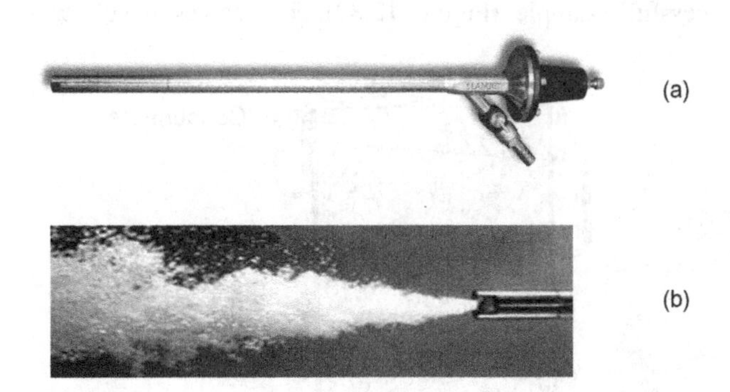

FIGURE 12.77 (a) SlamJet® sparger, and (b) in operation *(Courtesy Eriez Manufacturing Co.).*

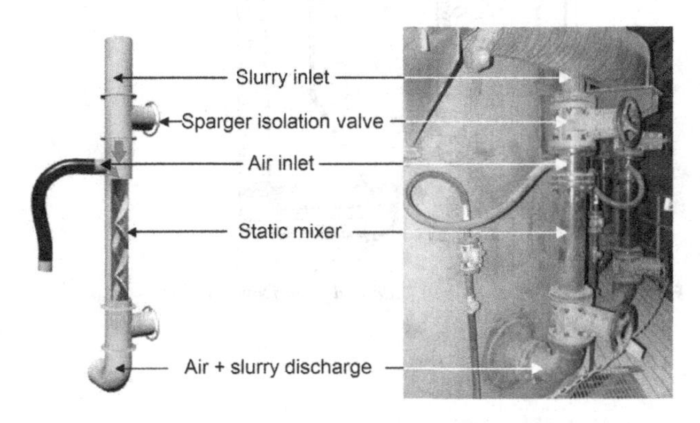

FIGURE 12.78 CISA/Microcel™ sparger *(Courtesy Metso).*

12.13.3 Reactor/Separator Flotation Machines

Conventional flotation machines house two functions in a single vessel: an intense mixing region where bubble–particle collision and attachment occurs, and a quiescent region where the bubble–particle aggregates separate from the slurry. The reactor/separator machines decouple these functions into two separate (or sometimes more) compartments. The cells are typically considered high-intensity machines due to the turbulent mixing in the reactor (see Section 12.9.5). The role of the separator is to allow sufficient time for mineralized bubbles to separate from the tailing stream which generally requires relatively short residence time (when compared to mechanical cells or columns).

Some of the earliest machine designs were of the reactor/separator-type. Figure 12.80 shows a design from a patent by Hebbard (1913). Feed slurry was mixed with entrained air in an agitation box (reactor) and flowed into the separation vessel where froth was collected as overflow. The design would be the basis for the Minerals Separation Corporation standard machine and early flotation cells used in the United States (Lynch et al., 2010).

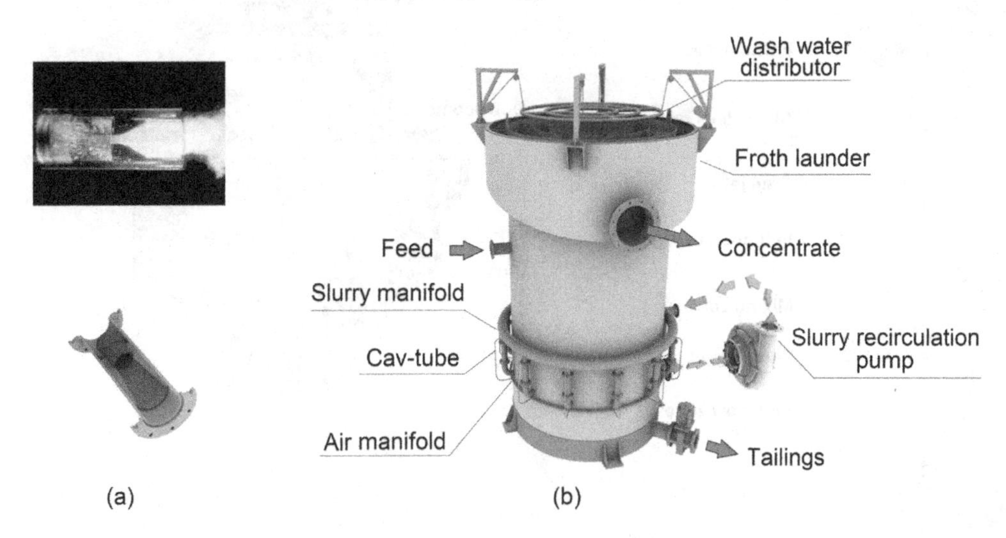

FIGURE 12.79 CavTube™: (a) principle of operation, bubbles form due to sudden pressure drop, and (b) general setup *(Courtesy Eriez Manufacturing Co.).*

The Davcra cell (Figure 12.81) was developed in the 1960s and is considered to be the first high-intensity machine. The cell could be thought of as a column or reactor/separator device. Air and feed slurry are contacted and injected into the tank through a cyclone-type dispersion nozzle, the energy of the jet of pulp being dissipated against a vertical baffle. Dispersion of air and collection of particles by bubbles occurs in the highly agitated region of the tank, confined by the baffle. The pulp flows over the baffle into a quiescent region designed for bubble–pulp disengagement. Although not widely used, Davcra cells replaced some mechanical cleaner machines at Chambishi copper mine in Zambia, with reported lower operating costs, reduced floor area, and improved metallurgical performance.

Several attempts have been made to develop more compact column-type devices, the Jameson cell (Jameson, 1990; Kennedy, 1990; Cowburn et al., 2005) being a successful example (Figure 12.82). The Jameson cell was

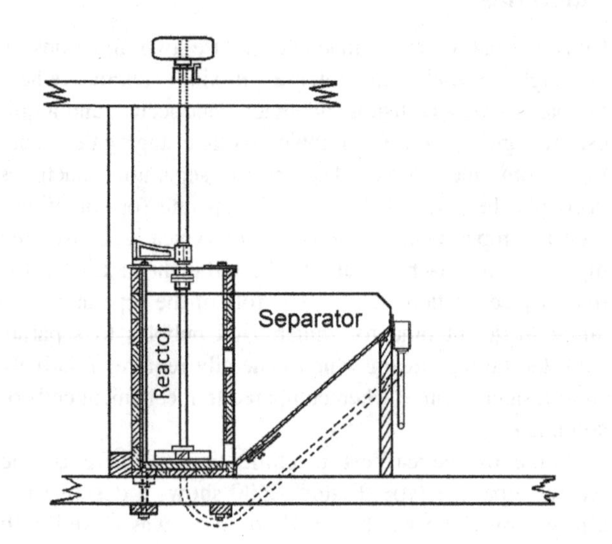

FIGURE 12.80 An early reactor/separator-type flotation machine *(Adapted from Hebbard (1913)).*

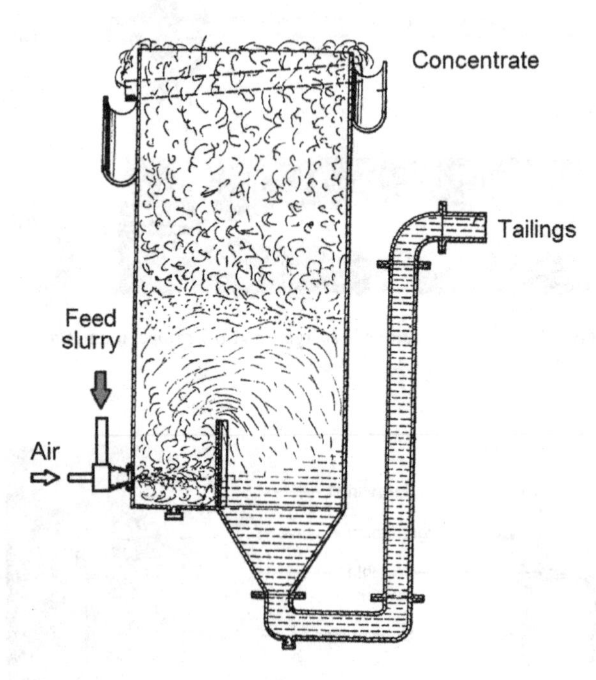

FIGURE 12.81 Davcra cell *(Adapted from Davis (1969)).*

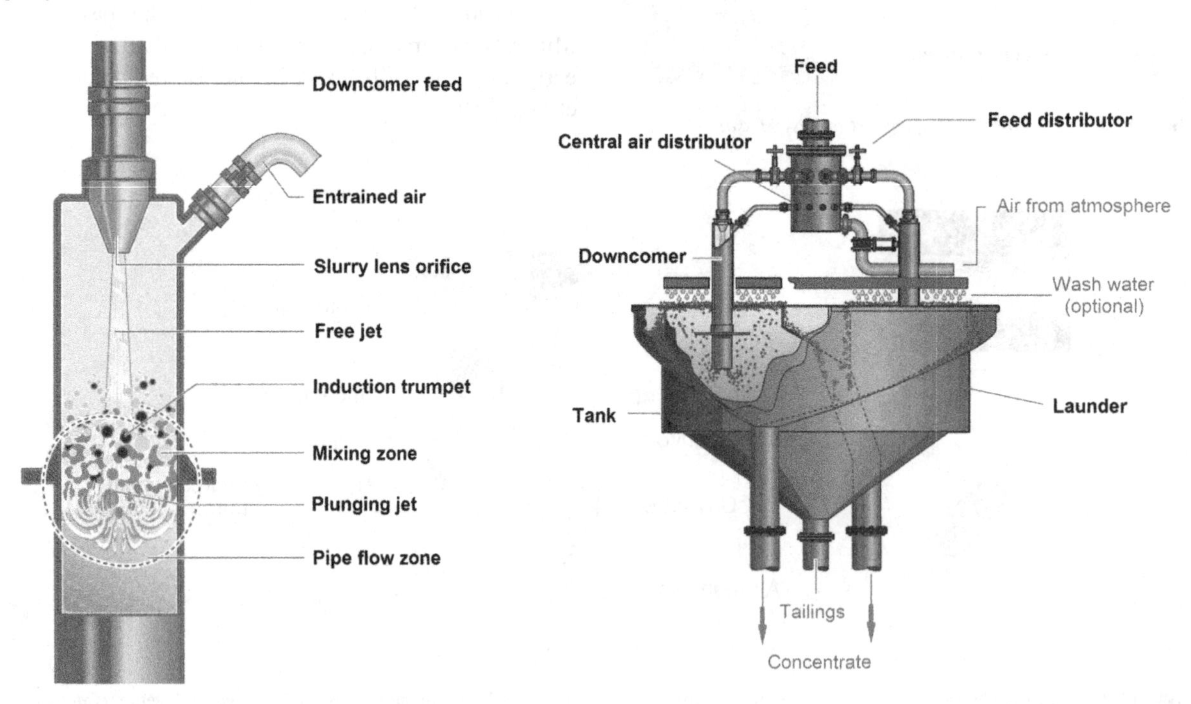

FIGURE 12.82 Principles of operation of the Jameson cell *(Courtesy Xstrata Technology).*

developed in the 1980s jointly by Mount Isa Mines Ltd and the University of Newcastle, Australia. The cell was first installed for cleaning duties in base metal operations (Clayton et al., 1991; Harbort et al., 1994), but it has also found use in coal plants and in roughing and preconcentrating duties. The original patent refers to the Jameson cell as a column method, but it can also be considered a reactor/separator machine: contact between the feed and the air stream is made using a plunging slurry jet in a vertical downcomer (the reactor), and the air–slurry mixture flows downwards to discharge and disengage into a shallow pool of pulp in the bottom of a short cylindrical tank (the separator). The disengaged bubbles rise to the top of the tank to overflow into a concentrate launder, while the tails are discharged from the bottom of the vessel. Air is self-aspirated (entrained) by the action of the plunging jet. The air rate is influenced by jet velocity and slurry density and level in the separator chamber.

The Jameson cell has been widely used in the coal industry in Australia since the 1990s. Figure 12.83 shows a typical cell layout where fine coal slurry feeds a central distributor which splits the stream to the downcomers. Clean coal is seen overflowing as concentrate from the separation vessel. The major advantage of the cell in this application is the ability to produce clean concentrates in one stage of operation by reducing entrainment, especially when wash water is used. It also has a novel application in copper solvent extraction/electrowinning circuits, where it is used to recover entrained organic droplets from electrolyte (Miller and Readett, 1992).

The Contact cell (Figure 12.84) was developed in the 1990s in Canada. The feed slurry is placed in direct contact with pressurized air in an external contactor which comprises a draft tube and an orifice plate. The slurry–air mixture is fed from the contactor to the column-type separation vessel, where mineralized bubbles rise to form froth. Contact cells employ froth washing similar to conventional flotation columns and Jameson cells. Contact cells have been implemented in operations in North America, Africa, and Europe.

The IMHOFLOT V-Cell (Figure 12.85(a)) was developed in the 1980–1990s and evolved from early pneumatic designs developed in Germany in the 1960–1970s (Imhof et al., 2005; Lynch et al., 2010). Conditioned feed pulp is mixed with air in an external self-aeration unit above the flotation cell. The air–slurry mixture descends a downcomer pipe and is introduced to the separation vessel via a distributor box and ring pipe with nozzles that

FIGURE 12.83 Jameson cell in coal flotation showing the downcomers and the clean coal concentrate being produced *(Courtesy JKMRC and JKTech Pty Ltd).*

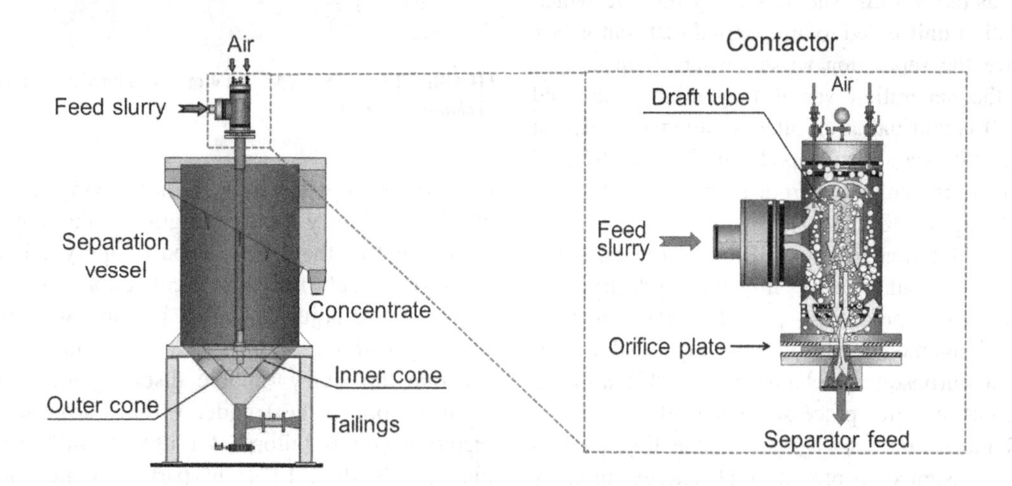

FIGURE 12.84 Contact cell and close-up of contactor (Note: downcomer is external to separation vessel) *(Courtesy Aminpro Ltd.).*

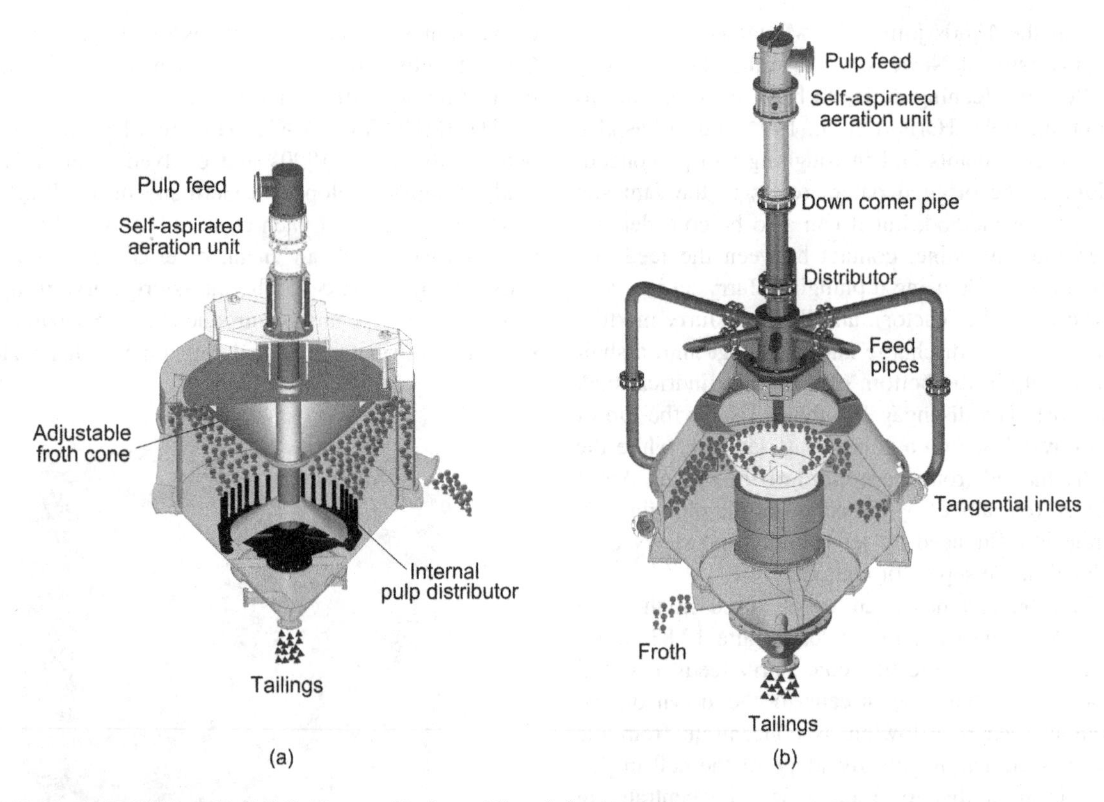

FIGURE 12.85 (a) V-Cell, and (b) G-Cell. *(Courtesy Maelgwyn Minerals Services Ltd.)*

redirect the flow upward in the cell. The separation vessel is fitted with an adjustable froth crowding cone which can be used to control mass pull. The concentrate overflows to an external froth launder, while the tailings stream exits at the base of the separation vessel. The V-Cell has been used to float sulfide and oxide ores with the largest operation being an iron ore application (Imhof et al., 2005).

The IMHOFLOT G-Cell (Figure 12.85(b)) was introduced in 2001 and employs the same external self-aerating unit as the V-Cell. The air–slurry mixture which exits the aeration unit is fed to an external distributor box (located above the separation vessel) where pulp is split and fed to the separation vessel tangentially via feed pipes. The cell is unusual as an internal launder located at the center of the vessel collects froth. The centrifugal motion of the slurry enhances froth separation with residence times being ca. 30 s.

The Staged Flotation Reactor (SFR) (Figure 12.86) is a recent development in the minerals industry. By sequencing the three processes—particle collection, bubble/slurry disengagement, and froth recovery—and assigning each to a purpose-built chamber, the SFR aims to optimize each of the three processes independently.

The SFR incorporates an agitator in the first (collection) chamber designed to provide high energy intensity (kW m^{-3}) and induce multiple particle passes through the high shear impeller zone, hence giving high collection efficiency. Slurry flows by gravity through the reactor stages, that is, there is no need to apply agitation to suspend solids, only for particle collection. As such, impeller speed can be adjusted online in correlation with desired recovery without sanding. The second tank is designed to deaerate the slurry (bubble disengagement) and rapidly recover froth to the launder without dropback. The froth recovery unit is tailored for use of wash water and for high solids flux. Efficient particle collection and high froth recovery translate into fewer, smaller cells, resulting

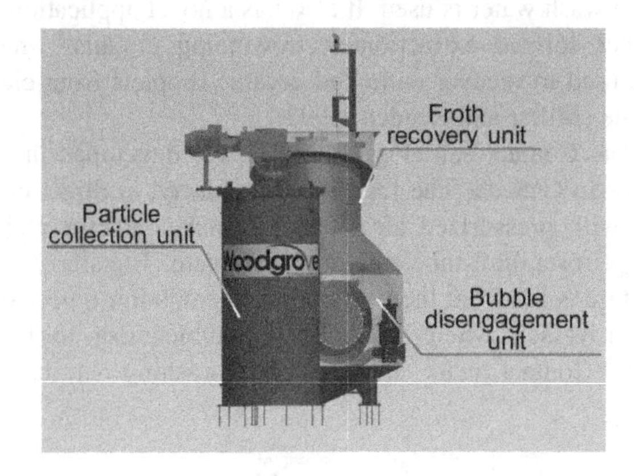

FIGURE 12.86 Staged Flotation Reactor. *(Courtesy Woodgrove Technologies Inc.)*

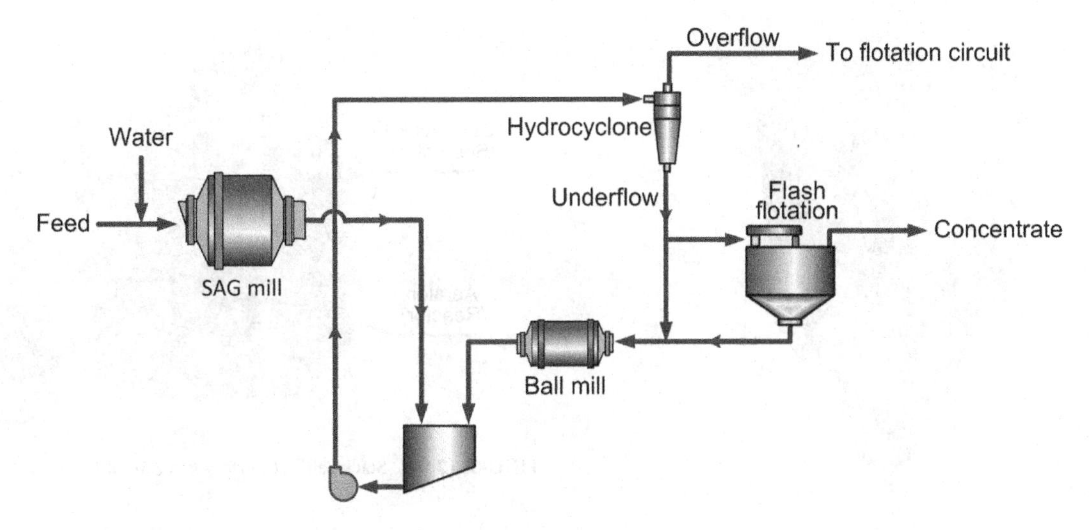

FIGURE 12.87 Typical flash flotation circuit.

in a smaller footprint and building height, with lower power consumption, and the potential for good selectivity in both roughing and cleaning applications.

12.13.4 Specialized Flotation Machines

Several flotation machines have been developed specifically to treat fine and coarse particle streams where kinetics are typically low (Section 12.10.1). This section highlights commercially available equipment that has made in-roads for specific applications along with some promising designs that are in the development stage.

Flash Flotation

The process aims to float coarse, high-grade particles early in the process flowsheet. Typically, a portion of the cyclone underflow stream from the ball mill circuit (i.e., the circulating load) comprises the feed to the flash flotation cell, with the flotation tails returning to the grinding mill (Figure 12.87). The product can be directed to a final concentrate (but see comments in Section 12.10.2) or a middlings stream, depending upon the grade. Flash cells are used in a variety of base metal operations where dense minerals accumulate in the cyclone underflow, but draw much attention for gold recovery as gold tends to accumulate in the recirculating load, not only due to its high density but also due to gold's malleability giving slow grinding kinetics (Banisi et al., 1991). The flash flotation cell is specifically designed with a sloped bottom and vertical discharge, which allows for by-pass of very coarse particles (Figure 12.88).

One of the earliest references to the concept was by Garrett (1933), who discussed experiments at North Broken Hill where ball mill discharge was directly floated in an eight-cell bank. Concentrate from the first three cells was directed to the final lead concentrate filters, with the remaining reporting to a middlings stream. In the early days of implementation, the machines were termed "unit cells." Flash cells have been widely used since the 1980s in new operations and retrofitted in older concentrators.

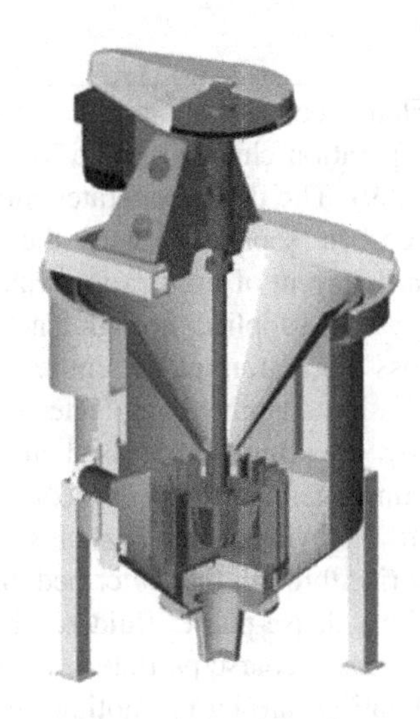

FIGURE 12.88 Flash flotation cell *(SkimAir® Courtesy Outotec).*

Hydrofloat™ Separator

This device was developed in the 1990s, with the first unit being installed for coarse particle recovery in the potash industry. The Hydrofloat™ separator has since been used to recover coarse phosphates, diamonds, spodumene, and vermiculite. Preclassified feed (typically >250 μm) is processed in the machine, which incorporates hindered settling and flotation principles that allow for recovery of coarse particles (when compared to conventional flotation systems).

FIGURE 12.89 Hydrofloat™ coarse particle separator *(Courtesy Eriez Manufacturing Co.).*

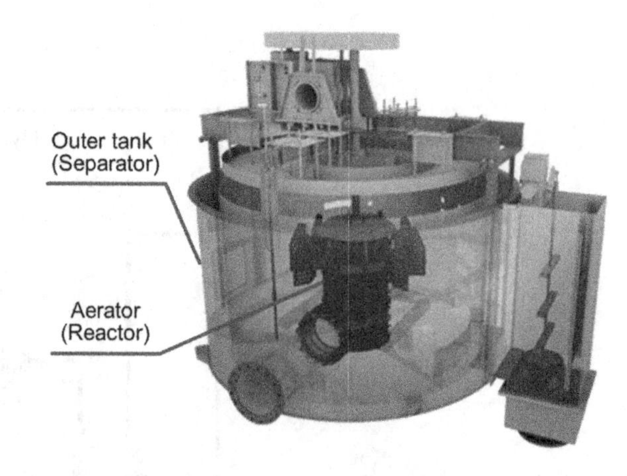

FIGURE 12.90 StackCell™ *(Courtesy Eriez Manufacturing Co.).*

The HydroFloat™ consists of a circular tank subdivided into an upper separation chamber and a lower dewatering cone (Figure 12.89). The device operates much like a traditional hindered-bed separator, with the feed settling against an upward current of fluidization water. The fluidization (teeter) water is supplied through a network of pipes that extend across the bottom of the entire cross-sectional area of the separation chamber. The teeter bed is continuously aerated by injecting compressed air and a small amount of frothing agent into the fluidization water. The bubbles attach to the hydrophobic particles, reducing their density till they rise through the teeter bed and are floated off. The use of the dense-phase, fluidized bed eliminates axial mixing, increases coarse particle residence time, and increases the flotation rate by promoting bubble–particle interactions. As a result, the rate of recovery is high for both coarse liberated and semiliberated particles. Hydrophilic particles that do not attach to the air bubbles continue to move down through the teeter bed and eventually settle into the dewatering cone. These particles are discharged as a high solids stream (e.g., 75% solids) through a control valve at the bottom of the separator. The valve is actuated in response to a signal provided by a pressure transducer mounted to the side of the separation chamber. This configuration allows a constant effective density to be maintained within the teeter bed.

StackCell™

This technology has recently been developed and implemented for fine particle flotation (typically treating particles <150 μm), primarily for coal, but also in iron ore and copper processing. In the StackCell™ (Figure 12.90), feed slurry is mixed with low-pressure air (i.e., preaeration) and the aerated slurry then travels to the high-shear aerator (reactor) located at the center of the cell. This arrangement provides efficient bubble–particle contacting, substantially shortening the residence time required for particle collection and thus reducing cell volume requirements. It has been found (Williams and Crane, 1983) that the flotation rate is proportional to bubble concentration, hydrophobic particle concentration, and energy (Section 12.9.5). With the shearing and mixing energy concentrated within the aerator and used for the sole purpose of creating bubbles and increasing bubble–particle collisions, the flotation rate offered by the StackCell™ is higher than in conventional cells. Once the slurry leaves the aerator and enters the outer tank, phase separation occurs between the mineralized bubbles (forming froth) and the pulp. A pulp level is maintained in the outer tank to provide a deep froth that can be washed to minimize the entrainment of fine material. The froth overflows into a launder, while the tailings are discharged through a control valve. The system is specifically designed to have both a small footprint and a gravity-driven feed system. This allows multiple units to be "stacked" in series on subsequent levels in the plant or easily placed ahead of existing (potentially overloaded) flotation circuits.

Recently, two devices have been proposed by developers at the University of Newcastle, Australia, for fine particle flotation: the *Concorde cell* (Jameson, 2006) and the *Reflux Flotation Cell (RFC)* (Dickinson and Galvin, 2014).

The Concorde cell (Figure 12.91) is specifically designed to float particles less than 20 μm. The cell resembles a Jameson cell and consists of two stages (i.e., reactor/separator-type machine). Small bubbles are first formed in a "blast tube" under pressure (Jameson, 2010). Similar to the Jameson cell, the feed enters as a vertical jet, where it is mixed with air, but in this case the air is fed under pressure. The bubble/slurry mixture then passes through a choke and is accelerated and ultimately exceeds

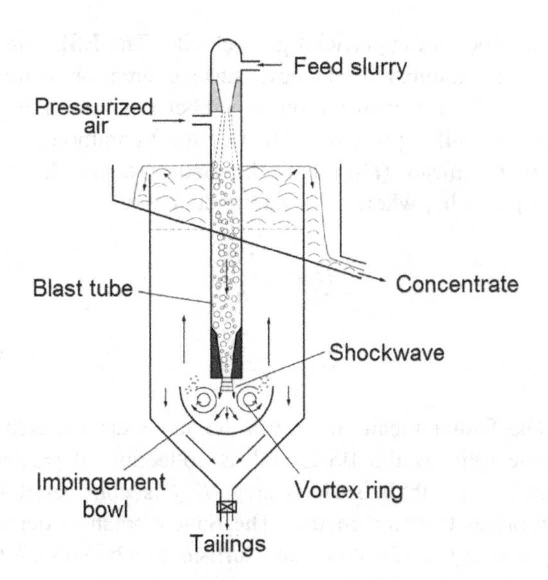

FIGURE 12.91 Schematic of the Concorde cell *(Used with permission, Jameson (2010). Copyright Elsevier).*

the speed of sound (travelling at ca. 20 m s^{-1}) (Jameson, 2010), which creates shock waves. The rapid pressure change in the choke results in fine bubble formation. The bubble/slurry mixture exits the choke and is fed to an impingement bowl, which redirects the flow upward into the separation vessel. Mineralized bubbles float to form froth and overflow, while the tailings stream exits via the bottom of the vessel. To date, pilot-scale tests have been undertaken (Jameson, 2010).

A diagram of the RFC is shown in Figure 12.92. Air is introduced vertically downward through a sintered sparger which is inserted into a vertical feed tube. Feed slurry is introduced to the tube and flows over the sparger, forming bubbles. The air/slurry mixture flows downward into the separation vessel. The separation vessel includes parallel inclined channels (PICs) toward the base, through which the tailings stream passes. The PICs increase the segregation of the bubbles from the tailings stream, which minimizes the loss of mineralized bubbles (Dickinson and Galvin, 2014). A feature of the RFC is that the top of the vessel is enclosed and froth is confined to a reduced area overflow outlet. Fluidization water is added via a distributor below the overflow. The RFC design allows for higher gas and wash water fluxes when compared to conventional machines.

12.14 FLOTATION CELL AND GAS DISPERSION CHARACTERIZATION

Improvements in flotation cell operation require parameters that can be measured and are related to the recovery and separation processes. The initial work adapted the approach taken to characterize mixers, leading to a series of scale-up criteria. Later work focused on the properties

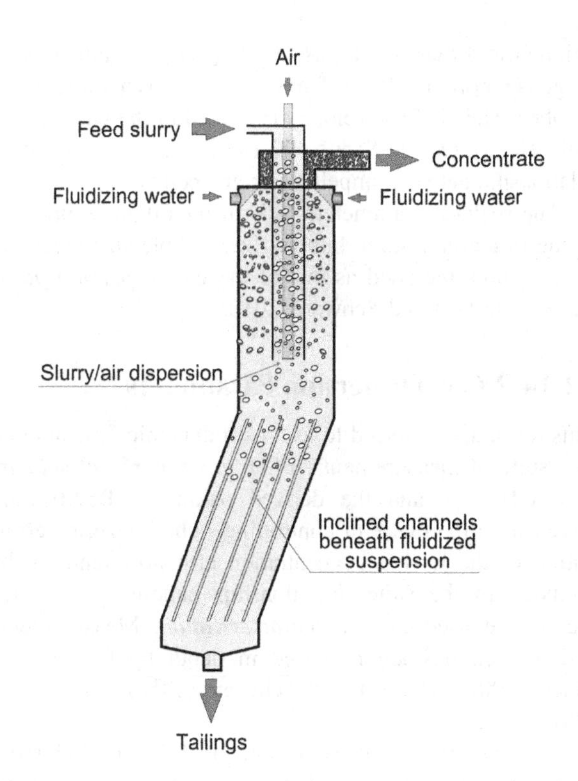

FIGURE 12.92 Schematic of the RFC *(Adapted from Dickinson and Galvin (2014)).*

of the air dispersed into bubbles. Both approaches are briefly reviewed.

12.14.1 Mechanical Cell Scale-Up Parameters

Partly based on fluid mixing and gas–liquid reactor design in chemical engineering (Oldshue, 1983), several scale-up numbers have been used in flotation cell design (Harris, 1976; Harris, 1986; Deglon et al., 2000). *Power intensity* is the net power per unit volume ($P_I = P_{net}/V$) with typical values 1–3 kW m^{-3} but the range can extend from 0.8–9.5 kW m^{-3} (Deglon et al., 2000). *Power number* is the ratio of net power to theoretical power ($N_P = P_{net}/\rho N^3 D^5$), where ρ is the effective fluid density, N is impeller speed (rps), and D is the diameter of the impeller. Typical values are 3.4–6.6. Note that introducing air lowers the effective density and the net power. *Impeller tip speed* ($S = \pi ND$) is typically 5–7 m s^{-1} but can go as high as 10 m s^{-1}. Recalling that power goes up as N^3 energy costs and motor size increase quickly with increasing N. *Air flow number* is the ratio between air flowrate and theoretical pumping rate ($N_Q = Q/ND^3$) where Q is volumetric air flowrate (m^3 s^{-1}). The range is quite large, 0.01–0.25, and thus N_Q is more a guideline than a scale-up criterion. *Air flow velocity* is the ability of the impeller to handle the volumetric flowrate of gas ($U_Q = Q/D^2$). Again a wide range is reported, 0.02–0.5 m s^{-1}, so not strictly a scale-up number. The air flow velocity is related to the gas superficial velocity (J_g see Section 12.14.2) as there is usually a relationship between impeller size and cell size.

The scale-up parameters lack an important component of the flotation system: bubble size. Bubble size measurement is now included as one of the *gas dispersion parameters* (Harbort and Schwarz, 2010).

12.14.2 Gas Dispersion Parameters

This term, also referred to as "hydrodynamics", is applied to a suite of measurements: BSD, *gas superficial velocity* (J_g), GH (ε_g), and the derived parameter BSAF (S_b). Given its importance in controlling bubble size, determination of the frother concentration also now tends to be included in the suite. Relationships among these parameters are used in *cell characterization*. Measurements and applications are reviewed in detail by Gomez and Finch (2007), Harbort and Schwarz (2010), and Nesset (2011).

An obstacle to measurement of BSD in industrial machines is the presence of slurry. The last 20 years have seen several measurement techniques developed for flotation systems (Randall et al., 1989; Yianatos et al., 2001; Gomez and Finch, 2002; Grau and Heiskanen, 2002; Rodrigues and Rubio, 2003; Miskovic and Luttrell, 2012). Most are variations on sampling bubbles to present to an imaging camera. The McGill Bubble Size Analyzer (Gomez and Finch, 2007), which has a viewing chamber with a sloped window to spread bubbles into a near monolayer and give an unambiguous focal plane (Figure 12.93), has gained quite wide acceptance (Harbort and Schwarz, 2010). A related device is the Anglo Platinum Bubble Sizer, which combines BSD with measurement of superficial gas velocity. The BSD can be output as number frequency, surface area or volume. Treating the full distribution is cumbersome and a single metric (mean) is preferred. The two most common are the arithmetic mean (D_{10}) and the Sauter mean diameter (D_{32}) given by, where d_i is bubble diameter:

$$D_{10} = \frac{\Sigma d_i}{n} \tag{12.55}$$

$$D_{32} = \frac{\Sigma d_i^3}{\Sigma d_i^2} \tag{12.56}$$

The Sauter mean size gives the same surface area to volume ratio as the BSD, and as collection of particles depends on bubble surface area, D_{32} is considered the appropriate flotation metric. The Sauter mean is derived by preservation of volume and surface area by solving the following system of two equations:

$$n_{32}(\pi D_{32}^3/6) = \Sigma(\pi\, d_i^3/6) \tag{12.57}$$

$$n_{32}(\pi\, D_{32}^2) = \Sigma(\pi\, d_i^2) \tag{12.58}$$

where n_{32} is the number of bubbles (note, not the actual number as we cannot satisfy three constraints simultaneously). The D_{32} is always greater than D_{10}; the closer they are the narrower the distribution. In most flotation cells, the bubble size range is approximately $0.5 < D_{32} < 2.5$ mm.

Gas superficial velocity (or just *air velocity*) is the volumetric air flowrate divided by cell cross-sectional area. The common unit is cm s^{-1}. With measurement of volumetric air rate to the cell, this definition works, provided the sectional area of the cell is well defined as, for example, in a flotation column. In many cases the area is difficult to ascertain because of baffles and launder inserts. Measurement is then made locally. This is achieved by a

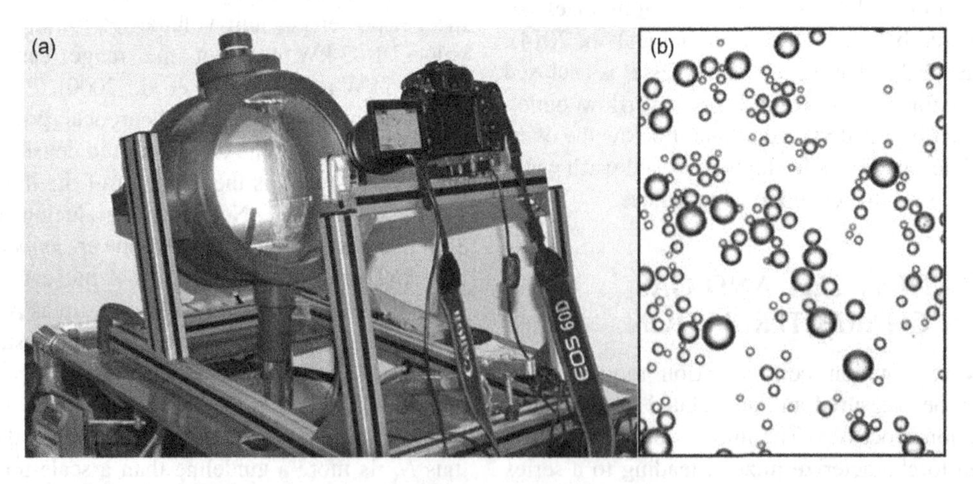

FIGURE 12.93 (a) McGill Bubble Size Analyzer in operation, and (b) example bubble image from an operating cell (background removed).

variety of methods, all based on collection of rising bubbles in a vertical tube filled with water. Measurement can be direct, using a flow meter on the exit of the tube to measure air rate and dividing by the tube cross-sectional area (Figure 12.94(a)); by measuring the rate of descent of the water in a transparent tube once the exit valve is closed (Figure 12.94(b)), which approximates the air velocity; or by measuring the rate of increase in pressure once the exit valve is closed (Figure 12.94(c)) where, with pressure in cm water, the rate of increase approximates the air velocity. Strictly, corrections to the latter two measurements are required for comparison between different locations (Gomez and Finch, 2007). The first and third methods give opportunity to record data; the second is the simplest (and is the technique used in the Anglo Platinum Bubble Sizer).

The air velocity depends on the local cell cross-sectional area, which will vary with inserts such as internal launders. The sensor is best located at some repeatable point in the cell, for example, the same position relative to the impeller shaft, and which is accessible to all cells to be measured, for example, in a bank. The J_g can be calibrated against the control room air flowrate in the case of cells with individual air flow measurement.

The third gas dispersion measure, GH, is the fraction of air in the air–slurry mix. It can be measured manually by trapping a sample of cell contents and allowing air to escape and recording the drop in slurry volume. Attention is required with this method to ensure the sampling chamber is initially filled with aerated slurry. Automatic methods use electrical conductivity or sonar-based technology (Figure 12.95). The conductivity technique uses two "flow conductivity cells," one measuring air–slurry conductivity and the other, excluding air using a "syphon cell", measuring slurry only conductivity, the ratio of the two signals being used to calculate GH from a fundamental model due to Maxwell (Gomez and Finch, 2007). The sonar-based technology uses the velocity of acoustic waves to determine GH (Chapter 3, Section 3.3.6).

GH increases as air flowrate to the cell is increased and increases by adding frother, which reduces bubble size and bubble rise velocity. The relationship between GH and air velocity can be used to define the *cell operating range*. An example is Figure 12.96: the operating range is the linear section. The trend is linear up to a certain air velocity, ca. 2.2 cm s^{-1} in this case, above which the cell "boils"; that is, the air rate exceeds the flow that can be dispersed by the mechanism and large bubbles ("slugs") form which disrupt the froth. In that situation the air dispersing mechanism is said to be "flooded." The upper air velocity limit is dictated by the GH. The maximum GH is typically about 15% (the dashed line on the figure). The 15% GH maximum may be reached at quite different air velocities, depending on bubble rise velocity: small bubbles rising slowly due to frother addition (or

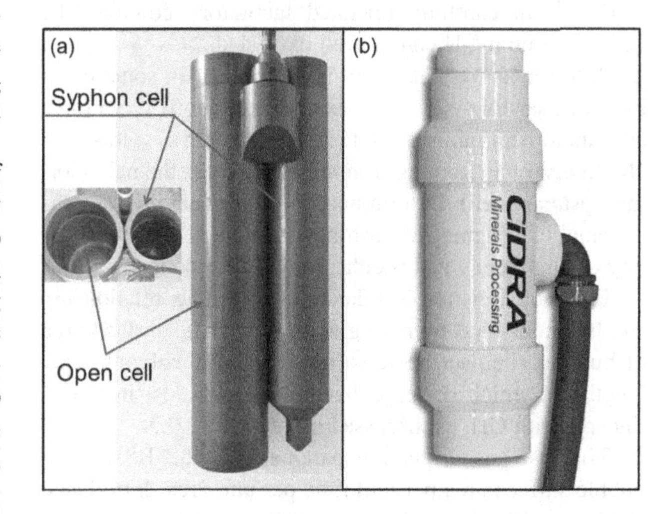

FIGURE 12.95 GH sensors: (a) conductivity-based, and (b) sonar-based *(Courtesy CiDRA Mineral Processing)*.

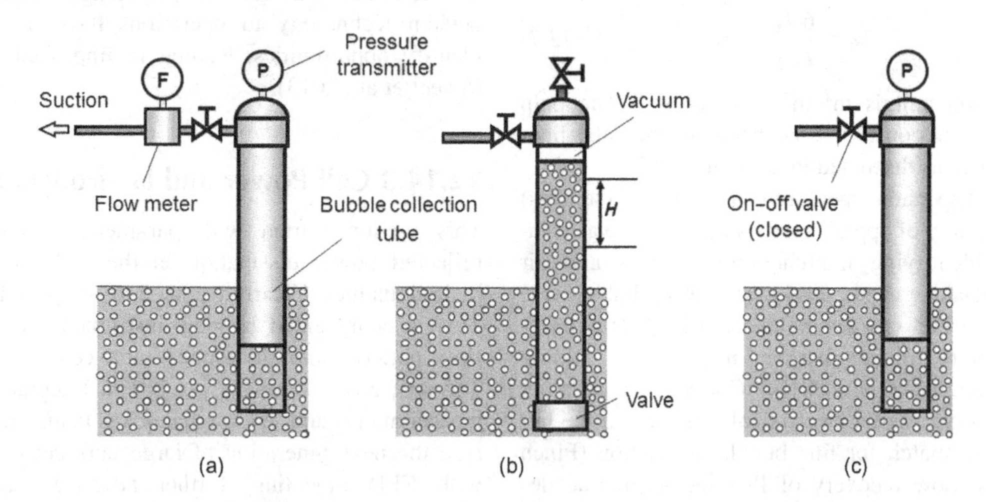

FIGURE 12.94 Methods of measuring air superficial velocity, J_g: (a) direct, (b) rate of descent of liquid, and (c) rate of increase in pressure.

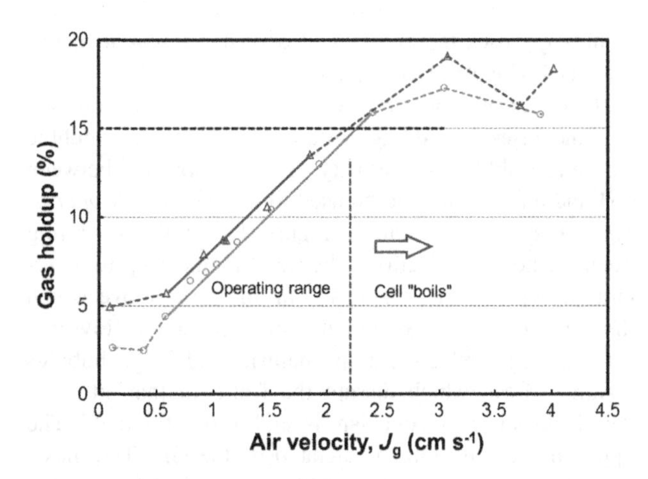

FIGURE 12.96 GH as a function of air velocity showing the operating range of the cell *(Adapted from Dahlke et al. (2005)).*

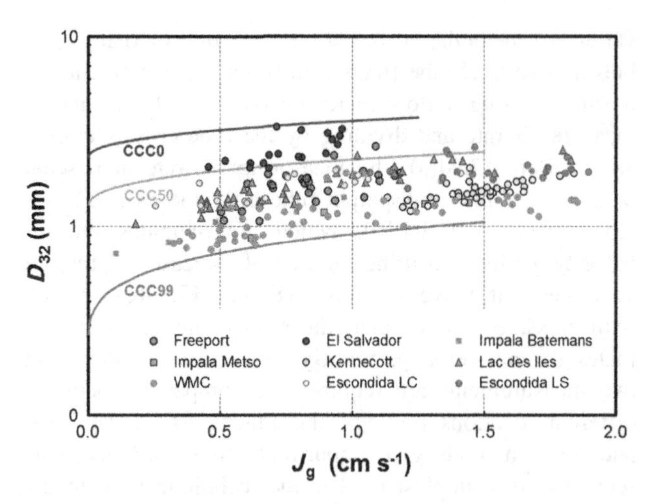

FIGURE 12.97 Model predicted bubble size as a function of gas velocity (rate) and dimensionless frother concentration (fraction of CCC) *(Adapted from Nesset et al. (2012)).*

high salt content) may result in GH reaching 15% at $J_g <$ 1 cm s^{-1}. In carefully operated laboratory columns the 15% maximum GH can be well exceeded.

A lower air velocity limit is determined in some cases by solids sanding out. In the example in Figure 12.96 there appears to be a minimum GH at $J_g \sim 0.5$ cm s^{-1}; this usually indicates air coming from upstream (e.g., the neighboring upstream cell). Determining the cell operating range is essential in designing air distribution strategies to avoid setting an air velocity that is either too high or too low.

Flotation kinetics is related to GH. The relationship can be understood by noting that the specific surface area of bubbles (i.e., surface area relative to air volume, units length^{-1}), which drives collection of particles, increases linearly with GH, as discussed in Section 12.9.5.

The last gas dispersion parameter is the BSAF, the bubble surface area per unit time per unit area. Introduced in Section 12.5.1, to remind it is calculated from (remembering to convert D_{32} to cm if J_g is in cm s^{-1}):

$$S_b = \frac{6\,J_g}{D_{32}} \tag{12.7}$$

The common unit is m^2 m^{-2} s^{-1} or s^{-1}. The pulp zone flotation rate constant k has been shown to be linearly related to S_b as discussed in Section 12.9.3.

The gas dispersion measurement devices (sensors) have seen a variety of applications. An early use was troubleshooting, identifying malfunctioning valves and air flow meters (Dahlke et al., 2001). Measuring bubble size given by Microcel™ spargers helped justify their replacement of perforated rubber spargers in a flotation column (Hernandez-Aguilar et al., 2010). Comparison of a jet sparger, Microcel™, and a mechanical cell showed the latter was hard to match for fine bubble production (Finch et al., 2007b). Low recovery of Pt minerals at Lac des Iles was traced to large bubbles in the downstream cells of the bank, which was corrected by addition of frother (Hernandez-Aguilar et al., 2006). That same work also noted the differences in bubble size between lab, pilot, and the full size units, a problem in scaling kinetic data. Derived from a bubble size database, a bubble size prediction model has been used to benchmark operations (Figure 12.97), for example, to identify possible opportunities for bubble size reduction by addition of frother (Nesset et al., 2012). Several groups have addressed the role of impeller speed on bubble size, as reviewed by Amini et al. (2013). It would seem that in commercial scale cells the effect of tip speed over a wide range, 5–9 m s^{-1}, does not have much impact on bubble size. Impeller speed will affect bubble–particle collision rates and thus have an impact on flotation kinetics, but it does mean that seeking the source of increased kinetics with impeller speed (or power) does not have to consider any role of bubble size. Training sessions to transfer gas dispersion technology to operations have been known to identify opportunities, leading to improved performance (Sweet et al., 2013).

12.14.3 Cell Power and Hydrodynamics

This section started with parameters, some of which reflected power dissipation in the cell, then moved to hydrodynamics. Clearly we need to integrate both aspects. This is being aided by new instrumentation to measure flow patterns and turbulence in a cell (e.g., Yang and Telionis, 2012; Amini et al., 2013). Integration of power measurements and hydrodynamics is being applied to analyze the new generation of large tank cells, and coupled with CFD modeling, further advances are envisaged (Grönstrand et al., 2012).

12.15 CONTROL OF FLOTATION PLANTS

Process control is increasingly used in flotation circuits, the strategies being almost as numerous as the number of plants involved. The key when considering the metallurgical performance (i.e., grade and recovery) is online chemical analysis (Chapter 3), which produces real-time analysis of the metal composition of process streams. This being said, the most fundamental objective in industrial practice is to achieve effective regulatory control, enabling a steady operation (e.g., flowrates, densities, froth depth), which is an absolute prerequisite to reach and maintain target key production indicators. It must be emphasized that regulatory control does not generally rely on on-stream chemical analysis.

Control strategies are implemented in distributed control systems (DCS) or programmable logic controllers (PLC), and sometimes in advanced process control systems (additional hardware external to the plant main control system), mainly for model-based predictive control (MPC) and fuzzy logic (FL) applications. There are many vendor-supplied solutions available.

Although several successful applications have been reported, in reality few if any plants can claim to be fully automatic in the sense of operating unattended over extended periods. This is despite the availability of robust instrumentation, a wide range of control algorithms (PID (proportional-integral-derivative)-based, MPC, and FL), and powerful computing assets. McKee (1991) reviewed some of the reasons explaining this gap, observations that remain pertinent today. The main problems have been in first stabilizing a complex process in a sustainable manner, and then developing process models that will define set-points and limits to accommodate changes in ore type, mineralogy, texture, chemical composition of the mine water, and contamination of the feed.

Process control practitioners Ruel (2007) and Bouchard et al. (2010) have more recently identified fundamental reasons why process control has been unsuccessful in many cases, among them:

- inadequate installation and maintenance of instrumentation and control elements,
- faulty control strategy design,
- suboptimal or inappropriate controller tuning.

Ruel (2007) presented astonishing figures that 30% of control valve installations are in poor mechanical condition, 85% of PID controllers exhibit inadequate tuning, and 85% of control loops are not performing according to the design objective. With all these commonly encountered problems, only 25% of control loops perform better in automatic mode than in manual mode. Fortunately, these housekeeping issues can be tackled.

12.15.1 Instrumentation

It is essential that all online instrumentation be regularly serviced and calibrated according to a scheduled program. However, this alone is not sufficient. Instruments must be installed following industry best practices to avoid problems such as systematic biases, premature wear, catastrophic failure, or lack of accuracy and precision. For instance, pH probes must not only remain clean, but they also have to be judiciously positioned. Factors to be considered include:

- mixing of the pH modifier reagent,
- distance between the addition point and the measuring point.

As the latter distance increases, more time is allocated for the reagent to react and the mixing tends to improve, thus increasing the likelihood of exposing the probe to a representative sample. On the other hand, increasing the distance between the points of addition and measurement also increases the time before the effect of the reagent on the pH can be detected; or in other words, there is a process time delay. This is detrimental for both control performance and robustness. The optimal location will establish a compromise between the effects of adequate mixing and minimizing the time delay. Other instruments will likewise require specific best setup arrangements.

A comprehensive control system thus involves investment that may seem significant, but that generally represents only 1–1.5% of the total project capital outlay. Moreover, the payback period of process control projects typically lies within the first year following commissioning. Figure 12.98 shows the instrumentation requirement for a simple feedforward strategy, which could assist in control of a sulfide rougher bank; and Figure 12.99 depicts the instrumentation used in the 1970s at Mount Isa copper flotation circuit in Queensland, Australia (Fewings et al., 1979).

Lynch et al. (1981) analyzed the cost of such installations. The majority of plants that installed instrumentation for manual or automatic control purposes reported improved metal recoveries, varying from 0.5% to 3.0%, sometimes with increased concentrate grades. Reductions in reagent consumptions in the 10–20% range have also been reported. These figures are still valid today, but it must be emphasized that monitoring capabilities alone (i.e., the instrumentation) are generally not enough, and to enable real and sustained savings requires the implementation of control loops.

12.15.2 Process Control Objectives

Implementation of high-level control strategies, involving grades and recoveries, at the plant design stage is challenging, partly because the most significant control variables are often not identified until plant operational experience has been gained, and partly because process control standards have not yet been well defined. Even following commissioning, the training of production and metallurgical staff in the principles and application of process control systems is not always straightforward. A shortage of skilled control engineers exacerbates the situation. The most successful applications have been those where the control room operator can interact with the plant control system when necessary to adjust set-points and limits. This allows taking advantage of the complementarity qualities of the human, able to cope with extraordinary situations, and a PLC or a DCS, constantly vigilant, not being affected by shift changeovers, coffee breaks, and other interruptions.

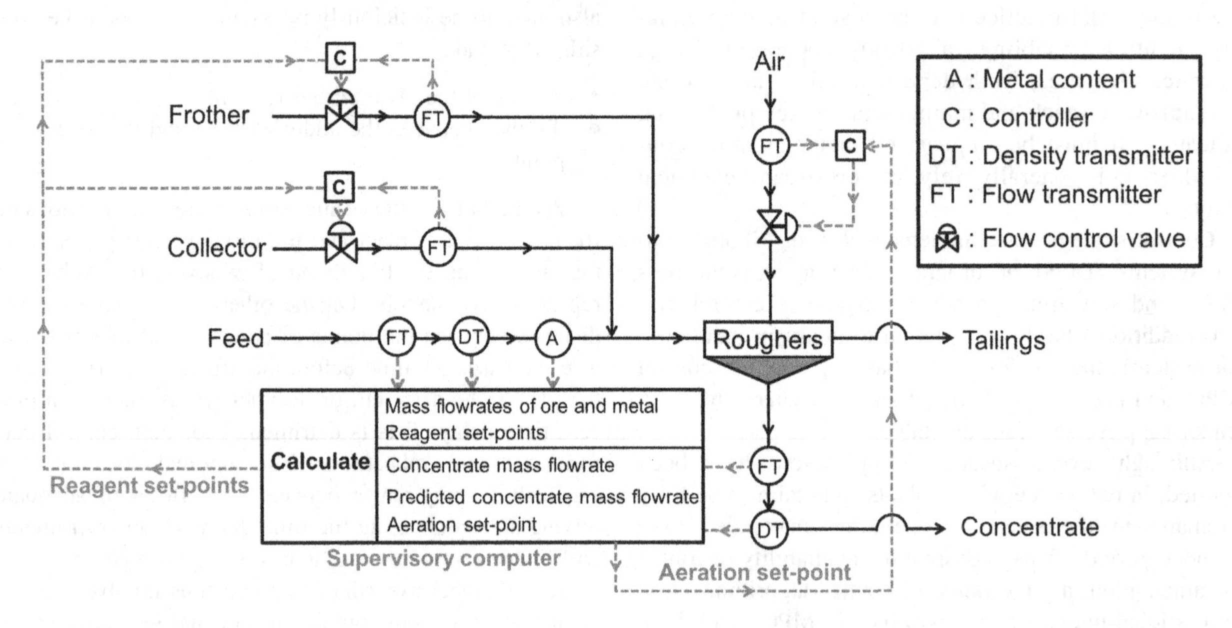

FIGURE 12.98 Instrumentation for rougher circuit control.

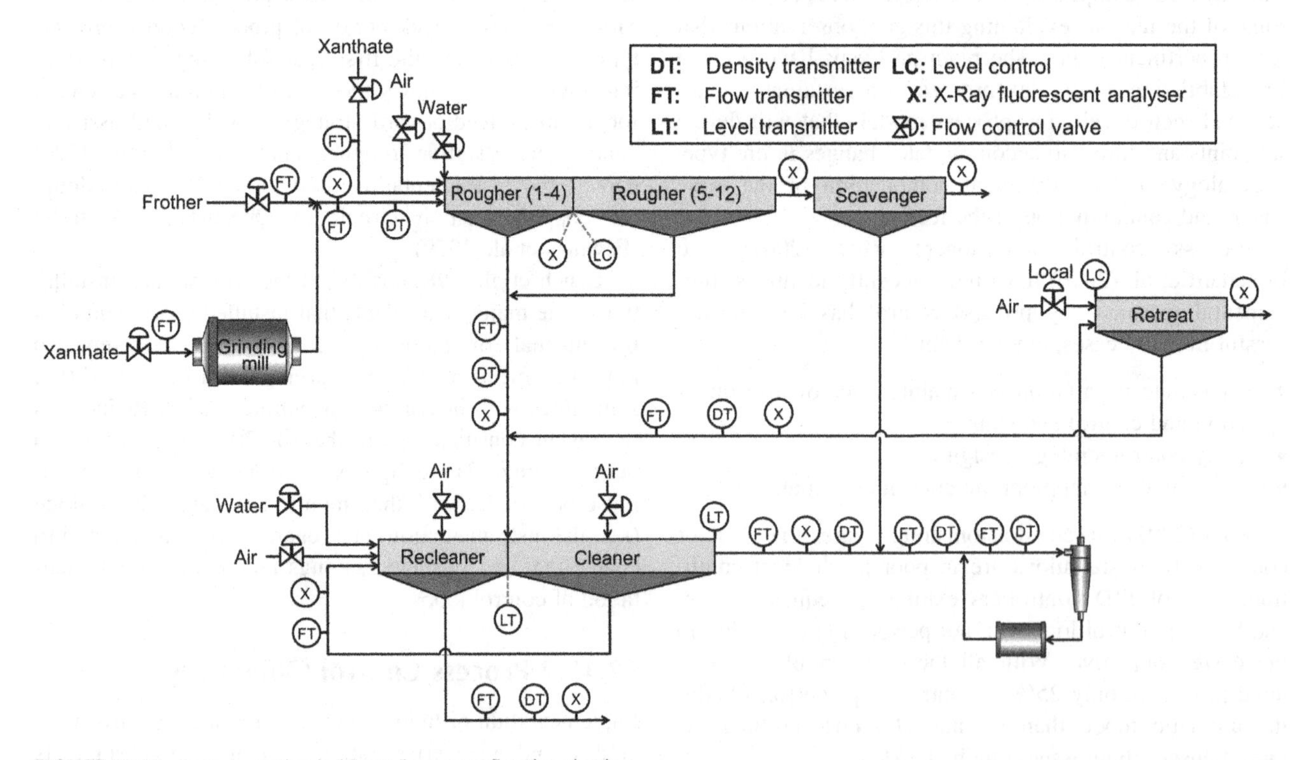

FIGURE 12.99 Instrumentation in Mount Isa copper flotation circuit.

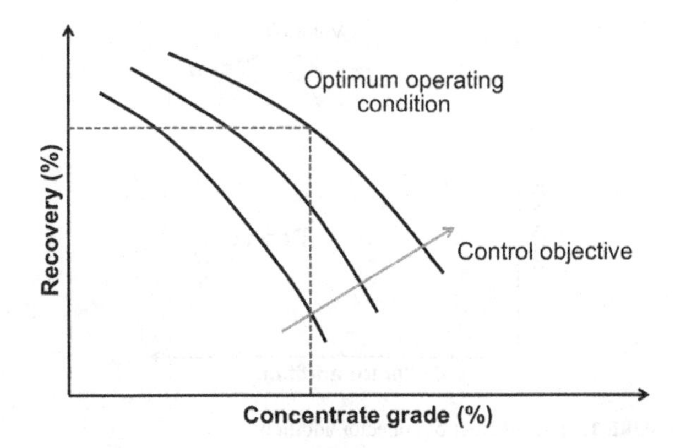

FIGURE 12.100 Flotation control objective.

The aim of high-level control strategies should be twofold:

1. improve metallurgical efficiency, that is, produce the best possible grade–recovery curve,
2. stabilize the process at the concentrate optimal grade.

Controlling the circuit at the optimal operating point will produce the most economic return based on throughput, as depicted in Figure 12.100, despite disturbances entering the circuit.

Disturbances caused by variations in feed rate, pulp density, and particle size distribution should be minimal if grinding circuit control is properly implemented. That being said, surge capacities, such as pump boxes and conditioning tanks, must be used to help maintain a steady throughput. Tight level control must then be avoided, as the slurry volume should be allowed to fluctuate within a certain operating range, thus allowing the buffering of disturbances, as long as complete cell drainage and pulp overflow are avoided.

The prime function of flotation control is to compensate for variations in mineralogy and floatability. Manually or automatically manipulated variables affecting this include: mass flowrates; reagent and air addition rates; pulp and froth levels; pH; and circulating loads through the control of cell-splits on selected banks. Best practice involves establishing basic control objectives, such as stabilizing control of pulp and sump levels, air, and reagent flows. More advanced stabilizing control can then be attempted, such as pH, reagent ratio control (based on plant input flows and assays), pulp flow, circulating load, concentrate grade, and recovery. Finally, true optimizing control can be developed, such as maximum recovery at a target grade. In any case, higher level optimizing control is generally not possible until stable operation has first been achieved (Bouchard et al., 2010). It is worth mentioning in this regard that simple, but properly implemented stabilizing control generally enables metallurgical gains deriving from the steadier operation.

12.15.3 Fundamental Controlled Variables: Pulp Level, Air Rate, pH, and Reagent Addition

The key variable to control is the pulp level in the cell, to ensure stable and efficient flotation performance. The pulp level can be measured by a number of different means. The most common and reliable system is a "float" resting at the froth/pulp interface, coupled with an ultrasonic sensor. The float is connected to a vertical shaft mounted with a plate target. The instrument measures the time sound waves take to reach the plate target and return to the source, and infers how far the float moves as the pulp level changes. A properly designed and maintained sensor of this kind is generally expected to be responsive (time constant less than a second), smooth, and nonoscillating. One problem that can occur is build-up of solids on the float altering its density.

Other systems such as conductivity probes, and differential pressure gauges are also in use in mineral processing plants. Conductivity probes register the difference in electrical conductivity between the froth and the pulp to determine the pulp level. Differential pressure cells are submerged in the flotation tank and measure the static head exerted by the slurry above. Image-based techniques have also been proposed (Jampana et al., 2009).

Control of pulp level is effected by dart valves or pinch valves. In older flotation plants, movable weirs are also used. In general, each bank of cells will have a level detection transducer (usually a float-based device) and the level is then controlled by a simple feedback PI loop which adjusts the valve on the bank tailings outlet based on a set-point either entered by the operator or determined by a higher-level control strategy responding to changes in grade, recovery, froth condition, or other criteria. Feedforward, in combination with feedback control, is often required to avoid disturbing interactions between different flotation banks. Feedforward control is based on feed flow measurement or inference (e.g., from a variable speed pump or preceding level controllers).

Level control can either be simple, as outlined above, or involve more complex interactions (Kämpjärvi and Jämsä-Jounela, 2003). Float-Star™, developed by Mintek in South Africa, is one example of an integrated package providing level control throughout a flotation circuit, and additional capabilities such as an algorithm to calculate optimum level set-points and/or aeration rates that aim to optimize the residence times, mass pulls, and circulating loads within a flotation circuit (Singh et al., 2003). Another, more current example, is Portage Froth

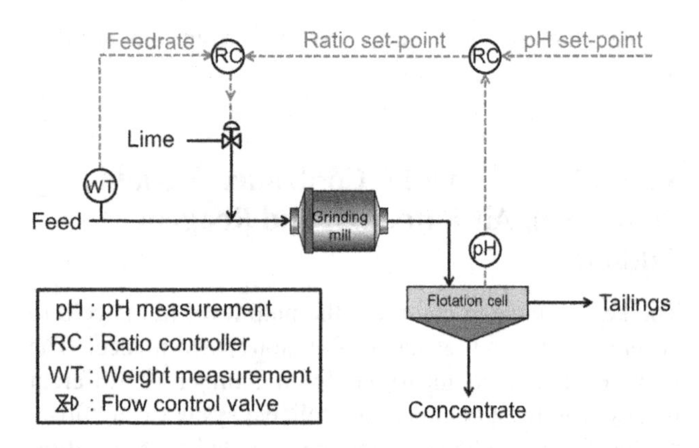

FIGURE 12.101 Control of pH in a flotation circuit.

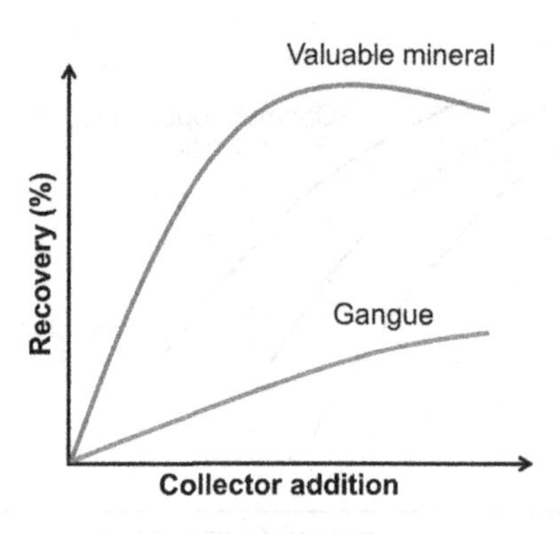

FIGURE 12.102 Effect of collector addition.

Characterization System® (PFC), from Portage Technologies in Canada, implemented in the control system (DCS or PLC) (Kewe et al., 2014), and allowing monitoring of froth characteristics (bubble size classification, velocity, stability, texture, and color) for process control purposes.

There is interest in monitoring and controlling the air flowrate to flotation cells which promise metallurgical gains (Section 12.10). In forced air flotation systems (e.g., tank cells and flotation columns), the available technology (flowrate sensors and butterfly valves) enables managing how much gas is introduced to the cell. Shean and Cilliers (2011) provide a comprehensive review on this topic. The case of self-aspirated cells is not as straightforward as the air addition rate is mainly dictated by the impeller speed and pulp level.

Control of slurry pH is an important requirement in many selective flotation circuits, the control loop generally being independent of the others, although in some cases the set-point is varied according to changes in flotation characteristics. Lime is often added to the grinding mills to minimize media corrosion and to precipitate heavy metal ions from solution. A typical pH control circuit is depicted in Figure 12.101. Lime addition is controlled by ratioing to the solids mass flow to the mill (feedforward control), with or without a bias term. A pH controller adjusts the ratio set-point from the difference between the pH measurement and a set-point established by the operator (feedback control). Lags are sufficient to allow appropriate mixing in the grinding mill.

Control of collector addition rate is sometimes performed by feedforward ratio control based on a linear response to assays or tonnage of valuable metal in the flotation feed. Typically, increase in collector dosage increases mineral recovery until a plateau is reached, beyond which further addition may either have no practical effect, or even a slight reduction in recovery may occur. The gangue recovery also increases with collector addition, such that beyond the plateau region selectivity is reduced, as illustrated in Figure 12.102. The operator can intervene to modify the ratio set-point (slope of the linear function) or bias (intercept of the linear function, see "reagent flowrate" equation below for an example) to respond to changing feed conditions.

The most common aim of collector control is to maintain the addition rate at the leading edge of the plateau, the main difficulty being in identifying this point, especially when the recovery–collector addition response changes due to changes in ore type, or interaction with other reagents. For this reason, automatic control using feedforward loops has rarely been successful in the long term. There are many cases of successful semiautomatic control, however, where the operator adjusts the set-point to accommodate changes in ore type, and the computer controls the reagent addition over fairly narrow limits of feed grade. For example, feedforward control of copper sulfate activator and xanthate to the zinc roughers has been used in the control strategy at Mattagami Lake Mines, Canada (Konigsmann et al., 1976). The reagents were varied in proportion to changes in feed grade according to a simple ratio/bias algorithm, which is a standard feature of all DCSs or PLCs:

$$\text{Reagent flow rate} = A + B \times \%\text{Zn in feed}$$

where A and B vary for different reagents. The operator may change the base amount A (bias) as different ore types are encountered. Almost four decades later, this kind of strategy has become part of standard practice in several plants. The Raglan Mine mill in Canada, for instance, controls the xanthate addition based on the combined feed grade of copper and nickel.

Although feedforward ratio control can provide a degree of stability, stabilization is more effective using feedback data. The time delay experienced with feedback loops utilizing tailings assays can be overcome to some extent by making use of the fact that the circuit begins to respond to changes in flotation characteristics immediately when the ore enters the bank. This can be detected by measurements in the first few cells. Controlling the rougher concentrate grade is a useful strategy, as this strongly influences the final cleaner concentrate grade.

The amount of frother added to the flotation system is an important variable, but automatic control has been unsuccessful in many cases, as the action of the frother is dependent on only very minor changes in addition rate and is much affected by intangible factors such as contamination of the feed, mine water chemistry, variation in particle composition, etc. At low addition rates, large bubbles are formed and the froth is unstable and recovery of minerals is low, whereas increasing the frother addition rate has a marked effect on increasing the flotation rate, increasing the mass pull and entrainment recovery thus reducing the grade of concentrate produced. The usual approach is to manually adjust the frother set-point, or less commonly to ratio the frother to the feed rate of solids and water.

12.15.4 Higher Level Controlled Variables: Mass Pull, Grade, and Recovery

From the perspective of controlling the metallurgical performance, frother dosage is not a controlled variable *per se*, but rather a manipulated variable as it highly impacts the flowrate of concentrate. Some systems capitalize on this feature. Cascade control can be used, where the concentrate grade controls the concentrate flowrate set-point, which in turn controls the frother addition set-point as illustrated in Figure 12.103.

Progress in froth imaging systems, with several off-the-shelf packages available today, opens new opportunities for froth velocity or mass pull control strategies manipulating the froth depth, gas (air) rate and/or the frother addition using standard feedback PID controllers.

Froth depth, gas rate, and frother addition directly affect recovery and can be used to control the concentrate grade, tailings grade, or mass flowrate of concentrate. Aeration and froth depth do not, however, have a lingering effect on subsequent operations (i.e., they provide local cell control), in contrast to frother (or other reagents), which are carried over from cell to cell and from previous stages. Aeration and froth depth are thus often used as primary manipulated variables. Flotation generally responds faster to changes in aeration than to changes in froth depth, and because of this, aeration is often a more effective control variable, especially where circulating loads have to be controlled. There is obviously interaction between frother addition, aeration, and froth depth, and where control loops are used, it is necessary to account for process dynamics and multivariable effects in the control system design. This can be achieved using fine/coarse control, where one variable (e.g., froth depth) is positioned (*coarse* control) in order to allow another (e.g., gas flowrate) to remain within a workable range to (*fine*) control the process variable of interest (e.g., concentrate grade). Some progress is reported in manipulating froth depth and air rate down a bank of cells, referred to as "profiling" (Section 12.11.2).

At Porgera in Papua New Guinea, Kewe et al. (2014) implemented a comprehensive system including vision cameras and on-stream analysis integrated with an expert system (AwaRE) strategy. The result was a stabilized circuit with reduced spillage and increased gold recovery by over 1.5%.

12.15.5 Advanced Control

The ultimate aim of a control system is to increase the economic efficiency of the process by seeking the optimal performance. There are several strategies that can be adopted to achieve this. Real-time optimization using quadratic programming has the potential for flotation

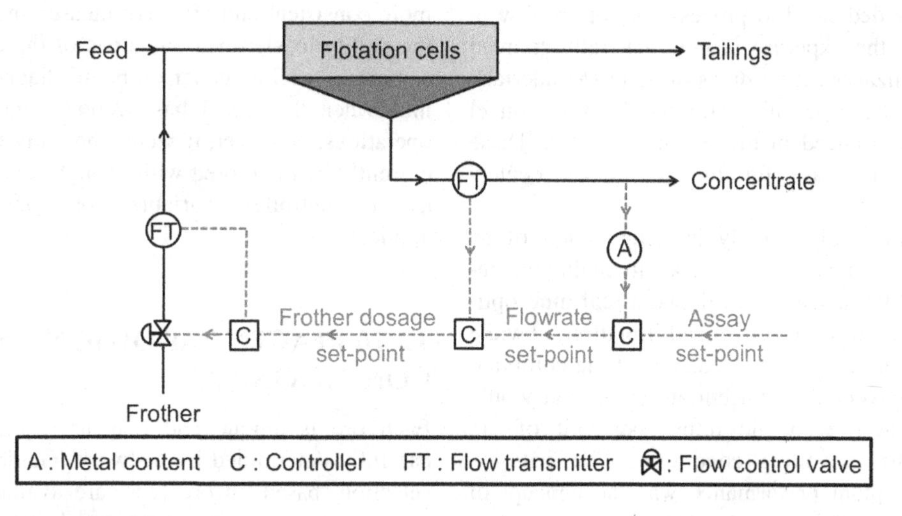

FIGURE 12.103 Cascade control of frother addition.

optimization, but has not yet been used. The method involves periodically adjusting the set-points of the controlled variables in order to minimize an objective function factoring in the economic efficiency. The optimization problem can be stated mathematically as (Edgar et al., 2001):

$$\begin{gathered}\operatorname{Min} f(x) \\ \text{subject to } a_i \le g_i(x) \le b_i \\ 1 \le x \le \mathrm{u}\end{gathered}$$

where x is a vector of n decision variables (i.e., the set-points) with lower and upper bounds l and u, f is the objective/cost function to be minimized, and g_i is the problem constraint (e.g., maximum tailings grade), with upper and lower limits a_i and b_i.

The set-points are then shifted slightly to move in the direction of the optimum, and the process is repeated until an optimum is encountered. Such methods cannot, however, be fully effective unless:

- satisfactory stabilization of plant performance can be achieved over long periods,
- normal operation does not rely on unmonitored/uncontrolled water or reagent addition manually adjusted in the field,
- maintenance issues (mechanical, instrumentation, etc.) are addressed in a timely manner.

McKee (1991) reviewed early attempts of process control applications based on metallurgical objectives, and a few examples from the 1980s were reported by practitioners (Thwaites, 1983; Twidle et al., 1985; Miettunen, 1983). Interest seems to have faded over the years, and plant practice today does not generally rely on process control systems explicitly based on economic criteria. Given how process control capabilities have leapt forward with technological progress in the last decades, this may seem paradoxical. However, considering that plant engineers exclusively dedicated to process control are few in number and that the expertise is now generally sourced outside the organization, it becomes difficult to undertake long term projects, especially with the high personnel turnover rate experienced in the mining industry. These problems make the case for remote control centers (Chapter 3).

A good example of an early implementation of an advanced control strategy is the Black Mountain concentrator in South Africa, which developed a real-time optimization application to control lead flotation (Twidle et al., 1985). Optimizing control calculated the combination of metal recovery and concentrate grade that would achieve the highest economic return per unit of ore treated under the prevailing conditions. The criterion used to evaluate plant performance was the concept of economic efficiency (Chapter 1), in this case defined as the ratio between the revenue derived per ton of ore at the achieved concentrate grade and recovery, and that derived at the target grade and recovery. Target concentrate grade and recovery were calculated from the operating grade−recovery curve, which was continuously updated based on a 24-h data bank, to allow for changes in the nature of the ore, quality of grinding, etc. Many factors influence the optimum combination of recovery and grade, such as commodity prices, reagent and treatment costs, transport costs, etc. The fundamental principle of real-time optimization is that online multivariable linear regression models can predict both the concentrate grade and recovery. The coefficients of the models are continuously updated from the 24-h data bank. Independent variables determining the grade and recovery can be reagent additions, grades of rougher concentrate and cleaner tailings, feed grade, and throughputs. Some independent variables are controllable whilst others are not.

In the last 25 years, adaptive control (Thornton, 1991), expert systems (Kittel et al., 2001; Kewe et al., 2014), and neural networks (Cubillos and Lima, 1997) have all been applied to flotation systems with varying degrees of success. In practice, the sustainability of these types of strategies is always challenging, the use of automatic controllers typically tending to decrease sharply following commissioning.

The texture, velocity, and color of flotation froths are diagnostic of the flotation condition. Skilled operators can use this information to adjust set-points, particularly air addition rates. This function has now been implemented in machine vision systems, which measure these properties online (van Olst et al., 2000; Holtham and Nguyen, 2002; Kewe et al., 2014), allowing control systems to make use of froth characteristics in optimizing performance (Kittel et al., 2001; Kewe et al., 2014). Innovation continues in the field of vision technology, resulting in more consistent and effective measurement. This includes the ability to view various areas of the cell with a single camera as well as better on-board diagnostics that recognize when the signal has become degraded and alarms operations. However, it seems that only froth velocity can currently be monitored with enough robustness to be used as a controlled variable for long-term industrial applications.

12.16 REAGENT ADDITION AND CONDITIONING

Each ore is unique and reagent requirements must be carefully determined by testwork. Guidelines for reagent selection, based on ore type, are available from reagent suppliers based on experience with many operations and with prior testwork programs. One vital requirement of a collector or frother is that it becomes totally dissolved or dispersed prior to use. Suitable emulsifiers must be used if this condition is not apparent.

Selection of reagents must be followed by careful consideration of the points of addition in the circuit. It is essential that reagents are fed smoothly and uniformly to the pulp, which requires close control of reagent feeding systems and pulp flowrate. When possible, frothers are added last, as they do not react chemically and only require dispersion in the pulp and long conditioning times are unnecessary. Adding frothers early can result in froth formation in conditioning tanks due to entrained air, which could cause pulp overflow and potentially cause uneven distribution of the collector.

In flotation, the amount of agitation and consequent dispersion are closely associated with the time required for physical and chemical reactions between the reagents and the mineral surfaces. Conditioning prior to flotation is considered standard practice. Effective conditioning can potentially result in decreased flotation time (i.e., increased rate constant), which is perhaps the most economical way of increasing the capacity of a flotation plant.

Although it is possible to condition in a flotation machine, the practice is generally not economic, although stage-addition of certain reagents is common practice. Many circuits add collector down a bank, particularly at the transition from rougher to scavenger collection. Agitated tanks, into which reagents can be fed, may be interposed between the grinding mills and the flotation circuit as surge capacity, which can stabilize feed rate and grade from the mills. Alternatively, reagents may be added to the grinding circuit in order to ensure optimum dispersion. Tumbling action in the ball mill is ideal for reagent mixing, especially for the case of oily collectors, which require emulsifying and long conditioning times. An added advantage of conditioning in the mill is that the collector is present as new "fresh" mineral surface is being formed, before oxidation can take place. The disadvantage is that control of reagent addition rate can be difficult due to continual minor feed grade fluctuations, and the mill may have a high circulating load which could result in over-conditioning. Where close control of conditioning time is essential, such as in the selective flotation of polymetallic ores, special conditioning tanks may be incorporated. Stage addition of reagents can yield higher recoveries at lower cost than if all the reagent is added at the head of the flotation bank or circuit (Bazin and Proulx, 2001). It is common practice to employ distribution boxes where slurry is split to feed parallel flotation banks. In this case, it is necessary to ensure equal splits between the banks to ensure proper reagent dosing if done after the split, which is not necessarily a simple task.

It is a common finding that the effectiveness of a separation may occur within a narrow pH window, in which case the key to success lies with the pH controller. This is especially true in selective flotation where separation pH may vary from one stage to the next. This, of course, makes it vitally important to regulate reagents to ensure conditions that promote separation. In some sulfide flotation systems control of the pulp potential is important.

Initial pH control is often undertaken by adding dry lime to the fine ore-bins, which tends to reduce oxidation of sulfides. Final close pH control may be carried out on the classifier overflow. Care must be taken to keep lime slurry moving, because otherwise it forms a hard cement within the pipelines. Reagents are typically added via either positive displacement metering pumps or automatically controlled valves, where reagents are added in frequent short bursts from a ring main or manifold. Where small dosages are required, peristaltic pumps can be used, where rollers squeeze a carrier tube seated in a curved track, thus displacing the reagent along the tube. Solid flotation reagents can be fed by rotating disc, vibro, and belt feeders, but more commonly reagents are added in liquid or slurry form. Data from the pumps and feeders are logged to enable remote monitoring, typically in the control room.

Insoluble liquids such as pine oil are often fed at full strength, whereas water-soluble reagents are made up to fixed solution strengths, normally about 10%, before addition. Reagent preparation is typically performed on day shifts in most mills, under close supervision, to produce a 24-h supply. Long storage times of reagent solutions should be avoided.

12.17 FLOTATION FLOWSHEETS AND PLANT PRACTICE

Although flotation is increasingly used for nonmetallic and oxidized minerals, the main tonnage is currently sulfide minerals of copper, lead, and zinc, often associated in complex ores. Comprehensive reviews of the complete range of sulfide, oxide, and nonmetallic flotation separations can be found elsewhere (Fuerstenau et al., 2007). The following presents example flotation flowsheets for a variety of ore types. The examples were selected to illustrate the variety of reagents, machine types, and circuit configurations encountered in industry. The flowsheets have been taken from recent sources but may not represent current configurations. The first five are examples of sulfide processing, the next three non-sulfides, and the final two energy minerals. All the flowsheets have been redrawn from the originals to try to maintain some uniformity in their presentation. The accompanying descriptions

TABLE 12.8 Top Producing Copper Mines in 2010 (ICSG, 2010; Schlesinger et al., 2011)

Mine	Country	Capacity (kt a^{-1} Cu)
Escondida	Chile	1,330
Codelco Norte	Chile	950
Grasberg	Indonesia	780
Collahuasi	Chile	518
El Teniente	Chile	457

of flotation practice should be read also with reference to Sections 12.4–12.8.

12.17.1 Copper Ores

In 2013, world copper production was over 17 Mt, with over 30% originating from Chile (USGS, 2014). Significant tonnages were also produced in China (9%), Peru (7%), the United States (7%), and Australia (6%) (USGS, 2014). In 2010, four of the five top producing copper mines were located in Chile (Table 12.8).

Several copper-bearing minerals are economically extracted (Appendix I), many of which may occur in the same deposit. Copper sulfides in the upper part of an ore body are often oxidized, resulting in the presence of a cap zone containing secondary copper minerals such as malachite ($CuCO_3 \cdot Cu(OH)_2$) and azurite ($(CuCO_3)_2 \cdot Cu(OH)_2$). Dissolved copper may also pass below the water table into reducing conditions, where high grade secondary sulfides (e.g., covellite (CuS) and chalcocite (Cu_2S)) may form.

The development of flotation, coupled with the introduction of vast tonnage open-pit mining methods, had an enormous impact on the copper industry. This made economical the processing of the huge low-grade copper sulfide deposits known as *porphyries*. Higher demand for copper, coupled with decreasing ore grades, has resulted in the expansion of these operations and prompted the increase in grinding mill and flotation cell size to enable concentrators to process tens of thousands of tons of ore per day.

The exact definition of copper porphyry has long been the subject of debate among geologists. Porphyries are essentially very large oval or pipe-shaped deposits (commonly referred to as disseminated) containing on average 140 Mt of ore, averaging about 0.8% Cu and 0.015% Mo, and a variable amount of pyrite (Sutolov, 1975).

Copper sulfide minerals are readily floatable and respond well to thiol collectors such as xanthates, notably amyl, isopropyl, and butyl. Alkaline circuits of pH 8.5–12 are generally used, with lime controlling the pH and used to depress pyrite. Frother usage has shifted away from natural reagents such as pine oil and cresylic acids, to synthetic frothers such as MIBC and polyglycol-types. Cleaning of rougher concentrates is usually necessary to achieve economic smelter grades (25–50% Cu depending on mineralogy), and rougher concentrates as well as middlings must often be reground for maximum recovery, which is usually between 80% and 90%. Typical flotation feed is ground to ca. 80% – 250 μm, rougher concentrates being reground to ca. 80% – 75 μm for further liberation. Reagent consumption is typically in the range 1–5 kg lime t^{-1} ore, 0.002–0.3 kg t^{-1} of xanthate, and 0.02–0.15 kg t^{-1} of frother.

All porphyry copper deposits contain at least traces of molybdenite (MoS_2), and in many cases molybdenum is an important by-product. Molybdenite, which is naturally floatable, is separated from the copper minerals after regrinding and cleaning of the copper rougher concentrates. Regrinding to promote optimum liberation requires careful control, as molybdenite is a soft mineral which slimes easily and whose floatability decreases as particles become finer (see Section 12.3). Bulk copper–molybdenum concentrates are thickened, after which the copper minerals are depressed, allowing molybdenite to float. Cleaning is important as molybdenite concentrates are heavily penalized by the smelter if they contain copper and other impurities, and the final copper content is often adjusted by leaching in sodium cyanide, which easily dissolves chalcocite and covellite and some other secondary copper minerals. Chalcopyrite, however, does not dissolve in cyanide, and in some cases is leached with hot ferric chloride.

Copper depression is achieved by the use of a variety of reagents, sometimes in conjunction with prior heat treatment. Heat treatment is used to destroy residual flotation reagents and is most commonly achieved by the use of steam injected into the slurry. Depression of chalcopyrite is usually effectively accomplished by the use of sodium sulfide or sodium hydrosulfide. Nitrogen is often used as the flotation gas, which helps avoid loss of reagent by side oxidation reactions and maintains low pulp potential that aids chalcopyrite depression. Other copper depressants are "Nokes Reagent," a product of the reaction of sodium hydroxide and phosphorus pentasulfide, arsenic Nokes (As_2O_3 dissolved in Na_2S), and thioglycolic acid. Replacement of these environmentally challenging reagents with more benign polymer-based depressants is on-going area of research now showing industrial promise (Section 12.6.5). Ye et al. (1990) have shown that ozone conditioning can also effectively depress copper minerals. The molybdenite is floated using a light fuel oil as collector. There can often be high circulating loads of molybdenite in the common C-CS type circuit (Section 12.11.3) using columns as cleaners and mechanical cells as cleaner–scavengers.

Highland Valley Copper (Teck Resources) located in British Columbia, Canada, is an example of a concentrator that produces copper and molybdenum concentrates. In 2013, the operation processed ca. 45 Mt of ore (ca. 125,000 t d^{-1}) with an average grade of 0.29% Cu (Teck, 2014) producing ca. 113,000 t of copper in concentrate and ca. 2,700 t of molybdenum in concentrate. The major copper-bearing minerals in the ore are chalcopyrite and bornite and the major molybdenum-bearing mineral is molybdenite.

A simplified flowsheet is shown in Figure 12.104 (Damjanović and Goode, 2000; Hernandez-Aguilar, 2010) with an equipment summary in Table 12.9. Crushed ore is divided among three SAG (A, B, and C) and two fully-AG (D and E) grinding lines. Each SAG mill feeds two ball mills, with each AG mill feeding a single ball mill. The AG/SAG mills are in closed-circuit with the AG mills employing pebble crushing. All ball mills are in closed-circuit with hydrocyclones.

Fuel oil (0.1 g t^{-1} ore), PAX (1.1 g t^{-1}), and pine oil (5.5 g t^{-1}) are added in the grinding circuit, which operates at pH 9.2 by lime addition. The hydrocyclone overflow (55% passing 150 μm) from each grinding line reports to one of three bulk Cu–Mo rougher–scavenger–cleaner circuits (modified R-S-C, which includes regrinding of the rougher concentrate, Table 12.7). The scavenger concentrates are recycled to the head of the roughers with the scavenger tailings reporting to tailings impoundment. Dowfroth 250 (7.2 g t^{-1}) and PAX (2.6 g t^{-1}) are stage-added in the bulk rougher–scavenger stages. A, B, and C rougher concentrates feed a regrind ball mill that operates in closed circuit with the cyclone overflow feeding the cleaner (two banks of six cells) and recleaner (two banks of five cells) section (all Denver 300 DR machines). Recleaner tails are fed to the head of the rougher circuit with the concentrate feeding Cu–Mo separation. The bulk cleaning stage pH is maintained at 10.5.

The D + E flotation circuit is operated in a slightly different manner to the A, B, C circuits, with the regrind cyclone overflow feeding two parallel flotation columns (2.13 m diameter). The circuit configuration also allows for tailings from one column to be reground in closed-circuit with the second column acting as a recleaner. For all circuits the column concentrate feeds Cu–Mo separation. The final bulk concentrate (from all bulk flotation circuits) is ca. 36% Cu and 0.7% Mo (Hernandez-Aguilar, 2010).

The bulk concentrate thickener produces an underflow stream ca. 60% solids which is fed to a Cu–Mo separation conditioning tank where sodium hydrosulfide (45 g t^{-1}) addition is used for copper sulfide depression. Fuel oil (0.2 g t^{-1}) is also added as a molybdenite collector. Denver 30 DR cells are configured into 5 roughers and 13 scavengers. Carbon dioxide is used in the first cell to maintain a pH of 9.0. The remaining cells utilize nitrogen as flotation gas to maintain reducing conditions and limit oxidation (loss) of the hydrosulfide. Scavenger tailings constitute the copper concentrate with the scavenger concentrate being recycled to the head of the copper–molybdenum separation circuit. Rougher concentrate is reground and cleaned in two 1 m diameter columns with the concentrate feeding the molybdenum leach plant. Column tailings are recycled to the head of Cu–Mo separation. (Note: current arrangement now employs tank cells and two columns, Hernandez-Aguilar (2010).) Column concentrate is typically 49% Mo and 2.5–3% Cu. Selective copper leaching using ferric chloride in the molybdenum leach plant reduces the final copper content to less than 0.25%.

Bulatovic et al. (1998), Castro and Henriquez (1998), and Amelunxen and Amelunxen (2009) have reviewed the various reagents, machine types, and circuits employed in copper–molybdenum flotation plants.

One of the largest copper concentrators is at the Freeport mine (Grasberg) in Indonesia (Table 12.8). The plant was progressively expanded since initial start-up in 1972 from 7,500 to 200,000 t d^{-1} to compensate for the lower grade ore encountered as the open pit deepened. The principal copper mineral in the porphyry deposit is chalcopyrite. Gold and silver are also present in the primary ore, which in 2013 graded 0.76% Cu, 0.69 g t^{-1} Au (Freeport-McMoran, 2013). The operation boasts the largest known gold and third largest copper reserve in the world.

The Grasberg flotation circuit is large (comprising four concentrators), but fairly simple. After primary grinding to produce a flotation feed grind size of 15% passing 212 μm, the ore is conditioned with lime, frother, and collector, before being fed to the rougher flotation circuit, which comprises four parallel banks of nine Wemco 127 m^3 cells. The cleaner circuit consists of 14 column cells for primary and secondary cleaning and 12 85 m^3 mechanical cleaner–scavenger cells (i.e., C-CS configuration). The concentrate produced from the columns reports to final concentrate, while the concentrate from the scavengers is recycled back to the cleaner feed. In 2006, the Grasberg operation produced 610,800 t of copper and 58,500 g of gold and 174,500 g of Ag (Anon, 2014a). Typical copper and gold recoveries are 86 and 76%, respectively.

The move to ever larger cells driven by the processing of these large tonnage low-grade porphyry copper ores is further illustrated at Chuquicamata (Codelco Norte), another of the world's large flotation plants (Table 12.8). The newer A1, A2 rougher lines comprise 160 m^3 Outotec tank cells, and testing of larger cells is on-going.

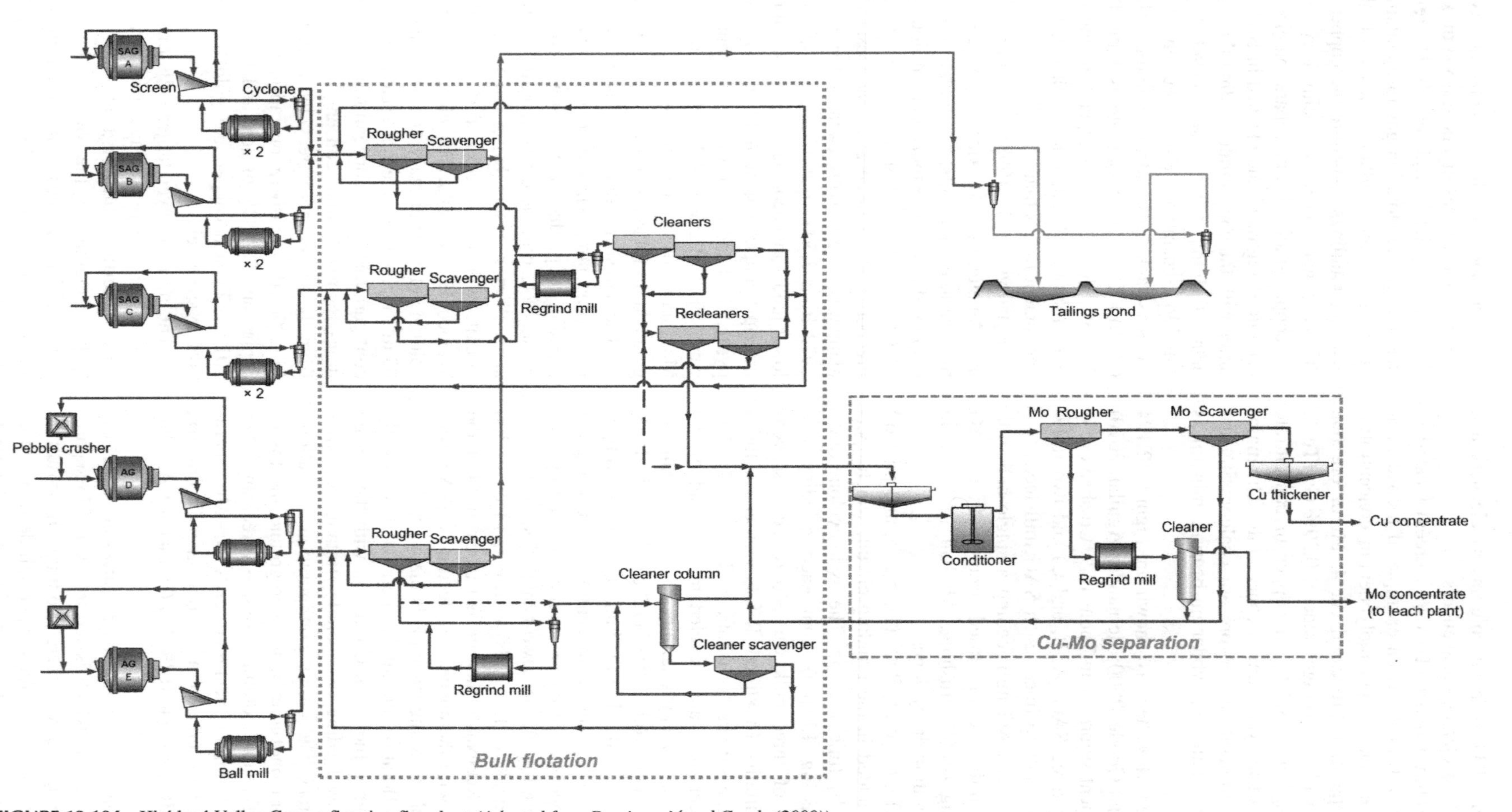

FIGURE 12.104 Highland Valley Copper flotation flowsheet *(Adapted from Damjanović and Goode (2000))*.

TABLE 12.9 Summary of HVC Bulk Flotation Circuits

Hydrocyclone Overflow from	No. Banks	Cell Type	No. Rougher Cells	No. Scavenger Cells
A + B	4	Denver 600 H-DR	8	14
C	4	Denver 1275 DR	3	5
D + E	4	Denver 300 DR	3	6

Note: Denver 600, 1275, and 300 refer to cell volume in ft^3.

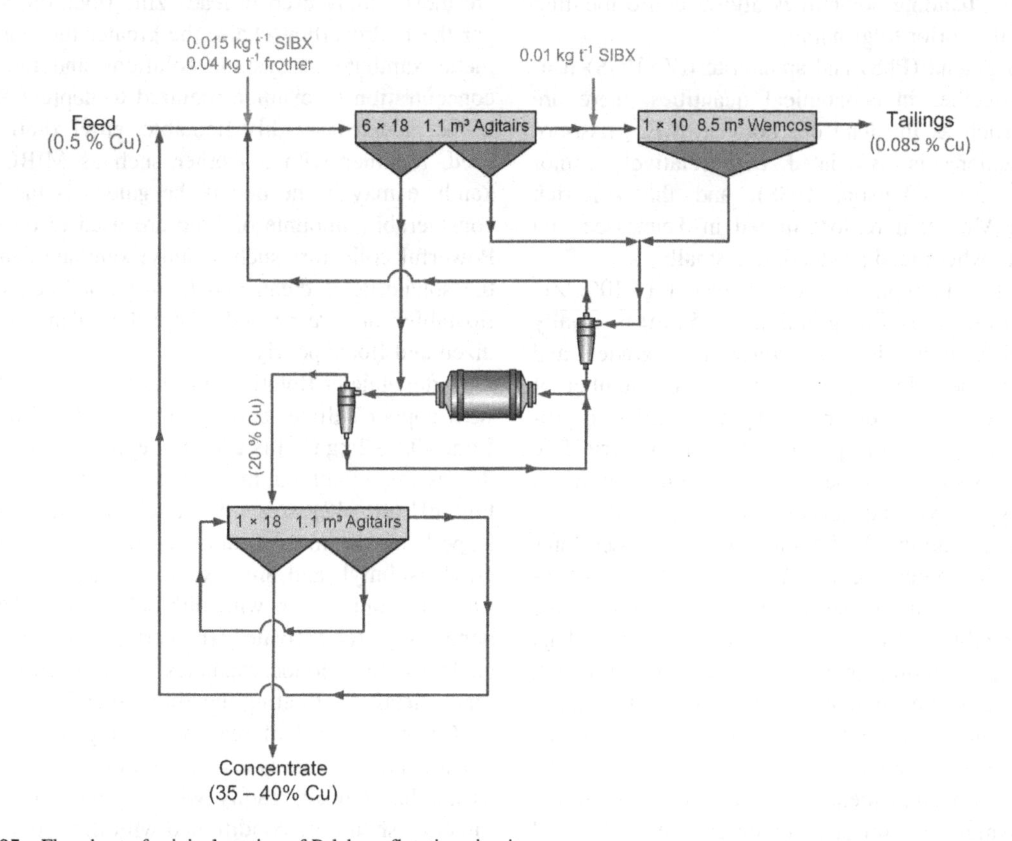

FIGURE 12.105 Flowsheet of original section of Palabora flotation circuit.

By-products play an important role in the economics of the Palabora Mining Co. in South Africa, which treats a complex carbonatite ore to recover copper, magnetite, uranium, and zirconium values. The ore assays ca. 0.5% Cu, the principal copper minerals being chalcopyrite and bornite, although chalcocite, cubanite ($CuFe_2S_3$), and other copper minerals are present in minor amounts. The flotation feed is coarse (80% − 300 μm) due to the high grinding resistance of the magnetite in the ore, which would increase grinding costs if ground to a finer size, plus the fact that the flotation tailings are treated by low-intensity magnetic separation to recover magnetite, and Reichert cone gravity concentration to recover uranothorite and baddeleyite both of which benefit from the coarser size.

The flotation circuit consists of eight separate sections, the last two sections being fed from an AG grinding circuit. The five parallel sections of the original Palabora flowsheet (Figure 12.105) each operate nominally at 385 t h^{-1}. Flotation feed is conditioned with sodium isobutyl xanthate and frother before being fed to the rougher flotation banks. The more readily floatable minerals, mainly liberated chalcopyrite and bornite, float in the first few cells. Collector is also added before the final scavenger cells to recover the less floatable minerals, such as cubanite and valleriite, the latter a copper–iron sulfide containing Mg and Al in the crystal lattice. Rougher and scavenger concentrates are reground to 90% −45 μm, before being fed to the cleaner circuit at a pulp density of 14% solids. This dilution is possible due to the rejection of magnetite and other heavy minerals into the tailings, and aids separation of the fine particle size produced after regrinding.

12.17.2 Lead–Zinc Ores

The bulk of the world's lead and zinc is supplied from deposits, which often occur as finely disseminated bands of galena and sphalerite, with varying amounts of pyrite, as replacements in various rocks, typically limestone or dolomite. This banding sometimes allows dense medium preconcentration prior to grinding.

Although galena (PbS) and sphalerite ((Zn,Fe)S) usually occur together in economical quantities, there are exceptions, such as the lead ore body in S.E. Missouri, where the galena is associated with relatively minor amounts of zinc (Watson, 1988), and the zinc-rich Appalachian Mountain region, mined in Tennessee and Pennsylvania, where lead production is small.

Feed grades are typically 1–5% Pb and 1–10% Zn, and although relatively fine grinding (<75 μm) is usually required, fairly high flotation concentrate grades and recoveries can be achieved. In an increasing number of cases, ultrafine grinding down to 10 μm is needed to produce acceptable flotation performance from very fine grained ores such as those at the Century mine in Australia. Typically, lead concentrates of 55–70% lead are produced containing 2–7% Zn, and zinc concentrates of 50–60% Zn containing 1–6% Pb. Sphalerite always contains some Fe in the lattice, which determines the maximum possible Zn concentrate grade. Although galena and sphalerite (including marmatite, sphalerite with high Fe) are the major ore minerals, cerussite ($PbCO_3$), anglesite ($PbSO_4$), and smithsonite ($ZnCO_3$) can also be significant and pose a challenge to flotation. In some deposits, the value of associated metals, such as silver, cadmium, gold, and bismuth, is almost as much as that of the lead and zinc, and lead–zinc ores are the largest sources of silver and cadmium.

Several processes have been developed for the separation of galena from zinc sulfides, but by far the most widely used method is that of two-stage selective flotation, where the zinc and iron minerals are depressed, allowing the galena to float, followed by the activation of the zinc minerals by copper sulfate addition in the lead tailings to allow a zinc float.

Lead (galena) flotation is usually performed at a pH of between 9 and 11. Low-cost lime is typically the preferred reagent for pH control. Not only does lime act as a strong depressant for pyrite, but it can also depress galena to some extent. Soda ash can be substituted in that case, especially when the pyrite content is relatively low.

Heavy metal ions, often present in process waters and derived from the ore, can accidentally activate sphalerite. Lime and soda ash addition in this case can play a deactivating role by precipitating (sequestering) the heavy metals. In most cases other depressants are also required, the most widely used being sodium cyanide (up to 0.15 kg t^{-1}) and zinc sulfate (up to 0.2 kg t^{-1}), either alone or in combination. These reagents are typically added to the grinding circuit, as well as to the lead flotation circuit, and their effectiveness depends on pulp alkalinity.

The effectiveness of depressants also depends on the concentration and selectivity of the collector. Xanthates are most widely used in lead–zinc flotation, and the longer the hydrocarbon chain, the greater the stability of the metal xanthate in cyanide solutions and the higher the concentration of cyanide required to depress the mineral. If the galena is readily floatable, ethyl xanthate may be used, together with a frother such as MIBC. Isopropyl xanthate may be needed if the galena is tarnished, or if considerable amounts of lime are used to depress pyrite. Powerful collectors such as amyl xanthate can be used if the sphalerite is clean and hydrophilic (i.e., highly nonfloatable) and are needed where the galena is highly oxidized and floats poorly.

After galena flotation, the tailings are usually treated with copper sulfate, between 0.2 to 1.6 g Cu per kg Zn. Lime (0.5–2 kg t^{-1}) is used to depress pyrite, as it has no depressing effect on the activated zinc minerals, and a high pH (10–12) is used in the circuit. Isopropyl xanthate is perhaps the most commonly used collector, although ethyl, isobutyl, and amyl xanthates are also used, sometimes in conjunction with dithiophosphate, depending on conditions. As activated sphalerite behaves similarly to chalcopyrite, thionocarbamates are also common collectors, selectively floating the zinc minerals from the pyrite.

Careful control of reagent feeding must be observed when copper sulfate is used in conjunction with xanthates, as xanthates react readily with copper ions. Ideally, the minerals should be conditioned with the activator separate from the collector, so that when the conditioned slurry enters the collector conditioner there is little residual copper sulfate in solution. Although the activation process is fairly rapid in acidic or neutral conditions, in practice it is usually carried out in an alkaline circuit used to depress pyrite flotation, and a conditioning time of ca. 10–15 min is required to make full use of the reagent. The alkali precipitates the copper sulfate as basic compounds which are sufficiently soluble to provide a reservoir of copper ions for the activation reaction. As noted in Section 12.6.1 some operations have switched, to add copper sulfate before final raising of pH.

The fine-grained nature and complexity of some lead–zinc ores has led to the need for ultrafine grinding. Flotation was undertaken at MacArthur River Mine in Australia on material ground to 12 μm to produce a bulk lead–zinc concentrate. At Mount Isa Mines rougher concentrates of lead and zinc are reground to 10 and 15 μm, respectively, prior to cleaner flotation (Young and Gao, 2000). At the Century Mine, zinc concentrates are reground to below 10 μm to effectively liberate fine-grained silicates (Burgess et al., 2003). Ultrafine grinding now commonly employs stirred mills which are more efficient than conventional tumbling mills for this duty (Chapter 7). Due to the high intensity of ultrafine grinding, inert grinding media is often used to prevent release of depressant iron oxy-hydroxyl complexes. The production of ultrafine concentrates usually results in tenacious froths, with pulping and material handling problems being common.

In some cases sphalerite is activated by copper ions released from copper minerals in the ore to such an extent that depression of sphalerite fails, even when the most powerful combinations of reagents, such as zinc sulfate and cyanide, are used. (An example where this problem was overcome by judicious combination of reagents is discussed below under "Cu–Zn ore" processing, Section 12.17.3.) Bulk flotation of lead and zinc minerals may in such cases have a number of economic advantages. Coarse primary grinding is often sufficient with bulk flotation, as the valuable minerals need be liberated only from the gangue, not from each other. The flotation circuit design is normally relatively simple. In contrast, selective flotation calls for finer primary grinding, in order to free the valuable minerals not only from the gangue, but also from each other. This increases grinding mill size and energy requirements.

However, the production of bulk lead–zinc concentrates is only reasonable if there are smelters which are equipped for such concentrates. The only smelting process available is the *Imperial Smelting Process* (ISP), which was developed at a time when most lead and zinc was recovered from low-pyrite ores. In recent years lead and zinc are increasingly being recovered from complex and highly pyritic ores which is problematic for the ISP process.

Bulk flotation followed by separation can sometimes be used, although subsequent depression of activated sphalerite and pyrite is difficult. Every attempt is made at plants using bulk flotation to use a minimum amount of collector. Bulk flotation followed by selective flotation is performed at the Zinkgruvan mine, Sweden (Anon., 2014b).

12.17.3 Copper–Zinc and Copper–Zinc–Lead Ores

Copper–Zinc Ores

In the flotation of copper–zinc ores, where lead is absent (or is not present in economic quantities), lime is almost universally used to control pH in the range 8–12. In certain instances, the addition of lime to the grinding mills and flotation circuit is sufficient to prevent the flotation of zinc minerals (by precipitating activating heavy metal ions), but in most cases supplementary depressants are required. Sodium cyanide is often added in small quantities (0.01–0.05 kg t^{-1}) to the grinding mills and cleaners. Zinc sulfate is also used in conjunction with cyanide, and in some cases sodium sulfite, bisulfite, or sulfur dioxide depressants are used.

After conditioning, the copper minerals are floated using either xanthates, or if the mineralogy allows, a selective copper collector such as thionocarbamate. Typically, copper concentrates contain 20–30% Cu and up to 5% Zn. Copper flotation tailings are activated with copper sulfate and zinc minerals floated as described above for Pb–Zn ores.

One example of a plant that uses cyanide as a depressant is the Canatuan Cu–Zn operation (TVI Resources) in the Philippines (Umipig et al., 2012). The operation began processing ore of 1.5–2% Cu with low levels of Zn (Cu:Zn ratio of > 5:1). The original plant produced acceptable copper grades by grinding to ca. 120 μm and flotation at a pH of 11.5–12 (using lime) with IPETC (45 g t^{-1}) as collector and Nasfroth HEL frother. No regrinding or cleaning stages were necessary.

In year three of operation, the ore transitioned to higher zinc, which prompted the construction of a Cu–Zn separation circuit. The high zinc grades, coupled with the presence of soluble secondary copper sulfides (67% of copper in the feed is present as chalcocite and 10% as covellite) which released Cu ions into the pulp, led to zinc "super activation." An extensive laboratory test program led to the development of a complex reagent addition scheme (Figure 12.106) to combat zinc activation. The inclusion of sodium cyanide as a reagent necessitated the implementation of a cyanide detoxification (destruction) circuit.

The new flowsheet (2,500 t d^{-1} ca. 1.2% Cu and 1.3% Zn) has bulk flotation at pH 11.8 with lime, again using IPETC, but with downstream addition of PAX. The bulk rougher tailings are discarded to the tailings pond with the bulk concentrate reporting to a regrind circuit (P_{80} ~ 45 μm), followed by Cu–Zn separation (Cu rougher) flotation. Sodium metabisulfite (SMBS) is added to the regrind mills, followed by aeration. The SMBS/aeration step at pH < 7 was key to the collector desorption necessary to effect Cu–Zn separation. The action appears to be due to formation of SO_2, which in combination with oxygen forms a powerful oxidizing environment. The Cu-rougher circuit operates at a pH of ca. 10.5, with sodium cyanide and zinc sulfate added to depress (deactivate) the sphalerite. Potassium ethyl xanthate (PEX) is used as a collector in the copper roughers with sodium sulfide stage added to control low pulp potential (−200 mV on SHE scale) and precipitate Cu ions. The Cu circuit concentrate assays ca. 24% Cu and 11% Zn.

The Cu rougher tailings feed the Zn flotation circuit, which consists of a rougher and four counter-current

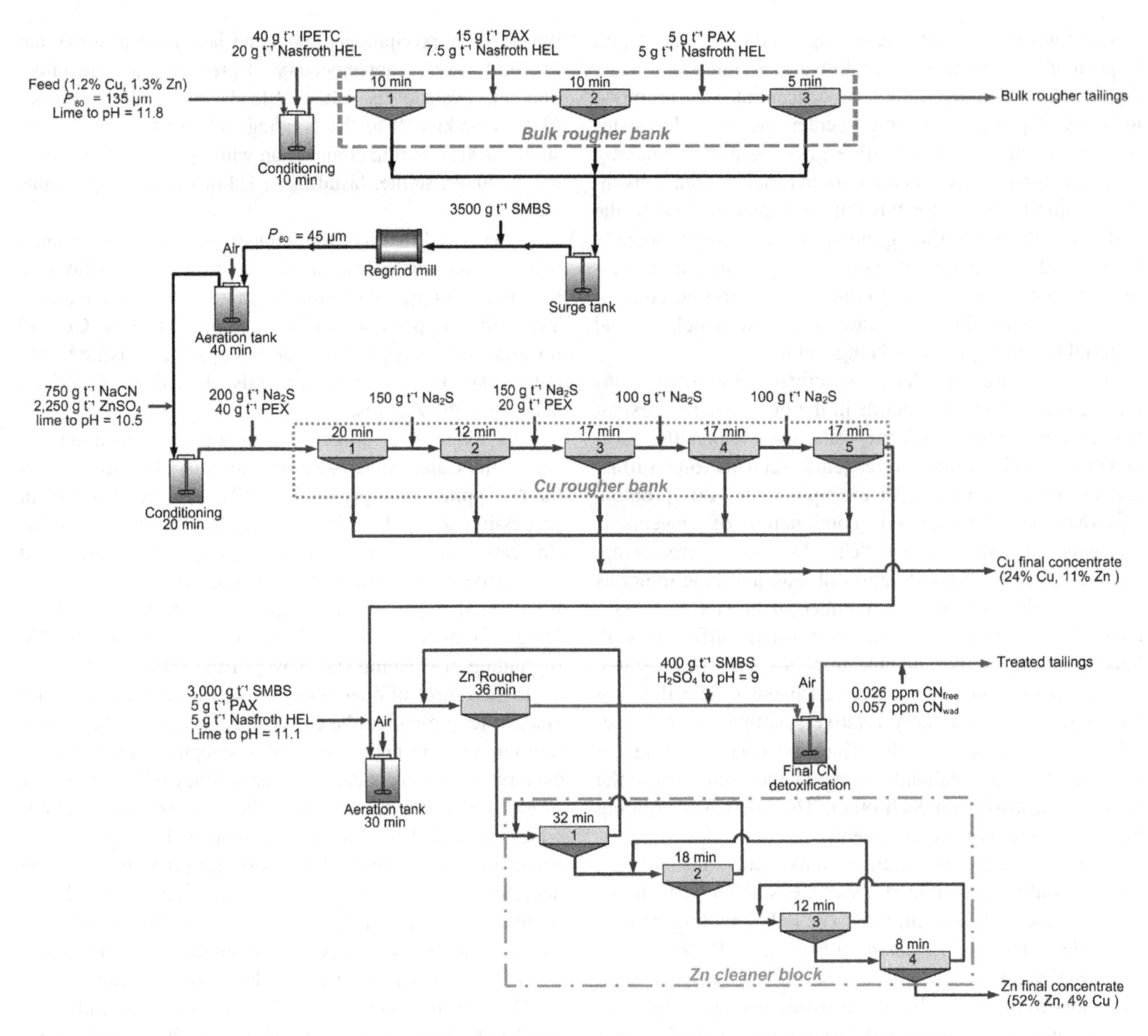

FIGURE 12.106 Canatuan flowsheet (note, IPETC is isopropyl ethyl thionocarbamate) *(Adapted from Umipig et al. (2012)).*

cleaner stages. The reagents (lime, SMBS, PAX, and Nasfroth HEL) are added ahead of an aerator which is held at pH 11. This aeration step effects partial destruction of the cyanide, again likely through SO_2/O_2 creating strong oxidizing conditions. Notably, copper sulfate addition was not required, the "natural" release of Cu ions providing the necessary sphalerite activation. The zinc rougher tailings undergo final cyanide detoxification (final discharge: CN_{free} ~ 0.026 ppm, CN_{wad} ~ 0.057 ppm) and then report to the tailings pond. The final zinc concentrate is ca. 52% Zn and 4% Cu.

Copper–Lead–Zinc Ores

The production of separate concentrates from copper–lead–zinc ores is complicated by the similar metallurgy of chalcopyrite and activated zinc minerals. The mineralogy of many of these ores is a complex assembly of finely disseminated and intimately associated chalcopyrite, galena, and sphalerite in a gangue consisting predominantly of pyrite or pyrrhotite (iron sulfides often constitute 80–90% of the ore), quartz, and carbonates. Such massive sulfide ores of volcanosedimentary origin are also a valuable source of silver and gold. Ore grades are 0.3–3% Cu, 0.3–3% Pb, 0.2–10% Zn, 3–100 g t^{-1} silver, and 0–10 g t^{-1} gold, on average.

The major processing problems encountered are related specifically to the mineralogy of the assemblies. Due to the extremely fine dissemination and interlocking of the minerals, extensive fine grinding is often required, usually below 75 μm. The New Brunswick deposits in Canada required grinding to 80% − 40 μm.

In most cases, concentrates are produced at relatively poor grades and recoveries, typical grades being:

	Cu (%)	Pb (%)	Zn (%)
Copper concentrates	20–30	1–10	2–10
Lead concentrates	0.8–5	35–65	2–20
Zinc concentrates	0.3–2	0.4–4	45–55

Recoveries of 40–60% for copper, 50–60% for lead, and 70–80% for zinc are typical for New Brunswick deposits. Smelting charges become excessive with contaminated concentrates, as very rarely is a metal paid for when it is not in its proper concentrate and penalties are often imposed for the presence of zinc and lead in copper concentrates. Silver and gold are well paid for in copper and lead concentrates, whereas payment in zinc concentrates is often zero.

A wide variety of flowsheets are in use involving sequential flotation or bulk flotation of copper and lead minerals followed by separation (Figure 12.52). Bulk flotation of all economic sulfides from pyrite has also been investigated.

The method most widely used to treat ores containing economic amounts of lead, copper, and zinc is to initially float a bulk lead–copper concentrate, while depressing the zinc and iron minerals. The zinc minerals are then activated and floated, while the bulk concentrate is treated by the depression of either the copper or lead minerals to produce separate concentrates.

The bulk float is performed in a moderately alkaline circuit, usually at pH 7.5–9.5, lime, in conjunction with depressants such as cyanide and zinc sulfate, being added to the grinding mills and bulk circuit. Depression of zinc and iron sulfides is sometimes supplemented by the addition of small amounts of sodium bisulfite or sulfur dioxide to the cleaning stages, although these reagents should be used sparingly as they can also depress galena.

The choice and dosage of collector used for bulk flotation are critical not only for the bulk flotation stage but also for the subsequent separation. Xanthates are commonly used, including combinations, and while a short-chain collector such as ethyl xanthate gives high selectivity in floating galena and chalcopyrite and permits efficient copper–lead separation, it does not allow high recoveries into the bulk concentrate, particularly of the galena. Much of the lost galena subsequently floats in the zinc circuit, contaminating the concentrate, as well as representing an economic loss. Because of this, a powerful collector (i.e., longer chained) such as amyl or isobutyl xanthate is commonly used, and close control of the dosage is required. Usually, fairly small collector additions, between 0.02 and 0.06 kg t^{-1}, are used, as an excess makes copper–lead separation difficult, and large amounts of depressant are required, which may depress the floating mineral.

Although the long-chain collectors improve bulk recovery, they are not as selective in rejecting zinc, and sometimes a compromise between selectivity and recovery is needed, and a collector such as isopropyl xanthate is chosen. Dithiophosphates, either alone or in conjunction with xanthates, are also used as bulk float collectors, and small amounts of thionocarbamate may be used to increase copper recovery.

The choice of the method for separating copper and lead minerals depends on the response of the minerals and their relative abundance. It is preferable to float the mineral present in least amount, for example, galena depression is usually performed when the ratio of lead to copper in the bulk concentrate is greater than unity.

Lead depression is also undertaken if economic amounts of chalcocite or covellite are present, as these minerals do not respond to depression by cyanide, or if the galena is oxidized or tarnished and does not float readily. It may also be necessary to depress the lead minerals if the concentration of copper ions in solution is high, as may be the case due to the presence of secondary copper minerals in the bulk concentrate. The standard copper depressant, sodium cyanide, combines with these ions to form complex cuprocyanides, thus reducing free cyanide ions available for copper depression. Increase in cyanide addition only serves to accelerate the dissolution of secondary copper minerals.

Depression of galena is achieved using sodium dichromate, sulfur dioxide, and starch in various combinations, whereas copper minerals are depressed using cyanide, or cyanide–zinc complexes. Methods of depression used at various concentrators can be found elsewhere (Wills, 1984).

In some plants, galena depression is aided by heating the slurry to about 40°C by steam injection. Kubota et al. (1975) showed that galena can be completely depressed, with no reagent additions, by raising the slurry temperature above 60°C. The xanthate adsorbed on the galena is removed, but that on the chalcopyrite surface remains. It is thought that preferential oxidation of the galena surface at high temperature is the mechanism for depression. At Woodlawn in Australia, the lead concentrate originally assayed 30% Pb, 12% Zn, 4% Cu, 300 ppm Ag, and 20% Fe, and received very unfavorable smelter terms (Burns et al., 1982). Heat treatment of the concentrate at 85°C for 5 min, followed by reverse flotation, gave a product containing 35% Pb, 15% Zn, 2.5% Cu, 350 ppm Ag, and 15% Fe, with improved smelter terms.

At the Brunswick Mining concentrator in Canada (McTavish, 1980; Damjanović and Goode, 2000), the bulk copper–lead concentrate is conditioned with 0.03 kg t^{-1} of a wheat dextrin–tannin extract mixture to depress the galena, and then the pH is lowered to 4.8 with liquid SO_2. The slurry is further conditioned for 20 min at this low pH, then 0.005 kg t^{-1} of thionocarbamate is added to float the copper minerals. The rougher concentrate is heated by steam injection to 40°C, and is then cleaned three times to produce a copper concentrate containing 23% Cu, 6% Pb, and 2% Zn. The lead concentrate produced is further upgraded by regrinding the copper separation tails, and then heating the slurry with steam to 85°C, and conditioning for 40 min. Xanthate and dithiophosphate collectors are then added to float pyrite. The rougher concentrate produced is reheated to 70°C and is cleaned once. The hot slurry from the lead upgrading tailings contains about 32.5% Pb, 13% Zn, and 0.6% Cu, and, after cooling, is further treated to float a lead–zinc concentrate, leaving a final lead concentrate of 36% Pb and 8% Zn. The Brunswick concentrator experienced a marked summer–winter change in metallurgical performance which correlated with changes in the thiosulfate levels in the recycle waters drawn from the tailings pond. Other operations in Canada, at least, note a similar shift in metallurgy with the season.

A simplified Brunswick flowsheet is shown in Figure 12.107 (Orford et al., 2005). It is similar to the type "c" circuit in Figure 12.52. Some specific features are: the Cu–Pb circuit includes a flotation stage after the SAG mill ("stage flotation") aimed at early removal of galena to reduce release of activating Pb ions; the Zn circuit is an example of the R-C-SC arrangement (Table 12.7), in this case with four cleaning stages (i.e., R-C_4-CS); the Pb upgrading circuit produces a pyrite concentrate primarily sent to tailings but with a side stream (indicated by dashed arrow) going to the Pb concentrate to maintain the pyrite "fuel" demanded by the Pb smelter. Although now closed (as of 2013), the experiences at the Brunswick Mine concentrator detailed in Orford et al. (2005), including the extensive use of the six sigma statistical toolbox, should prove invaluable to future Cu–Pb–Zn operations.

In general, where the ratio of lead to copper in the bulk concentrate is less than unity, depression of the copper minerals by sodium cyanide may be preferred. Where standard cyanide solution may cause unacceptable dissolution of precious metals and small amounts of secondary copper minerals, a cyanide–zinc complex can sometimes be used to reduce these losses. At Morococha in Peru (Pazour, 1979), a mixture of sodium cyanide, zinc oxide, and zinc sulfate has been used, allowing a recovery of 75% of the silver in the ore (Ag head grade ca. 120 g t^{-1}). Zinc oxide can prove superior to zinc sulfate for sphalerite depression, but is not necessarily the economic choice (Umipig et al., 2012).

Close alkalinity control is necessary when using cyanides, a pH of between 7.5 and 9.5 being common, although the optimum value may be higher, dependent on the ore. Cyanide depression is not used if economic quantities of chalcocite or covellite are present in the bulk concentrate, since it has little depressing action on these minerals. As cyanide is a very effective sphalerite depressant, most of the zinc reporting to the bulk concentrate is depressed into the copper concentrate, which may incur smelter penalties. Cyanide, however, has little action on galena, allowing effective flotation of the galena from the chalcopyrite, and hence a low lead copper concentrate. Lead is never paid for in a copper concentrate and is often penalized.

In a few cases, adequate metallurgical performance cannot be achieved by bulk flotation, and sequential

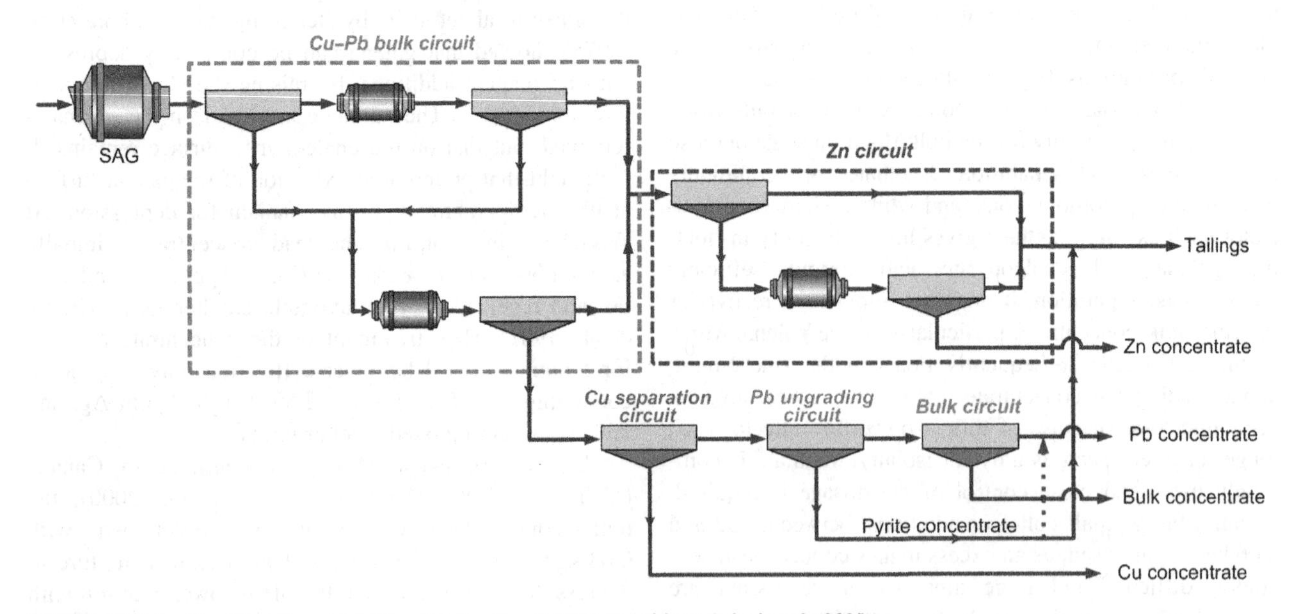

FIGURE 12.107 Simplified Brunswick Mine flotation flowsheet *(Adapted from Orford et al. (2005)).*

selective flotation must be performed. This necessarily increases capital and operating costs, as the bulk of the ore—often iron sulfide gangue minerals—is present at each stage in the separation, but it allows use of selective reagents to suit the mineralogy at each stage. The general flowsheet for sequential flotation involves conditioning the slurry with SO_2 at pH 5–7, and using a selective collector such as ethyl xanthate, dithiophosphate, or thionocarbamate, which allows a copper concentrate that is relatively low in lead to be floated. The copper tailings are conditioned with lime or soda ash, xanthate, sodium cyanide, and/or zinc sulfate, after which a lead concentrate is produced, the tailings being treated with copper sulfate prior to zinc flotation.

Sequential separation is required where there is a marked difference in floatability between the copper and lead minerals, which makes bulk rougher flotation and subsequent separation of the minerals in the bulk concentrate difficult, as at the Black Mountain concentrator in South Africa (Beck and Chamart, 1980). Metallurgical development at Woodlawn in Australia was an ongoing process. The original circuit, designed to depress lead with dichromate, was never effective for various reasons, and a combination of bulk and sequential flotation was then used (Roberts et al., 1980; Burns et al., 1982). The feed, containing roughly 1.3% Cu, 5.5% Pb, and 13% Zn, was conditioned with SO_2, starch, SMBS and a dithiophosphate collector, after which a copper concentrate was produced, which was cleaned twice. The copper tailings were conditioned with lime, NaCN, starch, and secondary butyl xanthate prior to flotation of a lead concentrate, which contained the less floatable copper minerals. This concentrate was reverse cleaned by steam heating to 85°C prior to flotation of the copper minerals with no further reagent addition. The floated copper minerals were pumped to the initial copper cleaning circuit. Lead rougher tailings fed the zinc roughing circuit.

12.17.4 Nickel Ores

Nickel is produced from two main sources: sulfidic ores and lateritic ores. Seventy percent of land-based nickel resources are contained in lateritic deposits, though the majority of the world's current production of nickel still derives from sulfidic sources (Bacon et al., 2002). The dominant nickel mineral in these deposits is pentlandite—$(NiFe)_9S_8$. However, many ores also have minor amounts of millerite (NiS) and violarite (Ni_2FeS_4). Nickel can also be found within the pyrrhotite (Fe_8S_9) lattice, substituting for iron. In some Sudbury area deposits of Canada, up to 10% of the nickel is in pyrrhotite (Kerr, 2002). Depending on the downstream smelting requirements, nickel flotation can occur by two processes: bulk sulfide flotation (e.g., in Western Australia's nickel operations) or separate Ni-mineral flotation (e.g., Canada's Sudbury area). In addition to iron sulfides, nickel often occurs with economic concentrations of copper (Sudbury), cobalt (Western Australia), and precious metals such as gold and platinum-group metals (e.g., Sudbury operations, the Noril'sk operation in northwest Siberia, and in the Bushveld Complex in South Africa). Kerr (2002) reviewed six of the major nickel flotation operations, which include those in Sudbury, Western Australia, and Russia.

Figure 12.108 shows an example nickel–copper processing flowsheet from Vale's Clarabelle mill (Doucet et al., 2010). Typical head grades are ca. 1.2% Ni (primarily pentlandite) and 1.3% Cu (primarily chalcopyrite), 20% pyrrhotite, and 75% NSG. The feed first undergoes magnetic separation. The magnetic stream (rich in pyrrhotite) is reground and feeds a flotation circuit in rougher–cleaner arrangement. Rougher tailings (largely comprising magnetic pyrrhotite) are discarded as final sulfide tailings, while the cleaner concentrate reports to final nickel concentrate. Cleaner tailings are recycled to the regrind mill. The non-magnetic stream reports to a rougher (A + B)-scavenger circuit. The rougher A concentrate comprises the fast floating copper minerals, which are upgraded in a copper rougher–scavenger–column cleaner arrangement (Cu circuit) to produce copper concentrate, and the copper scavenger tailings reporting to the final nickel concentrate. (The Cu circuit includes recycle of column tails to the scavenger, making it an R-S-C-CS circuit.) The rougher B and the scavenger have their own cleaning stages in a modified $(R\text{-}C\text{-}CS)_2$ type circuit. The rougher B concentrate is reground and cleaned (cleaner B) with concentrate reporting to final nickel concentrate and tails to the scavenger/cleaner. The scavenger concentrate is thickened and cleaned (scavenger/cleaner), with concentrate joining rougher B concentrate for regrinding and the tails reporting to final sulfide tailings. Scavenger tails is the final non-sulfide tailings, the scavenger acting to divide the rougher B tails into sulfide and non-sulfide fractions. The non-sulfide tailings are used to build tailings dams, while the sulfide tailings are impounded underwater at the center of the dams (Kerr, 2002). In this manner, oxidation of the sulfides to produce acid run-off (acid rock drainage) is contained (see Chapter 16).

PAX is used as the collector and Unifroth 250 C (a PPG methyl ether) as frother (Kerr, 2002). Kerr (2002) discusses the importance of pyrrhotite rejection in the flowsheet. The presence of nickel and copper ions in solution can result in pyrrhotite activation. The combination of TETA with sodium sulfite is used to minimize activation.

Vale's Sudbury operation also includes a Matte Separation plant, a mineral processing plant in the

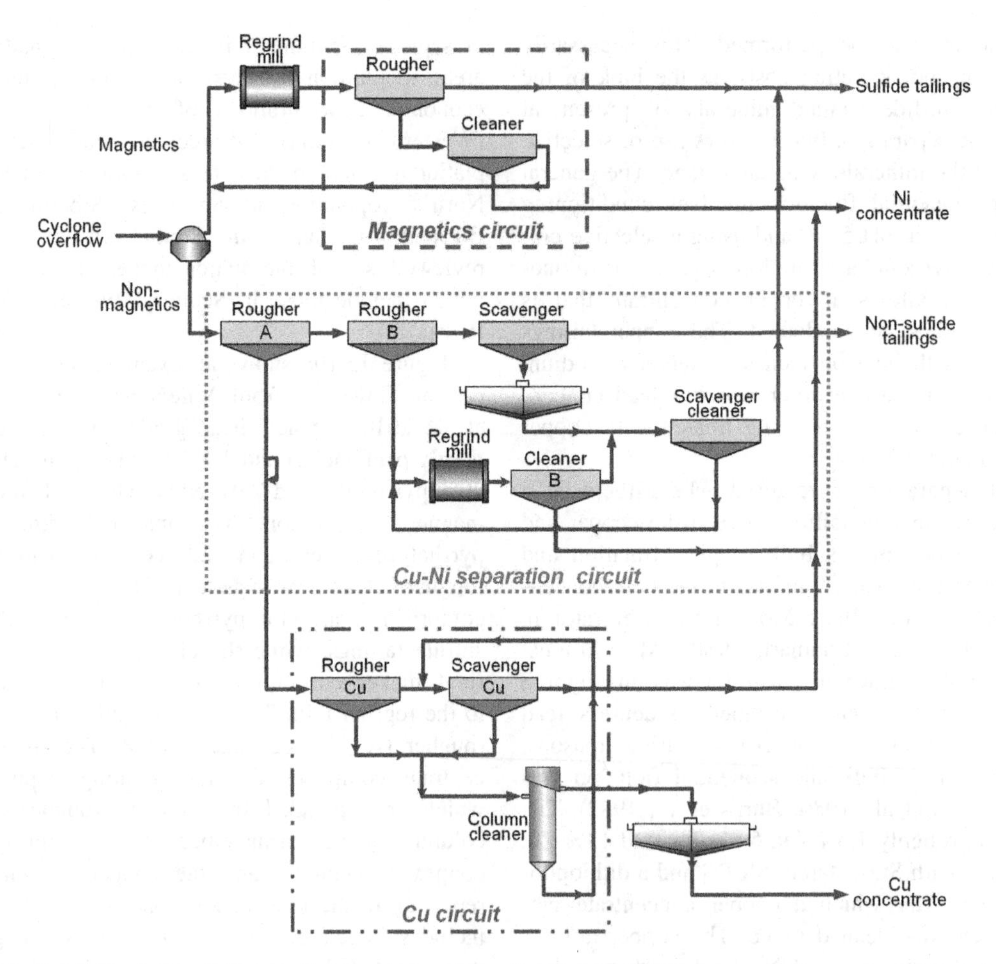

FIGURE 12.108 Vale's Clarabelle mill flotation flowsheet *(Adapted from Doucet et al. (2010)).*

smelter. Matte is slowly cooled to enlarge the mineral grains, ground, passed over magnets to remove a Cu/Ni alloy, then floated to produce Cu concentrate (chalcocite) and Ni concentrate (heazelwoodite), which are sent to their respective smelters.

12.17.5 Platinum Ores

Platinum is one of the Platinum Group Metals (PGMs), which also include palladium, iridium, osmium, rhodium, and ruthenium. They are generally found together in ores; 90% of PGM production comes from South Africa and Russia. In 2004, 44% of platinum was used in catalysts for motor vehicle emission control, and 33% in jewelry. The PGMs are classed with gold and silver as precious metals.

There are three main types of PGM deposit: PGM-dominant (e.g., the Bushveld Igneous Complex in South Africa), Ni–Cu dominant (e.g., Sudbury in Canada and Noril'sk in Russia), and miscellaneous. PGMs are usually recovered by flotation as a bulk low-grade sulfide concentrate, followed by smelting and refining.

There are over 100 known PGM-containing minerals, which include sulfides, tellurides, antimonides, arsenides, and alloys. Each of these has a unique metallurgical behavior, and the mode of occurrence and grain size varies considerably according to location (Corrans et al., 1982). The gangue minerals present specific challenges to flotation that affect downstream processing, notably talc (Shortridge et al., 2000) and chromite (Wesseldijk et al., 1999). Typical reagent suites include: thiol collectors (xanthate, sometimes with co-collectors dithiophosphate or dithiocarbamate); in some cases, copper sulfate as an activator; and polymeric depressants such as guar or carboxymethyl cellulose to inhibit recovery of naturally floatable talcaceous gangue (Wiese et al., 2005).

The wide range of valuable mineral densities in PGM ores presents problems in conventional classification in grinding circuits (see Chapter 9), so the South African flotation concentrators sometimes employ combined milling and flotation circuits without classification (Snodgrass et al., 1994). Flash flotation and preconcentration by DMS or gravity are also used.

12.17.6 Iron Ores

Iron ore minerals such as goethite (FeO(OH)) and hematite (Fe_2O_3) are floated by amine, oleate, sulfonate, or sulfate collectors. Processing involves preconcentration by gravity or magnetic separation, followed by flotation. Iron ore flotation has increased in importance due to market requirements for higher grade products. This requires the reverse flotation of silicate impurities. Amines are commercially used for the flotation of silica in many parts of the world (Das et al., 2005).

The requirement for higher grade product has seen an increase in the use of flotation columns in iron ore treatment. In Brazil, all new iron ore concentration circuits commissioned since the 1990s have consisted of rougher–cleaner–scavenger column-only configurations (Araujo et al., 2005b).

12.17.7 Niobium Ores

Niobec mine, located in the Saguenay region of Quebec, Canada, has been in operation since 1976 processing a niobium-bearing carbonatite ore. The minerals of economic interest are pyrochlore, $(Na,Ca)Nb_2O_6F$, and columbite, $(Fe,Mn)(Nb,Ta)_2O_6$. In 2012, the operation produced 4.7 million kg of niobium (Iamgold, 2014). The process is an example of non-sulfide flotation that can become quite complex. The plant employs desliming, carbonate reverse flotation, magnetic separation, pyrochlore flotation, and pyrite reverse flotation, followed by acid leaching to produce the niobium concentrate.

The Niobec mill flowsheet (ca. 1989) is shown in Figure 12.109 (Biss and Ayotte, 1989). Run-of mine ore is crushed in three stages to 100% − 20 mm, which feeds an open-circuit rod mill followed by a closed-circuit ball mill (classification performed by a screen and screw classifier). The ball mill circuit product (95% − 200 μm) is deslimed with the − 10 μm material, which has a deleterious effect on flotation, being discarded.

The fine (10–40 μm) and coarse (40–200 μm) fractions are processed in separate carbonate flotation circuits. Flotation is at natural pH (ca. 8) using emulsified fatty acid collector and sodium silicate as pyrochlore depressant. Concentrates (i.e., carbonate concentrate) from both circuits are deslimed, with the fines being sent to the tailings pond and the coarse fraction recycled to the head of coarse carbonate flotation. The coarse fraction of the coarse carbonate concentrate moves on to magnetic separation. The desliming stage also replaces process water with potable water (Rao et al., 1988).

The magnetic separation stage employs two drum separators in series to remove magnetite. The magnetics stream, which contains ca. 68% Fe, 0.8% SiO_2, and 0.08% Nb, is sent to the tailings pond with the nonmagnetic fraction continuing to pyrochlore flotation.

The pyrochlore flotation stage consists of a rougher and six cleaner stages and upgrades the feed from 1–1.2% Nb_2O_5 to 40–50% Nb_2O_5. Flotation is undertaken at gradually decreasing pH from 6.5 to 2.8 using hydrofluosilicic acid and carboxylic acid. The circuit utilizes secondary amine salts as collector, which in combination with the gradual pH decrease and the addition of modifiers (tapioca starch and sodium silicate) allows for a selective separation.

The pyrochlore cleaner concentrate comprises pyrochlores and pyrite (and small amounts of pyrrhotite) in roughly equal amounts, along with small concentrations of carbonates and apatite. A reverse flotation circuit using xanthate at ca. pH 10.5 produces a pyrite concentrate that is sent to the tailings pond and a pyrochlore tailings stream that undergoes hydrochloric acid leaching at ambient temperature to dissolve carbonates, apatite and some iron sulfide minerals. The leach residue undergoes a second pyrite flotation with the tailings stream being the final niobium concentrate, which is filtered and dried. Typical marketable concentrate quality requirements are: >55% Nb_2O_5, <4.0% SiO_2, <0.23% P_2O_5, and <0.15% S.

12.17.8 Phosphate Ores

Phosphate ores are widely used in the production of fertilizer and phosphoric acid. In 2012, world phosphate production was estimated to be 181 Mt (USGS, 2012). Roughly 80% of world reserves are sedimentary phosphates, with the remainder being from igneous sources. The world's largest reserves are found in the United States, Morocco, and China and are sedimentary deposits with major phosphate minerals francolite (carbonate rich apatite) and cellophane ($Ca_5(PO_4,CO_3)_3F$). Gangue minerals typically include clays, silica, calcite, and dolomite.

Figure 12.110 shows a simplified flowsheet for the IMC Four Corners operation in Florida, USA (Kawatra and Carlson, 2014). Feed from the washing plant is sized using screens and hydrocyclones into three size fractions which feed separate circuits:

Size Fraction	Stream	Circuit
−16 +24 mesh	Spiral feed	Spiral and scavenger flotation
−24 +35 mesh	Coarse flotation feed	Crago process
−35 +150 mesh	Fine flotation feed	Crago process

The coarse size fraction (0.7–1.2 mm), conditioned with a combination of fatty acid and fuel oil collectors, feeds a spiral film flotation circuit (Wiegel, 1999) (see

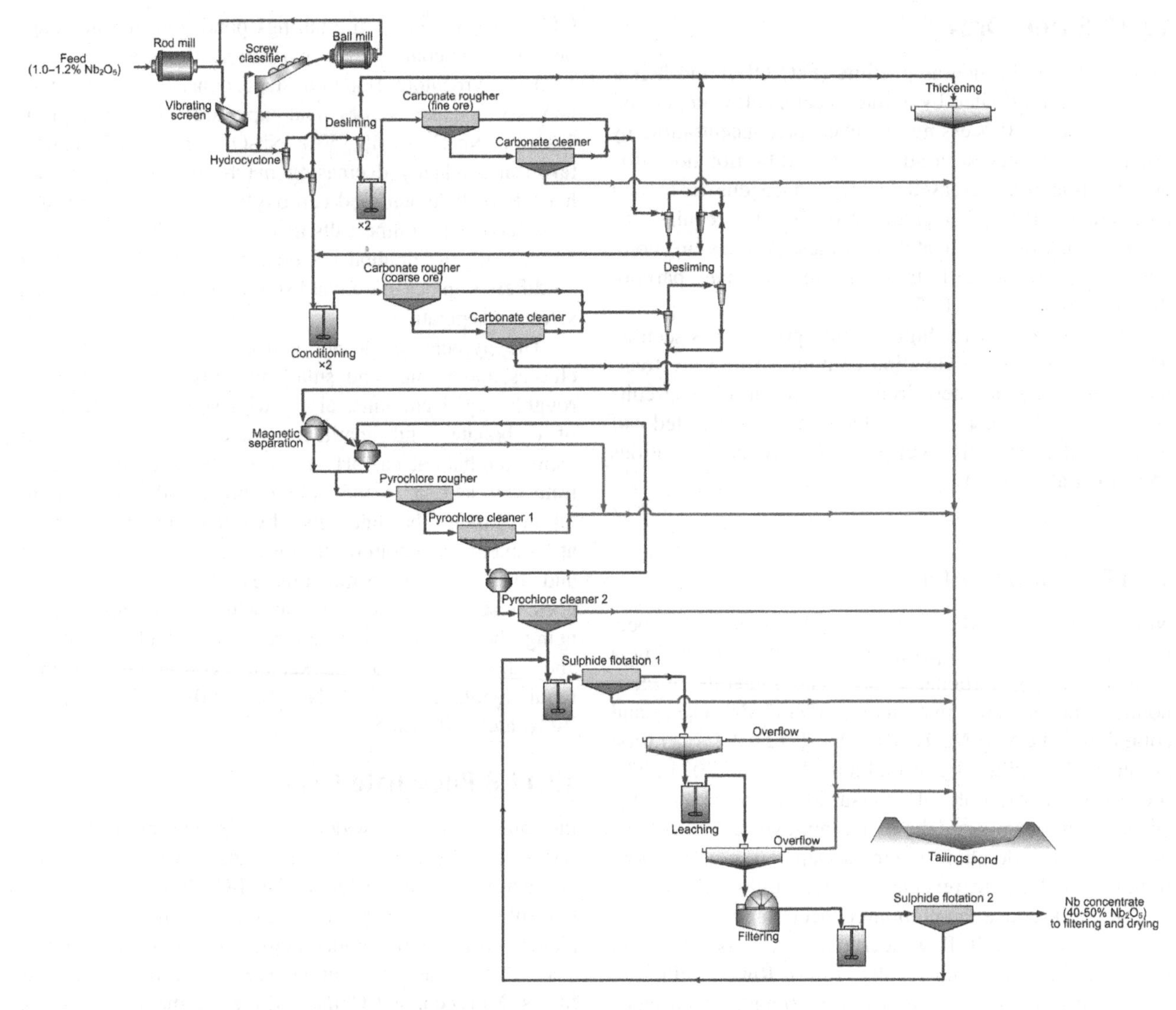

FIGURE 12.109 Niobec concentrator flowsheet *(Adapted from Biss and Ayotte (1989)).*

Section 12.18.6). The spiral concentrate is of high-enough phosphate quality to report to final concentrate. The spiral tailings feed a scavenger flotation circuit that produces final phosphate concentrate and (mainly) silica tails, which are sent to the tailings pond.

The coarse and fine flotation feed streams report to separate rougher flotation stages. Both streams undergo the *Crago double float* process, which entails a first stage rougher phosphate flotation using fatty acid/fuel oil collectors at pH 9.0–9.5 (typically controlled using soda ash) and a second stage amine reverse silica flotation of the rougher concentrate. Prior to silica flotation, the slurry undergoes an acid scrub to remove adsorbed collector. (There is also a reverse Crago process where fine silica is first floated using amine collector, followed by phosphate flotation using fatty acid/fuel oil collector (Kawatra and Carlson, 2014).)

12.17.9 Coal

Unlike most metalliferous flotation, where all the feed is treated by flotation, in coal processing only a portion is treated. This is typically 10–25% of the feed tonnage and represents the fines fraction, usually below 250 μm, but sometimes up to 1 mm. Mining methods, in particular the increased use of longwall mining, have resulted in an increase in fines production and made the flotation of coal fines more important. In many countries, environmental legislation has limited the amount of coal fines that can be sent to tailings ponds, with flotation being the only effective recovery method.

Coal flotation circuits are relatively simple roughing and scavenging stages, with sometimes roughing alone being adequate. The mass recovery in coal flotation is high (up to 70%) and frother usage rates can be high as

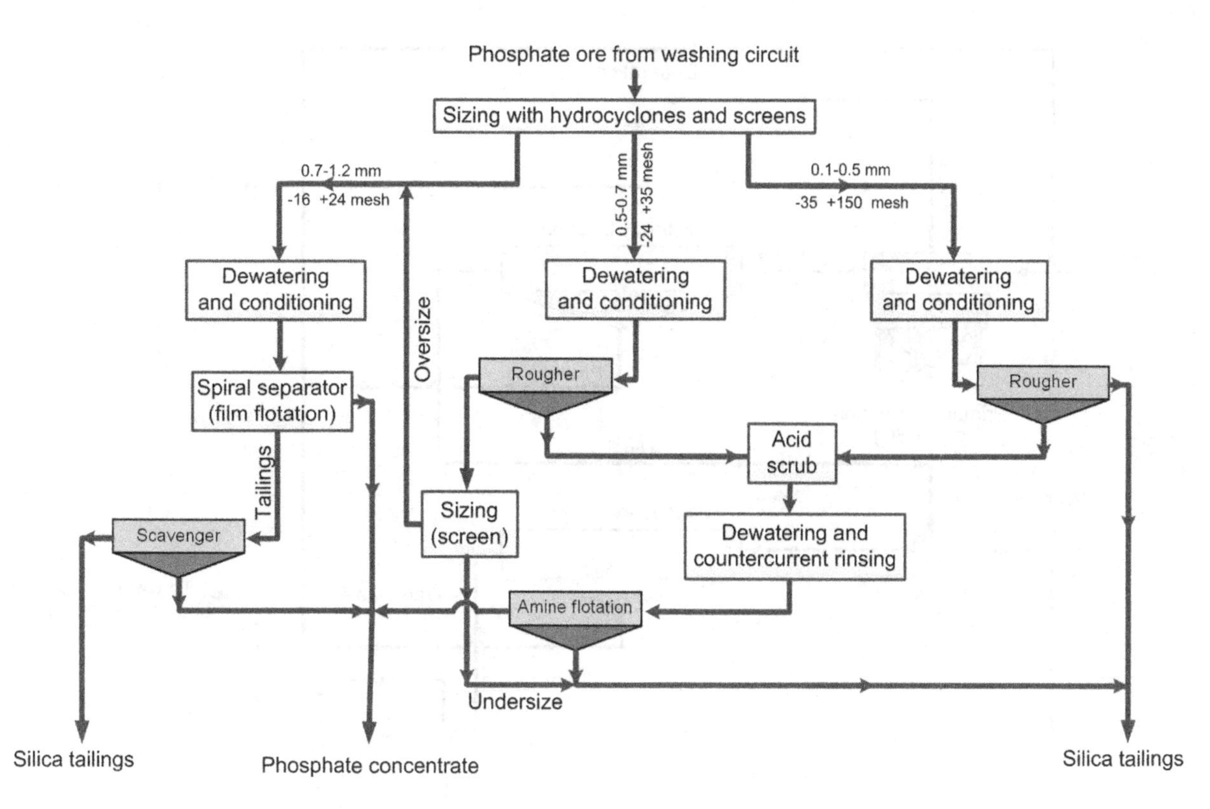

FIGURE 12.110 Example phosphate processing flowsheet *(Adapted from Kawatra and Carlson (2014)).*

some is consumed (adsorbed) by the coal. Many flotation circuits use mechanical paddles to physically remove the heavy froth from the flotation cells. Petrochemical products are usually used as collectors, with the most common being diesel oil, liquid paraffin, and kerosene.

Coal operations can produce one of two products, depending on the quality of coal mined, either high value coking coal for pyrometallurgical industries or lower value thermal coal for power generation. Coking coal demands few impurities and the ash content (noncombustible content) is typically between 5% and 8%. Given the lack of cleaning stage and this demand for low impurity levels, coking coals often require froth washing and this has seen increasing use of Jameson cells and flotation columns. Flotation concentrates for thermal coals range from 8% to 14% ash content, which can often be achieved without froth washing and mechanical flotation cells are still common (Nicol, 2000).

12.17.10 Oil Sands—Bitumen

Oil sands (also referred to as tar sands or bituminous sands) are unconsolidated sand deposits that contain viscous petroleum, commonly termed bitumen (Masliyah et al., 2004; Rao and Liu, 2013). The two largest sources of bitumen are found in Canada and Venezuela. It is estimated in Alberta, Canada, that there are ca. 300 billion barrels of recoverable oil (using today's technology).

Oil sands are composed of a mixture of bitumen and water-enveloped clays and silica (Takamura, 1982; Czarnecki et al., 2005). Pioneering work during the 1960s led to the development of the first commercially successful processing plant employing the Clark Hot Water Extraction (CHWE) process (Masliyah et al., 2004). Several water-based extraction processes now exist (Masliyah et al., 2004). The CHWE process uses hot water (50–80°C) and caustic soda (NaOH) to "liberate" the bitumen and this also releases natural surfactants which aid the flotation process (e.g., fine bubbles are "naturally" produced by the released surfactants so frothers are not required).

Over the years, oil sands processing has come to share many unit operations with mineral processing (Ritson and Ward, 2009; Nesset, 2011). Figure 12.111 shows a simplified flowsheet of the Muskeg river operation (Shell Canada). Crushers, tumblers, breakers, mixing boxes, and/or stirred tanks are used to break down the ore. Prior to the extraction plant the feed is conditioned in a hydrotransport pipeline into which warm water, reagents, and air are added. The conditioned slurry is fed to a large primary separation cell (PSC) (also referred to as PSV, primary separation vessel), unique to the oil sands industry, which acts as a gravity separator where bitumen buoyancy is

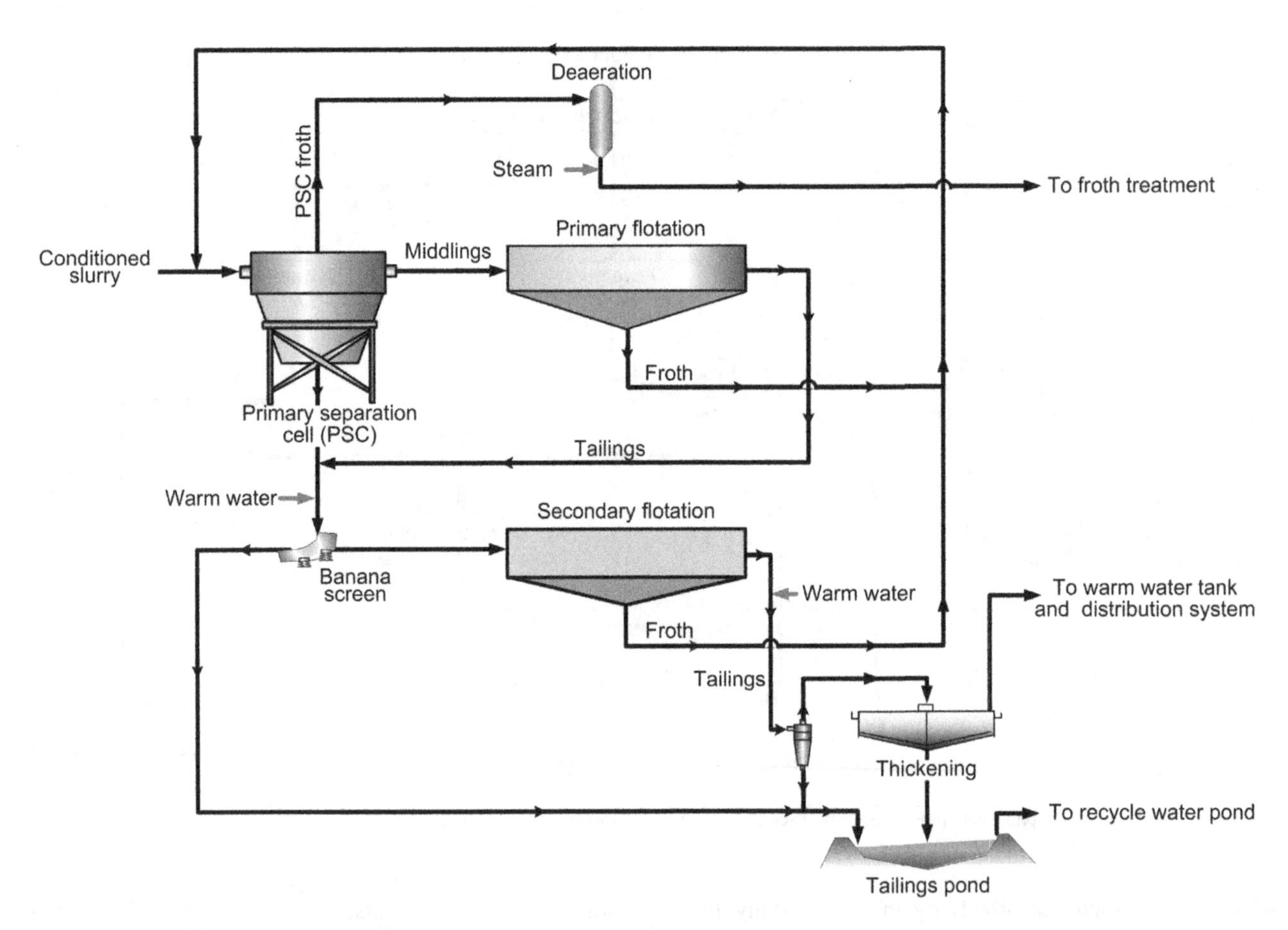

FIGURE 12.111 Muskeg River Mine process flowsheet *(Adapted from Nassif et al. (2014).)*

augmented by attached air bubbles to rise and form froth. The PSC treats close to 10,000 t h^{-1}. The PSC froth overflow is deaerated and feeds the froth treatment stage. Middlings from the PSC feeds a "primary" flotation stage comprising two parallel banks of four Dorr-Oliver Eimco 160 m^3 cells. The froth product is recycled to the PSC. The tailings are combined with PSC underflow and screened with the coarse fraction sent to tailings. The fine fraction goes to a "secondary" flotation stage comprising three parallel banks of four Dorr-Oliver Eimco 160 m^3 cells. The secondary froth product is recycled to the PSC and the secondary tailings are classified with the coarse fraction sent to final tailings and the fine fraction first thickened before disposal to final tailings. Overall bitumen recovery typically ranges from 88% to 95%, with the PSC froth product containing ca. 60% bitumen, 30% water, and 10% solids (Masliyah et al., 2004).

12.18 OTHER SURFACE CHEMISTRY AND FLOTATION-BASED SEPARATION SYSTEMS

12.18.1 Selective Flocculation

Recovery of fine particles in flotation is hampered by slow flotation kinetics. One option is *selective flocculation*; that is, to aggregate just the desired mineral particles (i.e., make them larger) and then separate (recover) the aggregates from the dispersed material. While attempts have been made to apply selective flocculation to a wide range of ore types, the bulk of the work has concerned the treatment of clays, iron, phosphate, and potash ores.

A prerequisite for the process is that the mineral mixture must be stably dispersed prior to introducing an aggregating reagent. Coagulating agents are a possibility, that is, manipulating surface charge to induce selective aggregation, but most work has been on using high molecular weight polymer flocculants which selectively adsorb on one of the constituents of the mixture (see Chapter 15 for further discussion on *coagulation* and *flocculation*). Selective flocculation is then followed by removal of the aggregates (*flocs*) from the dispersion by either flotation or sedimentation. The process of selective flocculation usually consists of four subprocesses: preparation of the material to be flocculated (particle dispersion, pH regulation, particle deactivation, etc.); selective adsorption of flocculant; floc formation and conditioning; and separation of flocs from nonflocculated material.

Selective flocculation has been successfully introduced in the treatment of fine-grained nonmagnetic oxidized taconites, which led to the development of Cleveland Cliffs Iron Company's (now Cliffs Natural Resources) operation in the United States. The finely intergrown ore is autogenously ground to 85% − 25 μm with caustic soda and sodium silicate, which act as dispersants for the fine silica. The ground pulp is then conditioned with a cornstarch flocculant which selectively flocculates the hematite. About one-third of the fine silica is removed in a deslime thickener, together with a loss of about 10% of the iron values. Most of the remaining coarse silica is removed from the flocculated underflow by reverse flotation, using an amine collector (Paananen and Turcotte, 1980; Siirak and Hancock, 1988).

In selective flotation of kaolin for removal of anatase impurity, one process includes dosing with sodium hexametaphosphate and sodium silicate to disperse the pulp and addition of high molecular weight ($>10^6$) PAM polymer to flocculate the anatase, which separates out in a thickener (Miller et al., 2002).

A related process is *carrier flotation*, where fine particles of the selected mineral attach to large particles of either the same or a different mineral added to the pulp, which are then removed with the attached fines. An application in kaolin processing was addition of ground limestone, conditioning with collector and flotation of the limestone with attached anatase. Rather than adding coarse particles followed by flotation, flocculation of fine magnetite selectively onto the target particle could render it (whether fine or coarse) recoverable by magnetic separation.

Selective flocculation remains a challenging technology, but an important one in seeking ways to treat the ever increasing volumes of fines as ore grades diminish and ore textural complexity increases. An attraction is the requirement of only small amounts of flocculant, usually several ppm. The most significant problem is the relatively low selectivity, with nontarget minerals entrapped in the flocs often requiring sequential dispersion and flocculation stages. Dewatering the flocs may also prove a challenge.

12.18.2 Oil-Assisted Separation Processes

The presence of a water-insoluble oil is used in several processes (Laskowski, 1992): to improve attachment of hydrophobic particles to bubbles (*emulsion flotation*), to increase size of fine particles to improve flotation rate (*agglomerate flotation*), or to allow separation by screening (*oil agglomeration*). In addition, there is also *oily bubble flotation*, where bubbles are precoated with oil to enhance attachment of particles (Su et al., 2006), and *liquid–liquid extraction*, where hydrophobic particles transfer to an oil phase (Lai and Fuerstenau, 1968). Outside of possible use in processing energy minerals coal and bitumen, the economics of employing oil (usually kerosene) seem questionable.

12.18.3 Gamma Flotation

In Section 12.3 the notion was introduced that high surface energy solids are wetted by water. This can be generalized to high surface energy solids are wetted by low surface energy liquids; and conversely, low surface energy solids are not wetted by high surface energy liquids (i.e., a contact angle forms). With water as the liquid, "high" and "low" energy refer to the surface energy (tension) of water, which for pure water at room temperature is about 72 mJ m^{-2}. Thus we see that low surface energy solids can be floated in water; that is, they are naturally hydrophobic. If the surface energy of water ($\gamma_{l/a}$) is lowered, then there is a value when the solid is just wetted; this is the *critical surface tension of wetting* (γ_c) of the solid, a concept first introduced by Zisman (1964). Since minerals will have different γ_c, it follows for two solids A and B that if $\gamma_{cA} < \gamma_{l/a} < \gamma_{cB}$ then solid A could be floated from solid B in water of surface energy $\gamma_{l/a}$. The technique has been demonstrated for separating naturally hydrophobic minerals (Kelebek and Smith, 1985) and plastics (Shent et al., 1999). Because it was surface tension being manipulated, the technique became known as "gamma (γ) flotation" (Buchan and Yarar, 1995).

Most of the literature focusses on measurement of γ_c. Surfactants can be used to control surface tension (Kelebek et al., 2001), but most work has used alcohol–water mixtures. The test procedure was either flotation in solutions of varying surface tension (Yarar and Kaoma, 1984), or *film flotation*, whereby particles placed on the surface of water would sink (be wetted) when the water surface tension was lowered below their γ_c (Fuerstenau and Williams, 1987).

While not a realistic practical option for mineral flotation, the concept has found application in characterizing naturally hydrophobic minerals by determining their γ_c. In the case of coals, which have variable composition, fractionating in solutions of controlled surface tension offers a way to describe the distribution in hydrophobicity (Hornsby and Leja, 1980).

While in principle the concept applies to high surface energy solids obviously the choice of liquid to determine their γ_c (let alone exploit for separation) is limiting. The addition of collector to high surface energy solids could be viewed as lowering the solid's γ_c to less than $\gamma_{w/a}$, and making the solid floatable. The γ_c of some collector-coated minerals has been determined (Finch and Smith, 1975; Kelebek et al., 1986).

12.18.4 Colloidal Gas Aphrons

The flotation of fine particles is considered to require correspondingly fine bubbles. There appears to be a limit to the fineness of bubbles that can be produced in conventional flotation systems, of the order of 500 μm (Section 12.5.1). Stable (i.e., noncoalescing) bubbles less than 100 μm can be created in the presence of surfactant when stirred at a very high speed, typically 8,000 rpm (Jauregi et al., 2000). Referred to as "colloidal gas aphrons" (CGAs), they can be defined as surfactant-stabilized micro-bubbles. Of interest in several fields because of the high surface area, in mineral recovery they remain a research idea (Waters et al., 2008).

12.18.5 Electroflotation

In this process, direct current is passed through the pulp within the cell, generating a stream of hydrogen and oxygen bubbles at the two electrodes. Considerable work has been done on factors affecting the bubble size on detachment from the electrodes, such as electrode potential, pH, surface tension, and contact angle of the bubble on the electrode. On detachment, the majority of bubbles are in the 10–60 μm range, and bubble concentration (gas holdup) can be controlled by current density. Electroflotation has been used for some time in waste-treatment applications to float solids from suspensions (Chen and Chen, 2010); it may have a future role in the treatment of fine mineral particles.

Some other factors have also been noted in addition to the fine bubbles. For example, the flotation of cassiterite is improved when electrolytic hydrogen is used for flotation. This may be due to nascent hydrogen reducing the surface of the cassiterite to tin (metal elements tend to be hydrophobic), allowing the bubbles to attach.

12.18.6 Agglomeration–Skin Flotation

In *agglomeration flotation*, the hydrophobic mineral particles are loosely bonded with small air bubbles to form agglomerates. When the agglomerates reach a free water surface, they are replaced by *skin-flotation* of individual particles. In skin (or film) flotation, surface tension forces result in holding the hydrophobic particles at the water surface, while the hydrophilic particles sink. The reagentized feed is fed to gravity concentration devices such as tables and spirals and the particles are floated off. It is used in treating coarse phosphate in some operations (Section 12.17.8) (Moudgil and Barnett, 1979).

12.18.7 Dissolved Air Flotation

In this process, air is dissolved in water and released as fine bubbles (<100 μm) by a sudden drop in pressure. Two versions are employed: *vacuum flotation*, where air is dissolved in water at atmospheric pressure and released under vacuum; and the more common *pressure flotation*, where air is dissolved under pressure and released at atmospheric pressure. The technology is used in treating municipal waste and industrial waste including oil removal (deoiling) from refinery effluent where natural gas can be substituted (and is preferred) for air. The flotation appears to be due to the fine bubbles nucleating on and clustering the hydrophobic matter (i.e., forming aggregates of matter and bubbles). Sometimes considered for fine mineral flotation, there are two disadvantages: one is the limited amount of air that can be introduced (not an issue with the low solid content municipal effluents); and two, the fine bubbles are too small to levitate the particles on their own (unless clustering can be induced). The CavTube™ used to disperse air in flotation columns (Section 12.13.2) exploits pressure release to produce fine bubbles, the aim being to nucleate the fine bubbles on the mineral surface to improve subsequent attachment of flotation-size bubbles.

12.18.8 Adsorptive Bubble Separation Processes

These processes take advantage of the attachment to bubbles of naturally or induced hydrophobic material other than minerals. *Foam fractionation* is the flotation of naturally hydrophobic material, ranging from bacteria to surfactants. *Ion flotation* is the removal of metal ions through addition of surfactants with appropriate functional groups to form a hydrophobic precipitate. Direct flotation of ions may be effected by first coating bubbles with extractant, such as in the air-assisted solvent extraction process (Tarkan et al., 2012). *Precipitate flotation* includes precipitates in general, both naturally hydrophobic and those induced through addition of surfactants. The processes employ both dispersed air and dissolved air machines. Adsorptive bubble and related processes are reviewed by Wang et al. (2007).

REFERENCES

Abramov, A.A., 2011. Design principles of selective collecting agents. J. Min. Sci. 47 (1), 109–121.

Adam, K., Iwasaki, I., 1984. Effects of polarisation on the surface properties of pyrrhotite. Miner. Metall. Process. 1 (Nov.), 246–253.

Adel, G.T., et al., 1991. Full-scale testing of microbubble column flotation. In: Agar, G.E., et al. (Eds.), Proceedings of Column'91: An International Conference on Column Flotation, vol. 1, Sudbury, ON, Canada, pp. 263–274.

Adkins, S.J., Pearse, M.J., 1992. The influences of collector chemistry on kinetics and selectivity in base-metal sulphide flotation. Miner. Eng. 5 (3-5), 295–310.

Agar, G.E., Kipkie, W.B., 1978. Predicting locked cycle flotation test results from batch data. CIM Bull. 71 (799), 119–125.

Agar, G.E., et al., 1980. Optimizing the design of flotation circuits. CIM Bull. 73 (824), 173–180.

Agar, G.E., et al. (Eds.), 1991. Proceedings of Column'91: An International Conference on Column Flotation, vols. 1 2, CIM, Sudbury, ON, Canada.

Ahmed, N., Jameson, G.J., 1989. Flotation kinetics. Miner. Process. Extr. Metall. Rev. 5 (1-4), 77–79.

Alexander, D., et al., 2000. The application of multi-component floatability models to full scale flotation circuits. Proceedings of the Seventh Mill Operators' Conference, AusIMM, Kalgoorlie, Australia, pp. 167–178.

Alexander, S., et al., 2012. Correlation of graphite flotation and gas holdup in saline solutions. In: Drelich, J. (Ed.), Proceedings of Water in Mineral Processing: First International Symposium, SME, Englewood, CO, USA, pp. 41–50.

Ali, S.S.M., et al., 2013. Introducing inverse gas chromatography as a method of determining surface heterogeneity of minerals for flotation. Powder Technol. 249, 373–377.

Allison, S.A., O'Connor, C.T., 2011. An investigation into the flotation behavior of pyrrhotite. Int. J. Miner. Process. 98 (3–4), 202–207.

Amelunxen, P., Amelunxen, R., 2009. Moly plant design considerations. Proceedings of SME Annual Meeting, SME, Denver, CO, USA. Preprint 09-136: 1-5.

Amelunxen, P.A., Rothman, R., 2009. The online determination of bubble surface area flux using the CiDRA GH-100 sonar gas holdup meter. Workshop on Automation in Mining, Minerals and Metals Industry (IFACMMM 2009), Vina del Mar, Chile.

Amelunxen, P., Runge, K., 2014. Innovations in froth flotation modeling. In: Anderson, C.G., et al., (Eds.), Mineral Processing and Extractive Metallurgy: 100 Years of Innovation. SME, Littleton, CO, USA, pp. 177–192.

Amelunxen, R.A., Sandoval, G., 2013. The column cell and contact cell, comparison, beliefs and operation. In: Álvarez, M. et al. (Eds.), Proceedings of 10th International Mineral Processing Conference (Procemin 2013), Gecamin, Santiago, Chile, pp. 249–259.

Amini, E., et al., 2013. Influence of turbulence kinetic energy on bubble size in different scale flotation cells. Miner. Eng. 45, 146–150.

Ananthapadmanabhan, K.P., Somasundaran, P., 1988. Acid-soap formation in aqueous oleate solutions. J. Colloid Interface Sci. 122 (1), 104–109.

Anon., 2014a. <http://en.wikipedia.org/wiki/Grasberg_mine>. Viewed Aug 8, 2014.

Anon., 2014b. <http://www.lundinmining.com/s/Zinkgruvan.asp>. Viewed July 31, 2014.

Araujo, A.C., et al., 2005a. Reagents in iron ores flotation. Miner. Eng. 18 (2), 219–224.

Araujo A.C., et al., 2005b. Flotation machines in Brazil: columns vs mechanical cells. In: Jameson, G. (Ed.), Proceedings of Centenary of Flotation Symposium, AusIMM, Brisbane, Australia, pp. 187–192.

Aravena, J.J., 1987. Column flotation applications at Chuquicamata's molybdenite flotation plant. In: Mular, A., et al. (Eds.), Proceedings of Copper'87, vol. 2, Universidad de Chile, Santiago, Chile, pp. 155–169.

Ata, S., 2012. Phenomena in the froth phase of flotation—a review. Int. J. Miner. Process. 102-103, 1–12.

Atrafi, A., et al., 2012. Frothing behavior of aqueous solutions of oleic acid. Miner. Eng. 36-38, 138–144.

Avotins, P.V., et al., 1994. Recent advances in sulfide collector development. In: Mulukutla, P.S. (Ed.), Reagents for Better Metallurgy. SME, Littleton, CO, USA, pp. 47–56.

Bacon, W.G., et al., 2002. Nickel Outlook—2000 to 2010. CIM Bull. 95 (1064), 47–52.

Bailey, M., et al., 2005. Development and application of an image analysis method for wide bubble size distributions. Miner. Eng. 18 (12), 1214–1221.

Baldauf, H., et al., 1985. Alkane dicarboxylic acids and aminoaphthol sulphonic acids a new reagent regime for cassiterite flotation. Int. J. Miner. Process. 15 (1-2), 117–133.

Banisi, S., et al., 1991. The behaviour of gold in Hemlo Mines Ltd. grinding circuit. CIM Bull. 84 (955), 72–78.

Barbery, G., 1984. Engineering aspects of flotation in the minerals industry: flotation machines, circuits and their simulation. In: Ives, K.J. (Ed.), The Scientific Basis of Flotation. Martinus Nijhoff Publishers/ Springer, Netherlands, pp. 289–348.

Barbery, G., 1986. Complex sulphide ores: processing options. In: Wills, B.A., Barley, R.W. (Eds.), Mineral Processing at a Crossroads—Problems and Prospects. Martinus Nijoff Publishers/Springer, Dordrecht/Netherlands, pp. 157–194.

Barbian, N., et al., 2003. Dynamic froth stability in froth flotation. Miner. Eng. 16 (11), 1111–1116.

Bartlett, H.E., Hawkins, D., 1987. Process evaluation. In: Stanley, G. G. (Ed.), The Extractive Metallurgy of Gold in South Africa, vol. 2. SAIMM, Johannesburg, South Africa, pp. 745–792 (Chapter 13).

Bazin, C., Proulx, M., 2001. Distribution of reagents down a flotation bank to improve the recovery of coarse particles. Int. J. Miner. Process. 61 (1), 1–12.

Bazin, C., et al., 1994. A method to predict metallurgical performances as a function of fineness of grind. Miner. Eng. 7 (10), 1243–1251.

Beck, R.D., Chamart, J.J., 1980. The Broken Hill concentrator of Black Mountain Mineral Development Co. (Pty) Ltd., South Africa. In: Jones, M.J. (Ed.), Complex Sulphide Ores. IMM, pp. 88–99.

Bergh, L.G., Yianatos, J.B., 2003. Flotation column automation: state of the art. Control Eng. Practice. 11 (1), 67–72.

Bhambhani, T., et al., 2012. Atypical grade-recovery curves: transport of Mg silicates to the concentrate explained by a novel phenomenological model. In: Young, C.A., Luttrell, G.H. (Eds.), Separation Technologies for Minerals, Coal, and Earth Resources. SME, Englewood, CO, USA, pp. 479–488.

Biss, R., 1984. Concentration of niobium-bearing minerals. Proceedings of the 16th Annual Meeting of the Canadian Mineral Processors Conference, CIM, Ottawa, Canada, pp. 198–214.

Biss, R., Ayotte, N., 1989. Beneficiation of carbonatite ore-bearing niobium at Niobec Mine. In: Dobby, G.S., Rao, S.R. (Eds.), Proceedings of Complex Ores. MetSoc/CIM, Pergamon Press, New York, NY, USA, pp. 497–506.

Bloom, F., Heindel, T.J., 2002. On the structure of collision and detachment frequencies in flotation models. Chem. Eng. Sci. 57 (13), 2467–2473.

Blonde, P., et al., 2013. Recovery profiling in a talc flotation roughing bank. Proceedings of the 23rd World Mining Congress, Paper No. 584, Montreal, QC, Canada.

Bouchard, J., et al., 2005. Recent advances in bias and froth depth control in flotation columns. Miner. Eng. 18 (7), 709–720.

Bouchard, J., et al., 2009. Column flotation simulation and control: an overview. Miner. Eng. 22 (6), 519–529.

Bouchard, J., et al., 2010. Asset optimisation through process control at Xstrata Alloys-Eland Platinum—optimal doesn't always rhyme with capital. Proceedings of the 42nd Annual Meeting of the Canadian

Mineral Processors Conference, CIM, ON, Ottawa, Canada, pp. 419–438.

Boulton, A., et al., 2003. Characterisation of sphalerite and pyrite flotation samples by XPS and ToF-SIMS. Int. J. Miner. Process. 70 (1-4), 205–219.

Boutin, P., Wheeler, D.A., 1967. Column flotation. World Mining. 20 (3), 47–50.

Box, G.E.P., et al., 2005. Statistics for Experimenters: Design Innovation and Discovery. second ed. Wiley-Interscience, Hoboken, NJ, USA.

Bozkurt, V., et al., 1994. Electrochemistry of chalcocite/heazelwoodite/sulfhydril collector systems. Can. Metall. Q. 33 (3), 175–183.

Bradshaw, D.J., O'Connor, C.T., 1996. Measurement of the sub-process of bubble loading in flotation. Miner. Eng. 9 (4), 443–448.

Bradshaw, D.J., Vos, F., 2013. The development of a small scale test for rapid characterization of flotation response (JKSMI). Proceedings of the 45th Annual Meeting of the Canadian Mineral Processors Conference, Ottawa, ON, Canada, pp. 43–57.

Brake, I.R., Eldridge, G., 1996. The Development of new Microcel™ column flotation circuit for BHP Australia coal's Peak Downs coal preparation plant. Proceedings of the 13th International Coal Preparation Conference, Lexington, KY, USA, pp. 237–251.

Broekaert, E., et al., 1984. New processes for cassiterite ore flotation. In: Jones, M.J., Gill, P. (Eds.), Mineral Processing and Extractive Metallurgy. IMM, London, UK, pp. 453–463.

Broman, P.G., et al., 1985. Experience from the use of SO_2 to increase the selectivity in complex sulphide ore flotation. In: Forssberg, K.S.E. (Ed.), Developments in Mineral Processing: Flotation of Sulphide Minerals. Elsevier Science Pub Co., Amsterdam, Netherlands, pp. 277–291.

Bruckard, W.J., et al., 2011. A review of the effects of grinding environment on the flotation of copper sulphides. Int. J. Miner. Process. 100 (1-2), 1–13.

Buchan, R., Yarar, B., 1995. Recovering plastics for recycling by mineral processing techniques. J. Metals (JOM). 47 (2), 52–55.

Buckley, A.N., Woods, R., 1997. Chemisorption—the thermodynamically favoured process in the interaction of thiol collectors with sulphide. Int. J. Miner. Process. 51 (1-4), 15–26.

Bulatovic, S.M., 1999. Use of organic polymers in the flotation of polymetallic ores: a review. Miner. Eng. 12 (4), 341–354.

Bulatovic, S.M., 2007. *Handbook of Flotation Reagents: Chemistry, Theory and Practice. Flotation of Sulfide Ores*, vol. 1. Elsevier, Amsterdam, Netherlands.

Bulatovic, S.M., 2010. *Handbook of Flotation Reagents: Chemistry, Theory and Practice. Flotation of Gold, PGM and Oxide Minerals*, vol. 2. Elsevier, Amsterdam, Netherlands.

Bulatovic, S.M., et al., 1998. Operating practice in the beneficiation of major porphory copper/molybdenum plants from Chile: innovated technology and opportunities: a review. Miner. Eng. 11 (4), 313–331.

Burdukova, E., Laskowski, J.S., 2009. Effect of insoluble amine on bubble surfaces on particle–bubble attachment in potash flotation. Can. J. Chem. Eng. 87 (3), 441–447.

Burgess, F., et al., 2003. Ramp up of the Pasminco Century Concentrator to 500 000 tpa zinc metal production in concentrate. Proceedings of the Eighth Mill Operators' Conference, AusIMM, Townsville, QLD, Australia, pp. 153–163.

Burns, C.J., et al., 1982. Process development and control at Woodlawn Mines. Proceedings of the 14th International Mineral Processing Congress, CIM, Toronto, ON, Canada, pp. IV18.1–IV18.14.

Cappuccitti, F., Finch, J.A., 2008. Development of new frothers through hydrodynamic characterization. Miner. Eng. 21 (12-14), 944–948.

Castro, S., 2012. Challenges in flotation of Cu–Mo sulphide ores in sea water. In: Drelich, J. (Ed.), Proceedings Water in Mineral Processing: First International Symposium, SME, Englewood, CO, USA, pp. 29–40.

Castro, H.S., Henriquez, C., 1998. By-product molybdenite recovery in Chile chemical factors in the selective flotation of molybdenite. Proceedings of the SME Annual Meeting Preprints, SME, Littleton, CO, USA, preprint 98-159: 1-9.

Castro, S., et al., 2012. Foaming properties of flotation frothers at high electrolyte concentration. In: Drelich, J. (Ed.), Proceedings of Water in Mineral Processing: First International Symposium, SME, Englewood, CO, USA, pp. 51–60.

Castro, S., et al., 2014. Depression of inherently hydrophobic minerals by hydrolysable metal cations: molybdenite depression in seaeater. Proceedings of the 27th International Mineral Processing Congress (IMPC) (Chapter 3), Paper: C0321, Santiago, Chile.

Chander, S., 1991. Electrochemistry of sulfide flotation: growth characteristics of surface coatings and their properties, with special reference to chalcopyrite and pyrite. Int. J.Miner. Process. 33 (1-4), 121–134.

Chander, S., 2003. A brief review of pulp potentials in sulfide flotation. Int. J. Miner. Process. 72 (1-4), 141–150.

Chandra, A.P., Gerson, A.R., 2009. A review of the fundamental studies of the copper activation mechanisms for selective flotation of the sulfide minerals, sphalerite and pyrite. Adv. Colloid Interface Sci. 145 (1-2), 97–110.

Chau, T.T., et al., 2009. A review of factors that affect contact angle and implications for flotation practice. Adv. Colloid Interface Sci. 150 (2), 106–115.

Chehreh Chelgani, S., et al., 2012a. Study of pyrochlore surface chemistry effects on collector adsorption by TOF-SIMS. Miner. Eng. 39, 71–76.

Chehreh Chelgani, S., et al., 2012b. Study of pyrochlore matrix composition effects on froth flotation by SEM–EDX. Miner. Eng. 30, 62–66.

Chehreh Chelgani, S., et al., 2013. A TOF-SIMS surface chemical analytical study of rare earth element minerals from micro-flotation tests products. Miner. Eng. 45, 32–40.

Chehreh Chelgani, S., Hart, B., 2014. TOF-SIMS studies of surface chemistry of minerals subjected to flotation separation—a review. Miner. Eng. 57, 1–11.

Chen, J., et al., 2010. Bulk flotation of auriferous pyrite and arsenopyrite by using tertiary dodecyl mercaptan as collector in weak alkaline pulp. Miner. Eng. 23 (11-13), 1070–1072.

Chen, X., Chen, G., 2010. Electroflotation. In: Comninellis, C., Chen, G. (Eds.), Electrochemistry for the Environment. Springer Science + Business Media Dordrecht, Technical Communications (Publishing) Ltd.,, Letchworth, England, pp. 263–278 (Chapter 11).

Cho, Y.S., Laskowski, J.S., 2002. Effect of flotation frothers on bubble size and foam stability. Int. J. Miner. Process. 64 (2-3), 69–80.

Chu, P., Finch, J.A., 2013. Frother and breakup in small bubble formation. In: Liu, Q. (Ed.), Proceedings of Materials Science & Technology 2013: Water and Energy in Mineral Processing, Montreal, QC, Canada, pp. 2034–2043.

Chu, P., et al., 2014. Quantifying particle pick up at a pendant bubble: a study of non-hydrophobic particle–bubble interaction. Miner. Eng. 55, 162–164.

Chudacek, M.W., 1991. EMDEE Microflot floatability test. Int. J. Miner. Process. 33 (1-4), 383–396.

Cienski, T., Coffin, V., 1981. Column flotation operation at Mines Gaspé molybdenum circuit. Can. Min. J. 102 (3), 28–33.

Clayton, R., et al., 1991. The development and application of the Jameson cell. Miner. Eng. 4 (7-11), 925–933.

Clift, R., et al., 2005. Bubble, Drops, and Particles. Dover Publications Inc., Mineola, NY, USA, pp. 169–202.

Coleman, R., Dixon, A., 2010. Tried, tested and proven-300 m^3 flotation cells in operation. Proceedings of the 25th International Mineral Processing Congress (IMPC), Brisbane, Queensland, Australia, pp. 3429–3440.

Cooper, M., et al., 2004. Impact of air distribution profile on banks in a Zn cleaning circuit. Proceedings of the 36th Annual Meeting of the Canadian Mineral Processors Conference, CIM, Ottawa, ON, Canada, pp. 525–540.

Corin, K.C., O'Connor, C.T., 2014. A proposal to use excess Gibbs energy rather than HLB number as an indicator of the hydrophilic-lipophilic behavior of surfactants. Miner. Eng. 58, 17–21.

Corrans, I.J., et al., 1982. The recovery of platinum group metals from ore of the UG2 Reef in the Bushveld Complex. In: Glen, H.W. (Ed.), Proceedings of the 12th Council of Mining and Metallurgical Institutions (CMMI) Congress, vol. 2, AusIMM, Johannesburg, South Africa, pp. 629–634.

Cowburn, J.A., et al., 2005. Design developments of the Jameson cell. In: Jameson, G. (Ed.), Proceedings Centenary of Flotation Symposium, AusIMM, Brisbane, Queensland, Australia, pp. 193–199.

Craig, V.S.J., et al., 1993. The effect of electrolytes on bubble coalescence in water. J. Phys. Chem. 97 (39), 10192–10197.

Crawford, R., Ralston, J., 1988. The influence of particle size and contact angle in mineral flotation. Int. J. Miner. Proc. 23 (1-2), 1–24.

Crozier, R.D., 1984. Plant reagents. Part 1: changing pattern in the supply of flotation reagents. Mining Mag. 151 (Sep.), 202–213.

Crozier, R.D., 1991. Sulphide collector mineral bonding and the mechanism of flotation. Miner. Eng. 4 (7-11), 839–858.

Crozier, R.D., 1992. Flotation: Theory, Reagents, and Ore Testing. Pergamon Press, Oxford, UK.

Crozier, R.D., Klimpel, R.R., 1989. Frothers: plant practice. Miner. Process. Extr. Metall. Rev. 5 (1-4), 257–279.

Cubillos, F.A., Lima, E.L., 1997. Identification and optimizing control of a rougher flotation circuit using an adaptable hybrid-neural model. Miner. Eng. 10 (7), 707–721.

Cunningham, R.R., Finch, J.A., 2009. Modification to foam volume measurements. Advances in mineral science and technology. In: Gomez, C.O., et al. (Eds.), Proceedings of the Seventh UBC-McGill-UA International Symposium on Fundamentals of Mineral Processing. COM 2009, Sudbury, ON, Canada, pp. 193–204.

Czarnecki, J., et al., 2005. On the nature of Anthabasca oil sands. Adv. Colloid Interface Sci. 114-115, 53–60.

Dahlke, R., et al., 2001. Trouble shooting flotation cell operation using gas velocity measurements. Proceedings of the 33rd Annual Meeting of the Canadian Mineral Processors Conference, CIM, Ottawa, ON, Canada, pp. 359–370.

Dahlke, R., et al., 2005. Operating range of a flotation cell from gas holdup vs. gas rate. Miner. Eng. 18 (9), 977–980.

Damjanović, B., Goode, J.R. (Eds.), 2000. Canadian Milling Practice, Special Volume 49. CIM, Montréal, QC, Canada.

Das, B., et al., 2005. Studies on the beneficiation of Indian iron ore slimes using the flotation technique. Proceedings of the Centenary of Flotation Symposium, AusIMM, Brisbane, Queensland, Australia, pp. 737–742.

Davis, W.J.N., 1969. Method and Apparatus for Froth Flotation. US Patent No. 3,446,353. United States Patent Office, pp. 1–22.

Deglon, D.A., et al., 2000. Review of hydrodynamics and gas dispersion in flotation cells on South African platinum concentrators. Miner. Eng. 13 (3), 235–244.

Dell, C.C., 1964. An improved release analysis procedure for determining coal washability. J. Inst. Fuel. 37, 149–150.

Dichmann, T.K., Finch, J.A., 2001. The role of copper ions in sphalerite-pyrite flotation selectivity. Miner. Eng. 14 (2), 217–225.

Dickinson, J.E., Galvin, K.P., 2014. Fluidized bed desliming in fine particle flotation—Part I. Chem. Eng. Sci. 108, 283–298.

Dixon, D.G., et al., 2007. Galvanox™—a novel galvanically-assisted atmospheric leaching technology for copper concentrates. Can. Metall. Q. 47 (3), 327–336.

Dobby, G., 2002. Column flotation. In: Mular, A.L., et al., (Eds.), Mineral Processing Plant Design, Practice and Control, vol. 1. Littleton, CO, USA, pp. 1239–1252.

Dobby, G.S., Finch, J.A., 1991. Column flotation: a selected review—part II. Miner. Eng. 4 (7-11), 911–923.

Dong, J., Xu, M., 2011. Evaluation of environmentally friendly collectors for xanthate replacement. Proceedings of the 43rd Annual Meeting of the Canadian Mineral Processors Conf., CIM, Ottawa, ON, Canada, pp. 289–302.

Doucet, J., et al., 2010. Evaluating the effect of operational changes at the Vale Canada Clarabelle Mill. Can. Metall. Q. 49 (4), 373–380.

Dreyer, J.P., 1976. Development of Agitair flotation machines. J. S. Afr. Inst. Min. Metall. 76 (11), 445–447.

Edgar, T.F., et al., 2001. Optimization of Chemical Processes. McGraw-Hill Chemical Engineering Series. McGraw-Hill, New York, NY, USA.

El-Ammouri, E., et al., 2002. Aggregation index and a methodology to study the role of magnesium in aggregation of sulphide slurries. CIM Bull. 95 (1066), 67–72.

Elmahdy, A.M., Finch, J.A., 2013. Effect of frother blends on hydrodynamic properties. Int. J. Miner. Process. 123, 60–63.

Elmahdy, A.M., et al., 2008. Zeta potential of air bubbles in presence of frothers. Int. J. Miner. Process. 89 (1-4), 40–43.

El-Shall, H., et al., 2000. Collector–frother interaction in column flotation of Florida phosphate. Int. J. Miner. Process. 58 (1-4), 187–199.

Espinosa, R., 2011. Operating practices at Peñoles concentrators-Mexico. Proceedings of the 43rd Annual Meeting of the Canadian Mineral Processors Conference, CIM, Ottawa, ON, Canada, pp. 91–105.

Espinosa-Gomez, R., et al., 1988. Coalescence and froth collapse in the presence of fatty acid. Colloids Surf. 32, 197–209.

Fahrenwald, A.W., 1922. Flotation Apparatus. US Patent No. 1,417,895. United States Patent Office, pp. 1–3.

Farrokhpay, S., 2011. The significance of froth stability in mineral flotation—a review. Adv. Colloid Interface Sci. 166 (1-2), 1–7.

Farrokhpay, S., 2012. The importance of rheology in mineral flotation: a review. Mining Eng. 36-38, 272–278.

Farrokhpay, S., Bradshaw, D.J., 2012. Effect of clay minerals on froth stability in mineral flotation: a review. In: Pradip, R. (Ed.), Proceedings of the 26th International Mineral Processing Congress (IMPC), New Delhi, India, pp. 4601–4611.

Fee, B.S., Klimpel, R.R., 1986. pH regulators. In: Malhotra, D., Riggs, W.F. (Eds.), Chemical Reagents in the Minerals Processing Industry. SME, Littleton, CO, USA, pp. 119–126 (Chapter 13).

Feteris, S.M., et al., 1987. Modelling the effect of froth depth in flotation. Int. J. Miner. Process. 20 (1-2), 121–135.

Fewings, J.H., et al., 1979. The dynamic behaviour and automatic control of the chalcopyrite flotation circuit at Mount Isa Mines Ltd. Proceedings of the 13th International Mineal Processing Congress (IMPC), vol. 2, Warsaw, Poland, pp. 405–432.

Finch, E., Riggs, W.F., 1986. Fatty acids—a selection guide. In: Malhotra, D., Riggs, W.F. (Eds.), Chemical Reagents in the Minerals Industry. SME, Littleton, CO, USA, pp. 95–98 (Chapter 10).

Finch, J.A., 1995. Column flotation: a selected review—part IV: novel flotation devices. Miner. Eng. 8 (6), 587–602.

Finch, J.A., 1998. Gaudin Lecture: fundamental fallout from column flotation. Mining Eng. 50 (12), 49–56.

Finch, J.A., Dobby, G.S., 1990. Column Flotation. Pergamon Press, Oxford, UK.

Finch, J.A., Smith, G.W., 1973. Dynamic surface tension of alkaline dodecylamine acetate solutions in oxide flotation. Trans. Inst. Min. Metall. Sec. C. 81, C213–C218.

Finch, J.A., Smith, G.W., 1975. Bubble–particle attachment as a function of bubble surface tension. Can. Metall. Q. 14 (1), 47–51.

Finch, J.A., Zhang, W., 2014. Frother function–structure relationship: dependence of CCC95 on HLB and the H-ratio. Miner. Eng. 61, 1–8.

Finch, J.A., et al., 1979. Laboratory simulation of a closed-circuit grind for a heterogeneous ore. CIM Bull. 72 (803), 198–200.

Finch, J.A., et al., 1996. Control of iron sulphides in mineral processing. In: Dutrizac, J.E., Harris, G.B. (Eds.), Iron Control and Disposal, Second International Symposium on Iron Control in Hydrometallurgy. CIM, Ottawa, ON, Canada, pp. 3–15.

Finch, J.A., et al., 2000. Gas dispersion properties: bubble surface area flux and gas holdup. Miner. Eng. 13 (4), 365–372.

Finch, J.A., et al., 2007a. Iron control in mineral processing. Proceedings of the 39th Annual Meeting of the Canadian Mineral Processors Conference, CIM, Ottawa, ON, Canada, pp. 365–386.

Finch, J.A., et al., 2007b. Column flotation. In: Fuerstenau, M.C., et al., (Eds.), Froth Flotation: A Century of Innovation. SME, Littleton, CO, USA, pp. 681–729.

Finch, J.A., et al., 2008. Role of frother on bubble production and behaviour in flotation. Miner. Eng. 21 (12-14), 949–957.

Finkelstein, N.P., 1997. The activation of sulphide minerals for flotation: a review. Int. J. Miner. Process. 52 (2-3), 81–120.

Fragomeni, D., et al., 2005. The use of end-members for grind/recovery modelling, tonnage prediction and flowsheet development at Raglan. Proceedings of the 37th Annual Meeting of the Canadian Mineral Processors Conference, Ottawa, ON, Canada, pp. 75–98.

Freeport-McMoran., 2013. Strength in Resources: 2013 Annual Report. View at December 2014: <www.fcx.com/ir/downloads/FCX_2012_10K-A.PDF>.

Fuerstenau, D.W., 2007. A century of developments in the chemistry of flotation processing. In: Fuerstenau, M.C., et al., (Eds.), Froth Flotation: A Century of Innovation. SME, Littleton CO, USA, pp. 3–64.

Fuerstenau, D.W., Williams, M.C., 1987. Characterization of the hydrophobicity of particles by film flotation. Colloids Surf. A. 22 (1), 87–91.

Fuerstenau, M.C., Palmer, B.R., 1976. Anionic flotation of oxides and silicates. In: Fuerstenau, M.C. (Ed.), Flotation A.M Gaudin Memorial Volume, vol. 1. SME, pp. 148–196 (Chapter 7).

Fuerstenau, M.C., Somasundaran, S., 2003. Flotation. In: Fuerstenau, M. C., Han, K.N. (Eds.), Principles of Mineral Processing. SME, Littleton, CO, USA, pp. 245–306 (Chapter 8).

Fuerstenau, M.C., et al., 1985. Chemistry of Flotation. AIMME, New York, NY, USA.

Fuerstenau, M.C., et al., 2007. Sulfide mineral flotation. In: Fuerstenau, M.C., et al., (Eds.), Froth Flotation: A Century of Innovation. SME, Littleton, CO, USA, pp. 425–464.

Garrett, A., 1933. Ore Concentrating Machine. US Patent No. 1,910,386. United States Patent Office, pp. 1–4.

Gélinas, S., Finch, J.A., 2007. Frother analysis: some plant experiences. Miner. Eng. 20 (14), 1303–1308.

Gerson, A., Napier-Munn, T., 2013. Integrated approaches for the study of real mineral flotation systems. Minerals. 3 (1), 1–5.

Ghobadi, P., et al., 2011. Optimization of the performance of flotation circuits using a genetic algorithm oriented by process-based rules. Int. J. Miner. Process. 98 (3-4), 174–181.

Gomez, C.O., Finch, J.A., 2002. Gas dispersion measurements in flotation machines. CIM Bull. 95 (1066), 73–78.

Gomez, C.O., Finch, J.A., 2007. Gas dispersion measurements in flotation cells. Int. J. Miner. Process. 84 (1-4), 51–58.

Gorain, B.K., 2005. Optimisation of flotation circuits with large flotation cells. Centenary of Flotation Symposium, Brisbane, QLD, Australia, pp. 843–851.

Gorain, B.K., et al., 1997. Studies on impeller type, impeller speed and air flow rate in an industrial flotation cell—Part 4: effect of bubble surface area flux on flotation performance. Miner. Eng. 10 (4), 367–379.

Gorain, B.K., et al., 1999. The empirical prediction of bubble surface area flux in mechanical flotation cells from cell design and operating data. Miner. Eng. 12 (3), 309–322.

Gorain, B.K., et al., 2007. Mechanical froth flotation cells. In: Fuerstenau, M.C., et al., (Eds.), Froth Flotation: A Century of Innovation. SME, Littleton, CO, USA, pp. 637–680.

Govender, D., et al., 2014. Large flotation cells in copper processing: experiences and considerations. Mining Eng. 66 (2), 24–32.

Grano, S.R., 2010. Chemical measurements during plant surveys and their interpretation. In: Greet, C.J. (Ed.), Flotation Plant Optimisation: A Metallurgical Guide to Identifying and Solving Problems in Flotation Plants. AusIMM, pp. 107–122.

Grano, S., et al., 2014. Innovations in flotation plant practice. In: Anderson, C.G., et al., (Eds.), Mineral Processing and Extractive Metallurgy: 100 Years of Innovation. SME, Englewood, CO, USA, pp. 193–207.

Grandon, F., Alvarez, J., 2014. Comparison of frother characterization method in two and three phase environments. 27th International Mineral Processing Congress (IMPC), Poster. Santiago, Chile.

Grau, R.A., Heiskanen, K., 2002. Visual technique for measuring bubble size in flotation machines. Miner. Eng. 15 (7), 507–513.

Gredelj, S., et al., 2009. Selective flotation of carbon in the Pb–Zn carbonaceous sulphide ores of Century Mine, Zinifex. Miner. Eng. 22 (3), 279–288.

Grubbs, F.E., 1969. Procedures for detecting outlying observations in samples. Techonometrics. 11 (1), 1–21.

Grönstrand, S., et al., 2012. Cell power input or hydrodynamics—which is more important in flotation? In: Young, C.A., Luttrell, G.H. (Eds.), Separation Technologies for Minerals, Coal, and Earth Resources. SME, Englewood, CO, USA, pp. 593–604.

Gu, G., et al., 2003. Effects of physical environment on induction time of air-bitumen attachment. Int. J. Miner. Process. 69 (1-4), 235–250.

Güler, T., et al., 2006. Adsorption of dithiophosphate and dithiophosphinate on chalcopyrite. Miner. Eng. 19 (1), 62–71.

Guo, B., et al., 2014. Cyanide chemistry and its effect on mineral flotation. Miner. Eng. 66-68, 25–32.

Gy, P.M., 1979. Sampling of Particulate Materials: Theory and Practice. Elsevier, Amsterdam, Netherlands.

Hadler, K., Cilliers, J.J., 2009. The relationship between the peak in air recovery and flotation bank performance. Miner. Eng. 22 (5), 451–455.

Hadler, K., et al., 2010. Recovery *vs.* mass pull: the link to air recovery. Miner. Eng. 23 (11-13), 994–1002.

Hadler, K., et al., 2012. The effect of froth depth on air recovery and flotation performance. Miner. Eng. 36-38, 248–253.

Han, M.Y., et al., 2004. The effect of divalent metal ions on the zeta potential of bubbles. Water Sci. Technol. 50 (8), 49–56.

Harbort, G.J., Schwarz, S., 2010. Characterisation measurements in industrial flotation cells. In: Greet, C.J. (Ed.), Flotation Plant Optimisation: A Metallurgical Guide to Identifying and Solving Problems in Flotation Plants. AusIMM, Carlton, VIC, Australia, pp. 95–106 (Spectrum series: No. 16), (Chapter 5).

Harbort, G.J., et al., 1994. Recent advances in Jameson flotation cell technology. Miner. Eng. 7 (2-3), 319–332.

Harbort, G., et al., 2003. Jameson cell fundamentals—a revised perspective. Miner. Eng. 16 (11), 1091–1101.

Hardie, C.A., et al., 1998. Application of mineral processing techniques to the recycling of wastepaper. Proceedings of the 30th Annual Meeting of the Canadian Mineral Processors Conference, CIM, Ottawa, ON, Canada, pp. 553–572.

Harris, C.C., 1976. Flotation machines. In: Fuerstenau, M.C. (Ed.), Flotation A.M. Gaudin Memorial Volume, vol. 2. AIME, New York, NY, USA, pp. 753–815 (Chapter 27).

Harris, C.C., 1986. Flotation machine design, scale-up and performance: database. In: Somasundaran, P. (Ed.), Advances in Mineral Processing. SME, Littleton, CO, USA, pp. 618–635 (Chapter 37).

Harris, M.C., et al., 2002. JKSimFloat as a practical tool for flotation process design and optimisation. In: Mular, A.L., et al., (Eds.), Mineral Processing Plant Design, Practice and Control, vol. 1. SME, Littleton, CO, USA, pp. 461–478.

Hebbard, J., 1913. Apparatus for Ore Concentration. US Patent No. 1,264,209A. United States Patent Office.

Herbst, J.A., Harris, M.C., 2007. Modeling and simulation of industrial flotation processes. In: Fuerstenau, M.C., et al., (Eds.), Froth Flotation: A Century of Innovation. SME, Littleton CO, USA, pp. 757–777.

Hernandez-Aguilar, J.R., 2010. Gas dispersion at Highland Valley Copper. Can. Metall. Q. 49 (4), 381–388.

Hernandez-Aguilar, J.R., 2011. On the role of bubble size in column flotation. Proceedings of the 43rd Annual Meeting of the Canadian Mineral Processors Conference, CIM, Ottawa, ON, Canada, pp. 269–287.

Hernández, H., et al., 2003. Gas dispersion and de-inking in a flotation column. Miner. Eng. 16 (8), 739–744.

Hernandez-Aguilar, J., et al., 2005. Testing the $k - S_b$ relationship at the micro-scale. Miner. Eng. 18 (6), 591–598.

Hernandez-Aguilar, J.R., et al., 2006. Experiences in using gas dispersion measurements to understand and modify metallurgical performance. Proceedings of the 38th Annual Meeting of the Canadian Mineral Processors Conference, CIM, Ottawa, ON, Canada, pp. 387–402.

Hernandez-Aguilar, J.R., et al., 2010. Improving column flotation operation in a copper/molybdenum separation circuit. CIM J. 1 (3), 165–175.

Holme, R.N., 1986. Sulphonate-type flotation reagents. In: Malhotra, D. W., Riggs, W.F. (Eds.), Chemical Reagents in the Minerals Processing Industry. SME, Littleton, CO, pp. 99–111 (Chapter 10).

Holtham, P.N., Nguyen, K.K., 2002. On-line analysis of froth surface in coal and mineral flotation using JKFrothCam. Int. J. Miner. Process. 64 (2-3), 163–180.

Hornsby, D.T., Leja, J., 1980. Critical surface tension and the selective separation of inherently hydrophobic solids. Colloids Surf. A. 1 (3-4), 425–429.

Houot, R., 1983. Beneficiation of iron ore by flotation—a review of industrial and potential applications. Int. J. Miner. Process. 10 (3), 183–204.

Houot, R., et al., 1985. Selective flotation of phosphatic ores having a siliceous and/or a carbonated gangue. Int. J. Miner. Process. 14 (4), 245–264.

Hu, J.S., et al., 1986. Characterization of adsorbed oleate species at the fluorite surface by FTIR spectroscopy. Int. J. Miner. Process. 18 (1-2), 73–84.

Huang, P., et al., 2012. Adsorption of chitosan on chalcopyrite and galena from aqueous suspensions. Colloids Surf. A. 409, 167–175.

Hunter, T.N., et al., 2008. The role of particles in stabilizing foams and emulsions. Adv. Colloid Interface Sci. 137 (2), 57–81.

Iamgold., 2014. Corporate website, <http://www.iamgold.com/English/Operations/Operating-Mines/Niobec-Niobium-Mine/> (accessed 25.07.14).

ICSG, 2010. ICSG Directory of Copper Mines and Plants 2008 to 201. International Copper Study Group, Lisbon, Portugal.

Imhof, R., et al., 2005. The successful application of pneumatic flotation technology for the removal of silica by reverse flotation at the iron ore Pellet Plant of Compañia Minera Huasco. In: Jameson, G. (Ed.), Proceedings of the Centenary of Flotation Symposium, AusIMM, Brisbane, QLD, Australia, pp. 1–9.

Iwasaki, I., et al., 1960. Flotation characteristics of goethite. Report of Investigation 5593 (US Bur. Min.).

Jameson, G.J., 1990. Column Flotation Method and Apparatus. US Patent No. 4,938,865. United States Patent Office.

Jameson, G.J., 1992. Flotation cell development. The AusIMM Annual Conference, Broken Hill, Australia, pp. 25–32.

Jameson, G.J., 2006. Method and Apparatus for Contacting Bubbles and Particles in a Flotation Separation System. International Application No.: PCT/AU2006/000123. Pub. No. WO 2006/081611.

Jameson, G.J., 2010. New directions in flotation machine design. Miner. Eng. 23 (11-13), 835–841.

Jameson, G.J., 2012. The effect of surface liberation and particle size on flotation rate constants. Miner. Eng. 36-38, 132–137.

Jampana, P.V., et al., 2009. Image-based level measurement in flotation cells using particle filters. In: Bergh, L.G. (Ed.), Automation in Mining, Mineral and Metal Processing, vol. 1. Venue, Chile, pp. 116–121.

Jauregi, P., et al., 2000. Colloidal gas aphrons (CGA): dispersion and structural features. AIChE J. 46 (1), 24–36.

Johnson, N.W., 1972. The Flotation Behaviour of Some Chalcopyrite Ores. PhD Thesis. University of Queensland, Brisbane, Australia.

Johnson, N.W., 1988. Application of electrochemical concepts to four sulphide flotation separations. In: Richardson, P., Woods, R. (Eds.), Proceedings of the Second International Symposium on Electrochemistry in Mineral and Metal Processing, The Electrochemical Society Pennington, NJ, USA, pp. 139–150.

Johnson, N.W., 2010. Existing methods for process analysis. In: Greet, C.J (Ed.), Flotation Plant Optimisation: A Metallurgical Guide to Identifying and Solving Problems in Flotation Plants, vol. 16. AusIMM, Australia, pp. 35–64. Spectrum Series.

Johnson, B.N.W., Munro, P.D., 2002. Overview of flotation technology and plant practice for complex sulphide ores. In: Mular, A.L., et al., (Eds.), Mineral Processing Plant Design Practice and Control, vol. 1. SME, Littleton CO, USA, pp. 1097–1123.

Johnson, N.W., et al., 1982. Oxidation–reduction effects in galena flotation: observations on Pb–Zn–Fe sulphides separation. Trans. Inst. Min. Metall. Sec. C. 91, C32–C37.

Jones, R.L., Horsley, R.R., 2000. Viscosity modifiers in the mining industry. Miner. Process. Extr. Metall. Rev. 20 (1), 215–223.

Jowett, A., 1966. Gangue mineral contamination of froth. Brit. Chem. Eng. II (5), 330–333.

Jowett, A., 1975. Formulae for the technical efficiency of mineral separations. Int. J. Miner. Process. 2 (4), 287–301.

Jowett, A., Sutherland, D.N., 1985. Some theoretical aspects of optimizing complex mineral separation systems. Int. J. Miner. Process. 14 (2), 85–109.

Kant, C., et al., 1994. Distribution of surface metal ions among the products of copper flotation. Miner. Eng. 7 (7), 905–916.

Kämpjärvi, P., Jämsä-Jounela, S.L., 2003. Level control strategies for flotation cells. Miner. Eng. 16 (11), 1061–1068.

Kawatra, S.K., Carlson, J.T., 2014. Beneficiation of Phosphate Ore. SME, Englewood, CO, USA.

Kaya, M., et al., 1990. Plant application of froth washing in mechanical cells. Proceedings of the 22nd Annual Meeting of the Canadian Mineral Processors Conference, CIM, Ottawa, ON, Canada, pp. 160–196.

Kelebek, S., Smith, G.W., 1985. Selective flotation of inherently hydrophobic minerals by controlling air/solution interfacial tension. Int. J. Miner. Process. 14 (4), 275–289.

Kelebek, S., Tukel, C., 1999. The effect of sodium metabisulphite and triethylenetetramine system on pentlandite–pyrrhotite separation. Int. J. Miner. Process. 57 (2), 135–152.

Kelebek, S., et al., 1986. Wettability and floatability of galena–xanthate systems as a function of solution surface tension. Colloids Surf. 20 (1-2), 89–100.

Kelebek, S., et al., 2001. Wetting behavior of molybdenite and talc in lignosulfonate/MIBC solutions and their separation by flotation. Sep. Sci. Tech. 36 (2), 145–157.

Kennedy, A., 1990. The Jameson flotation cell. Mining Mag. 163 (Oct.), 281–285.

Kerr, A., 2002. An overview of recent developments in flotation technology and plant practice for nickel ores. In: Mular, A.L., et al., (Eds.), Mineral Processing Plant Design, Practice and Control, vol. 1. SME, Littleton, CO, USA, pp. 1142–1158.

Kewe, T., et al., 2014. Porgera flotation circuit upgrade and expert system installation. Proceedings of the 12th AusIMM Mill Operators' Conference, pp. 345–356.

Kirkwood, B., et al., 2014. Low pH trial in the Ernest Henry Mining Copper rougher flotation circuit. Proceedings of the 46th Annual Meeting of the Canadian Mineral Processors, Ottawa, ON, Canada, pp. 343–350.

Kitchener, J.A., 1984. The froth flotation process: past, present and future—in brief. In: Ives, K.J. (Ed.), The Scientific Basis of Flotation. Martinus Nijhofff Publishers, The Hague, Netherlands, pp. 3–51. (NATO ASI Series, Series E: Applied Sciences: No. 75).

Kittel, S., et al., 2001. Rougher automation in Escondida flotation plant. SME Annual Meeting. SME, Denver, CO, USA. Preprint 01-053, pp. 1–7.

Klimpel, R.R., 1980. Selection of chemical reagents for flotation. In: Mular, A.L., Bhappu, R.B. (Eds.), Mineral Processing Plant Design, second ed. SME, New York, NY, USA, pp. 907–934. (Chapter 45).

Klimpel, R.R., 1984. Use of chemical reagents in flotation. Chem. Eng. 91 (18), 75–79.

Klimpel, R.R., 1995. The influence of frother structure on industrial coal flotation. In: Kawatra (Ed.), High-Efficiency Coal Preparation. SME, Littleton, CO, USA, pp. 141–151. (Chapter 12).

Klimpel, R., Isherwood, S., 1991. Some industrial implications of changing frother chemical structure. Int. J. Miner. Process. 33 (1-4), 369–381.

Konigsmann, K.V., 1985. Flotation techniques for complex ores. Zunkel, A.D. (Ed.), Complex Sulphide: Processing of Ores, Concentrates and By-Products. Pennslyvania, PA, USA, pp. 5–19.

Konigsmann, K.V., et al., 1976. Computer control of flotation at Mattagami Lake Mines. CIM Bull. 69 (767), 117–121.

Kowalczuk, P.B., 2013. Determination of critical coalescence concentration and bubble size for surfactants used as flotation frothers. Ind. Eng. Chem. Res., (I&EC). 52, 11752–11757.

Kuan, S.H., Finch, J.A., 2010. Impact of talc on pulp and froth properties in F150 and 1-pentanol frother systems. Miner. Eng. 23 (11-13), 1003–1009.

Kubota, T., et al., 1975. A new method for copper–lead separation by raising pulp temperature of the bulk float. Proceedings of the 11th International Mineral Processing Congress, Cagliari, Italy, pp. 623–637.

Labonté, G., Finch, J.A., 1990. Behaviour of redox electrodes during flotation and relationship to mineral floatabilities. Miner. Metall. Process. 7 (2), 106–109.

Lai, R.W.M., Fuerstenau, D.W., 1968. Liquid–liquid extraction of ultrafine particles. Trans. AIME. 241, 549–555.

Lascelles, D., Finch, J.A., 2002. Quantifying accidental activation: Part I Cu ion production. Miner. Eng. 15 (8), 567–571.

Laskowski, J.S., 1992. Oil assisted fine particle processing. In: Laskowski, J.S., Ralston, J. (Eds.), Colloid Chemistry in Mineral Processing. Elsevier, Amsterdam, Netherlands(Developments in Mineral Processing Series, Vol. 12), (Chapter 12), pp. 361–394.

Laskowski, J.S., 1998. Frothers and frothing. In: Laskowski, J.S., Woodburn, E.T. (Eds.), Frothing in Flotation II. Gordon and Breach Science Publishers, Amsterdam, Netherlands, pp. 1–49. (Chapter 1).

Laskowski, J.S., 2001. Flotation machines. Coal Flotation and Fine Coal Utilization. Elsevier, Amsterdam, Netherlands (Developments in Mineral Processing series. Vol. 14), (Chapter 8), pp. 225–262.

Laskowski, J.S., 2003. Fundamental properties of flotation frothers. In: Lorenzen, L., Bradshaw, D.J. (Eds.), Proceedings of the 22nd International Mineral Processing Congress (IMPC.), vol. 2, SAIMM, Cape Town, South Africa, pp. 788–797.

Laskowski, J.S., et al., 2003. Fundamental properties of polyoxypropylene alkyl ether flotation frothers. Int. J. Miner. Process. 72 (1-4), 289–299.

Lauder, D.W., 1992. The recycle mechanism in recirculation separation systems. Miner. Eng. 5 (6), 631–647.

Lawrence, G.A., et al., 1984. Flotation Separation Apparatus and Method. US Patent 4,425,232, United States Patent Office.

Lee, K., et al., 2009. Flotation of mixed copper oxide and sulphide minerals with xanthate and hydroxamate collectors. Miner. Eng. 22 (4), 395–401.

Lelinski, D., et al., 2005. Important considerations in the design of mechanical flotation machines. In: Jameson, G. (Ed.), Proceedings of Centenary of Flotation Symposium, AusIMM, Brisbane, QLD, Australia, pp. 217–223.

Lelinski, D., et al., 2013. Development of the largest flotation machine: 600 series SuperCell™ from FLSmidth. Proceedings of 10th International Mineral Processing Conference (Procemin 2013), Santiago, Chile, pp. 233–233.

Leroux, M., et al., 1994. Continuous minicell test of collectorless flotation at Mattabi Mines Ltd. CIM Bull. 87 (985), 53–57.

Lotter, N.O., 1995a. A Quality Control Model for the Development of High-Confidence Flotation Test Data. M.Sc.(Chem. Eng.) Thesis, University of Cape Town, South Africa.

Lotter, N.O., 1995b. Review of evaluation models for the representative sampling of ore. J. S. Afr. Inst. Min. Metall. 4, 149–156.

Lotter, N.O., 1995c. A quality control model for the development of high-confidence flotation test data. SME Annual Meeting, SME, Denver, CO, USA. Preprint: 95-40.

Lotter, N.O., Bradshaw, D.J., 2010. The formulation and use of mixed collectors in sulphide flotation. Miner. Eng. 23 (11-13), 945–951.

Lotter, N.O., Fragomeni, D., 2010. High confidence flotation testing at Xstrata Process Support. Miner. Metall. Process. 27 (1), 47–54.

Lotter, N.O., Oliveira, J.F., 2011. Sampling—a practical primer (short course). 43rd Annual Meeting of the Canadian Mineral Processors Conference, Ottawa, ON, Canada.

Lotter, N.O., Oliveira, J.F., 2013. Sampling—A Practical Primer (Short Course). University of Utah, Salt Lake City, UT, USA.

Lotter, N.O., et al., 2003. Sampling and flotation testing of Sudbury Basin drill core for process mineralogy modelling. Miner. Eng. 16 (9), 857–864.

Lotter, N.O., et al., 2011. Modern process mineralogy—two case studies. Miner. Eng. 24 (7), 638–650.

Lotter, N.O., et al., 2013. Flowsheet development for the Kamoa project—a case study. Miner. Eng. 52, 8–20.

Luttrell, G.H., Yoon, R.-H., 1984. The collectorless flotation of chalcopyrite ores using sodium sulphide. Int. J. Miner. Process. 13 (4), 271–283.

Lynch, A.J., et al., 1981.). Mineral and Coal Flotation Circuits: Their Simulation and Control. Elsevier Scientific Publishing Co., Amsterdam, Netherlands (Developments in mineral processing Series).

Lynch, A.J., et al., 2010. History of Flotation. AusIMM, Carlton, Melbourne, Australia (Spectrum Series: issue 18).

McTavish, S., 1980. Flotation practice at Brunswick Mining. CIM Bull. 73 (814), 115–120.

McTavish, S., 1985. Goldstream concentrator design and operation. Proceedings of the 17th Annual Meeting of the Canadian Mineral Processors Conference, CIM, Ottawa, ON, Canada, pp. 60–79.

Maldonado, M., et al., 2010. On-line estimation of frother concentration for flotation processes. Can. Metall. Q. 49 (4), 435–446.

Maldonado, M., et al., 2011. Optimizing flotation bank performance by recovery profiling. Miner. Eng. 24 (8), 939–943.

Maldonado, M., et al., 2012. An overview of optimizing strategies for flotation banks. Proceedings of the 44th Annual Meeting of the Canadian Mineral Processors Conference, CIM, Ottawa, ON, Canada, pp. 211–224.

Maldonado, M., et al., 2013. An experimental study examining the relationship between bubble shape and rise velocity. Chem. Eng. Sci. 98, 7–11.

Malghan, S.G., 1986. Role of sodium sulphide in the flotation of oxidised copper, lead, and zinc ores. Miner. Metall. Process. 3 (3), 158–163.

Malysiak, V., et al., 2004. An investigation into the floatability of a pentlandite–pyroxene system. Int. J. Miner. Process. 74 (1-4), 251–262.

Marchese, M.M., et al., 1992. Measurement of gas holdup in a three-phase concurrent downflow column. Chem. Eng. Sci. 47 (13-14), 3475–3482.

Margetts, I.R., Fagergren, W., 1920. Flotation Machine. US Patent No. 1,361,342. United States Patent Office.

Marticorena, M.A., et al., 1994. Inco develops new pyrrhotite depressant. In: Yalcin, T. (Ed.), Proceedings of Innovations in Mineral Processing Conference, Laurentian University, Sudbury, ON, Canada, pp. 15–33.

Martin, C.J., et al., 1989. Complex sulphide ore processing with pyrite flotation by nitrogen. Int. J. Miner. Process. 26 (1-2), 95–110.

Martin, J., et al., 2000. Noranda Inc.-Brunswick Mine. In: Damjanović, B., Goode, J.R. (Eds.), Canadian Milling Practice, Special vol. 49. CIM, Montreal, Quebec, Canada, pp. 185–193.

Masliyah, J., et al., 2004. Understanding water-based bitumen extraction from Anthabasca oil sands. Can. J. Chem. Eng. 82 (4), 628–654.

Mavros, P., 1992. Mixing and hydrodynamics in flotation cells. In: Mavros, P., Matis, K.A. (Eds.), Innovations in Flotation Technology, vol. 208. Kluwer Academic Publishers, Netherlands, pp. 211–234.

McKee, D.J., 1991. Automatic flotation control—a review of the last 20 years of effort. Miner. Eng. 4 (7-11), 653–666.

Mendez, D.A., et al., 2009. State of the art in conceptual design of flotation circuits. Int. J. Miner. Process. 90 (1-4), 1–15.

Miettunen, J., 1983. The Pyhäsalmi concentrator: 13 years of computer control. Proceedings of the Fourth IFAC Symposium on Automation in Mining, Mineral and Metal Processing, Helsinki, Finland, pp. 391–402.

Miller, G., Readett, D.J., 1992. The Mount Isa Mines Limited copper solvent extraction and electrowinning plant. Miner. Eng. 5 (10-12), 1335–1343.

Miller, J., et al., 2002. Nonsulfide flotation technology and plant practice. In: Mular, A.L., et al., (Eds.), Mineral Processing Plant Design, Practice and Control, vol. 1. SME, Littleton, CO, USA, pp. 1159–1178.

Miller, J.D., et al., 2006. Pyrite activation in amyl xanthate flotation with nitrogen. Miner. Eng. 19 (6-8), 659–665.

Miller, J.D., et al., 2007. Flotation chemistry and technology of nonsulfide minerals. In: Fuerstenau, M.C., et al., (Eds.), Froth Flotation: A Century of Innovation. SME, Littleton, CO, USA, pp. 465–554.

Mirnezami, M., et al. (2012). Comparison of electrokinetic behaviour of pentlandite from various sources. In: Pradip, R. (Ed.), Proceedings of the 26th International Mineral Processing Congress (IMPC), New Delhi, India, Paper No. 987: 3412–3418.

Miskovic, S., Luttrell, G., 2012. Comparison of two bubble sizing methods for performance evaluation of mechanical flotation cells. In: Young, C.A., Luttrell, G.H (Eds.), Separation Technologies for Mineral, Coal, and Earth Resources. SME, Englewood, CO, USA, pp. 563–574.

Moreno, P.A., et al., 2011. The use of seawater as process water at Las Luces copper–molybdenum beneficiation plant in Taltal (Chile). Miner. Eng. 24 (8), 852–858.

Mouat, J., 1996. The development of the flotation process: technological change and the genesis of modern mining: 1898–1911. Aust. Econ. Hist. Rev. 36 (1), 3–31.

Moudgil, B.M., Barnett, D.H., 1979. Agglomeration-skin flotation of coarse phosphate rock. Mining Eng. 31 (3), 283–289.

Muganda, S., et al., 2012. Benchmarking the flotation performance of ores. Miner. Eng. 26, 70–79.

Murdock, D.J., Wyslouzil, H.E., 1991. Large-diameter column flotation cells take hold. Eng. Min. J. 192 (8), 40–42.

Murphy, B., 2012. Less float bank for your buck—TankCell® e500. Outotec SEAP e-Newsletter. 30 (June), 1–3, <www.outotec.com/ImageVaultFiles/id.../2012_Issue30_TankCell.PDF>.

Nagaraj, D.R., 1994. A critical assessment of flotation agents. In: Mulukutla, P.S. (Ed.), Reagents for Better Metallurgy. SME, Littleton, CO, USA, pp. 81–90. (Chapter 10).

Nagaraj, D.R., Farinato, R.S., 2014a. Major innovations in the evolution of flotation reagents. In: Anderson, C.G., et al., (Eds.), Mineral Processing and Extractive Metallurgy: 100 Years of Innovation. SME, Englewood, CO, USA, pp. 159–175.

Nagaraj, D.R., Farinato, R., 2014b. Chemical factors effects in saline and hypersaline waters in the flotation of Cu and Cu and Cu–Mo ores. Proceedings of the 27th International Mineral Processing Congress (IMPC) (Chapter 3), Paper: C0315, Santiago, Chile.

Nagaraj, D.R., Ravishankar, S.A., 2007. Flotation reagents—a critical overview from an industry perspective. In: Fuerstenau, M.C., et al., (Eds.), Froth Flotation: A Century of Innovation. SME, Littleton, CO, USA, pp. 375–423.

Nagaraj, D.R., et al., 1986. Structure–activity relationships for copper depressants. Trans. Inst. Min. Metall. Sec C. 95 (Mar.), C17–C26.

Nakazawa, H., Iwasaki, I., 1985. Effect of pyrite–pyrrhotite contact on their floatabilities. Miner. Metall. Process. 2 (4), 206–211.

Napier-Munn, T.J., 2012. Statistical methods to compare batch flotation grade–recovery curves and rate constants. Miner. Eng. 34, 70–77.

Nassif, M., et al., 2013. Developing critical coalescence concentration curves for industrial process waters using dilution. Miner. Eng. 50-51, 64–68.

Nassif, M., et al., 2014. Determining frother-like properties of process water in bitumen flotation. Miner. Eng. 56, 121–128.

Ndlovu, B., et al., 2014. A preliminary rheological classification of phyllosilicate group minerals. Miner. Eng. 55, 190–200.

Nelson, M.G., et al., 2009. Design and operation of mechanical flotation machines. In: Malhotra, D., et al., (Eds.), Recent Advances in Mineral Processing Plant Design. SME, Littleton, CO, USA, pp. 168–189.

Nesset, J.E., 1988. The application of residence time distributions to flotation and mixing circuits. CIM Bull. 81 (919), 75–83.

Nesset, J.E., 2011. Significant Canadian developments in mineral processing technology—1961 to 2011. In: Kapusta, J., et al., (Eds.), The Canadian Metallurgical & Materials Landscape 1960 to 2011. MetSoc, CIM, Westmount, Montreal, Quebec, Canada, pp. 241–293.

Nesset, J.E., Finch, J.A., 2013. Correcting bubble size measurement for frother concentration in the McGill Bubble Size Analyzer. In: Liu, Q. (Ed.), Proceedings of the Materials Science & Technology (MS&T) 2013, Water and Energy in Mineral Processing. Montreal, QC, Canada, pp. 1824–1840.

Nesset, J.E., et al., 1998. The effect of soda ash and lime as pH modifiers in sphalerite flotation. Proceedings of the 30th Annual Meeting of the Canadian Mineral Processors Conference, CIM, Ottawa, ON, Canada, pp. 459–482.

Nesset, J.E., et al., 2007. Operating variables affecting the bubble size in forced-air mechanical flotation machines. Proceedings of the 9th Mill Operators' Conference, AusIMM, Fremantle, Australia, pp. 55–65.

Nesset, J.E., et al., 2012. A benchmarking tool for assessing flotation cell performance. Proceedings of the 44th Annual Meeting of the Canadian Mineral Processors Conference, CIM, Ottawa, ON, Canada, pp. 183–210.

Neethling, S.J., et al., 2006. The use of FrothSim to optimise the water addition to a column flotation cell. Miner. Eng. 19 (6-8), 816–823.

Newell, A.J.H., et al., 2007. Restoring the flotatability of oxidized sulfides using sulfidisation. Int. J. Miner. Process. 84 (1-4), 108–117.

Nguyen, A.V., Schulze, H.J. (Eds.), 2004. Colloidal Science of Flotation, vol. 118. Marcel Dekker/CRC Press, New York, NY, USA (Surfactant Science Series).

Ngobeni, W., Hangone, G., 2013. The effect of using pure thiol collectors on the froth flotation of pentlandite containing ore. S. Afr. J. Chem. Eng. 18 (1), 41–50.

Nicol, S.K., 2000. In: Swanson, A.R., Partridge, A.C. (Eds.), Fine Coal Beneficiation, vol. 4. Australian Coal Preparation Society (Advanced Monograph Series).

Niitti, T., Tarvainen, M., 1982. Experiences with large Outokumpu flotation machines. Proceedings of the 14th International Mineral Processing Congress, Paper No. VI-7, CIM, Toronto, ON, Canada, pp. V-I7.1–VI-7.12.

Noble, A., Luttrell, G.H., 2014. The matrix reduction algorithm for solving separation circuits. Miner. Eng. 64, 97–108.

Nooshabadi, A.J., et al., 2013. Formation of hydrogen peroxide by pyrite and its influence on flotation. Miner. Eng. 49, 128–134.

Oldshue, J.Y., 1983. Fluid Mixing Technology. Chemical Engineering (Publisher), McGraw Hill Pub. Co.

Orford, I., et al., 2005. The approach to process improvements at Brunswick Mine. Proceedings of the 37th Annual Meeting of the Canadian Mineral Processors Conference, CIM, Ottawa, ON, Canada, pp. 203–226.

Outotec, 2014. The first Outotec TankCell® e500 will be delivered to First Quantum Minerals in Finland. *Outotec Product News*.

Paananen, A.D., Turcotte, W.A., 1980. Factors influencing selective flocculation-desliming practice at the Tilden Mine. Mining Eng. 32 (8), 1244–1247.

Panayotov, V., Panayotova, M., 2013. Mineral Processing, Applied Electrochemistry and Physical Chemistry for Environmentally Friendly Extraction of Metals. University of Mining and Geology, Sofia, Bulgaria, Publishing House "St. Ivan Rilski".

Parks, G.A., 1965. The isoelectric points of solid oxides, solid hydroxides, and aqueous hydroxo complex systems. Chem. Rev. 65 (2), 177–198.

Partridge, A.C., Smith, G.W., 1971. Small-sample flotation testing: a new cell. Trans. Inst. Min. Metall., Sec. C. 80 (778), C199–C200.

Patwardhan, A., Honaker, R.Q., 2000. Development of a carrying capacity model for column froth flotation. Int. J. Miner. Process. 59 (4), 275–293.

Pazour, D.A., 1979. Morococha-five product mine shows no sign of dying. World Mining. 32 (12), 56–58.

Pease, J.D., et al., 2006. Designing flotation circuits for high fines recovery. Miner. Eng. 19 (6-8), 831–840.

Piantadosi, C., Smart, R.S.C., 2002. Statistical comparison of hydrophobic and hydrophilic species on galena and pyrite particles in flotation concentrates and tails from TOF-SIMS evidence. Int. J. Miner. Process. 64 (1), 43–54.

Poling, G.W., 1980. Selection and sizing of flotation machines. In: Mular, A.L., Bhappu, R.B. (Eds.), Mineral Processing Plant Design, second ed. SME, New York, NY, USA, pp. 887–906 (Chapter 44).

Poorkani, M., Banisi, S., 2005. Industrial use of nitrogen in flotation of molybdenite at the Sarcheshmeh copper complex. Miner. Eng. 18 (7), 735–738.

Pradip, Fuerstenau, D.W., 2013. Design and development of novel flotation reagents for the beneficiation of Mountain Pass rare-earth ore. Miner. Metall. Process. 30 (1), 1–9.

Pradip, Rai, B., 2003. Molecular modeling and rational design of flotation reagents. Int. J. Miner. Process. 72 (1-4), 95–110.

Pratt, A.R., et al., 1994. X-ray photoelectron spectroscopy and Auger studies of pyrrhotite and mechanism of air oxidation. Geochim. Cosmochim. Acta. 58 (2), 827–841.

Pratten, S.J., et al., 1989. An evaluation of the flotation response of coals. Int. J. Miner. Process. 27 (3-4), 243–262.

Pugh, R.J., 1986. The role of solution chemistry of dodecylamine and oleic acid collectors in the flotation of fluorite. Colloids Surf. 18 (1), 19–41.

Pugh, R.J., 1989. Macromolecular organic depressants in sulphide flotation—a review, 1. Principles, types and applications. Int. J. Miner. Process. 25 (1-2), 101–130.

Pugh, R.J., 2007. The physics and chemistry of frothers. In: Fuerstenau, M.C., et al., (Eds.), Froth Flotation: A Century of Innovation. SME, Littleton, CO, USA, pp. 259–281.

Quinn, J.J., Finch, J.A., 2012. On the origin of bi-modal bubble size distributions in absence of frother. Miner. Eng. 36-38, 237–241.

Quinn, J.J., et al., 2007. Comparing the effects of salts and frother (MIBC) on gas dispersion and froth properties. Miner. Eng. 20 (14), 1296–1302.

Quinn, J.J., et al., 2014. Critical coalescence concentration of inorganic salt solutions. Miner. Eng. 58, 1–6.

Rahal, K., et al., 2000. Flotation plant modelling and simulation using the floatability characterisation test rig (FCTR). Proceedings of the International Congress on Mineral Processing and Extractive Metallurgy (Minprex 2000), vol. 5, AusIMM, Melbourne, Australia, pp. 339–344.

Rahman, R.M., et al., 2012. The effect of flotation variables on the recovery of different particle size fractions in the froth and the pulp. Int. J. Miner. Process. 106-109, 70–77.

Ralston, J., 1991. E_h and its consequences in sulphide mineral flotation. Miner. Eng. 4 (7-11), 859–878.

Ralston, J., Healy, T.W., 1980. Activation of zinc sulphide with CuII, CdII and PbII: II. Activation in neutral and weakly alkaline media. Int. J. Miner. Process. 7 (3), 203–217.

Ralston, J., et al., 2001. The hydrophobic force in flotation—a critique. Colloids & Surf. A. 192 (1-3), 39–51.

Ramirez-Castro, J., Finch, J.A., 1980. Simulation of a grinding circuit change to reduce lead sliming. CIM Bull. 73 (822), 132–139.

Randall, E.W., et al., 1989. A method for measuring the sizes of bubbles in two and three-phase systems. J. Phys. E: Sci. Instr. 22 (10), 827–833.

Ranney, M.W. (Ed.), 1980. Flotation Agents and Processes: Technology and Applications. Noyes Data Corp., New Jersey, USA.

Rao, F., Liu, Q., 2013. Froth treatment in Anthabasca oil sands bitumen recovery process: a review. Energy Fuels. 27 (12), 7199–7207.

Rao, K.H., et al., 1991. Mechanism of oleate adsorption on salt-type minerals: IV. Adsorption, electrokinetic, and diffuse reflectance FT-IR studies of natural fluorite in the presence of sodium oleate. J. Colloid Interface. Sci. 145 (2), 314–329.

Rao, K.H., Forssberg, K.S.E., 2007. Chemistry of iron oxide flotation. In: Fuerstenau, M.C., et al., (Eds.), Froth Flotation: A Century of Innovation. SME, Littleton, CO, USA, pp. 498–513.

Rao, S.R., Leja, J., 2004. Surface Chemistry of Froth Flotation. second ed. Kluwer Academic/Plenum Publishers, New York, NY, USA.

Rao, S.R., et al., 1988. Effects of water chemistry on the flotation of pyrochlore and silicate minerals. Miner. Eng. 1 (3), 189–202.

Rao, S.R., et al., 2011. Activation of sphalerite by Cu ions produced by cyanide action on chalcopyrite. Miner. Eng. 24 (9), 1025–1027.

Raschi, F., Finch, J.A., 2006. Deactivation of Pb-contaminated sphalerite by polyphosphate. Colloids Surf., A. 276 (1-3), 87–94.

Ravichandran, V., et al., 2013. Gas dispersion characteristics of flotation reagents. Powder Technol. 235, 329–335.

Rawlins, C.H., 2009. Flotation of fine oil droplets in petroleum production circuits. In: Malhotra, D., et al., (Eds.), Recent Advances in Mineral Processing Plant Design. SME, Littleton, CO, USA, pp. 232–246.

Ritson, G., Ward, C., 2009. An overview of the oil sands process, comparing it to conventional mineral processing. Proceedings of the 41st Annual Meeting of the Canadian Mineral Processors Conference, CIM, Ottawa, ON, Canada, pp. 163–190.

Roberts, A.N., et al., 1980. Metallurgical development at Woodlawn Mines, Australia. In: Jones, M.J. (Ed.), Complex Sulphide Ores. IMM, pp. 128–134.

Rodrigues, R.T., Rubio, J., 2003. New basis for measuring the size distribution of bubbles. Miner. Eng. 16 (8), 757–765.

Rosas, J., et al., 2012. Update of Chilean mining projects and technological trends. Proceedings of the 44th Annual Meeting of the Canadian Mineral Processors Conference, CIM, Ottawa, ON, Canada, pp. 3–12.

Ross, V.E., 1990. Flotation and entrainment of particles during batch flotation tests. Miner. Eng. 3 (3-4), 245–256.

Ruel, M., 2007. Managing assets using performance supervision. Proceedings of the Sixth International Copper/Cobre Conference, vol. 7, Toronto, ON, Canada, pp. 73–84.

Runge, K., 2010. Laboratory flotation testing—an essential tool for ore characterization. In: Greet, C.J. (Ed.), Flotation Plant Optimisation: A Metallurgical Guide to Identifying and Solving Problems in Flotation Plants (Spectrum Series), vol. 16. AusIMM, Carlton, Australia, pp. 155–174 (Chapter 9).

Safari, M., et al., 2013. The effect of energy input on the flotation kinetics of galena in an oscillating grid flotation cell. Flotation'13, Session 1, Cape Town, South Africa, pp. 1–10.

Sandoval-Zambrano, G., Montes-Atenas, G., 2012. Errors in estimation of size-by-liberation flotation rate constants. Miner. Eng. 27-28, 1–10.

Sastry, K.V.S., 1988. Column Flotation'88. SME, Littleton, CO, USA.

Schlesinger, M.E., et al., 2011. Extractive Metallurgy of Copper. fifth ed. Elsevier, Amsterdam, Netherlands.

Schubert, H., Bischofberger, C., 1998. On the microprocesses air dispersion and particle-bubble attachment in flotation machines as well as consequences for the scale-up of macroprocesses. Int. J. Min. Process. 52 (4), 245–259.

Seaman, D.R., et al., 2004. Bubble load measurement in the pulp zone of industrial flotation machines—a new device for determining the froth recovery of attached particles. Int. J. Miner. Process. 74 (1-4), 1–13.

Seke, M.D., Pistorius, P.C., 2006. Effect of cuprous cyanide, dry and wet milling on the slective flotation of galena and sphalerite. Miner. Eng. 19 (1), 1–11.

Senior, G.D., et al., 1994. The flotation of pentlandite from pyrrhotite with particular reference to the effects of particle size. Int. J. Miner. Process. 42 (3-4), 169–190.

Senior, G.D., Thomas, S.A., 2005. Development and implementation of a new flowsheet for the flotation of a low grade nickel ore. Int. J. Miner. Process. 78 (1), 49–61.

Shackelton, N.J., et al., 2003. The use of amine complexes in managing inadvertent activation in a pentlandite-pyroxene flotation system. Miner. Eng. 16 (9), 849–856.

Shackleton, N.J., et al., 2007. Surface characteristics and flotation behaviour of platinum and palladium arsenides. Int. J. Miner. Process. 85 (1-3), 25–40.

Shannon, E.R., et al., 1993. Back to basics—the road to recovery: milling practice at Brunswick Mining. Proceedings of the 25th Annual Meeting Canadian Mineral Processors Conference, CIM, Ottawa, ON, Canada, pp. 2–17.

Shaw, D.R., 1981. Dodecyl mercaptan: a superior collector for sulphide ores. Mining Eng. 33 (5), 686–692.

Shean, B.J., Cilliers, J.J., 2011. A review of froth flotation control. Int. J. Miner. Process. 100 (3-4), 57–71.

Shen, Z., et al., 2010. Research on the design and processing characteristics of 320 m^3 air-forced mechanical flotation cell. Proceedings of the 25th International Mineral Processing Congress (IMPC), Brisbane, Australia, pp. 3481–3487.

Shent, H., et al., 1999. A review of plastics waste recycling and flotation of plastics. Resources, Conservation and Recycling. 25 (2), 85–109.

Shergold, H.L., 1984. Flotation in mineral processing. In: Ives, K.J. (Ed.), The Scientific Basis of Flotation, vol. 75. Martinus Nijhofff Publishers, The Hague, Netherlands, pp. 229–288. (NATO Advanced Study Institute Series: Series E: Applied Sciences).

Shortridge, P.G., et al., 2000. The effect of chemical composition and molecular weight of polysaccharide depressants on the flotation of talc. Int. J. Miner. Process. 59 (3), 215–224.

Siirak, J., Hancock, B.A., 1988. Progress in developing a flotation phosphorous reduction process at the Tilden iron ore mine. In: Forssberg, K.S.E., (Ed.), Proceedings of the 16th International Mineral Processing Congress, Stockholm, Sweden, pp. 1393–1404.

Simmons, G.L., 1997. Flotation of auriferous pyrite using Santa Fe Gold's N2TEC process. SME, Denver, CO, USA. Preprint, 97-27:1–8.

Singh, A., et al., 2003. Flotation stabilization and optimization. J. S. Afr. Inst. Min. Metall. 103 (9), 581–588.

Singh, N., Finch, J.A., 2014. Bank profiling and separation efficiency. Miner. Eng. 66-68, 191–196.

Smart, R.S.C., et al., 1999. XPS of sulfide mineral surfaces: metal-deficient, polysulfides, defects and elemental sulphur. Surf. Interface Anal. 28 (1), 101–105.

Smart, R.S.C., et al., 2003. Surface analytical studies of oxidation and collector adsorption in sulfide mineral flotation. In: Wandelt, K., Thurgate, S. (Eds.), Solid/Liquid Interfaces: Macroscopic Phenomena-Microscopic Understanding, vol. 85. Verlag, Berlin, Germany, pp. 3–60. Topics in Applied Physics.

Smart, R.St.C., et al., 2007. Surface characterization and new tools for research. In: Fuerstenau, M.C., et al., (Eds.), Froth Flotation: A Century of Innovation. SME, Littleton, CO, USA, pp. 283–337.

Smart, R.St.C., et al., 2014. Innovations in measurement of mineral structure and surface chemistry in flotation: past, present, and future. In: Anderson, C.G., et al., (Eds.), Mineral Processing and Extractive Metallurgy: 100 Years of Innovation. SME, Englewood, CO, USA, pp. 577–602.

Smith, C., et al., 2008. Air-rate profile optimization: from simulation to bank improvement. Miner. Eng. 21 (12-14), 973–981.

Smith, P.G., Warren, L.J., 1989. Entrainment of particles into flotation froths. Miner. Process. Extr. Metall. Rev. 5 (1-4), 123–145.

Somasundaran, P., Moudgil, B.M., 1987. Reagents in Mineral Technology, vol. 2. CRC Press, New York, NY, USA (Surfactant Science Series).

Somasundaran, P., Sivakumar, A., 1988. Advances in understanding flotation mechanisms. Miner. Metall. Process. 5, 97–103.

Somasundaran, P., Wang, D., 2006. In: Wills, B.A. (Ed.), *Solution Chemistry:* Minerals and Reagents, vol. 17. Elsevier, Amsterdam, New York, NY, USA (Developments in Mineral Processing Series).

Sorensen, T.C., 1982. Large agitair flotation design and operation. Proceedings of the 14th International Mineral Processing Congress (IMPC), vol. 9, CIM, Toronto, ON, Canada, pp. 1–10.

Snodgrass, R.A., et al., 1994. Process development and design of the Northam Merensky concentrator. Proceedings of the 15th Commonwealth Mining and Metallurgical Congress, vol. 2, SAIMM, Johannesburg, South Africa, pp. 341–357.

Stowe, K.G., et al., 1995. Mapping of composition of mineral surfaces by TOF-SIMS. Miner. Eng. 8 (4-5), 421–430.

Su, L., et al., 2006. Role of oily bubbles in enhancing bitumen flotation. Miner. Eng. 19 (6-8), 641–650.

Sui, C.C., et al., 1999. Comparison of the activation of sphalerite by copper and lead ions. Miner. Metall. Process. 16 (2), 53–61.

Sutherland, K.L., Wark, I.W., 1955. Principles of Flotation. AusIMM, Melbourne, Australia.

Sutolov, A., 1975. Copper Porphyries. World Mining Books, M. Freeman Publications.

Suttill, K.R., 1991. A technical buyer's guide to mining chemicals. Eng. Min. J. 192 (8), 23–34.

Sweet, J.A., et al., 2013. The AGDP in 2012—nine years of exceptional graduate training. In: Cilliers, J., et al. (Eds.), Minerals Industry: Education and Training. National Metallurgical Laboratory, The Indian Institute of Mineral Engineers (IIME), Jamshedpur, India, pp. 131–149.

Takamura, K., 1982. Microscopic structure of Anthabasca oil sand. Can. J. Chem. Eng. 60 (4), 538–545.

Tan, S.N., et al., 2005. Foaming of polypropylene glycols and glycol/MIBC mixtures. Miner. Eng. 18 (2), 179–188.

Tarkan, H.M., et al., 2012. Studies on air-assisted solvent extraction. In: Young, C.A., Luttrell, G.H. (Eds.), Separation Technologies, for Minerals Coal and Earth Resources. SME, Englewood, CO, USA, pp. 317–324.

Teck., 2014. Diversified Mining, Highland Valley Copper. Visited October 3, 2014. <https://www.teck.com/Generic.aspx?PAGE=Teck+Site%2fDiversified+Mining+Pages%2fCopper+Pages%2fHighland+Valley+Copper+Pages%2fProduction&portalName=tc>.

Thornton, A.J., 1991. Cautious adaptive control of an industrial flotation circuit. Miner. Eng. 4 (12), 1227–1242.

Thwaites, P., 1983. Continued development of copper flotation control at the Kidd Creek concentrator. CIM Bull. 76 (860), 41–46.

Tomiyama, A., et al., 2002. Terminal velocity of single bubbles in surface tension force dominant regime. Int. J. Multiphase Flow. 28 (9), 1497–1519.

Trahar, W.J., 1976. The selective flotation of galena from sphalerite with special reference to the effects of particle size. Int. J. Miner. Process. 3 (2), 151–166.

Trahar, W.J., 1981. A rational interpretation of the role of particle size in flotation. Int. J. Miner. Process. 8 (4), 289–327.

Trahar, W.J., Warren, L.J., 1976. The floatability of very fine particles—a review. Int. J. Miner. Process. 3 (2), 103–131.

Trombino, S., et al., 2012. Cellulose and dextran antioxidant polymers for biomedical applications. In: Cirillo, G., Iemma, F. (Eds.), Antioxidant Polymers: Synthesis, Properties, and Applications. John Wiley & Sons, Inc., Hoboken, NJ, USA, pp. 133–151 (Chapter 6).

Tsatouhas, G., et al., 2006. Case studies on the performance and characterisation of the froth phase in industrial flotation circuits. Miner. Eng. 19 (6-8), 774–783.

Tukel, C., Kelebek, S., 2010. Modulation of xanthate action by sulphite ion in pyrrhotite deactivation/depression. Int. J. Miner. Process. 95 (1-4), 47–52.

Türkel, N., 2011. Stability of metal chelates of some hydroxamic acid ligands. J. Chem. Eng. Data. 56 (5), 2337--2342.

Twidle, T.R., et al., 1985. Optimising control of lead flotation at Black Mountain. Proceedings of the 15th International Mineral Processing Congress, vol. 3, Cannes, French Riviera, France, pp. 189–198.

Uddin, S., et al., 2011. Processing an ultramafic ore using fibre disintegration by acid attack. Int. J. Miner. Process. 102-103, 38–44.

Uddin, S., et al., 2012. Effect of particles on the electrical charge of gas bubbles in flotation. Miner. Eng. 36-38, 160–167.

Umipig, C.V., et al., 2012. Canatuan Cu/Zn flotation metallurgy—dealing with zinc pre-activation. Proceedings of the 44th Annual Meeting of the Canadian Mineral Processors Conference, CIM, Ottawa, ON, Canada, pp. 155–168.

USGS., 2012. Minerals commodities summaries: phosphate rock. <http://minerals.usgs.gov/minerals/pubs/commodity/phosphate_rock/>.

USGS., 2014. Copper: U.S. Geological Survey. Mineral Commodity Summaries. <http://minerals.usgs.gov/minerals/pubs/commodity/copper/>.

Van Olst, M., et al., 2000. Improving flotation plant performance at Cadia by controlling and optimising the rate of froth recovery using Outokumpu FrothMaster™. Proceedings 7th Mill Operators' Conference, Kalgoorlie, Australia, pp. 127–135.

Vera, M.A., et al., 1999. Simultaneous determination of collection zone rate constant and froth zone recovery in a mechanical flotation environment. Miner. Eng. 12 (10), 1163–1176.

Wang, B., Peng, Y., 2013. The behaviour of mineral matter in fine coal flotation using saline water. Fuel. 109, 309–315.

Wang, L.K., et al.,2007. *Advanced Physicochemical Treatment Technologies. Handbook of Environmental Engineering*, vol. 5. Humana Press, NJ, USA.

Wang, X., et al., 1989a. The aqueous and surface chemistry of activation in the flotation of sulphide minerals—a review, Part I: an electrochemical model. Miner. Process. Extr. Metall. Rev. 4 (3-4), 135–165.

Wang, X., et al., 1989b. The aqueous and surface chemistry of activation in the flotation of sulphide minerals—review Part II: a surface precipitation model. Miner. Process. Extr. Metall. Rev. 4 (3-4), 167–199.

Waters, K.E., et al., 2008. The flotation of fine particles using charged microbubbles. Miner. Eng. 21 (12-14), 918–923.

Watson, J.L., 1988. South East Missouri lead belt—a review 1987. Miner. Eng. 1 (2), 151–156.

Watters, T.J., Sandy,W., 1983. Lead Flotation Practice and Control at North Broken Hill Limited. Proc AusIMM. Conference, Broken Hill, New South Wales, Australia, pp. 277–285.

Weiss, N.L. (Ed.), 1985. Flotation Kinetics. SME Mineral Processing Handbook, vol. 1. SME, Kingsport Press, Kingsport, TN, USA, pp. 94–98 (Chapter 5).

Welsby, S.D.D., 2014. Pilot-scale froth testing at Highland Valley Copper. Proceedings of the 46th Annual Meeting of the Canadian Mineral Processors Conference, CIM, Ottawa, ON, Canada, pp. 301–314.

Welsby, S.D.D., et al., 2010. Assigning physical significance to floatability components. Int. J. Miner. Process. 97 (1-4), 59–67.

Wesseldijk, Q.I., et al., 1999. The flotation behaviour of chromite with respect to the beneficiation of UG2 ore. Miner. Eng. 12 (10), 1177–1184.

Wiegel, R.L., 1999. Phosphate rock beneficiation practice. In: Parekh, B. K., Miller, J.D. (Eds.), Advances in Flotation Technology. SME, Littleton, CO, USA, pp. 213–218.

Wiese, J., et al., 2005. The influence of the reagent suite on the flotation of ores from the Merensky reef. Miner. Eng. 18 (2), 189–198.

Wiese, J.G., et al., 2010. The effect of increased frother dosage on froth stability at high depressant dosages. Miner. Eng. 23 (11-13), 1010–1017.

Wilkinson, E.W., et al., 1926. Flotation Machine. US Patent No. 1,588,077. United States Patent Office, pp. 1–10.

Williams, J.J.E., Crane, R.I., 1983. Particle collision rate in turbulent flow. Int. J. Multiphase Flow. 9 (4), 421–435.

Williams, M.C., Meloy, T.P., 2007. Optimal designs for homogeneous countercurrent flotation processing networks. In: Fuerstenau, M.C., et al., (Eds.), Froth Flotation: A Century of Innovation. SME, Littleton, CO, USA, pp. 739–756.

Wills, B.A., 1984. The separation by flotation of copper–lead–zinc sulphides. Mining Mag. 15 (1), 36–41.

Wood, K.R., 2002. Flotation equipment selection and plant layout. In: Mular, A.L., et al., (Eds.), Mineral Processing Plant Design, Practice, and Control. SME, Littleton, CO, USA, pp. 1204–1238.

Woods, R., 2010. Electrochemical aspects of sulfide mineral flotation. In: Greet, C.J. (Ed.), Flotation Plant Optimisation: A Metallurgical Guide to Identifying and Solving Problems in Flotation Plants. AusIMM, Carlton, VIC, Australia, pp. 123–136. (Spectrum series: No. 16), (Chapter 7).

Wrobel, S.A., 1970. Economic flotation of minerals. Mining Mag. 122 (4), 281–282.

Wyslouzil, H., et al., 2013. Optimization of the flotation process through development of purpose specific flotation machines. In: Alvarez, M., et al. (Eds.), Proceedings of the 10th International Mineral Processing Conference, (Procemin 2013), Gecamin, Chile, pp. 295–303.

Xu, M., et al., 1991. Maximum gas and bubble surface rates in flotation columns. Int. J. Miner. Process. 32 (3-4), 233–250.

Xu, M., et al., 1995. Reverse flotation of pyrite from a zinc-concentrate using nitrogen. Miner. Eng. 8 (10), 1159–1173.

Yalamanchili, M.R., et al., 1993. Adsorption of collector colloids in the flotation of alkali halide particles. Int. J. Miner. Process. 39 (1-2), 137–153.

Yang, Y., Telionis, D., 2012. Turbulence measurements in a flotation cell using fast-response probe. In: Young, C.A., Luttrell, G.H. (Eds.), Separation Technologies for Minerals, Coal, and Earth Resources. SME, Englewood, CO, USA, pp. 583–591.

Yang, S., et al., 2012. Nanoparticle flotation collectors for pentlandite. Proceedings of the 44th Annual Meeting of the Canadian Mineral Processors Conference, CIM, Ottawa, ON, Canada, pp. 225–236.

Yarar, B., Kaoma, J., 1984. Estimation of critical surface tension of wetting of hydrophobic solids by flotation. Colloids Surf. 11 (3-4), 429–436.

Ye, Y., et al., 1990. Molybdenite flotation from copper/molybdenum concentrates by ozone conditioning. Miner. Metall. Process. 11, 173–179.

Yeomans, T., 2008. Copper concentrate quality improvements at Myra Falls. Proceedings of the 40th Annual Meeting of the Canadian Mineral Processors Conference, CIM, Ottawa, ON, Canada, pp. 283–296.

Yianatos, J.B., et al., 1988. Effect of column height on flotation column performance. Miner. Metall. Process. 5 (Feb.), 11–14.

Yianatos, J.B., 2007. Fluid flow and kinetic modeling in flotation related processes: columns and mechanically agitated cells—a review. Chem. Eng. Res. Des. 85 (12), 1591–1603.

Yianatos, J.B., Henriquez, F., 2007. Boundary conditions for bubble size at the pulp-froth interface in flotation equipment. Miner. Eng. 20 (6), 625–628.

Yianatos, J., et al., 2001. Hydrodynamic and metallurgical characterization of industrial flotation banks for control purposes. Miner. Eng. 14 (9), 1033–1046.

Yianatos, J.B., et al., 2008. Froth recovery of industrial flotation cells. Miner. Eng. 21 (12-14), 817–825.

Yin, X., Miller, J.D., 2012. Wettability of kaolinite basal planes based on surface force measurements using atomic force microscopy. Miner. Metall. Process. 29 (1), 13–19.

Yoon, R.H., 1981. Collectorless flotation of chalcopyrite and sphalerite ores by using sodium sulfide. Int. J. Miner. Process. 8 (1), 31–48.

Yoon, R.-H., 1993. Microbubble flotation. Miner. Eng. 6 (6), 619–630.

Yoon, R.-H., 2000. The role of hydrodynamic and surface forces in bubble–particle interaction. Int. J. Miner. Process. 58 (1-4), 129–143.

Yoon, R.-H., et al., 1992. Beneficiation of kaolin clay by froth flotation using hydroxamate collectors. Miner. Eng. 5 (3-5), 457–467.

Young, M.F., Gao, M., 2000. Performance of the IsaMills in the George Fisher flowsheet. Proceedings of the Seventh Mill Operators Conference, AusIMM, Kalgoorlie, Australia, pp. 12–14.

Young, P., 1982. Flotation machines. Min. Mag. 146 (Jan.), 35.

Zangooi A., et al., 2014. Determining frother distribution in flotation circuits. Proceedings of the 27th International Mineral Processing Congress (IMPC), Paper: C0603, Santiago, Chile, (Chapter 6).

Zanin, M., et al., 2009. Quantifying contributions to froth stability in porphyry copper plants. Int. J. Miner. Process. 91 (1-2), 19–27.

Zhang, W., Poling, G.W., 1991. Sulphidization-promoting effects of ammonium sulphate on sulphidized xanthate flotation of malachite. Proceedings of the 17th International Mineral Processing Congress, vol. IV, Dresden, Germany, pp. 187–197.

Zhang, W., et al., 2009. Use of frother with sampling-for-imaging bubble sizing technique. Miner. Eng. 22 (5), 513–515.

Zhang, W., et al., 2010. Water recovery and bubble surface area flux in flotation. Can. Met. Q. 49 (4), 353–362.

Zhang, W., et al., 2012a. Determining independent control of dual-frother systems—gas holdup, bubble size and water overflow rate. Miner. Eng. 39, 106–116.

Zhang, W., et al., 2012b. Characterizing frothers through critical coalescence concentration (CCC95)-hydrophilic-lipophilic balance (HLB). Minerals. 2 (3), 208–227.

Zheng, X., et al., 2006. An evaluation of different models of water recovery in flotation. Miner. Eng. 19 (9), 871–882.

Zhou, Z.A., et al., 1994. On the role of cavitation in particle collection during flotation—a critical review. Miner. Eng. 7 (9), 1073–1084.

Zhou, Z.A., et al., 1997. Role of hydrodynamic cavitation in fine particle flotation. Int. J. Miner. Process. 51 (1-4), 139–149.

Zhou, Z., et al., 2000. Effect of natural surfactants released from Athabasca oil sands on air holdup in a water column. Can. J. Chem. Eng. 78 (4), 617–624.

Zhou, Z.A., et al., 2009. On the role of cavitation in particle collection in flotation—a critical review, II. Miner. Eng. 22 (5), 419–433.

Zieliński, P.A., et al., 2000. Preferential deportment of low-iron sphalerite to lead concentrate. Miner. Eng. 13 (4), 357–363.

Zisman, W.A., 1964. Relation of equilibrium contact angle to liquid and solid constitution. In: Gould, R.F (Ed.), Advances in Chemistry Series 43. ACS, Washington, DC, USA, pp. 1–51.

Chapter 13

Magnetic and Electrical Separation

13.1 INTRODUCTION

Magnetic and electrical separators are being considered in the same chapter, as there is often an overlap in the application of the two processes. A classic example of this is the processing of heavy mineral sands in which both magnetic and electrostatic separation are crucial to achieve separation.

13.2 MAGNETISM IN MINERALS

Magnetic separators exploit the difference in magnetic properties between the minerals in a deposit and are used to concentrate a valuable mineral that is magnetic (e.g., magnetite from quartz), to remove magnetic contaminants, or to separate mixtures of magnetic and nonmagnetic valuable minerals. An example of the latter is the tin-bearing mineral cassiterite, which is often associated with traces of the valuable minerals magnetite or wolframite, which can be removed by magnetic separators.

This text will only briefly introduce the concepts associated with magnetism in mineral separation. For those interested in further details there are a number of other sources (Jiles, 1990; Oberteuffer, 1974; Svoboda, 1987; Svoboda and Fujita, 2003).

All materials are affected in some way when placed in a magnetic field, although with many substances the effect is too slight to be easily detected. For the purposes of mineral processing, materials may be classified into two broad groups, according to whether they are attracted or repelled by a magnet:

1. Diamagnetic materials are repelled along the lines of magnetic force to a point where the field intensity is smaller. The forces involved here are very small and diamagnetic substances are often referred to as "nonmagnetic", although this is not strictly correct. Diamagnetic minerals will report to the nonmagnetic product ("non-mags") of a magnetic separator as they do not experience a magnetic attractive force.
2. Paramagnetic materials are attracted along the lines of magnetic force to points of greater field intensity. Paramagnetic materials report to the "magnetic" product ("mags") of a magnetic separator due to attractive magnetic forces. Examples of paramagnetic minerals which are separated in commercial magnetic separators are ilmenite ($FeTiO_3$), rutile (TiO_2), wolframite ($(Fe, Mn)WO_4$), monazite ($(Ce, La, Nd, Th)PO_4$), xenotime (YPO_4), siderite ($FeCO_3$), chromite ($FeCr_2O_4$), and manganese minerals.

Paramagnetism in a material originates due to the presence of unpaired electrons which create magnetic dipoles. When these magnetic dipoles are aligned by an externally applied magnetic field, the resultant magnetic moment causes the material to become magnetized and experience a magnetic force along the lines of the applied magnetic field. Certain elements have electron configurations with many unpaired electrons, but the magnetic response of a given mineral depends on the structure of the mineral as well as its constituent atoms. For example, pyrite (FeS_2) is very slightly paramagnetic, but the chemically similar pyrrhotite ($Fe_{1-x}S$) in the monoclinic structural form is actually strongly magnetic, referred to as ferromagnetic.

Ferromagnetism can be regarded as a special case of paramagnetism in which the magnetic dipoles of a material undergo exchange coupling so that they can more rapidly align themselves with an applied magnetic field. Examples of diamagnetic, paramagnetic, and ferromagnetic behavior are shown in Figures 13.1 and 13.2 represented as magnetization (density of magnetic dipoles) versus applied magnetic field strength. The slope of these curves represents the dimensionless magnetic susceptibility of the material. Figure 13.1 shows the paramagnetic susceptibility (shown by a positive linear slope) of chromite and the diamagnetic susceptibility (negative linear slope) of quartz, while Figure 13.2 shows the ferromagnetic trend of magnetite. Thanks to exchange coupling, ferromagnetic materials will have very high initial susceptibility to magnetic forces until all of the exchange coupled magnetic moments have aligned with the applied magnetic force. This results in a rapidly decreasing value of susceptibility with increased applied magnetic field (Figure 13.2, points 1–3). Once this alignment has occurred the material is said to have

Wills' Mineral Processing Technology.

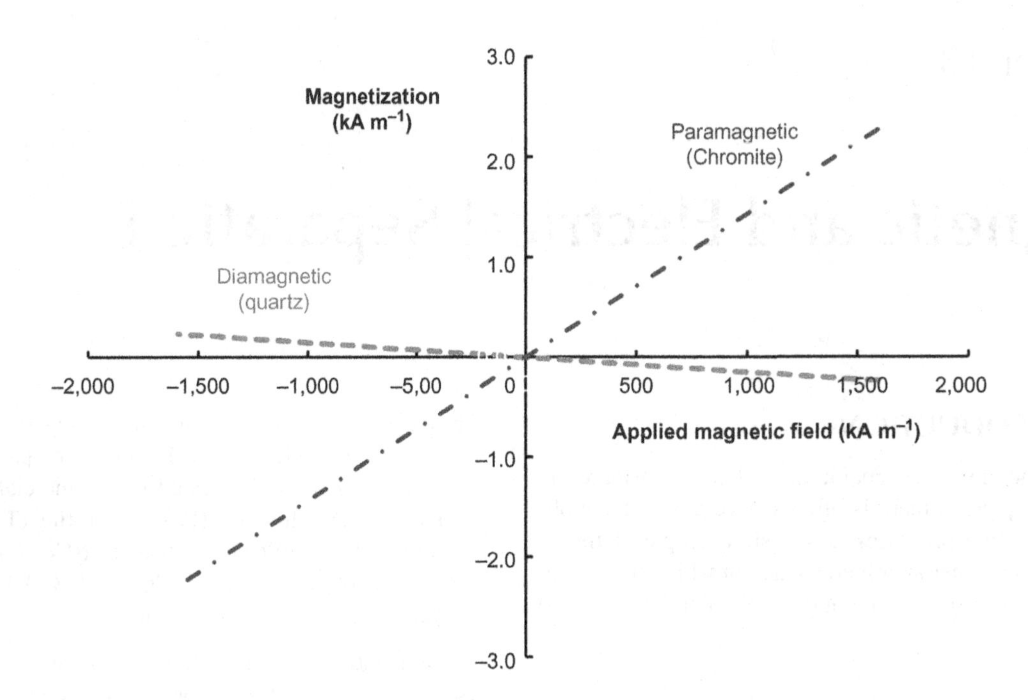

FIGURE 13.1 Magnetization versus applied magnetic field strength for idealized paramagnetic and diamagnetic minerals.

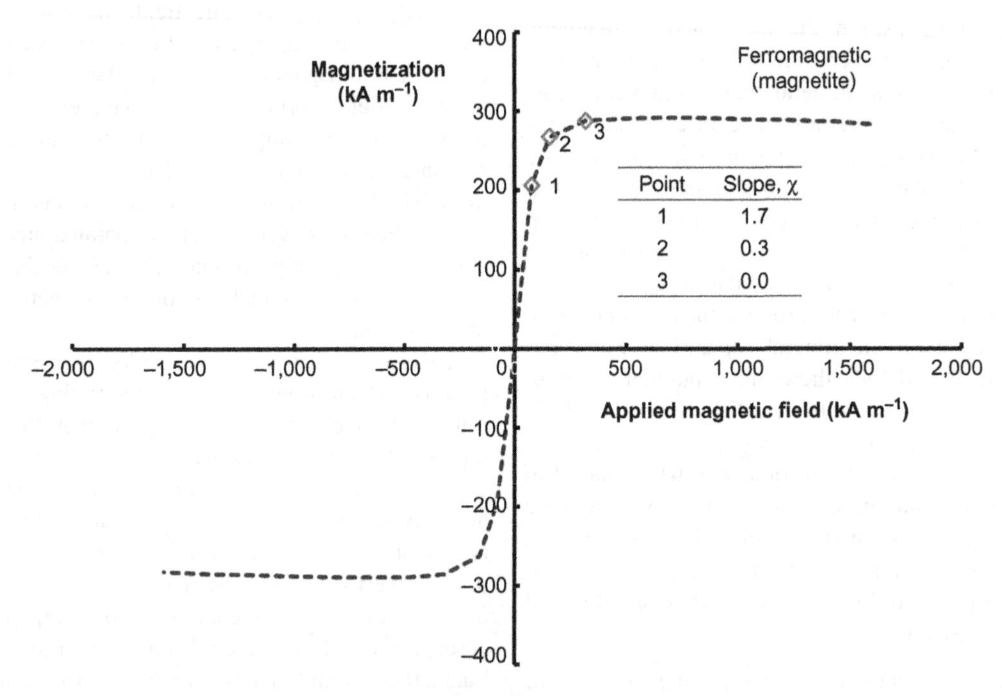

FIGURE 13.2 Magnetization versus applied magnetic field strength for a ferromagnetic mineral.

reached its *saturation magnetization* (a characteristic value of the material shown in Figure 13.2 as a plateau in magnetization), and any further increase in applied magnetic field will not be accompanied by a further increase in magnetization.

Compared to paramagnetic materials, which need high-intensity (high magnetic field) magnetic separators to report to the magnetic product, ferromagnetic materials are recovered in low-intensity magnetic separators.

Measuring Magnetic Properties

The magnetic properties of a material may be measured directly via a vibrating sample magnetometer (used to

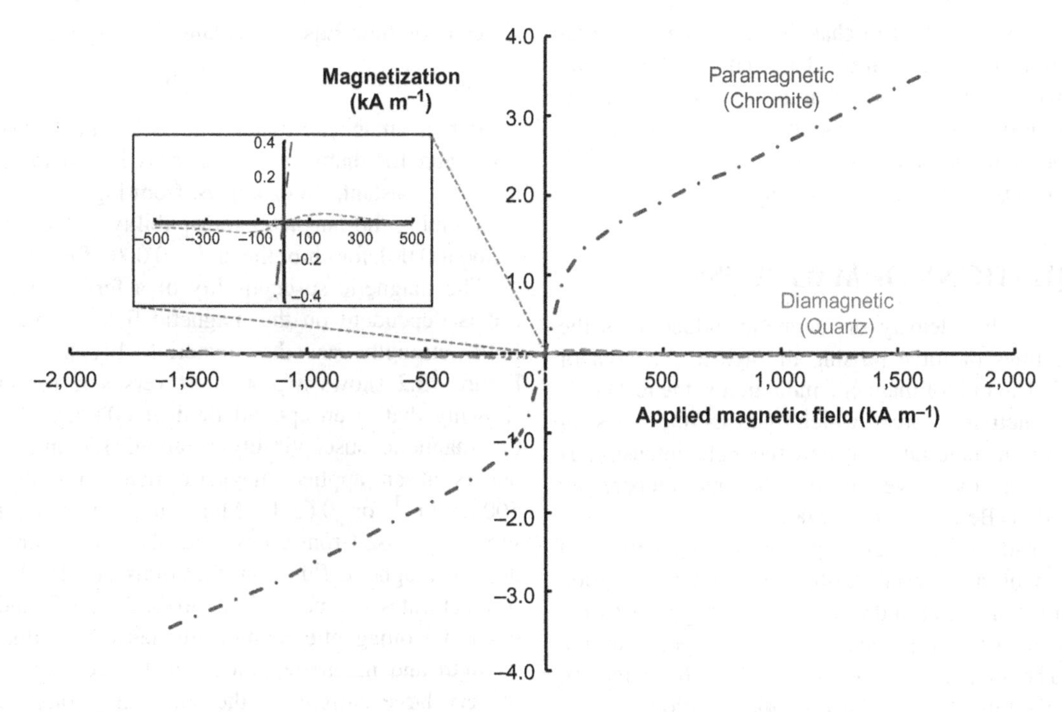

FIGURE 13.3 Magnetization versus applied magnetic field strength for natural samples.

obtain the data for Figures 13.1 and 13.2) which is specifically designed to capture the variation of the magnetic properties of a material as a function of applied magnetic field strength. Empirical data for paramagnetic mineral samples often includes both paramagnetic and ferromagnetic characteristics due to the presence of impurities in the sample, as mineral grains, or in the mineral's crystal structure. Similarly, measurements of a predominantly diamagnetic sample may show signs of a paramagnetic impurity. In Figure 13.3, the results are shown for the samples from Figure 13.1 prior to data processing to isolate the paramagnetic or diamagnetic trend.

Vibrating sample magnetometers are expensive instruments and require specialized personnel to operate. As such, they are typically found only in research settings, such as universities. A more practical tool in mineral processing labs for determining magnetic properties is the Frantz Isodynamic Separator, shown diagrammatically in Figure 13.4. In the Frantz separator, mineral particles are fed down a vibrating chute inclined at an angle θ_2 which is also inclined in the transverse direction at an angle θ_1. The force of gravity is opposed by the magnetic force generated by the electromagnetic coil through which the chute passes. The Frantz is referred to as an isodynamic separator due to the fact that the magnetic force felt by a particle of constant magnetic susceptibility and orientation remains constant throughout the length of the separator (McAndrew, 1957). When the attractive magnetic force is stronger than the force of gravity on a mineral

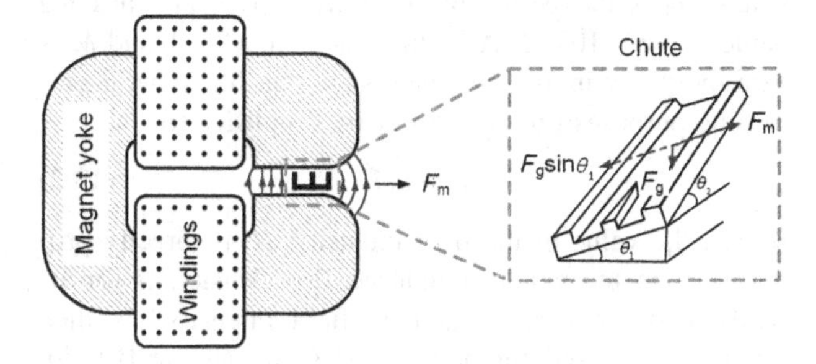

FIGURE 13.4 Diagram of Frantz Isodynamic Separator.

particle it will report to the chute on the right side of Figure 13.4, and when the magnetic force is insufficient to overcome gravity the mineral particle exits the separator on the left side of the chute. As the current through the electromagnetic coil and both θ_1 and θ_2 may be varied across a wide range, this separator is able to separate minerals of varying magnetic properties. It may even be used to concentrate diamagnetic minerals, in which case the side slope is moved past horizontal such that the chute exit on the right of Figure 13.4 becomes the down slope exit for particles where the diamagnetic force (repulsive magnetic force) is insufficient to overcome the force of gravity. The Frantz may also be used to determine the magnetic susceptibility of a given mineral, provided other materials of known susceptibility are available for proper calibration (McAndrew, 1957). Normally operated dry, it can be modified to operate wet (Todd and Finch, 1984).

The Franz is also used to characterize materials, separating fractions that can then be identified. These data can form the basis of predicting separation in full size magnetic separators. For ferromagnetic materials, the Davis tube is more suitable and is the common tool for characterizing magnetic iron ores (Davis, 1921).

13.3 EQUATIONS OF MAGNETISM

The magnetic flux density or magnetic induction is the number of lines of force passing through a unit area of material, B. The unit of magnetic induction is the tesla (T).

The magnetizing force, which induces the lines of force through a material, is called the field intensity, H (or H-field), and by convention has the units ampere per meter ($A\,m^{-1}$) (Bennett et al., 1978).

The intensity of magnetization or the magnetization (M, $A\,m^{-1}$) of a material relates to the magnetization induced in the material and can also be thought of as the volumetric density of induced magnetic dipoles in the material. The magnetic induction, B, field intensity, H, and magnetization, M, are related by the equation:

$$B = \mu_0(H + M) \quad (13.1)$$

where μ_0 is the permeability of free space and has the value of $4\pi \times 10^{-7}\,N\,A^{-2}$. In a vacuum, $M = 0$, and M is extremely low in air and water, such that for mineral processing purposes Eq. (13.1) may be simplified to:

$$B = \mu_0 H \quad (13.2)$$

so that the value of the field intensity, H, is directly proportional to the value of induced flux density, B (or B-field), and the term "magnetic field intensity" is then often loosely used for both the H-field and the B-field. However, when dealing with the magnetic field inside materials, particularly ferromagnetic materials that concentrate the lines of force, the value of the induced flux density will be much higher than the field intensity. This relationship is used in high-gradient magnetic separation (discussed further in Section 13.4.1). For clarity it must be specified which field is being referred to.

Magnetic susceptibility (χ) is the ratio of the intensity of magnetization produced in the material over the applied magnetic field that produces the magnetization:

$$\chi = \frac{M}{H} \quad (13.3)$$

Combining Eqs. (13.1) and (13.3) we get:

$$B = \mu_0 H(1 + \chi) \quad (13.4)$$

If we then define the dimensionless relative permeability, μ, as:

$$\mu = 1 + \chi \quad (13.5)$$

we can combine Eqs. (13.4) and (13.5) to yield:

$$B = \mu\mu_0 H \quad (13.6)$$

For paramagnetic materials, χ is a small positive constant, and for diamagnetic materials it is a much smaller negative constant. As examples, from Figure 13.1 the slope representing the magnetic susceptibility of the material, χ, is about 0.001 for chromite and -0.0001 for quartz.

The magnetic susceptibility of a ferromagnetic material is dependent on the magnetic field, decreasing with field strength as the material becomes saturated. Figure 13.2 shows a plot of M versus H for magnetite, showing that at an applied field of $80\,kA\,m^{-1}$, or 0.1 T, the magnetic susceptibility is about 1.7, and saturation occurs at an applied magnetic field strength of about $500\,kA\,m^{-1}$ or 0.63 T. Many high-intensity magnetic separators use iron cores and frames to produce the desired magnetic flux concentrations and field strengths. Iron saturates magnetically at about 2–2.5 T, and its nonlinear ferromagnetic relationship between inducing field strength and magnetization intensity necessitates the use of very large currents in the energizing coils, sometimes up to hundreds of amperes.

The magnetic force felt by a mineral particle is dependent not only on the value of the field intensity, but also on the field gradient (the rate at which the field intensity increases across the particle toward the magnet surface). As paramagnetic minerals have higher (relative) magnetic permeabilities than the surrounding media, usually air or water, they concentrate the lines of force of an external magnetic field. The higher the magnetic susceptibility, the higher the induced field density in the particle and the greater is the attraction up the field gradient toward increasing field strength. Diamagnetic minerals have lower magnetic susceptibility than their surrounding medium and hence expel the lines of force of the external field. This causes their expulsion down the gradient of the field in the direction of the decreasing field strength.

The equation for the magnetic force on a particle in a magnetic separator depends on the magnetic susceptibility of the particle and fluid medium, the applied magnetic field and the magnetic field gradient. This equation, when considered in only the x-direction, may be expressed as (Oberteuffer, 1974):

$$F_x = V(\chi_p - \chi_m)H\frac{dB}{dx} \quad (13.7)$$

where F_x is the magnetic force on the particle (N), V the particle volume (m^3), χ_p the magnetic susceptibility of the particle, χ_m the magnetic susceptibility of the fluid medium, H the applied magnetic field strength ($A\,m^{-1}$), and dB/dx the magnetic field gradient ($T\,m^{-1} = N\,A^{-1}\,m^{-2}$). The product of H and dB/dx is sometimes referred to as the "force factor."

Production of a high field gradient as well as high intensity is therefore an important aspect of separator design. To generate a given attractive force, there are an infinite number of combinations of field and gradient which will give the same effect. Another important factor is the particle size, as the magnetic force experienced by a particle must compete with various other forces such as hydrodynamic drag (in wet magnetic separations) and the force of gravity. In one example, considering only these two competing forces, Oberteuffer (1974) has shown that the range of particle size where the magnetic force predominates is from about 5 μm to 1 mm.

13.4 MAGNETIC SEPARATOR DESIGN

13.4.1 Magnetic Field Gradient

Certain elements of design are incorporated in all magnetic separators, whether they are wet or dry separators with low- or high-intensity magnetic fields. The prime requirement is the provision of a high-intensity field in which there is a steep field strength gradient. In a field of uniform magnetic flux, such as in Figure 13.5(a), magnetic particles will orient, but will not move along the lines of flux. The most straightforward method for producing a converging field is by providing a V-shaped pole above a flat pole, as in Figure 13.5(b). The tapering of the upper pole concentrates the magnetic flux into a very small area, giving high intensity. The lower flat pole has the same total magnetic flux distributed over a larger area. Thus, there is a steep field gradient across the gap by virtue of the different field intensity levels. Another method of producing a high field gradient is by using a pole which is constructed of alternating magnetic and nonmagnetic laminations (Figure 13.6).

The design of field gradient in magnetic separators may be divided into two types: open-gradient magnetic separators (OGMSs) and high-gradient magnetic separators (HGMSs). In an OGMS design, the magnetic gradient is created by the poles of the magnets themselves, and as a result the gradient is relatively weak (Kopp, 1991). These type of separators include free-fall separators and drum separators where particles passing by the separator are deflected into different streams based on their magnetic properties (Kopp, 1991). In HGMS, a ferromagnetic matrix element is introduced into the applied magnetic field to create many points of high field gradient with the intention of capturing the magnetic particles and allowing nonmagnetic particles to flow through the separator.

The introduction of particles with high magnetic susceptibility into a magnetic field will concentrate the lines of force so that they pass through the particles themselves (Figure 13.7). Since the lines of force converge to the particles, a high field gradient is produced, which causes the particles themselves to behave as magnets, thus attracting each other. Flocculation, or agglomeration, of the particles can occur if they are small and highly susceptible and if the field is intense. This is important as these magnetic "flocs" can entrain gangue mineral particles as well as bridge the gaps between magnetic poles, reducing the efficiency of separation.

Much of the optimization of high-intensity separators is based on providing as many sites of high field gradient as possible to improve the magnetic particle carrying capacity of the separator. Wet high-intensity magnetic

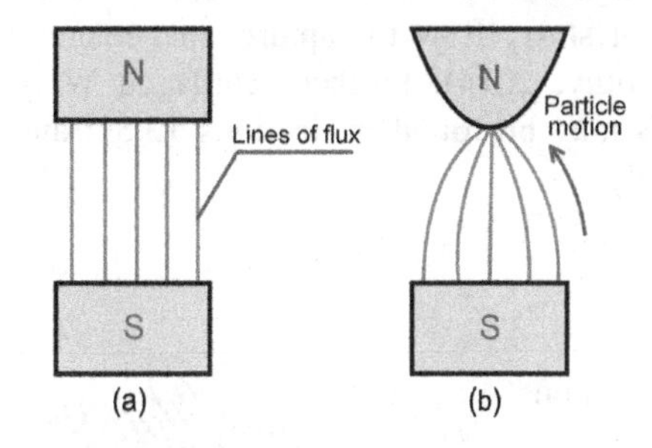

FIGURE 13.5 (a) Field of uniform flux, and (b) converging field generating force on particle.

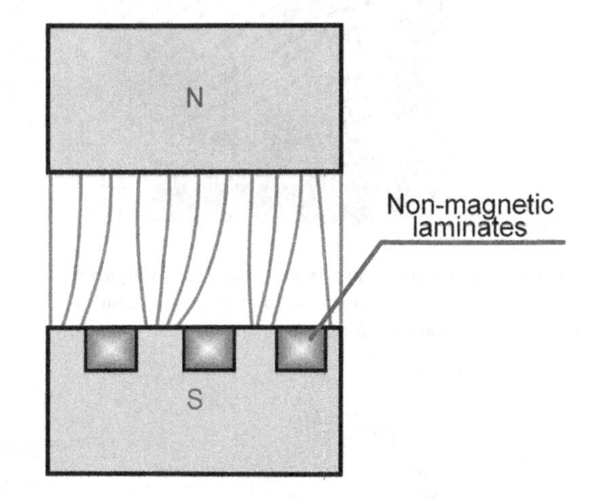

FIGURE 13.6 Production of field gradient by laminated pole.

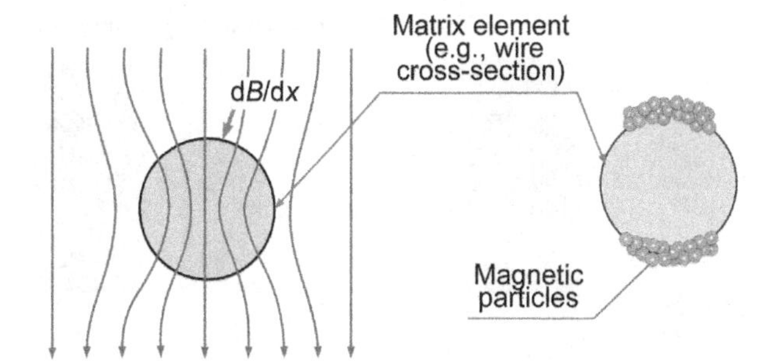

FIGURE 13.7 Production of filed gradient by strongly magnetic matrix material, and consequent capture and buildup of magnetic particles.

separators (WHIMSs) will often use a ferromagnetic matrix material to achieve this, such as those shown in Figure 13.8. The recently developed Outotec SLon vertically pulsating high-gradient magnetic separator (VPHGMS) offers improvements in magnetic matrix design via the use of steel rods.

A visual comparison of the effects of different matrix materials on magnetic flux may be seen in Figure 13.9. The design of the matrix can be further optimized by tapering the size and spacing of the rods throughout the matrix (Figure 13.10) so that coarse magnetic particles are trapped first near the slurry inlet, with additional points of high field gradient introduced further along the direction of slurry flow to capture finer magnetic particles (Novotny, 2014). Further details on WHIMS and VPHGMS may be found in Sections 13.5.2 and 13.5.3, respectively.

13.4.2 Magnetic Field Intensity

Provision must be incorporated in the separator for regulating the intensity of the magnetic field in order to deal with various types of material. This is easily achieved in electromagnetic separators by varying the current, while with permanent magnets the interpole distance can be varied. In the case of laboratory separators, this can also be achieved by interchanging the permanent magnets for magnets of higher magnetic field intensity. A special class of magnetic separators, known as superconducting separators, may be used when very high field intensities are required. Additional information on superconducting separators may be found in Section 13.5.4.

It is important to note that increasing field intensity does not necessarily lead to an improved separation. Work by Svoboda (1994) with HGMS has shown that the

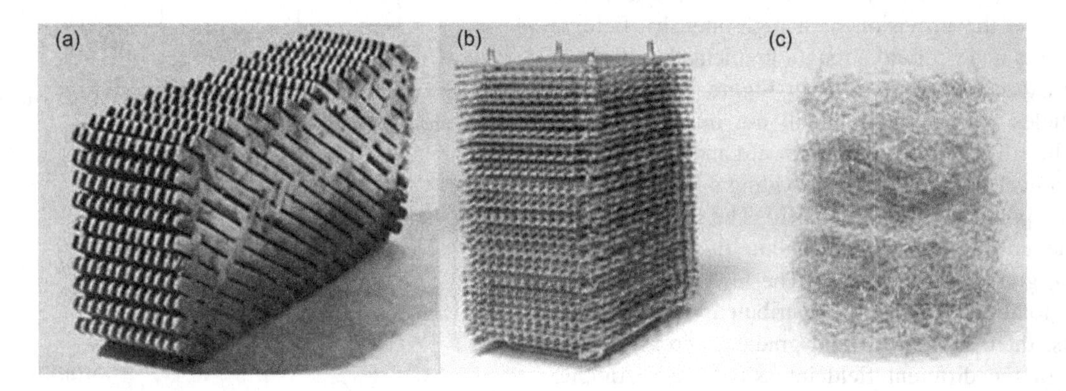

FIGURE 13.8 Examples of matrix materials used in high-intensity separators: (a) section through Boxmag-Rapid grid assembly showing matrix of stainless steel bars, (b) grid of expanded ferromagnetic stainless steel used for coarse particle sizes, and (c) ferromagnetic stainless steel wool used for fine particle sizes *(Courtesy Metso)*.

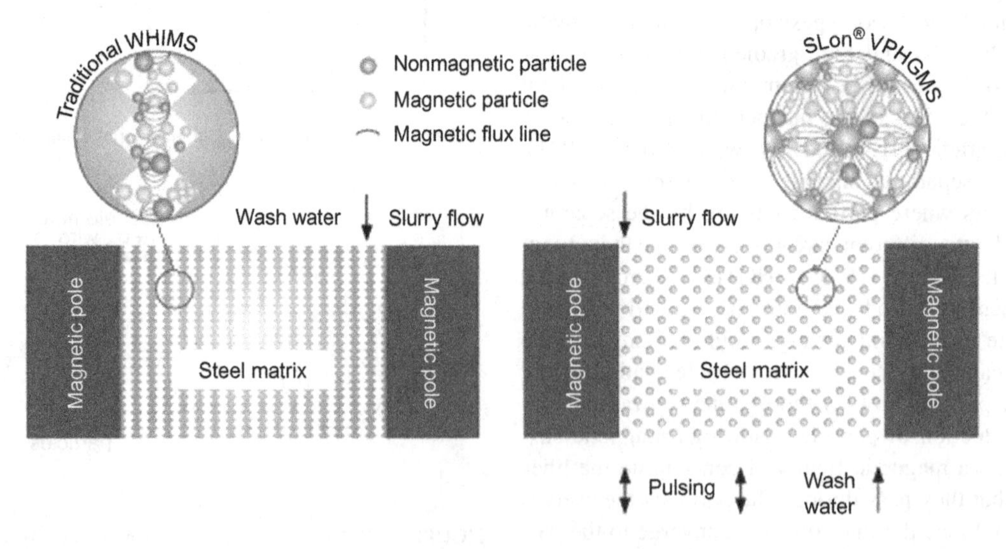

FIGURE 13.9 Comparison of effects of magnetic matrix design on magnetic flux in traditional WHIMS and SLon VPHGMS *(Courtesy Outotec)*.

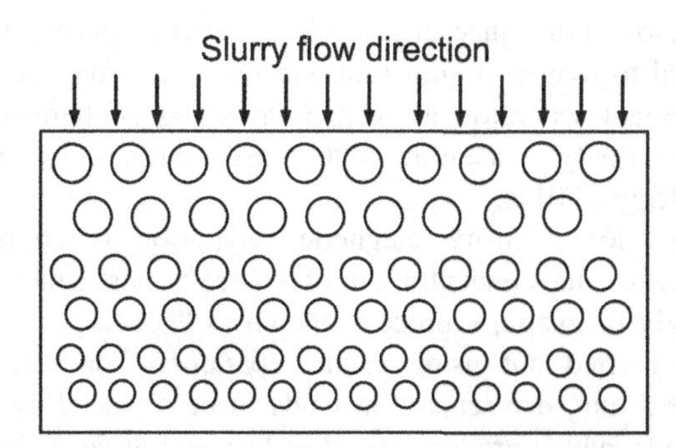

FIGURE 13.10 Design of magnetic matrix to optimize capacity as well as maximize recovery of fine magnetic particles *(Adapted from Novotny (2014)).*

magnetic field strength should be carefully selected according to the application, as higher field strengths may lead to increased capture of weakly magnetic gangue particles. The field should be sufficient to ensure that a particle which collides with a matrix element will remain fixed to that element; any further increase serves only to retain particles with weaker magnetic properties. Another negative impact of high field strength is that for particles exhibiting some degree of magnetic ordering, a relatively common situation in mineral processing, increased field strength actually serves to decrease the magnetic susceptibility (Svoboda, 1994). Work by Shao et al. (1996) to measure the magnetic susceptibility of iron minerals at varying field strengths showed that from 0.4 to 0.9 T the susceptibility of a hematite sample decreased by more than 50%. A similar result is seen in Figure 13.2, where the susceptibility of magnetite decreases from 1.7 to 0.1 as the applied magnetic field strength is increased from 0.1 to 0.4 T (80–320 kA m^{-1}). While such a decrease in magnetic susceptibility is significant, it must be considered in the context of the increasing applied magnetic field strength, as both H and χ affect the force experienced by a mineral particle (Eq. (13.7)). The product of the two, χH, should be calculated to capture the effect of both the decreasing magnetic susceptibility and the increasing applied magnetic field strength. An example of such a calculation for the mineral from Figure 13.2 is given in Table 13.1.

As magnetic field strength is increased, the magnetic field gradient in the separator will also change; this is not considered in the calculations in Table 13.1, although it will have a direct effect on the force experienced by the mineral particles. Excess applied field strength may actually decrease the field gradient in a given separator (Section 13.4.1) (Svoboda, 1994). Since the magnetic force on a particle is directly proportional to the magnetic susceptibility of the particle as well as the magnetic field gradient in the separator, as seen in Eq. (13.7), the net effect of increased field strength can actually be a decrease in the magnetic force experienced by the particle. It is therefore crucial that the appropriate magnetic field is applied for a given separation.

TABLE 13.1 Comparison of Product of Magnetic Susceptibility and Applied Magnetic Field Between Points Along the Curve in Figure 13.2

Point	Applied Magnetic Field (kA m^{-1})	χH (kA m^{-1})
1	80	134
2	160	49
3	320	23

13.4.3 Material Transport in Magnetic Separators

Commercial magnetic separators are continuous-process machines, and separation is carried out on a moving stream of particles passing into and through the magnetic field. Close control of the speed of passage of the particles through the field is essential, which typically rules out free fall as a means of feeding. Belts or drums are very often used to transport the feed through the field.

As discussed in Section 13.4.1, flocculation of magnetic particles is a concern in magnetic separators, especially with dry separators processing fine material. If the ore can be fed through the field in a monolayer, this effect is much less serious, but, of course, the capacity of the machine is drastically reduced. Flocculation is often minimized by passing the material through consecutive magnetic fields, which are usually arranged with successive reversals of the polarity. This causes the particles to turn through 180°, each reversal tending to free the entrained gangue particles. The main disadvantage of this method is that flux tends to leak from pole to pole, reducing the effective field intensity.

Provision for collection of the magnetic and nonmagnetic fractions must be incorporated into the design of the separator. Rather than allow the magnetics to contact the pole-pieces, which then requires their detachment, most separators are designed so that the magnetics are attracted to the pole-pieces, but come into contact with some form of conveying device, which carries them out of the influence of the field, into a bin or a belt. Nonmagnetic disposal presents no problems; free fall from a conveyor into a bin is often used. Middlings are readily produced by using a more intense field after the removal of the highly magnetic fraction.

13.5 TYPES OF MAGNETIC SEPARATOR

Magnetic separators are generally classified into low- and high-intensity machines, but here we include high-gradient and superconducting devices.

13.5.1 Low-Intensity Magnetic Separators

Low-intensity separators are used to treat ferromagnetic materials and some highly paramagnetic minerals.

As shown in Figure 13.2, minerals with ferromagnetic properties have high susceptibility at low applied field strengths and can therefore be concentrated in low intensity ($<\sim 0.3$ T) magnetic separators. For low-intensity drum separators (Figure 13.11) used in the iron ore industry, the standard field, for a separator with ferrite-based magnets, is 0.12 T at a distance of 50 mm from the drum surface (Novotny, 2014). Work by Murariu and Svoboda (2003) has also shown that such separators have maximum field strengths on the drum surface of less than 0.3 T. The principal ferromagnetic mineral concentrated in mineral processing is magnetite (Fe_3O_4), although hematite (Fe_2O_3) and siderite ($FeCO_3$) can be roasted to produce magnetite and hence give good separation in low-intensity machines.

The removal of "tramp" iron from feed belts can also be regarded as a form of low-intensity magnetic separation. However, tramp iron removal is usually accomplished by means of a magnetic pulley at the end of an ore conveyor (Figure 13.12) or by a guard magnet suspended over the conveyor belt (see Chapter 2). Tramp iron removal is important prior to crushing and in certain cases removal of the iron produced from grinding media wear can be important for downstream processing. A common example of the latter is processing gold ores, where the use of magnetic separation (typically a drum separator) in advance of centrifugal gravity concentration is used to remove tramp iron and prevent damage to the centrifugal separator, as well as avoiding contamination of the gravity concentrate with dense iron particles (Bird and Briggs, 2011).

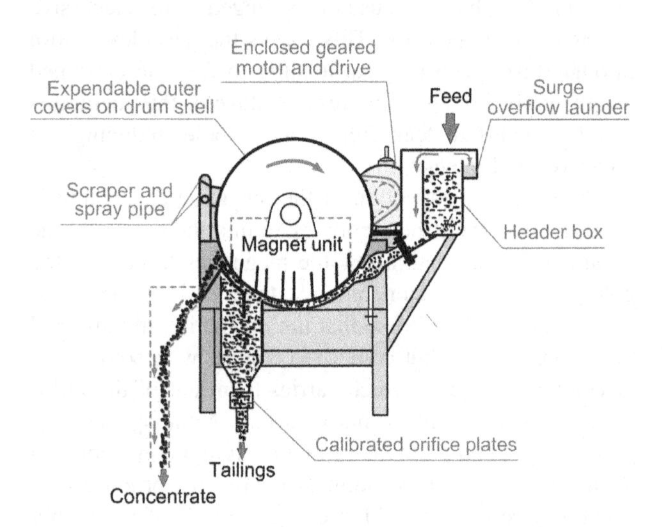

FIGURE 13.11 Diagram of a typical drum separator.

Dry low-intensity magnetic separation is confined mainly to the concentration of coarse sands which are strongly magnetic, a process known as "cobbing," and is often carried out using drum separators. For particles below 5 mm, dry separation tends to be replaced by wet methods, which produce less dust loss and usually yield a cleaner product. Low-intensity wet separation is widely used for recycling (and cleaning) magnetic media in dense medium separation (DMS) processes (see Chapter 11) and for the processing of ferromagnetic sands.

The general design of drum separators is a rotating, hollow, nonmagnetic drum containing multiple stationary magnets of alternating polarity. The medium-intensity Permos separator uses many small magnet blocks, whose direction of magnetization changes in small steps. This is said to generate a very even magnetic field, requiring less magnetic material (Wasmuth and Unkelbach, 1991). The spatial arrangement of the magnets within a drum separator may be varied depending on the specific requirements of the application. This is illustrated by the two variants of magnet configuration offered for Metso's low-intensity drum separator for DMS applications. The two configurations (Figure 13.13) demonstrate the trade-off between increasing magnetic loading capacity (Figure 13.13(a)) to capture more particles and increasing field gradient (Figure 13.13(b)) to capture finer or less susceptible particles. The high-capacity arrangement has fewer, larger poles, which results in a lower field gradient but a higher magnetic flux of 0.12 T at a distance of 50 mm from the roll surface (Metso, 2014a). The high gradient variant has more, smaller poles, resulting in a higher field gradient to better capture fine magnetic particles, but at reduced capacity (magnetic flux of only 0.06 T at a distance of 50 mm from the roll surface) (Metso, 2014a).

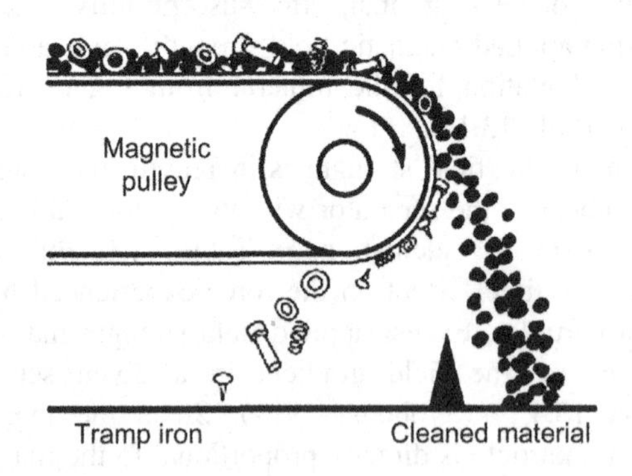

FIGURE 13.12 Example of magnetic pulley used to remove tramp iron from an ore prior to further processing *(Courtesy Eriez).*

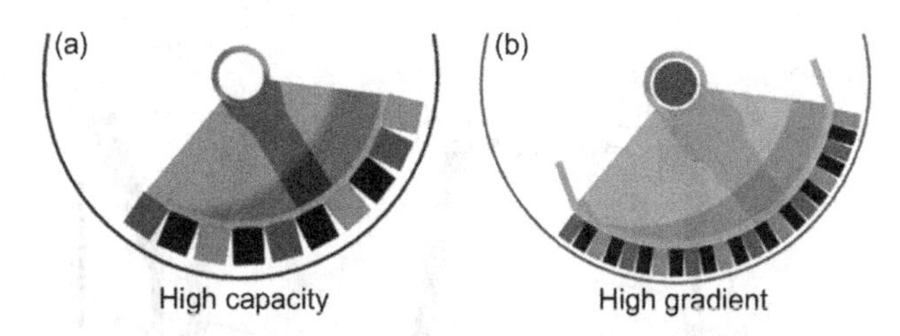

FIGURE 13.13 Alternate magnet configurations for a wet drum separator. (a) High-capacity arrangement, and (b) high-gradient arrangement *(Courtesy Metso)*.

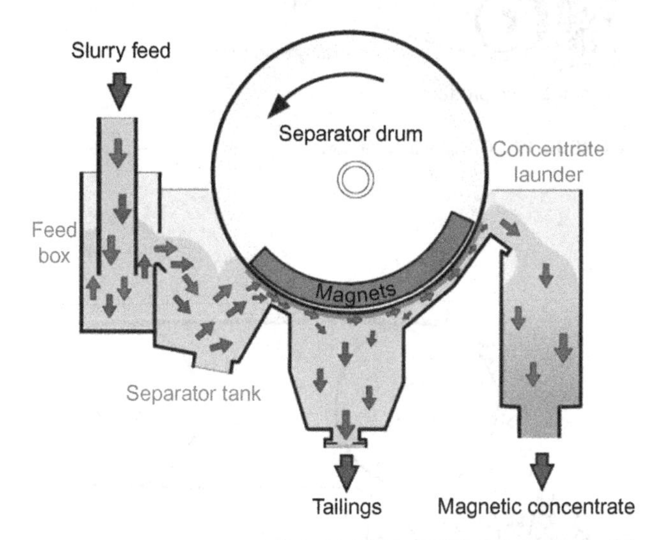

FIGURE 13.14 Concurrent configuration of a wet drum separator *(Courtesy Metso)*.

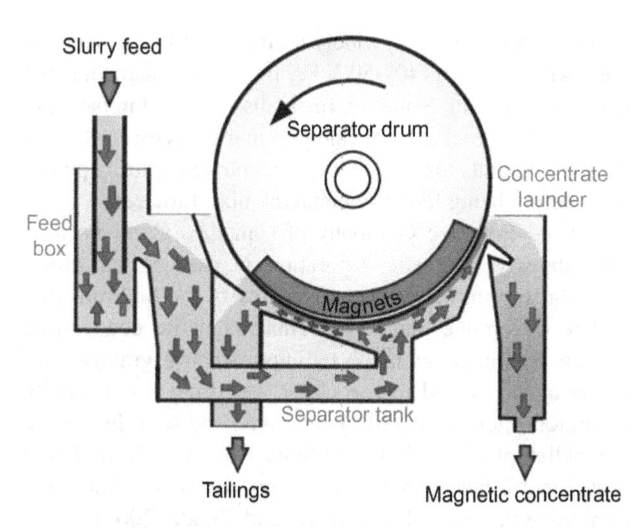

FIGURE 13.15 Counter-current configuration of a wet drum separator *(Courtesy Metso)*.

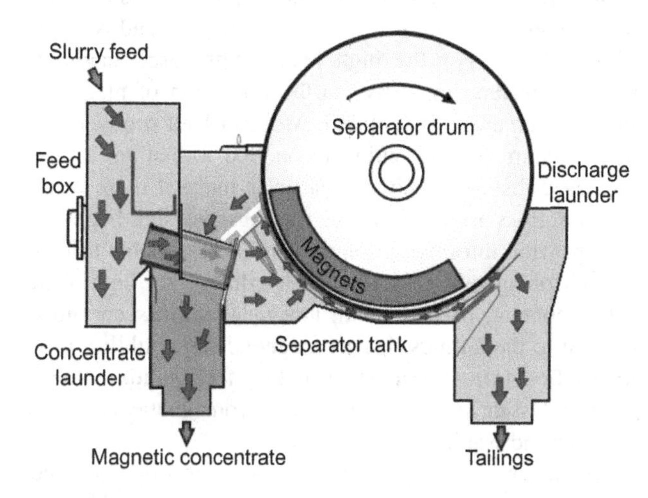

FIGURE 13.16 Counter-rotation configuration of a wet drum separator *(Courtesy Metso)*.

Although drum separators initially employed electromagnets, permanent magnets are used in modern devices, utilizing ceramic or rare earth magnetic alloys, which retain their intensity for an indefinite period (Norrgran and Marin, 1994). Separation in a drum separator occurs by the "pick-up" principle, wherein magnetic particles are lifted by the magnets and pinned to the drum and are conveyed out of the field, leaving the nonmagnetics (usually the gangue) in the tailings compartment. Water is introduced to provide flow, which keeps the pulp in suspension. Field intensities of up to 0.7 T at the pole surfaces can be obtained in this type of separator.

The drum separators shown in Figures 13.11 and 13.14 are of the concurrent type (as shown by the separator tank flow pattern in Figure 13.14), whereby the concentrate is carried forward by the drum and passes through a gap, where it is compressed and dewatered before leaving the separator. This design is most effective for producing a clean magnetic concentrate from relatively coarse feeds (up to 6–8 mm) and is widely used in dense medium recovery systems. In addition to the concurrent arrangement (Figure 13.14), drum separators may also be configured with counter-current and counter-rotation arrangements (Figures 13.15 and 13.16).

In a counter-current separator, the tailings are forced to travel in the opposite direction to the drum rotation and are discharged into the tailings chute. This type of separator is designed for finishing operations on relatively fine material, of particle size less than about 800 μm. Pulp densities in this type of separator are typically lower than in the concurrent configuration.

The third possible configuration is the counter-rotation type, where the feed flows in the opposite direction to the rotation. This type is used in roughing operations, where occasional surges in feed must be handled, and where magnetic material losses are to be held to a minimum when high solids loading is encountered, while an extremely clean concentrate is not required.

Drum separators are widely used to treat low-grade iron ores, which contain 40–50% Fe, mainly as magnetite, but in some areas with hematite, finely disseminated in bands in hard siliceous rocks. Very fine grinding is necessary to free the iron minerals that produce a concentrate requiring pelletizing before being fed to steelmaking blast furnaces.

At the Iron Ore Company of Canada's Carol Project, low-intensity magnetic separation is used as an initial cobbing step on the cyclone overflow of the ball mill discharge to remove magnetite. This magnetite concentrate is then combined with the tailings of spiral gravity concentrators to be fed to a rougher wet drum low-intensity separator where magnetite is removed and sent directly to the pellet plant, with the tailings from the drum being sent for further gravity processing to concentrate any remaining hematite (Damjanović and Goode, 2000).

The mechanism by which ferromagnetic particles are captured by low-intensity drum separators has been investigated for both high solids content (10–17% solids by weight) and low solids content (2%) (Rayner and Napier-Munn, 2000). For the high feed solids case, magnetic recovery occurs primarily via the formation of magnetic flocs, which are then captured. At lower feed solids, magnetic capture is not contingent on floc formation, as the increased distances between particles make it more difficult for flocs to form. These findings are significant as they provide information on the dominant variable affecting ferromagnetic losses to the tailings. At high feed solids content, particles with low magnetic susceptibility are lost to the tailings (higher magnetic susceptibility promotes floc formation), while at low feed solids content tailings losses are primarily fine ferromagnetic particles (Rayner and Napier-Munn, 2000).

At Palabora, the tailings from copper flotation (see Section 12.17.1) are deslimed, after which the +105 μm material is treated by wet low-intensity drum separators to recover 95% of the magnetite at a grade of 62% Fe.

Another example of wet low-intensity magnetic separation is the treatment of flotation tailings at the Niobec mine in Quebec, Canada (Section 12.17.7). The Niobec mine employs multiple flotation stages to produce a pyrochlore concentrate including carbonate, pyrochlore, and sulfide flotation circuits. The deslimed tails from the carbonate flotation bank are fed to low-intensity drum magnetic separators to remove approximately $1\ t\ h^{-1}$ of magnetite assaying 68.30% Fe, 0.08% Nb_2O_5, 0.80% SiO_2, and 0.16% P_2O_5. The nonmagnetic product (containing the bulk of the Nb) is then sent to the pyrochlore flotation circuit for upgrading (Biss and Ayotte, 1989).

The cross-belt separator (Figure 13.17) and disc separators, once widely used in the mineral sands industry, are now considered obsolete. They are being replaced with rare earth roll magnetic separators and rare earth drum magnetic separators (Arvidson, 2001).

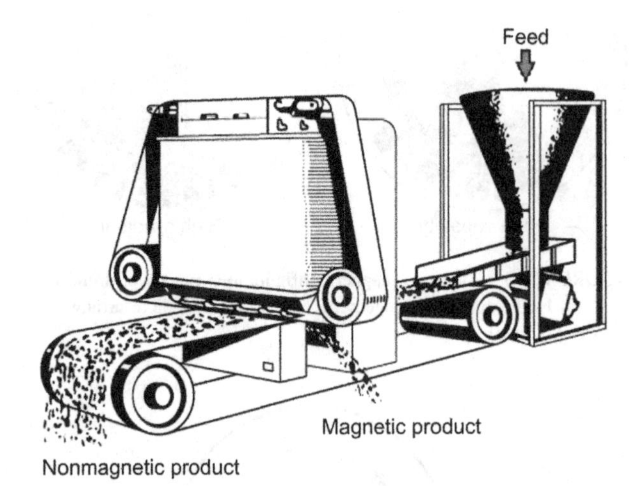

FIGURE 13.17 Cross-belt separator.

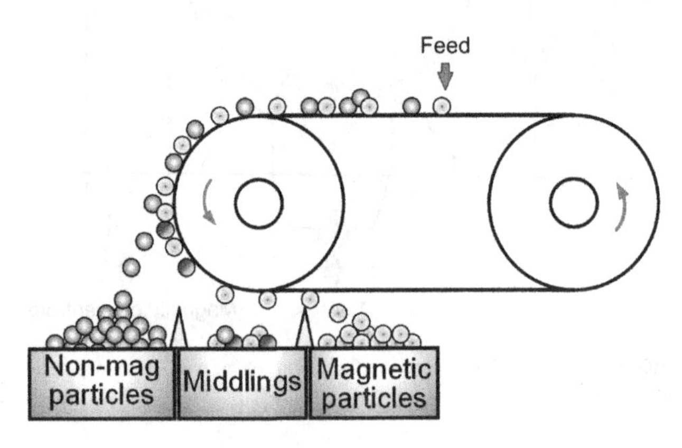

FIGURE 13.18 Schematic of rare earth roll separator *(Adapted from Dobbins et al. (2009)).*

Rare earth roll separators use alternate magnetic and nonmagnetic laminations (like those illustrated in Figure 13.6). Feed is carried onto the magnetic roll by a belt as shown in Figure 13.18 to limit bouncing or scattering of particles and to ensure they all enter the magnetic zone with the same horizontal velocity. In a rare earth roll separator, the variables affecting separation are the magnetic field strength, the feed rate, the linear speed of the roll surface, and the particle size of the material (Eriez, 2003). Most importantly, the centrifugal force applied to the mineral particles by the roll surface must be optimized to achieve a sharp separation (Eriez, 2003). To control the centrifugal force, roll speed can be adjusted over a wide range, allowing the product quality to be "dialed in."

Dry rare earth drum separators provide a "fan" of separated particles, which can often be seen as distinct streams (Figure 13.18). The fan can be separated into various grades of magnetic product and a nonmagnetic tailing. In some mineral sands applications, drum separators have been integrated with one or more rare earth rolls, arranged to treat the middlings particles from the drum. In any dry magnetic separator, the careful control of feed moisture is critical to avoid smaller particles sticking to larger particles (Oberteuffer, 1974). While increasing particle size increases the acceptable moisture limits, even at a particle size of 90% passing 20 mm, the recommended moisture limit for Metso's dry drum separators is only 3% (Metso, 2014b).

13.5.2 High-Intensity Magnetic Separators

Weakly paramagnetic minerals can only be effectively recovered using high-intensity (B-fields of 2 T or greater) magnetic separators (Svoboda, 1994). Until the 1960s, high-intensity separation was confined solely to dry ore, having been used commercially since about 1908. This is no longer the case, as many new technologies have been developed to treat slurried feeds.

Induced roll magnetic (IRM) separators (Figure 13.19) are widely used to treat beach sands, wolframite and tin ores, glass sands, and phosphate rock. They have also been used to treat weakly magnetic iron ores, principally in Europe. The roll, onto which the ore is fed, is composed of phosphated steel laminates compressed together on a non-magnetic stainless steel shaft. By using two sizes of laminations, differing slightly in outer diameter, the roll is given a serrated profile, which promotes the high field intensity and gradient required. Field strengths of up to 2.2 T are attainable in the gap between feed pole and roll. Nonmagnetic particles are thrown off the roll into the tailings compartment, whereas magnetics are held, carried out of the influence of the field and deposited into the magnetics compartment. The gap between the feed pole and rotor is adjustable and is usually decreased from pole to pole (to create a higher effective magnetic field strength) to take off successively more weakly magnetic products.

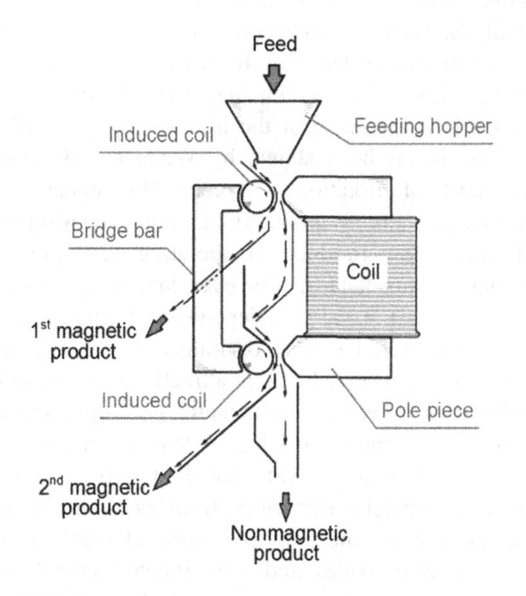

FIGURE 13.19 Schematic of an induced roll separator.

The primary variables affecting separation using an IRM separator are the magnetic susceptibility of the mineral particles, the applied magnetic field intensity, the size of the particles, and the speed of the roll (Singh et al., 2013).The setting of the splitter plates cutting into the trajectory of the discharged material is also of importance.

In most cases, IRM separators have been replaced by the more recently developed (*circa* 1980) rare earth drum and roll separators, which are capable of field intensities of up to 0.7 and 2.1 T, respectively (Norrgran and Marin, 1994). The advantages of rare earth roll separators over IRM separators include: lower operating costs due to decreased energy requirements, less weight leading to lower construction and installation costs, higher throughput, fewer required stages, and increased flexibility in roll configuration which allows for improved separation at various size ranges (Dobbins and Sherrell, 2010).

Dry high-intensity separation is largely restricted to ores containing little, if any, material finer than about 75 μm. The effectiveness of separation on such fine material is severely reduced by the effects of air currents, particle–particle adhesion, and particle–rotor adhesion.

Without doubt, the greatest advance in the field of magnetic separation was the development of continuous WHIMSs (Lawver and Hopstock, 1974). These devices have reduced the minimum particle size for efficient magnetic separation compared to dry high-intensity methods. In some flowsheets, expensive drying operations, necessary prior to a dry separation, can be eliminated by using an entirely wet concentration system.

Perhaps the most well-known WHIMS machine is the Jones separator, the design principle of which is utilized in many other types of wet separators found today. The machine has a strong main frame (Figure 13.20(a)) made of structural steel. The magnet yokes are welded to this frame, with the electromagnetic coils enclosed in air-cooled cases. The separation takes place in the plate boxes, which are on the periphery of the one or two rotors attached to the central roller shaft and carried into and out of the magnetic field in a carousel (Figure 13.20(b)). The feed, which is thoroughly mixed slurry, flows through the plate boxes via fitted pipes and launders into the plate boxes (Figure 13.21), which are grooved to concentrate the magnetic field at the tip of the ridges. Feeding is continuous due to the rotation of the plate boxes on the rotors and the feed points are at the leading edges of the magnetic fields (Figure 13.20(b)). Each rotor has two feed points diametrically opposed to one another.

The weakly magnetic particles are held by the plates, whereas the remaining nonmagnetic particle slurry passes through the plate boxes and is collected in a launder. Before leaving the field any entrained nonmagnetics are

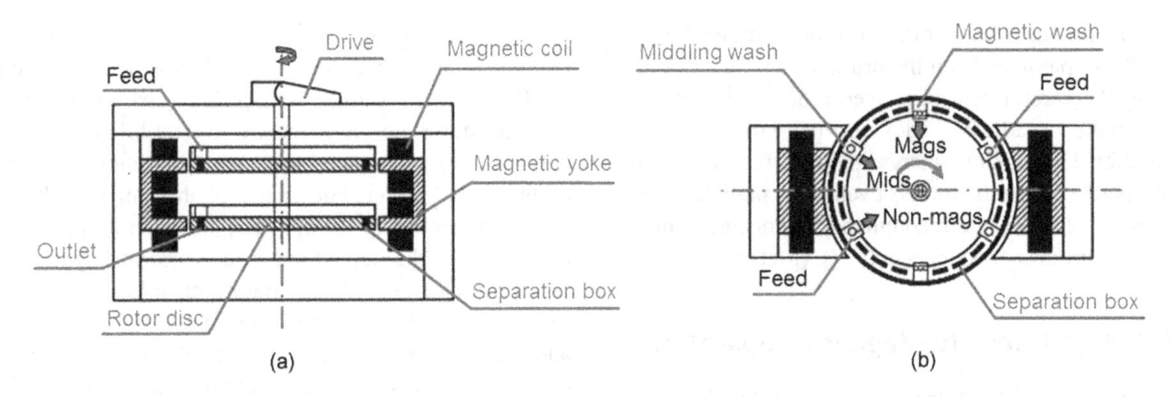

FIGURE 13.20 The Jones high-intensity wet magnetic separator in cross section: (a) plan view, and (b) top view.

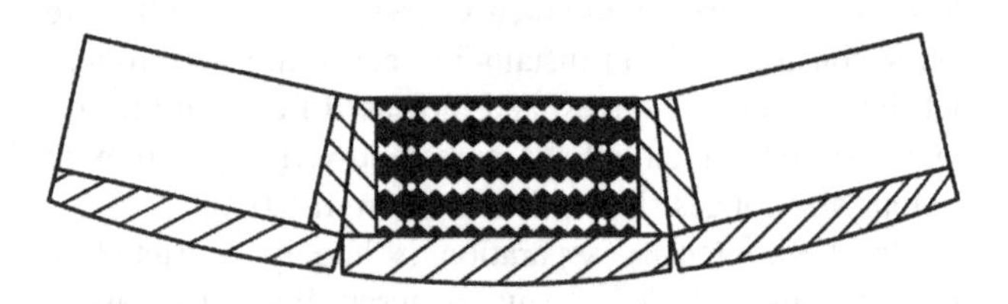

FIGURE 13.21 Jones plate box showing grooved plates and spacer bars.

washed out by low-pressure water and are collected as a middlings product.

When the plate boxes reach a point midway between the two magnetic poles, where the magnetic field is essentially zero, the magnetic particles are washed out using high-pressure scour water sprays operating at up to 5 bar. Field intensities of over 2 T can be produced in these machines, although the applied magnetic field strength should be carefully selected depending on the application (see Section 13.4.2). The production of a 1.5 T field requires electric power consumption in the coils of 16 kW per pole.

There are currently two types of WHIMS machines, one that uses electromagnetic coils to generate the required field strength, the other that employs rare earth permanent magnets. They are used in different applications; the weaker magnetic field strength produced by rare earth permanent magnets may be insufficient to concentrate some weakly paramagnetic minerals. The variables to consider before installing a traditional horizontal carousel WHIMS include: the feed characteristics (slurry density, feed rate, particle size, magnetic susceptibility of the target magnetic mineral), the product requirements (volume of solids to be removed, required grade of products), and the cost of power (Eriez, 2008). From these considerations the design and operation of the separator can be tailored by changing the following: the magnetic field intensity and/or configuration, the speed of the carousel, the setting of the middling splitter, the pressure/volume of wash water, and the type of matrix material (Eriez, 2008). The selection of matrix type has a direct impact on the magnetic field gradient present in the separation chamber. As explained in Section 13.4.2, increasing magnetic field can in some applications actually cause decreased performance of the magnetic separation step and it is for this reason that improvements in the separation of paramagnetic materials focus largely on achieving a high magnetic field gradient. The Eriez model SSS-I WHIMS employs the basic principles of WHIMS with improvements in the matrix material (to generate a high field gradient) as well as the slurry feeding and washing steps (to improve separation efficiency) (Eriez and Gzrinm, 2014). While this separator is referred to as a WHIMS, it is in fact more similar to the SLon VPHGMS mentioned in Sections 13.4.1 and 13.5.3. Further discussion on high-gradient magnetic separation (HGMS) may be found in Section 13.5.3.

Wet high-intensity magnetic separation has its greatest use in the concentration of low-grade iron ores containing hematite, where they are an alternative to flotation or gravity methods. The decision to select magnetic separation for the concentration of hematite from iron ore must balance the relative ease with which hematite may be concentrated in such a separator against the high capital cost of such separators. It has been shown by White (1978) that the capital cost of flotation equipment for concentrating weakly magnetic ore is about 20% that of a Jones separator installation, although flotation operating costs are about three times higher (and may be even higher if water treatment is required). Total cost depends on terms for capital depreciation; over 10 years or longer the high-intensity magnetic separator may be more attractive than flotation.

In addition to recovery of hematite (and other iron oxides such as goethite), wet high-intensity separators are now in operation for a wide range of duties, including removal of magnetic impurities from cassiterite concentrates, removal of fine magnetic material from asbestos, removal of iron oxides and ferrosilicate minerals from industrial minerals such as quartz and clay, concentration

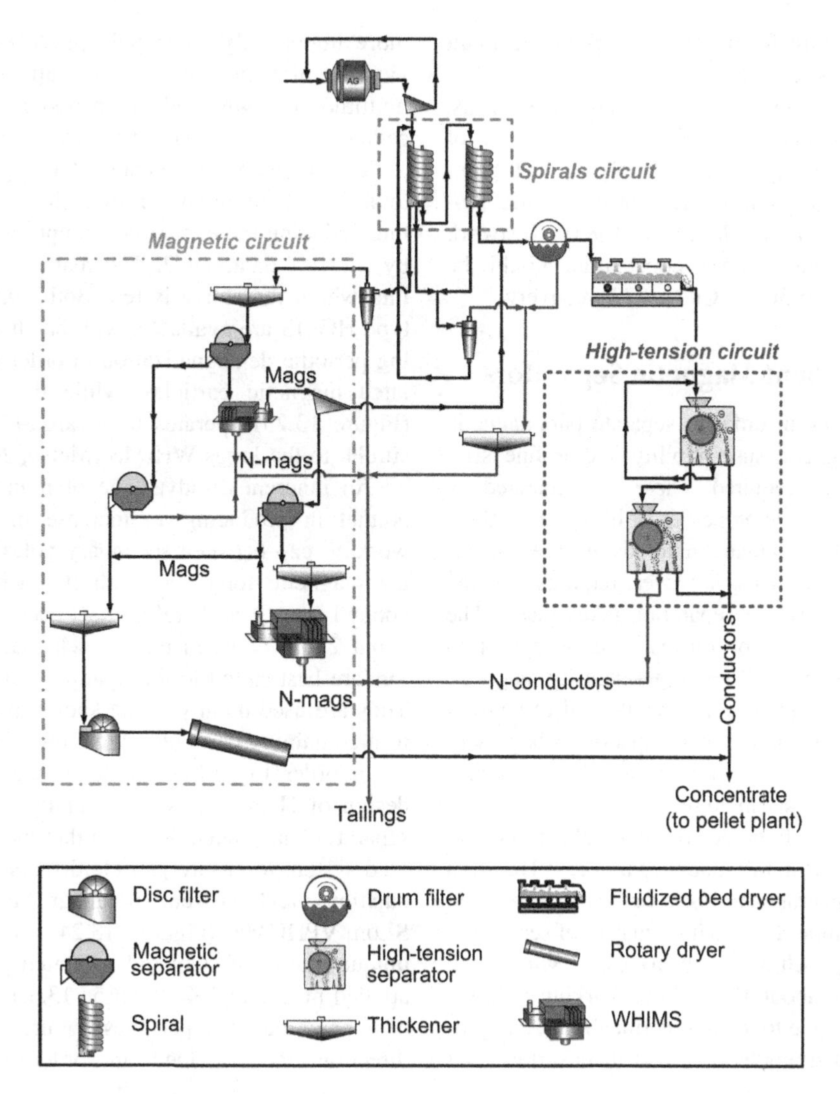

FIGURE 13.22 Flowsheet from Cliffs–Wabush iron ore mine *(Adapted from Damjanović and Goode (2000)).*

of ilmenite, wolframite, and chromite, removal of magnetic impurities from scheelite concentrates, purification of talc, the recovery of non-sulfide molybdenum-bearing minerals from flotation tailings, and the removal of Fe-oxides and Fe–Ti-oxides from zircon and rutile in heavy mineral beach sands (Corrans and Svoboda, 1985; Eriez, 2008). In the PGM-bearing Merensky Reef (South Africa), WHIMS has been used to remove much of the strongly paramagnetic orthopyroxene gangue from the PGM-containing chromite (Corrans and Svoboda, 1985). WHIMS has also been successfully used for the recovery of gold and uranium from cyanidation residues in South Africa (Corrans, 1984). Magnetic separation can be used to recover some of the free gold, and much of the silicate-locked gold, due to the presence of iron impurities and coatings. In the case of uranium leaching, small amounts of iron (from milling) may act as reducing agents and negatively affect the oxidation of U^{4+} to U^{6+}; treatment via WHIMS can reduce the consumption of oxidizing agents by removing a large portion of this iron prior to leaching (Corrans and Svoboda, 1985).

At the Cliffs–Wabush iron ore mine in Labrador, Canada (Figure 13.22), the cyclone overflow from the tailings of a rougher spiral bank is sent to a magnetic scavenger circuit utilizing both low-intensity drum separation and WHIMS. This circuit employs the low-intensity (0.07 T) drum separators to remove fine magnetite particles lost during the spiral gravity concentration step, followed by a WHIMS step using $100\ t\ h^{-1}$ Jones separators which are operated at field strengths of 1 T to concentrate fine hematite. Cleaning of only the gravity tailings by magnetic separation is preferred, as relatively small amounts of magnetic concentrate have to be handled, the bulk of the material being essentially unaffected by the magnetic field. The concentrate produced from this magnetic scavenging step is eventually recombined with the spiral concentrate before feeding to the pelletizing plant (Damjanović and Goode, 2000).

The paramagnetic properties of some sulfide minerals, such as chalcopyrite and marmatite (high Fe form of sphalerite), have been exploited by applying wet high-intensity magnetic separation to augment differential flotation processes (Tawil and Morales, 1985). Testwork showed that a Chilean copper concentrate could be upgraded from 23.8% to 30.2% Cu, at 87% recovery.

13.5.3 High-Gradient Magnetic Separators

As noted in Eq. (13.7), in order to separate paramagnetic minerals of low magnetic susceptibility and/or fine size, high field gradients are required. These are generated by exploiting the ferromagnetic properties of iron to generate a high B-field (induced field) many hundreds of times greater than the applied H-field. This, however, requires that the iron be in the volume where separation takes place. The steel plates in a Jones separator, for example, occupy up to 60% of the process volume. Thus, high-intensity magnetic separators using conventional iron circuits tend to be very massive and heavy in relation to their capacity. A large separator may contain over 200 t of iron to carry the flux, hence capital and installation costs are high.

Instead of using one large convergent field in the gap of a magnetic circuit, as in the Jones separator, in HGMS a solenoid is used to generate a uniform field with a solenoid core, or working volume, filled with a matrix of secondary ferromagnetic poles, such as ball bearings, or wire wool, the latter filling only about 10% of the working volume. Each secondary pole, due to its high permeability, can produce maximum field strengths of 2 T at their surface, but more importantly, each pole produces, in its immediate vicinity, high field gradients of up to 14 T/mm. Thus, a multitude of high gradients across numerous small gaps, centered on each of the secondary poles, is achieved.

The solenoid can be clad externally with an iron frame to form a continuous return path for the magnetic flux, thus reducing the energy consumption for driving the coil by a factor of about 2. The matrix is held in a canister into which the slurry is fed. Both continuous and batch-type HGMS are available, with batch-type HGMS requiring periodic demagnetization in order to remove accumulated magnetic particles, while the continuous HGMS (Figure 13.23) operates in a carousel-type configuration similar to the Jones WHIMS (Metso, 2014c,d).

An inherent disadvantage of high-gradient separators is that in producing an increase in field gradient, the working gap between secondary poles is reduced, so that the magnetic force has a short reach of no more than about 1 mm. It is therefore necessary to use gaps of only about 2 mm between poles, such that the matrix separators are best suited to the treatment of very fine particles. They are used mainly in the kaolin industry for removing iron-containing particles which lower brightness.

In order to address some of the deficiencies in the design of HGMS, new horizontally fed vertical carousel separators have been designed that incorporate a pulsating feed system to ensure particle dispersion (i.e., avoid flocculation) and prevent nonmagnetic entrainment. The SLon VPHGMS (Figures 13.24 and 13.25) employs a unique matrix of steel rods oriented perpendicular to the applied magnetic field (Section 13.4.1) as well as flushing of trapped magnetic particles (Figure 13.26) in the reverse direction to the feed in order to reduce particle

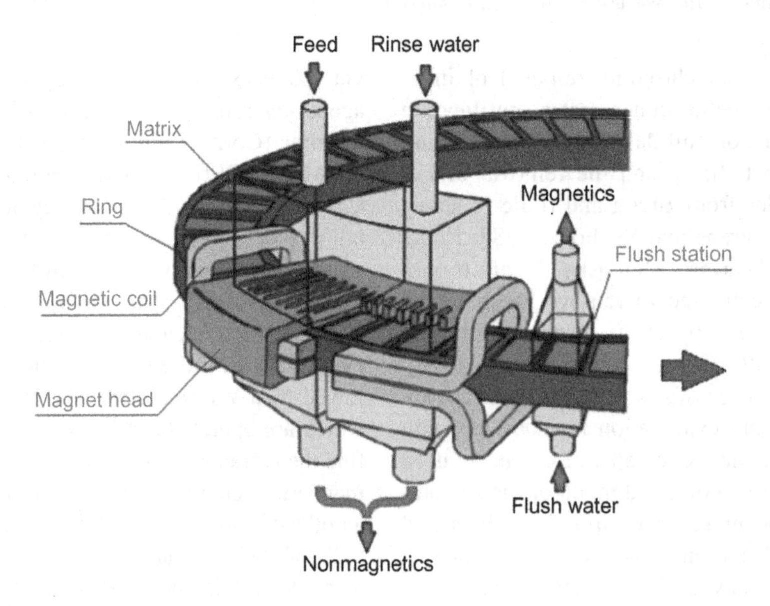

FIGURE 13.23 Metso HGMS operating principle *(Courtesy Metso)*.

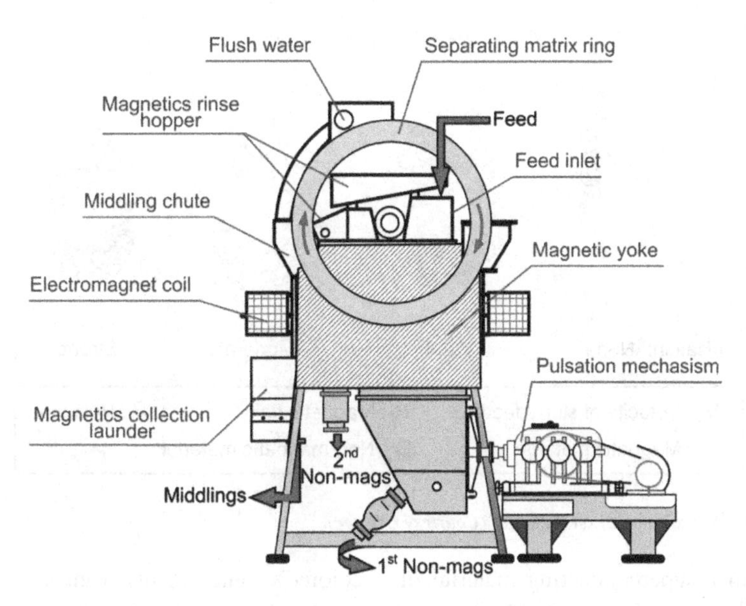

FIGURE 13.24 SLon VPHGMS *(Courtesy Outotec).*

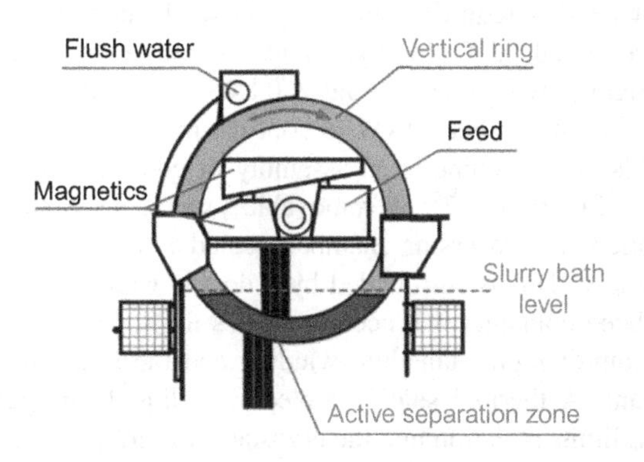

FIGURE 13.25 Plan view of separation zone in SLon VPHGMS *(Courtesy Outotec).*

momentum, maximize particle trapping, and improve separation (Outotec, 2013). The rod diameter in the matrix may be tailored for the given application to vary the maximum particle size that can pass through the separator from 0.6 up to 3.0 mm (Outotec, 2013). The averaged magnetic field intensity across the entire VPHGMS is no greater than 1.3 T; however, as the steel rod matrix becomes saturated, intensities up to 1.8 T can be achieved at the matrix surface with an applied magnetic field of only 1 T (Outotec, 2013). The SLon separator has been applied in the concentration of fine particles such as hematite and ilmenite, and for desulfurization and dephosporization of iron ore feeds prior to steelmaking (Xiong, 1994, 2004). Eriez also offers a vertical carousel-type WHIMS with similar innovations to the SLon VPHGMS, such as pulsating feed and high capacity due to improved matrix washing (Eriez and Gzrinm, 2014). The recently developed version of the Eriez separator is somewhat confusingly referred to as WHIMS, while using a rod matrix to produce high magnetic field gradients in a manner similar to HGMS (Eriez and Gzrinm, 2014). The Eriez separator has been successfully applied to the following: the concentration of Fe-bearing minerals (hematite, limonite, siderite, chromite), the cleaning of nonferrous ores (quartz, cassiterite, garnet), the recovery of rare earth minerals, and the purification of nonmetallic ores (quartz, feldspar, kaolin, alusite, kyanite) (Eriez and Gzrinm, 2014).

13.5.4 Superconducting Separators

Future developments and applications of magnetic separation in the mineral industry lie in the creation and use of increasingly higher product of field and field gradient, that is, the "force factor." Matrix separators with very high field gradients and multiple small working gaps can draw little benefit from field strengths above the saturation levels of the secondary poles (~2 T for an iron/steel matrix material). As discussed in Section 13.4.1, the alternative to HGMS is OGMS, where separators with large working volumes deflect coarser particles at high capacity, rather than capture particles, as in HGMS. As the gradient in OGMS is relatively low, these separators need to use the highest possible field strengths to generate the high magnetic forces required to treat weakly paramagnetic particles. Field strengths in excess of 2 T can only be generated economically by the use of superconducting magnets (Kopp, 1991; Watson, 1994).

Certain alloys have the property of presenting no resistance to electric currents at extremely low temperatures. An example is niobium–titanium at 4.2 K, the temperature of liquid helium. Once a current is established

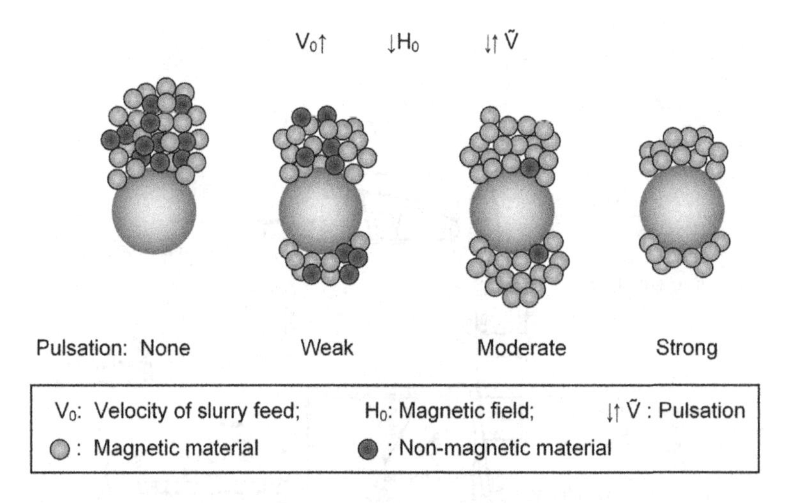

FIGURE 13.26 Buildup on the matrix in the SLon separator *(Courtesy Outotec).*

through a coil made from a superconducting material, it will continue to flow without being connected to a power source, and the coil will become, in effect, a permanent magnet. Superconducting magnets can produce extremely intense and uniform magnetic fields, of up to 15 T. The main problem, of course, is in maintaining the extremely low temperatures. In 1986, a Ba/La/Cu oxide composite was made superconductive at 35 K, promoting a race to prepare ceramic oxides with much higher superconducting temperatures (Malati, 1990). Unfortunately, these materials are of a highly complex crystal structure, making them difficult to fabricate into wires. They also have a low current-carrying capacity, so it is likely that for the foreseeable future superconducting magnets will be made from ductile niobium alloys, embedded in a copper matrix.

The main advantage of superconducting separators is that elevated magnetic field strength increases the maximum feed slurry velocity with a corresponding increase in capacity (Kopp, 1991). In order to fully utilize this capacity, downtime for removal of accumulated magnetic particles from the working volume of the separator must be minimized through the use of a reciprocating or continuously cycling matrix (Kopp, 1991). Another advantage of these separators is the reduced weight of the separators (smaller coils and windings along with much less iron required compared to the heavy frames and matrix materials used in HGMS) (Gillet and Diot, 1999). The factors limiting the adoption of superconducting separators are the difficulties in maintaining the very low temperatures necessary for the material to retain its superconducting properties against heat leaks, and the high energy costs associated with maintaining this refrigeration (Kopp, 1991). Superconducting magnets are generally only viable when large field volumes and magnetic fields greater than 2 T are required (Kopp, 1991).

In 1986, a superconducting HGMS was designed and built by Eriez Magnetics to remove magnetic (and colored) contaminants from kaolinite clay for operations in the United States. This machine used only about 0.007 kW in producing 5 T of flux, the ancillary equipment needed requiring another 20 kW. In comparison, a conventional 2 T high-gradient separator of similar throughput would need about 250 kW to produce the flux, and at least another 30 kW to cool the magnet windings.

The 5 T machine is an assembly of concentric components (Figure 13.27). A removable processing canister is installed in a processing chamber located at the center of the assembly. This is surrounded by a double-walled, vacuum-insulated container that accommodates the superconductive niobium/titanium–tantalum winding and the liquid helium coolant. A thermal shield, cooled with liquid nitrogen to 77 K, limits radiation into the cryostat. In operation, the supply of slurry is periodically cut off, the magnetic field is shut down, and the canister backwashed with water to clear out accumulated magnetic contaminants.

A picture of a superconducting magnetic separator in a horizontal arrangement installed in a plant is shown in Figure 13.28.

An open-gradient drum magnetic separator with a superconducting magnet system has been operating commercially since the 1980s (Unkelbach and Kellerwessel, 1985; Wasmuth and Unkelbach, 1991). Although separation is identical to that in conventional drum separators, the magnetic flux density at the drum surface can reach over 4 T.

The development of HGMS and superconducting separators capable of concentrating very fine or very weakly magnetic mineral particles has prompted the application of magnetic separation techniques to treat many waste streams from mineral processing operations. Fine ($<10\ \mu m$), weakly magnetic hematite and limonite have been recovered by a combination of selective flocculation using sodium oleate and kerosene followed by HGMS (Song et al., 2002). HGMS has been used to

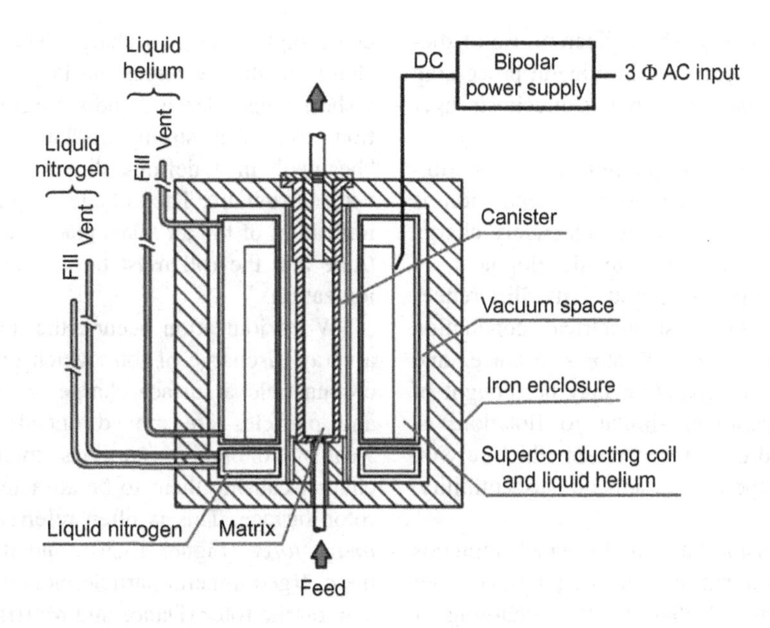

FIGURE 13.27 Diagram of a superconducting magnetic separator.

FIGURE 13.28 Superconducting magnetic separator *(Courtesy Imerys).*

recover fine gold-bearing leach residues from uranium processing, and fine Pb minerals containing V and Zn from a mining waste dump (Watson and Beharrell, 2006). A single-stage extraction of ilmenite from highly magnetic gangue minerals has been developed using a superconducting HGMS system (the difference in magnetic susceptibility between ilmenite and gangue is only significant at very high magnetic field strength). However, this process is still faced with the typical challenges associated with an industrial installation of a superconducting separator (Watson and Beharrell, 2006). Another interesting, and potentially significant, application of HGMS is in the treatment of wastewater streams containing heavy metal ions. Multiple authors have developed processes where the metal ions to be removed are coprecipitated with Fe ions to form a fine, dispersed magnetite phase which can be easily extracted through the use of HGMS (Gillet et al., 1999; Karapinar, 2003).

13.6 ELECTRICAL SEPARATION

Electrical separation exploits the differences in electrical conductivity between different minerals in a feed. Since almost all minerals show some difference in conductivity, it would appear to represent the universal concentrating method. In practice, however, the method has fairly limited applications due to the required processing conditions (notably a perfectly dry feed), and its greatest use is in separating some of the minerals found in heavy mineral sands from beach or stream placer deposits (Dance and Morrison, 1992). Electrical separation also suffers from a similar disadvantage to dry magnetic separation—the capacity is very small for finely divided material. For most efficient operation, the feed to most electrical separators should be in a layer, one particle deep, which reduces the throughput if the particle size is small (<75 μm).

There are two distinct forces which may be considered in the context of electrical separation. The electrophoretic force is the force experienced by a charged particle under the influence of an electric field, and the dielectrophoretic force is the force experienced by a neutral particle in a fluid when subjected to a nonuniform electric field. The dielectrophoretic force is somewhat analogous to magnetic force as it relies on the polarization of a neutral particle into an electric dipole as well as a nonuniform applied field (Lockhart, 1984). The deliberate use of dielectrophoresis is almost nonexistent in mineral processing however, as the electrophoretic force is much stronger (Lockhart, 1984).

In order to exploit the electrophoretic force for mineral separation, a treatment step prior to separation is required in all electrical separators to selectively charge the mineral particles. This selective development of charges on particles relies on conductivity differences between the minerals. As most electrical conduction occurs in the surface layers of atoms (Dance and Morrison, 1992), electrical separation may be thought of as a surface-based separation, similar to flotation, as opposed to magnetic and gravity separation which rely on differences in bulk properties (magnetic susceptibility, specific gravity).

There are three main mechanisms by which minerals are charged: ion bombardment (corona charging), conductive induction, and frictional charging (tribocharging or contact electrification). Each of these three mechanisms has a corresponding separator type, the details of which are described in the following sections. To understand electrical separation methods, knowledge of the electrical properties of materials is required. Introduction to the relevant concepts, as they apply to mineral processing, along with detailed descriptions of many industrial separators, may be found in the comprehensive reviews by Kelly and Spottiswood (1989a–c) and Manouchehri et al. (2000).

13.6.1 Ion Bombardment

Charging via ion bombardment occurs as a high voltage is applied between two electrodes so that the gas near the electrodes ionizes and forms a corona discharge, a continuous flow of gaseous ions. Mineral particles passing through this corona are bombarded with the flow of ions and develop a charge. A similar mechanism of charge application is employed in electrostatic precipitators used to remove fine particulate matter from flowing gas streams. In mineral separation applications, different conductivities of the charged mineral particles then result in different rates of charge decay and correspondingly different forces experienced by the particles.

The typical separator relying on corona charging is the high-tension roll (HTR) separator (Figure 13.29). In this separator the feed, a mixture of ore minerals of varying susceptibilities to surface charging, is fed to a rotating drum made from mild steel, or some other conducting material, which is grounded through its support bearings. An electrode assembly, comprising a brass tube in front of which is supported a length of fine wire, spans the complete length of the roll and is supplied with a fully rectified DC supply of up to 50 kV, usually of negative polarity. Together these two electrodes act to create a dense high-voltage discharge. The fine wire tends to discharge readily, whereas the large electrode tends to have a short range, dense, nondischarging field. This combination creates a strong discharge pattern that may be "beamed" in a definite direction and concentrated to a very narrow arc. The voltage supplied should be such that ionization of the air takes place. Arcing between the electrode and the roll must be avoided, as this destroys the ionization.

When ionization occurs, the mineral particles receive a spray discharge of ions which gives all particles in the corona field a surface charge. As the HTR drum rotates and particles are moved outside of the corona field, weakly conductive particles maintain a high surface charge, causing them to be attracted to and pinned to the rotor surface. This is often referred to as pinning by the *image force* (Figure 13.30), and it may be explained by the charged mineral particle inducing a charge of opposite sign on the rotor (Dance and Morrison, 1992). Pinned particles are removed from the rotor surface either through the eventual decay of their surface charge or mechanically by means of a brush.

Particles of relatively high conductivity lose their surface charge as the charge rapidly dissipates to the earthed rotor. The centrifugal force of the rotor, along with

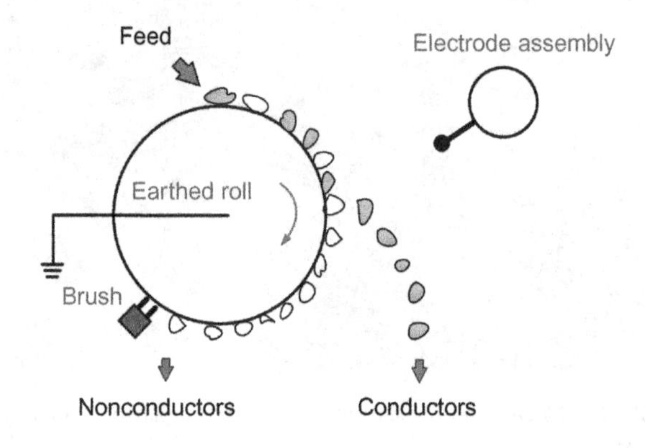

FIGURE 13.29 Diagram of HTR separator.

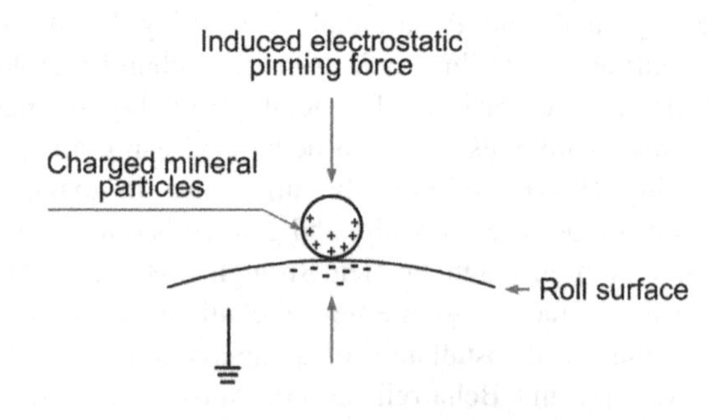

FIGURE 13.30 Representation of pinning force experienced by nonconducting particle in a HTR separator *(Courtesy OreKinetics).*

TABLE 13.2 Typical Conductivity and Behavior of Minerals in a High-Tension Separator

Nonconductive Minerals (Pinned)	Conductive Minerals (Thrown)
Apatite	Cassiterite
Barite	Chromite
Calcite	Diamond
Coal	Feldspar
Corundum	Galena
Garnet	Gold
Gypsum	Hematite
Kyanite	Ilmenite
Monazite	Limonite
Quartz	Magnetite
Scheelite	Pyrite
Sillimonite	Rutile
Spinel	Sphalerite
Tourmaline	Stibnite
Zircon	Tantalite
	Wolframite

gravitational and frictional forces, is then able to throw these particles from the roll and away from the relatively low-conductivity particles that remain pinned so that two streams of particles develop which may be collected separately through the use of a splitter. The separation can be optimized by varying the splitter position. However, predicting particle trajectories from an HTR separator is challenging, as Edward et al. (1995) have shown that particles do not instantaneously accelerate to the roll speed, due to slip on the rotor surface.

The primary industrial use of HTR separators is in the processing of heavy mineral sands (Dance and Morrison, 1992). Other uses include coal cleaning (Butcher and Rowson, 1995), and recycling metals from plastic waste (Dascalescu et al., 1993). Table 13.2 shows typical minerals which are either pinned to or thrown from the rotor during HTR separation.

A combination of pinning and lifting can be created by using a third "static" electrode following the corona discharge electrodes with a diameter large enough to preclude corona discharge. The conducting particles, which are thrown from the rotor, are attracted to this third electrode, and the combined process produces a very wide and distinct separation between conductive particles (lifted from the rotor surface) and nonconductive particles (pinned to the rotor surface).

High-tension separators operate on feeds containing particle sizes of 60–500 μm. Particle size influences separation behavior, as surface charges on a coarse particle are lower in relation to its mass than on a fine particle. Thus, a coarse particle is more readily thrown from the roll surface, and the conducting fraction (particles thrown from the rotor) often contains a small proportion of coarse nonconductors. Similarly, the finer the particles the more they are influenced by the surface charge, and the nonconducting fraction often contains some fine conducting particles. This cross-contamination may also be interpreted in terms of the interplay between the centrifugal force on a particle and the image force acting to pin a charged particle to a grounded surface. The centrifugal force varies with particle mass, while the image force varies with surface area (as charge is accumulated on the particle surface) so, consequently, the centrifugal force is dominant at coarse particle sizes (Dance and Morrison, 1992; Svoboda, 1993).

Some machine factors affecting the operation of an HTR separator include: geometry of the electrode assembly, electrode voltage and polarity, rotor speed, rotor diameter, and splitter position (Dance and Morrison, 1992). Larger rotor diameters help to increase recovery, while a smaller rotor diameter improves the grade of the conducting fraction (Svoboda, 1993). A similar dependence exists for particle density, rotor speed, and the coefficient of friction between the particle and rotor surface, so that separation selectivity is maximized at low particle density, small rotor diameter, high rotor speed, and high coefficient of friction (Svoboda, 1993). The effect of rotor speed on separation is complex and dependent on the conductivity of a given particle, as the act of increasing rotor speed decreases the time available for charge decay. In this way, increased rotor speed increases the chance that a conductive mineral particle will report to the nonconductor fraction, while high rotor speeds will also increase the centrifugal force on a nonconductive particle so that it is more likely to incorrectly report to the conductive fraction (Svoboda, 1993). Stated another way, increased rotor speed simultaneously increases the minimum particle size necessary for a conductive particle to be thrown from the rotor while decreasing the maximum nonconductive particle size that will be pinned to the rotor (Svoboda, 1993).

While HTR separation primarily exploits the differences in conductivities between minerals, an equally important criteria for successful operation is the presence of at least one strongly conductive (on an absolute basis) mineral species in the separator feed. It has been shown by Svoboda (1993) that very large differences in mineral conductivities (up to an order of magnitude) will not result in a sharp separation if both minerals are weak conductors.

Conversely, two strongly conducting minerals can be separated with only a small difference in their conductivities.

HTR separators have been one of the mainstays of the mineral sands industry for decades. Very little development of the machines has occurred in that period; their generally poor single pass separation has been tolerated, and overcome by using multiple machines and multiple recycle streams. However, in the last few years innovative new designs have started to appear, from new as well as established manufacturers. Roche Mining (MT) have developed the Carara HTR separator, which incorporates an additional insulated plate static electrode to help deflect the path of conductive particles thrown from the rotor (Germain et al., 2003). Outokumpu Technology developed the eForce HTR separator, which also incorporates additional static electrodes, as well as an electrostatic feed classifier (Elder and Yan, 2003).

OreKinetics has introduced the new CoronaStat machine (Figure 13.31), which is a significant improvement on existing HTR designs as it employs additional static electrodes to improve the efficiency of separation. Unlike existing machines, the static electrodes are not exposed, making the machines much safer to operate. The key improvement in the CoronaStat design relative to traditional HTR separators is the presence of induction electrodes, which simultaneously increase the pinning force on nonconducting particles and increase the rate of charge decay for conductive particles (Figure 13.32). This results in a larger distance between the two particle streams and therefore an improved separation.

13.6.2 Conductive Induction

The second charging mechanism used in electrical separators is conductive induction, in which polarization of a mineral particle occurs upon exposure to an electric field. Similar to charge decay in HTR separators, the ability of the mineral particle to respond to this induced polarization is directly related to its conductivity. Polarization results when an uncharged particle develops an opposite charge, relative to the electrode creating the electric field, at the surface closest to the electrode and a corresponding like charge to the electrode on the particle surface furthest from the electrode. Conductive particles are able to redistribute these induced charges across the particle surface, while nonconductive particles are unable to redistribute these charges and will remain polarized. The electric force on a polarized particle is a function of the degree to which it polarizes, which is in turn affected by both the size and shape of the particle (Manouchehri et al., 2000). When a polarized particle contacts a conductive surface it may conduct charge of one polarity to the surface, leaving a net charge on the particle. In such a situation nonconductive particles (with no net charge) will experience no attraction from an applied electric field, whereas

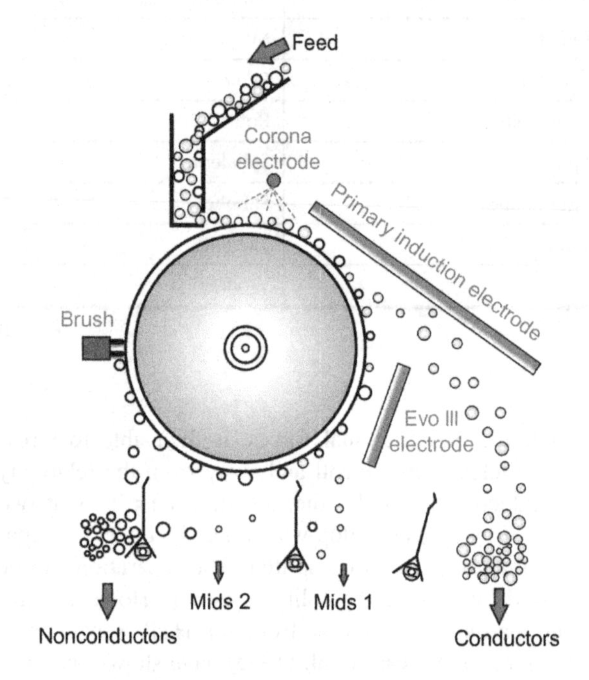

FIGURE 13.31 Diagram of CoronaStat separator *(Courtesy OreKinetics)*.

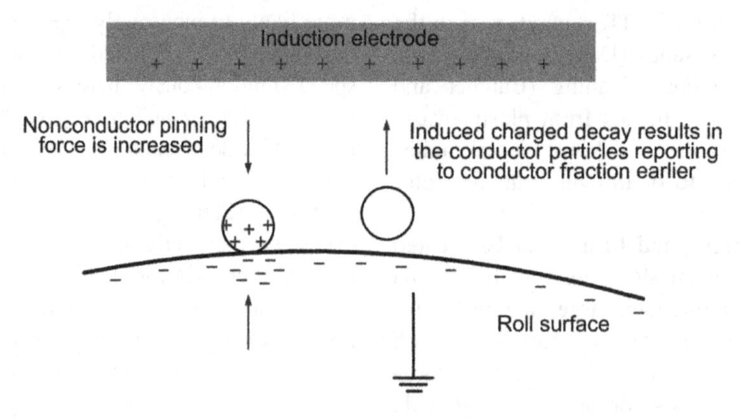

FIGURE 13.32 Effect of induction electrode in CoronaStat separator *(Courtesy OreKinetics)*.

conductive particles will be attracted to an oppositely charged electrode (Kelly and Spottiswood, 1989a). Conductive induction can therefore be thought of as a process in which charges are induced on uncharged conductive particles, leaving nonconductive particles with no net charge. A graphical representation of conductive induction may be seen in Figure 13.33.

Separators exploiting this charging mechanism are typically used to separate strongly conductive particles from weakly conductive particles and employ static electrodes to "lift" charged conductive particles from a grounded surface while nonconductive particles remain pinned to that surface. The most common such separator is the electrostatic plate (ESP) separator. In an ESP separator, material is gravity fed through the separator and the force on the charged particles acts to counteract the force of gravity. In contrast to HTR separators, coarse particles will tend to report to the nonconducting fraction, which is why final cleaning of the products of HTR separation is often carried out in purely electrostatic separators.

Modern electrostatic separators are of the plate or screen type (Figure 13.34), the former being used to clean small amounts of nonconductors from a predominantly conducting feed (Figure 13.34(a)), while the screen separators remove small amounts of conductors from a mainly nonconducting feed (Figure 13.34(b)). The principle of operation is the same for both types of separator. The feed particles gravitate down a sloping, grounded plate into an electrostatic field induced by a large, oval-shaped, high-voltage electrode. Fine particles are most affected by the lifting force, and so fine conductive particles are preferentially lifted to the electrode, whereas coarse nonconductors are most efficiently rejected. Machine parameters affecting ESP separators include: electrode geometry, electrode voltage and polarity, plate curvature, and position of the splitters (Dance and Morrison, 1992). For both HTR and ESP separators, system humidity is intentionally kept low, as excess moisture may alter the conductivity of the fluid medium of the separator (the air) as well as affecting the conductivity of the particle surface through the dual effects of water molecules themselves and dissolved ions in the water (Kelly and Spottiswood, 1989b).

Similar to the CoronaStat for HTR separation, OreKinetics has also developed an improved ESP separator known as the UltraStat separator (Figure 13.35). The primary improvements in this separator are different geometries of the electrode and particle feed path, the presence of secondary induction electrodes to further increase the lifting force on charged conductive particles as well as a secondary roll to clean the primary roll surface.

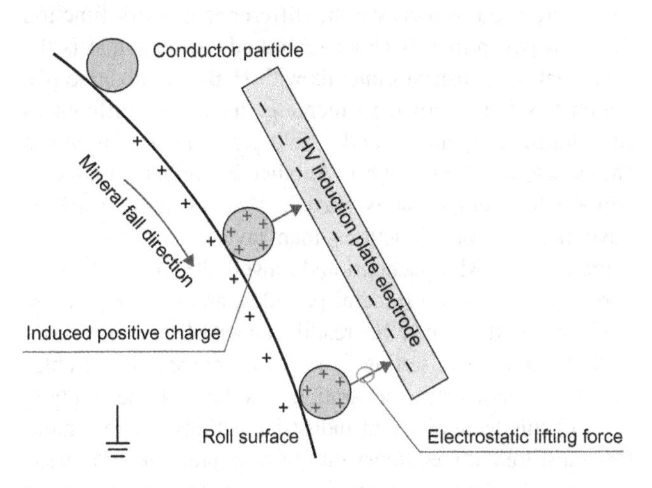

FIGURE 13.33 Representation of conductive induction *(Courtesy OreKinetics).*

13.6.3 Triboelectric Charging

The final charging mechanism used in mineral processing is triboelectrification, or contact electrification, in which two materials of dissimilar electrical properties exchange electrons upon coming into contact with one another. As most minerals are semi-conductors, with volume conductivities between 10^5 and $10^{-8}\ \Omega\ m^{-1}$ (Manouchehri et al., 2000), the charge acquired by two minerals after contacting one another may be predicted by the relative Fermi levels (energy level at which 50% of the energy states in a material are occupied by electrons) of the two minerals (Kelly and Spottiswood, 1989c). An alternative measure also used to predict triboelectric charging behavior is the

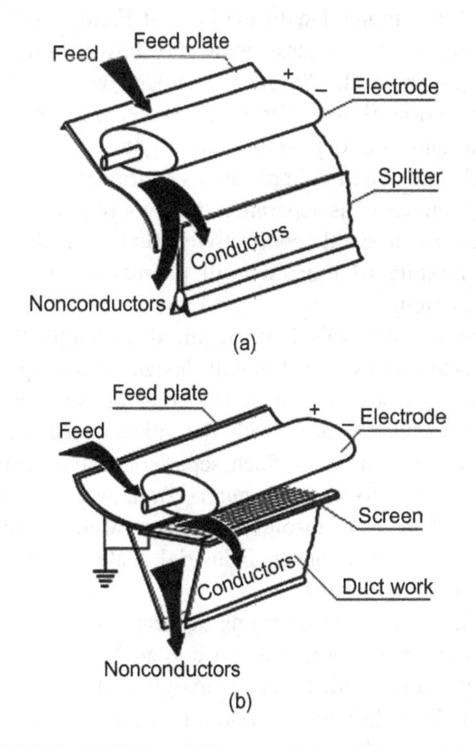

FIGURE 13.34 (a) Plate, and (b) screen electrostatic separators.

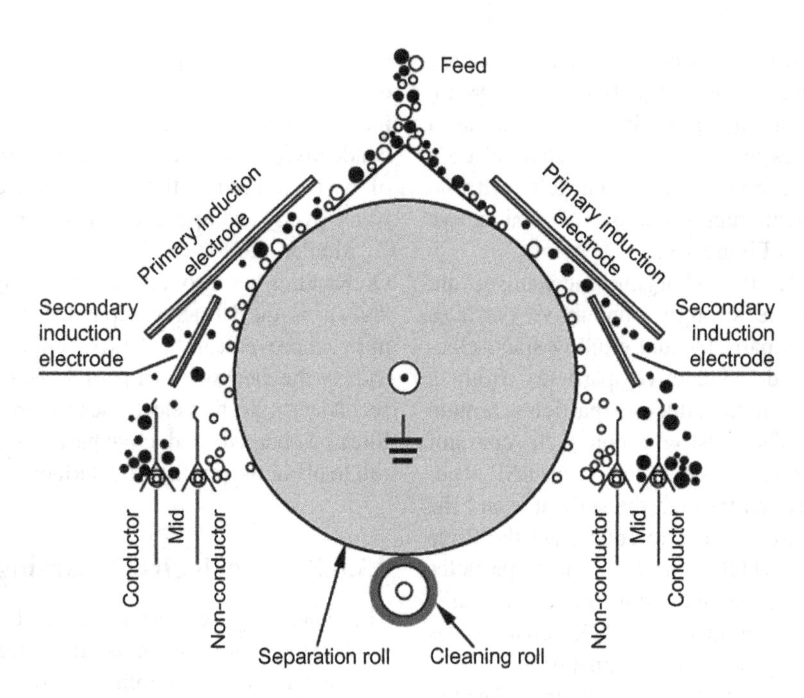

FIGURE 13.35 Diagram of UltraStat separator *(Courtesy OreKinetics).*

work function of a material, which is a measure of the energy required to bring an electron from the Fermi level of a given material to a free electron state (Kelly and Spottiswood, 1989c). A mineral with a low Fermi level must therefore have a higher work function than a mineral with a higher Fermi level. When two mineral particles come into frictional contact their Fermi levels will equalize, with the mineral with the highest Fermi level losing electrons to the mineral with the lower Fermi level (Manouchehri et al., 2000). The mineral with the highest work function (lower Fermi level) becomes negatively charged and the opposing mineral becomes positively charged. The potential applications of triboelectric separation are immense, as separation does not require large differences in mineral conductivity and virtually every binary mixture of minerals will possess a difference in work function.

Once the minerals have acquired a charge, they are often separated using a free-fall design consisting of two charged electrodes which deflect mineral particles based on their surface charge with the mineral particles collected in different bins. Such separators have been used on a lab scale to separate quartz from wollastonite and calcite, calcite from insoluble silicates (Manouchehri and Fawell, 2002), and on an industrial scale to beneficiate potash (Lockhart, 1984).

In all triboelectric charging devices, mineral particles come into contact with not just one another, but also the material from which the conveying device is constructed. It is therefore important to take this into consideration, as different materials, such as brass and Teflon, have large differences in work function and will therefore produce corresponding differences in the charges produced on a given mineral particle (Dwari et al., 2009). Even if the charge induced on a mineral surface using different charging materials is of the same sign, the amount of the charge transfer between two materials is dependent on the differences between the Fermi levels of the two materials.

Charge acquisition of mineral particles in triboelectric separation may also be controlled through the use of a surface treatment prior to tribocharging such as: surface cleaning, chemical pretreatment, thermal pretreatment, irradiation, changes in the atmospheric humidity, or surface doping (Manouchehri, 2010). Such surface treatments are used to increase the difference in work function between two minerals to be separated. An example is the treatment of industrial minerals with H_3BO_3 at alkaline pH, which has been shown to increase the charge differences of feldspar–quartz and feldspar–calcite mixtures (Manouchehri et al., 1999). Another important variable in triboelectric separation is particle size, as small particles have higher work functions than coarse particles of the same mineral (Manouchehri and Fawell, 2002). While measuring the charge on mineral particles, and even separating a binary mixture, can be readily accomplished in a controlled laboratory setting, the wide range of variables affecting triboelectric separation has limited the applications of this technology in industrial settings where separators must treat a feed consisting of multiple mineral types.

One of the explanations for continued research interest in triboelectric separation is the minimal effect of gravity

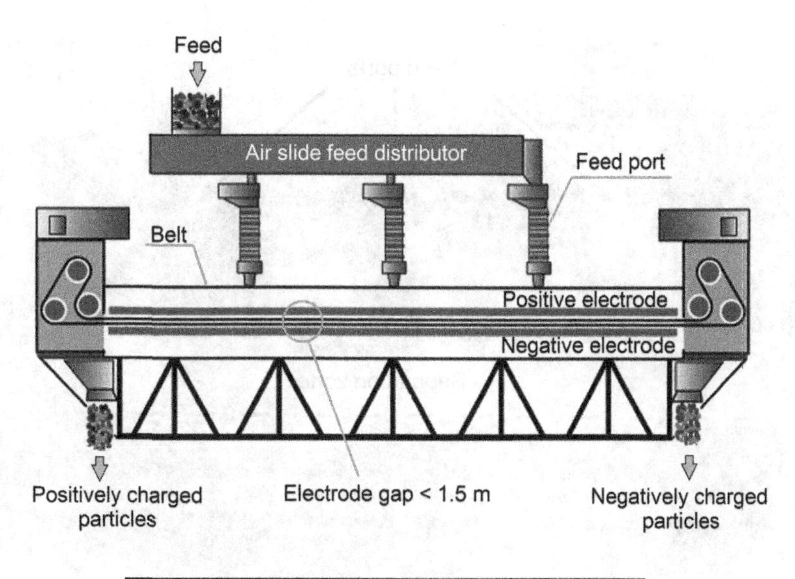

FIGURE 13.36 Diagram of ST separator *(Adapted from Bittner et al. (2014)).*

on the separation process, which may in the future be very beneficial in developing extraterrestrial or lunar mining operations (Li et al., 1999). In one such study, focused on the beneficiation of ilmenite from a synthetic lunar soil, ilmenite was found to report to both positive and negative electrodes in binary mixture separations depending on the gangue mineral chosen for the feed mixture (Li et al., 1999). When the full synthetic ore, four different gangue minerals along with the ilmenite, was processed through the triboelectric separation unit ilmenite was found to be concentrated (by a factor of 2–3) in the neutral particle collection bin, evidently acquiring little net charge due to the presence of gangue minerals with both higher and lower work functions than the ilmenite (Li et al., 1999). This finding is illustrative of the inherent difficulties in predicting mineral behavior through triboelectric separations in an industrial setting.

Recently a new triboelectric separator, the ST separator, has been developed by Separation Technologies which employs conventional interparticle contact to tribocharge mineral particles and a continuous loop open-mesh belt that travels at high speeds (5–20 m s^{-1}) between positive and negative electrodes for particle separation (Figure 13.36) (Bittner et al., 2014). Feed enters from the top of the unit (at feed rates of up to 40 t h^{-1}) with positive and negative charged particles exiting from opposite ends of the separator (Figure 13.37). The separation occurs within a narrow gap (<1.5 cm) between the electrodes. The top and bottom sections of the belt move in opposite directions, setting up a high particle number density, counter current flow within the electrode gap. Particles must travel across only a small fraction of the electrode gap (across the zone of high shear and lower velocities) under electrostatic forces to be separated into the oppositely flowing streams (Figure 13.38). The counter current highly turbulent flow enables multiple stages of separation to occur within a single pass through the separator, increasing both grade and recovery of the product streams (Bittner et al., 2014). This multistage separation zone requires that the particles maintain their charge, which is made possible due to the high degree of interparticle contacts occurring throughout the separation zone (Bittner et al., 2014). This separator can process particles from 1 to 300 μm, which is much smaller than conventional free-fall and HTR separators. It has been widely employed industrially in removing unburned coal char from fly ash (10–20 μm median diameter) generated by coal-fired power plants (Bittner et al., 2014). On a pilot plant scale (3–6 t h^{-1}), it has also been shown to be effective at beneficiating industrial minerals such as separating quartz from calcite (89% recovery, 99% grade) and magnesite from talc (77% recovery, 95% grade) (Bittner et al., 2014).

13.6.4 Example Flowsheets

Earlier in this chapter the possibility of combined magnetic and electrical separation was noted, particularly in the processing of heavy mineral sand deposits. Table 13.3

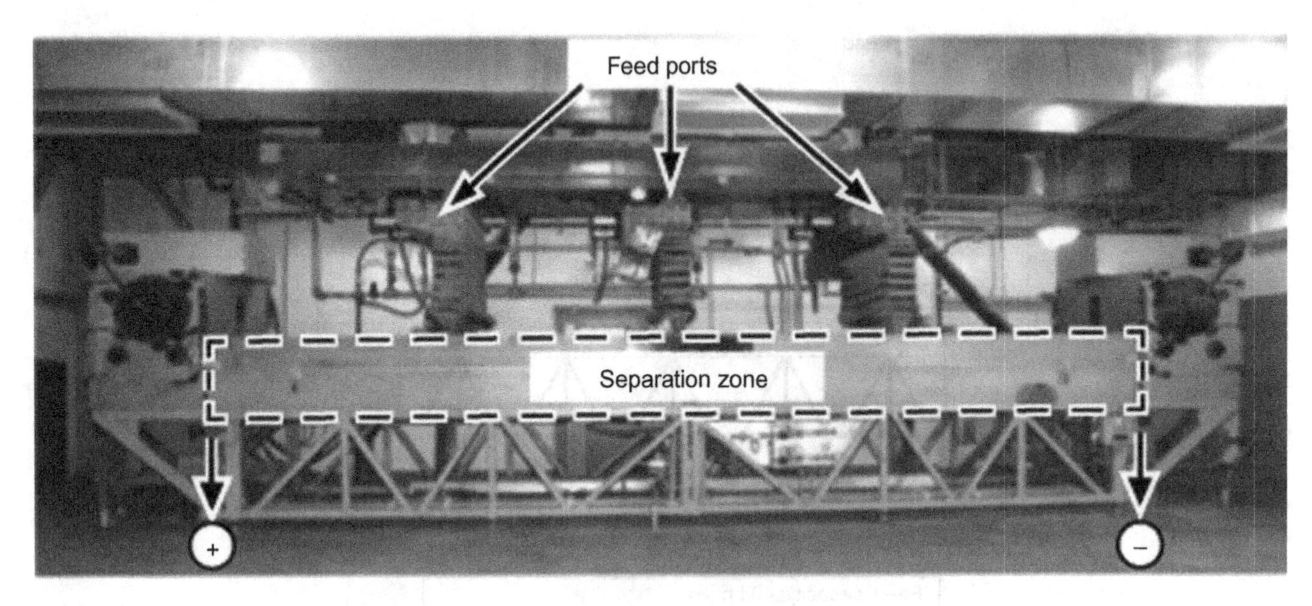

FIGURE 13.37 Industrial installation of ST separator *(Adapted from Bittner et al. (2014)).*

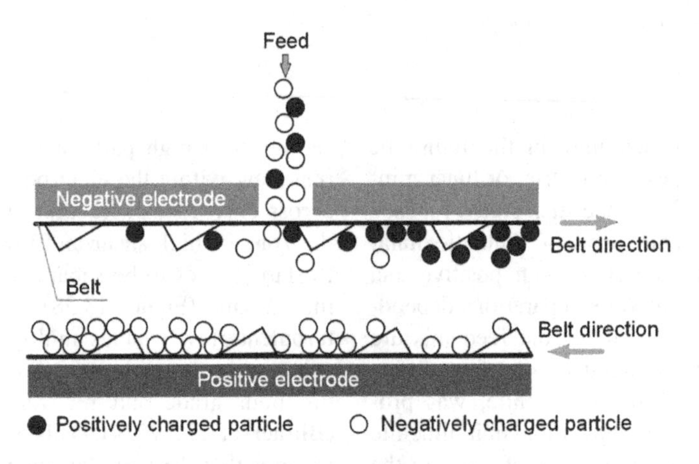

FIGURE 13.38 Separation zone of ST separator *(Adapted from Bittner et al. (2014)).*

TABLE 13.3 Magnetic and Electrical Behavior of Typical Heavy Mineral Sands Components

Magnetics	Magnetite—C	Ilmenite—C	Garnet—NC	Monazite—NC
Nonmagnetics	Rutile—C	Zircon—NC	Quartz—NC	

C, conductor; NC, nonconductor.

shows some of the common minerals present in such alluvial deposits, along with their properties, related to magnetic and electrical separation. Mineral sands are commonly mined by floating dredges, feeding floating concentrators at up to 2,000 $t\,h^{-1}$. Such concentrators, consisting of a complex circuit of sluices, spirals, or Reichert cones, upgrade the heavy mineral content to around 90%, the feed grades varying from less than 2%, up to 20% heavy mineral in some cases. The gravity preconcentrate is then transferred to the separation plant for recovery of heavy minerals by a combination of gravity, magnetic, and electrical (typically HTR) separation.

Mineral sands flowsheets vary according to the properties of the minerals present, wet magnetic separation often preceding high-tension separation where magnetic ilmenite is the dominant mineral, for example.

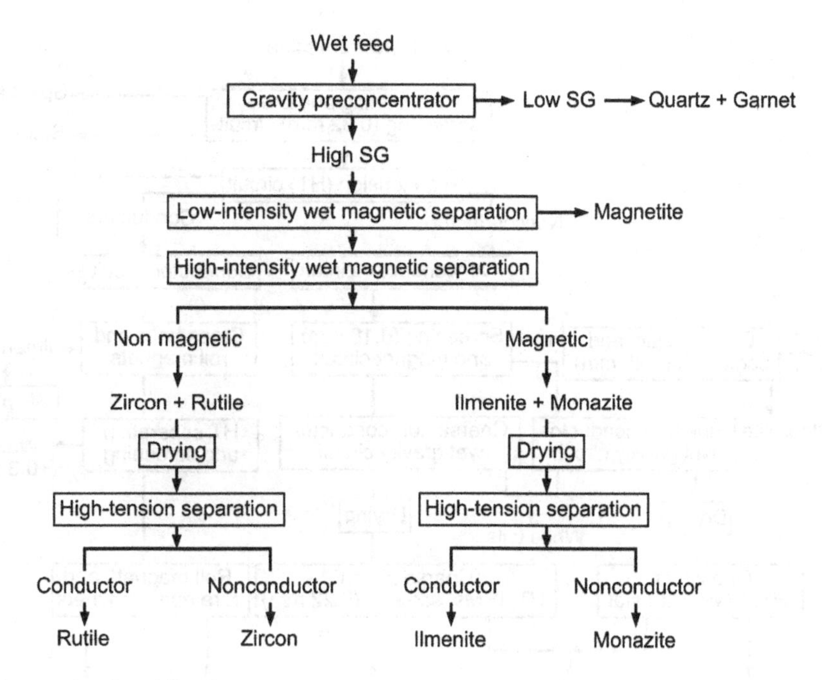

FIGURE 13.39 Typical heavy mineral sand flowsheet.

A generalized flowsheet is shown in Figure 13.39. Low-intensity drum separators remove any magnetite from the feed, after which high-intensity wet magnetic separators separate the monazite and ilmenite from the zircon and rutile. Drying of these two fractions is followed by HTR separation to produce final products, although further cleaning is sometimes carried out by ESP separators. For example, screen electrostatic separators (Figure 13.34(b)) may be used to clean the zircon and monazite concentrates, removing fine conducting particles from these fractions. Similarly, plate electrostatic separators (Figure 13.34(a)) could be used to reject coarse nonconducting particles from the rutile and ilmenite concentrates.

Figure 13.40 shows a simplified circuit used to process heavy minerals, on the west coast of Australia (Benson et al., 2001).

The heavy mineral concentrate is first divided into conductive and nonconductive streams using HTR separators. The conductors are treated using cross-belt and roll magnetic separators to remove the ilmenite as a magnetic product. The nonmagnetic stream is cleaned with high-intensity roll and rare earth magnets to separate the weakly paramagnetic leucoxene from diamagnetic rutile. The nonconductors undergo another stage of wet gravity separation to remove quartz and other low-specific-gravity contaminants, before sizing and cleaning using HTR, ESP and Ultrastat separators to produce fine and coarse zircon products. Similar flowsheets are used in South-East Asia for the treatment of alluvial cassiterite deposits, which are also sources of minerals such as ilmenite, monazite, and zircon.

In the case of the Cliffs–Wabush mine (discussed in Section 13.5.2), the gravity concentrate from the spirals bank is cleaned by a series of HTR separators (Damjanović and Goode, 2000). The spiral concentrate is first filtered and dried before being fed to 54 primary Carpco HTR separators (with a total of 288 rotors), with the tailings from the rougher HTR separators fed to six scavenger HTR separators (total of 24 rotors). Each rotor is 10 ft long with a roll diameter of 14 in. and is operated at a rotor speed of 100 rpm, electrode voltage of 23–25 kV, and a feed rate of $2.54\ t\ h^{-1}$. Through this separation, the nonconductive quartz gangue is pinned to the roll, with the valuable iron oxide mineral thrown from the roll. Cleaning of the gravity concentrate with HTR separators is preferred as relatively little material is pinned to the rotor in the high-tension treatment of the gravity concentrate (Fe-oxides are conductive). The rougher bank of HTR separators produces a final concentrate, a middlings stream that is recycled to the HTR feed and a tailings stream sent to the HTR scavenger circuit. The scavenger concentrate goes to the final concentrate (65.50% Fe, 2.55% SiO_2, 1.95% Mn), with the scavenger tailings sent to the final tailings. The HTR concentrate is then blended with the magnetic concentrate prior to being sent to pelletization.

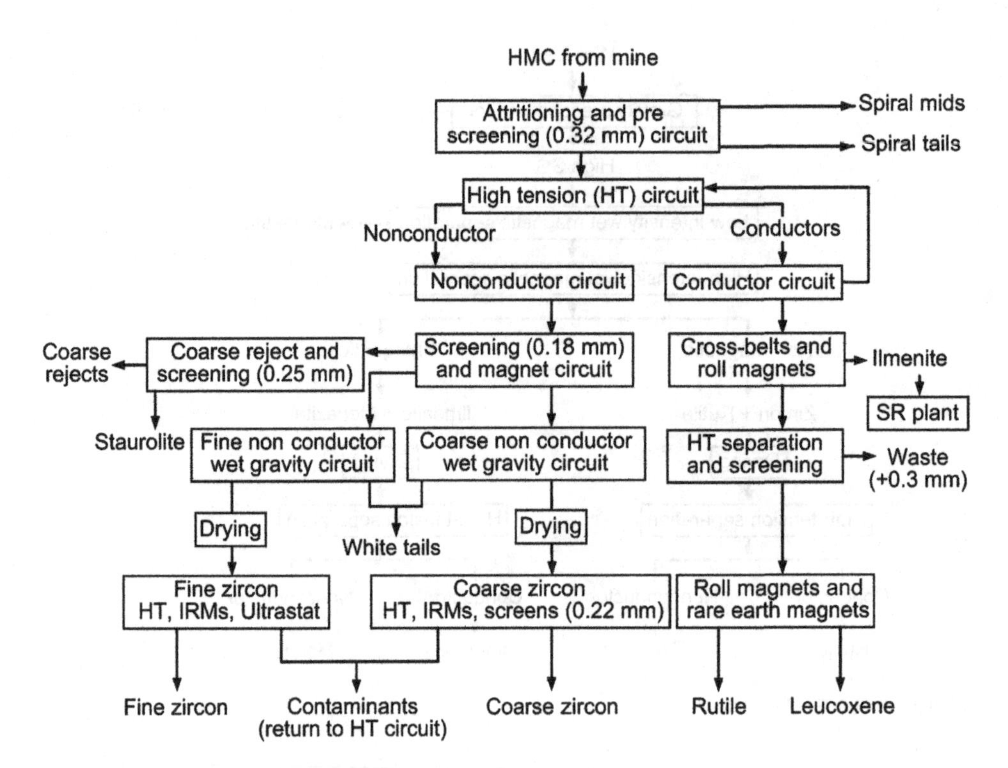

FIGURE 13.40 Simplified mineral sands circuit used by Tronox Limited (formerly Tiwest Joint Venture) at the Chandala processing plant, Western Australia *(Adapted from Benson et al. (2001)).*

REFERENCES

Arvidson, B.R., 2001. The many uses of rare-earth magnetic separators for heavy minerals sands processing. Proceedings of the International Heavy Minerals Conference. AusIMM, Perth, Australia, pp. 131–136.

Bennett, L.H., et al., 1978. Comments on units in magnetism. J. Res. Natl. Bur. Stand. 83 (1), 9–12.

Benson, S., et al., 2001. Quantitative and process mineralogy at tiwest. Proceedings of the International Heavy Minerals Confernce. AusIMM, Fremantle, Australia, pp. 59–68.

Bird, A., Briggs, M., 2011. Recent improvements to the gravity gold circuit at Marvel Loch. Proceedings of Metallurgical Plant Design and Operating Strategies (MetPlant) Conference. AusIMM, Perth, Australia, pp. 115–137.

Biss, R., Ayotte, N., 1989. Beneficiation of carbonatite ore-bearing niobium at Niobec Mine. In: Dobby, G.S., Rao, S.R. (Eds.), Proceedings of the International Symposium on Processing of Complex Ores. CIM, Halifax, Canada, pp. 497–506.

Bittner, J.D., et al., 2014. Triboelectric belt separator for beneficiation of fine minerals. Procedia Eng. 83, 122–129.

Butcher, D.A., Rowson, N.A., 1995. Electrostatic separation of pyrite from coal. Mag. Elect. Sep. 9, 19–30.

Corrans, I.J., 1984. The performance of an industrial wet high-intensity magnetic separator for the recovery of gold and uranium. J. S. Afr. Inst. Min. Metall. 84 (3), 57–63.

Corrans, I.J., Svoboda, J., 1985. Magnetic separation in South Africa. Mag. Sep. News. 1, 205–232.

Damjanović, B., Goode, J.R., 2000. Canadian Milling Practice, Special Vol. 49. CIM, Montréal, QC, Canada.

Dance, A.D., Morrison, R.D., 1992. Quantifying a black art: the electrostatic separation of mineral sands. Miner. Eng. 5 (7), 751–765.

Dascalescu, L., et al., 1993. Corona-electrostatic separation: an efficient technique for the recovery of metals and plastics from industrial wastes. Mag. Elect. Sep. 4, 241–255.

Davis, E.W., 1921. Magnetic Concentrator. US Patent No. 1474624.

Dobbins, M., Sherrell, I., 2010. Significant developments in dry rare-earth magnetic separation. Mining Eng. 62 (1), 49–54.

Dobbins, M., et al., 2009. Recent advances in magnetic separator designs and applications. Proceedings of the Seventh International Heavy Minerals Conference. South African Institute of Mining and Metallurgy, Johannesburg, South Africa, pp. 63–70.

Dwari, R.K., et al., 2009. Characterisation of particle tribo-charging and electron transfer with reference to electrostatic dry coal cleaning. Int. J. Miner. Process. 91, 100–110.

Edward, D., et al., 1995. The motion of mineral sand particles on the roll in high tension separators. Mag. Elect. Sep. 6, 69–85.

Elder, J., Yan, E., 2003. eForce—Newest generation of electrostatic separator for the mineral sands industry. Proceedings of Heavy Minerals 2003 Conference. South African Institute of Mining and Metallurgy, Johannesburg, South Africa, pp. 63–70.

Eriez, 2003. Rare earth roll magnetic separators. Eriez Mag.1–8.

Eriez, 2008. WHIMS Separators. Eriez Manufacturing Company, Erie, PA, USA, pp. 1–8.

Eriez, Gzrinm, 2014. Reliable WHIMS with Maximum Recovery. Eriez Manufacturing Company, Erie, PA, USA, pp. 1–12.

Germain, M., et al., 2003. The application of new design concepts in high tension electrostatic separation to the processing of mineral sands concentrates. Proceedings of the International Heavy Minerals Conference. South African Institute of Mining and Metallurgy, Johannesburg, South Africa, pp. 101–106.

Gillet, G., Diot, F., 1999. Technology of superconducting magnetic separation in mineral and environmental processing. Miner. Metall. Process. 16 (3), 1–7.

Gillet, G., et al., 1999. Removal of heavy metal ions by superconducting magnetic separation. Sep. Sci. Technol. 34 (10), 2023–2037.

Jiles, D., 1990. Introduction to Magnetism and Magnetic Materials. Chapman & Hall, London, UK.

Karapinar, N., 2003. Magnetic separation of ferrihydrite from wastewater by magnetic seeding and high-gradient magnetic separation. Int. J. Miner. Process. 71, 45–54.

Kelly, E.G., Spottiswood, D.J., 1989a. The theory of electrostatic separations: a review—part II. Particle charging. Miner. Eng. 2 (2), 193–205.

Kelly, E.G., Spottiswood, D.J., 1989b. The theory of electrostatic separations: a review—part III. The separation of particles. Miner. Eng. 2 (3), 337–349.

Kelly, E.G., Spottiswood, D.J., 1989c. The theory of electrostatic separations: a review—part I. Fundamentals. Miner. Eng. 2 (1), 33–46.

Kopp, J., 1991. Superconducting magnetic separators. Mag. Elect. Sep. 3, 17–32.

Lawver, J.E., Hopstock, D.M., 1974. Wet magnetic separation of weakly magnetic minerals. Miner. Sci. Eng. 6 (3), 154–172.

Li, T.X., et al., 1999. Dry triboelectrostatic separation of mineral particles: a potential application in space exploration. J. Electrostat. 47, 133–142.

Lockhart, N.C., 1984. Dry beneficiation of coal. Powder Technol. 40, 17–42.

Malati, M.A., 1990. Ceramic superconductors. Mining Mag. 163 (Dec.), 427–431.

Manouchehri, H.R., 2010. Triboelectric charge and separation characteristics of industrial minerals. Proceedings of the 25th International Mineral Processing Congress (IMPC), Brisbane, Australia, pp. 1009–1021.

Manouchehri, H.R., Fawell, S., 2002. Electrophysical properties, triboelectric charge characteristics and separation behaviour of calcite, quartz and wollastonite minerals. Proceedings of the SME Annual Meeting. SME, Phoenix, AZ, USA, pp. 1–10.

Manouchehri, H.R., et al., 1999. Changing potential for the electrical beneficiation of minerals by chemical pretreatment. Miner. Metall. Process. 16 (3), 14–22.

Manouchehri, H.R., et al., 2000. Review of electrical separation methods. Part 1: fundamental aspects. Miner. Metall. Process. 17 (1), 23–36.

McAndrew, J., 1957. Calibration of a Frantz isodynamic separator and its application to mineral separation. Proc. Aust. Inst. Min. Met. 181, 59–73.

Metso, 2014a. Magnetic Separators for Dense Media Recovery. Metso Minerals, Sweden, pp. 1–8.

Metso, 2014b. Drum Separators. Metso Minerals, Sweden, pp. 1–4.

Metso, 2014c. High Gradient Magnetic Separators—HGMS Cyclic. Metso Minerals, Sweden, pp. 1–5.

Metso, 2014d. High Gradient Magnetic Separators—HGMS Continuous. Metso Minerals, Sweden, pp. 10–4.

Murariu, V., Svoboda, J., 2003. The applicability of Davis tube tests to ore separation by drum magnetic separators. Phys. Sep. Sci. Eng. 12 (1), 1–11.

Norrgran, D.A., Marin, J.A., 1994. Rare earth permanent magnet separators and their applications in mineral processing. Miner. Metall. Process. 11 (1), 41–45.

Novotny, D., 2014. Technological advances in magnetic separation—Presented as part of the CMP 2014 "Iron Ore Processing in Canada" short course. The 46th Annual Meeting of the Canadian Mineral Processors Conference, Ottawa, ON, Canada.

Oberteuffer, J., 1974. Magnetic separation: a review of principles, devices, and applications. IEEE Trans. Magn. 10 (2), 223–238.

Outotec, 2013. SLon vertically pulsating high-gradient magnetic separator. 1–4. Outotec. <www.outotec.com/>.

Rayner, J.G., Napier-Munn, T.J., 2000. The mechanism of magnetics capture in the wet drum magnetic separator. Miner. Eng. 13 (3), 277–285.

Shao, Y., et al., 1996. Wet high intensity magnetic separation of iron minerals. Magn. Elect. Sep. 8 (1), 41–51.

Singh, V., et al., 2013. Particle flow modeling of dry induced roll magnetic separator. Powder Technol. 244, 85–92.

Song, S., et al., 2002. Magnetic separation of hematite and limonite fines as hydrophobic flocs from iron ores. Miner. Eng. 15, 415–422.

Svoboda, J., 1987. Magnetic Methods for the Treatment of Minerals. Elsevier, Amsterdam, North Holland, Netherlands.

Svoboda, J., 1993. Separation of particles in the corona-discharge field. Magn. Elect. Sep. 4, 173–192.

Svoboda, J., 1994. The effect of magnetic field strength on the efficiency of magnetic separation. Miner. Eng. 7 (5-6), 747–757.

Svoboda, J., Fujita, T., 2003. Recent developments in magnetic methods of material separation. Miner. Eng. 16 (9), 785–792.

Tawil, M.M.E., Morales, M.M., 1985. Application of wet high intensity magnetic separation to sulphide mineral beneficiation. In: Zunkel, A.D. (Ed.), Complex Sulfides: Processing of Ores, Concentrates and By-products. SEM, Pennsylvania, PA, USA, pp. 507–524.

Todd, I.A., Finch, J.A., 1984. Measurement of susceptibility using a wet modification to the Frantz Isodynamic Separator. Can. Met. Quart. 23 (4), 475–477.

Unkelbach, K.-H., Kellerwessel, H., 1985. A superconductive drum type magnetic separator for the beneficiation of ores and minerals. Congrès international de minéralurgie, Cannes, France, pp. 371–380.

Wasmuth, H.-D., Unkelbach, K.-H., 1991. Recent developments in magnetic separation of feebly magnetic minerals. Miner. Eng. 4 (7-11), 825–837.

Watson, J.H.P., 1994. Status of superconducting magnetic separation in the minerals industry. Miner. Eng. 7 (5-6), 737–746.

Watson, J.H.P., Beharrell, P.A., 2006. Extracting values from mine dumps and tailings. Miner. Eng. 19, 1580–1587.

White, L., 1978. Swedish symposium offers iron ore industry an overview of ore dressing developments. Eng. Min. J. 179 (4), 71–77.

Xiong, D.-H., 1994. New development of the SLon vertical ring and pulsation HGMS separator. Magn. Elect. Sep. 5, 211–222.

Xiong, D.-H., 2004. SLon magnetic separators applied in the ilmenite processing industry. Phys. Sep. Sci. Eng. 13 (3-4), 119–126.

Chapter 14

Sensor-based Ore Sorting

14.1 INTRODUCTION

The term "sensor-based (ore) sorting" (SBS) is introduced as an umbrella term for all applications where particles are singularly detected by a sensor technique and ejected by an amplified mechanical, hydraulic, or pneumatic process (Wotruba and Harbeck, 2012). Analogous terms include ore sorting, electronic sorting or automated sorting. A variety of sensor types are available and see use in the minerals, recycling, and food industries. SBS can be implemented at various positions in the mineral processing flowsheet:

1. Pre-concentration

   Pre-concentration is defined as a physical separation stage in mineral beneficiation, where a fraction of high grade coarse particles can be separated from the run-of-mine material to produce a final concentrate, prior to downstream processing of particles below ca. 5 mm.
2. Waste rejection

   Waste rejection is defined as a beneficiation stage where coarse non-valuable waste particles are separated from the run-of-mine ore.
3. Concentration

   Concentration using SBS is the creation of a final marketable product.
4. Ore-type diversion

Separation of one or more ore types that are fed alternately as batches into the same plant or parallel into multiple plant lines for specialized treatment.

Note that SBS is almost always used in diamond processing flowsheets where the terminology differs slightly: waste rejection is referred to as "concentration" and concentration is referred to as "recovery."

SBS refers to a concentration stage which identifies certain physical or chemical characteristics of individual rock particles and separates them from the process stream via a physical mechanism (Arvidson, 1987). Ore sorting is commonly undertaken after primary or secondary crushing, after sufficient liberation is achieved. Applications show that many mines have about 30 wt% barren waste liberated in the size range 10–100 mm, which allows material to be discarded without significant loss of value. Pre-concentration by sorting is seen as a method of improving the sustainability of mineral processing operations by reducing specific materials handling requirements, minimizing energy consumption and water in grinding and concentration, and achieving more benign tailings disposal (Cutmore and Ebehardt, 2002; Lessard et al., 2014). The ultimate goal is to minimize specific investment and processing costs while reducing the environmental footprint of an operation. Sensor-based sorting can be applied to a waste rejection stage (e.g., base and precious metals) or concentration (i.e., the production of an intermediate or final product, e.g., industrial minerals, ferrous metals, and gemstones). Sorting has also been used to upgrade previously mined/processed waste-rock material prior to re-processing (von Ketelhodt, 2009; Wotruba and Harbeck, 2012).

SBS is the automation of hand sorting, which is now extended by the use of additional sensing/detection technologies (i.e., techniques are not limited to optical sensors). Hand sorting is the original mineral concentration process, having been used by the earliest workers several thousand years ago. The practice was recorded by Agricola (1556) (Figure 14.1).

Hand sorting involves the visual assessment of individual ore particles and the rejection of those particles that do not warrant further treatment. Figure 14.2 shows hand sorting in the early days of operation at the Sullivan Mine (Cominco), British Columbia, Canada, which began operation in 1909 (Ednie, 2006).

Hand sorting has declined in importance due to the need to treat large quantities of low-grade ore which can require extremely fine grinding. Hand sorting of some kind, however, is still practiced (e.g., the removal of large pieces of timber or tramp iron from the run-of-mine ore). It is also still applied in countries with low labor pay rates.

Wills' Mineral Processing Technology.

FIGURE 14.1 Early reference to hand sorting by Agricola (1556) *(Used with permission Dover Publications, Inc., New York, 1950).*

FIGURE 14.2 Hand sorting at Sullivan Mine ca. 1915 *(Used with permission Columbia Basin Image Bank).*

Early workers identified ore sorting as a key variable which affected the economics of a mine. Rickard (1905) in the book "The Economics of Mining" stated "whether to (hand) sort or not, is a question of vital importance to the economics of a mine; it may mean the choice between a small yield of high-grade material or a large output of low-grade, which immediately affects all the operations carried on at the surface, as well as underground...."

SBS is reviewed here, but several other waste elimination methods exist, which are primarily gravity-based processes (Chapters 10 and 11) that could be employed underground or at the mine. The techniques are not currently employed underground due to the need for complex water handling and retreatment systems (Hughes and Cormack, 2008; Murphy et al., 2012). Complete SBS installations are relatively compact (small volume to throughput ratio), especially when operated dry. Thus, SBS is commonly employed using a semi-mobile operation. About 30% of today's machines being installed are containerized systems. This being said, there is great potential for SBS systems integration underground (near-to-face sorting) especially where stoping and cut-and-fill methods are used (Schindler, 2003; Dammers et al., 2013; Robben, 2014).

SBS was first introduced in the late 1940s, and although its application is fairly limited, it is an important technique for the processing of certain minerals (e.g., diamonds, uranium, limestone, magnesite, gemstones) (Sassos, 1985; Salter and Wyatt, 1991; Sivamohan and Forssberg, 1991; Collins and Bonney, 1998; Arvidson, 2002). Arvidson (1987) stated that vein-type, layered, brecciated or pebbly mineralizations were typically good candidate ores for sorting. There is a general misconception that sorters cannot be used in massive ores, but full liberation is not necessarily required for successful waste elimination (Chapter 1). Complete liberation may be required in certain instances for concentration, such as when limestone is required for filler purposes (Arvidson, 1987).

14.2 SENSOR-BASED SORTING PRINCIPLES

Many mineral properties have been used as the basis of sensor-based sorting, including reflectance and color in visible light (magnesite, limestone, base metal and gold ores, phosphates, talc, coal), ultraviolet (scheelite), natural gamma radiation (uranium ore), magnetism (iron ore), conductivity (sulfides), X-ray fluorescence (base metals), and X-ray luminescence (diamonds). Infrared, Raman, microwave attenuation, and other properties have also been tested. Table 14.1 gives examples of the properties that can be exploited using commercially-available sensor technologies along with example industrial applications. Reviews on historical developments, the various sensor types, and applications are given elsewhere (Salter and Wyatt, 1991; Wotruba, 2006; Wotruba and Harbeck, 2012).

SBS systems inspect particles to determine the value of some property using contactless and real-time measurements that obtain both location and material properties. The data are processed and the information (e.g., visible light reflectance) creates a basis for ejection (or retention) of those particles which meet some criterion (e.g., light vs. dark particles). Therefore, a distinct difference in the required physical property must exist between the valuable minerals and the gangue. Either valuables or waste may be selected for ejection. Automated sorters consist of

TABLE 14.1 Industrial Sensor Technologies and Applications

Sensor Type	Material Property	Example Applications
Radiometric	Natural gamma radiation	Uranium
X-ray Transmission	Atomic density	Base metals, coal
X-ray Fluorescence	Compositional analysis	Base/precious metals
X-ray Luminescence	Visual fluorescence	Diamonds
Color	Reflectance, brightness, transparency	Base/precious metals
Photometric	Monochromatic reflection/absorption	Gemstones
Near Infrared Spectrometry	Reflectance/absorption	Industrial minerals
Thermal Infrared Camera Electromagnetic	Differential heating Conductivity	Base/precious metals Base metals

Modified From Salter and Wyatt, 1991; Robben et al., 2013

four basic subsystems (Arvidson, 1987; Salter and Wyatt, 1991; von Ketelhodt, 2012; Robben et al., 2013):

1. Particle presentation
2. Sensing (particle examination/detection)
3. Electronic processing (data analysis)
4. Separation

Robben et al. (2013) include a material conditioning stage prior to particle presentation. Most important for successful operation is the presentation of a carefully selected and screened particle size range that shows both liberation and a minimum amount of fine material which is detrimental for high availability. For photometric sorting, particle surfaces sometimes must be moistened or washed, so that blurring of the signal by a covering layer does not occur. The upper size limit is technically 350 mm, but nonetheless liberation of barren waste is often experienced below 100 mm. The lower size limit is technically 0.5 mm for most detection technologies, but as the operating costs are inversely proportional to the average particle size (and weight), the economically feasible lower size limit is often in the range of 10–20 mm. Separation efficiency decreases when a wide range of particle sizes is fed to a single machine, the ratio between maximum and minimum particle size (*size range coefficient*) typically should not exceed three (Robben et al., 2013).

Particle presentation can be achieved via two system types: chute or belt-type systems (shown in Figure 14.3). The chute system senses the particles as they free-fall after being guided on a high-incline chute. For the belt system, the sensor is mounted above or below the conveyor belt, which feeds a monolayer of particles. The ore must be fed in a monolayer, as individual particles must be displayed to the sorting device for effective separation.

Various sensing technologies that are available and common industrial sensors are listed in Table 14.1. Electronic processing involves analysis of the data acquired by the detector. A wide range of site-specific algorithms are used, depending on the sensor type and ore characteristics. Physical separation is typically achieved using an array of about 200 high speed air valves. Mechanical ejectors are installed in low throughput units, for example, single particle XRL (X-ray luminescence) diamond or XRF (X-ray fluorescence) sorters. Water jets have been discussed and trialed, but have not found their way into today's industrial scale sorters (Fickling, 2011; Robben et al., 2013).

14.3 HISTORICAL DEVELOPMENT

First patents for SBS technology originated in the 1920s (Sweet, 1928). During the 1950s Kelly and Hunter (K + H) developed the Model 6 (M6) photometric sorter, which was subsequently installed at the Mary Kathleen uranium mine in Australia (Salter and Wyatt, 1991; Stewart, 1967). In 1966, Gold Fields of South Africa undertook a joint project with Rio Tinto-Zinc (RTZ) (through their subsidiary Ore Sorters) to develop sorting technology for use in South African gold-mining operations (Barton and Peverett, 1980). The project culminated with the installation of a Model 13 prototype photometric sorter at the Doornfontein Gold Mine in 1972 (Keys et al., 1974; Barton and Peverett, 1980). The sorters were the first to use laser technology and are considered the first high tonnage sorters (Salter and Wyatt, 1991). The Gunson's Sortex MP80 machine was probably the first sorter to employ microprocessor technology (Anon., 1980). The sorter handled minerals in the size range 10–150 mm at feed rates of up to 150 t h^{-1}.

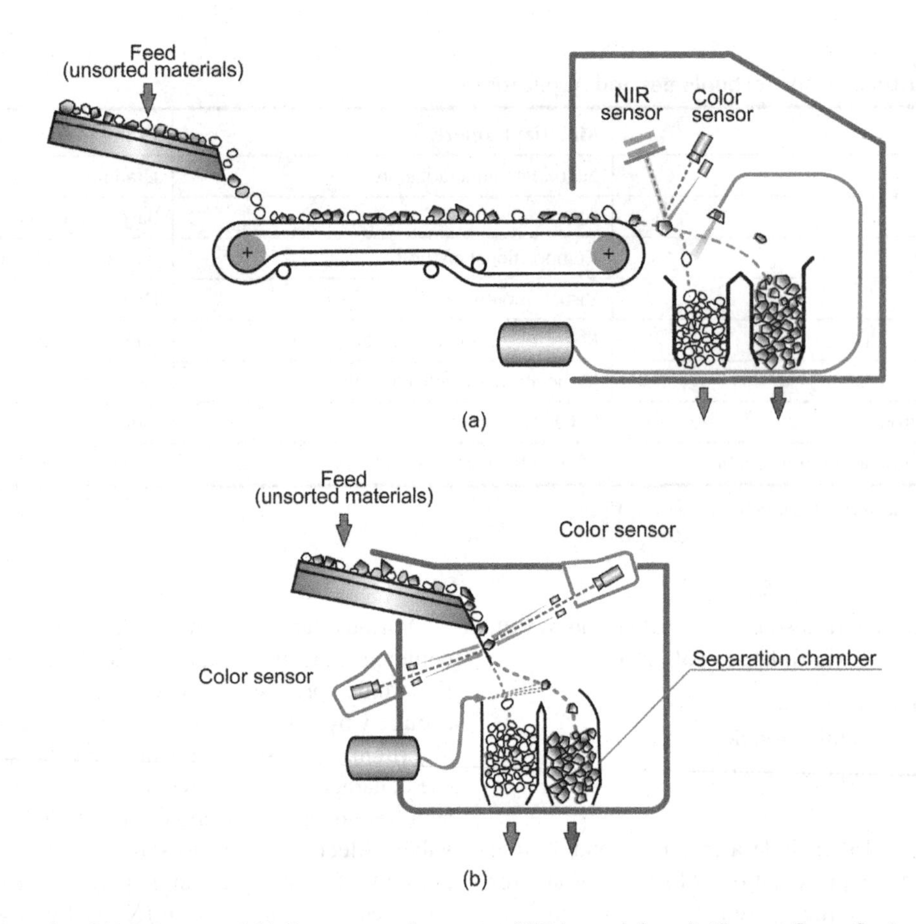

FIGURE 14.3 Schematics of (a) Belt-type, and (b) Chute-type sorting systems (NIR is near infra-red) *(Courtesy Tomra Sorting Solutions)*.

Rocks having white or gray quartz pebbles in a darker matrix were accepted, while quartzite ranging from light green through olive green to black, were rejected. Most of the gold occurred in rocks which reported to the "accept" category. Uniform distribution of the ore entering the sorter was achieved by the use of tandem vibrating feeders and the ore was washed on a second feeder to remove slimes which could affect light-reflecting qualities. The successful implementation of the Model 13 sorter led to the development of the RTZ Ore Sorters Model 16 photometric sorter, which has been used since 1976 on a wide range of ore types (e.g., magnesite, wolframite, gold) (Anon., 1981a; Barton and Peverett, 1980).

A subsequent development to the RTZ Ore Sorters Model 16 is the Ultrasort UFS120 photometric sorter, which is used in the processing of magnesite, feldspar, limestone, and talc. Ore passes from a vibrating feeder to high pressure water sprays and counterweight feeder where water is removed and the rocks are accelerated to form a monolayer. They drop onto a short conveyor moving at 2 m s^{-1} where they pass via a high speed 5 m s^{-1} "slinger" conveyor into free fall, now well separated. The rock layer, 0.8–1.2 m wide, is scanned by a laser beam at up to 4,000 times per second, and the reflection analyzed in less than 0.25 μs by photomultiplier tubes and high speed parallel processors. One or more of 120 air ejectors are fired to divert the value or waste past a cutter and into the accept/reject bins. As the position of the rock is accurately identified, and the ejector firing duration is less than 1 ms, the sorter can operate very selectively.

SBS has been employed in diamond recovery since the 1960s, initially using simple optical sorters and more recently machines based on the fact that diamonds luminesce when irradiated by X-rays (Anon, 1971; Rylatt and Popplewell, 1999a,b; Damjanović and Goode, 2000). X-ray luminescence sorters are used in almost all diamond operations for the final stages of recovery after the ore has been concentrated by DMS (Chapter 11). They replace grease separation (Taggart, 1945), which exploits the natural hydrophobicity (oleophilicity) of diamonds and is now used only in rare cases where the diamonds luminesce weakly or to audit the X-ray sorter tailings. Luminescence is a more consistent diamond property than oleophilicity, and sorters are more secure than grease belts or tables.

Figure 14.4 shows an early dry X-ray sorter, in which the DMS concentrates are exposed to a beam of X-rays in

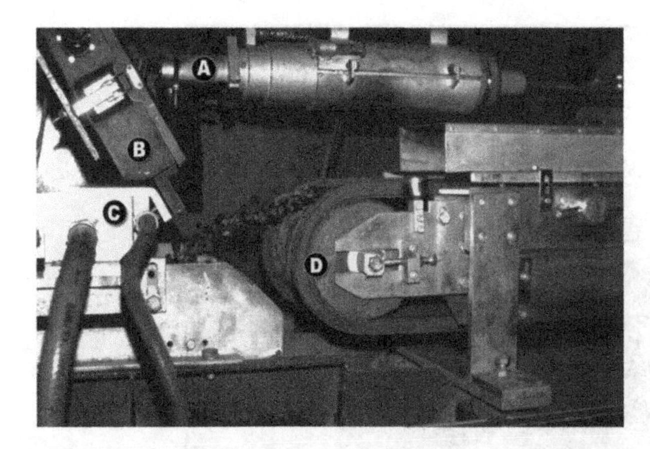

FIGURE 14.4 Early diamond sorter. A: X-Ray generator; B: Photomultiplier tubes; C: Air ejectors; D: Feed belt *(Courtesy JKMRC and JKTech Pty Ltd).*

free fall from a conveyor belt, the luminescence detected by photomultiplier tubes and the diamonds ejected by air ejectors. Both dry and wet X-ray machines are now available, and the process is usually multistage to ensure efficient rejection of waste with very high diamond recoveries.

The "Lapointe picker" was probably the earliest radiometric sorter, developed in Canada and used in the 1940s (Lapointe and Wilmot, 1952; Bettens and Lapointe, 1955; Salter and Wyatt, 1991). Radiometric sorting has since been used to pre-concentrate uranium ore in South Africa (Anon., 1981b), Namibia, Australia (Bibby, 1982), and Canada (IAEA, 1967, 1980, 1993). A sorter installed at the Rössing Mine in Namibia (Gordon and Heuer, 2000) detected gamma radiation from the higher grade ore pieces using scintillation counters comprising NaI crystals and photomultiplier tubes mounted under the belt. Lead shielding was used to achieve improved resolution of detection. A laser-based optical system similar to that used in photometric sorters was used to determine rock position and size for ejection, and could be adapted to determine additional optical characteristics of the rocks. Some uranium operations undertake selective mining practices, which could be considered ore sorting (IAEA, 1980, 1993). Radiometric readings are taken on truckloads of ore. Based on the readings, ore is placed in high-grade, low-grade, or waste stockpiles. The ore can then be appropriately blended or processed separately based on grade.

Several other physical properties of ores and minerals have been exploited in a range of sorting machines. Neutron absorption (or activation) separation has been used for the sorting of boron minerals (Mokrousov et al., 1975). The ore is delivered by a conveyor belt between a slow neutron source and a scintillation neutron detector. The neutron flux attenuation by the ore particles is detected and used as the means of sorting. The method is most applicable in the size range 25–150 mm. Boron minerals are easy to sort by neutron absorption since the neutron capture cross section of the boron atom is very large compared with those of common associated elements and thus the neutron absorption is almost proportional to the boron content of the particles. The technique has also been tested on gold ores (Uken et al., 1966, 1968).

Near infra-red (NIR), which has seen wide use in recycling applications, has recently been investigated for use in the mining industry. It has been successfully implemented for waste elimination from boron minerals (colemanite, ulexite) (Dehler et al., 2012). Development work has shown promise in separating talc from carbonate and quartz (von Ketelhodt and Bartram, 2009), processing porphyry copper samples (Dalm et al., 2014), and for waste rock removal in the diamond industry (von Ketelhodt and Bartram, 2014).

Photoneutron separation (gamma activation) is recommended for the sorting of beryllium ores. When a beryllium isotope in the mineral is exposed to gamma radiation of a certain energy, a photoneutron is released and this may be detected by scintillation or by a gas counter.

The RTZ Ore Sorters Model 19 sorter measured conductivity and magnetic properties and had application to a wide variety of ores including sulfides, oxides, and native metals (Anon., 1981a). The machine treated 25–150 mm rocks at up to $120\,\mathrm{t\,h^{-1}}$. Such systems employ a tuned coil under the belt which is influenced by the conductivity and/or magnetic susceptibility of the rocks in its proximity. Phase shift and amplitude are used to decide on acceptance or rejection.

Outokumpu developed the "Precon" sorter which was installed at the Hammaslahti copper mine (now closed) (Kennedy, 1985). The Precon sorter used gamma scattering analysis to evaluate the total metal content, and had a capacity of $7\,\mathrm{t\,h^{-1}}$ for 35 mm lumps rising to $40\,\mathrm{t\,h^{-1}}$ for 150 mm lumps. Primary crushed ore was pre-concentrated, rejecting about 25% as waste grading 0.2% copper, compared with an average feed grade of 1.2%.

RADOS XRF sorting technology has been used (primarily in Russia) over the past two decades at 49 operational sites for over 20 commodities (Fickling, 2011; RADOS, 2014). The sensor directly identifies ore elemental composition using X-ray fluorescence technology (able to detect elements with an atomic number >20) and sorts accordingly. Figure 14.5 shows the internal arrangement of the RADOS XRF chute-type sorter and a row of units operating at a copper/zinc mine in Russia. Particles free-fall from the chute past an X-ray source and detector. Particles are separated via a mechanical ejector into

FIGURE 14.5 Left: Internals of a RADOS XRF sorter (modified from Fickling, 2011), Right: Rados XRF sorter installation at Ural Mountains Mining Company's Svyatogor Cu/Zn Mine *(Courtesy RADOS International Technologies).*

concentrate and discard streams. Typically, feed grade must exceed 0.1 wt% for effective detection, although the system can also use a matrix of elements as the criteria for separation in low grade ore systems (e.g., gold, uranium or PGM) (RADOS, 2014).

Microwave attenuation has been used to sort diamond-bearing kimberlite from waste rock (Salter et al., 1989). The development was notable for the first use of high speed pulsed water ejectors. Equipment to sort asbestos ore has also been developed (Collier, 1972). The detection technique was based on the low thermal conductivity of asbestos fibers and used sequential heating and infrared scanning to detect the asbestos seams. The use of microwave heating coupled with infrared-thermography has been tested on a wide range of minerals (e.g., copper–molybdenum ore, iron ore) for sorting purposes (Sivamohan and Forssberg, 1991; van Weert et al., 2009; Ghosh et al., 2013).

A machine was installed at King Island Scheelite in Tasmania, where the scheelite was sensed by its fluorescence under ultraviolet radiation. XRT (X-ray transmission) has proven to be superior to UV fluorescence. Mittersill Mine currently operates XRT sorters for waste elimination where ca. 25% of the 130 $t\,h^{-1}$ run-of-mine stream grading 0.03% WO_3 is rejected before milling and sold as aggregate (Mosser and Gruber, 2010).

Rapid developments in sensing and computing technology have increased the capabilities of today's SBS machines significantly. In addition, industrial food processing and recycling industries have openly adopted the technology and tens of thousands of sorters are installed for numerous separation tasks. The mining industry is now benefiting from those developments, with proven detection and data processing being available for application on mining proof machines.

14.4 EXAMPLE FLOWSHEET AND ECONOMIC DRIVERS

SBS installations typically consist of: crusher, screen, sorter, and compressor. Because of the often large particle sizes and low concentrations of valuables, it is advised to install mechanical sampling, size reduction and splitting equipment to enable determination of plant performance.

Figure 14.6 shows an example flowsheet for a multi-stage, multi-machine stage operation. Machines are commonly operated in parallel for high throughput operations. Cascading circuit arrangements enhance separation efficiency and allow for redundancy and the potential for multiple detection technologies to be combined.

Capital and operating costs are typically roughly half the costs of dense-media systems. Since SBS is a single particle technology, the throughput is inversely proportional to the average particle size fed to the machine. Both types of specific costs are thus inversely dependent upon the average particle size. The upper particle size is determined by the liberation characteristics of the ore and the lower particle size limit by the economically viable specific costs per ton of product.

Typically, the highest economic benefit is obtained when eliminating deleterious waste that hinders downstream processes. In such cases, the overall recovery

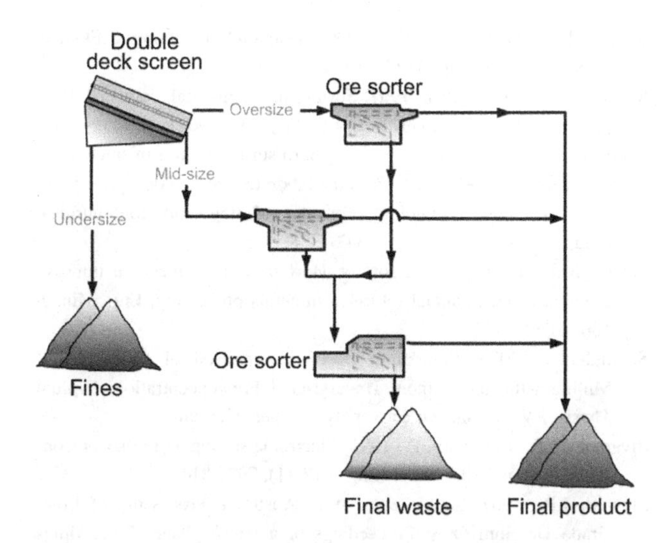

FIGURE 14.6 Example rougher-scavenger flowsheet for waste elimination from base and precious metals run-of-mine ore.

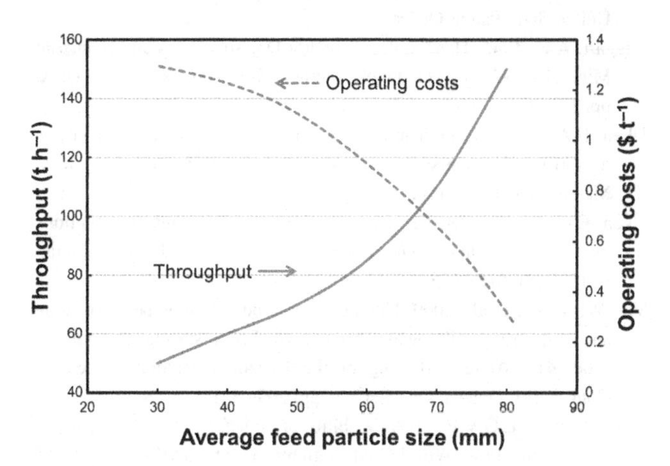

FIGURE 14.7 Estimation of throughput and operating costs depending on the average particle feed size. *(Courtesy C. Robben).*

(and potentially product quality) increases while reducing the specific costs per ton processed, and reducing the comminution and processing of waste material. Eliminating waste from the ROM stream also allows for higher dilution in the mine, which may allow for processing of marginal waste dumps or mining blocks (i.e., allowing for decreased cut-off grades and extended mine life) (Bamber et al., 2004; Pretz et al., 2011; Robben, 2013) (Figure 14.7).

In summary, SBS is a relatively low cost and versatile technology that has the potential for wide applicability in the minerals industry, if liberation of either waste or product is sufficient in coarse particle sizes. The currently available suite of detection technologies allows for discrimination of single particles based on a pre-determined discrimination criteria (e.g., NIR spectrum, color, conductivity, atomic density). Today's mining machines integrate state of the art technology into mechanical designs. Increased pressure to decrease processing costs and environmental impact will undoubtedly further the use of SBS in the mining industry.

REFERENCES

Agricola, G., 1556 De Re Metallica Libri XII. Wiebaden, Fourier Verlag GmbH 2003.

Anon, 1971. New generation of diamond recovery machines developed in South Africa. S. Afr. Min. Eng. J.(May), 17–18.

Anon, 1980. Micro-processor speeds optical sorting of industrial minerals. Mine and Quarry. 9 (Mar), 48–49.

Anon, 1981a. Photometric ore sorting. World Mining. 34 (4), 42–47.

Anon, 1981b. New ore sorting system. Mining J. 296-297 (Dec.), 446–447.

Arvidson, B., 2002. Photometric ore sorting. In: Mular, A.L., et al., (Eds.), Mineral Processing Plant Design, Practice and Control, vol. 2. SME, Littleton, CO, USA, pp. 1033–1048.

Arvidson, B., 1987. Economics and Technical Features of Preconcentration Using Sorting. *Proc. SME Annual Meeting*. SME, Denver, CO, USA, Preprinter: 87-125: 1-22.

Bamber, A., et al., 2004. Reducing selectivity in narrow-vein mining through the integration of underground pre-concentration. Proc. 15th International Symposium on Narrow Vein Mining Techniques, CANMET, Canada, pp. 1–12.

Barton, P.J., Peverett, N.F., 1980. Automated sorting on a South African gold mine. J. S. Afr. Inst. Min. Metall. 80 (3), 103–111.

Bettens, A.H., Lapointe, C.M., 1955. Electronic concentration of low grade ores with the Lapointe picker. Department of Mines Branch Technical Surveys (No. NP-5508; TR-123/54), Queen's Printer, Ottawa, ON, Canada.

Bibby, P.A., 1982. Preconcentration by radiometric ore sorting. Proc. 1st Mill Operators' Conf., AusIMM, Melbourne, North West Queensland, Australia, pp. 193–201.

Collier, D., 1972. Ore sorters for asbestos and scheelite. Proc. International Conference on Mining and Metallurgy, vol. 19, No. 72, London, UK, pp. 1007–1022.

Collins, D.N., Bonney, C.F., 1998. Separation of coarse particles (> 1 mm). Int. Min. Miner., (IM&M). 1-2 (Apr.), 104–112.

Cutmore, N.G., Ebehardt, J.E., 2002. The future of ore sorting in sustainable processing. *Proc. Green Process*. AusIMM, Cairns, Australia, pp. 287–289.

Dalm, M., et al., 2014. Application of near-infrared spectroscopy to sensor based sorting of a porphyry copper ore. Miner. Eng. 58, 7–16.

Damjanović, B., Goode, J.R. (Eds.), 2000. *Canadian Milling Practice*. Special, vol. 49. CIM, Montreal, QC, Canada.

Dammers, M., et al., 2013. Synergies and potentials of near-to-face processing–An integrated study on the effects of mining processes and primary resource efficiency. *Proc. Aachen* International Mining Symposia (AIMS), Aachen, Germany, pp. 283–297.

Dehler, M., et al., 2012. NIR versus color sorting of industrial minerals. *Proc. Sensor-Based Sorting Conf.*, Aachen, Germany.

Ednie, H., 2006. Celebrating a century of excellence: Teck Cominco's trail operations. CIM Mag. 1 (3), 35–43.

Fickling, R.S., 2011. An introduction to the RADOS XRF ore sorter. Proc. 6th South Africa Base Metals Conf., SAIMM, Phalaborwa, South Africa, pp. 99–110.

Ghosh, A., et al., 2013. A non-invasive technique for sorting alumina-rich iron ores. Miner. Eng. 45, 55–58.

Gordon, H.P., Heuer, T., 2000. New age radiometric sorting-the elegant solution. In: Ozberk, E., Oliver, A.J. (Eds.), Proc. Uranium 2000: Int. Symposium on the Process Metallurgy of Uranium. CIM, Saskatoon, SK, Canada. 35(43): 323–338.

Hughes, T., Cormack, G., 2008. Potential benefits of underground processing for the gold sector-conceptual process design and cost benefits. Proc. 1st International Future Mining Conf., Sydney, Australia, pp. 135–142.

IAEA (International Atomic Energy Agency)., 1967. *Processing of Low-Grade Uranium Ores:* Proceedings of a Panel Held in Vienna, Austria.

IAEA (International Atomic Energy Agency)., 1980. Significance of Mineralogy in the Development of Flowsheets for Processing Uranium Ores: Technical report. Series No. 196, Vienna, Austria.

IAEA (International Atomic Energy Agency)., 1993. Uranium Extraction Technology: Technical report Series No. 359, Vienna, Austria.

Kennedy, A., 1985. Mineral processing developments at Hammaslahti, Finland. Mining Mag. 152 (2), 122–125.

Keys, N.J., et al., 1974. Photometric sorting of ore on a South African gold mine. J. S. Afr. Inst. Min. Metall. 75 (2), 13–21.

Lapointe, C.M., Wilmot, R.D., 1952. Electronic concentration of ores with the Lapointe belt picker. Department of Mines and Technical Surveys, Memo No. 123 (No. NP-4384). Canada.

Lessard, J., et al., 2014. Developments of ore sorting and its impact on mineral processing economics. Miner. Eng. 65, 88–97.

Mokrousov, V.A., et al., 1975. Neutron-radiometric processes for ore beneficiation. Proc. 11th International Mineral Processing Congr., Cagliari, Italy.

Mosser, A., Gruber, H., 2010. Operational experience with a sensor based sorting system. Proc. Sensor-Based Sorting Conf., Aachen, Gesellschaft für Bergbau, metallurgie, Rostoff- und Umweltechnik.

Murphy, B., et al., 2012. Underground preconcentration by ore sorting and coarse gravity separation. Proc. Narrow Vein Mining Conf., Perth, West Australia, Australia, pp. 237–244.

Pretz, T., et al., 2011. Applications of Sensor-based Sorting in Raw Material Industry. Shaker Verlag GmbH (Publisher), Germany.

Rados, 2014. RADOS XRF Sorters. <http://www.radosxrf.com/index.htm>. Retrieved Feb. 2014.

Rickard, T.A. (Ed.), 1905. The Economics of Mining. 1st Edn., Eng. Min. J., New York, NY, USA, pp. 8–10.

Robben, C., et al., 2013. Potential of sensor-based sorting for the gold mining industry. CIM J. 4 (3), 191–200.

Robben, C., 2014. Characteristics of Sensor-based Sorting Technology and Implementation in Mining. Shaker verlag GmbH (Publisher), Aachen, Germany.

Rylatt, M.G., Popplewell, G.M., 1999a. Ekati diamond mine-background and development. Mining Eng. 51 (1), 37–43.

Rylatt, M.G., Popplewell, G.M., 1999b. Diamond processing at Ekati in Canada. Mining Eng. 51 (2), 19–25.

Salter, J.D., Wyatt, N.P.G., 1991. Sorting in the minerals industry: Past, present and future. Miner. Eng. 4 (7-11), 779–796.

Salter, J.D., et al., 1989. Kimberlite-gabbro sorting by use of microwave attenuation: Development from the laboratory to a 100t/h pilot plant. Proc. Today's Technology for the Mining and Metallurgical Industries. Kyoto, Japan, pp. 347–358.

Sassos, M.P., 1985. Mineral sorters: How these machines can improve efficiency and financial return in minerals processing. Eng. Min. J. 186 (6), 68–72.

Schindler, I., 2003. Simulation Based Comparison of Cut and Fill Mining with and without Underground Pre-concentration. Diplom Thesis, RWTH Aachen University, Aachen, Germany.

Sivamohan, R., Forssberg, E., 1991. Electronic sorting and other preconcentration methods. Miner. Eng. 4 (7-11), 797–814.

Stewart, J.R., 1967. Status report from Australia. Processing of Low-Grade Uranium Ores: Proceedings of a Panel. (Panel Proceedings Series), International Atomic Energy Agency, Vienna, Austria, pp. 3–8.

Sweet, A.T., 1928. Metallurgical Separator. Patent No. 1678884 A. United State Patent Office.

Taggart, A.F., 1945. Handbook of Mineral Dressing: Ores and Industrial Minerals. Wiley Handbook Series, New York, NY, USA, pp. 12–50.

Uken, E.A., et al., 1966. The application of neutron activation analysis to sorting Witwatersrand gold-bearing ores. J. S. Afr. Inst. Min. Metall. 67 (3), 99–114.

Uken, E.A., et al., 1968. The determination of gold in ores and solutions by activation analysis with a neutron generator. Int. J. Appl. Radiat. Isot. 19 (8), 615–623.

Van Weert, G., et al., 2009. Upgrading molybdenite ores between mine and mill using microwave/infrared (MW/IR) sorting technology. Proc. 41st Annual Meeting of the Canadian Mineral Processors Conf., CIM, Ottawa, ON, Canada, pp. 509–521.

von Ketelhodt, L.G.V.F., 2009. Viability of optical sorting of gold waste rock dumps. Proc. World Gold Conference 2009. SAIMM, Cradle of Humankind, Gauteng, South Africa, pp. 271–278.

von Ketelhodt, L.G.V.F., 2012. Beneficiation of Witwatersrand Type Gold Ores by Means of Optical Sorting. M.Sc. Thesis, University of Witwatersrand, South Africa.

von Ketelhodt, L., Bartram, K., 2009. New Developments in Sensor-based Sorting. In: Malhotra, D., et al., (Eds.), Recent Advances in Mineral Processing Plant Design. SME, Littleton, CO, USA, pp. 476–489.

von Ketelhodt, L., Bartram, K., 2014. Diamond processing using sensor based sorting. GSSA Kimberley Diamond Symposium & Trade Show, Kimberley, South Africa, pp. 1–5.

Wotruba, H., 2006. Sensor sorting technology-is the minerals industry missing a chance. (Plenary Lecture). Proc. 23rd International Mineral Processing Congr., Istanbul, Turkey, pp. 21–30.

Wotruba, H., Harbeck, H., 2012. Sensor-based sorting. Ullmann's Encyclopedia of Industrial Chemistry. Wiley-VCH Verlag GmbH & Co. KgaA, Weinheim, Germany.

Chapter 15

Dewatering

15.1 INTRODUCTION

With few exceptions, most mineral-separation processes involve the use of substantial quantities of water and the final concentrate has to be separated from a pulp in which the water–solids ratio may be high. Dewatering, or solid–liquid separation, produces a relatively dry concentrate for shipment. Partial dewatering is also performed at various stages in the concentrator, so as to prepare the feed for subsequent processes. Sometimes tailings are also de-watered (Chapter 16).

Dewatering methods can be broadly classified into three groups: sedimentation (gravity and centrifugal), filtration, and thermal drying. Dewatering in mineral processing is normally a combination of these methods, an example being Figure 15.1. The bulk of the water is first removed by thickening, which produces a thickened pulp of perhaps 55–65% solids by weight. Up to 80% of the water can be separated at this stage. Filtration of the thickened pulp then produces a moist filter cake of between 80% and 90% solids, which may require thermal drying to produce a final product of about 95% solids by weight.

The principles of solid–liquid separation, testing, equipment sizing, and operation are covered in detail by Concha (2014) and Mular et al. (2002); innovations are discussed by McCaslin et al. (2014), and specifically in coal preparation (washing) plants by Luttrell (2014).

15.2 GRAVITATIONAL SEDIMENTATION

Gravity sedimentation or *thickening* is the most widely applied dewatering technique in mineral processing, and it is a relatively cheap, high-capacity process. The *thickener* is used to increase the concentration of the suspension by sedimentation, accompanied by the formation of a clear liquid (*supernatant*). The principal type of thickener consists of a cylindrical, largely open tank from which the clear liquid is taken off at the top (thickener "overflow"), and the suspension is transported by rotating rakes to discharge at the bottom ("underflow") (Schoenbrunn and Laros, 2002). These are *conventional thickeners* and their variants, and they form the main content of this section of the chapter. Other sedimentation devices, including centrifuges, are briefly covered at the end of the section.

15.2.1 Sedimentation of Particles

Sedimentation is most efficient when there is a large density difference between liquid and solid. This is always the case in mineral processing where the carrier liquid is water. Sedimentation cannot always be applied in hydrometallurgical processes, however, because in some cases the carrier liquid may be a high-grade leach liquor having a density approaching that of the solids.

Settling of solid particles in a liquid produces a clarified liquid which can be decanted, leaving a thickened slurry. The process is illustrated in Figure 15.2 using the common laboratory test, that is, *batch sedimentation* in a graduated cylinder (*cylinder test* or *jar test*).

The settling rates of particles in a fluid are governed by Stokes' or Newton's laws (Chapter 9). Factors that affect sedimentation include: particles size and shape, weight and volume content of solids, fluid viscosity, and specific gravity of solids and liquid. Very fine particles, of only a few micrometers diameter, settle extremely slowly by gravity alone, and centrifugal sedimentation may have to be performed. Alternatively, the particles may be aggregated or *flocculated*, into relatively large clumps, called *flocs*, that settle out more rapidly (Section 15.2.2).

Batch settling tests are undertaken to size thickeners, which operate continuously, and to test performance of operating thickeners. The data for analysis are presented as solid–liquid interface height versus time (Figure 15.3).

The start of the settling curve represents the interface (or *mud-line*) between zone A, clear water (clarified *supernatant*), and zone B at the initial concentration; the settling rate is nearly constant. The *critical sedimentation point* indicates the loss of zone B and the interface now corresponds to that between zone A and the variable concentration zone C. The settling rate now decreases as the concentration in zone C increases, hindering the settling process. When zone

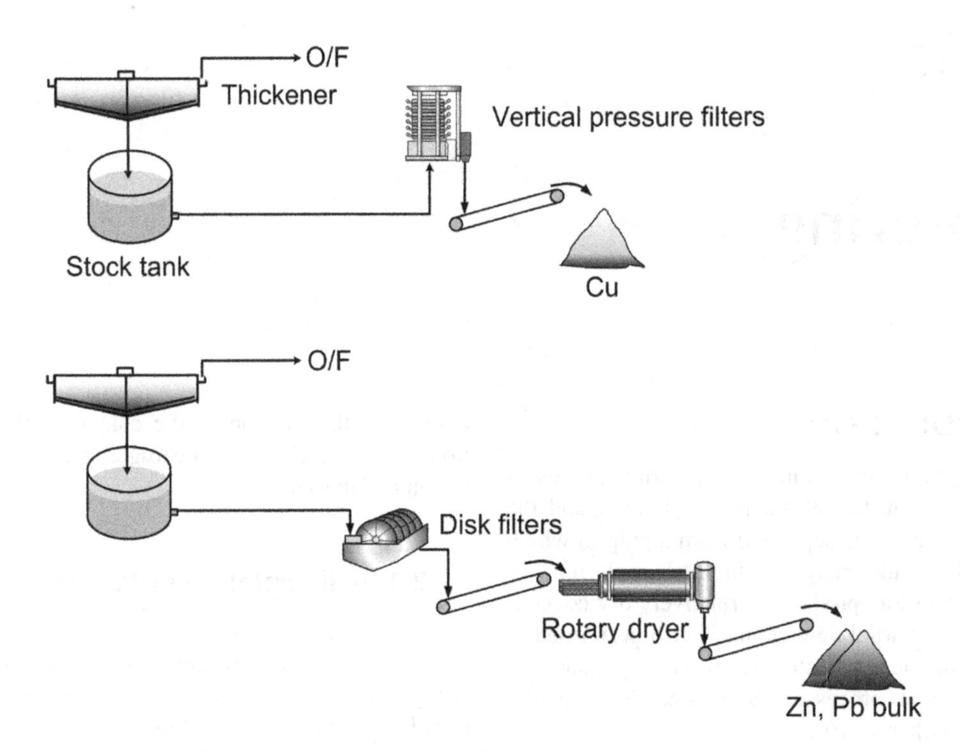

FIGURE 15.1 Example of thickening, filtering, and drying (Brunswick Mine concentrator).

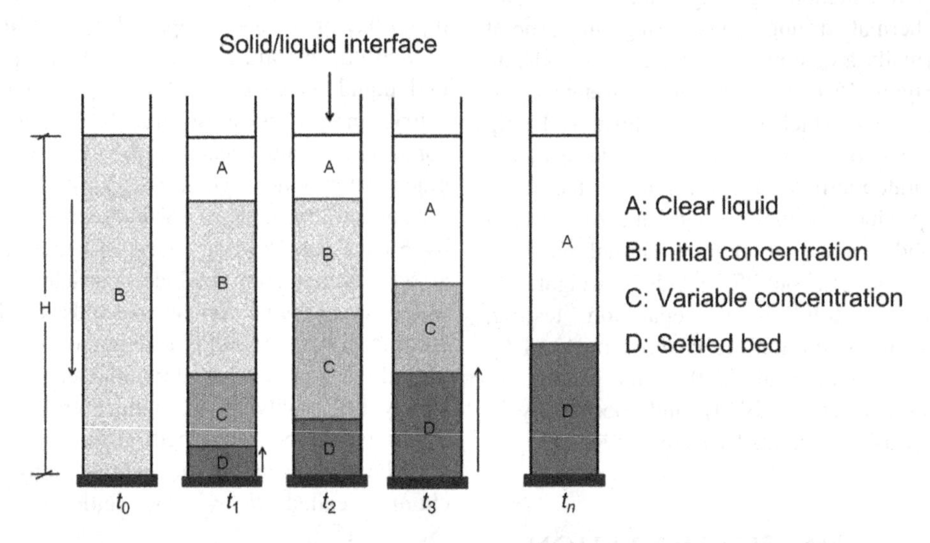

FIGURE 15.2 Batch settling process (t_i is time).

C is lost, the settling rate approaches zero, the interface only slowly decreasing as water is squeezed out by the weight of the particles compressing the bed. The settling curve should be independent of vessel geometry, avoiding, for example, wall effects by using a too small diameter cylinder. Concha (2014) describes a setup to measure settling characteristics in five cylinders simultaneously (*SediRack*).

Additional detail on suspension settling behavior can be obtained by measuring local solids concentration in zones C and D using, for example, mobile radiation sources and detectors (X-rays, gamma-rays) with appropriate calibrations (Owen et al., 2002; Kurt, 2006). These data can reveal channeling in the settling bed. In the case of flocculated suspensions, other parameters that can be determined include *compressibility* and the *gel point*, defined as the concentration at which flocs come into contact and start to form a self-supporting network structure. These parameters are used in theories of flocculated suspensions (Buscall and White, 1987). An array of conductivity sensors along a settling cylinder provides data on the concentration profile of the settled bed (Concha, 2014).

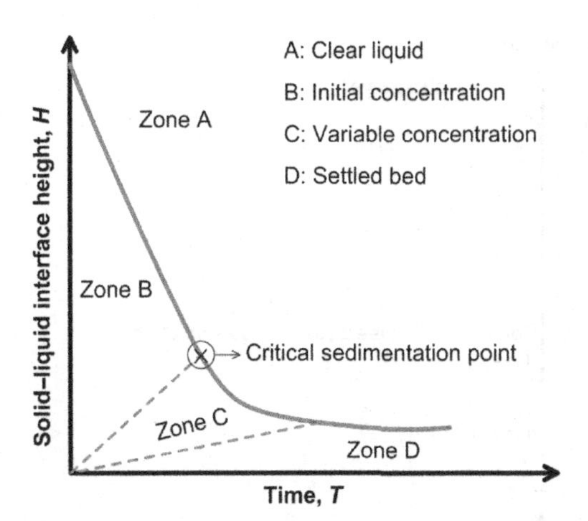

FIGURE 15.3 Example batch settling curve (interface or mud-line vs. time).

15.2.2 Particle Aggregation: Coagulation and Flocculation

Fine particles, say $-10\ \mu m$, settle slowly under gravity. All particles exert mutual attractive forces, known as *London-Van der Waals'* forces. Normally these attractive forces are opposed by the charge on the particle surface that originates from a variety of mechanisms (one was depicted in Chapter 12, Figure 12.10). *Coagulation* and *flocculation* refer to processes that cause particles to aggregate or agglomerate (i.e., adhere or cluster together) to increase settling rates. The cluster is variously called an aggregate, agglomerate, or more commonly a *floc*. Often used synonymously, coagulation is associated with modification of particle surface charge to cause aggregation, while flocculation involves addition of long-chain polymers that bind ("bridge") particles together. This difference leads to two classes of aggregating reagents: *coagulants* and *flocculants*.

Surface Charge

Particles in water always exhibit a surface charge. In a given system the electrical charge on the particle surfaces will be of the same sign and this causes mutual repulsion, which slows settling by keeping particles apart and in constant motion. In mineral processing most aqueous systems are alkaline and the charge is usually taken to be negative; a negative charge will be assumed in the discussion that follows. The charge at the particle surface affects the distribution of solute ions nearby (Figure 15.4 (a)). The solute ions form two layers: an inner layer that comprises an excess of cations (assuming the surface is negatively charged) more or less bound to the surface, the *bound layer* (or *Stern layer*); and a second layer that contains more loosely attracted ions, the *diffuse layer*. This depiction is referred to as the *electrical double layer model*. The charge is normally expressed as a potential. Figure 15.4(a) illustrates the distribution of ions (upper) and the corresponding electrical potential versus distance from the particle surface (lower).

The potential decays with distance from the surface (*surface potential*) to reach zero at the bulk solution. Since suspension behavior involves motion of the particle relative to the water, a layer of water is associated with the moving particle. The relevant potential then is not the surface potential, but that at the plane of shear (the boundary between the water layer carried by the particle and the bulk water), termed the *zeta potential*. The plane of shear is sometimes taken to coincide with the boundary between the Stern and diffuse layers. The zeta potential is measured by the common methods that track the motion of the particle under the influence of an electrical field (*electrophoresis*).

Zeta potential values range up to about 200 mV. A zeta potential of 20 mV or more (negative or positive) will tend to keep fine particles dispersed by electrostatic repulsion. There are two ways the charge can be manipulated to cause aggregation: reducing the zeta potential to near zero to remove the repulsive force (*charge neutralization*); or creating conditions to cause electrostatic attraction (*electrostatic coagulation*).

Charge Neutralization

In Chapter 12 we learned that zeta potential can be reduced to zero by adjusting pH to the iso-electric point (IEP). In fact, determining the pH giving maximum settling rate can be used to identify the IEP for some minerals. Altering pH is not usually a practical option and charge neutralization is achieved instead by adding multivalent cations, such as Fe^{3+}, Al^{3+}, and Ca^{2+} to neutralize the charge. The common reagents, known as *inorganic coagulants*, are aluminum sulfate (Alum) ($Al_2(SO_4)_3$), ferric sulfate ($Fe_2(SO_4)_3$), ferric chloride ($FeCl_3$), and lime ($Ca(OH)_2$).

Figure 15.4(b) illustrates the effect on the rate of potential decay by introducing multivalent ions, using Ca^{2+} ions as the example. The double charge on the cation causes the rate of potential decay to increase compared to monovalent ions, and thus the zeta potential is decreased. At sufficient Ca^{2+} concentration the zeta potential reduces to zero and electrostatic repulsion is lost and the system aggregates (in practice a value $< \pm 20$ mV usually suffices). This concentration is the called the *critical coagulation concentration* (CCC). The effect on increasing rate of decay of the potential is referred to as *double layer compression*. Sufficiently high concentration of monovalent ions can achieve double layer compression and cause coagulation, a situation that may be encountered in plants using seawater or bore water.

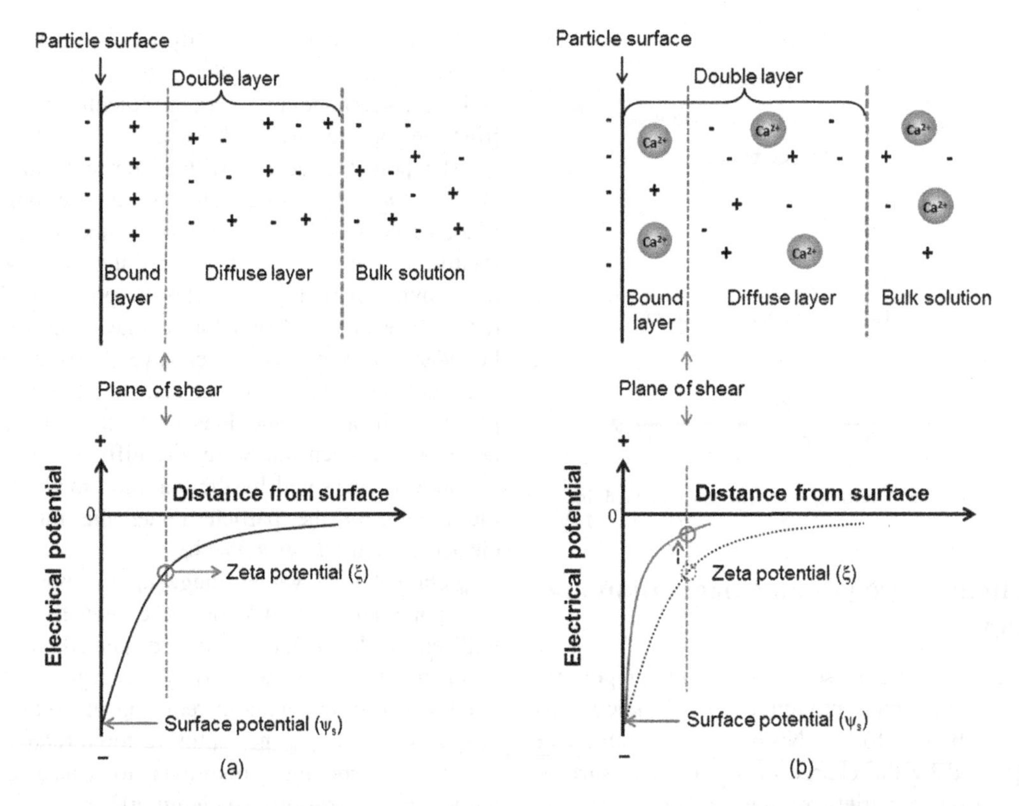

FIGURE 15.4 Distribution of ions at a charged surface (upper) and potential as a function of distance from negatively charged surface (lower): (a) dilute solution of monovalent ions, and (b) after adding Ca^{2+} ions (as lime).

The CCC (not to be confused with the previous definition in Chapter 12) is strongly dependent on the charge on the ion, thus Fe^{3+} and Al^{3+} ions are more effective (have lower CCCs) than Ca^{2+} which in turn is more effective than, say, Na^{+}. The action of ferric and aluminum salts, however, also involves formation of hydroxide precipitates (at pH $>$ 4–5) that collect particles as they settle, known as *sweep flocculation*. The Fe^{3+} and Al^{3+} coagulants are best employed when suspended solids is low. Somasundaran and Wang (2006) list the CCC of some coagulants. The aggregation mechanism is reversible; a Ca^{2+} concentration well above the CCC will create positive surface charge (positive zeta potential) and cause re-dispersion. (Charge reversal was illustrated in Chapter 12, Figure 12.27.)

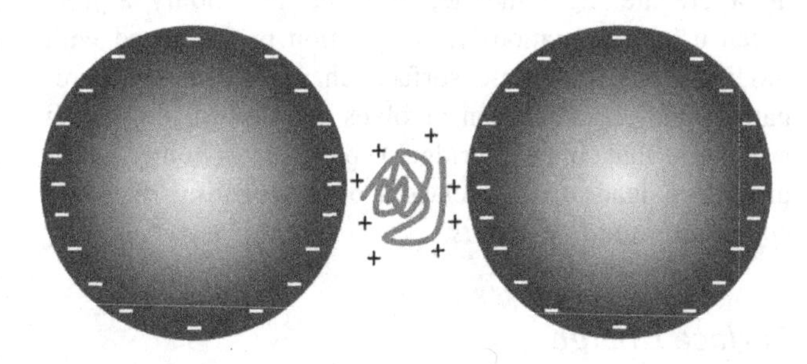

FIGURE 15.5 Coagulation through electrostatic attraction (electrostatic coagulation).

Electrostatic Coagulation

Coagulation by electrostatic attraction occurs when two particle types have opposite charge and there is mutual attraction. An example of silica and iron oxide minerals was discussed in Chapter 12. Talc presents an example of a single mineral subject to electrostatic coagulation, as the charge on the basal plane (face) is usually negative and the charge on the edge (at least below about pH 10) is positive and aggregation by a face-edge ("house-of-cards") arrangement occurs. Electrostatic coagulation in the present case is achieved by adding *organic coagulants*. These are polymers of low molecular weight (3,000 to 1 million) with cationic functional groups. Adsorption creates regions of positive charge ("patches") that promote electrostatic attraction with the polymer-free negative regions on other particles, as depicted in Figure 15.5. Also known as *patch flocculation*, like inorganic coagulants, over-dosing can cause re-dispersion, as the patches spread to cover the surface and the charge becomes uniformly positive.

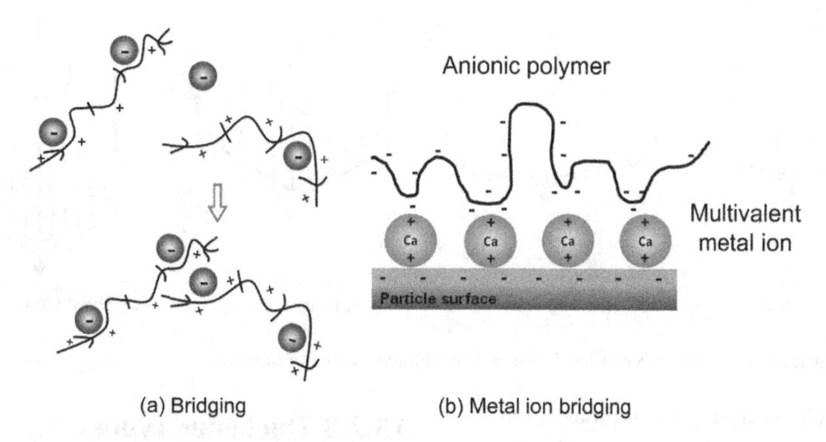

FIGURE 15.6 Flocculation mechanisms using organic polymers: (a) bridging by high MW cationic polymer, and (b) co-use of Ca^{2+} adsorption and bridging by anionic polymer.

Three families of organic coagulants are primarily used: polyamines, polyDADMAC, and dicyandiamide. More complete lists are given by Somasundaran and Wang (2006) and Bulatovic (2007).

Flocculation

Creating a more open structure than coagulation, flocculation involves the use of long-chain organic polymers of high molecular weight (>1 million) to form molecular bridges between particles (Hogg, 2000; Pearse, 2003; Tripathy and De, 2006; Usher et al., 2009). Formerly natural products such as starch derivatives and polysaccharides, they are now increasingly synthetic materials, loosely termed *polyelectrolytes*. Bridging is illustrated in Figure 15.6(a) for a cationic polymer. In practice, many such inter-particle bridges are formed, linking a number of particles together. Figure 15.6(b) shows the co-use of coagulant (Ca^{2+}) and an anionic polymer. The length of a completely uncoiled polymer is about 1 μm, up to perhaps a few tens of μm for the longest chains. Bridging with other particles stops when the floc size reaches about 10 mm (Rushton et al., 2000).

It would be expected that, since most suspensions encountered in the minerals industry contain negatively charged particles, cationic polyelectrolytes would be most suitable. Although this gives some level of charge neutralization, and aids attraction of the polymer to the particle surface, it is not necessarily true for the bridging role of the flocculant. For bridging, the polymer must be strongly attached, and this is promoted by chemical adsorption (formation of chemical bonds) through, for example, amide groups (e.g., $CONH_2$). In other words, the charge on the polymer is less important and the majority of commercially available polyelectrolytes are anionic, since these tend to be of higher molecular weight than the cationics, and are less expensive.

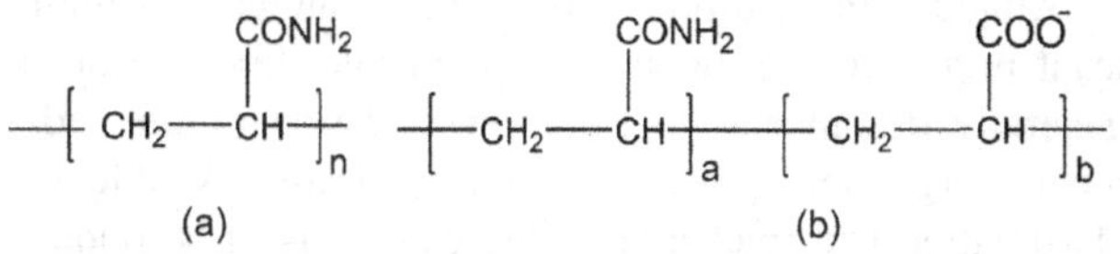

$$\left[CH_2 - \underset{\displaystyle |}{\overset{CONH_2}{CH}} \right]_n \qquad \left[CH_2 - \overset{CONH_2}{CH} \right]_a \left[CH_2 - \overset{COO^-}{CH} \right]_b$$

(a) (b)

FIGURE 15.7 Polyacrylamide flocculants: (a) uncharged polymer comprising n acrylamide monomers, and (b) an anionic polymer made by co-polymerization with acrylic acid.

Bridging requires that the polymer be strongly bonded to one particle, but have other bonding sites available for other particles. Excess polymer tends to adsorb on one particle and this can promote re-dispersion. The optimum polymer concentration and pH requires laboratory testing. Physical conditions, such as agitation and pumping, can also cause re-dispersion by breaking down the flocs. This is one reason why flocculants are generally not used with hydrocyclones and have limited application in centrifugal de-watering. In comparison, aggregates produced by coagulation will reform after disruption.

The polyacrylamides (abbreviated as PAM) are the most common flocculants. They can be manufactured to have non-ionic, anionic, or cationic character. Polyacrylamide formed from the acrylamide monomer ($-CH_2CHCONH_2$) is non-ionic (Figure 15.7(a)). (Being made of only one monomer type it is a *homopolymer*.) An anionic polymer can be made by hydrolysis of the non-ionic PAM, but more usually by co-polymerizing acrylamide with acrylic acid ($CH_2{=}CHCOOH$) monomer (i.e., making a *co-polymer*), where some of the side chains lose a proton to become negatively charged (Figure 15.7(b)). A cationic polymer is made by co-polymerizing with a cationic monomer.

The manufacture gives a certain degree of ionic character or *charge density*, which refers to the percentage of the monomer segments that carry a charge. It is now possible to obtain water soluble PAM products with a wide range of ionic character, varying from 100% cationic content through non-ionic to 100% anionic content

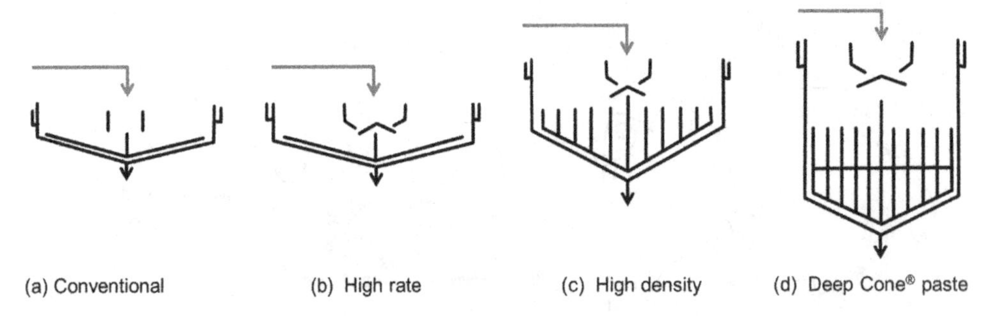

FIGURE 15.8 Types of thickener equipment (Deep Cone® is a registered trademark of FLSmidth).

and with molecular weights from several thousand to over 10 million, the highest among the synthesized polymers. Flocculants other than PAM (e.g., polyethylene—imines, polyamides—amines) are used under special conditions.

Although the addition of flocculants can lead to significant improvements in sedimentation rate, flocculation is generally detrimental to final consolidation of the sediment. Large flocs promote settling and are desirable for clarification and thickening. Floc density is of secondary importance in these processes. Conversely, dense flocs are most appropriate for consolidation of the sediment, and size is of lesser importance in this stage. Therefore the optimization of solid–liquid separation processes requires careful control of floc size and structure. If thickening is followed by filtration, the choice of flocculant may be important. Flocculants are widely used as *filter aids*. However, the specific requirements of a flocculant used to promote sedimentation are not necessarily the same as for one used as a filter aid; for example, flocs formed with high molecular weight products are relatively large, trapping water within the structure and increasing the final moisture content of the filter cake.

Laboratory batch cylinder tests are commonly used to assess the effectiveness of flocculants. Reproducibility of such tests is often poor and there is almost always conflict in determining an optimum dosage of flocculant between full-scale operations and laboratory tests (Scales et al., 2015). Methods of improving reproducibility and better approximating conditions in the thickener, such as shear rate, continue to be developed (Farrow and Swift, 1996; Scales et al., 2015; Parsapour et al., 2014).

In practice, polyelectrolytes are normally made up of stock solutions of about 0.5–1%, which are diluted to about 0.1% (maximum) before adding to the slurry. The diluted solution must be added at enough points in the stream to ensure its contact with every portion of the system. Agitation is essential at the addition points, and shortly thereafter, to assist in flocculant dispersion in the process stream. Care should be taken to avoid severe agitation after the flocs have been formed. The age of the stock solution can have a significant effect on flocculant performance (Owen et al., 2002).

15.2.3 Thickener Types

There are four types of thickener that have a more or less open tank design (Concha, 2014), shown in Figure 15.8: conventional (a), high rate (b), high density (c), and paste (d) thickeners. High rate (or *high capacity*) is the term applied to thickeners processing very high throughput by optimization of flocculation. In that regard, it may be more appropriate to talk of different thickener operation rather than thickener type. *High density* and *paste thickeners* are similar to conventional thickeners, but with steeper cone angles and higher sided tanks. The extra height increases the pressure on the sediment bed and thus gives higher density underflow. Applications of paste thickeners include mine backfill and tailings disposal (Chapter 16). The *clarifier* is similar in design, but is less robust, handling suspensions of much lower solid content than the thickener and designed for removal of solids rather than their compaction (Seifert and Bowersox, 1990). Given the basic similarity in design, just the conventional thickener and its variants will be described in any detail.

The *conventional thickener* consists of a cylindrical tank, the diameter ranging from about 2 to 200 m in diameter, and of depth 1–7 m. The clarified liquid overflows a peripheral launder, while the solids, which settle over the entire bottom of the tank, are withdrawn as a thickened pulp from an outlet at the center. The zones in the thickener Figure 15.9 mirror those recognized in the batch cylinder sedimentation test (Figure 15.2).

Pulp is fed into the center via a *feedwell* placed up to 1 m below the surface. The feedwell is a small concentric cylinder with several key functions, including (Loan et al., 2009; Owen et al., 2009; Lake and Summerhays, 2012): controlling momentum dissipation, de-aerating feed slurry, diluting feed (if required), optimizing flocculation, and ensuring even distribution of the feed stream into the thickener. Within the tank are one or more rotating radial arms, from each of which are suspended a series of blades, shaped so as to rake the settled solids toward the central outlet. On most thickeners today these arms rise automatically if the torque exceeds a certain value, thus preventing damage due to overloading. The blades also assist the

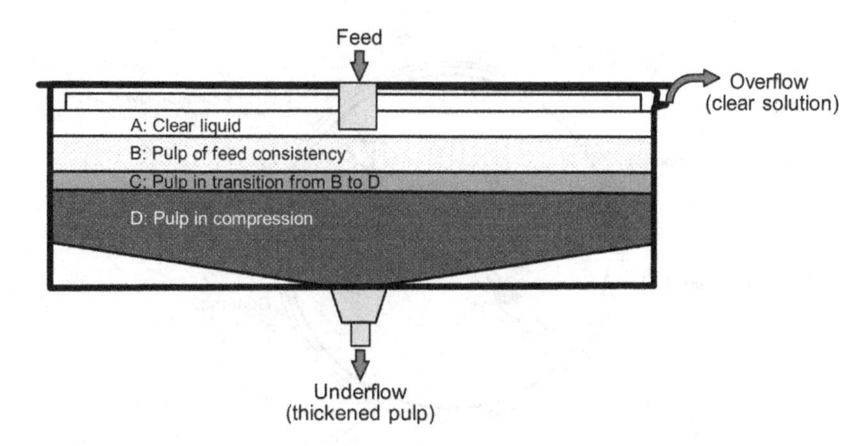

FIGURE 15.9 Concentration zones in a thickener.

compaction of the settled particles and produce a thicker underflow than can be achieved by simple settling by assisting the removal of water. In paste thickeners the high yield stress of the suspension can lead to a phenomenon known as "rotating beds," "doughnuts," or "islands", which should be avoided (Arbuthnot et al., 2005). This is associated with the presence of large aggregates that form ahead of and are compacted by the rotating blades. The solids in the thickener move continuously downwards, and then inwards toward the thickened underflow outlet, while the liquid moves upwards and radially outwards. In general, there is no region of constant composition in the thickener.

Thickener tanks are constructed of steel, concrete, or a combination of both, steel being most economical in sizes of less than 25 m in diameter. The tank bottom is often flat (e.g., Figure 15.10), while the mechanism arms are sloped toward the central discharge. With this design, settled solids must "bed-in" to form a false sloping floor. Steel floors are rarely sloped to conform to the rake arms because of expense. Concrete bases and sides become more common in the larger sized tanks. In many cases the settled solids, because of particle size, tend to slump and will not form a false bottom. In these cases the floor should be concrete and poured to match the slope of the arms. Tanks may also be constructed with sloping concrete floors and steel sides. Earth bottom thickeners are also in use, which are generally considered to be the lowest cost solution for thickener bottom construction (Hsia and Reinmiller, 1977).

The method of supporting the raking mechanism depends primarily on the tank diameter. In thickeners of diameter less than about 45 m, the drive head is usually supported on a superstructure spanning the tank, with the arms being attached to the drive shaft. Such machines are referred to as *bridge* or *beam* thickeners (Figure 15.10). The underflow is usually drawn from the apex of a cone located at the center of the sloping bottom.

A common arrangement for larger thickeners is to support the drive mechanism on a stationary steel or concrete center column. In most cases, the rake arms are attached to a drive cage, surrounding the central column, which is connected to the drive mechanism. The thickened solids are discharged through an annular trench encircling the center column (Figure 15.11). Figure 15.12 shows a thickener of this type in operation.

In the *traction thickener*, a single long arm is mounted with one end on the central support column, while to the other end are fixed traction wheels that run on a rail on top of the tank wall. The wheels are driven by motors that are mounted on the end of the arm and which therefore travel around with it. This is an efficient and economical design since the torque is transmitted through a long lever arm by a simple drive. They are manufactured in sizes ranging up to 200 m in diameter.

Cable thickeners have a hinged rake arm fastened to the bottom of the drive cage or center shaft. The rake arm is pulled by the cables connected to a torque or drive arm structure, which is rigidly connected to the center shaft at a point just below the liquid level. The hinge allows the rake to automatically lift when torque rises, which enables the rake arm to find its own efficient working level in the sludge, where the torque balances the rake weight. A feature of the design is the relatively small surface area of the raking mechanism.

In all thickeners, the speed of the raking mechanism is normally about 8 m min^{-1} at the perimeter, which corresponds to about 10 rev h^{-1} for a 15 m diameter thickener. Energy consumption is thus extremely low, such that even a 60 m unit may require only a 10 kW motor. Wear and maintenance costs are correspondingly low.

The underflow is usually withdrawn by pumping, although in clarifiers the material may be discharged under the hydrostatic head in the tank. The underflow is usually collected in a sludge-well in the center of the tank bottom, from where it is removed via piping through an

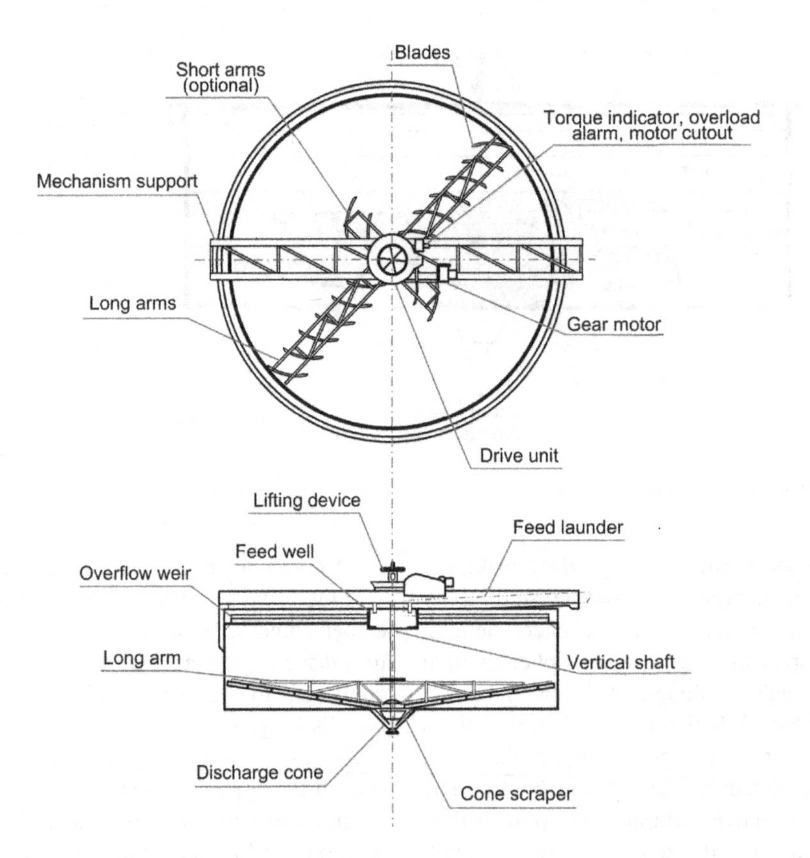

FIGURE 15.10 Thickener with mechanism supported by superstructure.

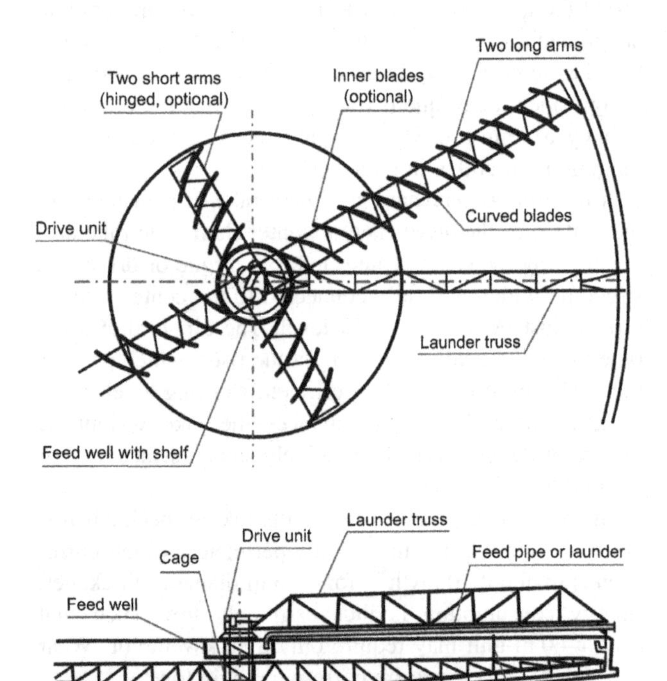

FIGURE 15.11 Thickener with rake mechanism supported by center column.

underflow tunnel. The underflow lines should be as short and as straight as possible to reduce the risk of choking, and this can be achieved, with large tanks, by taking them up from the sludge-well through the center column to pumps placed on top, or by placing the pumps in the base of the column and pumping up from the bottom. This has the advantage of dispensing with the expensive underflow tunnel. A development of this is the *caisson thickener*, in which the rake assembly is supported on hydrostatic bearings and the center column is enlarged sufficiently to house a central control room; the pumps are located in the bottom of the column, which also contains the mechanism drive heads, motors, control panel, underflow suction, and discharge lines. The caisson concept has lifted the possible ceiling on thickener sizes.

Underflow pumps are often of the *diaphragm* type. These are positive action pumps for medium heads and volumes, and are suited to the handling of thick viscous fluids. They can be driven by an electric motor through a crank mechanism, or by directly acting compressed air. A flexible diaphragm is oscillated to provide suction and discharge through non-return valves, and variable speed can be achieved by changing either the oscillating frequency or the stroke. In some plants, variable-speed pumps are connected to nucleonic density gauges on the thickener

FIGURE 15.12 A center-column supported thickener in operation *(Courtesy Outotec)*.

underflow lines, which control the rate of pumping to maintain a constant underflow density. The thickened underflow is commonly pumped to filters for further dewatering.

Thickeners often incorporate substantial storage capacity so that, for instance, if the filtration section is shut down for maintenance, the concentrator can continue to feed material to the dewatering section. During such periods the thickened underflow should be recirculated into the thickener feedwell. At no time should the underflow cease to be pumped, as choking of the discharge cone rapidly occurs.

15.2.4 Thickener Operation and Control

The two primary functions of the thickener are the production of a clarified overflow and a thickened underflow of the required concentration. For a given throughput, the clarifying capacity is determined by the thickener diameter, since the surface area must be large enough so that the upward velocity of liquid is at all times lower than the settling velocity of the slowest settling particle that is to be recovered. The degree of thickening produced is controlled by the residence time of the particles and hence by the thickener depth.

The solids concentration in a thickener varies from that of the clear overflow to that of the thickened underflow being discharged (Figure 15.9). When materials settle with a definite interface between the suspension and the clear liquid, as is the case with most flocculated mineral pulps, the solids-handling capacity determines the surface area. Solids-handling capacity is defined as the capacity of a material of given dilution to reach a condition such that the mass rate of solids leaving a region is equal to or greater than the mass rate of solids entering the region. The attainment of this condition with a specific dilution depends on the mass subsidence rate being equal to or greater than the corresponding rise rate of displaced liquid. A properly sized thickener containing material of many different dilutions, ranging from the feed to the underflow solids contents, has adequate area such that the rise rate of displaced liquid at any region never exceeds the subsidence rate.

Operation of the thickener to provide clarified overflow depends upon the existence of a clear-liquid zone at the top. If the zone is too shallow, some of the smaller particles may escape in the overflow. The volumetric rate of flow upwards is equal to the difference between the rate of feed of liquid and the rate of removal in the underflow. Hence the required concentration of solids in the underflow, as well as the throughput, determines the conditions in the clarification zone. Although not normally a problem with clarifiers, a stable froth bed can sometimes form on the surface of thickeners which may hinder operation (overflow is from under the froth bed in these cases). *Islands* of solids may also form.

Control of thickeners requires certain measurements. Concha (2014) describes novel instrumentation for operations to provide estimates of settling velocity and "solids stress" on the compacted solids. For on-line measurements, the following are considered among the most important.

Bed Level

The ability to monitor the "bed level" in a thickener is crucial in enhancing efficiency (Ferrar, 2014). Incorrect measurements can lead to problems such as excess water reporting to the underflow, sludge spillage into the overflow, or incorrect feedback in control of flocculation. Various techniques are employed depending on the application, including: calculated bed level based on density and hydrostatic pressure; ultrasound transducers to sense reflections from the solid bed; buoyancy-based electromechanical system; and conductivity-based probes.

Feed Mass Flow Rate

This is important to control clarity of the overflow water. Throughput can be optimized by combining mass flow measurement with ratio control of the flocculant dosage.

Flocculant Dosage Rate

As flocculants are expensive, keeping dosage to a minimum consistent with target performance is a priority and is one key to minimizing operating costs (Ferrar, 2014).

15.2.5 Thickener Sizing

Sizing methods include by experience, cylinder (or jar) settling tests, and pilot plant testwork (McIntosh, 2009). Experience (or "rule of thumb") is used if there is no sample, or to prepare a rough first draft for sizing equipment for budget purposes at the beginning of the project. Pilot plant testwork is the most reliable. A number of pilot plant units are available in various sizes and configurations. These units are small scale versions of commercial thickeners, with a feedwell, flocculant addition facility, underflow pumps, and rake mechanisms to duplicate the full-scale thickening process.

Batch settling tests, described above (Section 15.2.1), are most commonly used. There are two well-known methods to design thickeners based on the settling curve, the Coe and Clevenger (1916) method and the Talmage and Fitch (1955) method. The first determines the thickener area as follows.

If X is the liquid-to-solids ratio by weight at any region within the thickener, U the liquid-to-solids ratio of the thickener discharge, and W (t h^{-1}) the dry solids feed rate to the thickener, then, assuming no solids leave with the overflow, $(X - U)W$ (t h^{-1}) mass of liquid moves upwards with velocity V (m h^{-1}) :

$$V = \frac{(X-U)W}{AS} \tag{15.1}$$

where A is the thickener area (m^2) and S the density of the liquid (t m^{-3}). Because this upward velocity must not exceed the settling rate of the solids in this region, in the limit:

$$R = \frac{(X-U)W}{AS} \tag{15.2}$$

where R is the settling rate (m h^{-1}).

The required thickener area (m^2) is, therefore:

$$A = \frac{(X-U)W}{RS} \tag{15.3}$$

From a set of R and X values, the area required for various dilutions may be found by recording the initial settling rate as a function of dilution ranging from that of the feed to the discharge. The dilution corresponding to the maximum value of A represents the minimum solids-handling capacity and is the *critical dilution*. A scale up safety factor of between 1.2 and 1.5 is applied to A (Dahlstrom, 2003).

Originally proposed for suspensions without flocculant, in the Coe-Clevenger method the error in determining the settling rate increases as the flocculant dosage is increased. Parsapour et al. (2014) have proposed a modified procedure to account for the addition of flocculant.

The Coe and Clevenger method requires multiple batch tests at different pulp densities before an acceptable unit area can be selected. The Kynch model (1952) offers a way of obtaining the required area from a single batch-settling curve and is the basis of several thickening theories, which have been comprehensively reviewed by Pearse (1977) and Concha (2014).

The Talmage and Fitch method (1955) applies Kynch's model to the problem of thickener design. They showed that by constructing a tangent to the curve at any point on the settling curve (Figure 15.13), then:

$$CH = C_0H_0 \tag{15.4}$$

where H cm is the interface height corresponding to a uniform slurry of concentration C (kg l^{-1}) at the point where the tangent was taken, C_0 (kg l^{-1}) is the original feed solids concentration, and H_0 cm the original interface height. Therefore, for any selected point on the settling curve, the local concentration can be obtained from Eq. (15.4), and the settling rate from the gradient of the tangent at that

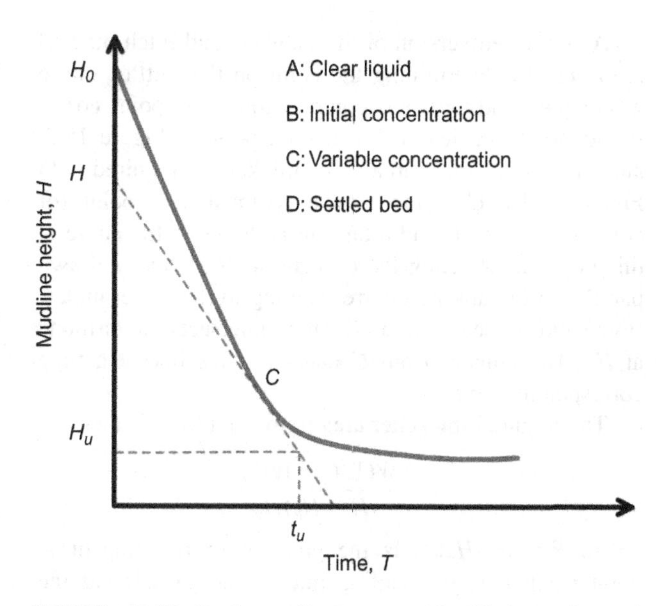

FIGURE 15.13 Batch settling curve (simplified version of Figure 15.2).

point. Thus a set of data of concentration against settling rate can be obtained from the single batch-settling curve.

To understand the approach we start by rewriting Eq. (15.3) in terms of concentration C. For a pulp of solids concentration C (kg l^{-1}), the volume occupied by the solids in 1 liter of pulp is C/d, where d (kg l^{-1}) is the density of dry solids.

Therefore the weight of water in 1 liter of pulp is

$$1 - \frac{C}{d} = \frac{d-C}{d}$$

and the water–solids ratio by weight becomes

$$= \frac{d-C}{dC}$$

For pulps of concentrations C kg l^{-1} of solids, and C_u kg l^{-1} of solids, the difference in water–solids ratio is

$$= \frac{d-C}{dC} - \frac{d-C_u}{dC_u}$$

$$= \frac{1}{C} - \frac{1}{C_u}$$

Therefore, the values of concentration obtained, C, and the settling rates, R, can be substituted in the Coe and Clevenger Eq. (15.3) to give:

$$A = \left(\frac{1}{C} - \frac{1}{C_u}\right)\frac{W}{R} \tag{15.5}$$

where C_u is the underflow solids concentration. (Note, to preserve A in m^2 C is now in t m^{-3} and S is eliminated.)

A simplified version of the Talmage and Fitch method is offered by determining the point on the settling curve where the solids go into compression. This point corresponds to the critical sedimentation point in Figure 15.2, and controls the area of thickener required. In Figure 15.13, C is the critical sedimentation point (or *compression point*) and a tangent is drawn to the curve at this point, intersecting the ordinate at H. A line is drawn parallel to the abscissa corresponding to the target underflow solids concentration C_u which intersects the ordinate at H_u. The tangent from C intersects this line at a time corresponding to t_u.

The required thickener area from Eq. (15.5) is then:

$$A = \frac{W(1/C - 1/C_u)}{(H - H_u)/t_u}$$

where $R = (H-H_u)/t_u$ is the gradient of the tangent at point C, that is, the settling rate of the particles at the compression point concentration. Since $CH = C_0H_0$, then:

$$A = \frac{W[(H/C_0H_0) - (H_u/C_0H_0)]}{(H - H_u)/t_u}$$

That is,

$$A = W\frac{t_u}{C_0H_0} \quad (15.6)$$

In most cases, the compression point concentration will be less than that of the underflow concentration. In cases where this is not so, the tangent construction is not necessary, and t_u is the point where the underflow line crosses the settling curve. The point of compression on the curve can be clear, but when this is not so, a variety of methods have been suggested for its determination (Fitch, 1977; Pearse, 1980; Laros et al., 2002).

The Coe and Clevenger and modified Talmage and Fitch methods are the most widely used in the metallurgical industry to predict thickener area requirements. Both methods have limitations (Waters and Galvin, 1991; Parsapour et al., 2014): the Talmage and Fitch technique relying critically on identifying a compression point, and both methods must be used in conjunction with empirical safety factors. However, the results of these methods are similar when the settling tests are carried out either on a single sample with the solids concentration of the thickener feed using the Talmage and Fitch method or on diluted samples using the Coe and Clevenger method (Parsapout et al., 2014).

Software has been developed for prediction of thickener area based on a phenomenological model of particle settling. The development of thickener models is reviewed by Concha and Burger (2003) and Concha (2014).

The mechanism of solids consolidation has been far less well expressed in mathematical terms than the corresponding clarifying mechanisms. The height of the thickener is, therefore, usually determined by experience.

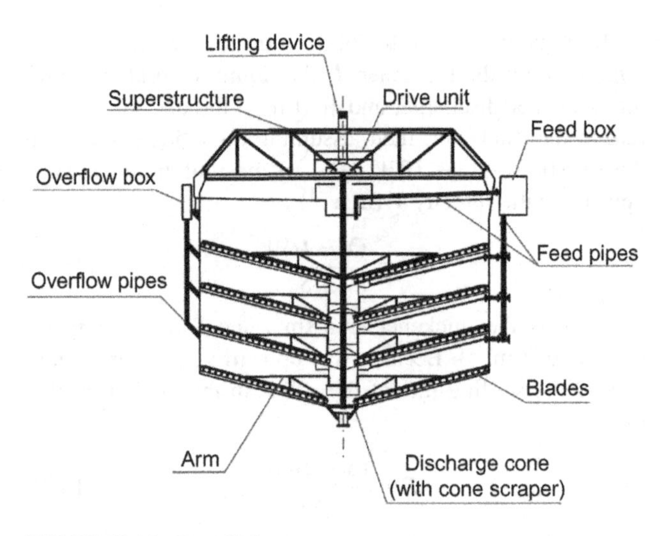

FIGURE 15.14 Tray thickener.

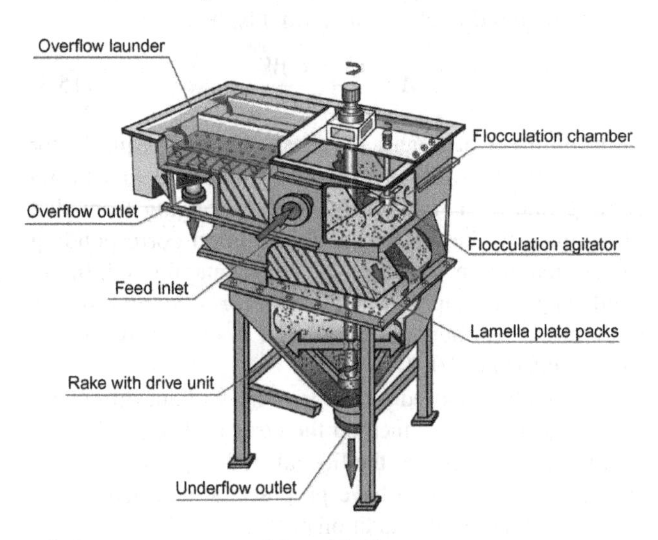

FIGURE 15.15 Lamella thickener/clarifier.

15.2.6 Other Gravity Sedimentation Devices

Tray Thickener

The diameter of a conventional thickener is usually large and therefore a large ground area is required. *Tray thickeners* (Figure 15.14) are sometimes installed to save space. In essence, a tray thickener is a series of unit thickeners mounted vertically above one another. They operate as separate units, but a common central shaft is utilized to drive the sets of rakes.

Lamella Thickener

Also known as an *inclined plate settler*, it has two main parts, an upper tank containing lamella plates inclined at 55° and a lower conical or cylindrical tank with a rake mechanism (Figure 15.15). The inclined plate gives a short distance for particle sedimentation, and with low friction, sliding down the plate increases the speed of separation (Anon., 2011). Clarification (clear supernatant) is achieved when the upstream liquid velocity is sufficiently low to allow solids to settle to the plate. Thickening is achieved through a combination of sedimentation onto the plate ("primary thickening") and conventional sedimentation in the lower tank.

15.3 CENTRIFUGAL SEDIMENTATION

Centrifugal separation can be regarded as an extension of gravity separation, as the settling rates of particles are increased under the influence of centrifugal force. It can, however, be used to separate emulsions which are normally stable in a gravity field.

Centrifugal separation can be performed either by hydrocyclones or centrifuges. The simplicity and cheapness of the hydrocyclone (Chapter 9) make it very attractive, although it suffers from restrictions with respect to the solids concentration that can be achieved and the relative proportions of overflow and underflow into which the feed may be split. Generally, the efficiency of even a small-diameter hydrocyclone falls off rapidly at very fine particle sizes and particles smaller than about 10 μm in diameter will invariably appear in the overflow, unless they have high density. Flocculation of such particles is limited, since the high shear forces due to the cyclonic action break up the agglomerates. The hydrocyclone is therefore inherently better suited to classification rather than thickening. Its dry counterpart, the (air) cyclone, is widely used for "dust" removal in a variety of industrial applications.

By comparison, centrifuges are much more costly and complex, but have a much greater clarifying power and are generally more flexible than hydrocyclones. Various types of centrifuge are used industrially (Bragg, 1983; Bershad et al., 1990; Leung, 2002), *the solid bowl centrifuge* (or *decanter*) having widest use in the minerals industry due to its versatility and ability to discharge the solids continuously.

The basic principles of a typical centrifuge are shown in Figure 15.16. It consists of a horizontal revolving shell or bowl, cylindroconical in form, inside which a screw conveyor of similar section rotates in the same direction at a slightly higher or lower speed. The feed pulp is admitted to the bowl through the center tube of the revolving-screw conveyor. On leaving the feed pipe, the slurry is immediately subjected to a high centrifugal force, causing the solids to settle on the inner surface of the bowl at a rate which depends on the rotational speed employed, this normally being between 1,600 and 8,500 rev min^{-1}. The separated solids are conveyed by the scroll out of the liquid and discharged through outlets at the smaller end of the bowl. The solids are continuously dewatered by

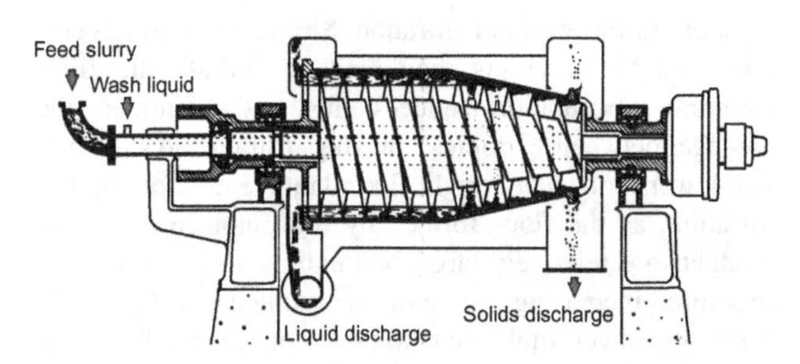

FIGURE 15.16 Continuous solid bowl centrifuge.

centrifugal force as they proceed from the liquid zone to the discharge. Excess entrained liquor drains away to the pond circumferentially through the particle bed. When the liquid reaches a predetermined level, it overflows through the discharge ports at the larger end of the bowl.

The actual size and geometry of these centrifuges vary according to the throughput required and the application. The length of the cylindrical section largely determines the clarifying power and is thus made a maximum where overflow clarity is of prime importance. The length of the conical section, or "beach," decides the residual moisture content of the solids, so that a long shallow cone is used where maximum dryness is required.

Centrifuges are manufactured with bowl diameters ranging from 15 to 150 cm, the length generally being about twice the diameter. Throughputs vary from about 0.5 to 50 $m^3 h^{-1}$ of liquid and from about 0.25 to 100 $t h^{-1}$ of solids, depending on the feed concentration, which may vary widely from 0.5% to 70% solids, and on the particle size, which may range from about 12 mm to as fine as 2 μm, or even less when flocculation is used. The application of flocculation is limited by the tendency of the scroll action to damage the flocs and thus redisperse the fine particles. The moisture content in the product varies widely, typically being in the range 5–20%.

15.4 FILTRATION

Filtration is the process of separating solids from liquid by means of a porous medium (the *filter*) which retains the solid but allows the liquid to pass. The most common filter type in mineral processing is *cake filtration*, where the liquid passes through the filter, called the *filtrate*, and the solids build-up on the filter is referred to as *filter cake*. The volume of filtrate collected per unit time is the *rate of filtration*. There are typically five steps in the process: cake formation, moisture reduction, cake washing (if required), cake discharge, and medium washing.

Filtration in mineral processing applications normally follows thickening. The thickened pulp may be fed to storage agitators from where it is drawn off at uniform rate to the filters. Flocculants are sometimes added to the agitators in order to aid filtration. Slimes have an adverse effect on filtration, as they tend to "blind" the filter medium; flocculation reduces this and increases the voidage between particles, making filtrate flow easier. The lower molecular weight flocculants tend to be used in filtration, as the flocs formed by high molecular weight products are relatively large, and entrain water within the structure, increasing the moisture content of the cake. With the lower molecular weight flocculants, the filter cake tends toward a uniform porous structure which allows rapid dewatering, while still preventing migration of fine particles through the cake (Moss, 1978). Other surfactant *filter aids* are used to reduce the liquid surface tension, or more likely, to modify particle surface properties to assist flow through the medium (Singh et al., 1998; Wang et al., 2010; Asmatulu and Yoon, 2012).

There is a large body of literature on types of filters, filtration principles, equipment selection, testing and sizing (Wakeman and Tarleton, 1999; Svarovsky, 2000; Rushton et al., 2000; Smith and Townsend, 2002; Cox and Traczyk, 2002; Welch, 2002; Dahlstrom, 2003; Stickland, 2008; Concha, 2014).

15.4.1 Brief Theory

Concha (2014) provides comprehensive theoretical treatment; the brief summary here based on Smith and Townsend (2002), Cox and Traczyk (2002), and Dahlstrom (2003) is just to introduce the main operational variables. Starting with the classical theory of Darcy and Poiseuille, the basic filtration equation can be written as:

$$v = \frac{1}{A}\frac{dV}{dt} = \frac{\Delta P}{\mu\left(\alpha w \frac{V}{A}\right)} \tag{15.7}$$

where v is filtrate flow rate (m s^{-1}), A area of filter (m^2), V filtrate volume (m^3), t time (s), ΔP pressure drop across the cake and medium (N m^{-2}), μ liquid viscosity (N.s m^{-2}), α specific cake resistance (m kg^{-1}), and w feed slurry concentration in terms of dry solids mass per unit of filtrate volume (kg m^{-3}). The equation may be modified by adding a resistance term for the filter medium.

Equation (15.7) shows that the filtration rate varies directly with the pressure drop across the filter and the area of the filter, and inversely with liquid viscosity, cake resistance (reciprocal of cake permeability), and slurry solids contents. These factors determine the necessary test variables in sizing filters.

The cake comprises a bundle of capillaries and water can be removed only when the applied pressure is greater than the capillary pressure P_C given by:

$$P_C = \frac{4\gamma_l \cos\theta}{d_C} \tag{15.8}$$

where γ_l is the liquid surface tension, θ the contact angle, and d_C is the diameter of the capillary. Equation (15.8) indicates why surfactant filter aids can assist, by decreasing surface tension and/or increasing contact angle (Asmatulu and Yoon, 2012). The equation also indicates that filtration becomes more difficult as particle size decreases, as the capillary (pore) diameter tends to decrease.

15.4.2 The Filter Medium

The choice of the filter medium is often the most important consideration in assuring efficient operation of a filter. Its function is generally to act as a support for the filter cake, while the initial layers of cake provide the true filter. The filter medium should be selected primarily for its ability to retain solids without blinding. It should be mechanically strong, corrosion resistant, and offer as little resistance to flow of filtrate as possible. Relatively coarse materials are normally used and clear filtrate is not obtained until the initial layers of cake are formed, the initial cloudy filtrate being recycled.

Filter media are manufactured from cotton, wool, linen, jute, silk, glass fiber, porous carbon, metals, rayon, nylon and other synthetics, ceramic, and miscellaneous materials such as porous rubber. Cotton fabrics are among the most common type of medium, primarily because of their low initial cost and availability in a wide variety of weaves.

15.4.3 Filtration Tests

It is not normally possible to forecast what may be accomplished in the filtration of an untested product, therefore preliminary tests have to be made on representative samples of pulp before the large-scale plant is designed. Bench scale testing of samples for specification of filtration equipment is described by Smith and Townsend (2002) and Tarleton and Wakeman (2006). Tests are also commonly carried out on pulps from existing plants, to assess the effect of changing operating conditions, filter aids, etc.

It is necessary to identify what are the objectives, for example, target moisture, and what are the candidate filter types, for example, pressure or vacuum? The slurry physical and chemical conditions that need to be replicated in the sample should include: solids concentration, particle density and size distribution, slurry pH, and chemical additives (e.g., flotation reagents and flocculants).

Vacuum Filtration Test

A simple vacuum *filter leaf* test setup is shown in Figure 15.17. The filter leaf, consisting of a section of the industrial filter medium, is connected to a filtrate receiver equipped with a vacuum gauge. A known weight of slurry

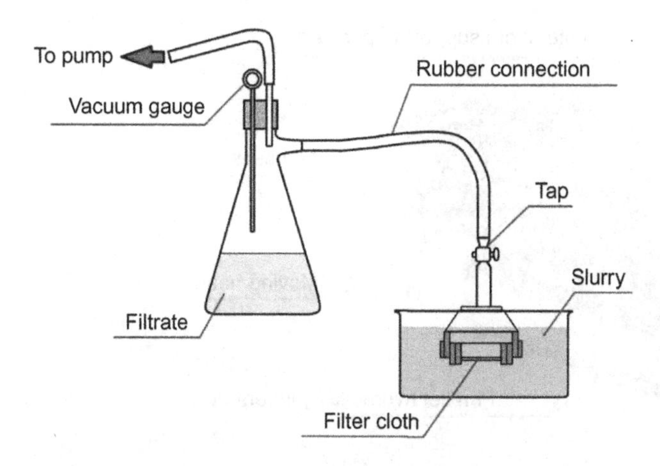

FIGURE 15.17 Laboratory test filter.

is introduced sufficient to approximate the target cake thickness. The receiver is connected to a vacuum pump. If the industrial filter is to be a continuous vacuum filter, this operation must be simulated in the test. The cycle is divided into three sections: cake formation (or "pick-up"), drying, and discharge. Sometimes pick-up is followed by a period of washing and the cake may also be subjected to compression during drying. While under vacuum, the test leaf is submerged for the pick-up period in the agitated pulp to be tested. The leaf is then removed and held with the drainpipe down for the allotted drying time.

At the end of drying, the filter cake is removed and the net weight and cake thickness are recorded. The sample is dried and weighed to determine moisture content. The daily filter capacity can then be determined by the dry weight of cake per unit area of test leaf multiplied by the daily number of cycles and the filter area. A range of conditions should be tested to cover the range of anticipated variables.

Pressure Filtration Test

Bench-scale pressure filtration testwork can be performed using a bomb device. A typical apparatus is a 250 mm length of 50 mm (outer-wall diameter) pipe capped with flanges. The lower flange supports the test filter cloth on a drainage grid above the filtrate collection port. The upper flange houses the air pressure connection, pressure gauge, and feed port. The test procedure is similar to that described above, with applied pressure being substituted for vacuum. A new pressure test procedure, *step pressure filtration*, suited to flocculated feeds, is described by Usher et al. (2001) and De Kretser et al. (2011).

15.4.4 Types of Filter

For particles coarse enough that capillary pressures are negligible (Eq. (15.8)), gravimetric dewatering can be employed, for example, dewatering screens in dense media recovery (Chapter 11). This is not usually the case and cake filters are the type most frequently used in mineral processing, where the recovery of large amounts of fine solids (typically <100 μm) from fairly concentrated slurries (50–60% solids) is the main requirement. Cake filters may be pressure or vacuum types, and operation may be batch or continuous (Cox and Traczyk, 2002). In pressure filters positive pressure is applied at the feed end and in vacuum filters there is a vacuum at the far side of the filter, the feed side being at atmospheric pressure. Dewatering is a combination of cake compression and air blow through.

Pressure Filters

Because of the increasing fineness of mineral concentrates (those of Cu, Pb, and Zn are commonly 80%<30 μm), coupled with shipping schedules calling for moisture contents 8–10 wt% on these fine concentrates, filtration under pressure has certain advantages over vacuum. (Given that many operations are at high altitude is an additional drawback for vacuum units.) Higher flow rates and better washing and drying result from the higher pressures that can be used. Pressure filters have become sufficiently large and reliable to handle the output of most concentrators and can produce low enough cake moisture to eliminate driers (Townsend, 2003). Thus, the trend is to pressure filtration (Cox and Traczyk, 2002).

The common pressure filters in mineral processing applications come in two basic forms, horizontal and vertical, defined either by the orientation of the filter plates (Concha, 2014), or the convention here, by the direction the pressure is applied: actuation either horizontally or vertically (Cox and Traczyk, 2002). They both represent more automated versions of *plate-and-frame* filters (Taggart, 1945; McCaslin et al., 2014).

Horizontal Pressure Filters

A typical horizontal filter is shown in Figure 15.18. In this arrangement, filter plates, usually made from lightweight polymer, are suspended vertically from a steel frame (hence the alternative name *vertical plate pressure filter*). Between the plates is hung filter cloth. The plates are held together and the press is opened and closed by a hydraulic piston. The slurry is pumped into the press to fill each chamber. Dewatering starts immediately the slurry enters, and the filtrate is removed. When the chamber is full of material, membranes on one side of the chamber are pressurized to hold the cake in place and squeeze out some water. At the end of this "feeding cycle", the main dewatering cycle is activated by forcing pressurized air through the cake, the "air dewatering cycle" with pressures up to 8 bar. A "washing cycle"

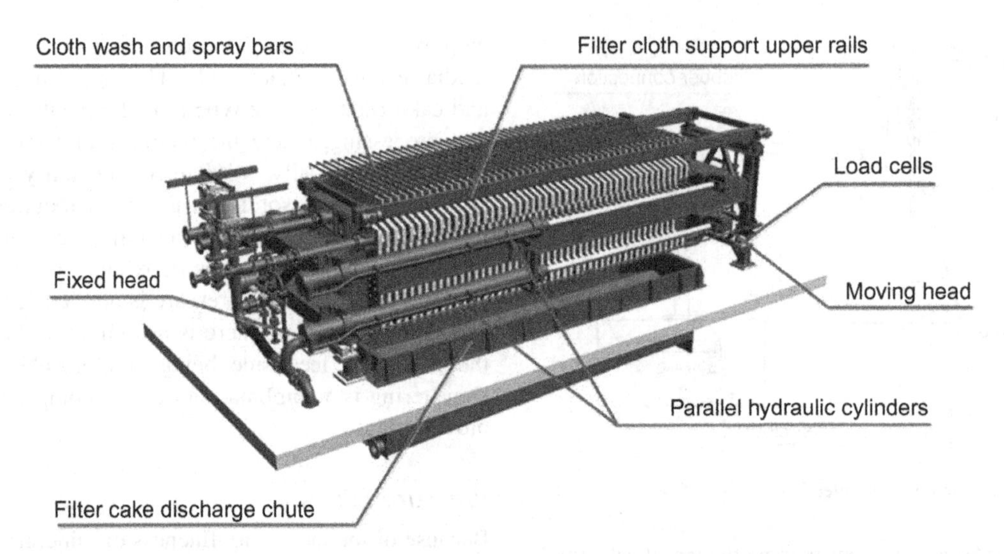

FIGURE 15.18 Horizontal pressure filter. *(Courtesy Metso).*

could be incorporated, the cake washed by replacing air by liquid with air pressure then re-applied to achieve final moisture.

At completion, the press is opened, the plates separate and the cake falls by gravity onto a conveyor. To finish, the filter cloth is cleaned by a combination of vibration and water sprays, the washings recycled to the feed tank, the chambers closed and the cycle repeated. While the operation is batch, with cycle times of ca. 10 min and a feed tank with sufficient storage, operation appears continuous.

Vertical Pressure Filters

These differ from the horizontal units by stacking the chambers on top of each other and rather than individual filter cloths between each chamber, the cloth in the vertical unit is continuous (Figure 15.19). The filtration cycle is similar to that for horizontal units. At completion, the press is opened and the filter cloth advanced to discharge the cake, followed by cloth washing. Both horizontal and vertical units are automated to control the various cycle times and may include sensors to monitor, for example, cloth condition (Townsend, 2003).

FIGURE 15.19 Vertical pressure filter *(Courtesy FLSmith).*

Tube Press

In some situations, dewatering of ultrafine ($<10\,\mu m$) material requires special equipment. Large capillary resistance forces demand greater air pressures (Eq. (15.8)) than in the units described above. By performing filtration in a tube, pressures up to 100 bar can be exploited. The *tube press* has been applied in a variety of difficult dewatering applications. It consists of a casing with a membrane at each end and a porous tube (or "candle") covered with cloth suspended inside. Feed is introduced under pressure to fill the casing and cake starts to form on the candle. Dewatering pressure is applied by the membrane and an air blow. Cake washing can also be incorporated. On completion, the membrane is retracted, and the candles lowered to discharge the cake, which can be aided by air blown behind the cloth. The candle is re-inserted in the casing and the cycle repeats. A related device is the "candle filter", which consists of a series of porous tubes (candles) inside a pressurized chamber (Concha, 2014) (Figure 15.20).

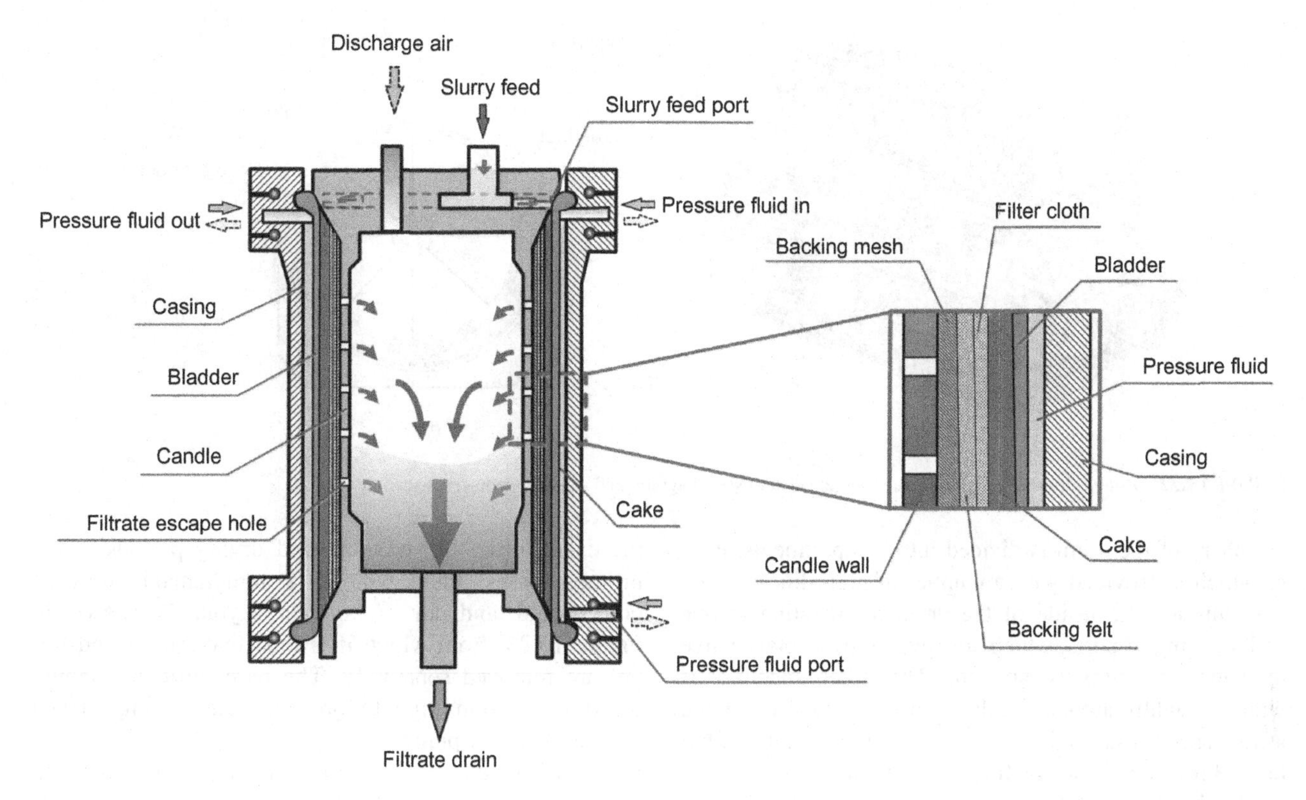

FIGURE 15.20 Operation of tube press.

Vacuum Filters

There are many different types of vacuum filter, but they all incorporate filter media suitably supported on a drainage system, beneath which the pressure is reduced by connection to a vacuum system.

Batch Vacuum Filters

There are two main types, the *vertical leaf filter* and the *horizontal leaf* or *tray filter*, which are similar except for the orientation of the leaf. The leaf consists of a metal framework or a grooved plate over which the filter cloth is fixed (Figure 15.21). Numerous holes are drilled in the pipe framework, so that when a vacuum is applied, a filter cake builds up on both sides of the leaf. A number of leaves are generally connected. For example, in the vertical leaf filter, the array is first immersed in slurry held in a slurry feed tank, removed, and then placed in a cake-receiving tank where the cake is removed by replacing the vacuum by air pressure. Although simple to operate, these filters require considerable floor space. They are now used only for clarification, that is, the removal of small amounts of suspended solids from liquors.

Continuous Vacuum Filters

These are the most widely used filters in mineral processing applications and fall into three classes: drums, discs, and horizontal filters.

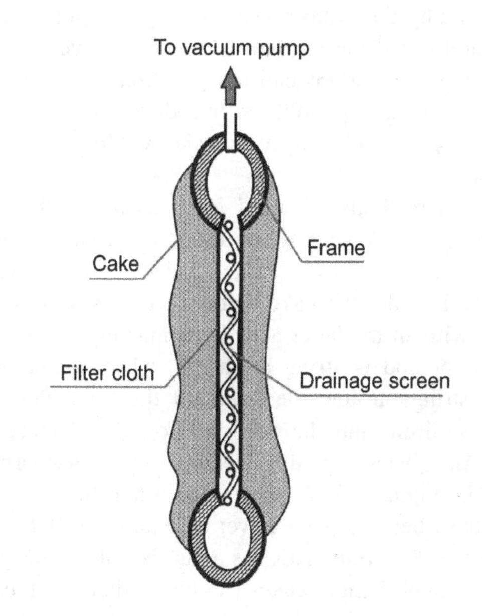

FIGURE 15.21 Cross section of typical leaf filter.

Rotary-Drum Filter

This is the most common, finding application both where cake washing is required and where it is unnecessary. The drum is mounted horizontally and is partially submerged in the filter trough, into which the feed slurry is fed and maintained in suspension by agitators (Figure 15.22). The

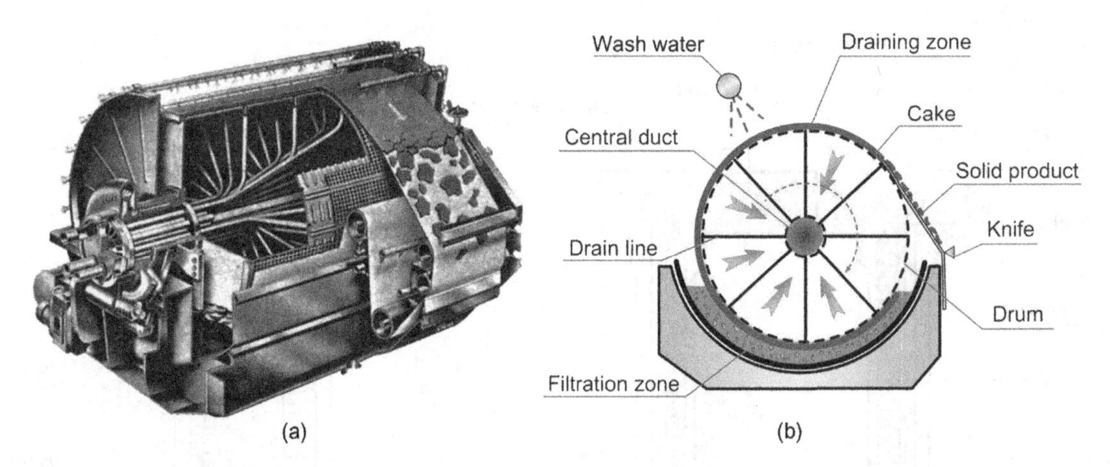

FIGURE 15.22 Rotary-drum filter with belt discharge: (a) cut-away diagram, and (b) schematic cross-section

periphery of the drum is divided into compartments, each of which is provided with a number of drain lines, which pass through the inside of the drum, terminating at one end as a ring of ports, which are covered by a rotary valve to which vacuum is applied. The filter medium is wrapped tightly around the drum surface which is rotated at low speed, usually in the range 0.1–0.3 rev min^{-1}, but up to 3 rev min^{-1} for fast filtering materials.

As the drum rotates, each compartment goes through the same cycle of operations, the duration of each being determined by the drum speed, the depth of submergence of the drum, and the arrangement of the valve. The normal cycle of operations consists of filtration, drying, and discharge, but it is possible to introduce other operations into the basic cycle, such as cake washing and cloth cleaning.

Various methods are used for discharging the solids from the drum, depending on the material being filtered. The most common form makes use of a reversed blast of air, which lifts the cake so that it can be removed by a knife, without the latter actually contacting the medium. Another method is string discharge, where a number of endless strings around the drum lift the filter cloth as it leaves the drum and the cake falls off. It is rarely used today. An advance on this method is belt discharge, as shown in Figure 15.22, where the filter medium itself leaves the filter and passes over an external roller, before returning to the drum. This has a number of advantages in that very much thinner cakes can be handled, with consequently increased filtration and draining rates and hence better washing and dryer products. At the same time, the cloth can be washed on both sides by means of sprays before it returns to the drum, thus minimizing the extent of blinding. Cake washing is usually carried out by means of sprays or weirs, which cover a fairly limited area at the top of the drum.

The capacity of the vacuum pump will be determined mainly by the amount of air sucked through the cake during the washing and drying periods when, in most cases, there will be a simultaneous flow of both liquid and air. A typical layout is shown in Figure 15.23, from which it is seen that the air and liquid are removed separately. The barometric leg should be at least 10 m high to prevent liquid being sucked into the vacuum pump.

Variations on standard drum filters to enable them to handle coarse, free-draining, quick-settling materials include top feed units where the material is distributed at between 90 and 180° from the feed point.

Disc Filter

The principle of operation of *disc filters* (Figure 15.24), is similar to that of rotary drum filters. The disc filter consists of sectors of cloth covered steel mounted on a central shaft that also connects a certain number of the sectors to vacuum. The solids cake is formed on both sides of the disc; the disc rotates and lifts the cake above the level of the slurry in the trough, whereupon the cake is suction-dried and is then removed by a pulsating air blow with the assistance of a scraper. Several discs are mounted along the shaft separated by about 30 cm and consequently a large filtration area can be accommodated in a small floor space. Cost per unit area is thus lower than for drum filters, but cake washing is virtually impossible with the disc filter.

Ceramic Disc Filter

A special type of disc filter uses micro-pore ceramic sectors rather than ones of steel covered with cloth. When submerged in the slurry pool, *capillary action* assists drawing liquid through the pores of the filter. This can be understood by reference to Eq. (15.8); for the filter (as opposed to the cake) we want the capillary pressure to be high, achieved by having small pore diameter and strongly hydrophilic material, that is, ceramic. The action

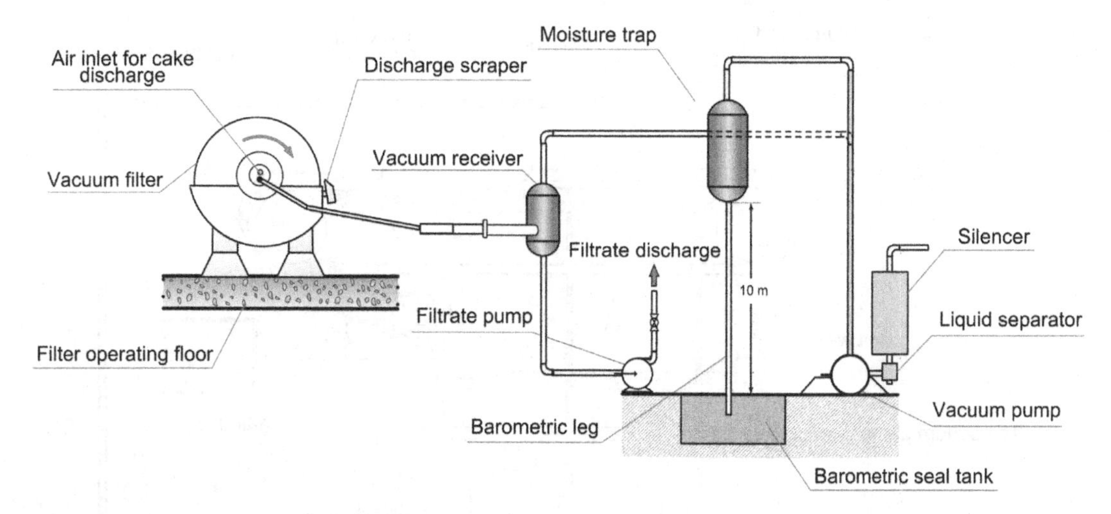

FIGURE 15.23 Typical rotary-drum filter system.

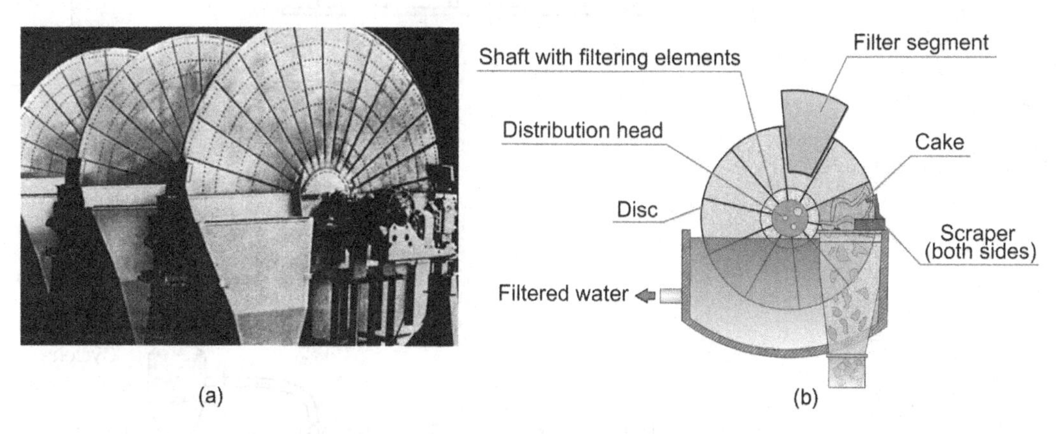

FIGURE 15.24 Rotary-disc filter: (a) general view, and (b) schematic cross-section.

reduces the size of vacuum pump required, resulting in reduced energy consumption. Referred to as *capillary filtration*, it can produce moisture contents that approach pressure filtration and has found application on mineral concentrates (Cox and Traczyk, 2002; Concha, 2014).

Horizontal Belt Filter

If ultimate cake moisture is not critical, but rather water recovery or production of solids that can be handled or stored is critical, then the *belt filter* may be suitable (Figure 15.25). Examples include tailings dewatering at operations with limited space or environmental restrictions on disposal in tailings ponds, and recovery of leach liquors in hydrometallurgical operations. It comprises an endless perforated rubber drainage deck supporting the filter cloth. Vacuum is applied by a series of suction boxes underneath the belt. Varying the number of the suction boxes gives control over the length of the filtration, drying, and washing (if included) stages. The cake is discharged as the belt reverses over a small diameter roller.

Pan Filter

This consists of a series of horizontal trays supporting the filter cloth rotating around a central vertical axis and connected to a common suction valve. The trays are trapezoidal in shape to accommodate the rotation and slightly tilted toward the center. Cake builds, can be washed, and is discharged by tipping the tray. Compared to a disc filter, which has two sides forming cake, the pan filter, with only one side, is lower capacity.

Hyperbaric Filters

By placing a conventional disc or drum vacuum filter inside a pressurized vessel, the available pressure drop can be increased, up to four bars or more (Bott et al., 2003; Concha, 2014). Cake moisture levels can be reduced from typically 15% in vacuum filters to ca. 8%. As with all pressure filters, cake discharge is a problem.

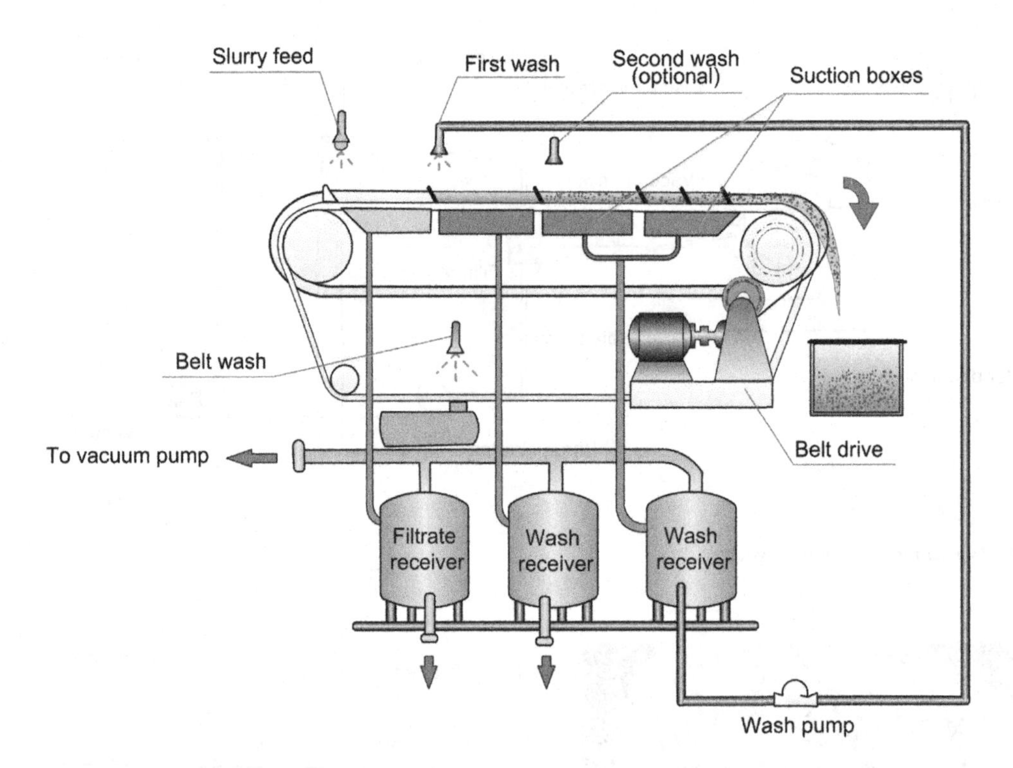

FIGURE 15.25 Horizontal belt filter.

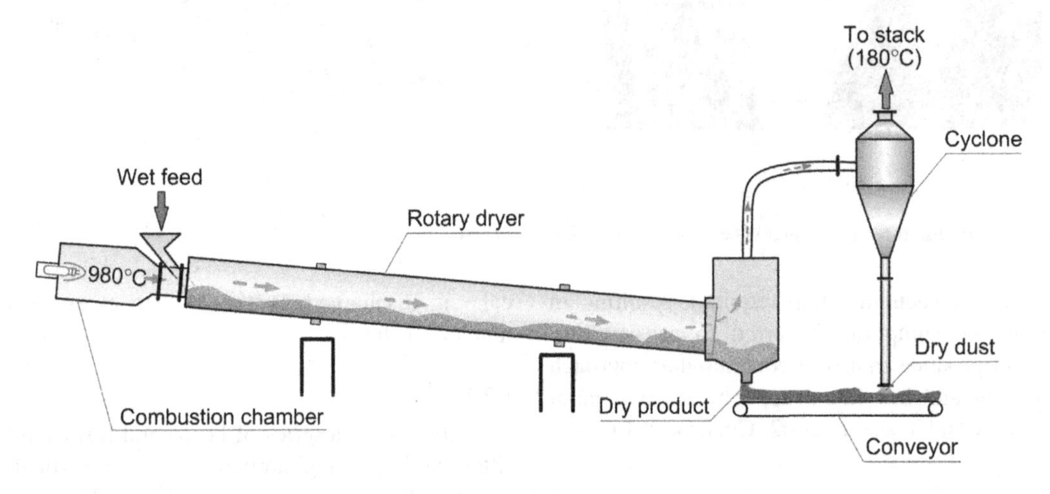

FIGURE 15.26 Direct fired, parallel flow rotary dryer *(Adapted from Kram (1980)).*

15.5 DRYING

The drying of concentrates prior to shipping, if done, is the last operation performed in the mineral processing plant. It reduces the cost of transport and is usually aimed at lowering the moisture content to about 5 wt%. Dust losses are often a problem if the moisture content is lower.

Prokesch (2002), Mujumdar (2006), and Kudra and Mujumdar (2009) review the types of drying equipment available and describe dryer selection based on the required duty. The dryer types include hearth, grate, shaft, fluidized bed, and flash. The one often used in concentrators is the rotary dryer.

Rotary Thermal Dryer

This unit consists of a relatively long cylindrical shell mounted on rollers and driven at a speed of up to 25 rev $\min^{-1}$. The shell is at a slight slope, so that material moves from the feed to discharge end under gravity. Hot gases, or air, are fed in either at the feed end to give parallel flow or at the discharge to give counter-current flow.

The method of heating may be either direct, in which case the hot gases pass through the material in the dryer, or indirect, where the material is in an inner shell, heated externally by hot gases. The direct-fired is the one most commonly used in the minerals industry, the indirect-fired type being used when the material must not contact the hot combustion gases. Parallel flow dryers (Figure 15.26) are used in the majority of operations because they are more fuel efficient and have greater capacity than counterflow types (Kram, 1980). Since heat is applied at the feed end, build-up of wet feed is avoided, and in general these units are designed to dry material to not less than 1% moisture. Since counter-flow dryers apply heat at the discharge end, a completely dry product can be achieved, but its use with heat-sensitive materials is limited because the dried material comes into direct contact with the heating medium at its highest temperature.

The product from the dryers is often stockpiled, before being loaded on to trucks or rail-cars as required for shipment. To control dust, containers may be closed, or the surface of the contents sprayed with various dust suppressing solutions (Kolthammer, 1978).

REFERENCES

Anon, 2011. Basics in Mineral Processing. eighth ed. Metso Corporation.

Arbuthnot, I., et al., 2005. Designing for paste thickening. *Proc. 37th Annual Meeting of Canadian Mineral Processors Conf.* CIM, Ottawa, ON, Canada, pp. 597–628.

Asmatulu, R., Yoon, R.H., 2012. Effects of surface forces on dewatering of fine particles. In: Young, C.A., Luttrell, G.H. (Eds.), Separation Technologies for Mineral, Coal, and Earth Resources. SME, Englewood, CO, USA, pp. 95–102.

Bershad, B.C., et al., 1990. Making centrifugation work for you. Chem. Eng. 97 (7-12), 84–89.

Bott, R., et al., 2003. Recent developments and results in continuous pressure and steam-pressure filtration. Aufbereitungs Techni. 44, 5–18.

Bragg, R., 1983. Filters and centrifuges (part 1- part 4). Mining Mag. Aug., 90–111.

Bulatovic, S.M., 2007. Handbook of Flotation Reagents, Theory and Practice. Flotation of Sulfide Ores, vol. 1. Elsevier, Amsterdam, Netherlands.

Buscall, R., White, R.L., 1987. The consolidation of concentrated suspensions. Part 1. The theory of sedimentation. J. Chem. Soc., Faraday Trans. 1: Phys. Chem. Condensed Phases. 83 (6), 873–891.

Coe, H.S., Clevenger, G.H., 1916. Methods for determining the capacities of slime-settling tanks. Trans. AIMME. 55, 356–384.

Concha, F., Burger, R., 2003. Thickening in the 20th century: a historical perspective. Miner. Metall. Process. 20 (2), 57–67.

Concha, F.A., 2014. *Solid-Liquid Separation in the* Mining *Industry*, vol. 105. Springer, Fluid Mechanics and Its Applications (Series).

Cox, C., Traczyk, F., 2002. Design features and types of filtration equipment. In: Mular, A.L., et al., (Eds.), Mineral Processing Plant Design, Practice and Control, vol. 2. SME, Littleton, CO, USA, pp. 1342–1357.

Dahlstrom, D.A., 2003. Liquid-solid separation. In: Fuerstenau, M.C., Han, K.N. (Eds.), Principles of Mineral Processing. SME, Littleton, CO, USA, pp. 307–362.

De Kretser, R.G, et al., 2011. Comprehensive characterisation of material properties for dewatering: how much is enough? FILTECH 2011 Congress Proc., vol. 1. Wiesbaden, Germany, pp. 383–390.

Farrow, J.B., Swift, J.D., 1996. A new procedure for assessing the performance of flocculants. Int. J. Miner. Process. 46 (3-4), 263–275.

Ferrar, G., 2014. Optimise your thickener efficiency for maximum profitability. <http://www.pacetoday.com.au/news/optimise-your-thickener-efficiency-for-maximum-pro> (viewed December 2014).

Fitch, E.B., 1977. Gravity separation equipment-clarification and thickening. In: Purchas, D.B., Wakeman, R.J. (Eds.), Solid-Liquid Separation Equipment Scale-up. Uplands Press, Croydon, UK.

Hogg, R., 2000. Flocculation and dewatering. Int. J. Miner. Process. 58 (1-4), 223–236.

Hsia, E.S., Reinmiller, F.W., 1977. How to design and construct earth bottom thickeners. Mining Eng. Aug., 36–39.

Kolthammer, K.W., 1978. Concentate drying, handling and storage. In: Mular, A.L., Bhappu, R.B. (Eds.), Mineral Processing Plant Design. AIMME, New York, NY, USA, pp. 601–617.

Kram, D.J., 1980. Drying, calcining, and agglomeration. Eng. Min. J. 181 (6), 134–151.

Kudra, T., Mujumdar, A.S., 2009. Advanced drying technologies. second ed. CRC Press, Taylor & Francis Group, Boca Raton, FL, USA.

Kurt, N., 2006. A Study of Channelling Behaviour in Batch Sedimentation. PhD thesis. School of Civil and Chemical Engineering, Royal Melbourne Institute of Technology University, Melbourne, Australia.

Kynch, G.J., 1952. A theory of sedimentation. Trans. Faraday Soc. 48, 166–176.

Lake, P., Summerhays, R.P., 2012. Thickener feed system design. Proc. 13th International Mineral Processing Symp., Bodrum, Turkey, pp. 197–204.

Laros, T., et al., 2002. Testing, sizing, and specifying sedimentation equipment. In: Mular, A.L., et al., (Eds.), Mineral Processing Plant Design, Practice and Control, vol. 2. SME, Littleton, CO, USA, pp. 1295–1312.

Leung, W., 2002. Centrifugal sedimentation and filtering for mineral processing. In: Mular, A.L., et al., (Eds.), Mineral Processing Plant Design, Practice and Control, vol. 2. SME, Littleton, CO, USA, pp. 1262–1288.

Loan, C., et al., 2009. Operational results from the Vane Feedwell-Cutting-Edge modeling turned into reality. Proc. 10th Mill Operators' Conf. AusIMM, Adelaide, SA, Australia, pp. 261–266.

Luttrell, G.H., 2014. Innovations in coal processing. In: Anderson, C.G., et al., (Eds.), Mineral Processing and Extractive Metallurgy: 100 Years of Innovation. SME, Englewood, CO, USA, pp. 277–296.

McCaslin, M.L., et al., 2014. Innovations in liquid/solid separation for metallurgical processing. In: Anderson, C.G., et al., (Eds.), Mineral Processing and Extractive Metallurgy: 100 Years of Innovation. SME, Englewood, CO, USA, pp. 333–343.

McIntosh, A., 2009. Thickener sizing and the importance of testwork. Output Aust. Minerals Metals Process. Solut. 23, 9–12.

Moss, N., 1978. Theory of flocculation. Mine Quarry. 7 (May), 57–61.

Mujumdar, A.S., 2006. Handbook of Industrial Drying. third ed. CRC press, Taylor & Francis Group, Boca Raton, FL, USA.

Mular, A.L., et al.,2002. Mineral Processing Plant Design, Practice, and Control, vol. 2. SME, Littleton, CO, USA (Chapter 9).

Owen, A.T., et al., 2002. The impact of polyacrylamide flocculant solution age on flocculation performance. Int. J. Miner. Process. 67 (1-4), 123–144.

Owen, T.A., et al., 2009. The effect of flocculant solution transport and addition conditions on feedwell performance in gravity thickeners. Int. J. Miner. Process. 93 (2), 115–127.

Parsapour, Gh.A., et al., 2014. Effect of settling test procedure on sizing thickeners. Sep. Purif. Technol. 122, 87–95.

Pearse, M.J., 1977. Gravity Thickening Theories: A Review. Department of Industry, Warren Spring Laboratory.

Pearse, M.J., 1980. Factors affecting the laboratory sizing of thickeners. Plumtree, A. (Ed.), Fine Particle Processing, vol. 2, pp. 1619–1642 (Chapter 81).

Pearse, M.J., 2003. Historical use and future development of chemicals for solid-liquid separation in the mineral processing industry. Miner. Eng. 16 (2), 103–108.

Prokesch, M.E., 2002. Selection and sizing of concentrate drying, handling and storage equipment. In: Mular, A.L., et al., (Eds.), Mineral Processing Plant Design, Practice and Control, vol. 2. SME, Littleton, CO, USA, pp. 1463–1477.

Rushton, A, et al., 2000. Solid-Liquid Filtration and Separation Technology. VCH, Weinheim, Federal Republic of Germany.

Scales, P.J., et al., 2015. Compressional dewatering of flocculated mineral suspension. Can. J. Chem. Eng. 93 (3), 549–552.

Schoenbrunn, F., Laros, T., 2002. Design features and types of sedimentation equipment. In: Mular, A.L., et al., (Eds.), Mineral Processing Plant Design, Practice and Control, vol. 2. SME, Littleton, CO, USA, pp. 1331–1341.

Seifert, J.A., Bowersox, J.P., 1990. Getting the most out of thickeners and clarifiers. Chem. Eng. 97 (7-12), 80–83.

Singh, B.P., et al., 1998. Use of surfactants to aid dewatering of fine clean coal. Fuel. 77 (12), 1349–1356.

Smith, C.B., Townsend, I.G., 2002. Testing, sizing and specifying of filtration equipment. In: Mular, A.L., et al., (Eds.), Mineral Processing Plant Design, Practice and Control, vol. 2. SME, Littleton, CO, USA, pp. 1313–1330.

Somasundaran, P., Wang, D., 2006. In: Wills, B.A. (Series Ed.), Developments in Mineral Processing. Solutions Chemistry: Minerals and Reagents, vol. 17. Elsevier, Amsterdam, Boston, USA.

Stickland, A.D., et al., 2008. Numerical modelling of flexible-membrane plate-and-frame filtration: formulation, validation and optimization. AIChE J. 54 (2), 464–474.

Svarovsky, L., 2000. Gravity clarification and thickening, Solid-Liquid Separation. fourth ed. Butterworth-Heinemann, Oxford, UK, pp. 166–190.

Taggart, A.F., 1945. Handbook of Mineral Dressing, Ore and Industrial Minerals. John Wiley, & Sons., Chapman & Hall, Ltd, London, UK.

Talmage, W.P., Fitch, E.B., 1955. Determining thickener unit areas. Ind. Eng. Chem. 47 (1), 38–41.

Tarleton, E.S., Wakeman, R.J., 2006. Solid/liquid Separation: Equipment, Selection and Process Design. Elsevier, Amsterdam, Netherlands.

Townsend, I., 2003. Automatic pressure filtration in mining and metallurgy. Miner. Eng. 16 (2), 165–173.

Tripathy, T., De, B.R., 2006. Flocculation: a new way to treat the waste water. J. Phys. Sci. 10, 93–127.

Usher, S.P., et al., 2001. Validation of a new filtration technique for dewaterability characterization. AIChE J. 47 (7), 1561–1570.

Usher, S.P., et al., 2009. Theoretical analysis of aggregate densification: impact on thickener performance. Chem. Eng. J. 151 (1-3), 202–208.

Wakeman, R.J., Tarleton, E.S., 1999. Filtration: Equipment Selection, Modelling and Process Simulation. first ed. Elsevier Advanced Technology, Elsevier Science Ltd, Oxford, UK.

Wang, X.Y, et al., 2010. Polymer aids for settling and filtration of oil sands tailings. Can. J. Chem. Eng. 88 (3), 403–410.

Waters, A.G., Galvin, K.P., 1991. Theory and application of thickener design. Filtr. Sep. 28 (2), 110–116.

Welch, D.G, 2002. Characterization of equipment based on filtration principals and theory. In: Mular, A.L., et al., (Eds.), Mineral Processing Plant Design, Practice and Control, vol. 2. SME, Littleton, CO, USA, pp. 1289–1294.

Chapter 16

Tailings Disposal

16.1 INTRODUCTION

The disposal of mill tailings is a major environmental problem, which is becoming more serious with the increasing exploration for metals and the working of lower-grade deposits. Apart from the visual effect on the landscape of tailings disposal, the major ecological effect is usually water pollution, arising from the discharge of water contaminated with solids, heavy metals, mill reagents, and sulfur compounds (Chalkley et al., 1989). Waste must therefore be disposed of in both an environmentally acceptable and, if possible, economically viable manner (Ritcey, 1989). Disposal is governed by legislation and may involve long-term rehabilitation of the site.

The nature of tailings varies widely; they are usually transported and disposed of as a slurry of high water content, but they may be composed of very coarse dry material, such as the float fraction from dense medium plants. Due to the lower costs of mining from open pits, ore from such locations is often of very low grade, resulting in the production of large amounts of fine tailings.

16.2 METHODS OF DISPOSAL OF TAILINGS

The methods used to dispose of tailings have developed due to environmental pressures, changing milling practice, and realization of profitable applications. Early methods included discharge of tailings into rivers and streams, which is still practiced at some mines, and the dumping of coarse dewatered tailings on to land. The many nineteenth-century tips seen in Cornwall and other parts of Britain are evidence of this latter method. Due to the damage caused by such methods, and the much finer grinding necessary on most modern ores, other techniques have been developed.

16.2.1 Tailings Dams

The design, construction, and operation of tailings dams is a major consideration for most new mining projects, as well as for many existing operations (Vick, 1984).

It is economically advantageous to site the impoundment close to the mine, but this imposes limits on site selection. The type of tailings embankment is generally determined by the local seismic activity, water clarification requirements, tailings properties and stability, tailings size distribution, foundation and hydrological conditions, and environmental factors (Azizli et al., 1995). The ground underlying the dam must be structurally sound and able to bear the weight of the impoundment. If such a site cannot be found close to the mine, it may be necessary to pump the tailings, at a high slurry density, to a suitable location.

Tailings dams may be built across river valleys, or as curved or multi-sided dam walls on valley sides, this latter design facilitating drainage. On flat, or gently sloping ground, lagoons are built with walls on all sides of the impoundment.

The disposal of tailings adds to the production costs, so it is essential to make disposal as cheap as possible. This requirement led initially to the development of the once commonly used *upstream method* of tailings dam construction, so named because the centerline of the dam moves upstream into the pond.

In this method, a small starter dam is placed at the extreme downstream point (Figure 16.1) and the dam wall is progressively raised on the upstream side. The tailings are discharged by spigoting off the top of the starter dike and, when the initial pond is nearly filled, the dike is raised and the cycle repeated. Various methods are used to raise the dam; material may be taken from the dried surface of the previously deposited tailings and the cycle repeated, or more commonly, the wall may be built from the coarse fraction of the tailings, separated out by cyclones, or spigots, the fines being directed into the pond (Figure 16.2 and 16.3).

The main advantages of the upstream construction are the low cost and the speed with which the dam can be raised by each successive dike increment.

The method suffers from the disadvantage that the dam wall is built on the top of previously deposited unconsolidated slimes retained behind the wall. There is a limiting height to which this type of dam can be built before failure occurs and the tailings flow out and,

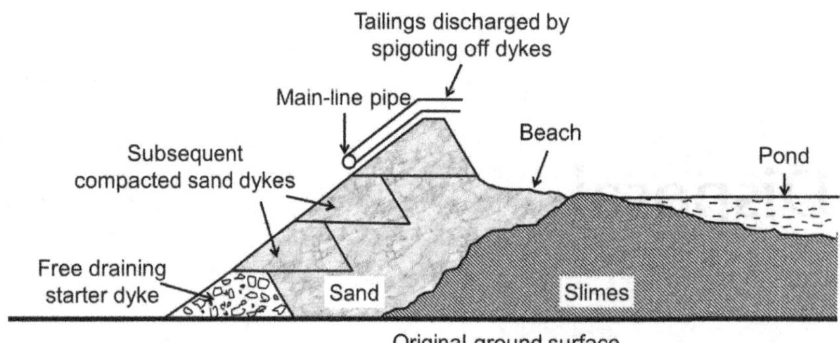

FIGURE 16.1 Upstream tailings dam.

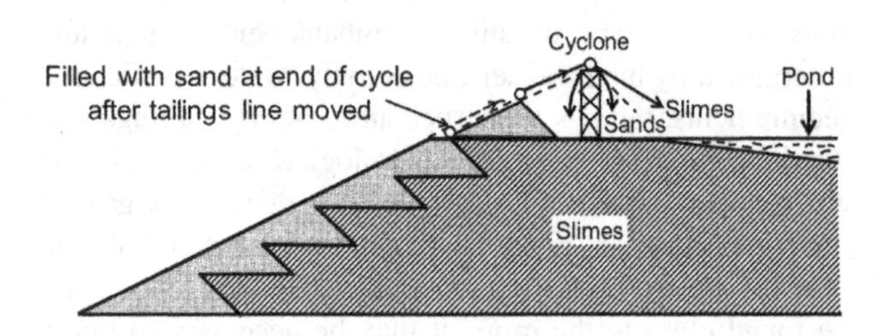

FIGURE 16.2 Construction of upstream tailings dam using cyclones.

because of this, the upstream method of construction is now less commonly used. Several major failures have involved tailings dams constructed with the upstream method (van Zyl, 2014).

The *downstream method* has evolved as a result of efforts to devise methods for constructing larger and safer tailings dams. This method produces safer dams both in terms of static and seismic loading (Azizli et al., 1995). It is essentially the reverse of the upstream method, in that as the dam wall is raised, the centerline shifts downstream, and the dam remains founded on coarse tailings (Figure 16.4). Most procedures involve the use of cyclones to produce sand for the dam construction.

Downstream dam building is the only method that permits design and construction of tailings dams to acceptable engineering standards. All tailings dams in seismic areas, and all major dams, regardless of their location, should be constructed using some form of the downstream method. The major disadvantage of the technique is the large amount of sand required to raise the dam wall. It may not be possible, especially in the early stages of operation, to produce sufficient sand volumes to maintain the crest of the tailings dam above the rising pond levels. In such cases, either a higher starter dam is required or the sand supply must be augmented with borrowed fill, such procedures increasing the cost of tailings disposal.

The *centerline method* (Figure 16.5) is a variation of that used to construct the downstream dam and the crest remains in the same horizontal position as the dam wall is raised. It has the advantage of requiring smaller volumes of sand-fill to raise the crest to any given height. The dam can thus be raised more quickly and there is less trouble keeping it ahead of the tailings pond during the early stages of construction. Care, however, must be exercised

FIGURE 16.3 Construction of tailings dam wall using cyclone underflow (spigoting) *(Courtesy Fraser Alexander)*.

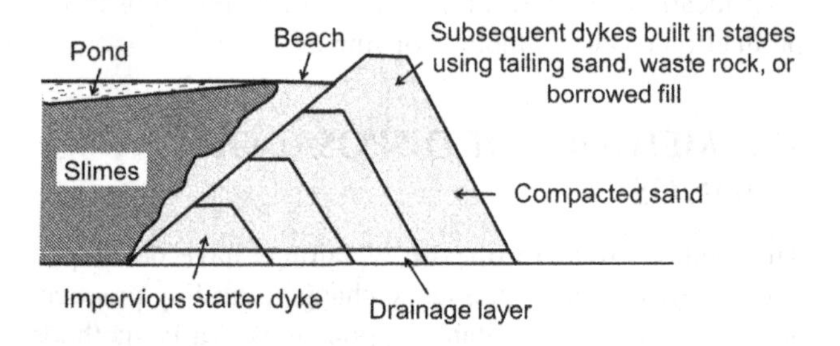

FIGURE 16.4 Downstream tailings dam.

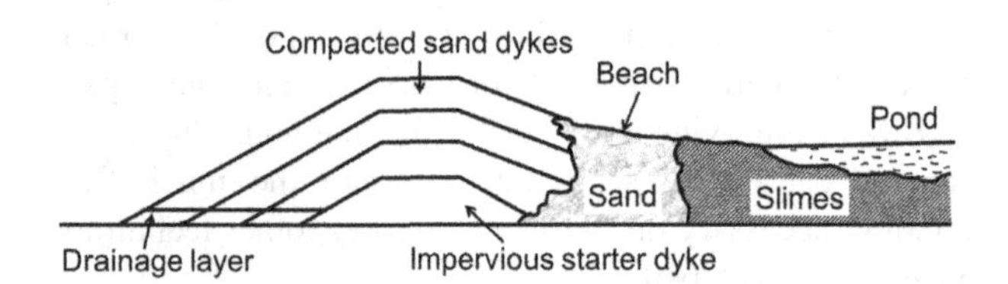

FIGURE 16.5 Centerline tailings dam.

in raising the upstream face of the dam to ensure that unstable slopes do not develop temporarily.

Very stable tailings dams can be constructed from open-pit over-burden, or waste rock, according to the local circumstances. An example is shown in Figure 16.6. Waste rock is considered an interesting construction material when available in sufficient quantities because of its ideal mechanical and geotechnical properties, such as well-graded particle size distribution, high internal friction angle, and high pore-water dissipation capacity (Julien and Kissiova, 2011). Since the tailings are not required for the dam construction, they may be fed into the pool without separation of the sands from the slimes. In some cases, the output of overburden may not be sufficient to keep the dam crest above the tailings pond, and it may be necessary to combine waste rock and tailings sand-fills to produce a safe economical dam.

Erosion of dams due to wind and rain can affect the stability and produce environmental problems. Many methods are used to combat this, such as vegetation of the dam banks (Hill and Nothard, 1973) and chemical stabilization to form an air- and water-resistant crust.

There is little doubt that tailings dams have visual impact. Perhaps the most conspicuous is the downstream type, whose outer wall is continually being extended, and cannot be re-vegetated until closure. There are, however, few reasons why dam walls should not be landscaped at some stage in their life, and many dams have been designed to permit early visual integration with the environment (Down and Stocks, 1977a). An example is the impoundment at Flambeau, North Wisconsin, USA (Shilling and May, 1977), where a rock-fill dam wall 18 m high, 24 m wide at the crest, and 111 m wide at the base was designed to minimize both visual and pollution effects (Figure 16.7). The wall consists of a clay core, with the downstream side faced with non-pyrite rock and covered with top-soil, permitting re-vegetation, and consequently, reduced visual impact.

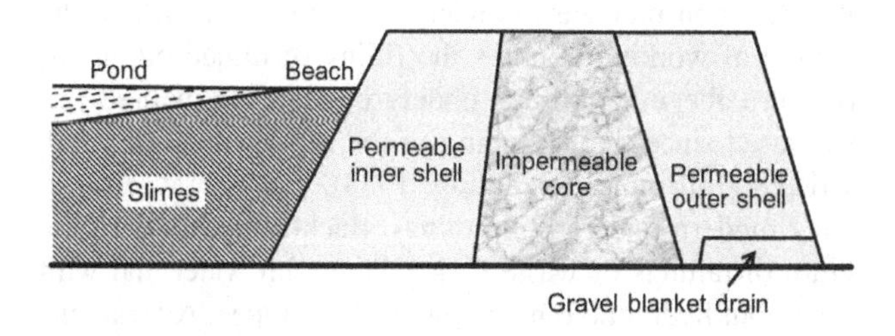

FIGURE 16.6 Dam constructed from overburden.

Figure 16.8 shows a generalized representation of water gain and loss at a tailings impoundment (Down and Stocks, 1977b). With the exception of precipitation and evaporation, the rates and volumes of the water can be controlled to a large extent. It is more satisfactory to attempt to prevent the contamination of natural waters rather than to treat them afterwards, and if surface run-off to the dam is substantial, then interception ditches should be installed. It is difficult to quantify the amount of water lost to groundwater, but this can be minimized by selecting a site with impervious foundations, or by sealing with an artificial layer of clay. Seepage through the dam wall is often minimized by an impervious slimes layer on the upstream face of the dam, but this is expensive, and many mines prefer to encourage free-drainage of the dam through pervious, chemically barren material. In the case of upstream dams, this can be a barren starter dike, while with downstream and centerline constructions, a free-draining gravel blanket can be used. A small seepage pond with impervious walls and floors situated below the main dam can collect this water, from where it can be pumped back into the tailings pond. If the dam wall is composed of metal-bearing rock, or sulfide tailings, the seepage is often highly contaminated due to its contact with the solid tailings, and may have to be treated separately.

The tailings are often treated with lime to neutralize acids and precipitate heavy metals as insoluble hydroxides before pumping to the dam. Such treated tailings may be

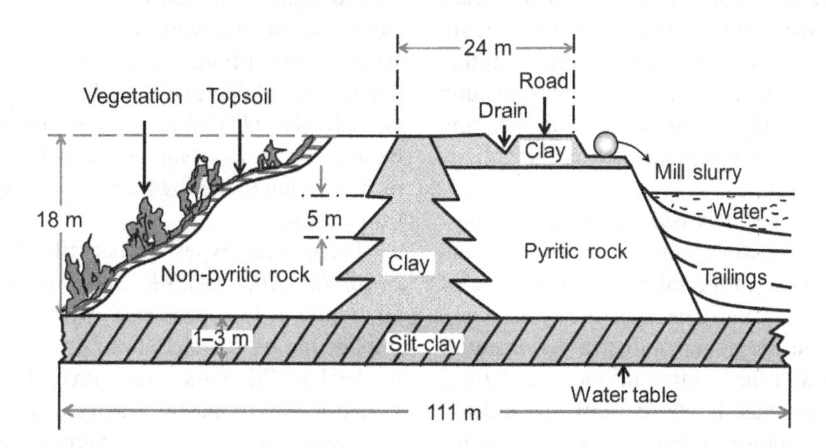

FIGURE 16.7 Flambeau impoundment.

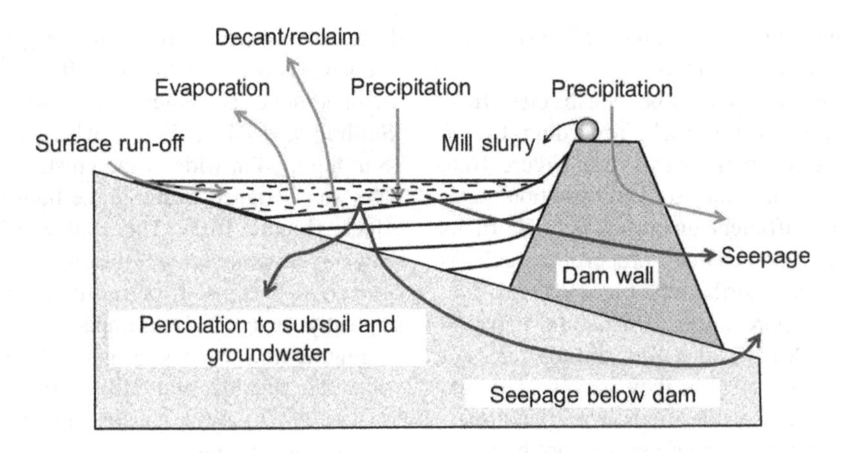

FIGURE 16.8 Water gain and loss in a typical tailings dam.

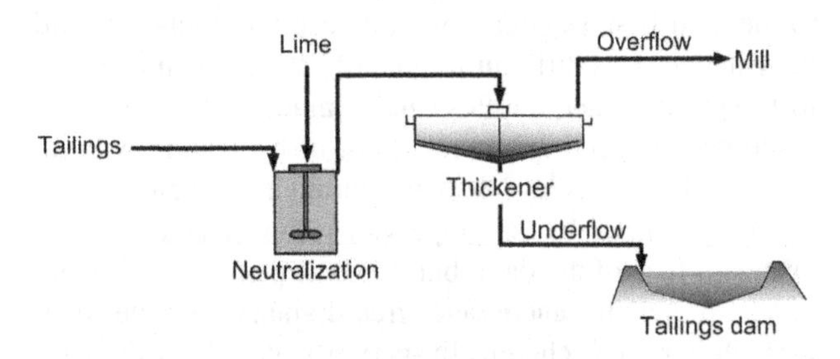

FIGURE 16.9 Treatment of tailings with lime.

thickened, and the overflow, free of heavy metals, returned to the mill (Figure 16.9), thus reducing the water and pollutant input to the tailings dam.

Assuming good control of the above inputs and outputs of dam water, the most important factor in achieving pollution control is the method used to remove surplus water from the dam. Decant facilities are required on all dams to allow excess free water to be removed. Inadequate decant design has caused major dam failures. Many older dams used decant towers with discharge lines running through the base of the dam to a downstream pump-house. Failures of such structures were common due to the high pressures exerted on the pipelines, leading to uncontrolled losses of fluids and tailings downstream. Floating, or movable, pump-houses situated in the tailings pond are now in common use.

Recycling of decant water is becoming more important. As much water as possible must be reclaimed from the tailings pond for reuse in the mill and the volume of fresh make-up water used must be kept to a minimum. The difference between the total volume of water entering the tailings pond and the volume of water reclaimed plus evaporation losses must be stored with the tailings in the dam. If that difference exceeds the volume of the voids in the stored tailings, there becomes a surplus of free water that can build up to tremendous quantities over the life of a mine. A typical dam-reclaim system is shown in Figure 16.10.

The main disadvantage of water reclamation is the recirculation of contaminants to the mill, which can interfere with processes such as flotation. Water treatment may overcome this, at little or no extra cost, as similar treatment would be required for the effluent discharge in any case.

16.2.2 Backfill

It is common practice in underground mines, in which the method of working requires the filling of mined-out areas, to return the mill tailings underground. This method has been used since the beginning of the 20th century in South Africa's gold mines (Stradling, 1988), and is now used in many modern underground mines. Backfilling involves the return of tailings or waste rock mixed with water and with or without hydraulic binders into empty stopes. After a curing period of a few months, the backfill can act as effective ground support and allow recovering ore from an adjacent stope, which otherwise would have been left as ground support. Backfilling worked-out stopes also reduces the volume of tailings that must be impounded on the surface; typically 40–60% of all tailings produced by the mine can be returned underground as backfill, therefore the footprint of the required tailings surface impoundment is significantly reduced.

Three main types of backfill are in use worldwide: *hydraulic fill*, *rockfill*, and *cemented paste backfill*. Hydraulic fill is produced with de-slimed tailings and implies the use of binders to provide mechanical strength to the backfill mass once cured. Hydraulic fill is transported underground by gravity or pumping as a 60–70% (by weight) solids slurry. Significant amounts of water and binder are recovered after deposition in the stope and

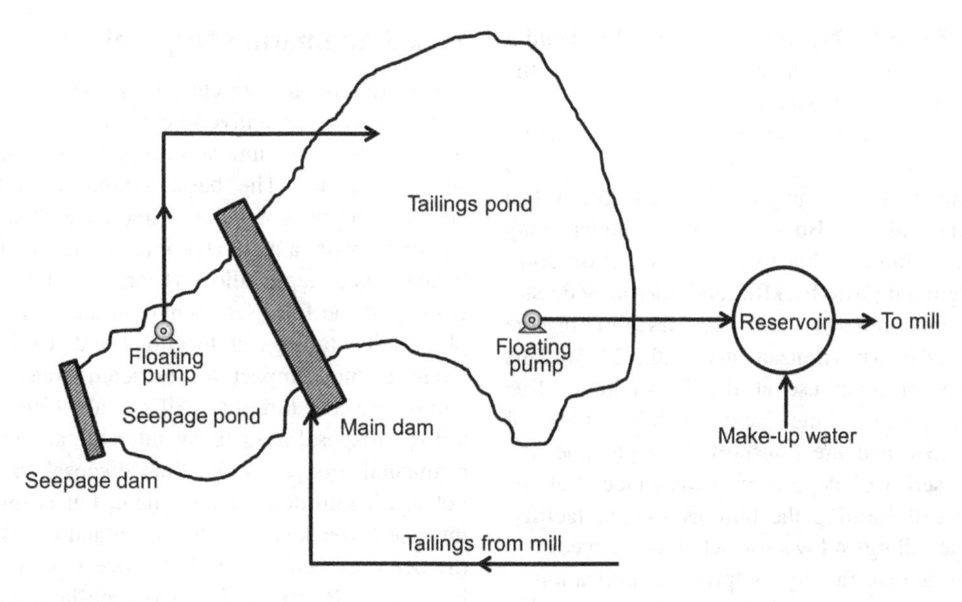

FIGURE 16.10 Water-reclamation system.

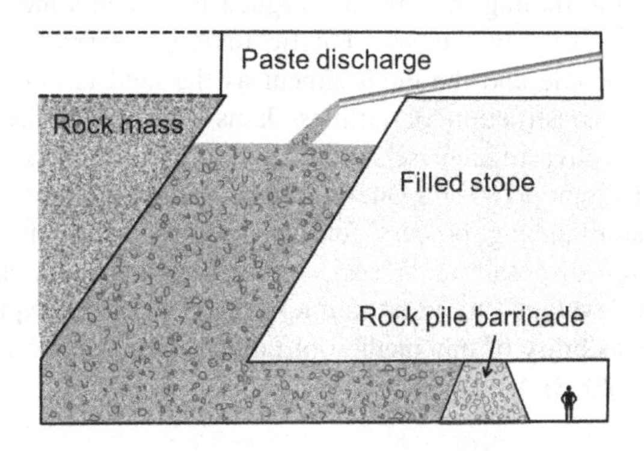

FIGURE 16.11 Schematic view of backfilled stope.

must be pumped to the surface. Rockfill is produced with waste rock distributed underground by truck or conveyor belts. In some cases, a binder slurry can be added to improve mechanical strength.

Cemented paste backfill has been applied since the 1990s and has rapidly gained popularity because of its advantages over hydraulic backfill, notably: use of the total tailings (no de-sliming required), improved mechanical strength, and possibility to return underground sulfidic tailings prone to acid rock drainage (ARD) (Belem and Benzaazoua, 2008). Cemented paste backfill consists of a mixture of tailings, water, and binder (e.g., Portland cement) that is usually pumped into underground stopes as a 70–80% solids paste, as shown in Figure 16.11. The choice and dosage of binder depends on the composition of tailings and water, and on the required mechanical strength, and ultimately on economics. Generally, the amount of binder used is between 3% and 7% of the dry tailings mass. Binder is the major expense in a paste backfill operation, therefore there is incentive to substitute part of the cement content with by-products from other industries with adequate properties, such as fly ash and blast-furnace slag (Peyronnard and Benzaazoua, 2011). Paste backfill recipes must be optimized for each site through laboratory testing and regular follow-up, since the tailings and water properties have an influence on strength acquisition.

16.2.3 Densified Tailings

Advances in the disposal of tailings using semidry or dry techniques offer a number of advantages over the wet disposal techniques. Collectively, these techniques are known as "densified tailings," and require that tailings be thickened or dewatered prior to disposal. The densified tailings techniques comprise *dry stacking* (or filtered tailings), *thickened tailings*, and *paste tailings*. These are all schemes that improve water and reagent recovery and decrease tailings volumes and footprint, which greatly assists site rehabilitation. Although disposal of densified tailings has benefits, these techniques require a detailed understanding of the rheology and transport of the densified tailings (Nguyen and Boger, 1998).

The thickened tailings technique involves the use of thickeners to bring the tailings to a solid content between 50% and 70%, to make the tailings mass more homogeneous and self-supporting, while reducing the size of the dams (Robinsky, 1999). Furthermore, thickened tailings disposal areas do not require settling ponds because of the improved water recovery at the thickening stage. This method of disposal was used at the Ecstall (Kidd Creek) operation at Texasgulf Canada Ltd. (Amsden, 1974). The tailings disposal area consists of 3,000 acres enclosed by a gravel dike. Mill tailings are thickened and pumped to a central spigoting location inside the dam. The system is designed to build a mountain of tailings in the central area and thus keep the height of the perimeter dike to a minimum.

Paste tailings are produced by thickening and filtering the tailings slurry up to 70–85% solids. The benefits of a high solids content are the improved hydrogeotechnical properties of the tailings (Bussière, 2007). Binders may be added to paste tailings, but in lower proportion compared to underground paste backfill, and may provide stabilization of contaminants (such as arsenic) in the cemented tailings matrix (Benzaazoua et al., 2004). This method of disposal is in use at the Bulyanhulu Mine (Tanzania). The paste tailings have a solids content of approximately 73% and are transported by pipeline and deposited by a series of deposition towers placed at the center of each cell forming the tailings storage facility. The use of paste tailings allows for better water recovery, reduced tailings storage facility footprint, minimization of contaminant transport, lower wind erosion, and possible progressive site rehabilitation (Theriault et al., 2003).

Dry stacking refers to the disposal of tailings having a solid content above 85%, usually obtained by thickening and filtering. At such low moisture content, the tailings have to be transported to the tailings storage facility by truck or conveyor belt. This method is particularly useful in arid climates, or when water handling issues are significant (Davies and Rice, 2001). For example, the Raglan Mine in Quebec is located in an arctic environment and uses dry stacking to recycle most of its process water and transports the tailings by truck to the impoundment area, where they are integrated into the permafrost (Bussière, 2007).

Densified tailings is expected to become more common in operating mines, particularly those with low ore grade and high throughput, which produce large volumes of tailings and require large impoundment areas. For those high tonnage operations, a reduction in tailings volume has a significant impact on dam size and the area of disturbed land.

16.2.4 In-pit Disposal

The disposal of tailings in mined-out open pits can be an alternative to constructing a new tailings storage facility, provided that the open pit and the tailings have the appropriate characteristics. The main risk associated with in-pit disposal is the contamination of the groundwater network by leachates from the tailings. This risk is minimized for open pits that act as water sinks, that is, groundwater flows toward the pit, and/or when the bedrock is mostly impervious (not fractured). Hydrogeological studies of the pit area must be undertaken to confirm the suitability of the pit to safely contain the tailings while maintaining the quality of the groundwater (Eary, 2011).

16.2.5 Submarine Disposal

For operations that are close to the sea, submarine tailings disposal is an alternative to conventional tailings disposal, provided the governmental regulations permit disposal in such a manner. The basic submarine tailings disposal design comprises a tailings line to a de-aeration/mixing chamber, with a seawater intake line, and discharge to location and depth allowing gravity flow of a coherent density to the final sedimentation area. Such systems can place mine tailings at locations and depths constraining environmental impact to restricted areas of the seabed and deep water turbidity (Ellis et al., 1995). This form of tailings disposal attracts considerable attention from environmental groups, as the final disposal of the tailings is not in a controlled impoundment, but is released directly into the lower levels of the ocean and can therefore affect the deep sea ecosystem. The process is increasingly used in the Asia-Pacific region where on-land disposal options are problematic. In comparison to tailing retention on land, the mining industry has argued that submarine tailings disposal in the Asia-Pacific region is safer for the local people and the environment as the land is unsuited to the construction of tailings dams due to the natural topography, regular seismic activity, and high rainfall (McKinnon, 2002). Due to the complexity of the decision-making process for the viability of submarine tailings disposal, tools such as an expert system have been developed to assist mining project planners explore the feasibility of this method of tailings disposal (Ganguli et al., 2002).

16.2.6 Reprocessing and Reuse of Tailings

The most satisfactory way of dealing with tailings is to make positive use of them, such as reprocessing in order to recover additional values, or to use them as a useful product in their own right, for example, the use of coarse (20–30 mm) DMS float as railway ballast and aggregate. Tailings can be used as a component of mortar for concrete or as fined-grained construction material. Depending on the jurisdiction, the reuse of tailings may be encouraged either on the mine site as construction material or for other usage outside the mine site. The incentive to strive for sustainable development should lead to new alternatives for tailings reuse; phytomining where plants, natural or genetically modified, are used to uptake metals is one emerging possibility (Moskvitch, 2014).

Reprocessing the tailings may in some cases become economically attractive, especially for older tailings produced when the processing methods were not as efficient as they are today. With the global drop in ore grades, yesterday's wastes may become tomorrow's resources. Mining and comminution costs are also significantly lower when reprocessing tailings than with typical ore (Muir et al., 2005).

16.3 ENVIRONMENTAL ISSUES

Apart from geotechnical stability issues, the most serious problem associated with the disposal of tailings is the release of polluted water, and this has been extensively investigated. The main effects of pollution are due to the effluent pH, which may cause ecological changes; dissolved heavy metals, such as copper, lead, zinc, etc., which can be lethal to fish-life if allowed to enter local water-courses; mill reagents, which are usually present in only very small quantities, but, nevertheless, may be harmful; and suspended solids, which should be minimal if the tailings have sufficient residence time in the dam, thus allowing the solids to settle and produce a clear decant.

16.3.1 Cyanide and Ammonia Management

Complexes of metals with cyanide and ammonia are especially prone to stabilization and solubilization in alkali solutions and may require special treatment other than straightforward neutralization by lime. Although natural degradation occurs to some extent, this is of little value in many cases during the winter months in some locales, when the tailings ponds may be ice-covered. Several processes have been developed to treat cyanide-bearing effluent, these occasionally being derived from cyanide used as a flotation reagent (see Section 12.17.3) but especially derived from cyanide use in gold extraction (Scott and Ingles, 1981). Alkaline chlorination, whereby cyanide is oxidized to cyanate, is one route (Eccles, 1977), but cyanides can also be effectively destroyed by oxidation with ozone (Jeffries and Tczap, 1978) or hydrogen peroxide, and by reactions with sulfur dioxide and air (Lewis, 1984), processes that are used industrially. Other methods include electrochemical treatment, ion exchange, and volatilization of hydrogen cyanide. In the latter method, which has been proven at full scale, the tailings are acidified to produce hydrogen cyanide which is volatilized by intensive air-sparging, while simultaneously recovering the evolved gas in a lime solution for recycling. The aerated, acidified barren solution is then re-neutralized to precipitate the metal ions.

Cyanide destruction may also cause other issues, such as production of thiocyanates and ammonia during SO_2-air treatment (Mudder et al., 2001; Gould et al., 2012), so it is imperative to test and identify potential problems before selecting a treatment option.

Ammonium ions enter the systems from the use of the blasting agent ammonium nitrate fuel oil (ANFO), from amines used as flotation reagents, and from certain hydrometallurgical operations. There is the possibility of biological control, using bacteria (*nitrifiers*) that gain their energy for growth from oxidation of ammonia (Kapoor et al., 2003).

16.3.2 Acid Rock Drainage

Acid rock drainage (ARD) (or acid mine drainage, AMD) is considered the most significant environmental issue related to mine waste management. It is produced through the natural oxidation of sulfidic minerals by air and water, accelerated by bacterial action (*thiobacillus*); thus exposed sulfide-bearing tailings (and waste rock) are prone to ARD generation. Pyrite and pyrrhotite are the main ARD generating sulfide minerals and are found in many deposits associated with base metals, gold and coal. The resulting acid leaches other heavy and toxic metals into the ARD (Ritcey, 1989). (Under other conditions, primarily moisture levels 3–8%, the same oxidation reactions can cause sulfide self-heating, a topic introduced in Chapter 2.)

The mineralogical nature of the tailings can provide some natural pollution control. For instance, the presence of alkaline gangue minerals such as limestone can render metals less soluble and neutralize oxidation products. Such ores thus present fewer problems than sulfide ores associated with neutral-acid gangues. Several testing methods have been developed to identify potentially acid-generating tailings, either based on the balance between acid-producing and neutralizing minerals (Sobek et al., 1978; Miller et al., 1991; Lawrence and Scheske, 1997) or on the reaction rates of acid generation and neutralization processes (Bussière et al., 1997; Morin and Hutt, 1997; Lapakko and White, 2000).

Once ARD generation has begun, it is difficult to control and stop. Therefore, several jurisdictions impose regulations on ARD prevention, usually by obligating mine operators to reclaim their tailings ponds at the end of mine-life, or preferably, during operation. To prevent ARD, the main approaches involve the removal of one of the components for sulfide oxidation, either air (oxygen), water, or the sulfide minerals themselves. The selection of the ARD prevention approach is based on tailings characterization, climatic conditions at the site, social and economic considerations, and desired performance. In humid climates, covers to limit oxygen transport are generally favored. Water covers, where tailings are deposited under a water layer in an impoundment built with impervious dikes, are efficient at limiting oxygen transport because oxygen solubility and diffusion in water are several orders of magnitude lower than in air (Davé et al., 1997). However, the long-term geotechnical stability risk of maintaining impervious dikes led to the development of alternative dry cover systems that induce lower pressures on the confining structures. Dry covers of various types can offer similar performance as water covers to limit oxygen transport when at least one cover layer maintains a high water content (Nicholson et al., 1989). Different types of materials can be used as dry covers: natural geomaterials (clay, silt, sand), industrial materials (neutral tailings, neutral waste rock, sludge), geomembranes and geocomposites (Aubertin et al., 1997; Yanful et al., 1999; Demers et al., 2008). Dry cover systems consist of single to multilayered covers; the selection of the appropriate cover system for a given site depends on material availability, site characteristics and required performance. In arid climates, water exclusion, by covering with different soils layers or with impervious geomembranes, is usually the most appropriate option (Williams et al., 1997; Zhan et al., 2001). In permafrost conditions, tailings may be integrated into the frozen soil, as the cold temperatures significantly reduce sulfide mineral oxidation rates (Elberling, 2001, 2005). However, global climate changes may affect the permafrost conditions and more robust ARD strategies must be developed for arctic tailings disposal and reclamation.

A novel approach to preventing ARD is the argument that since oxidative processes cause ARD, then reductive processes will counteract the oxidation. There are examples where reductive reactions have occurred naturally and eliminated acid drainage. They appear to be related to the formation of biofilms hosting oxygen-consuming bacteria (*acidophilic heterotrophs*), which under the right conditions can outcompete the bacteria promoting oxidation (Kalin et al., 2012). This is an example of the application of *ecological engineering* principles to solving the ARD problem (Kalin, 2005).

When ARD has begun, chemical treatment of acid effluents is essential, neutralization by lime usually being performed, which precipitates the heavy metals, and promotes flocculation as well as reducing acidity. Furthermore, a number of wastewater treatment techniques are available, such as physical adsorption methods using active carbon, bentonite clay or mineral slimes, biological oxidation of organics, biological sulfate reduction for metal precipitation as sulfides, removal of ionic species by ion exchange resins, and reverse osmosis (Rao and Finch, 1989; Younger et al., 2002; Lawrence et al., 2003). Passive treatment methods, such as wetlands and limestone drains, are proposed as alternatives to chemical treatment and are considered a more suitable solution for long-term treatment of ARD (Bernier et al., 2002; Neculita et al., 2007).

Rather than just treat, there may be options for recovery of heavy metals (and other resources) (Rao, 2006). For instance, it has been established that various microorganisms that may be recovered by flotation are able to abstract heavy metal ions from aqueous solutions, thus making it possible not only to solve some industrial environmental problems, but also to recover a currently wasted product (Veneu et al., 2014).

As a specific case, a major concern is control of selenium in discharge waters, especially from the mining of coal (Harrison et al., 2013). Several techniques are employed: ion exchange, zero-valent iron (ZVI), and active biological treatment. Ion exchange removes selenium by exchanging selenate for ions on a resin backbone. ZVI refers to using elemental iron to reduce the oxidized forms of selenium (selenite, selenate) to elemental selenium. The ZVI can be in powder, granular, or fiber forms. The elemental selenium is insoluble and exists as nanoparticles embedded in the iron and is recovered. In active biological treatment, certain bacteria gain their energy from reduction of selenites and selenates to selenium. One engineered system is to use a fluidized bed reactor (FBR). Solid-liquid separation is required to recover the biomass with the selenium. Two full-scale FBR-based treatment plants are scheduled for operation in 2014.

It is evident that there is much research potential in these areas. Particular attention is being given to the modification of mineral processing operations to mitigate environmental impact. The ultimate way of avoiding water-based environmental impact is to operate dry mineral processes, and consideration is being given to such options, particularly in arid areas.

REFERENCES

Amsden, M.P., 1974. Ecstall concentrator. CIM Bull. 67 (745), 105–115.

Aubertin, M., et al., 1997. Construction and instrumentation of in situ test plots to evaluate covers built with clean tailings. Proc. 4th International Conference on Acid Rock Drainage (ICARD), Vancouver, BC, Canada, pp. 717–730.

Azizli, K.A.M., et al., 1995. Design of the Lohan tailings dam, Mamut Copper Mining Sdn. Bhd., Malaysia. Miner. Eng. 8 (6), 705–712.

Belem, T., Benzaazoua, M., 2008. Design and application of underground mine paste backfill technology. Geotech. Geol. Eng. 26 (2), 147–174.

Benzaazoua, M., et al., 2004. A laboratory study of the behaviour of surface paste disposal. Proc. 8th International Symposium on Mining with Backfill. Nonferrous Metals Society of China, Beijing, China, pp. 180–192.

Bernier, L., et al., 2002. On the use of limestone drains in the passive treatment of acid mine drainage (AMD). Proc. Symposium 2002 sur l'Environnement et les Mines, Rouyn-Noranda, Canada.

Bussière, B., 2007. Colloquium 2004: hydrogeotechnical properties of hard rock tailings from metal mines and emerging geoenvironmental disposal approaches. Can. Geotech. J. 44 (9), 1019–1052.

Bussière, B., et al., 1997. Effectiveness of covers built with desulphurized tailings: column tests investigation. Proc. 4th International Conference on Acid Rock Drainage (ICARD), Vancouver, Canada, pp. 763–778.

Chalkley, M.E., et al. (Eds.), 1989. Tailings and effluent management. Proc. International Symposium on Tailings and Effluent Management, Halifax, NS, Canada.

Davé, N.K., et al., 1997. Water cover on reactive tailings and wasterock: laboratory studies of oxidation and metal release characteristics. Proc. 4th International Conference on Acid Rock Drainage (ICARD), Vancouver, Canada. pp. 779–794.

Davies, M.P., Rice, S., 2001. An alternative to conventional tailings management – "dry stack" filtered tailings. Proc. 8th International Conference on Tailings and Mine Wastes '01, Fort Collins, CO, USA. pp. 411–420.

Demers, I., et al., 2008. Column test investigation on the performance of monolayer covers made of desulphurized tailings to prevent acid mine drainage. Miner. Eng. 21 (4), 317–329.

Down, C.G., Stocks, J., 1977a. Methods of tailings disposal. Min. Mag. 136 (5), 345–359.

Down, C.G., Stocks, J., 1977b. Environmental problems of tailings disposal. Min. Mag. 137 (1), 25–33.

Eary, L.E., 2011. A review of closure strategies for pit lakes at hard rock metal mines in the United States. Proc. Symposium 2011 sur l'Environnement et les Mines, Rouyn-Noranda, Canada.

Eccles, A.G., 1977. Pollution control at Western Mines Myra Falls operations. CIM Bull. 70 (785), 141–147.

Elberling, B., 2001. Environmental controls of the seasonal variation in oxygen uptake in sulfidic tailings deposited in a permafrost-affected area. Water Resour. Res. 37 (1), 99–107.

Elberling, B., 2005. Temperature and oxygen control on pyrite oxidation in frozen mine tailings. Cold Regions Sci. Technol. 41 (2), 121–133.

Ellis, D.V., et al., 1995. Submarine tailings disposal (STD) for mines: an introduction. Mar. Georesour. Geotechnol. 13 (1-2), 3–18.

Ganguli, R., et al., 2002. STADES: an expert system for marine disposal of mine tailings. Mining Eng. 54 (4), 16–21.

Gould, D.W., et al., 2012. A critical review on destruction of thiocyanate in mining effluents. Miner. Eng. 34 (0), 38–47.

Harrison, T., et al., 2013. Selenium treatment for mine water discharge compliance. In: Craynon, J.R. (Ed.), Environmental Considerations in Energy Productio. SME, Englewood, CO, USA, pp. 443–449.

Hill, J.R.C., Nothard, W.F., 1973. The Rhodesian approach to the vegetating of slimes dams. J. S. Afr. Inst. Min. Metall. 73, 197–208.

Jeffries, L.F., Tczap, A., 1978. Homestake's Grizzly Gulch tailings disposal project. Min. Cong. J. 64 (11), 23–28.

Julien, M., Kissiova, M., 2011. Benefits and challenges of using tailings as foundation and construction material to increase capacity of tailings storage facilities. Proc. Symposium 2011 sur l'Environnement et les Mines, Rouyn-Noranda, Canada.

Kalin, M., 2005. Ecological engineering in Northern Ontario: tailor made solutions for an acid generating mine site. Canadian Reclamation. Canadian Land Reclamation Association, Calgary, Alberta, Canada. Special Conference Edition, (Winter/Spring), pp. 13–17.

Kalin, M., et al., 2012. Phosphate mining wastes reduces microbial oxidation of sulphidic minerals: a proposed mechanism. Proc. 9th International Conference. on Acid Rock Drainage (ICARD), Ottawa, ON, Canada, pp. 585–593.

Kapoor, A., et al., 2003. Use of a rotating biological contactor for removal of ammonium from mining effluents. Eur. J. Min. Proc. Environ. Prot. 3 (1), 88 100.

Lapakko, K.A., White, W.W., 2000. Modification of the ASTM 5744-96 Kinetic Test. Proc. 5th International Conference on Acid Rock Drainage (ICARD), vol. 1. Denver, CO, USA, pp. 631–639.

Lawrence, R.W., Scheske, M., 1997. A method to calculate the neutralization potential of mining wastes. Environ. Geol. 32 (2), 100–106.

Lawrence, R.W., et al., 2003. ARD treatment for selective metal recovery and environmental control using biological reduction technology-commercial case studies. Proc. 35th Annual Meeting of the Canadian Mineral Processors Conf. CIM, Ottawa, ON, Canada, pp. 123–136.

Lewis, A., 1984. New Inco Tech process attacks toxic cyanides. Eng. Min. J. 185 (7), 52–54.

McKinnon, E., 2002. The environmental effects of mining waste disposal at Lihir Gold Mine, Papua New Guinea. J. Rural Remote Environ. Health. 1 (2), 40–50.

Miller, S.D., et al., 1991. Use and misuse of the acid-base account for "AMD" prediction. Proc. 2nd International Conference on the Abatement of Acidic Drainage, vol. 3. Montreal, Canada, pp. 489–506.

Morin, K.A., Hutt, N.M., 1997. Environmental Geochemistry of Minesite Drainage: Practical Theory and Case Studies. Minesite Drainage Assessment Group (MDAG) Publishing.

Moskvitch, K., 2014. Field of dreams: the plants that love heavy metal. New Scientist. 221 (2961), 46–49.

Mudder, T.I., et al., 2001. The Chemistry and Treatment of Cyanidation Wastes. second ed. Mining Journal Books Ltd, London, UK.

Muir, A., et al., 2005. A practical guide to re-treatment of gold processing residues. Miner. Eng. 18 (8), 811–824.

Neculita, C.M., et al., 2007. Passive treatment of acid mine drainage in bioreactors using sulfate-reducing bacteria: critical review and research needs. J. Environ. Qual. 36 (1), 1–16.

Nguyen, Q.D., Boger, D.V., 1998. Application of rheology to solving tailings disposal problems. Int. J. Miner. Process. 54 (3-4), 217–233.

Nicholson, R.V., et al., 1989. Reduction of acid generation in mine tailings through the use of moisture-retaining cover layers as oxygen barriers. Can. Geotech. J. 26 (1), 1–8.

Peyronnard, O., Benzaazoua, M., 2011. Estimation of the cementitious properties of various industrial by-products for applications requiring low mechanical strength. Resour. Conserv. Recy. 56 (1), 22–33.

Rao, S.R., 2006. Resource Recovery and Recycling from Metallurgical Wastes. Waste Management Series, vol. 7. Elsevier.

Rao, S.R., Finch, J.A., 1989. A review of water re-use in flotation. Miner. Eng. 2 (1), 65–70.

Ritcey, G.M., 1989. Tailings Management: Problems and Solutions in the Mining Industry. Elsevier, Amsterdam, New York, NY, USA.

Robinsky, E.I., 1999. Thickened Tailings Disposal in the Mining Industry. E.I. Robinsky Associates Ltd (Publisher), Toronto, ON, Canada.

Scott, J.S., Ingles, J.C., 1981. Removal of Cyanide from Gold Mill Effluents. Mining and Metallurgical Division, Water Pollution Control Directorate, Environment Canada (Publisher), Ottawa, ON, Canada.

Shilling, R.W., May, E.R., 1977. Case study of environmental Impact-Flambeau project. Min. Cong. J. 63 (1), 39–44.

Sobek, A.A., et al., 1978. Field and Laboratory Methods Applicable to Overburdens and Minesoils (Research report). Industrial Environmental Research Laboratory, Office of Research and

Development. US Environmental Protection Agency, Cincinnati, Ohio, USA, EPA-600/2-78-054.

Stradling, A.W., 1988. Backfill in South Africa: developments to classification systems for plant residues. Miner. Eng. 1 (1), 31–40.

Theriault, J.A., et al., 2003. Surface disposal of paste tailings at the Bulyanhulu gold mine, Tanzania. Proc. 28th Annual Meeting of the Canadian Land Reclamation Association, Mining and the Environment, Sudbury, ON, Canada, pp. 265–269.

van Zyl, D., 2014. A perspective of innovations in tailings management. In: Anderson, C., et al., (Eds.), Mineral Processing and Extractive Metallurgy: 100 years of Innovation. SME, Englewood, CO, USA, pp. 629–637.

Veneu, D., et al., 2014. Sorptive flotation of heavy metals by microorganisms. Proc. 27th International Mineral Processing Congress (IMPC), Ch. 12, paper C1221, Santiago, Chile, pp. 1–10.

Vick, S.G., 1984. A systematic approach to tailings impoundment siting. Mining Sci. Technol. 1 (4), 285–297.

Williams, D.J., et al., 1997. A cover system for a potentially acid forming waste rock dump in a dry climate. Proc. 4th International Conference. on Tailings and Mine Waste '97, Fort Collins, CO, USA, pp. 231–235.

Yanful, E.K., et al., 1999. Soil covers for controlling acid generation in mine tailings: a laboratory evaluation of the physics and geochemistry. Water Air Soil Pollut. 114 (3-4), 347–375.

Younger, P.L., et al., 2002. Mine Water: Hydrology, Pollution, Remediation. Kluwer Academic publishers, USA, Springer.

Zhan, G., et al., 2001. Capillary cover design for leach pad closure. SME Trans. 310 (1), 104–110.

Chapter 17

Modeling and Characterization

17.1 INTRODUCTION

In this chapter, following on from Chapter 3, some more of the important tools used by the mineral process engineer are briefly described. Under modeling is included computer simulation for circuit design, computational fluid dynamics (CFD) and discrete element method (DEM) for equipment design, and design of experiments (DOE) for empirical model building. Under characterization, geometallurgy and applied mineralogy are the topics; another characterization tool, surface analysis, is covered in Chapter 12, Section 12.4.4.

17.2 CIRCUIT DESIGN AND OPTIMIZATION BY COMPUTER SIMULATION

Computer simulation has become an important tool in the design and optimization of mineral processing plants. The capital and operating costs of mineral processing circuits are high and in order to reduce them consistent with desired metallurgical performance, the design engineer must be able to predict the metallurgical performance, relate performance to costs, and select the circuit for detailed design based on these data. Simulation techniques are suitable for this purpose, provided that the unit models are valid, and considerable progress continues to be made in this area.

Computer simulation is intimately associated with mathematical modeling and realistic simulation relies heavily on the availability of accurate and physically meaningful models. King (2012) provides an excellent review of unit modeling and circuit simulation methods. Comminution and classification models are well developed and in routine use (Napier-Munn et al., 1996; Sbábarao and Del Villar, 2010). Kinetic models of flotation have been used for many years (Lynch et al., 1981), and new, more powerful approaches to modeling flotation are now available for practical use (Alexander et al., 2005; Fuerstenau et al., 2007; Collins et al., 2009). Models of gravity and dense medium processes are available, although they are not as widely used (Napier-Munn and Lynch, 1992), and the modeling of liberation has progressed, though is not yet routine (Gay, 2004a, b). Advances in modeling and simulation of dewatering systems can be found in Concha (2014).

Several commercial simulators for mineral processing are available, including: JKSimMet/Float, USimPac (Brochot et al., 2002), Modsim (King, 2012), and Plant Designer (Hedvall and Nordin, 2002). The first two simulators also provide data analysis and model calibration capabilities. Limn (Leroux and Hardie, 2003) is a flowsheet solution package implemented as a layer on top of Microsoft Excel with simulation capability.

Figure 17.1 illustrates the procedure for using simulation to predict process performance for design or optimization. The key inputs are the material characteristics (e.g., grindability or floatability) and the parameters of the process models. The latter are obtained by estimation from operating data (preferred) or from a parameter library.

Compared to laborious and expensive plant trials, computer simulation offers clear advantages in assessing alternative circuits, optimizing design, and estimating flow rates of process streams, which can be used to size material handling equipment (conveyors, pumps, and pipelines). However, the dangers of computer simulation also come from its computational power and relative ease of use, which encourage searching the "what-if" space. It is always necessary to respect the operating range over which the models are valid, as well as the realistic limits which must be placed on equipment operation, such as pumping capacity. In addition, it is worth remembering that good simulation models combined with poor data or poor model parameter estimates can produce highly plausible looking nonsense. Simulation studies are a powerful and useful tool, complementary to sound metallurgical judgment and familiarity with the circuit being simulated and its metallurgical objectives.

17.3 MACHINE DESIGN

To improve particle size reduction and particle separation systems and thus increase productivity of mineral

Wills' Mineral Processing Technology.

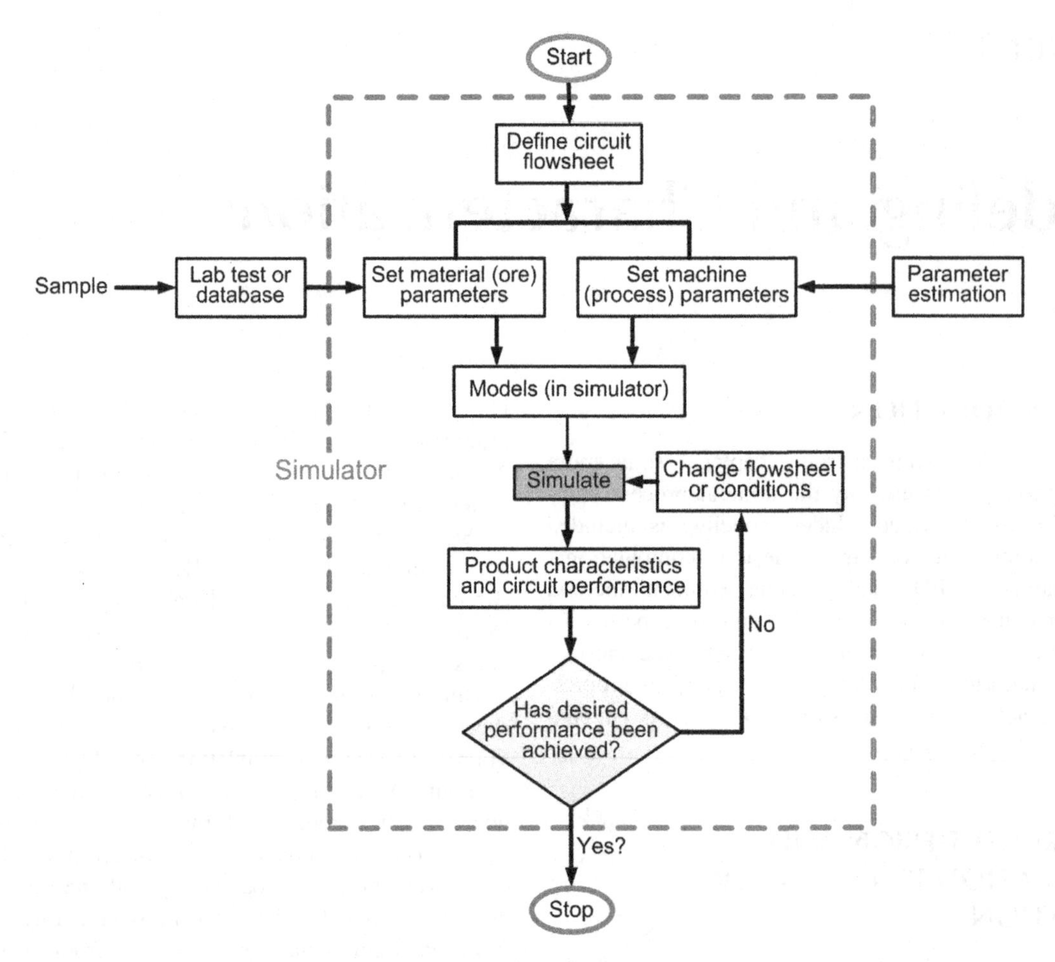

FIGURE 17.1 Procedure for simulation to optimize design or performance *(From Napier-Munn et al., 1996; Courtesy JKMRC, The University of Queensland).*

processing, new devices of varied designs and operating principles are constantly being developed. Mineral processing has always invited the talented innovator, as the patent literature makes evident. An important cost consideration in the pursuit of the optimal design is the number of iterations in the evolution of the design from the original concept.

Still today much testing is empirical, which can mean years to reach a conclusion straining budgets and financial backers' patience. Employing mathematical modeling tools can significantly reduce the time and cost involved. Advances in computational power enable multiple simulated iterations of a device's operation. *Computational Fluid Dynamics* and *Discrete Element Method* are two important groups of modeling techniques that are beginning to pay dividends in the development of new mineral processing devices and the re-design of existing units. This section describes the general principle of these two modeling approaches, and stresses the need to validate their outcomes.

17.3.1 Computational Fluid Dynamics

Computational fluid dynamics (CFD) is the application of algorithm and numerical techniques to solve fluid flow problems (Versteeg and Malalasekera, 2007). In CFD, the fluid body (i.e., the interior shape containing the fluid) is divided into small fluid elements called cells (usually with tetrahedral or hexahedral shape). Algebraic variables are attributed to each flow characteristic of each cell (e.g., mass, pressure, velocity, temperature). The interfaces of the fluid body are used as boundary conditions where some of those flow characteristics are known. The interfaces can be of many types such as: walls, inlets/outlets with imposed flowrate, and openings with fixed pressure drop. The known characteristics at the boundaries are attributed to the corresponding cells and variables. The conservation principle of mass, momentum and energy is applied to each cell with consideration of the neighboring cells. This creates a system of algebraic equations with the variables representing each cell's characteristics (some known from the boundary conditions,

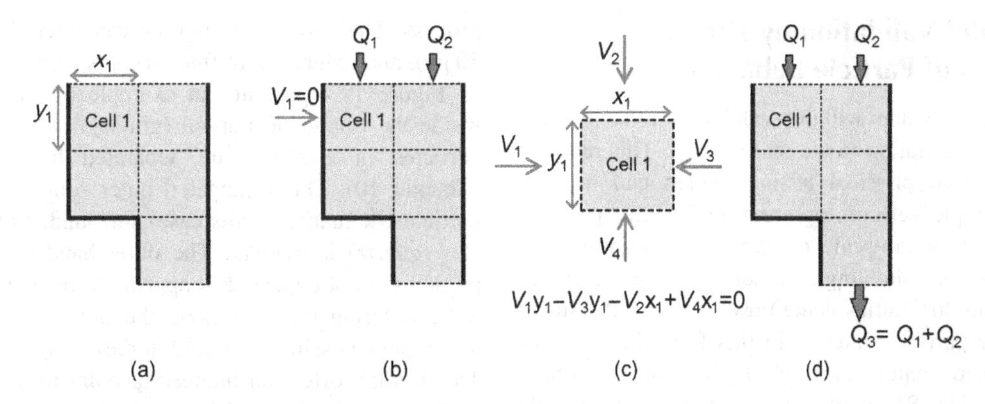

FIGURE 17.2 Simplified steps of a CFD study: (a) discretization, (b) boundary conditions, (c) algebraic equations, and (d) solution.

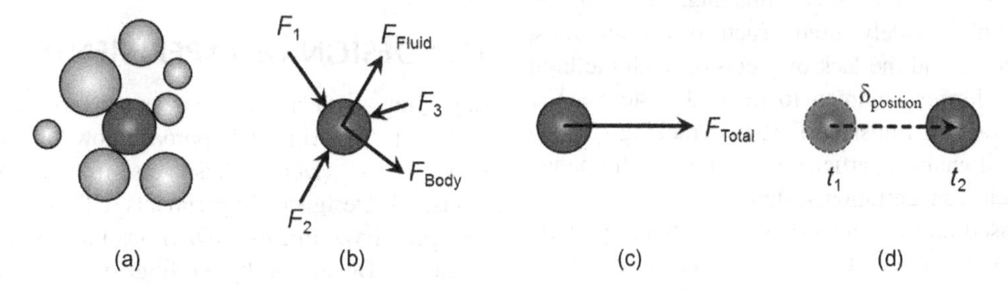

FIGURE 17.3 Example of a DEM iteration on a particle: (a) contacts identification, (b) forces application, (c) forces summation, and (d) particle displacement computation.

some unknown). This system of equations is solved with an iterative technique by the CFD code. The solutions are the flow characteristics for every cell. Knowing the discrete fluid behavior of each cell then enables one to determine the whole fluid body behavior (e.g., pressure, velocity, flowrate). Figure 17.2 shows a simplified application of the CFD technique to a 2D flow problem.

CFD is now applied to complex fluid flow problems with a high degree of confidence in the retrieved solution, even in the case of a mixture of two fluids like air and water. A sound knowledge of fluid dynamics, however, remains central to comprehending the simulations.

An important point to remember when applying CFD techniques to mineral processing problems is that slurries do not behave like water. This is due to the influence of variable viscosity and density depending on the local solids concentration and particle size. The model needs to take account of those slurry properties. Adequately incorporating slurry properties is a major focus of particle technology research.

17.3.2 Discrete Element Method

The discrete element method (DEM) is a numerical technique to simulate the behavior of a population of independent particles (Cundall and Strack, 1979). In this technique, each particle is represented numerically and is identified with its specific properties (e.g., shape, size, material properties, initial velocity). The interior shape of the vessel containing the particle is used as the domain of the simulation and is separated into a grid to identify the particle's position. Particles are then subjected to a small motion (based on Newton's laws) over a small time interval (iteration). The small motion will cause some particles to contact other particles or the domain boundaries. Each of these contacts is monitored and produces discrete reaction forces on each particle. The magnitude of the contacting forces is determined by a contact model (e.g., spring-dashpot model). The summation of the total force on each particle is then computed and forces created by external factors (e.g., fluid drag, magnetism, buoyancy, acceleration) can be added to this balance at this point. Newton's laws of motion are then used to determine the motion parameters of each particle (e.g., acceleration, velocity, displacement) over the small time interval.

An example of this process is shown in Figure 17.3. The new position of the particles is computed and the process of contact detection can restart for the next iteration. After computation of every time step, every particle's behavior is known over the total simulation time, hence the bulk behavior is known.

Combinations of CFD and DEM have been used to describe the behavior of particles moving and colliding inside a flowing fluid. In mineral processing where the ore is mostly processed wet, these CFD-DEM coupled approaches are of interest as they promise to optimize equipment design. However, it should be noted that these coupled simulations require large computational power and advanced technical knowledge.

17.3.3 Model Validation by Direct Observation of Particle Behavior

A model has no value without validation to assess the fidelity of the simulation to the real process. This requires a certain level of physical testing, which can be performed on simple set-ups representing specific parts of the process, or on lab-scale equipment representing the whole process. In validating a mineral processing operation, one of the difficulties is the observation of the slurry flow and how particles interact in this flow. Few techniques are able to quantitatively track and display particle or fluid behavior. Some of the techniques are: particle image velocimetry (PIV), laser Doppler anemometry (LDA), dye injection, high-speed imaging, and acoustic monitoring. While widely used, each has limitations, mostly the opacity and the lack of precision with the high solids content slurries common to mineral systems. The technique of Positron Emission Particle Tracking (PEPT) is attractive as it enables particles to be tracked in opaque and high particle concentration systems.

PEPT is based on tracking a radioactive tracer particle that is mixed with the feed to a unit (Parker et al., 1993). The tracer particle can be taken from the bulk to be irradiated and tracked, an advantage of the technique as the surface and body properties will be the same as the particles processed. By recycling the tracer, a picture of the particle motion is established over time while it flows, tumbles, collides and floats (in the case of flotation), and builds a picture of how a particle reacts to the system design, and what is its velocity, acceleration, residence time, etc. In mineral separation, knowing the motion of gangue and valuable mineral (at present tracked individually) will help to refine geometrical features of the separators and to validate competing models of the separation process. The current technology can track down to about 50 μm, and extending to finer sizes is a current challenge.

Figure 17.4 presents an example of particle tracking inside the trough of a spiral (gravity) concentrator where particles in a slurry are separated based on density (Chapter 10). The anticipated inner zone of high density particles (hematite in this case) surrounded by light particles (quartz) is evident. The outer band of high density particles is not expected; it appears to be due to these particles entering the trough on the outside and the turbulence (and possibly Bagnold force) keeping them there. This finding offers an interesting point to the observation of Bazin et al. (2014) (Chapter 10) that some coarse heavy particles are misplaced to the lights product.

17.4 DESIGN OF EXPERIMENTS

Beginning with its earliest incarnation called factorial design, the experimental approach now known as *Design of Experiments* (DOE) (also as Experimental Design, Industrial Designed Experiments (IDE) and Statistically Designed Experiments (SDE)) has been around for almost a century. Developed by the English statistician and evolutionary biologist Sir Ronald Fisher (1890–1962) as a method for testing and assessing crop yields at agricultural stations in England (Fisher, 1926), the method was slow to gain acceptance in other industries until the 1960s. It then became a mainstay of the continuous improvement revolution that launched the resurgence of Japanese manufacturing. With the advent of computer software to simplify the experimental design and data analysis (e.g., Stat Ease™, Minitab™), the approach is now gaining acceptance across engineering disciplines. It is particularly well suited for processing and manufacturing industries where many variables may interact in complex ways.

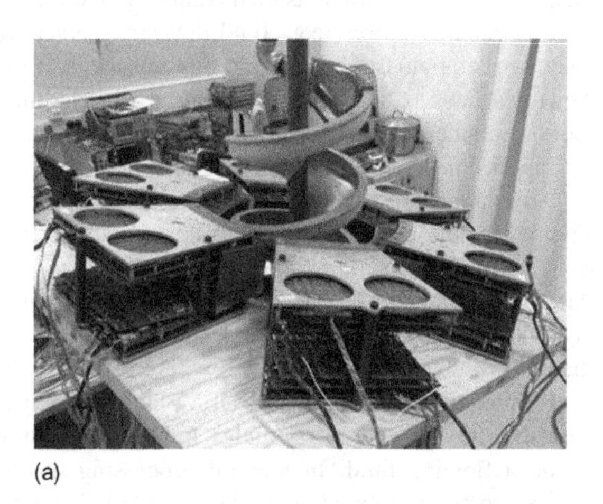
(a)

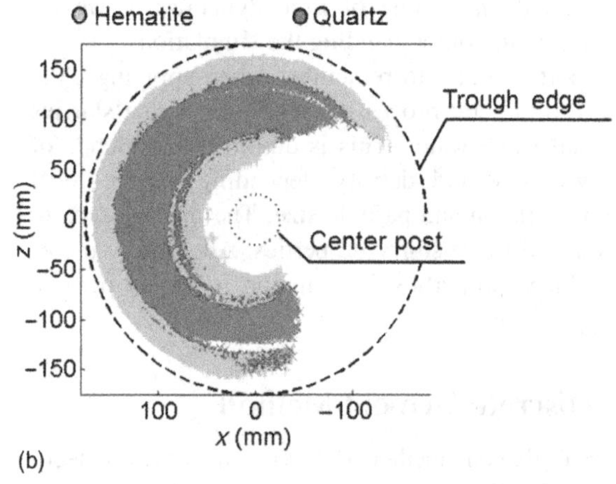

(b)

FIGURE 17.4 (a) Modular positron emission particle tracking detector mounted around a spiral concentrator to observe a slice of the separation process, and (b) Slice top view showing the combination of many trajectories of a hematite tracer (inner and outer zone) and a quartz (middle zone) tracer (particle diameters ≈ 1.5 mm) in an iron ore slurry *(Courtesy D. Boucher)*.

Experimental design is a structured process for investigating the relationship between input variables and output effects, *factors and responses*, in a process. Multiple input factors are considered and controlled simultaneously to ensure that the effects on the output responses are causal and statistically significant. Each variable is given equal weight across its full range by coding the variables in the regression analysis (e.g., max = 1, min = − 1), thus yielding the relative magnitude of the effect of each factor and a measure of the interaction between factors. DOE, therefore, represents a large improvement over the traditional *one-factor-at-a-time* (OFAT) experimental approach, by providing statistical information on the significance and magnitude of each factor and their interactions, through the method of *ANOVA* (analysis of variance) developed by Fisher (note: in doing so Fisher also developed the applied statistics concepts of the F-test and z distribution). The ANOVA provides equations (*transfer functions*) for each response in terms of the input factors exhibiting the desired level of significance (e.g., 90% or 95% confidence levels). Process models developed by DOE methods fall into the realm of empirical modeling; that is, they are not based on a physical or chemical representation of the process, but rather a mathematical regression that should only be applied within the range of the input factors used in its development. The DOE method, once learned, has the potential to provide the maximum information for the least cost in terms of time and resources. Montgomery (2012) provides an excellent reference for the subject. Somewhat easier to digest for the beginner is the reference by Schmidt and Launsby (1994).

Modern experimental design has evolved from the full factorial designs of Fisher, which had the serious limitation of being constrained by the number of factors that could be reasonably investigated. The following relationship illustrates the point. If k factors are to be tested at L levels, the number of tests (*runs*) N is given by:

$$N = L^k \quad (17.1)$$

At 3 levels (low, high, mid-point), a 3 factor design would require 27 tests, a 4 factor design, 81 tests, and for 5 factors, 243 tests. If only high and low levels are tested, the test numbers become a more reasonable 8, 16, and 32, respectively. Full factorial designs are therefore typically conducted at low and high levels only and are suitable for up to, at most, 5 (=32 runs) or 6 (=64 runs) factors. *Fractional factorial designs* were introduced in the 1930s by Yates (1937) to extend the number of factors that could be reasonably tested. The number of tests now required becomes:

$$N = L^{k-q} \quad (17.2)$$

where q is 1,2,3... etc., and represents a half, quarter, eighth...etc. fractional design. Because not all conditions are tested for each variable against all other variables, some information about factor interactions is indiscernible from other factor interactions. This is known as *aliasing* and is a limitation of fractional designs. Fractional designs that involve even fewer runs for a given number of variables have been developed (Plackett and Burman, 1946; Box and Behnken, 1960; Taguchi, 1987), but in these designs the aliasing patterns are more severe than in basic fractional designs and care and understanding of the process factors under test need to be exercised. Factors in DOE can be either quantitative or qualitative, a feature which enhances their applicability to mineral processing where not all factors are continuous (e.g., ON-OFF, or Chemical A vs. Chemical B).

Modern DOE has divided the overall design process into two stages. In the first stage many factors are tested to determine those that are critical to the process, and is called the *factorial testing* or *screening stage*. In the second stage these critical factors are re-tested at multiple levels to establish a more robust mathematical relationship that is better suited for process optimization. This stage is referred to as *response surface modeling* (RSM) (because the resulting equations may be non-linear) or *optimization* stage. Figure 17.5 schematically represents the concept. If the key process factors are known then testing can proceed directly to the optimization stage.

A favored experimental design for optimization is the Central Composite Design (CCD), shown in Figure 17.6(a),

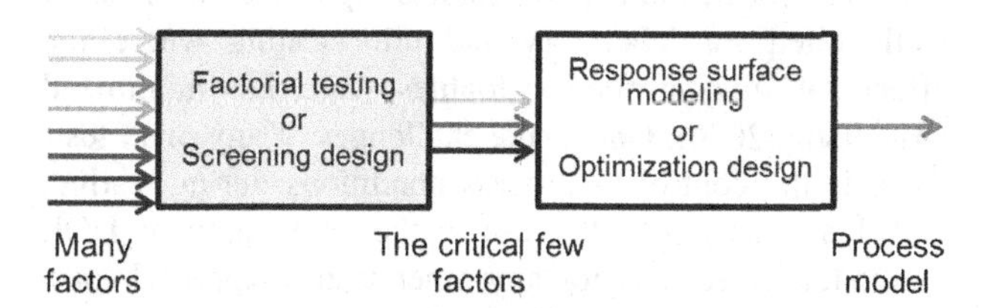

FIGURE 17.5 Two-step process for modern experimental design.

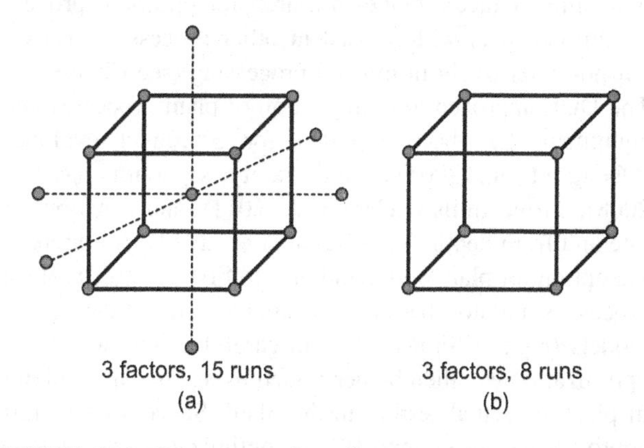

FIGURE 17.6 3D representation of: (a) central composite design (CCD) having 3 factors (x, y and z axes), and (b) full factorial design.

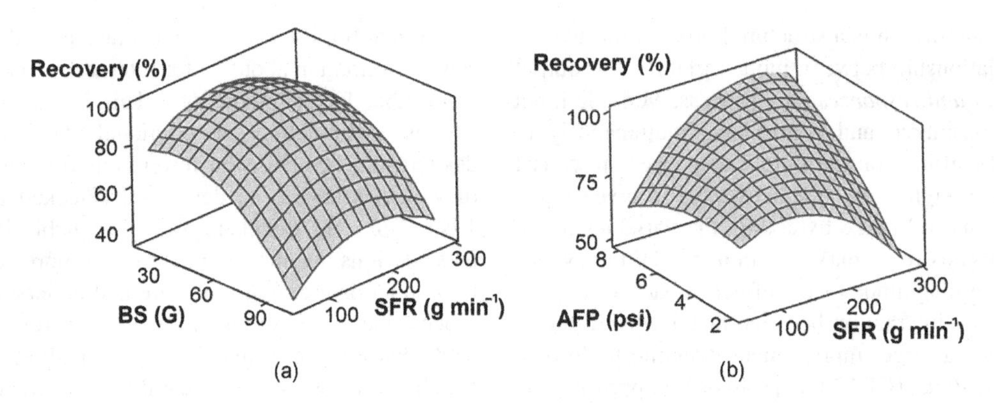

FIGURE 17.7 Examples of response surface plot from a CCD designed DOE for tungsten recovery from silica in a 3 inch laboratory Knelson gravity separator run under dry conditions: (a) BS = bowl speed (G), SFR = solids feed rate (g min^{-1}), and (b) AFP = air fluidizing pressure (psi) *(Adapted from Kökkılıç et al., 2015). (The curved response surfaces indicate a strong non-linear interaction for the 3 factors).*

which tests factors at 5 levels and requires 15 tests (not counting replicates) for 3 factors and 25 tests for 4 factors. Also shown for comparison purposes in Figure 17.6(b) is the 3D representation of a 3 factor, 2 level (2^3) design. Note that in the CCD, each factor is tested at 5 levels which greatly enhances the response definition over the 2-level factorial design. An example response surface using CCD to identify optimum conditions for recovery of tungsten from silica in a dry Knelson concentrator is shown in Figure 17.7. As with any experimental testwork, once optimum conditions have been identified, a set of *confirmation* runs should be performed to validate results.

The rigor of the experimental design method makes it well suited for laboratory and pilot testing where the effect of uncontrolled variables can be minimized (Anderson, 2006). One of the challenges of any plant testwork is the control of process conditions during testing, and this is as much the case in the application of DOE methodology as it is for any other testing approach in a plant (Napier-Munn, 2010). The EVOP (evolutionary operation) approach to process optimization was introduced by Box (1957) and uses simple factorial designs with limited factor range to search for optimum process conditions. It is widely used in other process industries, but not extensively in mineral processing (see Chapter 3). The DOE approach to optimization of plant processes and equipment has been reported for 3-product cyclones (Obeng et al., 2005), multi-gravity separators (Aslan, 2008), stirred mills (Celep et al., 2011), and flotation air rate and froth depth (Venkatesan et al., 2014). An interesting option for plant optimization is to first develop a robust process simulator based on available phenomenological models (e.g., JKSimMet™) with careful calibration. DOE optimization can then be performed using such a simulator in place of actual testing in the plant. Success using this approach has been reported for optimization of cyclone variables in a grinding circuit (Leroux et al., 1999), and for mill liner design in a SAG mill using discrete element modeling (Radziszewski et al., 2003).

17.5 GEOMETALLURGY

The traditional approach to plant design involved the extensive testing of a single large composite sample or a small number of composite samples that are reputed to represent the ore body. It is accepted that laboratory tests can accurately measure factors such as the grindability, floatability, or other process parameter of the sample by a technique representing that to be used in the plant. The size of equipment required to achieve a specified throughput and product quality is then calculated from one of a variety of models that have been developed over the years, with some examples being given in previous chapters. Since these tests and models are tried and tested, they are accepted as reasonably precise. During operation of the resulting plant, the design is sometimes found to be inadequate. It is then suspected that the flaw in the design process lies in the samples not being sufficiently representative of the ore body, since using only a single or small number of composite samples does not recognize the variability of the ore, nor does it allow for the lack of precision in the value of the metallurgical parameter used in the design.

A geometallurgical approach uses a design procedure suited to an ore body that is described by a geostatistical analysis of a reasonably large number of small samples of drill core. The analysis requires the identification of the location of the sample points and a geological plan of the ore body, together with a mine plan of the blocks to be mined during the proposed life of the mine. The plant design can then be made using the estimated metallurgical parameter of each individual block. A statistical error can be assigned to the estimated parameter for each block and for each production period such that the final design includes risk estimates and safety factors based on the possible errors that arise from having a limited amount of sample data (Kosick et al., 2002).

Consequently, we can define "geometallurgy" as:

- The geologically informed selection of a number of samples for the determination of metallurgical parameters
- The distribution of these parameters across the blocks of the ore body by some accepted geostatistical technique, where the distribution is usually influenced by the geology because lithology/alteration/texture has an effect on the parameters
- The subsequent use of the distributed data in metallurgical process models to generate economic parameters such as throughput, grind size, grade and recovery for each mine block for plant design and production forecasting that can be used in mine planning.

This "geometallurgical approach" is needed because:

- Ore bodies are variable in both grade and metallurgical response
- The variability is a source of uncertainty that affects plant design, results (both metallurgical and financial) and capital investment decisions
- Deposits are becoming lower grade and more complex
- Throughputs are necessarily increasing and profit margins reduced; the financial risks are escalating
- Mining industry risk must be more carefully managed for projects to attract the necessary finance.

A "geometallurgical project" takes a step-by-step approach:

- Geologically informed selection of a number of variability samples (i.e., samples selected to reveal variability)
- Determination of relevant metallurgical parameters for those samples
- Populating a spatial model of the ore body by geostatistical distribution of those parameters
- Use of the distributed dataset of parameters in metallurgical models for design and forecasting
- Estimation of the uncertainties in the knowledge of the ore body to calculate lack of precision in the results
- Managing risk by adding safety factors to designs and calculating error bars for the forecasts.

17.5.1 Variability Sampling

It is important to realize the interrelated technologies involved in a mining operation and to include all departments in the design and production forecasting project:

- Metallurgy determines expected results for throughput and recovery of the plant by:
  - The use of tests on drill core samples to generate metallurgical parameters for the ore: for example, grindability, flotation kinetics
  - The use of these parameters in process models.
- Mining determines temporal variability in the production sequence; consequently, the metallurgical parameters vary from time to time in the plant
- Geology affects geographic variability in the mineral assemblage of the ore body; consequently, the metallurgical parameters vary from place to place
- Geostatistics use the geological information to make an estimate of the metallurgical parameters for each block and get an idea of the estimation errors (i.e., the uncertainty). It is self-evident that the more samples that are tested and the better they are selected, the greater the certainty in the data used for forecasting and the lower the project risk.

Effective sampling of an ore body is both difficult and expensive, but important to reduce the greater cost from the financial risk of failure to meet the expected results in production (Chapter 3). The sample requirements and subsequent risk analysis vary with:

- The stage of the project: that is, preliminary, prefeasibility, or full feasibility
- The size of the ore body, and complexity of the geology, and resulting metallurgy.

The number of samples depends on the project stage, plant throughput, and ore variability.

Some guidelines for sample selection of drill core are:

- Always consult with the Geology and Mine Planning departments
- Try to include the variability of the ore types, that is, lithology, alteration, mineral occurrence of both values and gangue minerals; use geochemical and structural information
- Choose a representative number of each ore type; validate the sample set against the resource population. Outliers or unusual ore types should not be over- or under-represented
- Fresh core is better. Old badly stored core can supply erroneous data. But near-surface weathered ore must be included for testing, as it will be included in the mine plan
- Full or half-core is better than "assay rejects"
- Space the samples to allow uncertainty to be calculated; some close together, but most to cover the complete area of interest between drill holes and down-hole. A random distribution is acceptable for preliminary stages, but use of the drilling grid is better for feasibility design and forecasts
- Select samples to match the mining method while still showing variability; for example, composite by length

equal to bench height but interrupt by ore type change where necessary
- Choose a relevant mining time period from the mine plan as a source of most of the samples; for example, most of them from the first 5 years of production for new plant design. Add more samples for testing each year for on-going production forecasting
- Identify the number of samples needed for the current stage of the project:
  - Preliminary stage of a large project may need 35 samples to demonstrate the variability of the ore body and allow estimate of equipment size (albeit imprecise)
  - Rule-of-thumb for pre-feasibility life-of-mine sampling is one sample per million tons of ore under evaluation, or 1 sample per 400,000 m^3 (100 m × 100 m × 40 m)
  - A full feasibility study will require more samples for a large ore body, but only the first 10 years of the mine plan are of immediate interest. The number of samples should be based on a statistical analysis to ensure that it meets the required level of confidence
  - Collect a sample set that is representative of the variability of the section of the deposit that is of most interest when considering the financial risk in the project—not a "representative sample" (since there is no such thing).

17.5.2 Metallurgical Testing

Variability samples must be tested for the relevant metallurgical parameters. Ball mill design requires a Bond work index, BWi, for ball mills at the correct passing size; SAG mill design requires an appropriate SAG test, for example, SPI (Chapter 5). Flotation design needs a valid measure of kinetics for each sample, including the maximum attainable recovery and rate constants for each mineral (Chapter 12). Take care to avoid unnecessary testing for inappropriate parameters, saving the available funds for more variability samples rather than more tests on few samples. Remember that it must be possible to use the measured values for the samples to estimate the metallurgical parameters for the mine blocks in order to describe the ore body, and these estimates will be used in process models to forecast results for the plant. Always include some basic mineralogical examination of each sample.

17.5.3 Populating the Mine Block Model

Understanding and using the measured metallurgical parameters of the whole ore body requires that the test data are distributed across all the blocks in the mine plan. This exercise involves the consideration of much more information about the mine than would traditionally be used in plant design, which includes:

- Location of each sample within the ore body in terms of co-ordinates and section of core used
- Geological description of the sample, for example, lithology, alteration, rock type, and perhaps metal grade
- Mine block plan with similar geological information
- Planned mining schedule for the mine blocks, for example, by year.

The objective of the analysis is the distribution of the metallurgical test data across the blocks in the mine plan, assigning each block (and each mining period) an estimated metallurgical parameter value (e.g., BWi), and a precision of each estimate.

A suitable method involves:

- Basic statistical study to identify *geometallurgical (geomet) domains*, that is, areas where it can be shown by an ANOVA that the geological description coincides with distinctly different metallurgical parameters. There must be a sufficient number of samples in a domain for only those samples to be used in estimation of blocks of similar geology
- Use of a geostatistical technique, such as Kriging, or regression analysis, for distribution of metallurgical parameters within each geometallurgical domain. Independent variables such as BWi that can be used in statistical analysis can be distributed directly to mine blocks by the distance weighting method known as Kriging. Variables that are dependent on other parameters, such as maximum attainable recovery that is dependent on head grade, may be distributed to blocks by regression analysis using the estimated block value for head grade, and the estimate of maximum recovery can be improved by Kriging of "residuals." Parameters that are not amenable to statistical analysis, such as flotation rate constant, must be recognized and handled by transposing to some variable that is suitable for Kriging or some other geostatistical technique
- Acceptance of the uncertainty in the estimated value of each block. Distribution of metallurgical parameters is best done using a method that involves a measure of the statistical error. The error can be included in simulations of plant performance to show the uncertainty in the forecast results.

A short description of Kriging is a pre-requisite to understanding the geometallurgical approach (Amelunxen et al., 2001). This involves the construction of geostatistical *variograms* for each metallurgical parameter (e.g., BWi) by plotting the semi-variance of differences in the value of pairs of samples of equal (within a range)

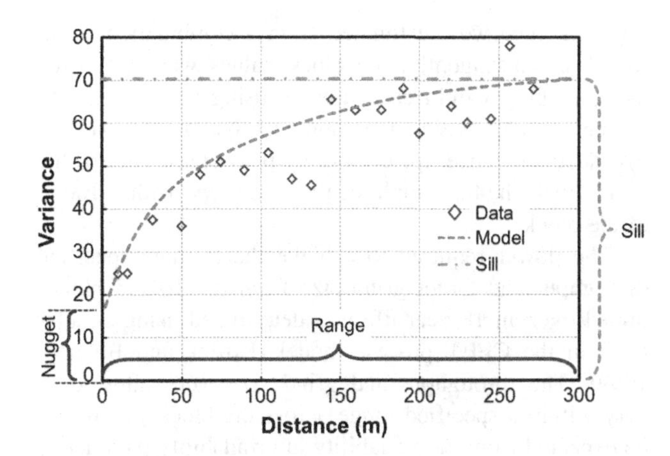

FIGURE 17.8 Typical variogram.

distance apart against that distance (or mid-range) – see Figure 17.8. It is evident that the variability between samples close together is less than for those further and further apart, till eventually there is no longer any influence on the variance. The establishment of a model (equation) for the variogram allows for the estimation (by interpolation) of the metallurgical parameter and the statistical error of each estimate to be made for each block by combination of the values of samples that are within the range of influence. That "range" is determined by the shortest distance apart for pairs having attained the maximum variance, known as the "sill." The value where the curve cuts the *y*-axis is referred to as the "nugget" (estimated from the variance of pairs of coincident samples) and is a measure of the inherent errors in sampling and measurement of individual datum points.

In practice, the estimate for a block is made from samples within an ellipsoid with dimensions selected after consideration of the geological structure, using the weighted average of a minimum of 3 samples that are not all in the same drill hole. If there are insufficient samples within those dimensions, a further ellipsoid is chosen with larger dimensions, and so on until there are sufficient samples. Enlarging the dimensions to use samples that are further away produces block estimates with larger standard errors. The use of data averaged from at least 3 samples results in a smoothing of variability in the block estimates. It is always instructive to compare the frequency distribution of parameters measured from the samples with that attributed to the blocks; excessive smoothing or shift in the mean indicates too few samples were tested.

Since blocks are normally identified by year on the mining plan, the geostatistical analysis also allows the determination of annual average parameter values and their statistical errors by the same Kriging technique. So, it is possible to design the plant to deliver a specified throughput and with optimum grade and recovery in each production year. It is also possible to extend the analysis to calculate how many more samples need to be tested from within the range of the blocks mined in a production year in order to improve the precision to any desired level.

17.5.4 Process Models for Plant Design and Forecasting

Metallurgical parameters (such as hardness work indices or mineral flotation kinetics) are estimated for each block in the mine model and all the blocks are used as an input dataset for process models. These process models are used for simulations to determine throughput, grade and recovery per block for new plant designs or for forecasting the results from existing operations (Bennett et al., 2004). The models must be capable of using the measured parameters and rapidly simulating many thousands of mine blocks to supply an optimal design or average forecast. Changing the plant or process design must be easy to accommodate within the model to allow various options to be considered.

It is important that the blocks are populated with the estimated metallurgical parameters. The results of the simulations in terms of throughput or recovery can then be assigned to each block. Remember that these results are specific to the plant design used in the model, and that changing the plant design will produce a new set of results for allocation to each block. Never try to populate the blocks with plant results that are determined by simulation from individual samples.

17.5.5 Estimating Uncertainty

Using a statistical approach to distribution of metallurgical parameters ensures that each block value is accompanied by a standard error, for example, Kriging or regression errors. This allows process models to be run as Monte Carlo simulations using the estimated value as the mean of a distribution of possible values that has a standard deviation based on the standard error. Hence the standard error of the resultant plant forecast can be determined, as a measure of uncertainty. Since the geostatistics also allows the estimation of annual average parameter values and their statistical errors, it is possible to calculate the uncertainty in the forecast for each production year.

17.5.6 Managing Risk

Error bars can be fitted to the production forecasts for blocks and for average annual production. The size of the bar is determined from standard error and the confidence that is to be applied. The more confidence we wish to place in the forecast, the larger the error bar to encompass the possible result. This concept may be used in design to choose a safety factor: enlarging the equipment size, for example, for grinding mills, will increase the forecast throughput to a point where the bottom of the error bar (minimum estimated throughput) will match the specification. Remember that there is always a statistical chance that the result will be below the error bar because of uncertainty in the estimated metallurgical parameter. The larger the acceptable risk, the lower the confidence limits required, and the smaller the grinding mill (in this example). The testing of more samples that are closer to the blocks will probably generate lower standard errors in the block estimates and narrower error bars, allowing for the design to involve a smaller safety factor (and smaller mill) for the same risk.

Adding a safety factor to equipment size may not reduce the risk in failing to achieve specified flotation recovery to an acceptable level, since maximum attainable recovery is a fixed limit. However, this approach quantifies the remaining risk, which can be invaluable to the financial viability of the project.

17.5.7 Case Study

In a study for the design of a SAG-Ball mill grinding plant (Bulled, 2007), 100 samples were selected, each comprising a 12 m bench intercept, from a total of 27 vertical drill holes. The samples were from six main lithologies further split into eleven sub-types. Minimum drill hole spacing was 50 m and minimum spacing within a hole was 12 m.

The samples were tested for SPI value for SAG design and BWi value for ball mill design. The variability in each case was slightly less than typical; standard deviation is normally about 20% of the mean value in terms of specific energy, kWh t^{-1}, for both grindability parameters.

The sample data were distributed across 10,903 mine blocks, representing approximately 130 million tons of ore to produce a dataset for the ore body on which to base the plant design.

Statistical analysis indicated significant grindability differences between lithologies but, with only 100 samples in total, there was insufficient data to do separate geostatistical analysis of each type. This difference indicated that lithology can be used as a guide to the hardness of any block, but the wide range of values within each domain suggested that there is some other factor (such as degree of alteration) that has a regional basis. The samples and mine blocks were grouped into three different ore types and there were sufficient samples in two of these to conduct a separate geostatistical analysis for each, allowing block values to be estimated from samples of the same ore type treated as domains. Blocks from the third ore type were estimated from a combination of all samples. Subsequently, the block values were estimated using Kriging within each domain using only the neighboring samples within that domain. Spacing of samples was adequate since, for each ore type, at least 75% of the mine blocks had a sample of the same type within 100 m of the block.

The power required to achieve the average specified throughput and target grind size from the 10,903 blocks mined over an 18-year-life was determined using simulations in the CEET process model (Kosick and Bennett, 1999). The throughput and grind size were allowed to vary within a specified range across the blocks, as would be expected from the variability in grindability parameters (SPI and BWi). Variations in both grindability parameters resulted in the bottleneck shifting between the SAG and ball mill; it is important to note that had the design been based on average grindability for each stage rather than the complete dataset, the mills would be about 5% too small to achieve target throughput due to the changing bottlenecks.

Examination of datasets of blocks grouped into annual production periods indicated years when throughput was below target due to limitations from either the SAG or the ball mill circuits. Model simulations indicated that this could be avoided to some extent by planning changes to the SAG discharge screen aperture. However, it was necessary to increase the SAG mill power requirement by 5% to ensure that target throughput was met on average during the year with hardest ore.

The statistical errors in average annual estimates of grindability were used in Monte Carlo simulations within the CEET simulator to determine uncertainty in the forecast throughput. Error bars were put on annual forecasts at 90% confidence limits, indicating that a safety factor of approximately 14% must be added to the power requirements for both mills to ensure that the specified throughput was achieved in every year of the mine life. This still left the 5% statistical chance that specified throughput will not be achieved on average in every year.

17.6 APPLIED MINERALOGY

Quantitative automated mineralogy (QEMSCAN: Quantitative Evaluation of Minerals by Scanning Electron Microscopy; MLA: Mineral Liberation Analyser; and recently TIMA–Tescan Integrated Mineral Analyser) is increasingly being applied to the study of ore deposits and the evaluation of mineral processing operations. The three instruments identified are all based on electron microscopes, although they may be operationally different, and form the main basis of this section. They provide information on polished sections, that is, 2D data, and have now been joined by X-ray microtomography, which provides 3D data, some details of which will be outlined later. The quantification and improved understanding of mineralogical parameters is making significant contributions to exploration, modeling of ore bodies to predict comminution and separation performance, and in diagnosing plant operations (Hoal et al., 2009; Miller and Lin, 2009; Evans et al., 2011; Lotter, 2011; Smythe et al., 2013).

Samples can be analyzed in many forms either as intact drill core, coarse reject material, and composite plant samples when evaluating mineral processing operations. Intact drill core samples are analyzed to determine rock-types and provide textural characterization. Intact core can be mounted on a polished section (or polished thin section) where chemical spectra are collected at a set interval within the field of view. Each field of view is then processed offline and a pseudo color image of the sample is produced, from which the modal mineralogy and texture of the sample can be extracted. ("Modal" refers to mineral proportion or weight% of mineral in a sample calculated taking into consideration the mineral specific gravity.) Data are acquired over the polished sections at a varied pixel size (e.g., 5 or 25 μm). Multiple mineral types can be identified and color coded for easy visual inspection.

The information is presented according to the questions being asked, as illustrated below. The emphasis is on quantitative data presented in a way to aid the mineral processor reach an informed decision.

17.6.1 Mineral Variability

Blended and homogenized coarse reject samples from, for example, 2-3 meter drill core intervals used for geochemical analyses, can be analyzed to determine mineral variability which might have an impact on the metallurgical response. These coarse reject samples are further ground and homogenized and a pre-defined number of particles are mapped at a selected resolution. Such studies are critical in defining the distribution of minerals in a deposit. As an illustration, Figure 17.9 shows the variation of rare earth minerals among the ore zones of the Nechalacho rare earth element (REE) deposit in NWT (Canada). The important observation is that the zones carry significantly different proportions of heavy (HREE) and light rare earth elements (LREE).

17.6.2 Grain Size, Liberation, and Association

Grain size distribution of the individual minerals can be extracted from an automated mineralogical analysis. This information illustrates the relationship of the grain size of the various phases within a sample. An example is shown in Chapter 4, Figure 4.15.

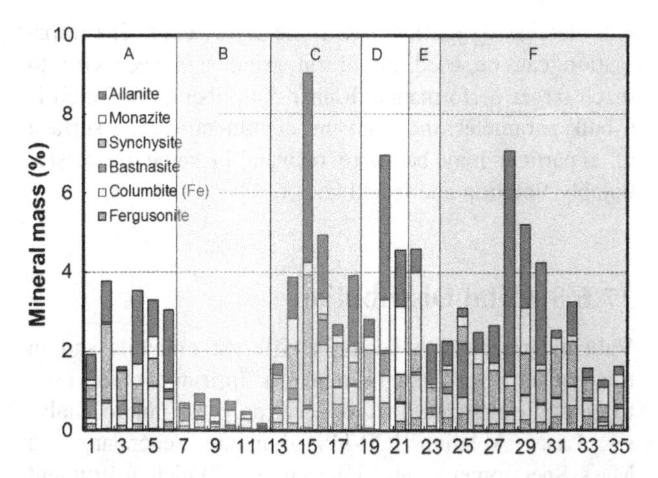

FIGURE 17.9 Variability of REE minerals in the Nechalacho REE Deposit, NWT (A-F refer to ore zones with specific geological and mineralogical characteristics). *(Courtesy T. Grammatikopoulos, SGS Canada Inc).*

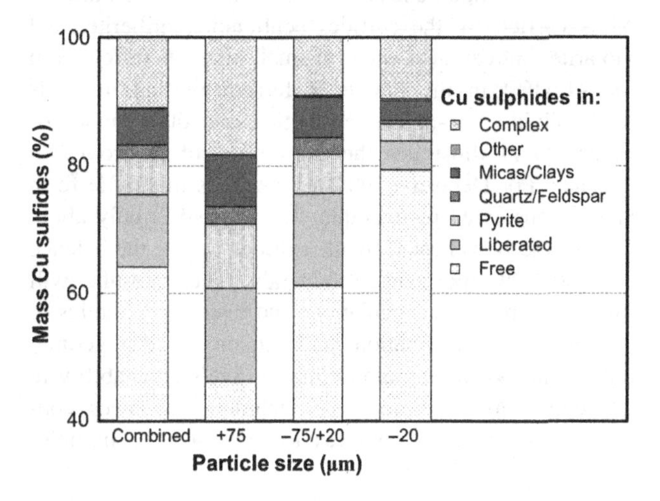

FIGURE 17.10 Liberation and association of copper sulfides (mass %) calculated for the sample (combined) and by size fraction. *(Courtesy T. Grammatikopoulos, SGS Canada Inc).*

Particles are classified into groups based on mineral-of-interest area percent: for example, free (≥95% of the total particle area), liberated (≥80%), and non-liberated (<80%). The non-liberated grains can be further classified according to association characteristics into binary and complex groups. Figure 17.10 illustrates an example of liberation and mineral associations from a Cu deposit. The analysis is conducted on sized samples then the data are combined to assess the whole sample. From Figure 17.10, in the combined, copper sulfides in the free plus liberated forms account for about 75% of the mass of the Cu sulfides. The Cu sulfides in associations are roughly evenly distributed among complex, micas/clays, and pyrite. The increase in liberation (liberated + free) with decreasing particle size is clearly evident. The information can be used to inform grind size necessary to reach target performance. Rather than liberation, which is a bulk parameter, the exposure of mineral on the surface of a particle may be more relevant in some processes, notably flotation and leaching.

17.6.3 Metal Distribution

Valuable metals can occur in different minerals and in trace quantities in gangue minerals. Instrumentation used to quantify elements include electron probe micro analysis, Laser Ablation ICP-MS, dynamic Secondary Ion Mass Spectrometry, and micro-pixie. (Which instrument to employ is mainly dependent on the ore type and the mineral assemblage.) Coupling with automated mineralogical analyses, the distribution of metals among the minerals can be quantified.

As an example, consider a Cu-Ni deposit. Commonly, Ni is carried by the sulfides pentlandite, millerite, and violarite, but can also occur in small amounts (a few ppm to ~1 wt%) in the lattice of ferromagnesian minerals (e.g., olivine), Fe-oxides pyrrhotite, and other minerals. Figure 17.11 illustrates the distribution of Ni from two deposits. For Deposit-1, 90% of the Ni is in sulfide form and is considered recoverable. For Deposit-2, only about 72% of the Ni is hosted by the sulfides, where the balance is hosted by the gangue minerals. These results will impact the resource calculations and overall economics of a project. This information can be incorporated in geometallurgical models to help define the ore's amenability to concentration. Instrument and technique advances continue to push these analyses to ever smaller particles (Brodusch et al., 2014).

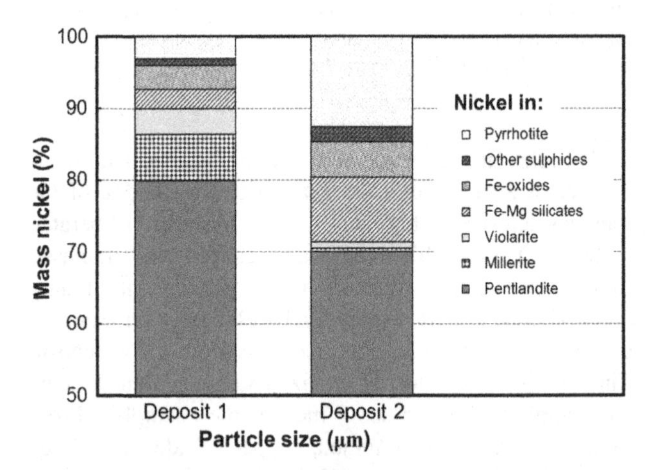

FIGURE 17.11 Distribution of Ni among sulfides, silicates and oxides. The Ni distribution is calculated based on the mineral mass estimated by the QEMSCAN analysis and electron probe micro analysis. *(Courtesy T. Grammatikopoulos, SGS Canada Inc).*

17.6.4 Liberation-limited Grade Recovery

Introduced in Chapter 1, the liberation-limited recovery is determined from the mineral composition of the particles presented to the separator. The mineralogical data are assembled to mimic a perfect separator, that is, one that accepts particles to the concentrate stream based purely on the mineral content, from highest liberated to least liberated. After allowing for mineral specific gravity to convert to a mass basis, the liberation-limited grade recovery curve is produced.

An example is given is Figure 17.12 which compares the liberation-limited curve determined by mineralogical analysis (MIN) with actual metallurgical test results (MET) from two rare earth element (REE) deposits. In the case of C-2, the metallurgical result approaches the theoretical; in case C-1, the metallurgical result falls well below the theoretical due to the loss of fines (slimes) rejected in the testwork.

Based on electron beam instruments, the mineralogical data is 2D. At one time the apparent need to correct to 3D was a major research activity (Barbery, 1991). At issue is that locked particles can be sectioned to appear liberated, meaning the degree of liberation is over-estimated. Correction procedures have ranged from simple to complex (Lin et al., 1999). In routine use, however, no corrections are usually made, but it has spurred development of 3D analysis using tomographic techniques. This cutting edge research is a fitting place to finish the chapter.

17.6.5 High-resolution X-ray Microtomography (HRXMT)

Cone beam X-ray microcomputed tomography (micro CT) systems were introduced commercially a decade ago.

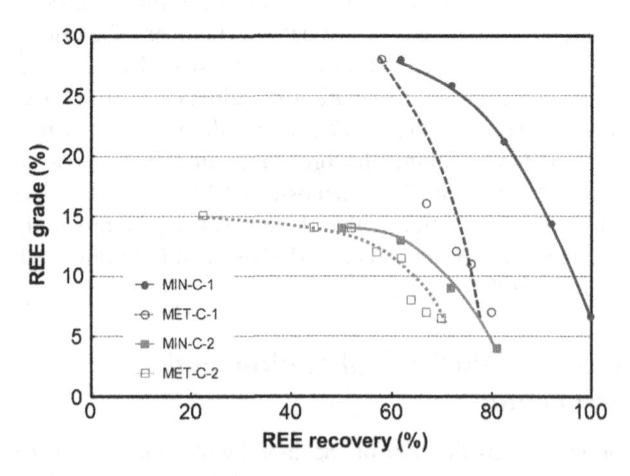

FIGURE 17.12 REE grade and recovery: liberation-limited curve calculated from mineralogical analysis (MIN) and compared to actual metallurgical results (MET) for two carbonatite samples, C-1 and C-2 *(Courtesy T. Grammatikopoulos, SGS Canada Inc).*

They allow for 3-dimensional visualization, characterization and analysis of multiphase systems at a voxel (value on a regular grid in three-dimensional space) resolution of 10 μm to generate three-dimensional images of particulate systems (Miller and Lin, 2009). An example application is described by Miller et al. (2009), who used the technique to diagnose separation performance on a phosphate project. They derived the limiting grade recovery curve from the 3D (volume) data and compared it with that generated from 2D (area) data and showed, as expected, that the 2D result over-estimated the liberation, evidenced by the grade being higher at any given recovery than shown by the 3D result. They also compared with plant data and found that the operating point (grade/recovery) fell on the 3D derived curve. As instrumentation advances and costs decline, this attractive direct measure of volume-based mineralogical data may become more widespread.

REFERENCES

Alexander, D.J., et al., 2005. Flotation performance improvement at Placer Dome Kanowna Belle Gold Mine. Proc. 37th Annual Meeting of the Canadian Mineral Processors Conf. CIM, Ottawa, ON, Canada, pp. 171–201.

Amelunxen, P., et al., 2001. Use of geostatistics to generate an ore body hardness dataset and to quantify the relationship between sample spacing and the precision of the throughput predictions. Proc. International Conference on Autogenous and Semiautogenous Grinding Technology (SAG) Conf., vol. 4. Vancouver, BC, Canada, pp. 207–220.

Anderson, C.G., 2006. The use of design of experimentation software in laboratory testing and plant optimization. Proc. 38th Annual Meeting of the Canadian Mineral Processors Conf., Ottawa, ON, Canada, pp. 577–606.

Aslan, N., 2008. Application of response surface methodology and central composite rotatable design for modeling and optimization of a multi-gravity separator for chromite concentration. Powder Techno. 185 (1), 80–86.

Barbery, G., 1991. Mineral Liberation, Measurement, Simulation and Practical Use in Mineral Processing. Les Editions GB, Quebec, Canada.

Bazin, C., et al., 2014. Simulation of an iron ore concentration circuit using mineral size recovery curves of industrial spirals. Proc. 46th Annual Meeting of the Canadian Mineral Processors Conf. CIM, Ottawa, ON, Canada, pp. 387–402.

Bennett, C., et al., 2004. Geometallurgical modeling: applied to project evaluation and plant design. Proc. 36th Annual Meeting of the Canadian Mineral Processors Conf. CIM, Ottawa, ON, Canada, pp. 227–240.

Box, G.E.P., 1957. Evolutionary operation: a method for increasing industrial productivity. J. Royal Statist. Soc. Series C (Applied Statistics). 6 (2), 81–101.

Box, G.E.P., Behnken, D.W., 1960. Some new three level design for the study of quantitative variables. Technometrics. 2 (4), 455–475.

Brochot, S., et al., 2002. USIM PAC3: design and optimization of mineral processing plants from crushing to refining, *Mineral Processing Plant Design, Practice* and Control, vol. 1. SME, Littleton, CO, USA, pp. 479–494.

Brodusch, N., et al., 2014. Ionic liquid-based observation technique for nonconductive materials in the scanning electron microscope: application to the characterization of a rare earth ore. Microsc. Res. Tech. 77 (3), 225–235.

Bulled, D., 2007. Grinding circuit design for Adanac Moly Corp using a geometallurgical approach. Proc. 37th Annual Meeting of the Canadian Mineral Processors Conf. CIM, Ottawa, ON, Canada, pp. 101–121.

Celep, O., et al., 2011. Optimization of some parameters of stirred mill for ultra-fine grinding of refractory Au/Ag ores. Powder Technol. 208 (1), 121–127.

Collins, D.A., et al., 2009. Designing modern flotation circuits using JKFIT and JKSimFloat. In: Malhotra, D., et al., (Eds.), Recent Advances in Mineral Processing Plant Design. SME, Littleton, CO, USA, pp. 197–203.

Concha, A.F., 2014. *Solid-Liquid Separation in the Mining Industry.* Fluid Mechanics and Its Applications (Series), vol. 105. Cham Heidelberg, New York, NY, USA, Dordrecht London Springer.

Cundall, P.A., Strack, O.D.L., 1979. A discrete numerical model for granular assemblies. Géotechnique. 29 (1), 47–65.

Evans, C.L., et al., 2011. Application of process mineralogy as a tool in sustainable processing. Miner. Eng. 24 (12), 1242–1248.

Fisher, R.A., 1926. The arrangement of field experiments. J. Ministry Agric. Great Brit. 33, 503–513.

Fuerstenau, M.C. (Ed.), 2007. *Froth Flotation: A Century of Innovation.* SME, Littleton, CO, USA.

Gay, S.L., 2004a. A liberation model for comminution based on probability theory. Miner. Eng. 17 (4), 525–534.

Gay, S.L., 2004b. Simple texture-based liberation modelling of ores. Miner. Eng. 17 (11-12), 1209–1216.

Hedvall, P., Nordin, M., 2002. PlantDesigner®: a crushing and screening modeling tool, *Mineral Processing Plant Design, Practice and Control*, vol. 1. SME, Littleton, CO, USA, pp. 421–441.

Hoal, K.O., et al., 2009. Research in quantitative mineralogy: examples from diverse applications. Miner. Eng. 22 (4), 402–408.

King, R.P., 2012. Modeling and Simulation of Mineral Processing Systems. second ed. SME, Englewood, CO, USA, Elsevier.

Kökkılıç, O., et al., 2015. A design of experiments investigation into dry separation using a Knelson concentrator. Miner. Eng. 72 (0), 73–86.

Kosick, G., Bennett, C., 1999. The value of orebody power requirement profiles for SAG circuit design. Proc. 31st Annual Meeting of the Canadian Mineral Processors Conf. CIM, Ottawa, ON, Canada, pp. 241–253.

Kosick, G., et al., 2002. Managing company risk by incorporating the Mine Resource Model into design and optimization of mineral processing plants. SGS Mineral Services, Technical Paper 21, Technical Bull. pp. 1–7.

Leroux, D., Hardie, C., 2003. Simulation of closed circuit mineral processing operations using Limn® Flowsheet Processing Software. Proc. 35th Annual Meeting of the Canadian Mineral Processors Conf. CIM, Ottawa, ON, Canada, pp. 543–558.

Leroux, D.L., et al., 1999. The application of simulation and Design-of-Experiments techniques to the optimization of grinding circuits at the Heath Steele Concentrator. Proc. 31st Annual Meeting of the Canadian Mineral Processors Conf. CIM, Ottawa, ON, Canada, pp. 227–240.

Lin, D., et al., 1999. Comparison of stereological correction procedures for liberation measurements by use of a standard material. Trans. Inst. Min. Metal. Sec C. 108, C127–C137.

Lotter, N.O., 2011. Modern process mineralogy: an integrated multi-disciplined approach to flowsheeting. Miner. Eng. 24 (12), 1229–1237.

Lynch, A.J., et al., 1981. *Mineral and Coal Flotation Circuits: Their Simulation and Control.* Developments in Mineral Processing, vol. 3. Elsevier Scientific Publishing Company, New York, NY, USA.

Miller, J.D., Lin, C.L., 2009. High resolution X-ray Micro CT (HRXMT)—advances in 3D particle characterization for mineral processing operations. In: Malhotra, D., et al., (Eds.), Recent Advances in Mineral processing Plant Design. SME, Littleton, CO, USA, pp. 48–59.

Miller, J.D., et al., 2009. Liberation-limited grade/recovery curves from X-ray micro CT analysis of feed material for the evaluation of separation efficiency. Int. J. Miner. Process. 93 (1), 48–53.

Montgomery, D.C., 2012. Design and Analysis of Experiments. eighth ed. John Wiley and Sons, New York, NY, USA.

Napier-Munn, T., 2010. Designing and analysing plant trials. In: Greet, C.J. (Ed.), Flotation Plant Optimization. AusIMM, pp. 175–190. Spectrum Series, No 15.

Napier-Munn, T.J., Lynch, A.J., 1992. The modelling and computer simulation of mineral treatment processes – current status and future trends. Miner. Eng. 5 (2), 143–167.

Napier-Munn, T.J., et al., 1996. Mineral Comminution Circuits: Their Operation and Optimisation. Julius Kruttschnitt Mineral Research Centre (JKMRC), University of Queensland, Brisbane, Australia.

Obeng, D.P., et al., 2005. Application of central composite rotatable design to modelling the effect of some operating variables on the performance of the three product cyclone. Int. J. Miner. Process. 76 (3), 181–192.

Parker, D.J., et al., 1993. Positron emission particle tracking-a technique for studying flow within engineering equipment. Nucl. Instrum. Methods Phys. Res. Sect. A. 326 (3), 592–607.

Plackett, R.L., Burman, J.P., 1946. The design of optimum multifactorial experiments. Biometrika. 33 (4), 305–325.

Radziszewski, P., et al., 2003. Lifter design using a DOE approach with a DEM charge motion model. Proc. 35th Annual Meeting of the Canadian Mineral Processors Conf. CIM, Ottawa, ON, Canada, pp. 527–542.

Sbárbarao, D., Del Villar, R. (Eds.), 2010. Advanced Control and Supervision of Mineral Processing Plants. Springer, London, Dordrecht Heidelberg, New York, NY, USA.

Schmidt, S.R., Launsby, R.G., 1994. In: Kiemele, M.J. (Ed.), Understanding Industrial Designed Experiments, fourth ed. Air Academy Press.

Smythe, D.M., et al., 2013. Rare earth element deportment studies utilising QEMSCAN technology. Miner. Eng. 52, 52–61.

Taguchi, G., 1987. System of Experimental Design: Engineering Methods to Optimize Quality and Minimize Costs. Kraus International Publications, White Plains, NY, USA.

Venkatesan, L., et al., 2014. Optimisation of air rate and froth depth in flotation using a CCRD factorial design-PGM case study. Miner. Eng. 66-68, 221–229.

Versteeg, H.K., Malalasekera, W., 2007. An Introduction to Computational Fluid Dynamics: The Finite Volume Method. second ed. Prentice Hall, Harlow, Essex, UK.

Yates, F., 1937. The design and analysis of factorial experiments (Technical report). Technical Communication No. 35 of the Commonwealth Bureau of Soils (alternatively attributed to the Imperial Bureau of Soil Science), Harpenden, UK.

Appendix I

Metallic Ore Minerals

Metal	Main Applications	Ore Minerals[a]	Formula	% Metal	Specific Gravity	Occurrence/Associations
ALUMINUM	Where requirements are lightness, high electrical and thermal conductivity, corrosion resistance, ease of fabrication. Forms high tensile strength alloys	BAUXITE	–	–	3.2–3.5	Bauxite, which occurs massive, is a mixture of minerals such as diaspore, gibbsite, and boehmite with iron oxides and silica. Occurs as residual earth from weathering and leaching of rocks in tropical climates
		Diaspore	AlO(OH)	45.0	3.2–3.5	
		Gibbsite	$Al(OH)_3$	34.6	2.38–2.42	
		Boehmite	AlO(OH)	45.0	3.2–3.5	
ANTIMONY	Flame-resistant properties of oxide used in textiles, fibers, and other materials. Alloyed with lead to increase strength for accumulator plates, sheet, and pipe. Important alloying element for bearing and type metals	STIBNITE	Sb_2S_3	71.8	4.5–4.6	Main ore mineral. Commonly in quartz veins and in limestone replacements. Associates with galena, pyrite, realgar, orpiment, and cinnabar
ARSENIC	Limited use in industry. Small amounts alloyed with copper and lead to toughen the metals. In oxide form, used as insecticide	Arsenopyrite	FeAsS	46.0	5.9–6.2	Widely distributed in mineral veins, with tin ores, tungsten, gold and silver, sphalerite, and pyrite. Since production of metal is in excess of demand, it is commonly regarded as gangue
		Realgar	AsS	70.1	3.5	Often associated in mineral veins in minor amounts
		Orpiment	As_2S_3	61	3.4–3.5	
BERYLLIUM	Up to 4% Be alloyed with copper to produce high tensile alloys with high fatigue, wear, and corrosion resistance, which are used to make springs, bearings, and valves, and spark-proof tools. Used for neutron absorption in nuclear industry. Used in electronics for speakers and styluses	BERYL	$Be_3Al_2Si_6O_{18}$	5	2.6–2.9	Only source of the metal. Often mined as gemstone—emerald, aquamarine. Commonly occurs as accessory mineral in coarse-grained granites (pegmatites) and other similar rocks. Also in calcite veins and mica schists. As similar density to gangue minerals difficult to separate other than by hand-sorting
BISMUTH	Pharmaceuticals; low-melting-point alloys for automatic safety devices, such as fire-sprinklers. Improves casting properties when alloyed with tin and lead	Native	Bi	100	9.7–9.8	Minor amounts in veins associated with silver, lead, zinc, and tin ores
		Bismuthinite	Bi_2S_3	81.2	6.8	Occurs in association with magnetite, pyrite, chalcopyrite, galena, and sphalerite, and with tin and tungsten ores. Majority of bismuth produced as by-product from smelting and refining of lead and copper
CADMIUM	Rust-proofing of steel, copper, and brass by electroplating and spraying; production of pigments; negative plate in alkali accumulators; plastic stabilizers	Greenockite	CdS	77.7	4.9–5.0	Found in association with lead and zinc ores, and in very small quantities with many other minerals. Due to volatility of the metal, mainly produced during smelting and refining of zinc, as a by-product

CESIUM	Low ionization potential utilized in photoelectric cells, photomultiplier tubes, spectro-photometers, infra-red detectors. Minor pharmaceutical use	POLLUCITE	$Cs_2Al_2Si_4O_{12}.2H_2O$	40.3	2.9	Occurs in pegmatites of complex mineralogical character. Rare mineral
		Lepidolite (Lithium mica)	$K(Li, Al)_3 (Si, Al)_4O_{10} (OH, F)_2$	–	2.8–2.9	Occurs in pegmatites, often in association with tourmaline and spodumene. Often carries traces of rubidium and cesium
CHROMIUM	Used mainly as alloying element in steels to give resistance to wear, corrosion, heat, and to increase hardness and toughness. Used for electroplating iron and steel. Chromite used as refractory with neutral characteristics. Used in production of bichromates and other salts in tanning, dyeing, and pigments	CHROMITE	$FeCr_2O_4$	46.2	4.1–5.1	Occurs in olivine and serpentine rocks, often concentrated sufficiently into layers or lenses to be worked. Due to its durability, it is sometimes found in alluvial sands and gravels
COBALT	Used as alloying element for production of high-temperature steels and magnetic alloys. Used as catalyst in chemical industry. Cobalt powder used as cement in sintered carbide cutting tools	Smaltite	$CoAs_2$	28.2	5.7–6.8	Smaltite and cobaltite occur in veins, often together with arsenopyrite, silver, calcite, and nickel minerals
		Cobaltite	$CoAsS$	35.5	6.0–6.3	
		Carrolite	$CuCo_2S_4$	38.0	4.8–5.0	Carrolite and linnaeite sometimes occur in small amounts in copper ores. Cobalt is usually only a minor constituent in ores such as lead, copper, and nickel and extracted as by-product
		Linnaeite	Co_3S_4	57.8	4.8–5.0	
COPPER	Used where high electrical or thermal conductivity is important. High corrosion resistance and easy to fabricate. Used in variety of alloys—brasses, bronzes, aluminum bronzes, etc.	CHALCOPYRITE	$CuFeS_2$	34.6	4.1–4.3	Main ore mineral. Most often in veins with other sulfides, such as galena, sphalerite, pyrrhotite, pyrite, and also cassiterite. Common gangue minerals quartz, calcite, dolomite. Disseminated with bornite and pyrite in porphyry copper deposits
		CHALCOCITE	Cu_2S	79.8	5.5–5.8	Often associated with cuprite and native copper. Constituent of matte in Vale's matte separation process
		BORNITE	Cu_5FeS_4	63.3	4.9–5.4	Associates with chalcopyrite and chalcocite in veins
		COVELLITE	CuS	66.5	4.6	Sometimes as primary sulfide in veins, but more commonly as secondary sulfide with chalcopyrite, chalcocite, and bornite
		CUPRITE	Cu_2O	88.8	5.9–6.2	Found in oxidized zone of deposits, with malachite, azurite, and chalcocite
		MALACHITE	$Cu_2CO_3(OH)_2$	57.5	3.6–4.0	Frequently associated with azurite, native copper, and cuprite in oxidized zone

Metal	Main Applications	Ore Minerals[a]	Formula	% Metal	Specific Gravity	Occurrence/Associations
		Native	Cu	100	8.9	Occurs in small amounts with other copper minerals
		Tennantite	$Cu_{12}As_4S_{13}$ Variable	51.5 Variable	4.4–4.6	Tennantite and tetrahedrite found in veins with silver, copper, lead, and zinc minerals
		Tetrahedrite	$Cu_{12}Sb_4S_{13}$ Variable	45.8 Variable	4.4–5.1	Tetrahedrite more widespread and common in lead–silver veins
		Azurite	$Cu_3(CO_3)_2(OH)_2$	55.3	3.8–3.9	Occurs in oxidized zone. Not as widespread as malachite
		Enargite	Cu_3AsS_4	48.4	4.4	Associates with chalcocite, bornite, covellite, pyrite, sphalerite, tetrahedrite, baryte, and quartz in near-surface deposits
GALLIUM	Electronics industry for production of light-emitting diodes. Used in electronic memories for computers	Occurs in some zinc ores, but no important ore minerals	–	–	–	About 90% of production is a direct by-product of alumina output. Also found in coal ash and flue dusts
GERMANIUM	Electronics industry	Argyrodite	Ag_8GeS_6	6.4	6.1	Occurs with sphalerite, siderite, and marcasite. No important ore minerals. Chief source is cadmium fume from sintering zinc concentrates
GOLD	Jewelry, monetary use, electronics, dentistry, decorative plating	NATIVE	Au	>85 (invariably alloyed with Ag and Cu, and other metals)	12–20	Disseminated in quartz grains, often with pyrite, chalcopyrite, galena, stibnite, and arsenopyrite. Also found alluvially in stream or other sediments. South African "banket" is consolidated alluvial deposit
		Sylvanite	$(Au,Ag)_2Te_4$	34.4	7.9–8.3	Sylvanite empirical formula $Au_{0.75}Ag_{0.25}Te_2$. Tellurides occur in Kalgoorlie gold ores of Western Australia
		Calaverite	$AuTe_2$	43.6	9.0	
HAFNIUM	Naval nuclear reactors, flashbulbs, ceramics, refractory alloys, and enamels	No ore minerals	–	–	–	Produced as co-product of zirconium sponge
INDIUM	Electronics, component of low-melting-point alloys and solders, protective coating on silverware and jewelry	Occurs as trace element in many ores	–	–	–	Recovered from residues and flue dusts from some zinc smelters

IRON	Iron and steel industry	HEMATITE	Fe_2O_3	69.9	5.3	Most important iron ore mineral. Occurs massive. Also in igneous rocks and veins, and as ooliths or cementing material in sedimentary rocks
		MAGNETITE	Fe_3O_4	72.3	5.1–5.2	The only ferromagnetic mineral. Widely distributed in several environments, including igneous and metamorphic rocks; and beach-sand deposits
		Goethite	FeO(OH)	62.9	4.0–4.4	Widespread occurrence, associated with hematite and limonite
		Limonite	$FeO(OH) \cdot nH_2O$	Variable	2.9–4.3	Natural rust, chief constituent being goethite. Often associates with hematite in weathered deposits
		Siderite	$FeCO_3$	48.3	3.7–3.9	Occurs massive in sedimentary rocks and as gangue mineral in veins carrying pyrite, chalcopyrite, galena
		Pyrrhotite	FeS Variable	63.6 Variable	4.6	In monoclinic form the only magnetic sulfide mineral. Occurs disseminated in igneous rocks, commonly with pyrite, chalcopyrite, and pentlandite. Usually regarded as gangue
		Pyrite	FeS_2	46.7	4.9–5.2	One of most widely distributed sulfide minerals. Used for production of sulfuric acid, but often regarded as gangue
LEAD	Batteries, corrosion resistant pipes and linings, alloys, pigments, radiation shielding	GALENA	PbS	86.6	7.4–7.6	Very widely distributed, and most important lead ore mineral. Occurs in veins, often with sphalerite, pyrite, chalcopyrite, tetrahedrite, and gangue minerals such as quartz, calcite, dolomite, baryte, and fluorite. Also in pegmatites, and as replacement bodies in limestone and dolomite rocks, with garnets, feldspar, diopside, rhodonite, and biotite. Often contains up to 0.5% Ag and is important source of this metal
		Cerussite	$PbCO_3$	77.5	6.5–6.6	In oxidized zone of lead veins, associated with galena, anglesite, smithsonite, and sphalerite
		Anglesite	$PbSO_4$	68.3	6.1–6.4	Occurs in oxidation zone of lead veins
		Jamesonite	$Pb_4FeSb_6S_{14}$	40.1	5.5–6.0	Rare mineral occurring in veins with galena, sphalerite, stibnite

Metal	Main Applications	Ore Minerals[a]	Formula	% Metal	Specific Gravity	Occurrence/Associations
LITHIUM	Lightest metal. Lithium carbide used in production of aluminum. Used as base in multipurpose greases; used in manufacture of lithium batteries. Large application in ceramics industry. Very little use in metallic form	SPODUMENE	$LiAlSi_2O_6$	3.7	3.0–3.2	Occurs in pegmatites with lepidolite, tourmaline, and beryl
		Amblygonite	$LiAlPO_4(OH,F)$	Variable	3.0–3.1	Rare mineral occurring in pegmatites with other lithium minerals
		Lepidolite	$K(Li,Al)_3\ (Si,Al)_4O_{10}\ (H,F)_2$	Variable	2.8–3.3	Mica occurring in pegmatites with other lithium minerals
		Tourmaline	Complex borosilicate of Al, Na, Mg, Fe, Li, Mn	–	3.0–3.2	Not a commercial source of metal. Some crystals used as gems. Occurs in granite pegmatites, schists, and gneisses
						50–70% world's Li reserves are estimated to be in Solar de Uyuni salt flat in Bolivia, now in process of being extracted
MAGNESIUM	Small amounts used in aluminum alloys to increase strength and corrosion resistance. Used to desulfur blast-furnace iron. Added to cast-iron to produce nodular iron. Used in cathodic protection, as a reagent in petrol processing and as reducing agent in titanium, and zirconium production. Structural uses where lightness required—magnesium die castings					Most magnesium extracted from brine, rather than ore minerals
		Dolomite	$MgCa(CO_3)_2$	13	2.8–2.9	Mineral used in manufacture of refractories. Occurs as gangue mineral in veins with galena and sphalerite. Also occurs widely as rock-forming mineral
		Magnesite	$MgCO_3$	28.6	3.0–3.2	Used mainly for cement and refractory bricks. Often associates with serpentinite
		Carnallite	$KMgCl_3 \cdot 6H_2O$	8.6	1.6	Occurs with halite and sylvite
		Brucite	$Mg(OH)_2$	41.4	2.4	Occurs in dolomitic limestones and veins with talc, calcite, and in serpentine
MANGANESE	Very important ferro-alloy. About 95% of output used in steel and foundry industry. Balance mainly in manufacture of dry cells and chemicals	PYROLUSITE	MnO_2	63.2	4.5–5.0	Often found in oxidized zone of ore deposits containing manganese. Also in quartz veins and manganese nodules
		Manganite	$MnO(OH)$	62.5	4.2–4.4	Occurs in association with baryte, pyrolusite, and goethite and in veins in granite
		Braunite	$3Mn_2O_3 \cdot MnSiO_3$	63.6	4.7–4.8	Occurs in veins with other manganese minerals
		Psilomelane	$(Ba,H_2O)_2\ Mn_5O_{10}$	38.5	3.3–4.7	Found with pyrolusite and limonite in sediments or quartz veins

MERCURY	Electrical apparatus, scientific instruments, manufacture of paint, electrolytic cells, solvent for gold, manufacture of drugs and chemicals	CINNABAR	HgS	86.2	8.0–8.2	Only important mercury mineral. Occurs in fractures in sedimentary rocks with pyrite, stibnite, and realgar. Common gangue minerals are quartz, calcite, baryte, and chalcedony
MOLYBDENUM	Main use as ferro-alloy. Metal used in manufacture of electrodes and furnace parts. Also used as catalyst corrosion inhibitor, additive to lubricants	MOLYBDENITE	MoS_2	60	4.7–4.8	Principal ore mineral in porphyry molybdenum deposits and an important ore mineral in porphyry Cu–Mo deposits. Occurs in granites and pegmatites with wolframite and cassiterite
		Powellite	$CaMoO_4$	48	4.25	Occurs in hydrothermal ore deposits of molybdenum within the near surface oxidized zones. Also appears as a rare mineral phase in pegmatite, tactite, and basalt
		Wulfenite	$PbMoO_4$	26.2	6.5–7.0	Found in oxidized zone of lead and molybdenum ores. Commonly with anglesite, cerrusite, and vanadinite
NICKEL	Important ferro-alloy due to its high corrosion resistance (stainless steels). Also alloyed with many non-ferrous metals—chromium, aluminum, manganese. Used for electroplating steels, as base for chromium plate. Pure metal corrosion resistant, and resists alkali attack. Is non-toxic and used for food handling and pharmaceutical equipment	PENTLANDITE	$(Fe,Ni)_9S_8$	34.2	4.6–5.0	Empirical formula, $Fe^{2+}{}_{4.5}Ni_{4.5}S_8$. Occurs invariably with chalcopyrite, and often intergrown with pyrrhotite, millerite, cobalt, selenium, silver, and platinum metals
		GARNIERITE	Hydrated Ni–Mg silicate	Variable	2.4	Often occurs massive or earthy, in decomposed serpentines, often with chromium ores, deposits being known as "lateritic"
		Niccolite	NiAs	44.1	7.3–7.7	Occurs in igneous rocks with chalcopyrite, pyrrhotite, and nickel sulfides. Also in veins with silver, silver–arsenic, and cobalt minerals
		Millerite	NiS	64.8	5.3–5.7	Occurs as needle-like radiating crystals in cavities and as replacement of other nickel minerals. Also in veins with other nickel minerals and sulfides
		Heazelwoodite	Ni_3S_2	73.3	5.8	Rare. Constituent of matte in Vale's matte separation process

Metal	Main Applications	Ore Minerals[a]	Formula	% Metal	Specific Gravity	Occurrence/Associations
NIOBIUM (Columbium)	Important ferro-alloy. Added to austenitic stainless steels to inhibit intergranular corrosion at high temperatures	PYROCHLORE (Microlite)	$(Na,Ca)_2\ Nb_2O_6\ (OH,F)$	52.5	4.2–6.4	Empirical formula, $Na_{1.5}Ca_{0.5}Nb_2O_6(OH)_{0.75}F_{0.25}$. Occurs in pegmatites associated with zircon and apatite. Pyrochlore is the name given to niobium-rich minerals, and microlite to tantalum-rich minerals
		COLUMBITE (Tantalite)	$(Fe,Mn)\ (Nb,Ta)_2O_6$	–	5.0–8.0	In granite pegmatites with cassiterite wolframite, spodumene, tourmaline, feldspar, and quartz. Columbite is name given to niobium-rich, and tantalite to tantalum rich-minerals in series
PLATINUM GROUP (Platinum Palladium Osmium, Iridium Rhodium Ruthenium)	Platinum and palladium have wide use in jewelry and dentistry. Platinum, due to its high melting point and corrosion resistance, is widely used for electrical contact material and in manufacture of chemical crucibles, etc. Also widely used as a catalyst. Iridium is also used in jewelry and in dental alloys and in electrical industry. Long life platinum–iridium electrodes used in helicopter spark-plugs. Rhodium used in thermocouples, and platinum–palladium–rhodium catalysts are used in control of automobile emissions					Platinum group metals occur together in nature as native metals or alloys
		NATIVE PLATINUM	Pt	45–86	21.5 Pure	Platinum alloyed with other platinum group metals, iron, and copper. Occurs disseminated in mafic and ultramafic igneous rocks, associates with chromite and copper ores. Found in lode and alluvial deposits
		SPERRYLITE	$PtAs_2$	56.6	10.6	Occurs in pyrrhotite deposits and in gold–quartz veins. Also with covellite and limonite
	Osmium, the heaviest metal known, with a melting point of 2200°C, and ruthenium have little commercial importance	Osmiridium	Alloy of Os–Ir	–	19.3–21.1	Found in small amounts in some gold and platinum ores, where it is recovered as by-product
RADIUM	Industrial radiography, treatment of cancer, and production of luminous paint	*See* URANIUM	–	–	–	Constituent of uranium minerals

RARE EARTHS	The cerium subgroup is the most important industrially. Rare earths used as catalysts in petroleum refining, iron–cerium alloys used as cigarette-lighter flints. Used in ceramics and glass industry, production of color televisions, and in magnets (Nd, Dy) which are being used extensively in wind turbines and hybrid cars	MONAZITE	$REEPO_4$	–	4.98–5.43	Source of Ce, La, Pr, and Nd. See also Thorium. Found throughout the world in placer deposits
		BASTNAESITE	$REE(CO_3)F$	–	4.9–5.2	Source of Ce, La, Pr, and Nd. Widespread, one of the more common rare-earth carbonates. Also found in carbonatite plutons, e.g., Bayan Obo, Mongolia, and Mountain Pass, California. Has replaced monazite as chief source of REEs
		XENOTIME	YPO_4	48.4	4.4–5.1	Rare but important source of yttrium, along with ion-adsorbed RE-bearing clays. Found in pegmatites and other igneous rocks
		Other potential RE minerals: parasite, synchysite, pyrochlore, fergusonite, allanite, eudialyte				
RHENIUM	Used as catalyst in production of low-lead petrol. Used as catalyst with platinum. Used extensively in thermocouples, temperature controls, and heating elements. Also used as filaments in electronic apparatus	Molybdenite	MoS_2	–	4.7–4.8	Rhenium occurs associated with molybdenite in porphyry Cu–Mo deposits, and recovered as by-product
RUBIDIUM	Rubidium and cesium largely interchangeable in properties and uses, although latter usually preferred to meet present small industrial demand	*See* CESIUM				Rubidium widely dispersed as minor constituent in major cesium minerals
SILICON	Used in steel industry and as heavy medium alloy as ferro-silicon. Also used to de-oxidize steels. Metal used as semi-conductor	QUARTZ	SiO_2	46.7	2.65	Commonest mineral, forming 12% of earth's crust. Essential constituent of many rocks, such as granite and sandstone, and virtually sole constituent of quartzite rock
SELENIUM	Used in manufacture of fade-resistant pigments, photo-electric apparatus, in glass production, and various chemical applications. Alloyed with copper and steel to improve machinability	Naumanite	Ag_2Se	26.8	8.0	Selenides occur associated with sulfides, and bulk of selenium recovered as by-product from copper sulfide ores
		Clausthalite	$PbSe$	27.6	8.0	
		Berzelianite	Cu_2Se	38.3	6.7	
SILVER	Sterling ware, jewelry, coinage, photographic and electronic products, mirrors, electroplate, and batteries. Silver nitrate solutions and other silver compounds used as disinfectants and microbiocides	ARGENTITE	Ag_2S	87.1	7.2–7.4	Closely associated with lead, zinc, and copper ores, and bulk of silver produced as by-product from smelting such ores
		ACANTHITE	Ag_2S			Stable polymorph of argentite below 180°C. Similar associations as argentite

Metal	Main Applications	Ore Minerals[a]	Formula	% Metal	Specific Gravity	Occurrence/Associations
		Native	Ag	Up to 100	10.1–11.1	Usually alloyed with copper, gold, etc., and occurs in upper part of silver sulfide deposits
		Cerargyrite	AgCl	75.3	5.8	Occurs in upper parts of silver veins together with native silver and cerussite
TANTALUM	Used in certain chemical and electrical processes due to extremely high corrosion resistance. Used in production of special steels used for medical instruments. Used for electrodes, and tantalum carbide used for cutting tools. Used in manufacture of capacitors	PYROCHLORE TANTALITE	*See* NIOBIUM			As well as ore minerals, certain tin slags are becoming important source of tantalum
TELLURIUM	Used in production of free machining steels, in copper alloys, rubber production, and as catalyst in synthetic fiber production	Sylvanite Calaverite	*See* GOLD			Produced with selenium as by-product of copper refining These metal tellurides, which are important gold ores, and other tellurides of bismuth and lead, are most important sources of tellurium
THALLIUM	Very poisonous, and finds limited outlet as fungicide and rat poison. Thallium salts used in Clerici solution, an important heavy liquid	Occurs in some zinc ores, but no ore minerals	–	–	–	By-product of zinc refining
THORIUM	Radioactive metal. Used in electrical apparatus, and in magnesium–thorium and other thorium alloys. Oxide of importance in manufacture of gas-mantles, and used in medicine	MONAZITE	$REEPO_4$	–	4.9–5.4	Although occurring in lode deposits in igneous rocks such as granites, the main deposits are alluvial, beach-sand deposits being the most important source. Occurs associated with ilmenite, rutile, zircon, garnets, etc.
		Thorianite	ThO_2	87.9	9.3	Occurs in some beach-sand deposits
TIN	Main use in manufacture of tin-plate, for production of cans, etc. Important alloy in production of solders, bearing-metals, bronze, type-metal, pewter, etc.	CASSITERITE	SnO_2	78.8	6.8–7.1	Found in lode and alluvial deposits. Lode deposits in association with wolfram, arsenopyrite, copper, and iron minerals. Alluvially, often associated with ilmenite, monazite, zircon, etc.

TITANIUM	Due to its high strength and corrosion resistance, about 80% of titanium used in aircraft and aerospace industries. Also used in power-station heat-exchanger tubing and in chemical and desalination plants	ILMENITE	$FeTiO_3$	31.6	4.5–5.0	Accessory mineral in igneous rocks especially gabbros and norites. Economically concentrated into alluvial sands, together with rutile, monazite, and zircon
		RUTILE	TiO_2	60	4.2	Accessory mineral in igneous rocks, but economic deposits found in alluvial beach-sand deposits
TUNGSTEN	Production of tungsten carbide for cutting, drilling, and wear-resistant applications. Used in lamp filaments, electronic parts, electrical contacts, etc. Important ferro-alloy, producing tool and high-speed steels	WOLFRAMITE	$(Fe,Mn)WO_4$	Variable	7.1–7.9	Occurs in veins in granite rocks, with minerals such as cassiterite, arsenopyrite, tourmaline, galena, sphalerite, scheelite, and quartz. Also found in some alluvial deposits
		SCHEELITE	$CaWO_4$	63.9	5.9–6.1	Occurs under same conditions as wolframite. Also occurs in contact with metamorphic deposits
URANIUM	Nuclear power plant fuel. Depleted uranium used in ammunition and in shielding for radioactive materials	PITCHBLENDE (URANINITE)	UO_2 (variable—partly oxidized to U_3O_8)	80–90	8–10	Most important uranium and radium ore. Much of the World's uranium comes from unconformity type deposits in which the uraninite is massive and hosted by sandstone a few tens of meters above a highly reducing metamorphic basement
		Carnotite	$K_2(UO_2)_2(VO_4)_2 \cdot 3H_2O$	52.8	3.7–4.7	Secondary mineral found in sedimentary rocks, also in pitchblende deposits. Source of radium
		Coffinite	$U(SiO_4)_{1-x}(OH)_{4x}$ Variable	72.6	5.1	Empirical formula, $U(SiO_4)_{0.9}(OH)_{0.4}$. Common secondary uranium mineral
		Torbernite	$Cu(UO_2)_2(PO_4)_2 \cdot 8\text{-}12H_2O$	48.0	3.2	Empirical formula, $Cu(UO_2)_2(PO_4)_2 \cdot 11(H_2O)$. Secondary uranium mineral
		Autunite	$Ca(UO_2)_2(PO_4)_2 \cdot 12H_2O$	48.3	3.1–3.2	Oxidation product of uranium minerals
VANADIUM	Important ferro-alloy. Used in manufacture of specialty steels, such as high-speed tool steels. Increases strength of structural steels—used for oil and gas pipelines. Vanadium–aluminum master alloys used in preparation of some titanium-based alloys. Vanadium compounds used in chemical and oil industries as catalysts. Also used as glass-coloring agent and in ceramics, and growing use in storage batteries	PATRONITE	VS_4	28.4	2.8	Occurs with nickel and molybdenum sulfides and asphaltic material
		CARNOTITE	*See* URANIUM			
		Roscoelite (Vanadium mica)	$K(V,Al,Mg)_2AlSi_3O_{10}(OH)_2$	9.9	3.0	Empirical formula, $KV^{5+}_{0.8}Al_{0.6}Mg_{0.4}AlSi_3O_{10}(OH)_2$. Found in epithermal Au–Ag–Te deposits and oxidized portions of sedimentary U-V ores
		Vanadinite	$Pb_5(VO_4)_3Cl$	Variable	6.6–7.1	Occurs in oxidation zone of lead, and lead–zinc deposits. Also with other vanadium minerals in sediments

Metal	Main Applications	Ore Minerals[a]	Formula	% Metal	Specific Gravity	Occurrence/Associations
ZINC	Corrosion protective coatings on iron and steel ("galvanizing"). Important alloying metal in brasses and zinc die-castings. About 50% consumed in form of compounds, e. g., zinc oxide as catalyst in manufacture of rubber, and white paint pigment; zinc sulfide as luminescent pigment	SPHALERITE	(Zn,Fe)S	67.1 Pure ZnS	3.9–4.1	Most common zinc ore mineral. Range in Fe content. Frequently associated with galena, and copper sulfides in vein deposits. Also occurs in limestone replacements, with pyrite, pyrrhotite, and magnetite
		Wurtzite	(Zn,Fe)S			High temperature polymorph of sphalerite
		Smithsonite (Calamine)	$ZnCO_3$	52	4.3–4.5	Mainly occurs in oxidized zone of ore deposits carrying zinc minerals. Commonly associated with sphalerite, galena, and calcite
		Hemimorphite (Calamine)	$Zn_4Si_2O_7(OH)_2 \cdot H_2O$	54.3	3.4–3.5	Found associated with smithsonite accompanying the sulfides of zinc, iron, and lead
		Marmatite	(Zn,Fe)S	46.5–56.9	3.9–4.1	High Fe content sphalerite
		Franklinite	Oxide of Zn,Fe,Mn	–	5.0–5.2	First identified at Franklin Mine
		Zincite	ZnO	80.3	5.4–5.7	Zincite often occurs with franklinite and willemite
		Willemite	Zn_2SiO_4	58.5	4.0–4.1	
ZIRCONIUM	Used, alloyed with iron, silicon, and tungsten, in nuclear reactors, and for removing oxides and nitrides from steel. Used in corrosion-resistant equipment in chemical plants	ZIRCON	$ZrSiO_4$	49.8	4.6–4.7	Widely distributed in igneous rocks, such as granites. Common constituent of residues of various sedimentary rocks, and occurs in beach sands associated with ilmenite, rutile, and monazite
		Baddeleyite	ZrO_2	74	5.5–6.0	Forms in igneous rocks low in silica. Found in rocks containing potassium feldspar and plagioclase. Associated minerals include ilmenite, apatite, fluorite, and pyrochlore

[a] *Uppercase is main mineral exploited.*

Appendix II

Common Nonmetallic Ores

Material	Main Applications	Main Ore Minerals	Formula	Specific Gravity	Occurrence
ANHYDRITE	Increasing importance as a fertilizer, and in manufacture of plasters, cements, sulfates, and sulfuric acid	ANHYDRITE	$CaSO_4$	2.95	An important gangue mineral in porphyry Cu–Mo deposits. Occurs with gypsum and halite as a saline residue. Occurs also in "cap rock" above salt domes, and as minor gangue mineral in hydrothermal metallic ore veins
APATITE	*See* Phosphates				
ASBESTOS	Heat-resistant materials, such as fire-proof fabrics and brake-linings. Also asbestos cement products, sheets for roofing and cladding, fire-proof paints, etc.				Group of six minerals with same habit, roughly 1:20 aspect ratio, thin fibrous crystals
		CHRYSOTILE (Serpentized asbestos)	$Mg_3Si_2O_5(OH)_4$	2.5–2.6	Fibrous serpentine occurring as small veins in massive serpentine
		CROCIDOLITE	$Na_2(Mg,Fe,Al)_5Si_8O_{22}(OH)_2$	3.4	Fibrous riebeckite, or blue asbestos, occurring as veins in bedded ironstones
		AMOSITE	$(Mg,Fe)_7Si_8O_{22}(OH)_2$	3.2	Fibrous grunerite, occurring as long fibers in certain metamorphic rocks
		ACTINOLITE	$Ca_2(Mg,Fe)_5Si_8O_{22}(OH)_2$	3.0–3.4	Occurring in schists and in some igneous rocks as alteration product of pyroxene
BADDELEYITE	Ceramics, abrasives, refractories, polishing powders, and manufacture of zirconium chemicals	BADDELEYITE	ZrO_2	5.4–6.0	Mainly found in gravels with zircon, tourmaline, corundum, ilmenite, and rare-earth minerals
BARYTES	Main use in oil- and gas-well drilling industry in finely ground state as drilling muds. Also in manufacture of barium chemicals, and as filler and extender in paint and rubber industries	BARYTE	$BaSO_4$	4.5	Most common barium mineral, occurring in vein deposits as gangue mineral with ores of lead, copper, zinc, together with fluorite, calcite, and quartz. Also as replacement deposit of limestone and in sedimentary deposits
BORATES	Used in manufacture of insulating fiberglass, as fluxes for manufacture of glasses and enamels. Borax used in soap and glue industries, in cloth manufacture and tanning. Also used as preservatives, antiseptics, and in paint driers	BORAX	$Na_2B_4O_5(OH)_4 \cdot 8H_2O$	1.7	An evaporate mineral, precipitated by the evaporation of water in saline lakes, together with halite, sulfates, carbonates, and other borates in arid regions
		KERNITE	$Na_2B_4O_6(OH)_2 \cdot 3H_2O$	1.95	Important source of borates. Occurrence as for borax
		COLEMANITE	$Ca_2B_6O_{11} \cdot 5H_2O$	2.4	In association with borax, but principally as a lining to cavities in sedimentary rocks
		ULEXITE	$NaCaB_5O_6(OH)_6 \cdot 5H_2O$	1.9	Occurs with borax in lake deposits. Also with gypsum and rock salt
		SASSOLITE	H_3BO_3	1.48	Occurs with sulfur in volcanoes and in hot lakes and lagoons
		BORACITE	$Mg_3B_7O_{13}Cl$	2.95	Occurs in saline deposits with rock-salt, gypsum, and anhydrite

CALCIUM CARBONATE	Many uses according to purity and character. Clayey variety used for cement, purer variety for lime. Marble for building and ornamental stones. Used as smelting flux, and in printing processes. Chalk and lime applied to soil as dressing. Transparent calcite (Iceland spar), used in construction of optical apparatus	CALCITE	$CaCO_3$	2.7	Common and widely distributed mineral, often occurring in veins, either as main constituent, or as gangue mineral with metallic ores. It is a rock-forming mineral, which is mainly quarried as the sedimentary rocks limestone and chalk, and metamorphic rock marble
CHINA CLAY	Manufacture of porcelain and china. Used as filler in manufacture of paper, rubber, and paint	KAOLINITE	$Al_2Si_2O_5(OH)_4$	2.6	A secondary mineral produced by the alteration of aluminous silicates, and particularly of alkali feldspars
CHROMITE	Used as refractory in steel-making furnaces	See CHROMIUM (Appendix I)			
CORUNDUM	Abrasive. Next to diamond, hardest known mineral. Colored variety used as gemstones	CORUNDUM	Al_2O_3	3.9–4.1	Original constituent of various igneous rocks, such as syenite. Also in metamorphic rocks such as marble, gneiss, and schist; and in pegmatites and alluvial deposits. Impure form is emery, containing much magnetic and hematite
CRYOLITE	Used as flux in manufacture of aluminum by electrolysis	CRYOLITE	Na_3AlF_6	3.0	Occurs in pegmatite veins in granite with siderite, quartz, galena, sphalerite, chalcopyrite, fluorite cassiterite, and other minerals. Only significant deposit in Greenland
DIAMOND	Gemstone. Used extensively for abrasive and cutting purposes—hardest known mineral. Used for tipping drills in mining and oil industry	DIAMOND	C	3.5	Distributed sporadically in kimberlite pipes. Also in alluvial beach and river deposits. Bort refers to shards of non-gem-grade diamonds, used industrially
DOLOMITE	Important building material. Also used for furnace linings and as flux in steel-making	DOLOMITE	$CaMg(CO_3)_2$	2.8–2.9	Rock-forming mineral. Occurs as gangue mineral in veins containing galena and sphalerite
EMERY	*See* CORUNDUM				
EPSOM SALTS	Medicine and tanning	EPSOMITE	$MgSO_4 \cdot 7H_2O$	1.7	Usually as encrusting masses on walls of caves or mine workings. Also in oxidized zone of pyrite deposits in arid regions
FELDSPAR	Used in manufacture of porcelain, pottery, and glass. Also in production of glazes on earthware, etc., and as mild abrasive	ORTHOCLASE (Isomorphous forms—Microline, Sanidine, and Adularia—the potassic feldspars)	$KAlSi_3O_8$	2.6	Most abundant of all minerals, and most important rock-forming mineral. Widely distributed, mainly in igneous, but also in metamorphic and sedimentary rocks
		ALBITE	$NaAlSi_3O_8$	2.6	

Material	Main Applications	Main Ore Minerals	Formula	Specific Gravity	Occurrence
		ANORTHITE	$CaAl_2Si_2O_8$	2.74	
		(*Plagioclase* feldspars form series having formulae ranging from $NaAlSi_3O_8$ to $CaAl_2Si_2O_8$, changing progressively from albite, through oligoclase, andesine, labradorite, and bytownite to anorthite)			
FLUORSPAR	Mainly as flux in steelmaking. Also for manufacture of specialized optical equipment, production of hydrofluoric acid, and fluorocarbons for aerosols. Color-banded variety known as *Blue-John* used as semi-precious stone	FLUORITE	CaF_2	3.2	Widely distributed, hydrothermal veins and replacement deposits, either alone, or with galena, sphalerite, barytes, calcite, and other minerals
GARNET	Mainly used as abrasive for sandblasting of aircraft components, and for wood polishing. Also certain varieties used as gemstones	PYROPE	$Mg_3Al_2(SiO_4)_3$	3.7	Widely distributed in metamorphic and some igneous rocks. Also commonly found as constituent of beach and river deposits. Andradite is an important mineral in skarns where it may be accompanied by chalcopyrite, spalerite, and galena
		ALMANDINE	$Fe_3Al_2(SiO_4)_3$	4.0	
		GROSSULAR	$Ca_3Al_2(SiO_4)_3$	3.5	
		ANDRADITE	$Ca_3Fe_2(SiO_4)_3$	3.8	
		SPESSARTITE	$Mn_3Al_2(SiO_4)_3$	4.2	
		UVAROVITE	$Ca_3Cr_2(SiO_4)_3$	3.4	
GRAPHITE (Plumbago)	Manufacture of foundry molds, crucibles, and paint; used as lubricant and as electric furnace electrodes	GRAPHITE	C	2.1–2.3	Occurs as disseminated flakes in metamorphic rocks derived from rocks with appreciable carbon content. Also as veins in igneous rocks and pegmatites
GYPSUM	Used in cement manufacture, as a fertilizer, and as filler in various materials such as paper, rubber, etc. Used to produce plaster of Paris	GYPSUM	$CaSO_4 \cdot 2H_2O$	2.3	Evaporate mineral, occurring with halite and anhydrite in bedded deposits. Can occur as a gangue mineral in volcanogenic massive sulfide deposits of Cu, Pb, and Zn
ILMENITE	About 90% of ilmenite produced is used for manufacture of titanium dioxide, a white compound used as a pigment and as a sunscreen agent	See TITANIUM (Appendix I)			
MAGNESITE	Used as refractory for steel furnace linings, and in production of carbon dioxide and magnesium salts	See MAGNESIUM (Appendix I)			
MICA	Used for insulating purposes in electrical apparatus. Ground mica used in production of roofing material, and in lubricants, wall-finishes artificial stone, etc. Powdered mica gives "frost" effect on Christmas cards and decorations	MUSCOVITE	$KAl_2(AlSi_3O_{10})(OH,F)_2$	2.8–2.9	Widely distributed in igneous rocks, such as granite and pegmatites. Also in metamorphic rocks—gneisses and schists. Also in sedimentary sandstones, clays, etc.
		PHLOGOPITE	$KMg_3(AlSi_3O_{10})(OH,F)_2$	2.8–2.85	Most commonly in metamorphosed limestones, also in igneous rocks rich in magnesia
		BIOTITE	$K(Mg,Fe)_3(AlSi_3O_{10})(OH,F)_2$	2.7–3.3	Widely distributed in granite, syenite and diorite. Common constituent of schists and gneisses and of contact metamorphic rocks

PHOSPHATES	Main use as fertilizers. Small amounts used in production of phosphorous chemicals	APATITE	$Ca_5(PO_4)_3(F, Cl, OH)$	3.1–3.3	Occurs as accessory mineral in wide range of igneous rocks, such as pegmatites. Also in metamorphic rocks, especially metamorphosed limestones and skarns. Principal constituent of fossil bones in sedimentary rocks. Other important phosphates are monazite and xenotime commonly associated with rare earth minerals and may be enriched in REE (*See* RARE EARTHS in Appendix I)
		PHOSPHATE ROCK	Complex phosphates of Ca, Fe, A1		Most extensive phosphate rock deposits associated with marine sediments, typically glauconite-bearing sandstones, limestones, and shales. *Guano* is an accumulation of excrement of sea-birds, found mainly on oceanic islands
POTASH	Used as fertilizers, and source of potassium salts. Nitre also used in explosives manufacture (*saltpetre*)	SYLVITE	KC1	2.0	Occurs in bedded evaporate deposits with halite and carnallite
		CARNALLITE	$KMgCl_3 \cdot 6H_2O$	1.6	In evaporate deposits with sylvite and halite
		ALUNITE	$KAl_3(SO_4)_2(OH)_6$	2.6–2.9	Secondary mineral found in areas where volcanic rocks containing potassic feldspars have been altered by acid solutions
		NITRE	KNO_3	2.1	Occurs in soils in arid regions, associated with gypsum, halite, and nitratine
QUARTZ	Building materials, glass making, pottery, silica bricks, ferro-silicon, etc. Used as abrasive in scouring soaps, sandpaper, toothpaste, etc. Due to its piezo-electric properties, quartz crystals widely used in electronics	*See* SILICON (Appendix I)			
ROCK SALT	Culinary and preservative uses. Wide use in chemical manufacturing processes	HALITE	NaCl	2.2	Occurs in extensive stratified evaporate deposits, formed by evaporation of land-locked seas in geological past. Associates with other water soluble minerals, such as sylvite, gypsum, and anhydrite
RUTILE	Production of welding rod coatings, and titanium dioxide. High refractive index makes it useful for manufacture of certain optical elements	*See* TITANIUM (Appendix I)			
SERPENTINE	Used as building stone and other ornamental work. Fibrous varieties source of asbestos (*See* ASBESTOS)	LIZARDITE	$Mg_3Si_2O_5(OH)_4$	2.55–2.6	The most common serpentine mineral. Typically a product of retrograde metamorphism, replacing olivine, orthopyroxene, or other minerals in ultramafic igneous rocks

Material	Main Applications	Main Ore Minerals	Formula	Specific Gravity	Occurrence
		ANTIGORITE	$(Mg,Fe)_3Si_2O_5(OH)_4$	2.5–2.6	Widespread, less common than lizardite. Found in regional and contact metamorphosed serpentinites. Associated with chromite, magnetite, chrysotile, olivine
		CHRYSOTILE	$Mg_3Si_2O_5(OH)_4$		Most commonly encountered form of asbestos. Naturally occurring fiber bundles
SILLIMANITE	Raw material for high-alumina refractories, for iron and steel industry, and other metal smelters. Also used in glass industry, and as insulating porcelains for spark-plugs, etc.	KYANITE (Disthene)	Al_2SiO_5	3.5–3.7	Typically in regionally metamorphosed schists and gneisses, together with garnet, mica, and quartz. Also in pegmatites and quartz veins associated with schists and gneisses
		ANDALUSITE	Al_2SiO_5	3.1–3.2	In metamorphosed rocks of clayey composition. Also as accessory mineral in some pegmatites, with corundum, tourmaline, and topaz
		SILLIMANITE	Al_2SiO_5	3.2–3.3	Typically in schists and gneisses produced by high-grade regional metamorphism
		MULLITE	$Al_6Si_2O_{13}$	3.2	Rarely found in nature, but synthetic mullite produced in many countries
SULFUR	Production of fertilizers, sulfuric acid, insecticides, gunpowder, sulfur dioxide, etc.	NATIVE SULFUR	S	2.0–2.1	In craters and crevices of volcanoes. In sedimentary rocks, mainly limestone in association with gypsum. Also in cap rock of salt domes, with anhydrite, gypsum, and calcite
		PYRITE	*See* IRON (Appendix I)		
TALC	As filler for paints, paper, rubber, etc. Used in plasters, lubricants, toilet powder, French chalk. Massive varieties used for sinks, laboratory tabletops, acid tanks, etc.	TALC	$Mg_3Si_4O_{10}(OH)_2$	2.6–2.8	Secondary mineral formed by alteration of olivine, pyroxene, and amphibole, and occurs along faults in magnesium-rich rocks. Also occurs in schists, in association with actinolite. Massive talc known as *soapstone* or *steatite*
VERMICULITE	Outstanding thermal and sound insulating properties, light, fire-resistant, and inert—used principally in building industry	VERMICULITE	$Mg_3(Al, Si)_4O_{10}(OH)_2 \cdot 4H_2O$	2.3–2.7	Occurs as an alteration product of magnesian micas, in association with carbonatites
WITHERITE	Source of barium salts. Small quantities used in pottery industry	WITHERITE	$BaCO_3$	4.3	Not of wide occurrence. Sometimes accompanies galena in hydrothermal veins, together with anglesite and barite
ZIRCON SAND	Used in foundries, refractories, ceramics, and abrasives, and in chemical production	See ZIRCONIUM (Appendix I)			

Appendix III

Technical Separation Efficiency: Definition and Derivation

Based on Jowett (1975) and Jowett and Sutherland (1985), as the basis for the definition of separation efficiency we take the following:

$$SE = \frac{\text{Amount perfectly separated}}{\text{Amount theoretically perfectly separable}}$$

A solution is derived from the cumulative recovery–cumulative total solids (or weight) recovery ($R-W$) method of presenting separation data. In this method of data presentation, the slope of any line represents a grade, or more precisely grade divided by the feed grade. This can be shown by noting:

$$\text{Slope} = \frac{\Delta R}{\Delta W} = \frac{\Delta(Cc)}{Ff} / \frac{\Delta C}{F} = \frac{c}{f}$$

This property is exploited in the derivation of SE. (As a consequence of this property, the plot is sometimes referred to as a *grade-gradient* plot.)

Figure A.III.1 shows a representative $R-W$ *operating line*, *OMQ*, with the necessary constructions: line *OQ* is the "no separation" line ($c = f$); *OPQ* represents perfect separation; and lines *J(I)M* and *KP* are parallel to the no separation line.

We can now define some features of the plot.

Weight recovery (or yield) at P, W_P:

By inspection, when metal recovery is 1 (100%) at pure mineral grade then the weight recovered, W_P, must be the weight of mineral contained in the feed, that is, the feed mineral grade. Using the same symbols as in Chapter 1, then if f is the feed metal grade and m is the content of metal in the mineral then the feed mineral grade is f/m.

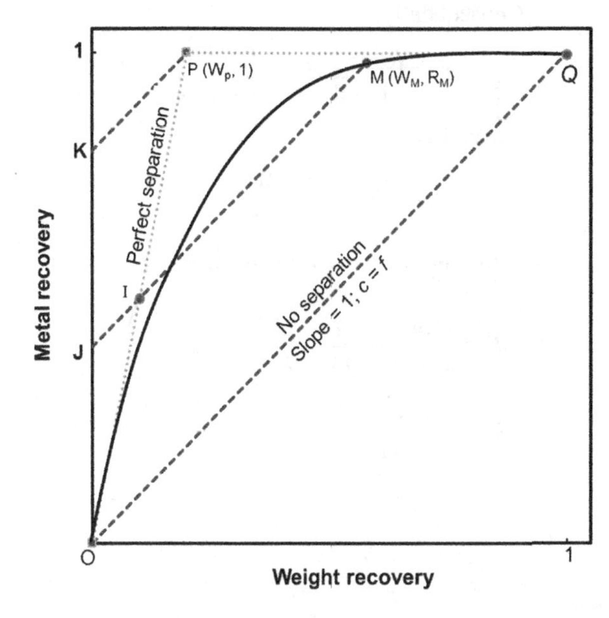

FIGURE A.III.1 Generic recovery–weight recovery ($R-W$) plot with constructions to determine separation efficiency.

Thus

$$W_P = \frac{f}{m}$$

Operating point M:

From the construction, we can consider point M to be made of part perfect separation, part feed by-passed to the concentrate, that is:

$$OM = OI + IM$$

where OI is the perfectly separated part, and IM is the by-passed feed part.

Separation efficiency:

Based on the definition above

$$SE = \frac{OI}{OP}$$

Derivation in terms of R, etc.

By similar triangles we can write

$$SE = \frac{OI}{OP} = \frac{OJ}{OK} = \frac{J}{K}$$

J and K:

$$\text{slope } JM = 1 = \frac{R_M - J}{W_M}$$

$$\text{thus } J = R_M - W_M$$

$$\text{slope } KP = 1 = \frac{1-K}{W_P}$$

$$\text{thus } K = 1 - W_P$$

Consequently:

$$SE = \frac{R_M - W_M}{1 - W_P} = \frac{R_M - W_M}{1 - f/m}$$

General case:

$$SE = \frac{R - W}{1 - f/m}$$

Substituting for $R = Cc/Ff$ and $W = C/F$

$$SE = \frac{C}{F}\frac{m}{f}\frac{(c-f)}{(m-f)} = \frac{C}{F}\left(\frac{c}{f} - \frac{m-c}{m-f}\right)$$

$$= \frac{C}{F}\frac{m}{f}\frac{(c-f)}{(m-f)}$$

The result, starting from the above definition of separation efficiency, is thus equivalent to Eq. (1.5), which in turn means SE is equal to difference in recoveries (Eq. (1.2)).

Maximum separation efficiency:

The maximum can be found by trial and error from the available recovery data for the two components, as illustrated in Chapter 12, Figure 12.58 (see also Appendix VI). It can also be found by inspection of Figure A.III.1, noting that the maximum SE will correspond to when OJ is maximum. This will occur when the line constructed through J parallel to the no separation line is tangent to the operating line; in other words, the maximum SE corresponds to the point on the operating line where the increment in grade (the tangent) is equal to the feed grade. This result has led to attempts to include the increment of grade concept in flowsheet design to optimize separation (see Chapters 11 and 12).

REFERENCES

Jowett, A., 1975. Formulae for the technical efficiency of mineral separations. Int. J. Miner. Process. 2 (4), 287–301.

Jowett, A., Sutherland, D.N., 1985. Some theoretical aspects of optimizing complex mineral separation systems. Int. J. Miner. Process. 14 (2), 85–109.

Appendix IV

Data Used in Figure 4.7

Particle Size Distribution: Effect of Grind Time in a Laboratory Ball Mill

Particle Size (μm)	Cumulative Finer (%)			
	Feed	Grind Time (min)		
		2	6	12
600	86.88	96.17	99.30	99.91
425	79.72	93.28	99.05	99.90
300	70.01	86.64	97.89	99.85
212	61.04	78.38	94.48	99.56
148	47.95	64.02	84.36	97.19
106	33.51	47.03	67.65	88.18
75	22.90	33.70	51.63	72.40
53	15.50	23.12	37.20	53.97
37	12.38	18.48	30.12	44.31

Appendix V

Derivation of Fully Mixed Recovery Equation (Chapter 12, Equation 12.28)

Figure A.V.1 represents a continuously operating flotation cell with retention time τ. For convenience, in this derivation it is easier to refer to the concentration of particles rather than mass, so we have a concentration C_0 entering the cell with steady-state concentration C in the cell, which is also the concentration in the exit (nonfloats) as the cell is perfectly mixed.

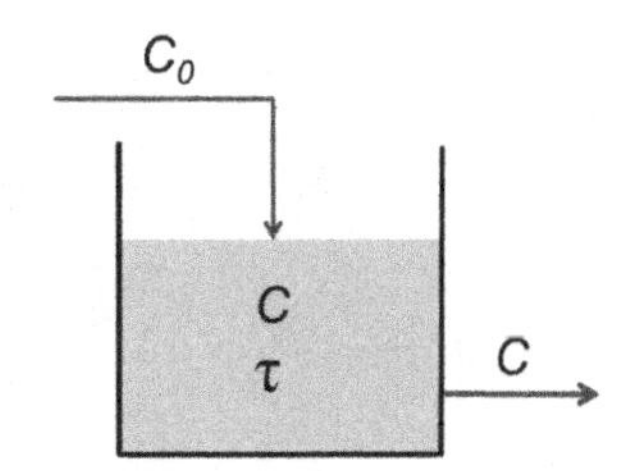

FIGURE A.V.1 Continuous operating cell; perfectly mixed, first-order kinetics.

The rate of recovery (by flotation) is then:

$$\frac{\mathrm{d}R}{\mathrm{d}t} = \frac{C_0 - C}{\tau}$$

From the assumption of first order we can equate:

$$\frac{C_0 - C}{\tau} = k\,C$$

From which we deduce:

$$\tau = \frac{C_0 - C}{k\,C} \rightarrow \frac{C_0}{C} = k\,\tau + 1 \rightarrow \frac{C}{C_0} = \frac{1}{1 + k\,\tau}$$

Since C/C_0 is the fraction left in the cell, then the recovery is:

$$R = 1 - \frac{C}{C_0} = 1 - \frac{1}{1 + k\,\tau} = \frac{k\,\tau}{1 + k\,\tau}$$

That is, the same as Eq. (12.28).

Appendix VI

Data and Computations to Determine Time Corresponding to Maximum Separation Efficiency (Figure 12.58), and to Determine Flotation Rate Constant (Figure 12.59)

1. Maximum separation efficiency (Figure 12.58)

1	2	3	4	5	6	7	8	9	10	11	12
Time (min)	Mass %	Assay (%)			Units (Mass % X % PbS)	Rec PbS (%)	Cum *R* PbS (%)	Units (Mass % X % G)	Rec G (%)	Cum *R* G (%)	CR PbS - CR G (%)
		Pb	PbS	G							
0.5	3.29	72.9	84.18	15.82	276.95	37.25	37.25	52.05	0.56	0.56	36.69
1	2.12	66.9	77.25	22.75	163.77	22.03	59.28	48.23	0.52	1.08	58.20
1.5	1.88	51.7	59.7	40.3	112.24	15.10	74.38	75.76	0.82	1.90	72.48
2	1.44	37.5	43.3	56.7	62.35	8.39	82.77	81.65	0.88	2.78	79.98
4	4.28	19.2	22.17	77.83	94.89	12.76	95.53	333.11	3.60	6.38	89.15
6.5	3.53	2.41	2.78	97.22	9.81	1.32	96.85	343.19	3.71	10.09	86.76
10	4.82	0.82	0.95	99.05	4.58	0.62	97.47	477.42	5.16	15.25	82.22
14	3.87	0.82	0.95	99.05	3.68	0.49	97.96	383.32	4.14	19.39	78.57
20	4.33	0.49	0.57	99.43	2.47	0.33	98.29	430.53	4.65	24.04	74.26
Tails	70.44	0.16	0.18	99.82	12.68	1.71	100.00	7031.32	75.96	100.00	0.00
Head	100.00	6.44	7.43	92.57	743.41	100.00		9257.00	100.00		

Computations:

Column 4: conversion of Pb assay to PbS

Column 5: determination of gangue (G) assay, 100-column 3b

Column 6: intermediate step in calculation for recovery of PbS, column 2 $\times$ column 4

Column 7: determination of increment PbS recovery, column 6/Σ column 6 ("head")

Column 8: cumulative recovery of PbS (or Pb)

Column 9: intermediate step in calculation for recovery of G, column 2 $\times$ column 5

Column 10: determination of increment G recovery, column 6/Σ column 9 ("head")

Column 11: cumulative recovery of G

Column 12: determination of separation efficiency, column 8 − column 11

Maximum separation efficiency is determined from:

a. plot of Column 12 (CR PbS − CR G) versus time; and

b. from plot of increment Pb grade (assay, %Pb, column 3) versus time, both given in Figure 12.58

2. Determination of flotation rate constant: disappearance plot (Figure 12.59)

1	2	3
Time (min)	Cum R PbS (%)	1 − Cum R/100
0.5	37.25	0.6275
1	59.28	0.4072
1.5	74.38	0.2562
2	82.77	0.1723
4	95.53	0.0447
6.5	96.85	0.0315
10	97.47	0.0253
14	97.96	0.0204
20	98.29	0.0171

Computations:

Column 2: same as column 8 above

Column 3: calculation of mass remaining, converted to fraction

From the plot of column 3 (1 − Cum R/100) versus time (the "disappearance plot," Figure 12.59), the first 2 min give good approximation to simple first order, with $k = 0.88\ \mathrm{min}^{-1}$.

In principle the same computations could be applied to gangue. In that case you will find an R_∞ is required to fit the data (i.e., Eq. (12.51)); and, of course, it assumes gangue recovery also follows first-order kinetics which is likely not the case, especially as entrainment tends to be an important factor in batch tests.

Index

Note: Page numbers followed by "*f*" and "*t*" refer to figures and tables, respectively.

D

I

J

K

L

M

N

O

T

U

V

W

X

Y

Z